THE ROUTLEDGE HANDBOOK OF GENDER AND COMMUNICATION

This volume provides an extensive overview of current research on the complex relationships between gender and communication. Featuring a broad variety of chapters written by leading and upcoming scholars, this edited collection uses diverse theoretical frameworks to provide insight into recent concerns regarding changing gender roles, representations, and resources in communication studies. Established research and new perspectives address vital themes in this comprehensive text, including the shifting politics of gender, ethical and technological trends in gendered media, and gender in daily life. Comprising 39 chapters by a team of international contributors, the *Handbook* is divided into six general themes:

- Gendered lives and identities
- Visualizing gender
- The politics of gender
- Gendered contexts and strategies
- Gendered violence and communication
- Gender advocacy in action

These sections examine central issues, debates, and problems, including the ethics and politics of gender as identity, impacts of media and technology, legal and legislative battlegrounds for gender inequality and LGBTQ+ human rights, changing institutional contexts, and recent research into gender violence and communication. The final section links academic research on gender and communication to activism and advocacy beyond the academy.

The Routledge Handbook of Gender and Communication will be an invaluable reference work for students and researchers working at the intersections of gender studies and communication studies. Its international perspectives and the variety of themes it covers make it an essential and pragmatic pedagogical resource.

Marnel Niles Goins (Ph.D., Howard University) is Interim Dean of the College of Sciences and Humanities and Professor of Communication at Marymount University in Arlington, Virginia.

Joan Faber McAlister (Ph.D., University of Iowa) is Associate Professor of Rhetoric, Media, & Social Change at Drake University in Des Moines, Iowa.

Bryant Keith Alexander (Ph.D., Southern Illinois University Carbondale) is Professor and Dean of the College of Communication and Fine Arts at Loyola Marymount University in Los Angeles, California.

THE ROUTLEDGE HANDBOOK OF GENDER AND COMMUNICATION

Edited by

Marnel Niles Goins, Joan Faber McAlister, and Bryant Keith Alexander

Routledge
Taylor & Francis Group

LONDON AND NEW YORK

First published 2021
by Routledge
2 Park Square, Milton Park, Abingdon, Oxon OX14 4RN

and by Routledge
52 Vanderbilt Avenue, New York, NY 10017

Routledge is an imprint of the Taylor & Francis Group, an informa business

British Library Cataloguing-in-Publication Data
A catalogue record for this book is available from the British Library

Library of Congress Cataloging-in-Publication Data
Names: Niles Goins, Marnel, editor. | Alexander, Bryant Keith, 1963– editor. | Faber McAlister, Joan, editor.
Title: The Routledge handbook of gender and communication / edited by Marnel Niles Goins, Bryant Keith Alexander, and Joan Faber McAlister.
Description: New York : Routledge, 2020. | Series: Handbooks to gender and sexuality | Includes bibliographical references and index. |
Identifiers: LCCN 2020025054 (print) | LCCN 2020025055 (ebook) | ISBN 9781138329188 (hardback) | ISBN 9780367622497 (paperback) | ISBN 9780429448317 (ebook) | ISBN 9780429827334 (adobe pdf) | ISBN 9780429827327 (epub) | ISBN 9780429827310 (mobi)
Subjects: LCSH: Communication–Social aspects. | Sex role.
Classification: LCC HM1206 .R68 2020 (print) | LCC HM1206 (ebook) | DDC 305.3–dc23
LC record available at https://lccn.loc.gov/2020025054
LC ebook record available at https://lccn.loc.gov/2020025055

ISBN: 978-1-138-32918-8 (hbk)
ISBN: 978-0-429-44831-7 (ebk)

Typeset in Bembo Std
by KnowledgeWorks Global Ltd.

CONTENTS

Contents

Contents

CONTRIBUTORS

Shadee Abdi (Ph.D., University of Denver) is Assistant Professor of Communication at San Francisco State University. Dr. Abdi's research emphasizes international and intercultural communication, gender, feminist and queer studies, performance studies, critical media studies, critical family communication, and performances of Iranian diaspora. Broadly, her work explores how conflicting narratives complicate and enhance our intersectional understandings of identity and power relative to race, culture, sexuality, gender, nationality, religion, ability, class, and family within familial and mediated contexts.

Fatima Zahrae Chrifi Alaoui (Ph.D., University of Denver) is Assistant Professor and Coordinator of Graduate Studies in the Department of Communication Studies at San Francisco State University. Her research engages critical rhetoric, political communication, new media, gender and sexuality studies, transnational feminism, and social change in a variety of contexts including social movements, political discourse, and pop culture. More particularly, Dr. Alaoui's scholarship considers how the often non-normative, un-institutionalized voices of resistance work to change their communities, and how normative or institutionalized discourses reinforce their ability to maintain power, with a specific focus on the Middle East and North Africa.

Bryant Keith Alexander (Ph.D., Southern Illinois University Carbondale) is Dean and Professor in the College of Communication and Fine Arts at Loyola Marymount University. He is the author or editor of three books, *Performance Theories in Education: Power, Pedagogy and the Politics of Identity* (2005, with Gary L. Anderson and Bernardo P. Gallegos), *Performing Black Masculinity: Race, Culture, and Queer Identity* (2006), and *The Performative Sustainability of Race: Reflections on Black Culture and the Politics of Identity* (2012). Dr. Alexander is the author of a wide range of essays and chapters, including contributions to the *Handbook of Critical and Indigenous Methodologies*, *Handbook of Performance Studies*, *Handbook of Qualitative Research* (Third Edition/Fifth Edition), *Handbook of Communication and Instruction*, *Handbook of Critical Intercultural Communication*, and the *Handbook of Autoethnography*. He is also Affiliate Faculty in the Educational Leadership for Social Justice, Ed.D Program in the School of Education.

Jace Allen (M.A., San Francisco State University) is an alumna from the Communication Studies department at San Francisco State University. His research focuses on sexuality and communication, representations of non-normative sexual identities in video games, new media, and sexual identity in cross-cultural settings with a focus on Asia and the United States.

Raisa F. Alvarado (Ph.D., University of Denver) is Assistant Professor in the Department of Communication Studies at Dixie State University. Broadly, her research focuses on the rhetoric of social movements, Chicana feminism, identity assemblages, and critical intercultural communication. Dr. Alvarado's most recent work explores the intersectional implications of girl-oriented social movements and the ways silenced histories are co-opted in adolescent literature.

Mariam Betlemidze (Ph.D., University of Utah) is Assistant Professor in the Department of Communication Studies at California State University, San Bernardino (CSUSB). Aside from teaching and research activities, she advises CSUSB's independent student newspaper, *The Coyote Chronicle*. Her areas of interest include new materialism, poststructural media theory, social change, and multimedia journalism. Before her graduate academic career, Dr. Betlemidze worked as a journalist in her home country of Georgia and neighboring countries of Armenia, Azerbaijan, Russia, and Turkey. Mariam Betlemidze is a cofounder of the non-profit media organization GO Group Media and JAM News, reporting stories primarily in the South Caucasus.

Sarah Jane Blithe (Ph.D., University of Colorado Boulder) is Associate Professor of Communication Studies at the University of Nevada, Reno. Her expertise is in organizational communication, with specific attention to intersectional gender, work-life balance, policy inequalities, and management learning. Dr. Blithe is the author of *Gender Equality and Work-Life Balance: Glass Handcuffs and Working Men in the U.S.* and *Sex and Stigma: Stories of Everyday Life in Nevada's Legal Brothels*, with Anna Wiederhold Wolfe and Breanna Mohr, both of which won multiple national book awards. She is also the recipient of the 2019 Thornton Peace Prize from the Nevada System of Higher Education, for her social justice advocacy.

Jaime Bochantin (Ph.D., Texas A&M University) is Associate Professor of Communication Studies and Organizational Science at the University of North Carolina, Charlotte. Her research examines the overall health and well-being of organizational members, including the interface of work and family, stress, emotion and burnout, and workplace mistreatment, including incivility and bullying.

Patrice M. Buzzanell (Ph.D., Purdue University) is Professor and Chair of the Department of Communication at the University of South Florida and Endowed Visiting Professor for the School of Media and Design at Shanghai Jiaotong University. Fellow and Past President of the International Communication Association (ICA), she served as President of the Council of Communication Associations and the Organization for the Study of Communication, Language and Gender. She is a Distinguished Scholar of the National Communication Association. Her research focuses on career, work-life policy, resilience, gender, and engineering design. She is co-editor of four books and more than 210 articles and chapters, plus proceedings in engineering education and other disciplines. She received ICA's Mentorship Award and the Provost Outstanding Mentor Award at Purdue, where she was University Distinguished Professor and Endowed Chair and Director of the Susan Bulkeley Butler Center for Leadership Excellence.

Bernadette Marie Calafell (Ph.D., University of North Carolina) is Professor and Chair of the Department of Critical Race and Ethnic Studies at Gonzaga University. She is Editor-Elect of the *Journal of International and Intercultural Communication* and Film Review Editor of *QED: A Journal in GLBTQ Worldmaking*. Additionally, she is co-editor of the Critical Intercultural Studies

Series with Thomas Nakayama and editor of the Horror and Monstrosity Studies series along with Kendall Phillips and Marina Levina. She is also author of *Monstrosity, Performance, and Race in Contemporary Culture*. Her work is centered in performance studies, monstrosity, queer of color theory, and Chicana feminisms.

David Carless (Ph.D., University of Bristol) is a researcher-writer-artist working on interdisciplinary projects across social science, health, and education using songwriting, music, storytelling, filmmaking, and live performance to create research that is relevant and accessible beyond academia. Research collaborations with Kitrina Douglas are available as performances, CDs, and film. His work has been published as journal articles and book chapters, and his latest book, *Doing Arts-Based Research*, will be published by Routledge. He is currently a Visiting Research Professor at Queen's University Belfast and University of Edinburgh.

Kundai Chirindo (Ph.D., University of Kansas) is Associate Professor in the Rhetoric and Media Studies department at Lewis & Clark College in Portland, Oregon. A rhetorical scholar interested in discourses that relate to the African continent, Kundai's work centers on discursive practices that contest, contribute to, and ultimately constitute ideas of Africa in American public life. Through exploring these themes, he contributes to scholarly conversations in rhetorical studies, environmental communication, African and African American Studies, and war and peace studies. His critical essays, commentaries, and book reviews have appeared in *Advances in the History of Rhetoric* (now named *Journal for the History of Rhetoric*), *Argumentation & Advocacy*, *Quarterly Journal of Speech*, *Rhetoric & Public Affairs*, *Women's Studies in Communication*, and in edited volumes.

KC Councilor (Ph.D., University of Wisconsin-Madison) is Assistant Professor in the Communication, Media, and Screen Studies Department at Southern Connecticut State University in New Haven. He is also a cartoonist who makes comics about the experiences of being transgender and transitioning. His work can be seen at www.kccouncilor.com.

Suzy D'Enbeau (Ph.D., Purdue University) is Associate Professor in the School of Communication Studies at Kent State University. Dr. D'Enbeau's research looks at how social change organizations navigate competing goals in a variety of contexts ranging from domestic violence prevention to transnational feminist organizing. Her work has appeared in leading journals such as *Communication Monographs*, *Human Relations*, *Journal of Applied Communication Research*, *Journal of Management Inquiry*, *Qualitative Inquiry*, and *Women's Studies in Communication*.

Phillippa C. Diedrichs (Ph.D., University of Queensland) is Professor of Psychology at the Centre for Appearance Research, University of the West of England, where she leads a team of researchers responsible for co-creating and evaluating body image interventions. She is an applied research psychologist with an international reputation for developing and evaluating evidence-based strategies to improve body image in community, education, corporate, and policy settings. She has co-created body image interventions that have been delivered through industry and community partnerships around the world. This work is underpinned by her research investigating predictors and outcomes of body image concerns among child and adult populations. Dr. Diedrichs has authored over 50 peer-reviewed papers and book chapters, and her research has received international media attention.

Marissa J. Doshi (Ph.D., Texas A&M University) is Associate Professor of Communication at Hope College. Her research draws on feminist perspectives to examine the creative and cultural dimensions of media and technology use. Doshi's secondary research interests include intercultural communication and media representation. Her scholarship has been published in journals such as *Communication Research, Journal of International & Intercultural Communication, International Journal of Communication,* and *Women's Studies in Communication.* At Hope College, she teaches courses on media writing, transnational feminisms, cultural studies, and social media activism.

Kitrina Douglas (Ph.D., University of Bristol) is a video/ethnographer/director, story-teller, musician, and narrative scholar/researcher whose research spans the arts, human-ities, and social sciences. Along with David Carless, she has carried out research for a variety of organizations including Department of Health, Addiction Recovery Agency, Royal British Legion, Women's Sports Foundation, UK Sport, local authority, and NHS Primary Mental Health Care Trusts. She is a professor of Narrative and Performance at the University of West London, a Visiting Professor at the University of Coimbra in Portugal, and has a fractional con-tract at Leeds Beckett University. She is Director of the Boomerang-Project.org.uk, an arts-based network for public engagement and performance of social science research, and co-produces the online qualitative research series "Qualitative Conversations."

Sean Eddington (Ph.D., Purdue University) is Assistant Professor of Communication Studies at Kansas State University. Sean's primary research interests exist at the intersections of organ-izational communication, new media technologies, gender, and organizing. Dr. Eddington's work has examined higher educational contexts including (1) the on-boarding experiences of new engineering faculty and (2) educational cultures that impact the professional formation of engineers (funded by the National Science Foundation). Both projects are published in the *Proceedings of the American Society of Engineering Education (ASEE).* Dr. Eddington's research on new media, gender, and organizing explores the intersections of gender organizing online and communicative construction of resilience in and around Reddit communities. This work has been published in *Social Media + Society* and *Journal of Applied Communication Research.*

Shinsuke Eguchi (Ph.D., Howard University) is Associate Professor in the Department of Communication and Journalism at the University of New Mexico. Their research interests focus on global and transcultural studies, queer of color critique, intersectionality and racialized gender politics, Asian/Pacific/American studies, and performance studies. Their most recent work will appear or has appeared for publication in *Women's Studies in Communication, Cultural Studies ↔ Critical Methodologies, Journal of Homosexuality, China Media Research,* and *QED: A Journal in GLBTQ Worldmaking.* They are a recipient of the 2019 Randy Majors Award—annually rec-ognizing an individual who has made outstanding contributions to lesbian, gay, bisexual, and/or transgender scholarship in communication studies—bestowed by the National Communication Association's Caucus on Gay, Lesbian, Bisexual, and Transgender Concerns.

Carolyn Elerding (Ph.D., Ohio State University) is an independent researcher. Her research on media, culture, and technology focuses on how technological change affects intersecting social differences like gender, race, and class, both onscreen and behind the scenes in produc-tion. Her work has appeared in numerous journals, including *Postmodern Culture, First Monday,* and *Communication, Culture & Critique.* She has also written for *Ms. Magazine.* Currently, she is working on a book titled *Animation's Others: Digitalizing Creative Labor.*

Jasmine Fardouly (Ph.D., UNSW Sydney) is a Research Fellow at the Centre for Emotional Health, Macquarie University, Australia. Her research interests include social influences on young people's mental and physical health. She is particularly interested in understanding how social media use may impact users' body image and is passionate about finding ways to reduce any negative impact of social media use for young people. She has won several awards for her contribution to the field and has published many highly cited peer-reviewed articles in leading psychology journals. Her research has gained substantial interest from global media, government, and the general public.

Deanna L. Fassett (Ph.D., Southern Illinois University, Carbondale) is Assistant Vice Provost for Faculty Development and Professor of Communication Studies at San José State University, which is located on land that is the traditional home of the Puichon Ohlone-speaking people and the present-day Muwekma Ohlone Tribe. She is the author, along with the late John T. Warren, of *Critical Communication Pedagogy, Communication: A Critical/Cultural Introduction* and *Coordinating the Communication Course: A Guidebook.* Her scholarship strives to create spaces for educators to engage in communication and/in instruction as inclusive, intentional, and attentive to structural inequality and justice. Her pronouns are she/her.

Dustin Bradley Goltz (Ph.D., Arizona State University) is a Professor of Performance Studies and a Vincent de Paul Professor in the College of Communication at DePaul University. He is the author of *Queer Temporalities in Gay Male Representation* (Routledge 2009), *Comic Performativities: Identity, Internet Outrage and the Aesthetics of Communication* (Routledge 2017), and co-author and editor of *Queer Praxis: Questions for LGBTQ Worldmaking* (Peter Lang, 2014).

Marnel Niles Goins (Ph.D., Howard University) is Interim Dean of the College of Sciences and Humanities and Professor of Communication at Marymount University in Arlington, Virginia. She teaches courses in Small Group Communication and Organizational Communication and has a special interest in gender and racial dynamics in organizational settings, particularly relating to organizational socialization and leadership. Dr. Niles Goins has written and participated in numerous publications. These include *Still Searching for Our Mothers' Gardens: Experiences of New, Tenure-Track Women of Color in 'Minority' Institutions,* a book about the experiences of women of color in the academy, as well as several articles and book chapters about Black women in organizational and group settings. She is on the editorial board for *Women's Studies in Communication.* At the time of this publication, she is President of the Western States Communication Association and the first Black president of the Association. She is immediate Past President of the Organization for Research on Women and Communication.

Nickesia S. Gordon (Ph.D., Howard University) is Associate Professor in the School of Communication at the Rochester Institute of Technology. Her research focuses on communication and gender, race, and nationality, as well as communication for social change with emphasis on the Caribbean region. Additionally, her research agenda includes examining how the communication curriculum in higher education can engage experiential learning practices and help foster civic/community engagement among college students. Dr. Gordon also holds a master's degree in English Literature from Clark University, Massachusetts.

Michele L. Hammers (Ph.D., Arizona State University) is a Professor of Communication Studies and Associate Dean of the College of Communication and Fine Arts at Loyola Marymount University. Her primary research interests are in contemporary rhetorical theory and criticism, and feminist media studies. She is also interested in ethnographic and autoethnographic explorations of identity, focusing on identity at the intersections of the public and private. Her work has appeared in regional, national, and international journals.

Anne Harris (Ph.D., Victoria University, Melbourne, Australia) is Associate Professor, Principal Research Fellow (RMIT University), and Australian Research Council Future Fellow at RMIT University in Melbourne, Australia. Dr. Harris writes and researches in the areas of critical autoethnography, gender, education, creativity, and creative methods. She is the director of Creative Agency research lab creativeresearchhub.com.

Kate Lockwood Harris (Ph.D., University of Colorado, Boulder) is Associate Professor of Organizational Communication at the University of Minnesota, Twin Cities. Her scholarship, which focuses on the relationship between communication and violence, has won accolades from regional, national, and international associations. Dr. Harris assumes that violence sustains inequity on the basis of difference, so she pays close attention to the intersections of gender, race, and sexuality. She is the author of *Beyond the Rapist: Title IX and Sexual Violence on US Campuses* (2019) and is at work on a second book about disorganizing systemic racism in higher education.

Serena Hawkins (B.A., Indiana University-Purdue University Indianapolis) is a Ph.D. student working toward her doctorate in American Politics at the University of North Carolina Chapel Hill. Her research interests include political psychology, electoral behavior, political biology, and survey research and methodology. In 2019 she was awarded a research grant to design a survey to help research the secondary effects of direct democracy and internal political efficacy.

Michelle A. Holling (Ph.D., Arizona State University) is Professor of Communication in the Department of Communication at California State University San Marcos. She identifies as a Chicana and womyn of color, which inform her roles as teacher-scholar-activist inside and outside of the classroom. Her commitment to social justice is apparent in her scholarly interests that examine racial and gendered violence; vernacular voices with specific emphasis on Chican@-Latin@ rhetoric; and academe as a site of inquiry about microaggressions, intersectionality, identity, and agency. She is co-editor, with Dreama Moon, of *Race(ing) Intercultural Communication: Racial Logics in a Colorblind Era* and co-editor, with Bernadette Calafell, of *Latina/o Discourse in Vernacular Spaces: Somos de Una Voz?* She has also published in an array of peer-reviewed journals.

Stacy Holman Jones (Ph.D., University of Texas at Austin) is Professor and Director of the Centre for Theatre and Performance at Monash University, Australia. Her research focuses broadly on how performance operates as a socially, culturally, and politically resistive and transformative activity. She specializes in critical qualitative methods, particularly critical autoethnography, performance, and feminist theory.

Kristina Horn Sheeler (Ph.D., Indiana University) is Professor of Communication Studies and the Executive Associate Dean of the Honors College at Indiana University-Purdue University Indianapolis. Her research is in the area of gender and political communication, studying the ways that political candidate identity is contested and constructed in popular media, political discourse, journalism, and punditry. Her book, *Woman President: Confronting Postfeminist Political Culture* (2013), co-authored with Karrin Vasby Anderson, won the Organization for the Study of Communication, Language, and Gender's outstanding book award in 2014 and the National Communication Association's 2014 James A. Winans–Herbert A. Wichelns Memorial Award for Distinguished Scholarship in Rhetoric and Public Address. She has published numerous essays and reviews in scholarly journals.

Yuhan Huang (Ph.D., Purdue University) is an Assistant Professor in Modern Languages and Cultures at Rochester Institute of Technology. Her graduate research focused on film studies, women's studies, and Chinese studies. Her dissertation "Remembrance and Rumination: The 1.5 Generation of the Chinese Cultural Revolution" examines the poetics and politics of reviving and representing childhood and adolescent memories of the Cultural Revolution by writers, filmmakers, and painters. Yuhan is the co-editor of *Mo Yan in Context* (2014), and a recipient of Purdue Research Foundation dissertation grant (2016–2017) and Lynn Fellowship (2013–2014).

Amber Johnson (Ph.D., Pennsylvania State University) is an award-winning Associate Professor of Communication and Social Justice at Saint Louis University and the creator of The Justice Fleet™, a mobile social justice museum that fosters healing through art, dialogue, and play. Dr. Johnson is a scholar/artist/activist whose research and activism focus on narratives of identity, protest, and social justice in digital media, popular media, and everyday lived experiences. As a polymath, their mixed-media artistry involves working with metals, recycled and reclaimed goods, photography, poetry, percussion, and paint to interrogate systems of oppression. Dr. Johnson's forthcoming book, *A Great Inheritance,* uses memoir and self-help education to highlight healthy forms of support for trans and nonbinary folks alongside visionary science fiction to speculate on gender futurity as a site of liberation.

Clare Johnson (Ph.D., Lancaster University) is Associate Professor in Art and Design at the University of the West of England, Bristol, UK, where she leads the Visual Culture Research Group. Dr. Johnson's primary research interests are feminist art history, visual culture, and maternal representations. She is the author of *Femininity, Time and Feminist Art* (2013) and has published on the work of Tracey Emin, Vanessa Beecroft, Eleanor Antin, Yoko Ono, Marina Abramović, Eti Wade, and Lena Simic, among others. She teaches in both undergraduate and postgraduate/research programs in visual culture.

Vani Kakar (M.Phil., National Institute of Educational Planning and Administration; M.S., University of Delhi) is a Ph.D. candidate at the Centre for Emotional Health, Department of Psychology, Macquarie University. Her research investigates beauty norms and body image and how it may be influenced by social media use in hundreds of adolescent girls living in diverse cultures around the world like Australia, India, Iran, and China. Her cross-cultural research focuses on body image and social media and is the perfect culmination of the confluence of interests she has developed over the last decade.

Elena Elías Krell (Ph.D., Northwestern University) is a songwriter and scholar based in the Hudson Valley of New York, 90 miles north of New York City. Their music project, Hen in the Foxhouse, draws from such wide-ranging influences as samba, Latin pop, dream pop, gypsy jazz, blues, and opera. Dr. Krell's doctoral work in Performance Studies was advised by E. Patrick Johnson. Dr. Krell is currently Assistant Professor of Women Studies at Vassar College in the Hudson Valley of New York. In addition to founding the Queer Studies minor, they teach courses in Women, Africana, Latino & Latin American, Music, and Media Studies. Their research centers on queer and trans of color critique, voice, and performance.

Benny LeMaster (Ph.D., Southern Illinois University, Carbondale) is Assistant Professor of Critical/Cultural Communication and Performance Studies at Arizona State University, which is located on the ancestral lands of the Tohono O'odham, Akimel O'otham, and Piipaash Peoples. Their scholarship engages the intersectional constitution of cultural difference with particular focus on queer and trans of color life, art, and embodiment. Their research can be found in academic anthologies as well as in peer-reviewed academic journals including, for example, *Women's*

Studies in Communication, Text and Performance Quarterly, Departures in Critical Qualitative Research, and *Qualitative Inquiry.* In addition, they joined Amber Johnson in co-editing the anthology *Gender Futurity, Intersectional Autoethnography: Embodied Theorizing from the Margins.* Their pronouns are they/them.

Kostia Lennes (M.S., University of Leuven, University of Lille) is a Ph.D. candidate in Anthropology (Université Libre de Bruxelles, Belgium) and Sociology (Université de Paris, France), and a fellow of the National Fund for Scientific Research (F.R.S.-FNRS, Belgium). He has previously conducted fieldwork in Mexico on queer migrants returning from the United States. His doctoral research explores the trajectories of men selling sex and intimacy in Paris. His main research interests include same-sex sexualities, masculinities, intimacies, sex work, borderlands, mobilities, online ethnography, and animal studies.

Nina M. Lozano (Ph.D., University of North Carolina at Chapel Hill) is a political consultant and Associate Professor of Communication Studies at Loyola Marymount University. Her areas of expertise include rhetoric, social movements, gender, and politics. Dr. Lozano earned her doctorate from the University of North Carolina at Chapel Hill. She is a former Carnegie Fellow and is a regular contributor to the Huffington Post and other political websites. Lozano is the author of *Not One More! Feminicidio on the Border* (2019), as part of the special series: "New Directions in Rhetoric and Materiality."

Joan Faber McAlister (Ph.D., University of Iowa) is Associate Professor of Rhetoric, Media, & Social Change at Drake University in Des Moines, Iowa. Her research focuses on visual, spatial, and material rhetorics in public culture as well as aesthetics and critical theory. The status, function, and material character of home, dwelling, and belonging are persistent themes running throughout her work. Dr. McAlister's articles have won multiple awards and she is also the recipient of two national awards for mentoring women and faculty in the discipline. She is a former editor of *Women's Studies in Communication* and co-editor of a special issue entitled "Placing New Materialities and Precarious Mobilities." Her work has appeared in numerous journals such as *Howard Journal of Communications, Critical Studies in Media Communication, Liminalities: A Journal of Performance Studies, Southern Journal of Communication, Communication and Critical/Cultural Studies, Rhetoric Society Quarterly,* and the *Quarterly Journal of Speech.* Dr. McAlister has also published chapters in edited collections including *Rhetoric, Materiality, and Politics; From Agamben to Žižek: Contemporary Critical Theorists; Communicating the City: Meanings, Practices, Interactions; DIY Utopias; Identity Politics and the Power of Representation: Adventures in Shondaland;* and *City Places, Country Spaces: Rhetorical Explorations of the Urban/Rural Divide.* Most recently, she authored the "Gender in Rhetorical Theory" and "Space in Rhetorical Theory" entries in the *Oxford Research Encyclopedia of Communication.*

Kimberly R. Moffitt (Ph.D., Howard University) is Director and Professor in the Language, Literacy & Culture Ph.D. program and affiliate professor in the Department of Africana Studies at University of Maryland Baltimore County (UMBC). Her research focuses on mediated representations of marginalized groups and the politicized nature of Black hair and the body. She has published five co-edited volumes, including *Gladiators in Suits: Race, Gender, and the Politics of Representation in Scandal* (2019), and *Blackberries and Redbones: Critical Articulations of Black Hair and Body Politics in Africana Communities* (2010), and published in academic journals and edited volumes. She is the co-creator of the blackhairsyllabus.com project and cohost of *the color treatment* podcast, exploring the effects of colorism within Africana communities.

Kesha Morant Williams (Ph.D., Howard University) is Associate Professor of Communication Arts & Sciences at Penn State University, Berks. She teaches a variety of courses including research methods, interpersonal communication, and health communication. Her research interest focuses on interpersonal communication, health communication, and popular media examined through a cultural lens. Her research has been published in peer review journals, book chapters, and the popular press. She is the author of the edited book *Reifying Women's Experiences with Invisible Illness: Illusions, Delusions, Reality* (Lexington Books, 2017).

Ashley Morgan (Ph.D., University of Sheffield) is a senior lecturer in the School of Art and Design, Cardiff Metropolitan University. Her current research interests are about masculinity, sex, and representations of the male body, especially but not limited to representations of masculinity in popular culture. Ashley has published on male geek identity, sexual asceticism as a viable form of male sexual behavior, and the presence of mediated toxic masculinity. She is especially interested in the intersections between masculine identity and clothing, and has published on men in skirts and the relationship between hegemonic masculinity and men's suits. Her paper on hybrid masculinities and men's clothing will be published in the forthcoming Intellect Journal, *Critical Studies in Men's Fashion*. She is currently working on book chapters about representation of female detectives in popular culture and digital discourses of masculinities.

J. Nautiyal (Ph.D., University of Texas at Austin) is a lecturer in the Communication Studies Department at Gonzaga University. She is interested in the study of quotidian rhetorics at the new materialist intersections of embodiment, aesthetics, affect, and queer theory. Dr. Nautiyal's work has been featured in journals such as *Women's Studies in Communication, Women & Language*, and *International Journal of Listening*, among others.

Mackenna Neal (M.A., University of Nevada, Reno) is Instructor of Communication Studies at Great Basin College. She is currently working on developing diversity initiatives and curriculum to help promote inclusivity in higher education. Throughout her academic career, her research has focused on gender, advocacy, and social justice. She hopes to pursue her Ph.D. and contribute meaningful research to help empower women who have faced oppression and marginalization.

Tracey Owens Patton (Ph.D., University of Utah) is Professor of Communication in the Department of Communication and Journalism, Adjunct Professor in African American and Diaspora Studies in the School of Culture, Gender, and Social Justice, and affiliate faculty in the Creative Writing MFA Program in the Department of Visual and Literary Arts at the University of Wyoming. Her areas of specialization are critical cultural communication, critical media studies, rhetorical studies, and transnational studies. She has authored a number of academic articles on topics involving the interdependence between race, gender, and power and how these issues interrelate culturally and rhetorically in education, media, memory, myth, and speeches. She published a co-authored book titled *Gender, Whiteness, and Power in Rodeo: Breaking Away from the Ties of Sexism and Racism* (2012) and is working on a second book involving race, memory, rejection, and World War II.

Craig O. Rich (Ph.D., University of Utah) is Associate Professor and Chair of the Department of Communication Studies at Loyola Marymount University, Los Angeles. His research explores the intersections of organization, communication, and gender and sexual diversity. His work has appeared in *TAMARA: The Journal of Critical Organization Inquiry, Communication Monographs,* and *Gender, Work, and Organization.*

Sage E. Russo (M.A., San Francisco State University; M.A., New York University) is a lecturer at multiple Bay Area academic institutions teaching critical communication studies. Their research and creative work, which has been published in multiple venues, weaves together aesthetic performance, (trans)gender studies, and queer theory.

Tammy Sanders Henderson (Ph.D., University of Maryland College Park) is lecturer in Africana Studies and affiliate faculty in Gender, Women's and Sexuality Studies at the University of Maryland Baltimore County. Dr. Henderson also serves on the faculty of the First Year Seminar Program teaching courses on race, maternity, and public policy. Her research focuses on African American women's history, motherhood, public policy, Black Feminist Thought, and mediated images of Black womanhood in popular culture. Her peer-reviewed publications including *Answering the Call: A Continued Response to Sprague's Call to Action to Instructional Communications Scholars and Beyond* and several conference presentations reflect her interdisciplinarity and vast expertise in many research areas. She often provides commentary on area radio programs and podcasts.

Clifton Scott (Ph.D., Arizona State University) is Professor of Communication Studies and Organizational Science at the University of North Carolina, Charlotte. His work examines the relationships among formal and informal communication processes, high reliability organizing, identity, and occupational safety. Much of his research focuses on how identity work among first responders enables and constrains their capacity to learn from emergency incidents and continuously improve the safety and reliability of their operations.

Rivka Neriya-Ben Shahar (Ph.D., The Hebrew University of Jerusalem) is a senior lecturer at Sapir Academic College in Sderot, Israel, teaching courses in research methods, communication, religion, and gender. Her dissertation at the Hebrew University of Jerusalem was entitled "Ultra-Orthodox Women and Mass Media in Israel: Exposure Patterns and Reading Strategies." She was a Fulbright post-doctoral fellow and a Scholar in Residence at the Hadassah-Brandeis Institute at Brandeis University and conducted a study that analyzes women's cultural-religious praxes. Dr. Neriya Ben-Shahar researches mass media from the perspectives of religion and gender. Her most recent research project addresses the tension between religious values and new technologies among Old Order Amish women and Jewish Ultra-Orthodox women.

Claire Sisco King (Ph.D., Indiana University) is Associate Professor of Communication Studies at Vanderbilt University, where she also teaches in the Cinema and Media Arts Program. A critical cultural scholar of media and visual rhetoric with a focus on the study of gender, she is the current Editor of *Women's Studies in Communication*. Dr. Sisco King is the author of *Washed in Blood: Male Sacrifice, Trauma, and the Cinema* (2011) and has been published in such journals as *Communication and Critical/Cultural Studies, Critical Studies in Media Communication, Feminist Media Studies, Quarterly Journal of Speech, Text and Performance Quarterly*, and *Women's Studies in Communication*. She has recently completed a manuscript about celebrity culture, metonymy, and theories of networks, and is pursuing new projects about the relationships between various modes of art/making, trauma, and changing notions of publicity.

Nancy Small (Ph.D., Texas Tech University) is Assistant Professor and Director of First Year Writing at the University of Wyoming. Prior to her current position, she spent six years on the liberal arts faculty of an international branch campus, Texas A&M at Qatar. Her research revolves around everyday storytelling and the rhetorical lifeworlds it generates, particularly with a critical eye toward gender and whiteness. Her current projects engage narrative methods to focus on feminist tactics in transnational spaces, ethics and transcultural research, and rhetorical impacts of public memory. Her work has been published in edited collections on globalization and education, as well as in journals of feminist scholarship and in journals in technical communication.

Mel Stanfill (Ph.D., University of Illinois, Urbana-Champaign) is Assistant in the College of Arts and Humanities at the University of Central Florida. Stanfill's recent publications include *Exploiting Fandom: How the Media Industry Seeks to Manipulate Fans* (2019) and *A Portrait of the Auteur as Fanboy: The Construction of Authorship in Transmedia Franchises* (with Anastasia Salter, fall 2020), as well as articles in *Porn Studies, Television & New Media, Social & Legal Studies, Journal of Film and Video*, and *Social Identities*. Recent presentations have included conferences such as Race and Intellectual Property, Association of Internet Researchers, Society for Cinema and Media Studies, and American Studies Association.

Aly Stetyick (B.A., University of Maryland Baltimore County) is a doctoral student in the Organizational Science program at the University of North Carolina, Charlotte. Her research focuses on the interplay between identity, perception, and organizational (dys)function. She is especially interested in high-risk and blue-collar work.

Sunera Thobani (Ph.D., Simon Fraser University) is Professor in the Department of Asian Studies at the University of British Columbia. Her scholarship is located at the intersection of the Social Sciences and Humanities and focuses on postcolonial, critical race, and transnational feminist theory; critical social theory, intersectionality, and social movements; colonialism, indigeneity, and racial violence; globalization, citizenship, and migration; South Asian women's, gender, and sexuality studies; representations of Islam and Muslims in South Asian and Western media; and Muslim Women, Islamophobia and the war on terror. The geographical areas of her research include Canada, the United States, South Asia, and the South Asian diaspora.

Tia C. M. Tyree (Ph.D., Howard University) is a Professor at Howard University in the Department of Strategic, Legal and Management Communications. She teaches graduate and undergraduate courses. Her research interests include hip-hop, rap, reality television, film, and social media, as well as African-American and female representations in mass media. She has several published book chapters and peer-reviewed articles in journals such as *Women and Language; Howard Journal of Communications; Journalism: Theory, Practice & Criticism; International Journal of Emergency Management* and *Journal of Black Studies*. She is the author of *The Interesting and Incredibly Long History of American Public Relations* and co-editor of *HBCU Experience—The Book; Social Media: Pedagogy and Practice,* as well as *Social Media: Culture and Identity*. She is also cofounder of the Social Media Technology Conference and Workshop.

Kallia O. Wright (Ph.D., Ohio University) is Associate Professor of Communication and Rhetorical Studies at Illinois College. She teaches a variety of courses including intercultural communication and health communication. She has been involved in several experiential-learning initiatives that focus on intercultural interactions. Originally from Jamaica, she is a qualitative researcher who examines narratives about chronic illness experiences and the impact of culture on those experiences. She has published work in *Health Communication*, communication encyclopedias, and chapters of edited books.

Eline van den Bossche (B.A., Indiana University-Purdue University) was born in Vichten, Luxembourg, and pursued her primary and secondary education in Luxembourg City, graduating in June of 2016. She moved to the United States at 16 to pursue her undergraduate degree. She is pursuing a J.D. in international and comparative law at the Maurer School of Law at Indiana University Bloomington.

Aimée Vega Montiel (Ph.D., Autonomous University of Barcelona) is a researcher at the National Autonomous University of Mexico, Center of Interdisciplinary Research in Sciences and Humanities. She is a specialist in feminism and communication studies and is particularly interested in women's human rights, media, and communication. She has published peer-reviewed articles and 8 single-authored/co-edited books. She is Vice-President of the International Association for Media and Communication Research (IAMCR). She has been President of the Mexican Association of Communication Researchers (AMIC) and Director of the Mexican Association of the Right to Information (AMEDI). She is currently preparing two books as author: *Media and Violence of Gender* and *Access and Participation of Women in Media Industries*.

Astrid M. Villamil (Ph.D., University of Kansas) is Assistant Professor in the Department of Communication at the University of Missouri-Columbia. Dr. Villamil's research focuses on minoritized voices in organizational contexts and the way structural and organizational policies affect the well-being of individuals from underrepresented groups. In addition, she has worked with a grant-funded leadership program that serves women from the Middle East, North Africa, and Central and South Asia regions.

Gust A. Yep (Ph.D., University of Southern California) is Professor of Communication Studies, Graduate Faculty of Sexuality Studies, and Faculty in the Ed. D. Program in Educational Leadership at San Francisco State University. His research examines communication at the intersections of culture, race, class, gender, sexuality, and nation, with a focus on sexual, gender, and ethnic minority communities. In addition to three books and a monograph, he has authored more than 100 articles in (inter)disciplinary journals and anthologies. He is recipient of numerous academic and community awards including the 2011 San Francisco State University Distinguished Faculty Award for Professional Achievement (Researcher of the Year), the 2015 Association for Education in Journalism and Mass Communication (AEJMC) Leroy F. Aarons Award for significant contributions to LGBT media education and research, and the 2017 National Communication Association (NCA) Outstanding Mentor in Master's Education Award.

Jason Zingsheim (Ph.D., Arizona State University) is a Professor of Communication and Chair of the Division of Arts and Letters in the College of Arts and Sciences at Governors State University. He is a co-author and editor of *Queer Praxis: Questions for LGBTQ Worldmaking* (2015). His scholarship has appeared in the *Journal of International and Intercultural Communication, Journal of Lesbian Studies,* and *Text and Performance Quarterly*, among other outlets.

INTRODUCTION

In the now infamous 2016 recorded interview of Donald Trump with *Access Hollywood* reporter Billy Bush, the world heard a United States presidential candidate boast about grabbing women "by the pussy."[1] In the days following this media event, more stories of Trump's disregard and disrespect for women in both the private and public domain emerged and his historical and ongoing media attacks on women critics gained more attention.[2] Scholars have connected these events and many others to resurgent rhetorics of bias and hate that continue to menace the lives of those who live outside the sheltered domain of social privilege and gender normativity. There are severe consequences of such an enacted rhetoric that also implicates, and maybe authorizes, white supremacist cis-hetero-masculinity. Despite declarations that feminist and LGBTQ + rights movements had achieved their primary goals by the second decade of the 21st century, the social scope and personal harms of misogyny and bigotry were evident in the disclosures of the "# MeToo" movement, the horrors of the Pulse Nightclub shooting in Orlando, Florida, and the ongoing murders of women, transgender people, and sexual minorities across the globe.

For these reasons and more, further intentional discussions about gender and communication are imperative. This handbook is not a response to any one person or event, but to persistent and insidious problems of inequity, exploitation, and abuse that are linked to hierarchies of gender and sexuality. It provides clear apprehensions of the problems as well as pathways to solutions. While the concerns the authors raise have been discussed elsewhere, their assembled studies foreground emergent topics and offer fresh approaches that produce unique and incisive analyses of urgent social issues. Overall, the researchers featured in this collection elucidate the many ways that established attitudes fomenting discrimination, sexism, racism, and transphobia are linked to other complex social problems with concrete impacts on international, intercultural, and interpersonal encounters shaping everyday life.

Harassment and hatred directed toward women and marginalized subjects demand a reconnoitering of the role of communication in forming, enforcing, and resisting social and cultural notions of difference. Any exploration of "gender and communication" brings together two powerful constructions that inform human engagement. *Gender* is a social construction of identity attributed by culture and relative to determinations of sexed and embodied possibilities. Gender, often assigned at birth via reductive and culturally established categories, becomes an ascription. It carries a set of expectations and anticipations linked to performative binaries of maleness/femaleness or masculinity/femininity that are entangled in asymmetrical vectors of power, worth, relationality, compensation, freedoms, and opportunities. Variously described, *communication* references the socially constructed means through which people, and increasingly recognized non-human subjects, share and transmit information of value that forms the collective milieu of culture and community. Communication influences not only interpersonal engagement but also institutions and knowledge systems molding the nature of what is perceived and what is accepted in any given society.

Many of the chapters in this book concern sexuality as well as gender, and their specific subjects are described and defined in different ways. Varied constructions such as LGBTIQCAPGNGFNBA+ (Lesbian, Gay, Bisexual, Transgender, Intersex, Queer, Curious, Asexual, Pansexual, Gender-non-conforming, Gender-Fluid, Non-binary, and Androgynous) are proliferating as part of an ongoing project to recognize and name identities and experiences once cloistered in social derision. Some object to these efforts to label these particularities and pluralities, arguing that increasing categories of being has the potential to foreground some at the expense of others. Efforts to include emergent identities can provoke concerns over diluted attention to subjects historically *and* currently marginalized and disenfranchised. Expanding attention to recently recognized groups can have the appearance of making some lives matter more and others matter less. We want to acknowledge this fear, but we also affirm that naming is an important tool of political activism that is critical to democratization and liberation. Social media activists employing #sayhername to mark and remember individual lives lost vividly illustrate that naming plays a vital role in social justice movements. It is a clarion call for all to be collectively involved in igniting an activism of radical inclusiveness, an activism designed to change the social constructions that uphold the politics of privilege. We hope this edited collection will contribute to efforts to call attention to diverse lived experiences, as this is essential to the change that we must not only *see* in the world but the change we must also *be* in the world. Thus, varying constructions and acronyms to highlight differences that matter will be used throughout the book.

When conjoined, "gender and communication" create a powerful alchemy that encompasses interpersonal relationality, systems of judgment, modes of becoming, performative contexts, means of identity, degrees of mobility in place and space, types of power over one's body, and sources of self-determination. Hence, invoking new insights and strategies related to "gender and communication" has the potential to break the cycle of culturation and enculturation into which we have all been inducted. To engage in deep critical conversation about interventionist strategies linked to gender and communication is to invest in a future in which existing and emerging gendered subjects might not only be able to articulate (communicate) their own realities but also promote new possibilities for others. That is, gendered subjects might be able to live in a future free from some harms of the existing politics of gender that regulate every aspect of our daily lives and influence the seats of national and world government, even in bastions of intellectual inquiry and academic freedom.

Communication scholars have long noted the roles that gender dynamics play in the inner sanctums of higher education and in the systems of evaluation and publication to which educators and researchers in the field must submit. In a 1994 *Quarterly Journal of Speech* article entitled "Disciplining the Feminine," Carole Blair, Julie R. Brown, and Leslie A. Baxter (in)famously exposed the gender politics of peer reviews that censor critical communication studies scholarship.[3] After detailing their own punishing journeys through the publishing process, the authors concluded that the conditions it created were inimical to feminist research:

Every ideological apparatus wielded by the initial referees would silence virtually any feminist statement. The silences would not be limited to those the referees consider extremist, Marxist, lesbian, and so forth—those they clearly do wish to squelch. All of feminism would be quarantined at the disciplinary border and ultimately deported. Such an ideology cannot be allowed to govern an academic field that so proudly espouses pluralism, diversity, and communication. Disagreement is to be kept in play in the dialogue of such a field, not silenced or kept cloistered behind the secretive curtain of blind peer reviewing.[4]

The ideals embraced in this passage are as relevant and urgent as ever for the scholarship assembled in this *Routledge Handbook of Gender and Communication*. The research it contains builds on these aims while refusing to be constrained by convention, even sparking disagreements over how the topics and terms deployed in early feminist scholarship (like the example above) may inadvertently enact exclusion, simplification, appropriation, or ableism across a range of gendered subjects.

This tension between the values and practices driving communication studies scholarship feeds the larger project to which the diverse research in this collection contributes, by offering fresh insights and questioning assumptions within and beyond the discipline. The authors in this volume evince a shared commitment to probing neglected topics and challenging existing perceptions through unique studies that nuance accounts of genders and sexualities and deepen understandings of how different bodies and subjects encounter gendered communication. To support this larger mission, the editors have invited and assembled important new research on the complex relationships between gender and communication. Featuring a broad variety of entries written by leading and upcoming scholars, this edited volume uses diverse theoretical frameworks to speak to recent concerns regarding changing gender roles, representations, and resources in communication studies. Established research and new perspectives address vital themes in this wide-ranging text including the shifting politics of gender, evolving trends in gendered media, and varying roles of gender in daily life.

This collection was designed to be useful to faculty, graduate students (master's and doctoral), advanced undergraduate students, and university libraries who are looking for a robust overview of current topics in gender and communication. In addition to providing a breadth of coverage in the discipline, the handbook offers international and intercultural perspectives as well as developing work on new technological, political, and economic contexts for gender and communication. The scope of themes covered makes it a pragmatic pedagogical resource. We have included chapters written by scholars experienced in or just entering communication studies, each of whom specialize in gender research, making this collection suitable for use as a textbook appropriate for a graduate seminar or advanced gender communication course.

This impressive array of important studies has been grouped into six sections: "Gendered lives and identities," "Visualizing gender," "The politics of gender," "Gendered contexts and strategies," "Gendered violence and communication," and "Gender advocacy in action." Each section has an introduction identifying salient themes, previewing individual studies, and drawing attention to larger conversations emerging from the research included. While these section introductions provide more detailed information, a general map of the content covered in the main segments of the volume follows below.

The research in Part I, "Gendered lives and identities," addresses the confrontation of gender expectation and embodied gender performativity in disparate and desperate situations. The chapters chart shifting effects and affects on both the performance of gender and the material conditions of bodies engaged in social negotiations and interpretations of social norms. Contributors in this section consider crucial sites of gendered practices and performances as well as issues of identification and disidentification, while also reflecting on recent debates regarding inclusion in feminist scholarship and challenging key assumptions of gender studies in diverse geographic, ethnic, and cultural contexts. The chapters tease at the intersections of communication, gender, and sexuality in the social domains of the self in society and build new arguments around the notion of gendered lives in locational contexts.

In Part II, "Visualizing gender," contributors focus on visual culture with particular attention to recent media studies scholarship and the impact of technologies on social relations. New research on global networks and transnational flows of influence join established literatures on media images, tropes, and consumption to engage the most salient topics on gender and communication in and through webs of visibility/invisibility, images, and ideologies. Authors demonstrate careful applications of and enhancements to central concepts mined from rich scholarly literatures on "controlling images," "the gaze," "media literacy," and "counterpublics" to consider complex processes for stigmatization, shame, discipline, or erasure via visual media. The studies in this section show that a wide variety of media forms offer new means of reinforcing or reinventing the scripts surrounding femininities, masculinities, and sexualities that have long concerned those producing feminist, queer, and critical-cultural analyses of communication.

The chapters in Part III, "The politics of gender," argue that gendered political progress is possible, in spite of current and past political tragedies. The section features profiles of recent research on the role of gender norms in political culture, as well as new legal and legislative battlegrounds over LGBTQ+ human rights and gender equality. In addition to examining political institutions and processes relating to gender and communication, this section examines broader political contexts for gendered relations between citizens and nations, gendered subjects, and collective bodies. A diverse set of transdisciplinary and transnational essays chart new paths in recent research and politics that avoid the gaps and assumptions shaping previous accounts of gender and communication. These chapters illustrate the complexity of gender in political systems worldwide, and also emphasize the importance of activism in reducing gendered political oppressions.

In Part IV, "Gendered contexts and strategies," contributors examine how gender and sexuality affect and are affected by their changing locations and situations. These contexts include male-dominated organizations, sports, and environments where health decisions are made, such as hospitals and hospice-care facilities. Homophobia, sexism, and racism are prevalent and powerful forces in these places. Further, narratives of what represents the "ideal" (worker, player, partner, patient, character, etc.) are framed around maleness, heterosexuality, youth, and whiteness. Using a variety of theories and frameworks, the authors discuss communication strategies used to manage gender imbalance in these settings and provide suggestions for coping with, and often resisting, dominant power structures.

Part V, "Gendered violence and communication," addresses recent research into communication implicated in and responding to assault, harassment, and abuse. The authors overview key developments in analyses of how media representations, rhetorics of masculinity, and interpersonal interactions shape sexual violence in particular cultural and political contexts. A variety of methods and theories are used to generate new approaches to understanding the role of communication in connecting violence to gender in harmful or hopeful ways. Through studies showcasing incisive interpretations of new discourses of gendered violence in film, journalism, social media, art, and television, the authors meticulously support original arguments about specific relationships between symbolic and material factors contributing to this social problem. These assembled studies also profile specific strategies employed to resist and reduce this dangerous dimension of social life.

The chapters in Part VI, "Gender advocacy in action," the final section of the handbook, link academic research on communication and gender to activism and advocacy beyond the academy. Entries explore practical pathways for scholars' critical insights to shape community outreach, coalitional movements, and pragmatic institutional intervention. Moreover, contributors in this closing segment challenge future researchers and current readers to reflect on the gendered dynamics and communicative ethics of their own practices in pedagogical, interpersonal, and communal encounters. The notion of "gender advocacy in action" that frames this section is about varying methods, modes, and modalities of overt partisan politics that promote the free expression of gendered identities toward a greater sense of social justice and thus, affirm the innate dignity of all human beings in and through their complex gendered identities.

In this volume, it was important to address urgent themes in communication and gender by highlighting research from international as well as U.S. scholars. However, the international scholarship in this book represents commitments and efforts of the editors that were difficult to maintain to our own satisfaction. The authors in this edited volume do indeed hail from many nations (including Australia, Canada, China, England, Georgia, India, Israel, Jamaica, Luxembourg, Mexico, the Republic of South Africa, and Wales), but many contributors are currently located in the United States. The absence of valuable voices and views initially included is due to complexities and disparities in academic labor that we want to mark here, as it speaks to the challenge of representation in communication scholarship.

Many researchers in underrepresented regions and communities do not have access to the necessary support and resources to publish their work in all chosen outlets, nor the capacity to adjust their other responsibilities to meet publisher demands and deadlines. For this reason, some invited scholars were unable to complete planned contributions to this collaborative project. It is our hope that a revised version of this handbook will include more studies conducted in East Asia, Africa, and a wide variety of places in the Global South. We are living in an increasingly globalized world, as the crisis of COVID-19—which intervened in the production of this volume—has painfully reminded us. However, it is also imperative to recognize multiple types of difference and diversity across a broad spectrum of the academy. These include the disparity in access to resources for faculty of color and women in private and public universities in the United States; the critical but sometimes cloistered voices of faculty at community colleges, two-year institutions, Historical Black Colleges and Universities (HBCUs), and Hispanic Serving Institutions (HSIs); and the politics of regionality within universities in the South, East Coast, Midwest, and West Coast. There are important differences in degrees of job security, compensation, and institutional support for marginalized scholars, and many workplace environments are hostile places for women and minorities, as well as lesbian, gay, bisexual, transgender, nonbinary, Two-Spirit, gender queer, queer, questioning, intersex, asexual, and ally (+) academics.

Although our featured authors are dispersed across ranks or stages of their careers, we are cognizant of the representational politics of the racial, ethnic, and gender balance in this volume. We must also acknowledge the ever-increasing gender disparities in research productivity due to the double duty and emotional labor that women encounter at disproportionately higher rates when local, national, and world demands dictate working from home (e.g., childcare, home schooling, and caring for sick family members). We hope that as critical discussions about gender and communication continue, increased attention is brought to inequities regarding who is (or is not) sitting at the table, or who is included in volumes of this nature.

It is also important to note that the researchers who are contributors to this collection are working with varying levels and types of institutional support and appreciation for their labor. As the editors who worked with these remarkable and dedicated writers, we want to acknowledge the many obvious and subtle variables—from our own locations in U.S. universities to the selection of a citation style—that disadvantaged some authors and advantaged others. We continue to hope that academic publishing becomes more inclusive and better compensated in the future, so that collections like this one can represent the outstanding research about gender and communication being generated in so many places.

We anticipate that our readers will find these studies, and the fascinating contrasts and comparisons they create between them, to be useful as starting points for their own projects. We look forward to future collections that showcase more of this work as it emerges. Like you, we are avid readers and students of research in gender and communication studies. We are grateful for this opportunity to share some new discoveries on this topic made by the scholars who contributed to this exciting collaborative project.

Notes

1 David A. Fahrenthold, "Trump Recorded Having Extremely Lewd Conversation about Women in 2005," *Washington Post*, October 8, 2016, www.washingtonpost.com/politics/trump-recorded-having-extremely-lewd-conversation-about-women-in-2005/2016/10/07/3b9ce776-8cb4-11e6-bf8a-3d26847eeed4_story.html

2 Tessa Berensen, "Donald Trump Personally Attacked the Women Accusing Him of Sexual Assault," *Time*, October 14, 2016, https://time.com/4531872/donald-trump-sexual-assault-accusers-attack/

3 Carole Blair, Julie R. Brown, and Leslie A. Baxter, "Disciplining the Feminine," *Quarterly Journal of Speech* 80, no. 4 (1994): 383–409.

4 Blair, Brown, and Baxter, "Disciplining the Feminine," 402.

PART I

Gendered lives and identities

Introduction to Part I

The idea of "gendered lives and identities" speaks to the interconnectivity of gender to how we define ourselves, and how we are thus socially constructed in the circulation of our daily lives as gendered beings. Maybe it begins with the now popular *gender reveal parties*, in which the colors pink or blue are still used as signifiers of sex, serving as the beginning of a gendering process. Within these rituals that vary across cultures, there is always an unspoken or maybe overly subscribed socialized desire in the determination, to which the ceremonial reveal and the evidenced response is always performatively enacted: either overly reacted or timidly tempered in the moment of public sharing. Each response speaks to the value of the presumed masculine in relation to the determined feminine. And so begins a socializing process of expectations and orientations of enforced binary fixities: male/masculine, female/feminine, and all that comes with those categorical containments, including heteronormativity. This process often establishes inequities of opportunity as well as restrictive options in the ever-emerging possibilities and potentialities of gendered being and becoming. Gendered beings, in their own time and manner, may then come to resist the social fixities and containments of their assigned and socially gendered determinations. And some actively begin to embody, and in some cases *trans-body*, their owned and emerging gender identities in a performatively enacted resistance of claiming their innate human rights and dignity.

Hence, the social investment in gendered identities also signals that issues of hierarchy and inequity inform and intervene in the connections between communication, gender, and culture. In her groundbreaking work on *Gendered Lives*, Julia T. Wood writes:

> If we don't want to be limited by the experiences, perspectives, and circumstances of people in other social positions ... realizing that inequality is socially constructed empowers us to be agents of change. We have a choice about whether to accept our culture's designation of who is valuable and who is not, who is normal and who is abnormal ... Instead, we can challenge social views that accord arbitrary and unequal value to people and that limit humans' opportunities and lives.[1]

In such a process of challenge, those of us who have been restricted by and captive to the social expectations of gender (which is most, if not all, of us outside the regimes of sustained power and

1

privilege) must stand up and speak out for the next generation of those who are *revealed* into a social system that has already predetermined their possibility.

This section continues and updates discussion of such issues by furthering vital debates relating to the ethics and politics of gender as identity and the social constructions and sharing of identity through communication. The chapters as a whole address the confrontation of gender expectation and embodied gender performativity in disparate and desperate situations, with shifting effects and affects on both the performance of gender and the material fact of bodies in social negotiations and interpretations. Contributors in this section consider crucial sites of gendered practices and performances as well as issues of identification and disidentification, while also reflecting recent debates regarding inclusion in feminist scholarship and challenging key assumptions of gender studies in diverse geographic, ethnic, and cultural contexts. The chapters tease at the intersections of communication, gender, and sexuality in the social domains of the self in society. In order, the chapters build new arguments on the notion of gendered lives in locational contexts.

In the first chapter, "Performing gender complaint as airport activism, or: don't get over it when it's not over," Stacy Holman Jones and Anne Harris offer a bracing auto/ethnographic documentation of experience—in which they situate their gendered and relational bodies in the public space of the airport. The auto/ethnographic, in this case, is literal to the articulated personal politics of lived experience by gendered beings, bodies in a cultural context, to which no one is left uncensored (with differing consequences). They explore how airport scanners have become a site of trans* and genderqueer anxiety and performativity much in the same way public toilets served those roles for generations of gender and sexuality warriors and resisters. They explore a new generation of LGBTQI assimilation that has not tempered the politics of visibility in such spaces for trans-masculine and femme-presenting women. If trans* has become hyper visible in an age of social media, femme queer sexuality and gender expression have become even more invisible. They argue that the now everyday experience of body scanning as a process of entry for airports and other "secure" areas both overtly and inadvertently categorizes and marginalizes gender outlaws. The chapter draws on critical autoethnographic narratives, in addition to feminist, trans*, and queer scholarship, as lenses through which to question the everyday politics of invisibility and security and to break down binaries in human and more-than-human intersubjectivity. It is the aim and hope of the authors that this work offers readers new and resistive ways of performing the rainbow of queer selves possible in contemporary communication scholarship.

Bryant Keith Alexander takes up the charge of Jones and Harris, as he in essence articulates his own *resistive ways* of being fully present in the breadth and depth of his personal and professional life as a Black queer man, amongst other things. In his chapter, "Dense particularities: race, spirituality, and queer/quare intersectionalities," Alexander uses the notion of dense particularities as a means of talking about our lived and imagined difference. Using theories of critical race theory and disidentification linked with queer/quare studies, queer color critique, and spirituality, the author provides strategies of knowing the gendered self in society through race and resistance; and maybe race as resistance. This knowing is with respect to prevailing notions of intersectionality, palimpsest, and theories of the flesh as co-informing tropes that are not conflated as the same, and yet, not *not* the same in the dailiness of living. Like Jones and Harris, Alexander then moves to recontextualize theory back into the body through a performatively autoethnographic piece that speaks to the intersections of his Black/male/queer-quare/Catholic gendered identity—working as an academic Dean in a Catholic Jesuit university—a positionality in which he refuses to compromise any aspect of the personal for the professionally political.

In the third chapter, "Gaysian fabulosity: quare(ing) the *normal* and *ordinary*," Shinsuke Eguchi queerly and powerful appropriates and claims the terms "gaysian" and "fabulosity." They theorize the terms in the empowerment of gay Asians, with fabulosity becoming not just a performatively aesthetic reference but a criticality of engagement and boldly owning the particularity of oneself in the world. The study uses performance and quare studies to interrogate how gaysian [gay + Asian] subjects self-represent their narratives via YouTube. The author argues specifically that gay anti-Asian racism reproduces the construct of gaysians as *feminine and submissive* Others. Simultaneously, gaysian identities, performances, and politics implicate a need to navigate the material fragmentations of places and spaces such as White gay sexual cultures and "Asian" cultures. Yet, gaysian narratives primarily draw from the hegemony of Asian-White interracial couplings as the relational norm. Eguchi explores three themes, "Gay Anti-Asian Racism," "Strategic Essentialism," and "Gaysian Critically, Needed" that represent the analysis of this engagement. They formalize the notion of fabulosity as both a performative presentation of self and a critical locational politic used as performative resistance to repressive regimes of the normal. Eguchi also uses fabulosity to normalize the construct of "gaysian" as a complex identity location that operates at the intersections of both gay and Asian culture to establish an emancipatory politic of possibility.

In an ever-increasing globalized world in which borders bleed in ways that are both figurative and literal, Astrid Villamil and Suzy D'Enbeau's chapter, "Communication, gender, and career in MENA countries: navigating the push and pull of empowerment and exclusion," explores how transnational discourses around gender take on multiple, shifting, and contested meanings. The fluidity of these terms appears especially complicated in geographies where local and cultural complexities do not mirror Western notions and ideals of gender, career, and family. The Middle East and North African (MENA) region presents a unique context to explore gender/ed dimensions at the intersections of family, career, and work. This extended literature review strives to unpack the ways of thinking about MENA countries reflected in academic research that informs underlying assumptions regarding choice, agency, and constrained obligation. The authors introduce transnational feminism to synthesize a snapshot of extant research in the contexts of education and career in MENA countries. They also summarize several strategies employed by women in the MENA region to navigate a push and pull between empowerment and exclusion as they pursue their education and career. The chapter concludes by suggesting several potential areas of future research.

In the fifth chapter, "Chicano masculinities," Kostia Lennes addresses in very nuanced ways the slippage in gender identities that are at times both particular and plural. Lennes offers a brief survey of key issues addressing Chicano masculinities in various disciplinary fields. Drawing on a critique of the "classification" of Chicano masculinities as marginalized masculinities, this chapter recalls that Chicano men, as all men, may embody a wide range of identities and cannot be considered as only marginal. Three fields of inquiry are then explored. First, the author highlights how the notion of machismo has been widely associated with Chicano men and addresses the question of performing machismo and violence through popular culture. Then, the author analyzes gay and queer Chicano masculinities through the lens of a case study: the changes in male same-sex practices among Mexicans in the United States reveal new definitions of Chicano gay masculinities (more specifically the male sexual role) since the early 1990s. What follows is an exploration of how Chicano masculinities are discussed in relation to the borderland identity and notions of hybridity. In the end, the author suggests an intersectional approach to investigating Chicano masculinities, as both particular and plural.

The final chapter of this section takes us back into an activist framework, one that offers us news ways of looking at, seeing, and recognizing gender injustice, with a particular strategy of intervention through media and new materialism. In "A new materialist framework for activism in the age of mediatization: the entanglement of bodies, objects, images, and affects," Mariam Betlemidze suggests and advocates for a new materialist approach to studying activism as gaining pertinence with the rapid growth of digital media technologies. The suggestion is made that the study of any social phenomenon, and especially social change, is no longer possible with a sole focus on humans, their agency, and their agenda. We are constantly surrounded by materialities that involve nonhuman and more-than-human elements that demand our attention. These are technologies and objects intended to satisfy our needs, yet seem to possess agency to change the way we think, act, and desire. Ranging from technological devices, garments, spaces, and statues to social media posts and images, non-human elements play crucial roles in the networks of activism. The chapter proposes media-activism-assemblage, a framework based on Bruno Latour's actor-network theory and Gilles Deleuze's and Felix Guattari's concept of assemblages. To better understand the co-constitutive nature of human and nonhuman elements in assemblages of media activism, the chapter contributes the concept of image-affects.

These chapters range in different interpretative, locational, and embodied politics linked with gendered lives and gender identities, each serving a form of activism against the social stigmatizing of difference. Jones and Harris, Alexander, and (through naming and claiming) even Eguchi put their gendered bodies on the line: a line variously situated at the invasive threshold of the airport scanner that, under a penetrating mechanistic and spectral gaze, makes visible not only the inter-subjectivity of personhood but the sociopolitical investment in questioning categories of gender realness within and outside issues of social safety; a line of social edicts that presumably disavows and then reclaims the confluence of race and queer identities and issues or spiritual yearning in which sexuality becomes a presumptive determination in the practice of faith; and the historically shifty line to which racism draws difference, and hence particular others must both resist and persist in their self-determination in queerness and personhood. Lennes and Betlemidze force us to see differently, through new ways of knowing plurality in the presumed particularity of gendered communities, and to find new ways of seeing how non-human elements play crucial roles in the networks of activism, including seeing the limitations and possibilities of gender performance.

Linked with notions of gender performativities, the chapters in this section, along with essays throughout the volume, can be recirculated through each other as both spectral and embodied modes of living gendered lives in place and space that have a potency of meaning in the encounter of self and others, and self as other. The idea of "gendered lives and identities" seems to pivot on the performative, a thing done and the iterative and presumed-to-be intrinsic qualities that assign meaning to the being and doing of gender. The essays in this section both do and signal the critical role of performance,

> crisscrossing lines of activity and analysis … as a work of *imagination*, as an *object of study*, as a *pragmatics of inquiry* (both as a model and method), as an optic and operation of research, as a *tactics of intervention*, and alternative space and mode of struggle

to which many of the chapters in this volume address—in ways that are both named and unnamed.[2] And maybe, the idea of gender lives and identities might also be decoupled to reference the "gendered lives" that many of us live by virtue of our social designations and our "gendered identities" that have yet to be realized in the fullness of our own processes of being and becoming. The chapters in this section just may inspire in ways that we never trusted to give thought or possibility.

Notes

1 Julia T. Wood, *Gendered Lives: Communication, Gender, & Culture.* 6th edn. Belmont, CA: Wadsworth/Thomson, 2012, 2–3.
2 Dwight Conquergood, "Performance Studies: Interventions and Radical Research." *The Drama Reviews*, 46, no. 2 (2002): 152.

1

PERFORMING GENDER COMPLAINT AS AIRPORT ACTIVISM

Or: don't get over it when it's not over

Stacy Holman Jones and Anne Harris

Feminism, hopeful resistance, and security

Sara Ahmed has taken an intersectional approach to contemporary feminism, tackling the persistent issues of institutionalized sexual harassment, the gendered nature and crippling inequality of academic work, and the pervasive and intertwined operations of trans* and homophobia, islamophobia, and racial politics. Since resigning and publicizing her experiences with gendered bullying, harassment, and failed efforts to create feminist solidarity at London's Goldsmith's University, Ahmed has used her blog *feminist killjoys* as a platform for publicly, theoretically, and politically confronting the powers that be. Some of this work has come together in her 2017 book *Living a Feminist Life*, which traces the ways in which feminism is an ethically responsible and relationally responsive way of living in the every day, rather than a mantle that one takes up (and puts down) at will. It is this kind of everyday embodiment of resistive politics that is, and must be, at work as we move through the world. Minoritarian subjects know this viscerally and powerfully, as we are ever aware of the constraints and openings that the politics of visibility and security deploy and offer us on a daily, hourly, and minute-by-minute basis. Ahmed opens the book with the idea and ideal of living a resistive life:

> Living a feminist life does not mean adopting a set of ideals or norms of conduct, although it might mean asking ethical questions about how to live better in an unjust and unequal world... how to create relationships with others that are more equal; how to find ways to support those who are not supported or less supported by social systems; how to keep coming up against histories that have become concrete, histories that have become as solid as walls...To live a feminist life is to make everything into something that is questionable.[1]

In this chapter we use contemporary feminist and queer theory, especially the ways in which Ahmed is deploying that scholarship in popularly political ways, to look again at the everyday transphobia and heteronormativity of airport scanners as a site of microaggressions, pervasive oppression, and public shaming. In particular, we use Ahmed's notion of "getting over it" as both a popular cultural and political imperative that has particularly been levelled at women, queer, and other minorities at times of seeming assimilation into the mainstream (as queer culture is experiencing through the same sex marriage movement).[2]

By looking together at trans* hypervisibility, coupled with persistent femme invisibility, we interrogate the power of non-binary discourses and practices. Both ends of the gender spectrum provide a range of constraints and affordances, and through this chapter we consider how non-binary subjectivities and practices in gender studies offer complaint as an approach to the still-unrealized possibilities of gender expression that creates what Ahmed calls "hope," which "is not at the expense of struggle but animates a struggle; hope give us a sense that there is a point to working things out, working things through."[3]

Where Ahmed's sense of hope "does not only or always point toward the future," our writing is purposefully future-focused, though not in a naïve or simple way. Rather, we write in the spirit of what José Esteban Muñoz has called "queer futurities," which work toward a changed future in full knowledge of and animated by the fact that the here and now, for queer people, is "simply not enough."[4] He writes, "Queerness should and could be about a desire for another way of being in both the world and time, a desire that resists mandates to accept that which is not enough."[5]

Like Ahmed's use of hope as the animating force behind struggling together, Muñoz's queer futurity is a relational and collective effort of "endurance and support."[6] Here, the hopeful resistance of feminist and queer work is a future-focused way of being and living in the present "not [as] an end but an opening or horizon, a modality of critique that speaks to quotidian gestures as laden with potentiality."[7]

Diversity workers

Ahmed's conclusion to *Living a Feminist Life* addresses those of us who are doing "diversity work"—work that struggles to change everyday oppressions, inequities, and inadequacies. In a world where queer, trans*, and other minoritarian subjects are often told to "get over it," diversity work says: "don't get over it when it's not over."[8] Diversity work says some things can and should be changed, modified, or opened up. Ahmed writes, "The modifications that are required for spaces to be opened to other bodies are often registered as impositions on those who were here first. Diversity workers end up challenging what gives security, warmth, place and position."[9]

When we think of secure places, we think of home. Places where we feel that we have a rightful place, warmth, openness, room to breathe. Ahmed sees the work of feminism as *home-work*—work we take up to modify and transform the "master's house" so that we might find refuge there. Even though the work of feminism is home-work, we might not feel at home in feminism or in the world. We might not feel safe and secure. The linking of home with the feeling of security asks us to think of the kinds of work that make home spaces safe—the work of *homeland security*, for example. It asks us to think about spaces of security that, rather than providing us with warmth, place, and position, feel like spaces of danger, not-belonging, and imposition. These are spaces of security that do not include us—that do not recognize our bodies. Or if they did, would require modification, change, and transformation. Take, for example, the airport security scanner.

Airport security scanners 1: threatening security and airport activism

There are multiple sites where policing gender and gendered embodiment takes place, starting with home, and expanding out to schools, workplaces, military organizations, sidewalks[10]. Two of the most notable—the most volatile spaces of danger, not-belonging, and imposition—in contemporary culture are public bathrooms and airport security scanners. These spaces are not, for many queer subjects, *safe* spaces (though they can be sites of pleasure and resistance). Airport security scanners have become a site of trans* and genderqueer anxiety and performativity in the ways that public toilets were in times past (and continue to be). They are also spaces where diversity work and the home-work of feminism start to come together.

*****

We have our patterns when we approach airport security. Our routines:

Usually Anne gets anxious, and Stacy starts to reassure.
Usually Stacy says, "Give it to me if you want, I'll take your bag through."

What Stacy means is that Anne is carrying a gender aid (a packer, a dick) in their suitcase, and is sometimes wearing one, too.

What Stacy means is that Anne has been stopped several times by airport security in some-times public and always humiliating ways. Anne is stopped and checked, to make sure that what they are carrying and embodying are not terrorist appendages.

They are stopped and Anne is subject to a pat-down, right there in the security line or, some-times, taken away and into a room for a private body search. Sometimes Anne requests this privacy, sometimes they are escorted into it, and sometimes Anne asks for it to be done in public, as a public pedagogy of gender fluidity.

What this means is that Anne now approaches airport security with anxiety. And that, at least in part, Anne's behavior has changed in response to these experiences; they don't always bring or wear their gender aid, despite that being what Anne wishes to do and feels most at home in/as/with.

*****

Trans* people use a range of material aids to give fuller expression to our chosen gender iden-tity. "Gender aids" are material tools that allow their wearers to feel more feminine (for example, by wearing silicone breast forms) or masculine (for example, by wearing a binder to compress the breasts or a "packer" or synthetic penis to fill out the silhouette of male pants and give the appearance of a male "bulge"). They are tools that help their wearers feel more safe, secure, and "at home" in their bodies and clothing. Some trans* men wear packers all the time; others wear them only in public. Packers are not the same as dildos, which are tools made for sexual penetra-tion. However, packers can sometimes serve both purposes, although dildos usually don't double as everyday wear (too uncomfortable!). Packers come in a range of shapes and sizes; whatever their manifestation, they bring joy to many pre- or non-surgical female-to-male transgender folks, and to their partners. In short, packers are sophisticated gender aids which serve a multitude of purposes, but which are often misunderstood by those outside trans* communities, particu-larly some airport security officers, who view them as a source of alarm.

*****

As we approach the airport security on this particular day, we both put our suitcases on the conveyor belt and walk toward the security scanners. There are usually two types: the old-school x-ray machine (which can be requested but always creates a scene of suspicion), and which does not detect or highlight or set off gender binary alarms in order to do its work. The second type is the more recently (read: post 9/11) installed magnetic imaging scanner, which, according to one manufacturer's website, enables full-body screening that reveals "metallic and non-metallic objects and other suspicious things on a passenger's body" without touching them.[11] Though for some of us, the screening is anything but hands-off, both in the sensory experience of being stripped of our clothes and examined, and in the embodied experience of being "touched" in the resulting security "pat-down," or a series of pat-downs, accompanied by being separated from your traveling companions, belongings, and security. The "unintended" or even well-intentioned "singling out" in the name of security isn't reserved for non-binary and trans* travelers. Just ask women of color who wear their hair natural. Or in braids. Or in twists. Or wear wigs. These women, too, are singled out for "hair pat-downs" (which involve security officers pulling them out of line and publicly running their fingers through their hair) because what they are wearing on their heads sets off false alarms.[12]

*****

You've been through them—these millimeter wave scanners—and have stood, with your feet apart and hands up while the scanner swooshes across your body, then exited and waited for the "all clear" sign before collecting your bags off the belt. Though have you turned around when you're on the other side of the scanner, waiting for your results, and looked at the high-technology gender binary buttons—blue for boys and pink for girls—that security personnel must choose and push before you enter the scanner? Here's how the US Transportation Security Administration (TSA) describes the process:

> When you enter the imaging portal, the TSA officer presses a button designating a gender (male/female) based on how you present yourself. The machine has software that looks at the anatomy of men and women differently. The equipment conducts a scan and indicates areas on the body warranting further inspection if necessary. If there is an alarm, TSA officers are trained to clear the alarm, not the individual.[13]

This "advancement" in technology is promoted by the TSA as part of an effort to not discriminate on the basis of gender (or sex, race, color, national origin, religion, or disability) in airport screening, because security officers are not viewing "actual" images of passenger's anatomy.[14] However, once a TSA officer presses the pink or blue button, any areas of "anomaly" that pop up in the scan mean that a passenger is taken aside for more intensive screening. In a public space that is already a site of anxiety, stress, and time poverty for non-gender-conforming people, this "technological sex/gender policing" is "intrusive, frightening and humiliating."[15] As Cary Gabriel Costello (2016), a trans* person who has written extensively about TSA security screening, describes it,

> The TSA says it does not discriminate on the basis of gender identity, and that travellers will be treated with respect as members of whatever gender they present themselves as. But in fact, the "Automated Target Recognition" software used by body scanners is set up to police both binary sex and cisgender bodily expectations. Bodies that vary from these expectations set off alarms and are treated as potential terrorist threats.

Typically, they pull you aside and they call loudly out for a pat-down by a security officer of the same (apparent) gender. There are more than a few problems with this, some obvious and some less obvious:

1. Gender identification doesn't always equal gender presentation. Not all women have vaginas and breasts; not all men have penises and testicles.

2. What happens to trans* travelers "after they are marked as 'anomalous' by body scanners is largely a matter of the training and attitude of the TSA agent called over for the public pat-down."[16]
3. If the officer "understands that the alarm has been set off by our trans anatomy and that the TSA is supposed to treat bodies with respect, we get a quick pat-down and are cleared… The trans passenger will experience a brief flash of adrenaline when their body sets off an alarm, and anxiety as their body is publicly palpated in an area likely to be one they do not wish to be scrutinized, but the interval of apprehension is brief and travel is not disrupted."[17]
4. When pat-downs confirm there is "*something there*," trans* passengers are pulled into rooms for further, more intrusive physical screening and questioning, which creates an extended experience of triggering and violation (Burns 2017) that also often delays travel.
5. Levels of trans* awareness and transphobia vary by community, geography, and culture and impact whether trans* people will face additional screening at some airports.[18]
6. Airport security don't like to be questioned, even though they are insufficiently trained in gender diversity and have little understanding of trans* people or their bodies.
7. No "amount of training can make up for a cis-sexist body screen system that's designed to detect passenger genitals."[19]
8. Though body parts and hair are not alarms and have nothing to do with airport security, a "non-conforming" passenger can run afoul of the law very easily in this situation, without meaning to.
9. Sometimes, though, one means to.
10. This might be called diversity home-work, or "airport activism."

*****

On this day, Anne fails the test and gets pulled aside on the "other" side of the security border. The TSA agents ask Anne to step aside and tell them that they'll need to call for an officer to pat them down. They are not sure who to call—a man or a woman. Anne says it is fine for a woman to pat her down. They call loudly for a "Female Pat-Down" and a woman comes.

On this day, two women officers come to pat Anne down. Stacy questions why they've sent *two* people to do the pat-down as she gathers the luggage. She says they should leave Anne alone—there is no indication of a threat, just an indeterminate gender presentation. The two female security officers look at each other. They are both young and it's becoming clear that they are inexperienced with trans* people. This often, if not always, means they prioritize alarm and threats to (national) security over and above the alarm and threat to individual security. They ask Anne if they prefer to have the pat-down in a private room, or if it is ok to examine them there, in the public security area. Anne says they want a private room. The officers take Anne to a private room 20 feet away from the border. They close the door. Then, they tell Anne to pull down their pants.

Anne says, "Do you want to see my gender aid?"

They say there is something in Anne's pants.

Anne says, "I know, it's a gender aid. Do you want me to take it out?"

They ask what a gender aid is.

"It's a dick," says Anne.

They get nervous. They look back and forth. They say, "Oh no, don't take it out." They call for reinforcement.

Eventually an older woman, a supervisor, comes. She closes the door and pats Anne down and says it's not necessary to take it out and they are not intending to make anyone uncomfortable.

Anne says: "But gender has nothing to do with national or airport security. So, yes, it is uncomfortable, and yes, it is unnecessary. I feel upset, and I am already a nervous flyer. This has been a bad experience."

The supervisor reassures Anne: "It'll all be okay once you have gender reassignment surgery—the machine won't pick it up anymore."

Anne says, "But I'm not going to have gender reassignment surgery."

The supervisor looks puzzled. "Why wouldn't you want to have the surgery?"

Anne says, "Not all trans* people want surgery. Does everyone here get training around trans* experience and treating trans* passengers with respect, given it's the United States and it's the twenty-first century?"

The supervisor replies: "The supervisors do, but not the other officers. And even the supervisors get a single session on transgender passengers. It's tricky."

Anne leaves the interrogation, and joins Stacy, who is livid and voices her anger to the security officers, who do not respond. Stacy's complaint, the anger of a cisgendered female and invisible femme lesbian (someone who "passes" as heteronormative) has no place. No home in the land of (false) alarms set off by body parts read as terrorist threats to national security. Her pink button, no pat-down, collect your bags, resume your journey while her partner is misgendered, stopped, taken away, searched, and questioned some more is no concern of the TSA. Unless she makes too much of a fuss, in which case the TSA will consider her, too, a threat, and subject her to the same searching.[20] On this day, her questions and complaints aren't too much of a threat to the security of the TSA. On this day, we collect our bags and ourselves and go to our gate.

Vomiting as a feminist act

In her writing on complaint, and the "time of complaint" we are currently living through, Ahmed counsels those who choose to complain that this work "takes time."[21] Despite the time it takes to lodge and document and recover from the act of complaining, feminist and diversity workers must not "get over it when it's not over." And in so many places—in airports and bathrooms and on university campuses—it is not over. Ahmed's complaints in recent years have been about harassment. Sexual harassment on university campuses, to be exact. And the thing about harassment is that if you complain about it, you are harassed. Harassment is a tactic to deal with the act of complaining about harassment. And in places of security—spaces where we are supposed to be safe and able to work or to get where we are going, like university campuses and airports—complaining about harassment almost always leads to more harassment. In fact, asking to not be put into a space where one might be harassed is viewed as an invitation to investigate because those who point out the instruments of discrimination are viewed as threats to the system. Ahmed writes, "If you try to stop harassment you come up against what enables harassment. The accusations [against you] are part of a system; a system works by making it costly to expose how a system works."

One of the costs of exposing how systems of oppression and harassment work is that the act of complaining is emotionally, intellectually, and physically *exhausting*. Home-work is like this: we find ourselves working overtime, pouring everything we have into it, only to do the same things over and over again. So, when Ahmed asks us to stay *in touch* with the corporeal harms that have been done to us, or to which we have been witness, even when our bodies are not "touched," she is returning us to the body as a political act.[22] She is also returning us to the body as a site of exhaustion; the embodiment of complaint is "not an abstraction." Everything pours out and we are left feeling shaken, our bodies flooded with adrenaline, and sometimes, because of this, physically sick. The body, Ahmed claims, can set and test limits where hegemonic culture refuses us.

When she resigns from her university post after Goldsmiths failed to address sexual harassment as an institutional problem, Ahmed writes:

> I feel sick. In my last proper day at work we had a departmental meeting. A caring and well-meaning colleague got my resignation on the agenda—he got sexual harassment on the agenda. It is mentioned along with another item. The other item is picked up. I hear it being passed over. I hear sexual harassment being passed over. I rush out of the room and I am sick in the toilets. I now think of that vomiting as a feminist act: all that came up, all that I refused to digest.[23]

The current culture of feminist complaint is a new "wave" of social media-informed and altered feminism that is very much focused on the crimes and transgressions of men (including Ahmed's work on sexual harassment). In addition, it takes aim at institutions and the kinds of systems that exclude women and female-presenting people. Yet perhaps one blind spot in the #MeToo feminism that is so visually and creatively galvanizing through the affordances of contemporary social media is the issue of gender policing itself. Trans* and queer people know that we carry our traumas in our bodies, and these traumas often make us sick. The question of visibility through social media, which can have such swift and effective policing effects on those (particularly men) who have harmed women, can also marginalize and be used against those of us who are visually unintelligible or unhegemonic. Social media affords a global platform to issues and events and identities that are visually clear, aesthetically beautiful, and simple. Its curation and circulation rely upon visual intelligibility, yet so many of us in the queer and trans* communities are not those things; indeed, often we take power and pride in our multiple liminalities and fuzziness. The hypervisibility of non-binary trans* people and the invisibility of femme or otherwise-passing queer women are not always suited to visual platforms like social media apps.

Ahmed says complaint comes into being through its affective emergence as "a bodily condition, an ailment, an illness," a bodily experience which (in yet another binary) "allows the institution to disappear."[24] We don't disagree, but also suggest instead that there are many kinds of "institutions," systems constructed of not only material matter like bricks but those constructed of belongings, of rage, of outsider or insiderness, institutions which seek to distract from the always-unfolding layers of oppression, exclusion, and gender bias and a reminder that these should be fought in every site at which they appear.

Airport security scanners 2: airport activism and feeling sick

Back at the airport again. Another trip home to the States. Anne is becoming fatigued by getting stopped and publicly humiliated at the security checkpoints in airports. They are sick of— and made sick by—the adrenaline rush that floods their body and makes their heart pound during these encounters and exhausted by the prospect of increased harassment if they, or Stacy, complains. Anne and Stacy both brace for what might, what will probably, happen. It goes like this:

In the line leading up to the security lines, Anne starts to fidget and scans the open security lines, trying to see if any are using the x-ray machine.

If there's an x-ray machine in use, Anne and Stacy angle, and sometimes defy the directions of TSA staff so that they get into the possible x-ray line.

Once there, Anne speeds and slows their pace at the baggage belt, trying to judge when and how to catch the security officer's eyes and maybe waved into the x-ray machine line, where they can relax and wait.

If it's unclear that Anne will be channeled into the x-ray line, they get anxious. When they reach the decision point—where some are directed to the magnetic imaging scanner—Anne asks to be sent into the x-ray line.

If the security officer denies the request (which they usually do), Anne knows that avoiding the magnetic imaging scanner means saying, "pat-down, please."

In the past, any passenger had a right to ask for a pat-down, instead of being scanned, for any reason. But now—today—complying with a passenger's request for a pat-down is at the discretion of the agent.[25]

When Anne asks for a pat-down, the security officer usually delays and denies. "Pat-down?"

"Yes, pat-down."

"Step through the scanner please."

Because "simply requesting to be scanned can be perceived as suspicious, most trans* passengers will [agree to] be body-scanned."[26]

Because Anne is not most trans* passengers, they persist: "I don't want to use this scanner. I want a pat-down," to which the security officer, in a change of tactics, usually replies, "this scanner is actually safer than the x-ray machine, don't be scared."

At this point, Anne will say, "I'm not scared. I want a pat-down."

Eventually, the security officer will scream out at the top of their lungs for a person to come do the pat-down. Usually the person is extremely embarrassed and unsure of what to do, but proceeds awkwardly.

They usually ask Anne if Anne has any weapons or things in their pants. Anne always says, "No."

Usually Anne will add, "I have a gender aid. It's a packer. It's a dick," while the security officer blinks with incomprehension, or embarrassment. Once they understand what Anne is saying, they sometimes abandon the mission, backing away and saying "That's fine, that's fine, that's fine."

Even when this happens, everyone else in the line is watching and listening with anxiety, sometimes disgust, and always extreme interest.

Other times, the security officer stops and says that Anne will need to be escorted into a private room for a more extensive search.

No matter which way it goes, the work of airport activism as the embodiment of complaint is so exhausting it almost always makes Anne feel sick.

*****

When exhaustion takes over and Anne doesn't wear their packer to the airport (and who could blame them for wanting some relief?), they might still "fail" the scanner test. In other words, even when they leave the packer at home or in the suitcase, their gender presentation sets off the performance of anomaly. Here's how *that* goes: The TSA worker presses the "blue-for-boy" button and Anne's cisgender body registers the airport scanner's criteria for the category of "woman." Anne has again scanned with alarm—signified by a big yellow "box"—around their genital area. On these days, we laugh uproariously about this: we problematize the haunting of a ghost "lesbian phallus."[27] Anne's penis—like so much about trans* and femme embodiment as complaint against heteronormativity and technological sex/gender policing—remains unintelligible. It shows up on the "trustworthy" airport body scanner, even when it is not there.

Complaint as not coping and hoping

Can the mantra that comes with performing gender as complaint—don't get over it when it's not over—be used to say something about how we can be simultaneously hypervisible and unintelligible trans* and femme subjects in the highly policed and harassing spaces of airport security lines and bathrooms? Sometimes, making yourself hypervisible by complaining about the unintelligibility of your gender preferences and identity—as trans* person and as a femme lesbian—is an act of not getting over it. An act of airport activism: complaint as keeping something going and refusing to "hold ourselves together" by "coping" (Ahmed 2018). Other times, there is humor, even pleasure in refusing to hold yourself together in the airport security line— for example, when a security officer pulls Anne's bag out of line for inspection. This makes Anne intensely embarrassed, but not Stacy. On these days, if Anne's bag is singled out for inspection, Stacy claims it—and Anne's gender aids—as her own. Stacy revels in claiming and talking about the dick. Usually it doesn't come to this, but when it does, Stacy is keeps things going, refusing to pass (too quickly) over what might otherwise turn the hypervisibility of a masculine body part into a something that doesn't belong in the suitcase of a femme-presenting person. This too is airport activism, because it refuses to keep things in—"when we let out, spill out, what we are supposed to contain"[28]: laughter, bodies, desires, the presence of something powerful, even if these things are not, technically, "there." Not yet. Performing gender as complaint is both a refusal to cope and a commitment to hope.

Airport security scanners 3: feminist and furious

Anne is at a US-based international conference they attend every year. Anne sees this conference and the community it generates as her tribe. At the conference, Anne performs a monologue about airport scanners. The monologue is both funny and a testimony about violation. Anne gets good feedback on their performance, but there are other currents of commentary. One person asks if Anne is feeling "fragile" after delivering the "exposing" monologue. "No," Anne says, "This is my life. It's not secret or shameful. It's what I experience. I have been performing my reality since I was 17 years old." Others share their own responses—"that made me angry" or "that made me feel afraid"—and advice: "just don't travel with those things." One colleague and friend in attendance asks if Anne will re-perform the monologue so that she can video record it and share it with her class back home. She is a woman of color, a respected scholar who does critical race theory and creative methods and a trusted friend. Anne agrees. They film the monologue in the ballroom of the hotel, back behind the book display. It's fun and funny, and Anne is gratified that her friend will use the video in her classes on diversity. The colleague asks Anne to Skype in to class after the students have watched the video and Anne agrees.

A few months later, Anne Skypes in. The class is quiet. Anne's sense is that the students don't know what to make of the monologue. After being prompted by her colleague, an African-American male student breaks the silence and engages in a dialogue with Anne after about intersectionality (queer, POC, critical scholarship). Anne feels gratified about the conversation and also squeamish, given this student—really none of the students—responded of their own volition or about the performance itself. Awkward!

Anne and the colleague/teacher talk afterwards, and she tells Anne that it was "transformative" for her students, even though Anne didn't really get that sense. As they sign off, Anne decides it's been a good and useful experience and moves on.

The following year, Anne and her colleague and friend attend the same conference, where the colleague gives a presentation which she framed as "coming out" as a (non-queer) scholar of something-else.

At dinner, Anne says,

Hey, I want to raise something with you because you are a critical scholar, and you're my friend. I felt uncomfortable with your use of "I'm outing myself" because it kind of feels like appropriation of queer culture—my culture. What do you think?

THE COLLEAGUE SAYS, "Ah yes, you are right! Totally inappropriate, Sorry! But also? I'm not as straight as I seem. I am queer-*ish*."

ANNE: "Oh, how so?"

COLLEAGUE: "I was attracted to women at one point, and I tried to tell my very traditional mother, but I just couldn't" (funny story ensues).

ANNE: "So how is that queerish? You are married to a man. You are enjoying heterosexual privilege. What is the relevance of you kissing or liking a girl in college?"

COLLEAGUE: "Well not everyone who is married is straight you know."

ANNE: "I know, but how are you queerish?"

Long pause.

COLLEAGUE: "Well, it's complicated. And I think now I want to tell you about something that happened with your monologue. Your airport scanner monologue."

Anne is intrigued.

COLLEAGUE: "Well I was visiting my mentor, and I decided to play it for her."

ANNE: "But I only gave you permission to play it for your class."

COLLEAGUE: "Yes, but this is interesting. My mentor watched it and was full of praise, absolutely full of praise for your performance, but she was quite upset by it too."

ANNE: "Oh really?"

COLLEAGUE: "Yes, because she felt that you were class-shaming the airport workers."

ANNE: "Excuse me?"

COLLEAGUE: "Yes, she felt like there was this middle-class academic passenger who was impatient with these people just trying their best to do their job, and the judgment in the monologue was thick."

Long pause.

ANNE: "But you yourself know that all of my work is about being working class!"

COLLEAGUE: "Oh yes, I know. I don't agree with her at all, but I just thought you should know."

ANNE: "Why? Why should I know? I've never even met this person. Why did you show her the video? You didn't have permission to do that."

COLLEAGUE: "Oh yes, I probably shouldn't have. I'm sorry! I just know how good it is, and I knew she would love it."

ANNE: "She didn't love it. She criticized me. As a white straight cisgendered heterosexual middle-class woman she felt privileged to criticize me, and you felt like you should pass that on. That's not ok. Please don't show it to anyone else."

COLLEAGUE: "Oh sure, absolutely. I'm so sorry if I've upset you. That was not my intention. I simply wanted to share your brilliant work, and absolutely I should have asked. But she has an interesting point, right?"

ANNE: "No. She does not."

The next day, the colleague returns home to her husband and posts many photos on social media of how happy they are to be reunited and he is the "best husband ever!" Queer-*ish*.

Anne has an email exchange with the white cisgendered heterosexual middle-class mentor. Rather than accepting that she had no right to see or invitation to critique the work, the mentor offers suggestions for improvement. She hopes they can meet and discuss the video and the strategies for improvement at the next conference. The next time Anne travels to the conference, they do not meet with the mentor. And they are wary of disclosing too much to their colleague and friend. Even before arriving at the conference, Anne responds to the exhaustion they feel around the hypervisibility and unintelligibility of gender complaint: they don't dress as they prefer; they pack the packer. And the next time, and the next time, and the next time. Queer fatigue sets in.[29] Anne feels that they are not coping, but as Ahmed suggests, "maybe not coping is an action… how we create a collective. That collective might be fragile but it is also feminist and furious."[30] And not coping might just give us the hope that animates our struggle to work things through.

*****

In this chapter, we have use queer and feminist scholarship to look at everyday gender oppressions, even within feminist and critical spaces and relationships, which nevertheless can easily become sites of microaggression and macroaggression and public shaming. In turning to Ahmed's notion of "getting over it," we remind ourselves that all spaces, even spaces of supposed security—like homes, airport security lines, and feminist academic relationships—are sites where diversity work can and should take place. Even when it might seem like the work is finished.

The Zen Buddhist nun Pema Chodron's book *Start Where You Are*[31] asserts the principle that we can only be where we are at, and a denial of that in the pursuit of an idealized (spiritual or other) state only makes us feel bad about ourselves. Accept where (not just who) we are. The airport scanner situation is an example of how we can all "start where we are." We can call out gender injustice in every nook and cranny we inhabit. Some of us can't get to marches. Some of us aren't the type to resist; even if we are, the work of resistance can be exhausting and can even make us sick. But we can call out injustice (no matter how quietly) in everyday spaces. If you see a brown person being singled out at airport security, you can say something. If you see gender non-conforming or gender-confusing passengers being humiliated by airport security workers by being publicly patted down, or escorted off to a private examination room, you can say something. Airport activism: Don't get over it, if it's not over.

Notes

1 Sarah Ahmed, "The Time of Complaint," *Feminist Killjoys* (blog), May 30, 2018, https://feministkilljoys. com/.
2 Ahmed (2018).
3 Ahmed (2018, x).
4 Jose Esteban Muñoz, *Cruising Utopia: The Then and There of Queer Futurity* (New York: NYU Press, 2009), 96.
5 Muñoz (2009, 96).
6 Muñoz (2009, 91).
7 Muñoz (2009, 91).
8 Sarah Ahmed, *Living a Feminist Life* (Durham: Duke University Press, 2017), x.
9 Ahmed (2017, x).
10 See, for example: Eng 2010; Halberstam 2005; Harris, Holman Jones, Faulker and Brook 2017; Peterson 2013; Ringrose 2012, among others; David L. Eng, *The Feeling of Kinship: Queer Liberalism and the Racialization of Intimacy* (Durham: Duke University Press, 2010); Judith Halberstam, *In a Queer Time and Place: Transgender Bodies, Subcultural Lives* (New York: NYU Press, 2005); Anne Harris, Stacy Holman Jones, Sandra Faulkner and Eloise Brook, *Queering Families, Schooling Publics: Keywords* (New York: Routledge, 2017); Cassie Peterson, "The Lies That Bind: Heteronormative Constructions of 'Family' in Social Work Discourse," *Journal of Gay and Lesbian Social Services* 25, no. 4 (2013): 486–508; Jessica Ringrose, *Postfeminist Education? Girls and the Sexual Politics of Schooling* (New York/London: Routledge, 2012).

11 Fox News (no author), "Magnetometers, X-Rays, and More: Airport Security Technology," December 29, 2009, www.foxnews.com/tech/magnetometers-x-rays-and-more-airport-security-technology.

12 Brenda Medina and Thomas Frank, "TSA Agents Say They're Not Discriminating Against Black Women, But Their Body Scanners Might Be," *ProPublica*, April 17, 2019, https://apple.news/Ad2b3W93yTWOTdCvxHjETzQ.

13 Transportation Security Administration, "Transgender Passengers," 2019. Online: "http://www.tsa.gov/transgender-passengers" www.tsa.gov/transgender-passengers.

14 Cary Gabriel Costello, "Traveling While Trans: The False Promise of Better Treatment," *TransAdvocate*, January 3, 2016, www.transadvocate.com/the-tsa-a-binary-body-system-in-practice_n_15540.htm. See also Transportation Security Administration (2019) "Civil Rights," www.tsa.gov/travel/pasenger-support/civil-rights.

15 Costello (2016); see also Katelyn Burns, "Traveling While Trans: Airport Security Sees Your Genitals as Cause for Alarm," *Rewire. News*, May 5, 2017, https://rewire.news/article/2017/05/05/traveling-trans-airport-security-sees-genitals-cause-alarm/.

16 Costello (2016).

17 Costello (2016).

18 Costello (2016).

19 Burns (2017).

20 Costello (2016).

21 Ahmed (2018).

22 Ahmed (2018).

23 Ahmed (2018).

24 Ahmed (2018).

25 Costello (2016).

26 Costello (2016).

27 Judith Butler, *Bodies That Matter: On the Discursive Limits of 'Sex'* (New York: Routledge, 1993). Butler writes about the 'lesbian phallus,' which is not 'the' or even 'a' penis, but instead, 'a displacement of the hegemonic symbolic of (heterosexist) sexual difference and the critical release of alternative imaginary schemas for constituting sites of erotogenic pleasure' (p. 91). And where Butler argues that "what is needed is not a new body part" (p. 91) in such displacements, it is precisely the introduction of 'new' or perhaps 'unexpected' and non-cis-conforming body parts that create the kinds if disruptions and displacements Butler imagines in airport and other 'secure' spaces.

28 Ahmed (2018).

29 Anne Harris and Stacy Holman Jones, "Feeling Fear, Feeling Queer: The Peril and Potential of Queer Terror," *Qualitative Inquiry* 23, no. 7 (2017): 561–568; Stacy Holman Jones and Anne M. Harris, *Queering Autoethography* (New York/London: Routledge, 2018).

30 Ahmed (2018).

31 Pema Chodron, *Start Where You Are: A Guide to Compassionate Living* (Boston: Shambhala Press, 2004).

Bibliography

Ahmed, Sara. "The time of complaint." *Feminist Killjoys* (blog), 30 May 2018. Online: https://feministkilljoys.com/

—. *Living a Feminist Life*. Durham: Duke University Press, 2017.

Burns, Katelyn. "Traveling while trans: Airport security sees your genitals as cause for alarm." *Rewire.News*, 5 May 2017. Online: https://rewire.news/article/2017/05/05/traveling-trans-airport-security-sees-genitals-cause-alarm/

Butler, Judith. *Bodies that matter: On the discursive limits of 'sex.'* New York: Routledge, 1993.

Chodron, Pema. *Start where you are: A guide to compassionate living*. Boston: Shambhala Press, 2004.

Costello, Cary Gabriel. "Traveling while trans: The false promise of better treatment." *TransAdvocate*, 3 January 2016. Online: www.transadvocate.com/the-tsa-a-binary-body-system-in-practice_n_15540.htm

Eng, David L. *The feeling of kinship: Queer liberalism and the racialization of intimacy*. Durham/London: Duke University Press, 2010.

Fox News. *Magnetometers, x-rays, and more: Airport security technology*, 29 December 2009. Online: www.foxnews.com/tech/magnetometers-x-rays-and-more-airport-security-technology

Harris, Anne, and Stacy Holman Jones. "Feeling fear, feeling queer: The peril and potential of queer terror." *Qualitative Inquiry* 23, no. 7 (2017): 561–568.

Halberstam, Judith. *In a queer time and place: Transgender bodies, subcultural lives*. New York: NYU Press, 2005.

Harris, Anne, Stacy Holman Jones, Sandra Faulkner and Eloise Brook. *Queering families, schooling publics: Keywords*. New York: Routledge, 2017.

Holman Jones, Stacy, and Anne M. Harris. *Queering autoethography*. New York/London: Routledge, 2018.

Medina, Brenda, and Thomas Frank. "TSA agents say they're not discriminating against Black women, but their body scanners might be." *ProPublica*, 17 April 2019. Online: https://apple.news/Ad2b3W93yTWOTdCvxHjETzQ

Muñoz, Jose Esteban. *Cruising utopia: The then and there of queer futurity*. New York: NYU Press, 2009.

Peterson, Cassie. "The lies that bind: Heteronormative constructions of 'family' in social work discourse." *Journal of Gay & Lesbian Social Services* 25, no. 4 (2013): 486–508.

Ringrose, Jessica. *Postfeminist Education? Girls and the sexual politics of schooling*. New York/London: Routledge, 2012.

Transportation Security Administration. "Civil Rights," 2019. Online: www.tsa.gov/travel/passenger-support/civil-rights

Transportation Security Administration. "Transgender Passengers," 2019. Online: www.tsa.gov/transgender-passengers

2
DENSE PARTICULARITIES
Race, spirituality, and queer/quare intersectionalities

Bryant Keith Alexander

Sataya P. Mohanty uses the construct of *dense particularities* to reference "our lived and imagined differences."[1] The construction for me, while suggestive of how "difference will be deployed, rendered, and positioned in regard to both the substance and process of learning," also references the concentrated complexity of charisms and characteristics that come into the confluence of identity.[2] Maybe this term also signals the notion of *intersectionality* as "core ideas of community organizing, identity politics, coalitional politics, interlocking oppressions, social justice" and potentialities that are at the crux of identity politics. In which case, there might be the possibility to actually build a greater sense of unification within/between individuals and communities of affiliation.[3] I am also intrigued with the notion of *palimpsest*, a site on which original writing has been effaced to make room for later writing, but of which traces of the original or preceding remain.

In which case, as palimpsest, and the histories of the body through genealogical tracking, biological manifestations, and lived/historical cultural experience might be said to have the sedimented effects of particular histories. These histories are particular orientations to experiences and different, but highly concentrated and layered, influences of culture that shape the intersections of identity, gender, and desire, all competing and comparative in a field of differences that are both particular and plural. Yet still, Cheri Moraga and Gloria Anzaldúa's *theories of the flesh* also speak to me about that complex location

> where the physical realities of our lives—our skin color, the land or concrete we grew up on, our sexual longings—all fuse to create a politic born of necessity. Here, we attempt to bridge the contradictions in our experience .... We do this bridging by naming ourselves and by telling our stories in our own words.[4]

This chapter uses the notion of dense particularities through intersecting discussions of race and a specific conceptualization of Critical Race Theory (CRT), disidentification linked with queer/quare studies, and spirituality as cosmologies of knowing gendered contexts and strategies. These intersecting discussions are done with respect to prevailing notions of intersectionality, palimpsest, and theories of the flesh as co-informing tropes that are not conflated as the same, and yet, not *not* the same. The chapter then moves to recontextualize the breadth and depth of these constructions back into the body—the body as a critical site of knowing, showing, and telling of lived experience and of communicating queer/quare identity. I then make an autoethnographic move to tell a story of theory through experience. In this case, autoethnography is engaged as an approach to research and writing that seeks to describe and systematically analyze (graphy) personal experience (auto) to understand cultural experience (ethno),[5] reinforcing the notion of communication and gender operating within a cultural context with the intent of articulating, sharing, and negotiating meaning across lived experiences.

Thus, the construct of queer/quare intersectionalities is the abiding and integrative trope of interest that synthesizes all these issues—to speak to radical perspectives of communicating gendered identity that operate in resistance to regimes of the normal to which queer theory often attends itself, and to which quare studies demands attention to occlusion of racialized experience,[6] all with the constant focus on the intervening and inextricable variables that shape gender identities within and through varying social formations.

Race and critical race theory as gendered variables[7]

I want to offer a brief foray into the notion of race. Why race? It seems to me that race (at least in the United States) is the conjectural assumption of identity that often precedes and yet informs notions of gender—and thus characterizations of heteronormative and homonormative gender performativity. Race is the categorized phenotypic materiality of the human body and signaled origins of people (e.g., Caucasoid, Mongoloid, and Negroid) and race is a social construct that marks difference—invoking imagined performativities that are psychological and physiological, as well as relational and imagined. This notion of race is exemplified in the constructions of Black sexuality that circulate in the white imagination as both desire and disdain, particularly the fixation on bodies and body parts like the size of the Black phallus or the made-spectacle-of Black female physique (see the history of Sara Baartman, the promoted "Hottentot Venus").[8] These constructions are marketed in the slave trade as commodity and spectacle of difference in an argument of hierarchical value. Thus, race is socially perpetuated and embodied through a set of relational practices and commitments that are historically and situationally derived. The effects of such distinctions are made manifest in forms of power, privilege, and propriety that both intervene and guide the politics of human social engagement: a relationality of value that always marks privilege over pathology with white (Caucasoid) being placed in the position of hierarchy over all others to name, and thus to claim, authority.

In establishing an argument of "Critical Race Theory and the Postracial Imaginary," Jamel K. Donnor and Gloria Ladson-Billings offer a particular frame that marks the enigma of race as an investment in material fact, thus affording and informing relational dynamics. They write:

> The challenge for social scientists working with race is that all social science disciplines (to some extent) use the concept "race" as it were a fact of nature despite the denial of its existence by natural scientists and social scientists. Anthropology is a discipline largely founded on the concept of race and racial hierarchy. Anthropology emerged after the age of western European exploration as 'the study of human,' and that study was almost always focused on the people in European colonies (Ladson-Billing, 2013). Thus, anthropology was conceived as a study of "other."[9]

Short of examining a history of race, the nature of the offering by Donnor and Ladson-Billings does signal the undergirding intention of CRT—which can be used to outline problematics related to race and thus race relations (e.g., racism).

In their book *Critical Race Theory: An Introduction*, Richard Delgado and Jean Stefancic outline basic tenets of CRT.[10] The framework states that racism, the enacted bias relative to the social importance of race, is ordinary, not aberrational. It is the usual way society does business and the common, everyday experience of most people of color in the United States. CRT conjectures that systems of white-over-color ascendancy serve important purposes, both psychic and material. Within the construction there is feature of ordinariness, which means that racism is difficult to cure or address because of the ubiquity of its presence. Also, the notion of color-blindness as an antiracism curative becomes problematic because it presumes an equality based in not seeing difference, thus seeking to remedy only the most blatant forms of discrimination. Conversely, this sometimes called "interest convergence," or material determinism, reifies and different dimension of racism. Because racism advances the interests of both white elites (materially) and working-class people (psychically), large segments of society have little incentive to eradicate it.

CRT posits that the social construction thesis holds that race and races are products of social thought and relations. Not objective, inherent, or fixed, they correspond to no biological or genetic reality; rather, races are categories that society invents, manipulates, or retires when convenient. People with common origins share certain physical traits, of course, such as skin color, physique, and hair texture. But these constitute only an extremely small portion of their genetic endowment, and have little or nothing to do with distinctly human, higher-order traits, such as personality, intelligence, and moral behavior. Through a construct of differential racialization, critical writers in law, as well as social science, have drawn attention to the ways the dominant society racializes different minority groups at different times, in response to shifting needs such as the labor market. Intersectionality and anti-essentialism in and resistance to the singularity of race as a dominant signifier suggests that no person has a single, easily stated, unitary identity.

CRT suggests that there is a unique voice of color, a culturally significant and self-reifying, self-inventive trope of storytelling or counter-telling that is critical to the survival of the signifying monkey. This construction coexists in a necessary tensiveness with arguments against anti-essentialism. The voice-of-color thesis holds that because of their different histories and experiences with oppression, Black, Indian, Asian, and Latino/a writers and thinkers may be able to communicate to their white counterparts matters that whites are unlikely to know. Minority status, in other words, brings with it a presumed competence to speak about race and racism, through the particular articulation of the regimes and results of racism, as well as the liberating and empowering ownership of particularity that is most often occluded in racist efforts to exclude the distinct experiences of people of color. The legal storytelling movement urges Black and Brown writers to recount their experiences with racism and the legal system and to apply their own unique perspectives to assess the master narratives of law that regulate human bodies relative to history markings of difference.[11] CRT has a further interest in the counternarrative to race pathology with a critique of white privilege that comes with being a member of the dominant race, and the social capital of whiteness as marketable commodity or the property functions of whiteness.[12]

Queer of color critique/disidentification/quare studies/intersectionality[13]

This brief overview of CRT is, as I believe CRT to be, not a counternarrative to discussing race and racism as actualizing principles, but in fact to talk about race through a prism of critique that seeks to dismantle as it articulates the features of its impact. In other words, it represents a resistive narration that establishes new templates of promoting politics and possibility of non-white subjects. Such an approach then aptly leads to a discussion that seeks to also de-reify the notion of a queer theory that excludes, as to disavow the experience of "queers of color" as yet another defining exemplar of racism's impact on the formulation of non-white genders. Gloria Anzaldúa explicitly argues,

> queer is used as a false unifying umbrella which all 'queers' of all races, ethnicities and classes are shored under. [Yet] even when we seek shelter under it, we must not forget that it homogenizes, and erases our difference which are always and already mean-ingful to our lived existence and oddly serves as both buffer and magnet in interracial constructions of desire and the colonization of queer bodies of color.[14]

And in this way Cathy Cohen echoes this notion when she says, "queer theorizing that calls for the elimination of fixed categories seems to ignore the ways in which some traditional identities and communal ties can, in fact, be important to one's survival."[15]

These communal ties cross borders of sexuality, but are established through histories of experi-ence, struggle, location, and displacement at the margins of nation and state, gender and sexu-ality, race and culture, and the promotion of particular ideologies of each, that promote and fuel identity politics around notions of queerness. These logics are particularly central in the work of Gaytri Gopinath as she writes: "When queer subjects register their refusal to abide by the demands placed on bodies to conform to sexual (as well as gendered racial) norms, they contest the logic and dominance of these regimes."[16]

In making a reference to how analogies are drawn comparing racism and sexism, Trina Grillo and Stephanie M. Wildman make a keen observation. Queer theory "perpetuate[s] patterns of racial domination by marginalizing and obscuring the different roles that race plays in the lives of people of color and of whites."[17] And while race is a contested term that "may be a whole cluster of strands including color, culture, [nation], identification and experience," it does offer points of unity and can become foundational elements in building community.[18] These are not "imagined communities" in the sense of how Benedict Anderson articulates the desire for community "only in the minds" of those who seek it.[19] These are realized communities that offer support, famil-iarity, and strategies for living.

In his keynote address at the Black Queer in the Millennium Conference, Phillip Brian Harper makes an insightful critique that queer theory bridges what is often a racial divide of inclusion and exclusion in the discussion of sex and sexuality.

> What is currently recognized as Queer Studies is unacceptably Euro-American in orientation, its purview effectively determined by the practically invisible, because of putatively non-existent bounds of racial whiteness. It encompasses as well, to continue for a moment with the topic of whiteness, the abiding failure of most supposedly queer critics to subject whiteness itself to sustained interrogation and thus to delineate its import in sexual terms, whether conceived in normative or non-normative modes.[20]

All of these constructions focus on the concerted exclusion of race, as an elemental tension in queer theory; criterion that fails to fully articulate the experience of queers of color—in what appears as either an intentional or unintentional act of racism in a project that has at its goals the notion of broad inclusivity. In the introduction to *Black Queer Studies: A Critical Anthology* edited by E. Patrick Johnson and Mae G. Henderson, which is, in part, a compilation of the projects presented at the Black Queer in the Millennium Conference in April 2000, Johnson and Henderson state:

> Despite its theoretical and political shortcoming, queer studies, like black studies, disrupts dominant and hegemonic discourses by consistently destabilizing fixed notions of identity by deconstructing binaries such as heterosexual/homosexual, gay/lesbian, and masculine/feminine as well as the concept of heteronormativity in general … Yet, as some theorists have noted, the deconstruction of binaries and the explicit 'unmarking' of difference (e.g. gender, race, class, region, able-bodiedness) have serious implications for those for whom these other differences 'matter.'[21]

Approaches to queer of color critique

José Esteban Muñoz' theory of disidentification as a form of queer of color critique defines "disidentification [as] the hermeneutical performance of decoding mass, high, or any other cultural field from the perspective of a minority subject who is disempowered in such a representational hierarchy."[22] As Roderick A. Ferguson reminds me, "Muñoz suggests queer of color critique decodes cultural fields not from a position outside those fields, but from within them, as those fields account for the queer of color subject's historicity."[23] So, I must fully embrace my positionality within the text of critique—as subject and object—and disavow an *inside and outside* status as a challenge *to distinguish one from the other.* In fact, the critical power of a queer of color perspective is that a positionality of being within gives credence to the limitations and pitfalls of queer theory relative to issues of a felt and experienced racial exclusion of homogenizing whiteness. And as such, a queer of color critique through disidentification can be seen as a standpoint theory, a method for analyzing intersubjective discourses rooted in individuals' knowledge and positionality,[24] with the valorization of a situated knowledge[25] that recognizes an individual's own perspectives is shaped by their social and political experiences. This situated knowledge also capitalizes on intersectionality as a more pronounced recognition of "the various ways in which race and gender intersect in shaping structural, political, and representational aspects" of not only women of color, but all raced and sexual minorities.[26]

Alexander Doty states,

> Queer reception doesn't stand outside personal and cultural histories; it is a part of the articulation of these histories. This is why, politically, queer reception (and production) practices can include everything from the reactionary to the radical to the indeterminate.[27]

And with this logic, queer reception does not stand outside critiques of things or people who are queer, but uses the reception space of queer theory to engage a meta-critique of homonormative paradigms, or use of the strategies of mainstream culture in minoritian populations with the same results: exclusion and domination. In this way, while I may claim a certain border existence within the rhetoric of queer theory, I do not willingly renounce my citizenship in the location that seeks to articulate the lived experiences of queers (gays, lesbians, transsexuals, bisexuals, twin-spirits, same-gender loving, and the multiple and varying ways in which people articulate their sexed, sexual, and gendered selves in the influence of race, nation, and state).

If queer theory seemingly promotes mostly white constructions of gay sexual identity, it most certainly is complicit in racial domination in the service of sexual specificity; a study of white queers at the exclusionary expense of all others. But herein may lay both the limits and possibilities of queer epistemology—especially when pushed by a queer of color critique that not only seeks to identify the "queer of color" in the gaze and spectrum of sexuality discourses but actually broadens the sphere of seeing that is more inclusive of both particular bodies and historicities of being that are eschewed in a more narrow focus. In *Aberrations in Black: Toward a Queer of Color Critique*, Roderick Ferguson outlines the premise of a queer of color critique. Ferguson draws his principles from and offers a critical reading of an essay by Chandan Reddy (1997) entitled and focused on "Home, Houses, Nonidentity: 'Paris is Burning.'" Ferguson cites a critical passage from the essay:

> Unaccounted for within both Marxist and liberal pluralist discussions of the home and nation, queers of color as people of color ... take up the critical task of both remembering and rejecting the model of the "home" offered in the United States in two ways: first by attending to the ways in which it was defined over and against people of color, and second, by expanding the locations and moments of that critique of the home to interrogate processes of group formation and self-formation from the experience of being expelled from their own dwellings and familiar for not conforming to the dictation of a demand for uniform gendered and sexual types.[28]

In his own voice Ferguson then postulates on Reddy's intent:

> By identifying the nation as the domain determined by racial difference and gender and sexual conformity, Reddy suggests that the decisive intervention of queer of color analysis is that racist practice articulates itself generally as gender and sexual regulation, and that gender and sexual difference variegate racial formations. This articulation, moreover, accounts for the social formations that composed liberal capitalism.[29]

Here I offer what I have teased out and framed as rough tenets of Ferguson's queer of color critique/analysis, tenets as guiding or undergirding principles that help to articulate and guide the approach of knowing and doing. Ferguson argues that queer of color analysis "interrogates social formations as the intersections of race, gender, sexuality, and class, with particular interest in how those formations correspond with and diverge from nationalist ideals and practices."[30]

Queer of color critique approaches culture as one site that compels identification with and antagonisms to the normative ideals promoted by state and capital. Queer of color analysis must examine how culture as a site of identification produces such odd bedfellows and how it—as the location of antagonisms—fosters unimagined alliances. Queer of color analysis, as an epistemological intervention, denotes an interest in materiality, but refuses ideologies of transparency and reflection, ideologies that have helped to constitute Marxism, revolutionary nationalism, and liberal pluralism. Queer of color analysis eschews the transparency of all these formulations and opts instead for an understanding of nation and capital as the outcome of manifold intersections that contradict the idea of the liberal nation-state and capitals as sites of resolution, perfection, progress, and confirmation.

Queer of color analysis presumes that liberal ideology occludes the intersecting saliency of race, gender, sexuality, and class in forming social practices. Approaching ideologies of transparency as formations that have worked to conceal those intersections means that queer of color analysis has to debunk the idea that race, class, gender, and sexuality are discrete formations, apparently insulated from one another. Queer of color critique challenges ideologies of discreteness. It attempts to disturb the idea that racial and national formations are disconnected. If the intersection of race, gender, sexuality, and class constitute liberal capitalism, then queer of color analysis obtains its genealogy within a variety of locations: women of color feminism names a crucial component of that genealogy as women of color theorists have historically theorized intersections as the basis of social formation. Queer of color analysis extends women of color feminism by investigating how intersecting facial, gender, and sexuality practices antagonize and/ or conspire with the normative investments of nation states and capital.

Queer of color analysis claims an interest in social formations, and locates itself within the mode of critique known as historical materialism. Since historical materialism has traditionally privileged class over other social relations, queer of color critique cannot take it up without revision, must not employ it without disidentification, if to disidentify means "[recycle] and [rethink] encoded meaning" and "to sue the code [of the majority] as raw material for representing a disempowered politics of positionality that has been rendered unthinkable by the dominant culture."[31] Queer of color analysis disidentifies with historical materialism to rethinking its categories and how they might conceal the materiality of race, gender, and sexuality.[32] Queer of color analysis is a heterogeneous enterprise made up of women of color feminism, materialist analysis, poststructuralist theory, and queer critique.[33]

Ferguson references his postulation as *queer of color critique* and *queer of color analysis*—the difference of which offers both a theoretical framework and methodological engagement similar to the argument made by D. Soyini Madison establishing critical ethnography in/as critical social theory with a focus on a method of doing: in this case, a method in this case that is both about a doing, observation of a doing, and the critical strategies of talking (writing) about a doing that makes manifest the undergirding and emergent theories of a queer of color critique/analysis that is theoretically embodied and astute.[34]

Quare studies: a counternarrative to queer theory

In his now germinal essay "Quare Studies, or (Almost) Everything I Know About Queer Studies I Learned from my Grandmother," E. Patrick Johnson offers the important construct of quare studies as both a counter theory and maybe more importantly a counternarrative to queer theory. Johnson argues that the dividing and discerning principle in queer theory is particular to whiteness and engages, with intention or not, the occlusion of "people of culture." What is key about the dimensions of counternarratives that Johnson proposes in his construction of "quare studies" is an embodied and intellectual resistance to master narratives, as well a rejection of hegemonic narratives that exclude the particularity of other lived experiences and ways of knowing—and indeed as experiencing bodies. This resistance performance plays out in Johnson's construction of quare studies—both as a theoretical pushback and in the manner in which his method of theorizing from the "little stories" of his grandmother. It provides space for the voicing of experience from a Black gay body that broadens the scope and inclusion of non-white bodies in the theorizing of queer theory. In particular, the "little story" that Johnson uses to spawn his theory (or his grandmother's theory) goes as such:

> I remembered how "queer" is used in my family. My grandmother, for example, used it often when I was a child and still uses it today. When she says the word, she does so in a thick, black, southern dialect: 'That sho'll is a quare chile.' The use of 'queer' is almost always nuanced. Still one might wonder, what if anything could a poor, black, eighty-something, southern, homophobic woman teach her educated, middle-class, thirty-something gay grandson about queer studies? Everything. Or *almost* everything. On the one hand, my grandmother uses 'quare' to denote something or someone who is odd, irregular, or slightly off kilter—definitions in keeping with traditional understandings

and uses of 'queer.' On the other hand, she also deploys 'quare' to connote something excessive—something that might philosophically translate into an excess of discursive and epistemological meanings grounded in African American cultural rituals and lived experience. Her knowing is not knowing vis-à-vis "quare" is predicated on her own "multiple and complex social, historical, and cultural positionality" (Henderson, p. 147). It is this culture-specific positionality that I find absent from the dominant and more conventional usage of "queer," particularly in its most recent theoretical reappropriation in the academy.[35]

I offer you this complete narrative for a number of reasons: first, to evidence or exemplify the "little story" in which the magnitude of meaning belies the referent. Second, it shows how Johnson uses an idiomatic Black cultural storytelling (oral history) as a trope of critical theorizing, both of his grandmother and his own academese. And third, the narrative shows the manner in which Johnson presents quare of color critique as an embodied praxis—a knowing that is both processed through lived experience but an articulation of doing that fuses the academic knowing back in a body that demands attending to the front lines of both academic theorizing and polit- ical activism as "an interventionist disciplinary project."[36] Johnson grounds a significant portion of quare studies in "theories of the flesh" that recognize the importance of particularity and lived experiences (plural) through a diversity of embodiments as a source of mounting theories of knowing and a politics of resistance.[37] Like Ferguson's queer of color critique, allow me to tease our tenets of quare studies that Johnson lay bare in his essay.

Johnson writes: "Because much of queer theory critically interrogated notions of selfhood, agency and experience, it is often unable to accommodate the issues faced by gays and lesbians of color who come from 'raced communities.'"[38] Quare not only speaks across identities, it articulates identities as well. Quare offers ways to critique stable notions of identity and, at the same time, to locate racialized and class knowledges. Quare studies is a theory of and for gays and lesbians of color. Quare studies closes the gap of queer theory's failure to acknowledge consistently and critically the intellectual, aesthetic, and political contributions of non-white and non-middle-class gays, bisexuals, lesbians, and transgendered people in the struggle against homophobia and oppression. Quare studies narrows the gap between theory and practice, per- formance and performativity to the extent that it pursues an epistemology rooted in the body. Quare studies focuses attention on the racialized bodies, experiences, and knowledge of trans- gendered people, lesbians, gays, and bisexuals of color. Quare studies grounds the discursive pro- cess of mediated identification and subjectivity in a political praxis that speaks to the material existence of colored bodies.

Johnson suggests that quare studies deploy theories of performance. Performance theory not only highlights the discursive effects of acts, it also points to how these acts are historically situated. Quare studies' theorizing of the social context of performance sutures the gap between discourse and lived experience by examining how quares use performance as a strategy of survival in their day-to-day experiences. Quare studies offers a more utilitarian theory of identity politics, focusing not just on performers and effects but also on contexts and historical situatedness. Quare studies encourages strategic coalition building around laws and politics that have the potential to affect across racial, sexual, and class divides. Quare theory incorporates under its rubric a praxis related to the sites of public policy, family, church, and community. Quare studies is specific and intentional in the dissemination and praxis of quare theory and committed to communicating and translating its political potentiality. Quare theory is "bi-directional": it theorizes from bottom to top and top to bottom (pun intended!). This dialogical/dialectical relationship between theory and practice, the lettered and unlettered, ivory tower and front porch is crucial to a joint and sustained critique of hegemonic systems of oppression. Quare studies cannot afford to dismiss, cavalierly, the role of the Black church in quare lives. However, it must never fail to critique the Black church's continual denial of gay and lesbian subjectivity.

And ultimately, for Johnson, quare studies addresses the concerns and needs of gay, lesbian, bisexual, and transgendered people across issues of race, gender, class, and other subject positions. Quare studies is committed to theorizing the practice of everyday life. Because we exist in material bodies, we need a theory that speaks to that reality."[39] Johnson's quare studies ask us, and maybe more so demands, that we "move beyond simply theorizing subjectivity and agency as discursively mediated to theorizing how that mediation may propel material bodies into action."[40]

Links between Johnson's quare studies and Muñoz' notion of disidentification

José Estaban Muñoz' construct of a critically applied method of disidentification has already been invoked as a form of queer of color critique. What is further important to note about the construct is that throughout the project he variously engages disidentification: as ambivalent structure of feeling, as anti-assimilationist, as breaking down political possibility, as camp, as counterperformativity, as resistant to dominant ideologies, as melancholia, as a mode of performance, as a paradigm of opposition reception, as a practice of freedom, as a reformulation of performativity, as representational protocols of identity, as a response to state and global power apparatus, as a strategy of resistance, as a survival strategy of minoritarian subjects, as a tactical misrecognition of self, and as theories of revisionary identification.

Each of these constructions invoke action and subversion through performance; performance as the strategic and critical re/enactment of identity, and the aesthetic re/fashioning of the self in relation/resistance to society in a publicly staged manner for public consumption, reflection, identification, and dis-identification. Muñoz is clear about the limitations and profundity of the method. In the beginning of the book he states:

> disidentification is *not always* an adequate strategy of resistance or survival for all minority subjects. At times, resistance needs to be pronounced and direct, on other occasions, queers of color and other minority subjects need to follow a conformist path if they hope to survive a hostile public sphere. But for some, disidentification is a survival strategy that works within and outside the dominant public sphere simultaneously.[41]

To this extent, while Johnson deploys theories of performance, Muñoz' theorizing of disidentification is grounded in a critical analysis of live performance, of queers of color who use the craftedness and craftiness of performance as a critical methodology of both performing politics and the performance of politics, and engages the embodiment of such politics through theories of their own flesh and the illumination of politics through performance, through the queer bodies of color. Muñoz writes:

> The cultural performers I am considering in this book must negotiate between a fixed identity disposition and the socially encoded roles that are available for such subjects. The essentialized understanding of identity (i.e., men are like this, Latinas are like that, queers are that way) by its very nature must reduce identities to lowest-common-denominator terms ... The version of identity politics that this book participates in imagines a reconstructed narrative of identity formation that locates the enacting of self at precisely the point where the discourses of essentialism and constructivism

short-circuit. Such identities use *and* are the fruits of a practice of disidentificatory reception and performance. The term *identities-in-difference* is a highly effective term for categorizing the identities that populate these pages.[42]

Muñoz also grounds his theory in the discourses of women of color who have helped to build up a politics of disidentification, particularly the work Gloria Anzaldúa and Cherrie Moraga. Muñoz' logic of (dis)identification becomes a contestation not only to dominant ideologies, but also to white feminism. He argues that "although the advancements of white feminists in integrating multiple sites of difference in their analytic approaches have not, in many cases, been significant, the anthology has proved invaluable to many feminists, lesbians, and gay male writers of color."[43]

Muñoz' deployment of disidentification becomes a critical performative praxis of queer worldmaking. Short of teasing out formal tenants, as was the case with Ferguson's queer of color critique/analysis or Johnson's quare studies, I draw from the last paragraph of Muñoz book, *Disidentifications*, and tease out what reads to me as his treatise or theorems of disidentification. Muñoz writes, "Disidentification is a point of departure, a process, a building. It is a mode of reading and performing and it is ultimately a form of building."[44] Building takes place in the future and in the present, which is to say that disidentificatory performance offers a utopian blueprint for a possible future while, at the same time, staging a new political formation in the present. Through the "burden of liveness," we are called to perform our liveness for elites who would keep us from realizing our place in a larger historical narrative. The minoritarian subject employs disidentification as a crucial practice of contesting social subordination through the project of worldmaking.[45]

The call for action seems indicative and descriptively responsive to a queer/quare of color critique. It is a call for action that is resistant to both the absences and occlusions of queer theory, not fixating in the space of merely critiquing queer theory, but marshalling a mobilization of cause and effort to a quare worldmaking to which queer theory is currently incapable of creating. To this extent, I have used the construct of a critical performative praxis of queer worldmaking. So drawing from previously outlined theories, perspectives, and methodologies allows me the comfort of offering a working definition.[46] Hence, a critical performative praxis of queer worldmaking (CPPQWM) takes as its charge an incisive examination of the creative, performative, intimate, publicly disruptive,[47] personal, and political everyday lives of queer identified folk (e.g., LGBTTQ and beyond). This is done with an express focus on their intersectional identities in relation to race, gender, ethnicity, sexuality, embodiedness, and all markers of their particularity. Along with the assumptions and commitments of recognizing struggle, amplifying voice, archiving memory, empowering transformation, and mobilizing social change as activist intention, CPPQWM engages the tools of equity while recognizing that indigenous rhetorics of the front porch, of "learning how to take our differences and make them strengths," makes particularity important. Particularlity, then, becomes a positionality that makes more salient, manifest, and present the lived experiences of queers of color in practice, performance, and protest as means of promoting equality across differences working to systematically dismantle the masters house.[48]

As I have written elsewhere, maybe CPPQWM is involved in establishing

> a queer decorum as a rhetorically reflexive process of critiquing the very foundations of what is assumed to be appropriate in any given situation. That is, it can be understood as an exploration that benefits others in finding all the available means of their engagement; as acts of social justice; and acts of undifferentiated and nonhierarchical humanity. Let us strive for a queer decorum that is not exclusively about LGBTTQ folks, but a queer decorum as a resurgent way of critiquing self in society.[49]

(For me, this call invokes, in a different way, the politics of intersectionality and noncompetent investments that plays within queer/quare communities.

Further notions of intersectionality

While I have invoked Kimberlee Crenshaw's notion of intersectionality, as both theory and method, even claiming its undergirding impulse as queer/quare—Gust Yep's move toward thick(er) intersectionalities might serve as a strong action step, a further call from queer/quare theorists to transgress the boundaries of stayed constructions of identity that might further reify and homogenize identities at the intersections. Yep calls for

> exploration of the complex particularities of individuals' lives and identities associated with their race, class, gender, sexuality, and national locations by understanding their history and personhood in concrete time and space, and the interplay between individual subjectivity, personal agency, systemic arrangements, and structural forces.[50]

He writes further and outlines the concept of thick(er) intersectionalites (TI):

> As such, this concept suggests that we need to attend to the lived experiences and biographies of the persons occupying a particular intersection, including how they inhabit and make sense of their own bodies and relate to the social world (Yep, 2013). TI features four defining characteristics associated with social identity in a neoliberal global world. First, it struggles against coherence and premature closure of identity. Second, it embraces the messiness of everyday experiences in the social world. Third, it focuses on the affective investments that people make in their identity performances. Fourth, it attempts to understand identities as embodied and lived by people within geopolitical historical contexts.[51]

Yep applies his analysis to the notion of a critical ethnography of multiple intersecting communities of Filipina trans women in San Francisco. It is a "study that focus[es] on the performances of identity, among other things, of several women" that "simultaneous def[ies] a number of U.S. cultural binaries, such as man/woman and heterosexual/homosexual, and reif[ies] a range of U.S. cultural normativities, such as physical beauty and femininity. In other words, their identity performances are complex and 'thick.'"[52]

Yep's construction towards a thick(er) intersectionality is most certainly reflected in the critical performance analysis complete by José Esteban Muñoz and is most certainly present in the oral history and performance-based project that E. Patrick Johnson engages with Black gay men from the South. It is also reflected in a project in ethnography in which Johnson is involved, not only in collecting oral narratives (which he re-performs live), but also in the book, *Sweet Tea*, a critical ethnographic analysis of the social and historical conditions associated with their race, class, gender, sexuality, cultural, and locational politics that effect their personhood, particularity and personal agency.[53] Johnson's work continues with a later focus on Black queer southern women.[54]

Gust Yep's construction takes me back to D. Soyini Madison's outlining of critical social theory at the service of a critical ethnography to which Yep is engaged. In particular, there is a critical necessity "to direct our attention to the critical expressions within different interpretive communities relative to their unique symbol systems, customs, and codes and to demystify the ubiquity and magnitude of power"[55]—to which I believe that queer/quare color of critique is most committed to in a critical praxis of thought and action.

Whether as particular backlash to queer theory or as a cultural conscious, community conscious, and race conscious critique for social transformation and empowerment—queer/quare of color critique and the still emerging interpretive queer methodologies embody, in more salient ways, the postcolonial move that should be at the core of queer theory: that is, queer theory lacks a focus on the complicated construals of queer identity across variables of race, class, nation, state, and geography, with a particular lack of focus on articulating experience and voice. In this sense, I want to echo an important construct offered in Jeffrey Q. McCune, Jr.'s analysis on the very queerness of blackness in the case of Michael Brown being murdered by Officer Darren Wilson in Ferguson, Missouri. He writes,

> Queerness of this black moment is also marked in the ways that suspension and suspicion cooperate, as folks engage with nonnormative bodies, sexualities, and genders as sometimes inside and sometimes outside. The oscillation here is also an allegory, for the life of nonnormative subjects within marginalized spaces—feeling at once free and trapped at sites of solidarity.[56]

McCune's work on African American men who have sex with men while maintaining a heterosexual lifestyle also highlights how the social expectations of Black masculinity intersect and complicate expressions of same-sex affection and desire.[57]

But this critique on the state of queer theory is not just an idiosyncratic bias. In *What's Queer About Queer Studies Now?*, David L. Eng, Judith Halberstam, and José Esteban Muñoz call for a renewed queer studies with a broadened consideration. They state,

> At such a historical juncture, it is crucial to insist yet again on the capacity of queer studies to mobilize a broad social critique of race, gender, class, nationality, and religion, as well as sexuality. Such a theoretical project demands that queer epistemologies not only rethink the relationship between intersectionality and normalization from multiple points of view, but also and equally important, consider how lesbian, gay and lesbian [bisexual and transgendered, and additional bodies] rights are being reconstituted as a type of reactionary (identity) politics of national and global consequence.[58]

It is a call for action toward the emergence of a critical interpretive queer methodology that addresses the concerns of both a nihilistic postcolonial perspective and homogenizing queer studies, thus suturing the pains and possibilities of each. It is a reimagined approach that works toward elaborating social action issues without simply replacing ills with additional harms but introducing new spaces of inquiry. This is, I believe, one of the most bracing qualities and intentions of postcolonial perspectives, in the caution and care of replacing an essentialist eurocentrism and reestablishing what Edward Said might reference as, building a "culture of resistance as a cultural enterprise."[59]

Moving in the vein of Johnson's quare studies that "would not only critique the concept of 'race' as historically contingent and socially and culturally constructed/performed, would also address the material effects in a white supremacist society" crossing or bleeding the borders of identity construction that effects the material practices of culture, gender/sexuality, and the socially delimited constructions of possibility.[60]

But I am taken with the voice of Jesus I. Valles-Morales and Benny LeMaster, when they write:

> To practice queer of color criticism is to live a life that is closer to a freedom, to liber-
> ation, to embracing the worlds we bring with us to the classroom, the worlds we may
> not be able to leave behind. To practice queer of color criticism is to allow the self to be
> porous to the hurt of others, to let our understandings be guided by what our bodies
> know, what our communities have taught us. So here, we breathe and we question, and
> in doing so, we begin.[61]

And voices like the queer praxis collective, Dustin Bradley Goltz, Aimee Carrillo Rowe, Meredith
M. Bagley, Kimberlee Perez, Raechel Tiff and Jason Zingsheim in their project, *Queer Praxis:
Questions for LGBTQ Worldmaking* work to

> centralize [queer] theory through and from a relational perspective. [They] generate
> and examine [queer praxis] in a manner that forefronts dialogic collectively, multiple
> points of access, and the living/shifting nature of critical queerness... *Queer Praxis* is
> resistant to notions of identity, for sure, but works to directly challenge and extend the
> logics of queer negativity [occlusion and exclusion] in favor of a queer sociality—of
> queer relationality.[62]

A spiritual interlude

Within the constructs of queer studies, and maybe more importantly quare studies and queer
color critique/analysis, there is what I experience as an explicit and implicit reference to spirituality. It is a tensive and emergent spirituality that resists religious and social edicts that sometime
quarantines, pathologizes, and make sinful same sex or nonnormative beings/desires. It establishes
a cosmology of connecting queer/quare desire with a force that acts both outside the regimen-
tation of sexual/cultural action and emanates from internal locations of natural intent. It is a
form of spirituality that thinks, believes, and yearns for a space, a practiced place of liberation and
possibility (e.g., queer/quare space, the gay night club, and even a liberated notion of church).
This spirituality is reflected in the pan-Indian notion of someone being two-spirited "to describe
Native people in their communities who fulfill a traditional third-gender (or other gender-
variant) ceremonial role in their cultures" and operates in part for me as a practiced place beyond
the divine.[63] The term seeks not only to identify difference, but to disrupt a unifying sense of
reductive gendered and sexed identities in a binary of social conformity, communally uplifting
the gender-variant in terms of a spiritual enlightenment with communal purpose. The term, as a
collective and particular cultural construction, is a form of disidentification, as a cultural practice
that differentiates not only bodies and desires but cultural ways of knowing and performing communal membership. This occurs with the potentialities of being within community while living
in one's particularity, plural or otherwise. It means creating a collectivity of gendered possibilities
that is recognized through the cosmologies of not only knowing, and believing, but being.
  Spirituality can be said to

> include a sense of connection to something bigger than ourselves, and it typically
> involves a search for meaning in life. As such, it is a universal human experience—
> something that touches us all. People may describe a spiritual experience as sacred or
> transcendent or simply a deep sense of aliveness and interconnectedness.[64]

When Ferguson discusses that "queer of color analysis has to debunk the idea that race, class, gender, and sexuality are discrete formations, apparently insulated from one another,"[65] I believe that he is advocating not just for a broadening and yet particularizing conceptualization of cultural critique. I believe that he is advocating for a liberating space, maybe even a spiritual space, of possibility that exists at the intersections; a liveable space where race, class, gender, and sexuality are discrete formations that also create a synergy of particularity that informs cosmologies of knowing and reknowing the self in the world.

I also believe that Ferguson's discussion connects with Muñoz's notion of a disidentification that tracks a utopian impulse to "use the stuff of the 'real world' to remake collective sense of 'worldness.'"[66] For me, that is created not only through his notion of "spectacles, performances, and willful enactments of the self for others," but engaging in spiritual acts as both performative resistance and worldmaking, which is an emergent performative act of spirituality. Which, invoking Johnson's tenants of quare studies, "cannot afford to dismiss, cavalierly, the role of the Black church in quare lives. However, it must never fail to critique the Black church's continual denial of gay and lesbian subjectivity."[67] This not only serves as an act of resistance against, but as disidentification, for most Black folks steeped in the religiosity of the church in general, and the Black church in particular. In which case, Johnson's charge is not a call to resist the church, but to not fail to critique the Black church as a means of relocating ourselves within the boundaries of a faith to which we are so intricately intersectionally connected—at the crossroads of community, culture, character, and identity.

And since the notions of gender and sexuality are performative dispositions, and race and class are all social manifestations that are embodied, performed, and witnessed (both as social recognitions and social assessments) allow me an embodied performative moment to speak to all these issues made manifest at the intersections of my personal and professional life.

On the intersections of (my) queer/quare and Catholic identity[68]

This is a partial rendition of a presentation given at what was touted as the first LGBTQ+ retreat at the Catholic Jesuit university in which I serve as an academic dean. The politics of place and space are important both in the narrative, and the broader context of the essay to which this entry serves as a critical praxis at the intersection of faith and sexuality in place and space.

On my desks in my home and work offices there is the same picture of my parents. It is my favorite picture of them. They are in a dancing embrace, only paused for the photographic moment. In the photo, my father has his signature sunshades on to protect his always sensitive eyes. And my mother, with a full smile, looks directly into the lens of the camera. I imagine that she is always looking at me, reminding me of the man that she wanted me to be.

I always look at that picture of my parents and particularly that image of my mother looking back at me—when I am on the phone or about to write that email that might shatter someone's reality. I look at my mother looking and smiling at me, and I pause. And while I still deliver the message that I need to communicate, from time-to-time I might soften the tone and edge of the delivery or invite them to a meeting, creating my process of performing my academic and administrative role with humanity. My mother taught me about care, conviction, and character. And through the photo and my memory of her, she reminds me every day about the man she wanted me to be. And in most cases, I imagine that she is proud. So, in reference to that daily ritual in the memorializing of my parents who are both long past, I offer you additional glimpses to things that my mother taught me throughout my childhood about the intersections of my queer and Catholic identity.

*****

My mother used to tell me that I was made in the image of God.

My mother used to tell me that she saw a light and joy in me that could give

comfort and care to others, and that it was my responsibility to do so.

My mother used to tell me to be consistently present in my whole self, in my whole person—but always give thanks to God for the many gifts and particularities that He has bestowed upon me.

My mother used to tell me to honor my own spirit, and to respect my mind and body.

My mother used to tell me that *God does not make mistakes.*

*****

So as far back as I can remember, my mother told me these things as lessons to live. She would tell me these things: sometimes triggered by things that I said or did, or that happened to me. Sometimes triggered by my hurt feelings from childhood name calling: fag, sissy, punk, queer. Names that were hurled at me with the intent to hurt and shame a still developing self-identity. Sometimes triggered by thuggish pro-masculine-heterosexual male bullying–when my interests turned to the gentle arts–fine, performing, literary, and communication–with *an unwavering commitment to social justice*, rather than playground brawls and bullying.[69]

And sometimes she would tell me these things after church when someone in robed authority would utter rigid religious rhetoric that attempted to delimit the importance of my being in the world. And once, in that not often discussed time when as an altar boy, that same hypocritical robed authority lured me into an inappropriate act that challenged my belief in God. I told my mother and she quickly reinforced to me that adults taking sexual advantage of children is abuse, not love, and that the robed authority was acting on the desire of man, not God. She then walked over to the church to confront and report the culprit. For you see, my mother knew of my gay identity before the happening. And in some ways my mother knew of my gay identity long before I could fully articulate it on my own terms. She loved me fully and sheltered me consistently. My mother's faith in God and the Church was not challenged, but her faith in those who did not practice what they preached was shaken.

I am the fourth boy of five boys, and I had another brother who was also gay. And it would take years before my older brother Nathaniel and I would talk about what we shared. The occasion of such discussion came only after he used one of the private mantras from my mother to comfort me. I repeated others to him. And we laughed. We laughed thinking that we were alone in our struggle but always in the comfort of our mother. And through that laughter, we recognized what our mother had been doing all of our lives: comforting, caring, and encouraging her two boys (who happen to be gay) to be fully present in their faith, and in their particularity, and encouraging us to find a mindful and respectful balance in each.

My brother and I came to understand that my mother was offering *lessons for being* in God's presence as we were reared in a Catholic household. She offered a series of lessons that supported our particularity without demonizing our character, spirit, and potentialities. She encouraged us to live fully alive, whole persons present. But we understood that her encouragement was not a blessing on our sexuality. It was a Catholic mother's love for her children, which was relentlessly supportive of each of us as children of God.

My mother was what many in my childhood would call a "good Catholic girl," and still what others would call "a God-fearing woman." But she would always resist the Old Testament notion of a wrathful God that we should fear because as she would say, "If your hearts are good, then God knows it." There is no need to fear Him. She revered the glory and majesty of God, of His care and compassion and celebrated all of His creations. And she taught us to practice the faith in relation to fulfilling our personal sense of being and humanity. She would tell us, "Always stay close to God." She had all of seven children go to catechism classes from kindergarten through high school. I was an altar boy and sang in the church choir. I taught catechism for years as a display of my care and passion of faith. I have been a lecturer and cantor in church. And I have been a Eucharistic Minister for most of my adult life. These are ways that I follow my mother's lessons of keeping God in my life. For both her and for myself. My mother would tell me that the nature of my faith is not vitiated or invalidated by my sexuality, though I should practice humility and respect in both. And despite some interpretive aspects of catholicity that might demonize my queer identity, my mother told me that a performed and enacted faith is not restricted by desire, but measured by the character and practice of that faith in its most empowering and altruistic features.

There is an old joke which is based in what I call *directional or misdirectional humor*. You might know the joke. It goes like this: Situated in New York City a tourist, or maybe a young musician, asks a passerby: "How do I get to Carnegie Hall (the famed music venue)?" And the answer is "Practice, Practice, Practice!" Such is the answer to the question that I have been asked to respond to in my personal life at this very particular retreat in this very particular place.

> Question: "How do you maintain your faith as a member of the LGBTQ+ community?"
>
> Answer: "Practice, Practice, Practice!"

For me, faith is not belief without proof. I believe that a credible faith is a rational faith that looks at the nature of being as evidence of knowing. And the commitment to faith is a rhetoric of engagement, a belief which we must embody and make manifest in our actions.[70]

So, when I say and tell others:[71]

> I believe in *God who is our loving Father and Creator. And that God's love is limitless and overflows into our hearts and lives. And that God has created us out of His love that sustains and supports us daily.* I then carry that with me in my daily life and in the choices that I make in my life. Being a member of the LGBTQ+ community does not offset that belief—in fact I hope that my faith emboldens my identity in all the things that I do.

So, when I say and tell others:

> I believe that my faith to *love God above all else and to love others* is manifested in a practice of forgiveness, mercy, and to care for the poor and helpless. That extends into my membership in the LGBTQ+ community and beyond. As a part of my faith, I practice and work for peace and justice in our world, which means that I am bringing God's kingdom of peace, unity, and love to a world faced with conflict, division, and strife through my embodied queer particularity, not despite of it.

So, when I say and tell others:

> My faith (LGBTQ+ embodied or otherwise) *recognizes the need for forgiveness and sees the sacrament of reconciliation as a means to receive this great gift of God's forgiveness.* That is not forgiveness for my queer identity but forgiveness for the many failures and foibles of my humanity at large.

33

My faith requires and recognizes the importance of *practiced prayer* and daily critical reflection—in both formal and informal ways: on my knees supplicating myself or negotiating the daily traffic behind the wheel of my car. An Examen, by any other name, is still prayerful contemplation.[72] And I tell people that my faith recognizes the importance of service to others. Believing that Jesus came to serve and not to be served, I too strive to be more Jesus-like in my daily living. Hence the answer to the question "How do you maintain your faith as a member of the LGBTQ+ community? is:"Practice, Practice, Practice!" Having faith means being engaged in faith practices, daily—in my scholarship, in my art, in my teaching, in my leadership, and the ways in which all of those meet in the confluence of everyday living.

But there is also another question that is asked of me:

How I do I maintain friends and relationships in my personal and professional life?

And I answer: Being consistent and fully present in myself, in my whole self, which includes my faith and queer identity amongst other things. It means making fully present all aspects of my identity, all aspects of my complex *intersectional* identity:[73]

My intersectional identity:

as a man

as Black/man

as a Black/gay/man

as a Black/gay/Catholic/man

as a Black/gay/Catholic/man/citizen

as a Black/gay/Catholic/man/citizen/teacher/artist/scholar/administrator

as a Black/gay/Catholic/man/citizen/teacher/artist/scholar/administrator/brother

as my mother's son and my father's boy

and always as a child of God.

It means making my intersectional self fully and consistently present to engage and help others (and the world) to recognize support and celebrate our joint humanity while promoting justice.

This is most appropriately done through practices of information, formation and transformation (for self and others).[74] And to do so with great PRIDE. But not just PRIDE as a political-activist move and a positional statement like, "We're here and we're queer!" which is always important. I also refer to pride as a feeling of deep pleasure or satisfaction in the possibilities derived from one's faith,[75] a faith through God to continually ask for the grace to stand with others (the poor and the suffering), to educate, to heal, to recognize the humanity of the stigmatized, and to work to liberate the oppressed (even as I am also oppressed).[76] From a secular place, as Adrienne Rich suggests, "Pride is often born in the place where we refuse to be victims, where we experience our own humanity under pressure, where we understand that we are not the hateful projections of others but intrinsically ourselves."[77] Such recognition of self-worth occurs in spite of social derision and to begin to thrive in our personhood is to assist others in their struggles.

I strongly believe in the importance of racial/ethnic/gender diversity within any academic and professional environment, as it contributes to not only the diversity of presence, but the diversity of ideas that enriches the human/humanizing experiences in everyday life. Such a stance is not only about an "open embrace and hospitality [as] a feature [that] distinguishes … the Catholic academy."[78] It is at the core of a performed catholicity—tolerance, welcome, and the recognition of each person's innate humanity. So, in my personal and professional life I have never had to overcome any perceived challenges of being accepted for any individual part of my intersectional identity—because if you're your truth, you cannot be called out of your reality. Though oppression, racism, and bias play out on all levels and are always and already collective to our whole identity, the fullness of my person lives and thrives where those places and elements of my intersectional identity not just meet, but overlap and co-inform each other. I draw strength from diverging and converging histories of resistance, where such histories meet to synergize, energize, and fortify my being. So, the struggle is not particular to one trait but plural to my whole self. And the survival of and within that struggle has established this *fabulous* person that my mother once described me to be.

The challenges are most often not my own (alone), but the perceptions of others who must negotiate their own values, who must negotiate their own sense of what is appropriate in the range of human social behavior, how they practice their humanity, and how they practice their own faith. Hopefully that is:

A faith that requires the practiced acceptance for human difference.

A faith that is celebratory of the diversity of all human beings.

A faith that recognizes the diversity of all God's creations.

A faith that respects the power of God's love made manifest in their practice.

And a faith that does justice. *(And there is no hate in a faith that does justice.)*

*****

So, my mother used to tell me that I was made in the image of God. My mother used to tell me that she saw a light and joy in me that could give comfort and care to others, and that it was my responsibility to do so. My mother used to tell me to be consistently present in my whole self, in my whole person—but always give thanks to God for the many gifts and particularities that He has bestowed upon me. My mother used to tell me to honor my own spirit and to respect my mind and body. My mother used to tell me that God does not make mistakes. My mother used to also tell me to keep God in my life, not just as refuge in times of strife but to practice faith as a commitment to the possibilities and potentialities of God's great promise of redemption. And that is what I attempt to do every day: be fully present in a faith that does justice, a faith that allows me to be fully present in the lives of others. And a faith that allows me to be fully present on my own personal journey of both being and becoming a child of God. I do not use my many gifts or my particularities to beat down, demean, or shame others in any aspects of my personal or professional life. Sometimes that is perceived as a weakness. My mother would tell me that it is a sign of God's strength in me. I am my mother's son. And all my life, and particularly in my realized/out gay identity, my mother would say, "*Keith is* _still_ *a relatively good Catholic boy.*" The "still" served as a mindful qualification of her own strict catholicity that still supported her gay-boy's commitment to his faith in relation to his particularity. It was a playful acknowledgement between the two of us. And we both smiled when she would say it.

On my desks in my home and work offices there is the same picture of my parents. It is my favorite picture of them. They are in a dancing embrace, only paused for the photographic moment. In the photo, my father has his signature sunshades on to protect his always sensitive eyes. And my mother, with a full smile, looks directly into the lens of the camera. I imagine that she is always looking at me and approving of the man that I have become. I have faith that all the things that my mother told me are true. I try to practice that faith daily in the fullness of my identity and authorizing others, particularly those with the same struggles, to do the same. This practice works against those who presume the privileges and comforts of naming and shaming others and who further marginalize the presumed other and create a schism in our joint humanity.

I tell my story through the critical practices of autoethnography, using my lived experience as a cultural terrain for close scrutiny and a critical reflexivity. Like the Pensieve of Harry Potter fame, the methodological approach is a technology of reviewing stored memories to see again what one has once seen (or felt) before and to revisit that experience again critically, with the exactitude of artistic practice, and then, with courage and conviction, share the process and product of that discovery to wider cultural, social, and political audiences as testimony and evidentiary knowledge. This approach is not for you to code on that which is presented, but to help you and others to see the technique and technology as a methodology so that you (they) can to begin your (their) own process of sense making. I practice religion as an act of faith; I perform faith as an act of generosity and restraint; and I strive to achieve my own possibility and potentiality of being and becoming, trying to assist others on their journey. These are all the same things that I wish for you, and more.

Not an "act of contrition": a postscript[79]

In the Catholic Church, confession is a sacrament, a sacrament of reconciliation that allows and invites those who have sinned to obtain forgiveness and reconcile with God and the Church. Often, as a particular penance of confession, the priest assigns the charge of deep reflection or discernment through a repetition of prayers—X number of Hail Marys, X number of "The Lord's Prayer," or an "Act of Contrition."

> O my God, I am heartily sorry for having offended thee,
>
> and I detest all my sins because I dread the loss of heaven and the pains of hell,
>
> But most of all because they have offended thee, my God, who art all good and deserving of all my love.
>
> I firmly resolve, with the help of thy grace,
>
> to confess my sins,
>
> to do penance,
>
> and to amend my life. Amen.

I am not a sinner, at least not related to my racial and gendered particularity.

And this is not an act of contrition.[80] "In my Father's house are many mansions … [He goes] to prepare a place for me" in my particular plurality that is also made in His image.[81] There is a place for me. There is a place for you.

God recognizes my heart and spirit. It is the motivations of man/society that demand the performance of guilt and forces choices between the particularity of belief and religious practice with a rigid margin of personal expression, at least as it relates to issues of sex, sexuality, and queer identity. I don't believe myself to be a sinner, at least not linked with my sexuality—and such is a statement that is part of a reconciliation with self and religion that my mother taught me.

Conclusion

Within the title of this chapter are multiple and interlocking notions that inform the broader discussion of race, spirituality, queer, and quare as performative aspects of gender identity meeting at complex intersections. I use the term "meeting," not like a carefully planned city map with planned intersections that direct the traffic of social investments with expected performances of directionality, turn-taking, and immanent domain but instead as the dense particularities of our identities demand a reckoning and reconnoitering of primacy in the unification of who we proport ourselves to be and thus, how we embody that sense of being in the sociocultural environs in which we circulate. How we perform that sense of being and how we thus communicate to the world our being as a positionality of purpose is a political act. A positionality of purpose in being must always influence our perception of self in the world and serve as our filter and reading of society, and also as the pivot point from which the politics of being and survival rotates in our communicative engagements in everyday life.

Those intersections are always in the process of informing and not disforming our sense of self in relation to how others construct the particular and not plural aspects of our identities—like my being a Black/gay/Catholic/administrator/teacher/scholar and more, in how I (and others) deploy our dense particularities. Again, the construct of queer/quare intersectionalities synthesizes all these issues and speaks to radical and not so radical perspectives of communicating gendered identity in the particularity and plurality of our being, including race as a self-affirming locational device linked with community and cosmologies of knowing, while creating a world worthy of living free, and fully alive.

Notes

1 Mohanty, 1989, 13.
2 Simon, 1995, 90.
3 Hill Collins, 2015, 10.
4 Moraga and Anzaldúa, 1981, 23.
5 Ellis, Adams, and Bochner, 2011.
6 See Warner, 1993.
7 This section is informed by Alexander, 2018.
8 See Crais and Scully, 2010; Holmes, 2007, 2008.
9 Donnor and Ladson-Billings, 2018, 199.
10 Delgado and Stefancic, 2008.
11 Delgado and Stefancic, 2008, 6–9.
12 Harris, 1993.
13 This section is partly informed by Alexander, 2018.
14 Anzaldúa, 1987, 250.
15 Cohen, 1999, 450.
16 Gopinath, 2005, 28.
17 Grillo and Wildman, 1995, 566.
18 Wildman and Davis, 1995, 578.
19 Anderson, 1991, 6.
20 Harper is cited in Alexander, 2002, 1288.
21 Johnson and Henderson, 2005, 5, cited in Braverman, 1997.
22 Muñoz, 1999, 25.

23 Ferguson, 2004, 4.
24 Harding, 1987, 1998, 2008.
25 Haraway, 1988, 1991.
26 Crenshaw, 1995, 358.
27 Doty, 1993, 15–16.
28 Reddy, 1997, 356–357.
29 Ferguson, 2003, 3.
30 Ferguson, 2003, 149.
31 Muñoz, 1995, 5.
32 Ferguson, 2004, 2–4.
33 Ferguson, 2004, 149.
34 Madison, 2005.
35 Johnson, 2001, 2.
36 Johnson, 2001, 20.
37 Moraga and Anzaldúa, 1983.
38 Johnson, 2001, 3.
39 Johnson, 2001, 3–20.
40 Johnson, 2001, 9.
41 Muñoz, 1999, 5.
42 Muñoz, 1999, 6.
43 Muñoz, 1999, 22.
44 Muñoz, 1999, 200.
45 Muñoz, 1999, 200.
46 Special emphasis on the works of Madison, 2005; Muñoz, 1999; Ferguson, 2004; Johnson, 2001; Morris and Nakayama, 2013; Berlant and Warner, 1998; and Goltz et al., 2015.
47 Drawn from Morris and Nakayama, 2003, vi.
48 Here I am taking great liberty with the important charge from Audre Lorde (2007), "The Master's Tools Will Never Dismantle the Master's House." In the essay she states:

> Those of us who stand outside the circle of this society's definition of acceptable women; those of us who have been forged in the crucibles of difference – those of us who are poor, who are lesbians, who are Black, who are older – know that survival is not an academic skill. It is learning how to take our differences and make them strengths. For the master's tools will never dismantle the master's house. They may allow us temporarily to beat him at his own game, but they will never enable us to bring about genuine change. And this fact is only threatening to those women who still define the master's house as their only source of support
>
> (110)

Lorde's work is of course germinal to the origins of a broader queer of color critique including *Sister Outsider* (1984) and *Burst of Light Essays* (1988).
49 Alexander, 2005, 207.
50 Yep, 2010, 173; 2016, 89.
51 Yep, 2016, 89.
52 Yep, 2016, 89–90.
53 Johnson, 2008.
54 Johnson, 2018.
55 Madison, 2005, 14.
56 McCune, 2015, 174.
57 McCune, 2014. Description drawn from the front materials of the book's promotional materials on Amazon.
58 Eng, Halberstam, and Muñoz, 2005, 4.
59 Said, 1990.
60 Johnson, 2001, 73.
61 Valles-Morales and LeMaster, 2015, 80.
62 Goltz and Zingsheim, 2015.
63 See: https://en.wikipedia.org/wiki/Two-spirit.

64 See: www.takingcharge.csh.umn.edu/what-spirituality.
65 Ferguson, 2004, 4.
66 Muñoz, 1999, 200.
67 Johnson, 2001, 19.
68 This presentation was originally delivered on the shared birthdate of my mother (Velma Ray Bell Alexander) and my brother (Nathaniel Patrick Alexander). Both are now deceased. My brother died in 1995 from HIV/AIDS. My mother retired from being a Pediatric Nursing Assistant to take care of my brother at home. After his death, my mother spoke at church functions about the relationship between faith, sexuality, and love, HIV/AIDS education in the Black community, and unmediated love of a mother for her child in faith.
69 I am also making an allusion to the mission statement of the Religious of the Sacred Heart of Mary—RSHM at Loyola Marymount University. https://academics.lmu.edu/marymount/aboutthemarymountinstitute/mission/.
70 See Austin, 2014.
71 For references in this section see, "What does it mean to be Catholic." www.bostoncatholic.org/Being-Catholic/Content.aspx?id=11316.
72 This is a reference to Ignatian Spirituality, with the Examen as a ritual process of meditation and reflection. See: https://jesuits.org/spirituality?PAGE=DTN-20130520125910.
73 Intersectionality is a theory which considers that the various aspects of humanity, such as class, race, sexual orientation, and gender do not exist separately from each other, but are complexly interwoven, and that their relationships are essential to an understanding of the human condition. https://en.wikipedia.org/wiki/Intersectionality. It is a sociological theory that describes multiple threats of discrimination when an individual's identities overlap. Most often attributed to Crenshaw, 1991.
74 Drawn from the interpretation of the "Education of the Whole Person" aspect of the LMU Mission Statement: http://mission.lmu.edu/missionstatement/.
75 A standard Google dictionary definition.
76 Drawn partially from the mission and statement and purpose of Sisters of St. Joseph of Orange (CSJ) at Loyola Marymount University: https://csjorange.org/about-us/our-mission/, and through the CSJ Center for Reconciliation and Justice: http://academics.lmu.edu/csjcenter/#.
77 Rich, 1994.
78 See Hornbeck II and Massingale, 2018.
79 See Confession: www.catholicscomehome.org/your-questions/what-is-the-sacrament-of-confession/.
80 Act of Contrition: https://en.wikipedia.org/wiki/Act_of_Contrition.
81 John 14: 2, King James version of the Holy Bible.

Bibliography

Agamben, Giorgio. *Potentialities*. Palo Alto, CA: Stanford University Press, 1999.

Alexander, Bryant K. "Bodies Yearning on the Borders of Becoming: A Performative Reflection on Three Embodied Axes of Social Difference." *Qualitative Inquiry* 20, no. 10 (2014): 1169–1178.

Alexander, Bryant K. "On Weddings, Resistance and Dicks: Autophotographic Responses to Questioning the Limits of Decorum." In *Queer Praxis: Questions for LGBTQ Worldmaking*, edited by Dustin B. Goltz and Jason Zingsheim, 199–211. New York: Peter Lang, 2015.

Alexander, Bryant K. "Performance of Race, Culture and Whiteness." In *Oxford Research Encyclopedia of Communication*. Oxford: Oxford University Press, 2018. doi: http://dx.doi.org/10.1093/acrefore/9780190228613.013.566.

Alexander, Bryant K. "Queer/Quare Theory: Worldmaking and Methodologies." In *Handbook of Qualitative Inquiry*, 5th edn, edited by Norman Denzin and Yvonna S. Lincoln, 254–307. Thousand Oaks, CA: Sage, 2018.

Alexander, Bryant K. "Queer(y)ing the Postcolonial Through the West(ern)." In *Handbook of Critical and Indigenous Methodologies*, edited by Norman Denzin and Yvonna S. Lincoln, 101–133. Thousand Oaks, CA: Sage Publications, 2008.

Alexander, Bryant K. "Reflections, Riffs and Remembrances: The Black Queer Studies in the Millennium Conference." *Callaloo* 23, no. 4 (2002): 1285–1302.

Anderson, Benedict. *Imagined Communities: Reflections on the Origin and Spread of Nationalism*. New York: Verso, 1991.

Anzaldúa, Gloria. *Borderlands/a Frontera: The New Mestiza*. San Francisco, CA: Spinsters/Aunt Lute, 1987.

Anzaldúa, Gloria, and Cherrie Moraga. *This Bridge Called My Back: Writing by Radical Women of Color*. New York: SUNY Press, 2015.

Austin, J.L. *How to Do Things with Words*. Oxford: Oxford University Press, 1975.

Austin, Michael W. "The Nature of Faith: Faith, Reason, and Evidence." www.psychologytoday.com/blog/ethics-everyone/201404/the-nature-faith.

Berlant, Lauren, and Michael Warner. "Sex in Public." *Critical Inquiry* 24, no. 2 (1998): 558.

Blasius, Mark. *Sexual Identities, Queer Politics*. Princeton, NJ: Princeton University Press, 2001.

Boal, Augusto. *Legislative Theatre: Using Performance to Make Politics*. London: Routledge, 1998.

Bradford, Mark. "In Conversation: Mark Bradford, A.L. Steiner and Wu Tsang." In *Scorched Earth, Mark Bradford*, edited by Connie Butler, 177–193. Los Angeles, CA: The Hammer Museum, 2015.

Braveman, Scott. *Queer Fictions of the Past: History, Culture, and Difference*. Cambridge: Cambridge University Press, 1997.

Butler, Judith. "Performative Acts and Gender Constitution: An Essay in Phenomenology and Feminist Theory." In *Performing Feminisms: Feminist Critical Theory and Theatre*, edited by Sue Ellen Case. Baltimore, MD: Johns Hopkins University Press, 1990.

Carrillo-Rowe, Aimee M. "Belonging: Toward a Feminist Politics of Relation." *NWSA Journal* 17, no. 2 (2005): 15–46.

Carrillo-Rowe, Aimee M., Raechel Tiffe, Dustin B. Goltz, Jason Zingsheim, Meredith M. Bagley, and Sheena Malhotra. "Queer Love: Queering Coalitional Politics." In *Queer Praxis: Questions for LGBTQ Worldmaking*, edited by Dustin B. Goltz, Jason Zingsheim, et al., 123–139. New York: Peter Lang, 2015.

Case, Sue Ellen. "Tracking the Vampire." *Differences* 3, no. 2 (1991): 1–20.

Cohen, Cathy. *The Boundaries of Blackness: AIDS and the Breakdown of Black Politics*. Chicago, IL: The University of Chicago Press, 1999.

Cohen, Cathy. "Punks, Bulldaggers, and Queens: The Radical Potential of Queer Politics." *GLQ: A Journal of Lesbian and Gay Studies* 3, no.1 (1997): 437–465.

Constantine-Simms, Delroy. *The Greatest Taboo: Homosexuality in Black Communities*. Los Angeles, CA: Alyson Books, 2000.

Crais, Clifton, and Pamela Scully. *Sara Baartman and the Hottentot Venus: A Ghost Story and a Biography*. Princeton, NJ: Princeton University Press, 2010.

Crenshaw, Kimberlee W. "Mapping the Margins: Intersectionality, Identity Politics, and Violence Against Women of Color." In *Critical Race Theory: The Key Writings that Formed the Movement*, edited by Kimberlee Crenshaw, Neil Gotanda, Gary Peller, and Kendell Thomas, 357–383. New York: The New Press, 1995.

Crenshaw, Kimberlee. "Mapping the Margins: Intersectionality, Identity Politics, and Violence against Women of Color." *Stanford Law Review* 43, no. 6 (1991): 1241–1299.

Davis, Dan-Ain, and Shaka McGlotten. *Black Genders and Sexualities*. London: Palgrave Macmillan, 2012.

Delgado, Richard, and Jean Stefancic. *Critical Race Theory: An Introduction*. New York: New York University, 2001.

Derrida, Jacques. *La Vérité en Peinture (Truth in Painting)*. Translated by Geoffrey Bennington and Ian McLeod, 9–12. Chicago, IL: The University of Chicago Press, 1987.

Denzin, Norman. *Interpretive Ethnography: Ethnographic Practices for the 21st Century*. Thousand Oaks, CA: Sage, 1997.

Donnor, Jamal K., and Gloria Ladson-Billings. "Critical Race Theory and the Postracial Imaginary." In *The Sage Handbook of Qualitative Inquiry*, edited by Norman K. Denzin and Yvonna S. Lincoln, 195–213. Thousand Oaks, CA: Sage, 2018.

Doty, Alexander. *Making Things Perfectly Queer: Interpreting Mass Culture*. Minneapolis, MN: University of Minnesota Press, 1993.

Duggan, Lisa. "Queering the State." In *Social Perspective in Lesbian and Gay Studies*, edited by Peter M. Nardi and Beth. E. Schneider, 564–572. New York: Routledge, 1998.

Eguchi, Shinsuke, Bernadette M. Calafell, and Nicole Files-Thompson. "Intersectionality and Quare Theory: Fantasizing African American Male Same-Sex Relationships in Noah's Arc: Jumping the Broom." *Communication, Culture & Critique* 7, no. 3 (2014): 371–389.

Ellis, Carolyn, Tony E. Adams, and Arthur Bochner. "Autoethnography: An Overview." *FQS Forum: Qualitative Social Research* 12, no. 1 (2011).

Eng, David L. "Transnational Adoption and Queer Disaporas." *Social Text* 21, no. 3 (2003): 1–37.

Eng, David L., Judith Halberstam, and Jose Esteban Muñoz. "What's Queer About Queer Studies Now?" *Social Text*, 84–85. Durham, NC: Duke University Press, 2005.

Ferguson, Roderick A. *Aberrations in Black: Toward a Queer of Color Critique*. Minneapolis, MN: University of Minnesota Press, 2004.

Freedman, Eric. "Producing (Queer) Communities: Public Access Cable TV in the USA." In *The Television Studies Book*, edited by Christine Geraghty and David Lusted. New York: St. Martin's Press, 1998.

Fuoss, Kirk W. "Performance as Contestation: An Agonistic Perspective on the Insurgent Assembly." *Text and Performance Quarterly* 13 (1993): 331–349.

Fuss, Diana. *Inside/Outside: Lesbian Theories, Gay Theories*. New York: Routledge, 1991.

Galvin, Mary E. *Queer Poetics: Five Modernist Women Writers*. Westport, CT: Praeger, 1999.

Gamson, Joshua. "Must Identity Movements Self-Destruct? A Queer Dilemma." In *Social Perspectives in Lesbian and Gay Studies*, edited by Peter M. Nardi and Beth E. Schneider, 589–604. New York: Routledge, 1998.

Gearhart, Sally M. "Foreword: My Trip to Queer." In *Queer Theory and Communication: From Disciplining Queers to Queering the Disciplines(s)*, edited by Gust A. Yep, Karen E. Lovass, and John P. Elia, xxi–xxx. New York: Harrington Park Press, 2003.

Geertz, Clifford. "Notes on the Balinese Cockfight." In *The Interpretation of Cultures*. New York: Basic Books, 1973.

Gilbert, Helen, and Joanne Tompkins. *Post-Colonial Drama: Theory, Practice, Politics*. New York, Routledge, 1996.

Giroux, Henry A., Colin Lankshear, Peter McLaren, and Michael Peters. *Counternarratives: Cultural Studies and Critical Pedagogies in Postmodern Spaces*. New York: Routledge, 1996.

Goltz, Dustin B., and Jason Zingsheim. *Queer Praxis: Questions for LGBTQ Worldmaking*. New York: Peter Lang, 2015.

Gopinath, Gayatri. *Impossible Desires: Queer Diasporas and South Asian Public Cultures*. Durham, NC: Duke, 2005.

Grillo, Trina, and Stephanie Wildman. "Obscuring the Importance of Race: The Implication of Making Comparisons Between Racism and Sexism (or Other-isms)." In *Critical Race Theory: The Cutting Edge*, edited by Richard Delgado, 564–572. Philadelphia, PA: Temple University Press, 1995.

Grossberg, Lawrence. "Introduction." In *Between Borders: Pedagogy and the Politics of Cultural Students*, edited by Henry A. Giroux and Peter McLaren. New York: Routledge, 1994.

Guba, Egons G., and Yvonna S. Lincoln. "Paradigmatic Controversies, Contradictions, and Emerging Confluences." In *The Sage Handbook of Qualitative Research*, 3rd edn, edited by Norman K. Denzin and Yvonna S. Lincoln, 191–215. Thousand Oaks, CA: Sage, 2005.

Hames-Garcia, Michael. "Queer Theory Revisited." In *Gay Latino Studies: A Critical Reader*, edited by Michael Hames-Garcia and Ernesto J. Martinez, 19–45. Durham, NC: Duke University Press, 2011.

Haraway, Donna J. "Situated Knowledges: The Science Question in Feminism as a Site of Discourse on the Privilege of Partial Perspectives." *Feminist Studies* 14 (1988): 575–599.

Haraway, Donna J. *Simiens, Cyborgs, and Women: The Invention of Nature*. London: Routledge, 1991.

Harding, Sandra. "Conclusion: Epistemological Questions: What Is 'Strong Objectivity?'" In *Feminism and Methodology*, edited by S. Harding, 181–190. Bloomington, IN: University of Indiana Press, 1987.

Harding, Sandra. *Is Science Multicultural? Postcolonialisms, Feminisms, and Epistemology*. Bloomington, IN: Indiana University Press, 1998.

Harding, Sandra. *Sciences from Below: Feminisms, Postcolonialitie, and Modernities*. Durham, NC: Duke University Press, 2008.

Harris, Cheryl I. "Whiteness as Property." *Harvard Law Review* 106, no. 8 (1993): 1707–1791.

Hawley, John C. *Postcolonial, Queer: Theoretical Intersections*. New York: State University of New York Press, 2001.

Hawley, John C. *Postcolonial and Queer Theories: Intersections and Essays*. Westport, CT: Greenwood Press, 2001.

Harper, Philip Brian. *Are We Not Men? Masculine Anxiety and the Problem of African American Identity*. Oxford: Oxford University Press, 1998.

Henderson, Mae G. *Borders, Boundaries, and Frames: Cultural Criticism and Cultural Studies*, 1–30. New York: Routledge, 1995.

Hill Collins, Patricia. "Intersectionality's definitional dilemmas," *The Annual Review of Sociology* 41 (2015): 10.

Holmes, Rachel. *African Queen: The Real Life of the Hottentot Venus*. New York: Random House, 2007.

Holmes, Rachel. *Hottentot Venus*. New York: Bloomsbury Publishing, 2008.

Hong, Grace K., and Roderick A. Ferguson. *Strange Affinities: The Gender and Sexual Politics of Comparative Racialization*. Durham, NC: Duke University Press, 2011.

Hornbeck II, J. Patrick and Rev. Bryan N. Massingale. "A Message of Concern and Care to Graduate Students in Theology." Association of Jesuit Colleges and Universities October 3, 2018: https://medium.com/jesuit-educated/a-message-of-concern-and-care-to-graduate-students-in-theology-6b2eeb517248.

Irvine, J. M. "A Place in the Rainbow: Theorizing Lesbian and Gay Culture." In *Social Perspectives in Lesbian and Gay Studies*, edited by Peter M. Nardi and Beth E. Schneider, 573–588. New York: Routledge, 1998.

Jacoby, Russell. "Marginal Returns: The Trouble with Post-Colonial Theory." *Lingua Franca* (1995): 30–37.

Jakobsen, J.R. "Queer is? Queer Does? Normativity and the Problem of Resistance." *GLQ: A Journal of Lesbian and Gay Studies* 4, no. 4 (1999): 511–536.

Johnson, E. Patrick. *Black. Queer. Southern. Women. An Oral History*. Chapel Hill, NC: University of North Carolina Press, 2018.

Johnson, E. Patrick. "Quare Studies, or (Almost) Everything I Know About Queer Studies I Learned from My Grandmother." *Text and Performance Quarterly* 21, no. 1 (2001): 1–25.

Johnson, E. Patrick. *Sweet Tea: Black Gay Men of the South, An Oral History*. Chapel Hill, NC: University of North Carolina Press, 2008.

Johnson, E. Patrick, and Mae G. Henderson (eds). *Black Queer Studies: A Critical Anthology*. Durham, NC: Duke University Press, 2005.

Kennedy, Dane. "Imperial History and Postcolonial Theory." *Journals of Imperial and Commonwealth History* 24, no. 3 (1996): 345–363.

Kinchloe, Joe L., and Peter McLaren. "Rethinking Critical Theory and Qualitative Research." In *Handbook of Qualitative Research*, edited by Norman K. Denzin and Yvonna S. Lincoln, 279–313. Thousand Oaks, CA: Sage, 2000.

King, Katie. "Global Gay Formations and Local Homosexualities." In *A Companion to Postcolonial Studies*, edited by Henry Schwartz and Sangeeta Ray, 508–519. Malden, MA: Blackwell Press, 2000.

Lorde, Audre. *Burst of Light Essays*. Ithaca, NY: Firebrand Books, 1988.

Lorde, Audre. "The Master's Tools Will Never Dismantle the Master's House." In *Sister Outsider: Essays and Speeches*, 110–114. Berkeley, CA: Crossing Press, 1984.

Lorde, Audre. *Sister Outsider: Essays and Speeches*. Trumansburg, NY: Crossing Press, 1984.

Madison, D. Soyini. *Critical Ethnography: Methods, Ethics and Performance*. Thousand Oaks, CA: Sage, 2005.

Madison, D. Soyini. "That Was My Occupation: Oral Narrative, Performance and Black Feminist Thought." *Text and Performance Quarterly* 13 (1993): 213–232.

Malhortra, Shenna, and Aimee Carrillo Rowe. *Silence, Feminism, Power: Reflections at the Edges of Sound*. New York: Palgrave, 2013.

Marcus, George E., and Michael M. Fischer. *Anthropology as Cultural Critique: An Experimental Moment in the Human Sciences*. Chicago, IL: The University of Chicago Press, 1985.

Marcus, Sharon. "Queer Theory for Everyone: A Review Essay." *Signs* 31, no. 1 (2005): 191–218.

McCune, Jeffrey, Q. "The Queerness of Blackness." *QED: A Journal in GLBTQ Worldmaking* 2, no. 2 (2015): 173–176.

McCune, Jeffrey, Q. *Sexual Discretion: Black Masculinity and the Politics of Passing*. Chicago, IL: The University of Chicago Press, 2014.

McLaren, Peter. *Schooling as a Ritual Performance: Towards a Political Economy of Education Symbols and Gestures*. New York: Routledge, 1986.

Minh-ha, Trinh T. *Framer Framed*. New York: Routledge, 1991.

Mohanty, Satya P. "Us and Them: On the Philosophical Bases of Political Criticism." *Yale Journal of Criticism* 2, no. 2 (1989): 1–31.

Moraga, Cherrie, and Gloria Anzaldúa (eds). "Entering the Lives of Others: Theory in the Flesh." In *This Bridge Called My Back: Writings by Radical Women of Color*, 23. San Francisco, CA: Aunt Lute Press, 1981.

Moraga, Cherrie, and Gloria Anzaldúa. *This Bridge Called My Back: Writings by Radical Women of Color*. New York: Kitchen Table, Women of Color Press, 1983.

Morris, Charles E. "Queer/Love/Yawp: Meditation Aloud." In *Queer Praxis: Questions for LGBTQ Worldmaking*, edited by Dustin B. Goltz and Jason Zingsheim, 107–115. New York: Peter Lang, 2015.

Morris, Charles E., and Thomas K. Nakayama. "Queer Editorial Overture." *QED: A Journal of GLBTQ Worldmaking* (2013): v–ix.

Muñoz, Jose Esteban. *Disidentification: Queers of Color and the Performance of Politics*. Minneapolis, MN: University of Minnesota Press, 1999.

Muñoz, Jose Esteban. "Stages: Queers, Punks, and the Utopian Performance." In *The Handbook of Performance Studies*, edited by D. Soyini Madison and Judith Hamera, 9–20. Thousand Oaks, CA: Sage, 2006.

Murillo, Eric G., Jr. "Mojado Crossing Along Neoliberal Borderlands." In *Postcritical Ethnography: An Introduction*, edited by George W. Noblit, Susana Y. Flores, and Eric G. Murrillo Jr., 155–179. Cresskill, NJ: Hampton, 2004.

Parry, Benita. "Directions and Ends in Postcolonial Studies." In *Relocating Postcolonialism*, edited by David Theo Goldberg and Ato Quayson, 67–81. Malden, MA: Blackwell Publishers, 2002.

Parry, Benita. "Resistance Theory/Theorizing Resistance or Two Cheers for Nativism." In *Colonial Discourse/Postcolonial Theory*, edited by Francis Barker, Peter Hulme, and Margaret Iversen, 172–196. New York: Manchester University Press, 1994.

Patton, Cindy, and Benigno Sánchez-Eppler. *Queer Diasporas*. Durham, NC: Duke University Press, 2000.

Pérez, Kimberlee, and Daniel C. Brouwer. "Potentialities and Ambivalences in the Performance of Queer Decorum." *Text & Performance Quarterly* 30 (2012): 317–323.

Phelan, Peggy. *Unmarked the Politics: The Politics of Performance*. New York: Routledge, 1993.

Plummer, Ken. "Critical Humanism and Queer Theory: Living with the Tensions, Postscript." In *The Sage Handbook of Qualitative Research*, edited by Norman K. Denzin and Yvonna S. Lincoln, 195–207. Thousand Oaks, CA: Sage, 2011.

Reddy, Chandan. *Freedom with Violence: Race, Sexuality, and the US State*. Minneapolis, MN: University of Minnesota Press, 2011.

Reddy, Chandan. "Home, Houses, Nonidentity: 'Paris is Burning'." In *Burning Down the House: Recycling Domesticity*, edited by Rosemary M. George, 355–379. Boulder, CO: Westview Press, 1997

Rich, A. "If Not with Others." In New Worlds of Literature: Writings from America's Many Cultures, edited by J. Beaty and J. P. Hunter, 786–791, 787. New York: P. Norton, 1994.

Riggs, Marlon T. "Black Macho Revisited: Reflections of a Snap! Queen." Originally published in *Black American Literature Forum* 25 (1991): 389–394. Republished in Connie Butler (ed.), *Scorched Earth, Mark Bradford*, 97–100. Los Angeles, CA: The Hammer Museum, 2015.

Roman, David. *Acts of Intervention: Performance, Gay Culture, and AIDS*. Bloomington, IN: Indiana University Press, 1998.

Said, Edward. *Orientalism*. New York: Pantheon, 1978.

Sedgwick, Eve K. *Epistemology of the Closet*. Berkeley, CA: University of California Press, 1990.

Sedgwick, Eve K. *Tendencies*. Durham, NC: Duke University Press, 1993.

Seidman, Steven. *Difference Troubles: Queering Social Theory and Sexual Politics*. Cambridge: Cambridge University Press, 1997.

Sigal, Pete. *Infamous Desire: Male Homosexuality in Colonial Latin America*. Chicago, IL: The University of Chicago Press, 2003.

Simon, Roger. "Face to Face with Alterity: Postmodern Jewish Identity and the Eros of Pedagogy." In *Pedagogy: The Question of Impersonation*, edited by Jane Gallop, 90-105. Bloomington, IN: Indiana University Press, 1995.

Somerville, Sionhan B. *Queering the Color Line: Race and the Invention of Homosexuality in American Culture*. Durham, NC: Duke University Press, 2000.

Stein, Arlene, and Ken Plummer. "'I Can't Even Think Straight': Queer Theory and the Missing Sexual Revolution in Sociology." *Sociological Theory* 12 (1994):178–187.

Stockton, Kathryn B. *Beautiful Bottom, Beautiful Shame: Where "Black" Meets "Queer"*. Durham, NC: Duke University Press, 2006.

Turner, William B. *A Genealogy of Queer Theory*. Philadelphia, PA: Temple University Press, 2000.

Valles-Morales, Jesus I., and Benny LeMaster. "On Queer of Color Criticism, Communication Studies, and Corporeality." *Kaleidoscope* 14 (2015): 77–81.

Warner, Michael. *Fear of a Queer Planet: Queer Politics and Social Theory*. Minneapolis, MN: University of Minnesota Press, 1993.

Wildman, Stephanie M., and Adrienne D. Davis. "Language and Silence: Making Systems of Privilege Visible." In *Critical Race Theory: The Cutting Edge*, 573–579. Philadelphia, PA: Temple University Press, 1995.

Yarbro-Bejarano, Yvonne. "Expanding the Categories of Race and Sexuality in Lesbian and Gay Studies." In *Professions of Desire: Lesbian and Gay Studies in Literature*, edited by George E. Haggerty and Bonnie Zimmerman. New York: MLA, 1995.

Yep, Gust A. "Queering/Quaring/Kauering/Crippin'/Transing 'Other Bodies' in Intercultural Communication." *Journal of International and Intercultural Communication* 6, no. 2 (2013): 118–126.

Yep, Gust A. "Toward the De-Subjugation of Racially Marked Knowledges in Communication." *Southern Communication Journal* 75, no. 2 (2010): 171–175.

Yep, Gust A. "Toward Thick(er) Intersectionalities: Theorizing, Researching, and Activating the Complexities of Communication and Identities." In *Globalizing Intercultural Communication: A Reader*, edited by Kathryn Sorrells and Sachi Sekimoto, 86–94. Thousand Oaks, CA: Sage, 2010.

Yep, Gust A., Karen E. Lovass, and John P. Elia (eds). "Introduction: Queering Communication: Starting the Conversation." In *Queer Theory and Communication: From Disciplining Queers to Queering the Disciplines(s)*. New York: Routledge, 2003.

3

GAYSIAN FABULOSITY

Quare(ing) the *normal* and *ordinary*

Shinsuke Eguchi

The centrality of whiteness and of White-on-White gay male relationships [is] a sense-making norm that fuels the logic by which we [people of color] ascribe value in the gay marketplace of desire.

Dwight A. McBride[1]

Queers of color are excluded or constructed in ways that further marginalize that identity construct, in the service of promoting heteronormative constructions of White masculinity—even in the presumed context of foregrounding queer identity

Bryant Keith Alexander[2]

Since moving from Japan to the United States in 2001, I have almost always observed, felt, and experienced the politics of racialized gender essentializing Asian men as *feminine and submissive Others* in and across gay sexual cultures. Whether I was in Orange County (2001–2003), San Francisco (2003–2005), New York (2006–2008), or Washington, DC (2008–2011) where Asian and Asian American populations are historically visible, the logics of whiteness that organize gay sexual cultures always already subordinate Asian men "as a subcultural category referencing the racialized fetishes of an older white male for the diminutive and effeminized Asian male."[3] In and across gay sexual cultural spaces such as clubs, pride parades, community events, and online dating sites I have frequented, youthful, healthy, able-bodied white men who look like buffed-out blue-eye gay Adonises represent the normative masculine power and ideal. As C. Winter Han argues, a catchphrase "*No Fat, No Femme, and No Asians*" that emerged from gay online dating and hookup websites mirrors how gay sexual cultures that naturalize and stabilize the youthful White masculine beauty standard continue to ostracize Asian men.[4]

This situation showcases what queer Asian American scholar David L. Eng has argued that the historical continuum of U.S. racial formation institutionally organized Asian men and masculinities as subordinate to white men and masculinities.[5] The active forces of White supremacy that normalize the power and privilege of white men and masculinities have thus far controlled the hetero-reproductive growth of the Asian population. In the early 1900s, the anti-Asian migration policy and legislation restricted women from nations in Asia to enter into the United States. So, large numbers of Asian men remained single and could not reproduce their off-spring. Additionally, the institution of White supremacy relegated Asian men to occupy the labor market such as domestic servants, hotel workers, and laundrymen largely considered as feminine jobs. Such material conditions of Asian men perpetuated the stereotypes of their physical traits (including body shape, height, and penis size) as smaller and therefore more feminine than white men, who signify and affirm the *normal* and *ordinary*. Because of Asian men historically seen as not man enough, Asian men have become to signify *undesirable feminine and submissive Others*.

This racialized gender politics has also affected the gay sexual cultural representations of Asian men. The prominent stereotype about gay + Asian (also known as gaysian) is that they are also smaller and more feminine than white men. This *feminine* stereotype often essentializes gaysians as sexual bottoms (meaning anal sex receivers).[6] Gaysians are not mostly expected to be masculine, active, and sexual tops (meaning anal sex penetrators) and/or to be sexually versatile even when the gay sexual cultural practices of top-bottom roles are much more fluid.[7] Indeed, a gay subcultural category "Asian" has been produced and materialized for the pornographic consumption of older men who desire youthful foreign Asian men as *feminine and submissive* Others. Hence, a stereotypical gaysian narrative that operates as the gay sexual cultural norm is that Asian men cannot easily find ideal sexual and romantic partners because they are "too" feminine and submissive.[8] Accordingly, gaysians are known to compete against each other to find their partners from the limited numbers of white men who desire Asian men. At the same time, it is ironic that gaysians are mostly regarded not to desire other Asian men and/or non-Asian men of color (e.g., Arab/Middle Eastern, Black, Latinix, Native, and Pacific Islander).[9] white men as the ideal and desirable partners occupy in the center of a stereotypical gaysian narrative. Thus, the politics of racialized gender that reify the essentialist, false White-Asian, West-East, and domestic-foreign binaries surround the performative rhetorics of gaysian.

However, I critique that the stereotypical gaysian narrative I have mentioned above solely depends on the discourse of oppression. Gaysian narratives are much more complex, contested, multiple, and fluid than a simply oppressed narrative. While the supremacy of whiteness as the normative gay masculine power and ideal can undeniably cultivate similar racialized gender experiences for the gaysian subjects, it should not entirely homogenize differences of gaysian narratives. Depending on locations where gaysian subjects are situated, they differently internalize, embody, and perform the intersectionality of gay and Asian. By intersectionality, I mean a simultaneous and multifarious operation of differences (such as race, ethnicity, gender, sexuality, class, language, citizenship, coloniality, and the body) that politicize, historicize, and contextualize one's experiences.[10] Accordingly, I am concerned with how the simplistic and highly limited gaysian narrative rooted in the discourse of oppression maintains the privilege of whiteness the normative gay masculine power. Thus, this chapter stands as a humble call to interrogate, complicate, and shift gaysian narratives to push the whiteness of gay sexual cultures to the margin. I am specifically interested in examining paradoxes between reconstitutions and fragmentations of the *normal* and *ordinary* embedded in the material realities of gaysian identities, performances, and politics.

To do so, I draw the genealogy of quare studies[11] as a critical interpretive methodology to analyze various YouTube vlogger-created gaysian contents. E. Patrick Johnson maintains, "'Quare' offers a way to critique stable notion of identity and, at the same time, to locate racialized and gendered knowledges."[12] Writing from this standpoint, I interrogate how gaysian subjects self-represent their internalizations of and/or resistances against the politics of racialized gender through their everyday performances. I also examine how such gaysian narratives ironically help sustain whiteness as the normative gay masculine power. To interrupt and break through such workings of whiteness, I reconsider possible ways in which the embodied performance of *gaysian fabulosity* works as a "quare" performative strategy. According to Fiona Buckland, fabulosity is an extra, over-the-top performance of fashion, beauty, and lifestyle emerged from gay bar/night club cultures, which develop counter-normative formations of sexual and gender identities, performances, and politics.[13] At the same time, Bryant Keith Alexander also reminds that the ways in which people of color perform (White) bourgeois characteristics, values, and attitudes allude to potential forms of resistance.[14] Elaborating Alexander's point, I propose that gaysian fabulosity is a political everyday performative in which gaysians present, articulate, and celebrate the intersectionality of their race/culture and queer sexuality as an ownership of particularity and pride. Such a strategically intentional performative is resistant to the social constructions of their identity and promotes a celebration of gaysian identity as worthy of a particular space in geo-political realms of queer identities, performativities, and desirabilities that encourages self-love. What follows is an explication of Johnson's notion of quare studies in the application to building a theory of gaysian identity in/as a politics of *Queer-of-Color critique*.

Quare studies as a methodology

There have been critiques against the field of queer studies as a site of reproducing the supremacy of whiteness.[15] Especially in light of the fact that the theoretical foundation of queer studies is supposedly about the plurality of agency and sexual freedom. A fluidity problematizes the essentialist paradigm of a straight and gay/lesbian binary rooted in the logics of cisheteronormativity. Queer studies are supposedly a call to destabilize and eliminate the fixed categories of gendered identity politics. People should be able to freely enjoy the particularity of sexualities and genders as they desire across a range of race/culture informed communities of engagement. However, such queer theorizing neglects to consider the complex roles of intersectionality—producing ways of knowing about sex, sexuality, and gender across categories of race/culture/class and situated origins.[16] Queer studies/theory as it exists erases, ignores, and marginalizes the culturally specific and nuanced ways in which people of color embody and perform their sexuality that intersects with race, ethnicity, gender, nationality, class, and the body. Differences are essentially lumped into one categorical paradigm—that is, queer. Consequently, Bryant Keith Alexander critiques, "Queer studies is unacceptably Euro-American in orientation, its purview effectively determined by the practically invisible, because putatively nonexistent—bounds of racial whiteness."[17]

To intervene the aforementioned practices and promotions of queer studies as circumferences of whiteness, performance studies scholar E. Patrick Johnson builds the notion of quare as a counter-narrative to queer theory that expands a space of inclusion for Queer-of-Color critique.[18] The goal of Queer-of-Color critique is for all non-White queers to articulate and rearticulate their particularized experiences to find a homeplace in an expanded sense of the *normal* and *ordinary*.[19] The offering of counter-narratives functions as both resistance and inclusion of gender and sexual minoritarian voices. In promotion of quare studies, Johnson is advancing a Queer-of-Color critique concerned with how the historical and contemporary realities of race intersecting with gender and class construct nuanced particularities of sexual knowledge among lesbian-gay-bisexual-transgender (LGBT) people of color. The culture-specific and text-specific knowledge embedded in the material realities of LGBT people of color explicate, elucidate, and elaborate the historical significances of relational and communal ties that has promoted collective resistances against White supremacy. The social constructions of racial category rooted in the politics of difference matter to LGBT people of color. While homophobia, transphobia, and related oppressions are prevalent in and across raced communities, LGBT people of color cannot easily cut off their ties with their counterparts who identify as heterosexual. Everyday operations of race and racial category develop shared dissatisfactions among raced subjects. At the same time, LGBT people of color experience the complex intersections among racism and classism when they enter into LGBT communities predominantly occupied by White sexual minoritarian [of this majority] counterparts. Thus, LGBT people of color's experiences are performative products of their intersectionality. They are neither just LGBT people nor just people of color, but both-and.

Jeffery Q. McCune Jr. exemplifies Johnson's theorizing of quare studies as he analyzes intersectionality of race, gender, sexuality, and class among Black men who engage in discreet same-sex sexual and romantic relations. These privately queer active men are also known as Black men on down low (DL). The term became popular in early 2000s through media.[20] However, McCune rejects the simplistic representation of these Black men as "closeted" gay men. Instead he offers,

> The DL may offer an alternative to the closet—a space where much more happens than Black men having sex with wives/girlfriends while having sex with other men—where men actively negotiate issues of race, gender, class, and sexuality.[21]

For instance, Black men are pressured to embody and perform a hetero-patriarchal Black hypermasculinity. Such racialized gender performances may work as the embodied methods for Black men to navigate and resist against the White control, discipline, and surveillance. At the same time, Black men are constructed as hypersexual. Particularly, the White sexual imagination has historically eroticized the Black male penis size as large. This was both as an actuality of being and as a promotional strategy during slavery to promote the relationality between chattel and the procreation of more slaves, as well as the notion of hypersexuality with deviance. Simultaneously, "homosexuality was effectively 'theorized' as a 'White disease' that had 'infected' the Black community."[22] Thus, they are not expected to engage in same-sex sexual and romantic relations because such sexual desires violate the norm of racialized gender.

Accordingly, Jose Esteban Muñoz's theorizing of disidentification is significant to the central development of quare studies. Disidentification helps identify how LGBT people of color perform ways of knowing, acting, and being within and against the majoritarian norm. Muñoz asserts, "Disidentification is the third mode of dealing with the dominant ideology, one that neither opts to assimilate within such a structure nor strictly opposes it; rather, disidentification is a strategy that works on and against dominant ideology."[23] Indeed, LGBT people of color must navigate back and forth between ideologically divided spaces of identity and difference. Accordingly, LGBT people of color must learn how to walk around material fragmentations of their intersectionalities as they both cope with and resist against the majoritarian codes of belonging. LGBT people of color are not structurally given privileges to either fully buy into or refuse dominant ideologies due to the lack of the cultural capital.

Such Quare theorizings have resulted its fruitful expansion of Johnson's initial proposal.[24] To explicitly borrow Johnson's quare studies, Wenshu Lee, a transnational Taiwanese womanist, proposes to "Kuare" queer studies.[25] Bringing a Chinese/Mandarin pronunciation of queer to the fore, Lee argues for accounting particular geopolitical places and spaces in which both local and transnational-level circulations of race, ethnicity, gender, sexuality, class, and nation occur. In fact, the contemporary Taiwanese productions and constitutions of sexualities implicate effects of both local politics and globalization. In addition, by centering the notion of dis/ability, Robert McRuer calls to "Crip" queer studies to interrogate assumptions about gender, sexuality, and sex through Disability Studies.[26] For instance, the historical medicalizations of homosexuality and transgenderism implicate how cisheterosexuality is discursively normalized as the healthy symbol of able-bodiedness in the United States At the same time, queer scholars such as Judith Halberstam, Railey Snorton, and Susan Stryker advocate for "Transing" the social systems and cultural processes of disciplining gender intersecting with race, ethnicity, sexuality, class, and the body.[27] Gender embodiments and performances implicate discursive and material effects of historical and existing power relations. These lines of methodological expansion help complicate and deepen the intellectual and political utilities of quare studies within and beyond Black queer studies from which quare studies has originally emerged from the site of original theorizing.

A theory of gaysian identity also elaborates such utilities of quare studies. Indeed, a number of gaysian scholars[28] examine how the historical production of gaysians as *feminine and submissive*. Other creates a glass ceiling for gaysians as they seek out dating and hookup in and across gay sexual cultures. At the same time, because of such structural constraints that normalize the White patriarchal phallic masculinity, the aforementioned gaysian scholars also argue that some gaysians both actively and strategically recycle their racialized gender images. By exaggerating their "femininity" and "submissiveness," they intend to attract men who desire feminine and submissive men. In fact, gay sexual cultures embrace and celebrate gaysians who perform as drag queens because they are assumed to be suitable to feminine gender roles.[29] Simultaneously, some gaysians put efforts to masculinize their body images to separate from their racialized gender images.[30] Still, because of the historical pervasiveness of Asian men as feminine and submissive, some gaysians borrow and imitate hypermasculine images portrayed onto other men of color to pass as masculine.[31] Thus, gaysians are not simply oppressed. They are both actively and strategically disidentifying with dominant ideologies as they simultaneously cope with and resist against the historical femininization and subordination of Asian men and masculinities.

Drawing from the genealogy of quare studies I have described above, in this essay I approach the conception of queerness as a fleeting moment of sexual and gender transgression politicized, historicized, and contextualized in and across the lines of differences. The major analytical significance of such queer transgression implicates a futurity. By a futurity, I mean a "a temporal arrangement in which the past is a field of possibility in which subjects can act in the present in the service of new futurity."[32] Performing queerness implicates an ongoing interplay of contradictory tensions between possibilities and impossibilities moving toward the future—that is, where the politics of difference maintaining dominant ideologies are no longer useful. Accordingly, queerness is indeed about failure that demonstrates missing holes of the material realities through which queers cannot be queers.[33] In this view, my methodological operationalization of queerness highlights what Bryant Keith Alexander has suggested is queerness as "a gendered identity location but as resistance to orthodoxy—expounding, elaborating, and promoting alternative ways of being, knowing, and narrating experience."[34] Thus, queerness is an intellectual and political paradigm of sexual and gender transgression.

YouTube and gaysian narratives

With the aforementioned methodological background, I engage in a close reading of YouTube vlogger-created contents about gaysian identities, performances, and politics. As Ivan Dylko and Michael McCluskey have argued, my reason for choosing YouTube as an analytical site is that it offers a space of self-representations in which each vlogger freely produces their own media contents.[35] Such self-representations are also never free from historical and existing power relations within which each vlogger is located.[36] Each vlogger actively reuses, remakes, and recycles local, national, and global circulations of power, culture, and ideology to narrate their everyday experiences. So, the viewers can consume the vlogger-created contents—relative to the personal intentionality of the vlogger or that of the viewer. Thus, I approach YouTube video contents as the media and cultural text through which gaysian narratives implicate the politics of racialized gender.

For this essay, I insert keywords such as "gay Asian" and "gay and Asian" in the search engine to select vlogger-created contents in YouTube. In so doing, I locate 23 videos uploaded from 2010 to 2018. To interrogate how gaysian subjects narrate the politics of racialized gender, I choose the contents of nine vloggers (e.g., calvinwoo, Collin Factor, DDCC, FlawlessKevin, OneWingedChris, Scottyarh, Tim Wong, Will from TerraNovaBoys, and Vladomir Lolinco). In particular, I am interested in analyzing how they address self-representations of their own gaysian narratives. To examine such contents through the lens of quare studies as a methodology, I also depend on my autoethnographic ways of knowing about gay and Asian. Queer performance studies scholars such as Bryant Keith Alexander, Bernadette Marie Calafell, and Amber Johnson have argued that the body plays a quintessential site of knowledge.[37] The researcher's subjective knowing is always a part of critical qualitative research design, analysis, and writing. Thus, I unapologetically use myself as a referencing point to explicate, elucidate, and elaborate an inter-subjective space through which the vlogger-created gaysian contents are closely read. Through these analytical processes, I organize my argument according to three themes: *Gay Anti-Asian Racism*, *Strategic Essentialism*, and *Gaysian Critically, Needed*. I begin sharing my analysis with the first theme *Gay Anti-Asian Racism* next.

Gay anti-Asian racism

The performative rhetorics of gaysian self-narratives in YouTube revolve around the material duality of "either gay or Asian." More precisely, the logics of whiteness almost always organize overall implications of being gay. Consequently, gaysians are structurally forced to adapt the logics of whiteness that subordinate their race/culture/class. At the same time, the White supremacy that promotes anti-Asian racism erases, ignores, and marginalizes queer sexualities associated with Asian men. The material fact of Asian men castrated as asexual according to the logics of phallic masculinity implicates the historical context of Asian men known as remaining single. However, I argue that both terms gay and Asian also implicate the stable, singular, and essentialist constructions of racialized gender politics. The plurality and fluidity of gay and Asian are not really accounted. As a result, the gaysian self-narratives exemplified by the YouTubers/YouTube personalities operate within the essentialist logic of an intersection between gay and Asian. As Maurice Kwong-Lai Poon has observed, "The [gaysian] personal narratives focus primarily on Asian men's interactions with White society, often in the form of relationships with white men, and the oppression that occurs as a result of the racism that exists within the gay community."[38] Overlooking intersectional approaches to their everyday experiences, the gaysian subjects mostly focus on white men to construct their narratives.

In the 6 minute and 13 second YouTube video "Being Gay and Asian!" published on March 9, 2018, FlawlessKevin politicizes his sexual and romantic encounters with gay men. He starts the video by arguing that the media representations of Asians are either Whitewashing or nerds. Then, he critiques media representations of gays as mostly White masculine men. To support his argument, he also names a couple of recently popular gay romantic movies such as *Call Me by Your Side* and *Love Simmons*. The following scripted skit of Kevin's dating experience represents the very typical feminized gaysian stereotypes imposed onto his body. Kevin's Asianness is almost always read according to the logics of gay subcultural categories such as twinkish and/or fetish. Or, no gay man is interested in dating him literarily because he is neither White nor masculine. Kevin says that he has seen multiple online dating app profiles such as Grindr suggesting "No blacks, No feminine guys, and No Asians." At the same time, Kevin as a feminine gaysian subject is "naturally" expected to be submissive and not to talk back to (white) men. Kevin is normally asked if he cross-dresses because he often uses makeup products such as foundation, lip gloss, and concealer to present himself. However, Shinsuke Eguchi and Hannah R. Long have suggested that an embodied performance of nonbinary femininity does not automatically indicate whether a gaysian subject is actually cross-dressing or not.[39] Thus, this video concludes that Kevin is never free of Asian male feminization.

Another YouTuber based in Canada, Wil normally produces vlogger-content videos with his white male partner Adam under the name of TerraNovaBoys and published his gaysian narrative on January 3, 2017. Wil starts off this 9 minute and 40 second video "Growing up Gay and Asian" by saying that his experience is different from Adam, who is benefited by the logics of whiteness. For instance, he was looking to make friends through gay social networking apps. Because Wil is an Asian male, white men whom he messaged often rejected him. Some of them immediately blocked him from messaging them further. Or they messaged back to Wil saying "not interested in Asians." Such blunt racism made him feel like he is "garbage." He was hurt especially because he expected that (white) gay people experiencing homophobia and heterosexism would be welcoming him. Thus, Wil says that he learned being "a dominant strong gay white guy" is a social requirement to fit into a gay world. Such strict policy of gay social dress code is the aesthetic production of homonormativity. By homonormativity, I mean the politics of respectability promoting the hetero-standardized beauty, style, and value of men and masculinities rooted in the whiteness.[40] As a consequence, Wil went to the therapy to work on his racialized gender struggle. Based on such experience, he questions why there are no gaysian male leads in media. There is always the effeminate Asian male who cooks, cleans, and does laundry. Asian men are not assigned to do "manly things." Yet Wil says, "There are Abercrombie & Fitch models with the six-pack abs, blond hairs, and scruff" in media and popular culture. A Queer-of-Color critic Dwight A. McBride also reminds, "Abercrombie codes for race and class without actually having to name it. … the A&F look is styled on a celebration of racial and cultural whiteness."[41] Thus, this video ends with Wil's questions and critiques of whiteness as the gay sexual cultural capital.

By closely reading these videos, I argue that gay anti-Asian racism remakes a structural (re) production of power that sustains the white male superiority. The material realities of gay male beauty, style, and value symbolize and affirm the logics of White/Western/U.S. American exceptionalism through which Asians and non-Asian people of color are marginalized.[42] At the same time, cisgender, able-bodied, and affluent white men are given privileges to easily perform *liberal outlooks of sexual freedom*. Yet Asian men are strategically lumped into one racialized gender—that is, *a feminine submissive foreigner*. Consequently, both Kevin and Wil, speaking English as their first languages without any foreign accents, share experiencing similar sexual and romantic rejections rooted in the simultaneous technology of gay anti-Asian racism intersecting with anti-femininity.

Hence, I maintain that our desires can never be simply just preferences. What we are sexually and romantically attracted to is an embodied product of dominant ideologies. For instance, the logics of whiteness that organize gay sexual cultures shape how some white men reject these gaysian subjects regardless of their (racialized) gender presentations and performances. In fact, Kevin's gender performance is very different from Wil. Kevin seems to be comfortable with his femininity. He also presents his face with campy makeups. Simultaneously, Wil is more of a masculine gaysian. He presents his body by wearing a baseball hat and a sports jersey jacket in the video I have mentioned above. Wil is not what I think of as a feminine gaysian whose body type is represented in the gay pornographic marketplace category of "Asian." At the same time, it is noteworthy to mention that (white) gay men unfairly devalue, discredit, and discount diverse gender presentations and performances of gaysians just because they are Asian.[43] A Louisiana-native YouTuber Scottyarh also introduces such anti-Asian racism in the 4 minute and 28 second video "Getting Rejected on Grindr!?" published on January 2, 2015. Scotty represents his body with the performative signs of masculinity including his outfit, yet he tells the viewers, "they [men in the Grinder] don't give me a chance because I am Asian." Some of them said to Scotty, "I am not into Asians." Thus, regardless of how each of them embodies and performs racialized gender, the logics of whiteness as the normative gay masculine power and ideal maintains the always-alreadyness of gaysians as *feminine* and *submissive* Others.

Strategic essentialism

Still these gaysian narratives overlook the complex roles of intersectionality in producing differences. Richard Fung reminds that the implications of nationality affect Western gay pornographic representations of Asian men.[44] To be specific, Filipino men may be imaged as sexually available while Japanese men are assumed to be sexually kinky. Nguyen Tan Hoang also demonstrates how the Western gay pornography give gaysian actors performing assimilated images more freedom and flexibility to be both sexual tops and bottoms than gaysian actors performing foreign-born images.[45] Considering these arguments, I must also acknowledge that my intersectionality such as citizenship (U.S.), nationality (Japanese), location of birth (foreign), language (accented English), and class and occupational status (international scholar and tenured university faculty) constructs my everyday gaysian experiences. This does not mean that I do not experience what Kevin, Scotty, and Wil have argued. I too experience the politics of racialized gender. However, being a gaysian U.S. citizen migrating from Japan, I have also encountered both Asian and non-Asian gay and bisexual men who are interested in Japanese and Japanese American cultures, histories, and politics. Driven by liberal capitalism, the globalization promotes rapid flows of Japanese cultures and commodities via business, media, social media, and technology. Some of them have traveled to and/or lived in Japan. So, I was not always feeling like "garbage" as I participated in gay sexual culture. The feeling of being desired by other men was *secretly* "ego" boosting despite the fact that the gay sexual logics of fetishism for Japanese might have driven their desires. My experience implicates the political, economic, and cultural hierarchy of Japan as the major Global North center. The contemporary realities of globalization give me an unearned privilege of being Japanese. Thus, I am reluctant to uncritically promote the essentialist version of gaysian narratives *as if* all gaysians are the same.

From this line of thought, I reframe gaysian narratives through the conceptual lens of strategic essentialism as a disidentificationary practice working on and against the dominant ideology. I borrow what Dwight A. McBride has witnessed is that "African American intellectuals often use to reclaim a racial essentialism based on experience that authorizes or legitimizes their speech in some very politically important ways."[46] Indeed, this racial politics is what a post-colonial theorist Gayatri Chakravorty Spivak also argued is strategic essentialism.[47] That is, an intellectual tactic through which minoritarians mobilize and push forward their shared identities to increase their group visibilities to achieve political goals. By borrowing such identity politics, I argue that strategic essentialism is compulsory for gaysian subjects to interrupt and break through the whiteness of gay sexual cultures and the cisheteronormativity of Asian ethnic cultures.

In the 4 minute and 25 second video "2 Short Grindr Stories: 'But Where Are You REALLY From?' published on December 8, 2016, OneWingedChris says that he is almost always assumed to be a migrant in Grindr; this as a consequence of his Asian appearance. When Chris said to a man that he is from Southern California, this man replied to Chris, "Not your pretentious answer" and "But, where are you really from?" The question "where are you from?" is indeed a discursive form of racism rooted in whiteness as a normative U.S. American point of origin. Asian Americans can never get to be considered as *just Americans* even when they were born and raised in the United States Thomas K. Nakayama has already argued the overall historical orientation of Asian Americans as *forever foreign* in the White racial imagination.[48] Thus, Chris' response could have been his resistance against anti-Asian racism or his sincere sense of U.S. citizenship.

This kind of racialized gender experience is also shared by Tim Wong in his 4 minute and 54 second video "Gay Asian American's 'Joy Luck Club' Conundrum" published on October 3, 2015. Tim starts this video by sharing that he always felt he was *too* Asian in the United States His body symbolizes always-alreadyness of Otherness. At the same time, Tim is *too* American in his motherland, that is, China. He says, "I have no idea of what I'd be like in China." He has never dated Asian and/or Chinese guys and cannot eat Chinese foods all the time. He is accustomed to date American guys and eat American foods. Here, I am reminded that Tim, as a gaysian subject, resides in an ambiguous state of in-betweenness requiring ongoing negotiations of contradictions complicated by his intersectional identity.

Such *in-betweenness* is further exemplified by both U.S. and Canada-based YouTubers such as calvinwoo, Vladomir Lolinco, and Wil from TerraNovaBoys. These gaysians talk about their experiences of coming out to their traditional, conservative, and tiger "Asian" parents. They were originally concerned that the parents and family would never understand why they publicly want to live as gaysians. The heteronormative path of passing on the family's name to next generation is significantly valued for cisgender males in and across "Asian" communities rooted in patriarchy. Let me explain how this is unique to various "Asian" communities, as some readers may not understand otherwise.

The paradigm of coming out of the closet, rooted in the logics of individualism, responsibility, choice, and merit, is indeed a White/Western/U.S. American organization of same-sex desire. Such individualistic sexual freedom is not often practiced in non-Western cultures. The organizations of same-sex desire have been different in and across "Asian" communities. In Chinese mainland, for instance, there is a culturally nuanced practice through which queer women and men get married to meet the relational, familial, and institutional expectation of marriage.[49] So, they can maintain their private spaces in which they engage in queer desire, intimacy, and relationality. The institution of marriage is rather a duty than a choice to fall in love.

Thus, gaysians coming out of the closet can be seen as wanting to be like *liberal* and *progressive* white men, who are publicly open and frank about their sexuality. Therefore, as Calvinwoo, Vladomir and Wil have demonstrated on YouTube, the central framing of gaysian narratives "is predicated on a liberal concept of individual autonomy and universal justice that permeates the political and social landscape of Western civil rights movements (particularly in the U.S. context)."[50] Accordingly, the politically important mechanisms of gaysian narratives are useful mostly only in Western contexts in which gaysians are structurally forced to coordinate the unbalanced fragmentations of their intersectionalities.

More precisely, gaysians are less likely to develop a sense of shared critical dissatisfaction without working on and against the centrality of whiteness and cisheteronormativity. Kevin is Vietnamese. Tim and Wil are Chinese. Scotty is Laotian. Vladomir is Filipino. Calvin and Chris do not explicitly elaborate their ethnic identifications in their narratives, while I can guess their ethnicities. Other aspects of difference such as gender performance, class, migration status, citizenship, age, and location also vary. However, these YouTubers represent the strategically essentialized version of racialized gender narratives as gay and Asian. So, other gaysians from diverse cultural and ethnic backgrounds make relatable and supportive comments and constructive feedback on YouTube gaysian vlogger-created content. Accordingly, they can collectivity dismantle the whiteness of gay sexual cultures and the cisheteronormativity of Asian ethnic cultures. Strategic essentialism is a racialized gender formation of gaysian as *a political identity*.

Gaysian criticality, needed

Simultaneously, I remain concerned with the gaysian self-narratives I have introduced above. They explicitly focus on the gay cultural value of sexual and intimate relations with (youthful, cisgender, healthy, and able-bodied) white men. More specifically, the politically important moves of gaysians self-narrating their experiences ironically re-center, re-secure, and readjust the White cismale power and privilege. However, gaysians indeed meet, make friends, and/or fall in and out of love with Asian men and non-Asian men of color in and across gay sexual cultures, for example. The same-sex color-to-color relations can be politically significant and revolutionary. Such same-sex couplings can possibly marginalize the superiority of whiteness from gaysian narratives. At the same time, these couplings can also mimic the logics of whiteness as the gay relational norm when the subjects uncritically participate in such relationships. The representations of racial categories, emerged from whiteness as a normative knowledge production, may affect how men of color make sense of one another. Thus, I maintain that gaysians must be critically reflexive of their same-sex desires (re)defining who they are, what they do, and how they make sense of what they do.

For instance, a gaysian YouTuber Collin Factor uploaded his six-minute and 33-second video "Not into Asians" on April 3, 2016 first. He has a follow-up nine-minute and 40-second video "Asians Not into Asians" published on May 6, 2018. In both video contents, Collin features his dialogues with his gaysian friend Marvin. Just like any other gaysian YouTubers, the first video of their production focuses on how Asian men unequally and unfairly experience dating and hook up in and across gay sexual cultures. Then in the second video they also respond to the critiques that the viewers have offered to the first video. Specifically, some viewers have critiqued that both Collin and Marvin center white men as the ideal sexual and romantic partners. In so doing, they do not elaborate Asian men who date Asian men and non-Asian men of color. In addition, some viewers have mentioned that both Collin and Marvin as the Filipino subjects are "their own enemies." They essentialize "Asians" as a single culture according to the White/ Western/U.S. American logics of racial category. So, both Collin and Marvin own up to their biases and/or shortcomings that directed their production of the first video. They say that they grew up in the social and cultural environments in which white people are overwhelmingly the majority. People of color were not as visible. As a consequence, they developed their sexual and romantic desires toward white men. Most significantly, they question if the material legacy of U.S. colonialism among Filipino/a/xs might have affected their sexual and romantic aspirations toward white men.

While Collin and Marvin's co-performance of self-reflection yet requires a critically in-depth interrogation, I appreciate how they publicly exemplify a referencing point for gaysians to evaluate their own same-sex desires. Through the lens of an ongoing interplay between privilege, power, and oppression, both Collin and Marvin demonstrate that gaysians actively turn their attentions to white men as the ideal dating partners. Indeed, gaysian desires always already implicate the simultaneous technology of whiteness intersecting with patriarchy, hetero/homonormativity, and capitalism. Thus, it is necessary for gaysians to critically problematize the strategically essentialized framings of their "oppressed" narratives. Adapting what Celine Parreñas Shimizu has suggested,[51] I maintain that all gaysians must begin recognizing and working with their unearned privileges within and beyond the historical feminization and subordination of Asian men and masculinities.

Consequently, I advocate for pushing forward intellectual and political spaces in which gaysian counter-narratives destabilize the hegemony of Asian-White interracial couplings. While some scholars have previously worked on gaysian color-to-color sexual encounters and romantic relations,[52] these collections of studies are not enough. Gaysian narratives continue to be overwhelmingly dominated by their rejections by white men. In addition, several gaysian YouTubers I have introduced also suggest there is a dating competition driven by the capitalist logic of gay sexual cultures. Gaysians are fighting against one another to earn attentions from the small pools of (white) men who desire Asian men. This creates a discursive phenomenon of what Collin and Marvin's second video's title says is "Asians not Into Asians." Simultaneously, I question if such competition also implicates political rivalries, economic hierarchies, and historical tensions among gaysian men. Thus, more and more gaysian counter-narratives need to come forward. So, the *collective* shared performances of gaysian criticality can be further cultivated for change.

Gaysian fabulosity as a "quare" performative strategy: a conclusion

In this essay, through a lens of quare studies I have attempted to showcase how YouTube vloggers self-represent their gaysian experiences. The performative rhetorics of gaysian narratives, being produced by these vloggers, explicitly explicate, elucidate, and elaborate the supremacy of whiteness as the gay sexual cultural value of desire. More specifically, gay anti-Asian racism produces the racialized gender representation of Asian men as *feminine and submissive* Others for sustaining the power and privilege of White patriarchal phallic masculinity. Gaysian narratives also implicate the complexities and contradictions of identities, performances, and politics as they simultaneously navigate the material fragmentations of places and spaces such as white gay sexual cultures and "Asian" cultures. Yet gaysian narratives primarily draw from the hegemony of Asian-White interracial couplings as the relational norm. Thus, self-representational productions of gaysian narratives require ongoing questions and critiques. The performative meanings of becoming and being gaysians must be complicated, problematized, and shifted further.

Before concluding this essay, I highlight queerness of gaysian identities, performances, and politics implicated by the YouTube vlogger-created contents. I want to recognize that the gaysian YouTubers are in search of self-presenting and self-representing their racialized gender aesthetics working *on and against* the simultaneous technology of anti-Asian racism intersecting with the supremacy of whiteness and cisheteronormativity. For example, the Australian-based gaysian couple YouTuber DDCC published their intraracial coupling images in the four-minute and 49-second video "Cute Gay Asian Couple 2018 Recap" on December 30, 2018. The images are composed of picture slides featuring David and Eri showing off their same-sex intimacies to each other. Specifically, one of the images is their couple ring while there is no description of ring such as promising, engagement, or wedding. Also, some of the images include their photo shoots gathering with their family members and friends. The overall framing of their coupling image is *fabulous*. I recognize that some readers may discount their video as "extra," "over the top," and "ostentatious." They may also say the couple mimics White, cisheteronormative, and capitalistic logics of love, sex, and relationship. The ways in which they showcase their gaysian same-sex love is not any different from a Hollywood (or West Hollywood, a gay subdivision) love story. However, as a gaysian subject, I highly value their vlogger-created contents. It does not at least recycle the hegemony of Asian-White interracial couplings as the *normal* and *ordinary*. In addition, gaysian intraracial images and representations are not often available, especially in the Western contexts such as Australia, Canada, the United States, and the United Kingdom. Accordingly, I maintain that David and Eri's co-performance of gaysian intraracial desire is very politically important. It is a queer subcultural production of gaysian fabulosity; that is, a collective intelligence of gaysians who fabulously perform the racialized gender aesthetics to counter white gay sexual cultures.

In closing, I end this essay by reiterating gaysian fabulosity as a "quare" performative strategy. Gaysian identities, performances, and politics are never simple, clear-cut, and linear. Intersectionality, a construction of messy texts of privilege, power, and oppression produces and constitutes the material realities of gaysians. The ways in which gaysians simultaneously identify and disidentify with the majoritarian codes of belonging, taste, and value are much more complex and contradictory. Thus, there remain intellectual and political spaces for quareing racialized, gendered, and classed knowledge embedded in the material realities of gaysian fabulousity further. Therefore, I invite the readers to engage in questions, critiques, and practices of gaysian fabulousity as "quare."

Notes

1 Dwight A. McBride, *Why I Hate Abercrombie & Fitch: Essays on Race and Sexuality* (New York: New York University Press, 2005), 125.
2 Bryant Keith Alexander, "Queer(y)ing the Postcolonial Through the Western," in *Handbook of Critical and Indigenous Methodologies*, eds Norman K. Denzin, Yvonna S. Lincoln, and Linda T. Smith (Thousand Oaks, CA: Sage, 2008), 113.
3 Eng-Beng Lim, *Brown Boys and Rice Queens: Spellbinding Performances in the Asias* (New York: New York University Press, 2014), 27.
4 See the detailed discussion on a catchphrase in Chong-suk Han, "No Fats, Femmes, or Asians: The Utility of Critical Race Theory in Examining the Role of Gay Stock Stories in the Marginalization of Gay Asian Men," *Contemporary Justice Review* 11, no. 1 (2008): 11–22.
5 David L. Eng, *Racial Castration: Managing Masculinity in Asian America* (Durham, NC: Duke University Press, 2001), 2.
6 See, for example, Shinsuke Eguchi, "Negotiating Sissyphobia: A Critical/Interpretive Analysis of One 'Femme' Gay Asian Body in the Heteronormative World," *Journal of Men's Studies* 19, no. 1 (2011): 37–56; C. Winter Han, *Geisha of a Different Kind: Race and Sexuality in Gaysian America* (New York: New York University Press, 2015); Lim, *Brown Boys and Rice Queens*; and Voon Chin Phua, "Contesting and Maintaining Hegemonic Masculinities: Gay Asian American Men in Mate Selection," *Sex Role* 57 (2007): 909–918.
7 Nguyen Tan Hoang, *A View from the Bottom: Asian American Masculinity and Sexual Representaion* (Durham, NC: Duke University Press, 2014), 6–7.
8 Maurice Kwong-Lai Poon and Peter Trung-Thu Ho, "Negotiating Social Stigma among Gay Asian Men," *Sexualities* 11, no. 2 (2008): 245–268.
9 See Eguchi, "Negotiating Sissyphobia".
10 Gust A. Yep, "Queering/Quaring/Kuaring/Crippin'/Transing 'Other Bodies' in Intercultural Communication," *Journal of International and Intercultural Communication* 6, no. 2 (2013): 118–126.
11 "Quare" is an African American vernacular sounding/pronunciation of "queer" having commonly uttered in the South. Johnson coopts this construction as uttered by his grandmother to build a theory of "quare studies/quare theory." See E. Patrick Johnson, "'Quare' Studies or (Almost) Everything I Know About Queer Studies I Learned from My Grandmother," *Text and Performance Quarterly* 21, no. 1 (2001): 1–25.
12 Johnson, "'Quare' Studies," 3.
13 Fiona Buckland, *Impossible Dance: Club Culture and Queer World-Making* (Middletown, CT: Wesleyan University Press, 2002), 36–37.
14 See Bryant K. Alexander, "Boojie! A Question of Authenticity," in *From Bourgeois to Boojie: Black Middle Class Performances*, ed Vershawn Ashanti Young with Bridget Harris Tsemo (Detroit, MI: Wayne State University Press, 2011), 309–330.
15 See Shinsuke Eguchi and Godfried Asante, "Disidentifications Revisited: Queer(y)ing Intercultural Communication Theory," *Communication Theory* 26, no. 2 (2016): 171–189.
16 See Cathy J. Cohen, "Punks, Bulldaggers, and Welfare Queens: The Real Radical Potential of Queer Politics?," *GLQ: A Journal of Lesbian and Gay Studies* 3, no. 4 (1997): 437–465.
17 Bryant Keith Alexander, "Reflections, Riffs and Remembrances: The Black Queer Studies Conference," *Callaloo*, 23 (2002): 1288.

18 See Johnson, "'Quare' Studies"; E. Patrick Johnson and Mae G. Henderson, eds, *Black Queer Studies: A Critical Anthology* (Durham, NC: Duke University Press, 2005); and E. Patrick Johnson, ed, *No Tea No Shade: New Writings in Black Queer Studies* (Durham, NC: Duke University Press, 2016).

19 See, for example, Richard A. Ferguson, *Toward a Queer of Color Critique: Aberrations in Black* (Minneapolis, MN: University of Minnesota Press, 2004).

20 Jeffery Q. McCune Jr., "Out in the Club: The Down Low, Hip-Hop, and the Architexture of Black Masculinity," *Text and Performance Quarterly* 28, no. 3 (2008): 299.

21 Ibid., 299.

22 Johnson and Henderson, *Black Queer Studies*, 4.

23 José Esteban Muñoz, *Disidentifications: Queers of Color and the Performance of Politics* (Minneapolis, MN: University of Minnesota Press, 1999), 11.

24 See Yep, "Queering/Quaring/Kuaring/Crippin'/Transing 'Other Bodies' in Intercultural Communication."

25 See Wenshu Lee, "Kuaring Queer Theory: My Autocritography and a Race-Conscious, Womanist, and Transnational Turn," *Journal of Homosexuality* 45, nos 2/3/4 (2003): 147–170.

26 See Robert McRuer, *Crip Theory: Cultural Signs of Queerness and Disability* (New York: New York University Press, 2006).

27 See, for example, Jack Halberstam, *Trans: A Quick and Quirky Account of Gender Variability* (Oakland, CA: University of California Press, 2018); Railey C. Snorton, *Black on Both Sides: A Racial History of Trans Identity* (Minneapolis, MN: University of Minnesota, 2017); and Susan Stryker, "(De)subjugated Knowledges: An Introduction to Transgender Studies," in *The Transgender Studies Reader*, eds Susan Stryker and Stephen Whittle (New York: Routledge, 2006), 1–17.

28 See, for example, Eguchi, "Negotiating Sissyphobia"; Han, *Geisha of a Different Kind*; Hoang, *A View from the Bottom*; Lim, *Brown Boys*; and Poon and Ho, *Sex Role*.

29 Han, *Geisha of a Different Kind*, 148.

30 See, for example, Shinsuke Eguchi, "Queerness as Strategic Whiteness: A Queer Asian American Critique of Peter Le," in *Interrogating the Communicative Power of Whiteness*, eds Dawn Marie D. McIntosh, Dreama G. Moon, and Thomas K. Nakayama (New York: Routledge, 2019), 29–44.

31 See an analysis of gaysian porn star Brendon Lee in Hoang, *A View from the Bottom*, 29–70.

32 José Esteban Muñoz, *Cruising Utopia: The Then and There of Queer Futurity* (New York: New York University Press, 2009), 16.

33 See Muñoz, *Cruising Utopia*.

34 Alexander, "Queer(y)ing the Postcolonial," 108.

35 Ivan Dylko and Michael McCluskey, "Media Effects in an Era of Rapid Technological Transformation: A Case of User-Generated Content and Political Participation," *Communication Theory* 22, no. 3 (2012): 250–278.

36 Lori Kido Lopez, "Blogging While Angry: The Sustainability of Emotional Labor in the Asian American Blogospheres," *Media, Culture and Society* 36, no. 4 (2014): 421–436.

37 See, for example, Bryant Keith Alexander, *The Performative Sustainability of Race: Reflections on Black Culture and the Politics of Identity* (New York: Peter Lang, 2012); Bernadette Marie Calafell, "Monstrous Femininity: Constructions of Women of Color in the Academy," *Journal of Communication Inquiry* 36, no. 2 (2012): 111–130; and Amber Johnson, "Confessions of A Video Vixen: My Autocritography of Sexuality, Desire, and Memory," *Text and Performance Quarterly* 34, no. 2 (2014): 182–200.

38 Maurice Kwong-Lai Poon, "The Discourse of Oppression in Contemporary Gay Asian Diasporal Literature: Liberation or Limitation?," *Sexuality & Culture* 10, no. 3 (2006): 34–35.

39 Shinsuke Eguchi and Hannah R. Long, "Queer Relationality as Family: Yas Fats! Yas Femmes! Yas Asians!" *Journal of Homosexuality* 66, no. 11 (2019): 1589–1606.

40 Lisa Duggan, "The New Homonormativity: The Sexual Politics of Neoliberalism," in *Materializing Democracy: Toward a Revitalized Cultural Politics*, eds Russ Castronovo and Dana D. Nelson (Durham, NC: Duke University Press, 2002), 175–194.

41 McBride, *Why I Hate Abercrombie & Fitch*, 71.

42 See, for example, David L. Eng, *The Feeling of Kinship: Queer Liberalism and the Racialization of Intimacy* (Durham, NC: Duke University Press, 2010); and Jasbir K. Puar, *Terrorist Assemblages: Homonationalsim in Queer Times* (Durham, NC: Duke University Press, 2007).

43 Han, *Geisha of a Different Kind*, 141–142.

44 See Richard Fung, "Looking for My Penis: The Eroticized Asian in Gay Video Porn," in *Popular Culture: A Reader*, eds Raiford Guins and Omayra Z. Cruz (London: Sage Publications, 2005), 338–348.

45 See Hoang, *A View from the Bottom.*

46 McBride, *Why I Hate Abercrombie & Fitch*, 164.

47 See, Gayatri Chakravorty Spivak, *Other Asia* (Malden, MA: Blackwell Publishing, 2008).

48 Thomas K. Nakayama, "Show/Down Rime: 'Race,' Gender, Sexuality, and Popular Culture," *Critical Studies in Media Communication* 11, no. 2 (1994): 162–179.

49 See Shuzen Huang and Daniel C. Brouwer, "Coming Out, Coming Home, and Coming With: Models of Queer Sexuality in Contemporary China," *Journal of International and Intercultural Communication* 11, no. 2 (2018): 97–116.

50 Poon, "The Discourse of Oppression," 43.

51 Celine Parreñas Shimizu, *Straight Sexualities: Unbinding Asian American Manhoods in the Movies* (Stanford, CA: Stanford University Press, 2012), 243.

52 See, for example, Shinsuke Eguchi, "Queer Intercultural Relationality: An Autoethnography of Asian-Black (Dis)connections in White Gay America," *Journal of International and Intercultural Communication*, 8, no. 1 (2015): 27–43; and Cynthia Wu, *Sticky Rice: A Politics of Intraracial Desires* (Philadelphia, PA: Temple University Press, 2018).

4

COMMUNICATION, GENDER, AND CAREER IN MENA COUNTRIES

Navigating the push and pull of empowerment and exclusion

Astrid M. Villamil and Suzy D'Enbeau

Women with professional aspirations throughout the world face a competitive, professional landscape that is gendered masculine, where the ideal worker is male with unlimited time to spend devoted to work and who possesses leadership skills and technical expertise.[1] In contrast, the ideal woman privileges her family's needs[2] and is able to seamlessly balance the demands of work and non-work commitments.[3] Thus, women with professional aspirations throughout the world face challenges in their attempts to navigate these competing gendered identities.[4]

Yet, despite the similarities of the global professional landscape, transnational discourses around gender and sexuality take on multiple, shifting, and contested meanings in an increasingly globalized world and differences emerge in how women from diverse parts of the world frame and manage these challenges.[5] These challenges are especially complicated in geographies where local and cultural complexities fail to mirror Western notions and ideals of gender and sexuality. In this chapter, we focus specifically on feminist, organizational, and management communication scholarship about women from the MENA region in their efforts to redress dominant gendered discourses.

The MENA region includes 19 Muslim countries: 14 countries in the Middle East (Bahrain, Iran, Iraq, Jordan, Kuwait, Lebanon, Oman, Palestine, Qatar, Turkey, Saudi Arabia, Syria, United Arab Emirates, and Yemen) and 5 countries in North Africa (Algeria, Egypt, Libya, Morocco, and Tunisia). We do not mean to paint this region as homogenous; indeed, there are marked differences in the everyday lived experiences of women from these parts of the world that are informed by class, religion, education, age, and other factors.[6] However, for the purposes of this chapter, we are interested in striking a balance between the similarities and differences among women from the MENA region, with an eye toward offering a stark contrast to most gender, career, and work-life scholarship that is focused primarily on the experiences of women in the West.

We begin our extended literature review by explicating two ways of thinking about MENA as reflected in academic research and everyday discourse that inform underlying assumptions with regard to choice, agency, responsibility, and constrained obligation. We then introduce a transnational feminist approach to synthesize a snapshot of extant research in the contexts of education and career. We summarize several strategies employed by women in the MENA region to navigate a pull between empowerment and exclusion as they pursue their education and career. We conclude by suggesting several potential areas for future research.

Transnational feminist theoretical assumptions

To lay the foundation for our transnational feminist approach, we first set the stage by describing the dominant ways of conceptualizing gender in the MENA region and the perpetuation of problematic stereotypes informed by binary-laden ideologies, most notably as a result of globalization. First, there are symbolic implications for how women in the MENA region are represented in scholarly and popular discourse. These representations are often informed by "binary logics … one form of discursive logic that reproduces male privilege in organizing practices"[7] and also positions Western ways of thinking as superior to non-Western approaches. For example, women's experiences and gendered performances in MENA countries are often described as oppressive, backwards, and static, while in contrast, the West is constructed as a superior, more dynamic, and accepting space for women.[8] This orientalist binary presents a narrow reading of gender in MENA countries and strips away any nuanced sense of women's agency. That is, binary logics structure gendered representations in a way that strips away women's ability to "fashion (or refashion) transformative identities and subjectivities in otherwise constrained conditions."[9]

Second, there are material implications that result from globalization in terms of women's experiences in the MENA region. Globalization has promulgated new patriarchal forms, including neo-patriarchy or patriarchal social structures combined with capitalist economic structures.[10] Sharabhi defines neo-patriarchy as the product between discourses of modernity and tradition that render an Arab society full of contradictions and unstable political, social, and cultural dynamics.[11] Neo-patriarchy developed as a product of Western discourses of independent economy, wealth expansion, and oil markets that have traveled through MENA countries. What is more, neo-patriarchal assumptions have clashed with radically different worldviews, including cultural and religious values that do not instinctively complement capitalist discourses but that have been forced to fabricate an amalgamation of these distinct ideologies. That is, "under the global sway of neoliberalism, with its celebration of consumerism and privatization, socials issues are recoded as private challenges and this compromises the very idea of political agency."[12] For example, Haghighat suggests that while discourses of economic independence have become more acceptable among women in MENA countries, many times the pursuit of financial independence brings more burden to women who have to experience exploitation, abuse, and unfair conditions in the workforce, all while still expecting to fulfill reproductive and familial obligations[13]. In this way, it is incumbent upon scholars to consider the power relations, also often constructed through problematic binaries, that surface as we problematize socially constructed divisions like public/private, local/global, and masculine/feminine.[14]

In concert, the dominance of binary and neo-patriarchal logics positions women of MENA countries in a shifting landscape that demands attention to how they understand and navigate this terrain. Transnational feminist theories are ideal in defining frameworks and problematizing extant feminist discourses that are aided by globalization, but anchored in distinct geopolitical and cultural spaces. In short, transnational feminism theorizing focuses on acknowledging that patriarchal processes are not unified but rather undertake different meanings depending on the cultural and political characteristics of a context.[15] More specifically, globalization has circulated Western discourses about gender and sexuality throughout much of the world, and transnational feminist theories invite critique of how these discourses play out in non-Western regions, in effect, exacerbating tensions between ideal woman and ideal worker.[16] The colonialist nature of Western discourses, particularly with regard to individualism, often obscure and further marginalize the voices of women from the MENA region and recommended strategies to navigate gendered dynamics that often do not take into consideration the specific structural and cultural constraints in this context.[17] Thus, while issues might be similar across cultural contexts, gendered experiences, problems, and strategies are multifaceted, nuanced, and specific. This transnational feminist approach to constructions of gender and sexuality in the MENA region underscores our commitment to radically revise and reformulate how we approach, understand, and analyze the experiences of women in diverse parts of the world.[18]

Thus, our transnational feminist approach allows us to describe the convoluted meanings of gender and sexuality in MENA countries; engage the recurring tensions between transnational, cultural, and local discourses around gender and sexuality; and describe the gendered paradoxes and attendant discursive strategies that MENA agents experience in navigating their own identities. Our reading of extant literature contextualizes and unpacks the ways gender/ed discourses surface in ironic and paradoxical ways in the areas of education and career. In doing so, we document how global discourses of gender have traveled and informed women's experiences in MENA countries with the understanding that these issues are present in many other geographies, including developed countries. We begin by describing the competing discourses of empowerment and exclusion as anchors of gendered experiences in the MENA region.

The push and pull of empowerment and exclusion in education and career

The gendered experiences of women in the MENA region are accentuated by a broader tension around empowerment and exclusion; that is, "a context of women's increasing mobilization and empowerment, on one hand, and enduring forces of gender inequality and exclusion, on the other."[19] We take a constitutive view of tension to highlight the stress, anxiety, and discomfort that are revealed as women respond to the push and pull of both asserting their agency through and being constrained by binary logics and neo-patriarchal and religious assumptions.[20]

The empowerment-exclusion tension is exacerbated by contested interpretations of the cultural and religious significance of Islam in light of discourses of modernity and capitalism. As the dominant religion, Islam is a recurrent variable in theorizing about gender in MENA countries. However, the interpretation of religious texts and its influence on women's experiences takes at least three strands. A first strand presents Islam as interchangeable with patriarchal discourses and therefore obstructing women's development.[21] A second strand advocates for Islam as the ultimate feminist practice. A third strand acknowledges the patriarchal co-optation of Islam, but also recognizes the central role of Islam in women's experiences while advocating for a progressive interpretation of the Qur'an that privileges gender egalitarianism.[22] As a case in point, although the Qur'an advocates for women's right to be educated and to work; concurrent interpretations also state that women may only work in certain industries as long as they preserve their and their family's safety and dignity.[23]

Thus, the empowerment-exclusion tension renders a complicated scenario: women's educational development is encouraged, but the utilization of their professional abilities often gets diluted by societal and familial gender/ed expectations. For instance, Beverly Metcalfe describes how interpretations of labor laws in the majority of MENA societies are often guided by the concept of *urf*, or the need to protect women in order to preserve a moral work environment, and *qiwama*, the expectation for men to preserve women's modesty and keep women's sexuality veiled.[24] Thus, these deeply ingrained and implicit values create trenches in the labor landscape that are unique for women to navigate. Western feminist assumptions might paint this dilemma as more contentious, but Metcalfe found that this gender separation for businesses is seen as a productive way to preserve the gendered social order.[25] With this empowerment-exclusion tension in mind, we explore extant research about education and career in different parts of the MENA region to highlight the complex and multiple ways in which this tension plays out.

In terms of the empowerment side of this tension, closing the gender gap on educational needs has been at the forefront of almost every country in the MENA region during the last decade.[26] Most countries have made great strides toward providing access to primary and secondary education. In fact, most countries have closed or are closing the gender gap in youth literacy, as well as primary, secondary, and tertiary education. For example, in Iran, the increase in education efforts has become a mobility vehicle for women into the workforce.[27] Women from rural areas in Iran have traveled away from their households and live independently from their immediate families, facilitating more agentic decisions about postponing marriage and career pursuits. Conversely, countries like Morocco and Yemen have struggled to overcome poverty, safety, and other cultural barriers that keep young women from accessing education.[28]

Despite these changes, the exclusion side of this tension showcases some of the cultural constraints in the area of education. In Saudi Arabia, for instance, education systems are deeply informed by Shari'a law that requires differentiated educational content based upon sex. As such, men are educated to use ambition to pursue career goals and women are taught about gender-segregated careers and their nurturing role as mothers and housewives.[29] In addition, logistical and material factors still make it difficult for women in Saudi Arabia to have access to education. For example, Saudi women face mobility barriers because transportation is limited and autonomous vehicle driving only recently became legal.[30] Nevertheless, some Saudi parents have become more open to supporting daughters as they study abroad and downplay the law that requires women to travel with a male companion.[31]

In sum, discourses that advocate for equal access to education for women have become normative in part because of government policies that present higher education as a universal right. However, these discourses come loaded with contradictions and conflicting messages. For example, James-Hawkins, Qutteina, and Yount found that women in Qatar are aware of the benefits of higher education but some have also pointed to the contradictory narratives when considering joining the workforce or postponing family development.[32] That is, women in Qatar receive conflicting messages from the government. These messages center around being primarily responsible for the family and privileging motherhood over employment; yet conversely, the government is actively trying to increase women's employment. In another example from Iran, access to secondary and tertiary education is still tied to socioeconomic factors; as such, women from middle and upper classes experience radically different gender discourses and female employment is predicated on neoliberal premises that advocate for free trade and the privatization of local industries and ignore the consequences these policies have on low-income women.[33] In short, discourses of women's empowerment can be problematic because they are often presented as an emancipatory, one-size-fits-all solution to a multilayered landscape; in doing so, these discourses can become constricting, exploitative, and unhelpful.

This empowerment-exclusion tension also informs understandings of and access to career options. Although efforts to increase gender equity in education systems in many of the MENA countries has yielded productive outcomes,[34] weak educational pipeline infrastructures and lack of career choice support remain a constant struggle. As a result, countries that have made advances to close the gap in primary and secondary education have not yet experienced similar outcomes when it comes to diversifying the workforce or other aspects of the public sphere.[35] In theory, efforts to close the educational gap should lead to a more diverse workforce and visibility in leadership, but there is still a clear disconnect between access to education and opportunities for paid employment and long-term careers.[36]

More specifically, career choices for women in the MENA region are informed by interlocking discourses of religion, family, and culture.[37] To begin, Islam is a robust foundation in determining workplace access and organizational policy and practice. For example, many universities and workplaces have incorporated leaves and breaks in their human resource policies to accommodate prayers and rituals[38] and some corporations have included deeply religious rhetoric in their institutional documents in order to build legitimacy and provoke moral validation among stakeholders.[39] Evidently, the institutionalization of Islam directly informs gender performances. On the one hand, it can serve as a facilitator for gender development within traditional frameworks, as women are more and encouraged to seek educational and professional opportunities and institutionalized pathways for these opportunities are more common; on the other hand, it can be co-opted by neo-patriarchal discourses and restrict opportunities for women's advancement by advancing binary discourses that often make women choose between career and family. Thus, despite various interpretations of Islam, it is intimately tied to women's workplace experiences, ranging from everyday interactions to formalized organizational policy and practice.

Furthermore, the influence of Islam on career also intersects with entrenched understandings of the role of family. The collectivistic nature of MENA countries has centered nuclear and extended families as a cornerstone of society. To this end, socialization processes are fueled with gendered discourses that charge women with responsibilities like caring for husbands, parents, and children.[40] A byproduct of this socialization cycle favors family over career- and income-related professions.[41] Indeed, the superiority of the family over career not only influences everyday dynamics but, as Afiouni explains, it is "enshrined in the constitutions of many Arab states."[42]

In another example, James-Hawkins and colleagues found that women in Qatar fear that normative changes regarding education and workforce have also brought shattering consequences to institutions like marriage. In this study, Qatari women perceived high divorce rates, even though this perception did not reflect public data, and reported increasing anxiety at the thought of their current or future spouses divorcing them. This uncertainty fueled their decisions to keep pursuing education and career goals as a way to mitigate the economic uncertainty that a prospective divorce might bring.[43] In other words, a latent fear of future divorce activated motivations to continue pursuing career goals as a preparedness strategy. Conversely, dilemmas about timing, approaches to motherhood, and making more decisions about family conception and contraception are starting to be more visible among women in MENA. Thus, the incorporation of technologies like in vitro fertilization (IVF) as alternative routes for family planning or emergency contraception to interrupt are becoming emergent.[44] In this way, the motivation to pursue education and a career was informed more by the potential instability of family structures than professional aspirations.

Comparably, cultural norms influence career access and options for women from the MENA region. For example, social capital, or *wasta*, is an important factor in determining successful career pathways for women. Metcalfe describes wasta as "the recognition that power in society is related to tribal and familial structures."[45] In this context, women whose families and personal circles are centrally located among those with more power have a better chance of career advancement. The concept of wasta might appear to parallel Western ideals of nepotism and social capital; however, Metcalfe argues that wasta is deeply ingrained in the teaching of the Qur'an and is part of a multilayered pursuit of balance and equilibrium in social and work relationships.[46] For instance, Tlaiss and Kauser reported how women in Lebanon used available social networks to seek career opportunities for recruitment, promotion, and advancement.[47] The tribal and collectivistic nature of MENA countries centers wasta as a key navigation tool for women who are entering the workforce and need to expand their networks.

Ironically, these networks often are limited and continue to perpetuate the scarcity of resources—social, political, and economical—necessary for women to expand their professional capacity. McElwee and Al-Riyami advocate for more networks of women in similar positions to have spaces to share information and seek career advice,[48] pointing to how wasta is seamlessly integrated into cultural norms around career. However, Metcalfe also acknowledges the perpetuation of inequalities for women whose families do not have the connections and influence necessary to facilitate career pathways. In this way, cultural norms such as wasta also heighten the influence of socio-economic and class-related dynamics. The predominant use of wasta and its subsequent favoring of women in more power-laden networks has brought consequences for women in places like Bangladesh, India, and Indonesia who have fled their own geopolitical struggles in search for a more promising future in an oil-boosting economy and often end at the periphery of these networks, working as nannies and housekeepers.[49] This is particularly problematic, as Western neo-patriarchal discourses that favor social capital and elitist connections have populated the way MENA countries conduct business. Furthermore, the intersectional nature of these power imbalances accentuates the struggle of individuals whose identities are not socially rewarded. As such, refugees, gender nonconforming individuals, and immigrants with other religious beliefs are not likely to have the same access to the job benefits and career status as many of the women in MENA who are positioned in spaces of privilege.[50]

In this section, we outlined how gendered discourses intersect with religious, familial, and cultural norms for women with professional aspirations in the MENA region. Our synthesis unpacks the different ways that women are both empowered to fully participate and excluded from full participation in the public world. In the next section, we document the innovative strategies employed by women to navigate this empowerment-exclusion tension.

Strategies to navigate the empowerment-exclusion tension

Theories of transnational feminism call for a revisioning of how we understand gender and sexuality in the MENA region, with an eye toward gendered transformative possibilities and potential.[51] The imminent influence of neo-patriarchal ideologies that globalization has circulated has made it possible for women to occupy alternative agentic spaces. In doing so, MENA women are resisting, contesting, and redefining oppressive patriarchal discourses.[52] These contested spaces have made two factors apparent. First, targets of resistance for MENA women are deeply dictated by their cultural, religious, and familial backgrounds. Second, they conceal, problematize, and resolve tensions about their experiences in ways that may seemingly look contradictory but are, in fact, carefully crafted to attend to the unique intersections MENA women encounter to advance both their careers and families without violating the multiple cultural and gender/ed expectations from society. In this section, we synthesize how extant literature has conceptualized strategies employed by everyday women in navigating the broader empowerment-exclusion tension in the contexts of education and career. These strategies coalesce around four main themes; the first three highlight the ways in which women in the MENA region symbolically perceive gender and the last theme focuses on structural changes.

First, some research has explored how women from the MENA region emphasize gender equality while retaining gender difference. For example, D'Enbeau and colleagues explain that the young women they interviewed privileged gender complementarity to explicate the equal, but different, mentality supported by Muslim societies. In doing so, women still honor religious norms and highlight the importance of gender differentiation as a tool for society advancement.[53] In a second example, Tlaiss and Kauser discovered that Lebanese women with managerial aspirations did not "believe that a successful manager needs to possess masculine traits" but rather they saw their feminine traits as a productive alternative that could be useful in the workplace.[54] In this way, this theme represents strategies that privilege gender difference and use that difference as a mechanism for career success.

The second theme encapsulates strategies employed by women in the MENA region that identify productive overlap between family and career success. D'Enbeau and colleagues highlight how young women envision professional and familial goals as contingent upon one another.[55] Tlaiss and Kauser explain that Lebanese female managers did not feel guilty about their career aspirations as long as their family responsibilities did not suffer. In most cases, successful female managers who were married and had children were also able to rely on domestic help so that their husbands did not feel the burden of their wife's career.[56] In this way, this strategy also points to the socioeconomic-based luxuries for some women in the MENA region to pursue professional aspirations. In brief, "traditional gender roles and the needs and desires of family and significant others will shape personal conceptualizations of career success for women in the region."[57]

The third theme underscores strategies that require MENA women to thread cultural and religious pride into their careers. D'Enbeau and colleagues describe how young women embrace their cultural and religious pride to overcome tensions between individual and collective goals. In fact, women from the MENA region decided to advocate for more gender equitable spaces– not to counteract their religious and cultural beliefs but *because* of their attachment to them.[58] Furthermore, this transcending mechanism served as a safe way to advocate for change, acknowledging that "change must be strategic but is not always about sweeping structural transformations."[59] Similarly, McIntosh and Islam argue that Bahraini women entrepreneurs who preserve religious and cultural traditions like wearing the hijab or having support from their family had better access to business networks.[60] Syed also suggests that women from MENA regions can lean into modesty to promote gender equality and achieve cooperation from men who might feel ambivalent about women joining the workforce.[61] Afiouni highlights women's agentic role in negotiating independent and interdependent commitments by enacting more traditionally interdependent gender scripts while generating structural change with their individual performances at work.[62] To be sure, Karam, Afiouni, and Nasr acknowledge the creative aspect of this strategy in terms of rethinking women's enactment of gender as a subversive act where women do not passively internalize gender roles but instead choose to eloquently "deploy a variety of strategic responses to bend these norms, while at the same time maintaining legitimacy."[63]

The final theme encompasses strategies that require changes to organizational culture and policy. For example, Tlaiss and Kauser explain that women often do not have access to wasta in the way that men do. The authors suggest that "human resource departments should set up policies that reduce the influence of wasta within an organization which may limit its potential influence to those who have met the minimum requirements in terms of qualifications and work experience."[64] In doing so, organizations can enact more transparent and equitable labor processes.

To this end, Metcalfe proposes four suggestions for Western corporations that can enhance the promotion of women in business in MENA regions: keep a close track of statistics to monitor and track women's progression and engagement in the workforce; tailor training and development programs to build upon competencies about women in business while at the same time being sensitive about Islam and cultural norms; emphasize Islam in cultural training and how it intersects with all aspects of women's lives; and design mentoring programs to promote positive role models among women.[65] In another example, James-Hawkins, Qutteina, and Yount suggest that "governments in highly patriarchal countries that wish to advance women in their societies need to encourage men and women to share more equally within the private sphere."[66] In this way, this theme points to the need for an attitudinal shift in organizations and society in general in order to recognize women as competent professionals and leaders.[67]

Taken together, these strategies constitute a cartography for women in MENA countries to carefully navigate tensions between discourses of empowerment and exclusion. Moreover, these strategies entertain the friction between the two poles and in doing so, provide an alternate space where women in MENA countries creatively operate as agents and pacers of their careers while still performing as interdependent family and society members. Women in MENA countries perceive gender equality as access to education and work, while at the same time acknowledging religious and neo-patriarchal norms by following occupational differences. And women simultaneously navigate collective gender roles and new career paths while also "adhering to the collective norms that constrained gender performances."[68]

Future directions

This review of the literature suggests a need for additional research. The empowerment-exclusion tension discussed in this chapter will likely keep shaping the discourses around gender and sexuality in MENA countries. As such, future research on gendered performances in the MENA region could consider the following five areas.

First, as demonstrated in this review, women from the MENA region experience dilemmas in navigating the contentious intersections of individual goals and collective gendered obligations, particularly in the contexts of education and career. We argue that these dilemmas coalesce around a broader tension of empowerment and exclusion. Extant research has identified these dilemmas, but future research could explore the underlying emotions of these dilemmas, an important endeavor given that most tension-navigation strategies assume a rational approach and downplay the significant influence of emotions in working through dilemmas.[69]

Second, and relatedly, exploration of the (lack of) effectiveness of organizational policy recommendations to accommodate women in the workforce is warranted. Although suggestions to human resource (HR) policies may default to the implementation of more consistent and merit-based rules and procedures, Baranik, Gorman, and Wales and Vendello remind us that these propositions rely too much on Western concepts and ignore the religious and cultural nexus in which MENA countries are positioned.[70] Notwithstanding, institutional change through the incorporation of sensitive and tailored polices can be a useful avenue for women to navigate the emerging tensions in the workplace.[71] Additionally, scholars can expand on the risks and benefits associated with MENA-based corporations that center Islamic values as organizational pillars, especially considering the neoliberal push to streamline processes and participate in free trade economies.[72]

Third, future research should contemplate the ecological consequences of traveling Western and neo-patriarchal discourses in MENA family structures. More specifically, research could examine alternative family formation processes like single motherhood, foster parenting and adoption, and family planning and infertility. For example, Newman exposes the paradoxes of single motherhood in Morocco and how embracing motherhood as dominant identity becomes the rescuing mechanism and point of entry for women to access public and political realms.[73] Further, the shunning and punishment of abortion, abandonment, and adoption alternatives dominate the MENA family landscape. Thus, scholars could expand on reproductive and contraceptive processes and the incorporation of technologies like in vitro fertilization (IVF) to kickstart conception or emergency contraception to interrupt it, and how these processes inform gendered dynamics in the region.[74] In short, more scholarship is needed to unpack the existing tensions around and shifting landscape of family constitution in MENA countries.

Fourth, future research should focus on the intersection of Muslim identities and other cultural categories such as race, nationality, class, and sexual orientation.[75] That is, gender should be theorized as constitutive with other identity-based categories that are inseparable from gender. In taking this intersectional approach, inquiries about how gender, race, class, and ability can create necessary spaces to think about tensions of equality and difference.[76] Specifically, MENA researchers could focus on exploring the tensions that LGBTQI individuals navigate in terms of gendered performances amidst cultural and religious traditions and norms. Nasser-Edin, Abu-Assab, and Greatick, for example, draw attention to the troublesome predicaments of people who identify as LGBTQI when facing geographical displacement or seeking asylum.[77]

Finally, as development policies around women in MENA continue to unfold, gendered constructions of masculinities will also shift. Future research could look at the way men negotiate their power and privilege in their roles as spouses, fathers, and coworkers. Specifically, researchers could explore the interplay between Islamic paternalism and neo-patriarchy and identify existing dimensions of masculine role negotiation in this range. Amar invites a way of theorizing about masculinities that steers away from totalizing discourses where masculinities are scrutinized and rendered immoral, fetishized, and in crisis, especially when contrasted with the also essentialized gaze toward women from MENA countries. Instead, Amar invites us to engage in "a more materialist approach that focuses on industries and institutions that are producing the particular subjects of masculinity who are seen as animating these crises."[78]

Conclusion

In this chapter, we present an overview of extant research about gender and sexuality in the MENA region in the contexts of education and career. Our transnational feminist approach to this systematic review allows us to accomplish several objectives. First, we delineate the way in which binary logics underscore representations of women in the MENA region and constrain everyday lived gendered experiences. We also articulate the role of neo-patriarchy's influence upon the gendered dynamics of the region. Binary logics and neo-patriarchal discourses create a broader system of empowerment and exclusion that reveals a push-and-pull dynamic; on one hand, women are empowered to access education and pursue their career aspirations, while on the other hand, they are also excluded from full participation in the public sphere without consequence. In pulling the literature together, we synthesize the strategies used to navigate the tensions outlined in extant research as women navigate this push and pull. Finally, we propose several avenues of future research that can productively extend our understanding of the complexities of gender in the MENA region.

Notes

1 Joan Acker, "Hierarchies, Jobs, Bodies: A Theory of Gendered Organizations," *Gender & Society* 4 (1990): 139–158, doi: 10.1177/089124390004002002; Patrice Buzzanell, "Gaining a Voice: Feminist Organizational Communication Theorizing," *Management Communication Quarterly* 7, no. 4 (1994): 339–383, doi: 10.1177/0893318994007004001; Beverly Metcalfe, "Women, Management and Globalization in the Middle East," *Journal of Business Ethics* 83 (2008): 85–100, doi: 10.1007/s10551-007-9654-3.

2 Fida Afiouni, "Women's Careers in the Middle East: Understanding Institutional Constraints to the Boundariless Career", *Career Development International* 19, no. 3 (2014): 314–336, doi: 10.1108/CDI-05-2013-0061; Patrice Buzzanell and Suzy D'Enbeau, "Stories of Caregiving: Intersections of Popular Gender, Academic Research, and One Woman's Experiences," *Qualitative Inquiry* 15 (2009): 1119–1224, doi: 10.1177/1077800409338025; Suzy D'Enbeau, Astrid Villamil, and Rose Helens-Hart, "Transcending Work–Life Tensions: A Transnational Feminist Analysis of Work and Gender in the Middle East, North Africa, and India," *Women's Studies in Communication* 38 (2015): 273–294, doi:10.1080/07491409.2015.1062838.

3 Patrice Buzzanell, Rebecca Meisenbach, Robyn Remke, Meina Liu, Vanessa Bowers, and Cindy Conn, "The Good Working Mother: Managerial Women's Sensemaking and Feelings about Work–Family Issues," *Communication Studies* 56 (2005): 261–285, doi: 10.1080/10510970500181389; Kirby Erika and Kathleen Krone, "The Policy Exists but You Can't Really Use It," *Journal of Applied Communication Research* 30, no. 1 (2002): 50–77, doi: 10.1080/00909880216577; Haifa Tlaiss and Saleema Kauser, "The Impact of Gender, Family, and Work in the Career Advancement of Lebanese Women Managers," *Gender in Management: An International Journal* 26, no. 1 (2011): 8–36, doi: 10.1108/17542411111109291.

4 Laurie James-Hawkins, Yara Qutteina, and Kathryn Yount, "The Patriarchal Bargain in a Context of Rapid Changes to Normative Gender Roles: Young Arab Women's Role Conflict in Qatar," *Sex Roles* 77, nos 3–4 (2017): 155–168, doi: 10.1007/s11199-0160708-9; John McIntosh and Samia Islam, "Beyond the Veil: The Influence of Islam on Female Entrepreneurship in a Conservative Muslim Context," *International Management Review* 6, no. 1 (2006): 101–109.

5 Suzy D'Enbeau, Astrid Villamil, and Rose Helens-Hart, "Transcending Tensions"; Fauzia Al-bakr, Elizabeth Bruce, Petrina Davidson, Edit Schlaer, and Ulrich Kropiunigg, "Empowered but Not Equal: Challenging the Traditional Gender Roles as Seen by University Students in Saudi Arabia," *FIRE: Forum for International Research in Education* 4, no. 1 (2017): 52–66, doi:10.18275/re201704011083.

6 Valentine Moghadam, *Modernizing Women: Gender and Social Change in the Middle East* (Boulder, CO: Lynne Rienner Publishers, 2003).

7 Debbie Dougherty and Marlo Goldstein Hode, "Binary Logics and the Discursive Interpretation of Organizational Policy: Making Meaning of Sexual Harassment Policy," *Human Relations* 69, no. 8 (2016): 1730, doi: 10.1177/0018726715624956.

8 Roksana Bahramitash, "Iranian Women during the Reform Era (1994–2004): A Focus on Employment," *Journal of Middle East Women's Studies* 3, no. 2 (2007): 86–109.

9 Nikki Townsley, "Love, Sex, and Tech in the Global Workplace," in *The Sage Handbook of Gender and Communication*, eds Bonnie Dow and Julia Wood (Thousand Oaks, CA: SAGE, 2006), 145.

10 Elhum Haghighat, "Social Status and Change: The Question of Access to Resources and Women's Empowerment in the Middle East and North Africa," *Journal of International Women's Studies* 14, no. 1 (2013): 273–299; Hisham Sharabi, *Neopatriarchy: A Theory of Distorted Change in Arab Society* (New York: Oxford University Press, 1988).

11 Hisham Sharabi, *Neopatriarchy.*

12 Radha Hegde, "Globalizing Feminist Research in Communication," in *The Sage Handbook of Gender and Communication*, eds Bonnie Dow and Julia Wood (Thousand Oaks, CA: SAGE, 2006), 434.

13 Elhum Haghighat, "Social Status and Change."

14 Nikki Townsley, "Love, Sex, and Tech."

15 Rebecca Dingo, *Networking Arguments: Rhetoric, Transnational Feminism, and Public Policy Writing* (Pittsburgh, PA: University of Pittsburgh Press, 2012); Beverly Melcafe, "Women Management, and Globalization," in *Globalizing Women: Transnational Feminist Networks*, ed Valentine Moghadam (Baltimore, MD: The Johns Hopkins University Press, 2005).

16 Suzy D'Enbeau, Astrid Villamil, and Rose Helens-Hart, "Transcending Tensions."

17 Rebecca Dingo, *Networking Arguments.*

18 Radha Hegde, "Globalizing Feminist Research."

19 Loubna Skalli, "Constructing Arab Female Leadership Lessons from the Moroccan Media," *Gender & Society* 25, no. 4 (2011): 478.

20 Linda Putnam, Gail Fairhurst, and Scott Banghart, "Contradictions, Dialectics, and Paradoxes in Organizations: A Constitutive Approach," *The Academy of Management Annals* 10, no. 1 (2016): 65–171.

21 Aysel Morin, "Victimization of Muslim Women in Submission," *Women's Studies in Communication* 32, no. 3 (2009): 380–408.

22 Aysel Morin, "Victimization of Muslim Women."

23 Mona Al Munajjed, *Women in Saudi Arabia Today* (London: Macmillan, 1997).

24 Beverly Metcalfe, "Exploring Cultural Dimensions of Gender and Management in the Middle East," *Thunderbird International Business Review* 48, no. 1 (2006): 97.

25 Beverly Metcalfe, "Exploring Cultural Dimensions."

26 UNICEF, *Regional Overview for the Middle East and North Africa: MENA Gender Equality Profile*, 2011.

27 Roksana Bahramitash, "Iranian Women during the Reform Era."

28 World Bank, *The Status & Progress of Women in the Middle East & North Africa* (World Bank, Washington, DC, 2009).

29 Roula Baki, "Gender-Segregated Education in Saudi Arabia: Its Impact on Social Norms and the Saudi Labor Market," *Education Policy Analysis Archives* 12, no. 28 (2004): 1–12.
30 Margaret Coker, "Saudi Women can Drive, but here's the Real Roadblock," *The New York Times*, December 15, 2018, www.nytimes.com/2018/06/22/world/middleeast/saudi-arabia-women-driving. html.
31 Eleanor Doumato, "Saudi Arabia," in *Women's rights in the Middle East and North Africa: Progress amid Resistance*, eds Sanja Kelly and Julia Breslin (New York, NY: Rowman & Littlefield, 2010), 425–458.
32 Laurie James-Hawkins, Yara Qutteina, and Kathryn Yount, "The Patriarchal Bargain."
33 Roksana Bahramitash, "Iranian Women during the Reform Era."
34 World Bank, *The Status & Progress of Women*, 2009.
35 Elhum Haghighat, "Social Status and Change."
36 Roksana Bahramitash, "Iranian Women during the Reform Era"; Laurie James-Hawkins, Yara Qutteina and Kathryn Yount, "The Patriarchal Bargain."
37 Fida Afiouni, "Women's Careers in the Middle East"; Suzy D'Enbeau, Astrid Villamil, and Rose Helens-Hart, "Transcending Tensions."
38 Fida Afiouni, "Women's Careers in the Middle East."
39 Yussuf Sidani and Sammy Showail, "Religious Discourse and Organizational Change Legitimizing the Stakeholder Perspective at a Saudi Conglomerate," *Discourse & Organizational Change* 26, no. 6 (2013): 931–947.
40 Haifa Tlaiss and Saleema Kauser, "The Impact of Gender, Family, and Work."
41 Fida Afiouni, "Women's Careers in the Middle East."
42 Fida Afiouni, "Women's Careers in the Middle East," 317.
43 Laurie James-Hawkins, Yara Qutteina, and Kathryn Yount, "The Patriarchal Bargain."
44 Ellen Amster, "Global IVF, Infertility, and Emergency Contraception in the Middle East and North Africa," *Journal of Middle East Women's Studies* 14, no. 3 (2018): 343–347, doi: 10.1215/15525864-7025455.
45 Beverly Metcalfe, "Gender and Human Resource Management in the Middle East," *The International Journal of Human Resource Management* 18, no. 1 (2007): 57, doi: 10.1080/09585190601068292.
46 Beverly Metcalfe, "Gender and Human Resource Management."
47 Haifa Tlaiss and Saleema Kauser, "The Impact of Gender, Family, and Work."
48 Gerard McElwee and Rahma Al-Riyami, "Women Entrepreneurs in Oman: Some Barriers to Success," *Career Development International* 8, no. 7 (2003): 339–346, doi: 10.1108/13620430310505296.
49 Beverly Metcalfe and Christopher Rees, "Gender, Globalization, and Organization: Exploring Power, Relations and Intersections," *Equality, Diversity and Inclusion* 24, no. 1 (2010): 5–22, doi: 10.1108/02610151011019183.
50 Joseph Vendello, "Do We Need a Psychology of Women in the Islamic World?," *Sex Roles* 75 (2016): 623–629, doi: 10.1007/s11199-016-0691-1.
51 Rebecca Dingo, *Networking Arguments.*
52 Loubna Skalli, "Constructing Arab Female Leadership."
53 Suzy D'Enbeau, Astrid Villamil, and Rose Helens-Hart. "Transcending Tensions."
54 Haifa Tlaiss and Saleema Kauser, "The Impact of Gender, Family, and Work," 18.
55 Suzy D'Enbeau, Astrid Villamil, and Rose Helens-Hart, "Transcending Tensions."
56 Haifa Tlaiss and Saleema Kauser, "The Impact of Gender, Family, and Work."
57 Fida Afiouni, "Women's Careers in the Middle East," 329.
58 Suzy D'Enbeau, Astrid Villamil, and Rose Helens-Hart, "Transcending Tensions."
59 Suzy D'Enbeau, Astrid Villamil, and Rose Helens-Hart, "Transcending Tensions," 287.
60 John McIntosh and Samia Islam, "Beyond the Veil."
61 Jawad Syed, "An Historical Perspective on Islamic Modesty and its Implications for Female Employment," *Equality, Diversity and Inclusion: An International Journal* 29, no. 2 (2010): 150–166, doi: 10.1108/02610151011024475.
62 Fida Afiouni, "Women's Careers in the Middle East."
63 Charlotte Karam, Fida Afiouni, and Nour Nasr, "Walking a Tightrope or Navigating a Web: Parameters of Balance within Perceived Institutional Realities," *Women's Studies International Forum* 40 (2013): 99.
64 Haifa Tlaiss and Saleema Kauser, "The Impact of Gender, Family, and Work," 29.
65 Beverly Metcalfe, "Exploring Cultural Dimensions of Gender."
66 Laurie James-Hawkins, Yara Qutteina, and Kathryn Yount, "The Patriarchal Bargain," 133.
67 Loubna Skalli, "Constructing Arab Female Leadership."
68 Suzy D'Enbeau, Astrid Villamil, and Rose Helens-Hart, "Transcending Tensions," 281.

69 Linda Putnam, Gail Fairhurst, and Scott Banghart, "Contradictions, Dialectics, and Paradoxes."
70 Lisa Baranik, Brandon Gorman, and William Wales, "What Makes Muslim Women Entrepreneurs Successful? A Field of Study Examining Religiosity and Social Capital in Tunisia," *Sex Roles* 78 (2018): 208–219, doi: 10.1007/s11199-017-0790-7; Joseph Vendello, "Do We Need a Psychology of Women in the Islamic World?," *Sex Roles* 75 (2016): 623–629, doi: 10.1007/s11199-016-0691-1.
71 Wafa Almobaireek and Tatiana Manolova, "Entrepreneurial Motivations Among Female University Youth in Saudi Arabia," *Journal if Business Economics and Management* 14, no. 1 (2013): 56–75, doi: 10.3846/16111699.2012.711364.
72 Sidani Yussuf and Sammy Showail, "Religious Discourse and Organizational Change."
73 Jessica Newman, "Aspirational Maternalism and the 'Reconstitution' of Single Mothers in Morocco," *Journal of Middle East Women's Studies* 14, no. 1 (2018): doi: 10.1215/15525864-4297006.
74 Ellen Amster, "Global IVF, Infertility, and Emergency Contraception."
75 Joseph Vendello, "Do We Need a Psychology of Women?"
76 Momin Rahman, "Queer as Intersectionality: Theorizing Gay Muslim Identities," *Sociology* 44, no. 5 (2010): 944–961.
77 Nof Nasser-Eddin, Nour Abu-Assab, and Aydan Greatick, "Reconceptualising and Contextualizing Sexual Rights: Beyond LGBTQI Categories," *Gender & Development* 26, no. 1 (2018): 173–189, doi: 10.1080/13552074.2018.1429101
78 Paul Amar, "Middle East Masculinity Studies: Discourses of 'Men in Crisis', Industries of Gender in Revolution," *Journal of Middle East Women's Studies* 7, no. 3 (2011): 36–70.

Bibliography

Acker, Joan. "Hierarchies, Jobs, Bodies: A Theory of Gendered Organizations," *Gender & Society*, 4 (1990): 139–158, doi: 10.1177/089124390004002002

Afiouna, Fida. "Women's Careers in the Middle East: Understanding Institutional Constraints to the Boundariless Career," *Career Development International*, 19, no. 3 (2014): 314–336, doi: 10.1108/CDI-05-2013-0061

Al-bakr, Fauzia, Elizabeth Bruce, Petrina Davidson, Edit Schlaer, and Ulrich Kropiunigg. "Empowered but Not Equal: Challenging the Traditional Gender Roles as Seen by University Students in Saudi Arabia," *FIRE: Forum for International Research in Education*, 4, no. 1 (2017): 52–66, doi: 10.18275/re201704011083

Almobaireek, Wafa, and Tatiana Manolova. "Entrepreneurial Motivations among Female University Youth in Saudi Arabia," *Journal of Business Economics and Management*, 14, no. 1 (2013): 56–75, doi: 10.3846/16111699.2012.711364

AlMunajjed, Mona. *Women in Saudi Arabia Today*. London: Macmillan, 1997.

Amar, Paul. "Middle East Masculinity Studies: Discourses of 'Men in Crisis,' Industries of Gender in Revolution," *Journal of Middle East Women's Studies*, 7, no. 3 (2011): 36–70.

Amster, Ellen. "Global IVF, Infertility, and Emergency Contraception in the Middle East and North Africa," *Journal of Middle East Women's Studies*, 14, no. 3 (2018): 343–347, doi: 10.1215/15525864-7025455

Bahramitash, Roksana. "Iranian Women during the Reform Era (1994–2004), a Focus on Employment," *Journal of Middle East Women's Studies*, 3, no. 2 (Spring 2007): 86–109.

Baki, Roula. "Gender-Segregated Education in Saudi Arabia: Its Impact on Social Norms and the Saudi Labor Market," *Education Policy Analysis Archives*, 12, no. 28 (2004): 1–12.

Baranik, Lisa, Brandon Gorman, and William Wales. "What Makes Muslim Women Entrepeneurs Successful? A Field of Study Examining Religiosity and Social Capital in Tunisia," *Sex Roles*, 78 (2018): 208–219, doi: 10.1007/s11199-017-0790-7

Buzzanell, Patrice. "Gaining a Voice: Feminist Organizational Communication Theorizing," *Management Communication Quarterly*, 7, no. 4 (1994): 339–383, doi: 10.1177/0893318994007004001

Buzzanell, Patrice, and Suzy D'Enbeau. "Stories of Caregiving: Intersections of Popular Gender, Academic Research, and one Woman's Experiences," *Qualitative Inquiry*, 15 (2009): 1119–1224, doi: 10.1177/1077800409338025

Buzzanell, Patrice, Rebecca Meisenbach, Robyn Remke, Meina Liu, Vanessa Bowers, and Cindy Conn. "The Good Working Mother: Managerial Women's Sensemaking and Feelings about Work–Family Issues," *Communication Studies*, 56 (2005): 261–285, doi: 10.1080/10510970500181389

Coker, Margaret. "Saudi Women Can Drive, but Here's the Real Roadblock." *The New York Times*. June 22, 2018. www.nytimes.com/2018/06/22/world/middleeast/saudi-arabia-women-driving.html

D'Enbeau, Suzy, Astrid Villami, and Rose Helens-Hart. "Transcending Work–life Tensions: A Transnational Feminist Analysis of Work and Gender in the Middle East, North Africa, and India," *Women's Studies in Communication*, 38, no. 3 (2015): 273–294, doi: 10.1080/07491409.2015.1062838

Dingo, Rebecca. *Networking Arguments: Rhetoric, Transnational Feminism, and Public Policy Writing*. Pittsburgh, PA: University of Pittsburgh Press, 2012.

Dougherty, Debbie, and Marlo Goldstein Hode. "Binary Logics and the Discursive Interpretation of Organizational Policy: Making Meaning of Sexual Harassment Policy," *Human Relations*, 69, no. 8 (2016): 1729–1755, doi: 10.1177/0018726715624956

Doumato, Eleanor. "Saudi Arabia," in *Women's Rights in the Middle East and North Africa: Progress amid Resistance*, eds Sanja Kelly and Julia Breslin (New York, NY: Rowman & Littlefield, 2010), pp. 425–458.

Haghighat, Elhum. "Social Status and Change: The Question of Access to Resources and Women's Empowerment in the Middle East and North Africa," *Journal of International Women's Studies*, 14, no. 1 (2013): 273–299.

James-Hawkins, Laurie, Yara Qutteina, and Kathryn M. Yount. "The Patriarchal Bargain in a Context of Rapid Changes to Normative Gender Roles: Young Arab Women's Role Conflict in Qatar," *Sex Roles*, 77, nos 3–4 (2017): 155–168, doi: 10.1007/s11199-016-0708-9

Karam, Charlotte, Fida Afiouni, and Nour Nasr. "Walking a Tightrope or Navigating a Web: Parameters of Balance within Perceived Institutional Realities," *Women's Studies International Forum*, 40 (2013): 87–101.

Kirby, Erika, and Kathleen Krone. "The Policy Exists but You Can't Really Use It," *Journal of Applied Communication Research*, 30, no. 1 (2002): 50–77, doi: 10.1080/00909880216577

McElwee, Gerard, and Rahma Al-Riyami. "Women Entrepreneurs in Oman: Some Barriers to Success," *Career Development International*, 8, no. 7 (2003): 339–346, doi: 10.1108/13620430310505296

McIntosh, John, and Samia Islam. "Beyond the Veil: The Influence of Islam on Female Entrepreneurship in a Conservative Muslim Context," *International Management Review*, 6, no. 1 (2006): 101–109.

Metcalfe, Beverly. "Exploring Cultural Dimensions of Gender and Management in the Middle East," *Thunderbird International Business Review*, 48, no. 1 (2006): pp. 93–107.

Metcalfe, Beverly. "Gender and Human Resource Management in the Middle East," *The International Journal of Human Resource Management*, 18, no. 1 (2007): 54–74, doi: 10.1080/09585190601068292

Metcalfe, Beverly. "Women, Management and Globalization in the Middle East," *Journal of Business Ethics*, 83 (2008): 85–100, doi: 10.1007/s10551-007-9654-3

Metcalfe, Beverly, and Christopher Rees. "Gender, Globalization, and Organization: Exploring Power, Relations and Intersections," *Equality, Diversity and Inclusion*, 24, no. 1 (2010): 5–22, doi: 10.1108/02610151011019183

Moghadam, Valentine. *Modernizing Women: Gender and Social Change in the Middle East*. Boulder, CO: Lynne Rienner Publishers, 2003.

Moghadam, Valentine. *Globalizing Women: Transnational Feminist Networks*. Baltimore, MD: The Johns Hopkins University Press, 2005.

Morin, Aysel. "Victimization of Muslim Women in Submission," *Women's Studies in Communication*, 32, no. 3 (2009): 380–408.

Nasser-Eddin, Nof, Nour Abu-Assab, and Aydan Greatick. "Reconceptualising and Contextualizing Sexual Rights: Beyond LGBTQI Categories," *Gender & Development*, 26, no. 1 (2018): 173–189, doi: 10.1080/13552074.2018.1429101

Newman, Jessica. "Aspirational Maternalism and the 'Reconstitution' of Single Mothers in Morocco," *Journal of Middle East Women's Studies*, 14, no. 1 (2018): doi: 10.1215/15525864-4297006

Putnam, Linda, Gail Fairhurst, and Scott Banghart. "Contradictions, Dialectics, and Paradoxes in Organizations: A Constitutive Approach," *The Academy of Management Annals*, 10, no. 1 (2016): 65–171.

Radha, Hegde. "Globalizing Feminist Research in Communication," *The Sage Handbook of Gender and Communication*, eds Bonnie Dow and Julia Wood (Thousand Oaks, CA: SAGE, 2006), pp. 433–450.

Rahman, Momin. "Queer as Intersectionality: Theorizing Gay Muslim Identities," *Sociology*, 44, no. 5 (2010): 944–961.

Sharabi, Hisham. *Neopatriarchy: A Theory of Distorted Change in Arab Society*. New York: Oxford University Press, 1988.

Sidani, Youssuf and Sammy Showail. "Religious Discourse and Organizational Change Legitimizing the Stakeholder Perspective at a Saudi Conglomerate," *Discourse & Organizational Change*, 26, no. 6 (2013): 931–947, 10.1108/JOCM-11-2012-0175

Skalli, Loubna. "Constructing Arab Female Leadership Lessons from the Moroccan Media," *Gender & Society*, 25, no. 4 (2011): 473–495.

Syed, Jawad. "An Historical Perspective on Islamic Modesty and Its Implications for Female Employment," *Equality, Diversity and Inclusion: An International Journal*, 29, no. 2 (2010): 150–166, doi: 10.1108/02610151011024475

Tlaiss, Haifa and Saleema Kauser. "The Impact of Gender, Family, and Work in the Career Advancement of Lebanese Women Managers," *Gender in Management: An International Journal*, 26, no. 1 (2011): 8–36, doi: 10.1108/17542411111109291

Townsley, Nikki. "Love, Sex, and Tech in the Global Workplace," in *The Sage Handbook of Gender and Communication*, eds Bonnie Dow and Julia Wood (Thousand Oaks, CA: SAGE, 2006), pp. 143–160.

UNICEF. Regional Overview for the Middle East and North Africa: MENA Gender Equality Profile, 2011.

Vendello, Joseph. "Do We Need a Psychology of Women in the Islamic World?," *Sex Roles*, 75 (2016): 623–629, doi: 10.1007/s11199-016-0691-1

World Bank. *The Status & Progress of Women in the Middle East & North Africa* (World Bank, Washington, DC, 2009).

5

CHICANO MASCULINITIES

Kostia Lennes

Introduction

This introductive section offers some general definitions of the main concepts that are used throughout this chapter as well as the theoretical framework it refers to.

Chicanos/as

The term Chicano/a carries different meanings. In its broadest sense, "Chicano/a" may be used a synonym of "Mexican American," which therefore would include a large part of the U.S. population. However, while many American citizens of Mexican descent may be identified as Chicanos/as, the term remains deeply associated with a chosen political and/or cultural identity, which does not necessarily rely on a specific point of origin. Thus, other Latin American people may self-identify as Chicanos/as whereas American citizens of Mexican descent may reject this label, preferring other terms such as Mexican American, Latinos, or *mestizos*. In this respect, Hurtado and Sinha (2016, 6) stated that

> when interacting with older family members such as grandparents, a US-born man of Mexican descent may communicate primarily in Spanish and identify as Mexicano while simultaneously identifying as Chicano when interacting in English with his Mexican-descent peers at the university. The same person, when interacting with members of out-groups, for example, whites, may identify as Latino in order to avoid confusion with those who lack familiarity with the meanings embedded in the label *Chicano*. [...] Because ethnic labels have historically, regionally, and culturally specific dimensions, the meaning imputed to the labels can vary based on who is involved in the interaction, the language being spoken, and geographical location.

Yet this chapter will focus on a more specific—though widely used in the academia—definition of the term, that is, the Chicano/a cultural identity, which is often associated with the working-class culture and cultural hybridity embodied by the borderland.

Masculinities and hegemonic masculinity

There are various ways of defining both *masculinity* and *masculinities*. In this chapter, I refer to *masculinity* and *masculinities* as relational concepts. In the singular, *masculinity* tends to put the emphasis on its opposition to femininity. As Connell (1995) stated, *masculinity* is largely defined as what it is not. Thus, *masculinity* can be considered as the way people think of themselves, conceptualize themselves, and self-identify *as men*. However, as most works in men's studies nowadays, I rather use the plural *masculinities* in this chapter as a relational concept again, in the sense that gender identities are not fixed, nor mutually exclusive. *Masculinities* are then relational between themselves (because they are hierarchized in specific contexts) but also when they interact and intersect with other social identities such as race, social class, sexual orientation, and nationality. Therefore, *masculinities* are not only personal, private, or intimate identities. More than this, they are political, cultural, and social markers that inherently shape other social identities and dynamics of power and domination.

Connell (1995, 2000) was one of the pioneering researchers to consider the diversity of masculinities coexisting in modern societies. She distinguishes four categories of masculinities: hegemonic masculinity, complicit masculinity, subordinated masculinity, and marginalized masculinity. According to Connell (1995, 76), "hegemonic masculinity is not a fixed character type, always and everywhere the same. It is, rather, the masculinity that occupies the hegemonic position in a given pattern of gender relations, a position always contestable." Hegemonic masculinity is then *hegemonic* only in specific contexts and always remains subject to changes and variations. Furthermore, Connell's categories are not permanent. Instead, she advocates for a relational approach between these four concepts.

Marginalized masculinities

Connell's account on marginalized masculinities is exemplified by the case of Black men in the United States. According to her, these men are marginalized because of their position in the American racial and economic orders. In the wake of Connell's theory (1995) on masculinities, other social scientists (Cheng 1999; Haywood and Johansson 2017; Jackson and Moshin 2013) have focused on marginalized masculinities in a wide range of approaches.

However, as Mirandé et al. pointed out (2010, 310), Connell's conceptualization of marginalized masculinities "reduces racial/ethnic groups to a broad 'marginalized' category that is ultimately grounded in the traditional white/black where Black men are relegate to a 'marginal' category and Mexican/Latino men are virtually invisible or non existent." Mirandé (1997) argues that Latino masculinities should not be regarded as marginalized only as they are "as complex and varied as Euro-American masculinities" (1997, 147). As Mirandé's position, other scholars (Aldama 2005; Almaguer 1991; Barriga 2001; Beattie 2002; Cantú 2001, 2009; Gutmann 1996, 2003; Hondagneu-Sotelo 1994) have underlined the plurality of Mexican and Chicano masculinities. Classifying Chicano masculinities within the "marginal" category would certainly lead us to the essentialization of identities, which both critical men's studies and ethnic studies should avoid. Therefore, Mirandé makes an interesting point. Are all colonized and racialized masculinities necessarily marginal? If they are probably marginal in some contexts, can they not be hegemonic or simply complicit (Connell 1995) in others? While reflecting on marginalized masculinities, Cheng argues (1999, 298–299) that

> we are simultaneously members of multiple groups, including dominant and marginalized groups. *One may be marginalized by a visible marker*, such as race, sex, or the display of behavior generally regarded as "gay," or wearing religious adornments

of a non-Christian group, but this *does not mean one is marginalized based on gender performance.* In fact, many members of these marginalized groups perform hegemonic masculinity in order to gain patriarchal privileges within their group, if not the larger society. Performing hegemonic masculinity by a marginalized person is seen as a passing behavior that distracts from her/his stigma.

Consequently, one must also answer the following question: why Chicano masculinities? Are they specific enough to be labeled in such an expression, even in the plural? This chapter attempts to go further the "zoo approach" (Cheng 1999), that is, the temptation to look at Chicano men as one ethnic/gendered group among other racialized masculinities. Rather, this contribution explores how a relational approach of masculinities may help us to grasp Chicano men *as men.*

Machismo and violence

Machismo and violence are probably some of the most discussed concepts when it comes to Latino men in general and Chicano men in particular. Machismo has been defined as a "cult of exaggerated masculinity" (Gutmann 1996), which manifests itself in particular by dominance over women (Baca Zinn 1980, 1982), non-heteronormative men, and the children within the family (Mirandé 1977). Machismo is then often described as a cultural pattern of the Chicano people, which tends to designate Chicano men as *others* within the American racial and gender orders.

Machismo as an "adaptive characteristic"

Early works on Chicano men *as men* (Baca Zinn 1975, 1980, 1982; Mirandé 1979, 1985, 1997) have looked at Chicano masculinities principally through the lens of machismo. However, without underestimating the reality and the effects of machismo as a patriarchal system, these researchers offered a critique of what might be called the culturalization of machismo as a properly Mexican/Chicano characteristic. For instance, Baca Zinn, one of the first social scientists to actually study Chicano masculinities, views Chicanos' machismo as an "adaptive characteristic" (Baca Zinn 1982). She makes the following statement (1982, 39):

> Being male is one sure way to acquire status when other roles are systematically denied by the workings of society. [...] To be "hombre" may be a reflection of both ethnic and gender components and may take on greater significance when other roles and sources of masculine identity are structurally blocked. Chicanos have been excluded from participation in the dominant society's political-economic system. Therefore, they have been denied resources and the accompanying authority accorded men in other social categories. [...] It may be worthwhile to consider some expressions of masculinity as attempts to gain some measure of control in a society that categorically denies or grants people control over significant realms of their lives.

Baca Zinn thus explains Chicanos' machismo as a response to racial and class oppressions. Mirandé offers a similar critique of the concept and stated that "native men develop an overly masculine and aggressive response in order to compensate for deeply felt feelings of powerlessness and weakness" (1997, 36). However, in a quantitative survey, Mirandé et al. (2010) demonstrated that machismo, and the negative—and sometimes positive—attitudes that are traditionally associated with, are actually "largely class-based rather than culturally based phenomena" (328). From the pioneering works on Chicano masculinities to recent investigations, there is now much evidence that machismo is *not* rooted in Chicano (or Mexican) culture. Furthermore, some scholars (Broughton 2008; Cantú 2001, 2009; Carrillo 2017; Hondagneu-Sotelo 1994; Ramirez 2011; Smith 2006) acknowledged the fact that migration to the United States impacts redefinitions of masculinity among Mexican men.

In one of the most important works within the field of Chicano/a studies, *Borderland/La Frontera*, Gloria Anzaldúa (1987) offers some insights on Chicano masculinities and machismo. Indeed, she underlines the domination of white hegemonic masculinity over Chicanos' marginalized masculinity (Connell 1995). Kynčlová (2014, 5) analyses one of Anzaldúa's poems (We Call Them Greasers), which tells the story of a rape, as follows:

> The poem narrates an incident in which a husband is forced to watch the spectacle of his wife's brutal rape and murder executed by a White Anglo. Because the Chicano husband … is tied to a mesquite tree … he is deprived of any sort of agency and is made to be passive, powerless onlooker of his wife's doom, and the subject of victimization carried out by a man who not only represents the colonizer's political, economic, and cultural domination, but also embodies hegemonic masculinity. (Connell and Messerschmidt)

Here, white hegemonic masculinity represents symbolic, economic, political, and physical domination both over Chicana women and Chicano men. As it has been suggested earlier, Chicanos would adopt these macho behaviors in response to white, middle-to-upper class, American, and masculine domination. In this respect, I argued elsewhere (Lennes 2016, 7) that "Chicano men would incorporate American supremacy and then perform machismo as a means to defend their 'injured' masculine identity."

Performing machismo and violence

Delgado's analysis (2000) of the Chicano rapper Kid Frost performing what he calls "Brown masculinity" offers an interesting example of the popular representations of Chicano men's machismo and violence. In his lyrics, Kid Frost portrays Chicano men as the representatives of the purest form of Chicano masculinity (note the singular). "Brown masculinity" is thus intimately related to machismo, of course, but also to working-class and gang cultures, rebellion against the institutions (more specifically the police), and so on. Delgado states (2000, 395) that

> we cannot overlook, however, the fact that Kid Frost's lyrical narratives do direct us to see how violence intertwines itself with the Chicano body and macho identity. The Chicano male is located within a bordered and policed environment that incubates anger, marginalization, violence, and virulent strains of machismo … Kid Frost, like any good gangster rapper, imposes himself on others through his ability to physically dominate and his ability to lyrically dismantle … Kid Frost performs a deconstruction of the forms of power and identity allocated to Chicano males in East Los Angeles. The game, such as it is, is relayed in lyrics that clearly allow Kid Frost to metaphorically perform his role as predator in search of prey—other rappers, gang bangers, women, the weak.

Kid Frost's performance of machismo and violence therefore exemplifies the phantasm of Chicano masculinity in American popular culture. By performing "Brown masculinity," Kid Frost somehow embodies many of the common research interests when it comes to Chicano men: machismo (Baca Zinn 1982; Mirandé et al. 2010), of course, but also gang culture (Flores and Hondagneu-Sotelo 2013), lowrider culture (Chavez 2013), or rap music (Baker-Kimmons and McFarland 2011).

Additionally, it is important to underline that many other cultural productions corroborate these traditional and somehow caricatural (because they are exaggerated and overacted) representations. Indeed, Baker-Kimmons and McFarland (2011) demonstrated that many Chicano rappers tend to offer performances of Brown hegemonic masculinities that challenge white American masculinity and its racism. Yet, as Baker-Kimmons and McFarland (2011) argue, most of the time these lyrics do not contest sexism and capitalism. Thus, they deal with gender identities by celebrating Chicano hypermasculinity in contrast to white hegemonic masculinity but they fail to include Chicana women (or even other working-class and racialized women) in their political claims. Likewise, these Chicano rappers assert the powerful potential of their working-class culture (also associated with gang culture and lowrider culture, for instance) but they rarely challenge the capitalist system constraining their lives.

Redefining queer sexualities and masculinities: a case study

As mentioned earlier, it is difficult to classify Chicano masculinities within the "marginal" category (Connell 1995) because Chicano male identities are multidimensional. Since the early 1990s, there is a significant body of literature (Almaguer 1991; Cantú 2001, 2009; Carrillo 2017; Carrillo and Fontdevilla 2014; Cuevas 2018; Pérez 2011; Rodríguez 2011; Thing 2009, 2010; Viego 1999) focusing on non-heteronormative Chicano identities in the United States. Although attention has been paid principally on gay, bisexual and queer men, Cuevas (2018) recently provided an interesting analysis of Chicana female masculinities. Thus, this section provides further insight on gay and queer Chicano masculinities. Various scholars working on Mexican queer immigrants in the United States (Cantú 2009; Carrillo 2017; Carrillo and Fontdevila 2014) have argued that gender and sexual identities are negotiated after relocation.

Changing same-sex sexualities, changing masculinities

Chicano/Mexican male same-sex sexualities have often been theorized as hybridized between a traditional definition of homosexuality, linked to the Mexican/Latino culture, and the American/Western model of gay sexuality. In Mexico, as in most Latin American countries (Vasquez del Aguila 2014), male same-sex sexuality has been perceived through a "gender-stratified activo/passivo model of homosexuality" (Thing 2010, 809) in which only the *pasivo* (passive, bottom, sexually penetrated) partner was considered as gay—and often repressed for this reason—whereas the *activo* (active, top, sexually penetrating) partner was usually not affected by the homosexual stigma. This model exists in contrast to the "object-choice gay model of homosexuality" (Thing 2010, 809), as is effective in most "Western countries." The "object-choice" model implies that only the sex of the partner matters, regardless of sexual roles.

Since the 1990s, it has been argued (Almaguer 1991; Cantú 2009; Carrillo 1999; Thing, 2009, 2010) that Mexican understanding of homosexuality is shifting from the traditional definition to the American gay model. In this regard, Cantú (2009) highlighted the tensions around the existence of a supposed Mexican gay identity that would distinguish itself from "Western"—more specifically American—gay identity and construct itself as authentically "Mexican." Furthermore, by the late 1990s, Carrillo (1999) pointed out the role played by the spread of the American culture and media in Mexico regarding this issue. The strong influence of the United States has stimulated new considerations about the gay subculture and LGBTQ people. Carrillo observed the emergence of a new model of homosexuality in Mexico, which is in his view the result of hybridization between both definitions ("object-choice" and "gender-stratified"). More recently, Thing (2009, 2010) pointed out a similar trend in the way queer Mexican men self-identify during their post-migration experience in the United States. Thus, rather than a direct shift from the traditional gender-stratified concept to the "modern" object-choice concept, the dominant perception of male homosexuality would now be a complex combination of both models. Finally, as Thing (2009, 2010) demonstrated, these influences impact queer Mexican men in different ways depending on their social class and geographical background.

Internacional: the Mexican American gay sexuality?

Therefore, the conflict around these two Mexican perceptions of male homosexuality also impacts the way queer men self-identify. The symbolic dichotomy of the "active/passive axis" (Almaguer 1991) marks a strong social boundary between *activos* and *pasivos* in the context of male same-sex intercourse. The following insight highlights the primacy of these categories within the Mexican gay subculture: while often referring to the terms *activo* and *pasivo* to specify one's sexual role, versatile gay men (those who perform both roles) are usually called and define themselves as *internacionales* (internationals). The use of this term (often shortened as *inter*) is quite telling. It underlines the supposed incompatibility of "performing both roles" with the Mexican traditional view of sexual dichotomy (penetrating or being penetrated). So-called *internacionales* are not necessarily foreigners as their label would suggest, but they are classified in a category located outside the normative system used by most Mexican gay and bisexual men. It could also be argued that this is now a third category that has emerged in the last decades.

Yet the term *internacional* designates gay men whose sexual practices have been reshaped by intercultural and/or international experiences, as they are viewed by many gay and bisexual men in Mexico (Carrier 1995). According to Cantú (2009), these supposedly new practices could partly be attributed to the rise of gay tourism (mainly from the United States) in Mexico since the 1980s. Consequently, the use of the word *internacional* to refer to sexual versatility reflects that the normative Mexican dichotomy between *activos* and *pasivos* has been blurred by the American experience. Finally, beyond queer Chicano masculinities, these tensions around the definition of homosexuality within the Mexican and Chicano culture also challenge the definition of masculinity itself. Indeed, behind the meaning of homosexuality for Mexicans who have been confronted with the American culture lies the definition of the male sexual role as well. Therefore, Chicano queer masculinities cannot be examined without paying attention to this confrontation with the American gay subculture.

Gay masculinities and the specter of borderland identity

The latter reflections on the meaning of male homosexuality in the Chicano culture are the continuity of Gloria Anzaldúa's (1987) theory on borderland identity: the hybrid and self-conscious identity embodied by the Chicanos living on both sides of the border. The *mestiza*[1] writer theorized cultural hybridity—more precisely what she calls *mestizaje*—as a major element of Chicano/a identity. As a queer Chicana researcher, Anzaldúa analyzes how this borderland identity affects Chicanos/as not only as people of color, but also as gendered and sexual beings. The Anzalduan concept of cross-border identity largely echoes the tensions around the elusive sexual identities that queer Chicanos/as constantly have to face. The difficulty of coming to terms with one's own identity while navigating two, sometimes contradictory, cultural paradigms or national frameworks, of which neither accommodate the embodied subjectivities and trajectories of queer Chicano men is another aspect of what Anzaldúa has referred to as borderland identity. Likewise, the academic debate surrounding the supposed emergence of *internacional* sexual versatility or the two opposed models of homosexuality—always through the comparative lens of the United States—is reminiscent of the borderland thinking. Anzaldúa addresses this issue with what she called the *mestiza* consciousness (*la conciencia de la mestiza*, in Spanish), which is an identity construction that goes beyond Western binary thinking of race, language, gender, and sexuality.

Chicano masculinities, hybridity, and the borderland

In the wake of Anzaldúa's border theory that was discussed previously, this section discusses the hybridity of Chicano masculinities. It has been argued (Carrillo 2013; Delgado 2000; Lennes 2016; Rivera 1997) that Chicano masculinities often deal with what Anzaldúa (1987) calls the borderland identity. This section provides further insight on borderland identity and subjectivity through the lens of the Anzalduan theory and explores their relationship with Chicano masculinities.

Cultural hybridity and borderland subjectivity

In her semiautobiographical work *Borderlands/La Frontera*, Anzaldúa introduced the concept of the "new *mestiza*" to illustrate Chicanos/as' borderland identity and thus counters the Western border ideology. The border here must be understood both as a physical delineation and a symbolic and mental reality. Anzaldúa states (1987, 25–26) the following:

> Borders are set up to define the places that are safe and unsafe, to distinguish us from them. A border is a dividing line, a narrow strip along a steep edge. A borderland is a vague and undetermined place created by the emotional residue of an unnatural boundary. It is a constant state of transition. The prohibited and forbidden are its inhabitants … Gringos in the U.S. Southwest consider the inhabitants of the borderlands transgressors, aliens—whether they possess documents or not, whether they're Chicanos, Indians or Blacks. Do not enter, trespassers will be raped, maimed, strangled, gassed, shot … Tension grips the inhabitants of the borderlands like a virus.

The borderland is here perceived as a consequence of the Western thought hierarchizing differences. Yet this dualism does not consider Chicanos/as' and Mexicans' lived experiences. As some scholars (Carrillo 2013; Kynčlová 2014) suggest, Anzaldúa's understanding of the border echoes Bhabha's concepts of "inbetween spaces" (Bhabha 1994) and the "unhomely" (Bhabha 1992). Indeed, according to Kynčlová (2014, 2), "it is these grey zones [the borderlands] that attract attention as they are spaces where meanings and identities are constantly in the process of negotiation, becoming, and struggle for recognition." In this respect, I have stated elsewhere (Lennes 2016, 5) that

> the Chicano borderland identity could be seen as this specific culture embodied by the borderland inhabitants—the Chicanos/as—whose condition is subjected to the borderland regime and authority. In other words, the borderland identity must be understood here as a set of intersecting identities that some Chicanos/as embody in a specific way. Their homeland is the borderland itself, that is, a hybridized and still undefined territory.

Then, drawing on the concept of borderland identity, borderland subjectivity can be defined as the "inbetweenness that goes beyond the reifying effects of national identity" (Ashcroft 2009, 20) or even "the ability to navigate in/between/among/within different cultures, languages, and epistemological systems, and to embody this hybridity consciously and constructively with respect to one's own racial/ethnic background, gender identity, class belonging, and reflected lived experience" (Kynčlová 2014, 2). Then, one must consider the borderland subjectivity as a productive political distinctiveness in the sense that it may allow Chicano/a people to strengthen their positionality in the American society.

Borderland masculinities

Yet, Chicano masculinities remain deeply rooted in cultural hybridity. Rivera pointed out (1997, 35) that "the question of origin, or of remembering one's origins, lies at the center of Chicano identity and is often expressed as an alternate to both American and Mexican cultural identities." I have defined "borderland masculinity" as a "concept to characterize the embedment of borderland subjectivity (*mestizaje*) in Chicanos' masculine identity" (Lennes 2016, 7). As I stated (Lennes 2016, 13),

> "borderland masculinity" in no case is a fixed category of masculinity, but rather a constantly negotiable gender identity, subject to changes and reinterpretations. Thus, if many Chicano males experience the trauma of the border and "carry" it with them, they do not manifest it in a univocal way. [...]. As an elastic—and somehow elusive—concept, the "borderland masculinity" does not capture one single masculine identity. Instead, it must be appreciated as a broad approach allowing a better understanding of the complex strategies used by Chicano men to negotiate their identity on the borderlands.

The concept of borderland masculinity is reflected in some cases studies focusing on Chicano men's performances of masculinity. For instance, in an article on working-class Chicano students based in southern Texas, near the United States-Mexico borderland, Carrillo (2013) demonstrates how these men are confronted to white hegemonic masculinity in the university context. To resist, they adopt what Carrillo calls "unhomely masculinities," that is, a hybrid identity that contrasts both with the "educated masculinity" expected by their American counterparts and their Chicano and working-class backgrounds. Carrillo defines (2013, 194) this concept as "the specific ways in which working-class Latino masculinities produce culturally situated notions of academic/intellectual manhood." In this case, the Chicano men have to face two different gender regimes (American university versus Chicano working-class family) while they navigate throughout both sides of the borderland. Therefore, the Chicano students have to produce new masculine identities (what Carrillo calls "unhomely masculinities") in order to go beyond this impossible dilemma. By doing so, these men genuinely perform *mestizaje*: in a context where gender, race, and class intersect to challenge their positionality within the educative system, they invent their own performance of identity to maneuver their lives on the borderland.

Concluding remarks: intersectionalizing Chicano masculinities?

To conclude this chapter, I propose in this final section some reflections on the potential "intersectionalization" of Chicano masculinities. Drawing on the case of African American women in the United States, Crenshaw (1989) defined intersectionality as the combination of simultaneous forms of systemic discrimination. It is nowadays a widely used concept when it comes to Chicana women. Anzaldúa herself, although she does not use the term in her work, · is often cited in intersectional Chicana studies. However, there are very few works focusing on Chicano masculinities through an intersectional approach.

As Sinatti noted (2014, 218), "given that in a situation of social inequality men are often implicitly assumed to be the ones holding privilege and power, research applying intersectionality to men is relatively rare." Indeed, only a few scholars (Coston and Kimmel 2012; Hearn 2009; Hurtado and Sinha 2008, 2016; Leek and Kimmel 2015; Perraudin 2014; Sinatti 2014) have proposed intersectional analyses of masculinities. Yet some of these works tend to essentialize masculinities by considering them as fixed categories of power that may be strengthened or weakened by other criteria such as disability, sexual orientation, or social class (Coston and Kimmel 2012), for instance. Instead, in her work on Otomi[2] migrants in Mexico City, Perraudin (2014) draws on the intersectional theory by analyzing how masculinity directly intersects with ethnicity and social class to produce exclusion. Thus, she highlights how associations assisting the Otomis in Mexico City tend to portray Otomi men as macho, violent, and alcoholic, unable to educate their children, while their spouses are seen as more responsible.

Therefore, if we really want to go *beyond machismo* (Hurtado and Sinha 2016), I argue that our intersectional approaches of Chicano masculinities should further explore how the plurality of male identities *directly intersect* with other social identities to produce new categories of power in specific contexts. The aim is obviously not to argue that men—even those who may embody marginalized masculinities—are generally disadvantaged in the global gender regime. Rather, I call for further research on the potential vulnerabilities of Chicano masculinities in a society where marginalized and racialized men are subject to hate speeches and fear that always somehow rely on their threatening masculinities.

Notes

1 *Mestizo/a* (usually not translated into English) is a Spanish word used in Mexico to designate people of mixed descent. In Mexico, the majority of the population is *mestizo/a*. However, in the wake of several Chicano/a intellectuals such as Gloría Anzaldúa, *mestizo/a* can also be related to a philosophical or political theory (e.g., *mestiza* consciousness) that counters white American hegemony.
2 The Otomi are an indigenous people in Mexico, mostly located in the center of the country.

Bibliography

Aldama, Frederick L. *Brown on Brown: Chicano/a Representations of Gender, Sexuality, and Ethnicity*. Austin: University of Texas Press, 2005.

Almaguer, Tomas. "Chicano Men: A Cartography of Homosexual Identity and Behavior." *Differences* 3 (1991): 75–100.

Anzaldúa, Gloria. *Borderlands/La Frontera: The New Mestiza*. San Francisco, CA: Aunt Lute Book, 1987.

Ashcroft, Bill. "Chicano Transnation." In *Imagined Transnationalism: U.S. Latino/a Literature, Culture, and Identity*, edited by Kevin Concannon, Francisco A. Lomeli, and Marc Priewe, 13–28. New York: Palgrave Macmillan, 2009.

Baca Zinn, Maxine. "Political Familism: Toward Sex Role Equality in Chicano Families." *International Journal of Chicano Studies Research* 6 (1975): 13–26.

Baca Zinn, Maxine. "Gender and Ethnic Identity among Chicanos." *Frontiers: A Journal of Women Studies* 5, no. 2 (1980): 18–24.

Baca Zinn, Maxine. "Chicano Men and Masculinities." *Journal of Ethnic Studies* 10, no. 2 (1982): 29–44.

Baker-Kimmons, Leslie and Pancho McFarland. "The Rap on Chicano and Black Masculinity: A Content Analysis of Gender Images in Rap Lyrics." *Race, Gender & Class* 18, nos 1/2 (2011): 331–344.

Barriga, Miguel D. "*Vergüenza* and Changing Chicano and Chicana Narratives." *Men and Masculinities* 3, no. 3 (2001): 278–296.

Beattie, Peter. "Beyond Machismos: Recent Examinations on Masculinities in Latin America." *Men and Masculinities* 4 (2002): 303–308.

Bhabha, Homi. "The World and the Home." *Social Text* 31/32 (1992): 141–153.

Bhabha, Homi. *The Location of Culture*. London: Routledge, 1994.

Broughton, Chad. "Migration as Engendered Practice: Mexican Men, Masculinity, and Northward Migration." *Gender & Society* 22, no. 5 (2008): 568–589.

Cantú, Lionel. "A Place Called Home: A Queer Political Economy of Mexican Immigrant Men's Family Experiences." In *Queer Families, Queer Politics: Challenging Culture and State*, edited by Mary Bernstein, and Renate Reimann, 112–136. New York: Columbia University Press, 2001.

Cantú, Lionel. *The Sexuality of Migration: Border Crossings and Mexican Immigrant Men*. New York: New York University Press, 2009.

Carrier, Joseph. *De los Otros: Intimacy and Homosexuality among Mexican Men*. New York: Columbia University Press, 1995.

Carrillo, Hector. "Cultural Change, Hybridity and Male Homosexuality in Mexico." *Culture, Health and Sexuality* 1, no. 3 (1999): 223–238.

Carrillo, Hector. *Pathways of Desire: The Sexual Migration of Mexican Gay Men*. Chicago, IL: The University of Chicago Press, 2017.

Carrillo, Hector and Jorge Fontdevila. "Border Crossings and Shifting Sexualities among Mexican Gay Immigrant Men: Beyond Monolithic Conceptions." *Sexualities* 17, no. 8 (2014): 919–938.

Carrillo, Juan F. "The Unhomely in Academic Success: Latino Males Navigating the Ghetto Nerd Borderlands." *Culture, Society and Masculinities* 5, no. 2 (2013): 193–207.

Chavez, Michael J. "The Performance of Chicano Masculinity in Lowrider Car Culture: The Erotic Triangle, Visual Sovereignty, and Rasquachismo." PhD diss., University of California Riverside, 2013.

Cheng, Cliff. "Marginalized Masculinities and Hegemonic Masculinity: An Introduction." *The Journal of Men's Studies* 7, no. 3 (1999): 295–315.

Connell, Raewyn. *Masculinities*. Berkeley, CA: University of California Press, 1995.

Connell, Raewyn. *The Men and the Boys*. Berkeley, CA: University of California Press, 2000.

Connell, Raewyn and James Messerschmidt. "Hegemonic Masculinity: Rethinking the Concept." *Gender and Society* 19, no. 6 (2005): 829–859.

Coston, Bethany M. and Michael Kimmel. "Seeing Privilege Where it Isn't: Marginalized Masculinities and the Intersectionality of Privilege." *Journal of Social Issues* 68, no. 1 (2012): 97–111.

Crenshaw, Kimberlé. "Demarginalizing the Intersection of Race and Sex: A Black Feminist Critique of Antidiscrimination Doctrine, Feminist Theory and Antiracist Politics." *University of Chicago Legal Forum* (1989): 139–167.

Cuevas, T. Jackie. *Post-Borderlandia. Chicana Literature and Gender Variant Critique*. New Brunswick, NJ: Rutgers University Press, 2018.

Delgado, Fernando. "All Along the Border: Kid Frost and the Performance of Brown Masculinity." *Text and Performance Quarterly* 20, no. 4 (2000): 388–401.

Flores, Edward O. and Pierrette Hondagneu-Sotelo. "Chicano Gang Members in Recovery: The Public Talk of Negotiating Chicano Masculinities." *Social Problems* 60, no. 4 (2013): 476–490.

Gutmann, Matthew. *The Meanings of Macho: Being a Man in Mexico City*. Berkeley, CA: University of California Press, 1996.

Gutmann, Matthew. *Changing Men and Masculinities in Latin America*. Durham, NC: Duke University Press, 2003.

Haywood, Chris and Thomas Johansson (eds). *Marginalized Masculinities: Contexts, Continuities and Change*. New York: Routledge, 2017.

Hearn, Jeff. "Neglected Intersectionalities in Studying Men: Age(ing), Virtuality, Transnationality." In *Framing Intersectionality: Debates on a Multifaceted Concept in Gender Studies*, edited by Helma Lutz, Herrera Vivar, Maria Teresa and Linda Supik, 89–104. London: Ashgate, 2011.

Holling, Michelle. "*El Sympático* Boxer: Underpinning Chicano Masculinity with a Rhetoric of *Familia* in *Ressurection Blvd*." *Wester Journal of Communication* 70, no. 2 (2006): 91–114.

Hondagneu-Sotelo, Pierrette. *Gendered Transitions: Mexican Experiences of Immigration*. Berkeley, CA: University of California Press, 1994.

Hurtado, Aída and Mrinal Sinha. "More than Men: Latino Feminist Masculinities and Intersectionality." *Sex Roles* 59 (2008): 337–349.

Hurtado, Aída and Mrinal Sinha. *Beyond Machismo: Intersectional Latino Studies*. Austin, TX: University of Texas Press, 2016.

Jackson, Ronald M. and Jamie E. Moshin (eds). *Communicating Marginalized Masculinities. Identity Politics in TV, Film and New Media*. New York: Routledge, 2013.

Kynčlová, Tereza. "Elastic, Yet Unyielding: The U.S.-Mexico Border and Anzaldúa's Oppositional Rearticulations of the Frontier." *European Journal of American Studies* 9, no. 3 (2014): 1–15.

Leek, Cliff and Michael Kimmel. "Conceptalizing Intersectionality in Superordination: Masculinities, Whitenesses, and Dominant Classes." In *Routledge International Handbook of Race, Class, and Gender*, edited by Shirley A. Jackson. New York: Routledge, 2015.

Lennes, Kostia. "Constructing, Negotiating, and Performing Chicano Manhood as a Borderland Masculinity." *Journal of Borderlands Studies* (2016): 1–16.

Mirandé, Alfredo. "The Chicano Family: A Reanalysis of Conflicting Views." *Journal of Marriage and the Family* 39 (1977): 747–756.

Mirandé, Alfredo. "A Reinterpretation of Male Dominance in the Chicano Family." *Family Coordinator* 28, no. 4 (1979): 473–497.

Mirandé, Alfredo. *The Chicano Experience: An Alternative Perspective*. Notre Dame, IN: University of Notre Dame Press, 1985.

Mirandé, Alfredo. *Hombres y Machos: Masculinity and Latino Culture*. New York: Routledge, 1997.

Mirandé, Alfredo, Pitones, Juan M., and Jesse Díaz. "Quien es el mas Macho? A Comparison of Day Laborers and Chicano." *Men and Masculinities* 14, no. 3 (2010): 309–334.

Pérez, Daniel Enrique. "Comment. Entre Machos y Maricones: (Re)Covering Chicano Gay Male (Hi) Stories." In *Gay Latino Studies. A Critical Reader*, edited by Michael Hames-García and Ernesto Javier Martínez. Durham, NC: Duke University Press, 2011.

Perraudin, Anna. "Migrar para afianzar las masculinidades.: La renegociación de las relaciones de género de la Ciudad de México a Estados Unidos: el caso de una población indígena." In *Género en movimiento*, edited by Rozée, Virginie, Cosio, Maria Eugenia Zavala, 333–357. Mexico City: El Colegio de México, 2014.

Ramirez, Hernan. "Masculinity in the Workplace: The Case of Mexican Immigrant Gardeners." *Men and Masculinities* 14, no. 1 (2011): 97–116.

Rivera, Andrew. "Remembrance and Forgetting: Chicano Masculinity on the Border." *Latino Studies Journal* 8 (1997): 35–55.

Rodríguez, Richard T. "Carnal Knowledge: Chicano Gay Men and the Dialectics of Being." In *Gay Latino Studies. A Critical Reader*, edited by Michael Hames-García and Ernesto Javier Martínez. Durham, NC: Duke University Press, 2011.

Sinatti, Giulia. "Masculinities and Intersectionality in Migration: Transnational Wolof Migrants Negotiating Manhood and Gendered Family Roles." In *Migration, Gender and Social Justice. Hexagon Series on Human and Environmental Security and Peace*, edited by Truong, T. D., Gasper D., Handmaker, J., and Bergh S. Berlin: Springer, 2014.

Smith, Robert. *Mexico en Nueva York*. Mexico City: Porrua, 2006.

Thing, James. "Entre Maricones, Machos y Gays. Globalization and the Construction of Sexual Identities among Queer Mexicans." PhD diss., University of Southern California, 2009.

Thing, James. "Gay, Mexican and Immigrant: Intersecting Identities among Gay Men in Los Angeles." *Social Identities* 16, no. 6 (2010): 809–831.

Vasquez, Fernando. "Negotiation Chicano Masculinities at Institutions of Higher Education: Voices of South Texas Chicano Men." PhD diss., University of Texas at Austin, 2006.

Vasquez del Aguila, Ernesto. *Being a Man in a Transnational World: The Masculinity and Sexuality of Migration*. New York: Routledge, 2014.

Viego, Antonio. "The Place of Gay Male Chicano Literature in Queer Chicana/o Cultural Work." *Discourse* 21, no. 3 (1999): 111–131.

6

A NEW MATERIALIST FRAMEWORK FOR ACTIVISM IN THE AGE OF MEDIATIZATION

The entanglement of bodies, objects, images, and affects

Mariam Betlemidze

Introduction

The "material turn" in feminism signifies a shift in focus to considering material elements of bodies and physical environments in which they operate.[1] This strand of scholarship radically rethinks the relation between human and nonhuman elements as co-constitutive. My own turn to new materialist feminism is motivated by its potential to explicate the question of activism and social change in the age of mediatization. Although controversial, new material feminist scholarship provides unique tools for examining how activists and their audiences are constituted by the materialities of their bodies and spaces in additional to various physical, electronic, and online elements as well as technological affordances. Emphasizing these complex factors requires communication scholars to revise existing poststructuralist, postmodernist, and cultural studies approaches and to begin examining the material and discursive relationship between the non-human, the posthuman, and the prehuman. There is an evolving need to conduct research in ways that will "more productively account for the agency, semiotic force, and dynamics of bodies and natures."[2]

From the point of view of new materialist feminism, linguistic, discursive, and/or representational approaches are reductive and ineffective. Extensive attention to the identity politics of the race, gender, and sexuality of the modern human subject lose their usual appeal and appear exclusive in light of new materialist feminism. Such dramatic changes in theory and criticism do not take place without creating tensions between scholars.

On feminist accounts of subjectivity and agency, the influential author of *Bodies that Matter*, Judith Butler[3] often becomes a scholar against whom new materialist feminists choose to contrast their own approaches.[4] Sara Ahmed finds this move unjust and challenges other scholars[5] for singling out Butler "as a primary example of a feminist who reduces matter to culture."[6] Ahmed criticizes new materialist scholars for their "rejection" of Butler as a means for solidifying "a new terrain." In other words, Ahmed contends that such scholars have opportunistically employed Butler to their own ends. New materialist scholars may appear opportunistic, to put it mildly, as creating new theoretical frameworks often requires the undoing of existing influences. Nonetheless, Ahmed's stance is somewhat reductive, as it is rooted in the anthropocentric understanding of hegemony, ideology, identity, power, and even affect. It is time to evolve past such a humanistic focus to encompass more-than-human entities and to transform discursive artifacts into material-discursive relations.

This chapter makes a case for the affirmative study of bodies and objects with their biological, technological, mediatized, physical, and lived experiences, which often defy expectations, treatments, hierarchies, and authorities—thereby rendering themselves recalcitrant. To accomplish this goal, the chapter provides an overview of the foundational influences within new materialist feminism and then weaves in the theory of affect as a way to illustrate the autonomy and recalcitrance of human and nonhuman elements to established ways of thinking and operating. I then propose media-activism assemblage, a framework based on Bruno Latour's Actor-Network Theory and Gilles Deleuze's and Felix Guattari's concept of assemblages. To better understand the co-constitutive nature of human and nonhuman elements in media-activism assemblages, the chapter outlines the concept of image-affect. This assemblage theorizes images as inseparable from the affects they create in relation to humans, other images, and media artifacts. Such an assemblage is especially pertinent to the study of the visual rhetoric of activist groups and their appeals for social change.

To illustrate the theoretical framework of media-activism assemblage and the concept of image-affect, the chapter provides examples that help account for agency within a vast web of human and nonhuman elements. These range from recalcitrant half-naked female bodies clashing with religious objects to thousands of children's shoes laying on the Capitol Hill signifying lives lost to gun violence in America to yellow vests in Paris protesting raised transportation costs and other activist events encompassing surrounding spaces, architecture, and media. In conclusion, the chapter once again brings together new materialist feminist thoughts and affect theory to emphasize the perspectives offered by this new approach to the study of mediatized activism.

Foundations

New materialist feminism is deeply influenced by Donna Haraway's concepts of the cyborg[7] and companion species[8] that blur the boundaries between traditionally opposing elements such as nature-culture, women-men, ideal-material, same-other, and similar binary categories. She proposes this new way of thinking through onto-epistemology and material-discursive terms, emphasizing the co-constitution of domains hitherto treated as separate. Kathryn Hayles likewise highlights the importance of "the mutually constitutive interactions between the components of a system rather than on message, signal, or information."[9]

In the opening of her seminal book *Meeting the Universe Halfway: Quantum Physics and the Entanglement of Matter and Meaning*, Karen Barad claims that matter and meaning are "inextricably fused together, and no event, no matter how energetic, can tear them asunder."[10] Delineating entanglements from entwinements, she suggests that there are no independent, self-contained entities. Formation of an individual is an endless process of entanglement and "intra-action" that has no beginning or culmination. This Baradian term "intra-action" is a notable one. As Vicky Kirby accurately observed,[11] Barad invented the word to cover for the insufficient connotations of the word interaction, just as Jacques Derrida came up with the word différance[12] to enrich the watered-down force of difference in its common usage and many applications.

Invention and hyphenated joining of words are characteristic to the new materialist feminist approach, as it is in a constant quest to expand boundaries and bridge divides to grasp the nuances of material-discursive world that is always in flux. Alaimo terms the productive mingling between environmental and corporeal theories "trans-corporeality," defined as "the time-space where human corporeality, in all its material fleshiness, is inseparable from 'nature' or 'environment.'"[13] Alaimo is interested in the crucial ethical and political possibilities that emerge from this "contact zone"[14] between the environmental and the corporeal. She suggests a way of thinking across bodies that would let us see the more-than-human world with its own needs, claims, and (often unpredictable) actions.

In her celebrated book *Vibrant Matter,* Jane Bennett furthers posthuman and new materialist scholarship by giving voice to nonhuman elements in public life. Influenced by Deleuze and Guattari's theorization of assemblages and Latour's concept of the "actant," she examines the "vitality" of matter. She is curious about the "capacity of things—edibles, commodities, storms, metals—not only to impede or block the will and designs of humans but also to act as quasi agents or forces with trajectories, propensities, or tendencies of their own."[15] This is no wonder, given Bennett's earlier writing about the *Enchantment of Modern Life*,[16] wherein she claims that all bodies and objects, inorganic or organic, are affective. Indeed, moving forward in complex symbolic-materialist analysis is impossible without engaging affect.

As Pedwell and Whitehead[17] note, affects can be most productive "for the possibilities they offer for thinking (and feeling) beyond what is already known and assumed."[18] This distrust of the preexisting or given is what affect theory and new materialist feminisms have in common. Guided by the principle of emergence, both strands of scholarship reject the often unquestionable validity of moral claims that operate within the binary of right versus wrong and always revolve around discrete and universalized human subjects.[19]

For good reasons, an affective turn aligns perfectly with new material feminisms and new empiricisms.[20] According to Patricia Clough, this shift to affective materialities is conditioned by changes in political, economic, and cultural fields all around the world. She justifies the affective turn by "changing global processes of accumulating capital and employing labor power through the deployment of technoscience to reach beyond the limitations of the human body, or what is called 'life itself.'"[21] What matters in this way of thinking is the realization of "transcorporeality"[22] and the intensity that runs through and across various human and nonhuman entities.

The promise of affect

Much has been written about affect as the intensity that passes through bodies and objects, exerting forces of reciprocal action.[23] Spinozian–Deleuzian affect is a pre-subjective sensemovement that destabilizes representational and physical understandings of human bodies as well as nonhuman elements that coexist with them.[24] In this manner, affect creates and/or becomes a tangle of possibilities for thinking *otherwise* across human and nonhuman materialities. These possibilities span biological, technological, environmental, and discursive plateaus of scholarly thought.

Deleuze and Guattari's concept of the body without organs (BwO)[25] helps to illustrate the intensity of affect. They explain "a BwO is made in such a way that it can only be populated by intensities. Only intensities pass and circulate."[26] This emphasis on circulation still remains popular and its utility may be heightened in an age of 24/7 circulation of media images and stories. Here, I would like to bring the capacities of the transmitting, touching, vibrating, and shimmering of affect into the discussion. I provide each of these characteristics of affect in gerund form to underline their continuous, in-flux nature to provide more operational tools in the new materialist toolbox.

Transmitting, for Teresa Brennan,[27] can be exemplified through the work of pheromones, chemicals that act similar to hormones associated with various social and sexual activity in humans and animals. Brennan proposes that we think about affect as a transfer and, in some cases, even a contagion. According to her, "the transmission of affect, conceptually, presupposes a horizontal line of transmission: the line of the heart."[28] Brennan acknowledges that these transfers do not take place in vacuums but does not elaborate on the conditions and situations surrounding such transfers, some of which cannot happen without connection, or a touch.

Touching and sensing aspects of affect are explicated movingly through the study of dance as we can see in Erin Manning's work.[29] She writes about how dancing bodies operate in relationality to (perhaps even "violently") produce power. "I cannot affect you violently: I affect us. To touch is to violently or gently encounter a surface, a contour."[30] She talks about affect as an interruption and expansion of one's own contours, charting new perimeters that happen in the moment of the touch.

Vibrating is as another innovative contribution to affect theory. In his fascinating article, "The vibrations of affect and their propagation on a night out on Kingston's dancehall scene," Julian Henriques proposes a new model for understanding affect. He "addresses the intensities of affect in terms of auditory amplitudes, as with sonic dominance; feelings as frequencies; and the distinctive meaning of affect as timbre."[31] Henriques (like Manning) remains in the realm of senses, rather than feelings, with this account. His writing about affect differs from discursive, representational, and emotional conceptualizations in this regard.

Shimmering is another gerund we encounter in relation to affect. Melissa Gregg and Gregory Seigworth use this word eloquently in the opening chapter of their *Affect Theory Reader*.[32] They cite Roland Barthes, who describes affect as a "microscopic fragment of emotion ... which implies an extreme changeability of affective moments, a rapid modification, into shimmer."[33] Following Barthes in his ongoing quest for passions of gradient elucidating continuous differences, Gregg and Seigworth open up the discussion of affect with an inventory of shimmers.

Incongruent strands within affect

While some of these theorizations of affect are popular and promising, they do render certain other strands ill at ease if not in contradiction with one another. One of the most notable tensions is between Deleuzian and psychoanalytic approaches. Ranjana Khanna laments, "it has become something of a commonplace in recent feminist affect studies to sideline the psychoanalytic and the deconstructive in favor of Deleuzian notions of affect." She describes affect as displacing the subject and replacing it with "becoming animal," warning against the potential dismissal of important feminist notions such as labor.

Other notable divergences with the Deleuzean way of theorizing affect as pre-subjective and autonomous appear in Sara Ahmed's and Lauren Berlant's works. Their intricate, elaborate, and thought-provoking writing utilizes affect interchangeably with emotion. Thus, they see affect as another mechanism for cultural disciplining of the human subjects. Such an emphasis on the cultural impedes new materialist feminist attempts to shift a/symmetries between human and nonhuman agency and underscore their capacity as co-constitutive of social change.

Contrary to Marxist and psychoanalytic theorizations of affect, Deleuzian affect colors the world outside the lines of the cultural, economic, and political grid—not to mention identity politics, representation, ideology, and hegemony. What we acquire from this shift is attunement with our material worldings and imperceptible but mighty nonhuman elements. Such an approach allows a thorough study of gender and social change via contingent encounters with technologies and matters. To me, affect can be imagined as a rhizome,[34] a vivacious, versatile, decentralized root system of a plant that grows based on the principles of rupture and connection.

Ways, modes, and styles of examining affect and matter

To proceed with a multifaceted, messy, immanently organized new materialist study of affective activism, I propose a media-activism assemblage approach. This framework helps to understand the processes that go into the co-constitution of activists, including their human bodies, non-human elements, images, stories, spaces, and the materialities they are immersed in offline and online. The idea of media-activism assemblages comes from Bruno Latour's Actor-Network Theory,[35] which highlights how human elements are often entangled with nonhumans. One of the key ideas of Actor-Network theory is expressed by the hyphenation between the two terms in its name, which renders them fully interchangeable. An actor may well be a network and vice versa. As Vicki Kirby in her new materialist feminist writing suggests, Latour helps us realize the "radical disjunction/inseparability is comprehensive—a fault line that runs throughout all of human nature. It articulates the nonlocal within the local, nature within culture, and human within nonhuman."[36] Accordingly, human actors cannot control nonhuman elements but can co-create with them. An example of such entanglement is a human body, which consists of billions of microbes carrying out their day-to-day activities despite our knowledge and instructions.

The obvious inspiration for this media-activism assemblage is Deleuze and Guattari's conceptualization of rhizomatic assemblage.[37] Assemblage is an interconnected decentering system that proceeds in a nonlinear fashion, ceaselessly establishing connections between various materials and discursive realities. According to Jane Bennett:

> Assemblages are ad hoc groupings of diverse elements, of vibrant materials of all sorts. Assemblages are living, throbbing confederations that are able to function despite the persistent presence of energies that confound them from within.[38]

What makes Bennett's utilization of Deleuze and Guattari so pertinent for my approach is her weaving of Latour's Actor-Network Theory, specifically the concept of actant, into an entity of human or nonhuman composition or origin igniting an action. This linking of mediatized assemblages to various entities does not demonstrate spontaneity as much as a contingent affirmation of particular flows vis-à-vis an interrelation of times, spaces, and processes. These parts of assemblages are wholes characterized by relations of exteriority and interiority.[39] Parts of the whole, unlike seamless totalities, are detachable from the assemblage and can be plugged into a different assemblage.[40]

The rhizomatic principles of connection and heterogeneity manifest daily in the mediatized assemblages: "any point of a rhizome [digital network] can be connected to anything other, and must be."[41] In this rhizomatic web, interactions are "overflowing in all directions" defying any hidden, structural force of a central, presupposed context.[42] The ideas of unpredictability, decentralization, multiplicity, and dissemination[43] are well represented in new media technologies as they continue to dissolve and destabilize the sovereignty and autonomy of the human subject (the so-called rhetor/orator of communication theory) into the network of nonhuman and more-than-human entities.

When we shift attention to nonhuman entities or objects, it is not their cultural force that matters but the relationality between these objects and other humans. In his book *Prince of Networks: Bruno Latour and Metaphysics*, Graham Harman explains that an object is an "actor trying to adjust or inflict its forces, not unlike Nietzsche's cosmic vision of the will to power."[44] Such articulation of the object resonates with Barad's notion of agency that surpasses objects, subjects, or their intentionality:

> Agency is a matter of intra-acting … Agency is about the possibilities and accountability entailed in reconfiguring material-discursive apparatuses of bodily production, including the boundary articulations and exclusions that are marked by those practices.[45]

Translating this into media-activism assemblage is not a matter of examining a human subject at an activist event but considering the transgressive intra-action with pertinent objects and spaces as they come together, co-creating a mediatized happening.

These intra-actions operate in ways similar to what Nietzschean traces as "joyous affirmation of the play of the world and of the innocence of becoming, ... without origin which is offered to an active interpretation."[46] In contemporary theory, the interplay occurs from the peripheries of multiple decentered knots, objects, subjects, human and otherwise, sending "nomadic waves or flows of deterritorialization" that then go from new peripheries to new centers and knots "falling back to the old center[s] and launching forth to the new [ones]."[47] Such movement illustrates the protest actions of activist groups that start on the margins and then permeate into centers of political discussions, warping movement across contexts, times, and spaces.

From this perspective, media-activism assemblage is concerned with speed and intensity, medium-specific features, affects, and effects of the actors it carries along. Looking at the multitude of digital threads (likes, shares, tweets, posts, and comments aggregated by smart algorithms), Deleuze and Guattari's proposition about the speed of this intensity becomes clearer:

> ...it is in the middle where things pick up speed. [The space] between things does not designate a localizable relation going from one thing to the other and back again, but a perpendicular direction, a transversal movement that sweeps one and the other away, a stream without beginning or end that undermines its banks and picks up speed in the middle.[48]

The media-activism assemblage enacts new possibilities that rest on their posthuman, contingent, and decentralized character. Such new materialistic entwinement of activism and media challenges the preconceived notions of morals, ethics, values, identities, pragmatics, and ideas about progress. Thinking in terms of media-activism assemblages troubles reductionist approaches, as it unveils an incredibly dynamic and contingent way of seeing transformations.

Image-affects

Material-discursive assemblages and networks are produced through heterogeneous alliances between activist bodies, objects, images, spaces, online discussion threads, and journalists, etc. Each of these nonhuman elements is an assemblage of its own kind that plugs into other assemblages, co-transforming and co-creating with others. Images are assemblages of their own, frequently becoming sources of unexpected actions, on their own and on behalf of others. They may even appear to have goals and desires of their own, disseminating and multiplying across the web of digital screens.[49]

Inspired by new materialist re-readings of Deleuzian affect theory, Jane Bennett's concept of "thing-power,"[50] and Kevin DeLuca's "image event,"[51] I find it helpful to think about provocative activist mediatized protests in terms of image-affects. This assemblage theorizes images as inseparable from the affects they create in relation to humans, other images, and media artifacts. Such an assemblage is most pertinent in the study of the visual rhetoric of activist groups and their appeals for social change. The concept of image-affect is composed of the human and nonhuman elements and assembled in a way that generates an autonomous source of power, a force. This power is similar to the one Bennett characterizes as "not transpersonal or intersubjective but impersonal, an affect intrinsic to forms that cannot be imagined (even ideally) as persons."[52] This power is a force of image-affect that attracts other human and nonhuman elements to it, propelling image-affect's movement across material and discursive plains, plugging in and out of other assemblages, producing polyvalent transformations.

In a world structured by information and communication technologies, it is not possible to spend a day or even an hour without a certain extent of visual meditation, whereby images cause sensations and act as references to various events or trends. DeLuca's "image event" describes a tactic of oppositional movements as they use visual rhetoric in the advancement of their political goals. However, it is not the particular power of images that causes social change, but the way their processes create affect and desires through which an image can acquire the agency of a "living being."[53] Images can be thought of as produced by humans, but their process of creation relies on and folds in nonhuman elements, thereby transforming both humans and nonhumans as they co-create images. Similar to Bennett's thing-power, image-affects "can aid or destroy, enrich or disable, ennoble or degrade us."[54]

In the current new media environment, patterns of perception become highly fragmented, distracted, and decentered.[55] In such a disposition of senses toward mediatized daily life, thought becomes even more dependent upon the contingency of an encounter.[56] These encounters could be of anyone viewing an online picture of a protest action with various human and non-human elements. Painted or printed slogans, bodies, body parts, cameras, digital screens, and urban landmarks within an image are encountering not only cameras capturing them for the viewers, but also they are encountering each other. These encounters are responding to each other without necessarily signifying anything but producing certain affective forces.[57]

Much has been written about affective encounters between bodies and images. One of the most hailed theorizations of such encounters is provided by Mark Hensen. He suggests that affectivity is an intrinsic element of an image: "affectivity thereby becomes the very medium of interface with the image."[58] According to Hensen, affect works as a force flipping the virtual-actual binary between images and bodies. As media technology and immersive content develop, we coevolve with new media and become more dependent on it. Hensen's argument becomes more pertinent as the actuality of the images increases and our bodies become more virtual. This process of flipping between images and bodies, virtual and actual, is one of the effects or symptoms of image-affect.

This symptom corresponds with the new materialist perspective of Elizabeth Grosz. In her 2017 book, she suggests that incorporeal conditions may well be exceeding corporeality.[59] This is an updated viewpoint from when Grosz's *Volatile Bodies: Toward a Corporeal Feminism* became one of the foundational texts for a new materialist turn in feminisms. In *The Incorporeal: Ontology, Ethics, and the Limits of Materialism*, this Deleuzian feminist provides a more nuanced exploration of corporeality in the contexts of extramaterialism or "the inherence of ideality, conceptuality, meaning, or orientation that persists in relation to and within materiality as its immaterial or incorporeal conditions."[60] I believe that within communication studies, new media technologies and affordances are the incorporeal conditions that frame and orient human and nonhuman entities into and within image-affects.

Ever-evolving and multiplying new media tools such as searchability[61] and spreadability[62] enable us to enter the era of "digitally afforded affect,"[63] in its networked, collective nature. Adding a new materialist perspective to this argument highlights the force of nonhuman entities along with their human counterparts. In their new materialist feminist writing, Vicky Kirby and Susan Heckman often cite and credit Bruno Latour for "folding humans and nonhumans into each other."[64] As Latour demonstrates in *Pandora's Hope*, "the mobilization of nonhumans inside the collective"[65] brings in fresh and unexpected resources, creating new hybrids who are actants that change the collective.

In an activist image-affect, nonhuman and human elements intra-act and create new hybrids, folding in aspects of sexuality, urbanity, and public space. The theme of sexualities in cities is particularly pertinent, as many of these mediatized protests take place in urban settings. Lighting, advertising screens, and the fast pace of cities create the expectation of sexuality.[66] These factors "effectively remind viewers that the city is a sexual marketplace."[67] A city as a sexual marketplace is often dominated by lively screens, small and big, showing bodies that conform to mass-mediated standards of attractive femininity. Researched by Erving Goffman[68] almost four decades ago, these standards remain largely unchanged: mass media advertising is still dominated by conventionally attractive, slim, mostly white, and docile/submissively withdrawn bodies.[69]

While women are expected to dominate city screens by serving as passive props for sexual imagery, public space remains a domain for masculine actions. Sennett,[70] tracing the histories of human bodies in relation to cities from ancient Greece to Medieval Europe, illustrated how public space has been a masculine arena, while femininity remained restricted to domesticity. With the new materialist feminist perspective, the reality appears more promising, as it creates opportunities for femininity to make likely, and often unlikely, alliances with nonhuman elements that are abundantly present in our material spaces. According to Henry Lefebvre, space "rejoins material production: the production of goods, things, objects of exchange—clothing, furnishings, houses or homes—a production which is dictated by necessity."[71] This material production dictated by necessity or design can intensify certain encounters, pains, gratifications, or contradictions. Human, nonhuman, and technologically enhanced hybrids of both can effectively contradict the necessities, design-intended uses, and expectations of public space, creating gendered dissonances of activist image-affects.

Illustrative examples

As I was writing this chapter, one particular image-affect kept reemerging in my thoughts: Femen activist Inna Shevchenko cutting down the crucifix in Kiev downtown. This was the protest I have analyzed frame by frame in the past to show how the image of her activism opened ruptures within the traditional Ukrainian and Eastern European Christian Orthodox community while creating connections with secular French and Dutch communities and transforming discourses pertaining to the place of women in culture, politics, and religion.[72]

Now that I look back at this analysis, I am curious about the degree to which I properly traced the affective intra-actions between human and nonhuman elements in this protest action. The wooden cross with a crucifix erected on a hilltop overlooking Kiev Euromaidan is easily recognizable. The reason for that is the highly recognizable landmark in the background— the Monument of Ukrainian Independence, a female deity from Slavic folklore and mother-protector, Bereginia.[73] The resemblance between the Femen activist cutting down the crucifix and Bereginia is striking. Both have ribbons in their hair, Inna Shevchenko has a flower wreath on her head, while Bereginia is holding an unfolded wreath of flowers above her head. How these nonhuman elements on a live activist's body and statue speak to each other remains imperceptible but still poignant. The larger issue at stake in this event is the clash that results from bringing together a half-naked woman and a religious icon. The expectations for Orthodox Christian femininity in public space, especially in the proximity of religious objects, are absolute and specifically require timidity, submissive trust, and worship. The Femen activist's appearance and actions enacted exactly the opposite of these expectations. "Female obedience, ethics of care, and the virtue of never exhausting patience"[74] remain pertinent in Slavic Orthodox communities. This being the case, a half-naked Ukrainian women with the slogan "free riot" painted across her chest cutting down the most sacred Christian symbol with a chainsaw was immanently materialized to create image-affects that continue to shimmer and vibrate across digital screens. From a new materialist perspective, chainsaw, cross, body paint, words, red shorts, combat boots, Statue of Independence, and cameras capturing the event (as well as other elements) co-created such image-affects.

Femen is just one example of the strategic entangling of human actors with nonhuman entities like garments (or the lack of thereof). Other illustrative examples of activism relying on nonhumans and similar objects include Guerrilla Girls with their subversive use of gorilla masks, Pussy Riot with their colorful balaclavas, the SlutWalks with short skirts, the women's march with pink pussy hats, #TimesUp supporters in all black dresses at the Golden Globes and President Trump's first State of the Union address, and, most recently, the movement of "yellow vests"[75] protesting fuel price hikes in France with its fluorescent yellow hazard vests. In these instances of activism, humans co-create within the assemblages of nonhuman elements (in the form of particular garments or accessories) and media affordances capturing, transmitting, and multiplying activism's image-affects across the networks of digital screens.

Similarly, one of the most prominent uses of garments as nonhuman elements enacting activism can be seen in protests utilizing empty shoes to visualize the void created by a loss of human lives. In March 2018, 7,000 pairs of shoes covered the lawn in front of the U.S. State Capitol building. The protest was organized in the name of #notonemore, protesting the loss of 7,000 children's lives to gun violence since 2012 and demanding gun control laws.[76] Sneakers, slippers, sandals, and boots (ranging in size from those worn by preschoolers to high schoolers) not only covered 10,000 square feet of one of the most venerated spaces in America, they created a material space of dissonance on that symbolic ground. In her video interview, Nell Greenberg, a campaign director at the global activist network Avaaz, specifically emphasizes the shoes displayed at the protest: "I saw roller skates behind me, I saw little ruby slippers, I saw little children's sneakers … and these should be the shoes worn by the children, running around, playing, jumping … and instead, they are empty soles." As she went on to say in the interview, the purpose of the protest was to "deliver to Congress' doorstep the true, vivid, heartbreaking cost of their inaction."[77]

As Greenberg and other people were expressing their thoughts in the interviews, the imagery in the background, foreground, and middle ground was doing its own thing. The sea of neatly arranged, colorful children's shoes subversively occupied the area frequented by the tourists and children on school trips for its iconic manifestation of democracy. Some of the shoes, sent by parents of the gun violence victims, came with messages and little toys stuffed inside them. Shoes displayed in one of the most venerated spaces of the United States materialized untold stories of children who are longer alive. The collective of the various human and nonhuman entities, consisting of protest organizers, journalists, by-passers, cameras, shoes, grass, the Capitol Building, wind, as well as media affordances came together in the assemblage of this image-affect. Similar political events utilizing shoes as protestors, acting for their former wearers, have also taken place in different parts of the world.

In November 2015, activists in Valladolid, Spain, arranged a total of 1,800 dyed-red shoes on one of the city's most iconic landmarks, Plaza Mayor. The shoes here, like those in the U.S. Capitol Hill protest, were symbolically materializing the void left by victims of violence against women. The protest aimed to raise awareness about the International Day for the Elimination of Violence Against Women.[78] Similar protests took place in several European countries to protest femicide. "The colour red, and particularly red shoes, have become a symbol of the struggle for women's rights in Italy," says caption on a header photo in Italian online media article "Italy's 'small, silent protest' against femicide."[79] Euro News's online story uses an imbedded tweet with a photo from the red shoes protest in Spain: "A man walks between red shoes displayed as part of a protest to highlight violence against women in Barcelona (AP Photo/Manu Fernandez)."[80] A close-up shot of shoes awkwardly pointed outward from the same protest in Barcelona open up the Guardian's call for more stories on the subject.[81] "How will you mark 16 days of activism against gender abuse?" asks the *Guardian* headline.[82] When studying stories with the shoes as protestors, I noticed that many journalistic stories tend to devote more attention to human protesters. Nevertheless, the sheer volume of the images with empty shoes protesting in various places of Europe engenders the emergence new, a more visceral way of seeing the issue of femicide and the protest.

To examine these nonhuman actors during women's shoe protests, I will discuss in more detail a 2018 protest that took place on a much smaller scale in my home country of Georgia, in a small town of Zugdidi. This small town is located a 30-minute drive away from the conflict zone of Abkhazia, a breakaway territory of Georgia still guarded by Russian military troops. In light of the unresolved ethnic-territorial conflict, political turbulence, and economic hardship, gender equality is hardly the first issue on the radar of Zugdidi city-dwellers. Nevertheless, the local branch representatives of the Equality Movement took this issue to a key public place at the heart of Zugdidi: the City Boulevard. In comparison to other large-scale protests, this one stands out with its utilization of printed human stories, attached to each pair of women's shoes. Examples of these short narratives include the following:

> I live in a little town. I cannot say aloud anywhere that I am a transgender woman. This has to remain hidden due to the transphobia in our society.
>
> (Lola, 40)[83]

> I cannot participate in this protest to defend women's right because I was killed by my husband in 2016.
>
> (Lika, 31)[84]

> I cannot be on this protest action because I cannot afford to miss my work. I work from 8 in the morning till midnight in a local supermarket and the earnings are barely enough to cover my basic needs.
>
> (Tamo, 23)[85]

These snippets of stories provide glimpses into precarious lives and deaths of women who remain mostly unnoticed due to their recalcitrance to the traditional norms and patriarchal macho views imbued by left-righteous Georgian Orthodox-Christian moralism.

An online news story with photos and an embedded video of the event explains that the empty shoes were placed in the city center to draw people's attention. The protest organizer, Tsabunia Vartagava, states that there should be women there *in* their shoes protesting, but women could not participate "because they have been taught by the society to remain silent."[86] The humans in the protest, like Vartagava, remained on the sidelines, as the epicenter of the action was 20 pairs of shoes along with their printed unique stories. They stood empty on the cobblestone road of the boulevard in the rainy afternoon, intra-acting with the passers-by and reporters and their storytelling gear. This image-affect illustrates Ron Burnett's point precisely: "the relationship developed with objects transforms all the partners in the exchange."[87] Burnett, analogously to Latour, calls this relationship a hybridization. In the example of this Zugdidi gender violence protest, women's empty shoes, rain, cobblestones, cameras, and stories printed on white paper carefully placed in transparent cellophane files, among other objects, found a common ground in a way that exceeds corporeal conditions, design-intended object-ives, and mundane necessities—thereby culminating its intra-action in the image-affect of the protest.

Conclusion

It is often said that change happens on the margins and that "marginal subjects produce different, more reliable knowledge."[88] But in light of the approach this chapter charted, it may well be the marginal objects and not the subjects that produce reliable knowledge and instigate social change. Illustrative examples provided at the end of the chapter show how images, technologies, clothes, shoes, spaces, and myriad other elements intra-act, co-creating image-affects.

The goal of this chapter was to bring together new materialist feminist thoughts and affect theory to create a unique approach for the study of mediatized activism. This approach focuses on the realm of physical, brute matter as well as its digital manifestations, which are acquiring new, autonomous lives in the networks of digital screens. Such analysis is beneficial for studying how physical protest actions transpire and continue living in the online realms.

To better study the contingency of (dis)connections between human, nonhuman, and other entities, the chapter drew on both Latour's Actor-Network Theory with Deleuze and Guattari's concept of assemblage. These thinkers have been and continue to be well-respected among new materialist feminists for good reasons: they help communication scholars to consider how affect plays a catalytic role in processes of material-digital social change. These forces operate between subjects and objects and work in all directions simultaneously, thereby allowing not only humans to affect matter but also to be affected by matter. Inspired by rereading these thinkers, specific-ally in the context of affect and mediatized social change, I offer visual and media scholars the concept of image-affect, which can be used to better understand the inseparable amalgamation of subject-objects and reciprocal processes of affect across and in between them as they come together to create images.

Any shift or change—especially a material one—comes at a cost. In this case, what may be lost are human-oriented avenues of discourse, performance, and identity. The overarching aim of new materialist approaches is based on ethics that resist taking a moral stance or making prescriptions as to what should be done and why. New materialisms can open up spaces for empowering and promising dialogues, but this will only happen through a detailed and thorough examination of phenomena via material-discursive routes. Such a careful and complex analytic lens makes unique contributions to the field of communication, especially to the study of contemporary social activism's connections to human bodies, technologies, media, and the environment. Future research can advance the conversation by examining how various human and nonhuman elem-ents touch, vibrate, and transmit collective, digitally afforded image-affects.

Through various theoretical entanglements, this chapter illustrates that the agency of non-human elements constitutes rich material milieus that cannot be separated from the heteroge-neous processes of which they are a part. In order to proceed with new materialist approaches in media studies, scholars have to sift through the abundances of digital nodes as well as physical and material traces in a nonlinear way that enables detours, deferrals, falls, jumps, and flights.

The lines between human and nonhuman, cultural and technological, virtual and actual, present and absent, as well as visual and nonvisual, are dissolving. What is at stake is the cap-acity to chart constant processes of technologically mediated transformations through rupture, connection, co-constitution, and co-creation. This heterogeneous mediatized process unhinges binary-based views, making pure non-mediatized modes of being unviable.

Notes

1 Stacy Alaimo and Susan Hekman, *Material Feminisms* (Bloomington, IN: Indiana University Press, 2008); Karen Barad, *Meeting the Universe Halfway: Quantum Physics and the Entanglement of Matter and Meaning* (London: Duke University Press, 2007); Jane Bennett, *Vibrant Matter* (Durham, NC: Duke University Press, 2010); Diana Coole and Samantha Frost, *New Materialisms: Ontology, Agency, and Politics* (Durham, NC: Duke University Press, 2010).

2 Alaimo and Hekman, *Material Feminisms*, 6.

3 Judith Butler, *Bodies that Matter: On the Discursive Limits of Sex* (London: Routledge, 1993).

4 Karen Barad, "Posthumanist Performativity: Toward an Understanding of How Matter Comes to Matter," *Signs* 28, no. 3 (2003): 801–831; Pheng Chea, "Mattering," *Diacritics* 26, no. 1 (1996): 108–39; Patricia Clough, "Introduction," in *The Affective Turn: Theorizing the Social*, eds P. Clough and J. Halley (Durham: Duke University Press, 2007), 1–33; Mariam Fraser, "What is the Matter of Feminist Criticism," *Economy and Society* 31, no. 4 (2002): 606–25; Alaimo and Hekman, *Material Feminisms*.

5 Barad, "Posthumanist Performativity," 801–831; Clough, "Introduction," 8; Chea, "Mattering," 108–139; Fraser, "What Is the Matter," 606–625.

6 Sara Ahmed, "Open Forum Imaginary Prohibitions: Some Preliminary Remarks on the Founding Gestures of the 'New Materialism,'" *European Journal of Women's Studies* 15, no. 1 (2008): 23–39, doi: 10.1177/1350506807084854, 33.

7 Donna Haraway, "Manifesto for Cyborgs: Science, Technology, and Socialist Feminism in the 1980s," *Socialist Review* no. 80 (1985): 65–108.

8 Donna Haraway, *Modest Witness@Second_Millennium. FemaleMan©Meets_OncoMouse: Feminism and Technoscience* (New York: Routledge, 1997).

9 Kathrine Hayles, *How We Became Posthuman: Virtual Bodies in Cybernetics, Literature, and Informatics* (Chicago: The University of Chicago Press), 11.

10 Barad, *Meeting the Universe Halfway*, 3.

11 Vicki Kirby, "Natural Convers(At)Ions: Or, What if Culture was Really Nature all Along?," in *Material Feminisms*, eds Stacy Alaimo and Susan Hekman (Bloomington, IN: Indiana University Press, 2008), 214–236.

12 Jacques Derrida, "Différance," in *Margins of Philosophy*, ed. Jacques Derrida (Chicago: The University of Chicago Press, 1982), 3–27.

13 Stacey Alaimo, "Trans-corporeal Feminisms and the Ethical Space of Nature," in *Material Feminisms*, eds Stacey Alaimo and Susan Hekman (Bloomington, IN: Indiana University Press, 2008), 238.

14 Alaimo, "Trans-corporeal Feminisms"; Mary Louise Pratt, "Arts of the Contact Zone," *Profession* (1991): 33–40.

15 Bennet, *Vibrant Matter*, viii.

16 Jane Bennett, *The Enchantment of Modern Life: Attachments, Crossings, and Ethics* (Princeton, NJ: Princeton University Press, 2001).

17 Carolyn Pedwell and Anne Whitehead, "Affecting Feminism: Questions of Feeling in Feminist Theory," *Feminist Theory* 13, no. 2 (2012): 115–129, doi:10.1177/1464700112442635.

18 Pedwell and Whitehead, "Affecting Feminism," 117.

19 Barad, *Meeting the Universe Halfway*, 23.

20 Patricia Clough, "The New Empiricism: Affect and Sociological Method," *European Journal of Social Theory* 12, no. 1 (2009): 43–61, doi:10.1177/1368431008099643.

21 Clough and Halley, *The Affective Turn: Theorizing the Social*, 117.

22 Alaimo, "Trans-corporeal Feminisms."

23 Melissa Gregg and Gregory Seigworth, "Inventory of Shimmers," in *The Affect Theory Reader*, eds Melissa Gregg and Gregory Seigworth (Durham, NC: Duke University Press, 2010), 1–28.

24 Brian Massumi, *The Principle of Unrest* (London: Open Humanities Press, 2017).

25 Gilles Deleuze and Félix Guattari, *A Thousand Plateaus: Capitalism and Schizophrenia* (London: Continuum, 1987).

26 Deleuze and Guattari, *A Thousand Plateaus*, 153.

27 Teresa Brennan, *The Transmission of Affect* (Ithaca, NY: Cornell University Press, 2004).

28 Brennan, *The Transmission of Affect*, 75.

29 Erin Manning, *Politics of Touch: Sense, Movement, Sovereignty* (Minneapolis, MN: University of Minnesota Press, 2007).

30 Manning, *Politics of Touch,* 13.

31 Julian Henriques, "The Vibrations of Affect and their Propagation on a Night Out on Kingston's Dancehall Scene," *Body & Society* 16, no. 1 (2010): 57–89, 57.

32 Gregg and Seigworth, *Affect Theory Reader.*

33 Gregg and Seigworth, *Affect Theory Reader,* 10.

34 Deleuze and Guattari, *A Thousand Plateaus.*

35 Bruno Latour, *The Pasteurization of France* (Cambridge, MA: Harvard University Press, 1993); Bruno Latour, "On Actor-Network Theory: A Few Clarifications," *Soziale Welt* 47 (1996): 369–381; Bruno Latour, *Reassembling the Social: An Introduction to Actor-Network Theory* (Oxford: Oxford University Press, 2005).

36 Kirby, "Natural Convers(at)ions," 233–234.

37 Deleuze and Guattari, *Thousand Plateaus.*

38 Bennett, *Vibrant Matter,* 23–24.

39 Manuel DeLanda, *A New Philosophy of Society: Assemblage Theory and Social Complexity* (New York: Continuum, 2006).

40 DeLanda, *A New Philosophy.*

41 Deleuze and Guattari, *Thousand Plateaus,* 7.

42 Latour, *Reassembling the Social,* 202.

43 Kevin DeLuca and Jennifer Peeples, "From Public Sphere to Public Screen: Democracy, Activism, and the 'Violence' of Seattle," *Critical Studies in Media Communication* 19, no 2 (2002): 125–151, doi: 10.1080/07393180216559; John Durham Peters, *Speaking in to the Air: A History of the Idea of Communication* (Chicago: The University of Chicago Press, 1999); Jacques Derrida, *Dissemination* (New York: Continuum, 1981/2005).

44 Graham Harman, *Prince of Networks: Bruno Latour and Metaphysics* (Prahran: re.press, 2009), 16.

45 Barad, *Meeting the Universe Halfway,* 214.

46 Jacques Derrida, *Writing and Difference* (Chicago: The University of Chicago Press, 1978), 368.

47 Deleuze and Guattari, *A Thousand Plateaus,* 44.

48 Deleuze and Guattari, *A Thousand Plateaus,* 25.

49 W. J. T. Mitchell, *What Do Pictures Want?: The Lives and Loves of Images* (Chicago: The University of Chicago Press, 2005).

50 Bennett, *Vibrant Matter.*

51 Kevin DeLuca, *Image Politics: The New Rhetoric of Environmental Activism* (New York: The Guilford Press, 1999).

52 Bennet, *Vibrant Matter,* xii.

53 Mitchell, *What Do Pictures Want?*

54 Bennet, *Vibrant Matter,* ix.

55 DeLuca, Kevin M., Sean Lawson and Ye Sun. "Occupy Wall Street on the Public Screens of Social Media: The Many Framings of the Birth of a Protest Movement," *Communication, Culture & Critique* 5, no. 4 (2012): 483–509.

56 Gilles Deleuze, *Difference and Repetition* (New York: Columbia University Press, 1994), 139.

57 Marco Abel, *Violent Affect: Literature, Cinema, and Critique after Representation* (Lincoln, NE: University of Nebraska, 2007); Brian Massumi, "The Autonomy of Affect," *Cultural Critique* (1995): 83–109.

58 Mark Hansen, "Affect as Medium, or the 'Digital-Facial-Image,'" *Journal of Visual Culture* 2, no. 2 (August 2003): 205–228, doi:10.1177/14704129030022004, 208.

59 Elizabeth Grosz, *The Incorporeal: Ontology, Ethics, and the Limits of Materialism* (New York: Columbia University Press, 2017).

60 Grosz, *The Incorporeal,* 5.

61 Danah Boyd, "Social Network Sites as Networked Publics: Affordances, Dynamics, and Implications," in *Networked Self: Identity, Community, and Culture on Social Network Sites,* ed. Zizi Papacharissi (London: Routledge, 2010), 39–58.

62 Henry Jenkins, Sam Ford and Joshua Green, *Spreadable Media: Creating Value and Meaning in a Networked Culture* (New York: New York University Press, 2013).

63 Papacharissi, Zizi. *Affective Publics: Sentiment, Technology, and Politics* (New York: Oxford University Press, 2015), 8.

64 Bruno Latour, *Pandora's Hope: Essays on the Reality of Science Studies* (Cambridge: Harvard University Press, 1999). 176.

65 Latour, *Pandora's Hope*, 194.
66 Phil Hubbard, *Cities and Sexualities* (New York: Routledge, 2013).
67 Hubbard, *Cities and Sexualities*, 10.
68 Erving Goffman, *Gender Advertisements* (Cambridge, MA: Harvard University Press, 1979).
69 Gregory Fouts and Kimberley Burggraf, "Television Situation Comedies: Female Weight, Male Negative Comments, and Audience Reactions," *Sex Roles* 42 (2000): 925, https://doi.org/10.1023/A:1007054618340; Jack Glascock, "Gender Roles on Prime-Time Network Television: Demographics and Behaviors," *Journal of Broadcasting & Electronic Media* 45 (2001): 656–669; Nancy Signorielli and Aaron Bacue, "Recognition and Respect: A Content Analysis of Prime-Time Television Characters across Three Decades," *Sex Roles* 40 (1999): 527–544.
70 Richard Sennett, *Flesh and Stone: The Body and the City in Western Civilization* (New York: W.W. Norton & Company, 1996).
71 Henri Lefebvre, *Production of Space* (Cambridge: Blackwell, 1991), 137.
72 Mariam Betlemidze, "Mediatized Controversies of Feminist Protest: FEMEN and Bodies as Affective Events," *Women's Studies in Communication* 38, no. 4 (2015): 374–379. doi: 10.1080/07491409.2015.1089103.
73 Ylia Buiskikh [Юлія Буйських], "'Bereginia': an attempt to deconstruct one 'office myth'" ["Берегиня": спроба деконструкції одного «кабінетного» міфа], Modern Ukraine, May 24, 2018, http://uamoderna.com/md/buyskykh-berehynia-myth.
74 Julia Khrebtan-Hörhager and Tanya Gordiyenko, "Clash of Cultural Capitals in European Mixed Marriages," *Connections: European Studies Annual Review* 39 (2012), 38–49.
75 Le Monde, "Movement of "Yellow Vests" [MOUVEMENT DES "GILETS JAUNES"]" (n.d.). Retrieved from www.lemonde.fr/mouvement-des-gilets-jaunes/; Vanessa Friedman, "The Power of the Yellow Vest," *The New York Times*, December 4, 2018, www.nytimes.com/2018/12/04/fashion/yellow-vests-france-protest-fashion.html.
76 Marilyn Icsman, "7,000 Pairs of Shoes Were Outside U.S. Capitol to Represent Children Killed by Guns," *USA Today*, March 13, 2018, www.usatoday.com/story/news/politics/onpolitics/2018/03/13/7-000-pairs-shoes-were-outside-u-s-capitol-represent-children-killed-guns/421433002/.
77 Ruptly, "USA: 7,000 Pairs of Shoes Cover Capitol Lawn to Protest Gun Violence," YouTube, March 13, 2018, www.youtube.com/watch?v=s7Mitu5FXno.
78 CBS News, "Gender Violence and Femicide Protest," November 25, 2015, www.cbsnews.com/pictures/what-am-i-seeing-curious-strange-photos-november-27-2015/16/.
79 The Local. "Italy's 'Small, Silent Protest' against Femicide," June 2, 2016, www.thelocal.it/20160602/italys-small-silent-protest-against-femicide-sara-di-pietrantonio.
80 Gill, Joanna, "Turkey to Buenos Aires: Worldwide Rallies to Protest Violence against Women," November 26, 2016, www.euronews.com/2016/11/26/turkey-to-buenos-aires-worldwide-rallies-to-protest-violence-against-women.
81 The Guardian, "How Will You Mark 16 Days of Activism Against Gender Abuse? Share Your Stories," November 25, 2016, www.theguardian.com/global-development/2016/nov/25/how-will-you-mark-16-days-of-activism-against-gender-abuse-share-your-stories.
82 The Guardian, "16 Days of Activism Against Gender Abuse."
83 Livepress.ge. "Stories Told by Women's Shoes [ფეხსაცმლით მოყოლილი ქალთა ისტორიები]," March 8, 2018, www.livepress.ge/ka/akhali-ambebi/article/22658fekhsacmlithmoyoliliqalthaistorieb-ifotovideo.html?fbclid=IwAR3jKGZn1VVeXxU5PKpigr6F4b4aQ2YDnZ%25E2%2580%25A6, para. 2.
84 Livepress.ge, "Stories Told by Women's Shoes [ფეხსაცმლით მოყოლილი ქალთა ისტორიები]," March 8, 2018, www.livepress.ge/ka/akhali-ambebi/article/22658fekhsacmlithmoyoliliqalthaistorieb-ifotovideo.html?fbclid=IwAR3jKGZn1VVeXxU5PKpigr6F4b4aQ2YDnZ%25E2%2580%25A6, para. 3.
85 Livepress.ge, "Stories Told by Women's Shoes [ფეხსაცმლით მოყოლილი ქალთა ისტორიები]," March 8, 2018, www.livepress.ge/ka/akhali-ambebi/article/22658fekhsacmlithmoyoliliqalthaistorieb-ifotovideo.html?fbclid=IwAR3jKGZn1VVeXxU5PKpigr6F4b4aQ2YDnZ%25E2%2580%25A6, para. 4.
86 Livepress.ge, "Stories Told by Women's Shoes [ფეხსაცმლით მოყოლილი ქალთა ისტორიები]," March 8, 2018, www.livepress.ge/ka/akhali-ambebi/article/22658fekhsacmlithmoyoliliqalthaistorieb-ifotovideo.html?fbclid=IwAR3jKGZn1VVeXxU5PKpigr6F4b4aQ2YDnZ%25E2%2580%25A6, para. 7.
87 Ron Burnett, *How Images Think* (Cambridge, MA: MIT Press, 2005), 172.
88 Clare Hemmings, "Affective Solidarity: Feminist Reflexivity and Political Transformation," *Feminist Theory* 13, no. 2 (August 2012): 147–61, doi: 10.1177/1464700112442643, 155.

Bibliography

Abel, Marco. *Violent Affect: Literature, Cinema, and Critique after Representation.* Lincoln, NE: University of Nebraska, 2007.

Ahmed, Sara. "Open forum imaginary prohibitions: some preliminary remarks on the founding gestures of the 'New Materialism'." *European Journal of Women's Studies*, 15, no. 1 (2008): 23–39. doi: 10.1177/1350506807084854, 33.

Alaimo, Stacy and Hekman, Susan, eds *Material Feminisms.* Bloomington, IN: Indiana University Press, 2008.

Alaimo, Stacy. "Trans-corporeal feminisms and the ethical space of nature." In *Material Feminisms*, eds Stacy Alaimo and Susan Hekman. Bloomington, IN: Indiana University Press, 2008.

Barad, Karen. "Posthumanist performativity: toward an understanding of how matter comes to matter." *Signs*, 28, no. 3 (2003): 801–31.

Barad, Karen. *Meeting the Universe Halfway: Quantum Physics and the Entanglement of Matter and Meaning.* Durham, NC: Duke University Press, 2007.

Bennett, Jane. *The Enchantment of Modern Life: Attachments, Crossings, and Ethics.* Princeton, NJ: Princeton University Press, 2001.

Bennett, Jane. *Vibrant Matter.* Durham, NC: Duke University Press, 2010.

Betlemidze, Mariam. "Mediatized controversies of feminist protest: FEMEN and bodies as affective events." *Women's Studies in Communication*, 38, no. 4 (2015): 374–379. doi: 10.1080/07491409.2015.1089103.

Boyd, Danah. "Social network sites as networked publics: affordances, dynamics, and implications." In *Networked Self: Identity, Community, and Culture on Social Network Sites*, ed. Zizi Papacharissi. London: Routledge, 2010.

Brennan, Teresa. *The Transmission of Affect.* Ithaca, NY: Cornell University Press, 2004.

Buiskikh, Ylia [Юлія Буйських], "'Bereginia': an attempt to deconstruct one 'office' myth" [«Берегиня»: спроба деконструкції одного «кабінетного» міфа], Modern Ukraine (May 24, 2018). Retrieved from http://uamoderna.com/md/buyskykh-berehynia-myth.

Burnett, Ron. *How Images Think.* Cambridge, MA: MIT Press, 2005.

Butler, Judith. *Bodies that Matter: On the Discursive Limits of Sex.* London: Routledge, 1993.

CBS News. Gender violence and femicide protest (November 25, 2015). Retrieved from www.cbsnews.com/pictures/what-am-i-seeing-curious-strange-photos-november-27-2015/16/.

Chea, Pheng. "Mattering." *Diacritics*, 26, no. 1 (1996): 108–139.

Clough, Patricia. "Introduction." In *The Affective Turn: Theorizing the Social*, eds Patricia Clough and Jean Halley. Durham, NC: Duke University Press, 2007: 1–33.

Clough, Patricia. "The new empiricism: affect and sociological method." *European Journal of Social Theory*, 12, no. 1 (2009): 43–61. doi: 10.1177/1368431008099643.

Coole, Diana and Frost, Samantha. *New Materialisms: Ontology, Agency, and Politics.* Durham, NC: Duke University Press, 2010.

Deleuze, Gilles and Guattari, Félix. *A Thousand Plateaus: Capitalism and Schizophrenia.* London: Continuum, 1987.

Deleuze, Gilles. *Difference and Repetition.* New York: Columbia University Press, 1994.

DeLanda, Manuel. *A New Philosophy of Society: Assemblage Theory and Social Complexity.* New York: Continuum, 2006.

DeLuca, Kevin. *Image Politics: The New Rhetoric of Environmental Activism.* New York: Routledge, 1999.

DeLuca, Kevin and Peeples, Jennifer. "From public sphere to public screen: democracy, activism, and the 'violence' of Seattle." *Critical Studies in Media Communication*, 19, no. 2 (2002): 125–151. doi: 10.1080/07393180216559.

DeLuca, Kevin, Lawson, Sean, and Sun, Ye. "Occupy Wall Street on the public screens of social media: the many framings of the birth of a protest movement." *Communication, Culture & Critique*, 5, no. 4 (2012): 483–509.

Derrida, Jacques. *Writing and Difference.* Chicago: The University of Chicago Press, 1978.

Derrida, Jacques. *Dissemination.* New York: Continuum, 1981/2005.

Derrida, Jacques. "Différance." In *Margins of Philosophy*, ed. Jacques Derrida. Chicago: The University of Chicago Press, 1982: 3–27.

Fouts, Gregory and Burggraf, Kimberley. "Television situation comedies: female weight, male negative comments, and audience reactions." *Sex Roles*, 42 (2000): https://doi.org/10.1023/A:1007054618340.

Fraser, Mariam. "What is the matter of feminist criticism." *Economy and Society*, 31, no. 4 (2002): 606–625.

Friedman, Vanessa. "The power of the yellow vest." *The New York Times*, December 4, 2018. Retrieved from www.nytimes.com/2018/12/04/fashion/yellow-vests-france-protest-fashion.html.

Gill, Joanna. Turkey to Buenos Aires: worldwide rallies to protest violence against women (November 26, 2016). Retrieved from www.euronews.com/2016/11/26/turkey-to-buenos-aires-worldwide-rallies-to-protest-violence-against-women.

Glascock, Jack. "Gender roles on prime-time network television: demographics and behaviors." *Journal of Broadcasting & Electronic Media*, 45 (2001): 656–669.

Goffman, Erving. *Gender Advertisements*. Cambridge, MA: Harvard University Press, 1979.

Gregg, Melissa and Seigworth, Gregory, eds *The Affect Theory Reader*. Durham, NC: Duke University Press, 2010.

Grozs, Elizabeth. *The Incorporeal: Ontology, Ethics, and the Limits of Materialism*. New York: Columbia University Press, 2017.

Hansen, Mark. "Affect as medium, or the 'digital-facial-image.'" *Journal of Visual Culture*, 2, no. 2 (2003): 205–228. doi: 10.1177/14704129030022004, 208.

Haraway, Donna. "Manifesto for cyborgs: science, technology, and socialist feminism in the 1980s." *Socialist Review*, no. 80 (1985): 65–108.

Haraway, Donna. *Modest Witness@Second_Millennium. FemaleMan©Meets_OncoMouse: Feminism and Technoscience*. New York: Routledge, 1997.

Hayles, Kathrine. *How We Became Posthuman: Virtual Bodies in Cybernetics, Literature, and Informatics*. Chicago: The University of Chicago Press, 2008.

Harman, Graham. *Prince of Networks: Bruno Latour and Metaphysics*. Prahran: re.press, 2009.

Hemmings, Clare. "Affective solidarity: feminist reflexivity and political transformation." *Feminist Theory*, 13, no. 2 (2012): 147–61. doi: 10.1177/1464700112442643, 155.

Henriques, Julian. "The vibrations of affect and their propagation on a night out on Kingston's dancehall scene." *Body & Society*, 16, no. 1 (2010): 57–89, 57.

Hubbard, Phil. *Cities and Sexualities*. New York: Routledge, 2013.

Icsman, Marilyn. "7,000 pairs of shoes were outside U.S. Capitol to represent children killed by guns." *USA Today*, March 13, 2018. Retrieved from www.usatoday.com/story/news/politics/onpolitics/2018/03/13/7-000-pairs-shoes-were-outside-u-s-capitol-represent-children-killed-guns/421433002/.

Jenkins, Henry, Ford, Sam, and Green, Joshua. *Spreadable Media: Creating Value and Meaning in a Networked Culture*. New York: New York University Press, 2013.

Kirby, Vicki. "Natural convers(at)ions: Or, what if culture was really nature all along?" In *Material Feminisms*, eds Stacy Alaimo and Susan Hekman. Bloomington, IN: Indiana University Press, 2008: 214–236.

Khrebtan-Hörhager, Julia and Gordiyenko, Tanya. "Clash of cultural capitals in European mixed marriages." *Connections: European Studies Annual Review* (2012): 38–49.

Latour, Bruno. *The Pasteurization of France*. Cambridge, MA: Harvard University Press, 1993.

Latour, Bruno. "On actor-network theory: a few clarifications." *Soziale welt*, 47 (1996): 369–381.

Latour, Bruno. *Pandora's Hope: Essays on the Reality of Science Studies*. Cambridge, MA: Harvard University Press, 1999.

Latour, Bruno. *Reassembling the Social: An Introduction to Actor-Network Theory*. Oxford: Oxford University Press, 2005.

Lefebvre, Henri. *Production of Space*. Cambridge: Blackwell, 1991.

Le Monde. Movement of "Yellow Vests" [MOUVEMENT DES "GILETS JAUNES"] (n.d.). Retrieved from www.lemonde.fr/mouvement-des-gilets-jaunes/.

Livepress.ge. Stories told by women's shoes [ფეხსაცმლით მოყოლილი ქალთა ისტორიები] (March 8, 2018). Retrieved from www.livepress.ge/ka/akhali-ambebi/article/22658fekhsacmlithmoyoliliqa lthaistoriebifotovideo.html?fbclid=IwAR3jKGZn1VVeXxU5PKpigr6F4b4aQ2YDnZ%25E2%2580 %25A6, para. 4.

Manning, Erin. *Politics of Touch: Sense, Movement, Sovereignty*. Minneapolis, MN: University of Minnesota Press, 2007.

Massumi, Brian. "The autonomy of affect." *Cultural Critique*, 31 (1995): 83–109.

Massumi, Brian. *The Principle of Unrest*. London: Open Humanities Press, 2017.

Mitchell, W. J. T. *What Do Pictures Want?: The Lives and Loves of Images*. Chicago: The University of Chicago Press, 2005.

Papacharissi, Zizi. *Affective Publics: Sentiment, Technology, and Politics*. New York: Oxford University Press, 2015.

Pedwell, Carolyn and Whitehead, Anne. "Affecting feminism: questions of feeling in feminist theory." *Feminist Theory*, 13, no. 2 (2012): 115–29. doi: 10.1177/1464700112442635.

Peters, John Durham. *Speaking in to the Air: A History of the Idea of Communication*. Chicago: The University of Chicago Press, 1999.

Pratt, Mary Louise. "Arts of the contact zone." *Profession*, 91 (1991): 33–40.

Ruptly, "USA: 7,000 pairs of shoes cover Capitol lawn to protest gun violence," YouTube, March 13, 2018. Retrieved from www.youtube.com/watch?v=s7Mitu5FXno.

Sennett, Richard. *Flesh and Stone: The Body and the City in Western Civilization*. New York: W.W. Norton & Company, 1996.

Signorielli, Nancy and Bacue, Aaron. "Recognition and respect: a content analysis of prime-time television characters across three decades." *Sex Roles*, 40 (1999): 527–544.

The Local. Italy's 'small, silent protest' against femicide (June 2, 2016). Retrieved from www.thelocal.it/20160602/italys-small-silent-protest-against-femicide-sara-di-pietrantonio.

The Guardian. How will you mark 16 days of activism against gender abuse? Share your stories (November 25, 2016). Retrieved from www.theguardian.com/global-development/2016/nov/25/how-will-you-mark-16-days-of-activism-against-gender-abuse-share-your-stories.

PART II

Visualizing gender

Introduction to Part II

Visual communication, while not the only force shaping ideologies of gender and gendered practices, plays formative and often hegemonic roles in establishing and reinforcing identities, values, and practices. These communicative interactions promote particular views of gender and sex that can fuel misogyny and sexism, heterosexism and transphobia, racialized hypersexualization, Orientalist objectification, and many other types of discrimination and dehumanization. As the following studies show, a wide variety of forms of media offer new means of reinforcing or reinventing the scripts surrounding femininities, masculinities, and sexualities that have long concerned those producing feminist, queer, and critical-cultural analyses of communication.

Popular media representations model aesthetic and behavioral norms that denigrate and demonize particular subjects and practices while elevating and idolizing others. Social media provide dynamic circuits for conveying images that reinforce and/or respond to such norms while interacting with, excluding, or even generating specific audiences. Cinema and other forms of visual art create singular and moving encounters with the cultural aesthetics and social norms at work in performances and expressions of gender and sexuality. These are the texts and topics taken up by the authors, who contributed fresh insights to this section.

The chapters profiled as projects in "Visualizing gender" engage with key theories concerning how gender differences are depicted and viewed. Authors demonstrate careful applications of and enhancements to central concepts mined from rich scholarly literatures on "controlling images," "the gaze," "media literacy," and "counterpublics" to consider complex processes for stigmatization, shame, discipline, or erasure via visual media. Examples are examined with innovative and nuanced versions of intersectional, aesthetic, and historiographic methods to produce discoveries showcasing the potential for apprehending the complex ways that bodies and bodies of work create connections through media characters and engender pleasures (as well as pains) for media consumers. These are forceful cultural channels for communicating the price and promise of mediated genders and sexualities.

The first chapter, "Interrogating the awkward Black girl: beyond controlling images of Black women in televised comedies" by Kimberly R. Moffitt and Tammy Sanders Henderson, is a tropological and textual engagement with new depictions of young Black femininity in popular situation comedies. The authors relate early work on "controlling images" in representations of

Black women to contemporary programs that have earned significant fandoms and accolades. Along the way, Moffitt and Henderson explore the problems and possibilities of a character termed the "awkward Black girl" through a detailed study of how this figure has shifted over time. Readers are invited to consider complex ethical tensions between and within mass media performances of young women and blackness, and also find signs of hope that the entrenched power of televisual imagery may yet be marshaled to make room for self-invention.

Claire Sisco King takes up another influential concept in visual studies in the chapter that follows, "The male gaze in visual culture." King not only provides a detailed overview of the emergence and impact of Laura Mulvey's work on the objectifying and commodifying perspective that captures female forms in cinema for an idealized heterosexual and male viewer, she explains important critiques and extensions of gaze theory in a variety of arenas. In exploring such challenges and revisions, King draws particular attention to the valuable contributions of work in intersectional analysis and disability studies that has produced extensive literatures containing startling insights and ethical interventions into conversations about gender and visuality.

The third contribution to this section, "*Vida*: anti-colonial queer and feminist Web TV and the gaze of allyship" by Carolyn Elerding, builds on the momentum of the previous chapter by connecting gaze theory to colonialist visions of self and other. With skilled appreciation for apparatuses of production and reception, Elerding accounts for the digital labor of content creators and active viewers that is opening up de-colonialist vistas through visual media. Such openings can counter legacies of whiteness, nationalism, and classism, as this study of an online program illustrates. At the same time, Elerding models more reflective and critical analyses of the ethical entailments of being an ally, especially the need to actively decenter privileged perspectives and eschew cultural appropriation to support the work of social change movements.

The next chapter, "Body image and global media," tackles a life-threatening problem of international scope. Jasmine Fardouly, Vani Kakar, and Phillippa C. Diedrichs demonstrate the global reach of Western social media in promoting unattainable and unhealthy ideals of beauty with the potential to inflict tremendous pain and suffering. Carefully grounded in current research, this comprehensive study synthesizes existing knowledge about relations between media images and a host of hazards to physical and psychological well-being. The authors also evaluate different modes of intervention (such as media literacy) and provide guidelines for productive measures to take to combat this crisis.

Inverting the dynamic explored in the previous chapter, Marissa J. Doshi demonstrates what can happen when the flow of media influence is reversed—challenging body shaming with digital provocation. "Blood, bodies, and shame: Indian artists combating menstrual stigma on Instagram" details online artistic activism portraying menstruating bodies for viewers within and beyond networks comprising South Asian diasporas. This striking study in cyberfeminism shows how social media can convey and connect embodied experiences and initiate radical departures from norms of Indian culture, with Doshi delivering insights that are relevant for research into many other places where leaking bodies are shamed and stigmatized. Although the creative digital works featured in this chapter are credited with enjoining counterpublics, the author also asks incisive questions about how such social media campaigns may exclude or even endanger bodies marginalized by ciscentric and caste-based social systems.

Two detailed studies of provocative films follow the chapters above: Bernadette Marie Calafell's "Monstrous erasure: quare femme (in)visibility in *Get Out*" and J. Nautiyal's "Queer aesthetics, playful politics, and ethical masculinities in Luca Guadagnino's filmic adaptation of André Aciman's *Call Me by Your Name*." Each author draws on very different sources of queer theory and cultural analysis to investigate seminal cinematic statements that broke generic conventions.

Both Calafell and Nautiyal work to alter the way that readers view and experience these memorable masterpieces directed by Jordan Peele and Luca Guadagnino in 2017.

Despite ample attention to portrayals of Black masculinity in Jordan Peele's horror film set in white affluent suburbia, Calafell homes in on how Black women are represented through the (seemingly) minor character of Georgina. Making the most of unique discoveries that emerge when queer theory meets feminist analysis guided by Afrofuturism, Calafell plumbs the depths of a complex cinematic text to find resistive readings of Black identity, femme performance, and non-normative desire within a filmic narrative that disavows queerness and reinscribes masculine agency.

Nautiyal's lush and enthralling encounter with *Call Me by Your Name*'s controversial Italian love story not only revels in unlikely couplings of theory but also performs methods of reading film and writing scholarship that overturn rhetorical criticism's long reliance on masculine rationality and disembodied critique. By joining John Dewey's account of aesthetics with the social theory of Gilles Deleuze and Félix Guattari, Nautiyal plucks the "artful rhizome" out of this unexpected mingling of American pragmatism and French philosophical psychoanalysis. Bringing this new concept to a fleshly engagement with the senses and sensibilities driving Guadagnino's film, Nautiyal proves that wondrous queer aesthetics and desirable modes of masculinity may be handled in playful and promising ways.

The final chapter in this section reminds readers that many forms of art often overlooked in media studies are nonetheless powerful players in visual culture. In this engaging collection of close readings of artworks, Clare Johnson explicates the possibilities of feminist and queer art as a vehicle for activism. Encompassing an impressive array of important works by talented artists from different times and places, Johnson gathers creative works under shared themes, affects, and ethics—all without displacing specialized attention to each medium. "Feminist and queer arts activism" takes us on an intellectual and political journey that travels from 1970s collectives in Los Angeles to 1980s performances in London, and from Viennese billboards at the start of the 21st century back to internationally exhibited sculptures from the end of the 20th century. Throughout, Johnson refuses to be constrained by canonical interpretations and established chronologies, creating intimate affinities and tensions between artists who are well worth extensive critical attention: Tanja Ostojić, Hannah Wilke, Adrian Howells, Gillian Wearing, and Louise Bourgeois. However, it is this author's unique attention to radical ways of apprehending the emotional resonances of feminist and queer art that makes this chapter so striking.

The communication scholarship collected in "Visualizing gender" provides important insights about how gender is represented and viewed, created and resisted. The implications of this research are far-reaching, as authors argue for significant changes in the texts communication scholars study, the theories they draw upon, and the methods and assumptions that inform their interpretations and evaluations. At the very least, readers will leave these chapters on the visibilities and invisibilities of gender and sexuality with new outlooks on the harms and hopes that cultural images and collective imaginations offer diverse audiences.

As the chapters previewed above illustrate, mediated visuality is crucial in the formation and reformation of norms and hierarchies governing social life. However, it is worth noting that these studies also feature analyses of how mediums make use of senses other than sight in the appeals and experiences they enact. Audible laughter and awkward silences provide vital cues in Moffitt and Henderson's study of Black women in television comedies, while soundtracks and sound effects are central in the films examined by Thobani and Calafell. Visceral associations with feelings and smells are important in the social media described by Doshi, and textures and tastes emerge in the cinematic encounters detailed by Nautiyal. Johnson's consideration of multimedia artists also underscores the materiality of art and its rich multi-modal resources. While visual culture merits the critical attention of scholars investigating gender and sexuality, the host of bodily senses and sensations that can be represented, resourced, or even created through communication deserve further analysis and commentary in future projects.

In a section addressing visuality and gender, it is also crucial to emphasize how transgender scholarship and activism is challenging hegemonic ciscentric and transphobic ways of imaging and viewing bodies. Although this is a significant point underscored by Doshi's study and related work is featured in multiple chapters of this volume it deserves explicit mention here. We encourage readers to note the theme of trans bodies and trans theories in chapters from other sections that are also relevant to visual culture, and hope they will seek out research on these themes published by *Handbook* authors in other venues. These works also contain bibliographies that will be of use when entering this transnational and interdisciplinary conversation. It is likely that much more of this important research will be published soon to move the conversation about gender's in/visibilities forward.

7

INTERROGATING THE AWKWARD BLACK GIRL

Beyond controlling images of Black women in televised comedies

Kimberly R. Moffitt and Tammy Sanders Henderson

> I was frustrated with the black female characters I was seeing in the media and I realized that this 'awkward' character hadn't really been explored or portrayed in anything I'd ever seen.
>
> Issa Rae[1]

"The Misadventures of Awkward Black Girl" was not only a successful YouTube web series (2011–2013), but it gave us the language to critically examine this unique representation of Black women. It must be noted that the "awkward Black girl" (ABG) was here before Issa Rae's popular online show, yet largely ignored and rarely discussed in entertainment sectors and academic realms. The static limitations of broadcast media have now given way to media innovations such as social platforms (e.g., Twitter, Instagram, Facebook) and other user-generated content (UGC) venues (e.g., YouTube) that allow for more opportunity to engage complex representations for specific audiences. A fan favorite because of her quirky sense of humor, relatability, and vulnerability, ABG is arguably, the only Black woman trope able to challenge the more restrictive and dated "controlling images." Although the ABG trope has been with us for a while, this chapter will explore the evolution of this representation over the last three decades. We argue that ABG is an extension of what Patricia Hill Collins would identify as a "controlling image" and consider the full range of this representation, some of which may escape the limits of a controlling image.[2]

Wanzo suggests an ABG is a Black woman character who appears to struggle with her seemingly incongruous identities of Black and awkward.[3] In her assessment of J, Issa Rae's character from her web series and the HBO series *Insecure*, Wanzo remarked that J has an

> inability to inhabit a black "cool pose," an identity mostly associated with men but one that clearly resonates in the everyday performance culture of African American women as well, [and] was embraced by black geeks as a revolutionary media representation.[4]

This chapter will add to the literature on controlling images and Black feminist thought by specifically highlighting the ABG image of Black femininities in televised media or streaming services. This analysis will include an exploration of four Black women characters on Black-oriented sitcoms who, we argue, embody the Black and awkward representation. We will consider two characters from popular television shows of the 1980s and 1990s and two contemporary television series. Analyzing the evolution of this representation may allow us to transcend static controlling images and work "toward self-definition," as suggested by Collins.[5] Only time will tell if television's ABG can be used to help us "break free" and open up spaces for new and more complex representations.

Controlling images

In Patricia Hill Collins' groundbreaking work, *Black Feminist Thought: Knowledge, Consciousness, and The Politics of Empowerment*, she explains several key examples of what she identifies as controlling images. These first formulations include the mammy, "the faithful, obedient domestic servant," the matriarch who is held "responsible for the failures of African American families;" the welfare mother, the "updated breeder woman from slavery and exploiter of unearned social resources;" the Black lady who is "educated, assertive, and emasculating;" and the Jezebel who is simply a "whore."[6] These images would later evolve into other iterations from Collins' follow-up work entitled *Black Sexual Politics: African Americans, Gender, and the New Racism* (2004). She identifies these as the Bitch who is "aggressive, loud, rude;" the bad Black mother who is "promiscuous and lazy"; modern mammies who affirm "job first, marriage second"; and the educated Black bitch whose focus is on "money, power and good jobs."[7]

Traditionally, Black women have been captured through the lens of these controlling images.[8] Even televised comedies employ images to tell us familiar stories about Black women. Such narratives generally revolve around a Black woman being the butt of jokes, unattractive, and rarely the intellectual authority (e.g., Aunt Esther on *Sanford and Son*). Even a Black woman playing herself like comedian Leslie Jones on the sketch comedy show *Saturday Night Live*, may use her undesirability as part of her comedy routine.[9] In fact, there have been very few sitcoms in which Black women have been the lead or a central character, with the exception of shows like *Julia* (1968–1971), *227* (1985–1990), *Sister, Sister* (1994–1999), *Girlfriends* (2000–2008), and *Half and Half* (2002–2006), programs on which they were allowed to think, love, and laugh along with the audience.

Of course, none of these controlling images exists outside of the pervasive narrative of respectability politics. E. Frances White delves into the specific ways in which scientific racism, nationalism, and the lack of critique taking sexuality into account have been the undoing of a sustainable and liberatory Black feminist movement.[10] Such a movement, in turn, would allow us a more holistic understanding of Black women's identity. Toward this end, younger Black women scholars who embrace "third wave feminism" have used memoir with a theoretical basis to explain their relationship to these, at times, controlling/liberatory images. Joan Morgan adds to that discussion with her own unique relationship with the hip-hop movement despite its many problematic narratives of gendered violence and misogyny.[11] Scholars like Stephens and Phillips and French, Lewis, and Neville help us to understand the socialization of Black girls and women into sexual politics that they negotiate in reality.[12] Similarly, Wanzo tackles how these social scripts have come to make up newer imagery of Black women in popular culture.[13] Cunningham examines how new media forms like YouTube or UGC have been responsible for the popularity of contemporary representations like that of the awkward girl.[14] In an effort to further contextualize our conversation, we have chosen to utilize Black feminist thought here in hopes of giving voice to Black female spectators, including the two authors.

Black feminist thought

Black feminists have theorized over the decades about the creation and construction of controlling images that limit and damage our understandings of Black women's reality. Patricia Hill Collins' foundational text *Black Feminist Thought* and her follow-up work *Black Sexual Politics* explore these over-exaggerated caricatures of Black women's identity. Similarly, Janell Hobson and Melissa Harris-Perry both tackle the impact of those controlling images in mass media.[15] With a backdrop of Black feminist thought, these and other works help us to understand the historical development and impact of these media representations. It is clear that their far-reaching tentacles come to inform how individuals are socialized and ultimately maneuver through life, especially within the context of U.S. popular culture. Black feminist thought is a useful framework for interrogating these representations of Black women because it makes visible the invisible and gives voice to the voiceless while realigning the object as subject. All of these dynamics are inextricably tied to race and gender, as conveyed by Smith-Shomade.[16] In order to interpret the controlling representations of Black women, we must engage with Black feminist thought and Black feminist spectatorship, which situates us in the historical epistemological constructions of the necessary element of self-identification in media. Specifically, Black feminist spectatorship "productively destabilizes the dominant gaze as standard" and allows our own intersectionality to be centered.[17]

While there are similarities in how the two authors came to this project, there are also distinctions in our experiences as spectators who are also Black women that color our understanding of this "new" representation of the "ABG. One of us claims a womanist positionality that seems to align with the very women about which this chapter is written: she embodies aspects of the ABG. She is simultaneously privileged and marginalized as a heterosexual woman of color raised in a middle-class environment and, by way of education, has worked to ensure that space. Evidently, there have been experiences that suggest she has benefitted from classism, colorism, and sizeism, but she is also cognizant of marginalized spaces she occupies as one who chooses to wear her hair in a natural style (locs). Numerous instances in the academy and other traditional spaces have marked her as different and unwilling to conform to certain expectations around hairstyles. The other coauthor of this chapter enters this project as a Black woman who identifies as both feminist and womanist. Her roots within a marginalized racial group inform her perspective as a cis woman raised in a lower working-class southern community. At the same time, she acknowledges that her academic background has afforded her a degree of privilege. Like the ABG, she navigates the world with her hair in its natural curl pattern, which looks more like Issa's on *Insecure* than Freddie's on *A Different World*. All of these intersections endear the ABG to her, as she also teeters between multiple positions simultaneously.

In order to ascertain further understanding of the "ABG and examine its potential to move us beyond controlling images, we analyzed four television sitcoms: *A Different World* (1987–1993), *Living Single* (1993–1997), *Black-ish* (2014–), and *Insecure* (2016–), all featuring Black women characters who embody aspects of Wanzo's description of the ABG.[18] The following exploration of the ABG follows the work of scholars like Griffin, Guerrero, and Harris, who analyze texts and contexts of cultural production to garner means of empowerment for women of color.[19] We begin with a synopsis of each female character considered. The first and last episodes of each season (a total of 36) in which the shows aired were reviewed to add breadth and depth to the characters, aiding in our careful consideration of the ABG representation and its potential to move us beyond long-established and studied controlling images.

Enacting the awkward Black girl: character synopsis

Freddie

Situated as a spinoff of *The Cosby Show, A Different World* ran from 1987 to 1993. It is set on the fictional historically Black college (HBCU) campus of Hillman College where Denise Huxtable, daughter of Cliff and Clair Huxtable, studies. The television series explores student life at an HBCU. After the actress playing Denise, Lisa Bonet, left the cast, actress Cree Summer joined the show as Winifred "Freddie" Brooks, the roommate of students Jaleesa Vinson and Kimberly Reese. Freddie, a native of New Mexico, is introduced to us in "Dr. War Is Hell" in season 2, episode 1. Like her hippie (white) mother, Freddie is "spiritual, energetic, and compassionate … a sort of 'free spirit,' who is "politically conscious and often engages in environmental, community, and charitable activities."[20] Her biracial background is, at times, raised to highlight the complexities of race, while also giving voice to her lived experience. Freddie evolves in many ways during her Hillman years, but in the end remains committed to social justice issues in her goal to become a lawyer.

Synclaire James-Jones

The ensemble cast of *Living Single* included four single Black women living and striving in New York City to achieve their professional and personal goals. Played by comedian Kim Coles, Synclaire James-Jones is the "perpetually perky but not too bright" cousin of Khadijah James, who is the owner and operator of *Flavor* magazine.[21] The cast also included two Black male characters, one of whom becomes Synclaire's love interest and eventual husband, Overton Jones. Synclaire, while not the dominant character among the four women, played significant roles of unifier, collaborator, and consoler, as well as the "butt of the joke." Her naive persona was often mocked, but at the same time, it was appreciated because of its sincerity and the love it expressed.

Rainbow Johnson

A hit comedy that first aired in 2014 as part of ABC's Tuesday night lineup, *Black-ish* stars comedians Anthony Anderson and Tracee Ellis Ross as a happily (on occasion) married couple raising five children in an upper-middle-class suburb of Los Angeles. Tracee Ellis Ross's character, Rainbow ("Bow") Johnson is an anesthesiologist, while her husband Andre ("Dre") is an advertising executive. At the start of the series, the four children are school age (ranging from middle to high school) and Bow becomes pregnant with a fifth child in season 3. Andre's divorced parents live with the family as well. She is described as "kind, funny, selfless and nice—usually, the person to put Dre and his craziness in his place."[22] Bow is also a biracial character with hippie parents (a Black mother and a white father). Her racial identity is often raised in episodes to explore the question of "what is Black?" and she is, at times, the "butt of the joke" because of her mixed-race identity.

Issa Dee

The half-hour comedy *Insecure* aired on the premium channel HBO in 2016, starring Issa Rae in the leading role with the same first name. With an Inglewood, California, backdrop, Issa on *Insecure* is socially awkward, having only a few close friends. She is single after having been in a five-year relationship with her boyfriend, Lawrence. Issa is college-educated, works for a non-profit organization, and often struggles financially. She is crafted as the average millennial, with average 30-something financial, employment, and dating issues. *Insecure* revels in its ability to capture through representation the regular lives of young, up-and-coming Black Americans as they transition through adulthood, just as we see their neighborhoods gentrifying.

Awkward Black girl themes

In our analysis of 36 television episodes, we developed the following themes to capture what are considered key elements of the ABG as illustrated by Freddie, Synclaire, Rainbow, and Issa. Facial expressions, quirky responses, sexual insecurity/naivete, being "schooled" on Black culture, and Black hair are all recurring themes. Although these may not be exhaustive, they are certainly consistent across the four Black women characters selected for this analysis.

Facial expressions

Undoubtedly, one theme of the ABG that extends across time is her use of comedic facial expressions. A telltale sign of the ABG character is her unique ability to visually convey her lack of understanding and all-around bewilderment at what others would consider commonsensical observations. Synclaire James (*Living Single*), played by actress Kim Coles, is a stand-up comic who generally uses her expressive face in her routines. It is quite common to see her large, round eyes broaden when she is startled, expressing concern, or adding emphasis to her declarations. In the "Judging by the Cover" episode, her bright eyes widen when she thinks her love interest, Overton, is going to touch a wire that may bring him harm.[23] In "The Shake-Up," Synclaire is again wide-eyed as she puffs on several cigarettes simultaneously when distraught over roommate Regine's decision to move out of their apartment.[24] Her facial expressions enable us to experience her bewilderment, shock, and even concern for those around her. Although Freddie does not use a wide-eyed expression often, she does use her eyes to convey that she is perplexed by a scenario in which she is unfamiliar or unsure. For example, in "Strangers on a Plane," Freddie remarks upon returning to Hillman from summer break that "Everyone in this school has changed but me."[25] Although she widens her eyes, she also uses them to sulk and convey her "outsider" status. Her eyes inform us, nonverbally, that she does not feel included or fit in, and that her status in the group is indeed marked as different and unusual. This is even more obvious in "There's no Place like Home" when many friends gather to say goodbye before departing for the summer. Freddie, while in the midst of the group, looks uncomfortable and downcast as the group takes pictures without her.[26] Even her exclamation that they will write to each other during the summer is dismissed and ignored, leaving Freddie with a perplexed facial expression and disappointed eyes.

Bow Johnson (*Black-ish*), played by actress Tracee Ellis Ross (who has naturally large almond-shaped eyes), is seen in several episodes with her enlarged eyes adding emphasis to her moments of disbelief and surprise. Bow effectively uses her eyes to stare directly at other characters, particularly during conflict: it is the signal used to convey she is ready for a fight. On the other hand, Issa's (*Insecure*) facial expressions most often reveal vulnerability and insecurity. Issa shies away from looking directly at people during uncomfortable moments. She tilts her head down and then looks up timidly, reminding the audience of her insecurity. She may even pout. The only time it is difficult to read Issa's emotions, which are typically revealed on her face, is when she interacts with her coworkers. Her face becomes steely, eyes vacuous, and mouth taut during these microaggressive or outright racist exchanges, as if she is willing herself to be as mentally and emotionally distanced as possible.

Quirky responses

The ABGs on these programs offer responses that often startle those around them because they are interpreted as weird, unusual, and at times unnecessary. The odd responses, which some may read as ridiculous or even stupid, make the characters quite lovable because their sincerity is conveyed to the audience.

Synclaire is most known for her quirky responses to a variety of social situations. Her "woo woo woo" expression became synonymous with her character and was often used to express her emotional reaction to any situation when she seemed stumped. More than a situation comedy catchphrase, actress Kim Coles shared that her mother often used this expression with her as a child when her mother seemed at a loss for words; yet the phrase could be used to show love and comfort or even offer distraction.[27] It became a part of the show as an impromptu, ad-lib moment while on the set that appeared to fit Synclaire's character and hence remained. It was the magical expression that seemed to make everything, and everyone, feel better, even though it was an odd, unexpected, and almost childish utterance. While "woo woo woo" is not used in every episode of the show, it is clearly a core aspect of Synclaire's identity and helps situate her as the "original awkward Black Girl," according to Uwumarogie.[28] Her responses come to be expected by all, even her love interest, Overton Jones. In "Come Back Little Diva" Overton says, "Sometimes I love your wide-eyed innocence," as he watches Synclaire put her cousin Khadijah's underwear over her eye as an eye patch to pretend to be a pirate.[29] Synclaire's response to this, "And sometimes you wish I'd shut up?" shows that she recognizes her quirkiness as well. Nonetheless, she seems comfortable in her own skin.

On the other hand, Freddie is not as comfortable in her skin. She deliberately chose to place herself in an environment foreign to her by attending Hillman College, which features a majority-Black student population. She awkwardly maneuvers through her college days fighting for social justice issues, but not quite certain of the "right" thing to say. For example, when bestowing a crystal on a chain to Whitley as a graduation gift, she informs her, "As you know I'm not very materialistic, but I got this for you. It's really powerful … very spiritual. It will give you so much powerful energy. Also, on the subway, it doubles as a blunt instrument."[30] Although this remark is played for laughs by the naïve southwestern girl, it also highlights Freddie's insinuation that Whitley will need to protect herself while on the New York subway, so this meaningful gift also has practical usage as well. And during an argument with her initial college nemesis, Ron, she retorts, "I am rubber. You are glue. Everything you say bounces off of me and sticks to you."[31] The remark, reminiscent of childhood play, causes Ron to raise his eyebrows, perplexed by another young adult using this expression during an argument.

In *Insecure*, Issa's quirky responses are often due to her inability to grasp social cues. In "Hella Great," the show opens with Issa and her white coworker giving a presentation to a class of Black and Brown middle-schoolers, introducing them to the tutoring program sponsored by Issa's non-profit job, "We Got Y'all."[32] The middle-schoolers quickly begin to question Issa's way of speaking, her outfit, her natural hairstyle, and even her romantic life. When questioned by a Black female student about why she talks like a white girl, Issa awkwardly responds, "You got me. I'm rocking blackface," and in turn, her white coworker and the student's Asian teacher look her way with uneasy smiles displayed on their faces. Issa, too, is flustered and her facial expressions let the audience know that she is in over her head when up against the more streetwise youth. In this scene, she looks embarrassed and begins to overshare with the students details about her age, why she chose her outfit, and even her relationship status. In "Familiar Like"[33] Issa constantly finds herself in conversations that she seems unable to navigate. These moments are funny, yet uncomfortable. For example, when she and her beau, Daniel, are meeting a celebrity rapper, Issa begins to make nonsensical comments that leave the people around her looking puzzled.

Like Issa, Bow (*Black-ish*) is socially awkward, but in order to downplay her inability to connect, she uses her position as a medical doctor to avoid developing friendships with the neighborhood moms. In a tense exchange with another mom at a pool party, Bow rambles on about how her job will not allow her to be more involved. In response to the other mother's questioning of Bow's absence, she asserts, "I was late for a funeral." The neighbor reminds her that she said she had to get to work. Bow replies that she had to get to a "work funeral … That's what happens when you don't have the life/work balance. You die at work."[34] Bow's awkwardness in social settings even includes Dre and his friends. Dre and his best friend Gigi (played by Tyra Banks) relate in ways that Bow finds difficult to share. Her quirky responses in these instances illustrate how out of step she is in these moments. For example, Gigi, Dre, and Bow are all watching a basketball game on television and while Gigi and Dre celebrate a play with hand slaps, Bow jumps to her feet and yells out "Yeah sports!" and actually gives herself a hand slap, prompting Gigi and Dre to give her an uncomfortable look.

Sexual insecurity/naivete

In addition to their quirky retorts, these four characters also exhibit an awkwardness around issues of sex and intimacy. While they all are in relationships and even married at various times, the manner in which they engage in discussions about sex and how they navigate their sexual relationships appears unnatural or naive.

Synclaire (*Living Single*) was enamored with neighbor and local handyman, Overton James. She seemed comfortable in expressing that love but was rarely cast as a sexual being openly exhibiting her sexual appeal on screen. She often served as the moral barometer for her roommates and friends. At the end of season 1, she raises doubts for Khadijah (who is considering a proposal to move in with her boyfriend), saying "Sure there's no beautiful white dress, big cake, and God doesn't recognize it, but still."[35] Or, when Khadijah's magazine, *Flavor*, features a cover story with the title, "Dating Married Men: The Sleeping Dog That Lies," Synclaire seems confused by the title and unsure of what it means, even though the episode focused on Regine's discovery that her prince charming was in fact a married man.[36] At the same time, Synclaire seems to present herself as knowledgeable about men, even though this is often read as naive. Examples of this include comments such as "Girl, it's raining men!" when Khadijah receives gifts from two men or when she quips at the end of an episode, "Did you ever stop to think about what the world would be with no men?"[37]

Freddie (*A Different World*) openly acknowledges her insecurity and naivete around sex. She is the one who admits her lack of sexual awareness when she confesses to Ron, "My parents are right. I'm all screwed up … [I'm] still waiting on my Prince Charming."[38] Even upon arriving at Hillman, Freddie finds herself swooning over Dwayne Wayne, a fellow student. After barely exchanging words with him, she asks her roommate Jalessa, "Do you think Dwayne wants children?"[39] Seasons later in "Everything Must Change,"[40] Whitley shares that Freddie's interest in Dwayne Wayne is futile because "Let's just say Dwayne brought back a souvenir that's not that easy to get rid of." To which Freddie asks, "You mean like a disease?" Dwayne had, in fact, returned from his trip to Japan with a new girlfriend. These responses are quite frequent for Freddie and only diminish toward the end of the television series when she finally engages in sexual encounters with Shazza, her boyfriend, and later, with Ron.

In "Hella Open" Issa (*Insecure*) is inept at asking for sex from men.[41] On two occasions throughout the episode, Issa fumbles with her words and her hands (when attempting to undress), laughs uncomfortably when she is being touched by her date, and inadvertently insults him. She goes to leave, and stumbles through the apartment looking for the exit. Then in "Hella Blows," Issa is working through her "hoetation," (single, dating, and choosing to have sex with whomever) even though she is now ready to be more assertive in asking for what she wants sexually.[42] Again, these efforts fall short, and in the end she and her on-again, off-again lover and friend,

Daniel, have an incredibly awkward sexual encounter when he unceremoniously ejaculates in her eye after she provides him with oral sex. Issa leaves embarrassed by what has just occurred but also with the realization that Daniel has, in fact, treated her like a "hoe."

"Schooled" on Black culture

The ABGs featured here represent characters who are biracial and characters who are not. Yet, a feature shared by them all is their inability to navigate an understanding of Black culture easily. Representing different regions of the country, each of these characters seem to struggle with nuances of Black culture, or even when they "perform" them, they are still challenged and/ or critiqued. In "What's Next?"[43] Synclaire suggests taking Overton to an "Avant Garde 70s film festival" to which her friends, without words, convey the ridiculousness of such an outing. Although subtle, it becomes clear to Synclaire that the group thinks she is odd and not investing in activities that align with what they consider Black culture.

Freddie (*A Different World*) is often at the center of such discussions about Black culture. Her friends appear to use her biracial background as a means of understanding her lack of knowledge, but it could also be because she was raised in New Mexico around a less diverse set of friends. After returning to school from summer vacation, Freddie, in deep thought, reflects, "I need to do something with my life. Maybe I'll take more courses. No, I'll write a book … [M]aybe the Marines. I could work with kids. I could join the Peace Corps or Rainbow Coalition. I met Jesse."[44] After the remark about Jesse (Jackson), Whitley walks away with her hands thrown up in the air to signal the hopelessness of Freddie's scatterbrained ideas, as well as her presumed connection to the Rainbow Coalition by casually meeting its founder Jesse Jackson. The next season when Freddie returns from summer vacation, she is heard telling Jaleesa, "a word of caution … stay Black." Visibly shocked by Freddie's new knowledge, Jaleesa responds, "I see someone discovered Malcolm X this summer."[45] These quips again suggest that Freddie is not and *cannot even become* knowledgeable about the Black experience because of her own background. While this serves as comic relief in many instances on the television series, it also raises questions about how her character evolves and matures as a woman attending an HBCU.

Several episodes of *Black-ish* feature Bow's husband, Dre, and/or mother-in-law, Ruby, schooling her on her lack of cultural understanding. In "Hope" the subject of police brutality is tackled.[46] Bow is "schooled" on the lack of justice for Black people who are forced to engage with the criminal justice system. The pair provides specific information to her to reveal the prevalence of unarmed Black people being killed by police and why Black people are not hopeful that indictments will be made on their behalf when they are victimized. Similarly, in an attempt to explain women's rights in the episode, "Johnson and Johnson," Bow reveals to her husband that she did not take his last name once they were married.[47] The humor in this 22-minute revelation revolves around the fact that both Bow and Dre already shared the same last name. Yet Bow makes it clear to her husband that as a self-identified feminist, she never planned to take her husband's last name. He explains to her what it means to a Black man for a wife to take her husband's name. While she continues to believe her stance was right and just, she comes to understand why his position is so important to him.

Issa, too, needs to be "educated" about dating and relationships from her male and female friends. In the pilot episode, while having dinner with her best friend, Molly, Issa seems to be oblivious to the dating challenges facing Black women: "If I have sex on the first date, they leave. If I wait to have sex, they say 'I didn't think you were interested.' If I don't have sex, [then it's] 'No! I didn't sign up for that shit.'"[48] Issa listens but still seems clueless about the reality faced by some Black women. This is even more evident in another episode where Issa makes fun of

Molly's dating situation by doing an impromptu rap she calls "Broken Pussy." Similarly, in the pilot episode, the Black middle schoolers who questioned and teased Issa about her hair, her clothing, and her speech were highlighting her inability to be culturally aware of what is acceptable in Black culture. In fact, she is "schooled" by one of the teenage girls who suggests she put tracks (hair extensions) in her hair, implying that weaves are more culturally acceptable than a *TWA* (teeny weenie afro).

Black hair

A major aspect of the ABG is not her mannerisms or behaviors, but how she styles her hair. Kobena Mercer calls hair "the most tangible sign of racial difference."[49] It also "functions as a key ethnic signifier because compared with bodily shape or facial features, it can be changed more easily."[50] Synclaire, Freddie, Rainbow, and Issa all have natural tresses. Their hair, although differently textured, does not appear to be chemically altered/treated, a process used to straighten Black women's hair. Synclaire and Issa both often wear braided hairstyles, while Freddie and Rainbow allow their less tightly coiled ringlets to frame their faces. Freddie and Rainbow flaunt their shoulder-length curls and stand out from the other female characters on their shows, whose hairstyles reflect fashionable wigs, weaves, and extensions. All except for Freddie have worn braided hairstyles reflecting contemporary trends like small individual braids, goddess braids, or cornrows during their appearances on their respective television series. For both Freddie and Rainbow, who are played by biracial actresses, there have been episodes that focus specifically on their hair being straightened and how this superficial change alters their acceptance into their respective communities. Although ABGs do not have to be adorned by natural hairstyles to be included in the representation, their aesthetics—whether in hair or clothing—have traditionally been makers of difference and just the decision to wear particular hairstyles becomes an "uber-political" statement.[51]

Hair becomes a signifier to those who interact with these characters and affects the manner in which they are perceived and treated. Freddie is called "Raggedy Ann" by Whitley, which is a direct reference to her busy, natural curls beaming with streaks of light golden colors that call to mind images of the stuffed toy doll.[52] And classmate Ron, who later becomes Freddie's beau, is heard denigrating the look of her hair further when in anger he refers to her as a "mismatched, Rainbow Coalition, Medusa-looking, freckled-face mutt."[53] Those remarks clearly characterize Freddie's skin hue and hair in specific ways, none of which are affirming. It is only when Freddie chooses more conservative hair and clothing styles that Ron acknowledges her in a more positive light in "Goodbye Tracy Chapman. Hello, Barbara Jordan."[54] The title of this episode is important to note, as it clearly aligns with the perceived evolution of Freddie from a more free-spirited "flower child" to a persona like the woman whose political acumen made her a force to be reckoned with in the U.S. Congress. Bow, too, finds herself only accepted by the white mothers in her neighborhood when she alters her hairstyle to a straightened look that lengthens her hair significantly.[55]

Issa, on the other hand, is the only one out of the four characters who wears a short and tightly coiled natural hairstyle. Audiences may watch *Insecure* as much for Issa's hair as they do any other part of her awkwardness.[56] By using Issa's natural hair pattern and casting an actress who is a darker-complected Black woman, a space is created for a certain type of ABG representation to exist that had not been seen often in prior American popular sitcoms. Hair is seemingly not an issue for the show or for Issa, even when comments are raised by white coworkers. Her hair is just that … hers and for her own consumption, suggesting a new and potentially resistive representation of Black women.

These five themes were most apparent across the four Black women characters considered for this exploration. There are additional distinct complexities to these depictions that have not been explored here. However, as the ABG evolves in media, other aspects of this representation will become more apparent and offer additional interpretations of the characters embodying it.

Conclusion

Through a detailed analysis of ABG televisual depictions, this chapter attempted to shed light on a phenomenon long accepted, yet rarely interrogated. The controlling images of mammy, matriarch, welfare mother, Black lady, bitch, bad Black mother, and educated Black bitch are widely known and perpetuated in various media forms.[57] Yet, in the 21st century, we have an opportunity to reflect on those tropes and consider new or different ones. With the advent of media innovations that enable audiences to garner their media content in a variety of venues, there may, in fact, be a new way of viewing the essence and complexity of Black women in popular culture. This chapter considered the roles of four Black women characters who embody aspects of Wanzo's understanding of an ABG as both Black and awkward, and explored ways in which this representation existed prior to Issa Rae's web series of the same name, evolving into its current formation.[58]

While acknowledging that these characters share the five themes of facial expressions, quirky responses, sexual insecurity/naivete, needing to be "schooled" on Black culture, and distinctive Black hairstyles, we also note that the evolution of the ABG has grown through the popular characters of Freddie and Synclaire and now Rainbow (*Black-ish*) and Issa (*Insecure*), marking a noticeable shift in the construction of Black women in comedy for media programming. None of these characters have been constructed to fit neatly into past traditional controlling images such as those long established and studied. For example, Bow's character does not depart from the burdens of "respectability politics" with attempts to prove to her white neighbors that her family is financially secure and can afford to live in their neighborhood or teaching her children how to eat breakfast properly when out in public. The audience is constantly reminded that for as much as she tries to fit in, her attempts often fail in private and public. Whether it is her children's discreet disdain of her presence in their lives or the opinions of the neighbors constantly reminding her of her inadequate attempts at juggling both mothering and a high-pressure career, Rainbow is no Claire Huxtable. But she is not trying to be … she has flaws and is embracing them.

Although Issa is the only millennial represented in this examination of ABG examples, it is this character that has spawned a new conversation about depicting Black people as they are: simply people. Specifically, Issa's aesthetics would traditionally have placed her along the spectrum of problematic controlling images of Black women. For instance, as recently as 2017, the *Straight Outta Compton* casting call identified "poor, out of shape and medium to dark skin" Black women as "D Girls," further illustrating this representation of both hair and a particular skin hue as degrading.[59] Yet, Issa's character, more so than the others, is what provides hope that representations of Black femininity on the small screen (and potentially the big screen as well) can transcend controlling images that continue to limit depictions of the range and complexity of Black women's lives and experiences.

Along the same lines, although Collins does not emphasize colorism in her exploration of controlling images, we acknowledge that the construction of the ABG may be impacted by past tropes surrounding skin hue and hair texture discussed by other scholars.[60] In the past, skin color was used to define Black women's desirability, likability, and even loyalty to Black culture typically under the guise of the trope of the "tragic mulatto." Donald Bogle expounds upon this in his seminal text, *Toms, Coons, Mulattoes, Mammies, and Bucks: An Interpretive History of Blacks in*

American Films, suggesting that "the mulatto is made likable—even sympathetic...."[61] Freddie (*A Different World*), Synclaire (*Living Single*), and Rainbow (*Black-ish*) are all lighter complected Black women, signaling a possible trend in the construction of these awkward characters. Were the writers suggesting an inherent *misfitting* due to biracialism or light skin? Is this a marker of awkwardness? Were they pulling on old stereotypes because of presumed audience recognition? The purpose of this examination was not to argue that any of this is the case, only to analyze the important similarities and disparities between recent ABGs on television.

What we do know is that the ABG leaves room for new interpretations of Black femininity as depicted on television and film. Only time will tell if this representation will move us past controlling images in significant ways, but with the new opportunities for Issa Rae to spearhead other HBO projects and Shonda Rhimes to join the streaming giant Netflix to work on new projects, there is a good chance that change is on the horizon. Because of the success of the YouTube series "The Misadventures of Awkward Black Girl," we have new language to critique past and present representations of Black women that had largely been ignored critically. ABG helps to push the boundaries set by long-standing controlling images because this newfound visibility is due in part to how Issa Rae reenvisioned a representation that had not been centered previously, enabling audiences to see her full humanity as both awkward and Black.

Notes

1 Phillip Lamarr Cunningham, *Get a Crew ...And Make It Happen: The Misadventures of Awkward Black Girl and New Media's Potential for Self-Definition* (Santa Barbara, CA: Praeger Publishers, 2013).
2 Patricia Hill Collins, *Controlling Images and Black Women's Oppression* in *Seeing Ourselves: Classic, Contemporary and Cross-Cultural Readings in Sociology*, eds John J. Macionis and Nijole V. Benokraitis (New York: Pearson, 2007).
3 Rebecca Wanzo, "Precarious-Girl Comedy: Issa Rae, Lena Dunham, and Abjection Aesthetics," *Camera Obscura: Feminism, Culture and Media Studies* 31, no. 2 (2016): 27–28.
4 Wanzo, "Precarious-Girl Comedy."
5 Patricia Hill Collins, *Black Feminist Thought: Knowledge, Consciousness, and the Politics of Empowerment* (Boston, MA: Unwin Hyman, 1990), 72–81.
6 Collins, *Black Feminist Thought.*
7 Patricia Hill Collins, *Black Sexual Politics: African Americans, Gender, and the New Racism* (New York: Routledge, 2004), 123–148.
8 Collins, *Black Sexual Politics.*
9 *Saturday Night Live,* season 39, episode 14, "Andrew Garfield/Coldplay," directed by Don Roy King (aired May 3, 2014 on NBC, New York: NY: Studio 8H).
10 E. Frances White, *Dark Continent of Our Bodies: Black Feminism & Politics of Respectability* (Philadelphia, PA: Temple University Press, 2001).
11 Joan Morgan, *When Chickenheads Come Home to Roost: A Hip-Hop Feminist Breaks It Down* (New York, NY: Simon and Schuster, 2000).
12 Dionne P. Stephens and Layli Phillips, "Integrating Black Feminist Thought into Conceptual Frameworks of African American Adolescent Women's Sexual Scripting Processes," *Sexualities, Evolution & Gender* 7, no. 1 (2005): 37–55; Bryana H. French, Jioni A. Lewis, and Helen A. Neville, "Naming and Reclaiming: An Interdisciplinary Analysis of Black Girls' and Women's Resistance Strategies," *Journal of African American Studies* 17, no. 1 (2013): 1–6.
13 Wanzo, "Precarious-Girl Comedy."
14 Cunningham, *Get a Crew ...And Make It Happen.*
15 Janell Hobson, *Venus in the Dark: Blackness and Beauty in Popular Culture* (New York: Routledge, 2018); Melissa V. Harris-Perry, *Sister Citizen: Shame, Stereotypes, and Black Women in America* (New Haven, CT: Yale University Press, 2013).
16 Beretta E. Smith-Shomade, *Shaded Lives: African-American Women and Television* (Rutgers, NJ: Rutgers University Press, 2002).

17 Rachel A. Griffin, "Push-ing into Precious: Black Women, Media Representation, and the Glare of the White Supremacist Capitalist Patriarchal Gaze," *Critical Studies in Media Communication* 31, no. 3 (2014): 184.

18 Wanzo, "Precarious-Girl Comedy."

19 Griffin, "Push-ing into Precious," 184; Lisa A. Guerrero, "Single Black Female: Representing the Modern Black Woman in Living Single," in *African Americans on Television: Race-ing for Ratings*, eds David Leonard and Lisa Guerrero (Santa Barbara, CA: Praeger, 2013), 177–90; Tina M. Harris, "Interrogating the Representation of African American Female Identity in the Films Waiting to Exhale and Set It Off," in *African American Communication and Identities: Essential Readings*, ed. Ronald L. Jackson (Thousand Oaks, CA: Sage, 2004), 189–98.

20 Fandom, "Freddie Brooks," accessed May 1, 2019, https://differentworld.fandom.com/wiki/Freddie_Brooks.

21 Fandom, "Living Single," accessed May 1, 2019, https://livingsingle.fandom.com/wiki/Living_Single_Wiki.

22 Fandom, "Rainbow Johnson," accessed May 1, 2019, https://blackish.fandom.com/wiki/Rainbow_Johnson.

23 *Living Single*, season 1, episode 1, "Judging by the Cover," directed by Tony Singletary (aired August 22, 1993 on Fox Network, Burbank, CA: Warner Brothers, 1995), 22 min.

24 *Living* Single, season 2, episode 27, "The Shake-Up," directed by Ellen Falcon (aired August 22, 1993 on Fox Network, Burbank, CA: Warner Brothers), 22 min.

25 *A Different World*, season 3, episode 1, "Strangers on a Plane," directed by Debbie Allen (aired September 28, 1989 on NBC, Universal City, CA: Universal), 22 min.

26 *A Different World*, season 2, episode 22, "There's No Place like Home," directed by Debbie Allen (aired May 4, 1989 on NBC, Universal City, CA: Universal), 22 min.

27 Saaret E. Yoseph, "Glad I Got My Girls," *The Root,* accessed May 1, 2019, www.theroot.com/glad-i-got-my-girls-1790900202.

28 Victoria Uwumarogie, "Synclaire James was the Original Awkward Black Girl," *Madame Noire*, April 23, 2018, https://madamenoire.com/1037319/living-single-synclaire/.

29 *Living Single*, season 3, episode 1, "Come Back Little Diva," directed by Tony Singletary (aired August 31, 1995 on Fox, Burbank, CA: Fox Network), 22 min.

30 *A Different World*, season 4, episode 25, "To Be Continued," directed by Michael Peters (aired May 2, 1991 Universal City, CA: National Broadcasting Company), 22 min.

31 *A Different World*, season 5, episode 5, "In the Eye of the Storm," directed by Debbie Allen (aired October 17, 1991 on NBC, Universal City, CA: Universal), 22 min.

32 *Insecure*, season 2, episode 1, "Hella Great," directed by Melina Matsoukas (aired July 23, 2017 on HBO, Los Angeles: Home Box Office), 31 min.

33 *Insecure*, season 3, episode 2, "Familiar-Like," directed by Pete Chatmon (aired August 19, 2018 on HBO, Los Angeles: Home Box Office), 31 min.

34 *Black-ish*, season 2, episode 14, "Sink or Swim," directed by Michael Schultz (aired February 10, 2016 on ABC, Pasadena, CA: ABC), 22 min.

35 *Living Single*, season 1, episode 27, "What's Next?," directed by Ellen Falcon (aired May 15, 1993 on Fox Network, Burbank, CA: Warner Brothers), 22 min.

36 *Living Single*, season 1, episode 1, "Judging by the Cover."

37 *Living Single*, season 1, episode 27, "What's Next?"

38 *A Different World*, season 5, episode 5, "In the Eye of the Storm."

39 *A Different World*, season 2, episode 1, "Dr. War Is Hell," directed by Debbie Allen (aired October 6, 1988 on NBC, Universal City, CA: Universal), 22 min.

40 *A Different World*, season 4, episode 1, "Everything Must Change," directed by Debbie Allen (aired September 20, 1990 on NBC, Universal City, CA: Universal), 22 min.

41 *Insecure*, season 2, episode 3, "Hella Open," directed by Marta Cunningham (aired August 6, 2017 on HBO, Los Angeles: Home Box Office,) 31 min.

42 *Insecure*, season 2, episode 6, "Hella Blows," directed by Pete Chatmon (aired August 27, 2017 on HBO, Los Angeles: Home Box Office), 31 min.

43 *Living Single*, season 1, episode 27, "What's Next?"

44 *A Different World*, season 3, episode 1, "Strangers on a Plane."

45 *A Different World*, season 4, episode 2, "Everything Must Change."

46 *Black-ish*, season 2, episode 6, "Hope," directed by Beth McCarthy-Miller (aired February 22, 2016 on ABC, Pasadena, CA: ABC), 22 min.

47 *Black-ish*, season 2, episode 20, "Johnson and Johnson," directed by Rob Hardy (aired April 13, 2016 on NBC, Pasadena, CA: American Broadcasting Company, 2016), 22 min.

48 *Insecure*, season 1, episode 1, "Insecure as Fuck," directed by Melina Matsoukas (aired October 9, 2016 on HBO, Los Angeles, CA: Home Box Office, 2016), 22 min.

49 Kobena Mercer, "Black Hair/Style Politics," in *Welcome to the Jungle: New Positions in Black Cultural Studies* (New York: Routledge, 2013), 97–128.

50 Mercer, 98.

51 Kimberly R. Moffitt and Tunisia Lumpkin, "The Tangled Weave We Wear: It Still Matters," *Media Res*, accessed January 10, 2019, /imr/2018/02/17/tangled-weave-we-wear-it-still-matters.

52 *A Different World*, season 2, episode 22, "There's No Place like Home."

53 *A Different World*, season 5, episode 5, "In the Eye of the Storm."

54 *A Different World*, season 6, episode 1, "Honeymoon in L.A. (Part 1)," directed by Debbie Allen (aired September 24, 1992 on NBC, Universal City, CA: Universal), 22 min.

55 *Black-ish*, season 4, episode 13, "Unkept Woman," directed by Pete Chatmon (aired Feb 6, 2018 on ABC, Pasadena, CA: American Broadcasting Company, 2018), 21 min.

56 Jessica Wilkins, "Issa Rae proves why she's the Queen of 4C natural hair on "Insecure," *Essence*, accessed May 1, 2019, www.essence.com/hair/issa-rae-insecure-natural-4c-hair/.

57 Donald Bogle, *Toms, Coons, Mulattoes, Mammies, and Bucks: An Interpretive History of Blacks in American Film* (New York, NY: Bloomsbury Publishing, 2016); Collins, *Black Sexual Politics*; Griffin, "Push-ing into Precious"; Harris, "Interrogating the Representation"; Harris-Perry, *Sister Citizen*.

58 Wanzo, "Precarious-Girl Comedy."

59 Danielle Cadet, "The 'Straight Outta Compton' Casting Call Is So Offensive It Will Make Your Jaw Drop," *HuffPost*, July 17, 2014, www.huffpost.com/entry/straight-out-of-compton-casting-call_n_5597010.

60 Collins, *Black Sexual Politics*; Ingrid Banks, *Hair Matters: Beauty, Power, and Black Women's Consciousness* (New York: New York University Press, 2000); Regina E. Spellers and Kimberly R. Moffitt, *Blackberries and Redbones: Critical Articulations of Black Hair/Body Politics in Africana Communities* (New York, NY: Hampton Press, 2010); Marita Golden, *Don't Play in the Sun: One Woman's Journey through the Color Complex* (New York: Anchor Books, 2007).

61 Bogle, *Toms, Coons, Mulattoes, Mammies, and Bucks.*

Bibliography

Allen, Debbie, dir. *A Different World*, season 2, episode 1, "Dr. War Is Hell." Aired October 6, 1988 on NBC, Universal City, CA: Universal.

Allen, Debbie, dir. *A Different World*, season 4, episode 1, "Everything Must Change." Aired September 20, 1990 on NBC, Universal City, CA: Universal.

Allen, Debbie, dir. *A Different World*, season 6, episode 1, "Honeymoon in L.A. (Part 1)." Aired September 24, 1992 on NBC, Universal City, CA: Universal.

Allen, Debbie, dir. *A Different World*, season 5, episode 5, "In the Eye of the Storm." Aired October 17, 1991 on NBC, Universal City, CA: Universal.

Allen, Debbie, dir. *A Different World*, season 3, episode 1, "Strangers on a Plane." Aired September 28, 1989 on NBC, Universal City, CA: Universal.

Allen, Debbie, dir. *A Different World*, season 2, episode 22, "There's No Place like Home." Aired May 4, 1989 on NBC, Universal City, CA: Universal.

Banks, Ingrid. *Hair Matters: Beauty, Power, and Black Women's Consciousness*. New York, NY: New York University Press, 2000.

Bogle, Donald. *Toms, Coons, Mulattoes, Mammies, and Bucks: An Interpretive History of Blacks in American Films*. New York, NY: Bloomsbury Publishing, 2016.

Cadet, Danielle. "The 'Straight Outta Compton' Casting Call Is so Offensive It Will Make Your Jaw Drop." *HuffPost*, Last modified July 17, 2014. www.huffpost.com/entry/straight-out-of-compton-casting-call_n_5597010.

Chatmon, Pete, dir. *Insecure*, season 3, episode 2, "Familiar-Like." Aired August 19, 2018 on HBO, Los Angeles: Home Box Office.

Chatmon, Pete, dir. *Insecure*, season 2, episode 6, "Hella Blows." Aired August 27, 2017 on HBO, Los Angeles: Home Box Office.

Chatmon, Pete, dir. *Black-ish*, season 4, episode 13, "Unkept Woman." Aired Feb 6, 2018 on ABC, Pasadena, CA: American Broadcasting Company.

Collins, Patricia Hill. *Black Feminist Thought: Knowledge, Consciousness, and the Politics of Empowerment*. Boston, MA: Unwin Hyman, 1990.

Collins, Patricia Hill. *Black Sexual Politics: African Americans, Gender, and the New Racism*. New York, NY: Routledge, 2004.

Collins, Patricia Hill. "Controlling Images and Black Women's Oppression." In *Seeing Ourselves: Classic, Contemporary, and Cross-Cultural Readings in Sociology*, edited by John J. Macionis and Nijole V. Benokraitis, 7th edn, 266–73. New York: Pearson, 2007.

Cunningham, Marta, dir. *Insecure*, season 2, episode 3, "Hella Open," Aired August 6, 2017 on HBO, Los Angeles: Home Box Office.

Cunningham, Phillip Lamarr. "'Get a Crew … And Make It Happen': The Misadventures of Awkward Black Girl and New Media's Potential for Self-Definition." In *African Americans on Television: Race-Ing for Ratings*, edited by David Leonard and Lisa A. Guerrero, 402–13. Santa Barbara, CA: Praeger Publishers, 2013.

Falcon, Ellen, dir. *Living Single*, season 2, episode 27, "The Shake-Up." Aired August 22, 1993 on Fox Network, Burbank, CA: Warner Brothers.

Falcon, Ellen, dir. *Living Single*, season 1, episode 27, "What's Next." Aired May 15, 1993 on Fox Network, Burbank, CA: Warner Brothers.

Fandom. "Freddie Brooks." Accessed May 1, 2019. https://differentworld.fandom.com/wiki/Freddie_Brooks.

Fandom. "Living Single." Accessed May 1, 2019. https://livingsingle.fandom.com/wiki/Living_Single_Wiki.

Fandom. "Rainbow Johnson." Accessed May 1, 2019. https://blackish.fandom.com/wiki/Rainbow_Johnson.

French, Bryana H., Lewis, Jioni A., and Neville, Helen A. "Naming and Reclaiming: An Interdisciplinary Analysis of Black Girls' and Women's Resistance Strategies." *Journal of African American Studies* 17, no. 1 (2013): 1–6.

Golden, Marita. *Don't Play in the Sun: One Woman's Journey through the Color Complex*. New York, NY: Anchor, 2007.

Griffin, Rachel Alicia. "Push-ing into Precious: Black Women, Media Representation, and the Glare of the White Supremacist Capitalist Patriarchal Gaze." *Critical Studies in Media Communication* 31, no. 3 (2014): 182–97.

Guerrero, Lisa A. "Single Black Female: Representing the Modern Black Woman in Living Single." In *African Americans on Television: Race-Ing for Ratings*, edited by David Leonard and Lisa A. Guerrero, 177–90. Santa Barbara, CA: Praeger, 2013.

Hardy, Rob. *Black-ish*, season 2, episode 20, "Johnson and Johnson." Aired April 13, 2016 on NBC, Pasadena, CA: American Broadcasting Company, 2016.

Harris, Tina M. "Interrogating the Representation of African American Female Identity in the Films Waiting to Exhale and Set It Off." In *African American Communication and Identities: Essential Readings*, edited by Ronald L. Jackson, 189–98. Thousand Oaks, CA: Sage, 2004.

Harris-Perry, Melissa V. *Sister Citizen: Shame, Stereotypes, and Black Women in America*. New Haven, CT: Yale University Press, 2013.

Hobson, Janell. *Venus in the Dark: Blackness and Beauty in Popular Culture*. London: Routledge, 2018.

King, Don Roy, dir. *Saturday Night Live*, season 39, episode 14, "Andrew Garfield/Coldplay." Aired May 3, 2014 on NBC, New York, NY: Studio 8H.

Matsoukas, Melina, dir. *Insecure*, season 2, episode 1, "Hella Great." Aired July 23, 2017, on HBO, Los Angeles, CA: Home Box Office.

Matsoukas, Melina, dir. *Insecure*, season 1, episode 1, "Insecure as Fuck." Aired October 9, 2016, on HBO, Los Angeles, CA: Home Box Office, 2016.

McCarthy-Miller, Beth, dir. *Black-ish*, season 2, episode 6. "Hope." Aired February 22, 2016 ABC, Pasadena, CA: ABC.

Mercer, Kobena. "Black Hair/Style Politics." In *Welcome to the Jungle: New Positions in Black Cultural Studies*, 97–128, New York, NY: Routledge, 2013.

Moffitt, Kimberly R., and Lumpkin, Tunisia. "The Tangled Weave We Wear: It Still Matters," *In Media Res*, 2018. /imr/2018/02/17/tangled-weave-we-wear-it-still-matters.

Morgan, Joan. *When Chickenheads Come Home to Roost: A Hip-Hop Feminist Breaks It Down.* New York, NY: Simon and Schuster, 2000.

Peters, Michael, dir. *A Different World*, season 4, episode 25. "To Be Continued." Aired on May 2, 1991 on NBC, Universal City, CA: Universal.

Schultz, Michael, dir. *Black-ish*, season 2, episode 14, "Sink or Swim." directed by Michael Schultz, Aired February 10, 2016 on ABC, Pasadena, CA: ABC.

Singletary, Tony, dir. *Living Single*, season 3, episode 1, "Come Back Little Diva." Aired August 31, 1995 on Fox, Burbank, CA: Fox Network.

Singletary, Tony, dir. *Living Single*, season 1, episode 1, "Judging by the Cover." Aired August 22, 1993 on Fox Network, Burbank, CA: Warner Brothers.

Smith-Shomade, Beretta E. *Shaded Lives: African-American Women and Television.* New Brunswick, NJ: Rutgers University Press, 2002.

Spellers, Regina E., and Moffitt, Kimberly R. *Blackberries and Redbones: Critical Articulations of Black Hair/Body Politics in Africana Communities.* New York, NY: Hampton Press, 2010.

Stephens, Dionne P., and Phillips, Layli. "Integrating Black Feminist Thought into Conceptual Frameworks of African American Adolescent Women's Sexual Scripting Processes." *Sexualities, Evolution & Gender* 7, no. 1 (April 1, 2005): 37–55. https://doi.org/10.1080/14616660500112725.

Uwumarogie, Victoria. "Synclaire James was the Original Awkward Black Girl MadameNoire." *Madame Noire,* April 23, 2018. https://madamenoire.com/1037319/living-single-synclaire/.

Wanzo, Rebecca. "Precarious-Girl Comedy: Issa Rae, Lena Dunham, and Abjection Aesthetics." *Camera Obscura: Feminism, Culture, and Media Studies* 31, no. 2 (September 1, 2016): 27–59. https://doi.org/10.1215/02705346-3592565.

White, E. Frances. *Dark Continent of Our Bodies: Black Feminism and Politics of Respectability.* Philadelphia, PA: Temple University Press, 2001.

Wilkins, Jessica. "Issa Rae Proves Why She's the Queen of 4C Natural Hair on 'Insecure.'" *Essence.* Accessed May 1, 2019. www.essence.com/hair/issa-rae-insecure-natural-4c-hair/.

Yoseph, Saaret E. "Glad I Got My Girls." *The Root.* Accessed May 1, 2019. www.theroot.com/glad-i-got-my-girls-1790900202.

8

THE MALE GAZE IN VISUAL CULTURE

Claire Sisco King

The *male gaze* remains one of the most canonical and contested concepts in feminist film and media studies. Laura's Mulvey's essay "Visual Pleasure and Narrative Cinema" contends that the camera in Hollywood films positions spectators to assume the perspective of heterosexual male viewers, objectifying and commodifying women's bodies toward sadistic ends. Since Mulvey introduced the phrase in 1975 in her study of Hollywood cinema, scholars have applied the concept in relation to not only film but also other forms of visual culture including television, advertising, art, theater, and museums. As Clifford Manlove notes, "Mulvey's thesis concerning the patriarchal structure of an 'active' male gaze has spread its influence far beyond feminist film critiques of Alfred Hitchcock and Hollywood, to film and cultural theory, and to theories of perception generally."[1] For this reason, Mulvey's theories have shaped scholarship in a diverse array of fields that includes but is not limited to English, communication studies, art history, theater, archaeology, and critical geography.

Mulvey's concept has also served as the foundation for considering multiple gazes that structure power relations through the acts of looking and being looked at, helping to instantiate what is often called "gaze theory."[2] Such expansions of Mulvey's theory have included attention to the female gaze, the white supremacist gaze, the colonialist gaze, the oppositional gaze,[3] and the ableist, or "normate," gaze[4]—among others. Mulvey's arguments have incited debates about the limitations of and potential for psychoanalytical film theories, as well as raising questions about the contours of spectatorial identification, desire, agency, and resistance. The vast and varied responses to Mulvey's work indicate, on the one hand, the centrality of visual culture to everyday life and the affordances of Mulvey's theory for thinking about power and relationality between subjects and objects and, on the other hand, some shortcomings of her theoretical approach. With this framing in mind, this essay will explicate Mulvey's theories and consider some of the critiques, revisions, and extensions of her figuration of the male gaze in studies of visual culture.

Defining the male gaze

Mulvey's essay, published in the British journal *Screen* in 1975, appeared in the context of the emergence of feminist film theory and criticism.[5] Just three years prior, the American journal *Women and Film* had dedicated itself to "taking up the struggle with women's image in film and women's roles in the film industry."[6] In 1973, Claire Johnston's *Notes on Women's Cinema* considered how films by women might counter the extent to which "male-dominated cinema" presented women as what they meant "for men."[7] And, Molly Haskell's 1974 *From Reverence to Rape* critiqued the cinema's contribution to cultural perceptions of "women's inferiority."[8]

Similar to John Berger's contemporaneous interest in cultural "ways of seeing,"[9] Mulvey's essay argues that Hollywood films foster particular structures for looking at women's bodies in line with the desires of heterosexual men.[10] Drawing on psychoanalytic theories, Mulvey posits the cinema as an apparatus that creates spectatorial positions and offers particular kinds of pleasures, including scopophilic ones (or, pleasures in the act of looking). Mulvey's interest in psychoanalysis is consonant with work being done at the time by a number of other scholars of film theory, including Christian Metz and Jean-Louis Baudry; but whereas Metz and Baudry presume masculine spectatorial perspectives, Mulvey's essay foregrounds attention to gender and gender difference, arguing that the "unconscious of patriarchal society has structured film form."[11] Mulvey explicitly positions her essay as a kind of rejoinder to other scholarship published in *Screen*, which deployed psychoanalytic frameworks for analyzing the cinema without, in her estimation, sufficiently addressing the "representation of the female form."[12]

Mulvey conceptualizes scopophilia in terms of an active, "controlling and curious gaze" that renders its targets as objects; and she identifies a gendered division of labor that is "split between active/male and passive/female."[13] These structures of looking create a metonymic reduction wherein women come to "connote *to-be-looked-at-ness*" and assume the exhibitionist role of show-stopping spectacle.[14] Mulvey figures this form of looking as having a tripartite structure: camera, spectators, and characters. As Teresa de Lauretis explains,

> The look of the camera (at the profilmic), the look of the spectator (at the film projected on the screen), and the intradiegetic look of each character (at other characters, objects, etc.) intersect, join, and relay one another in a complex system.[15]

While the camera helps to create such an imbalance in looking, Mulvey argues that narrative structure in Hollywood film reinforces this split, positioning onscreen men as sites for spectatorial identification and as causal agents in the story world, in contrast to representations of women as objects and impediments to the "development of a story line."[16] The work of camera and narrative align when, for example, Hollywood films encourage spectators to identify with men as driving narrative forces and bearers of the look, reduplicating the camera's objectification of women's bodies with onscreen depictions of men gazing at women. Mulvey likens this process of identification to Lacan's description of the mirror stage as a process of misrecognition, in which a child mistakes a mirrored image of the self as proof of his/her/their wholeness, such that spectatorial identifications with active male figures in the cinema reinforce fantasies of masterful, whole subjectivity and disavow the lack that Lacanian psychoanalysis understands as constitutive of all human subjects.[17]

Although Mulvey locates these ways of looking in the potential pleasures of the cinema, she also argues that images of women can generate "unpleasure."[18] Insofar as Freudian psychoanalysis understands women as signifiers of sexual difference, or lack, they also come to imply the threat of castration, which the patriarchal unconscious desires to control. Within Hollywood cinema, the patriarchal unconscious manages this imagined threat through two particular scopic mechanisms: voyeurism and fetishism. First, the cinema encourages a figuratively voyeuristic experience by operating as a "hermetically sealed world" that feigns indifference to or ignorance of the presence of the audience.[19] For example, characters in Hollywood films rarely acknowledge the camera or a film's imagined viewers. Writing in the context of theatrical reception of films prior to the advent of playback technologies, Mulvey also cites the phenomenological structures of watching a film silently in a darkened theater as contributing to the voyeuristic sensibilities of the cinema.[20] As women become the eroticized target of both the camera and men represented on screen, often without express consent, such a sadistic mode of looking presumes that women are guilty and deserving of punishment, hence the cruelty of the voyeuristic look.

Second, Mulvey introduces the notion of "fetishistic scopophilia," which positions women as "reassuring rather than dangerous" figures by fully rendering them as objects or "perfect product[s]."[21] Fetishism, Mulvey argues, "builds up the physical beauty of the object, transforming it into something satisfying in itself."[22] Cinematography and editing, particularly in the form of the close-up, operate in tandem to fragment the female form into a fetish object. The fixation on eroticized parts of women's bodies—legs, breasts, lips—reduces the female form to these parts, thereby nullifying women's potential power and disavowing the threat of castration.[23] Such fetishistic representations of women's bodies might be likened to a violent process of dismemberment, which often gets literalized in violent genres such as horror.

To develop her theoretical treatise, Mulvey offers close readings of films by Josef von Sternberg and Alfred Hitchcock as respective illustrations of the scopic operations of voyeurism and fetishism, but her essay makes clear that these mechanisms are not unique to these particular filmmakers. Rather, she reads them as illustrative of a patriarchal unconscious that structures all Hollywood film culture. To demonstrate, I will turn to a Hollywood film not addressed directly by Mulvey, *The Seven Year Itch* (directed by Billy Wilder, 1955), to help explicate her claims. In this film, Marilyn Monroe plays "The Girl," a commercial actress and model who becomes the object of affection for a married heterosexual man, Richard Sherman (played by Tom Ewell). While his wife and family vacation for the summer, Sherman pursues Monroe's character, who has moved into his apartment building. In one scene, Sherman and The Girl attend a screening of the monster movie *The Creature from the Black Lagoon* (directed by Jack Arnold, 1954) and stop on the sidewalk outside the theater on their way home as they discuss the film.

This scene offers clear examples of both voyeurism and fetishism, as Mulvey defines them. First, in addition to the voyeuristic sensibility that drives the illusion of the film itself as a "private world" into which viewers can peer, the scene depicts Sherman ogling The Girl, desirously taking in her form in an act of looking she seems not to notice. The film's structures of identification, Mulvey's "Visual Pleasure" essay might suggest, encourage spectators not only to take pleasure in seeing Monroe's body through the look of the camera but also to identify with Sherman's active, controlling gaze. In this conversation, The Girl also lapses rather unthinkingly into a recitation of lines she has delivered for a toothpaste commercial, reinforcing her position as the object of the gaze and encouraging the imagined viewers of *The Seven Year Itch* to identify even further with this other set of implied viewers.

Second, cinematography in this scene renders Monroe's body a fragmented object that performs an interruptive function within the film's narrative. Wearing a now-famous white halter dress, The Girl stands atop a subway grate as a train speeds underneath her, not once but twice. In both moments, the camera offers a medium close-up of Monroe's bare legs as the wind produced by the train blows up the hem of her dress. The Girl stands bemusedly still as this event happens, while Sherman watches with delight, and the passing of the train literally suspends their walk home, grinding the narrative to a temporary halt. Just as the train distracts The Girl and facilitates Sherman's manipulation of her (so that he may trick her into kissing him), the fetishistic shots of Monroe's body offer reassurance to the film's presumed heterosexual male viewers that she is a nonthreatening object of pleasure. Rhetorically speaking, this fetishization synecdochally reduces Monroe to her eroticized parts, denying her any semblance of full personhood. The fact that Monroe's character has no specific name but instead a generic (gendered) moniker further reinforces this objectification and de-subjectification.

Beyond the male gaze

The above application of Mulvey's theories to *The Seven Year Itch* typifies a great deal of scholarly work on the male gaze, in which critics search texts, artifacts, and practices for evidence of the gaze. For example, as recently as 2018 scholars have drawn from Mulvey's theorizations as the basis for their critical discussions of such topics as U.S. celebrity Alyssa Milano's social media images of breastfeeding, representations of women in contemporary James Bond movies, and the effects of sexually objectifying music videos on their viewers.[24] The influence of Mulvey's concept of the male gaze has not, however, shielded it from critique and/or revision. As Manlove argues,

> While most respondents accept Mulvey's argument that there is a "gaze" at work in cultural and power relations, invariably these respondents also seek to redefine Mulvey's gaze, depending on whether their primary object of study (or attack) is feminism, film, or psychoanalysis.

In addition, notes Manlove, some critics "seek not only to reject Mulvey's theory of the gaze but also, in some cases, to reject the use of psychoanalysis, 'feminist film theory,' or other critical/interpretive approaches to film."[25]

Operating within the latter category of critique, some film scholars have responded with objections to the use of psychoanalysis in film theory and criticism. For example, in the issue of *Screen* following the one in which "Visual Pleasure and Narrative Cinema" appeared, four members of the journal's editorial board wrote an essay describing psychoanalysis as a problematic paradigm for film studies given its "account of women" and lack of a clear "interpretive method."[26] A year later these same authors—Edward Buscombe, Christine Gledhill, Alan Lovell, and Christopher Williams—would write another statement announcing their resignation from the *Screen* editorial board because of the journal's continued interest in psychoanalytic theory, which they characterized as "unnecessarily obscure and inaccessible" and "intellectually unsound and unproductive" in its politics.[27] Other works in film studies have also decried psychoanalysis, both within and beyond Mulvey's work, as having tendencies toward ahistorical and universalizing logics.[28] For example, David Bordwell and Noël Carroll's *Post-Theory* objects to what they perceived as the hegemony of Lacanian psychoanalysis and gaze theory within film studies, lamenting the purported quest for a "Grand Theory" and calling instead for more empiricist and "problem-driven research."[29]

Other scholars have drawn from and pushed back against Mulvey's work to complicate or expand her figuration of the gaze, often addressing, as Joan Faber McAlister notes, "issues of subjectivity, agency, representation, and identity."[30] Indeed, multiple feminist film scholars have raised questions about the possibility of female spectatorship and women's agency, including Mary Ann Doane, Tania Modleski, Annette Kuhn, E. Ann Kaplan, and Teresa de Lauretis.[31] As Jackie Stacey explains it, "If feminist film criticism in the 1970s was characterized by debates about the male gaze, debates in the 1980s were characterized by their emphasis on female spectatorship."[32]

For example, Doane asks, "even if it is admitted that the woman is frequently the object of the voyeuristic or fetishistic gaze in the cinema, what is there to prevent her from reversing the relation and appropriating the gaze for her own pleasure?" Exploring this possibility, Doane's analysis attends to cinematic examples of masquerade, or the "flaunting" or "hyperbolization of the accoutrements of femininity," in ways that effect a denaturalization of the feminine. She also considers films that emphasize women's capacity as spectators, as is often literalized by having these characters wear glasses.[33] Although Doane does not write about *The Seven Year Itch*, her arguments apply fittingly to this film. For example, when leaving the screening of *The Creature from the Black Lagoon*, The Girl has not only assumed the literal role of spectator in the film, but she also describes identifying with and feeling empathy toward the film's creature such that Sherman acknowledges her "interesting point of view." Likewise, the fetishization of Monroe's body in the white halter dress, which has itself become a signifier of her larger image culture, might be read as so excessive, or hyperbolic, as to call critical attention to the exhibitionist role often assigned to women whose femininity is staged as a form of drag.

In an example drawn from within the field of communication studies, Brenda Cooper draws on Lorraine Gamman's work to argue for the feminist potential of films that center women through the registers of both production and content.[34] Offering a close reading of the film *Thelma & Louise* (directed by Callie Khouri, 1991), Cooper identifies three strategies through which this film may critique and even appropriate the male gaze: "resistance to male objectification and dominance" through the women's mockery of men's performances of hegemonic masculinity, "'returning the look' by making men spectacles for women's attention," and celebrating relationships between women.[35] Cooper contends that while "female and male gazes may comprise a filmic text," these competing gazes invite feminist reception of them, affording new opportunities for pleasure for women viewers.[36] It should be noted that Cooper's use of Mulvey's theory largely dispenses with its psychoanalytic underpinnings, which is a tactic common to many extensions of and variations on the concept of the gaze.

Attention to the potential for a female gaze have also invited considerations of the possibility of lesbian spectatorial desire, including the work of Jackie Stacey and Judith Mayne. While Stacey notes the importance of attention to sexual diversity in spectatorship, she cautions against overgeneralizations about lesbian spectators. She writes, "There is likely to be a whole set of desires and identifications with different configurations at stake which cannot necessarily be fixed according to the conscious sexual identities of the cinematic spectator."[37] Considerations of the female gaze have also inspired attention to the objectification of men's bodies onscreen. In a case study of film star Rudolph Valentino, Miriam Hansen asks, "If a man is made to occupy the place of erotic object, how does this affect the organization of vision?"[38] Hansen reads a figure such as Valentino, who was the object of intense affective investment by women fans in the early 20th century, as blurring the boundaries between looking/being looked at, activity/passivity, masculinity/femininity, and sadism/masochism. She posits Valentino as a potential object of desire for women spectators and an ambivalent figure for identification for them, noting that women viewers might identify with Valentino as an object of the gaze and also aspire to his potential to be a bearer of the look by virtue of his masculinity.[39] Likewise, scholars such as Steve Neale and Kaja Silverman call for greater attention to spectacles of masculinity as the objects of a masochistic, rather than sadistic, male gaze.[40] Silverman maintains that the "spectatorial inscription" associated with arguments like Mulvey's risks putting "masculinity under erasure" and reinforcing the metonymic links between masculinity and "aggression and sadism" at the expense of considering other iterations of masculine desire and pleasure.[41]

Beyond its neglect of female desire and agency and its inattention to the spectacularization of masculinity (for the pleasures of male and female viewers), "Visual Pleasure and Narrative Cinema" has engendered multiple critiques for its failure to consider the intersecting factors of race and class. Confronting the presumptive whiteness of Mulvey's theories, bell hooks laments the lack of attention to "black female spectatorship" in works of film theory by white feminists. Noting that Black women are subjected to the controlling and interarticulated gazes of white supremacy and colonialism, hooks argues that the gaze is always raced; she further contends that even the most sadistic gazes can be answered or returned in acts of resistance. Such forms of looking back, hooks notes, constitute an "oppositional gaze." She writes, "Looking and looking back, black women involve ourselves in a process whereby we see our history as counter-memory, using it as a way to know the present and invent the future."[42] Accordingly, hooks understands the gaze as always multiple—which is to say, she understands multiple gazes to be at work at once in any given encounter. She also calls attention to the agency of those who are looked at, drawing from the experiences of Black women to note that one can be both the object of an other's gaze and a subject with agency at the same time.

Akin to hooks's attention to the multiplicity and contradictoriness of the gaze, Jane Gaines argues for the importance of analyses that address the multiple "structures of oppression" that affect women, including not only race but also class.[43] Gaines proffers that too much focus on a film's investment in "male pleasure" may facilitate its strategic obfuscation of other power relations and imbalances, including racial discrimination and class-based inequities.[44] That is, attention to the (presumptively white) male gaze may exacerbate the erasure of non-privileged women. Thus, Gaines's call for an intersectional perspective reminds white feminist critics that the choice to see gender without seeing race and/or class becomes complicit in the perpetuation of inequality.[45] In response to a film such as *The Seven Year Itch*, for instance, Gaines might contend that attention to the film's objectifying treatment of white characters like The Girl should not obscure critical assessment of the film's depiction of Native Americans, who are played by white actors as childlike and "uncivilized" caricatures.

Writing about the sexualization of men's bodies in various image cultures, including advertising, Rosalind Gill also emphasizes the importance of an intersectional approach. She contends that, while men's bodies seem increasingly subject to sexualizing imagery in the 21st century, critics must read these images carefully to attend to their specific representational strategies and norms. For example, ageism and "patterns of racialization" affect which men's bodies appear in sexualized ways and how they take shape with varying degrees of objectification.[46] Such images, she argues, also address multiple audiences and encourage different modes of looking from such spectators as gay men, heterosexual women, and heterosexual men.[47]

Intersectional theories of the gaze have addressed not only race and class-based discrimination but also ableism. For instance, critical disability scholars, including Rosemarie Garland Thomson, have advocated for analysis of the ableist or "normate" gaze, or the gaze to which people with disabilities are subject.[48] Similarly, Johnson Cheu introduces the intersectional category of the "normative gaze," which he defines as the "white, male, heterosexual, able-bodied cinematic gaze."[49] While calling for increased scholarship on the disabled gaze, or the agency of disabled people to look back at those who might objectify them,[50] Cheu also warns that the conflation of agency with the capacity to look risks further marginalization of those for whom sight is limited or inaccessible, reinforcing figurations of the blind as dependent and helpless. Thus, considerations of nonnormative gazes should be attentive to the complex ways in which all bodies and subjects may enact agency and resistance in response to and outside of the gaze.[51]

The notion of "looking back" has also been extended to suggest that the act of spectatorship may not generate mastery but may instead position the viewer as subject to the returned gaze of the object. Some such arguments hinge on critical appraisals of Mulvey's use of Lacanian psycho-analysis, including the work of Gaylyn Studlar, Joan Copjec, and Todd McGowan. For example, Copjec accuses Mulvey of a "kind of 'Foucauldinization' of Lacanian theory," which assumes that the desire to look at another is equivalent to the desire for power over that other.[52] This perspective argues that rather than seeking mastery or the reaffirmation of the ego, the spectator may desire the shattering or transcending of subjectivity that can arise from encounters with and views of an other.[53] Further, McGowan explains, "The gaze is not the look of the subject at the object, but the point at which the object looks back."[54] From this perspective, the gaze does not assume or assert a masterful position but instead illustrates vulnerability.[55]

Peter Willemen and Wheeler Winston Dixon share this interest in the film's capacity to gaze at the spectator, suggesting that the possibility of the viewer's self-awareness about their own act of looking may invite attention to the possibility of being looked at. Willemen writes, "When the scopic drive is brought into focus, the viewer also runs the risk of becoming the object of the look, of being overlooked in the act of looking."[56] Willemen dubs the awareness of being seen in the act of looking as the "fourth look," which he adds to the tripartite structure of Mulvey's theory. This fourth look is imagined by the viewer and projected onto the film itself. As Dixon explains, "The film acts upon us, addressing us, viewing us as we view it, until the film itself becomes a gaze, rather than an object to be gazed upon."[57] These arguments complicate the distinction between activity and passivity, suggesting the capacity of the image (or an object) to affect and act on the spectator.

Meenakshi Gigi Durham articulates Willemen's and Dixon's interest in the returned gaze of the object with attention to race and national identity. Specifically, she links Willemen's and Dixon's theories to arguments in critical race scholarship, not unlike that of hooks, which emphasize the capacity of racially or ethnically othered subjects to look back at those who would objectify them. Reading images of South Asian women in mainstream Western media, Durham argues, "The viewing of images involves multiple processes and multiple "looks, which is to say that constituting the viewer as active subject and the viewed as passive object is a simplistic and heuristically limited formulation."[58] Mediated representations of South Asian women for audiences presumed to be/positioned as white may reduplicate the capacity of both the text and the subjects it represents to look back at spectators, who may become more aware of not only their spectatorial status but also their racialized positions.[59]

The possibility of the returned gaze matters not only to film studies but also to such fields as critical geography and museum studies. For example, building from Hilde Hein's contention that, absent explicitly feminist practices, museum displays often subject representations of women to a fetishistic male gaze (along with a host of other potential gazes, including white supremacist and colonialist ones),[60] McAlister suggests that museum displays can problematize the presumed relationships of looking/being looked at, activity/passivity. Offering close analysis of the nontraditional curatorial practices and "material aesthetics" of the Women's Jail Museum in Johannesburg, McAlister argues that this museum's installations "make use of a relational property of visuality," recognizing that the "gaze need not be conceptualized as unidirectional and all-powerful."[61] Specifically, the museum's design animates the capacity of the "sensory, experiential, and embodied practice of looking" to emphasize the vulnerability of the "looker" and to enable "agency to flow in both directions of the engagement."[62]

In other words, all of these revisions and extensions of Mulvey's work make clear that no visual encounter is limited to a singular gaze, nor is any gaze all-powerful or all-perceiving. Multiple, often conflicting, gazes operate in any given situation, and attention to the acts of looking and being looked at should be both intersectional and mindful of context. The act of looking at an other may or may not be aimed at achieving mastery for the looker, but it can neither guarantee masterful subjectivity (which is itself an impossibility) nor secure the passivity of its target (because objects themselves have agency). Such refinements to gaze theory are contributions Mulvey herself has acknowledged. Identifying her 1975 article as a "polemic" and "manifesto" that was necessarily hyperbolic in its "attack,"[63] Mulvey has suggested that the gaze should be reimagined as multifarious and subject to opposition. In "The Pleasure Principle," an essay written to commemorate the 40th anniversary of "Visual Pleasure," Mulvey concludes:

> Needless to say, modes of spectatorship were always more complex than the 'Visual Pleasure…' essay allowed and the 'male gaze' could always be transgressed by anyone who cared to assert their own sexual identity or proclivity, which might have involved reading against the grain or intuiting a subversive instinct already there on the screen.[64]

Technology and the gaze

Although scholars have applied the concept of the gaze to a wide array of texts, artifacts, and practices well beyond the cinema, others have insisted that Mulvey's focus on the filmmaking norms of a particular era of Hollywood cannot be disarticulated from the scopic mechanisms she identifies. Likewise, beyond the specific style of filmmaking "Visual Pleasure and Narrative Cinema" addresses, her theories also presume a theatrical exhibition context and the specific norms of reception that such a viewing experience entails. Given these factors, media scholars interested in the politics of looking beyond and outside of the cinema have questioned the viability of the gaze as the best, or only, paradigm for thinking about spectatorship, desire, and power. As Jeffrey Bennett notes, "While film studies has an extensive history exploring the 'gaze,' television scholars often employ the 'glance' to explain small screen reception."[65]

From this perspective, broadcast television is a medium encountered in states of distraction or even disengagement and therefore does not invite the kind of sustained attention that the cinema, particularly in theatrical settings, encourages. The gaze, for John Ellis, suggests a "concentration of the spectator's activity into that of looking," while the "glance implies that no extraordinary effort is being invested in the act of looking."[66] This mode of looking may involve frequent interruptions, including commercial breaks and myriad other possible interferences within the viewing space that lead viewers to, in Chris Chesher's terms, "drift" from the screen or even leave the television set entirely.[67] Likewise, television is often experienced on a "much smaller screen" in relatively familiar or "mundane" settings that bear little resemblance to the dreamlike spectacle of watching films on a big screen with surround sound in a darkened and otherwise quiet theater. In other words, home or mobile viewing environments seem decidedly removed from the "hermetically sealed" world of the cinema where Mulvey locates voyeuristic sensibilities.

Even this figuration of the glance, however, seems constrained by a particular model of televisuality that is no longer the only, or even the primary, way in which viewers engage with media texts. For example, Chesher notes the insufficiency of both concepts of the "gaze" and the "glance" to describe console gaming practices, which he makes clear are "not television" and "not cinema." Video games, Chesher argues, might be better understand as inviting a "glaze" from the audience due to their immersive sensibilities (in which players' eyes "glaze over" through intense involvement), interactive features (in which games have a kind of "stickiness" that promotes player investment), and mimetic qualities (in which games often have a specular, or reflective, quality).[68] While the immersive intensity of video games harkens to the voyeurism of the cinema, they also typically appear on television sets in potentially distracting domestic spaces; thus, Chesher draws from both theorizations of the gaze and the glance to suggest the particularities of the glaze for game studies.

Likewise, in the context of digital media and convergence culture, the concepts of medium specificity and medium-specific modes of looking become increasingly complex to parse. Streaming media and the binge-watching reception practices they encourage, along with the advent of cinematic television stylistics, suggest that neither the theatrical model of the gaze nor the televisual model of the glance will suffice. Mulvey herself has argued as much, noting that, among other things, the interactive qualities of digital media increase opportunities for active, rather than passive, spectatorship. She notes that in the context of digital mediation, the act of "inventing and imagining one's own mode of spectatorship is a matter of a simple push of the button."[69] She further acknowledges that the presumption of a theatrical spectatorial experience no longer holds, and she frames her later scholarship as a corrective on this point.

Mulvey writes, "It was partly to underline the irrelevance of 'Visual Pleasure and Narrative Cinema' to contemporary modes of spectatorship that I wrote my book *Death 24x a Second: Stillness and the Moving Image*."[70] In this book, Mulvey considers how playback technologies and the digitization of media afford more agency to spectators who have the capacity to stop moving images and, in these experiences of stillness, contemplate both the indexicality of cinematic images and their relationship to time.[71] Such changing modes of spectatorship (wherein viewers may pause, rewind, fast forward, zoom, etc. what they watch) may also diminish the spellbinding capacity of narrative and redistribute viewer "attention to detail and gesture" on screen. With such possibilities, the spectator—whom Mulvey dubs the "fetishistic spectator"—now "controls the image to dissolve voyeurism and reconfigure the power relation between spectator, camera and screen."[72]

In her discussion of this changing landscape of spectator, camera, and screen, Mulvey ponders "whether new practices of spectatorship have effectively erased the difficulty of sexual difference and the representation of gender in Hollywood cinema." She suggests that the changing power dynamics may also reconfigure relations between "male and female," but she does *not* contend that questions of power are displaced entirely. The nature of the look and the terms of activity/passivity may be changed by more interactive modes of spectatorship, but Mulvey implies that both circuits of power and pleasure remain present in all viewing experiences. With this in mind, her declaration that digital media render "Visual Pleasure and Narrative Cinema" irrelevant seems rather hyperbolic; perhaps she was again being polemical. In any case, Mulvey's newer scholarship on contemporary forms of mediation and spectatorship, along with the work of other scholars, reminds us that considerations of pleasure in visual culture must be attentive not only to the various and intersecting lines of power at work in any given situation but also to the specific iterations of visuality and mediation at play. Thus, while Mulvey's 1975 concept has traveled far and wide in the study of visual culture and while her theories have offered insights into a variety of visual artifacts and practices, scholars should always look closely and curiously at the particular scopic capacities and contexts they study.

Notes

1 Clifford Manlove, "Visual 'Drive' and Cinematic Narrative: Reading Gaze Theory in Lacan, Hitchcock, and Mulvey," *Cinema Journal* 46, no. 3 (Spring 2007): 83.
2 Manlove, "Visual 'Drive,'" 83.
3 bell hooks, *Black Looks: Race and Representation* (Boston, MA: South End Press, 1992).
4 Rosemary Garland Thompson, *Extraordinary Bodies: Figuring Physical Disability in American Culture and Literature* (New York: Columbia University Press, 2017/1997), 25–26.
5 Mulvey has indicated that her primary focus in writing the essay was extending the work of the women's liberation movement, with the study of film as a secondary interest. See Laura Mulvey, "The Pleasure Principle," *Sight and Sound* 25, no. 6 (2015): 50–51.
6 Siew-Hwa Beh and Saunie Salyer, "Overview," *Women in Film* 1, no. 1 (1972): 5.
7 Claire Johnston, "Women's Cinema as Counter-Cinema," in *Notes on Women's Cinema*, ed. Claire Johnston (London: Society for Education in Film and Television, 1973), 25.

8 Molly Haskell, *From Reverence to Rape: The Treatment of Women in the Movies* (Chicago and London: The University of Chicago Press, 1974), 1.

9 Berger's text *Ways of Seeing* was also produced as a television series, which consisted of four 30-minute segments and was broadcast on BBC Two. The adaptation of Berger's book for television illustrates its cultural resonance and its reach beyond strictly academic audiences or the field of art history.

10 John Berger, *Ways of Seeing* (London: BBC Books, 1972).

11 Laura Mulvey, "Visual Pleasure and Narrative Cinema," *Screen* 16, no. 3 (1975): 6.

12 Ibid., x.

13 Ibid., 8, 11.

14 Ibid., 11.

15 Teresa de Lauretis, *Alice Doesn't: Feminism, Semiotics, Cinema* (Bloomington, IN: Indiana University Press, 1984), 138.

16 de Lauretis, *Alice Doesn't*, 11.

17 For Lacan, the mirror stage occurs in babies who have not yet acquired language. Upon recognizing an image of themselves reflected in the mirror, children misrecognize themselves as coherent and whole subjects. Lacan understands the mirror stage as not only a discrete process that happens during early childhood development but as a kind of structural condition for humans who continually struggle between the misapprehension of themselves as whole and the fraught recognition of their lack and fragmentation.

18 Mulvey, "Visual Pleasure," 13.

19 Ibid., 9.

20 Ibid...

21 Ibid., 14.

22 Ibid..

23 Ibid..

24 See Raeann Ritland, "Visual Pleasure from Motherhood: Alyssa Milano Challenging the Male Gaze," *Media, Culture, & Society* 40, no. 8 (November 2018): 1281–1291; Lisa Funnell, "Reworking the Bond Girl Concept in the Craig Era," *Journal of Popular Film & Television* 46, no. 1 (January–March 2018): 11–21; and Kathrin Karsay, Jörg Matthes, Phillip Platzer, and Myrna Plinke, "Adopting the Objectifying Gaze: Exposure to Sexually Objectifying Music Videos and Subsequent Gazing Behavior," *Media Psychology* 21, no. 1 (January–March 2018): 27–49.

25 Manlove, "Visual 'Drive,'" 85.

26 Edward Buscombe, Christine Gledhill, Alan Lovell, and Christopher Williams, "Statement: Psychoanalysis and Film," *Screen* 16, no. 4 (1975): 119–130.

27 Edward Buscombe, Christine Gledhill, Alan Lovell, and Christopher Williams, "Statement: Why We Have Resigned from the Board of *Screen*," *Screen* 17, no. 2 (July 1976): 107, 108.

28 Manlove, "Visual 'Drive,'" 87. See also David Bordwell and Noël Carroll, eds, *Post-Theory: Reconstructing Film Studies* (Madison, WI: University of Wisconsin Press, 1996).

29 David Bordwell and Noël Carroll, eds, *Post-Theory: Reconstructing Film Studies* (Madison, WI: University of Wisconsin Press, 1996), xvii.

30 Joan Faber McAlister, "Collecting the Gaze: Memory, Agency, and Kinship in the Women's Jail Museum, Johannesburg," *Women's Studies in Communication* 36, no. 1 (2013): 4.

31 See Mary Ann Doane, "Film and the Masquerade: Theorising the Female Spectator," *Screen* 23, nos 3–4 (September 1982): 74–88; Tania Modleski, *The Women Who Knew Too Much: Hitchcock and Feminist Theory* (New York: Metheun, 1988); Annette Kuhn, *Women's Pictures: Feminism and Cinema* (London and New York: Routledge and Kegan Paul, 1982); E. Ann Kaplan, *Women and Film: Both Sides of the Camera* (New York and London: Metheun, 1983); and Teresa de Lauretis, *Alice Doesn't: Feminism, Semiotics, Cinema* (Bloomington, IN: Indiana University Press, 1984).

32 Jackie Stacey, *Star Gazing: Hollywood Cinema and Female Spectatorship* (London and New York: Routledge, 1994), 22.

33 Doane, "Film and the Masquerade," 67.

34 Lorraine Gamman, "Watching the Detectives: The Enigma of the Female Gaze," in *The Female Gaze: Women as Viewers of Popular Culture*, eds Lorraine Gamman and Margaret Marshment (Seattle: Real Comet Press, 1989), 8–26.

35 Brenda Cooper, "'Chick Flicks' as Feminist Texts: The Appropriation of the Male Gaze in *Thelma & Louise*," *Women's Studies in Communication,* 23, no. 3 (Fall 2000): 285.

36 Ibid., 301.

37 Jackie Stacey, "Desperately Seeking Difference: Desire Between Women in Narrative Cinema," *Screen*, 28, no. 1 (January 1987): 48–61.

38 Miriam Hansen, "Pleasure, Ambivalence, Identification: Valentino and Female Spectatorship," *Cinema Journal* 25, no. 4 (1986), 10.

39 Ibid., 6–32. Susan Bordo also writes about the eroticization of male bodies in her book *The Male Body: A New Look at Men in Public and Private* (New York: Farrar, Straus, and Giroux, 2000).

40 See Steve Neale, "Masculinity as Spectacle: Reflections on Men and Mainstream Cinema," *Screen* 24, no. 6 (1983): 2–16; Kaja Silverman, "Masochism and Male Subjectivity," *Camera Obscura* 6, no. 2 (1988): 30–67.

41 Silverman, "Masochism and Male Subjectivity," 50.

42 bell hooks, *Black Looks*, 131.

43 Jane Gaines, "White Privilege and Looking Relations: Race and Gender in Feminist Film Theory," *Cultural Critique* no. 4 (Autumn 1986): 59–79.

44 Gaines, "White Privilege," 72.

45 Gaines's essay does not use the term "intersectionality," which Kimberlé Crenshaw would coin in just a few years. See Crenshaw, "Mapping the Margins: Intersectionality, Identity Politics, and Violence Against Women of Color," *Stanford Law Review* 43, no. 6 (1991): 1241–129.

46 Rosalind Gill, "Beyond the Sexualization of Culture Thesis: An Intersectional Analysis of 'Sixpacks,' 'Midriffs' and 'Hot Lesbians' in Advertising," *Sexualities* 12, no. 2 (2009): 145.

47 Gill, "Beyond the Sexualization," 146.

48 Rosemarie Garland Thomson coined the term "normate" to refer to a subject position of those who imagine themselves to be "definitive human beings," and their gaze at others aims at establishing their "position of authority" over those understand as bearing markers of difference or deviance from the norm (8). See Gill, *Extraordinary Bodies*.

49 Johnson Cheu, "Seeing Blindness of Screen: The Cinematic Gaze of Blind Female Protagonists," *The Journal of Popular Culture* 42, no. 3 (2009): 483.

50 Johnson Cheu, "Performing Disability, Problematizing Cure," in *Bodies in Commotion: Disability & Performance*, eds Carrie Sandahl and Philip Auslander (Ann Arbor, MI: The University of Michigan Press, 2005).

51 Cheu, "Seeing Blindness on Screen," 493.

52 Joan Copjec, *Read My Desire: Lacan against the Historicists* (Cambridge: The MIT Press, 1994), 19.

53 See also Steven Shaviro, *Cinematic Body* (Minneapolis, MN: University of Minnesota Press, 1993); Claire Sisco King and Joshua Gunn, "On a Violence Unseen: The Womanly Object and Sacrificed Man," *Quarterly Journal of Speech* 99, no. 2 (May 2013): 200–208.

54 Todd McGowan, "Looking for the Gaze: Lacanian Film Theory and its Vicissitudes," *Cinema Journal* 42 (2003): 27–28.

55 In the field of rhetorical studies, Debra Hawhee also addresses questions about the authority or vulnerability of the gaze. In her work on ancient Greek rhetoric about the body, for example, Hawhee argues that vision neither assumes a position of active mastery nor passive reception but is figured as a site of connection that establishes a relationship between the one who looks and the one who is looked at. See Debra Hawhee, *Bodily Arts: Rhetoric and Athletics in Ancient Greece* (Austin, TX: University of Texas Press, 2004), 178.

56 Peter Willemen, *Looks and Frictions: Essays in Cultural Studies and Film Theory* (London: British Film Institute, 1994), 114.

57 Winston Wheeler Dixon, *It Looks at You: The Returned Gaze of the Cinema* (Albany, NY: State University of New York Press, 1995), 2.

58 Meenakshi Gigi Durham, "Displaced Persons: Symbols of South Asian Femininity and the Returned Gaze in U.S. Media Culture," *Communication Theory* 11, no. 2 (May 2001): 208.

59 Durham, "Displaced Persons," 209.

60 Hilde Hein, "Looking at Museums from a Feminist Perspective," in *Gender, Sexualities, and Museums: A Routledge Reader*, ed. Amy K. Levin (New York and London: Routledge, 2010), Kindle.

61 McAlister, "Collecting the Gaze," 17.

62 Ibid.

63 Roberta Sassatelli, "Interview with Laura Mulvey: Gender, Gaze and Technology in Film Culture," *Theory, Culture, & Society* 28, no. 5 (2011): 128.

64 Mulvey, "The Pleasure Principle," 51.

65 Jeffrey A. Bennett, "In Defense of Gaydar: Reality Television and the Politics of the Glance," *Critical Studies in Media Communication* 23, no. 5 (December 2006): 413.

66 John Ellis, *Visible Fictions: Cinema, Television, Video* (London: Routledge & Kegan Paul, 1982), 37.
67 Chris Chesher, "Neither Gaze nor Glance, but Glaze: Relating to Console Game Screens," *SCAN: Journal of Media Arts and Culture* 4, no. 2 (August 2007), http://scan.net.au/scan/journal/print.php?j_id=11&journal_id=19.
68 Chesher," Neither Gaze nor Glance."
69 Mulvey, "Pleasure Principle," 51.
70 Ibid., 50.
71 Laura Mulvey, *Death 24x a Second: Stillness and the Moving Image* (London: Reaktion Books, 2006), 7.
72 Mulvey, *Death 24x a Second*, 167.

Bibliography

Beh, Siew-Hwa and Salyer, Saunie. "Overview." *Women in Film* 1, no. 1 (1972): 5.

Bennett, Jeffrey A. "In Defense of Gaydar: Reality Television and the Politics of the Glance." *Critical Studies in Media Communication* 23, no. 5 (December 2006): 408–425.

Berger, John. *Ways of Seeing*. London: BBC Books, 1972.

Bordo, Susan. *The Male Body: A New Look at Men in Public and Private*. New York: Farrar, Straus, and Giroux, 2000.

Bordwell, David and Carroll, Noël, eds. *Post-Theory: Reconstructing Film Studies*. Madison, WI: University of Wisconsin Press, 1996.

Buscombe, Edward, Gledhill, Christine, Lovell, Alan, and Williams, Christopher. "Statement: Psychoanalysis and Film." *Screen* 16, no. 4 (1975): 119–130.

—. "Statement: Why We Have Resigned from the Board of *Screen*." *Screen* 17, no. 2 (July 1976): 106–109.

Chesher, Chris. "Neither Gaze nor Glance, but Glaze: Relating to Console Game Screens." *SCAN: Journal of Media Arts and Culture* 4, no. 2 (August 2007): http://scan.net.au/scan/journal/print.php?j_id=11&journal_id=19.

Cheu, Johnson. "Performing Disability, Problematizing Cure." In *Bodies in Commotion: Disability & Performance*, edited by Carrie Sandahl and Philip Auslander. Ann Arbor, MI: The University of Michigan Press, 2005.

—. "Seeing Blindness of Screen: The Cinematic Gaze of Blind Female Protagonists." *The Journal of Popular Culture* 42, no. 3 (2009): 480–496.

Cooper, Brenda. "'Chick Flicks' as Feminist Texts: The Appropriation of the Male Gaze in *Thelma & Louise*." *Women's Studies in Communication* 23, no. 3 (Fall 2000): 277–306.

Copjec, Joan. *Read My Desire: Lacan Against the Historicists*. Cambridge: The MIT Press, 1994.

Crenshaw, Kimberlé. "Mapping the Margins: Intersectionality, Identity Politics, and Violence Against Women of Color." *Stanford Law Review* 43, no. 6 (1991): 1241–129.

de Lauretis, Teresa. *Alice Doesn't: Feminism, Semiotics, Cinema*. Bloomington: Indiana University Press, 1984.

Dixon, Winston Wheeler. *It Looks at You: The Returned Gaze of the Cinema*. Albany, NY: State University of New York Press, 1995.

Doane, Mary Ann. "Film and the Masquerade: Theorising the Female Spectator." *Screen* 23, nos 3–4 (September 1982): 74–88.

Durham, Meenakshi Gigi. "Displaced Persons: Symbols of South Asian Femininity and the Returned Gaze in U.S. Media Culture." *Communication Theory* 11, no. 2 (May 2001): 201–217.

Ellis, John. *Visible Fictions: Cinema, Television, Video*. London: Routledge & Kegan Paul, 1982.

Funnell, Lisa. "Reworking the Bond Girl Concept in the Craig Era." *Journal of Popular Film & Television* 46, no. 1 (January–March 2018): 11–21.

Gaines, Jane. "White Privilege and Looking Relations: Race and Gender in Feminist Film Theory." *Cultural Critique*, no. 4 (Autumn 1986), 59–79.

Gamman, Lorraine. "Watching the Detectives: The Enigma of the Female Gaze." In *The Female Gaze: Women as Viewers of Popular Culture*, edited by Lorraine Gamman and Margaret Marshment, 8–26. Seattle: Real Comet Press, 1989.

Gill, Rosalind. "Beyond the Sexualization of Culture Thesis: An Intersectional Analysis of 'Sixpacks,' 'Midriffs' and 'Hot Lesbians' in Advertising." *Sexualities* 12, no. 2 (2009): 137–160.

Hansen, Miriam. "Pleasure, Ambivalence, Identification: Valentino and Female Spectatorship." *Cinema Journal* 25, no. 4 (1986): 6–32.

Haskell, Molly. *From Reverence to Rape: The Treatment of Women in the Movies*. Chicago and London: University of Chicago Press, 1974.

Hawhee, Debra. *Bodily Arts: Rhetoric and Athletics in Ancient Greece*. Austin, TX: University of Texas Press, 2004.

Hein, Hilde. "Looking at Museums from a Feminist Perspective." In *Gender, Sexualities, and Museums: A Routledge Reader*, edited by Amy K. Levin. New York and London: Routledge, 2010. Kindle.

hooks, bell. *Black Looks: Race and Representation*. Boston, MA: South End Press, 1992.

Johnston, Claire. "Women's Cinema as Counter-Cinema." In *Notes on Women's Cinema*, edited by Claire Johnston, 24–31. London: Society for Education in Film and Television, 1973.

Kaplan, E. Ann. *Women and Film: Both Sides of the Camera*. New York and London: Metheun, 1983.

Karsay, Kathrin, Matthes, Jörg, Platzer, Phillip, and Plinke, Myrna. "Adopting the Objectifying Gaze: Exposure to Sexually Objectifying Music Videos and Subsequent Gazing Behavior." *Media Psychology* 21, no. 1 (January–March 2018): 27–49.

King, Claire Sisco and Gunn, Joshua. "On a Violence Unseen: The Womanly Object and Sacrificed Man." *Quarterly Journal of Speech* 99, no. 2 (May 2013): 200–208.

Kuhn, Annette. *Women's Pictures: Feminism and Cinema*. London and New York: Routledge and Kegan Paul, 1982.

Manlove, Clifford. "Visual 'Drive' and Cinematic Narrative: Reading Gaze Theory in Lacan, Hitchcock, and Mulvey." *Cinema Journal* 46, no. 3 (Spring 2007): 83–108.

McAlister, Joan Faber. "Collecting the Gaze: Memory, Agency, and Kinship in the Women's Jail Museum, Johannesburg." *Women's Studies in Communication* 36, no. 1 (2013): 1–27.

McGowan, Todd. "Looking for the Gaze: Lacanian Film Theory and its Vicissitudes." *Cinema Journal* 42 (2003): 27–47.

Modleski, Tania. *The Women Who Knew Too Much: Hitchcock and Feminist Theory*. New York: Metheun, 1988.

Mulvey, Laura. *Death 24x a Second: Stillness and the Moving Image*. London: Reaktion Books, 2006.

—. "The Pleasure Principle." *Sight and Sound* 25, no. 6 (2015): 50–51.

—. "Visual Pleasure and Narrative Cinema." *Screen* 16, no. 3 (1975): 6–18.

Neale, Steve. "Masculinity as Spectacle: Reflections on Men and Mainstream Cinema." *Screen* 24, no. 6 (1983): 2–16.

Ritland, Raeann. "Visual Pleasure from Motherhood: Alyssa Milano Challenging the Male Gaze." *Media, Culture, & Society* 40, no. 8 (November 2018): 1281–1291.

Sassatelli, Roberta. "Interview with Laura Mulvey: Gender, Gaze and Technology in Film Culture." *Theory, Culture, & Society* 28, no. 5 (2011): 123–143.

Shaviro, Steven. *Cinematic Body*. Minneapolis, MN: University of Minnesota Press, 1993.

Silverman, Kaja. "Masochism and Male Subjectivity." *Camera Obscura* 6, no. 2 (1988): 30–67.

Stacey, Jackie. "Desperately Seeking Difference: Desire between Women in Narrative Cinema." *Screen* 28, no. 1 (January 1987): 48–61.

—. *Star Gazing: Hollywood Cinema and Female Spectatorship*. London and New York: Routledge, 1994.

Thomson, Rosemarie Garland. *Extraordinary Bodies: Figuring Physical Disability in American Culture and Literature*. New York: Columbia University Press, 2017/1997.

Willemen, Peter. *Looks and Frictions: Essays in Cultural Studies and Film Theory*. London: British Film Institute, 1994.

9

VIDA

Anti-colonial queer and feminist Web TV and the gaze of allyship

Carolyn Elerding

Tanya Saracho's Web TV program *Vida* (2018–) resists stereotypes and models organized opposition to inequality through its empowering portrayals of under- and misrepresented people.[1] *Vida* explores how material and cultural appropriation correlate with numerous dimensions of social identity including class, race, ethnicity, language, region, gender, and sexual orientation. *Vida* also challenges privileged audiences to cultivate a gaze of allyship as a means of leveraging their digital labor as viewers.[2] Instead of sexualizing, stereotyping, or appropriating, such a gaze is characterized by well-informed respect for those who have been colonized, marginalized, and oppressed in the past and present. It strives to direct attention to the agency of women, LGBT★ people, and people of color as "subjects of power."[3] The purpose of this essay is to show how *Vida* and the gaze of allyship oppose myriad forms of objectification with critical subjectivity that can support and encourage political action.[4]

The number of politicized Web TV programs with intersectional queer and feminist themes is growing, as is the variety of their production processes and distribution platforms. Independently produced short-form programs such as *Brujos* (2016–) and *Her Story* (2015–) utilize YouTube and Vimeo for free-of-charge display. These series thereby avoid excluding potential viewers due to subscription costs, and as independent productions they also sidestep corporate oversight that might curtail social critique. However, despite *Vida*'s status as a commercial production, it features a notably detailed engagement with a large number of interrelated critical themes, and not just by comparison with other commercial TV series. Its content is also radical by comparison with other shows on the Starz network, which has positioned itself among the vanguard of diversity-oriented (and therefore, arguably, to some extent politically progressive) programming. *Vida* is also distinguished by rapidly rising popularity. As of late September 2019, *Vida*'s rating on Rotten Tomatoes was "100% fresh." It had also won the GLADD Media Award for Outstanding Comedy, as well as an award at the 2018 Outfest: Los Angeles Gay & Lesbian Film Festival. At that time, the second season had been available since late May, with a third season planned for production.[5]

As a program about intersectional gender issues in a rapidly changing working-class neighborhood, *Vida* invites close and critical examination of neocolonialism in relation to cultural production. By analyzing *Vida*, this essay demonstrates ways in which digital popular culture is related to the physical displacement of poor and working-class neighborhoods in which people of color, women, LGBT★ people, and immigrants are overrepresented. Drawing upon influential theorists from multiple queer, feminist, and anti-colonial currents and social movements, the conceptual framework for the analysis, like *Vida*, centers the experiences of Latinx people in relation to colonialism, gentrification, political economy, and whiteness.[6] This tapestry of resources is intended to demonstrate an approach to honoring differences while promoting coalition-building on a theoretical level.

For example, drawing upon the "U.S. Third World" feminism of the late 20th century, this essay makes frequent reference to Gloria Anzaldúa, a queer and working-class Chicana from near the border of Texas and Mexico, as well as Chandra Talpade Mohanty, an immigrant feminist of color from India. Since the publication of Anzaldúa's *Borderlands/La frontera* in 1986 and Mohanty's (2003) collection of previously published essays called *Feminism Without Borders* calls for anti-colonization and ethical allyship have acquired new urgency with the intensification of what today is known as gentrification.[7] Although the word "gentrification" is used only infrequently in the theoretical writings of Anzaldúa and Mohanty's generation, they mapped in thorough detail many of the issues at stake in this social problem. Among the other theorists mentioned are Minnie Bruce Pratt, Mia McKenzie, and Marxist social reproduction feminists in the field of digital media studies.

The term "gentrification" is used critically in this essay in a dual sense, as it is in *Vida*, to acknowledge the harmful effects of "urban renewal" on the racialized poor. First, in accordance with the classic definition formulated by Ruth Glass in 1964, gentrification refers to the displacement of low-income urban communities and the concomitant dismantling of residents' networks of mutual support when old, affordable residential and commercial buildings are replaced by new, prohibitively expensive ones.[8] Second, mentioning gentrification invokes the history of settler colonialism, whereby indigenous territory has become occupied and "owned" by nonindigenous people, as natives were violently estranged from the land as well as the social relationships that sustained them.[9] For this reason, gentrification can also be understood as an example of what Nayan Shah describes as colonial and neocolonial "estrangement" in the ongoing history of the displacement of queer people of color from their places of belonging.[10] Gentrification is additionally a contemporary mode of what Marxists have called *primitive accumulation*, which David Harvey has defined expansively as "accumulation by dispossession."[11] From these points of view, gentrification exists on a continuum of enslavement, displacement, objectification, and appropriation.[12] Today, as the neoliberal downsizing of government-sponsored social programming leaves economically vulnerable individuals, families, and communities unprotected, land-based relations of mutual assistance through neighborhood and kinship acquire heightened significance because those who have already suffered most from colonialism face new cycles of colonization in the form of gentrification. In *Vida*, the issue of gentrification is represented onscreen in a critical light, inviting the construction of a particular point of view on the audience's part.

Cinema studies have explored a number of "gazes," the socially conditioned perspectives from which viewers consume onscreen spectacles. In 1975, Laura Mulvey's foundational essay critiqued the viewing pleasures obtained through the sexually objectifying patriarchal male gaze.[13] Subsequent research and criticism has explored other oppressive forms of *le regard*—Western, imperial, and colonial, to name only a few. As well, bell hooks has conceptualized a Black feminist "oppositional gaze" that promotes agency and resistance to challenge objectification.[14] The present essay envisions a related perspective in the form of a gaze of allyship that learns to promote social equality while wrestling with the psychological and social implications of its own unearned privilege, as well as the ocularcentric—that is, acquisitive, rationalizing, scrutinizing, alienating, and quantifying—perspective often conferred by such advantages.[15] Accustomed to unproblematized ease and belonging, many would-be allies have much to learn about navigating the rough terrain of political awakening, which Chandra Talpade Mohanty describes in terms of "its tentativeness, its consisting of fits and starts, and the absence of linear progress toward a visible end."[16] Viewing in allyship affords opportunities for such learning and practice.

Written in allyship with *Vida*'s production community as well as with anti-gentrification activism, the close reading of *Vida* in the present essay explores some of the program's social and technological contexts as well as its content. The analysis shows how *Vida*'s depiction of gender, sexuality, and the threat of displacement in a working-class Latinx neighborhood is actively anti-colonial, establishing groundwork for bridging sociocultural differences to increase understanding in ways that may build political coalition. The essay explores *Vida*'s production and distribution processes and then turns to the show's onscreen material. Asking how viewers involved in allyship might best engage politically with the anti-colonial queer and feminist Web TV programming made available by the experimental and less constraining environments of "over-the-top" (webstreaming) delivery, the analysis explores how a gaze of allyship can contribute materially to the growing social movement that *Vida* expresses and inspires.[17]

Web TV and the post-auteurism of the showrunner

Due largely to the development of new communications technologies such as Web TV platforms, U.S. popular culture in the late 20th and early 21st centuries has seen some improvement in the representation of women, LGBT★ people, and people of color.[18] In television, indications of progress toward equality have surfaced behind the scenes in the industry as well as onscreen. The year 2015 was record-breaking in terms of gender equity in American TV, despite an impending turn toward racism and misogyny in national politics the following year. Women won numerous TV industry awards and several female-centered programs premiered simultaneously. African-American showrunner Shonda Rhimes produced three concurrent hits, and Jenji Kohan's unprecedentedly diverse *Orange Is the New Black* was renewed for a third year. Despite these notable advances, objectionable characterizations of women, LGBT★ people, and people of color have persisted, as has the overrepresentation of young, straight, cisgendered, and white people.[19]

Women, people of color, and LGBT★ people have gained only a small share of material power over television, one of the most significant forms of contemporary cultural production.[20] Overall, the industry also remains white-, cis/hetero-, and male-dominated in terms of hiring and salary equity.[21] Behind-the-scenes personnel disparities in the newer digital forms of distribution and production are only slightly alleviated compared with broadcast TV.[22] This underrepresentation of diversity offscreen continues despite the much celebrated "disruption" that cable, satellite, and especially Internet distribution have wrought upon television since the 1990s by splintering its audiences into niches that often correspond to social identities.[23] At the same time, by comparison with broadcast TV, the new forms of distribution allow for greater exploration of sexual and political content, often bundling the two together.[24] As a result, "narrowcasting" has normalized diversity-oriented programming that would seem overly risky to a broadcasting network executive catering to a "family" audience long presumed to be white and middle-class.[25]

Concurrent with these technological and representational shifts, a significant new role has emerged in TV production: the showrunner. Going beyond the multiple responsibilities conventionally attributed to a writer-producer or head writer, a showrunner juggles an even greater number of tasks to coordinate the creative focus and material logistics of a television series.[26] Miranda Banks writes that the showrunner's role

> demands the skills of a visionary: someone who can hold the entire narrative of the series in their head; who is the gatekeeper of language, tone, and aesthetics on the set and behind the scenes; who knows where the series has been and [has] a sense, if not a plan, for its future.[27]

Although showrunners often receive no official screen credit, they are acknowledged in the industry and among fans as the major impetus behind the most successful TV programs. Increasing numbers of showrunners are women, queer, or of color. These numbers are doubly significant given that industry research indicates strongly that, for example, the presence of a woman showrunner greatly increases opportunities for women writers and actors. As television critic Joy Press writes, "It's an ever-expanding circle in which powerful female showrunners can enable others to create cultural images and narratives that inspire the next generation of powerful women."[28]

The successes of women, LGBT★ people, and people of color as showrunners suggest a nascent redistribution of privilege as well as new uses for old terminology. The term *auteur*, for instance, has been rightfully criticized in cinema and media studies for its individualistic elitism and consequent erasure of the collaborative nature of the production process.[29] However, in the context of Web TV's emerging equitability, it auteurism can be appropriated and used restoratively in a new form. Press suggests that showrunners are more deserving of the title of auteur than were film directors.[30] This seems especially appropriate given the increasing diversity of showrunners as well as the length of a television series by comparison with a feature film. Hence, *post-auteurism* celebrates the concentration of agency and resources now wielded by the growing number of showrunners from marginalized social backgrounds.[31] The figure of the *post-auteur* therefore evokes a reversal of objectification and a reinforcement of empowerment.

Latinx showrunner Tanya Saracho's post-auteurism vividly exemplifies the opportunities and challenges of diverse and politicized Web TV. Saracho gained experience by working with other woman showrunners on Lena Dunham's *Girls* (2012–2017) and on *How to Get Away with Murder* (2014–), for which Rhimes was executive producer. Saracho used her power as showrunner in inventive ways to facilitate a notably collective and multicultural approach to numerous aspects of *Vida*'s creation.[32] Telling the stories of queer, feminist, and working-class Latinx characters resisting patriarchy, racism, and homophobia, as well as the gentrification of their Los Angeles neighborhood due to rising rents and taxes as well as exploitative lending. Saracho's *Vida* illuminates possibilities for active opposition to the displacement of poor and working-class minorities. Saracho explains that the show intentionally foregrounds Latinxs, women, and LGBT★ people behind the scenes as well as onscreen: the "entire writer's room is Latinx. Half of it is queer. Our editors are all women. Our cinematographer is Latina. Our composer is a Latina."[33] Using an architectural metaphor suggestive of the story's rapidly gentrifying urban setting, Saracho—a resident of the Pilsen district of Chicago, a gentrification hot zone—says that "building the story" in this way "with a lot of brown females" results in "no need for translation because you have that cultural shorthand." In the same interview, Saracho also suggests obliquely that the series encourages viewing in allyship, emphasizing that "*Vida* is going to be a lot of things to a lot of different people."

Situated and subjectifying: a post-realism from the margins

Saracho's post-auterial vision and practice offers a sustained critique of gentrification from a situated, queer, and feminist Latinx perspective that empowers and subjectifies the cast, crew, community, and audience through sophisticated characterization of both the people and the place in which they live.[34] In a series of online interviews, *Vida*'s cast members discuss their excitement at the opportunity to play Latinx characters with such an unusual degree of complexity for traditional TV.[35] Saracho works to create non-stereotypical—in other words, realistic—Latinx characters.[36] Even the relatively minor figure of Nelson, an unscrupulous mortgage broker who grew up in the neighborhood, expresses a detailed web of tensions. Another example is the character of Vidalia, whose forceful posthumous presence is the center around which the narrative orbits. Vidalia's years of shame as a closeted lesbian scarred her queer daughter Emma, another complexly figured character. Emma, a successful businessperson living in Chicago, now faces responsibility for her late mother's struggling bar, one of the only queer Latinx spaces in the Boyle Heights neighborhood. Each character's details are as contradictory as the multiple worlds in which they constantly cross borders to live.

Saracho attributes her capacity for character development to her experience as a playwright. Conceptualizing *Vida*'s first six episodes in the form of a play, Saracho framed the entire season as roughly equivalent to a "three-hour pilot."[37] This approach allowed her the duration and continuity to create "Latinx characters that aren't one-dimensional."[38] *Vida*'s heightened degree of detail differs, however, from most literary and cinematic realism because it is a collective self-portrait in which marginalized people, rather than privileged outsiders, represent themselves. Like post-auteurism, such realism from the margins suggests a subversion of previous eras' "death of the subject" critiques.[39] Rather than solely deconstructing privileged subjectivity, *Vida* enhances the subjectivities of marginalized people, implicitly proposing a redistribution of cultural and, by extension, material resources to less advantaged communities.

In addition to its realistic characterizations, *Vida* simulates the temporality of "real life" experience. In interviews, cast members discuss two of Saracho's unconventional practices that contribute to *Vida*'s feeling of immediacy. First, Saracho extends the writing to overlap into production. Hence, rather than separating the writing and filming processes into a conventionally linear sequence, Saracho makes them concurrent and mutually recursive. Second, rather than delivering scripts to cast members well in advance of shooting as is customary, she distributes lines only shortly before.[40] Due to both of these techniques, the resulting performances cyclically inform the ongoing process of writing, creating an intensely naturalistic style of acting grounded in a close approximation of experience in continuous real time.[41] The experience of the cast members during filming is more spontaneous, less fully predetermined, and therefore more agential and empowering than in more conventional TV production.[42]

Vida's critiques of gentrification are situated geographically as well as socially. Describing the setting itself as a character, members of the cast and crew mention how greatly they value the experience of shooting on location, which has become rare in Los Angeles and in California more generally.[43] In the same set of interviews, cast members also emphasize how Saracho maintains high standards for respectful interaction with residents as well as learning and honoring the history of the Boyle Heights neighborhood, without shying away from criticizing what could be improved—all practices that exemplify allyship. For example, as the character Mari points out, a racist anachronism is produced by the name of Vidalia's bar, La Chinita ("the little Chinese girl"—which is questionable enough already), combined with the large image of a stereotyped Japanese geisha on its vintage sign. As Mari suggests, it is curious that contemporary neighborhood residents fail to notice that the familiar sign is offensive, let alone protest it. This fleeting moment in the story serves as a reminder that the neighborhood's many former Japanese American residents faced racism and stereotyping and ultimately were displaced from their homes and communities during the Second World War. Boyle Heights' Japanese American community was not disrupted by gentrification, although many did indeed lose property and become too impoverished to return, but rather by internment. By referring to a deeper and broader local history that involves Asian Americans as well as Latinxs, *Vida* shows how those who benefit from allyship can also extend it to others. Through such complex characterization of the neighborhood, Saracho creates a realistic portrayal demonstrating allyship that gestures toward coalition building across social and cultural differences.

Digital labor and the gaze of allyship

Research in audience and fan studies has closely examined viewers' agency in shaping the cultural production that influences their lives.[44] While it is certain that viewers have gained increasing creative control, particularly through their fandom on the Internet and social media in the course of consuming politicized and diverse Web TV such as *Vida*, audience agency can furthermore become a basis for political resistance and allyship. Feminist theories of digital labor prove helpful for considering how this is so. Through their digitally mediated cultural practices, Web TV viewers, much like users of social media, produce significant material value while constructing webs of mutual care and political alliance. The incidental data that social media users generate as they upload, rate, and comment upon content gets monetized through aggregated purchasing data culled by marketing firms, and in this sense social media can be understood as exploitative of users. Similarly, Web TV viewers provide valuable information to the platforms distributing their programming, and this could be considered a form of exploitation as well.[45] However, as Kylie Jarrett argues, belaboring social media also leaves users with what Marxian analysis refers to as "inalienable use-values," such as relationships, empowering feelings, or critical knowledge. Among these inalienable use-values obtained from participating in and valorizing social media, it is reasonable to include lessons in allyship, and the same could be said of Web TV viewing. With respect to watching Web TV—certainly the active fandom carried out online, but also the mere process of viewing—feminist theories of digital labor and "networked social reproduction" that explain economies of online sharing and caring offer an additional starting point for tracing the concrete implications of audience agency for activism and social movements.[46]

Arguably, theories of digital labor and networked social reproduction are perhaps even more applicable to Web TV than social media, since they rely on Dallas Smythe's classic formulation of the commodification of broadcast TV viewers' labor-time, a concept later revitalized by theorists of Internet economy.[47] The digital labor of the viewer produces economic value that enriches the socioeconomic system overall as well as the corporate interests that underwrite gentrification.[48] In this sense, as Smythe argued, viewers are objectified and exploited as an audience rendered commodity and ultimately through this process made more vulnerable both economically and ideologically. Yet the content of the programming can reinforce viewers' critical subjectivity, constituting an inalienable use-value, and furthermore leading to greater mutual well-being and potentially to social change. Hence, seeing the viewing process as networked social reproduction also suggests that a gaze of allyship can, to a limited extent and in a relatively indirect manner, promote socioeconomic equality.

For Web TV viewers in allyship to commit their digital labor and networked social reproduction to learning to avoid cultural appropriation can be a valuable contribution to anti-colonial projects such as resisting the gentrification of low-income urban neighborhoods. Cultural appropriation is likely the form of colonial objectification most familiar to this volume's readership. It receives sustained attention on college campuses in the United States, particularly in relation to Halloween costumes that borrow haphazardly from non-Western cultures. Gloria Anzaldúa connects appropriation explicitly with colonialism, as do Mohanty and Biddy Martin in their exploration of Minnie Bruce Pratt's discussion of "cultural impersonation": white women denying their guilt and implication in social inequalities by imitating "the identity of the other in order to avoid not only guilt but pain and self-hatred."[49] Culturally underwriting this dialectic of appropriation is the problematic but common notion of whiteness as an unmarked absence of identity, a stance Mohanty and Martin describe as similarly grounded in privileged denial motivated by fear of the "loss that accompanies change" (such as losing one's position in one's community or family by failing to convincingly perform belonging).[50] Anzaldúa counters claims that whiteness has no culture by arguing that "this whole country and the dominant culture is their culture. What white people watch on television is their culture."[51] Anzaldúa further argues that, to avoid appropriating the cultural roots of others, white people should explore their own folk heritages—without romanticizing.[52]

Given this complexity, for viewers aspiring to allyship to consume a diverse and politicized program such as *Vida* in a non-appropriative manner can present a significant challenge due to the blind spots produced by privilege. Yet Anzaldúa does not deny the potential for such undertakings. Anzaldúa describes "alliance work" as "the attempt to shift positions, change positions, reposition ourselves regarding our individual and collective identities … we are confronted with the problem of how we share or don't share space."[53] Anzaldúa argues that, by crossing borders to less advantaged forms of experience, the subjectivities of the privileged can (to some extent) become liminal, hybrid, or mixed. Proceeding in a well-informed, aware, and respectful manner, white allies can practice what Anzaldúa frames as cultural, intellectual, affective, and even spiritual mestizaje.[54] Anzaldúa describes this "new mestiza" as a "liminal subject who lives in borderlands between cultures, races, languages, and genders. In this state of in-betweenness the mestiza can mediate, translate, nego-tiate, and navigate these different locations."[55] Such "bridge work" requires reflection. Allyship necessitates formulating difficult questions of oneself, such as: "What can I do with my privileges? … Am I one more white woman ripping off yet another culture? Am I one more white woman bringing her guilt and wanting to be exonerated?"[56] Grounded in such reflexivity, consuming anti-colonial Web TV can support the critical self-reflection needed to engage in substantive allyship.

Numerous African-American writers, including award-winning and best-selling authors Ta-Nehisi Coates and Roxane Gay, have written persuasively and with great clarity about the term "ally."[57] Gay insists, "[t]he problem with allyship is that good intentions are not enough." One must "support other marginalized people—making their fights [one's] own because that's the only way forward."[58] According to Gay's view, without sufficient efforts that lead to less appropriative, less oppressive practices away from the screen, an ally's gaze would fall short of opposing inequality, even when viewing politicized Web TV—unless such viewing leads to action. For similar reasons, Mia McKenzie, founder of *Black Girl Dangerous*, a widely read anti-racist, queer, and feminist website, eschews the term "ally" altogether in favor of "allyship."[59] McKenzie frames allyship as ongoing and specific action: "currently operating in solidarity with …," "showing support for …," "operating with intentionality around …," "using [one's] priv-ilege to help by …," "demonstrating [one's] commitment to ending [insert oppression system] by …," and "showing up for [insert marginalized group] in the following ways …" Taking Gay and McKenzie's criticisms into account, the gaze of allyship must be *actively* anti-colonial.[60] More than a static identity, allyship is a social process—both "a promise and a way of life."[61]

McKenzie furthermore argues that the often well-intentioned role of the ally is frequently co-opted and gets frozen into an identity or pose rather than enacted as a practice and process: "It's supposed to be a way of living your life that *doesn't* reinforce the same oppressive behaviors you're claiming to be against." As McKenzie further explains, the role of the ally often gets reduced to a misguided or insincere "performance" centering the emotions, desires, and status of the socially advantaged ally rather than those subjected to historical and ongoing societal mar-ginalization. Similarly, Mohanty and Martin write in the early 1980s about their frustration with middle-class white feminists' "self-paralyzing guilt and/or defensiveness," as well their needless dependence upon women of color for the multicultural education they could obtain through other means.[62] Anzaldúa also mentions the difficulty of dealing with white women's self-entitled focus on their own emotions, which leads them to seek "reassurance, acceptance, and valid-ation from women of color."[63] If viewing Web TV is passive by comparison with other political practices, it is also true that watching programs such as *Vida* can provide some of the education and emotional processing that allies might otherwise seek from those they wish to support. In this sense, the digital labor of viewing, when conducted through a gaze of allyship, equates to time and energy saved on behalf of activists of color.

Parallel to the aforementioned post-auteurism and realism from the margins enabled by the politicized Web TV showrunner, but analyzing *Vida* at the point of consumption rather than production, certain claims begin to take shape. First, through their digital labor and networked social reproduction, viewers can meet *Vida*'s non-stereotypical representations halfway by actively developing a less appropriative, less oppressive gaze of allyship. Second, even watching TV in private and for entertainment, but from an oppositional point of view, can become a small act of anti-colonial queer and feminist alliance that meaningfully exceeds merely supporting a series as a consumer. This is particularly so if the program being consumed is as diverse and politicized as *Vida*. As McKenzie observes, "allyship is an every day practice" due to the constant operation of systems of oppression, and therefore "the work of an ally is never ceasing."[64] McKenzie writes that "[i]t's a creative thing that must be done over and over again, in the largest and smallest ways, *every day*."[65] Viewing Web TV with a gaze of allyship can provide training that can reorient other activities toward allyship.

By appealing to aspiring allies as well as to the Latinx communities it represents on- and offscreen, offering training in non-objectification and non-appropriation, *Vida* contributes to a cultural basis for building coalition across social differences. As the first season of *Vida* progresses, viewers enter the interrelated and contradictory tensions comprising daily experience in Boyle Heights. For example, even though Emma is at first dismissive of the neighborhood while Mari struggles actively to protect her community, it is difficult to argue with Emma's criticisms of Mari's aggressive behavior toward anglos, given that locally owned businesses in Boyle Heights struggle financially. Through such subplots, audiences can see that, just as there are no stereotypes in *Vida*'s realistic representation of Chicanx life, there is no pure dualism of powerless victim and powerful oppressor, thus leaving room in the portrayal for realistic agency and resistance. This is just one example of the many lessons that *Vida*'s viewers in allyship can begin to learn by performing the digital labor and networked social reproduction of watching the show.

Conclusion: leaving home through the eyes of allyship

The digital labor and networked social reproduction involved in the anti-colonial Web TV gaze of allyship are by no means negligible in scope, scale, or material significance, despite typically occurring during leisure time and in private. The gaze of allyship demanded by *Vida*'s situated and subjectifying characterization can train viewers in the critical self-reflexivity that leads away from objectification or appropriation and instead toward understanding and active resistance in coalition. This is not to claim that, by itself, watching a web series constitutes allyship's full political potential. After all, it is possible to consume even the most critical programming in a passive, unthinking, or ill-informed—in other words, appropriative—manner. Rather, viewing programs such as *Vida* with ethical intentionality can become one element among many in the process of "currently operating in solidarity with ..." However, for the anti-colonial digital labor and networked social reproduction of allyship to become effective resistance in material terms, its force must concretely surpass the economic value of the data it produces, and that force must be protected and expanded by the community engagement it generates.

There are aspects of anti-colonial cultural production and networked social reproduction that allies should not or cannot do, out of concern for avoiding appropriation. However, there are also contributions that *only* allies can—and must—make: namely, to become other than what they currently are, without abdicating responsibility for the privilege of unearned advantages. Viewing, like anything else, should therefore be approached with maximum intentionality if it is to lead to action.

The self-reflexivity required for effective anti-gentrification allyship is easily obstructed by unexamined desires for safety and comfort. The feelings associated with being at home involve a sense of belonging that is rarely recognized as relying on the exclusion of others, and this oversight provides a cultural buttress for gentrification. With an anti-colonial gaze of allyship, however, viewers can gain experience with the painful but necessary self-criticism that leads to political struggle, toward more effective allyship through better, more thoughtful, and more equitable material relations with others. As Mohanty writes:

> Illusions of home are always undercut by the discovery of the hidden demographics of particular places, as demography also carries the weight of histories of struggle .... Each of us carries around those growing-up places, the institutions, a sort of backdrop, a stage set. So often we act out the present against the backdrop of the past, within a frame of perception that is so familiar, so safe that it is terrifying to risk changing it even when we know our perceptions are distorted, limited, constricted by that old view.[66]

In addition to rigorously forswearing objectification and appropriation, as well as practicing critical self-reflection, effective allyship ultimately renounces entitlement to the comforts of safety and belonging, to focus instead on the oppositional project of the marginalized. The privileged viewer adopting a gaze of allyship must leave home and never look back.

Notes

1 Content note for teaching purposes: *Vida* contains harsh language and explicit sexuality.
2 The phrase "gaze of allyship" is the author's contribution to the theories of cinematic gazes first conceptualized by Laura Mulvey in her essay theorizing the male gaze, "Visual Pleasure and Narrative Cinema" first published in 1975. The author wishes to acknowledge valuable discussions of this idea with colleagues at HASTAC 2019 at the University of British Columbia in Vancouver and at Persistence of Vision: Reframing Animation 2019 at Goldsmiths University in London.
3 Chandra Talpade Mohanty, *Feminism Without Borders: Decolonizing Theory, Practicing Solidarity* (Durham: Duke University Press, 2003), 37. "LGBT" stands for lesbian, gay, bisexual, and transgender. The asterisk has been added to signal the complexity and ongoing negotiation of identities and differences among non-heterosexual and non-cisgendered people.
4 Donna Haraway, "Situated Knowledges: The Science Question in Feminism and the Privilege of Partial Perspective," *Feminist Studies* 14, no. 3 (1988): 575–599. The close reading presented in this essay is "situated" in the sense that Haraway conceptualizes as producing knowledge from a particular embodied, socially embedded, historically specific location. The author of the present essay is attempting to write in anticolonial allyship, working to decenter their white privilege and resist the ways they have been socialized to appropriate and objectify. Their interpretation of *Vida* is not meant to represent the perspective of a universal viewer, nor a universal gaze of allyship, though it does suggest some features that could be common to many such gazes, as well as some useful viewing practices.
5 Mishel Prada, interviewed by Erika Abad, "*Vida*: Family, Love, & Identity," April 22, 2019, panel at ClexaCon, April 11–15, 2019. www.youtube.com/watch?v=mhIVosC6Ce8&fbclid=IwAR1Vt1xxyO ctJwfA8Dah2-h7V146kqNdE90hgMYuEHdU2oBGD2ehxGgODPc.
6 The author is grateful to her interlocutors at HASTAC 2019 for their comments on this approach. While the primary intended audience for *Vida* may be Latinx, a central goal of this essay is to reach students and other readers interested in allyship, regardless of their social identities. As a person of white privilege, in writing this article the author hopes that the value of their intellectual and political contribution outweighs its flaws.

7 Gloria Anzaldúa, *Borderlands/La frontera: The New Mestiza* (San Francisco, CA: Spinters/Aunt Lute Books, 1987).

8 Ruth Glass, *London: Aspects of Change* (London: MacGibbon & K33, 1964), cited in Neil Smith, *The New Urban Frontier: Gentrification and the Revanchist City* (London and New York: Routledge, 1996). On page 32, Smith writes that gentrification is the process "by which poor and working-class neighborhoods in the inner city are refurbished via an influx of private capital and middle-class homebuyers and renters— neighborhoods that had previously experienced disinvestment and a middle-class exodus."

9 For an overview of anticolonial queer studies, see Nayan Shah, "Queer of Color Estrangement and Belonging," in *The Routledge History of Queer America*, ed. John Romesburg (New York: Routledge, 2018). Shah's definition of gentrification as a form of colonialism is also supported by the following entry in the *Oxford English Dictionary Online*: "To occupy (land) as a bona-fide settler." For an expansive critique of the use of "de-colonizing" to describe nonindigenous social justice projects, see Eve Tuck and K. Wayne Yang, "Decolonization Is Not a Metaphor," *Decolonization: Indigeneity, Education and Society* 1 (2012): 1–40. In accordance with Tuck and Yang's criticism, the term "anticolonial" is preferred over "de-colonial" in the present essay.

10 Shah, "Queer of Color Estrangement and Belonging." See also Sandra Harding, "Latin American Decolonial Studies of Scientific Knowledge: Alliances and Tensions," *Science, Technology, & Human Values* 41, no. 6 (2016): 1063–1087; as well as María Lugones, "Heterosexualism and the Modern/Colonial Gender System," *Hypatia* 22, no. 1 (Winter 2007): 186–209. For a study of LGBT★ communities in relation to gentrification, see Christina B. Hanhardt, *Safe Space: Gay Neighborhood History and the Politics of Violence* (Durham and London: Duke University Press, 2013).

11 David Harvey, *The New Imperialism* (Oxford and New York: Oxford University Press, 2003), 137–182. See also Karl Marx, *Capital*, Chap. 31, Vol. I (Moscow: Progress Publishers, 1965).

12 Jared Sexton, "The *Vel* of Slavery: Tracking the Figure of the Unsovereign," *Critical Sociology* 42, nos 4–5 (2016): 1–15.

13 Laura Mulvey, "Visual Pleasure and Narrative Cinema," in *Film Theory and Criticism: Introductory Readings*, 7th edn, eds Leo and Marshall Cohen (New York: Oxford University Press, 2009), 803–816.

14 bell hooks, "The Oppositional Gaze: Black Female Spectator," in *The Feminism and Visual Culture Reader*, ed. Amelia Jones (New York: Routledge, 2003), 94–105.

15 On the colonizing Western gaze in relation to epistemology, see Gloria Anzaldúa, "Haciendo caras, una entrada," in *The Gloria Anzaldúa Reader*, ed. AnaLouise Keating (Durham, NC: Duke University Press, 2009), 131. See also Haraway, "Situated Knowledges," 575–599.

16 Mohanty, *Feminism Without Borders*, 100.

17 Prada, "*Vida*."

18 "Pride in Progress: Nielsen's TV Panel Enhancement Better Identifies Same Gender Spouse and Partner Audiences," *Nielsen*, accessed September 22, 2019, www.nielsen.com/us/en/press-room/2018/pride-in-progress-nielsens-tv-panel-enhancement-better-identifies-gender-spouse-partner-audiences.html.

19 For an overview of gender studies on representation in various media, see Carolyn M. Byerly, "The Geography of Women and Media Scholarship," in *The Handbook of Gender, Sex, and Media*, ed. Karen Ross (Hoboken, NJ: Wiley-Blackwell, 2012), 3–19. On the history of LGBT★ representation in TV, see Katherine Sender's documentaries, *Off the Straight & Narrow* (Northampton, MA: Media Education Foundation, 1999) and *Further Off the Straight & Narrow* (Northampton, MA: Media Education Foundation, 2006).

20 Stacy L. Smith, March Choueiti, and Katherine Pieper, *Inclusion or Invisibility? Comprehensive Annenberg Report on Diversity in Entertainment* (Los Angeles: IDEA, 2016), 12–16; Joy Press, *Stealing the Show: How Women Are Revolutionizing Television* (New York: Atria/Simon & Schuster, 2018), 2–3.

21 Press, *Stealing the Show*, 3.

22 Amanda Lotz, *The Television Will Be Revolutionized*, 2nd edn (New York: New York University Press, 2014), 233–262; Press, *Stealing the Show*, 3.

23 Lotz, *The Television Will Be Revolutionized*, 40–46.

24 Smith, Choueiti, and Pieper, *Inclusion or Invisibility?*, 12–16.

25 Lotz, *The Television Will Be Revolutionized*, 167–205.

26 Miranda J. Banks, "*I Love Lucy*: The Writer-Producer," in *How to Watch Television*, eds Ethan Thompson and Jason Mittell (New York: New York University Press, 2013), 245.

27 Banks, "*I Love Lucy*," 245.

28 Press, *Stealing the Show*, 3.

29 By most accounts, the phrase "auteur theory" was originated by Andrew Sarris in "Notes on the Auteur Theory in 1962," *Film Culture* 27 (Winter 1962–1963): 1–8.

30 Press, *Stealing the Show*, 9.

31 The terms and concepts "post-auteurism" and "post-auteur" are new terms developed by the author of the present essay.

32 The directors of the episodes are: 101, Alonso Ruizpalacios; 102, So Yong Kim; 103, Rashaad Ernesto Green; 104, Rose Troche; 105, Catalina Aguilar Mastretta; 106, Rose Troche. Using different directors for various episodes in a series is not unusual in contemporary TV, but Saracho's degree of attention to diversity among directors is.

33 Starz, "World of *Vida*," accessed September 22, 2019, promotional video, "World of *Vida*," www.starz.com/video/a3424e5d-9773-400f-b021-f003c4abf2ff. The video is located on the *Vida* page of the Starz website under the tab labeled "Extras" ("Season 1 Extras").

34 The concept of situated knowledge and the manner in which this essay addresses subject-object relations are derived from Haraway, "Situated Knowledges," 575–599.

35 Ben Kenber, "The Ultimate Rabbit," interviews with cast, accessed September 22, 2019, https://theultimaterabbit.com.

36 Starz, "World of *Vida*."

37 Starz, "World of *Vida*."

38 Kenber, "Ultimate Rabbit." The quote is from Chelsea Rendon, who played the role of Mari.

39 Haraway, "Situated Knowledges," 585–586.

40 Kenber, "Ultimate Rabbit."

41 Kenber, "Ultimate Rabbit." Melissa Barrera (Lyn) uses the term "natural."

42 For a classic exposition of the concepts of structure and agency, see Anthony Giddens, *The Constitution of Society* (Cambridge: Polity Press, 1984).

43 Kenber, "Ultimate Rabbit."

44 Henry Jenkins, *Fans, Bloggers, and Gamers: Exploring Participatory Culture* (New York: New York University Press, 2006).

45 On algorithmic platforms in Web TV, see Tricia Jenkins, "Netflix's Geek Chic: How One Company Leveraged Its Big Data to Change the Entertainment Industry," *Jump Cut* 57 (2016): n.p., www.ejumpcut.org/archive/jc57.2016/-JenkinsNetflix/. For more on queer and feminist Web TV in relation to digital labor, see Carolyn Elerding, "The Digital Labor of Queering Feminist Web TV," *First Monday* 23, nos 3–5 (March 2018): n.p., http://dx.doi.org/10.5210/fm.v23i3.8289.

46 The phrase "networked social reproduction" was first used by Elise Thorburn in "Networked Social Reproduction: Crises in the Integrated Circuit," *TripleC: Communication, Capitalism & Critique* 14, no. 2 (2016): 380–396, https://doi.org/10.31269/triplec.v14i2.708.

47 Christian Fuchs, "Google Capitalism," *TripleC: Communication, Capitalism & Critique* 10, no. 1 (2012): 42–48, https://doi.org/10.31269/triplec.v10i1.304; Ursula Huws, "Eating Us Out of House and Home: The Dynamics of Commodification and Decommodification of Reproductive Labour in the Formation of Virtual Work," *International Journal of Media and Cultural Politics* 14, no. 1 (2018): 111–18; Tiziana Terranova, "Free Labor: Producing Culture for the Digital Economy," *Social Text* 18, no. 2 (2000): 33–58.

48 For an analysis of this tension in the context of social media, see Kylie Jarrett, *Feminism, Labour, and Digital Media: The Digital Housewife* (New York and London: Routledge, 2016).

49 Mohanty, *Feminism without Borders*, 102.

50 Mohanty, *Feminism without Borders*, 102–103.

51 Gloria Anzaldúa, "The New Mestiza Nation," in *The Gloria Anzaldúa Reader*, ed. AnaLouise Keating (Durham and London: Duke University Press, 2009), 214.

52 Anzaldúa, "New Mestiza Nation." 214.

53 Anzaldúa, "New Mestiza Nation." 143.

54 Anzaldúa, "New Mestiza Nation." 192–203.

55 Anzaldúa, "New Mestiza Nation." 209.

56 Anzaldúa, "New Mestiza Nation." 213.

57 Ta-Nehisi Coates, *Between the World and Me* (New York: Spiegel & Grau, 2015).

58 Roxane Gay, "On Making Black Lives Matter," *marie claire*, July 11, 2016, www.marieclaire.com/culture/a21423/roxane-gay-philando-castile-alton-sterling/.

59 Mia McKenzie, "No More 'Allies'," *Black Girl Dangerous*, September 30, 2013, www.bgdblog.org/2013/09/no-more-allies/.

60 Mohanty, *Feminism without Borders*, 100.
61 Mohanty, *Feminism without Borders*, 93.
62 Anzaldúa, "Haciendo caras," 130.
63 Becky Thompson, *A Promise and a Way of Life: White Antiracist Activism* (Minneapolis and London: University of Minnesota Press, 2001).
64 Thanks to Erika Abad for her provocations on this topic.
65 Original emphasis.
66 Mohanty, *Feminism without Borders*, 90.

Bibliography

Anzaldúa, Gloria. *Borderlands/La Frontera: The New Mestiza*. San Francisco, CA: Spinsters/Aunt Lute Books, 1987.

Anzaldúa, Gloria. "Haciendo caras, una entrada." In *The Gloria Anzaldúa Reader*, edited by AnaLouise Keating, 124–139. Durham, NC: Duke University Press, 2009.

Anzaldúa, Gloria. "The New Mestiza Nation: A Multicultural Movement." In *The Gloria Anzaldúa Reader*, edited by AnaLouise Keating, 203–216. Durham, NC: Duke University Press, 2009.

Banks, Miranda J. "*I Love Lucy*: The Writer-Producer." In *How to Watch Television*, edited by Ethan Thompson and Jason Mittell, 244–252. New York: New York University Press, 2013.

Byerly, Carolyn M. "The Geography of Women and Media Scholarship." In *The Handbook of Gender, Sex, and Media*, edited by Karen Ross, 3–19. Hoboken, NJ: Wiley-Blackwell, 2012.

Coates, Ta-Nehisi. *Between the World and Me*. New York: Spiegel & Grau, 2015.

Elerding, Carolyn. "The Digital Labor of Queering Feminist Web TV." *First Monday* 23, nos 3–5 (March 2018). http://dx.doi.org/10.5210/fm.v23i3.8289.

Fuchs, Christian. "Google Capitalism." *TripleC: Communication, Capitalism & Critique* 10, no. 1 (2012): 42–48. https://doi.org/10.31269/triplec.v10i1.304.

Gay, Roxane. "On Making Black Lives Matter." *marie claire*, July 11, 2016. www.marieclaire.com/culture/a21423/roxane-gay-philando-castile-alton-sterling/.

Giddens, Anthony. *The Constitution of Society*. Cambridge: Polity Press, 1984.

Glass, Ruth. *London: Aspects of Change*. London: MacGibbon & Kee, 1964.

Hanhardt, Christina B. *Safe Space: Gay Neighborhood History and the Politics of Violence*. Durham, NC and London: Duke University Press, 2013.

Haraway, Donna. "Situated Knowledges: The Science Question in Feminism and the Privilege of Partial Perspective." *Feminist Studies* 14, no. 3 (1988): 575–599.

Harding, Sandra. "Latin American Decolonial Studies of Scientific Knowledge: Alliances and Tensions." *Science, Technology, & Human Values* 41, no. 6 (2016): 1063–1087.

Harvey, David. *The New Imperialism*. Oxford and New York: Oxford University Press, 2003.

hooks, bell. "The Oppositional Gaze: Black Female Spectator." In *The Feminism and Visual Culture Reader*, edited by Amelia Jones, 94–105. New York: Routledge, 2003.

Huws, Ursula. "Eating Us Out of House and Home: The Dynamics of Commodification and Decommodification of Reproductive Labour in the Formation of Virtual Work." *International Journal of Media and Cultural Politics* 14, no. 1 (2018): 111–118.

Jarrett, Kylie. *Feminism, Labour, and Digital Media: The Digital Housewife*. New York and London: Routledge, 2016.

Jenkins, Henry. *Fans, Bloggers, and Gamers: Exploring Participatory Culture*. New York: New York University Press, 2006.

Jenkins, Tricia. "Netflix's Geek Chic: How One Company Leveraged Its Big Data to Change the Entertainment Industry." *Jump Cut* 57 (2016). www.ejumpcut.org/archive/jc57.2016/-JenkinsNetflix/.

Keating, AnaLouise, ed. *The Gloria Anzaldúa Reader*. Durham, NC: Duke University Press, 2009.

Kenber, Ben. "The Ultimate Rabbit." Accessed September 22, 2019. Video interviews. https://theultimaterabbit.com.

Lotz, Amanda. *The Television Will Be Revolutionized*. 2nd edn. New York: New York University Press, 2014.

Lugones, María. "Heterosexualism and the Modern/Colonial Gender System." *Hypatia* 22, no. 1 (Winter 2007): 186–209.

Marx, Karl. *Capital Volume I*. Moscow: Progress Publishers, 1965.

McKenzie, Mia. "No More 'Allies'." *Black Girl Dangerous*, September 30, 2013. Blog. www.bgdblog. org/2013/09/no-more-allies/.

Mohanty, Chandra Talpade. *Feminism without Borders: Decolonizing Theory, Practicing Solidarity*. Durham, NC: Duke University Press, 2003.

Mulvey, Laura. "Visual Pleasure and Narrative Cinema." In *Film Theory and Criticism: Introductory Readings*, 7th edn, edited by Leo Braudy and Marshall Cohen, 803–816. New York: Oxford University Press, 2009.

Nielsen Media Research. "Pride in Progress: Nielsen's TV Panel Enhancement Better Identifies Same Gender Spouse and Partner Audiences." Nielsen Holdings Plc. October 25, 2018. Press Release. www.nielsen.com/us/en/press-room/2018/pride-in-progress-nielsens-tv-panel-enhancement-better-identifies-gender-spouse-partner-audiences.html.

Mishel Prada, interviewed by Erika Abad. "*Vida*: Family, Love, & Identity." April 22, 2019. Panel at ClexaCon, April 11–15, 2019. www.youtube.com/watch?v=mhIVosC6Ce8&fbclid=IwAR1Vt1xxyOctJwfA8 Dah2-h7V146kqNdE90hgMYuEHdU2oBGD2ehxGgODPc.

Press, Joy. *Stealing the Show: How Women Are Revolutionizing Television*. New York: Atria/Simon & Schuster, 2018.

Sarris, Andrew. "Notes on the Auteur Theory in 1962." *Film Culture* 27 (Winter 1962–1963): 1–8.

Sender, Katherine E., dir. *Further Off the Straight & Narrow*. Northampton, MA: Media Education Foundation, 2006.

—, dir. *Off the Straight & Narrow*. Northampton, MA: Media Education Foundation, 1999.

Sexton, Jared. "The *Vel* of Slavery: Tracking the Figure of the Unsovereign." *Critical Sociology* 42, nos 4–5 (2016): 1–15.

Shah, Nayan. "Queer of Color Estrangement and Belonging." In *The Routledge History of Queer America*, edited by Don Romesburg. New York: Routledge, 2018.

Smith, Neil. *The New Urban Frontier: Gentrification and the Revanchist City*. London and New York: Routledge, 1996.

Smith, Stacy L., Choueiti, Marc, and Pieper, Katherine. *Inclusion or Invisibility? Comprehensive Annenberg Report on Diversity in Entertainment*. Media, Diversity & Social Change Initiative of the Institute for Diversity and Empowerment at Annenberg (IDEA). Los Angeles: USC Annenberg, 2016. https://annenberg. usc.edu/sites/default/files/2017/04/07/MDSCI_CARD_Report_FINAL_Exec_Summary.pdf.

Smythe, Dallas. "On the Audience Commodity and Its Work." In *Media and Cultural Studies: Keyworks*, edited by Meenakshi Durham and Douglas Kellner, 230–256. Malden, MA: Blackwell, 2001.

Starz. "World of *Vida*." Accessed September 22, 2019. Promotional video. www.starz.com/video/ a3424e5d-9773-400f-b021-f003c4abf2ff.

Terranova, Tiziana. "Free Labor: Producing Culture for the Digital Economy." *Social Text* 18, no. 2 (2000): 33–58.

Thompson, Becky. *A Promise and a Way of Life: White Antiracist Activism*. Minneapolis and London: University of Minnesota Press, 2001.

Thorburn, Elise. "Networked Social Reproduction: Crises in the Integrated Circuit." *TripleC: Communication, Capitalism & Critique* 14, no. 2 (2016): 380–396. https://doi.org/10.31269/triplec.v14i2.708.

Tuck, Eve, and Yang, K. Wayne. "Decolonization Is Not a Metaphor." *Decolonization: Indigeneity, Education and Society* 1 (2012): 1–40.

10

BODY IMAGE AND GLOBAL MEDIA

Jasmine Fardouly, Vani Kakar, and Phillippa C. Diedrichs

Media plays an important role in the development and maintenance of people's body image concerns.[1] Traditionally, body image research has focused on mass media, such as television, magazines, and films. However, given the rising popularity of social media, particularly among young people,[2] increasing attention is now being paid to how such new forms of communication impact body image. In this chapter, we will review the existing literature on global media and body image, with a particular focus on social media. We will also discuss research on the impact of Western media in different cultures. Researchers, government bodies, and the public are becoming increasingly concerned about the negative effects of idealized, homogenous media images on young people. Individual and macro-level interventions have been developed to improve media literacy, and to change the media's narrow and unrealistic portrayals of beauty. These interventions will be reviewed and critiqued. Finally, recommendations will be made for researchers, government bodies, businesses, and individuals looking to improve the media landscape to foster positive body image.

Body image: a global public health concern

Body image refers to one's perceptions, thoughts, and feelings about the way they look.[3] Body dissatisfaction encompasses the negative and dysfunctional thoughts and feelings associated with one's body image,[4] such as body shame and discontent with one's appearance. Body dissatisfaction can develop from a young age[5] and has been found across people of differing body sizes, genders, and among various cultures around the world.[6] For example, a recent large-scale cross-cultural study[7] with 10,500 girls and women aged 10–60 years old from 13 different countries (e.g., the United Kingdom, China, India, Russia, South Africa) found that body dissatisfaction is prevalent, regardless of age and culture. On average, 54% of girls across the world reported that they did not have high body esteem, with rates of dissatisfaction as high as 93% in Japan, 61% in the United Kingdom, and 52% in Brazil. Body dissatisfaction has traditionally been thought of as an issue primarily affecting women, which has resulted in a female-focused research field.[8] Historically, women have reported being more concerned about their appearance than men.[9] However, body image concerns are becoming more recognized in men,[10] particularly when concern about muscularity is being considered.[11] There is now a growing literature on men, and increased awareness of the importance of body image across genders. However, research in the body image field tends

146

to take a binary approach to gender and there is a lack of research on transgender and nonbinary people. Existing research suggests that the prevalence of eating disorders may be substantially higher among transgender and gender diverse people than cisgender people and that levels of body image concerns may vary based on gender identity.[12] Thus, further research is needed to better understand the prevalence, predictors, and implications of body image concerns among diverse genders.

Body dissatisfaction is associated with a host of negative outcomes across key areas of people's lives. It is one of the most consistent and robust modifiable risk and maintenance factors for eating disorders[13] and is a significant predictor of poor quality of life,[14] low self-esteem, and depression.[15] People who are more dissatisfied with their appearance are also more likely to engage in unhealthy diet and exercise regimes,[16] and to use steroids.[17] Beyond impacts on physical and mental health, research suggests that females who are more dissatisfied with their appearance are less likely to assert their opinions[18] and tend to have poorer academic performance.[19] Given the diversity and severity of these problems, the high prevalence of body dissatisfaction around the world is increasingly recognized as a global public health concern.

Societal beauty ideals

Sociocultural influences are a common focus in research seeking to understand what predicts body image concerns. From a young age, people, particularly girls, learn to place high importance on their physical appearance and to believe that physical attractiveness is synonymous with success and happiness.[20] Although a range of body shapes and sizes exist naturally among genders, the range of bodies reflected in beauty ideals, particularly in Western society, is narrow, with one body shape for women and one for men generally being promoted as ideal for physical attractiveness, and an absence of androgyny and non-normative gender presentations. These homogenous gendered beauty ideals are problematic because they are not representative of the general population and are biologically unachievable for most people,[21] potentially resulting in (or at least contributing to) the high prevalence of body dissatisfaction worldwide.[22]

Beauty ideals for women

Since the 1950s, the ideal body shape for women in Western society has changed, with the portrayal of attractive women becoming thinner in adult and children's media, such as magazines[23] and television.[24] For example, one study found that women featured on the cover of four popular American women's magazines (*Vogue, Mademoiselle, Cosmopolitan,* and *Glamour*) published between 1959 and 1999 became thinner over time.[25] In addition to the ideal focusing on thinness, which has been pervasive for decades, there is now an added focus on women being fit and toned[26] and/or curvaceous with a large buttocks and thin waist.[27] This is particularly evident on social media with the considerable number of "fitspiration" (an amalgamation of the words "fitness" and "inspiration," sometimes referred to as "fitspo") images, which are purportedly designed to inspire people to exercise and eat healthily to obtain an attractively toned body.[28] Both ideals are also evident in the popularity of social media celebrities and influencers, such as the Kardashians, who post sexualized images of their curvaceous, yet thin, bodies.[29] In addition to specific standards for overall body shape, other norms for female beauty in Western society include being young, tall, blemish-free, and having large breasts, a symmetrical face, big eyes, straight white teeth, light skin that appears tanned, and long flowing hair,[30] all of which increase the unattainability of the female beauty ideal for most women without surgical intervention, digital retouching, and extensive beauty and body work.

Beauty ideals for men

The ideal body shape for men has also changed in the past 50 years, with the portrayal of attractive men becoming more muscular.[31] For example, one study found that every body part of toy action figures (e.g., G.I. Joe, Batman, Superman) except the waist became significantly larger and more muscular from the 1970s to the early 2000s.[32] Similarly, another study found that the male centerfold models in *Playgirl* magazines published between 1973 and 1997 became more "dense" and muscular over time.[33] Current evidence of the male muscular ideal can be found in the many fitspiration images containing muscular men on social media,[34] and the rising sales of muscle enhancing supplements.[35] Other aspects of the male beauty ideal include being young, tall, and having "chiseled" facial features, a large penis, and a full head of hair.[36] All of these expectations also increase the unattainability of the male beauty ideal for most men.

Culture and beauty ideals

Culture may contribute to the development and maintenance of body dissatisfaction. Although the majority of body image research has been conducted in Western cultures and high-income countries,[37] there is an emerging literature examining cross-cultural differences in beauty ideals and body image. For example, studies suggest that there is variability in the desired skin color among different cultures. In South Asian countries, lighter skin is desired over darker skin,[38] whereas in Western countries, light skin that appears tanned is more desirable than light skin without a tan.[39] There is also evidence that the ideal female and male body shape may vary between cultures. For example, the ideal female and male body is more slender in Chinese and South Korean cultures,[40] and the ideal female and male body is larger in South African cultures.[41] More recent research, however, points to a homogenization of the female beauty ideal around the world, especially among countries with high socioeconomic status.[42] This is partly attributed to a globalization of Western media. One study, for example, found an increase in dieting and unhealthy weight loss practices among adolescent girls in Fiji after the introduction of television and access to Western media in their country.[43] Similar changes in body size ideals have also been noted in Nicaragua following the introduction of electricity and television.[44] Further research is needed on female and male beauty ideals in more diverse cultures to provide a more comprehensive understanding of the influences of media and cultural changes on body image.

Predictors of body dissatisfaction

Sociocultural models suggest that societal pressures to conform to beauty ideals can lead to body dissatisfaction and eating disturbances. The most widely accepted model in the literature is the Tripartite Model of Influence,[45] which suggests that pressures to conform to the societal ideals of appearance come from three main sources: peers, parents, and the media. According to this model, appearance-related pressure from those sources lead to body dissatisfaction and in turn eating disturbance through two pathways: internalization of the societal ideals and a tendency to make appearance comparisons to others. Specifically, people who subscribe to appearance-related pressures and cognitively "buy into" the societal ideals of appearance, but who fail to achieve those ideals, are likely to become dissatisfied with their appearance. Furthermore, one of the ways in which women can determine whether or not they conform to the societal ideal is by comparing their appearance to that of other women, and appearance comparisons to women perceived to be more attractive than oneself can lead to body dissatisfaction.[46] Thus, appearance comparisons can be seen as a mechanism through which pressures to conform to societal beauty norms and internalization of the ideals lead to body dissatisfaction and eating disturbance.[47]

Evidence in support of sociocultural models of body dissatisfaction has been found in a variety of different samples, including female and male university students,[48] adolescents,[49] preadolescent girls,[50] midlife women,[51] lesbian women,[52] and young people from a variety of different cultures (e.g., France, Hungary, Japan).[53] Thus, there is substantial evidence for the importance of internalization of beauty ideals and appearance comparisons as mechanisms responsible for societal appearance-related pressures on body image.

Media and body image

Mass media has been identified as one of the most influential and pervasive causes of body dissatisfaction.[54] Given the proliferation and popularity of social media around the globe and the focus on appearance in many platforms (e.g., 300 million images are posted on Facebook and 95 million images are posted on Instagram each day),[55] there has also been tremendous momentum in recent years to investigate its impact on users' body images. Studying the role and contribution of social media is now at the center stage of understanding how people view and feel about their own bodies. In this section, we will discuss the academic literature that has documented the potential implications and ill effects of exposure to traditional media and social media on women's and men's perceptions of their own bodies.

Traditional media and body image

Traditional forms of media, such as magazines, films, and television, are strong promoters of homogenized and narrowly defined beauty ideals, and they provide regular opportunities for people to make appearance comparisons. Not only do these media platforms contain models and celebrities who match these ideals, but the images and videos are often edited to enhance appearance through lighting, makeup, styling, and computer manipulation techniques.[56] This ultimately makes beauty ideals even less attainable for the average man or woman. Thus, when comparing their own appearance to these edited media images, people continually come up short (i.e., make upward appearance comparisons).[57]

The effects of traditional forms of media (e.g., magazines, television) on body image have been extensively studied since the 1980s, and several meta-analyses of this research have consistently found small-to-moderate adverse effects from short-term exposure to idealized media images on young women's and men's appearance concerns.[58] Correlational research has found a positive association between the consumption of fashion and fitness magazines,[59] and television programs and music videos containing characters who match the ideals[60] and body image concerns among female and male adolescents and adults. Longitudinal research on traditional media usage and body dissatisfaction has primarily been conducted in female samples (e.g., preadolescent girls, female university students, Caucasian women, Latino women) and has found mixed results.[61] For example, Harrison and Hefner[62] found that television viewing but not magazine usage among preadolescent girls predicted greater eating disturbance and a thinner future body ideal one year later. Further, Stice et al.[63] found no difference in body dissatisfaction or negative mood between adolescent girls who were provided with a 15-month fashion magazine subscription and those who were not over a 20-month period. They did, however, find that for vulnerable girls (e.g., those with high body dissatisfaction, elevated pressures to be thin, and a lack of social support), exposure to thin ideal images resulted in more body dissatisfaction and negative mood over time. Thus, some people are more likely to be vulnerable to the impact of media usage than others.

Numerous experimental studies have examined the causal impact of media usage on the body image of women and men. Typically, these studies randomly assign participants to view media images (or videos) that contain people who meet the beauty ideals or appearance neutral control images (e.g., images of objects, such as furniture, or images of plus-size models). Participants are subsequently asked to report on their body satisfaction. Research has repeatedly shown that exposure to ideal images in magazines,[64] television,[65] and music videos[66] leads to greater body dissatisfaction among young women and men than does exposure to neutral control images.[67] There are, however, mixed results in the field, with other studies finding no association between exposure to idealized media imagery and body image.[68] One meta-analysis examining both published and unpublished research in the field suggested that there was stronger evidence for the negative impact of idealized imagery in media on body image for women who were already more dissatisfied with their appearance.[69] Thus, there is a growing consensus that media may not impact all people equally, and that some people are more vulnerable than others.

Overall, results from correlational, longitudinal, and experimental research suggest that traditional media usage can negatively affect women's and men's body image. However, consumer and market research suggests that the popularity of these media types has been overtaken by the popularity and availability of more interactive and online media, such as the Internet and social networking sites, particularly among adolescents. For example, magazine usage is in decline[70] and young people are now multitasking and browsing the Internet while watching television.[71] Therefore, the focus in recent research has shifted to the effects of newer forms of media, such as social media, on body image concerns.

Social media and body image

The use of social media is pervasive worldwide and has grown rapidly in popularity since the early 2000s. Social media use is particularly popular among young people, with around 90% of youth in the Western world having at least one social media account,[72] which they browse for several hours each day.[73] Social media is popular among the populations of low-, middle-, and high-income countries. For example, in India it is estimated that 258 million people were using social media in 2019.[74] There are a variety of different social media platforms, and they all slightly differ in their functions and features. Image-based platforms, such as Instagram, Snapchat, and YouTube, are currently among the most popular platforms used by youth,[75] although Facebook has the highest number of users, with over 2.23 billion active users worldwide.[76] Most research in the field has focused on Facebook, but there are an increasing number of studies focused on Instagram. More research is needed on other platforms, such as Snapchat, Pinterest, YouTube, WhatsApp, and Reddit. Females are larger users of social media than males[77]; however, the impact of social media use on body image appears to be similar for girls or women and boys or men.[78]

Each day, more than 300 million images are posted on Facebook[79] and 95 million images are posted on Instagram.[80] Thus, social media provide users with forums for frequently engaging in appearance comparisons. Social media allow users to create public or semi-public personal profiles, and to customize their pages with photos and information about themselves. Like magazine images, which are edited and often "enhanced" before publication, social media users are able to edit images of themselves before uploading them. They are also able to closely refine and edit their self-presentation in order to portray an idealized or "hoped for possible" version of the self.[81] For example, social media users can enhance images through the use of lighting, makeup, camera angles, image filters, or image manipulation programs/applications. Therefore, the images that people view and are available for comparison on social media may not depict natural (unenhanced) or attainable beauty. Importantly, just as with exposure to idealized images in more traditional media, viewing and comparing to one's own or other people's idealized images and profiles on social media could put pressure on users to internalize beauty ideals and may also have a negative impact on their self-evaluations and overall well-being.

A unique aspect of social media is the different types of people presented. Unlike more traditional forms of media (such as magazines and television), which predominantly feature images of models and celebrities, social media contains images of a variety of different types of individuals. This can include images of models and celebrities through advertisements, fan pages, and other commercial sites, but also people who are known to the user (i.e., "friends"). Furthermore, these known others can vary in relational closeness to the user, including family members, close friends, and acquaintances. Sociocultural models of body image, such as the Tripartite Model of Influence,[82] make specific reference to the impact of appearance-related pressures from different sources, including one's family, peers, and the media. Given that social media contains all three sources of pressure (family members, peers, models/celebrities), it may be a particularly strong purveyor of appearance ideals. This potential has prompted growing concerns about the influence of social media on the body image of young people.

Correlational research

The majority of research on social media and body image has utilized correlational designs with self-report questionnaires. Early research focused on how the amount of time spent on social media relates to body image, and differences between the body image of users and nonusers of certain platforms. Users of social media platforms such as Facebook and Instagram tend to report being more concerned about their appearance than nonusers.[83] Further, some studies have found a link between more time spent on platforms such as Facebook and Instagram, and higher body image concerns among adolescent and adult women and men.[84] However, others have found no association between these variables.[85] The mixed results in the field are perhaps not surprising given that social media relies on user generated platforms, allowing users to choose whose content they access, what they post themselves, and how they interact with others.[86] For this reason, each user will be exposed to different content on social media. Therefore, browsing social media is only likely to influence users' body image if they seek out and/or engage in harmful appearance-focused activities or make negative comparisons to others. For example, research has found that the link between time spent on social media and body dissatisfaction is partly accounted for (or mediated by) the amount of appearance comparisons users make while online.[87]

More consistent links have been found between users' body image and specific types of activities they engage in on social media. With the worldwide adoption of smartphones, people can take a large number of photos of themselves (i.e., "selfies") quickly, cheaply, and easily, and are able to instantly share those images with others online. Selfie activities have gained increasing attention in the literature. Research suggests that engaging in more photo-based activities, such as being invested in and manipulating selfies on social media, is associated with more appearance concerns.[88] Another aspect of social media that has garnered increasing attention from researchers is the impact of exposure to "fitspiration" images. These images are purportedly designed to motivate users to exercise and eat healthily in order to maintain a healthy lifestyle. However, content analyses of fitspiration posts have found that they are highly appearance focused (e.g., images that focus on aspects of the body like the stomach and biceps, and images in which people are passively posing in gym clothes) and primarily feature people who match gendered and narrow beauty ideals.[89] Correlational research has found that the more women and men view fitspiration content on social media and/or the more they post fitspiration images online, the more likely they are to experience appearance concerns.[90] This relationship is mediated and explained by people making appearance comparisons with the people in the fitspiration images and internalizing the gendered beauty ideals as a result of viewing those images.[91] Taken together, correlational research to date suggests that viewing appearance-focused content on social media, in addition to posting and editing images of oneself online, is linked to more body image concerns among women and men. However, because this research is correlational, the direction of the relationship cannot be determined. Further, the relationship between these variables may be bidirectional. For example, engaging in more appearance-based activities on social media may increase users' appearance concerns, and those with heightened appearance concerns may choose to engage in more appearance-based activities on social media. More experimental and longitudinal research is needed to determine the complex interactions between social media and body image.

Experimental research

Like correlational research in the field, experimental studies have taken different approaches to examine the impact of social media on body image. The majority of this research has focused on young, usually undergraduate, women in Western countries. Some studies have examined the impact of scrolling through one's own social media newsfeed (i.e., stream of posts from people they follow) on women's body image. Those studies found that browsing social media does not impact all women equally. Furthermore, this type of activity among women who had a high tendency to compare their appearance to others seems to have a negative influence on concerns about their facial features but not their bodies.[92] This research adds further evidence to suggest that media, both social and traditional, may be more harmful for women who are already more concerned about their appearance. It also suggests that social media use may be particularly harmful for women's facial appearance concerns, perhaps because of the large number of portrait images and selfies posted online that primarily focus on the face.[93]

Other experimental studies have examined the impact of specific content found on social media, such as idealized portraits of peers and celebrities or fitspiration pictures. Similar to traditional media images, research has found that exposure to idealized depictions of women on social media, regardless of whether they are an unknown peer or a celebrity, can increase women's body dissatisfaction.[94] Viewing fitspiration content taken from social media has also been found to increase women's body dissatisfaction,[95] although one study found that fitspiration posts had no impact on body image.[96] Further, the impact of fitspiration posts on women's body image was partly accounted for by women comparing their own appearance to the idealized appearance of people in the posts.[97] In addition to viewing others' content, recent research has experimentally examined the impact of posting and editing images on social media. Women reported being less physically attractive and having a more negative mood after being instructed to post an image of themselves on their social media profile, regardless of whether they were able to edit and enhance the image or not.[98] Together, this research suggests that both viewing and posting images on social media can increase women's appearance concerns.

Social media provide interactive platforms and users can not only post and view content but can also engage with each other's posts through "likes" and comments. Receiving appearance-related feedback and endorsements from others on social media could increase the salience of idealized images and likewise increase the appearance focus of the social media environment. Few studies have considered the influence of social media user feedback on appearance concerns. One recent study found that viewing idealized images of women with positive appearance-related comments increased women's body dissatisfaction compared to viewing those images with appearance-neutral comments.[99] In contrast, another study found that the number of likes images received had no impact on how women feel about their body, but viewing images with more likes made women *more* happy with their facial appearance.[100] Thus, the impact of likes and comments on body image seems to be complex, and more research is needed to better understand why different types of feedback can disparately influence how women feel about certain aspects of their appearance.

Longitudinal research

Several longitudinal studies have investigated the influence of social media on users' body image over time. Some studies have focused on time spent on social media, while others have focused on social comparisons and specific aspects of the social media environment. One study with adolescent girls and boys in the Netherlands found that spending more time on social media was associated with high levels of body dissatisfaction 18 months later, but body dissatisfaction did not predict greater social media use over time.[101] In contrast, another study with adolescent girls in Australia found that time spent on social media did not predict increases in body image concerns over a 2-year period, nor did body image predict time spent on social media.[102] They did, however, find that having a higher number of "friends" on social media predicted greater drive for thinness and internalization of the thin beauty ideal, but drive for thinness and internalization also predicted an increased number of social media friends. Consistent with correlational and experimental research, there are mixed findings on the impact of hours spent on social media and body image over time, and having a higher number of friends on social media may be predictive of users' appearance-related concerns.

Studies examining the influence of online social comparisons over time have also found mixed results. One study with adolescent girls and boys in Belgium found that body dissatisfaction predicted greater social media appearance comparisons 6 months later but comparisons did not predict body dissatisfaction.[103] Another study with Australian women found that the relationship between social media comparisons and body dissatisfaction was bidirectional over a 1-month period.[104] Other studies in the United States have also found that maladaptive social media use (including making negative comparisons to others) among young women was associated with more body dissatisfaction and eating pathology over a 1-month period.[105] Seeking and receiving negative feedback on social media has also been found to predict higher eating pathology 1 month later among young women and men.[106] Taken together, current results in the field indicate that the relationship between social media comparisons and body image concerns may be bidirectional. Further longitudinal research is needed in diverse cultures and over longer periods of time.

Media-focused body image interventions

Research documenting the negative impact of media on body image for women and men has led to various initiatives by researchers, schools, governments, and the general public to improve body image. Individual-level approaches have included school-based media literacy programs. Macro-level interventions have included the introduction of government policies to include disclaimer labels on digitally edited media images, minimum body mass index (BMI) standards for fashion models, and promoting body positivity via social media posts and campaigns. In this section, we will discuss and critique these individual and macro-level interventions.

Individual-level interventions

One of the most common approaches to individual level interventions has been the development of school-based media literacy programs for body dissatisfaction and eating disorders. Media literacy refers to an individual's ability to access, critically evaluate, and create media.[107] Media literacy programs are designed to encourage individuals to critique images presented in the media by deconstructing their idealized and edited nature, and thinking about the intentions behind the images, such as selling products.[108] These programs are based on the premise that thinking critically about idealized media images will reduce their harm to body image. To date, they have primarily focused on media literacy in relation to traditional media formats (e.g., celebrity images, magazines, television). There is limited correlational evidence for the association between media literacy and body image. However, one experimental study found that women with high media literacy were less influenced by exposure to idealized images.[109] In their systematic review, McLean et al.[110] found that a number of media literacy interventions had positive outcomes for weight and shape concerns, thin-ideal internalization, and risk for disordered eating, some with sustained improvements after 3 months. Overall, there is evidence of the efficacy of media literacy programs for improving body image; however, the utility and sustainability of these traditional media literacy programs is being questioned as newer styles of media develop. Furthermore, by not addressing the root cause of the problem (i.e., the media images themselves), media literacy interventions may not be sustainable and may place the burden of responsibility on consumers.

There is now emerging interest in the development of social media literacy programs. Social media literacy has been defined as having the knowledge and skills to analyze, evaluate, produce, and participate in social media.[111] In their recent study, Tamplin et al.[112] found that social media literacy, and specifically commercial literacy on social media, protected against the negative impact of exposure to idealized social media images on women's, but not men's, body image. A recent pilot study for a social media literacy intervention also found positive results for body image in a sample of Australian adolescent girls.[113] Furthermore, some interventions have sought to address media literacy in relation to both traditional and social media.[114] Thus, preliminary research suggests that media literacy may be an effective body image intervention strategy. However, more research across genders is needed in this area.

Macro-level interventions

In addition to changing individuals' reactions to idealized images, there is also increasing pressure to change the media landscape on a macro-level. The idealized nature of media images and their potential harm to body image has come to the attention of governments around the world, who have sought strategies to change the media environment. One such strategy, adopted by governments in France, Israel, and Australia, was to attach disclaimer labels to advertisements that have been digitally altered. The aim of the disclaimers was to educate viewers on the edited and unrealistic appearance of the models, which would presumably reduce the number of comparisons by preventing the models from being seen as relevant comparison targets.[115] Although logical, a growing body of research has found that disclaimer labels on media advertisements are ineffective at improving women's body image and do not reduce the comparisons made to the models they contain.[116] Alarmingly, research suggests that these labels can actually *increase* women's body image concerns if they specify the aspects of the image that have been edited, because they may attract women's attention to those aspects of their own appearance.[117] Similar results have been found for social media self-disclaimers, in which users describe the unrealistic and edited aspects of their own images in the captions of their idealized posts.[118] Together, the literature on social media and traditional media disclaimers suggests that they are not an effective body image intervention, perhaps because viewers are still exposed to the idealized image itself.

Researchers have argued that a more effective intervention strategy may be to reduce the number of idealized images in both social media and traditional media. Indeed, several interventions have been introduced to change the media's narrow and unattainable beauty ideals. One such intervention was the introduction of a minimum BMI of 18 (i.e., within the "normal" weight range) for fashion models, which was implemented in France, Israel, Madrid, and Milan.[119] The aim of this intervention was not only to reduce eating pathologies among models themselves but also to move the female beauty ideal into a "healthy" weight range. This strategy, however, has not been well accepted within the modeling industry,[120] and research is needed to test the effectiveness of this intervention on both models and women more generally.

Another approach has been to include models with larger bodies in media advertisements and on fashion runways. Research shows that including average-sized women and men in advertisements is an effective way to improve body image and is equally as effective at selling products as is using thin or muscular models.[121] There has been increasing social pressure on companies to include models with diverse body shapes and sizes in their advertisements.[122] This approach has been adopted by the brand Dove for decades but is also now being adopted by other companies, such as H&M and Revlon. Not only is the inclusion of larger models in advertisements likely to improve body image, but there is also evidence to suggest that it increases the company's revenue.[123] Thus, including people with more diverse body shapes and sizes in the media may be an effective macro-level intervention.

The inclusion of more diverse bodies in advertisements is partly attributed to the body positivity movement on social media. With its origins in fat acceptance activism in the 1960s, the body positive movement today primarily involves social media users challenging societal beauty ideals, promoting body acceptance, and encouraging the inclusion of natural and realistic images on both social and traditional media. Body positive content has become increasingly popular on social media platforms, particularly on Instagram. A recent search of the hashtag #bodypositive on Instagram elicited over 8,233,727 posts (Instagram, January 2019). However, little research has examined the impact of viewing body positive content on social media. One experimental study found that viewing body positive posts taken from Instagram increased women's body satisfaction and mood, but it also made women more focused on their appearance.[124] The body positivity movement has been criticized for its focus on appearance and beauty, with researchers suggesting that it may be more effective to focus attention toward positive non-appearance aspects of oneself in order to improve body image and reduce appearance investment. Further research is needed to test the impact of body positive posts in users' everyday lives and to examine the effectiveness of those posts compared to appearance-neutral content.

Conclusions

For decades, narrowly defined and homogenous beauty ideals have been promoted in traditional media formats, resulting in increased body image concerns, particularly for those who are already more dissatisfied with and conscious of their appearance. Social media can also perpetuate beauty ideals, and engaging in appearance-related activities on social media (such as viewing idealized images of others, posting and editing images, and viewing appearance-related comments) can increase body dissatisfaction among users.

Most research examining media and body image has focused on women in Western societies. More research is needed with other groups, such as men and diverse cultures around the world, particularly due to the globalization of Western media. Furthermore, it will be important for future research to pay more attention to how body image and media relates to race, class, gender, and sexuality.[125] Historically, body image research has focused on, and privileged, the opinions and experiences of white, able-bodied, cisgendered, heterosexual adolescents and young adults from Western high-income countries. This continues to be a limitation of the field, and researchers have called for more studies investigating diverse populations and social identities.[126]

Media literacy programs may aid individuals with the skills needed to reduce the negative impact of idealized images. However, more research is needed to test the effectiveness and sustainability of these programs, particularly social media literacy programs, over longer periods of time. Effective macro-level interventions employ the inclusion of more diverse models in media advertisements and promote more natural (unaltered) images on social media, but these are yet to be widely implemented and there is a paucity of research examining their effectiveness. Due in part to the rise in social media use, there is increasing social pressure to change narrow societal beauty ideals and promote body positivity and acceptance. Changing the media landscape could substantially improve the body image of people around the world. Therefore, focusing on both individual- and macro-level interventions remains a priority for the field of body image research.

Notes

1 Shelly Grabe, L. Monique Ward, and Janet Shibley Hyde, "The Role of the Media in Body Image Concerns among Women: A Meta-Analysis of Experimental and Correlational Studies," *Psychological Bulletin* 134, no. 3 (2008): 460–476; Lisa Groesz, Michael Levine, and Sarah Murnen, "The Effect of Experimental Presentation of Thin Media Images on Body Satisfaction: A Meta-Analytic Review," *International Journal of Eating Disorders* 31, no. 1 (2002): 1–16; Christopher Barlett, Christopher Vowels, and Donald Saucier, "Meta-Analyses of the Effects of Media Images on Men's Body-Image Concerns," *Journal of Social and Clinical Psychology* 27, no. 3 (2008): 279–310.
2 Office for National Statistics. "Internet Access – Households and Individuals: 2017," www.ons.gov. uk/peoplepopulationandcommunity/householdcharacteristics/homeinternetandsocialmediausage/ publications?filter=bulletin; Monica Anderson and Jingjing Jiang, "Teens, Social Media & Technology 2018," www.pewinternet.org/2018/05/31/teens-social-media-technology-2018/; Australian Bureau of Statistics. "Household Use of Information Technology, Australia, 2014–15," www.abs.gov.au/ausstats/ abs@.nsf/0/ACC2D18CC958BC7BCA2568A9001393AE?Opendocument.
3 Sarah Grogan, *Body Image: Understanding Body Dissatisfaction in Men, Women, and Children* (New York, NY: Routledge, 1999).
4 David Garner, "Body Image and Anorexia Nervosa," in *Body Image: A Handbook of Theory, Research, and Clinical Practice*, eds Thomas Cash and Thomas Pruzinsky (New York, NY: Guilford, 2002), 295–303.
5 Hayley Dohnt and Marika Tiggemann, "Body Image Concerns in Young Girls: The Role of Peers and Media Prior to Adolescence," *Journal of Youth and Adolescence* 35, no. 2 (2006): 135.
6 Viren Swami, David Frederick, Tovio Aavik, Lidia Alcalay, Juri Allik, Donna Anderson, Sonny Andrianto, et al. "The Attractive Female Body Weight and Female Body Dissatisfaction in 26 Countries across 10 World Regions: Results of the International Body Project," *Personality and Social Psychology Bulletin* 36, no. 3 (2010): 309–325; Shelly Grabe and Janet Shibley Hyde, "Ethnicity and Body Dissatisfaction among Women in the United States: A Meta-Analysis," *Psychological Bulletin* 132, no. 4 (2006): 622–640.
7 Dove, "New Dove Research Finds Beauty Pressures Up, and Women and Girls Calling for Change," *PR Newswire*, www.prnewswire.com/news-releases/new-dove-research-finds-beauty-pressures-up-and-women-and-girls-calling-for-change-583743391.html; Dove, "Girls on Beauty: New Dove Research Finds Low Beauty Confidence Driving 8 in 10 Girls to Opt out of Future Opportunities," *PR Newswire*, www.prnewswire.com/news-releases/girls-on-beauty-new-dove-research-finds-low-beauty-confidence-driving-8-in-10-girls-to-opt-out-of-future-opportunities-649549253.html.
8 Barlett et al., "Men's Body-Image Concerns," 281.
9 Marika Tiggemann and Barbara Pennington, "The Development of Gender Differences in Body-Size Dissatisfaction," *Australian Psychologist* 25, no. 3 (1990): 306–313; Sarah Kate Bearman, Katherine Presnell, Erin Martinez, and Eric Stice, "The Skinny on Body Dissatisfaction: A Longitudinal Study of Adolescent Girls and Boys," *Journal of Youth and Adolescence* 35, no. 2 (2006): 229–241.

10 Deborah Mitchison, Phillipa Hay, Shameran Slewa-Younan, and Jonathan Mond, "The Changing Demographic Profile of Eating Disorder Behaviors in the Community," *BMC Public Health* 14, no. 1 (2014): 943.

11 Bryan Karazsia, Sarah Murnen, and Tracy Tylka, "Is Body Dissatisfaction Changing across Time? A Cross-Temporal Meta-Analysis," *Psychological Bulletin* 143, no. 3 (2017): 293–320.

12 Elizabeth Diemer, Julia Grant, Melissa Munn-Chernoff, David Patterson, and Alexis Duncan, "Gender Identity, Sexual Orientation, and Eating-Related Pathology in a National Sample of College Students," *Journal of Adolescent Health* 57, no. 2 (2015): 144–149; Penelope Strauss, Ashleigh Lin, Sam Winter, Angus Cook, Vanessa Watson, and Dani Wright Toussaint, "Trans Pathways: The Mental Health Experiences and Care Pathways of Trans Young People," in *Summary of Results* (Perth, Australia: Telethon Kids Institute, 2017).

13 Eric Stice and Mark Van Ryzin, "A Prospective Test of the Temporal Sequencing of Risk Factor Emergence in the Dual Pathway Model of Eating Disorders," *Journal of Abnormal Psychology* 128, no. 2 (2018): 119–128; Eric Stice, "Risk and Maintenance Factors for Eating Pathology: A Meta-Analytic Review," *Psychological Bulletin* 128 (2002): 825–848.

14 Tufan Nayir, Ersin Uskun, Mustafa Volkan Yürekli, Hacer Devran, Ayşe Çelik, and Ramazan Azim Okyay, "Does Body Image Affect Quality of Life?: A Population Based Study," *Plos One* 11, no. 9 (2016): e0163290.

15 Susan Paxton, Dianne Neumark-Sztainer, Peter Hannan, and Maria Eisenberg, "Body Dissatisfaction Prospectively Predicts Depressive Mood and Low Self-Esteem in Adolescent Girls and Boys," *Journal of Clinical Child and Adolescent Psychology* 35 (2006): 539–549.

16 Dianne Neumark-Sztainer, Susan Paxton, Peter Hannan, Jess Haines, and Mary Story, "Does Body Satisfaction Matter? Five-Year Longitudinal Associations between Body Satisfaction and Health Behaviors in Adolescent Females and Males," *Journal of Adolescent Health* 39, no. 2 (2006): 244–251.

17 Gen Kanayama, Steven Barry, James Hudson, and Harrison Pope, "Body Image and Attitudes toward Male Roles in Anabolic-Androgenic Steroid Users," *American Journal of Psychiatry* 163, no. 4 (2006): 697–703.

18 Dove, "Beauty Pressures Up".

19 Todd Florin, Justine Shults, and Nicolas Stettler, "Perception of Overweight Is Associated with Poor Academic Performance in US Adolescents," *Journal of School Health* 81, no. 11 (2011): 663–670.

20 Vanessa Buote, Anne Wilson, Erin Strahan, Stephanie Gazzola, and Fiona Papps, "Setting the Bar: Divergent Sociocultural Norms for Women's and Men's Ideal Appearance in Real-World Contexts," *Body Image* 8, no. 4 (2011): 322–334; Sharlene Hesse-Biber, Patricia Leavy, Courtney Quinn, and Julia Zoino, "The Mass Marketing of Disordered Eating and Eating Disorders: The Social Psychology of Women, Thinness and Culture," *Women's Studies International Forum* 29, no. 2 (2006): 208–224.

21 Joel Thompson, Lauren Schaefer, and Jessie Menzel, "Internalization of Thin-Ideal and Muscular-Ideal," in *Encyclopedia of Body Image and Human Appearance*, ed. Thomas Cash (Oxford: Academic Press, 2012), 499–504; Kelly Brownell, "Dieting and the Search for the Perfect Body: Where Physiology and Culture Collide," *Behavior Therapy* 22 (1991): 1–12.

22 Dove, "Beauty Pressures Up".

23 Mia Foley Sypeck, James Gray, and Anthony Ahrens, "No Longer Just a Pretty Face: Fashion Magazines' Depictions of Ideal Female Beauty from 1959 to 1999," *International Journal of Eating Disorders* 36, no. 3 (2004): 342–347; Tim Seifert, "Anthropomorphic Characteristics of Centerfold Models: Trends Towards Slender Figures over Time," *International Journal of Eating Disorders* 37, no. 3 (2005): 271–274.

24 Gregory Fouts and Kimberley Burggraf, "Television Situation Comedies: Female Weight, Male Negative Comments, and Audience Reactions," *Sex Roles* 42, no. 9 (2000): 925–932; Sylvia Herbozo, Stacey Tantleff-Dunn, Jessica Gokee-Larose, and J. Kevin Thompson, "Beauty and Thinness Messages in Children's Media: A Content Analysis," *Eating Disorders* 12, no. 1 (2004): 21–34.

25 Sypeck et al., "No Longer Pretty Face," 345.

26 Kristin Homan, Erin McHugh, Daniel Wells, Corrinne Watson, and Carolyn King, "The Effect of Viewing Ultra-Fit Images on College Women's Body Dissatisfaction," *Body Image* 9, no. 1 (2012): 50–56.

27 Diana Betz and Laura Ramsey, "Should Women Be "All about that Bass?": Diverse Body-Ideal Messages and Women's Body Image," *Body Image* 22 (2017): 18–31.

28 Leah Boepple, Rheanna Ata, Ruba Rum, and J. Kevin Thompson, "Strong Is the New Skinny: A Content Analysis of Fitspiration Websites," *Body Image* 17 (2016): 132–135; Marika Tiggemann and Mia Zaccardo, "'Strong Is the New Skinny': A Content Analysis of #Fitspiration Images on Instagram," *Journal of Health Psychology* 23, no. 8 (2016): 1003–1011.

29 Hannah Meyer, "Kardashian Curvy or Rousey Strong: It's Time for Women to Fight Back against Body Shaming," www.abc.net.au/news/2016-01-21/kim-kardashian-photos-rhonda-rousey-video-war-on-body-shaming/7100998.

30 Buote et al., "Setting the Bar," 326; Guy Cafri, J. Kevin Thompson, Megan Roehrig, Patricia van den Berg, Paul Jacobsen, and Stephen Stark, "An Investigation of Appearance Motives for Tanning: The Development and Evaluation of the Physical Appearance Reasons for Tanning Scale (Parts) and Its Relation to Sunbathing and Indoor Tanning Intentions," *Body Image* 3, no. 3 (2006): 199–209; Lora Jacobi and Thomas F Cash, "In Pursuit of the Perfect Appearance: Discrepancies among Self-Ideal Percepts of Multiple Physical Attributes," *Journal of Applied Social Psychology* 24, no. 5 (1994): 379–396.

31 J. Kevin Thompson and Guy Cafri, "The Muscular Ideal: Psychological, Social, and Medical Perspectives," in *The Muscular Ideal: Psychological, Social, and Medical Perspectives*, eds J. Kevin Thompson and Guy Cafri (Washington, DC: American Psychological Association, 2007).

32 Timothy Baghurst, Daniel Hollander, Beth Nardella, and Gregory Haff, "Change in Sociocultural Ideal Male Physique: An Examination of Past and Present Action Figures," *Body Image* 3, no. 1 (2006): 87–91.

33 Richard Leit, Harrison Pope Jr., and James Gray, "Cultural Expectations of Muscularity in Men: The Evolution of Playgirl Centerfolds," *International Journal of Eating Disorders* 29, no. 1 (2001): 90–93.

34 Tiggemann and Zaccardo, "Strong Is New Skinny," 1005.

35 New Hope Network, "Supplement Business Report 2016," www.newhope.com/sites/newhope360.com/files/2016%20NBJ%20Supplement%20Business%20report_lowres_TOC.pdf.

36 Marika Tiggemann, Yolanda Martins, and Libby Churchett, "Beyond Muscles: Unexplored Parts of Men's Body Image," *Journal of Health Psychology* 13, no. 8 (2008): 1163–1172.

37 Grabe et al., "Role of Media," 461.

38 Nazia Hussein, "Colour of Life Achievements: Historical and Media Influence of Identity Formation Based on Skin Colour in South Asia," *Journal of Intercultural Studies* 31, no. 4 (2010): 403–424.

39 Ivanka Prichard, Anna Kneebone, Amanda Hutchinson, and Carlene Wilson, "The Relationship between Skin Tone Dissatisfaction and Sun Tanning Behaviour," *Australian Journal of Psychology* 66, no. 3 (2014): 168–174.

40 Jaehee Jung and Gordon Forbes, "Body Dissatisfaction and Disordered Eating among College Women in China, South Korea, and the United States: Contrasting Predictions from Sociocultural and Feminist Theories," *Psychology of Women Quarterly* 31, no. 4 (2007): 381–393.

41 Vinet Coetzee and David Perrett, "African and Caucasian Body Ideals in South Africa and the United States," *Eating Behaviors* 12, no. 1 (2011): 72–74.

42 Swami et al., "Attractive Female Body," 322; Dove, "Beauty Pressures Up".

43 Anne Becker, Rebecca Burwell, David Herzog, Paul Hamburg, and Stephen Gilman, "Eating Behaviours and Attitudes Following Prolonged Exposure to Television among Ethnic Fijian Adolescent Girls," *British Journal of Psychiatry* 180, no. 6 (2002): 509–514.

44 Lynda Boothroyd, Jean-Luc Jucker, Tracey Thornborrow, Mark Jamieson, D. Michael Burt, Robert Barton, Elizabeth Evans, and Martin Tovee, "Television Exposure Predicts Body Size Ideals in Rural Nicaragua," *British Journal of Psychology* 107, no. 4 (2016): 752–767.

45 Patricia van den Berg, J. Kevin Thompson, Karen Obremski-Brandon, and Michael Coovert, "The Tripartite Influence Model of Body Image and Eating Disturbance: A Covariance Structure Modeling Investigation Testing the Mediational Role of Appearance Comparison," *Journal of Psychosomatic Research* 53 (2002): 1007–1020; J. Kevin Thompson, Leslie Heinberg, Madeline Altabe, and Stacey Tantleff-Dunn, *Exacting Beauty: Theory, Assessment, and Treatment of Body Image Disturbance* (Washington, DC: American Psychological Association, 1999).

46 Taryn Myers and Janis Crowther, "Social Comparison as a Predictor of Body Dissatisfaction: A Meta-Analytic Review," *Journal of Abnormal Psychology* 118 (2009): 683–698.

47 Rachel Rodgers, Sian McLean, and Susan Paxton, "Longitudinal Relationships among Internalization of the Media Ideal, Peer Social Comparison, and Body Dissatisfaction: Implications for the Tripartite Influence Model," *Developmental Psychology* 51, no. 5 (2015): 706–713.

48 van den Berg et al., "Tripartite Influence Model," 117; Tracy Tylka, "Refinement of the Tripartite Influence Model for Men: Dual Body Image Pathways to Body Change Behaviors," *Body Image* 8, no. 3 (2011): 199–207.

49 Helene Keery, Patricia van den Berg, and J. Kevin Thompson, "An Evaluation of the Tripartite Influence Model of Body Dissatisfaction and Eating Disturbance with Adolescent Girls," *Body Image* 1 (2004): 237–251; Rachel Rodgers, Camille Ganchou, Debra Franko, and Henri Chabrol, "Drive for Muscularity and Disordered Eating among French Adolescent Boys: A Sociocultural Model," *Body Image* 9, no. 3 (2012):

318–323; Emma Halliwell and Martin Harvey, "Examination of a Sociocultural Model of Disordered Eating among Male and Female Adolescents," *British Journal of Health Psychology* 11 (2006): 235–248.

50 Elizabeth Evans, Martin Tovée, Lynda Boothroyd, and Robert Drewett, "Body Dissatisfaction and Disordered Eating Attitudes in 7- to 11-Year-Old Girls: Testing a Sociocultural Model," *Body Image* 10, no. 1 (2013): 8–15.

51 Julie Helen Slevec and Marika Tiggemann, "Predictors of Body Dissatisfaction and Disordered Eating in Middle-Aged Women," *Clinical Psychology Review* 31, no. 4 (2011): 515–524.

52 Caroline Huxley, Emma Halliwell, and Victoria Clarke, "An Examination of the Tripartite Influence Model of Body Image: Does Women's Sexual Identity Make a Difference?," *Psychology of Women Quarterly* 39, no. 3 (2015): 337–348.

53 Yuko Yamamiya, Hemal Shroff, and J. Kevin Thompson, "The Tripartite Influence Model of Body Image and Eating Disturbance: A Replication with a Japanese Sample," *International Journal of Eating Disorders* 41, no. 1 (2008): 88–91; Rachel Rodgers, Henri Chabrol, and Susan Paxton, "An Exploration of the Tripartite Influence Model of Body Dissatisfaction and Disordered Eating among Australian and French College Women," *Body Image* 8, no. 3 (2011): 208–215; Ildikó Papp, Róbert Urban, Edit Czegledi, Bernadett Babusa, and Ferenc Tury, "Testing the Tripartite Influence Model of Body Image and Eating Disturbance among Hungarian Adolescents," *Body Image* 10, no. 2 (2013): 232–242.

54 Marika Tiggemann, "Sociocultural Perspectives on Human Appearance and Body Image," in *Body Image: A Handbook of Science, Practice, and Prevention*, eds Thomas Cash and Linda Smolak (New York, NY: Guilford Press, 2011), 12–19.

55 Reuters, "Instagram's User Base Grows to More Than 500 Million," www.reuters.com/article/us-facebook-instagram-users/instagrams-user-base-grows-to-more-than-500-million-idUSKCN0Z71LN; Adriana Manago, Michael Graham, Patricia Greenfield, and Goldie Salimkhan, "Self-Presentation and Gender on Myspace," *Journal of Applied Developmental Psychology* 29, no. 6 (2008): 446–458.

56 Shiela Reaves, Jacqueline Bush Hitchon, Sung Yeon Park, and Gi Woong Yun, "If Looks Could Kill: Digital Manipulation of Fashion Models," *Journal of Mass Media Ethics* 19, no. 1 (2004): 56–71.

57 Myers and Crowther, "Social Comparison Predictor," 690.

58 Grabe et al., "Role of Media," 461; Groesz et al., "Effect of Experimental Presentation," 14; Barlett et al., "Men's Body-Image Concerns," 303.

59 Renée Botta, "For Your Health? The Relationship between Magazine Reading and Adolescents' Body Image and Eating Disturbances," *Sex Roles* 48, no. 9 (2003): 389–399; Marika Tiggemann, "Media Exposure, Body Dissatisfaction and Disordered Eating: Television and Magazines Are Not the Same!," *European Eating Disorders Review* 11, no. 5 (2003): 418–430.

60 Marika Tiggemann, "Television and Adolescent Body Image: The Role of Program Content and Viewing Motivation," *Journal of Social and Clinical Psychology* 24, no. 3 (2005): 361–381; Kimberly Bissell and Peiqin Zhou, "Must See TV or Espn: Entertainment and Sports Media Exposure and Body Image Distortion in College Women," *Journal of Communication* 54, no. 1 (2004): 5–21.

61 Hayley Dohnt and Marika Tiggemann, "The Contribution of Peer and Media Influences to the Development of Body Satisfaction and Self-Esteem in Young Girls: A Prospective Study," *Developmental Psychology* 42, no. 5 (2006): 929–936; Kristen Harrison and Veronica Hefner, "Media Exposure, Current and Future Body Ideals, and Disordered Eating among Preadolescent Girls: A Longitudinal Panel Study," *Journal of Youth and Adolescence* 35 (2006): 153–163; Eric Stice, Diane Spangler, and W. Stewart Agras, "Exposure to Media-Portrayed Thin-Ideal Images Adversely Affects Vulnerable Girls: A Longitudinal Experiment," *Journal of Social and Clinical Psychology* 20, no. 3 (2001): 270–288; Marika Tiggemann, "The Role of Media Exposure in Adolescent Girls' Body Dissatisfaction and Drive for Thinness: Prospective Results," *Journal of Social and Clinical Psychology* 25, no. 5 (2006): 523–541; Deborah Schooler, "Real Women Have Curves: A Longitudinal Investigation of TV and the Body Image Development of Latina Adolescents," *Journal of Adolescent Research* 23, no. 2 (2008): 132–153.

62 Harrison and Hefner, "Longitudinal Panel Study," 159.

63 Stice et al., "Vulnerable Girls," 284.

64 Brit Harper and Marika Tiggemann, "The Effect of Thin Ideal Media Images on Women's Self-Objectification, Mood, and Body Image," *Sex Roles* 58 (2008): 649–657; Emma Halliwell and Helga Dittmar, "Does Size Matter? The Impact of Model's Body Size on Women's Body-Focused Anxiety and Advertising Effectiveness," *Journal of Social and Clinical Psychology* 23, no. 1 (2004): 104–122.

65 Daniel Agliata and Stacey Tantleff-Dunn, "The Impact of Media Exposure on Males' Body Image," *Journal of Social and Clinical Psychology* 23, no. 1 (2004): 7–22; Leslie Heinberg and J. Kevin Thompson, "Body

Image and Televised Images of Thinness and Attractiveness: A Controlled Laboratory Investigation," *Journal of Social and Clinical Psychology* 14, no. 4 (1995): 325–338.

66 Marika Tiggemann and Amy Slater, "Thin Ideals in Music Television: A Source of Social Comparison and Body Dissatisfaction," *International Journal of Eating Disorders* 35, no. 1 (2004): 48–58.

67 Grabe et al., "Role of Media," 461; Groesz et al., "Effect of Experimental Presentation," 14; Barlett et al., "Men's Body-Image Concerns," 303.

68 Emma Halliwell, Helga Dittmar, and Jessica Howe, "The Impact of Advertisements Featuring Ultra-Thin or Average-Size Models on Women with a History of Eating Disorders," *Journal of Community & Applied Social Psychology* 15, no. 5 (2005): 406–413.

69 Christopher Ferguson, "In the Eye of the Beholder: Thin-Ideal Media Affects Some, but Not Most, Viewers in a Meta-Analytic Review of Body Dissatisfaction in Women and Men," *Psychology of Popular Media Culture* 2, no. 1 (2013): 20–37.

70 Statistica, "U.S. Magazine Industry - Statistics & Facts," www.statista.com/topics/1265/magazines/.

71 Hilde Voorveld, Claire Segijn, Paul Ketelaar, and Edith Smit, "Investigating the Prevalence and Predictors of Media Multitasking across Countries," *International Journal of Communication* 8 (2014).

72 Office for National Statistics, "Households and Individuals"; Anderson and Jiang, "Teens Social Media"; Australian Bureau of Statistics, "Information Technology Australia."

73 Common Sense Media, "The Common Sense Census: Media Use by Tweens and Teen," www.commonsensemedia.org/research/the-common-sense-census-media-use-by-tweens-and-teens; Rachel Cohen, Toby Newton-John, and Amy Slater, "'Selfie'-Objectification: The Role of Selfies in Self-Objectification and Disordered Eating in Young Women," *Computers in Human Behavior* 79 (2018): 68–74.

74 Statistica, "Number of Social Network Users in India from 2015 to 2022 (in Millions)," www.statista.com/statistics/278407/number-of-social-network-users-in-india/.

75 Hannah Whyte-Smith, Emmeline Boreham, and Vanessa Patel, "Opinions of Generation Z's Ambitions and Priorities Differ Greatly between the Generations," www.ipsos.com/ipsos-mori/en-uk/opinions-generation-zs-ambitions-and-priorities-differ-greatly-between-generations.

76 Statistica, "Number of Monthly Active Facebook Users Worldwide as of 2nd Quarter 2018 (in Millions)," www.statista.com/statistics/264810/number-of-monthly-active-facebook-users-worldwide/.

77 Amanda Kimbrough, Rosanna Guadagno, Nicole Muscanell, and Janeann Dill, "Gender Differences in Mediated Communication: Women Connect More Than Do Men," *Computers in Human Behavior* 29, no. 3 (2013): 896–900; Samantha Stronge, Lara Greaves, Petar Milojev, Tim West-Newman, Fiona Kate Barlow, and Chris Sibley, "Facebook Is Linked to Body Dissatisfaction: Comparing Users and Non-Users," *Sex Roles* 73, nos 5–6 (2015): 200–213.

78 Grace Holland and Marika Tiggemann, "A Systematic Review of the Impact of the Use of Social Networking Sites on Body Image and Disordered Eating Outcomes," *Body Image* 17 (2016): 100–110.

79 Zephoria, "The Top 20 Valuable Facebook Statistics – Updated December 2018," https://zephoria.com/top-15-valuable-facebook-statistics/.

80 Reuters, "Instagram's User Base Grows to More Than 500 Million," www.reuters.com/article/us-facebook-instagram-users/instagrams-user-base-grows-to-more-than-500-million-idUSKCN0Z71LN.

81 Adriana Manago, Michael Graham, Patricia Greenfield, and Goldie Salimkhan, "Self-Presentation and Gender on Myspace," *Journal of Applied Developmental Psychology* 29, no. 6 (2008): 446–458; Shanyang Zhao, Sherri Grasmuck, and Jason Martin, "Identity Construction on Facebook: Digital Empowerment in Anchored Relationships," *Computers in Human Behavior* 24, no. 5 (2008): 1816–1836; Kathrynn Pounders, Christine Kowalczyk, and Kirsten Stowers, "Insight into the Motivation of Selfie Postings: Impression Management and Self-Esteem," *European Journal of Marketing* 50, nos 9/10 (2016): 1879–1892.

82 van den Berg et al., "Tripartite Influence Model," 117.

83 Marika Tiggemann and Amy Slater, "Nettweens: The Internet and Body Image Concerns in Preteenage Girls," *Journal of Early Adolescence* 34 (2014): 606–620; Marika Tiggemann and Amy Slater, "Netgirls: The Internet, Facebook, and Body Image Concern in Adolescent Girls," *International Journal of Eating Disorders* 46 (2013): 630–633; Rachel Cohen, Toby Newton-John, and Amy Slater, "The Relationship between Facebook and Instagram Appearance-Focused Activities and Body Image Concerns in Young Women," *Body Image* 23 (2017): 183–187.

84 Jasmine Fardouly, Natasha Magson, Carly Johnco, Ella Oar, and Ronald Rapee, "Parental Control of the Time Preadolescents Spend on Social Media: Links with Preadolescents' Social Media Appearance Comparisons and Mental Health," *Journal of Youth and Adolescence* 47, no. 7 (2018): 1456–1468; Jasmine

Fardouly, Brydie Willburger, and Lenny Vartanian, "Instagram Use and Young Women's Body Image Concerns and Self-Objectification: Testing Mediational Pathways," *New Media & Society* 20, no. 4 (2018): 1380–1395.

85 Ji Won Kim and T. Makana Chock, "Body Image 2.0: Associations between Social Grooming on Facebook and Body Image Concerns," *Computers in Human Behavior* 48 (2015): 331–339; Evelyn Meier and James Gray, "Facebook Photo Activity Associated with Body Image Disturbance in Adolescent Girls," *Cyberpsychology, Behavior, and Social Networking* 17 (2014): 199–206.

86 Richard Perloff, "Social Media Effects on Young Women's Body Image Concerns: Theoretical Perspectives and an Agenda for Research," *Sex Roles* 71, no. 11 (2014): 363–377.

87 Jasmine Fardouly and Lenny Vartanian, "Negative Comparisons About One's Appearance Mediate the Relationship between Facebook Usage and Body Image Concerns," *Body Image* 12 (2015): 82–88.

88 Sian McLean, Susan Paxton, Eleanor Wertheim, and Jennifer Masters, "Photoshopping the Selfie: Self Photo Editing and Photo Investment Are Associated with Body Dissatisfaction in Adolescent Girls," *International Journal of Eating Disorders* 48, no. 8 (2015): 1132–1140; Alexandra Rhodes Lonergan, Kay Bussey, Jonathan Mond, Olivia Brown, Scott Griffiths, Stuart Murray, and Deborah Mitchison, "Me, My Selfie, and I: The Relationship between Editing and Posting Selfies and Body Dissatisfaction in Men and Women," *Body Image* 28 (2019): 39–43.

89 Tiggemann and Zaccardo, "Strong Is New Skinny," 1005; Courtney Simpson and Suzanne Mazzeo, "Skinny Is Not Enough: A Content Analysis of Fitspiration on Pinterest," *Health Communication* 32, no. 5 (2017): 560–567.

90 Fardouly et al., "Testing Mediational Pathways," 1390; Scott Fatt, Jasmine Fardouly, and Ronald Rapee, "#Malefitspo: Links between Viewing Fitspiration Posts, Muscular-Ideal Internalisation, Appearance Comparisons, Body Satisfaction, and Exercise Motivation in Men," *New Media & Society* 21, no. 6 (2019): 1311–1325; Grace Holland and Marika Tiggemann, "'Strong Beats Skinny Every Time': Disordered Eating and Compulsive Exercise in Women Who Post Fitspiration on Instagram," *International Journal of Eating Disorders* 50, no. 1 (2017): 76–79.

91 Fardouly et al., "Testing Mediational Pathways," 1390.

92 Annalise Mabe, K. Jean Forney, and Pamela Keel, "Do You "Like" My Photo? Facebook Use Maintains Eating Disorder Risk," *International Journal of Eating Disorders* 47 (2014): 516–523; Jasmine Fardouly, Phillipa Diedrichs, Lenny Vartanian, and Emma Halliwell, "Social Comparisons on Social Media: The Impact of Facebook on Young Women's Body Image Concerns and Mood," *Body Image* 13 (2015): 38–45.

93 Nina Haferkamp, Sabrina Eimler, Anna-Margarita Papadakis, and Jana Vanessa Kruck, "Men Are from Mars, Women Are from Venus? Examining Gender Differences in Self-Presentation on Social Networking Sites," *Cyberpsychology, Behavior, and Social Networking* 15, no. 2 (2012): 91–98.

94 Nina Haferkamp and Nicole C. Kramer, "Social Comparison 2.0: Examining the Effects of Online Profiles on Social-Networking Sites," *Cyberpsychology Behavior and Social Networking* 14, no. 5 (2011): 309–314; Zoe Brown and Marika Tiggemann, "Attractive Celebrity and Peer Images on Instagram: Effect on Women's Mood and Body Image," *Body Image* 19 (2016): 37–43; Rachel Cohen and Alex Blaszczynski, "Comparative Effects of Facebook and Conventional Media on Body Image Dissatisfaction," *Journal of Eating Disorders* 3 (2015): 23; Megan Vendemia and David DeAndrea, "The Effects of Viewing Thin, Sexualized Selfies on Instagram: Investigating the Role of Image Source and Awareness of Photo Editing Practices," *Body Image* 27 (2018): 118–127.

95 Marika Tiggemann and Mia Zaccardo, "'Exercise to Be Fit, Not Skinny': The Effect of Fitspiration Imagery on Women's Body Image," *Body Image* 15 (2015): 61–67; Lily Robinson, Ivanka Prichard, Alyssa Nikolaidis, Claire Drummond, Murray Drummond, and Marika Tiggemann, "Idealised Media Images: The Effect of Fitspiration Imagery on Body Satisfaction and Exercise Behaviour," *Body Image* 22 (2017): 65–71; Ivanka Prichard, Annabel C. McLachlan, Tiffany Lavis, and Marika Tiggemann, "The Impact of Different Forms of #Fitspiration Imagery on Body Image, Mood, and Self-Objectification among Young Women," *Sex Roles* 78, no. 11 (2018): 789–798.

96 Amy Slater, Neesha Varsani, and Phillippa Diedrichs, "#Fitspo or #Loveyourself? The Impact of Fitspiration and Self-Compassion Instagram Images on Women's Body Image, Self-Compassion, and Mood," *Body Image* 22 (2017): 87–96.

97 Tiggemann and Zaccardo, "Exercise to be Fit," 64.

98 Jennifer Mills, Sarah Musto, Lindsay Williams, and Marika Tiggemann, "'Selfie' Harm: Effects on Mood and Body Image in Young Women," *Body Image* 27 (2018): 86–92.

99 Marika Tiggemann and Isabella Barbato, "'You Look Great!': The Effect of Viewing Appearance-Related Instagram Comments on Women's Body Image," *Body Image* 27 (2018): 61–66.

100 Marika Tiggemann, Susannah Hayden, Zoe Brown, and Jolanda Veldhuis, "The Effect of Instagram 'Likes' on Women's Social Comparison and Body Dissatisfaction," *Body Image* 26 (2018): 90–97.

101 Dian de Vries, Jochen Peter, Hanneke de Graaf, and Peter Nikken, "Adolescents' Social Network Site Use, Peer Appearance-Related Feedback, and Body Dissatisfaction: Testing a Mediation Model," *Journal of Youth and Adolescence* 45, no. 1 (2016): 211–224.

102 Marika Tiggemann and Amy Slater, "Facebook and Body Image Concern in Adolescent Girls: A Prospective Study," *International Journal of Eating Disorders* 50, no. 1 (2017): 80–83.

103 Ann Rousseau, Steven Eggermont, and Eline Frison, "The Reciprocal and Indirect Relationships between Passive Facebook Use, Comparison on Facebook, and Adolescents' Body Dissatisfaction," *Computers in Human Behavior* 73 (2017): 336–344.

104 Francis Puccio, Fiona Kalathas, Matthew Fuller-Tyszkiewicz, and Isabel Krug, "A Revised Examination of the Dual Pathway Model for Bulimic Symptoms: The Importance of Social Comparisons Made on Facebook and Sociotropy," *Computers in Human Behavior* 65 (2016): 142–150.

105 April Smith, Jennifer Hames, and Thomas Joiner Jr., "Status Update: Maladaptive Facebook Usage Predicts Increases in Body Dissatisfaction and Bulimic Symptoms," *Journal of Affect Disorders* 149, no. 1–3 (2013): 235–240.

106 Alexandra Hummel and April Smith, "Ask and You Shall Receive: Desire and Receipt of Feedback via Facebook Predicts Disordered Eating Concerns," *International Journal of Eating Disorders* 48, no. 4 (2015): 436–442.

107 Patricia Aufderheide, "Media Literacy: A Report of the National Leadership Conference on Media Literacy," Paper presented at the National Leadership Conference on Media Literacy, Washington, DC, 1993.

108 Sian McLean, Susan Paxton, and Eleanor Wertheim, "The Role of Media Literacy in Body Dissatisfaction and Disordered Eating: A Systematic Review," *Body Image* 19 (2016): 9–23.

109 Sian McLean, Susan Paxton, and Eleanor Wertheim, "Does Media Literacy Mitigate Risk for Reduced Body Satisfaction Following Exposure to Thin-Ideal Media?," *Journal of Youth and Adolescence* 45, no. 8 (2016): 1678–1695.

110 McLean et al, "Role of Media Literacy," 18.

111 Natalie Tamplin, Siân McLean, and Susan Paxton, "Social Media Literacy Protects against the Negative Impact of Exposure to Appearance Ideal Social Media Images in Young Adult Women but Not Men," *Body Image* 26 (2018): 29–37.

112 Tamplin et al., "Social Media Literacy Protects," 34.

113 Sian McLean, Eleanor Wertheim, Jennifer Masters, and Susan Paxton, "A Pilot Evaluation of a Social Media Literacy Intervention to Reduce Risk Factors for Eating Disorders," *International Journal of Eating Disorders* 50, no. 7 (2017): 847–851.

114 Phillippa Diedrichs, Melissa Atkinson, Rebecca Steer, Kirsty Garbett, Nichola Rumsey, and Emma Halliwell, "Effectiveness of a Brief School-Based Body Image Intervention 'Dove Confident Me: Single Session' When Delivered by Teachers and Researchers: Results from a Cluster Randomised Controlled Trial," *Behaviour Research and Therapy* 74 (2015): 94–104.

115 Nicole Paraskeva, Helena Lewis-Smith, and Phillippa Diedrichs, "Consumer Opinion on Social Policy Approaches to Promoting Positive Body Image: Airbrushed Media Images and Disclaimer Labels," *Journal of Health Psychology* 22, no. 2 (2017): 164–175; Belinda Bury, Marika Tiggemann, and Amy Slater, "Disclaimer Labels on Fashion Magazine Advertisements: Does Timing of Digital Alteration Information Matter?," *Eating Behaviors* 25 (2017): 18–22.

116 Belinda Bury, Marika Tiggemann, and Amy Slater, "The Effect of Digital Alteration Disclaimer Labels on Social Comparison and Body Image: Instructions and Individual Differences," *Body Image* 17 (2016): 136–142; Marika Tiggemann, Amy Slater, Belinda Bury, Kimberley Hawkins, and Bonny Firth, "Disclaimer Labels on Fashion Magazine Advertisements: Effects on Social Comparison and Body Dissatisfaction," *Body Image* 10, no. 1 (2013): 45–53; Marika Tiggemann, Amy Slater, and Veronica Smyth, "'Retouch Free': The Effect of Labelling Media Images as Not Digitally Altered on Women's Body Dissatisfaction," *Body Image* 11, no. 1 (2014): 85–88.

117 Bury et al., "Individual Differences," 140.

118 Jasmine Fardouly and Elise Holland, "Social Media Is Not Real Life: The Effect of Attaching Disclaimer-Type Labels to Idealized Social Media Images on Women's Body Image and Mood," *New Media & Society* 20, no. 11 (2018): 4311–4328.

119 Katherine Record and S. Bryn Austin, "'Paris Thin': A Call to Regulate Life-Threatening Starvation of Runway Models in the Us Fashion Industry," *American Journal of Public Health* 106, no. 2 (2016): 205–206.

120 Rachel Rodgers, Sara Ziff, Alice Lowy, Kimberly Yu, and S. Bryn Austin, "Results of a Strategic Science Study to Inform Policies Targeting Extreme Thinness Standards in the Fashion Industry," *International Journal of Eating Disorders* 50, no. 3 (2017): 284–292.

121 Phillippa Diedrichs and Christina Lee, "Waif Goodbye! Average-Size Female Models Promote Positive Body Image and Appeal to Consumers," *Psychology & Health* 26, no. 10 (2011): 1273–1291; Phillippa Diedrichs and Christina Lee, "Gi Joe or Average Joe? The Impact of Average-Size and Muscular Male Fashion Models on Men's and Women's Body Image and Advertisement Effectiveness." *Body Image* 7, no. 3 (2010): 218–226.

122 Larry Woodard, "Advertisers Embrace a Plus-Size Reality," https://abcnews.go.com/Business/advertisers-curvier-models-stay/story?id=11313808; Emma Bazilian, "Why More Brands Are Embracing Plus-Size Models," www.adweek.com/brand-marketing/why-more-brands-are-embracing-plus-size-models-170984/.

123 Emily Shugerman, "The Reason You're Seeing More Plus-Size Models in Ads Isn't Body Positivity," www.revelist.com/us-news/plus-size-models-raise-sales/6707.

124 Rachel Cohen, Jasmine Fardouly, Toby Newton-John, and Amy Slater, "#Bopo on Instagram: An Experimental Investigation of the Effects of Viewing Body Positive Content on Young Women's Mood and Body Image," *New Media & Society* 21, no. 7 (2019): 1546–1564.

125 Marika Tiggemann, "Considerations of Positive Body Image across Various Social Identities and Special Populations," *Body Image* 14 (2015): 168–176.

126 Tiggemann, "Considerations of Positive Body Image," 170.

Bibliography

Agliata, Daniel, and Tantleff-Dunn, Stacey. "The Impact of Media Exposure on Males' Body Image." *Journal of Social and Clinical Psychology* 23, no. 1 (2004): 7–22.

Anderson, Monica, and Jiang, Jingjing. "Teens, Social Media & Technology 2018." www.pewinternet.org/2018/05/31/teens-social-media-technology-2018/.

Aufderheide, Patricia. "Media Literacy. A Report of the National Leadership Conference on Media Literacy." Paper presented at the National Leadership Conference on Media Literacy, Washington, DC, 1993.

Australian Bureau of Statistics. "Household Use of Information Technology, Australia, 2014–2015." www.abs.gov.au/ausstats/abs@.nsf/0/ACC2D18CC958BC7BCA2568A9001393AE?Opendocument.

Baghurst, Timothy, Hollander, Daniel B., Nardella, Beth, and Haff, G. Gregory. "Change in Sociocultural Ideal Male Physique: An Examination of Past and Present Action Figures." *Body Image* 3, no. 1 (2006): 87–91.

Barlett, Christopher P., Vowels, Christopher L., and Saucier, Donald A. "Meta-Analyses of the Effects of Media Images on Men's Body-Image Concerns." *Journal of Social and Clinical Psychology* 27, no. 3 (2008): 279–310.

Bazilian, Emma. "Why More Brands Are Embracing Plus-Size Models." April 24, 2016. www.adweek.com/brand-marketing/why-more-brands-are-embracing-plus-size-models-170984/.

Bearman, Sarah Kate, Presnell, Katherine, Martinez, Erin, and Stice, Eric. "The Skinny on Body Dissatisfaction: A Longitudinal Study of Adolescent Girls and Boys." *Journal of Youth and Adolescence* 35, no. 2 (2006): 229–241.

Becker, Anne E., Burwell, Rebecca A., Herzog, David B., Hamburg, Paul, and Gilman, Stephen E. "Eating Behaviours and Attitudes Following Prolonged Exposure to Television among Ethnic Fijian Adolescent Girls." *British Journal of Psychiatry* 180, no. 6 (2002): 509–514.

Betz, Diana E., and Ramsey, Laura R. "Should Women Be "All About That Bass?": Diverse Body-Ideal Messages and Women's Body Image." *Body Image* 22 (2017): 18–31.

Bissell, Kimberly L., and Zhou, Peiqin. "Must See TV or ESPN: Entertainment and Sports Media Exposure and Body Image Distortion in College Women." *Journal of Communication* 54, no. 1 (2004): 5–21.

Boepple, Leah, Ata, Rheanna N., Rum, Ruba, and Thompson, J. Kevin. "Strong Is the New Skinny: A Content Analysis of Fitspiration Websites." *Body Image* 17 (2016): 132–135.

Boothroyd, Lynda G., Jucker, Jean-Luc, Thornborrow, Tracey, Jamieson, Mark A., Burt, D. Michael, Barton, Robert A., Evans, Elizabeth H., and Tovee, Martin J. "Television Exposure Predicts Body Size Ideals in Rural Nicaragua." *British Journal of Psychology* 107, no. 4 (2016): 752–767.

Botta, Renée A. "For Your Health? The Relationship between Magazine Reading and Adolescents' Body Image and Eating Disturbances." *Sex Roles* 48, no. 9 (2003): 389–399.

Brown, Zoe, and Tiggemann, Marika. "Attractive Celebrity and Peer Images on Instagram: Effect on Women's Mood and Body Image." *Body Image* 19 (2016): 37–43.

Brownell, Kelly D. "Dieting and the Search for the Perfect Body: Where Physiology and Culture Collide." *Behavior Therapy* 22 (1991): 1–12.

Buote, Vanessa M., Wilson, Anne E., Strahan, Erin J., Gazzola, Stephanie B., and Papps, Fiona. "Setting the Bar: Divergent Sociocultural Norms for Women's and Men's Ideal Appearance in Real-World Contexts." *Body Image* 8, no. 4 (2011): 322–34.

Bury, Belinda, Tiggemann, Marika, and Slater, Amy. "Disclaimer Labels on Fashion Magazine Advertisements: Does Timing of Digital Alteration Information Matter?." *Eating Behaviors* 25 (2017): 18–22.

Bury, Belinda, Tiggemann, Marika, and Slater, Amy. "The Effect of Digital Alteration Disclaimer Labels on Social Comparison and Body Image: Instructions and Individual Differences." *Body Image* 17 (2016): 136–142.

Cafri, Guy, Thompson, J. Kevin, Roehrig, Megan, van den Berg, Patricia, Jacobsen, Paul B., and Stark, Stephen. "An Investigation of Appearance Motives for Tanning: The Development and Evaluation of the Physical Appearance Reasons for Tanning Scale (Parts) and Its Relation to Sunbathing and Indoor Tanning Intentions." *Body Image* 3, no. 3 (2006): 199–209.

Coetzee, Vinet, and Perrett, David I. "African and Caucasian Body Ideals in South Africa and the United States." *Eating Behaviors* 12, no. 1 (2011): 72–74.

Cohen, Rachel, and Blaszczynski, A. "Comparative Effects of Facebook and Conventional Media on Body Image Dissatisfaction." *Journal of Eating Disorders* 3 (2015): 23.

Cohen, Rachel, Fardouly, Jasmine, Newton-John, Toby, and Slater, Amy. "#Bopo on Instagram: An Experimental Investigation of the Effects of Viewing Body Positive Content on Young Women's Mood and Body Image." *New Media & Society* 21 (2019): 1546–1564.

Cohen, Rachel, Newton-John, Toby, and Slater, Amy. "The Relationship between Facebook and Instagram Appearance-Focused Activities and Body Image Concerns in Young Women." *Body Image* 23 (2017): 183–187.

Cohen, Rachel, Newton-John, Toby, and Slater, Amy. "'Selfie'-Objectification: The Role of Selfies in Self-Objectification and Disordered Eating in Young Women." *Computers in Human Behavior* 79 (2018): 68–74.

Common Sense Media. "The Common Sense Census: Media Use by Tweens and Teens." 2015. www.commonsensemedia.org/research/the-common-sense-census-media-use-by-tweens-and-teens.

de Vries, Dian A., Peter, Jochen, de Graaf, Hanneke, and Nikken, Peter. "Adolescents' Social Network Site Use, Peer Appearance-Related Feedback, and Body Dissatisfaction: Testing a Mediation Model." *Journal of Youth and Adolescence* 45, no. 1 (2016): 211–224.

Diedrichs, Phillippa C., Atkinson, Melissa J., Steer, Rebecca J., Garbett, Kirsty M., Rumsey, Nichola, and Halliwell, Emma. "Effectiveness of a Brief School-Based Body Image Intervention 'Dove Confident Me: Single Session' When Delivered by Teachers and Researchers: Results from a Cluster Randomised Controlled Trial." *Behaviour Research and Therapy* 74 (2015): 94–104.

Diedrichs, Phillippa C., and Lee, Christina. "GI Joe or Average Joe? The Impact of Average-Size and Muscular Male Fashion Models on Men's and Women's Body Image and Advertisement Effectiveness." *Body Image* 7, no. 3 (2010): 218–226.

Diedrichs, Phillippa C., and Lee, Christina. "Waif Goodbye! Average-Size Female Models Promote Positive Body Image and Appeal to Consumers." *Psychology & Health* 26, no. 10 (2011): 1273–1291.

Diemer, Elizabeth W., Grant, Julia D., Munn-Chernoff, Melissa A., Patterson, David A., and Duncan, Alexis E.. "Gender Identity, Sexual Orientation, and Eating-Related Pathology in a National Sample of College Students." *Journal of Adolescent Health* 57, no. 2 (2015): 144–149.

Dohnt, Hayley, and Tiggemann, Marika. "The Contribution of Peer and Media Influences to the Development of Body Satisfaction and Self-Esteem in Young Girls: A Prospective Study." *Developmental Psychology* 42, no. 5 (2006): 929–936.

Dohnt, Hayley K., and Tiggemann, Marika. "Body Image Concerns in Young Girls: The Role of Peers and Media Prior to Adolescence." *Journal of Youth and Adolescence* 35, no. 2 (2006): 135.

Dove. "Girls on Beauty: New Dove Research Finds Low Beauty Confidence Driving 8 in 10 Girls to Opt out of Future Opportunities." *PR Newswire*, October 5, 2017. www.prnewswire.com/news-releases/girls-on-beauty-new-dove-research-finds-low-beauty-confidence-driving-8-in-10-girls-to-opt-out-of-future-opportunities-649549253.html.

Dove. "New Dove Research Finds Beauty Pressures up, and Women and Girls Calling for Change." *PR Newswire*, June 21, 2016. www.prnewswire.com/news-releases/new-dove-research-finds-beauty-pressures-up-and-women-and-girls-calling-for-change-583743391.html.

Evans, Elizabeth H., Tovée, Martin J., Boothroyd, Lynda G., and Drewett, Robert F. "Body Dissatisfaction and Disordered Eating Attitudes in 7- to 11-Year-Old Girls: Testing a Sociocultural Model." *Body Image* 10, no. 1 (2013): 8–15.

Fardouly, Jasmine, Diedrichs, Phillipa C., Vartanian, Lenny R., and Halliwell, Emma. "Social Comparisons on Social Media: The Impact of Facebook on Young Women's Body Image Concerns and Mood." *Body Image* 13 (2015): 38–45.

Fardouly, Jasmine, and Holland, Elise. "Social Media Is Not Real Life: The Effect of Attaching Disclaimer-Type Labels to Idealized Social Media Images on Women's Body Image and Mood." *New Media & Society* 20, no. 11 (2018): 4311–4328.

Fardouly, Jasmine, Magson, Natasha R., Johnco, Carly J., Oar, Ella L., and Rapee, Ronald M. "Parental Control of the Time Preadolescents Spend on Social Media: Links with Preadolescents' Social Media Appearance Comparisons and Mental Health." *Journal of Youth and Adolescence* 47, no. 7 (2018): 1456–1468.

Fardouly, Jasmine, and Vartanian, Lenny R. "Negative Comparisons About One's Appearance Mediate the Relationship between Facebook Usage and Body Image Concerns." *Body Image* 12 (2015): 82–88.

Fardouly, Jasmine, Willburger, Brydie K., and Vartanian, Lenny R. "Instagram Use and Young Women's Body Image Concerns and Self-Objectification: Testing Mediational Pathways." *New Media & Society* 20, no. 4 (2018): 1380–1395.

Fatt, Scott J., Fardouly, Jasmine, and Rapee, Ronald M. "#Malefitspo: Links between Viewing Fitspiration Posts, Muscular-Ideal Internalisation, Appearance Comparisons, Body Satisfaction, and Exercise Motivation in Men." *New Media & Society* 21, no. 6 (2019): 1311–1325.

Ferguson, Christopher J. "In the Eye of the Beholder: Thin-Ideal Media Affects Some, but Not Most, Viewers in a Meta-Analytic Review of Body Dissatisfaction in Women and Men." *Psychology of Popular Media Culture* 2, no. 1 (2013): 20–37.

Florin, Todd A., Shults, Justine, and Stettler, Nicolas. "Perception of Overweight Is Associated with Poor Academic Performance in US Adolescents." *Journal of School Health* 81, no. 11 (2011): 663–670.

Fouts, Gregory, and Burggraf, Kimberley. "Television Situation Comedies: Female Weight, Male Negative Comments, and Audience Reactions." *Sex Roles* 42, no. 9 (2000): 925–932.

Garner, David M. "Body Image and Anorexia Nervosa." In *Body Image: A Handbook of Theory, Research, and Clinical Practice*, edited by Thomas Cash and Thomas Pruzinsky, 295–303, New York, NY: Guilford, 2002.

Grabe, Shelly, and Hyde, Janet Shibley. "Ethnicity and Body Dissatisfaction among Women in the United States: A Meta-Analysis." *Psychological Bulletin* 132, no. 4 (2006): 622–640.

Grabe, Shelly, Ward, L. Monique, and Hyde, Janet Shibley. "The Role of the Media in Body Image Concerns among Women: A Meta-Analysis of Experimental and Correlational Studies." *Psychological Bulletin* 134, no. 3 (2008): 460–476.

Groesz, Lisa M., Levine, Michael P., and Murnen, Sarah K. "The Effect of Experimental Presentation of Thin Media Images on Body Satisfaction: A Meta-Analytic Review." *International Journal of Eating Disorders* 31, no. 1 (2002): 1–16.

Grogan, Sarah. *Body Image: Understanding Body Dissatisfaction in Men, Women, and Children.* New York, NY: Routledge, 1999.

Haferkamp, Nina, Eimler, Sabrina C., Papadakis, Anna-Margarita, and Kruck, Jana Vanessa. "Men Are from Mars, Women Are from Venus? Examining Gender Differences in Self-Presentation on Social Networking Sites." *Cyberpsychology, Behavior, and Social Networking* 15, no. 2 (2012): 91–98.

Haferkamp, Nina, and Kramer, Nicole C. "Social Comparison 2.0: Examining the Effects of Online Profiles on Social-Networking Sites." *Cyberpsychology Behavior and Social Networking* 14, no. 5 (2011): 309–314.

Halliwell, Emma, and Dittmar, Helga. "Does Size Matter? The Impact of Model's Body Size on Women's Body-Focused Anxiety and Advertising Effectiveness." *Journal of Social and Clinical Psychology* 23, no. 1 (2004): 104–122.

Halliwell, Emma, Dittmar, Helga, and Howe, Jessica. "The Impact of Advertisements Featuring Ultra-Thin or Average-Size Models on Women with a History of Eating Disorders." *Journal of Community & Applied Social Psychology* 15, no. 5 (2005): 406–413.

Halliwell, Emma, and Harvey, Martin. "Examination of a Sociocultural Model of Disordered Eating among Male and Female Adolescents." *British Journal of Health Psychology* 11 (2006): 235–248.

Harper, Brit, and Tiggemann, Marika. "The Effect of Thin Ideal Media Images on Women's Self-Objectification, Mood, and Body Image." *Sex Roles* 58 (2008): 649–657.

Harrison, Kristen, and Hefner, Veronica. "Media Exposure, Current and Future Body Ideals, and Disordered Eating among Preadolescent Girls: A Longitudinal Panel Study." *Journal of Youth and Adolescence* 35 (2006): 153–163.

Heinberg, Leslie J., and Thompson, J. Kevin. "Body Image and Televised Images of Thinness and Attractiveness: A Controlled Laboratory Investigation." *Journal of Social and Clinical Psychology* 14, no. 4 (1995): 325–338.

Herbozo, Sylvia, Tantleff-Dunn, Stacey, Gokee-Larose, Jessica, and Thompson, J. Kevin. "Beauty and Thinness Messages in Children's Media: A Content Analysis." *Eating Disorders* 12, no. 1 (2004): 21–34.

Hesse-Biber, Sharlene, Leavy, Patricia, Quinn, Courtney E., and Zoino, Julia. "The Mass Marketing of Disordered Eating and Eating Disorders: The Social Psychology of Women, Thinness and Culture." *Women's Studies International Forum* 29, no. 2 (2006): 208–224.

Holland, Grace, and Tiggemann, Marika. "'Strong Beats Skinny Every Time': Disordered Eating and Compulsive Exercise in Women Who Post Fitspiration on Instagram." *International Journal of Eating Disorders* 50, no. 1 (2017): 76–79.

Holland, Grace, and Tiggemann, Marika. "A Systematic Review of the Impact of the Use of Social Networking Sites on Body Image and Disordered Eating Outcomes." *Body Image* 17 (2016): 100–110.

Homan, Kristin, McHugh, Erin, Wells, Daniel, Watson, Corrinne, and King, Carolyn. "The Effect of Viewing Ultra-Fit Images on College Women's Body Dissatisfaction." *Body Image* 9, no. 1 (2012): 50–56.

Hummel, Alexander C., and Smith, April R. "Ask and You Shall Receive: Desire and Receipt of Feedback Via Facebook Predicts Disordered Eating Concerns." *International Journal of Eating Disorders* 48, no. 4 (2015): 436–442.

Hussein, Nazia. "Colour of Life Achievements: Historical and Media Influence of Identity Formation Based on Skin Colour in South Asia." *Journal of Intercultural Studies* 31, no. 4 (2010): 403–424.

Huxley, Caroline J., Halliwell, Emma, and Clarke, Victoria. "An Examination of the Tripartite Influence Model of Body Image: Does Women's Sexual Identity Make a Difference?" *Psychology of Women Quarterly* 39, no. 3 (2015): 337–348.

Jacobi, Lora, and Cash, Thomas F. "In Pursuit of the Perfect Appearance: Discrepancies among Self-Ideal Percepts of Multiple Physical Attributes." *Journal of Applied Social Psychology* 24, no. 5 (1994): 379–396.

Jung, Jaehee, and Forbes, Gordon B. "Body Dissatisfaction and Disordered Eating among College Women in China, South Korea, and the United States: Contrasting Predictions from Sociocultural and Feminist Theories." *Psychology of Women Quarterly* 31, no. 4 (2007): 381–393.

Kanayama, Gen, Barry, Steven, Hudson, James I., and Pope, Harrison G. "Body Image and Attitudes toward Male Roles in Anabolic-Androgenic Steroid Users." *American Journal of Psychiatry* 163, no. 4 (2006): 697–703.

Karazsia, Bryan T., Murnen, Sarah K., and Tylka, Tracy L. "Is Body Dissatisfaction Changing across Time? A Cross-Temporal Meta-Analysis." *Psychological Bulletin* 143, no. 3 (2017): 293–320.

Keery, Helene, van den Berg, Patricia, and Thompson, J. Kevin. "An Evaluation of the Tripartite Influence Model of Body Dissatisfaction and Eating Disturbance with Adolescent Girls." *Body Image* 1 (2004): 237–251.

Kim, Ji Won, and Chock, T. Makana. "Body Image 2.0: Associations between Social Grooming on Facebook and Body Image Concerns." *Computers in Human Behavior* 48 (2015): 331–339.

Kimbrough, Amanda M., Guadagno, Rosanna E., Muscanell, Nicole L., and Dill, Janeann. "Gender Differences in Mediated Communication: Women Connect More Than Do Men." *Computers in Human Behavior* 29, no. 3 (2013): 896–900.

Leit, Richard A., Pope, Harrison G., Jr., and Gray, James J. "Cultural Expectations of Muscularity in Men: The Evolution of Playgirl Centerfolds." *International Journal of Eating Disorders* 29, no. 1 (2001): 90–93.

Lonergan, Alexandra Rhodes, Bussey, Kay, Mond, Jonathan, Brown, Olivia, Griffiths, Scott, Murray, Stuart B., and Mitchison, Deborah. "Me, My Selfie, and I: The Relationship between Editing and Posting Selfies and Body Dissatisfaction in Men and Women." *Body Image* 28 (2019): 39–43.

Mabe, Annalise G., Forney, K. Jean, and Keel, Pamela K. "Do You "Like" My Photo? Facebook Use Maintains Eating Disorder Risk." *International Journal of Eating Disorders* 47 (2014): 516–523.

Manago, Adriana M., Graham, Michael B., Greenfield, Patricia M., and Salimkhan, Goldie. "Self-Presentation and Gender on Myspace." *Journal of Applied Developmental Psychology* 29, no. 6 (2008): 446–458.

McLean, Sian A., Paxton, Susan J., and Wertheim, Eleanor H. "Does Media Literacy Mitigate Risk for Reduced Body Satisfaction Following Exposure to Thin-Ideal Media?" *Journal of Youth and Adolescence* 45, no. 8 (2016): 1678–1695.

McLean, Sian A., Paxton, Susan J., and Wertheim, Eleanor H. "The Role of Media Literacy in Body Dissatisfaction and Disordered Eating: A Systematic Review." *Body Image* 19 (2016): 9–23.

McLean, Sian A., Paxton, Susan J., Wertheim, Eleanor H., and Masters, Jennifer. "Photoshopping the Selfie: Self Photo Editing and Photo Investment Are Associated with Body Dissatisfaction in Adolescent Girls." *International Journal of Eating Disorders* 48, no. 8 (2015): 1132–1140.

McLean, Sian A., Wertheim, Eleanor H., Masters, Jennifer, and Paxton, Susan J. "A Pilot Evaluation of a Social Media Literacy Intervention to Reduce Risk Factors for Eating Disorders." *International Journal of Eating Disorders* 50, no. 7 (2017): 847–851.

Meier, Evelyn P., and Gray, James. "Facebook Photo Activity Associated with Body Image Disturbance in Adolescent Girls." *Cyberpsychology, Behavior, and Social Networking* 17 (2014): 199–206.

Meyer, Hannah. "Kardashian Curvy or Rousey Strong: It's Time for Women to Fight Back against Body Shaming." January 20, 2016. www.abc.net.au/news/2016-01-21/kim-kardashian-photos-rhonda-rousey-video-war-on-body-shaming/7100998.

Mills, Jennifer S., Musto, Sarah, Williams, Lindsay, and Tiggemann, Marika. "'Selfie' Harm: Effects on Mood and Body Image in Young Women." *Body Image* 27 (2018): 86–92.

Mitchison, Deborah, Hay, Phillipa, Slewa-Younan, Shameran, and Mond, Jonathan. "The Changing Demographic Profile of Eating Disorder Behaviors in the Community." *BMC Public Health* 14, no. 1 (2014): 943.

Myers, Taryn A., and Crowther, Janis H. "Social Comparison as a Predictor of Body Dissatisfaction: A Meta-Analytic Review." *Journal of Abnormal Psychology* 118 (2009): 683–698.

Nayir, Tufan, Uskun, Ersin, Yürekli, Mustafa Volkan, Devran, Hacer, Çelik, Ayşe, and Okyay, Ramazan Azim. "Does Body Image Affect Quality of Life?: A Population Based Study." *Plos One* 11, no. 9 (2016): e0163290.

Neumark-Sztainer, Dianne, Paxton, Susan J., Hannan, Peter J., Haines, Jess, and Story, Mary. "Does Body Satisfaction Matter? Five-Year Longitudinal Associations between Body Satisfaction and Health Behaviors in Adolescent Females and Males." *Journal of Adolescent Health* 39, no. 2 (2006): 244–251.

New Hope Network. "Supplement Business Report 2016." www.newhope.com/sites/newhope360.com/files/2016%20NBJ%20Supplement%20Business%20report_lowres_TOC.pdf.

Office for National Statistics. "Internet Access – Households and Individuals: 2017." www.ons.gov.uk/peoplepopulationandcommunity/householdcharacteristics/homeinternetandsocialmediausage/publications?filter=bulletin.

Papp, Ildikó, Urban, Róbert, Czegledi, Edit, Babusa, Bernadett, and Tury, Ferenc. "Testing the Tripartite Influence Model of Body Image and Eating Disturbance among Hungarian Adolescents." *Body Image* 10, no. 2 (2013): 232–242.

Paraskeva, Nicole, Lewis-Smith, Helena, and Diedrichs, Phillippa C. "Consumer Opinion on Social Policy Approaches to Promoting Positive Body Image: Airbrushed Media Images and Disclaimer Labels." *Journal of Health Psychology* 22, no. 2 (2017): 164–175.

Paxton, Susan J., Neumark-Sztainer, Dianne, Hannan, Peter J., and Eisenberg, Maria E. "Body Dissatisfaction Prospectively Predicts Depressive Mood and Low Self-Esteem in Adolescent Girls and Boys." *Journal of Clinical Child and Adolescent Psychology* 35 (2006): 539–549.

Perloff, Richard M. "Social Media Effects on Young Women's Body Image Concerns: Theoretical Perspectives and an Agenda for Research." *Sex Roles* 71, no. 11 (2014): 363–377.

Pounders, Kathrynn, Kowalczyk, Christine M., and Stowers, Kirsten. "Insight into the Motivation of Selfie Postings: Impression Management and Self-Esteem." *European Journal of Marketing* 50, nos 9/10 (2016): 1879–1892.

Prichard, Ivanka, Kneebone, Anna, Hutchinson, Amanda D., and Wilson, Carlene. "The Relationship between Skin Tone Dissatisfaction and Sun Tanning Behaviour." *Australian Journal of Psychology* 66, no. 3 (2014): 168–174.

Prichard, Ivanka, McLachlan, Annabel C., Lavis, Tiffany, and Tiggemann, Marika. "The Impact of Different Forms of #Fitspiration Imagery on Body Image, Mood, and Self-Objectification among Young Women." *Sex Roles* 78, no. 11 (2018): 789–798.

Puccio, Francis, Kalathas, Fiona, Fuller-Tyszkiewicz, Matthew, and Krug, Isabel. "A Revised Examination of the Dual Pathway Model for Bulimic Symptoms: The Importance of Social Comparisons Made on Facebook and Sociotropy." *Computers in Human Behavior* 65 (2016): 142–150.

Reaves, Shiela, Hitchon, Jacqueline Bush, Park, SungYeon, andYun, Gi Woong. "If Looks Could Kill: Digital Manipulation of Fashion Models." *Journal of Mass Media Ethics* 19, no. 1 (2004): 56–71.

Record, Katherine L., and Austin, S. Bryn. "'Paris Thin': A Call to Regulate Life-Threatening Starvation of Runway Models in the Us Fashion Industry." *American Journal of Public Health* 106, no. 2 (2016): 205–206.

Reuters. "Instagram's User Base Grows to More Than 500 Million." June 21, 2016. www.reuters.com/article/us-facebook-instagram-users/instagrams-user-base-grows-to-more-than-500-million-idUSKCN0Z71LN.

Robinson, Lily, Prichard, Ivanka, Nikolaidis, Alyssa, Drummond, Claire, Drummond, Murray, and Tiggemann, Marika. "Idealised Media Images: The Effect of Fitspiration Imagery on Body Satisfaction and Exercise Behaviour." *Body Image* 22 (2017): 65–71.

Rodgers, Rachel F., Chabrol, Henri, and Paxton, Susan J. "An Exploration of the Tripartite Influence Model of Body Dissatisfaction and Disordered Eating among Australian and French College Women." *Body Image* 8, no. 3 (2011): 208–215.

Rodgers, Rachel F., Ganchou, Camille, Franko, Debra L., and Chabrol, Henri. "Drive for Muscularity and Disordered Eating among French Adolescent Boys: A Sociocultural Model." *Body Image* 9, no. 3 (2012): 318–323.

Rodgers, Rachel, F., McLean, S.A., and Paxton, S.J. "Longitudinal Relationships among Internalization of the Media Ideal, Peer Social Comparison, and Body Dissatisfaction: Implications for the Tripartite Influence Model." *Developmental Psychology* 51, no. 5 (2015): 706–713.

Rodgers, Rachel F., Ziff, Sara, Lowy, Alice S., Yu, Kimberly, and Austin, S. Bryn. "Results of a Strategic Science Study to Inform Policies Targeting Extreme Thinness Standards in the Fashion Industry." *International Journal of Eating Disorders* 50, no. 3 (2017): 284–292.

Rousseau, Ann, Eggermont, Steven, and Frison, Eline. "The Reciprocal and Indirect Relationships between Passive Facebook Use, Comparison on Facebook, and Adolescents' Body Dissatisfaction." *Computers in Human Behavior* 73 (2017): 336–344.

Schooler, Deborah. "Real Women Have Curves: A Longitudinal Investigation of TV and the Body Image Development of Latina Adolescents." *Journal of Adolescent Research* 23, no. 2 (2008): 132–153.

Seifert, Tim. "Anthropomorphic Characteristics of Centerfold Models: Trends Towards Slender Figures over Time." *International Journal of Eating Disorders* 37, no. 3 (2005): 271–274.

Shugerman, Emily. "The Reason You're Seeing More Plus-Size Models in Ads Isn't Body Positivity." January 31, 2017. www.revelist.com/us-news/plus-size-models-raise-sales/6707.

Simpson, Courtney C., and Mazzeo, Suzanne E. "Skinny Is Not Enough: A Content Analysis of Fitspiration on Pinterest." *Health Communication* 32, no. 5 (2017): 560–567.

Slater, Amy, Varsani, Neesha, and Diedrichs, Phillippa C. "#Fitspo or #Loveyourself? The Impact of Fitspiration and Self-Compassion Instagram Images on Women's Body Image, Self-Compassion, and Mood." *Body Image* 22 (2017): 87–96.

Slevec, Julie Helen, and Tiggemann, Marika. "Predictors of Body Dissatisfaction and Disordered Eating in Middle-Aged Women." *Clinical Psychology Review* 31, no. 4 (2011): 515–524.

Smith, A. R., Hames, J. L., and Joiner, T. E., Jr. "Status Update: Maladaptive Facebook Usage Predicts Increases in Body Dissatisfaction and Bulimic Symptoms." *Journal of Affective Disorders* 149, no. 1–3 (2013): 235–240.

Statistica. "Number of Monthly Active Facebook Users Worldwide as of 2nd Quarter 2018 (in Millions)." www.statista.com/statistics/264810/number-of-monthly-active-facebook-users-worldwide/.

Statistica. "Number of Social Network Users in India from 2015 to 2022 (in Millions)." www.statista.com/statistics/278407/number-of-social-network-users-in-india/.

Statistica. "U.S. Magazine Industry—Statistics & Facts." August 27, 2019. www.statista.com/topics/1265/magazines/.

Stice, Eric. "Risk and Maintenance Factors for Eating Pathology: A Meta-Analytic Review." *Psychological Bulletin* 128 (2002): 825–848.

Stice, Eric, Spangler, Diane, and Agras, W. Stewart. "Exposure to Media-Portrayed Thin-Ideal Images Adversely Affects Vulnerable Girls: A Longitudinal Experiment." *Journal of Social and Clinical Psychology* 20, no. 3 (2001): 270–288.

Stice, Eric, and Van Ryzin, Mark J. "A Prospective Test of the Temporal Sequencing of Risk Factor Emergence in the Dual Pathway Model of Eating Disorders." *Journal of Abnormal Psychology* 128, no. 2 (2018): 119–128.

Strauss, Penelope, Lin, Ashleigh, Winter, Sam, Cook, Angus, Watson, Vanessa, and Toussaint, Dani Wright. 2017. "Trans Pathways: The Mental Health Experiences and Care Pathways of Trans Young People." In *Summary of Results*. Perth, Australia: Telethon Kids Institute.

Stronge, Samantha, Greaves, Lara M., Milojev, Petar, West-Newman, Tim, Barlow, Fiona Kate, and Sibley, Chris G. "Facebook Is Linked to Body Dissatisfaction: Comparing Users and Non-Users." *Sex Roles* 73, nos 5–6 (2015): 200–213.

Swami, Viren, Frederick, David. A., Aavik, Tovio, Alcalay, Lidia, Allik, Juri, Anderson, Donna, Andrianto, Sonny, et al. "The Attractive Female Body Weight and Female Body Dissatisfaction in 26 Countries across 10 World Regions: Results of the International Body Project." *Personality and Social Psychology Bulletin* 36, no. 3 (2010): 309–325.

Sypeck, Mia Foley, Gray, James J., and Ahrens, Anthony H. "No Longer Just a Pretty Face: Fashion Magazines' Depictions of Ideal Female Beauty from 1959 to 1999." *International Journal of Eating Disorders* 36, no. 3 (2004): 342–347.

Tamplin, Natalie C., McLean, Siân A., and Paxton, Susan J. "Social Media Literacy Protects against the Negative Impact of Exposure to Appearance Ideal Social Media Images in Young Adult Women but Not Men." *Body Image* 26 (2018): 29–37.

Thompson, J. Kevin, and Cafri, Guy. *The Muscular Ideal: Psychological, Social, and Medical Perspectives*. The Muscular Ideal: Psychological, Social, and Medical Perspectives, edited by J. Kevin Thompson and Guy Cafri. Washington, DC: American Psychological Association, 2007.

Thompson, J. Kevin, Heinberg, Leslie J., Altabe, Madeline, and Tantleff-Dunn, Stacey. *Exacting Beauty: Theory, Assessment, and Treatment of Body Image Disturbance*. Washington, DC: American Psychological Association, 1999.

Thompson, Joel K., Schaefer, Lauren M., and Menzel, Jessie E. "Internalization of Thin-Ideal and Muscular-Ideal." In *Encyclopedia of Body Image and Human Appearance*, edited by Thomas Cash, 499–504. Oxford: Academic Press, 2012.

Tiggemann, Marika. "Considerations of Positive Body Image across Various Social Identities and Special Populations." *Body Image* 14 (2015): 168–176.

Tiggemann, Marika. "Media Exposure, Body Dissatisfaction and Disordered Eating: Television and Magazines are not the Same!" *European Eating Disorders Review* 11, no. 5 (2003): 418–430.

Tiggemann, Marika. "The Role of Media Exposure in Adolescent Girls' Body Dissatisfaction and Drive for Thinness: Prospective Results." *Journal of Social and Clinical Psychology* 25, no. 5 (2006): 523–541.

Tiggemann, Marika. "Sociocultural Perspectives on Human Appearance and Body Image." In *Body Image: A Handbook of Science, Practice, and Prevention*, edited by Thomas Cash and Linda Smolak, 12–19, New York: Guilford Press, 2011.

Tiggemann, Marika. "Television and Adolescent Body Image: The Role of Program Content and Viewing Motivation." *Journal of Social and Clinical Psychology* 24, no. 3 (2005): 361–381.

Tiggemann, Marika, and Barbato, Isabella. "'You Look Great!': The Effect of Viewing Appearance-Related Instagram Comments on Women's Body Image." *Body Image* 27 (2018): 61–66.

Tiggemann, Marika, Hayden, Susannah, Brown, Zoe, and Veldhuis, Jolanda. "The Effect of Instagram "Likes" on Women's Social Comparison and Body Dissatisfaction." *Body Image* 26 (2018): 90–97.

Tiggemann, Marika, Martins, Yolanda, and Churchett, Libby. "Beyond Muscles: Unexplored Parts of Men's Body Image." *Journal of Health Psychology* 13, no. 8 (2008): 1163–1172.

Tiggemann, Marika, and Pennington, Barbara. "The Development of Gender Differences in Body-Size Dissatisfaction." *Australian Psychologist* 25, no. 3 (1990): 306–313.

Tiggemann, Marika, and Slater, Amy. "Facebook and Body Image Concern in Adolescent Girls: A Prospective Study." *International Journal of Eating Disorders* 50, no. 1 (2017): 80–83.

Tiggemann, Marika, and Slater, Amy. "Netgirls: The Internet, Facebook, and Body Image Concern in Adolescent Girls." *International Journal of Eating Disorders* 46 (2013): 630–633.

Tiggemann, Marika, and Slater, Amy. "Nettweens: The Internet and Body Image Concerns in Preteenage Girls." *Journal of Early Adolescence* 34 (2014): 606–620.

Tiggemann, Marika, and Slater, Amy. "Thin Ideals in Music Television: A Source of Social Comparison and Body Dissatisfaction." *International Journal of Eating Disorders* 35, no. 1 (2004): 48–58.

Tiggemann, Marika, Slater, Amy, Bury, Belinda, Hawkins, Kimberley, and Firth, Bonny. "Disclaimer Labels on Fashion Magazine Advertisements: Effects on Social Comparison and Body Dissatisfaction." *Body Image* 10, no. 1 (2013): 45–53.

Tiggemann, Marika, Slater, Amy, and Smyth, Veronica. "'Retouch Free': The Effect of Labelling Media Images as Not Digitally Altered on Women's Body Dissatisfaction." *Body Image* 11, no. 1 (2014): 85–88.

Tiggemann, Marika, and Zaccardo, Mia. "'Exercise to Be Fit, Not Skinny': The Effect of Fitspiration Imagery on Women's Body Image." *Body Image* 15 (2015): 61–67.

Tiggemann, Marika, and Zaccardo, Mia. "'Strong Is the New Skinny': A Content Analysis of #Fitspiration Images on Instagram." *Journal of Health Psychology* 23, no. 8 (2016): 1003–1011.

Tylka, Tracy L. "Refinement of the Tripartite Influence Model for Men: Dual Body Image Pathways to Body Change Behaviors." *Body Image* 8, no. 3 (2011): 199–207.

van den Berg, Patricia, Thompson, J. Kevin, Obremski-Brandon, Karen, and Coovert, Michael. "The Tripartite Influence Model of Body Image and Eating Disturbance: A Covariance Structure Modeling Investigation Testing the Mediational Role of Appearance Comparison." *Journal of Psychosomatic Research* 53 (2002): 1007–1020.

Vendemia, Megan A., and DeAndrea, David C. "The Effects of Viewing Thin, Sexualized Selfies on Instagram: Investigating the Role of Image Source and Awareness of Photo Editing Practices." *Body Image* 27 (2018): 118–127.

Voorveld, Hilde A.M., Segijn, Claire M., Ketelaar, Paul E., and Smit, Edith G. "Investigating the Prevalence and Predictors of Media Multitasking across Countries." *International Journal of Communication* 8 (2014): 2755–2777.

Whyte-Smith, Hannah, Boreham, Emmeline, and Patel, Vanessa. "Opinions of Generation Z's Ambitions and Priorities Differ Greatly between the Generations." www.ipsos.com/ipsos-mori/en-uk/opinions-generation-zs-ambitions-and-priorities-differ-greatly-between-generations.

Woodard, Larry D. "Advertisers Embrace a Plus-Size Reality." April 20, 2010. https://abcnews.go.com/Business/advertisers-curvier-models-stay/story?id=11313808.

Yamamiya, Y., Shroff, H., and Thompson, J.K. "The Tripartite Influence Model of Body Image and Eating Disturbance: A Replication with a Japanese Sample." *International Journal of Eating Disorders* 41, no. 1 (2008): 88–91.

Zephoria. "The Top 20 Valuable Facebook Statistics – Updated December 2018." https://zephoria.com/top-15-valuable-facebook-statistics/.

Zhao, Shanyang, Grasmuck, Sherri, and Martin, Jason. "Identity Construction on Facebook: Digital Empowerment in Anchored Relationships." *Computers in Human Behavior* 24, no. 5 (2008): 1816–1836.

11

BLOOD, BODIES, AND SHAME

Indian artists combating menstrual stigma on Instagram

Marissa J. Doshi

In early 2015, Rupi Kaur, a Canadian of Indian heritage who is now a famous poet with a sizable Instagram following, posted an image that showed her lying on a bed with menstrual blood seeping out of her gray pajamas and on her faded, floral-print sheets. In the caption accompanying the photo, Kaur explained that it was part of a series she developed for a college course on visual rhetoric in order to normalize menstruation and challenge societal norms that coded visible signs of women menstruating as offensive but found images that objectified women's bodies permissible. Soon after, similar images from the series were posted and then Instagram removed them, stating that they violated community standards, which, in many ways, bolstered Kaur's argument. The images were subsequently made available on the social media site after Kaur challenged their removal. The incident received widespread media coverage and the posts subsequently went viral.[1]

I begin with this event because it marks a significant moment in contemporary digital activism aimed at combating menstrual stigma. First, Kaur's Instagram images showcase the potential and limitations of digital spaces for such challenges. Second, this moment highlights the participation of a young woman of Indian heritage in digital activism to end norms of shame and secrecy surrounding menstruation. This study builds on these points by examining the menstruation-related digital art of two female Indian artists that is available on Instagram. I argue that in their menstruation-related work, these artists use feminist strategies such as recovering embodiment and consciousness raising to end taboos of concealment and silence with regard to menstruation. Examining the online visual work of these artists reveals how such strategies facilitate the development of feminist digital counterpublics.

This study focuses on female Indian artists for a number of reasons. Examining the activism of female Indian artists stalls hegemonic Western narratives that frame women from India as passive victims.[2] Western narratives often reinforce the white savior trope by denying Indian women agency and homogenizing their experiences. In this way, Western narratives often affirm Western white superiority. By focusing on art produced by Indian women to end menstrual stigma, this study emphasizes Indian women's acts of self-representation and activism, challenging Western white savior narratives. Thus, this study takes seriously the creative strategies being used and developed by women from India, and in doing so challenges discourses that delegitimize knowledge produced from the Global South.[3] Further, by examining the potential and limitations of such interventions and highlighting how internal social hierarchies in Indian society play out in digital spaces, this study seeks to disrupt and complicate homogenizing discourses about India, Indian feminist activism, and more specifically, feminist digital activism in India. This approach is in line with the cyberfeminist framework advanced by Radhika Gajjala,[4] which asks researchers to acknowledge both the possibilities and limitations of digital spaces for activism. Specifically, Gajjala emphasizes that any online research about countries in the Global South (like India) must attend to the complexities of life for women and other marginalized groups in the Global South. In keeping with prior cyberfeminist research, I seek to unpack the politics of art-based digital advocacy efforts and reveal how internal social hierarchies are implicated in emerging feminist digital counterpublics.[5]

I begin this chapter by providing an overview of menstrual stigma in India. Next, I examine contemporary feminist efforts to challenge stigma in India, while focusing on the role of digital activism in this regard. Finally, I analyze the Instagram art created by two female Indian artists that challenges menstrual stigma and end by evaluating the strategic potential of their efforts for countering menstrual stigma and building a feminist digital counterpublic.

Menstrual stigma in India

Globally, menstruation continues to be widely stigmatized, with research showing that menstrual stigma is a public health issue in many places, negatively affecting quality of life and contributing to lower social status for menstruators.[6] Although menstrual stigma is experienced by all people who menstruate,[7] research has typically focused on its impact on women and young girls. In India, menstrual stigma is maintained by framing menstruation as dirty and a pollutant.[8] This framing of menstruation as "impure" becomes the rationale for prohibiting menstruating women's entry into religious spaces, kitchens, and other areas. Further, this negative framing means that menstruation is perceived as a source of shame, validating norms of concealment and silence surrounding it. Thus, as Yagnik[9] explains, menstrual stigma is maintained through three strategies: (1) Concealment: menstruators are encouraged to conceal the bleeding as well as any other signs of menstruation. Media also play a role in validating concealment via advertisements of menstrual hygiene products, which depict menstruation as undesirable, dirty, and something that needs to be hidden. (2) Restricted mobility: this refers to social norms that prohibit menstruators from accessing certain public and private spaces. In some cases, interpersonal contact is also prohibited, leading to social isolation. For example, menstruating women might be prohibited entry into the worship room and kitchen (and prohibited from cooking food) in the home and, in some cases, they might be made to sleep in a separate room or even outside the house.[10] Additionally, menstruating women might not only be denied entry into religious spaces, but they might also be prohibited from participating in religious events. (3) Silence: this norm codes direct (and especially descriptive) speech about menstruation as socially undesirable and unacceptable.[11] Consequently, euphemisms are used to reference menstruation during conversations, and on the rare occasions that the topic is addressed, it is discussed in hushed tones. Thus, as many Indian activists and researchers have argued, menstrual stigma needs to end because it operates as form of violent social control by perpetuating gender hierarchies and discrimination while also curtailing the legal rights of menstruators.[12]

There is a large body of public health research focused on the structural challenges that Indian menstruators, particularly young women, face with regard to maintaining menstrual hygiene (including lack of access to adequate clean water, privacy, and disposal options).[13]

They are also frequently unable to access or afford safe menstrual hygiene products, such as sanitary pads. Additionally, young women rarely use insertable menstrual hygiene products (tampons, for example) because their use might be incorrectly equated with masturbation and loss of virginity, both of which are taboo.[14] Young people also often face information deficits and social support challenges. For example, multiple studies have shown that adolescent girls in India lack basic information about menstruation and puberty.[15] Research has also documented that young girls are made to drop out of school upon the onset of menarche, despite legal protections for education of girls and women.[16]

Notably, the experience of menstrual stigma varies with menstruators' social location. For example, in large cities like Mumbai (which lack affordable housing and have high costs of living), poor urban women face heightened risks to their menstrual health because of lack of access to basic amenities.[17] A number of government programs have been started to educate rural and urban poor female adolescents about menstruation, but they often target adolescents and leave out mothers, who are usually the primary source of information in a family and might face challenges of their own with regard to menstruation.[18]

Religious and caste identities also play a role in how menstruation is experienced. For example, Guterman, Mehta, and Gibbs document religion-specific taboos pertaining to menstruation in major Indian religions: Hinduism, Islam, and Christianity.[19] Caste also influences how menstrual stigma is experienced. Within the Hindu caste system, Brahmins are considered upper caste, with Dalits considered low caste. This caste hierarchy is maintained through various mechanisms of social, cultural, and political control, including by policing notions of purity and pollution. Within the caste hierarchy, Brahmins are considered "pure," while Dalits and others at the lowest ends of the caste hierarchy are considered "polluting" and "untouchable." During menstruation, Brahmin women are considered temporarily impure, and their activities and interactions are heavily curtailed as a way of preventing them from "polluting" other members and spaces of the household. Thus, menstrual stigma is "crucial in sustaining notions of Brahmin-hood within their homes and among other Brahmins,"[20] which is in "sharp contrast to the eternal untouchable condition that the actual untouchables suffer."[21] Further, low-caste men and women, who are often employed as household help, are expected to clean the rags used by menstruating Brahmin women or dispose of used sanitary pads. By having low-caste men engage in cleaning the "dirt" of Brahmin women and engage in the otherwise feminized activity of washing clothes, menstrual stigma serves also to emasculate low-caste men.[22] Through practices such as giving low-caste women upper-caste girl's clothes stained during her first period, low-caste women become "receivers of the pollution in order to protect the purity of the upper-caste households."[23] Thus, menstrual stigma plays a significant role in upholding violently unjust caste hierarchies.

Finally, almost all scholarly research I encountered while reviewing the literature on menstrual stigma in India focused on how it affects women and young girls. By limiting inquiry to the female gender, efforts to end menstrual stigma often reinscribe and uphold the gender binary by erasing the menstruation experiences of trans men and non-binary people.[24]

Challenging menstrual stigma

Activism to end menstrual stigma has a long history globally, with women and legal activists playing an important role in this regard. However, an exhaustive review of such activism is beyond the scope of this chapter. Instead, I will focus on activism to end menstrual stigma in India specifically. Further, in keeping with the focus of this chapter, this section will examine contemporary digital activism by Indian women to end menstrual stigma rather than public health efforts enacted by the Indian government.

Destabilizing stigma involves creating alternative framings that transgress boundaries of visibility, mobility, and speech regarding menstruation. However, participating in such transgressions can lead to ostracism, bullying, censorship, and even violence. One example of such violence occurred on January 2019, when two women of menstruating age entered the Sabarimala temple in the state of Kerala, which refused entry to all women of menstruating age despite a court ruling that permitted all women to enter. Women who protested this ban faced widespread condemnation from conservative Hindu groups who accused them of not respecting religious norms and tradition. Further, protesters were also sprayed with tear gas.[25]

These women are part of a larger movement working to reclaim the public sphere by protesting various menstruation taboos in a variety of ways.[26] Indeed, activism advocating for allowing women entry into religious spaces has been occurring in the form of legal challenges as well as social media campaigns.[27] The #HappytoBleed campaign, for example, was started by 20-year-old Nikita Azad in 2015 in response to a comment that the chief priest of the Sabarimala temple made stating that women would be allowed into the temple only after a scanner is developed to determine that they are not currently menstruating. Those who participated in the campaign posted images of themselves holding posters or sanitary pads with the words "happy to bleed." Women have similarly protested bans on entering other temples and religious places while menstruating, such as the Shani Shingnapur Temple and Haji Ali Dargah in Maharashtra.[28] Another example of digital activism regarding menstruation is the #Red Alert: You've Got a Napkin campaign, which began after women employed by a company in Kerala were strip-searched after a used sanitary pad was found in the bathroom. Indian women who participated in the campaign mailed used and unused sanitary napkins to the company managers as a form of protest.[29] When the government instituted a tax on sanitary pads by categorizing them as luxury products, SheSays (an Indian nongovernmental organization) collaborated with Indian social media influencers to start an online campaign titled #LahukaLagaan (translation: tax on blood). They also launched legal challenges against the tax.[30] The Hindi movie industry has also begun to participate in efforts to end menstrual stigma, most notably through movies such as *Padman*, which showcased the efforts of Arunachalam Muruganantham in his quest to provide low-cost sanitary pads for rural women. The movie prompted the #padmanchallenge in which participants sought to break the silence around menstruation by posing with sanitary pads on social media.[31] These efforts to end menstrual stigma in India show the ways in which offline and online actions are connected.

However, a number of these contemporary advocacy efforts have been critiqued because they "tend to skew toward the opinions of youthful tech-savvy participants and reflect the views of elites in education, caste, and class."[32] Critics of such efforts point out that these protests persist in ignoring the ways in which menstrual stigma is a strategy for maintaining the caste system.[33] For example, while middle-class urban women, who have been the loudest voices in contemporary digital protests, have advocated for bins and better disposal systems in workplaces, they have ignored that the majority of sanitation workers are low caste and face significant health risks and social stigma as a result of engaging in sanitation work.[34] Further, while advocacy to end the luxury tax on sanitary pads mobilized a number of urban women and men, these efforts seemed to ignore that sanitary pads, even when not taxed, remain prohibitively expensive for the majority of Indian menstruators.[35] Finally, contemporary activism to end menstrual stigmas has been critiqued for framing menstruation as central to womanhood and valuing it as a symbol of reproductive potential. Doing so conflates gender and sex, and this move further fixes motherhood and fertility as central to womanhood. Consequently, such efforts reinforce heteronormativity while also contributing to the marginalization of women who might choose to or not be able to have biological children.[36]

Cyberfeminism and feminist digital counterpublics

When analyzed through the works of Dona Haraway[37] and Judy Wajcman,[38] cyberfeminists seem to share a commitment to understanding how technology and gender are co-constitutive. Thus, cyberfeminists are invested in exploring the emancipatory potential of technology even while being attentive to its challenges and limitations. Further, cyberfeminists reject digital dualism, and have shown that offline and online lives are connected rather than separate. Finally, cyberfeminists are attentive to how internal hierarchies manifest in digital spaces and practices as well as how these internal social hierarchies get reconfigured or reified within global flows of information and capital.[39] Given that this chapter aims to understand how digital artists are using an online space, Instagram, to challenge norms that affect the lived experiences of Indian menstruators, cyberfeminism is a productive lens for examining this topic.

A number of studies have interrogated how gender structures cyberculture. These studies range from understanding the role of access in engendering or limiting women's empowerment[40] to exploring the value of networked communities for women.[41] Others have examined how digital spaces are structured by gendered neoliberal forces that shape labor markets for women.[42] A robust body of research has also focused on the role of digital spaces in shaping feminist practice, ranging from efforts to change policy to engaging in consciousness raising and organizing intersectional protests around issues that center gender injustices.[43] In short, researchers have begun to take note of and theorize how women are using digital tools and spaces to challenge prevailing discourses of gendered oppression.

One body of research that looks at these issues focuses on the dynamics within feminist digital counterpublics. This research is grounded in Rita Felski and Joseph Felski's[44] and Nancy Fraser's[45] critique of Jurgen Habermas'[46] conceptualization of the public sphere as one that ignores the ways in which women have been excluded from democratic debate. In response to this erasure, they advance the idea of counterpublics, which are conceptualized as being in opposition to the masculinized public sphere and centering feminist discourse. Such counterpublics exist not only in ideological opposition to the dominant public but are constituted by discourses that "in other contexts would be regarded with hostility or with a sense of indecorousness."[47] Thus, unpacking the contours of the discourse shaping counterpublics is important if we are to understand their oppositional nature and ideological limits.

Recent scholarship has adopted a critical and intersectional[48] stance when studying feminist counterpublics to point out the pitfalls of assuming shared experiences of womanhood. For example, Sarah Jackson and Sonia Banaszczyk's research on feminist counterpublics on Twitter shows that

> even as Twitter works as a space where the historically marginalized standpoints of women can be elevated through virality and collective advocacy, the technological architecture of the platform's trending, retweeting, and mentioning functions, along with the ways mainstream and elite individuals and outlets legitimate the "popular," reproduce the marginalization of intersectional experiences.[49]

Similarly, Radhika Gajjala has highlighted how prominent Indian feminists dismissed #LoSha, an online curated a list of sexual harassers in Indian academia that was developed by queer, Dalit feminist Raya Sarkar.[50]

Research on feminist digital counterpublics has also explored how traditional strategies of feminist organizing, such as focusing on the possibilities and challenges for recovering embodiment and consciousness raising, manifest in digital spaces. Feminist digital counterpublics often facilitate community building and sharing of lived experiences. In doing so, they pose a challenge to popular framing of digital spaces as anonymous and disembodied.[51] Feminists also use digital spaces to normalize embodied practices such breastfeeding though the posting of brelfies (photos that show breastfeeding) on social media.[52] While digital feminist counterpublics offer opportunities for recovering embodiment, Hester Baer's[53] analysis of a number of transnational feminist digital protests shows that many fail to account for local body politics. Given that menstruation is an embodied practice that is taboo, unpacking whether digital spaces help to alleviate stigma associated with it is important.

Many studies on the role of feminist digital counterpublics highlight their role in building and sustaining community. The potential of these communities to generate social action has been discussed through the concept of consciousness raising. As part of contemporary consciousness raising, "feminists share their stories, listen to others' stories, consume popular culture in ways that they find empowering, and create new vocabularies to enhance their own lives."[54] Urszula Pruchniewska emphasizes that the digital

> allows a level of sharing similar to traditional face-to-face [consciousness raising] while adding shareability (a circulation function) and storage (an archiving function). Thus, the Internet does not change the essential nature of [consciousness-raising] as a feminist practice, but simply changes the ways in which it is done, using new tools to make the practice more efficient and beneficial to a larger number of people.[55]

Although the move to large-scale emancipation has typically been understood as central to consciousness raising, online communities might not always achieve that aspect of societal change. Others, however, emphasize the ways that consciousness raising in digital spaces may promote a form of micropolitics that allows one to understand the gendered dimensions of one's own experiences of oppression, which can then help to forge a connection with feminist politics.[56]

A note about method

In the pages that follow, I build on scholarship on feminist digital activism by conducting a comparative analysis of challenges to menstrual stigma in digital art created by @wallflowergirlsays and @doodleodrama, two female Indian artists. These two artists are not the only Indian artists challenging menstrual stigma. I chose to focus on the art of these two illustrators in this chapter because they are both primarily digital artists and the affordances of digital platforms like Instagram (for example, creating slideshows, adding hashtags, and allowing for reposting) potentially plays a role in shaping their activism. Thus, their art can help with understanding strategies used in digital advocacy. Both artists also share some similarities in that their art regularly addresses feminist concerns that center the body, such as fat positivity, beauty standards, and colorism. At the same time, the two artists are distinguished by interesting differences: While @doodleodrama has a number of posts addressing menstrual stigma specifically, @wallflowergirlsays addresses menstrual stigma as part of a larger set of social issues and has only two posts that tackle the issue directly.

Creating a case study comparing the menstruation-related art of two digital artists is useful because it allows for unpacking similarities as well as differences.[57] Further, in keeping with the case study approach, I have conducted an in-depth analysis of the contributions of each artist, and consequently, this method helps advance a nuanced understanding of the strategies being used by individual influential actors within India's digital counterpublics. By analyzing social media activism in this focused way, I am able to contextualize each post challenging menstrual stigma within the whole body of Instagram art produced by each artist, while remaining attentive to the larger sociopolitical context within which these posts exist. This comparative case study is also useful for contrasting the strategies of digital activism along multiple axes, such as aesthetics, storytelling, and platform affordances. Thus, this approach is useful for addressing my research question, which concerns understanding how female Indian artists challenge menstrual stigma in digital spaces.

To answer this question, I first identify key strategies used by these two artists and then discuss the ways in which these strategies facilitate the development of a feminist digital counterpublic. This analysis builds on my previous research on the feminist politics of digital embodiment[58] and how internal social hierarchies manifest within the Indian mediasphere.[59]

Analysis

In my analysis of the digital art of @doodleodrama and @wallflowergirlsays that addresses menstrual stigma, I pay particular attention to the following: aesthetics, storytelling, and platform affordances. In the section that follows, I first address each artist's work separately and then present a comparison in order to identify similarities and differences.

Digital art of @wallflowergirlsays

@wallflowergirlsays is the Instagram handle of Mumbai-based Kaviya Ilango. Ilango says she began illustrating in order to express her frustration with hypocritical and restrictive social norms. On June 6, 2017, she started a project titled #100daysofdirtylaudry with a caption that read,

> I hope to cover everything unholy, uncomfortable, cringe-worthy—be it complicated love, indefinable sexuality, masturbation, periods, over-gloried travel, manic materialism, alcohol binges, smelly farts, burps, ugly scars, anything & everything I am guilty about, ashamed to share in public. This project is going to be uncomfortable to do. But such dirt is a part of who I am. Who we all are. Just that some acknowledge it, some speak about it but most people hide it. I am just going to draw and rant about it.[60]

As part of Ilango's series, there were two posts that addressed menstruation: the first post (appearing on Day 18) addressed menstruation indirectly as part of a larger set of bodily functions that are considered taboo, while the second post (appearing on Day 40) challenged menstrual stigma directly. Analysis of these two images reveals that the overarching strategy in Ilango's art involves making menstruation hypervisible. As the subsequent section demonstrates, this strategy recovers embodiment in menstruation and encourages viewers to confront their own biases regarding this bodily process, thereby acting as a form of consciousness raising. Thus, Ilango uses hypervisiblity and candid discussion to break the silence and concealment taboos associated with menstrual stigma.

The first post that tackles menstruation is the Day 18 post captioned "ode to my humble toilet."[61] It depicts a girl wearing a pink blouse sitting on a toilet (with a handheld bidet beside it) with her underwear around her ankles and a blue towel covering her thighs. Hair loose, wearing pink high heels, and adorned with a matha patti (also known as maang tikka or sirbandi, which is jewelry worn along the hairline and often includes a piece resting on the forehand) and a pink bindi, the girl sits on the toilet, gazing straight ahead, nonchalantly, one hand cupping her chin. In the caption, Illango explains why this image is part of her series by listing the numerous milestones that have happened while on the toilet, one of those being that her toilet was "witness to the first blood that fell out & the shocked tears that followed."[62]

In this image, we are invited to both gaze upon a woman in a typically private space—the toilet—and also to meet her gaze. Her nonchalant expression reminds us that she is engaged in an ordinary act—nothing to see here! And yet, as we look at her, we are forced to question why viewing the image feels uncomfortable and transgressive. Arguably, the image derives its power through our discomfort. By making the female body visible in an act that is socially coded as private, this image returns our attention to the various visceral embodied experiences that might occur there, including menstruation. One way in which menstrual stigma is maintained by using disembodied or euphemistic language to create distance from its corporal nature. Ilango counters this tactic by making the menstruating body not only visible but also the focus of the post. The words about menstruation in the caption (i.e., "my toilet was witness to the first blood that fell out & the shocked tears that followed") remind us that for many, their first period is shocking and scary, perhaps because of the lack of information provided to adolescents about menstruation. The caption also reminds us that these feelings are often dealt with alone, perhaps while on a toilet seat.

Finally, there are subtle reminders of caste and class: the woman depicted in the post is adorned with gold jewelry, which implies a woman of upper caste who also has class privilege, given that women of lower castes and classes are either not permitted or expected to wear jewelry or might not have the resources to own such ornaments.[63] Unlike the squatting toilet found in many rural homes as well as in poor urban homes, the sitting-style toilet (colloquially referred to as a "western" toilet) is found in middle-class urban homes. The presence of the handheld bidet further marks the middle- to upper-class status of the setting because it implies water security, which is not a given in the homes of rural or urban poor folk. Indeed, lack of indoor toilets is a major issue in urban India that places poor urban women at risk for sexual violence.[64] Finally, the fashion choices in the image—the easy mixing of traditional jewelry with Western-style high-heeled shoes—invokes the breezy cosmopolitanism of privileged, urban Indians.

The second post from Ilango addressing menstruation is from Day 40. It is titled "Whispers and murmurs,"[65] and tackles silence and shame explicitly. The image depicts a girl lying in a pink bathtub filled with blue water. Her arms are crossed covering her chest so we can only see pink straps on her shoulders. She stares straight ahead, the lower half of her body submerged in bathwater with a pool of blood seeping out from the point where her legs cross and appear above the water. As in the previous image, she is wearing a matha patti and pink bindi, but here, her hair is braided, and she is holding a lollipop near her lips. The woman depicted in this second image is also economically secure: the presence of a bathtub that is filled with water suggests middle-class and comfortable urban living, as does the presence of jewelry.

Unlike the image from Day 18, menstrual blood is clearly visible in the Day 40 image. The red of the blood stands in stark contrast to the blue of the water, which is also the color used to euphemistically depict menstrual blood in commercials that include demonstrations of the absorbency of sanitary products like pads and tampons.[66] By using contrasting colors, Ilango's image makes menstrual bleeding hypervisible. The blood seeping out of the woman's body pushes us to confront the embodied nature of menstruation. The lollipop in the subject's hand can be read as a nod to the craving for sweets, often a side effect of menstruation. Even a cursory glance at the post reveals that it contests the assertion that menstruation is a clinical process that should be understood and discussed in a sanitized, detached manner. Instead, as the image shows, menstruation is messy, emotional, and connected with a living body. Perhaps most interesting is the way in which this image, as in Day 18, evokes the trope of the "good Indian woman": the gold jewelry and bindi are evocative of the upper-caste Hindu woman who is the template for appropriate Indian femininity, one who is docile and follows social norms.[67] In both images, this trope is evoked only to be subsequently destroyed. As our eyes travel downward, this "good Indian woman," we realize, is likely unclothed and in a very private space in the midst of an extremely private practice. Her direct gaze challenges and fixes interpretation of the image so that the person portrayed becomes what Sara Ahmed terms, "a willful subject"—one who refuses to adjust in an unjust world.[68] In the second image, the willfulness of the woman lies not only in bleeding but in bleeding unabashedly and in her invitation to the audience to view and acknowledge menstrual blood. These are the "actions" that mark her as willful, undisciplined, and unafraid. She transgresses the norms of silence and privacy regarding menstruation, and in doing so, rejects the social control enacted by menstrual stigma.

The caption accompanying this second image provides some additional insight into the artist's intentions: in it, Ilango references her annoyance with the use of blue to depict menstrual blood in advertisements saying,

> I wanted to colour the stain under the legs blue, just like an innocuous ink blot, just like ones in those sanitary pad ads. Red does look weird I agree, even when I look down every time, 4 days a month. Wish I didn't have to bleed blue only for the Indian cricket team.[69]

By acknowledging the "weirdness" of her period even while embracing it, Ilango provides a glimpse into the complicated relationship women might have with menstruation. Discussing her feelings openly is a strategy for countering stigma, specifically because hegemonic discourse tries to fix women's relationship with their periods by casting it as a shameful experience that women must experience silently. In contrast, Ilango boldly states that she "hates" her period:

> I wouldn't have hated it with such vengeance if all it did was appear coyly 4 days a month & quietly made its way out. But no, it needs drama, just like every woman u say? So it plots to make me sulk, weep over rom-coms 3 days prior, bitterly cry for no reason 2 days prior & transform me into Dracarys, the dragon a day prior. Basically, do not disturb me when I'm PMSing or else I will find u & burn u.[70]

She goes on to explain how periods take up a quarter of a woman's life and thus, rather than an inconvenient aberration, they are a routine occurrence. She also expresses her frustration at having to discuss menstruation in hushed tones in workplaces as if it is something shameful that requires secrecy. Later in the post, she references controversies around emerging workplace policies about giving working women a day off if they have their periods, demonstrating how in the case of menstrual stigma, the personal is indeed political.

In summary, in Ilango's artwork, a number of strategies are used to counter menstrual stigma: the bold colors used are in contrast to the hushed silence expected when discussing menstruation; the hypervisibility of the bleeding, menstrual body counters the disembodied language and imagery that is used to portray menstruation in commercial media; and captions allude to the complex relationship menstruators might have with their periods as well as the ways in which workplaces need to acknowledge the needs of menstruators in the workforce. All of these elements both emphasize the importance of, and actually enact, openly discussing menstruation. The hashtag #normalizeperiods that accompanies the second image is apt, in that it succinctly states the overarching goal of normalization that Ilango wants to achieve in order to counter prevailing patriarchal viewpoints that code periods as dirty, an aberration, and taboo.

Although Ilango's art is to be commended for advocating normalizing periods, it is also important to note that it centers the concerns of upper-caste, cis women regarding menstruation.

Indeed, by using markers of upper-caste and middle- or upper-class status, I argue that Ilango forecloses possibilities for her art to portray the concerns of a more expansive set of Indian women. By highlighting the centering of upper-caste, cis women's daily lives in @wallflowergirlsays' art, my intention is to show the limits of existing digital advocacy discourse to end menstrual stigma and point out the need for complexity within feminist digital counterpublics in India. As this analysis reveals, offline social hierarchies can and do manifest in online spaces as well. Recovering complexity in feminist digital advocacy to end menstrual stigma can start with noting the ways in which caste and class privilege buffer how stigma is experienced while also showcasing how stigma persists despite these privileges. Presenting this type of complicated narrative regarding the experience of menstrual stigma might ultimately lead to activism that is not only more inclusive but also more radical.

Digital art of @doodleodrama

@doodleodrama is the Instagram handle of Mounica Tata, an Indian digital artist who uses cartoons to advocate for a number of feminist causes and portrays everyday experiences in ways that are often humorous, witty, and bold. On Instagram, she discusses her craft, posts photos of her life, and also shares her art. Her visual work addresses a variety of themes, with menstrual stigma being one topic that is repeatedly featured through a variety of panel-style comics.

Tata often draws on her personal experiences to challenge menstrual stigma. Her signature style of drawing "rounded" cartoon figures is both playful and deceptively simple, allowing serious issues like menstrual stigma to be addressed with wit and humor. As the analysis that follows will show, Tata's art challenges menstrual stigma by engaging in social critique while providing public education. Specifically, through humorous storytelling that centers embodiment and prioritizes consciousness raising with information about menstruation, Tata counters menstrual stigma via her art on the topic.

In a panel published on February 10, 2017, titled "women, sanitary pads, and workplaces" Tata offers social critique by addressing the effort women must undertake to make sure that coworkers don't know they are menstruating.[71] In this comic, Tata illustrates the minutiae involved in a woman figuring out where to hide her sanitary pad as she walks from her desk to the bathroom. The post is humorous but also depicts an experience that resonates with many (as evidenced by the comments generated by the post). Here, Tata uses humor to show how concealment taboos associated with menstruation have a negative impact on the everyday lives of menstruators. While her other artwork that tackles menstruation is a deep pink, this comic is in blue, perhaps as a nod to the blue used to depict menstrual blood in commercials.[72] Tata engages in further social critique in the accompanying caption that addresses women, challenging them to reject the norm of concealment associated with menstruation, and reminding readers that there is nothing shameful about this bodily process. She then goes on to exhort organizations to have adequate facilities if they have female employees, such as a bin for disposing of sanitary pads. In this way, similar to Ilango, Tata moves discussions about menstruation from the private sphere to the public while also showing how menstruation is an issue with consequences extending beyond the personal and reflecting systemic failures.

Another example of Tata's visual social critique is a panel on sanitary pads titled "How not to sell sanitary pads—a tiny guide," which pokes fun at commercials that use outlandish persuasive appeals to market their products.[73] The protagonist here is a cartoon "talking" vagina who exaggerates the selling points common to sanitary pad commercials. For example, many commercials emphasize that their pads are perfumed, have wings, and are organic and soft.

In Tata's art, we see the cartoon vagina holding onto a pad and mockingly exclaiming, "Guys this smells like unicorns and dreams," "This one's made of the soft-est, fresh-est, sun-kissed clouds. 100% organic!" and "check out these real angel wings."[74]

The "talking" vagina is humorous, but it is also an example of how Tata recovers the embodied nature of menstruation in her art. Moreover, illustrating a vagina is a bold move, given that in popular culture, more often than not, vaginas are associated with sexual objectification. Furthermore, digital platforms often censor images depicting vaginas. Tata's humorous and daring depiction of a vagina is an alternative, desexualized, functional framing of this bodily organ. Her cartoon vagina is also a reminder that menstruation is an embodied process and that challenging menstrual stigma involves openly discussing bodies as well as body parts that are conventionally considered taboo and kept hidden.

In other panels, we see Tata combine humous storytelling and embodiment with consciousness raising as a strategy for busting myths about menstruation. For example, in a comic posted on October 27, 2017, Tata illustrates a conversation between a girl and her uterus, during which the girl is pleading with her uterus to delay menstruation because she has a beach vacation planned.[75] In the end the uterus wins, and the last panel shows the girl sitting in a pool of her menstrual blood, tears spilling from her eyes. While the outlandishness of negotiating with one's uterus forms the premise of this humorous illustration, the comic also acts as a form of consciousness raising: it provides a glimpse into the biology undergirding menstruation and thus works to dismantle patriarchal superstitions about menstruation that might be internalized, such as it being a curse or the result of some sort of "sinful" behavior. Just like in the cartoon vagina panel, the organs involved in menstruation are made visible. Additionally, the emotional relationship the girl has with her body is illustrated. In this way, through her art, Tata once again emphasizes the embodied nature of menstruation by making organs and emotions involved in the physiological and psychological experience of the process hypervisible while simultaneously providing information.

As is evident in the uterus conversation and similar to Ilnango, Tata often discusses emotions associated with her period. For example, the caption appearing with the abovementioned illustration reads, "My periods are bloody painful and I think I'm done crying about it so that's why I make comics about them now. This is my idea of black humor. Rather red humor. Teeheee no? Too much?"[76] By discussing the pain she experiences while on her period on Instagram not only does Tata challenge the social norm of not discussing periods in a "public" setting, she also resists the trope of the "happy" period that is often portrayed in advertising media.

In addition to the illustrations analyzed above, Tata educates her audience about painful periods in multiple posts. For example, in a post on March 5, 2017, she discusses her struggles with polycystic ovarian syndrome and explains how it can lead to painful periods.[77] This personal account thus provides further education on menstruation, which is a key issue in reproductive health that is rarely addressed in social forums. Another comic called "Period cramps be like." provides visceral illustrations of Tata's experience with painful periods: in one panel she depicts an elephant sitting on top of her, flattening her into the ground, and in another we panel we see her stomach being punched by a boxing glove.[78]

The theme of consciousness raising through education is most prominently seen in a comic titled "#menstrualcups and me!"[79] In this post, Tata begins by illustrating the fears and qualms she had about using a menstrual cup. In the next panel, she demonstrates methods for inserting the cup and identifies the one that worked for her. Finally, the third panel shows the reasons why she has now come to prefer the menstrual cup, which include benefits such as comfort and the eco-friendly nature of the product.

However, as with Ilango, Tata's art on menstruation centers the concerns of urban middle-class women. For example, although Tata's critique of the marketing of sanitary pads[80] is warranted, the primary concern for most menstruators in India is access to and affordability of menstrual hygiene products.[81] Similarly, while the menstrual cup is an ecologically sensitive choice, its use is predicated on having access to clean water for sanitizing it, and clean water remains scarce for many rural and poor urban women.

It is important to note that one illustration by Tata posted on February 8, 2018, titled "It's natural. Period" tackles issues of access and affordability of menstrual hygiene products by depicting three women hugging and holding sanitary pads.[82] Below the image are the words, "Charity begins at home. Buy your house help a pack of sanitary pads every month." In the caption, Tata suggests the depicted strategy as an alternative to the #padmanchallenge that was trending in India at the time. The #padmanchallenge was critiqued for being a form of slacktivism[83] because it involved celebrities and others posting a photo on social media of themselves holding up a pad, to purportedly lessen the concealment taboo around menstruation. This challenge was also criticized for being insensitive and elitist given that it involved wastage of pads—which seemed to be in opposition to one of the purported goals of the movement: to increase access to menstrual hygiene products.

Tata's solution, as depicted in her art, counters such wastefulness by suggesting that her Instagram audience give their household help a pack of pads every month.[84] Although the idea has some merit, it reveals some assumptions: First, Tata assumes her audience is composed of the demographic most able and likely to have hired household help. Consequently, it is no surprise that much of her art tackling menstrual stigma confronts the concern of menstruators within a privileged class bracket. Second, when viewed through the lens of feminist solidarity, the act of providing sanitary pads does not really address root causes undergirding the current lack of access, such as the affordability of sanitary pads, inadequate sanitation, or low wages. Instead, such seemingly benevolent practices keep social hierarchies intact by limiting the agency of poor menstruators in making choices about their health while increasing their dependence on employers.

Comparison: similarities, differences, and limitations

The artwork of both digital artists profiled in this case study shares some important similarities and differences in approaches and appeals. Through visceral, explicit illustrations, both artists break silence and concealment taboos about menstruation. By making menstrual blood and menstruating bodies visible in their art, both artists emphasize the embodied nature of menstruation. Further, both artists also address the complex relationship menstruating women have with their bodies in their Instagram posts by tackling issues such as period pain, inconvenience, and polycystic ovarian syndrome. In doing so, they displace shame, which is the emotion prescribed by hegemonic discourses about menstruation, with a variety of emotions, including anger, pain, and frustration, and engage in consciousness raising. Unlike some media narratives that celebrate menstruation as a sign of reproductive potential, neither artist frames menstruation in that way in their art.[85] By rejecting shame and celebration as the only two permissible accounts of menstruation and replacing them with complex feelings rooted in their lived experiences, both artists are able to frame menstruation in pragmatic terms that embrace nuance and ambivalence. Tata's illustrations go a step further by using humorous storytelling to educate her audiences about menstrual hygiene and the process of menstruation. Her illustrations are informative on topics ranging from the biological organs at work in menstruation to how to use a menstrual cup.

In terms of exploiting the affordances of Instagram, both artists use popular hashtags related to reducing menstrual stigma to signal that they are part of a larger online conversation around the topic. The use of public Instagram accounts, as on other social media platforms, allows users to tag others in posts, extending their reach, circulation, and visibility. Commenting features encourage engagement, while the platform's constraints around easy reposting and sharing allow the artists relatively more control over how their posts are framed in online conversations. Although both artists use Instagram to share their artwork pertaining to menstruation, Tata frequently reposts or creates art related to menstrual stigma, suggesting that advocating an end to menstrual stigma is a specific (although not singular) concern for her. In contrast, Ilango has created only two posts tackling menstrual stigma and they were part of a larger project (#100daysofdirtylaundry) suggesting that for her, menstrual stigma is part of a bigger conversation about behaviors and practices that are shamed in Indian society.

Neither artist addresses the restriction of activity that is enforced during menstruation, and in their art, both emphasize the concerns of urban middle-class women. The struggles of rural and poor women, particularly those of low caste, are conspicuously absent from their advocacy efforts—as are considerations of transgender community members who menstruate. Because both artists depict themselves as urban middle-class cisgender women drawing on their own personal experiences, some might argue that such critique is unwarranted. However, the point of this critique is not to dismiss the concerns of fairly privileged Indian women. Rather, it is aimed at pointing out the importance of incorporating reflexivity when using lived experiences to advocate for social causes. Reflexivity involves critical self-introspection. It allows for a radical contextualization of social experiences that is attentive to the complex webs of privilege and marginalization in which all experiences are enmeshed.[86] Therefore, a reflexive approach to storytelling can act as a corrective to common oversimplifications of personal narratives. Expanding the scope of issues addressed might ultimately help nurture empathy; cultivate caste, class, and gender consciousness; and educate audiences about the multiple ways in which menstrual stigma is implicated in other systems of domination such as casteism, classism, and heteronormativity. Indeed, in the absence of such reflexivity, digital advocacy to dismantle menstrual stigma advances a narrow, elitist vision that lacks intersectional complexity and nuance. The lack of direct confrontation with how menstrual stigma is a route to maintaining caste hierarchies acts as a form of representational and epistemic violence that allows such systems of power to persist in everyday discussions and representations.

Concluding thoughts

The past few years have seen an explosion of advocacy in India, online and offline, to challenge menstrual stigma. My goal in this chapter has been to identify the strategies used by two female Indian digital artists to engage in such advocacy via their Instagram posts. In this section, I put my analysis in conversation with existing ideas about digital feminist counterpublics.

Mohanty's question, "What happens to the key feminist construct of 'the personal is political' when the political (the collective public domain of politics) is reduced to the personal?"[87] is productive for unpacking the politics of digital feminist counterpublics, given that these counterpublics aim to intervene into the dominant public sphere. This case study of Instagram posts that challenge menstrual stigma shows that digital spaces can be successfully used to politicize the personal, with personal storytelling playing an important role in this regard. More specifically, when private experiences are communicated via Instagram posts (as images or images in conjunction with text), this format allows taboo topics like menstruation to be illustrated and discussed. As in the case of the art analyzed in this study, menstruation-related body parts, practices, and emotions can be made visible in these posts. Indeed, by being posted on public Instagram accounts, this art can provoke conversation among commenters and build networks via the use of hashtags and tagging. Thus, these posts can potentially become generative spaces for discussion and deliberation regarding menstrual stigma. Consequently, such personal storytelling via Instagram posts challenges the discourses of disembodiment that undergird the concealment and silence taboos regarding menstruation. Moreover, this type of Instagram art aligns closely with the traditional feminist activist strategy of consciousness raising. Hashtags, for example, signal connection with others engaging in similar efforts to end menstrual stigma, and thus, Ilango and Tata's art on the topic become part of an archive of Instagram posts on menstruation. Because their posts validate a range of emotions and provide an alternative framing for the various intimate and rarely discussed aspects of menstruation, the posts can engage a wider audience. Additionally, these posts and the discussions they provoke online can include non-menstruators who might not be able to participate in offline discussions about menstruation because of prevailing stigma. Although this study did not focus on the discussions generated by these posts, subsequent research might examine these discussions in order to understand if and how the possibilities for consciousness raising promoted by these posts are realized.

The strategy advanced by the menstruation-related art examined in this study, however, is not without its limitations. When personal stories showcased in digital spaces lack reflexivity, they limit the possibilities for coalition building. Importantly, this strategy reflects a neoliberal ethos that focuses on individual experiences and emancipation without attention to how these experiences are made possible or challenged by social systems. Although these storytelling practices are rooted in feminist politics of valuing lived experiences, they reinforce problematic notions of shared womanhood that are predicated on the erasure of those on the margins of Indian society.[88] For these reasons, such strategies can accelerate fragmentation within already precarious feminist digital counterpublics. Ultimately, this approach limits the possibilities for developing intersectional feminist digital counterpublics even if it provides an entry point into feminist politics for privileged Indian women.

Notes

1. Heather, Saul, "Menstruation-themed Photo Series Artist 'Censored by Instagram' Says Images Are to Demystify Taboos Around Periods," Independent.co.uk, last modified March 30, 2015, www.independent.co.uk/arts-entertainment/art/menstruation-themed-photo-series-artist-censored-by-instagram-says-images-are-to-demystify-taboos-10144331.html.
2. Chandra Talpade Mohanty, *Feminism without Borders: Decolonizing Theory, Practicing Solidarity* (Durham, NC: Duke University Press, 2003).
3. Mohanty, *Feminism without Borders.*
4. Radhika Gajjala, "'Third World' Perspectives on Cyberfeminism." *Development in Practice,* 9, no. 5 (1999): 616–619.
5. Radhika Gajjala, "Snapshots from Sari Trails: Cyborgs Old and New." *Social Identities,* 17, no. 3 (2011): 393–408; Deepa Fadnis, "Feminist Activists Protest Tax on Sanitary Pads: Attempts to Normalize Conversations About Menstruation in India Using Hashtag Activism." *Feminist Media Studies,* 17, no. 6 (2017): 1111–1114.
6. Marni Sommer, Jennifer S. Hirsch, Constance Nathanson, and Richard G. Parker, "Comfortably, Safely, and without Shame: Defining Menstrual Hygiene Management as a Public Health Issue." *American Journal of Public Health,* 105, no. 7 (2015): 1302–1311.
7. Sian Ferguson, "4 Ways to Make Your Period-Positivity More Inclusive." *Everyday Feminism* last modified January, 24, 2016, https://everydayfeminism.com/2016/01/inclusive-period-positivity/
8. Arpan Shailesh Yagnik, "Reframing Menstruation in India: Metamorphosis of the Menstrual Taboo with the Changing Media Coverage." *Health Care for Women International,* 35, no. 6 (2014): 617–633.
9. Yagnik, "Reframing Menstruation." 617–633.

10. Manju Kaundal and Bhopesh Thakur, "A Dialogue on Menstrual Taboo." *Indian Journal of Community Health,* 26, no. 2 (2014): 192–195.
11. Yagnik, "Reframing Menstruation." 617–633.
12. Yagnik, "Reframing Menstruation." 617–633; Fadnis, "Feminist Activists Protest Tax." 1111–1114; Haripriya Narasimhan, "Adjusting Distances: Menstrual Pollution among Tamil Brahmins." *Contributions to Indian Sociology,* 45, no. 2 (2011): 243–268; Chitra Karunakaran Prasanna, "Claiming the Public Sphere: Menstrual Taboos and the Rising Dissent in India." *Agenda,* 30, no. 3 (2016): 91–95.
13. Elizabeth R. MacRae, Thomas Clasen, Munmun Dasmohapatra, and Bethany A. Caruso. "'It's Like A Burden on The Head': Redefining Adequate Menstrual Hygiene Management Throughout Women's Varied Life Stages in Odisha, India." *PloS One,* 14, no. 8 (2019): e0220114
14. Anna Maria van Eijk et al., "Menstrual Hygiene Management Among Adolescent Girls in India: A Systematic Review and Meta-Analysis." *BMJ Open* 6, no. 3 (2016): e010290.
15. Kaundal and Thakur, "A Dialogue on Menstrual Taboo." 192–195.
16. Suneela Garg, and Tanu Anand. "Menstruation Related Myths in India: Strategies for Combating It." *Journal of Family Medicine and Primary Care,* 4, no. 2 (2015): 184–186.
17. Arundati Muralidharan, "Constrained Choices? Menstrual Health and Hygiene Needs among Adolescents in Mumbai Slums." *Indian Journal of Gender Studies,* 26, no. 1–2 (2019): 12–39.
18. Muralidharan, "Constrained Choices?" 12–39.
19. Mark Guterman, Payal Mehta, and Margaret Gibbs. "Menstrual Taboos among Major Religions." *The Internet Journal of World Health and Societal Politics,* 5, no. 2 (2008): 1–7.
20. Narasimhan, "Adjusting Distances." 243.
21. Tamalapakula Sowjanya, "Critique on Contemporary Debates on Menstrual Taboo in India: Through Caste Lens." *Pramana Research Journal,* 9, no. 8 (2019): 254.
22. Sowjanya, "Critique on Contemporary Debates," 253–258.
23. Sowjanya, "Critique on Contemporary Debates," 256.
24. Nupur Alok, "Menstruation and the Need for Inclusivity in the 'Period Positive' Movement," Feminism in India, last modified August 30, 2016, https://feminisminindia.com/2016/08/30/menstruation-period-positive-movement/.
25. Pokharel Sugam and Nikhil Kumar, "First Women to Enter India Temple in Centuries Now in Hiding as Protests Rage," CNN, last modified January 3, 2019, www.cnn.com/2019/01/03/asia/india-kerala-temple-intl/index.html.
26. Prasanna, "Claiming the Public Sphere." 91–95.
27. Prasanna, "Claiming the Public Sphere." 91–95.
28. Prasanna, "Claiming the Public Sphere." 91–95.
29. Aswathy Gopala Gopalakrishnan, "Red Alert: Kerala Activists Send Sanitary Pads to Factory Where Menstruating Women Were Insulted," OneIndia.com, last modified January 1, 2015, www.oneindia.com/india/red-alert-kerala-activists-send-sanitary-pads-factory-menstruating-women-insulted-1609044.html.
30. Fadnis, "Feminist Activists Protest Tax." 1111–1114.
31. Akshita Prasad, "Why the PadMan Challenge Does Not Really Combat Menstruation Stigma," Feminism in India, last modified February 8, 2018, https://feminisminindia.com/2018/02/08/pad-man-acquisitive-marketing/.
32. Fadnis, "Feminist Activists Protest Tax." 1113.
33. Sowjanya, "Critique on Contemporary Debates." 253–258.
34. Sowjanya, "Critique on Contemporary Debates." 253–258.
35. Fadnis, "Feminist Activists Protest Tax." 1111–1114.
36. Alok, "Menstruation and the Need for Inclusivity."
37. Donna Haraway, "A Cyborg Manifesto: Science, Technology, and Socialist-Feminism in the Late 20th Century," in *Simians, Cyborgs and Women: The Reinvention of Nature,* ed. Donna Haraway (New York and London: Routledge, 1991), 117–158.
38. Judy Wajcman, *Technofeminism* (Cambridge: Polity, 2004).
39. Radhika Gajjala, "South Asian Digital Diasporas and Cyberfeminist Webs: Negotiating Globalization, Nation, Gender and Information Technology Design." *Contemporary South Asia,* 12, no. 1 (2003): 41–56; Radhika Gajjala and Yeon Ju Oh, *Cyberfeminism 2.0* (New York: Peter Lang Publishing, 2012).
40. Leda Cooks and Kirsten Isgro, "The" Cyber Summit" and Women: Incorporating Gender into Information and Communication Technology UN Policies." *Frontiers: A Journal of Women Studies,* 26, no.1 (2005): 71–89.

41. Gajjala, "South Asian Digital Diasporas," 41–56; Radha S. Hegde, "Food Blogs and the Digital Reimagination of South Asian Diasporic Publics." *South Asian Diaspora,* 6, no. 1 (2014): 89–103.

42. Alison Adam and Eileen Green, "Gender, Agency, Location and the New Information Society," in *Cyberspace Divide,* ed. Brian Loader (New York: Routledge, 2004) 97–112; Jack Bratich, "The Digital Touch: Craft-Work as Immaterial Labour and Ontological Accumulation." *Ephemera: Theory & Politics in Organization,* 10, no. 3/4 (2010): 303–318.

43. Shahrzad Mojab, "The Politics of 'Cyberfeminism' in The Middle East: The Case of Kurdish Women." *Race, Gender & Class* (2001): 42–61; Ann Mojab, "Parallel Subaltern Feminist Counterpublics in Cyberspace." *Sociological Perspectives,* 46, no. 2 (2003): 223–237; Aristea Fotopoulou, "Digital and Networked by Default? Women's Organisations and the Social Imaginary of Networked Feminism." *New Media & Society,* 18, no. 6 (2016): 989–1005; Sarah J. Jackson and Sonia Banaszczyk, "Digital Standpoints: Debating Gendered Violence and Racial Exclusions in the Feminist Counterpublic." *Journal of Communication Inquiry,* 40, no. 4 (2016): 391–407.

44. Rita Felski and Joseph Felski, *Beyond Feminist Aesthetics: Feminist Literature and Social Change* (Cambridge: Harvard University Press, 1989).

45. Nancy Fraser, "Rethinking the Public Sphere: A Contribution to the Critique of Actually Existing Democracy." *Social Text,* 25/26 (1990): 56–80.

46. Jürgen Habermas, *The Structural Transformation of the Public Sphere: An Inquiry into a Category of Bourgeois Society* (Cambridge: MIT Press, 1991).

47. Michael Warner, "Publics and Counterpublics." *Public Culture,* 14, no. 1 (2002): 86.

48. Kimberlé Crenshaw, "Mapping the Margins: Intersectionality, Identity Politics, and Violence Against Women of Color." *Stanford Law Review,* 43 (1991): 1241–1299; Mohanty, *Feminism without Borders.*

49. Jackson and Banaszczyk, "Digital Standpoints." 403–404.

50. Radhika Gajjala, "When an Indian Whisper Network Went Digital." *Communication Culture & Critique,* 11, no. 3 (2018): 489–493.

51. Ann Travers, "Parallel Subaltern Feminist Counterpublics in Cyberspace." *Sociological Perspectives,* 46, no. 2 (2003): 223–237.

52. Elisabetta Locatelli, "Images of Breastfeeding on Instagram: Self-Representation, Publicness, and Privacy Management." *Social Media+ Society,* 3, no. 2 (2017): 2056305117707190.

53. Hester Baer, "Redoing Feminism: Digital Activism, Body Politics, and Neoliberalism." *Feminist Media Studies,* 16, no. 1 (2016): 17–34.

54. Stacey K. Sowards and Valerie R. Renegar, "The Rhetorical Functions of Consciousness–Raising in Third Wave Feminism." *Communication Studies,* 55, no. 4 (2004): 548.

55. Urszula Maria Pruchniewska, "Working Across Difference in the Digital Era: Riding the Waves to Feminist Solidarity." *Feminist Media Studies,* 16, no. 4 (2016): 738.

56. Shelley Budgeon, "Emergent Feminist (?) Identities: Young Women and the Practice of Micropolitics." *European Journal of Women's Studies,* 8, no. 1 (2001): 7–28.

57. Robert K. Yin, *Case Study Research and Applications: Design and Methods* (Thousand Oaks, CA: Sage, 2017).

58. Marissa J. Doshi, "Barbies, Goddesses, and Entrepreneurs: Discourses of Gendered Digital Embodiment in Women's Health Apps." *Women's Studies in Communication,* 41, no. 2 (2018): 183–203.

59. Marissa J. Doshi, "Hybridizing National Identity: Reflections on the Media Consumption of Middle-Class Catholic Women in Urban India," in *Media and Power in International Contexts: Perspectives on Agency and Identity,* eds Apryl Williams, Ruth Tsuria, and Laura Robinson (Bingley: Emerald Publishing, 2018): 101–131.

60. Kaviya Ilango (@wallflowergirlsays), "I am a millennial," Instagram post, June 6, 2017, www.instagram.com/p/BU_5mWBh5Zg/.

61. Kaviya Ilango (@wallflowergirlsays), "#100daysofdirtylaundry Day 18 - An ode to my humble toilet," Instagram post, June 27, 2017, www.instagram.com/p/BV3CQvLhlAr/.

62. Ilango, "#100daysofdirtylaundry Day 18."

63. Ashwaq Masoodi, "The Changing Fabric of Dalit Life." Livemint.com, last modified April 21, 2017, www.livemint.com/Leisure/avsrwntNuBHG3THdAb5aMP/The-changing-fabric-of-Dalit-life.html.

64. Muralidharan, "Constrained Choices?," 12–39.

65. Kaviya Ilango (@wallflowergirlsays), "#100daysofdirtylaundry Day 40 - Whispers & murmurs," Instagram post, July 29, 2017, www.instagram.com/p/BXJJyeehOgT/.

66. Arpan Shailesh Yagnik, "Mis-Fitting Menstrual Hygiene Products: An Examination of Advertisements to Identify Gaps in the Diffusion of Innovation." *IAFOR Journal of Psychology & the Behavioral Sciences*, 2, no. 3 (2016): 33–45.
67. Sushmita Chatterjee, "'English Vinglish' and Bollywood: What Is 'New' About the 'New Woman'?." *Gender, Place & Culture,* 23, no. 8 (2016): 1179–1192.
68. Sara Ahmed, *Living a Feminist Life* (Durham, NC: Duke University Press, 2017), 84.
69. Kaviya Ilango (@wallflowergirlsays), "#100daysofdirtylaundry Day 40 - Whispers & murmurs," Instagram post, July 29, 2017, www.instagram.com/p/BXJJyeehOgT/.
70. Ilango, "#100daysofdirtylaundry Day 40."
71. Mounica Tata (@doodleodrama), "Ok! Firstly, women, you DON'T need to go all undercover with your sanitary pad," Instagram post, February 10, 2017, www.instagram.com/p/BQVXsAAFoE3/.
72. Yagnik, "Mis-fitting Menstrual Hygiene Products." 33–45.
73. Mounica Tata (@doodleodrama), "How not to sell sanitary pads-a tiny guide," Instagram post, July 23, 2017, www.instagram.com/p/BW4fYU0BjZ3/.
74. Tata, "How not to sell sanitary pads."
75. Mounica Tata (@doodleodrama), "I was working on the #realhorror series for a period tracking app when this idea struck," Instagram post, October 27, 2017, www.instagram.com/p/BawXHsvHeD9/.
76. Tata, "I was working on the #realhorror."
77. Mounica Tata (@doodleodrama), "When people call me a rebel, I laugh it off because they haven't met my ovaries yet," Instagram post, March 5, 2017, www.instagram.com/p/BRPtKe4BVCH/.
78. Mounica Tata (@doodleodrama), "Period cramps be like," Instagram post, November 21, 2017, www.instagram.com/p/BbwsD1HH0_c/.
79. Mounica Tata (@doodleodrama), "#menstrualcups and me!" Instagram post, April 16, 2018, www.instagram.com/p/BhokynQno9b/.
80. Tata, "How not to sell sanitary pads."
81. Fadnis, "Feminist Activists Protest Tax." 1111–1114.
82. Mounica Tata (@doodleodrama), "There's a lot of hullabaloo on the internet around the #padmanchallenge," Instagram post, February 8, 2018, www.instagram.com/p/Be8BE8knrzj/.
83. Evgeny Morozov, "The brave new world of slacktivism," Foreign Policy, last modified May 19, 2009. https://foreignpolicy.com/2009/05/19/the-brave-new-world-of-slacktivism/.
84. Tata, "There's a Lot of Hullabaloo."
85. Lauren Rosewarne, *Periods in Pop Culture: Menstruation in Film and Television* (Lanham, MD: Lexington Books, 2012), 178.
86. Linda Finlay, "Negotiating the Swamp: The Opportunity and Challenge of Reflexivity in Research Practice." *Qualitative Research,* 2, no. 2 (2002): 209–230.
87. Chandra Talpade Mohanty, "Transnational Feminist Crossings: On Neoliberalism and Radical Critique." *Signs: Journal of Women in Culture and Society,* 38, no. 4 (2013): 971.
88. Mohanty, *Feminism without Borders.*

Bibliography

Adam, Alison, and Green, Eileen. "Gender, Agency, Location and the New Information Society." In *Cyberspace Divide,* edited by Brian Loader, 97–112. New York: Routledge, 2004.
Ahmed, Sara. *Living a Feminist Life.* Durham, NC: Duke University Press, 2017.
Alok, Nupur. "Menstruation and the Need for Inclusivity in the 'Period Positive' Movement." *Feminism in India.* Last modified August 30, 2016. https://feminisminindia.com/2016/08/30/menstruation-period-positive-movement/.
Baer, Hester. "Redoing Feminism: Digital Activism, Body Politics, and Neoliberalism." *Feminist Media Studies* 16, no. 1 (2016): 17–34.
Bratich, Jack. "The Digital Touch: Craft-Work as Immaterial Labour and Ontological Accumulation." *Ephemera: Theory & Politics in Organization* 10, no. 3/4 (2010): 303–318.
Budgeon, Shelley. "Emergent Feminist (?) Identities: Young Women and the Practice of Micropolitics." *European Journal of Women's Studies* 8, no. 1 (2001): 7–28.
Chatterjee, Sushmita. "'English Vinglish' and Bollywood: What Is 'New' About the 'New Woman'?." *Gender, Place & Culture* 23, no. 8 (2016): 1179–1192.

Cooks, Leda, and Isgro, Kirsten. "The "Cyber Summit" and Women: Incorporating Gender into Information and Communication Technology UN Policies." *Frontiers: A Journal of Women Studies* 26, no. 1 (2005): 71–89.

Crenshaw, Kimberlé. "Mapping the Margins: Intersectionality, Identity Politics, and Violence Against Women of Color." *Stanford Law Review* 43 (1991): 1241–1299.

Doshi, Marissa J. "Barbies, Goddesses, and Entrepreneurs: Discourses of Gendered Digital Embodiment in Women's Health Apps." *Women's Studies in Communication* 41, no. 2 (2018): 183–203.

Doshi, Marissa J. "Hybridizing National Identity: Reflections on the Media Consumption of Middle-Class Catholic Women in Urban India." In *Media and Power in International Contexts: Perspectives on Agency and Identity*, edited by Apryl Williams, Ruth Tsuria, and Laura Robinson, 101–131. Bingley, UK: Emerald Publishing, 2018.

Fadnis, Deepa. "Feminist Activists Protest Tax on Sanitary Pads: Attempts to Normalize Conversations about Menstruation in India Using Hashtag Activism." *Feminist Media Studies* 17, no. 6 (2017): 1111–1114.

Felski, Rita, and Felski, Joseph. *Beyond Feminist Aesthetics: Feminist Literature and Social Change*. Cambridge, MA: Harvard University Press, 1989.

Ferguson, Sian. "4 Ways to Make Your Period-Positivity More Inclusive." *Everyday Feminism*. January 24, 2016. Last modified https://everydayfeminism.com/2016/01/inclusive-period-positivity/.

Finlay, Linda. "Negotiating the Swamp: The Opportunity and Challenge of Reflexivity in Research Practice." *Qualitative Research* 2, no. 2 (2002): 209–230.

Fraser, Nancy. "Rethinking the Public Sphere: A Contribution to the Critique of Actually Existing Democracy." *Social Text* 25/26 (1990): 56–80.

Fotopoulou, Aristea. "Digital and Networked by Default? Women's Organisations and the Social Imaginary of Networked Feminism." *New Media & Society* 18, no. 6 (2016): 989–1005.

Gajjala, Radhika. "'Third World' Perspectives on Cyberfeminism." *Development in Practice* 9, no. 5 (1999): 616–619.

Gajjala, Radhika. "Snapshots from Sari Trails: Cyborgs Old and New." *Social Identities* 17, no. 3 (2011): 393–408.

Gajjala, Radhika. "South Asian Digital Diasporas and Cyberfeminist Webs: Negotiating Globalization, Nation, Gender and Information Technology Design." *Contemporary South Asia* 12, no. 1 (2003): 41–56.

Gajjala, Radhika. "When an Indian Whisper Network Went Digital." *Communication Culture & Critique* 11, no. 3 (2018): 489–493.

Gajjala, Radhika, and Oh, Yeon Ju. *Cyberfeminism 2.0*. New York: Peter Lang Publishing, 2012.

Garg, Suneela, and Anand, Tanu. "Menstruation Related Myths in India: Strategies for Combating It." *Journal of Family Medicine and Primary Care* 4, no. 2 (2015): 184–186.

Gopalakrishnan, Aswathy Gopala. "Red Alert: Kerala Activists Send Sanitary Pads to Factory Where Menstruating Women Were Insulted." OneIndia.com. January 1, 2015. www.oneindia.com/india/red-alert-kerala-activists-send-sanitary-pads-factory-menstruating-women-insulted-1609044.html.

Guterman, Mark, Mehta, Payal, and Gibbs, Margaret. "Menstrual Taboos Among Major Religions." *The Internet Journal of World Health and Societal Politics* 5, no. 2 (2008): 1–7.

Habermas, Jürgen. *The Structural Transformation of the Public Sphere: An Inquiry into a Category of Bourgeois Society*. Cambridge: MIT Press, 1991.

Haraway, Donna. "A Cyborg Manifesto: Science, Technology, and Socialist-Feminism in the Late 20th Century." In *Simians, Cyborgs and Women: The Reinvention of Nature*, edited by Donna Haraway, 117–158. New York and London: Routledge, 1991.

Hegde, Radha S. "Food Blogs and the Digital Reimagination of South Asian Diasporic Publics." *South Asian Diaspora* 6, no. 1 (2014): 89–103.

Ilango, Kaviya (@wallflowergirlsays). "I am a millennial." Instagram post, June 6, 2017. www.instagram.com/p/BU_5mWBh5Zg/.

Ilango, Kaviya (@wallflowergirlsays). "#100daysofdirtylaundry Day 18 - An ode to my humble toilet." Instagram post, June 27, 2017. www.instagram.com/p/BV3CQvLhlAr/.

Ilango, Kaviya (@wallflowergirlsays). "#100daysofdirtylaundry Day 40 - Whispers & murmurs." Instagram post, July 29, 2017. www.instagram.com/p/BXJJyeehOgT/.

Jackson, Sarah J., and Banaszczyk, Sonia. "Digital Standpoints: Debating Gendered Violence and Racial Exclusions in the Feminist Counterpublic." *Journal of Communication Inquiry* 40, no. 4 (2016): 391–407.

Kaundal, Manju, and Thakur, Bhopesh. "A Dialogue on Menstrual Taboo." *Indian Journal of Community Health* 26, no. 2 (2014): 192–195.

Locatelli, Elisabetta. "Images of Breastfeeding on Instagram: Self-Representation, Publicness, and Privacy Management." *Social Media+ Society* 3, no. 2 (2017): 2056305117707190.

MacRae, Elizabeth R., Clasen, Thomas, Dasmohapatra, Munmun, and Caruso, Bethany A. "'It's Like a Burden on the Head': Redefining Adequate Menstrual Hygiene Management Throughout Women's Varied Life Stages in Odisha, India." *PloS One* 14, no. 8 (2019): e0220114.

Masoodi, Ashwaq. "The Changing Fabric of Dalit Life." Livemint.com. Last modified April 21, 2017. www. livemint.com/Leisure/avsrwntNuBHG3THdAb5aMP/The-changing-fabric-of-Dalit-life.html.

Mohanty, Chandra Talpade. *Feminism without Borders: Decolonizing Theory, Practicing Solidarity*. Durham, NC: Duke University Press, 2003.

Mohanty, Chandra Talpade. "Transnational Feminist Crossings: On Neoliberalism and Radical Critique." *Signs: Journal of Women in Culture and Society* 38, no. 4 (2013): 967–991.

Mojab, Shahrzad. "The Politics of 'Cyberfeminism' in the Middle East: The Case of Kurdish Women." *Race, Gender & Class* (2001): 42–61.

Morozov, Evgeny. "The Brave New World of Slacktivism." Foreign Policy. Last modified May 19, 2009. https://foreignpolicy.com/2009/05/19/the-brave-new-world-of-slacktivism/.

Muralidharan, Arundati. "Constrained Choices? Menstrual Health and Hygiene Needs among Adolescents in Mumbai Slums." *Indian Journal of Gender Studies* 26, nos 1–2 (2019): 12–39.

Narasimhan, Haripriya. "Adjusting Distances: Menstrual Pollution among Tamil Brahmins." *Contributions to Indian sociology* 45, no. 2 (2011): 243–268.

Prasad, Akshita "Why the Padman Challenge Does Not Really Combat Menstruation Stigma." Feminism in India. Last modified February 8, 2018. https://feminisminindia.com/2018/02/08/pad-man-acquisitive-marketing/.

Prasanna, Chitra Karunakaran. "Claiming the Public Sphere: Menstrual Taboos and The Rising Dissent in India." *Agenda* 30, no. 3 (2016): 91–95.

Pruchniewska, Urszula Maria. "Working Across Difference in the Digital Era: Riding the Waves to Feminist Solidarity." *Feminist Media Studies* 16, no. 4 (2016): 737–741.

Pokharel Sugam and Nikhil Kumar. "First Women to Enter India Temple in Centuries Now in Hiding as Protests Rage." CNN. Last modified January 3, 2019. www.cnn.com/2019/01/03/asia/india-kerala-temple-intl/index.html.

Rosewarne, Lauren. *Periods in Pop Culture: Menstruation in Film and Television*. Lanham, MD: Lexington Books, 2012.

Saul, Heather. "Menstruation-Themed Photo Series Artist 'Censored by Instagram' Says Images Are to Demystify Taboos Around Periods." Independent.co.uk. Last modified March 30, 2015. www.independent.co.uk/arts-entertainment/art/menstruation-themed-photo-series-artist-censored-by-instagram-says-images-are-to-demystify-taboos-10144331.html.

Sommer, Marni, Hirsch, Jennifer S., Nathanson, Constance, and Parker, Richard G. "Comfortably, Safely, and Without Shame: Defining Menstrual Hygiene Management as A Public Health Issue." *American Journal of Public Health* 105, no. 7 (2015): 1302–1311.

Sowards, Stacey K., and Renegar, Valerie R. "The Rhetorical Functions of Consciousness-Raising in Third Wave Feminism." *Communication Studies* 55, no. 4 (2004): 535–552.

Sowjanya, Tamalapakula. "Critique on Contemporary Debates on Menstrual Taboo in India: Through Caste Lens." *Pramana Research Journal* 9, no. 8 (2019): 253–258.

Tata, Mounica (@doodleodrama). "Ok! Firstly, women, you DON'T need to go all undercover with your sanitary pad." Instagram post, February 10, 2017. www.instagram.com/p/BQVXsAAFoE3/.

Tata, Mounica (@doodleodrama). "How not to sell sanitary pads - a tiny guide." Instagram post, July 23, 2017. www.instagram.com/p/BW4fYU0BjZ3/.

Tata, Mounica (@doodleodrama). "I was working on the #realhorror series for a period tracking app when this idea struck." Instagram, October 27, 2017. www.instagram.com/p/BawXHsvHeD9/.

Tata, Mounica (@doodleodrama). "When people call me a rebel, I laugh it off because they haven't met my ovaries yet." Instagram post, March 5, 2017. www.instagram.com/p/BRPtKe4BVCH/.

Tata, Mounica (@doodleodrama). "Period cramps be like." Instagram post, November 21, 2017. www.instagram.com/p/BbwsD1HH0_c/.

Tata, Mounica (@doodleodrama). "#menstrualcups and me!" Instagram post, April 16, 2018. www.instagram.com/p/BhokynQno9b/.

Tata, Mounica (@doodleodrama). "There's a lot of hullabaloo on the internet around the #padmanchallenge." Instagram post, February 8, 2018. www.instagram.com/p/Be8BE8knrzj/.

Travers, Ann. "Parallel Subaltern Feminist Counterpublics in Cyberspace." *Sociological Perspectives* 46, no. 2 (2003): 223–237.

van Eijk, Anna Maria, Sivakami, M., Thakkar, Mamita Bora, Bauman, Ashley, Laserson, Kayla F., Coates, Susanne, and Phillips-Howard, Penelope A. "Menstrual Hygiene Management among Adolescent Girls in India: A Systematic Review and Meta-analysis." *BMJ Open* 6, no. 3 (2016): e010290.

Wajcman, Judy. *Technofeminism*. Cambridge: Polity, 2004.

Warner, Michael. "Publics and Counterpublics." *Public Culture* 14, no. 1 (2002): 49–90.

Yagnik, Arpan Shailesh. "Reframing Menstruation in India: Metamorphosis of the Menstrual Taboo with the Changing Media Coverage." *Health Care for Women International* 35, no. 6 (2014): 617–633.

Yagnik, Arpan Shailesh. "Mis-Fitting Menstrual Hygiene Products: An Examination of Advertisements to Identify Gaps in the Diffusion of Innovation." *IAFOR Journal of Psychology & the Behavioral Sciences* 2, no. 3 (2016): 33–45.

Yin, Robert K. *Case Study Research and Applications: Design and Methods*. Thousand Oaks, CA: Sage, 2017.

12

MONSTROUS ERASURE

Quare femme (in)visibility in *Get Out*

Bernadette Marie Calafell

Recent studies have shown that a "majority of whites say discrimination against them exists in America today"[1] and is as big of a problem as discrimination against racial minorities.[2] Indeed, researchers argue there is a growing sense among whites that white men have been denied "rightful" opportunities because of a perceived increased focus on diversity.[3] Furthermore, we see white women using their votes to support white patriarchy and racism, as evidenced by the large number of white women voting for Donald Trump, a candidate who openly espoused racist and sexist views, during the 2016 presidential election.[4] The previously mentioned studies can be understood in relationship to key events, such as the Abigail Fischer Supreme Court case touting discourses of reverse racism and attacks against affirmative action and the continual statesanctioned murders of African Americans by the police.[5] These studies must also be placed in conversation with discourses of post-racialism after the 2008 election of President Barack Obama and his subsequent reelection in 2012.

As Ta-Nehisi Coates suggests, "Obama, his family, and his administration were a walking advertisement for the ease with which black people could be fully integrated into the unthreatening mainstream of American culture, politics, and myth. And that was always the problem."[6] Obama's presidency "assaulted the most deeply rooted notions of white supremacy and instilled fear in its adherents and beneficiaries."[7] It became clear, as Coates notes, that "race is not simply a portion of the Obama story. It is the lens through which many Americans view all his politics."[8] However, Eric King Watts argues that this racialized lens also relies on a post-racial dream indebted to white masculine supremacy:

> Fantasies of the postracial depend on and reanimate the affective economies concurrent to the abjection of blackness. This is so because the postracial is not a space-time of racial absence or of racial transcendence; it does not signify the dissolution of the brutality of racism; it does not mark the beginning of a universal humanity. And yet it is *not* an absurd fiction either: rather it registers a reoccurring traumatic condition triggered by ruptures or fractures … of a society, felt as threats to the imaginary and symbolic status of masculine whiteness.[9]

This imagined threat to white masculinity allows white men to simultaneously rhetorically occupy a space of victimage and privilege.[10]

Certainly, the façade of the post-racial was publicly ripped away for many liberal whites by the election of Donald Trump in 2016, as well as the increased visibility and mainstream acceptance of white supremacist groups under the banner of the alt right or white nationalism. The rhetoric of "forgotten people" or victimized whiteness rematerialized even more strongly in such groups. To analyze these dynamics, Alison Reed draws on Robyn Wiegman's insights about "the white liberal tendency toward a kind of victimized whiteness born out of a class-based solidarity or historical patterns of immigration and racialization."[11] As Reed further notes:

> This leads to fallacious claims to prewhite injury, in part motivated by affirmative action backlash, and set into motion by the guise of an originary discursive blackness that simultaneously particularizes and dis-identifies with the political power of white skin. This "discursive blackness" is guilty of "participating in—indeed actively forging a counterwhiteness whose primary characteristic is its disaffiliation from white supremacist practices." This strategic alignment of discursive blackness and deracinated whiteness reproduces white supremacy under the banner of progressivism.[12]

Such white liberal deracialization of whiteness ironically recenters whiteness and white supremacy through discourses of social justice. The performance of this faux white liberal progressivism, that appropriates the traumas and imagined power of people of color as its own, is a compelling context in which to understand and situate Jordan Peele's 2017 film *Get Out,* which revisits the *Guess Who's Coming to Dinner* (1967) narrative through the lens of horror.

Peele's film was released on February 24, 2017. Although it was reportedly made with an estimated budget of $5,000,000, it grossed over $176,000,000 in the United States as of February 8, 2018, and over $254,000,000 worldwide as of January 2018.[13] Written by Peele and his first film as a solo director, *Get Out* garnered a great deal of accolades, including three Academy Award nominations for Best Motion Picture of the Year, Best Performance by an Actor in a Leading Role, and Best Achievement in Directing, and a win for Best Original Screenplay.[14]

The genre of the film, horror, is significant. Scott Poole argues,

> American monsters are born out of American history. They emerge out of the central anxieties and obsessions that have been part of the United States from colonial times to the present and from the structures and processes where these obsessions found historical expression.[15]

Monstrous identities provide spaces to discuss public anxieties and "register our national traumas."[16] Additionally, Marina Levina and Diem-My Bui suggest that monsters "can be read as a response to a rapidly changing social, cultural, political, economic, and moral landscape."[17] Likewise, Jeffrey Jerome Cohen proposes a monster as "an embodiment of a certain cultural moment—of a time, a feeling, and a place."[18] Functioning in this way, monsters "ask us to reevaluate our cultural assumptions about race, gender, sexuality, our perception of difference, our tolerance toward its expression. They ask us why we have created them."[19]

Given the ability of monsters to act as cultural barometers, *Get Out* provides us with an important text that seeks to disrupt particular discourses relevant to its historical-cultural context: those relating to post-racialism. I unpack the representations of Black women in the film, with a specific focus on Georgina, the Armitage family's housekeeper, to consider her positioning in relationship to both whiteness and Black masculinity. Following dominant cultural logics that frame blackness through masculinity and womanhood through whiteness, I explore how the film dismisses—or even erases—the possibilities of femme quareness. Instead, the film favors a critique that focuses on the violence enacted against Black masculinity and heterosexuality that echoes toward a narrow vision of Black nationalism. The film unfolds during a weekend visit that Chris, a young African American photographer, makes to meet his white girlfriend Rose's family. During this stay in a wealthy suburb, Chris discovers that Rose's family uses a process they call the "Coagula" procedure to place parts of their brains into the bodies of African Americans, rendering them as zombies or vessels for white consciousness. Rose's mother, Missy, uses hypnosis to trap the consciousness of her victims in what she terms "the sunken place." Chris uncovers the true motives of his girlfriend's family when they violently attempt to perform the procedure on him after he has been auctioned off to the highest white bidder (in a scene that harkens back to slave auctions). At the end of the film, Chris realizes that the white girlfriend who pretended to love him was in on the plot (and has many other victims). He only escapes with the help of his friend Rod, a TSA agent who long suspected that Chris was in danger from this white family.

While many will rightly focus on the character of Chris and what his representation means in this contemporary moment, I am particularly interested in the way Black women are depicted in the film, especially as they are positioned in relationship to whiteness and Black masculinity. Kinitra Brooks marks the lack of criticism focused on representations of Black women within horror.[20] She insists that when present, "Black women horror characters" appear "plagued by their construction as a mistreated tool used to further the more careful and considerate constructions of other characters."[21] Addressing identity and politics, Jared Sexton asks us to consider shifting from "a narrow emphasis on the attack on Black masculine empowerment—what used to be called 'manhood rights'—to a far more expansive formulation of assault on Black reproductive freedom." He asks, "what might it mean to stand on its head the patriarchal assumption that Black women's suffering is a disgrace to Black men and rethink the gendered violence against Black men as a component of gendered violence against Black women?"[22] Similarly, Patrice Douglas asks how "Black feminist politics account for the dead and dying inclusive of all genders?"[23] Informed by these scholars, much like Kara Keeling's analysis of the Black lesbian femme character, Ursula, in *Set It Off* (1996), I am interested in what we can learn from a relatively minor character in the film's plot.[24] We are introduced to Georgina as a woman who is working for Rose's family, yet we come to find out that Georgina is actually housing the consciousness of Rose's grandmother.

Like Douglas, I am not interested in reifying gender stratifications; rather, I want to use a Black feminist lens to be inclusive of all genders. Georgina exists in what Sharon Holland terms a "queer place," which is not necessarily based on acts in which individuals engage, but instead "the coercive norms that place their desires into a position of conflict with the present order."[25] Furthermore, I am reminded by Jennifer Nash that "to talk about black sexualities is always to be engaged in a political project"[26] as "sexual world-making and political world-making go hand in hand" since "Black political and Black sexual freedom are one and the same."[27] Thus, I argue that centering and unpacking the quare femme in *Get Out* allows us to read the resistive possibilities of this figure and how it helps us to consider and imagine radical potentialities, utopias, or futures.

Discourses of post-racialism and even pre-white injury pervade the film, including Rose's father, Dean, sharing with Chris the prototypical white liberal comment: that he would have voted for Obama for a third time, if possible. These discourses stand in relation to performances of neoliberal or inferential racism in the film, such as discussions about the physical superiority and social advantages of African Americans. For example, as Dean gives Chris a tour of the house, remarking on the privilege of experiencing another's culture through travel, he tells the story of his father, Roman, who was beaten by Jesse Owens in the qualifying round for the Berlin Olympics in 1936. Dean notes that Owens would go on to win in front of Hitler, suggesting that his father "almost got over it." The story proves to be important when we are introduced to the Coagula procedure, which in a video, Roman refers to as a "man-made miracle" perfected by his own flesh and blood. The symbolic wounding of white masculinity, through the narrative of Roman's loss, comes back around as patriarchy bites back in a twisted Eugenics perspective influenced by discourses of multiculturalism.

The film's fictional plot must be contextualized within the real racialized history of science, such as, "The Tuskegee experiment, in which innocent Black men were injected with syphilis for scientific study, or the use of the immortal cells of Henrietta Lacks."[28] Science fiction has also taken up narratives of race and science, such as in the 1971 film *The Omega Man,* which presents us with Charlton Heston's character, Neville, a white man whose "blood is the serum that can cure the adolescent Black male and all the others who have not yet 'gone over'" into monstrosity.[29] Neville's white blood serves as a "racial metaphor, with white blood presented as a means to cure and repopulate a diseased and dying world."[30] *Get Out* offers an interesting point of discussion in these narratives of science and race.

Quaring and transing *Get Out*

It would be a natural move to simply consider the film in terms of Afropessimism to theorize the Black characters as "positioned outside the scope of humanness," the position of the unthought, or man.[31] Likewise, we might reflect upon Hortense Spillers' discussion of the captive body—"a theft of the body—a willful and violent severing…from its motive will, and its active desire."[32] As M. Shadee Malaklou and Tiffany Willoughby-Herard remind us, "the freedom of all oppressed people depends entirely on the freedom of Black women, which we might extrapolate to include Black queer and trans* persons, who are exceptionally made 'it' in humanist discourse."[33] Resisting this cultural current, E. Patrick Johnson offers quare theory as a means to attest to the experiences of queers of color. He suggests quare "speaks across" and "articulates identities."[34] Furthermore,

> Quareness is born out of the specific historical context of African Americans. This specificity is important as it locates the material effects of U.S. (emphasizing Southern) history on queer African Americans, honoring the importance of identity in a queer perspective that often holds anti-identity politics at its center, but making sure that identity does not become so generalized that it is rendered meaningless.[35]

Further pushing Black queer or quare studies, Kai Green encourages us to "mine trans for its use as a method or optic" as it "articulates a unique relation between two or more identity categories where one marks the limits and excess of the other, simultaneously deconstructing and reconstructing or reimagining new possible ways of being and doing."[36] Green elaborates on this approach in the following passage:

> A Trans* method asks that we not be so invested in what follows Black is or Black ain't but rather that we be attuned to the ways in which Black is present or not, when, where, how, why, and most important, in relation to what. A Trans* method requires that we be more attuned to difference rather than sameness…A Trans* method shows us how people become representable as things, categories, and names because it shows us the excess as a perpetual challenge to containment.[37]

Moreover, a trans* method calls us to listen to the "fullness embedded in the silences and gaps" and "further names the work of charting the present absence in multiples sites of intersection for demanding a moment of critical presence."[38] When applied in these ways, a trans* method "allows us to see certain things that might not normally be seen. It also enables us to understand how the seeing is being shaped."[39] This kind of trans* method informs my own unpacking of the film by marking limits of the representations of blackness, specifically in terms of gender and sexuality. It points to excess or the excessive. It marks a starting point to begin to imagine other possibilities for the quare femme that are linked to the feminist underpinnings of Afrofuturism and utopic possibilities.

"You ruined my house!" Heteronormative subservience

When Chris and Rose arrive at the family home, we are introduced to Georgina, who we are told worked for Rose's grandparents. Georgina, though a young woman, dresses in matronly, modest clothes and wears her hair (which we come to find out is a wig) in a style traditionally associated with older women. As the film progresses and we learn of the Coagula procedure, it is clear that Georgina, who was a victim of Rose's romantic entrapment, has had her body taken over by the grandmother's consciousness. Her quareness, her experiences as a Black lesbian, are erased as she becomes symbolically and heteronormatively paired with Walter's Grandpa. In other words, her sexuality is a form of monstrosity that must be tamed in order for her to perform the politics of respectability as Grandma.

Georgina, who occupies a space of subservience, is referred to by Dean as a servant—echoing histories of Black female subservience in film, including the 1940 Best Supporting Actress Oscar win by Hattie McDaniel, who received her award for her portrayal of Mammy in the film *Gone With the Wind*.[40] Georgina's subservience is further manifested when she over-pours tea, which harkens back to the Black queer vernacular of "spilling the tea," telling the tales, or sharing gossip. This happens when Georgina is instructed by Missy, Rose's mother (who we do not yet know is her daughter-in-law) to pour the tea.

Though Georgina is the unwilling captive of the grandmother's consciousness, symbolically she is the matriarch of the family. While she must play a subservient role for Chris to maintain the ruse, her blackness and femininity consistently deny her the ability to exercise power in the white family, especially given the violence she has experienced. She is positioned as "the faithful, obedient domestic servant" who "may be well loved" but knows "her place" and has "accepted her subordination."[41] She literally echoes the expectation that Holland describes, "that black women will cease a connection with their own families in order to respond to the needs of white persons."[42]

As Chris settles in with the family, he starts to notice strange things. For example, he suspects that Georgina unplugged his charging cell phone. He suggests that she is unhappy that he and Rose are together because, "It's a thing." Rose responds, "So you are so sexy that people are just unpluggin' your phone?" This conversation is significant as it places Georgina immediately within the realm of heteronormativity as Chris drudges up a tired old script that positions "good" Black men as rare commodities that must be protected by and for Black women at all costs. Furthermore, Rose's response fails to consider her white privilege and a history of tension between white women and women of color because of white women's inability to confront their racial privilege. Georgina (and the other Black women she represents by default) is accused of unplugging Chris' phone because of jealousy and positioned simply as hater, devoid of historical context and relations of power between white women, Black women, and Black men.

However, the audience gets a glimmer of the "real" Georgina, while she apologizes to Chris for unplugging his cell phone. As he shares that he gets nervous "if there's too many white people," her smile fades and her breathing changes. As a tear runs down her check, she laughs saying,

"That's not my experience. Not at all. The Armitages are so good to us. They treat us like family." Georgina marks the limit of her relationship with the Armitages—"*like* family"—which can further reflect the lack of humanity she is granted in her role. As she leaves, Chris concludes, "This bitch is crazy." What does it mean that the quare Black femme is never able to break free of the "sunken place" and is just a crazy bitch from the perspective of the main character?

Moreover, within the constraints of the genre, Georgina is unable to embody Carol Clover's[43] final girl—the masculine performing virginal woman who survives until the end of the horror film and defeats the monster because of her virtues. While Brooks suggest that Black women are rendered masculine in the public imagination, which is a key aspect of the final girl, they rarely occupy such positions. In the case of Georgina, the embodied performance of white feminine respectability as grandma renders her further incapable of being Clover's final girl. Not only is Georgina unable to fulfill this role, she is also unable to break free of the label of monstrous in some ways. Though we see her cry as the real Georgina seeks to get out, she never breaks through, and instead dies inflicting violence upon Chris in the name of white supremacy; in the name of her (Grandma's) house, symbolic of the white family. Georgina, now in a hybrid body like that of Frankenstein's creation, comes to represent the danger and lack of trust that is often associated with hybridity or ambiguity. Furthermore, her performance of respectability and civility will not save her, particularly because women of color are consistently viewed through the lens of monstrosity.[44] For example, the framing of women of color as monstrous permeates popular culture through archetypes such as the welfare queen[45] or representations of excessive sexuality.[46]

Michael Lacy draws on Elizabeth Young's discussion of Black Frankenstein, in which the "Black Frankenstein metaphor has come from liberal or progressive critics (Black and white) who viewed the government-sanctioned political economies and practices (e.g., slavery) as the monster-maker that created, enslaved, and oppressed Blacks."[47] Lacy suggests that "Black Frankenstein monsters and metaphors critique the horrors of hegemonic oppression experienced by black people, and white people's fear of revolting blacks."[48] Lacy argues these "Black Frankenstein narrative form and figures reemerge in contemporary American films to both critique and redeem racial neoliberalism."[49] The narrative of black Frankenstein Lacy describes is at work in *Get Out* as well; however, it is imperative to consider how it is feminized and queered, especially in relationship to the practices of identification many queer viewers have with monsters.[50] Some argue that Peele is playing with the zombie narrative that is often de-raced from African connections in contemporary representations, but it is important to note that Frankenstein has long been seen as a figure connected to racial histories of blackness.[51]

The Armitages, beneficiaries of a white supremacist capitalist patriarchy, further exploit the system to literally use science in an attempt to create monsters. Historically, Frankenstein's monster has come to represent Otherness, specifically blackness, through scenes of the vigilante mob pursuing him with lit torches.[52] Within a historical context that includes *Birth of a Nation* (1915), lynchings, the eugenics movement, and cultural anxieties about interracial sex, this meaning becomes even clearer. Frankenstein's monster has also come to represent racial revolt, as Young observes in connection to slave revolts.[53] However, Young notes that in *The Bride of Frankenstein* (1935), the monster's bride is coded as white when her rejection of him signals the threat of Black masculinity to white femininity. What does this all mean for Georgina, who has been made monstrous by the Coagula procedure and is never able to get out of the "sunken place"? What are the meanings associated with quare femininity in this representation? Young, writing of a female version of Frankenstein's monster, argues that the horror of a body that is already infantilized becomes even further "victimized and laughable" when a woman occupies this role because her strangeness and Otherness intensifies the lack of intelligibility of the monster.[54]

Yet her unique monstrous threat is that she has the "capacity to reproduce": through the frame of Black Frankenstein, all Black women become "monster-makers."[55]

Throughout the film, Georgina is placed within the family home in contrast to Walter, an African American man who we are told works for the family and who houses Roman's consciousness. We first see Walter when Rose and Chris arrive to visit the family. Later, Chris sees him running in the night. Like Georgina, we get glimpses of him throughout the film. As the film progresses, viewers discover that Walter is also one of Rose's victims. Walter is consistently positioned outside of the home, running or chopping wood. His physicality is highlighted, which recalls the Armitage family story about the Olympic trials—as well as the racist figure of the buck, a common depiction with a long history in U.S. public culture wherein a Black man is oversexualized and positioned as a threat to white women, such as in the racist film *Birth of a Nation*. The symbolism of the buck figures throughout the film, as Rose hits a deer on the drive to her parent's home and Chris watches the animal die, bringing a sense of foreboding as we see him look deeply into the buck's eyes, as if he is having a moment of recognition with it. Later, Rose's father Dean says he wishes he could kill them all as they are nuisances. Thus, it is powerful and symbolic when Chris later uses the horns of a mounted buck's head against Dean to escape.

As the unwilling host body for the family patriarch, Roman, Walter occupies a more social or public position than Georgina, and he is seen happily greeting guests at the Armitage's party. Though located on the outside of the family, Walter seems to still maintain some of Roman's patriarchal power. This is further evidenced at the end of the film when he asks Rose to give him a gun, and she does so without hesitation.

In contrast, Georgina is only seen outside of the home when Chris tries to escape. In many cases throughout the film, she is positioned as framed in a window as she stares out or we see her reflection in a mirror, symbolizing her captivity. After hitting her with a car, Chris begins thinking of his mother, who he believes died because of his inaction (the memory used to trap him in his own "sunken place"). Not wanting to make the same mistake, Chris drags her into the car with him. Awaking from her momentary loss of consciousness, she begins lunging at him, screaming, "You ruined my house!" Chris has revealed and stopped the ruse, as well as ended the white family's lineage/legacy and racist scheme—their contribution to white supremacy. As a result, they crash into a tree, Georgina is killed, and Chris's escape is foiled.

The very different positioning of Walter and Georgina is significant. Walter has a sense of agency as he strikes back against Rose by shooting her as he is able to momentary leave the sunken place with the help of Chris. Georgina, on the other hand, is never quite able to leave the sunken place. Instead she demonstrates dismay over her house being destroyed, which potentially frames her within the tired trope of the house negro who is complicit in the maintenance of white supremacy. Furthermore, Patricia Hill Collins argues the image of the mammy was used to deny the violence of slavery, including rape and gendered violence. Collins's analysis of the hegemonic representations, "controlling images," that sustain white supremacy has echoes in Georgina. Rod, Chris's best friend (who is outside of the immediate situation and thus is concerned about Chris' safety), makes a joke throughout the film that white people are using African Americans as sex slaves. However, this seemingly hyperbolic conjecture is precisely the violence that is literally happening. Whites have *actually taken over Black bodies* to do with what they wish, in ways that echoes toward the necropolitics that govern the livelihood (or lack thereof) of Black lives, and what bell hooks[56] terms "eating the Other," simultaneously.

It is important to remember that the film was released in 2017 and should be understood within the context of #BlackLivesMatter; a hashtag and activist movement that was borne after George Zimmerman was acquitted of the murder of 17-year-old African American teenager Trayvon Martin. We must further place the character of Georgina in relation to the fact that

Black queer women, such as Alicia Garza, are intentionally at the forefront of Black Lives Matter, one of the most important movements for social justice of our time. Acknowledging histories of the contributions of LGBTQ people to social movements going unrecognized (and of them being pushed out of organizations and movements entirely), Black Lives Matter intentionally places "those at the margins closer to the center."[57] Brittany Cooper, Susana Morris, and Robin Boylorn write,

> To be a Black feminist in this moment is to have to continually negotiate the problem of rampant police brutality against Black communities, the regular killing of unarmed young Black men, while vigilantly proclaiming that 'Black women's lives matter too.' And when we say Black women's lives matter, we don't just mean cisgender Black women. In 2015 more than twenty trans women of color were murdered in the United States.[58]

Furthermore, research on voting patterns evidence that Black women continue to show up and be more progressive in politics, while white women tend to more often vote against their own interests.[59] How do we reconcile the violence against Black women and their radical leadership with representations in *Get Out* that position them as upholding white supremacy and patriarchy? How do these representations symbolically throw Black (quare) women under the bus in favor of defaulting to framing resistance to white supremacy as led by Black (heterosexual) masculinity? How does this narrative construct quare women as cultural outsiders who can't be trusted, reifying associations of queerness with whiteness and antithetical to blackness? These are important questions to explore. My unpacking of the film and the figure of Georgina is less a traditional film analysis and instead a rumination on the visibility and erasure of the quare Black femme within the contemporary moment. It points to the continual struggles for representation and the reductionism of blackness to heterosexuality and manhood, particularly in this moment when Black quare women continue to be at the head of social justice movements in the United States.

A radical elsewhere?

While audiences are familiar with cultural narratives and reverberations of heterosexual interracial relationships between Black men and white women, such as the relationship between Chris and Rose in *Get Out*, the revelation of the relationship between Georgina and Rose as Chris looks through pictures of Rose's many victims in her closet disrupts its heteronormative logics as well as butch/femme pairings. The irony of Chris finding Georgina in the closet (and then quickly putting her back in) is significant, as it can be seen as a symbolic gesture toward the historical erasure of Black quare women both in mainstream women's rights movements and Black nationalist movements, both of which often relied on a strategic essentialism that was premised on centering a dominant narrative of (white) womanhood or (male) blackness. Furthermore, from a quare perspective we might consider part of Georgina's fate as a form of conversion therapy, as she has been cast into a heterosexual pairing in her new role as grandma. Rose and Georgina's relationship is also placed in the lens of exception because it stands out as the only same-sex relationship in Rose's history of entrapment. Keeling argues that hegemonic common sense does not recognize a feminine Black woman, let alone a "feminine black lesbian."[60] Therefore, the femme-femme pairing of Georgina and Rose disrupts the dominant butch-femme framing of lesbianism in the popular imaginary.

So what are the possibilities of the femme? Elizabeth Galewski notes that historically femmes have been "implicitly conflated with weakness, passivity, and even complicity in the face of oppression," particularly in relationship to the butch-femme pairing.[61] Femmes may be viewed at times with suspicion similar to the ways bisexuals are viewed both in and out of queer communities. Reflecting on this history, Lisa Duggan and Kathleen McHugh offer the tongue-in-cheek assessment that, "you can never trust a fem(me)."[62] Kaila Adia Story goes further in stating,

> Black and Brown women's performances of femme of color identities are rarely seen as radical due to this socialized racist and heteronormative logic that is imbedded within American culture. It is also this logic … that often allows outsiders to interpret femme identities of color as though they are not powerful, radical, or even revolutionary.[63]

However, it is important to recognize that Black femme identity can embody a "resistive femininity."[64]

Keeling unpacks the character Ursula in *Set It Off*, who is the Black femme partner of Cleo, a Black butch woman. She argues,

> the only shot in the film where Ursula occupies the frame alone, her presence in the frame itself testifies to a more disturbing presence, one which cannot even be said to exist, but rather to "insist" or "subsist," a more radical Elsewhere outside homogenous space and time.[65]

In the fleeting moment we see the snapshot of Rose and Georgina, which signals Georgina's quareness, we are left to wonder about her story. It is a story we don't hear, a sequence of events that may be present in the glimmer in her eyes as the tears form during her chat with Chris. Georgina, as a quare Black femme, is in a "queer place" that "places" her "desires into a position of conflict with the present order."[66] The story we don't see, which occurs before the start of the film, is a radical possibility that is also taken away by her queerness made visible, only to be quickly dismissed in order to shift the narrative back to a focus on race, and in particular, Black *masculinity* and *heterosexuality*. This shift in some ways echoes nationalist ideologies that might subjugate femininity and queerness in favor of a larger focus on race, which comes to align itself commonsensically to masculinity.

> "Racism," Holland suggests, "consistently embeds us in a 'past' that we would rather not remember, where time stretches back toward the future, curtailing the revolutionary possibilities of queer transgression."[67] Similarly, "the black femme, while a product of" butch-femme relationality, "also might be a portal to a reality that does not operate according to the dictates of the visible and the epistemological, ethical, and political logics of visibility."[68]

Of femmes, Duggan and McHugh write, "Seemingly 'normal,' she responds to 'normal' expectations with a sucker punch—she occupies normality abnormally."[69] Similarly, Kathryn Hobson suggests that femme is a complicated identity, "because those who identify as femme have the supposed advantage of fitting into what appears to be normative, stereotypical femininity …. Queer femmes are viewed as inauthentic and conforming queers, which is then not viewed as queer at all."[70] Certainly, there is the potential to read Georgina as weak or normative, but it doesn't have to be that simple. Ulrika Dahl describes the "femme as a figuration [that] collects both hopes and fears, possibilities and dangers, the brazen and the injured."[71] Like Dahl,

Keeling suggests, "even beyond cinematic reality itself, the black femme might restore a critical belief in the world by revealing that alternatives persist within it."[72] Collins similarly declares that loving Black people in a society that hates blackness is a revolutionary act.[73] Similarly, loving (Black) quare femmes in a society that hates femininity is revolutionary.

While I have demonstrated how Georgina's story functions in the film to further a critique focused on centering the violence against Black masculinity, as part of a trans* analytic, I want to consider what the representation could mean in terms of imagining other possibilities. The possibilities the femme may signal lead us down a path similar to that taken by José Esteban Muñoz in his work on utopian performance.[74] Writing from an Afrofuturism perspective, Reynaldo Anderson notes that there has been "a tendency by scholars to criticize utopian frameworks as escapist"; however, these critiques ignore "the need and desire for peoples to share states of being in forms of interchange and expression that seek to recover an impulse or longing in various cultural forms."[75] Like utopias, writing about the promise and potential of Afrofuturism, particularly as a feminist tactic, Ytasha Womack argues, "imagination is the key to progress, and it's the imagination that is all too often smothered in the name of conformity and community standards."[76]

Given Womack's comments, where is the space of imagination for the quare femme? Muñoz writes, "Queerness is not yet here. Queerness is an ideality. Put another way, we are not yet queer ....We must strive, in the face of the here and now's totalizing rendering of reality, to think and feel a *then and there.*"[77] How can a trans* analytic push us toward not only a queer or quare time, but quare and femme potentiality? How can marking absence start to render presence rather than simply erasure? Perhaps this requires a move from pessimism to futurism, which is driven by a feminist impetus, or even to what Justin Louis Mann terms pessimistic futurism grounded in a Black female experience that "couches the prospects of tomorrow in the uncertainties conditioned by the past and present."[78] Of hope, Muñoz's shares,

> Practicing educated hope is the enactment of a critique function. It is not about announcing the way things ought to be, but, instead, imagining what things could be. It is thinking beyond the narrative of what stands for the world today by seeing it as not enough.[79]

Tiffany Willoughby-Herard writes that in the United States we are predisposed to "view Black Women as everything, but objects of human sympathy."[80] Inspired by Aisha Durham, I share her words as I end. To quare Black femmes, "I do see you. You have not been forgotten. You are loved.".[81]

Notes

1 Don Gonyea, "Majority of White Americans Say They Believe Whites Face Discrimination," NPR, October 24, 2017, www.npr.org/2017/10/24/559604836/ majority-of-white-americans-think-theyre-discriminated-against.
2 Ryan Struyk, "Blacks and Whites See Racism in the United States Very, Very Differently," CNN, August 18, 2017, www.cnn.com/2017/08/16/politics/blacks-white-racism-united-states-polls/index.html.
3 Jaime Paldin and Armando Diaz, "EY Studies Race, Gender, and Exclusion: The Top Takeaways," EY, October 6, 2017, www.ey.com/us/en/newsroom/news-releases/news-ey-studies-race-gender-and-exclusion-the-top-takeaways.
4 For further elaboration see Sarah Jaffe, "Why Did a Majority of White Women Vote for Trump?" *New Labor Forum* 27, no. 1 (2018): 18–26.
5 For elaboration please see Jia Tolentino, "All the Greedy Abigail Fishers and Me," Jezebel, June 28, 2016. https://jezebel.com/all-the-greedy-young-abigail-fishers-and-me-1782508801.

6 Ta-Nehisi Coates, *We Were Eight Years in Power: An American Tragedy* (Large Print ed.) (New York: Random House, 2017), xviii.

7 Ibid., xx.

8 Ibid., 206.

9 Eric King Watts, "Postracial Fantasies, Blackness, and Zombies," *Communication and Critical/Cultural Studies* 14, no. 4 (2017): 318.

10 Casey Ryan Kelly, "The Wounded Man: Foxcatcher and the Incoherence of White Masculine Victimhood," *Communication and Critical/Cultural Studies* 15, no. 2 (2018): 161–178.

11 Alison Reed, "The Whiter the Bread, the Quicker You're Dead: Spectacular Absence and Post-Racialized Blackness in (White) Queer Theory," in *No Tea, No Shade: New Writings in Black Queer Studies*, ed. E. Patrick Johnson (Durham, NC: Duke University Press, 2016), 52.

12 Ibid.

13 "Get Out." *Internet Movie Database,* Accessed February 9, 2018, www.imdb.com/title/tt5052448/?ref_=nv_sr_1.

14 "Get Out: Awards." *Internet Movie Database,* Accessed February 9, 2018, www.imdb.com/title/tt5052448/awards?ref_=tt_awd.

15 W. Scott Poole, *Monsters in America: Our Historical Obsession with the Hideous and Haunting* (Waco, TX: Baylor University Press, 2011), 4.

16 Ibid., 23.

17 Marina Levina and Diem-My T. Bui. eds, "Introduction: Toward a Comprehensive Monster Theory in the 21st Century," in *Monster Culture in the 21st Century: A Reader* (London: Bloomsbury Press, 2013), 2.

18 Jeffery Jerome Cohen, ed, "Monster Culture (Seven Theses)," in *Monster Theory: Reading Culture* (Minneapolis, MN: University of Minnesota Press, 1996), 4.

19 Ibid., 20.

20 Kinitra Brooks, *Searching for Sycorax: Black Women's Hauntings of Contemporary Horror* (New Brunswick, NJ: Rutgers University Press, 2018), 23.

21 Ibid., 8.

22 Jared Sexton, "Unbearable Blackness," *Cultural Critique*, 90 (2015): 169.

23 Patrice D. Douglas, "Black Feminist Theory for the Dead and Dying," *Theory & Event* 21, no. 1 (2018): 109.

24 Kara Keeling, *The Witch's Flight: The Cinematic, the Black Femme, and the Image of Common Sense* (Durham, NC: Duke University Press, 2007).

25 Sharon Patricia Holland, *The Erotic Life of Racism* (Durham, NC: Duke University Press, 2012), 9.

26 Jennifer C. Nash, "Black Sexualities," *Feminist Theory* 19, no. 1 (2017): 3.

27 Ibid., 4.

28 Ytasha L. Womack, *Afrofuturism: The World of Black Sci-Fi and Fantasy Culture* (Chicago, IL: Lawrence Hill Books, 2013), 36.

29 Adilifu Nama, *Black Space: Imagining Race in Science Fiction Film* (Austin, TX: University of Texas Press, 2008), 49.

30 Ibid., 49.

31 Douglas, "Black Feminist Theory," 115.

32 Hortense Spillers, "Mama's Baby, Papa's Maybe: An American Grammar Book," *Diacritics* 17, no. 2 (1987): 67.

33 M. Shadee Malaklou and Tiffany Willoughby-Herard, "Notes from the Kitchen, the Crossroads, and Everywhere Else, Too: Ruptures of Though, Word, and Deed in the 'Arbiters of Blackness Itself,'" *Theory & Event* 21, no. 1 (2018): 10.

34 E. Patrick Johnson, "'Quare' Studies, or (Almost) Everything I Know About Queer Studies I Learned From My Grandmother," *Text and Performance Quarterly* 21, no. 1 (2001): 3.

35 Shinsuke Eguchi, Bernadette Marie Calafell, and Nicole Files-Thompson, "Intersectionality and Quare Theory: Fantasizing African American Male Same-Sex Relationships in Noah's Arc: Jumping the Broom," *Communication, Culture, and Critique* 7 (2014): 373.

36 Kai Green, "Troubling the Waters: Mobilizing a Trans* Analytic," in *No Tea, No Shade: New Writings in Black Queer Studies*, ed. E. Patrick Johnson (Durham, NC: Duke University Press, 2016), 66.

37 Ibid., 79.

38 Ibid., 80.

39 Ibid..

40 "Hattie McDaniel." *Biography,* Accessed September 20, 2019, www.biography.com/actor/hattie-mcdaniel.

41 Patricia Hill Collins, *Black Feminist Thought: Knowledge, Consciousness, and the Politics of Empowerment* (New York: Routledge, 2008), 80.
42 Sharon Patricia Holland, *The Erotic Life of Racism* (Durham, NC: Duke University Press, 2012), 2.
43 Carol Clover, *Men, Women, and Chain Saws: Gender in the Modern Horror Film* (Updated ed.) (Princeton, NJ: Princeton University Press, 2015).
44 Bernadette Marie Calafell, *Monstrosity, Performance, and Race in Contemporary Culture* (New York: Peter Lang, 2015).
45 Patricia Hill Collins, *Black Sexual Politics: African Americans, Gender, and the New Racism* (New York: Routledge, 2004).
46 Isabel Molina-Guzmán, *Dangerous Curves: Latina Bodies in the Media* (New York: New York University Press, 2010).
47 Michael G. Lacy, "Black Frankenstein and Racial Neoliberalism in Contemporary American Cinema: Reanimating Racial Monsters in *Changing Lanes,*" in *The Routledge Companion to Popular Culture,* ed. Toby Miller (New York: Routledge, 2014), 230.
48 Ibid., 231.
49 Ibid..
50 Harry M. Benshoff, *Monsters in the Closet: Homosexuality and the Horror Film* (Manchester: Manchester University Press, 1997); Calafell; Susan Stryker, "My Words to Victor Frankenstein above the Village of Chamounix: Performing Transgender Rage," *GLQ* 1 (1994): 237–254.
51 Poole, *Monsters in America.*
52 Ibid.
53 Elizabeth Young, *Black Frankenstein: The Making of an American Metaphor* (New York: New York University Press, 2008).
54 Ibid., 59.
55 Ibid.
56 bell hooks, "Eating the Other: Desire and Resistance," in *Black Looks: Race and Representation* (Boston, MA: South End Press, 1992), 21–39.
57 "Herstory." *Black Lives Matter,* Accessed February 9, 2018, https://blacklivesmatter.com/about/herstory/.
58 Brittney C. Cooper, Susana M. Morris, and Robin M. Boylorn, "Introduction: Race and Racism: All Black Lives Matter," in *The Crunk Feminist Collection,* eds Brittney C. Cooper, Susana M. Morris, and Robin M. Boylorn (New York: Feminist Press, 2017), 42.
59 Kerra Bolton, "How Black Women Saved Alabama – and Democracy," CNN, December 14, 2017, www.cnn.com/2017/12/13/opinions/black-women-voters-white-feminism-bolton-opinion/index.html.
60 Kara Keeling, "'Ghetto Heaven': *Set It Off* and the Valorization of Black Femme-Butch Sociality," *The Black Scholar* 33, no. 1 (2003): 41.
61 Elizabeth Galewski, "Figuring the Feminist Femme," *Women's Studies in Communication* 28, no. 2 (2005): 186.
62 Lisa Duggan and Kathleen McHugh, "A Fem(me)inist Manifesto," *Women and Performance* 8, no. 2 (1996): 154.
63 Kaila Adia Story, "Fear of a Black Femme: The Existential Conundrum of Embodying a Black Femme Identity While Being a Professor of Black, Queer, and Feminist Studies," *Journal of Lesbian Studies* 21, no. 4 (2017): 412.
64 Ibid.
65 Keeling, *The Witch's Flight,* 137. Keeling cites Gilles Deleuze's formulation of 'the powers of the false' from *Cinema 2,* Chapter 6, 17.
66 Holland, *Erotic Life,* 9.
67 Ibid., 44.
68 Keeling, *The Witch's Flight,* 143.
69 Duggan and McHugh, "Fem(me)inist Manifesto," 155.
70 Kathryn Hobson, "Sue Sylvester, Coach Beiste, Santana Lopez, and Unique Adams," in *Glee and New Directions for Social Change,* eds Brian C. Johnson and Daniel K. Faill (Rotterdam: Sense Publishers, 2015), 102.
71 Ulrika Dahl, "Femmebodiment: Notes on Queer Feminine Shapes of Vulnerability," *Feminist Theory* 18, no. 1 (2017): 36.
72 Keeling, *The Witch's Flight,* 145.
73 Collins, *Black Sexual Politics.*

74 José Esteban Muñoz, *Cruising Utopia: The Then and There of Queer Futurity* (New York: New York University Press, 2009).

75 Reynaldo Anderson, "Fabulous: Sylvester James, Black Queer Afrofuturism, and the Black Fantastic," *Dancecult: Journal of Electronic Dance Music Culture* 5, no. 2 (2013), Accessed February 7, 2018, https://dj.dancecult.net/index.php/dancecult/article/view/394/413.

76 Womack, *Afrofuturism,* 191.

77 Muñoz, *Cruising Utopia,* 1.

78 Justin Louis Mann, "Pessimistic Futurism: Survival and Reproduction in Octavia Butler's Dawn," *Feminist Theory* 19, no. 1 (2018): 62.

79 Lisa Duggan and José Esteban Muñoz, "Hope and Hopelessness: A Dialogue," *Women and Performance: A Journal of Feminist Theory* 19, no. 2 (2009): 277.

80 Tiffany Willloughby-Herard, "(Political) Anesthesia or (Political) Memory: The Combahee River Collective and the Death of Black Women in Custody," *Theory & Event* 21, no. 1 (2018): 264.

81 Aisha Durham, "Do We Still Need a Body Count to Count? Notes on the Serial Murders of Black Women," in *The Crunk Feminist Collection*, eds Brittney C. Cooper, Susana M. Morris, and Robin M. Boylorn (New York: Feminist Press, 2017), 24.

Bibliography

Anderson, Reynaldo. "Fabulous: Sylvester James, Black Queer Afrofuturism, and the Black Fantastic." *Dancecult: Journal of Electronic Dance Music Culture* 5, no. 2 (2013): n.p. Accessed February 7, 2018. https://dj.dancecult.net/index.php/dancecult/article/view/394/413.

Benshoff, Harry M. *Monsters in the Closet: Homosexuality and the Horror Film.* Manchester: Manchester University Press, 1997.

Bolton, Kerra. "How Black Women Saved Alabama—and Democracy." CNN. Last modified December 14, 2017. www.cnn.com/2017/12/13/opinions/black-women-voters-white-feminism-bolton-opinion/index.html.

Brooks, Kinitra. *Searching for Sycorax: Black Women's Hauntings of Contemporary Horror.* New Brunswick, NJ: Rutgers University Press, 2018.

Calafell, Bernadette Marie. *Monstrosity, Performance, and Race in Contemporary Culture.* New York: Peter Lang, 2017.

Clover, Carol. *Men, Women, and Chain Saws: Gender in the Modern Horror Film.* Princeton, NJ: Princeton University Press, 2015.

Coates, Ta-Nehisi. *We Were Eight Years in Power: An American Tragedy* (Large Print ed.). New York: Random House, 2017.

Cohen, Jeffrey Jerome. "Monster Culture (Seven Theses)." In *Monster Theory: Reading Culture*, edited by Jeffrey Jerome Cohen, 3–25. Minneapolis, MN: University of Minnesota Press, 1996.

Collins, Patricia Hill. *Black Feminist Thought: Knowledge, Consciousness, and the Politics of Empowerment.* New York: Routledge, 2008.

Collins, Patricia Hill. *Black Sexual Politics: African Americans, Gender, and the New Racism.* New York: Routledge, 2004.

Cooper, Brittney C., Morris, Susana M., and Boylorn, Robin M. "Introduction: Race and Racism: All Black Lives Matter." In *The Crunk Feminist Collection*, edited by Brittney C. Cooper, Susana M. Morris, and Robin M. Boylorn, 41–44. New York: Feminist Press, 2017.

Dahl, Ulrika. "Femmebodiment: Notes on Queer Feminine Shapes of Vulnerability." *Feminist Theory* 18, no. 1 (2017): 35–53.

Douglas, Patrice D. "Black Feminist Theory for the Dead and Dying." *Theory & Event* 21, no. 1 (2018): 106–123.

Duggan, Lisa, and McHugh, Kathleen. "A Fem(me)inist Manifesto." *Women and Performance* 8, no. 2 (1996): 153–159.

Duggan, Lisa, and Muñoz, José Esteban. "Hope and Hopelessness: A Dialogue." *Women and Performance: A Journal of Feminist Theory* 19, no. 2 (2009): 275–283.

Durham, Aisha. "Do We Still Need a Body Count to Count? Notes on the Serial Murders of Black Women." In *The Crunk Feminist Collection*, edited by Brittney C. Cooper, Susana M. Morris, and Robin M. Boylorn, 22–24. New York: Feminist Press, 2017.

Eguchi, Shinsuke, Calafell, Bernadette Marie, and Files-Thompson, Nicole. "Intersectionality and Quare Theory: Fantasizing African American Male Same-Sex Relationships in *Noah's Arc: Jumping the Broom*." *Communication, Culture, and Critique* 7 (2014): 371–389.

Galewski, Elizabeth. "Figuring the Feminist Femme." *Women's Studies in Communication* 28, no. 2 (2005): 183–206.

"Get Out." *Internet Movie Database*, accessed February 9, 2018. www.imdb.com/title/tt5052448/?ref_=nv_sr_1.

"Get Out: Awards." *Internet Movie Database*, accessed February 9, 2018. www.imdb.com/title/tt5052448/awards?ref_=tt_awd.

Green, Kai. "Troubling the Waters: Mobilizing a Trans* Analytic." In *No Tea, No Shade: New Writings in Black Queer Studies*, edited by E. Patrick Johnson, 65–82. Durham, NC: Duke University Press, 2016.

Gonyea, Don. "Majority of White Americans Say They Believe Whites Face Discrimination." NPR, last modified October 24, 2017. www.npr.org/2017/10/24/559604836/majority-of-white-americans-think-theyre-discriminated-against.

"Herstory." *Black Lives Matter*, https://blacklivesmatter.com/about/herstory/. Accessed February 9, 2018.

Hobson, Kathryn. "Sue Sylvester, Coach Beiste, Santana Lopez, and Unique Adams." In *Glee and New Directions for Social Change*, edited by Brian C. Johnson and Daniel K. Faill, 95–107. Rotterdam: Sense Publishers, 2015.

Holland, Sharon Patricia. *The Erotic Life of Racism*. Durham, NC: Duke University Press, 2012.

Johnson, E. Patrick. "'Quare' Studies, or (Almost) Everything I Know about Queer Studies I Learned from My Grandmother." *Text and Performance Quarterly* 21, no. 1 (2001): 1–25.

Keeling, Kara. "'Ghetto Heaven': *Set It Off* and the Valorization of Black Femme-Butch Sociality." *The Black Scholar* 33, no. 1 (2003): 33–46.

Keeling, Kara. *The Witch's Flight: The Cinematic, the Black Femme, and the Image of Common Sense*. Durham, NC: Duke University Press, 2007.

Kelly, Casey Ryan. "The Wounded Man: Foxcatcher and the Incoherence of White Masculine Victimhood." *Communication and Critical/Cultural Studies* 15, no. 2 (2018): 161–178.

Lacy, Michael G. "Black Frankenstein and Racial Neoliberalism in Contemporary American Cinema: Reanimating Racial Monsters in *Changing Lanes*." In *The Routledge Companion to Popular Culture*, edited by Toby Miller, 229–243. New York: Routledge, 2014.

Levina, Marina, and Bui, Diem-My T. "Introduction: Toward a Comprehensive Monster Theory in the 21st Century." In *Monster Culture in the 21st Century: A Reader*, edited by Marina Levina and Diem-My T. Biu, 1–13. London: Bloomsbury Press, 2013.

Malaklou, M. Shadee, and Willoughby-Herard, Tiffany. "Notes from the Kitchen, the Crossroads, and Everywhere Else, Too: Ruptures of Though, Word, and Deed in the 'Arbiters of Blackness Itself.'" *Theory & Event* 21, no. 1 (2018): 2–67.

Mann, Justin Louis. "Pessimistic Futurism: Survival and Reproduction in Octavia Butler's." *Feminist Theory* 19, no. 1 (2018): 61–76.

Molina-Guzmán, Isabel. *Dangerous Curves: Latina Bodies in the Media*. New York: New York University Press, 2010.

Muñoz, José Esteban. *Cruising Utopia: The Then and There of Queer Futurity*. New York: New York University Press, 2009.

Nama, Adilifu. *Black Space: Imagining Race in Science Fiction Film*. Austin, TX: University of Texas Press, 2008.

Nash, Jennifer C. "Black Sexualities." *Feminist Theory* 19, no. 1 (2017): 3–5.

Paldin, Jaime, and Diaz, Armando. "EY Studies Race, Gender, and Exclusion: The Top Takeaways." EY, last modified October 6, 2017, www.ey.com/us/en/newsroom/news-releases/news-ey-studies-race-gender-and-exclusion-the-top-takeaways.

Poole, W. Scott. *Monsters in America: Our Historical Obsession With the Hideous and Haunting*. Waco, TX: Baylor University Press, 2011.

Reed, Alison. "The Whiter the Bread, the Quicker You're Dead: Spectacular Absence and Post-Racialized Blackness in (White) Queer Theory." In *No Tea, No Shade: New Writings in Black Queer Studies*, edited by E. Patrick Johnson, 48–64. Durham, NC: Duke University Press, 2016.

Sexton, Jared. "Unbearable Blackness." *Cultural Critique* 90 (2015): 159–178.

Spillers, Hortense J. "Mama's Baby, Papa's Maybe: An American Grammar Book." *Diacritics* 17, no. 2 (1987): 65–81.

Story, Kaila Adia. "Fear of a Black Femme: The Existential Conundrum of Embodying a Black Femme Identity While Being a Professor of Black, Queer, and Feminist Studies." *Journal of Lesbian Studies* 21, no. 4 (2017): 407–419.

Struyk, Ryan. "Blacks and Whites See Racism in the United States Very, Very Differently." CNN. Last modified August 18, 2017. www.cnn.com/2017/08/16/politics/blacks-white-racism-united-states-polls/index.html.

Stryker, Susan. "My Words to Victor Frankenstein above the Village of Chamounix: Performing Transgender Rage." *GLQ* 1 (1994): 237–254.

Watts, Eric King. "Postracial Fantasies, Blackness, and Zombies." *Communication and Critical/Cultural Studies* 14, no. 4 (2017): 317–333.

Willloughby-Herard, Tiffany. "(Political) Anesthesia or (Political) Memory: The Combahee River Collective and the Death of Black Women in Custody." *Theory & Event* 21, no. 1 (2018): 259–281.

Womack, Ytasha L. *Afrofuturism: The World of Black Sci-Fi and Fantasy Culture*. Chicago, IL: Lawrence Hill Books, 2013.

Young, Elizabeth. *Black Frankenstein: The Making of An American Metaphor*. New York: New York University Press, 2008.

13

QUEER AESTHETICS, PLAYFUL POLITICS, AND ETHICAL MASCULINITIES IN LUCA GUADAGNINO'S FILMIC ADAPTATION OF ANDRÉ ACIMAN'S *CALL ME BY YOUR NAME*

J. Nautiyal

Introduction

Against the resplendent backdrop of Luca Guadagnino's mediated entanglement (2017) adapted from André Aciman's 2007 novel *Call Me by Your Name* (henceforth *CMBYN*), I propose an embodied study of communication and gender through an attention to the body in its aesthetically queer sensitivities. *CMBYN's* sensual politics inspire an idea I call an *artful rhizome* in this book chapter. It is based on the works of American pragmatist John Dewey's theory of aesthetics from his *Art as Experience* (1934) and French philosophy-psychoanalysis duo Deleuze and Guattari's nonconforming mode of connecting with the world, the "rhizome." I tease out the Deweyan notions of artfulness as a playful attitude and embodied practice of attentive absorption in the heightened vitality of the everyday.[1] Artfulness is also the delightful perception of an aesthetic experience all of which for Dewey is grounded in communicative practices.[2] The rhizome, on the other hand, is a subterranean, lateral, anti-unity, nonbinary, and primarily queer modality of communicating with and engaging the world as a human, bestial, vegetal, and machinic multiple.[3] Put together, an artful rhizome is not a culmination or a resolution of differences between Deweyan aesthetics and affect theory. Instead, an artful rhizome is a *communicative impasse* formed within a collaborative oddity, one that offers scholars competing and consonant somatic approaches to unpacking the gendered dimensions of communication with aesthetic and affective inflections in other words. The impasse is a transformational bloom space for a playful deadlock and loss of an autonomous self (as if) through the haptic interweaving of interdisciplinary bodies in my work (much like the multiple agents and material scenes in *CMBYN*). With Dewey, Deleuze, and Guattari in concert, I utilize *CMBYN's* filmic sensorium to demonstrate the ways in which an artful rhizome underscores queer habits of attention and space-inhabiting practices involving more-than-human envelopments.

At the outset of this chapter's theoretical orientation which I elaborate later, the problematic queer visibility afforded to *CMBYN's* predominantly white and elitist cast of characters with their "empty, sanitized intimacy," and a pressing concern from some critics about its predatory themes are quite clear.[4] What is so remarkable (yet again) about a wealthy, white cosmopolitan family summering away somewhere in Italy while waxing eloquent about the vagaries and vicissitudes of love in multiple languages?[5] What is so captivating (once more with feeling) about a prototypically handsome older white figure who is charming in his intellectual demeanor and desirable in his bodily disposition enough to get a precocious 17-year-old's heart racing and aching?[6] Granting such justified critiques and cynicisms, I focus on Guadagnino's filmic adaptation with its unique and cosmopolitan romance featuring two young Jewish men. The film's unfolding in its narrative's historical context of the 1980s' AIDS panic combines with the amalgamated precarity of the main characters' queer desire on account of their Jewish identities to offer us a fascinatingly rich communicative artifact that warrants an immersive analysis of its sensual politics.[7]

Guadagnino's rendition stands out politically and eloquently on account of its *queer sensorium*, which artfully misplaces the sensory indices of desire in heterocentrist naturecultures.[8] What do I mean by this? For example, heterocentrist networks of family, education, religion, and mediated environments all spell out and fix "appropriate" objects of desire (or what Sara Ahmed would call "happy objects") for an individual from the very start, which then circulate as the happy gospel of the good life.[9] One is told repeatedly where to look, what to see, when to touch, how to taste, why to listen, whom to smell—courtesy of the heterocentrist genius to "[...] subjugate sexuality to the reproductive model."[10] *CMBYN* takes this subjugating sensory predetermination to task and playfully disrupts the toxic illusion of its goodness through sensual politics that enact libidinal rearrangements of desire.

The film's queer sensorium emerges through sense-enveloping material scratchings in normative aesthetic economies. Such economies are built on the social capital ("positive affective value") of compulsory heterosexuality as it manifests through the heteronormative circulation of sight, sound, taste, touch, and smell.[11] Within this fictional sensorium, *CMBYN's* queer registers perform a peculiar politics of freeing and mobilizing the desire to love from the exclusionary and immobilizing demands of heterocentrism.[12] Although I find other predominantly Western and queer representations (including *Brokeback Mountain, Boys Don't Cry, and Moonlight*) to offer pathbreaking narratives, I find *CMBYN's* sensibilities to be significant in its solidarity with and departure from these films.[13]

While such cinematic media have helped expand the communicative spectra of sexuality, race, class, and gender, *CMBYN's* queer sensorium can enrich the affective-aesthetic politics of embodiment and gender in communication studies in unique ways. Most importantly, I submit that Guadagnino's seemingly generous attention to sensory rearrangements amid more-than-human material actants in his narrative craft makes way for what Katie Stewart calls "a new regime of sensation."[14] *CMBYN's* sensory regime offers "[...] a sharpening of attention to the expressivity of something coming into existence," which is precisely a communicative remodeling of limiting discourses around masculinities, family life, and LGBTQIA-Z+ lived experiences.[15] In fact, *CMBYN* is a multisensory communicative glimpse into thoughtful practices of ethical masculinity that loosen the hegemonic noose of toxic, homophobic, and violent gendering habits in naturecultures. On account of *CMBYN's* keen attunement to the body in its surroundings, senses are not preconditioned to land on a predetermined object of desire. They wander. They hesitate. They take risks. They give in to the unknown: "All the world is a bloom space now [...] Anything can be a bloom space."[16] Like the bloom spaces of weird theoretical entanglements. Like the bloom spaces of *CMBYN*.

Strange theoretical encounters and dramatic contingencies within a queer sensorium

I turn toward the collaborative theoretical oddity of Dewey, Deleuze, and Guattari because of their unexpected appeal in hinging issues of everyday embodiment and materiality with aesthetically queer sensibilities and ethical masculinities. Furthermore, communication scholars can make use of new oppositional vocabularies emerging from this strange theoretical encounter through which to politicize the aesthetic bedrock of identity, intimacy, desire, and love. In fact, the encounter of aesthetics and affect is what Ahmed would call a "drama of contingency," created by a terrific, terrifying, messy, and inconveniently delightful contact that transpires in contact zones of proximity between two bodies.[17] Together and apart and within an interlocking impasse, Dewey, Deleuze and Guattari build what Lauren Berlant calls a temporary dwelling place for apprehending such events, or what I call an artful rhizome.[18]

"An impasse. So much the better."[19]

On his end, for instance, Dewey needs no introduction to the field of communication. A foray into the early archives of the field's history from early 20th century (1914–1915) reveals the disciplinary motivation to base rhetorical practices (called speech education at the time) on Dewey's scientific approach to communication and psychological theories of social adjustment.[20] The field's goal at the time, as Herman Cohen describes, was the attainment of a "well-adjusted personality," a quest that factored heavily into speech's inextricable connection with students' emotionally balanced and overall mature personality (called mental/speech hygiene).[21] What is perhaps one of the more prevailing markers of speech hygiene later in the field between 1930 and 1945 is the perceived binarization and opposing faculties of feeling/emotion and reason.[22]

In the field's historical desire for scientific rigor in well-adjusted speech practices that imply a privileging of rationality over emotion, the body ironically suffered an infantilizing compartmentalization in the foundational years of the discipline. Interestingly, the disciplinary history does not point explicitly to nor even include Dewey's aesthetic theory from the 1930s, although it offers a more inclusive upgrade for communicative practices based on community, mindful engagement, and the body. Therefore, my goal in turning to Deweyan aesthetics in this chapter is *reparative* in spirit, serving to redress a disciplinary tendency for somatic compartmentalization. Of course, next comes the concern for a capacious range of bodies and their feelings that might be represented in attunement with more representative (i.e., non-normative, nonrational, atypical, extra-human, Cyborgian, queer, disabled, and racially self-determined) desires and demands of today's democratic and more politically proliferated global circuits.

In making the current argument in 2020, I simply cannot assume the stability and/or even the universality of the democratic subject (predominantly cisgendered, straight, able-bodied, neurotypical, and white male) around whom Dewey built his aesthetic theory of communication in the early 1930s. This is why it is at this delicate juncture that Dewey's historically rich work needs pruning, grafting, and updating. This is why I turn to Deleuze and Guattari, whose geopolitical (read: underground) philosophy of the rhizome makes room for more daring alliances and nonrepresentational worldings with the body.[23] My hope here is to gently exfoliate Dewey's pragmatic yet fairly dated and anthropocentric theory of aesthetics.

What is artful about the impasse, or this playful theoretical deadlock, is a rhizomatic decoupling of desire from its reductive signification within sensory libidinal economies and the symbolic sedimentation of linguistic meaning:

the rhizome … is a liberation of sexuality not only from reproduction but also from genitality […] What is at question in a rhizome is a relation to sexuality—but also to the animal, the vegetal, the world, politics, the book, things natural and artificial—that is totally different from the arborescent relation: all manner of "becomings."[24]

Instead of considering transcendental or vertical lineage structures prevalent in Western thought that confine human sexuality to reproductive structures, Deleuze and Guattari consider lateral mode of kinships and queer world-making forays (much like a rhizome) without discriminating between organic, inorganic, machinic, human, and non-modes of experience, or becomings in their articulations.

What is artful about the impasse is a perceptual freeing up of sensual matter to rearrange itself as the "immanence of the flesh" in multi-body "circuits of inter-involvements"—vital matter, inanimate matter, seasonal rhythms of love in naturecultures; What doesn't matter in all manner of becomings?[25] This manner of becoming also implies that the works of my theorists dance with each other at the tensions where the interplay of ideas happens in what Diana Coole and Samantha Frost elegantly call "choreographies of becoming."[26] In these Deweyan-Deleuzo-Guattarian choreographies, artful and rhizomatic pirouettes of sensual "capacities and poten-cies" emerge, assemble, organize, forge ties, and dissipate as an intersensory *pas-de-deux*. Their inter-animations are similar to the simmering, summering, usurping space in which the multiple actants in *CMBYN* come to experience love in the artfully destabilizing call of their own names addressing the Other. But there is yet another impasse in translating these affective-aesthetic "choreographies of becoming" into words. Writing about something as elusive yet piercing as sense-overlapping experiences cannot be done with the surgical scalpel of clinical precision. Needless to say, even from a methodological standpoint, the academic libidinal economies of scholarly writing fail me with quiet pride in this space.

If an artful rhizome is about the freeing up of sensual matter to rearrange itself in accord with the queer registers and sensual politics of intersensory experiences in *CMBYN*, my writing feels ethically beholden to expressing itself in a similar fashion. Without tuning into ambient and performative resources of criticism, my writing simply cannot access the sensuousness of experi-ence such that the "… imbrications between embodiment, language, disposition, perception, and mood are always in operation."[27] Without sounding out an atmospheric refrain in writing, *CMBYN*'s queer choreographies of becoming remain adrift in the "dense entanglement of affect, attention, the senses, and matter."[28] Therefore, as my theoretical engagements with *CMBYN* call me deeper into rhythmic folds of inter-involvements, a politically poetic frames the many moods of the text's multi-agentic encounters. This ambient mode of criticism imbues the story with emotional accents ranging from the tumultuous trepidation of desire, the fragile vulnerability of anticipation, the thalassic warmth of love, and the lachrymose desolation of heartbreak in *CMBYN*.[29] Similarly, I tune into the apocrine circuits and inhale the heart notes of the headiest milieu—that of love—when being called by an artful rhizome in all manner of becomings. And now into the thick of the artful rhizome itself within *CMBYN*'s queer sensorium.

The becoming of an artful rhizome in *CMBYN*

To burrow into a fairly dense theoretical bloom space, I attempt to follow *CMBYN*'s subterranean canal as an artful rhizome featuring the beautifully painstaking vivacity of the devastating nectar that is love and heartbreak. I hesitate to categorize the experiences of the film's protagonists as "firsts" in the germinal series of their many love encounters. Is it birth or is it death each time that love calls? The tendency to rank experiences as first or last is misplaced from within an artful rhizome's locational politics because ranking fixes bodies and desires. Ranking forecloses on and/or sutures the possibility of loving and living life ethically, "with an open wound," regardless of what transpires at love's altar—irrespective of first or last moments for vital beating matter.[30] Moving on, *CMBYN* is a queer, playfully political, and sensual assembly of everyday aesthetics and affects that performs the sensuousness of ordinary experience, deterritorializes the aesthetic terrain of human achievement and masculinist mastery, and shatters the illusion of organic unity especially with/in matters of the heart.

The rhizomatic liberation that Deleuze and Guattari mention is queer in *CMBYN*, and not only because it decouples sexuality from culturally determined reproductive economies by artfully overriding them with the libidinal celebration of loving whom one loves. For communication and gender scholars, the queerness of the artful rhizome is its peculiarly absorbed, playful, and powerful repurposing of material sensory arrangements in everyday sense-scapes. This implies the rearrangement of sensory experiences and attunements that one is conditioned to associate with normatively aesthetic embodiments of love and desire in everyday life. Such queerness implies libidinal explorations of detours rather than investments in expected returns.[31] Libidinal detours usurp the body from the claws of language and its social categories, away from the fixed destinations at which language so ferociously demands the arrival of its subjects. These detours allow the body to reconfigure new experiences between the not-yet-explored horizons of naturecultures. For example, *CMBYN's* evocative advocacy of a queer aesthetic through its sensual politics around ambient spaces resonates more with stories such as Barry Jenkins' 2016 film, *Moonlight*. *Moonlight's* oceanic atmospheres, black and deep, offer the main character (*Chiron*) a bottomless refuge of kinship to dissolve his lifelong tensions pertaining to black hypermasculinity, queerness, systemic abuse, delinquency, and self-determination.[32] *CMBYN's* multi-agentic queer elements find more kinship with Jenkins' *Moonlight* than Ang Lee's 2005 film *Brokeback Mountain (BM)*, whose atmospheric elements recenter heterocentrism. The protagonists in *BM* consummate their desire for each other by validating their "heteromasculinity," as reflected in the masculine ruggedness of the wilderness. Catriona Mortimer-Sandilands and Bruce Erickson consider the portrayal of wilderness in *BM* as "a vast field of homoerotic possibility," that separates same-sex encounters from the queerness of the protagonists.[33] Here, the wildness of the ecological space, that is the "lush hanging river valley surrounded by trees and dramatic, snow-striped peaks," abets, affirms, and enshrines same-sex desire as "masculine, rural, virile."[34]

Comparisons aside, the warm lushness of *CMBYN's* North Italian summer, bathed in the fruit bearing succulent and anti-teflonic vegetal life abuzz with its sticky flies affirm the *kinetic potentiality of desire* over the cultural fixity of love objects.[35] Archeological floats from the nearby river with "not a straight body" in them and cool summer waters all invite libidinal movement without cultural sedimentation.[36] It is the ambient (and rather privileged) sphere that encourages the corporeally enmeshed bodies to lean into that bodily abyss of which they are unaware, the one whose vistas one will never fully know and feel.[37] Even though the protagonists can be seen wrestling with their feelings, I offer that the North Italian ecology and seasonal rhythms as depicted in the film are largely erotically charged and libidinally pluripotent. As I dive deeper into my analysis, I argue that the ecology co-participates in nourishing what I consider queer aesthetics as part of the film's intersensory rhetoric. This queer aesthetic's rearrangements playfully disorient and powerfully dissolve the heteronormatively sticky promises of "happy objects" pertaining to the social norms/goods of desire, sex, sexuality, culture, and nature.[38] All in all *CMBYN* exemplifies the impasse-as-artful rhizome, a quotidian yet unique queer alliance of aesthetic and affective experiences that I illustrate with the help of cinematic encounters in the following sections.

The encounter

"*L'usupateur.*" The usurper. These are the first few words in *CMBYN* that the precociously charming and refreshingly awkward Elio utters upon the arrival of Oliver, the confident and strapping American PhD student.[39] Oliver, who is joining the Perlmans during a warm north Italian summer is the usurper for whom Elio begrudgingly vacates his room to adjoining quarters that share a doorway.[40] As it turns out, both stand evicted when called home to love, the usurper of usurpers whose ego-destroying and life-affirming encounter with Elio and Oliver unfolds as a grand larceny of hearts, minds, bodies, libidos, words, and tongues. Any former sense of solid ground is completely awash and submerged in the unspecified intensity of a theft so rich in its sensations that they would be abundant in their strangeness to themselves. For as long as both shall live, they will only have died each time in calling one by the other's name. *(P)each time in the best way.*

Love, the usurper, summered a deep bond between multiple agents, bodies, ecologies and among all the things that once had nothing to do with each other. This is why everything—a multiplicity rendered speechless in the glory of the usurper, love. They basked in the after-glow of the aider and abettor, love, the mystery of existence, itself. This would be the summer of their lives in which both would experience the things that really mattered: the exquisite agony of desire, sensuality, jealousy, affection, thrill, joy, separation, heartbreak, and sweet surrender into the lonely arms of uncertainty. In opening up to love's devastating and trans-formative call through the power of another's name, *CMBYN* shows us how an artful rhizome liberates sexuality from its reproductive apparatuses and desire from its sensory predetermin-ation embedded in aesthetic economies of normative kinships. The freedom (if at all, apropos of Nietzsche) is in "cultivating joyful modes of confronting the overwhelming intensity of *bios-zoē,*" *zoē* being the "absolute" threatening vivaciousness of life's abundance itself.[41] What threatens and hurts more profoundly with its vitality than the call of love whenever, however, and whomever it calls? For instance, in an epochal scene from the film Elio and Oliver finally admit to their feelings for each other and find ways to decenter their identities associated with their culturally fixed names and all senses associated with the body through the following entanglement: *Call me by your name and I will call you by mine.*[42] What this means is that we will both come to experience life and love it from a "nonanthropocentric view" that "depersonalize[s] subjective life-and-death.": "This is just one life, not my life. The life in 'me' does not answer to my name: 'I' is just passing."[43] My name is just passing. Your name is just passing. Freedom is in being called mutually to concede the first and last territories marked by *bios-zoē*: the givenness of our names.

Call me by your name and I will call you by mine. Baptized in the name of the salt of ecstasy and heartbreak he tasted on his skin, in the name of the touch he tasted while holding the peach in his hand, in the name of the lips he felt in tracing over the sculptural outline of the archaeological float they found at sea, in the name of the taste that enveloped him in the way he consumed the very last drop of his apricot juice, in the name of his gaze that was irresistibly drawn to the egg yolk oozing out of the egg onto his lips, in the name of his fragrance that lingered in his tousled hair after they had made love. The pronouns stopped mattering once the things that really mattered baptized Elio and Oliver unto freedom. He who came to usurp. He who was usurped. He was just passing. In the name of the father, the son, and the holy spirit. All dead and resurrected. I was just passing.

"Feel my feet above the ground, hand of god deliver me ..."[44]

Queer aesthetics and playful politics in *CMBYN*

The following sections now unpack and focus on the various artful and rhizomatic rhythms unfurling in the *CMBYN* ecology. I understand rhythm through Henri Lefebvre's account of vital and repetitive investments/expenditures toward temporal and spatial interactions of bodies and environments in the context of everyday life: "Everywhere where there is interaction between a place, a time and an expenditure of energy, there is rhythm."[45] For Dewey also, rhythmic structure is key to artfulness as an everyday experiential orientation, interactive tendency, and mindful habit of embodied attention. Artfulness as a mind-body composite of such dispositions interacts with art as lived experience (and vice versa) to produce the emphatically delightful perception, or the rhythmic doing and undergoing of energy, which Dewey calls an aesthetic experience.[46] Conversely, rhythm in the Deleuzo-Guattarian sense has less to do with artful craft as a rhythmic balance of energetic investments:

> From chaos, *Milieus* and *Rhythms* are born ....The notion of the milieu is not unitary; not only does the living thing continually pass from one milieu to another, but the milieus pass into one another; they are essentially communicating. The milieus are open to chaos, which threatens them with exhaustion or intrusion. Rhythm is the milieus' answer to chaos. What chaos and rhythm have in common is the in-between—between two milieus rhythm-chaos or the chaosmos ....Action occurs in a milieu, whereas rhythm is located between two milieus, or between two inter milieus ....To change milieus, taking them as you find them: such is the rhythm. Landing, splashdown, takeoff ...[47]

In other words, rhizomatic rhythms have more to do with the liminality of chaos whose middle-place or *milieu* is rhythm's artful passageway featuring a free orientation to sexuality, decentered, in situ modes of multiple relationalities, and vanishing points. In all manner of becomings, I highlight two instances of such multi-agentic encounters within *CMBYN*'s queer sensorium, whose inter-animating sensual politics artfully destabilize hegemonic narratives around the aesthetic experience of love itself. First, I discuss in more vivid detail how the aestival rhythms of the North Italian vistas aid and abet the attraction and desire Oliver and Elio feel toward each other. These rhythms include the lushness of the environment and recreational activities such as music and food. Second, I focus on the Perlmans as a peculiar familial interactant within the *CMBYN* sensorium. The Perlmans' rhizomatic artfulness toward life as evident through their linguistic, educational, and cultural inter-milieus marks them as what I consider a *nomadic family and a happy abject*. They enact the embodied absorption and in-between rhythmicity of an artful rhizome that speaks to a liberated queer identity. Such an identity diffuses as it slants the material sensations and redistributes as it intensifies delightful perceptions associated with normative aesthetic experiences of familial togetherness, masculinity, and love.

Multi-agentic rhythms in CMBYN ecology (environment and music)

Guadagnino's artful direction showers abundant attention on the rhizomatic landscapes of a vibrant Italian summer. This kind of storytelling is frustratingly faithful to the fruitful and fateful season of a transformative encounter between Oliver and Elio. The film's luscious narrative foregrounds the interactive tendencies and mutations of experience between disparate milieus becoming intimate with one another.[48] In between these inter-milieus, fruits and bodies, trees and bodies, machines and bodies, water and bodies, food and bodies, music and bodies form critical rhythm-chaos with each other each sans a serial and linear march toward "dogmatic" logics.[49] This chaosmos is *CMYBN*'s milieus of all milieus passing into one another—communicating with one another.[50] The lush chaos of the environment in which the blues and greens of the surrounding waters intertwine, interweave, and mingle to morph into green grass affirm and assist the rhythmic inter-milieus of their burgeoning desire. *Landing. Splashdown. Takeoff.*[51] The environment welcomes a misplaced blurring, rerouting, and amplification of senses that Connolly attributes to the intersensory or "interwoven" aspects of perception itself, an intersensory queering of senses, or a libidinal estrangement from normative sensory economies.[52]

"Oh to see without my eyes, the first time that you kissed me, boundless by the time I cried, I built your walls around me"[53]

Stevens' *Mystery of Love*'s aforementioned lyrical expressivity helps us comprehend that in such an aphrodisiacal atmosphere, love's sight cannot see what its taste has already imagined and perceived hearing, touching, and smelling in the juice of apricots, peaches, and cherries. Such is the North Italian aestival rhythm that words ripened with the succulence of youth and desire overflow in the stickiness of summer warmth. Trees laden with a juicy gamut of fruits burst with the constant buzz of desiring flies. For example, in one scene foregrounding Oliver's official introduction to the Perlmans' over breakfast, Oliver's mundane act of cracking an egg open to savor the warmth of the yolk draws in Elio's gaze. *Over easy. Overdetermined.* It is such tiny quirks of intersensory immersion, amplification, and dispersion that beckon the destabilizing stirrings of love transfiguring (and transfigured by) its arterial landscape at the speed of lightening. The oscillations between desire and disgust, hospitality and hostility, usurper and usurped, night and day—such are the palpitating rhythms of desiring bodies without organs and without libidinal predetermination.[54] Elio is both mesmerized and mildly disgusted with Oliver's attention to food: on the fence about his feelings … but transfixed and perched, nonetheless.

Similarly, *CMBYN*'s use of music (its accompaniments, and renditions) as an integral part of everyday summertime activities invites further speculation. For example, in an outdoorsy scene in the Perlmans' juice-filled garden of summer flirtations, a shirtless Elio gently reproduces Bach's symphony on his guitar and draws Oliver's attention.[55] Upon Oliver's request to replay the musical version, Elio moves to their living room and turns to his piano to perform the same symphony with playful theatrics.[56] Each time, Elio's reprisals are artful, absorbed, and playful in attending to the rhythms of classical music. From a Deweyan orientation, playfulness is a key and rather paradoxical attitude mobilizing the everyday practice of artfulness. One embodies a playful attitude in the doing and undergoing of something artful for Dewey which follows in his own words: "Play remains as an attitude of freedom from subordination to an end imposed by external necessity; as opposed that is to labor; but it is transformed into work in that activity is subordinated to *production* of an objective result."[57] To clarify: for Dewey, artfulness entails a playful and ethical attitude that freely absorbs the immediacy of the means and does not let the aesthetic end dominate the moment of free-play. In that sense, artfulness remains a playful attitude without becoming tedious like labor. Paradoxically, artfulness as a playful attitude requires that the *means* of free-play "serve the purpose of a developing experience," such that the aesthetic end is produced as a *work* of ongoing and transformative processes to ensure the "complete merging of playfulness and seriousness."[58]

Keeping the playfulness of the moment between Elio and Oliver in context, there is something more important happening in the above scene that decenters masculinist modes of aesthetic mastery. Each time Elio plays the symphony, he alters it a little from the guitar version to both irritate and draw in Oliver further. Elio's constant deferral of great maestros' musical ownership in his own renditions is a provocative (rhizomatic) move in *CMBYN's* queer sensorium and sensual politics. In other words, Elio's reprisal of Bach's symphony from Listz' orientation and Listz' symphony in Buzoni's style underscores and troubles Dewey's approach to playfulness. Elio's performative deferrals absorb the immediacy of his musical means without letting the aesthetic end (i.e., great works by musical artists) dominate his moment of free-play. At the same time, Elio's promiscuous playfulness toward his activity does not serve the developing purpose of reproducing Bach's great work as its own aesthetic and masterful end. In fact, Elio's flirtatious and irreverent dislocation of Bach's symphony to charm Oliver is a rhizomatic rhythm emerging in-between the rapidly changing artistic inter-milieus of Bach, Listz, and Buzoni. Such is the nonanthropocentric rhythm of rhizomatic deterritorialization that dethrones Bach's aesthetic mastery of sensory arrangements within classical music and retunes formal expectations of pleasure associated with the genre.[59]

"Have I offended you?"[60]

Bach did not write the symphony for the guitar or ever imagine that a young fool about to fall in love would prevent Bach, Listz, and Buzoni from claiming their own works of art. These aesthetic deferrals are precisely the alchemical detours of desire in Elio's renditions for Oliver just so he can usurp Oliver's attention and presence a second longer. The encounter's musical inter-milieus thus affirm the Deweyan idea of serious play while shattering the culturally bounded pleasure and aesthetic purity associated with the dogmatic meter and bodily responses to classical works of art. Lefebvre notes that Bach's musical work is *logogenic* in that it aligns with the production of meaning as opposed to being pathogenic or emotion-producing work.[61] However, Elio's Bachian rendition to impress Oliver with his precocious talents rearranges Bach's sensory apparatus into *pathogenic* playfulness by rerouting the great artist's musical rhythm through the foolish and emotionally rich detours of desire. The musical refrains of this non-masterful and playful encounter resonate with the slippery territory of language rhythms whose multilingual interconnections further flesh out *CMBYN's* queer sensorium.

Nomadic families and happy abjects: the polyglottal and polymorphously perverse Perlmans

The artfully rhizomatic liberation of identity within *CMBYN's* queer sensorium is also in its portrayal of what can only be the *polyglottal stop* of what families could be together in their everyday interactions around music, art, poetry, food, reading, writing, sexuality, and sensualities. *CMBYN's* Perlman family enters playgrounds of love, pain, heartbreak, and repair—the artful juiciness of all that is sweet and tart in the poetry of being fully alive. I emphasize: *the polyglottal stop, not the glottal stop*. Not discounting their socioeconomic privileged flair, The Perlmans feature as the polyglot family in the *CMBYN* ecology, communicating across several tongues all at once (in English, German, French, Greek, and Latin).[62] But they are also precarious in terms of their Jewish identity, which is why Elio tells Oliver that his mother reminds the formers that they are "Jews of discretion."[63] Oliver reciprocates by saying he understands what it means to be the odd Jew out in a small American town.[64] Taking their privilege and precarity hand in hand, the family affirms the wretched playfulness and blessed mystery of love and love alone. They do so in each familial polyglottal stop, that is the consonantal sounds of not one breathtaking tongue but a lively babble of multiple breathtaking tongues.

For instance, on one rainy evening in the film, the Perlmans are huddled together reading a Germanic text about a young knight's inability to bring up the subject of love to his paramour and wonder aloud in English when the knight finally has the courage to ask the question: is it better to speak or to die?

> When Elio responds: "I will never have the courage to ask a question like that," his father affirms him with what I imagine can only be the most attentive and liberating thing to hear from a parent: *"I doubt that. Hey Elly-Belly, you know you can always talk to us."*[65]

The subtext of all their polyglottal stops in all their foreshadowing tongues seems to be along the following patterns of experience. We will honor the absorption of experience you are about to have with Oliver. You can always talk to us because we will not strip you of that experience by naming and claiming the call to which you heart and body has responded with utter confusion, anxiety, and boundless joy. You can always talk to us because we will not strip you of your vitality by redirecting you toward culturally determined objects of desire that subjugate sexuality to reproductive apparatuses within toxic masculinities. You can talk to us because we will affirm unequivocally that yes, you will suffer from the exquisite pain of falling in love with someone so unique and unexpected, and having your heart usurped from your body, but you will live. You will breathe at the acme of joy and in the deepest depths of despair. You will live unconditionally and not unrequitedly. You will live freely in the ironies, absurdities, miscalculations, and missteps of love, loss, and everything critical that transpires in-between their rhythm-chaos.

> "Oh will wonders ever cease? blessed be the mystery of love."[66]

What is so singular about this particular familial exchange? In my reading, it is really the Perlmans' playful vivacity toward the "singularity, force, movement, through assemblages or webs of interconnections with all that lives" in-between several languages that affirms their nomadic-rhizomatic orientation to lived experience.[67] For Braidotti, the polyglot is indeed a nomad by way of language.[68] The polyglot's manifold mixes in multiple languages only affirm the "incongruities, the repetitions, the arbitrariness of the languages he deals with."[69] The figure of the nomad relates with the rhizomatic rhythm of connecting with the world, marking and unmarking those worldings by uprooting and replanting fixed notions of identity—particularly as they come to be materialized, disciplined, and normalized in language. The Perlmanesque-nomadic-polyglottery sets the artful rhythm for several conversational topics all resounding the slipperiness of language. Their multitongued provocations affirm the blooming of bodily desire sans identity fixtures within language's slippery vocal folds—that is, its queer registers.

For instance, in one scene from the film, Oliver is helping out Professor Perlman in archiving and cataloging some information when Annella (Elio's mother and Professor Perlman's spouse) brings over a pitcher of fresh apricot juice. While Oliver is gulping down and relishing the last dregs of the juice, the air in the room turns playful and intellectual. Professor Perlman launches into philological and etymological modes in tracing the history of the word "apricot":

> The word apricot comes from the Arabic—it's like the words "algebra," "alchemy," and "alcohol." It derives from an Arabic noun combined with the Arabic article 'al-' before it. The origin of our Italian "albicocca" was "al-barquq"...

> It's amazing that today in Israel and many Arab countries the fruit is referred to by a totally different name: "mishmish".[70]

After listening to Professor's Perlman's provocation, Oliver jumps in with a rebuttal and replants the slippery roots of the apricot:

OLIVER: The word is not actually an Arabic word.

PERLMAN: How so?

OLIVER: It's a long story, so bear with me, Prof. Many Latin words are derived from the Greek. In the case of "apricot," however, it's the other way around. Here the Greek takes over from Latin. The Latin word was praecoquum, from pre-coquere, precook, to ripen early, as in precocious, meaning premature. The Byzantines—to go on—borrowed praecox, and it became prekokkia or berikokki, which is finally how the Arabs must have inherited it as al-barquq.[71]

This rhythmic exchange is merely Professor Perlman's aptitude test for Oliver, one that he passes with true colors. However, it is the artful rhizomaticity of the exchange that is of interest to my argument. I return to Braidotti's writings on the nomadic polyglot to enmesh the artful rhizome further: "The polyglot is a specialist of the treacherous nature of language. Words have a way of not standing still, of following their own ways. They come and go, pursuing preset semantic trails, leaving behind acoustic, graphic, or unconscious traces."[72] The polyglottal and rhythmic exchange into which Professor Perlman invites Oliver is less about the latter's intellectual capacity to trace the lineage of the apricot's root. The exchange is more about this treacherous play of language, of its rhythmic inter-milieus foreshadowing the story's prime milieu of milieus, that of stripping the disciplinary limitations of languages and following their in-between sensory flows of desire wherever they take the characters. The plot's bloom space-to-come is about resetting preset semantic trails, leaving behind acoustic, graphic, or unconscious traces on an inter-milieued rhythm-chaosmos of skins and surfaces: fruits, water, trees, and bodies passing into one another, multiplying the narrative's sense of space-time-territory. Rewinding back to that moment of philological tracing, Oliver unconsciously colocates Elio's precocious embodiment with the germinal traces of his desire-to-come for Oliver (and vice versa) in all the cosmopolitan tongues such as Latin, Greek, and Arabic. The incongruities, the repetitions, and the arbitrariness of the word *apricot's* signification across the seven seas catalyze the preripened, tart juiciness of desire that ironically blooms prior to and alongside the identity-fixing confines of its transnational markers. Peachy!

Furthermore, Oliver's hearty and absorbed attention to the last vestige of the apricot juice in his breathless intake of the fluid intrigues and amuses Elio. Among several other moments in the film, this moment of intrigue rearranges and mobilizes what might otherwise be limited as heterosexed sensory objects of desire—opening into feeling before calling and naming the inexplicable, desire-bending-love-breaking-heart-barking-love. The collective embodiment of Perlmans' familial exchanges foreground their orientation "to live intensely and be alive to the nth degree [pushing them] to the extreme edge of mortality" that Braidotti calls the affirmative ethic of "living with an open wound."[73] The Perlmans' affirmative mode of togetherness places my reading in the proximity of another queer register of ordinary sensibilities, one that I consider as the kinship space of *happy abjects*.

Happy abjects

The Perlmans embody and shatter the fantasy of the familial line of reproduction that Ahmed indicts as the self-perpetuating economy of the happy object.[74] As a polyglottal family, The Perlmans reinstall the hap, the contingency in the happiness of normative happy objects thereby acting as the radical and refreshing acetone in which the enduring stickiness of jaded, disciplining familial affects dissolves. The Perlmans embody their organic position as an intersensory familial acetone in the *CMBYN* ecology. The Perlmans are what I would delight in calling a *happy abject* as opposed to a happy object because they all understand the ethical importance and risks of embracing those experiences and desires that surprise, fray, and deterritorialize the reproduction of existing lines of happiness.

In their conversations around experiences pertaining to love, the Perlmans would rather take up space in the world as the *happy abject* of families, an oblique un-model affirming the wretched openness to not knowing in loving and loving anyway rather than letting socially determined frames of reference decide the fate and line of feelings (i.e., their orientation toward love objects). The Perlmans focus instead is on the cruel contingencies of both joy and sorrow within which the drama of embodied beings unfurls, undulates, and unfolds. Happens. Instead of defending their son's experiences from the contingency of happiness, the family affirms the vulnerability of courage in sounding out one's feelings. Instead of wishing Elio a normative sense of "I want you to be happy," often coded as a desire to insulate the beloved from a life marked by queerness, the family encourages an openness to the vivacious threats of feeling, of life's abundance itself (*bios-zoē*).[75] The Perlmans do not want to foreclose on Elio's experience just because the stigma of queerness threatens the promise of normative happy economies of heterosexual familial edifices. Better to be an unhappy queer or a happily undecided becoming that has tried to feel the ecstasy and eviction of love than to seek a straight and poker-faced existence that circulates and accrues value in society simply because it follows the hetero-patriarchal family's script.

Conclusion

To mark my return to the readers' attention that I hope I usurped to some extent during this meditation on one vivid heartbeat of love, I turn to Professor Perlman's conversation with his son, Elio. The professor delivers a piquant and thoughtful monologue to comfort his vulnerable son after Oliver departs and Elio is visibly frayed. The monologue that follows affirms the jouissance of being threateningly alive to the mysteries of love:

> We rip out so much of ourselves to be cured of things faster that we go bankrupt by the age of 30 and have less to offer each time we start with someone new. But to make yourself feel nothing so as not to feel anything—what a waste! Just remember, our hearts and our bodies are given to us only once, and before you know it, your heart's worn out. And as for your body, there comes a point when no one looks at it, much less wants to come near it. Right now, there's sorrow, pain; don't kill it, and with it, the joy you've felt.[76]

Professor Perlman's powerful and soothing monologue in response to his precocious son's heart-ache is both artful and rhizomatic in its call. From an artful perspective, Professor Perlman invites his son to be open to the searing experience of heartbreak and not attempt to evade or sedate the pain. Through the character of Elio's father, *CMBYN* articulates an understanding of the reasons why people stop engaging with life in all of its juicy, flavorful, and charred intensifications. It is because we anaesthetize, compartmentalize, and stamp out our perceptual capacities to experience the emotional depths breath has to offer in the desire for quick cures. Of course, those emotional depths are heightened during disorienting moments of queer desire. What happens when breath quickens and comes alive in sensations that we did not know even existed precisely because we internalized those moments as sensory outcasts whose call we must not hear, whose touch we must not feel, whose taste we must not savor, and whose existence we must deny? Such spaces and moments duly threaten a person's internalized glory of society's happy objects in all manner of becomings. They expose the corrupt arbitrariness of hegemonic gendering practices and their absolute social status to jam what Ahmed considers heterocentrism's "straightening device."[77]

Professor Perlman's words are the words I wish that every distressed young adult, who has felt something for another, would hear from their caregiver regardless of where they fall on the sex-gender-sexual orientation-sexuality spectrums. More than that, Perlman's monologue offers the viewers a hopeful glimpse of an ethical masculinity as opposed to its dominant heteropatriarchal forms equipped with toxic, homophobic, and vicious straightening devices. *CMBYN's* professorial parent's emotionally attuned masculinity embodies a mentally wholesome orientation. Through this orientation the film's father figure artfully encourages his son to understand the devastating price of foreclosing on potential queer encounters by saying yes to the fragile fantasy of heterosexuality in each lived experience.[78] Better to be a happy abject than a happy object whose "happiness" is predetermined and vastly limited. The fantasy is not worth it if it means starting with the fractured premise that the promise of disorienting queer moments in love and loss is off the table! What a waste!

Dewey's approach to artfulness helps us understand Professor Perlman's emotionally attentive tone to his son's despair in this important scene. As deep as the rhizomatic tendrils connect with the interlocking impasse, they make way for a new queer identity. This queer identity is less about mastering a manicured aesthetic balance one gets from reading Dewey and more about emergent, precarious rearrangements, and redistributions of delightful perceptions and throbbing pains among multi-agentic, rhythm-chaosmotic encounters. Such queerness emerges on account of an aparallel evolution through "transversal communications of heterogeneous populations" somewhere in North Italy in the *CMBYN* queer sensorium. Everything desires. Everything in the desire of a bloom space: glistening bodies, musical and dancing bodies, eating bodies, peaches, apricots, cherries, fresh water from mountain springs, water bodies, flies, sticky juices, juicy matter, books, the cobble stones of somewhere in North Italy, the bicycles, the green grass, two bodies … making the multiple.

The artful aparallel evolution of such multi-body encounters among the milieus of milieus is precisely the rhizomatic production, the "creative involution" of a new aesthetic-affective unconscious "between and beneath" the intercultural meeting of the two bodies.[79] This new aesthetic-affective unconscious, with its new queer identity, becomes "communicative or contagious" in the dissolution of old aesthetic-affective forms, liberation of "times and speeds," and the issuance of "new statements, different desires."[80] It offers less adjustment to empty happy objects and their straightening devices and more attunement to becoming *happy abjects* via unexpected connections that jolt and surprise breath with the vivaciously threatening limits of *zoē*.

For all of these mysteries of love, I argue that *CMBYN's queer sensorium* embodies the playful aesthetics and everyday politics of an artful rhizome. The film's special affirmation of the sweet and profoundly painful rhythms of desire and attachment particularly shatter the ideal of organic unity of romantic love as a perfectly curated aesthetic experience in which separated lovers triumph all odds and obstacles to be reunited eventually. On the contrary, the film ends in the milieu of all milieus: arresting heartbreak evidenced by an unceasing pain, a silent deluge glistening on Elio's face in the light of the bright fire, quietly accepting what he cannot fight: his burning love for Oliver and the fact that Oliver is engaged to marry another. I cannot think of a better accompaniment to this visual punctum than Hélène Cixous's vulnerary words in the *Book of Promethea*:

> In love not all is love. injustice, anger, hunger, delicate hunger and raging hunger, innocent hungers and cruel hungers proud of being so and ashamed of being so and even prouder … because in love not all is love.[81]

The milieus of love, desire, and attachment are open to chaos that threatens them with exhaustion or intrusion. With destruction. Heartbreak. Disorientation. Rupture. Rhythm is the milieus' answer to chaos. The rhythm of becoming queer, becoming polyglottal, becoming polymorphously perverse, becoming and living as an aerobic wound, becoming nomadic, becoming a happy abject whirling like the dervish worlding with rhythm. Rhythm is landing. Splashdown. Takeoff. Elsewhere. Elsewhen. Rhythm is allthetimewhen everywhere is a bloom space. I back the profundity of *CMBYN's* call to name love by its blithe, burdensome, and incalculable rhythms by citing the final verse of Stevens' *Mystery of Love.* In all manner of becomings:

> *How much sorrow can I take? Blackbird on my shoulder*
>
> *And what difference does it make when this love is over*
>
> *Shall I sleep within your bed, river of unhappiness, Hold my hands upon my head*
>
> *Till I breathe my last breath*
>
> *Oh Woe is me… The last time that you touched me.*
>
> *Oh will wonders ever cease? Blessed be the mystery of love.*[82]

Notes

1 John Dewey, *Art as Experience* (New York: Perigee, 1934), 3–291.
2 Dewey, *Art,* 3–291.
3 Gilles Deleuze and Félix Guattari, *A Thousand Plateaus: Capitalism and Schizophrenia,* trans. Brian Massumi (London: Athlone Press, 1988), 9.
4 Richard Brody, "The Empty, Sanitized Intimacy of 'Call Me by Your Name,'" *The New Yorker,* November 28, 2017.
5 "Call Me by Your Name: Screenplay by James Ivory (based on the novel by André Aciman)," Sony Classics, accessed September 22, 2019, https://sonyclassics.com/awards-information/screenplays/callmebyyourname_screenplay-20171206.pdf
6 "Call Me," Sony Classics.
7 Moshe Sakal, "Sodom and Diaspora—Jewish Identity in 'Call Me by Your Name,'" accessed July 2, 2019, www.intomore.com/you/sodom-and-diasporajewish-identity-in-call-me-by-your-name
8 Donna J. Haraway, "A Cyborg Manifesto: Science, Technology, and Socialist Feminism in the Late Twentieth Century," in *Simians, Cyborgs and Women: The Reinvention of Nature* (New York: Routledge, 1991), 150; Donna J. Haraway, *The Companion Species Manifesto: Dogs, People, and Significant Otherness* (Chicago, IL: Prickly Paradigm Press, 2003), 3.
9 Sara Ahmed, *The Promise of Happiness* (Durham, NC: Duke University Press, 2010), 22; Ahmed's work on *queer phenomenology* and the *politics of good feeling* has gained much traction in the field of communication studies as the field's interest in embodiment and materiality has grown through the critical study of affect and emotions.
10 Deleuze and Guattari, *A Thousand Plateaus,* 18.
11 Karl Schoonover and Rosalind Galt, *Queer Cinema in the World* (Durham, NC: Duke University Press, 2016), 213; Ahmed, *The Promise,* 22.
12 Schoonover and Galt, *Queer,* 213; Ahmed, *The Promise,* 22.
13 Omise'eke Natasha Tinsley, "Black Atlantic, Queer Atlantic: Queer Imaginings of the Middle Passage," *GLQ: A Journal of Lesbian and Gay Studies* 14, nos 2–3 (2008): 199, https://doi.org/10.1215/10642684-2007-030; Catriona Mortimer-Sandilands and Bruce Erickson, eds, *Queer Ecologies: Sex, Nature, Politics, Desire* (Bloomington, IN: Indiana University Press, 2010), 3; Judith Halberstam, "The 'Boys Don't Cry' Debate: The Transgender Gaze in 'Boys Don't Cry' (From Film Director Kimberly Peirce)," *Screen* 42, no. 3 (2001): 294–296.
14 Katie Stewart, "Afterword: Worlding Refrains," in *The Affect Theory Reader,* eds Melissa Gregg and Gregory J. Seigworth (Durham, NC: Duke University Press, 2010), 340.

15 Stewart, "Afterword," 340.

16 Stewart, "Afterword," 340–341.

17 Sara Ahmed, *Queer Phenomenology: Orientations, Objects, Others* (Durham, NC: Duke University Press, 2006), 124.

18 Lauren Gail Berlant, *Cruel Optimism* (Durham, NC: Duke University Press, 2011), 5.

19 Berlant, *Cruel*, 5.

20 Pat J. Gehrke and William M. Keith, eds, *A Century of Communication Studies: The Unfinished Business* (New York: Routledge, 2015), 20.

21 Herman Cohen, *The History of Speech Communication: The Emergence of a Discipline, 1914–1945* (Annandale: Speech Communication Association, 1994), 119–120; Earl Emery Fleischman, "Speech and Progressive Education," *Quarterly Journal of Speech* 27 (1941): 513; Charles H. Woolbert, "The Problem in Pragmatism," *Quarterly Journal of Public Speaking* 2, no. 3 (1916): 264; F. H. Lane, "Action and Emotion in Speaking," *Quarterly Journal of Public Speaking* 2, no. 3 (1916): 228.

22 Pat J. Gehrke, *The Ethics and Politics of Speech: Communication and Rhetoric in the Twentieth Century* (Carbondale, IL: Southern Illinois University Press, 2009), 18–22.

23 A New Materialist reference to the inter-animations of human and non-human world making entanglements; Stewart, "Afterword," 340–44; Helen Palmer and Vicky Hunter, "—Worlding," New Materialism Almanac, March 16, 2018, https://newmaterialism.eu/almanac/w/worlding.html

24 Rosi Braidotti, *Nomadic Subjects: Embodiment and Sexual Difference in Contemporary Feminist Theory* (New York: Columbia University Press, 2011), 109; Deleuze and Guattari, *A Thousand Plateaus*, 18.

25 Connolly, "Materialities," 183.

26 Diana Coole and Samantha Frost, eds, *New Materialisms: Ontology, Agency, and Politics* (Durham, NC: Duke University Press, 2010), 10.

27 Connolly, "Materialities," 182; Thomas Rickert. *Ambient Rhetoric: The Attunement of Rhetorical Beings* (Pittsburgh, PA: University of Pittsburgh Press, 2013), 5–16.

28 Stewart, "Afterword," 340.

29 Lyrics from American singer-songwriter Sufjan Stevens' *Mystery of love* back my play all through this ambient mode of criticism; Rickert, *Ambient*, 16.

30 Rosi Braidotti, "From the Politics of 'Life Itself' and New Ways of Dying," in *New Materialisms: Ontology, Agency, and Politics*, eds Diana Coole and Samantha Frost (Durham, NC: Duke University Press, 2010), 213.

31 Jaishikha Nautiyal, "Becoming a *Detour de Force*: Dehierarchizing Directionality and Mobility in Research," *Women's Studies in Communication* 41 no. 4 (2018): 431. doi: 10.1080/07491409.2018.1551983

32 Tinsley, "Black Atlantic," 199.

33 Mortimer-Sandilands and Erickson, eds, *Queer*, 3.

34 Mortimer-Sandilands and Erickson, eds, 2–3.

35 "Call Me," Sony Classics.

36 "Call Me," Sony Classics.

37 Benedict Spinoza, *Ethics; On the Correction of Understanding*, trans. Andrew Boyle (London: Everyman's Library, 1959), 87; Melissa Gregg and Gregory J. Seigworth, eds, *The Affect Theory Reader* (Durham, NC: Duke University Press, 2010), 3.

38 Ahmed, *The Promise,* 2; Sara Ahmed, "Happy Objects," in *The Affect Theory Reader*, eds, Melissa Gregg and Gregory J. Seigworth (Durham, NC: Duke University Press, 2010), 30.

39 *Call Me by Your Name*, directed by Luca Guadagnino, Sony Pictures Classics, 2018, film.

40 *Call Me,* Guadagnino.

41 Braidotti, "From the Politics," 210.

42 "Call Me," Sony Classics.

43 "Call Me," Sony Classics.

44 Sufjan Stevens, "Mystery of Love," in *Call Me by Your Name*, Sony Classical, 2017, http: spotify.com

45 Henri Lefebvre, *Rhythmanalysis: Space, Time, and Everyday Life*, trans. Stuart Elden and Gerald Moore (New York: Bloomsbury Academic, 2013), 22–27.

46 Dewey, *Art*, 3–4.

47 Deleuze and Guattari, *A Thousand Plateaus*, 313–14.

48 Lefebvre, *Rhythmanalysis*, 41; Deleuze and Guattari, *A Thousand Plateaus*, 313.

49 Deleuze and Guattari, *A Thousand Plateaus*, 313.

50 Deleuze and Guattari, *A Thousand Plateaus*, 314.

51 Deleuze and Guattari, *A Thousand Plateaus*, 314.

52 Connolly, "Materialities," 182.
53 Stevens, "Mystery."
54 Deleuze and Guattari, *A Thousand Plateaus*, 314.
55 "Call Me," Sony Classics.
56 "Call Me," Sony Classics.
57 Dewey, *Art*, 291.
58 Dewey, *Art*, 291.
59 Deleuze and Guattari, *A Thousand Plateaus*, 315–317.
60 "Call Me," Sony Classics.
61 Lefebvre, *Rhythmanalysis*, 68.
62 Anthony Lane, "'Call Me by Your Name: An Erotic Triumph,'" *The New Yorker*, December 4, 2017, www.newyorker.com/magazine/2017/12/04/call-me-by-your-name-an-erotic-triumph
63 "Call Me," Sony Classics.
64 "Call Me," Sony Classics.
65 "Call Me," Sony Classics.
66 Stevens, "Mystery."
67 Braidotti, "From the Politics," 210.
68 Braidotti, *Nomadic*, 29.
69 Braidotti, *Nomadic*, 29.
70 "Call Me," Sony Classics.
71 "Call Me," Sony Classics.
72 Braidotti, *Nomadic*, 29.
73 Braidotti, "From the Politics," 210–213.
74 Ahmed, "Happy," 30.
75 Ahmed, "Happy," 30.
76 "Call Me," Sony Classics.
77 Sara Ahmed, "Orientations: Toward a Queer Phenomenology," *GLQ: A Journal of Lesbian and Gay Studies* 12, no. 4 (2006): 562.
78 Ahmed, "Orientations," 558–560.
79 Deleuze and Guattari, *A Thousand Plateaus*, 18–267.
80 Deleuze and Guattari, *A Thousand Plateaus*, 238–239.
81 Hélène Cixous, *Le Livre de Promethea* [*The Book of Promethea*], trans. Betsy Wing (Lincoln: University of Nebraska Press), 65–66; A reference to Roland Barthes' *Camera Lucida*—*punctum* is the visual aspect of a photograph/cinematic shot that stings and bruises the preceptor (as opposed to the stadium which offers the preceptor a general sense of the visual moment without having a piercing effect like the punctum), 26–27.
82 Stevens, "Mystery."

Bibliography

Ahmed, Sara. "Happy Objects." In *The Affect Theory Reader*, edited by Melissa Gregg and Gregory J. Seigworth, 30, Durham, NC: Duke University Press, 2010.

Ahmed, Sara. "Orientations: Toward a Queer Phenomenology." *GLQ: A Journal of Lesbian and Gay Studies* 12, no. 4 (2006): 543–574. Project MUSE.

Ahmed, Sara. *The Promise of Happiness*. Durham, NC: Duke University Press, 2010.

Ahmed, Sara. *Queer Phenomenology: Orientations, Objects, Others*. Durham, NC: Duke University Press, 2006.

Barthes, Roland. *Camera Lucida: Reflections on Photography*, translated by Richard Howard, New York: Hill and Wang, 1981.

Berlant, Lauren Gail. *Cruel Optimism*. Durham, NC: Duke University Press, 2011.

Braidotti, Rosi. "From the Politics of "Life Itself" and New Ways of Dying." In *New Materialisms: Ontology, Agency, and Politics*, edited by Diana Coole and Samantha Frost, 213, Durham, NC: Duke University Press, 2010.

Braidotti, Rosi. *Nomadic Subjects: Embodiment and Sexual Difference in Contemporary Feminist Theory*. New York: Columbia University Press, 2011.

Brody, Richard. "The Empty, Sanitized Intimacy of "Call Me by Your Name." *The New Yorker*, November 28, 2017.

Call Me by Your Name, directed by Luca Guadagnino, New York: Sony Pictures Classics, 2018.

Cixous, Hélène. *Le Livre de Promethea [The Book of Promethea]*, translated by Betsy Wing, Lincoln: University of Nebraska Press, 1991.

Cohen, Herman. *The History of Speech Communication: The Emergence of a Discipline, 1914–1945*. Annandale: Speech Communication Association, 1994.

Connolly, William E. "Materialities of Experience." In *New Materialisms: Ontology, Agency, and Politics*, edited by Diana Coole and Samantha Frost, 183, Durham, NC: Duke University Press, 2010.

Coole, Diana, and Frost, Samantha, eds. *New Materialisms: Ontology, Agency, and Politics*. Durham, NC: Duke University Press, 2010.

Deleuze, Gilles, and Félix, Guattari. *A Thousand Plateaus: Capitalism and Schizophrenia*, translated by Brian Massumi, London: Athlone Press, 1988.

Dewey, John. *Art as Experience*. New York: Perigee, 1934.

Gehrke, Pat J. *The Ethics and Politics of Speech: Communication and Rhetoric in the Twentieth Century*. Carbondale, IL: Southern Illinois University Press, 2009.

Gehrke, Pat J., and Keith, William M, eds. *A Century of Communication Studies: The Unfinished Business*. New York: Routledge, 2015.

Gregg, Melissa, and Seigworth, Gregory J, eds. *The Affect Theory Reader*. Durham, NC: Duke University Press, 2010.

Halberstam, J. "The 'Boys Don't Cry' Debate: The Transgender Gaze in 'Boys Don't Cry' (From Film Director Kimberly Peirce)." *Screen* 42, no. 3 (2001): 294–98.

Haraway, Donna J. "A Cyborg Manifesto: Science, Technology, and Socialist Feminism in the Late Twentieth Century." In *Simians, Cyborgs and Women: The Reinvention of Nature*, 150, New York: Routledge, 1991.

Haraway, Donna J. *The Companion Species Manifesto: Dogs, People, and Significant Otherness*. Chicago, IL: Prickly Paradigm Press, 2003.

Lane, Anthony. "Call Me by Your Name: An Erotic Triumph." *The New Yorker*. December 4, 2017. www.newyorker.com/magazine/2017/12/04/call-me-by-your-name-an-erotic-triumph

Lane, F. H. "Action and Emotion in Speaking." *Quarterly Journal of Public Speaking* 2, no. 3 (1916): 221–228.

Lefebvre, Henri. *Rhythmanalysis: Space, Time, and Everyday Life*, translated by Stuart Elden and Gerald Moore, New York: Bloomsbury Academic, 2013.

Mortimer-Sandilands, Catriona, and Erickson, Bruce, eds. *Queer Ecologies: Sex, Nature, Politics, Desire*. Bloomington, IN: Indiana University Press, 2010.

Nautiyal, Jaishikha. "Becoming a *Detour de Force*: Dehierarchizing Directionality and Mobility in Research." *Women's Studies in Communication* 41, no. 4 (2018): 430–440. doi: 10.1080/07491409.2018.1551983

Palmer, Helen, and Hunter, Vicky. "—Worlding." New Materialist Almanac. March 16, 2018. https://newmaterialism.eu/almanac/w/worlding.html

Rickert, Thomas. *Ambient Rhetoric: The Attunement of Rhetorical Beings*. Pittsburgh, PA: University of Pittsburgh Press, 2013.

Sakal, Moshe. Sodom and Diaspora—Jewish Identity in 'Call Me by Your Name'. Accessed July 2, 2019. www.intomore.com/you/sodom-and-diasporajewish-identity-in-call-me-by-your-name

Schoonover, Karl, and Galt, Rosalind. *Queer Cinema in the World*. Durham, NC: Duke University Press, 2016.

Sony Classics. "Call Me by Your Name: Screenplay by James Ivory (based on the novel by André Aciman)." Accessed September 22, 2019. https://sonyclassics.com/awards-information/screenplays/callmebyyourname_screenplay-20171206.pdf.

Spinoza, Benedict. *Ethics; On the Correction of Understanding*, translated by Andrew Boyle, London: Everyman's Library, 1959.

Stevens, Sufjan. "Mystery of Love." In *Call Me by Your Name: Original Motion Picture Soundtrack*. Sony Classical, 2017, http://spotify.com.

Stewart, Katie. "Afterworld: Worlding Refrains." In *The Affect Theory Reader*, edited by Melissa Gregg and Gregory J. Seigworth, 340–344, Durham, NC: Duke University Press, 2010.

Tinsley, Omise'eke Natasha. "Black Atlantic, Queer Atlantic: Queer Imaginings of the Middle Passage." *GLQ: A Journal of Lesbian and Gay Studies* 14, no. 2–3 (2008): 191–215. doi: https://doi.org/10.1215/10642684-2007-030.

Woolbert, Charles H. "The Problem in Pragmatism." *Quarterly Journal of Public Speaking* 2, no. 3 (1916): 264–274.

14

FEMINIST AND QUEER ARTS ACTIVISM

Clare Johnson

In the summer of 2016, the Guerrilla Girls, a group of anonymous feminist activists who formed in 1985, wrote to the directors of 383 museums and galleries across Europe to ask them to respond to 14 questions about diversity in their collections and exhibition programs. The questions concerned the number of artists included in recent exhibitions who are women, gender nonconforming, or from Africa, Asia, South Asia, and South America.[1] The responses formed the basis of a campaign exhibited at the Whitechapel Gallery in London entitled "Is It Even Worse in Europe?" The Guerrilla Girls are well known for their use of humor and bold information graphics, and this campaign featured a banner on the outside of the gallery declaring that only one quarter of the museum directors responded. It invited people to come into the gallery to find out more, including the names of the institutions that did not respond. The ongoing work of the Guerrilla Girls makes clear the need not only to challenge the exclusion of particular groups from the power brokers of the artworld but also to find structurally inclusive alternatives to the canon, which celebrates the importance of some artists and shamelessly ignores the contributions of others.

The story told of feminist thought and its relation to art is often linear and chronological, a succession of practices that leads us to an enlightened present. This chimes with what Clare Hemmings describes as a persistent narrative of progress or loss in accounts of Western second wave feminist theory.[2] Indeed, in the early stages of drafting this chapter, I was tempted to present queer approaches to artmaking as the endpoint in a narrative of progress from Brechtian-inspired feminist strategies of the 1970s and 1980s, such as distanciation and defamiliarization, to conceptually sophisticated queer strategies of the 1990s and beyond, characterized by an understanding of identities as multiple, fluid, and provisional. However, this would have been problematic for a number of reasons. It would suggest a knowable topology of types of art making, which imposes order on the rich messiness of feminist and queer art practices at any one moment in time. It would also risk de-contextualizing strategies of art making that seem problematic, or even naïve, when read through contemporary concerns, but that were crucial at the time. These include references to a shared female biology, such as Judy Chicago and Miriam Schapiro's celebration of "central core" imagery in the 1970s. This was art about female subjectivity in which motifs of orifices were used to articulate an essence shared by women.[3] This form of artmaking was heavily critiqued for its apparent essentialism and polarization of sexual difference long before queer theory challenged binaries such as female/male and the alignment of these terms with feminine/masculine characteristics.[4] Furthermore, it would anchor artworks in their historical moment of production, which limits potential interpretations and strategic uses.

Instead, I argue for a historiographic approach to thinking about feminist and queer arts activism. This is less concerned with masculinist notions of precedent, origin, and influence and more to do with overlapping temporalities and provisional affiliations. I do not, therefore, offer a chronology of feminist and queer arts/activist practices, preferring instead to signal affinities between works produced in different contexts and historical moments. This includes the possibility that feminist strategies employed in earlier works can be reimagined through the lens of contemporary works in an approach akin to what Miele Bal calls "preposterous history."[5] Indeed, to embrace the entanglement of past-present practices and a nonlinear approach is to queer historical notions of sequence and precedent, where "to queer" is understood as a verb rather than a noun.[6] This matters because the naturalization of linear history as "common sense" has resulted in a canon that has routinely excluded work by women and LGBTQI artists among others.

The role of feminist and queer artists is crucial to disrupting the canon. Indeed, the "movement" associated with feminist art during the 1970s and 1980s grew out of an urgent need to challenge the male-dominated art historical canon and its associated critical discourses. My working definition of "feminist art" includes those practices that challenge the production of woman as commodity, patriarchal expectations of gendered behavior and affective repertoire, but does not prescribe what this looks like in the making of art. In detaching this identification from specific artists, as if an artist is either feminist or not, and allowing it to stick to practices, a more diverse range of arts activity can be understood as feminist in sensibility, if not obviously in subject matter. The very categorization needed to make sense of the field is itself called into question by some of the artworks associated with it. Similarly, to "define" queer art may seem antithetical to a set of practices that purposely embrace ambiguity as a radical force. Nevertheless, this is work that disrupts the naturalization of heteronormative categories, rituals, and assumptions. It is work that values alterity, protects otherness, and uses the threat that this poses to shift the social fabric of everyday life. In this sense, queer is understood as a *position* rather than an identity and is not only aligned with artists who identify as queer.

The idea of agitating for change is central, but the field is characterized by a necessary refusal to settle on established ways of thinking, acting, and making work. Indeed, the premises upon which feminist-identified and queer-identified arts activism are based are not always aligned. There is a tension between a feminist art making that takes its cues from the notion of identity through, for example, denaturalizing the construction of woman in visual images and the need to address the myriad ways in which women are excluded from art historical discourses, and a queer politics premised on the disruption of singular and stable forms of identity. From a queer perspective of gender fluidity and what Amelia Jones calls "radical undecidability," there is a danger that feminism reinstates essentialism, despite a raft of art practices of the 1970s and 1980s that aimed to expose the social and cultural production of women, because it necessarily relies on *identifying as a woman*.[7] Here I do not intend to smooth over such tensions. Indeed, the complexity of the context in which feminist and queer art is both produced and consumed needs to be acknowledged.

In what follows, I adopt an intergenerational approach to the work of Tanja Ostojić and Hannah Wilke, and then to works by Adrian Howells, Gillian Wearing, and Louise Bourgeois. The method I employ is intended to signal the queering of canonical linear histories and to keep alive the rich array of activist tactics used in arts practice. Critical arts practice shares with communication research a desire to interrogate institutions that shape our cultural consciousness.

It aims to intervene to expose the social, political, and economic mechanisms that reproduce unethical power relations. Indeed, communication research is part of an interdisciplinary debate about the construction and politics of knowledge, which overlaps with the subversive impulse of critical arts practice, in particular an imperative to work towards cultural change. The specific strategies I argue for are *provocation, critical mimicry,* and *emotional ambiguity*. Using these strategies, the artworks and performances discussed in this chapter aim to transform attitudes, behaviors, politics, and imagination on a number of issues including nationhood, immigration, intimacy, domesticity, and maternal ambivalence. My choice of artworks reflects what I consider to be some of the most pressing issues for feminist and queer art making at this point in time. In the context of widespread right-wing populism in Europe and the United States, for example, Ostojić's refusal to accept the entrenching of nationalistic and paternalistic imperatives in the artworld matters a great deal. The need to embrace uncertainty and ambiguity in a time of dangerous absolutism in international politics motivates my choice of Howell's performance work and Wearing's film. In both cases the stability of a fixed emotional state is challenged along with the assumption that stability is even desirable. The equation of political strength with an unchanging vision has been satirized by British artist Jeremy Deller. In 2017 typographic posters designed by Deller appeared across London reading "Strong and stable my arse," which refers to a phrase used repeatedly by former Conservative party leader and Prime Minister Theresa May. Wearing and Howells' work predates May's government, but the tyranny of certainty ridiculed by Deller is a reminder of the importance of artworks that refuse to equate political strength with absolute certainty. The choice to pair Ostojić's work with Wilke's, and Wearing's with Bourgeois', is designed to rebuke the tendency to deny contemporary female artists their historiographic affinities with earlier women artists. In the context of #MeToo and #Time'sUp, it is essential that women feel connected, not by shared biology or by assumptions about how they live their lives, but via an expanded historical awareness of feminist struggle including resonances between women artists of different generations.

The artworks I focus on differ in many ways. Some are designed to be seen by many, others are intended for an audience of one. They are made using different media (performance, photography, film, sculpture) and do not all take place within a gallery setting. The means of communication is not incidental or supplementary, but central to the artworks' capacity (or not) to challenge the reproduction of social realities. The artworks were produced at different moments in time and do not all share the same kind of identification with feminist and/or queer politics. What they do share, however, is a commitment to change and a desire to realize a transformative politics of visuality. I have deliberately chosen artworks that do not espouse a direct message, as if there is an artist who can see the light and a viewer who needs to be enlightened. In this respect, I am arguing for the significance of artworks that do not wear their critical credentials as a badge on their sleeve. Amelia Jones has argued that feminist art criticism was for a long time dominated by Brechtian avant-gardist strategies designed to increase the agency of the viewer by creating critical distance; a critique of realism in which the spectator's political faculties are heightened by a refusal to seduce the viewer.[8] Such tactics can be empowering, but they are problematic in that they assume that some people have the political awareness to see beyond the illusions of representation and others are in need of critical help. It is a hierarchical form of feminism, which ultimately does little to change the structure of power relations. The works discussed here are far less didactic in their approach, preferring to create an atmosphere in which change can take place by engaging the viewer/participant as an embodied agent. This strategy is exemplified by the work of Tanja Ostojić, which is discussed in the next section. As an approach it packs its political punch by seducing, rather than distancing, the viewer, only to then subvert their expectations of the field of vision.

Provocation and the politics of exclusion

In December 2005, a poster artwork called *Untitled/After Courbet* by Yugoslavian-born feminist artist and activist Tanja Ostojić was displayed on rotating billboards in Vienna as part of an exhibition called EuroPart (Figure 14.1). The artwork, originally made in 2004, had been chosen by curators for the public exhibition, which was timed to coincide with the Austrian prime minister taking over the presidency of the EU. However, the poster was removed by the exhibition curators two days later in response to a media scandal about the work, along with another artwork by Spanish artist Carlos Aires. Once the artworks had been removed from several billboards in Vienna, a larger poster was erected on the facade of the Forum StadtPark in Graz in defiance of censorship and in solidarity with the artist. The poster measures 3.5 × 4 meters and is a photograph of a reclining women's torso posed and cropped to reference Gustav Courbet's *L'origine du Monde* (The Origin of the World, 1866). Where Courbet's nude leads the viewer's gaze to the model's pubic area, in Ostojić's poster the woman is wearing blue underwear depicting the 12 stars of the European Union flag. The particular shade of blue is unmistakably that of the EU flag, and the stars are positioned at the center of the viewer's line of sight. The work is a critical commentary on the EU's immigration policies, which often require women

Figure 14.1 Tanja Ostojić, *Untitled/After Courbet (L'origine du monde)* (2004)

Colour photo, 46 × 55 cm
Photo: David Rych
© Ostojić/Rych

from southeastern Europe to marry an EU citizen to gain entry. The poster is also a disturbing reminder of the sexual economy of trafficking women from Eastern Europe and elsewhere who are forced to enter Austria and other EU countries illegally to work in Western Europe as sex workers, domestic servants, and slaves. Far from celebrating Austria's EU presidency, the poster critically examines immigration policies, border regimes, and European biopolitics from a radical feminist position grounded in the desire to subvert structural power relations.

Ostojić has a history of challenging what she understands as the arrogance of EU policies with respect to southeastern European women. In 2000 to 2005, she developed a complex five-year-long project involving an online performance called *Looking for a Husband with an EU Passport*. Again, employing the language of advertising, she posted an image of herself online, naked and shaven, bringing to mind troubling images of prison camp inmates, and asked for potential suitors to contact her. This action resulted in a legal marriage to fellow artist Klemens Golf, from whom she later divorced, and a marriage visa that enabled her to reside in Germany. It was explicitly a marriage of convenience designed to circumvent difficulties in obtaining a visa to live in Western Europe, where her work was increasingly gaining recognition. Working in a range of media forms, Ostojić operates from the point of view of the migrant woman to critique hierarchies of power within the Western artworld and the position of women within contemporary power structures.

Untitled/After Courbet was met with public outrage and accusations that it offended Austrian public morality. However, if the Austrian authorities thought that removing the poster would deny it visibility, they were sorely mistaken. The media furor surrounding the work's censorship resulted in over 100 articles being written and more than 1,000 reader comments on the situation.[9] Without needing any text to anchor the image, Ostojić exposed the hypocrisy of an aesthetic economy in which this poster is denounced as pornographic and media images that reveal far more of a woman's body pass without comment.[10] It is not the woman's body on its own that causes offense so much as the visualization of a libidinal economics of nationhood and trafficking. Bojana Videkanić argues that *After Courbet*'s formal characteristics, such as its size, cropping of the body, and art historical reference, are highly affective.[11] Drawing on Brian Massumi's work on affect, Videkanić argues that these elements "create zones of intensity that, in turn, function in our own in-between spaces, spaces where our cognitive side has not yet understood what the body has already absorbed."[12]

This affective intensity is due, in part, not just to the size of the work but to its scale. Billboard posters are designed to be seen from the car, their mode of consumption linked to notions of mobility and the movement of people and goods. However, Ostojić's poster is connected to movement that is illegal, enforced, and disempowering. The artist presents her own body as a commodity, inviting access to her fragmented part-body, only to thwart the transaction both visually and politically through her use of the underwear. Ostojić controls access to her body in a critical revisioning of the biopolitics that renders many Eastern European women subservient to the demands of traffickers. As a visual image, she refuses the breaching of her bodily border and evokes the politics of exclusion within EU immigration policies and those of the international artworld.

In the context of her wider body of work, Ostojić's practice relates the exclusion of women from positions of power within the artworld to the transnational issues of maltreatment and subjugation examined in *After Courbet*. Using the internationalization of the artworld, in the form of art fairs and biennials, as a landscape for intervention, Ostojić refuses to let the artworld establishment absolve itself of its responsibility for an aesthetic economy in which women are routinely denied privilege. In a previous work entitled *I'll Be Your Angel* (2001–2002), Ostojić accompanied the curator Harold Szeemann during the opening days of the 49th Venice Biennale, effectively

Figure 14.2   Tanja Ostojić, *I'll Be Your Angel* (2001)

Four-day performance with Harald Szeemann, *Plato of Humankind*, 49th Venice Biennale.
Photo: Borut Krajnc.
Courtesy: Tanja Ostojić

escorting him to cocktail parties, dinners, and press conferences (Figure 14.2). The Biennale is an international art fair attended by artists, curators, tourists, and financiers, and by performing as his younger female artist/muse, Ostojić shone a light on Szeemann's elevated position within the global artworld. The primary form of communication used here is behavioral. Without asserting a message or trying to persuade anyone of her position, Ostojić's acquiescence offered Szeemann up as a powerful figure and, consequently, highlighted gendered inequities in the process. Hers is an art of provocation and irony, integrated into the processes of everyday life.

Ostojić's strategy includes an ongoing dynamic between hiding and exposing. Szeemann refused to allow the publication of Ostojić's diary of the event, censoring part of her text inside the official Venice Biennale catalogue in 2001. The artist did, however, publish this subsequently in her books *Venice Diary* (2002) and *Strategies of Success/Curator Series* (2004) in addition to her video/video installations *I'll Be Your Angel* and *Strategies of Success*. Szeemann could not, however, disallow her commentary on his decision. *I'll Be Your Angel* included the artist's own decision to hide a conceptual artwork called *Black Square on White* (2001), in which her pubic hair was trimmed into a square. It was a reference to the work of Russian avant-garde artist Kazimir Malevich and remained hidden from sight for all except Szeemann. The Austrian authority's attempt to hide *After Courbet* from public view resonates with the treatment of the Courbet painting on which it is based. It is thought that *L'origine du Monde* was commissioned by Turkish-Egyptian diplomat Khalil-Bey, but it is not known what happened to the painting before its acquisition by the Musée d'Orsay, Paris, in 1995, at which point it was owned by the French psychoanalyst Jacques Lacan.[13] It was an audacious representation of a female nude, but despite its notoriety was rarely seen. Like Ostojić's recasting of it in relation to European border politics, its lack of visibility had the effect of amplifying its impact.

This section has focused on a reading of Ostojić's work, which emphasizes the disarming and provocative ways in which her practice interrogates a transnational politics of visibility. Ultimately, Ostojić asks who is allowed to officially exist, what they must do to gain recognition, and the price of their visibility. Her work chimes with the imperative of communications research to reveal the conditions of existence within which some voices are empowered, others are silenced, and some are rendered unimaginable. In the next section I position the tactics employed by Ostojić in relation to an earlier feminist artist, Hannah Wilke. My intention is to historicize Ostojić's practice, not by establishing a sequential line of precedence, but by tracing a thread of criticality shared across generations.

Critical mimicry and performance

Ostojić uses her *performance* of being pleasant, anodyne, and pleasing to strategic effect, mimicking the mannerisms of the flirtatious muse who never leaves the side of her older, more powerful master. The photographs of this performance show her in a low-cut dress gazing adorably at Szeemann, taking his arm or looking up at him while he speaks. The irony of the work is that by presenting herself so deliberately in this light—objectified, fetishized, and as an adjunct to his power—she turns Szeemann into the object of our gaze. There is a distinct difference in body language between the two with her ease and appearance of pliability emphasizing his discomfort. In this section I situate Ostojić's approach in relation to works of the mid-1970s by American artist Hannah Wilke, specifically focusing on the cross-generational use of anti-essentialist strategies such as performance and mimicry, which have gained critical visibility since the 1990s through queer theory.

The satirical performance of femininity in *I'll Be Your Angel* chimes with earlier feminist artworks such as Hannah Wilke's performances and performance photographs of the mid-1970s. In a number of works including *Starification Object Series* (1974–1975), *Hello Boys* (1975) and *Hannah Wilke Through the Large Glass* (1976), Wilke uses the poses and gestures of compliant sexuality to expose deeply entrenched inequalities. Like Ostojić, in *Through the Large Glass* Wilke critiqued both the commodification of woman as objectified muse/model and the art institution that perpetuates this power structure. In the video of this performance Wilke appears dressed in a white suit and fedora and enacts a striptease standing behind Marcel Duchamp's cracked sculpture *The Bride Stripped Bare by Her Bachelors, Even (The Large Glass)*. The performance took place at the Philadelphia Museum of Art in 1976, and *Through the Large Glass* documents this act.

In the mid-1970s Wilke made a number of works that she called performalist self-portraits. *Starification Object Series* (1974–1975) includes a set of black-and-white photographs featuring Wilke posing in various states of undress to mimic the representation of a range of feminine types. In some of the photographs she uses props such as hair rollers, a cowboy hat, and a men's tie, and in each there are small chewing gum sculptures carefully placed on her body. Multiple forms of semiotic communication are at play, including the signifying potential of gestures, poses, and props. However, this is combined with an embodied approach in which viewers are implicated in different forms of exchange both visually and literally. The work was originally a performance in which Wilke asked audience members to chew the gum, which she then molded into small one-fold sculptures and attached to her body. As "scars" that interrupt the surface of her body, the gum sculptures punctuate each photograph and thwart the scopophilic transaction that the images appear to promise. Where Ostojić comments on the breaching of both bodily and national borders in *After Courbet*, Wilke uses gum to deny the viewer unbridled access to her body. The photographs are the well-known element of this multidimensional work, which has existed as a performance and installation including vitrines containing the gum sculptures. As "performalist" self-portraits, they are works in and of themselves (individually and as a group), as opposed to stills from a filmed performance.

Critical commentary on *S.O.S.* has often focused on Wilke's beauty. The critic Lucy Lippard famously derided Wilke for the "confusion of her roles as beautiful woman and artist."[14] Since Wilke's death from lymphoma in 1993, her performalist self-portraits have been reevaluated in the light of her late works, such as the *Intra-Venus Project*, in which she continued to focus on her body as it gradually succumbed to disease. In the mid-1970s critics could not read Wilke's use of her own body as strategic and she was considered too flirtatious to be feminist. The idea that the job of feminism was to *transform* femininity rendered artworks that operated within the parameters of normative sexualized femininity too pleasurable to be political. However, in more recent readings, which are often influenced by queer theory's focus on anti-essentialist perform- ance, the transformative potential of the pose takes center stage.[15] Read through Ostojić's tactical use of compliant femininity in *I'll Be Your Angel*, Wilke's flirtations with the camera can clearly be identified as performances. The two works share an emphasis on the *practicing* of gestures and poses that constitute the feminine. By drawing attention to the work involved in practicing this body language, they expose the myth of "natural" femininity. Furthermore, by reiterating and rehearsing hetero-sexualized femininity only to expose its existence as performance, they enact an Irigarayan critical mimicry in which subordination is turned into activism. The Belgian-born French feminist, philosopher, linguist, and psychoanalyst Luce Irigaray argued, "One must assume the feminine role deliberately. Which means already to convert a form of subordination into an affirmation, and thus to begin to thwart it."[16] It is a disarming strategy because it so closely resembles the normative femininity that it ultimately distorts.[17]

In considering *I'll Be Your Angel* and *S.O.S.* alongside one another—as affinity rather than precedent—I cannot help but wonder how Ostojić managed to sustain her performance of adoration for Szeemann when encountering people in the unstructured spaces of the Venice Biennale. The mutual consideration of the works throws into sharp relief the labor involved in performing in unpredictable environments. In her wider body of work Wilke employed a number of queer strategies, which are recast by Ostojić in relation to EU border politics and artworld power relationships, such as the tactical significance of performing rather than being, a fluid approach to identification (flirt, feminist, artist, American, Jew, daughter), and the radical potential of seduction and desire.

Both Wilke and Ostojić developed practices that are queer in position, if not in identification. Further, the tactics used by Ostojić have a history within women's art practice. Looking back at Wilke's work, through the lens of Ostojić's provocative performance, enables a historical relation of affinity rather than succession. The relationship is not one of canon formation, but cross- generational resonance, which lends each historical significance without the masculinist burdens of influence and precedence.

In the next section, I discuss a different kind of performance work, which picks up on queer theory's critique of dichotomous logic in order to create affective vulnerability, thus refusing the certainty of artworks that declare themselves as resistant. Where Ostojić's work creates an emo- tional connection in which to draw a viewer/participant, only to thwart their expectations using a strategy of critical mimesis, the works discussed in the final section use a different strategy to shift relations of power, namely the construction of emotional ambiguity.

Emotional ambiguity and the difficulty of intimacy

Here I read works by Adrian Howells, Gillian Wearing, and Louise Bourgeois in terms of emotional fragility and argue that their political power lies not in delivering a message, but in producing feelings of uncertainty and ambiguity, which cannot easily be ignored. The works are made using different media (performance, film, and sculpture, respectively) but share a concern with fragile intimacy. As such, their primary mode of communication is affective, often provoking uncomfortable feelings without necessarily resolving any resulting tension. Viewers and participants are implicated in ways that risk their own unease and feelings of discomfort.

In her brilliant book on difficulty and emotion in contemporary art, Jennifer Doyle discusses a number of artworks that can make people feel uncomfortable in ways that are deeply personal and potentially intimate.[18] In the opening pages she confronts her own feelings about missing an appointment to experience a performance-for-one with the British performance artist Adrian Howells. As a proponent of "intimate theatre" and one-on-one performance encounters, Howells' work dealt with deeply personal, confessional, and sometimes autobiographical experiences. This is a form of performance art in which there is no possibility of voyeuristic detachment. As Dee Heddon argues: "In this form of performance practice—intimate, personal and interactive—the boundary between performer and spectator dissolves in the process of exchange, an exchange that asks for a very committed, and at times vulnerable, sort of spectatorship."[19]

Howells produced many works that embodied vulnerability and care until his untimely death in 2014. Participants for these performance works were invited to share in ritualized and caring encounters such as having their feet washed and massaged (*Foot Washing for the Sole,* 2010), being bathed, fed and cradled (*The Pleasure of Being: Washing, Feeding, Holding,* 2011) and having their dirty laundry cleaned, both literally and metaphorically, by Howells' quasi-drag performance persona Adrienne (*Adrienne's Dirty Laundry Experience,* 2003). The performance that Doyle discusses is called *Held* and was staged in Glasgow, Scotland, in 2006 and in London in 2007. Howells met his audience of one in an apartment and interacted with them in three scenarios. During the first encounter the artist and his guest held hands, drank tea, and talked while sitting at a kitchen table. In the second encounter they sat on the sofa together, held hands, and watched television. Finally, they went to a bedroom, lay down on a bed, and spooned in a physical demonstration of intimacy. Doyle recounts subconsciously sabotaging her opportunity to participate in *Held* by booking a hair appointment on the same day knowing that she had not left enough time to cross London and get to the venue for the encounter. Her shame is exacerbated by canceling so late that nobody else could take her place and then traveling to the location to apologize to the artist in the hope that he would forgive her. In reflecting on this experience, Doyle articulates the relationship between intimacy and control: "I managed to extract the caretaking that Howells offered within the boundaries of *Held* but outside the boundaries of the event. I insisted on getting what the artist had promised me, but on my own terms."[20]

Reading Doyle's account of this experience I am reminded of an event at the Arnolfini gallery in Bristol, UK, in 2007 when Doyle was in conversation with the Italian-born performance artist Franko B. During the event Franko B described at length his piece *I Miss You,* which was performed at Tate Modern, London, in 2003 as part of the Tate's "Live Culture" program. The performance featured the artist walking down a catwalk covered in white fabric and illuminated by strobe lights, his naked body covered in white paint. As Franko walked, blood dripped from cannulas positioned in both elbows until the fabric was splattered with his blood. During his description of *I Miss You,* I started to feel unwell, finding the mental image of spilt blood and physical endurance nauseating. Feeling trapped in the middle of a row of seats in the auditorium, I stuck it out because to leave would have felt like an admission of failure that I could not stomach difficult performance art, even in description only. Doyle describes her stamina for this kind of difficulty in experiencing the queer performances of artists such as Ron Athey and Bob Flanagan, as well as Franko B, which often involve violence to the body and overt sexualization. However, the private, domestic sphere that Howells deploys in his performances turns out to be utterly disarming in ways that are, for Doyle, far more disconcerting and difficult than more visceral and explicit forms of queer performance art.

Doyle's failure to attend her appointment with Howells concerns the difficulty of intimacy, the idea that tenderness and care cannot be guaranteed. Howells worked hard to create a feeling of safety for his participants, but the risk that this may not endure in a one-to-one relationship has been explored by other artists working in different media forms. The fragility of care is the subject of Gillian Wearing's short film *Sacha and Mum* (1996). Running at 4 minutes 30 seconds, Wearing's film depicts the emotionally charged relationship between a mother and her adult daughter. The film is shot in black and white and, unlike Howells' use of non-art spaces, is displayed as a large projection or on a monitor within a gallery space. Both parts are played by actresses who perform a psychological and physical struggle, which ebbs and flows throughout the piece. In the domestic setting of a bedroom, the mother and daughter initially smile at each other and embrace in a demonstration of intimacy, tenderness and care. There is, however, something strange about the encounter and the daughter appears peculiarly vulnerable. The power imbalance in this maternal relation is implicit from the outset because the daughter wears only her underwear. She is positioned as an adult-child in contrast to her mother's conservative attire. The feeling of unease that this creates is justified as a struggle ensues and the mother pulls the daughter's hair, moving her head back and forth in a display of maternal aggression. At points, the daughter is kneeling with her head pushed to the floor. She appears strangely complicit given that her mother's movements are less agile than her own. The viewer is left wondering why she does not break free. The feeling that things are not what they initially seem is heightened by the use of sound, which is amplified speech and played in reverse. This is, like the imagery, difficult to comprehend and is exacerbated by slightly speeded up video and the circling motion of the camera, which further heightens the sense of entanglement and confusion.

Wearing's film shares the connection between intimacy and control experienced by Doyle in her failure to participate in *Held*. However, the emotional repertoire of *Sacha and Mum* is more extreme. The maternal relation in this work hovers on the border of love and hate. It is a highly unusual visualization of what Rozsika Parker calls "maternal ambivalence" in which motherhood is experienced as both pleasure and pain.[21] Parker challenges the shame attached to such feelings by exposing the cultural invisibility of maternal ambivalence. Furthermore, she argues that the coexistence of loving and hating maternal feelings has a productive outcome because it sharpens a mother's understanding of her relationship with her child. Her argument revised psychoanalytic readings of maternal ambivalence and contributed to feminist research methodologies by studying this issue from the perspective of the mother rather than the child. Parker's use of the word "hate," as opposed to softer emotions, such as frustration or even anger, is particularly emotive. In an interview with Melissa Ben, she explained that she did consider other terms, but ultimately "nothing quite seemed to capture the raw feelings that so many parents have as 'hate.'"[22] It is the possibility of maternal hatred, rather than dislike, that the viewer is asked to confront in *Sacha and Mum*. The dynamic within Wearing's film is both affectionate and cruel, loving and violent. Furthermore, the rhythm of the film is such that the transitions flow both ways not only from affection to violence but back again in a circling of emotion and dependency.

The conflicting emotions of Wearing's film resonate with Louise Bourgeois' maternal works in particular *Maman* (1999), despite the use of an entirely different medium and scale. The story of Bourgeois' inclusion into the art historical canon is as fascinating as it is problematic. She was finally offered a retrospective at New York's Museum of Modern Art in 1982 when she was 70 years old. The catalogue that accompanied this exhibition was the first monograph detailing her extensive oeuvre. In 1993, Bourgeois was selected to represent the United States at the Venice Biennale, and by 2000, she had been selected to show at the opening of Tate Modern, London. Bourgeois produced many paintings, drawings, and prints in the 1940s and appeared in group shows along with Jackson Pollock, Adolph Gottlieb, Robert Motherwell, Mark Rothko, and others. However, it wasn't until the early 1970s that her work found a favorable context in the women's art movement, which emerged from the rise of second-wave feminism. By the late 1970s, a renewed interest in content and meaning, rather than formal properties, increased the appeal of her references to memory and family life. The emergence of revisionist histories of women's art, as well as interest in psychoanalytical readings of art in the 1980s, helped to increase the visibility of Bourgeois' practice. Deborah Wye points out that during the 1950s, 1960s, and 1970s Bourgeois was known to the New York art audience but not to a wider public.[23] Lucy Lippard put it more forcefully when she said in 1975 that:

> Despite her apparent fragility, Louise Bourgeois is an artist, and a woman artist, who has survived almost 40 years of discrimination, struggle, intermittent success and neglect, in New York's gladiatorial art arenas. The tensions which make her work unique are forged between just those poles of tenacity and vulnerability.[24]

The appalling exclusion of Bourgeois' work in art historical discourses for so many years is in stark contrast to Wearing's recognition at an earlier stage in her career as part of the Young British Artists movement. Wearing won the Turner prize in 1997 for *Signs That Say What You Want Them to Say and Not Signs That Say What Someone Else Wants You to Say* (1992–1993), a work in which she asked people to write what they were feeling on a card and photographed them holding this up. However, the two artists share a strategy of depicting emotional ambiguity, which crosses generations, techniques, and materials. A chronological account of artists engaging with feminist issues (if not necessarily of feminist artists) would place Bourgeois before Wearing, one generation following the other. This line of descent would not, however, give us the whole story because Wearing made *Sacha and Mum* before Bourgeois produced some of her late maternal works. Bourgeois returned to the themes of pregnancy and fertility in her 80s and 90s, producing *Maman* when she was 88 years old. Within Bourgeois' practice, there are multiple returns to earlier life stages and a playful attitude toward temporality. In her discussion of Bourgeois' work on the maternal subject, Rosemary Betterton argues that "Bourgeois' art is constantly informed by returns to her past in repeated and contradictory ways that refute the concept of a whole and singular self-bound by a chronological narrative."[25] In this respect a chronology of artworks, rather than artists, may be more helpful.

Maman stands at 36 feet high and is a sculpture of a spider constructed from steel and marble (Figure 14.3). The enormity of Bourgeois' sculpture means it can only be installed outside or in particularly large industrial buildings. It was part of the inaugural exhibition at Tate Modern's Turbine Hall in 2000 and was later displayed outside the gallery as part of a major retrospective of Bourgeois' work in 2007. *Maman* translates as "mummy" and for Bourgeois the spider is a maternal figure. The viewer can walk underneath the sculpture and look up at the egg sac, which contains 17 white and gray marble eggs. The spider's eight steel legs are simultaneously enormous and spindly. They are substantial and thick toward the top but pointy at the ends, as if the spider is walking on tiptoes. The sculpture is made of hard materials yet represents a fiercely protective form of nurturing. The spider as mother is terrifying yet caring, strong but vulnerable.

Figure 14.3   Louise Bourgeois, *Maman* (1999)

© The Easton Foundation/VAGA at ARS, NY, and DACS, London 2020

The emotional push and pull of Wearing's *Sacha and Mum* brings into focus the uncomfortable tension in *Maman* between the maternal spider as menacing, frightening, overwhelming (at this scale), overbearing, and hard on the one hand and enveloping, protective, precariously balanced, and fragile on the other. Elizabeth Manchester writes of the ambiguity of the maternal metaphor in this work when she says,

> Encountering *Maman* always from the perspective of the child looking up from below, the viewer may experience the sculpture as an expression of anxiety about a mother who is universal—powerful and terrifying, beautiful and, without eyes to look or a head to think, curiously indifferent.[26]

Maman was made with reference to the artist's own mother and like much of Bourgeois' practice evokes painful childhood memories of familial distress. It shares with *Sacha and Mum* the psychological childhood fear of a murderous mother; the anxiety that a mother's enveloping embrace can turn into something else. At various points in *Sacha and Mum,* the daughter's face is covered with a towel and the viewer is left unsure if this act is loving or cruel. The mother's actions hover between trying to calm a disturbed child and attempting to suffocate her. The activism of *Maman* and *Sacha and Mum* lies in their powerful depiction of emotional ambiguity and refusal to comply with binary forms of understanding: love *and* hate rather than love *or* hate. They are a reminder that care can so quickly turn to abuse and that the incredible strength of a parental safety net is fragile and can disappear in an instant.

By positioning Wearing in relation to Bourgeois, my intention has been to demonstrate the value of queering art historical discourse too long obsessed with patrilineal chronology and descent. Positioning women artists of different generations alongside each other, rather than in chronological sequence, matters if we are to find alternative ways to talk about which practitioners and practices are valuable and why. These artists are not often discussed alongside each other, working as they did in different geographic, political, and aesthetic spaces, yet *Maman* and *Sacha and Mum* share with *Held* a preoccupation with emotional vulnerability, intensity, and difficulty. Artworks communicate in a myriad of ways through which our subjectivities are constituted, sometimes fleetingly and at other times permanently. In this sense, communication is productive of who we are rather than reflective of a preformed subjectivity. Furthermore, by making their primary register that of emotions, and departing from the need to deliver a message, these artworks shift the debate about feminist and queer activisms from ideological resistance to productive ambiguity.

Conclusion

Throughout this chapter I have been careful not to identify specific artists as feminist or queer, as if these identifications are secure attachments. Nor do I want to suggest that only some types of practice can be understood in these terms. I have deliberately chosen to include the work of artists who have not routinely been identified as feminist or queer, for example, Gillian Wearing, as well as others who have, such as Tanja Ostojić and Adrian Howells. It would make little sense to replace a patriarchal art historical canon with another that fixes the positions of those deemed sufficiently critical to be included. Instead, my intention has been to queer art historical discourse by challenging the boundaries of inclusion, and to argue that feminist and queer can be understood as sensibilities rather than categories of practice. There is no universal story of feminist and queer arts activism. To attempt such an exercise would be to smooth the edges of a field characterized by disruption, multiplicity, and instability.

An overarching concern has been to identify and argue for the significance of *provocation, critical mimicry*, and *emotional ambiguity* as embodied strategies of feminist and queer arts activism. These approaches differ from now orthodox ideas of critical distance and defamiliarization. They work with, rather than speaking to, their viewer/participants in their attempts to encourage behavioral and attitudinal change in the arenas of migration, biopolitics, intimacy, and maternal ambivalence. This requires, and in some cases demands, the viewer to relinquish control and the certainty of a secure viewing position. The care offered by Howells, for example, is not guaranteed, and the maternal comfort enacted in the works by Wearing and Bourgeois is at best precarious and at worst terrifying. Indeed, to engage with these works is to risk disappointment, frustration, even anger, but includes the possibility of intimacy and care. In my view, however, the emotional precarity instigated by these works is precisely where their feminist and/or queer arts activism lies. It is a form of criticality that refuses the certainty of a message, and with it the comfort of a secured place within feminist and/or queer art histories, but which connects on an affective level to our hopes and fears about intimacy, control, and care.

Notes

1 "Guerrilla Girls: Is it Even Worse in Europe?," Whitechapel Art Gallery, accessed December 10, 2018, www.whitechapelgallery.org/exhibitions/guerrilla-girls/.
2 Clare Hemmings, "Telling Feminist Stories," *Feminist Media Studies* 6, no. 2 (2005): 115.

3 See Judy Chicago and Miriam Schapiro, "Female Imagery," *Womanspace Journal* (1973).

4 The biological determinism of central core imagery was challenged by Mary Kelly and Griselda Pollock, among others, who drew on semiotic, Marxist, and psychoanalytic theories to argue for the cultural and ideological, rather than biological, construction of woman. See, for example, Griselda Pollock, *Vision and Difference: Femininity, Feminism and the Histories of Art* (London: Routledge, 1988).

5 Mieke Bal, *Quoting Caravaggio: Contemporary Art, Preposterous History* (Chicago and London: The University of Chicago Press, 1999), 7–10.

6 For a discussion of the development of this position see Nikki Sullivan, *A Critical Introduction to Queer Theory* (Edinburgh: Edinburgh University Press, 2003), 50.

7 Amelia Jones and Erin Silver, eds, *Otherwise: Imagining Queer Feminist Art Histories* (Manchester: Manchester University Press, 2016), 3.

8 Amelia Jones, *Body Art: Performing the Subject* (Minneapolis, MN: University of Minnesota Press, 1998), 173.

9 "After Courbet." Tanja Ostojić, accessed November 12, 2018, http://southasastateofmind.com/article/courbet/.

10 Many second-wave feminist artworks addressed this hypocrisy including VALIE EXPORT's *Tap and Touch Cinema* performance (1968) in which the artist constructed a small movie theatre around her naked chest. She went out on to the street and invited people to touch her body inside this box. In so doing she challenged people to engage with a real woman's body as opposed to an image on screen. EXPORT called this approach "expanded cinema," film without celluloid in which the artist's body activates the watching process.

11 Bojana Videkanić, "Tanja Ostojić's Aesthetics of Affect and Postidentity." *Art Margins* (2009), accessed November 12, 2018, www.artmargins.com/index.php/featured-articles/414-tanja-ostojis-aesthetics-of-affect-and-postidentity-series-qnew-critical-approachesq.

12 Videkanic. "Tanja Ostojić's Aesthetics of Affect and Postidentity."

13 Musée d'Orsay. "Gustave Courbet, The Origin of the World," accessed November 13, 2018, www.musee-orsay.fr/en/collections/works-in-focus/search/commentaire/commentaire_id/the-origin-of-the-world-3122.html.

14 Helena Reckitt and Peggy Phelan, eds, *Art and Feminism* (London: Phaidon, 2001), 214.

15 See Amelia Jones, "The Rhetoric of the Pose: Hannah Wilke and the Radical Narcissism of Feminist Body Art," in *Body Art: Performing the Subject* (Minneapolis, MN: University of Minnesota Press, 1998).

16 Luce Irigaray, *This Sex which Is Not One* (New York: Cornell University Press, 1985), 76.

17 For a detailed discussion of this, see Clare Johnson, *Femininity, Time and Feminist Art* (Basingstoke, UK: Palgrave Macmillan, 2013), 77–114.

18 Jennifer Doyle, *Hold It Against Me: Difficulty and Emotion in Contemporary Art* (Durham, NC: Duke University Press, 2013).

19 Dee Heddon and Adrian Howells, "From Talking to Silence: A Confessional Journey," *PAJ: A Journal of Performance and Art* 33, no. 1 (2011): 1–12.

20 Doyle, *Hold It Against Me*, 3.

21 Rozsika Parker, *Torn in Two: The Experience of Maternal Ambivalence* (London: Virago, 2005).

22 Melissa Benn, "Deep Maternal Alienation," *The Guardian*, October 28, 2006, www.theguardian.com/lifeandstyle/2006/oct/28/familyandrelationships.family2.

23 Deborah Wye and Carol Smith, *The Prints of Louise Bourgeois* (New York: MOMA, 1994).

24 Lucy Lippard, "From the Inside Out," in *From the Center: Feminist Essays on Women's Art* (New York: Dutton, 1976), 249.

25 Rosemary Betterton, *Maternal Bodies in the Visual Arts* (Manchester and New York: Manchester University Press, 2014), 162.

26 Elizabeth Manchester, "Louise Bourgeois: Maman," *Tate*, December 2009, www.tate.org.uk/art/artworks/bourgeois-maman-t12625.

Bibliography

Bal, Mieke. *Quoting Caravaggio: Contemporary Art, Preposterous History*. Chicago and London: The University of Chicago Press, 1999.

Benn, Melissa. "Deep Maternal Alienation." *The Guardian* online (October 28, 2006). www.theguardian.com/lifeandstyle/2006/oct/28/familyandrelationships.family2.

Betterton, Rosemary. *Maternal Bodies in the Visual Arts*. Manchester and New York: Manchester University Press, 2014.

Chicago, Judy, and Schapiro, Miriam. "Female Imagery," *Womanspace Journal* 1, no. 3 (Summer 1973): 11–14.

Doyle, Jennifer. *Hold It Against Me: Difficulty and Emotion in Contemporary Art*. Durham, NC: Duke University Press, 2013.

Heddon, D., and Howells, Adrian. "From Talking to Silence: A Confessional Journey." *PAJ: A Journal of Performance and Art* 33, no. 1 (2011): 1–12.

Hemmings, Clare. "Telling Feminist Stories." *Feminist Media Studies* 6, no. 2 (2005): 115–139.

Irigaray, Luce. *This Sex which Is Not One*. New York: Cornell University Press, 1985.

Johnson, Clare. *Femininity, Time and Feminist Art*. Basingstoke, UK: Palgrave Macmillan, 2013.

Jones, Amelia. *Body Art: Performing the Subject*. Minneapolis, MN: University of Minnesota Press, 1998.

Jones, Amelia, and Silver, Erin, eds. *Otherwise: Imagining Queer Feminist Art Histories*. Manchester: Manchester University Press, 2016.

Lippard, Lucy. "From the Inside Out." In *From the Center: Feminist Essays on Women's Art*. New York: Dutton, 1976.

Manchester, Elizabeth. "Louise Bourgeois: Maman." Tate online (December 2009), www.tate.org.uk/art/artworks/bourgeois-maman-t12625.

Musée d'Orsay. 2018. "Gustave Courbet, The Origin of the World." Accessed November 13, 2018. www.musee-orsay.fr/en/collections/works-in-focus/search/commentaire/commentaire_id/the-origin-of-the-world-3122.html.

Ostojić, Tanja. 2015. "After Courbet." Accessed November 12, 2018. http://southasastateofmind.com/article/courbet/.

Parker, Rozsika. *Torn in Two: The Experience of Maternal Ambivalence*. London: Virago, 2005.

Pollock, Griselda. *Vision and Difference: Femininity, Feminism and the Histories of Art*. London: Routledge, 1988.

Reckitt, Helena, and Phelan, Peggy, eds. *Art and Feminism*. London: Phaidon, 2001.

Sullivan, Nikki. *A Critical Introduction to Queer Theory*. Edinburgh: Edinburgh University Press, 2003.

Videkanić, Bojana. 2009. "Tanja Ostojić's Aesthetics of Affect and Postidentity". *Art Margins*. Accessed November 12, 2018. www.artmargins.com/index.php/featured-articles/414-tanja-ostojis-aesthetics-of-affect-and-postidentity-series-qnew-critical-approachesq.

Whitechapel Art Gallery. "Guerrilla Girls: Is it Even Worse in Europe?" Accessed December 10, 2018. www.whitechapelgallery.org/exhibitions/guerrilla-girls/.

Wye, Deborah, and Smith, Carol. *The Prints of Louise Bourgeois*. New York: MOMA, 1994.

PART III

The politics of gender

Introduction to Part III

Gender is political. Discussions of gender cannot exist without also discussing the intersectionality of gendered politics, including race, ethnicity, sexuality, religion, nationality, and citizenship. The chapters in this section, "The politics of gender," argue that gendered political progress is possible, in spite of current and past political tragedies. It features profiles of recent communication research on the role of gender norms in political culture, as well as new legal and legislative battlegrounds over LGBTQ human rights and gender equality. In addition to examining political institutions and processes relating to gender and communication, this section analyzes broader political contexts for gendered relations between citizens and nations, religion, and gendered subjects and collective bodies. A diverse set of transdisciplinary and transnational essays chart new paths in recent research and politics that avoid the gaps and assumptions shaping previous accounts of gender and communication.

To exist in a gendered and raced world means that the rhetorical strategies and outcomes of elections, laws, and bills marginalize and exclude groups that do not perform and conform to traditionally white and masculine roles. Throughout the world, historically marginalized groups are threatened in varying contexts and processes because of their existence, and even more so because of their activism. KC Councilor, for example, discusses the suffering endured by trans★ people due to recent legislation in the United States, including bathroom bills in states such as Texas and North Carolina. Kristina Horn Sheeler, Serena Hawkins, and Eline van den Bossche offer additional examples in their chapter, finding that women who run for political office also suffer, notably because they are caught in double binds when they do not embody traditional masculinity.

This section also argues that the notion of *tradition* extends beyond masculinity in gendered politics, though whiteness is still upheld as ideal. For example, Tracey Owens Patton and Nancy Small's chapter about Congresswoman Maxine Waters details the death threats she received for performing authentic womanhood and blackness. Similarly, Fatima Zahrae Chrifi Alaoui and Shadee Abdi discuss the rhetoric of victimization that is used to discuss Arab and Muslim womanhood, which is contrasted against what is deemed a progressive and free white American feminism. Acceptable femininity also threatens gendered political progress, especially at the intersections of race, ethnicity, nationality, and citizenship.

The first chapter in this section, "Making waves: Maxine Waters' Black Feminist and Womanist rebuke of supremacist hegemony," notes the significant exclusion of Black women from the feminist movement and points out that Black women have long been at the forefront of feminist struggles in the United States. Activists such as Ida B. Wells, Harriet Tubman, and Anna Cooper, to name a few, fought for the right to be Black and woman. The authors, Tracey Owens Patton and Nancy Small, provide a thoughtful review of Alice Walker's Womanism and the Combahee River Collective's Black Feminism. From here, they argue that U.S. Congresswoman Maxine Waters performs Womanism and Black Feminism. That is, she demonstrates Black Feminist discourse, struggle, and authenticity, and actively aligns herself with LGBTQ allies. Though Waters experiences emotional and physical threats, her verbal resistance embodies Womanism and Black Feminism, particularly as she continues to fight against interlocking systems of domination.

Yet in spite of the progress that is being made through varying forms of resistance, the movement of gendered freedom persists. Michele L. Hammers, Nina M. Lozano, and Craig O. Rich's chapter, "One step forward … gender, communication, and the fragility of gender(ed) political progress," provides an overview of the #MeToo movement and highlights the communication discipline as an area that can make positive change toward gender(ed) legal and political progress. Specifically, they discuss U.S. President Donald Trump and consider his 2016 presidential campaign as the "perfect storm of gendered politics," particularly as it led to reductions of sex protection under the law and an increase in antiprotest rhetoric. Hammers, Lozano, and Rich urge communication scholars to contribute to cultural movements in order to improve gender(ed) legal and political protections.

KC Councilor's chapter, "The specter of trans bodies: public and political discourse about 'Bathroom Bills,'" furthers the discussion about gendered legal and political protections and argues that public space, and who has access to it, is an important component of political culture. The author analyzes bathroom bills in two states, North Carolina and Texas, due to the high-profile nature of the bills, the usage of the states' arguments by other states, and the significance of the states in LGBTQIA+ policies. Councilor examines the framing of bathroom bills by conservatives and finds similarities between both bathroom advocates and opponents approaches and those used in previous movements for social equality, with current policies ultimately portraying trans★ people as deceivers. This is exemplified in social media protests that ignore gender fluidity and affirm gender(ed) stereotypes. Ultimately, Councilor maintains that the legal system must not be allowed to define trans★ people.

Kristina Horn Sheeler, Serena Hawkins, and Eline van den Bossche thoughtfully continue the conversation on gender and politics in their chapter, "Research on gender and political rhetoric: masculinity, ingenuity, and the double bind." They provide an extensive literature review of feminist scholarship in rhetorical studies and highlight intersecting areas within politics. First, they discuss masculinity and bias in U.S. presidential elections and cite numerous examples of barriers faced by presidential candidates who lack traditional masculinity. Then, they provide case studies of strategies used by women who ran for President of the United States, including resistance against obstacles imposed by politicians and other voters. The authors conclude with a discussion of the stereotypes and hypermasculinity associated with presidency. This new study shows that U.S. politics is still filled with double binds that assume traditional ideals and that result in the exclusion of those whose values are not shared and performed.

Fatima Zahrae Chrifi Alaoui and Shadee Abdi's chapter, "Resisting Orientalist/Islamophobic feminisms: (re)framing the politics of difference," discusses the exclusion of those who do not fulfill stereotypical roles within white feminist movements. The authors argue that the relationship between "Muslim," "Arab," and "Feminism" that has been deemed antonymic is currently a contested zone due to the resurgence of the popular Orientalist and Islamophobic stereotypical racial discourses on Arab and Muslim subjectivities. They argue that the liberal deployment of a rhetoric that conflates Arab and Muslim womanhood with victimization also positions a mainstream "American" feminist as the feminist subject par excellence. The authors begin their chapter by providing an overview on the literature on Orientalist/Islamophobic feminism that has "shaped" Arab and Muslim feminism as oxymoronic and nonexistent, emphasizing the influence of the events of 9/11. Then, the authors draw on theories of women of color to demonstrate how they forged an inclusive visibility for Arab and Muslim feminism. They finalize their chapter by theorizing Muslim feminism as a movement that embarks to new spaces to understand the multiplicity and specificity of subjectivities within Muslim and Arab communities that redefine feminism. Ultimately, they argue that Muslim feminists create a middle space "in between" the binaries to produce alternate discourses that challenge and disrupt the knowledge, representations, and discourses associated with them.

The next chapter in this section furthers the discussion of spaces that are created and produced to understand feminism. Rivka Neriya-Ben Shahar's chapter, "Negative spaces in the triangle of gender, religion, and new media: a case study of the Ultra-Orthodox community in Israel," focuses on the multiple and complex relationships between gender, religion, and new media through a case study of the Ultra-Orthodox community in Israel. She argues that gender, religion, and new media are a triangle comprising a system in which each part influences the others in multiple ways. This argument contends that a contemporary study of any one aspect should use an intersectional analysis that takes the others and their interactions into account. Moreover, analyzing intersections of gender, religion, and new media requires attention to the relations, continuums, and negative spaces that both create and emerge from the pairing of the ostensibly oppositional categories of women/men, religion/secularization, new/traditional media. The Ultra-Orthodox community in Israel provides a case study demonstrating how the intersections between gender, religion, and new media necessarily entail changing boundaries between women and men, religion and secularization, and traditional and new media.

Drawing on Miranda Fricker's concept of epistemic injustice, Kundai Chirindo's chapter, "Invisible in/humanity: feminist epistemic ethics and rhetorical studies," analyzes epistemic practices in communication and rhetorical studies. The author argues that the heuristic schemas that dominate in rhetorical studies tend to be masculinist, hetero- and cis-, and Euro- and U.S. American-centered. To the extent that these dominant paradigms enable neglecting broad representations of human communicative and rhetorical practices, they consign the humanity of others to epistemic oblivion. This is because from the perspective of in/justice, epistemic practices are morally ambivalent, capable of effecting both vice and virtue. Chirindo focuses on two types of practices that effect epistemic injustice: testimonial injustice and hermeneutical injustice. The chapter illustrates the invisible inhumanity of these practices through a brief reception study of the rise and presidency of Barack Obama in rhetorical studies. It concludes by suggesting why scholars in communication and rhetorical studies should strive to make their respective subdisciplines more human through the pursuit of epistemic justice.

The chapters in this section illustrate the complexity of gender in political systems worldwide and also emphasize the importance of activism in reducing gendered political oppressions, which essentially affects and is affected by all gender inequity. Those citizens who are excluded are left out not only because they lack traditional masculinity but also because they do not embody traditional femininity or white Western feminism. Communication scholars are uniquely positioned to examine these gendered complexities, particularly as they intersect with other types of oppressions.

Embedded throughout the chapters in this section is a consistent call for resistance. The strategy of political resistance varies and can be practical, as Maxine Waters embodies Womanism and Black Feminism, or as other women politicians resisted, and continue to resist, obstacles during their election processes. Political resistance may also occur at a theoretical level, with calls being made for the study of intersectional work, such as with gender, religion, and new media, as well as for research that pursues epistemic justice. Ultimately, these strategies of resistance are necessary to combat gendered political injustice and continue gendered political progress.

15

MAKING WAVES

Maxine Waters's Black Feminist and Womanist rebuke of supremacist hegemony

Tracey Owens Patton and Nancy Small

Understanding the varieties and complexities of feminism is an important prerequisite in evaluating dominant narratives surrounding gender-centered norms and roles. Feminist scholar bell hooks defined feminism as the

> struggle to end sexist oppression. Its aim is not to benefit solely any specific group of women, any particular race or class of women. It does not privilege women over men. It has the power to transform in a meaningful way all our lives. Most importantly, feminism is neither a lifestyle nor a ready-made identity or role one can step into.[1]

The dominant narrative of "feminism" in the United States typically focuses on white women's suffrage and white women's fight for equal opportunity. Within that narrative, the Black feminist movement is often portrayed as an offshoot of white feminism, but in fact, the story of Black feminisms in the United States is more complex and nuanced than that. Black women's relationships with feminist movements in the United States have been complicated, even ambivalent. Fighting for their rights as both women and as People of Color has meant being estranged both from the (white) feminist movement because of its racial oppression and from the Black liberation movement because of its patriarchal foundations. Topics within conversations regarding race and gender—for example, traditional gender roles, opinions of homosexuality, the roles of religion—were at the heart of why different Black feminisms historically overlapped and diverged.

This chapter briefly reviews the history of Black women's early activism in the feminist movement, then pauses to consider how Alice Walker's Womanism and the Combahee River Collective's stance on Black feminism are examples of specific perspectives emerging during the second wave. The two perspectives shared a common foundation even as they took slightly divergent means to the same end, and they serve as representative examples of the multiple strains of second-wave Black feminist voices. Those overlapping, yet distinct, approaches would eventually be (re)united through the contemporary theory of intersectionality. Maxine Waters is a beautiful embodiment of this (re)integrated Black Feminist/Womanist positionality, so we use her work to illustrate how the strains of Black feminist history remain a key rhythm driving contemporary rhetorics. Waters performs an intersectional public rhetoric in order to unite the disenfranchised, fight patriarchy, and continually reclaim the capability of Black women as leaders in resisting white supremacy.

From intersectional roots grow branches of perspective

Black women have been at the heart and forefront of feminist struggles in the United States since the beginning, because first-wave Black feminists' connections to slavery meant that they were already concerned with facing the double damnation of both racial and gender-based oppression. Sojourner Truth, Harriet Tubman, Ida B. Wells, Mary Church Terrell, Josephine St. Pierre Ruffin, Fannie Barrier Williams, Victoria Earle Matthews, Anna Cooper, Francis Ellen Watkins Harper, and others publicly appealed for their rights from the intersection of being Black—often former slaves—and women.[2] In her iconic speech, "Ain't I a Woman?" (May 29, 1851), Truth did what few rhetoricians did back then, or do even today: she called out the complicity of white women for their supremacist views while at the same time impugning Black men in their willingness to accept the crumbs and benefits of patriarchal privilege when doled out to them.[3] Sojourner Truth was one of our first Black Feminists/Womanists.

The efforts of first-wave Black feminists were muted further and eventually silenced by the passing of the 19th amendment in 1920, by the effects of war, and by the emergence of Jim Crow laws. Black women were betrayed by white women who turned their backs on notions of "sisterhood" and took up support of the white patriarchy. As hooks explained, most "white women did not use their voting privileges to support women's issues: they voted as their husbands, fathers, or brothers voted."[4] In other words, winning the vote was seen as "a victory for racist principles."[5] Feminism, as it had been popularly articulated by white, educated, middle-class women, was exclusive and inadequate for addressing overlapping oppressions. On the heels of World War I, Jim Crow apartheid emerged and became a race-based threat more significant than gender-based oppression. As a result, Black women "ceased to struggle over women's rights issues and concentrated their energies on resisting racism."[6] Following decades of relative silence from feminist activists, the end of World War II reignited questions of women's roles as men returned to the home front, and increasing resistance to systematic discrimination under Jim Crow laws sparked the emergence of the civil rights movement and renewed questions over women's roles in the public sphere.[7]

The double bind of being woman and being Black reemerged with stark clarity when the second-wave rights movement took hold in the United States in the 1960s. As white women returned their attention to the Equal Rights Amendment, Black women were estranged from the feminist movement as well as enmeshed in the struggle for civil rights.[8] Central to fighting the dehumanizing effects of Jim Crow apartheid were efforts led by Black men to assimilate into white middle-class culture. However, as Black people fought for their basic freedoms and dignity based on their race, Black women were asked by the men leading the liberation movement to take a back seat because of their gender. For both Black and white women, fitting in meant women adopting idealized notions of femininity: being "pretty" (light-skinned or white), keeping a tidy home, focusing on child-rearing, and being in service to the man of the house. These expectations forced women firmly back into their private spheres. Adding racist fuel to the situation, the Moynihan report—*The Negro Family: The Case for National Action*—painted Black families as dominated by women, suggesting that "the negative effects of racist oppression of black people could be eliminated if black females were more passive, subservient and supportive of patriarchy."[9] A Black woman "acting out" damaged how her race *and* her gender were perceived, and in fact, likely enhanced white, patriarchal threats to the very safety of the Black community.

From the simmering frustration over the racism of the Jim Crow laws, the sexism of the Black liberation movement, and the continued exclusionary agenda of the white feminist movement, Black feminisms arose. The plurality of Black feminist efforts was a result of both their grassroots rekindling and the nuanced differences of their perspectives. The movement was not centralized and efforts to establish a unified Black feminist agenda were not successful.[10] Among the voices,

Alice Walker's Womanism and the CRC's Black Feminism were two strains that would prove enduring, as both still resonate in contemporary activist work over half a century later.[11] Although Walker and the CRC produced two branches on the same tree, understanding more about their positional divergences and overlaps is useful. These overlaps not only illustrate some of the nuanced tensions in the second-wave Black feminist movement, but also lay the foundation for demonstrating how contemporary rhetors such as Congresswoman Waters use a remixed Womanist/Black Feminist perspective in their continued resistance against white patriarchy and white supremacy. We begin by briefly defining features of Womanism and CRC Black Feminism as two overlapping perspectives growing from the same position of defiance.

Although Black feminist activism reemerged through numerous community and intellectual groups in the 1960s and 1970s, the "Combahee River Collective Statement" is the best-known manifesto from that time. The CRC, located in the Boston metropolitan area, formed in 1973 as a splinter group wishing to separate and distinguish itself from the National Black Feminist Organization (NBFO) because founding CRC members "had serious disagreements with NBFO's bourgeois-feminist stance and their lack of a clear political focus."[12] Specifically, the NBFO was affiliated with NOW (National Organization for Women), which was seen as a white women's feminist organization, connected with second-wave feminism and not seen as an organization fighting against oppression of Women of Color. In juxtaposition to the NBFO, the CRC embarked upon a "revolutionary task" involving multiple lines of resistance.[13] The CRC was "actively committed to struggling against racial, sexual, heterosexual, and class oppression."[14] Underpinning the focus on interlocking systems of oppression was a commitment to "identity politics," which expanded "the personal is political" to emphasize the particular and embodied nature of the struggle faced by Women of Color. As an exemplar of Black feminist concerns within the contexts of the Black liberation and civil rights fronts, the CRC's statement laid out specific commitments and positions. The CRC declared "the most profound and potentially most radical politics come directly out of our own identity, as opposed to working to end somebody else's oppression."[15] Although they rejected white feminism and the patriarchy of Black liberation, the CRC's statement established that their concerns were with *all* facets of Black women's lives, particularly those related to economics and material conditions. The CRC sought to be inclusive, collective, and nonhierarchical, as well as "committed to a continual examination of our politics" through reflexive thinking.[16] "Black women have always embodied, if only in their physical manifestation, an adversary stance to white male rule," which meant that their Black feminism was "the outgrowth of countless generations of personal sacrifice, militancy, and work by our mothers and sisters."[17] In other words, the CRC as a Black feminist organization considered their work a continuation of women such as Truth.

"Womanism" also emerged through rejection of white feminism, concerns over Black liberation, and the need for Black women to reassert their voices. Alice Walker created the term "Womanist," which appeared first in her 1979 short story "Coming Apart" and later was developed through a four-part definition in her 1983 volume of essays, *In Search of Our Mothers' Gardens.*[18] Like Black Feminist thought as defined by the CRC, Womanism—or the perspective from which a Womanist proceeds—focused on the intersection of gender and racial discrimination. A "Womanist" was "a black feminist or feminist of color" who may display "outrageous, audacious, courageous, or *willful* behavior" because she is "Responsible. In charge. *Serious.*"[19] In other words, Womanism was defined in opposition to silly, girlish frivolity. Walker's Womanism did not reject men and was not in overt opposition to traditional definitions of family and gender roles. Grounded in the feminine, a Womanist "sometimes love[d] individual men" even as she remained loyal to the "survival and wholeness of entire people, male *and* female."[20]

Table 15.1 Comparing Womanist and Black Feminist perspectives

	CRC Black Feminism	Walker's Womanism
On Labels	Rejects white feminist agenda as racist but embraces the politically charged "feminist" label	Rejects "feminist" label
On Men	Overtly critiques and resists Black patriarchy[1]	Remains inclusive of Black men
On Gender Roles	Takes a firmly anti-essentialist stance[2]	Implies strategic essentialism
On Sexuality	Declares a lesbian-centric critical theory, confronts homophobia, and embraces "identity politics"[3]	Embraces all kinds of love but doesn't overtly attend to the politics of homophobia
On Afrocentrism	Rejects Afrocentrism as an extension of male thought[4]	Underpins perspective (e.g., *mati*-ism) but reflexive in application[5]
On Religion	Remains respectful of but directly tied to religion	Extends from (radical) spirituality[6]
Primary Focus	Economic systems/socialism	Relationships, relationality
Emphasis	Critical analysis and reflexive militancy	Recuperation of community
Language	Resistance/"War" metaphors	Harmonizing/"Mother" metaphors
Perspective	Issue based (economics, politics, systems)	Holistic (interdependent, family/community oriented) and wholistic (of body, spirit, and mind)

[1] The CRC statement calls out Black men's "notoriously negative" response to feminism (279) but is in "solidarity with progressive Black men" because of shared racist oppression (275).

[2] The CRC statement calls feminism "very threatening to the majority of Black people because it calls into question some of the most basic assumptions about our existence" (278). One of those "most basic assumptions" are family and gender roles within the heteronormative community.

[3] Smith, xiv.

[4] Combahee River Collective, 273. The CRC emphasizes "Afro-*American* women" over Afrocentrism.

[5] Maparyan, 19. For example, Walker's Womanist love between women can be interpreted as an illustration of Afrocentric *mati*-ism, a group-centered perspective based in female friendships, including sexual relations with both men and women.

[6] Walker grounds Womanism in love of the moon and of "the Spirit" (19). As Womanism has continued to evolve it has adopted a more radical spiritual connection. See Maparyan, Chapter 1.

Table 15.1 summarizes a selection of differences between Walker's Womanist and CRC Black Feminist perspectives. Although the table admittedly is limited and reductive, considering variations within Black feminist perspectives assists in a deeper understanding of the conversations influencing the range of feminisms. Just as we pause to temporarily pull these perspectives apart, contemporary rhetors such as Patricia Hill Collins and Congresswoman Waters continue to integrate them back into a transcendent whole.

While neither Womanism nor CRC Black Feminism specifically called for gender segregation, Womanism remained more open to gender integrated antiracist efforts. As a Black middle class emerged and the Black liberation movement called for women to take their places back in the domestic sphere, Black men were called out for not making enough money to support a single-income lifestyle, fueling "black female contempt for black men."[21] Of course, such tensions only served to support the oppression of Black communities as they bought into the sham of the "idealized" white middle-class myth. But stresses in the home were bound to shape the agendas of Black feminisms, potentially dividing communities into men who were for or against the movement. The CRC called out Black men's "notoriously negative" response to feminism, but left the door in gender division cracked open in order to be in "solidarity with progressive Black men" because of shared racist oppression.[22] Nevertheless, the myth persisted that Black feminism—by its agenda as well as its association in name with white feminism—was "nothing but man-hating."[23] In contrast, Womanism was oriented toward a "pluralist version" of Black empowerment, "premised not on individual assimilation but on *group* integration."[24] Womanism overtly rejected feminist "separatism" and focused on the "survival and wholeness of entire people, male and female."[25] As Womanism has continued to develop in contemporary circles, it has maintained such inclusiveness as a "family-oriented type of communalism."[26]

CRC Black Feminists openly embraced lesbian sisterhood, while Womanists engaged a love of feminine culture and a strategically essential approach to traditional gender roles. Lesbianism was seen by some members of the Black activist community as "counterrevolutionary," at odds with the patriarchal assimilation promoted by the Black liberation movement.[27] Founding member of the CRC, Barbara Smith outlined five "myths" perpetuated by racist people and systems and designed to reinforce Black women's oppression. One of these is that Black feminists are "nothing but Lesbians," a lie serving to prevent heterosexual women from joining the Black feminist movement out of fear their sexual identities would be called into question.[28] Smith called this "the most pernicious myth of all" because it "reduced Lesbians to a category of beings deserving of only the most violent attack, a category totally alien from 'decent' Black folks."[29] Standing up for Black women with a range of gender and sexual identities, the CRC explicitly welcomed differences of identity within women's communities, rejecting a universal femininity and "any type of biological determinism."[30] Through this position, they overtly appealed to building alliances by making space for differences in "female" identity. By comparison, Womanists were generally silent "on the 'taboo' sexuality of lesbianism."[31] Walker's Womanism did not directly deny lesbian identity, but instead moved beyond the homosexual/heterosexual binary all together by embracing a range of sexualities. Such a fluid approach to gender roles functioned deeply in Womanism. As part of a broader community view, a Womanist might purposefully engage her feminine energies for harmonizing and healing tensions and wounds inflamed by ongoing oppression. In that way, her femininity and her biological role as "mother" could be deployed in a strategically essential fashion.[32] The strength and capability of being a Womanist came from understanding that choosing to adopt particular roles is a sign of being an adaptable remixer. A Womanist might engage her "queenhood" in the opportune space and time, a move that could be considered outrageous or audacious, even if it was particularly courageous.

Both Womanists and CRC Black Feminists amplified the capability, adaptability, and creative strength of Black women. The CRC statement declared Black women as "inherently valuable" and deserving of "autonomy" fighting against stereotypes of the day such the "mammy, matriarch...[or] whore."[33] Arguing for basic human gender equality, the CRC overtly rejected "pedestals, queenhood, and walking ten paces behind."[34] Womanists considered Black women talented samplers or remixers, making them "capable generators" who engaged each other's rhythms" and "[brought] a confidence in evaluating our own experiences, dictating our own discourses, and defining our own terms."[35] As CRC member and Black feminist leader Barbara Smith proclaimed, "Black women as a group have never been fools. We couldn't afford to be."[36] Walker's frequently cited, "Womanist is to feminist as purple is to lavender" analogously foregrounds the strength and intensity of Black women.[37] Both Womanism and CRC Black Feminism confronted the oppressive effects of the patriarchal systems into which the Black liberation movement sought to assimilate, and both sought to highlight the historical as well as ongoing *capability* of Black women, whose contributions and leadership were being persistently suppressed or erased by men and by white feminists.

Both CRC Black Feminism and Womanism also sought to apply pragmatic "home truths" to critique racism and sexism as well as to improve of women's lives, and they ultimately shared a strong motivation in healing the wounds of interlocking oppressions as well as continuing to fight against their hegemonic sources.[38] Beginning in the 1980s and continuing on to become a crucial part of critical race work into the 21st century, "intersectionality" as both standpoint and analytical tool emerged as a direct result of CRC Black Feminism, Womanism, and other voices of second-wave Black feminisms.[39] A cornerstone of the third-wave movement, intersectionality acknowledges multiple axes of oppression affecting people's lives and the complexity of working for social justice within specific contexts. Much more than a lens or identity, intersectionality is a tool for solving complex problems, a source of agency, and a means of forming strategic alliances.[40] To illustrate how strains of Womanism and Black Feminism intersect and interweave in contemporary rhetorics, we turn to the audacious, courageous, remixing, and alliance-building public performances of Congresswoman Waters.

"Auntie Maxine" as a Womanish Black Feminist

Congresswoman Maxine Waters is one of the most powerful women in U.S. politics and is known as a champion for LGBTQ, People of Color, the poor, and women. Elected in 2018 to her 15th term in the U.S. House of Representatives in the 43rd Congressional District of California, Waters is the Ranking Member of the House Committee on Financial Services, a member of the Congressional Progressive Caucus, and the past chair of the Congressional Black Caucus. Waters has been a public servant for the people of California for nearly four decades beginning in 1980, when she served on the Democratic National Committee (DNC).[41] "She was a key leader in five presidential campaigns: Sen. Edward Kennedy (1980), Rev. Jesse Jackson (1984 & 1988), and President Bill Clinton (1992 & 1996)."[42] This national work for the DNC provided her the leverage and experience she needed to run for the House of Representatives in 1990, where she

> already attracted national attention for her no-nonsense, no-holds-barred style of politics… She was responsible for some of the boldest legislation California has ever seen: the largest divestment of state pension funds from South Africa; landmark affirmative action legislation; the nation's first statewide Child Abuse Prevention Training Program; the prohibition of police strip searches for nonviolent misdemeanors; and the introduction of the nation's first plant closure law.[43]

In other words, in taking both a Black Feminist and Womanist approach, she courageously did her job representing the people of California—even if the issue was not her personal politics—thus challenging the binary divisiveness that has a stranglehold of politics today with its draconian alliance to party first and people second.

Popularly known as "Auntie Maxine," Waters's supporters love her performance as an "impeccably dressed older woman" who shares "her opinion, even if it might be painfully honest."[44] Waters embraces the role of "Auntie," which is used in this context as a positive term. During U.S. enslavement of Black people, however, "Auntie" invoked a white supremacist hierarchy with "an auntie" seen as someone "less than" and more akin to a mammy; docile, dark-skinned, deferent, servile, and silent. White children often referred to the Black women who took care of them as "Auntie" or "Mammy." Auntie, even in the progressive circles in which Waters runs, is problematic given the recent and racist historical context of the term. However, Auntie is typically used by Black diasporic populations in a positive way that highlights one's goodness, power, and strength. Waters is a smart, strong, sassy, tell-it-like-it-is woman who is endearing and embodies the positive definition behind the word "Auntie." More than a badass, an Auntie is an elder who embarks on a tough love, tough talking trajectory and is someone you trust to tell you

the truth no matter how raw, because it is for your own good. There may or may not be a biological connection with an Auntie, and such a connection is not necessary because if an Auntie
advocates for you and tells you the unvarnished truth, it is because she loves you. It is because she
wants you to be better, to be smarter, and to be more powerful and empowered. The very role of
being "Auntie" Maxine demonstrates how Waters inhabits a space where Womanism and Black
Feminism overlap. She is a motherly figure who loves and seeks harmony, while at the same time,
she is brave, stands up in public, and can militantly defend her wide and broad "family." She talks
tough, tells home truths, believes in the ultimate capability of herself and her people, and serves
as a role model for both activists and for audiences seeking to learn more about the damages
inflicted by oppressive systems.

Waters says things that other people are afraid to say and has been seen as a leader of the resistance following the 2016 Presidential election. Echoing Sojourner Truth, Waters is open about
her audacity when she says, "I'm a strong Black woman, and I cannot be intimidated. I cannot be
undermined."[45] In this statement Waters invokes not only why she is called "Auntie Maxine" by
some, but also why she invokes Black Feminism; she is Black and a woman and a woman who
cannot be intimidated or undermined. On top of this gender and race narrative, Waters combines
power and politics and actively shows she is not someone who is easily handled, embracing tenets
of both Womanism and Black Feminism. Her leadership and history reinforce the often-held
notion in Black Feminism and Womanism that politics for People of Color involves both the
personal and professional life as Waters's gendered identity and life are simultaneously politically
performative.[46] It is this intersectional politic of being that is manifested in her public rhetoric
and what we explore here.

Black Feminist/Womanist positionality—or performance—is exemplified through Waters's
"fiery speech," published on YouTube as "U. S. Representative Maxine Waters Speaks at HRC
Los Angeles Dinner." Her speech is nearly 25 minutes long and uses an intersectional approach
to garner audience support, as well as talk back to an administration that is failing the needs of its
people. The speech also highlights her use of Black Feminist and Womanist appeals to continue
to fight for social justice. As of this writing, it is one of Waters's most recent and longest speeches.
Although she speaks prolifically to the media, these are often "sound bites" of 30 seconds to 4 to
5 minutes in length.

Waters was asked to be the keynote speaker for the Human Rights Campaign (HRC) Los
Angeles dinner to an audience of more than 1,000 active members and supporters in California.
This fundraising event was held on March 10, 2018, with equality as it foundational message and
raising "crucial funds in the fight for LGBTQ equality" as the central theme for the evening.[47]
Waters spoke just hours after U.S. President Trump had given a rally speech in Moon Township,
Pennsylvania, where he said, "And Maxine Waters, a very low IQ individual. She's a low IQ
individual. She can't help it."[48] As a result, the exigency of Waters's speech conceivably may have
changed from a traditional fundraising dinner speech to a speech intended to respond to President
Trump's personal attack.

"Love is Struggle". Love is Power: foregrounding intersectional allyship

Congresswoman Waters followed in the footsteps of foremothers such as Truth in taking an
intersecting—Black Feminist and Womanist—approach in her HRC speech. Waters began her
speech by thanking those who came before her. Often a rhetorical strategy used in African
diasporic speaking, Waters acknowledged that her social justice advocacy came not only from
living in a gendered and raced body, but by learning from those and valuing the life experiences
that do not play out in her performative positionality. From this intersectional politic of

being, she began by stating, "I'm going to take what time has been allotted to try and continue with you my description of what I think is going on in Washington, DC, and with your President."[49] Waters's rhetorical strategy here is similar to that of iconic orator and first-wave feminist, Frederick Douglass, in his 1852 speech, "What to the Slave is the Fourth of July." By using "yours" and not "my," Waters widens the gulf between her and President Trump, but this political position, far from distancing herself from her audience, instead endears her more to her LBGTQ-supporting listeners—Trump is not their president either, as their laughter and applause confirms. In this rhetorical move of "yours" not "my," Waters shows that just because she is a member of Congress does not mean that she follows the regime. She will neither tolerate nor walk ten steps behind a fool. According to communication scholar Wen Shu Lee, "a womanist, or feminist of color, then, is a person who is aware of and makes a conscious effort to engage in the multifaceted struggles (i.e., race [*and*] gender together) for human equality."[50] Waters demonstrated the multifaceted struggle through a Black Feminist lens in the first five minutes of her speech noting how she herself had learned from and working with a diverse group of people in social justice advocacy roles, including civil rights activists, entertainers, human rights activists, and political strategists, all of whom are men and most of some of whom are also members of LGBTQ communities.

Waters's performance shows that while she is actively part of a Black Feminist discourse that destabilizes hegemonic hierarchies, she also puts her efforts into building alliances across gender and sexual identities, working in Womanist ways to harmonize communities disillusioned by current political leadership. Waters emphasizes collective agency and autonomy in the LGBTQ community, echoing claims of inherent value and capability that Black feminists have been voicing for centuries:

> No one has done more for you than you have done for yourself. It is the time, the energy, the money, the blood sweat and tears that each of you put into this movement that advanced LGBTQ equality in ways some never thought was possible.[51]

Waters also demonstrates a Womanist discourse emphasizing authenticity and action-oriented allyship. Although her position is one of political national prominence, she amplifies her connections to community. Through love of those relationships, she has been sustained. She continues,

> I have a lot of friends in the LGBTQ community and have been very active in this movement for many years. I want to take time to remind you of those who took time to educate me, to work with me, to nurture me over the years. And I want to tell you it is that kind of love and support that has helped me to become who I am.[52]

In showing the work Waters had done in order to build her ethos as not only an ally but also an advocate, she embraces her own privilege that she has as a high-ranking, powerful political figure. As Lee articulated about white women feminists, their "failures to confront women's different privileges head-on in established feminist discourse" allowed hierarchies within a movement designed to challenge such patriarchal norms to remain intact.[53] In other words, white feminists' push to unify all women through their essentialized characteristics—both biological and socially assigned—only further strengthened male-dominated society by reinforcing simplified and stereotypical gender-based divisions. Lee comments, "I come to realize that the more privilege a woman is born into, the fewer forms of oppression she is 'forced' to fight against."[54] As a wealthy, educated, cisgender, straight African American woman, Waters deftly navigates complicated terrain, not by speaking for groups to which she does not have automatic, and sometimes unwanted, membership, but rather by learning, listening, then acting when appropriate. Waters echoes Womanism's holistic embracing of others, wanting to know more and in greater depth to contribute to unification.

"'You acting womanish,' i.e., like a woman": inhabiting and valuing the space between Black Feminism and Womanism's strengths

From standing on the shoulders of those who came before her, Waters's 2018 speech further built her ethos by noting what she has done to stand on her own as a formidable opponent to injustice. She is serious, responsible, and accountable to the community. Keeping her audience in mind, Waters focused on achievements that directly impact them: she fought against the Defense of Marriage Act (DOMA), created the minority AIDS initiative federal program to reduce the spread of HIV in Black, Latinx, and other Communities of Color, and fought to protect housing opportunities for persons living with HIV (HOPWA).[55] She stated, "We need to remain vigilant in protecting programs that we know work, including HOPWA, for example. And I, for one, will never stop advocating for the resources that LGBTQ people need to succeed."[56] In using the inclusive term, "we," coupled with a desire for the group to engage in action and be "vigilant," in the face of corruption, divisiveness, and a common enemy yet to be named (Trump, and in particular the Republican party and White House staff), Waters set up a rhetorical battle between good and evil: "We have been bombarded with numerous policies from the White House rescinding important rights and protections for LGBTQ people across various programs and sectors."[57] Harkening back to Alice Walker's definition of Womanism, Lee's womanist statement is apt: Waters "has a strong, positive, courageous, and inclusionary (but critical) attitude toward life and people."[58]

In a Black Feminist move calling out Black men as complicit in hegemonic destruction, Waters then took aim at Ben Carson, the only remaining African American serving in the White House as head of the Department of Housing and Urban Development (HUD). Carson, a trained pediatric brain surgeon, has no experience leading an organization with an overt advocacy role fighting for those who are oppressed. Waters dismantled Carson's record with humor: "That's Ben Carson I'm talking about. That's the one they say is so brilliant. Brain surgeon. Send him back to the operating room!"[59] Whereas Womanism seeks harmony between Black men and women, Black Feminism overtly resists and critiques Black patriarchy. Waters questioned Carson's ethos and called out the mistaken assumption that his medical credentials—earned through the white patriarchal U.S. education system—qualify him to work as an advocate for marginalized and disadvantaged groups. Railing against the hegemonic establishment that erroneously equates advanced degrees and professional status with knowing what is "best" for those of "lesser" education and status, she offered a plain ("home") truth: what makes Carson "brilliant" in one area renders him insufficient in another. Rather than taking the Womanist route of suggesting what Carson's training might provide or by suggesting his strengths, as well as potential areas for learning and improvement, Waters demonstrated a Black Feminist approach—rejecting him because he is a contributor to the problem.

Waters's critique of Carson also amplified both Womanist and Black Feminist concerns over economic, class, and racially biased systems. Balancing humor with facts, Waters noted that under Carson's leadership, HUD has violated civil rights laws and nondiscrimination rules by making LGBTQ people homeless and less safe in homeless shelters. "It is totally unacceptable that HUD has backed away from its responsibility to provide homeless service providers with the tools and guidance they need to ensure compliance with civil rights laws and HUDs non-discrimination rules."[60]

Waters further called out Carson for trying to change HUD's social justice mission: "[Carson] announced a proposed change to HUD's mission statement and that would remove text promising inclusive and discrimination-free communities. This is just the latest in a series of hostile overtures the Trump administration has made against the LGBTQ community."[61] In recounting Carson's complicity with Trump and his willful violation of civil rights laws, Waters refused to keep LGBTQ people's lives invisible and erased. Her appeals to "responsibility" in support of homeless service providers speak from the Womanist perspective seeking to serve larger, broader communities rather than address race- and gender-based injustice. In the same breath, her direct calling out of "hostile overtures" discriminating against LBGTQ people was in more of a militant, Black Feminist style. Waters's combined Womanist/Black Feminist approach highlighted the multiple oppressions perpetuated by the White House administration. Rhetorically, her focused critique then prepared her to directly take on her primary opponent, President Trump. She labeled him "dishonorable, deceitful, and despicable;" in other words, a direct antithesis to Black Feminist values centered on confronting injustice and to Womanist values centered on harmonizing home truths. [62]

"Audacious, courageous or willful*": fighting back*

Waters showed her diverse audience there is strength in numbers, and equality, if they work together. As bell hooks noted,

> Aware feminist activists have insisted that the anti-racist struggle is best advanced by theory that speaks about the importance of acknowledging the way positive recognition and acceptance of difference is a necessary starting point as we work to eradicate white supremacy.[63]

hooks calls out the unnatural alliance people have to the naturalized binary divisions of race; the same can be said about gender rights: "A wealthy, well-educated, straight, white woman becomes a traitor to her class, race, and sexuality by fighting for the advancement of poor black, brown, red, yellow, and white women of different sexual orientations."[64] Using Black Feminism, Waters challenged raced and gendered binaries (because even oppressed people can be classist, racist, sexist, and anti-transgender), by focusing on the tie that binds them all together—Donald Trump. She recounted 29 diverse unpatriotic acts by Trump, including banning transgender people from military service; denying application of the 1964 Civil Rights Act to LBGTQ people; supporting white nationalists; insulting the people of Puerto Rico following the devastation of Hurricane Maria; lying, insulting, and assaulting women; and labeling immigrants as terrorists and rapists. In demonstrating why she is referred to as "Auntie Maxine," her "home truth" invective took up eight minutes of her speech.[65] In the climax of her speech, Waters invoked patriotism and Republican complicity: "I don't know how long his Republican friends on the opposite side of the aisle are going to stand with him; you know these are the ones who claim that they're more patriotic than anybody else."[66] Through calling out the division and wrongfulness of the other side, Waters reinforced for her audience that they belonged together, on the side of justice, love, and inclusion.

To further unite her diverse audience together as one, as one group ready to fight against a disparate number of oppressions, Waters used pathos appeal to claim that they, in fact, are the true patriots. She rhetorically encouraged her audience to move to political action and points out that they have already been doing just that, Womanist acts of community recuperation. She told her audience,

You have taken the abuse, you have loved your country, you have worked with your country; you have done everything to strengthen this democracy. That's what real patriotism is all about. It is not about what is happening now with Republicans who are intimidated and afraid to stand up against this man...You supported this democracy, and you've refused to allow anybody to separate you from your right to be a legitimate part of everything that goes on.[67]

Again hearkening back to how Black people have been enslaved, excluded, segregated, ignored, and denied their rights as full citizens of the nation, Waters reassured her audience that they are visible, that their agency and autonomy matter, and that they are rightful citizens of the American community. Her Womanist approach highlighted the efforts of the diverse group to encourage their continued unified fight, which completed her battle analogy as related to her audience. Waters is experienced with having to fight, to defend herself against racist and sexist insults. Throughout her tenure in the national spotlight, the embodied nature of her politics has become crystal clear. Beyond deflecting racist and gendered insults, when her personal safety has been threatened—her response has been, "I ain't scared."[68] Congresswoman Waters, like the Black feminist women who preceded her and who fight alongside her, must constantly defend her right to exist, to have a voice, and to contribute to (or *lead*) in the public sphere.

"Responsible. In charge. Serious." Reclaiming my time

To reclaim is to "recover" or to return to a state of cultivation. Black women have lived a history of reclaiming their bodies, their voices, and their presence in the face of intersectional oppressions. In another unforgettable moment of public performance, Waters's reclamation of her own presence and agency—represented as her "time" to speak as a Congresswoman—went viral over social, as well as traditional, media.

The phrase "reclaiming my time" was Waters's response to Treasury Secretary Steve Mnuchin's attempts to silence her. During a House Financial Services Committee hearing, on the Congressional House Floor, in front of stoic white men, Waters, a lone African American woman, asked why Mnuchin's office had failed to respond to her inquiries regarding the investigations into Trump's ties to Russia. Because of Mnuchin's failure to respond to Waters's letter, a deadline to impose sanctions against Russia, as ordered by the U.S. Congress, was missed. Instead of immediately responding to Waters's question during this committee hearing, Mnuchin tried to use up the time she was allotted to ask her question by offering Waters "platitudes and compliments."[69] His attempts to stall for time did not work. In response, "Waters shut down his rambling and redirected him to her question again and again with the phrase, 'Reclaiming my time,' a stone-faced invocation of House procedural rules."[70] She said, "We don't want to take my time up with how great I am" and "when you're on my time I can reclaim it."[71] And with those words, a reclamation rallying crying began.

Indicative of both Black Feminism and Womanism, People of Color often have to reclaim their time. Reclaim their agency. Reclaim their bodies. Reclaim their voices. Waters's Womanist insistence on having her portion of the allotted process, rather than being obedient and subservient, was cheered because it was an "outrageous, audacious, courageous"[72] act in the face of white patriarchal silencing. It was an overt political move, aimed at stopping a micro-moment of economic injustice (where time, as a resource, was not fairly distributed). Waters found herself in an unbearable situation. Rules were there for white men to access and enact, not for African American women to access and use. Waters would not be made to feel grateful and flattered for merely being in the presence of the committee. It was all too easy for Mnuchin to hide from his white supremacist strategy of flattery, thinking that a compliment was all that was needed to distract "Auntie Maxine."

"Reclaiming my time" was also a way for Waters to envelope and include the other social justice demonstrations that have resonated with American lives; she wanted her audience to be reminded that rights marches did not begin and end in the 1950s, 1960s, or 1970s, but rather they have the right to march, they have the right to be angry, and they have the right to be heard. Anger, as bell hooks reminds us, is a powerful emotion that should be embraced. In her HRC speech, Waters reassured the audience that "they had a right to be angry about the election of Trump [because] this president is not normal. But… all is not lost because each of us has a power to set this country back on track."[73] Amplified by the personal also being political, such anger informs CRC-style Black Feminist militant attitudes and actions designed to resist and disrupt oppressive regimes. Just as the generations of Black Feminists and Womanists before her, Waters has been and continues to advocate for power and justice beyond the scope of what would benefit her as a Black woman.

Waters has shown, whether it be through her agency with "reclaiming her time" to subvert Mnuchin's tactics or in her HRC speech, that she has been "punished." In her HRC speech, she described retaliatory efforts to defame her character:

> They're lying about me forging documents. The racist billboards that they putting up. The racist smears and the dirty tricks to try and defeat me in this election. I'm in this fight. I am NOT going to back down. And I believe that all of us deserve better than Donald Trump. I believe this country deserves better than Donald Trump.[74]

The covert "racist smears" that Waters alludes to are deeply entrenched in historical racism against African Americans. Trump's most common racist denigration of Waters is that she is "a very low IQ," "a low IQ individual. [She] can't help it" and "[Waters] should take an IQ test."[75] And in 2018 at a rally in Ohio before an ecstatic crowd of supporters Trump referred to Waters as "a seriously low-IQ person."[76] While Trump insults political rivals or people he believes have wronged him, he reserves insults against one's intelligence for people who are Black/African American. In 2018 alone, Trump referred to Waters as having a "low IQ" seven times.[77] CNN journalist Don Lemon said, "referring to African Americans as dumb is one of the oldest canards of America's racist past and present: that Black people are of inferior intelligence. This president constantly denigrates people of color and women."[78] According to Peter Wehner, a veteran of three Republican administrations, "Trump's made the same criticism of black athletes, black journalists and black Members of Congress. He attacks their intelligence. His racist appeals aren't even disguised anymore."[79] As political science professor Michael Cornfield stated, "The strategic value is obvious: Waters is a quadruple-category demon to please Trump's base, being black, female, leftist, and aggressive in her own rhetoric."[80]

Waters's body as a woman and more significantly as a Black woman has been wounded, threatened, and insulted, yet she has refused to give up. Her body in this instance becomes the visual signifier for how Black feminism differs from white feminism. As Audre Lorde famously said, "Black feminism is not white feminism in blackface" and the use of racism by her political adversaries points to the racist wound they tried to inflict upon Waters in an effort to damage her ethos as well as her power to speak out.[81] In the HRC speech, Waters's love of country shined through; she was being willful and would not lie down and be silent or silenced. Through her body as sacrifice, Waters's rhetoric sought to awaken her supporters' unified Womanist and Black Feminist consciousness, inspiring brave persistence in an inclusive progressive movement:

The resistance is a progressive movement against this President composed of everybody woman, man, trans, gay, white, black, Latino, Muslim, Asian, young, and old who have been energized and activated by all of the terrible things Trump is trying to do. I'm so honored to have been adopted by them and the millennials who view me as their Auntie in the resistance.[82]

Auntie Maxine ended her fiery HRC speech with a call to action and a call to work together for the greater good. This call to action was broader and deeper than the white feminist movement, because Waters was engaging in Black Feminist and Womanist perspectives. Her positioning and call not only acknowledged interlocking systems of domination like gender, race, and LGBTQ status, but also fought against the white supremacist, patriarchal hegemonic order—something with which first and second waves of feminism did not ultimately break ranks. As our literature review detailed, in each moment when white women could have promoted all women in the name of sisterhood, they instead chose to stand with the white men in their lives. White feminist organizations turned their backs on marginalized and disenfranchised communities in favor of the maintenance of white supremacist hegemonic hierarchy.

Reunification through Black Feminism and Womanism: a conclusion

Whether harmonizing the connections between activism in the LBGTQ and Black rights movements or defending her right to be present and speak, "Auntie Maxine" engages an embodied identity politics, unafraid and "acting out" to fight for transformative change through the lens of integrated Black Feminism/Womanism. Waters embraces both the militant warrior resistance of the CRC Black Feminists and the inclusive, harmonizing role of the Womanist. In doing so, she reunifies the Black feminist perspectives that splintered during the second wave.

Waters's innovative approach demonstrates how contemporary Black feminism outpaces white feminism by fighting through the lens of interlocking systems of domination, also referred to as intersectionality. By "interlocking systems of domination," we mean that Waters goes beyond fighting on behalf of those whom she directly represents as a woman and a Black woman. She engages other aspects of marginalization and disenfranchisement through her allyship with People of Color, the LBGTQ community, people who are homeless and impoverished, people who are living with long-term health and disabilities, and more. Whereas white feminism has focused primarily on gender-only issues, Black feminisms have not had that luxury. They have been fighting an intersectional fight all along.

Waters shows us how to use an interlocking systems or intersectional approach to seek justice across a range of intersecting identities. Through her rhetorical positioning and strategies, she reveals that traditional white feminism is outdated. If the focus is on gender only, then all of the other intersecting identities that exist on the margins are erased. If third-wave "intersectionality" has provided the theory and the method to understanding interlocking oppressions, then Waters's intersectional positioning and rhetoric enact how to speak home truths and build alliances that resist, subvert, and work to dismantle the people and systems that continue to marginalize, erase, and oppress.

Notes

1 hooks, *Feminist Theory*, 28.
2 hooks, *Ain't I*, 159–172.

3 We now know the famously circulated version of Truth's speech was written by a white woman over a decade after the comments were delivered. However, the rhetorical situation—Truth speaking to a white woman's rights convention and calling out oppressive systems—is persistent in both the inaccurate version and a more accurate version Marius Robinson recorded and worked with Truth to review. See Podell.

4 hooks, *Ain't I*, 171.

5 hooks, *Ain't I*, 171.

6 hooks, *Ain't I*, 173.

7 See Roth, *Separate Roads to Feminism,* 80-86 and hooks *Ain't I,* 177–184.

8 hooks provides a moving personal first-hand account of the disillusioned frustration felt my women of color (*Ain't I,* 188–191).

9 hooks, *Ain't I,* 179–181.

10 These included The Black Women's Liberation Group of Mount Vernon/New Rochelle, New York; the Third World Women's Alliance; the National Black Feminist Organization; the National Alliance of Black Feminists; Black Women Organized for Action; and the Combahee River Collective (see Roth, *Separate Roads to Feminism,* Chapter 3).

11 Capitalization of Womanist and Black Feminist are intentional to identify them as particular strains of thought tracing back to Alice Walker and to the CRC, respectively. When a more general collective of perspectives is meant, "Black feminism" is used.

12 Combahee River Collective, "Combahee River Collective Statement," 279.

13 Combahee River Collective, "Combahee River Collective Statement," 281.

14 Combahee River Collective, "Combahee River Collective Statement," 72.

15 Combahee River Collective, "Combahee River Collective Statement," 275.

16 Combahee River Collective, "Combahee River Collective Statement," 281.

17 Combahee River Collective, "Combahee River Collective Statement," 273.

18 Layli Marparyan explains that two other Womanist theories emerged during the same general time period as Walker's. Chikwenya Okonjo, a Nigerian literary critic, developed her theory of African Womanism while US African Studies scholar Clenora Hudson-Weems introduced Africana Womanism (22–28).

19 Walker, "Womanist," 19.

20 Helen (charles) questions the feasibility of the Womanist movement, finding it potentially exclusive even as it seeks to provide an overly inclusive–and strategically essential–umbrella term that both embraces and departs from feminism. Helen concludes that Walker's "womanist ideology can be seen to describe more the spirit of black women, which also has a place, as opposed to a fiercely political identity" (367).

21 hooks, *Ain't I,* 178.

22 Combahee River Collective, "Combahee River Collective Statement," 279, 275.

23 Smith, "Introduction," xxvii.

24 Collins, "Sisters and Brothers," 60.

25 Walker, "Womanist," 19.

26 Maparyan, *Womanist Idea,* 322.

27 Roth, *Separate Roads to Feminism,* 98.

28 Smith, "Introduction," xxix.

29 Smith, "Introduction," xxix-xxx.

30 Combahee River Collective, "Combahee River Collective Statement," 277.

31 Collins, "Sisters and Brothers," 61.

32 The essentialism/anti-essentialism debates are too complex to address here. "Essentialism" assumes that some foundational quality of an identity exists—for example, that being a biological woman is equated with specific feminine qualities such as emotionality, empathy, and nurturing. *Strategic* essentialism is the purposeful engagement of essential identities for the political purposes. "Anti-essentialism," as the CRC asserted, denies any such essence as part of identity. Post-colonial theorist Gayatri Spivak is credited with introducing the concept, although she also backed away from it later in her work. See Danius, Jonsson, and Spivak.

33 Combahee River Collective, "Combahee River Collective Statement," 274–275.

34 Combahee River Collective, "Combahee River Collective Statement," 275.

35 Phillips and McCaskill, "Daughters and Sons," 93.

36 Smith, "Introduction," xxv.

37 Walker, "Womanist," 19.

38 Smith, "Introduction," xxxv.
39 Patricia Hill Collins and Sirma Bilge's chapter, "Getting the History of Intersectionality Straight?," traces the multiple, intertwined influences that led to the emergence of this important contemporary theory and tool. The CRC and Walker are both acknowledged as contributing to the process.
40 Collins and Bilge, *Intersectionality*, 5.
41 "About Maxine."
42 "About Maxine."
43 "About Maxine."
44 Wire, "Maxine Waters Became Auntie Maxine."
45 Wire, "Maxine Waters Became Auntie Maxine."
46 "About Maxine."
47 Parshall, "Resistance Reclaims Its Time,".
48 Cillizza, "The 64 Most Outrageous Line,".
49 Human Rights Campaign, 00:22-00.35.
50 Lee, "One Whiteness Veils Three Uglinesses," 281.
51 Human Rights Campaign, 01:54-02.14.
52 Human Rights Campaign, 02:16-02:44.
53 Lee, "One Whiteness Veils Three Uglinesses," 281.
54 Lee, "One Whiteness Veils Three Uglinesses," 281.
55 Human Rights Campaign, 06:01-07:28.
56 Human Rights Campaign, 07:32-07:47.
57 Human Rights Campaign, 08:04-08:16.
58 Lee, "One Whiteness Veils Three Uglinesses," 281.
59 Human Rights Campaign, 08:20-08:35.
60 Human Rights Campaign, 09:23-09:39.
61 Human Rights Campaign, 09:53-10:14.
62 Human Rights Campaign, 11:20-11:23
63 hooks, *Black Looks*, 13.
64 Lee, "One Whiteness Veils Three Uglinesses," 282.
65 Human Rights Campaign, 10:16-18:02
66 Human Rights Campaign, 18:02-18:13.
67 Human Rights Campaign, 18:16-19:13.
68 In October 2018, Waters was sent a bomb in the mail—part of a larger plot by a Trump follower seeking to extract revenge on those who would resist the President's agenda—and her response on being asked about her reaction to the bomb was "I ain't scared" (Sonmez).
69 Emba, "Reclaiming My Time,".
70 Emba, "Reclaiming My Time,".
71 Live On-Air News, 1:19, 1:35-1:38, 2:02-2:06.
72 Walker, "Womanist," 19.
73 Human Rights Campaign, 22:40-22:59.
74 Human Rights Campaign, 19:26-19:54.
75 Ruiz, "Trump Again Questions Maxine Waters' Intelligence," para. 1, 4, 6.
76 Smith "Trump's Tactic to Attack Black People and Women,", para. 5.
77 Smith, "Trump's Tactic to Attack Black People and Women," para. 5.
78 Smith, "Trump's Tactic to Attack Black People and Women," para. 13.
79 Smith, "Trump's Tactic to Attack Black People and Women," para. 4.
80 Smith, "Trump's Tactic to Attack Black People and Women," para. 9.
81 Lorde, "Sexism," 44.
82 Human Rights Campaign, 19:55-20:19.

Bibliography

"About Maxine: Biography," *Congresswoman Maxine Waters*. Accessed October 19, 2018. https://waters.house.gov/about-maxine/biography.
(charles), Helen. "Harmony, Hegemony, or Healing?" In *The Womanist Reader*, edited by Layli Phillips, 361–378. New York: Routledge, 2006.

Cillizza, Chris. "The 64 Most Outrageous Line from Donald Trump's Untethered Pennsylvania Speech." *The Point with Chris Cillizza*. March 12, 2018. https://edition.cnn.com/2018/03/11/politics/trump-speech-pennsylvania/index.html.

Collins, Patricia Hill. "Sisters and Brothers: Black Feminists on Womanism." In *The Womanist Reader*, edited by Layli Phillips, 57–67. New York: Routledge, 2006.

Collins, Patricia Hill and Sirma Bilge. *Intersectionality*. Malden, MA: Polity Press, 2016.

Combahee River Collective. "The Combahee River Collective Statement." In *Home Girls: A Black Feminist Anthology*, edited by Barbara Smith, 272–282. New Brunswick, NJ: Rutgers University Press, 1983.

Danius, Sara, Stefan Jonsson and Gayatri Chakravorty Spivak. "An Interview with Gayatri Chakravorty Spivak." *Boundary* 20, no. 2 (1993): 24–50.

Emba, Christine. "'Reclaiming My Time' is Bigger Than Maxine Waters." *The Washington Post*. August 1, 2017. www.washingtonpost.com/blogs/post-partisan/wp/2017/08/01/reclaiming-my-time-is-bigger-than-maxine-waters/?noredirect=on&utm_term=.5878a6e251c3.

hooks, bell. *Ain't I a Woman: Black Women and Feminism*. Boston, MA: South End Press, 1981.

hooks, bell. *Black Looks: Race and Representation*. Boston, MA: South End, 1992.

hooks, bell. *Feminist Theory: From Margin to Center*, 2nd edn. Cambridge, MA: South End Press, 2000.

Human Rights Campaign. "U.S. Representative Maxine Waters Speaks at HRC Los Angeles Dinner." YouTube video. 24:34. Posted March 11, 2018. www.youtube.com/watch?v=ioiU3mrR_Ro.

Lee, Wen Shu. "One Whiteness Veils Three Uglinesses: From Border-Crossing to a Womanist Interrogation of Gendered Colorism." In *Whiteness: The Communication of Social Identity*, edited by Thomas K. Nakayama and Judith N. Martin, 279–298. Thousand Oaks, CA: Sage, 1999.

Live On-Air News, "Maxine Waters Grills Steve Mnuchin On Russia & Trump 7/27/17," YouTube video. 11.44. Posted July 27, 2017. www.youtube.com/watch?v=jFJOJUIGEwY.

Lorde, Audre. "Sexism: An American Disease in Blackface." In *I Am Your Sister: Collected and Unpublished Writings of Audre Lorde (Transgressing Boundaries: Studies in Black Politics and Black Communities)*, edited by Rudolph P. Byrd, Johnnetta Betsch Cole, and Beverly Guy-Sheftall, 44. Oxford, UK: Oxford University Press, 2011.

Maparyan, Layli. *The Womanist Idea*. New York: Routledge, 2012.

Parshall, Helen. "The Resistance Reclaims Its Time at the HRC Los Angeles Dinner." *Human Rights Campaign*. March 14, 2018. www.hrc.org/blog/the-resistance-reclaims-its-time-at-the-hrc-los-angeles-dinner.

Phillips, Layli and Barbara McCaskill. "Daughters and Sons: The Birth of Womanist Identity." In *The Womanist Reader*, edited by Layli Phillips, 85–95. New York: Routledge, 2006.

Podell, Leslie. "Compare the Speeches." *The Sojourner Truth Project*. Accessed February 8, 2019. www.thesojournertruthproject.com/compare-the-speeches/.

Roth, Benita. *Separate Roads to Feminism: Black, Chicana, and White Feminist Movements in America's Second Wave*. Cambridge, UK: Cambridge University Press, 2004.

Ruiz, Joe. "Trump Again Questions Maxine Waters' Intelligence, Says She's 'Very Low IQ'." March 11, 2018. www.cnn.com/2018/03/10/politics/trump-waters-low-iq-individual/index.html.

Smith, Barbara. "Introduction." In *Home Girls: A Black Feminist Anthology*, edited by Barbara Smith, xix–lvi. New Brunswick, NJ: Rutgers University Press, 1983.

Smith, David. "Trump's Tactic to Attack Black People and Women: Insult Their Intelligence." August 10, 2018. www.theguardian.com/us-news/2018/aug/10/trump-attacks-twitter-black-people-women.

Sonmez, Felicia. "Rep. Maxine Waters on Being Targeted by Mail Bombs: 'I Ain't Scared'." *Washington Post*. October 25, 2018. www.washingtonpost.com/politics/rep-maxine-waters-on-being-targeted-by-mail-bombs-i-aint-scared/2018/10/25/6c8e46b8-d8a7-11e8-a10f-b51546b10756_story.html.

Walker, Alice. *In Search of Our Mothers' Gardens: Womanist Prose*. New York: Harcourt Brace Jovanovich, 1983.

Walker, Alice. "Womanist." In *The Womanist Reader*, edited by Layli Phillips, 19. New York: Routledge, 2006.

Wire, Sarah. "How Maxine Waters Became 'Auntie Maxine' in the Age of Trump." *LA Times*. April 30, 2017. www.latimes.com/politics/la-pol-ca-maxine-waters-20170430-htmlstory.html.

16

ONE STEP FORWARD ...

Gender, communication and the fragility of gender(ed) political progress

Michele L. Hammers, Nina M. Lozano, and Craig O. Rich[1]

On October 10, 2017, Ronan Farrow's article in *The New Yorker* detailed a story of systematic sexual harassment by entertainment mogul Harvey Weinstein.[2] Farrow's article came just days after *The New York Times* published its own story, by Jodi Kantor and Meghan Twohey, on Weinstein.[3] These articles would share the 2018 Pulitzer Prize for Public Service.

Originally founded in 2006 by Tarana Burke on the MySpace platform in an effort to empower women of color who had experienced sexual abuse,[4] the *#MeToo* hashtag leapt into high profile usage across social media platforms in 2017 following the Weinstein coverage. *#MeToo* has served as a powerful tool for exposing sexual harassment as a pernicious problem—and not just in the entertainment industry. *#MeToo* also sparked discussion of harassment in the sciences, academics, and politics. In addition to providing a platform for the sharing of stories, *#MeToo* has drawn attention to the importance of supporting survivors who come forward, particularly in the face of uneven power dynamics, organizational and professional cultures that turn a blind eye to questionable practices, and the larger social and political forces at the heart of it all. Between *#MeToo's* founding in 2006 and its sudden social media explosion in 2017, there was a similar social media initiative; in 2014 *#yesallwomen* rose to prominence after a shooting spree in Isla Vista, California, which was (partially) attributed to the shooter's misogyny.[5] As with *#MeToo*, the grassroots *#yesallwomen* movement saw an outpouring of women's experiences with harassment, assault, and other forms of violence.[6]

Of course, with something as power-laden as gender and sex, responses to *#yesallwomen* and *#MeToo* have been complex—if not controversial. The *#yesallwomen* campaign was criticized for unfairly implying that all men are predatory or unsafe. Efforts within the Twitter discourse to differentiate between the claim that "all women have at some point felt unsafe" and the claim that "all men are predators" were fraught. The woman who originated the *#yesallwomen* tag faced rape and death threats.[7] In addition to challenges from those who equated the tag with unfair misandry, the tag also drew critique from women of color who used the tag *#YesAllWhiteWomen* to draw attention to the ways in which the original tag—like many aspects of traditional feminism—erased their voices and experiences.

#MeToo and *#yesallwomen* provide a snapshot of some of the ways in which matters of gender, sex, and power play out in the general arena of our everyday lived experiences and in our efforts to fully participate in particular spaces of publicness.[8] The battlelines are, perhaps, more sharply drawn with increased effort going into expanding inclusivity, targeting specific points of

intersectionality, and tackling particular legal contexts. However, at heart we are still engaged in the same fundamental battle—the fight for legal protection and full, public inclusion for all individuals regardless of sex, gender, sexual orientation, or the expression of any of these.

The current U.S. political climate, specifically under the Trump administration, has seen these fundamental battles reinvigorated by force of necessity. From Supreme Court appointments with serious implications for reproductive rights and same-sex marriage to policy-by-tweet attacks on transgender rights, the Trump administration has destabilized the progress we thought we'd made. Brett Kavanaugh was confirmed to the U.S. Supreme Court in 2019, despite credible allegations of sexual assault by Dr. Christine Blasey Ford. As the confirmation process played out, Dr. Ford faced what she (and we) all knew to be the inevitable backlash against those who bring forward allegations against powerful men—efforts to discredit her memory, accusations of political opportunism, mockery, and death threats.

Millions of women—and men—rallied to support Dr. Ford and, through her, all survivors of assault who feel silenced by the power structures that dominate our private, corporate, and civic cultures. Alternatively, Kavanaugh defenders relied upon the all too common defenses of "boys will be boys" and "his life shouldn't be ruined over a mistake he made years ago." The Kavanaugh confirmation demonstrates the stark realities of gender, power, gender(ed) power, and the ongoing, all-out war of Trump's *#MAGA* administration on our sexed/gendered political rights and legal protections.[9]

This overview just begins to touch upon the ways in which gender(ed) identity and expression continue to play dominant roles in United States' national, as well as international, political (in)action. While we have, in some cases, seen gender(ed) and sex(ed) identities more broadly included in social discourse and legal protections, under the Trump administration, we also have seen numerous legal rollbacks of these same rights. Additionally, our professional and public spheres continue to be constrained by regressive cultural constructions of sex, gender, and sexual orientation, as well as restrictive articulations of the potential intersections among them.

As a discipline, Communication Studies has the potential to both explain and intervene in the gendered legal and political processes that dominate our contemporary experiences. We explore this potential by presenting three close, critical sub-disciplinary analyses of gendered politics and practices. First, public sphere and media studies scholarship provides critical insight into the gendered hierarchy that continues to define how female politicians have their identities and participation constructed as part of the political process. Next, organizational communication scholarship further highlights the ongoing relevance of conservative, sex/gender-based definitions of identity and behavior by examining (or leaving unexamined) questions of Title XII and Title IV protections. Finally, critical analysis of Trump's anti-protest rhetoric and his rollback of sex/gender-based rights—particularly trans rights—provides insight into the importance of public protest as a tool for social change. Ultimately, we contend that future social change depends upon increased emphasis on the broader potentialities of gender(ed) identities and that substantial, lasting legal change toward increased protections and inclusion must be driven by theoretical and political practices informed by social change from below.

Gender, politics, and the media: grabbing pussy and the double-bind

In 2016, the United States saw a woman receive a major party nomination for the office of President for the first time. Former Secretary of State Hillary Clinton won the Democratic Party's nomination and landed in a campaign against Republican nominee Donald Trump. Throughout the campaign we saw overt misogynistic themes play out in Trump's campaign rhetoric, the rhetoric of Trump supporters, and in more subtle ways, media coverage of Clinton's campaign efforts.[10]

The 2016 campaign was a perfect storm of gender(ed) politics. Trump's comments about getting away with grabbing women "by the pussy"[11] caused outrage among many, but failed to register as definitively problematic with the majority of his supporters.[12] Trump, buoyed by his supporters, once publicly declared that he could "stand in the middle of 5th Avenue and shoot somebody and not lose any voters."[13] But while Trump basked in the untouchable-ness of his overtly misogynistic, toxic masculinity, at the other end of the spectrum, Clinton, under public scrutiny for both past political actions and (most problematically) for gendered identity traits, faced the damned- if-you-do and damned-if-you-don't double-bind of daring to be female in public.

This double-bind, grounded in the mind/body dichotomy, goes back to classical philosophy and has wound its way through Western thought. The "mind," associated with rationality and publicness, also represents the transcendence of the always lesser "body." The body—associated with the home—also carries with it the tincture of excessive emotion, illness, and death. In classical philosophy, the pursuit of knowledge and (T)ruth required the transcendence of the mundane, ephemeral, subject-to-decay realm of the body.[14] Because of their association with transcendent values, questions of civic importance—as opposed to questions of the personal or familial—required the triumph of mind over body. Thus, the mind/body duality was paired with a parallel public/private duality.

These binaries also carried gendered associations; the realm of mind, along with the public realm, was exclusively male. The body and the realm of the private were then, and have continued to be, coded as female. These binaries and their extended associations are value-laden, with the body/private/female construction being devalued vis-à-vis the mind/public/male triumvirate. Just as the transcendent, rational mind trumps the weakness of the decaying, emotional body, the public sphere (with its public interests) trumps the private realm (with private concerns), and the performative-masculine trumps the always, already devalued performance of the feminine.[15] In this hierarchical, gendered dichotomy, women are always, already defined by their bodies— bodies that are, by default, inappropriate for public life.[16]

The mind/body duality has been even more narrowly articulated by others studying women and leadership. Dustin Harp, Jaime Loke, and Ingrid Bachman invoke the dichotomy between brain and womb in their analysis of media coverage of Hillary Clinton during the hearing related to her involvement in the 2012 Benghazi incident. They explain that

> to be a woman (i.e., to have a womb) and to perform feminine qualities conflicts with perceptions of competence. Power is understood to be a masculine domain, and within the framework of double bind theory, to be feminine, then, is to lack power.[17]

Women in (or seeking) positions of leadership, particularly would-be political leadership at the level of the U.S. presidency,[18] confront this double-bind as they navigate the practical demands of being female-in-public. Further, the mind/body double-bind consistently plays out in an overt—and undue—emphasis on female bodies *as bodies* in the public sphere. For example, Eva Flicker's study of visual representations of high-level, female politicians globally affirms: "Women in exposed professional positions are often judged not only by their professional achievements but also on the basis of their physical attributes…physiognomy and wardrobe."[19] During the 2016 presidential elections, we saw this emphasis on the devalued female-body-in-public in the fact that references to and attacks on Clinton-as-candidate were often grounded in her body—her age, health, and appearance.[20] As Bock, Byrd-Craven, and Burkley note, both Clinton and Carly Fiorina—a Republican Presidential candidate—frequently had "their personhood... reduced to appearances."[21]

But reduction to the body *per se* is not the only way in which Clinton and other female political actors are reduced to the devalued feminine. We have to remember that the body is associated with emotionality—so when it's not the body itself that is under scrutiny, it's the body-as-emotional that draws fire "because emotions in general are seen as needing to be controlled (consider expressions like "losing control," "emotional outburst," or "feeling upset"), the woman-emotion pair further conveys women's lack of reason and competence to lead or hold positions of power."[22]

For instance, coverage of Clinton during the Benghazi hearings, while including a narrative addressing Clinton's political competence, offset that seemingly positive framing with other narratives that reinforced traditional gender stereotypes and assumptions/prescriptions related to public participation. Specifically, coverage focused on Clinton's portrayal of emotion—both anger and sadness—during the hearings.[23] Later, during her 2016 bid for the presidency, Clinton was repeatedly accused of not having the right temperament or general personality traits for the role.[24]

Whether facing undue emphasis on their bodies or heightened focus on their (lack of) emotionality, female politicians, like Clinton, have to make the impossible choice between being perceived as competent and unfeminine or being seen as feminine and incompetent. One might think that choosing competence would be the obvious choice—except that women who are viewed as unfeminine often suffer from the damaging consequences of being considered unlikeable.[25]

Burdened by a centuries-old hierarchical binary, the female body disallows full participation in publicness. From Elizabeth Dole in 2000 to Hillary Clinton nearly 20 years later, female bids for the U.S. presidency demonstrate the fact that traditional gendered stereotypes continue to dominate how we respond to—and whether we ultimately vote for—female political candidates.[26]

Unfortunately for Clinton and for the rest of us, there is (still) no good way to be female-in-public. The mind/body, rational/emotional, public/private, male/female string of hierarchical binaries constrains and contorts women's public performance of political (or any other kind of) leadership. Even more unfortunately, these ages old social constructions are not only deeply embedded in how we "do" sexed/gendered politics—they are just as deeply embedded in how we construct and understand sexed/gendered identities when it comes to the laws we create, interpret, and enforce through our political processes.

Dis/organizing sex: protections (?) under the law

The regressive devaluation of the female body in the public sphere, resulting from its hierarchical binary construction, also includes women in organizational settings. In many ways, organizations are gendered in their central substance—whether as structures that reproduce masculinity and advantage its embodiment[27] or as cultures that problematize women's bodies and their performance of professionalism.[28] Through this lens, organizations, consistent with our larger political culture, reify the female-body-as-problem and devalue femaleness and femininity as less than its ideal masculine counterpart.

As we confront and re-confront this gendered organization of power relations that continue to oppress and marginalize women, we must also push our work further—both in terms of the law and the role of heteronormativity in galvanizing this organization of sex and gender. Set against #*MeToo* and Hillary Clinton's double-binded presidential campaign, *The New York Times* reported on October 21, 2018, that the Department of Health and Human Services (DHHS) is organizing an effort among other federal agencies to (re)define sex, having significant implications for federal protections offered through Titles VII and IX and the LGBTQ community.[29] In drawing together *#MeToo* and *#WontBeErased*, this section draws attention to the dis/organization of sex (as well as its complex connections with gender and sexuality) in order to underscore a deeper challenge for both movements, namely the politics of the binary in constructing legal protections offered by Titles VII and IX.

As federal laws, both Titles VII and IX prohibit discrimination within employment, educational, and public contexts. In particular, Title VII of the Civil Rights Act of 1964 outlaws discriminatory practice or unequal treatment "because of such individual's race, color, religion, sex, or national origin."[30] For instance, sexual harassment is a legally actionable form of sex discrimination under Title VII by virtue of the fact that it is a form of discrimination "because of...sex," or rather an individual's sex is the basis for inequitable treatment or harassment. Comparatively, Title IX of the Education Amendments of 1972 bans sex-based discrimination in educational activities and institutions.[31] According to Title IX, "No person in the United States shall, on the basis of sex, be excluded from participation in, be denied the benefits of, or be subjected to discrimination under any education program or activity receiving Federal financial assistance." Collectively, both Titles VII and IX provide legally sanctioned protection against sex discrimination across a variety of organizational and public contexts; however, "because...of sex" is heavily contested, and our scholarly apprehension of it tends to be deeply flawed and ironically exclusionary.

While sex discrimination has received attention from a variety of our subfields in Communication Studies, organizational communication studies of sexual harassment are particularly illustrative in shedding light on how our scholarly discourses have constructed deeply flawed understandings of sex discrimination. For example, the vast bulk of scholarship on sexual harassment focuses on female victims, and although notable exceptions exist that study the sexual harassment of men,[32] they are constructed as doubtful through the frequent use of "allegedly"[33] or as an exotic, "interesting" spectacle.[34] By and large, Communication Studies scholarship assumes and reproduces a gendered model of sex discrimination in which—always already—women are victims and men are aggressors.[35]

The gendered model of sex discrimination is further problematized in its coverage of gay and lesbian victims. In the rare case that they are studied, the sexual harassment of gays and lesbians was constructed as either a "minor" problem, too inconsequential perhaps to warrant further study,[36] or a "minority" problem, too unique to the gay and lesbian community in light of the larger problem faced by our heterosexual counterparts.[37] Through this lens, our model of sexual harassment is deeply heteronormative, implanting a normatively heterosexual desire between the woman/victim and man/aggressor,[38] and when coupled with the deafening silence surrounding trans* persons and sexual harassment, it is fair to assume that this model is cisgendered.

Perhaps most troubling is the harmony between the Communication Studies model of sexual harassment and the proposed redefinition of sex spearheaded by the Trump administration's DHHS.[39] According to a memo obtained by *The Times*, sex would be defined as "a person's status as male or female based on immutable biological traits identifiable by or before birth," and definitive evidence of one's sex would include what is "...listed on a person's birth certificate, as originally issued,...unless rebutted by reliable genetic evidence."[40] Through its efforts to coordinate a standardized, uniform, and alarmingly restrictive definition of sex among federal agencies, the DHHS is ultimately organizing "because...of sex" under Titles VII and IX with the result only erasing trans* persons, while also problematizing would-be federal protects for LGBTQ individuals.

In many ways, the DHHS' efforts to codify a binary model of sex exists in tension with a growing recognition of the complex and heavily fluid nature sex and gender. For instance, members of the scientific community recognize that sex, far from a neat-and-tidy binary organization, is heavily varied at both the chromosomal and anatomical level. To posit that sex is "either/or" grossly ignores abundant examples of "both/and" in human genetics and physiology, such as individuals who are intersexed or have chromosomal variations beyond XX or XY.[41]

Furthermore, legal precedence also attests to how Titles VII and IX's "because...of sex" includes forms of sex discrimination centered around gender performance and expression. In particular, the Supreme Court case of *Price Waterhouse v. Hopkins* expanded sex discrimination to include gender stereotyping.[42] In the case, Ann Hopkins sued her employer Price Waterhouse for denying her promotion because of her gender performance, citing how colleagues suggested that she should "walk more femininely, talk more femininely, dress more femininely, wear make-up, have her hair styled, and wear jewelry."[43] The Supreme Court held that Title VII's provision against sex discrimination included forms of gender stereotyping, or rather discrimination, rooted in expectations for "appropriate" gendered behavior aligned with sex. Through this lens, *Price Waterhouse v. Hopkins* acknowledges a larger, more fluid performance of gender uncoupled from sex, and forms of discrimination rooted in stereotypical alignments of sex and gender are illegal and actionable.

Collectively, the growing recognition of sex and gender's fluidity served as the foundation and rationale for the Obama administration's expansion of gender protections under Titles VII and IX, particularly their extension to include gender identity. However, the Trump administration has steadily rolled back those efforts, ranging from rescinding guidelines that protect transgender students to attempts to formalize a transgender military ban. In many ways, the DHHS' work to (re)define sex is consistent with prior actions of the Trump administration; however, the DHHS efforts point to a reductionist return to a flawed, binary-based model of sex and gender. Such a model, painfully consistent with our own construction of sex discrimination, largely erases trans* and intersex persons while problematizing a larger project aimed at creating equity for both members of the LGBTQ community, as well as cisgendered women.

When seen in this light, we—as an international community of Communication Studies scholars, practitioners, educators, and activists—must embrace a project of dis/organization, namely one that critically questions and disrupts our own latent assumptions that continue to reify heteronormative, cisgendered assumptions of sex discrimination. Such an endeavor should also be firmly allied with fostering a dialogic engagement with the field of law while also forming coalitional partnerships across differences. Ultimately, "because...of sex" is "up for grabs," and to the extent we remain stymied in flawed frameworks of sex, gender, and sexuality, we lose out on the radical potential to reshape or organize a more equitable future.

A rejection of civility to protest groups' erasures

As noted above, in 2018 the Trump administration's DHHS was exploring policy that would legally define gender as solely based upon an immutable biological categorization defined by the individual's genitalia at birth. In addition to the implications for Title VII and Title IX protections, this move by President Trump is a rollback of trans* rights gained under the Obama administration that included legislative protections such as bathroom bills, college dormitory policies, and single-sex programs.[44] Implications for trans* individuals are far-reaching and warrant particular attention as we consider the potential for communication scholars and activists to engage in meaningful projects for social change.

Before discussing how social change can be made, we further interrogate the current political climate. The dominant discourse, coming from the Trump administration, calls for groups who oppose current political decisions to engage in rhetorics of civility. Trump, on October 23, 2018, stated, "The media also has a responsibility to set a civil tone and to stop the endless hostility and constant negative and oftentimes false attacks and stories."[45] But as scholars of social justice and social movements have pointed out, historically, and today, discourses of civility are used to protect the current social order. Work by Itagaki reminds us, "Civility has been about making sure that the status quo, the hierarchy of the status quo at the moment, which means racial inequality, gender inequality, class inequality, stays permanent." [46] Further, As Lozano-Reich and Cloud[47] have argued, appeals to rhetorics of civility have been used historically to silence the most historically and currently oppressed and marginalized groups. Indeed, any social justice movement has always and already operated outside the boundaries of civility and decorum. Mayo's[48] work, similarly, argues that "civility is a form of social discrimination, for it is predicated on making distinctions that support accepted practices and values, and entails enacting those distinctions to the detriment of the purportedly uncivil." Indigenous groups had to be civilized. LGBTQ non-normative enactments of desire operate outside the norms of civility. People of color are continually depicted as "angry uncivil" groups. In other words, protest is inherently uncivil and is used as a silencing strategy for communication projects of dissent. As such, we call for an "uncivil tongue"[49] when attempting to challenge inequalities and injustice. Further, we call for scholars engaging these justice projects to enact the subject position of the scholar-activist.[50]

Communication Studies scholarship provides insight into the repressive role of "civility" as a dissent-limiting political tool. This insight is particularly relevant to an understanding of the political context under the Trump administration, which has taken an aggressive stance against dissenters who choose to engage in strategies and tactics of protest. Within this context it has become increasingly difficult for protestors to engage safely, and without fear of rebuke, in places and spaces of resistance. When analyzing the rhetoric of President Trump in relation to individuals and groups who engage in strategies and tactics of protest, we see that a clear pattern of attempts to intimidate and quell protest, under the guise of calls for civility, currently exists.

President Trump espouses an explicit and bold pattern of anti-protest rhetoric. For example, as part of the *#MeToo* movement's protests enacted at the confirmation hearings of now Supreme Court Justice Brett M. Kavanaugh, President Trump, in an interview with the *Daily Caller*, stated,

> I don't know why they don't take care of a situation like that. I think it's embarrassing for the country to allow protestors. You don't even know what side the protestors are on. In the old days, we used to throw them out. Today, I guess they just keep screaming.[51]

In reality, the anti-protest climate began well before the Kavanaugh hearings. For example, in July of 2018, prior to President Trump's travels to Britain, when it was announced that massive demonstrations would be greeting the President in London, alongside a 20-foot-tall blow-up blimp effigy of Donald Trump depicted as a baby in a diaper, the President stated, "I guess when they put out blimps to make me feel unwelcome, no reason for me to go to London."[52] In addition, in September 2018, President Trump called upon National Football League owners to fire contract athletes who, following the lead of Colin Kaepernick, chose to "take a knee" during the national anthem to protest systematic police brutality and support the *#Blacklivesmatter* movement. Trump stated, "Wouldn't you love to see one of these NFL owners when somebody disrespects our flag, to say, 'Get that son of a bitch off the field right now, out. He's fired. He's fired!'"[53] President Trump's patterned rhetoric has created a clear climate of anti-protest in the United States of America.

With this context in mind, we can look more closely at the particular political dangers and activist efforts surrounding trans* populations.[54] We focus here on the trans* population both as an extension of our discussion of the regressive reach of traditional gender and sexed-based binaries and as a population facing particular political and social threats at this time.[55]

In 2016, trans* advocacy groups documented at least 23 deaths of transgender individuals in the United States. While the material circumstances around these deaths certainly vary, it is clear that "fatal violence disproportionally affects transgender women of color and that the intersections of racism, sexism, homophobia, and transphobia conspire to deprive them of employment, housing, healthcare, and other necessities,"[56] which make them a particularly vulnerable population. In the year 2017, for instance, at least 25 transgender people were killed,[57] and in 2018, 22 trans* deaths in the United States have been documented.[58]

The term "femicide" was first coined by Russell[59] to describe the killing of women and girls based upon their gender. Subsequently, work by Lozano[60] argues that gendered killings must be attended to in all of their material, cultural, political, and socioeconomic contexts and locations. What we are witnessing in this current historical moment merits its own theoretical term: "transcide." Transcide urges scholars and activists to examine the material, political, and symbolic exigencies surrounding the killing of trans* folk.

Knowing the very real threats facing the trans* population, we now turn briefly to the types of communicative and embodied acts of resistance for trans* rights. These acts of resistance, of course, take place in relation to the current political climate, which increasingly insists upon civility as the solution for individuals engaging in forms of dissent. Presently protests against the Trump administration are taking place under the hashtag *#WontBeErased*. For example, on October 22, 2018, demonstrators gathered outside of the White House to demand that transgender people not be erased from society. In addition to protests taking place outside of the White House, rallies for transgender rights also took place in New York, Washington, Chicago, Los Angeles, as well as other cities. Transgender advocate Chelsea Manning tweeted: "Laws don't determine our existence—*we* determine our existence."[61] Notably, trans* activists, belonging to the TransLatin Coalition, attempting to take advantage of a viewership of 14.3 million viewers during the Dodgers and Boston Red Sox game five of the World Series,[62] unfurled a huge banner, down the left-field line of Dodger Stadium, on the top deck, that stated, "Trans People Deserve to Live." Although the activist tactic of the banner drop made international news outlets and went viral on social media, the banner was never shown during the Fox Sports Broadcast.[63]

The aforementioned examples are part of a larger societal crack-down on individuals and groups' right to dissent. In 2017, the National Lawyers Guild reported that conservative-led anti-dissent pieces of legislation have already doubled since 2016. In 2018, the number of anti-protest bills peaked to 58 in 31 states, with no hint of ending anytime soon. Eight bills have already become law, including the states of North Carolina, North Dakota, South Dakota, Oklahoma, and Tennessee. Moreover, 28 bills are currently under review in state legislatures. These pieces of legislation coincide with the crackdowns of free speech on college campuses, as well as the growing phenomenon of professor watch lists. These crackdowns on dissent are not a coincidence; the Koch-funded policy organizations are the same interest groups behind these limits on dissent.[64]

The consequences of anti-dissent pieces of legislation, as well as the President calling to make protest illegal, cannot be understated. For those engaged in the fights for justice and liberation movements, the current exigence to counter this chilling climate—both politically and legally, against dissent, is dire. *#MeToo, #Blacklivesmatter, #WontBeErased*, alongside other movements, are matters of life and death. Hate crimes against women, trans* people, Blacks, and other minority groups have increased 17% in 2017, compared to 2016. Of the more that 7,000 hate crimes reported in 2017, almost three out of five were documented as motivated by race and ethnicity. Religion and sexual orientation were the other two primary motivators of the hate crimes.[65] Consequently, with hate crimes on the rise, and the ability for individuals and groups to dissent decreasing at an alarming rate, continuing to resist Trump's calls for civility, hate speech, and anti-dissent legislation is urgent. Continuing to demand justice, is not just a political project, it is a life and death project. And aren't those the same?

Conclusion

Hierarchical gendered/sexed binaries—like the mind/body split—unrealistically burden female participation in public and professional contexts. These deeply embedded binaries also contribute to the regressive resistance to broader acceptance and legal protection for gender/sexual fluidity. Ultimately, those burdened with the weight of differently gendered/sexed bodies face barriers to full, effective, and even safe participation in political and professional spheres. In today's current climate of *#MeToo* and *#WontBeErased,* these barriers are more ripe for analysis and action than ever before.

Hillary Clinton may not have won the 2016 presidential election, despite winning the popular vote, but in 2018 we saw a record number of women win seats in the U.S. House of Representatives, and other firsts—such as the first Native American women and Muslim woman elected to Congress, the first openly gay man elected governor, and the first female U.S. senators from Tennessee and Arizona.[66] It was not the "Blue Wave" that many hoped for, but small inroads were made toward more full and diverse representation in political office. Whether those elected to office will be able to effectively outmaneuver and outlast the stranglehold of "politics as usual" remains to be seen.

While some work to change the political status quo through the process of voting for, and hopefully electing, candidates who are potentially change agents, others seek change through challenging the very system itself. For example, despite the anti-protest climate created by the Trump administration, radical activists are increasingly making bold efforts to prevent the legal erasure of trans* individuals, amongst the other aforementioned current social movements. What Trump bemoans as a "problem"—the rise of well-organized, persistent, public protests—may be one of our greatest strengths at this time. Indeed, many activists are calling for more radical change, as exemplified in the dominant slogan, "we are unstoppable, another world is possible."

To help imagine and create this other possible world, we call upon our field to undertake a reparative reworking of our own scholarly assumptions of sex and gender, especially our latent, cisgendered and heteronormative assumptions. Our field also has the ability to push boundaries by seeking out increased interdisciplinary engagement, particularly with legal scholarship—and its practical application in both legislative and judicial arenas. Finally, we reiterate Lozano's call for communication scholars to occupy, as much as possible, the position of the scholar-activist in one (or all) of the many variations of that role.

There is much concern regarding gender, politics, and the law, especially in an era when the President believes it is appropriate to grab women "by the pussy." As scholars, educators, professionals in particular fields, or public advocates, we each have the opportunity to contribute to broader, more inclusive conversations and cultural movements. We must remain vigilant with regard to the sexed/gendered legal protections we have won over the years. We also must never forget that even our most generous protections fail to provide full justice to many. And as long as even one individual remains unprotected by our laws and vulnerable to cultural bigotry, none of us are really safe.

Notes

1 All authors are equal contributors. Authors' names are listed alphabetically.
2 Ronan Farrow, "From Aggressive Overtures to Sexual Assault: Harvey Weinstein's Accusers Tell Their Stories," *The New Yorker*, October 10, 2017, www.newyorker.com/news/news-desk/from-aggressive-overtures-to-sexual-assault-harvey-weinsteins-accusers-tell-their-stories.
3 Jodi Kantor and Megan Twohey, "Harvey Weinstein Paid Off Sexual Harassment Accusers for Decades," *The New York Times*, October 5, 2017, www.nytimes.com/2017/10/05/us/harvey-weinstein-harassment-allegations.html.
4 "Me Too." Accessed February 27, 2020, https://metoomvmt.org. It should be noted that actress Alysa Milano began using the *#Me Too* hashtag in response to the Weinstein accusations helping spur its 2017 rise; she later acknowledged Burke's prior use of the tag.
5 The shooter had posted a highly misogynistic "manifesto" online, fueling attributions that his shooting spree—which appeared to target women—was grounded in that misogyny. However, there was speculation that he suffered from mental health problems and that the shootings were linked more to mental illness than to misogyny.
6 Sasha Weiss, "The Power of #YesAllWomen," *The New Yorker*, May 26, 2014, www.newyorker.com/culture/culture-desk/the-power-of-yesallwomen.
7 Kaye M., "On #YesAllwomen One Year Later," *The Toast*, May 26, 2015, http://the-toast.net/2015/05/26/yesallwomen-one-year-later/.
8 There is one more movement that's worth reflecting upon. Almost two decades before *#Me Too* and *#yesallwomen* flooded the social media sphere, another movement was taking on these same fundamental issues. The *V-Day* movement, grounded in Eve Ensler's play *The Vagina Monologues*, began as a small groundswell of performances across the United States determined to draw attention to—and end—women's experiences of harassment, assault, and abuse. While flawed, the *V-Day* movement ignited both controversy and conversation—particularly on college campuses, where student groups continue to perform the play as a means of raising funds and awareness. Importantly, the *V-Day* movement has evolved in its two decades. Over time the movement received criticism for a narrow definition of "woman" and for not being inclusive of transwomen. The movement now explicitly defines its purpose as "eliminating violence against all women and girls (cisgender, transgender, and gender non-conforming)." What these movements, originating 20 years apart, tell us about our current social and political situation is that while we have obtained expanded legal rights in some areas related to gender, sex, and sexual orientation, really, at a basic level not that much has changed. "V-Day." Accessed February 27, 2020, www.vday.org/homepage.html.
9 Garrett Epps, "The Subtext of Kavanaugh's Nomination Bursts into the Open," *The Atlantic,* September 16, 2018, www.scribd.com/article/388733957/The-Subtext-Of-Kavanaugh-S-Nomination-Bursts-Into-The-Open.
10 Jarrod Bock, Jennifer Byrd-Craven, and Melissa Burkley, "The Role of Sexism in Voting in the 2016 Presidential Election," *Personality and Individual Differences* 119 (2017): 189–193; and Nina Lozano Reich, "Sexism: Alive and Well in 2016 Presidential Campaign," *Huffpost*, February 8, 2016, www.huffingtonpost.com/nina-m-lozanoreich-phd/sexism-alive-and-well-in-2016-presidential-campaign_b_9172186.html.
11 "Transcript: Donald Trump's Taped Comments About Women," *The New York Times*, October 8, 2016, www.nytimes.com/2016/10/08/us/donald-trump-tape-transcript.html.
12 Bonnie Dow argues that we should consider the possibility that Trump's supporters liked the general impulses he stood for, while not necessarily believing he would take the extreme actions and positions he espoused. Thus, they may not have supported his racism, xenophobia, and misogyny—but they were able (and willing) to look past it. Bonnie Dow, "Taking Trump Seriously: Persona and Presidential Politics," *Women's Studies in Communication* 40, no. 2 (2017): 136–139.
13 "Donald Trump, '"Fifth Avenue Comment,'" Snopes. Accessed October 2, 2018, www.snopes.com/fact-check/donald-trump-fifth-avenue-comment/.
14 Susan Bordo, *Unbearable Weight: Feminism, Western Culture, and the Body* (Berkeley, CA: University of California Press, 1995).
15 Dustin Harp, Jaime Loke, and Ingrid Bachmann, "Hillary Clinton's Benghazi Hearing Coverage: Political Competence, Authenticity, and the Persistence of the Double Bind," *Women's Studies in Communication* 39, no. 2 (2016): 193.

16 Harp, Loke, and Bachmann, "Hillary Clinton's Benghazi Hearing Coverage."

17 Harp, Loke, and Bachmann, "Hillary Clinton's Benghazi Hearing Coverage," 195.

18 Caroline Heldman, Susan Carroll, and Stephanie Olson, "'She Brought Only a Skirt': Print Media Coverage of Elizabeth Dole's Bid for the Republican Presidential Nomination," *Political Communication* 22 (2005): 316.

19 Eva Flicker, "Fasionable (Dis-)order in Politics: Gender, Power and the Dilemma of the Suit," *International of Media & Cultural Politics* 9, no. 2 (2013): 202.

20 Bock, Byrd-Craven, and Burkley, "The Role of Sexism," 189. In addition, Jayeon Lee and Young-shin Lim found that Clinton's own social media communications emphasized personality traits, particularly her masculine traits, as more Trump did in his social media communication. "Gendered Campaign Tweets: The Cases of Hilary Clinton and Donald Trump," *Public Relations Review* 42, no. 5 (2016): 849–855. Finally, in "She Brought Only A Skirt," Heldman, Carroll and Olson examine Elizabeth Dole's 2000 bid for the presidency and find similar emphasis on Dole's appearance, along with her personality and family.

21 Bock, Byrd-Craven, and Burkley, "The Role of Sexism," 189.

22 Harp, Loke, and Bachmann, "Benghazi Hearing Coverage," 196.

23 Harp, Loke, and Bachmann, "Benghazi Hearing Coverage."

24 Bock, Byrd-Craven, and Burkley, "The Role of Sexism," 189.

25 Karrin Vasby Anderson, "Presidential Pioneer or Campaign Queen? Hillary Clinton and the First-timer/Frontrunner Double Bind," *Rhetoric & Public Affairs* 20, no. 3 (2017): 526–527.

26 Bock, Byrd-Craven, and Burkley, "The Role of Sexism," 192.

27 Joan Acker, "Hierarchies, Jobs, Bodies: A Theory of Gendered Organizations," *Gender & Society* 4, no. 2 (1990): 139–158.

28 Angela Trethewey, "Disciplined Bodies: Women's Embodied Identities at Work," *Organization Studies* 20, no. 3 (1999): 423–450.

29 Erica Green, Katie Benner, and Robert Pear, "'Transgender' Could Be Defined Out of Existence under Trump Administration," *The New York Times*, October 21, 2018, www.nytimes.com/2018/10/21/us/ politics/transgender-trump-administration-sex-definition.html.

30 United States, "Civil Rights Act of 1964," Title VII, Sec. 701.

31 United States, "Education Amendments," Title IX, Sec. 1681.

32 Robin P. Clair, "Resistance and Oppression as a Self-contained Opposite: An Organizational Communication Analysis of One Man's Story of Sexual Harassment," *Western Journal of Communication* 58, no. 4 (1994): 235–262.

33 Clair, "Resistance and Oppression."

34 Jennifer A. Scarduzio and Patricia Geist-Martin, "Accounting for Victimization: Male Professors' Ideological Positioning in Stories of Sexual Harassment," *Management Communication Quarterly* 24, no. 3 (2010): 419–445.

35 Joanna Brewis, "Foucault, Politics and Organizations: (Re)-Constructing Sexual Harassment," *Gender, Work and Organization* 8, no. 1 (2001): 37–60.

36 Denise H. Solomon and Mary L. Miller Williams, "Perceptions of Social-Sexual Communication at Work as Sexually Harassing," *Management Communication Quarterly* 11, no. 2 (1997): 159.

37 Diana K. Ivy and Stephen Hamlet, "College Students and Sexual Dynamics: Two Studies of Peer Sexual Harassment," *Communication Education* 45, no. 2 (1996): 155, 158.

38 Craig O. Rich, "The Longings of Labor and Working Identity: Gender, Sexuality, and the Organization of Hairstylists and Barbers." (PhD diss., University of Utah, 2009).

39 Silvia Gherardi, *Gender, Symbolism and Organizational Cultures* (London: Sage, 1995). Gherardi observes an unwitting allyship between organizational discourses that construct organizations as asexual and feminist discourses aimed at eradicating coercive sexuality in the workplace; the eerily combined union produces "a form of social control which regards sex with prudish distaste" within a larger goal of organizational desexualization.

40 Green, Benner, and Pear, "'Transgender' Could Be Defined," 5.

41 Anne Fausto-Sterling, *Sexing the Body: Gender Politics and the Construction of Sexuality* (New York, NY: Basic Books, 2000).

42 *Price Waterhouse v. Hopkins*, 490 U.S. 228 (1989).

43 *Price Waterhouse v. Hopkins*, 490 U.S. 228 (1989), 4.

44 Green, Benner, and Pear, "'Transgender' Could Be Defined."

45 John Wagner, "Trump Doubles Down on Blaming Media as Suspicious Packages Continue to Surface," *Washington Post*, October 25, 2018, www.washingtonpost.com/politics/trump-doubles-down-on-blaming-media-as-suspicious-packages-continue-to-surface/2018/10/25/507aeec2-d848-11e8-a10f-b51546b10756_story.html.

46 Leila Fadel, quoting an interview with Lynn Itagaki, "In These Divided Times, is Civility Under Siege?" *NPR*, March 12, 2019, www.npr.org/2019/03/12/702011061/in-these-divided-times-is-civility-under-siege.

47 Nina Lozano-Reich and Dana Cloud, "The Uncivil Tongue: Invitational Rhetoric and the Problem of Inequality," *Western Journal of Communication* 73, no. 2 (2009): 220–226.

48 Cris Mayo, "The Binds That Tie: Civility and Social Difference," *Educational Theory* 52, no. 2 (2002): 169–186.

49 Kyra Pearson and Nina Lozano-Reich, "Cultivating Queer Publics with an Uncivil Tongue: Queer Eye's Critical Performances of Desire," *Text and Performance Quarterly* 29, no. 4 (2009): 383–402.

50 Nina Maria Lozano, *Not One More!: Feminicidio on the Border* (Columbus, OH: The Ohio State University Press, 2019).

51 Felicia Sonmez, "Trump Suggests that Protesting Should Be Illegal," *Washington Post*, September 05, 2018, www.washingtonpost.com/politics/trump-suggests-protesting-should-be-illegal/2018/09/04/11cfd9be-b0a0-11e8-aed9-001309990777_story.html?utm_term=.bfb780467314/.

52 Paul Sandle, "Snarling Orange 'Trump Baby' Blimp Flies Outside British Parliament," *Reuters*, July 13, 2018, www.reuters.com/article/us-usa-trump-britain-blimp/snarling-orange-trump-baby-blimp-flies-outside-british-parliament-idUSKBN1K30Z3/.

53 Jeremy Gottlieb and Mark Maske, "Roger Goodell Responds to Trump's Call to 'Fire' NFL Players Protesting During National Anthem," *Washington Post*, September 23, 2017, www.washingtonpost.com/news/early-lead/wp/2017/09/22/donald-trump-profanely-implores-nfl-owners-to-fire-players-protesting-national-anthem/?utm_term=.afb004e6d3af/.

54 We recognize that transgender individuals may self-identify in many ways, including gender non-binary, gender fluid, gender-non-conforming, etc. Indeed, the HRC tracking of transgender violence, one category is "non-binary/other/identity unclear." However, for the purpose of tracking transgender violence, we use this umbrella term throughout.

55 While we focus on trans* populations, we should be mindful not to rank oppression, but view subject positions of discrimination as "interlocking systems of oppression." Patricia Hill Collins, *Black Feminist Thought: Knowledge, Consciousness, and the Politics of Empowerment*. 1st edition (London: Routledge, 2008). All movements must be viewed in an intersectional fashion. Kimberle Crenshaw, "Demarginalizing the Intersection of Race and Sex: A Black Feminist Critique of Antidiscrimination Doctrine, Feminist Theory and Antiracist Politics," *University of Chicago Legal Forum* 140 (1989): 139–167.

56 Human Rights Campaign, "Violence Against the Transgender Community in 2017," Accessed February 27, 2020, www.hrc.org/resources/violence-against-the-transgender-community-in-2017.

57 Jackie Flynn Mogensen, "2017 Was the Deadliest Year for Trans People in at Least a Decade," *Mother Jones*, November 17, 2017, www.motherjones.com/crime-justice/2017/11/its-2017-and-trans-people-are-dying-violent-deaths-in-record-numbers/.

58 Human Rights Campaign, "Violence Against the Transgender Community in 2018," Accessed February 27, 2020. www.hrc.org/resources/violence-against-the-transgender-community-in-2018.

59 Diana E. Russell, *Femicide in Global Perspective* (Columbia: Teachers College, 2001).

60 Lozano, *Not One More!: Feminicidio on the Border*.

61 Sarah Mervosh and Christine Hauser, "At Rallies and Online, Transgender People Say They #WontBeErased," *The New York Times*, October 22, 2018, www.nytimes.com/2018/10/22/us/transgender-reaction-rally.html.

62 Joe Otterson, "World Series Rating Fall 23% from 2017," *Variety*, October 29, 2018, https://variety.com/2018/tv/news/world-series-ratings-2018-1202994171/.

63 Cindy Boren, "Transgender Rights Activist Unfurl Giant Banner During World Series: 'Trans People Deserve to Live,'" *Washington Post*, October 29, 2018, www.washingtonpost.com/sports/2018/10/29/transgender-rights-activists-unfurl-giant-banner-during-world-series-trans-people-deserve-live/?utm_term=.b2e756c603fc.

64 Traci Yoder, "Conservative-led Anti-Protest Legislation Already Doubled Since Last Year," *National Lawyers Guild*, February 15, 2018, www.nlg.org/conservative-led-anti-protest-legislation-already-doubled-since-last-year/.

65 John Eligon, "Hate Crimes Increase for the Third Consecutive Year, F.B.I. Reports," *The New York Times*, November 13, 2018, www.nytimes.com/2018/11/13/us/hate-crimes-fbi-2017.html.
66 Eli Watkins, "Women and LGBTQ Candidates Make History in 2018 Midterms," *CNN Politics*, November 7, 2018, www.cnn.com/2018/11/07/politics/historic-firsts-midterms/index.html.

Bibliography

Acker, Joan. "Hierarchies, Jobs, Bodies: A Theory of Gendered Organizations." *Gender & Society* 4, no. 2 (1990): 139–158.

Anderson, Karrin Vasby. "Presidential Pioneer or Campaign Queen? Hillary Clinton and the First-timer/Frontrunner Double Bind." *Rhetoric & Public Affairs* 20, no. 3 (2017): 525–538.

Bock, Jarrod, Byrd-Craven, Jennifer, and Burkley, Melissa. "The Role of Sexism in Voting in the 2016 Presidential Election." *Personality and Individual Differences* 119 (2017): 189–193.

Bordo, Susan. *Unbearable Weight: Feminism, Western Culture, and the Body*. Berkeley, CA: University of California Press, 1995.

Boren, Cindy. "Transgender Rights Activist Unfurl Giant Banner During World Series: 'Trans People Deserve to Live.'" *Washington Post*, October 29, 2018. www.washingtonpost.com/sports/2018/10/29/transgender-rights-activists-unfurl-giant-banner-during-world-series-trans-people-deserve-live/?utm_term=.b2e756c603fc.

Brewis, Joanna. "Foucault, Politics and Organizations: (Re)-Constructing Sexual Harassment." *Gender, Work and Organization* 8, no. 1 (2001): 37–60.

Clair, Robin P. "Resistance and Oppression as a Self-contained Opposite: An Organizational Communication Analysis of One Man's Story of Sexual Harassment." *Western Journal of Communication* 58, no. 4 (1994): 235–262.

Crenshaw, Kimberle. "Demarginalizing the Intersection of Race and Sex: A Black Feminist Critique of Antidiscrimination Doctrine, Feminist Theory and Antiracist Politics." *University of Chicago Legal Forum* 140 (1989): 139–167.

"Donald Trump 'Fifth Avenue' Comment." Snopes. Accessed October 2, 2018. www.snopes.com/fact-check/donald-trump-fifth-avenue-comment/.

Dow, Bonnie. "Taking Trump Seriously: Persona and Presidential Politics." *Women's Studies in Communication* 40, no. 2 (2017): 136–139.

Eligon, John. "Hate Crimes Increase for the Third Consecutive Year, F.B.I. Reports." *The New York Times*, November 13, 2018. www.nytimes.com/2018/11/13/us/hate-crimes-fbi-2017.html.

Epps, Garrett. "The Subtext of Kavanaugh's Nomination Bursts into the Open." *The Atlantic*, September 16, 2018. www.scribd.com/article/388733957/The-Subtext-Of-Kavanaugh-S-Nomination-Bursts-Into-The-Open.

Fadel, Leila. "In These Divided Times, Is Civility Under Siege?" *NPR*, March 12, 2019. www.npr.org/2019/03/12/702011061/in-these-divided-times-is-civility-under-siege.

Farrow, Ronan. "From Aggressive Overtures to Sexual Assault: Harvey Weinstein's Accusers Tell Their Stories." *The New Yorker*, October 10, 2017. www.newyorker.com/news/news-desk/from-aggressive-overtures-to-sexual-assault-harvey-weinsteins-accusers-tell-their-stories.

Fausto-Sterling, Anne. *Sexing the Body: Gender Politics and the Construction of Sexuality*. New York, NY: Basic Books, 2000.

Flicker, Eva. "Fashionable (Dis-)order in Politics: Gender, Power, and the Dilemma of the Suit." *International of Media & Cultural Politics* 9, no. 2 (2013): 201–219.

Gherardi, Silvia. *Gender, Symbolism and Organizational Cultures*. London: Sage, 1995.

Gottlieb, Jeremy and Maske, Mark. "Roger Goodell Responds to Trump's Call to 'Fire' NFL Players Protesting During National Anthem." *Washington Post*, September 23, 2017. www.washingtonpost.com/news/early-lead/wp/2017/09/22/donald-trump-profanely-implores-nfl-owners-to-fire-players-protesting-national-anthem/?utm_term=.afb004e6d3af/.

Green, Erica, Benner, Katie, and Pear, Robert. "'Transgender' Could Be Defined Out of Existence Under Trump Administration." *The New York Times*, October 21, 2018. www.nytimes.com/2018/10/21/us/politics/transgender-trump-administration-sex-definition.html.

Grosz, Elizabeth. *Volatile Bodies: Toward a Corporeal Feminism*. Bloomington, IN: Indiana University Press, 1994.

Harp, Dustin, Loke, Jaime, and Bachmann, Ingrid. "Hillary Clinton's Benghazi Hearing Coverage: Political Competence, Authenticity, and the Persistence of the Double Bind." *Women's Studies in Communication* 39, no. 2 (2016): 193–210.

Heldman, Caroline, Carroll, Susan J., and Olson, Stephanie. "'She Brought Only a Skirt': Print Media Coverage of Elizabeth Dole's Bid for the Republican Presidential Nomination." *Political Communication* 22 (2005): 315–335.

Hill Collins, Patricia. *Black Feminist Thought: Knowledge, Consciousness, and the Politics of Empowerment.* 1st edition. London: Routledge, 2008.

Human Rights Campaign. "Violence Against the Transgender Community in 2017." Accessed February 27, 2020. www.hrc.org/resources/violence-against-the-transgender-community-in-2017.

Human Rights Campaign. "Violence Against the Transgender Community in 2018." Accessed February 27, 2020. www.hrc.org/resources/violence-against-the-transgender-community-in-2018.

Ivy, Diana K. and Hamlet, Stephen. "College Students and Sexual Dynamics: Two Studies of Peer Sexual Harassment." *Communication Education* 45, no. 2 (1996): 149–166.

Kantor, Jodi and Twohey, Megan. "Harvey Weinstein Paid Off Sexual Harassment Accusers for Decades." *The New York Times*, October 5, 2017. www.nytimes.com/2017/10/05/us/harvey-weinstein-harassment-allegations.html.

Lee, Jayeon and Lim, Young-shin. "Gendered Campaign Tweets: The Cases of Hillary Clinton and Donald Trump." *Public Relations Review* 42, no. 5 (2016): 849–855.

Lozano, Nina Maria. *Not One More!: Feminicidio on the Border.* Columbus, OH: The Ohio State University Press, 2019.

Lozano-Reich, Nina. "Sexism: Alive and Well in 2016 Presidential Campaign." *Huffpost*, February 8, 2016. www.huffingtonpost.com/nina-m-lozanoreich-phd/sexism-alive-and-well-in-2016-presidential-campaign_b_9172186.html.

Lozano-Reich, Nina and Cloud, Dana. "The Uncivil Tongue: Invitational Rhetoric and the Problem of Inequality." *Western Journal of Communication* 73, no. 2 (2009): 220–226.

M., Kaye. "On #YesAllwomen One Year Later." *The Toast*, May 26, 2015. http://the-toast.net/2015/05/26/yesallwomen-one-year-later/.

Mayo, Cris. "The Binds That Tie: Civility and Social Difference." *Educational Theory* 52, no. 2 (2002): 169–186.

"Me Too." Accessed February 27, 2020. https://metoomvmt.org.

Mervosh, Sarah and Hauser, Christine. "At Rallies and Online, Transgender People Say They #WontBeErased." *The New York Times*, October 22, 2018. www.nytimes.com/2018/10/22/us/transgender-reaction-rally.html.

Mogensen, Jackie Flynn. "2017 Was the Deadliest Year for Trans People in at Least a Decade." *Mother Jones*, November 17, 2017. www.motherjones.com/crime-justice/2017/11/its-2017-and-trans-people-are-dying-violent-deaths-in-record-numbers/.

Otterson, Joe. "World Series Rating Fall 23% from 2017." *Variety*, October 29, 2018. https://variety.com/2018/tv/news/world-series-ratings-2018-1202994171/.

Pearson, Kyra and Lozano-Reich, Nina. "Cultivating Queer Publics with an Uncivil Tongue: Queer Eye's Critical Performances of Desire." *Text and Performance Quarterly* 29, no. 4 (2009): 383–402.

Rich, Craig O. "The Longings of Labor and Working Identity: Gender, Sexuality, and the Organization of Hairstylists and Barbers." (PhD diss., University of Utah, 2009).

Russell, Diana E. H. *Femicide in Global Perspective.* New York: Teachers College Press, 2001.

Sandle, Paul. "Snarling Orange 'Trump Baby' Blimp Flies Outside British Parliament.' *Reuters*, July 13, 2018. www.reuters.com/article/us-usa-trump-britain-blimp/snarling-orange-trump-baby-blimp-flies-outside-british-parliament-idUSKBN1K30Z3/.

Scarduzio, Jennifer A. and Geist-Martin, Patricia. "Accounting for Victimization: Male Professors' Ideological Positioning in Stories of Sexual Harassment." *Management Communication Quarterly* 24, no. 3 (2010): 419–445.

Solomon, Denise H. and Miller Williams, Mary L. "Perceptions of Social-Sexual Communication at Work as Sexually Harassing." *Management Communication Quarterly* 11, no. 2 (1997): 147–184.

Sonmez, Felicia. "Trump Suggests that Protesting Should be Illegal." *Washington Post*, September 05, 2018. www.washingtonpost.com/politics/trump-suggests-protesting-should-be-illegal/2018/09/04/11cfd9be-b0a0-11e8-aed9-001309990777_story.html?utm_term=.bfb780467314/.

"Transcript: Donald Trump's Taped Comments About Women." *The New York Times*, October 8, 2016. www.nytimes.com/2016/10/08/us/donald-trump-tape-transcript.html.

Trethewey, Angela. "Disciplined Bodies: Women's Embodied Identities at Work." *Organization Studies* 20, no. 3 (1999): 423–450.

"V-Day." Accessed February 27, 2020. www.vday.org/homepage.html.

Wagner, John. "Trump Doubles Down on Blaming Media as Suspicious Packages Continue to Surface." *Washington Post*, October 25, 2018. www.washingtonpost.com/politics/trump-doubles-down-on-blaming-media-as-suspicious-packages-continue-to-surface/2018/10/25/507aeec2-d848-11e8-a10f-b51546b10756_story.html.

Watkins, Eli. "Women and LGBTQ Candidates Make History in 2018 Midterms." *CNN Politics*, November 7, 2018. www.cnn.com/2018/11/07/politics/historic-firsts-midterms/index.html.

Weiss, Sasha. "The Power of #YesAllWomen." *The New Yorker*, May 26, 2014. www.newyorker.com/culture/culture-desk/the-power-of-yesallwomen.

Yoder, Traci. "Conservative-led Anti-Protest Legislation Already Doubled Since Last Year." *National Lawyer's Guild*, February 15, 2018. www.nlg.org/conservative-led-anti-protest-legislation-already-doubled-since-last-year/.

17

THE SPECTER OF TRANS BODIES

Public and political discourse about "Bathroom Bills"

KC Councilor

Beginning in 2015, laws that require people to use the bathroom matching the biological sex on their birth certificates rather than their gender identities, so-called transgender "bathroom bills," have become "a flashpoint in the U.S. culture wars."[1] Of the many issues that transgender people face (from discrimination in employment and housing and health care to transphobic violence, poverty, and exclusion), legislation concerning public bathrooms might seem like the least of our problems. After all, bathrooms are already fraught spaces for trans and gender-nonconforming people. How much worse could these laws make things? How could bathrooms and locker rooms possibly be policed in practice?

However, access to public space, including bathrooms, has long been central to struggles over political power. The laws are symbolic measures used to demonize trans people as predators and consolidate conservative opposition to LGBTQIA+ political protections. Such legal documents have serious and wide-ranging effects. This essay analyzes public discourse around bathroom bills, particularly in North Carolina and Texas between 2015 and 2018. I argue that the vulnerability inherent in bathroom spaces, where people have to do private things in public places, creates an affectively charged site for mobilizing fear around transgender populations. At the same time, some of the tactics employed by some transgender activists normalize passable trans bodies and center white, able-bodied norms reinforcing stereotypes of how men and women "should" look, demonstrating the limits of trans inclusion within liberal rights paradigms.

Although there have been bathroom bills in a number of states, my analysis here is focused on deliberations surrounding the North Carolina and Texas bills for a few reasons. First, they were the highest-profile cases, drawing the most media attention and public commentary. Second, these case studies represent the broad themes present in debates over similar measures in other places—at least 19 states considered bathroom bills in 2016[2]—and are likely to reappear in future proposed legislation and public discourse. Third, North Carolina and Texas are important states in terms of LGBTQIA+ legislation and policy. The landmark case *Lawrence v. Texas*, decided by the Supreme Court in 2003, struck down Texas' sodomy law, and in doing so, invalidated similar statutes in 13 additional states.[3] North Carolina is a southern state that is largely conservative, but with several large liberal cities, like Charlotte. For this reason, it is an arena for contests between liberal municipalities and conservative state government, a dynamic apparent in many U.S. states.[4] The fact that activists and organizers in North Carolina mobilized a significant countermovement also makes it worthy of close attention. In what follows, I will speak to method and trans* positionality before moving into analysis of the cases themselves. Finally, I offer a critique of trans-normative activism and media narratives.

On trans* positionality

First a note on terminology: I am choosing to use the term "trans*" in this chapter because it is a more expansive category that allows for flexibility in not being limited to a single identity category. Trans* includes gender nonconforming and nonbinary people and those whose gender presentation makes them suspect in public spaces like bathrooms, not all of whom identify as transgender. Avery Tompkins states that using trans* with an asterisk "signals greater inclusivity of new gender identities and expressions and better represents a broader community of individuals."[5] In his 2018 book entitled *Trans**, Jack Halberstam builds on the work of Eva Haywood and Jami Weinstein,[6] explaining that he "embraces the nonspecificity of the term 'trans' and uses it to open the term up to a shifting set of conditions and possibilities rather than attach it only to the life narratives of a specific group of people."[7] Kai M. Green pushes this further, using trans* as an analytic: "as a decolonial demand; a question of how, when, and where one sees and knows; a reading practice that might help readers gain a reorientation to orientation."[8] Debates over terminology used to describe gender identity and sexual orientation are often contentious and have shifted rapidly over the last half-century as people grapple with language and definitions to describe their embodied experiences. I use the term trans* to be purposefully expansive, yet specific, and in an attempt to meet Green's call for an analytic that allows for creative resistance and new coalitions to challenge systems of power.

Gender is a highly administrated category from birth to death. On nearly every of the many official forms we fill out (or others complete on our behalf), there is a line where people must mark M or F. Every institutional space we encounter, from public bathrooms to schools, jails, immigration prisons, hospitals, shelters, religious institutions and so on, are sex-segregated. For trans* people, these bureaucratic encounters are often painful, dangerous, and fraught with anxiety. Being in the world necessitates navigating gender. As Dean Spade writes, "For trans people, administrative gender classification and the problems it creates for those who are difficult to classify or are misclassified is a major vector of violence and diminished life chances and life spans."[9] For trans* people who are also seeking asylum, or who are incarcerated, harms are intensified exponentially.[10] In the history of the United States, Black women's gender and sexuality has been a primary site of regulation and social control, as Dorothy Roberts has thoroughly argued.[11] The rule of law always exceeds its formal enactment and thus requires rhetorical critics to study the context of a legal culture that enables its existence, including, according to Marouf Hasian Jr., "extrajudicial" artifacts, spaces where laws are enforced, and "all documents associated with identity formation."[12] In this chapter, I assemble this context by drawing together a wide range of sources, from tweets to press coverage to the texts of the laws themselves.

Before I proceed, I should note that positionality matters a great deal in trans* scholarship. There is certainly important work on trans* issues by scholars who do not identify as trans*.[13] At the same time, some cisgender scholars have taken up writing about trans* people as a new "hot topic" without being in community with trans* people. This concern was taken up in a 2018 forum in *Women & Language*. In their provocative essay that generated the forum, G Patterson writes, "the institutionalization of transgender studies has perhaps enabled cisgender allies to gain access to and personally (and unevenly) profit from trans spaces, people, and perspectives."[14] Their critique does not argue one has to identify as trans* to write about anything trans*-related, but rather makes the point that trans* scholarship by trans* scholars is not getting enough attention and scholarship by allies too often does not benefit trans* communities. Lisa Flores has issued a similar demand about rhetorical scholars and scholarship on race: "intellectually, morally, and politically, scholars of color engaged in racial rhetorical criticism cannot be marginalized."[15] I write this study as a transmasculine person who now has passing privilege and uses men's restrooms, but who still has significant anxieties around using public facilities—and who experienced years of harassment (and avoidance) as a butch lesbian in women's restrooms. I mark my positionality specifically here simply to say that I am intimately familiar with what is at stake in bathroom bills, and I write in solidarity with calls for more trans* scholarship to be done by trans* scholars. As an author who is also a white, able-bodied trans* person, I am mindful of demands from Treva Ellison, Kai M. Green, Matt Richardson, and C. Riley Snorton for Black feminist contributions to scholarship to be recognized, and for Black trans* women scholars to be the creators (and not simply subjects) of trans* scholarship.[16] To be *of* trans* experience is to exist and move within a broad spectrum of positionalities. Too often, transgender is understood as a fixed identity category within privileged social norms such as whiteness.[17]

The impacts and implications of bathroom bills

Stereotype threat is a significant impact of bathroom bills, even when they are not passed. This threat causes fear, however unfounded, of a trans* person fulfilling a stereotype and being outed. This often results in coping with that threat through behavioral response: increased avoidance of public bathroom facilities. Avoiding public facilities has a number of physical, social, and emotional impacts, not the least of which is decreased participation in public and civic life. Rob Kitchin and Robin Law coined the metaphor of the "bladder's leash" to describe the limitations of a lack of accessible bathroom accommodations for people with disabilities.[18] The same holds true for trans* people who are denied access to facilities by law or by stereotype threat: they can only travel as far as their bladder will allow before they need to return home again.[19] Rachel McKinnon writes that

> trans* persons are regularly harassed and are even the subjects of violence merely for wanting to use the bathroom like anyone else. And even for trans* persons who "pass," in that it's not easily discernible that they're trans*, the risk is still present of being "found out" in such situations. This leads to anxiety and stress, and in some cases, situational avoidance.[20]

Like all trans* and gender-nonconforming people, I can attest to the stresses of public bathrooms. I spent years structuring my days, my interactions, and my liquid intake around bathroom access, and frankly, I still do. As a sometimes-passing transmasculine person using men's bathrooms in 2016, I was on heightened alert because of people's increased attention to who was in the bathroom with them—people were on the lookout for transgender people in their bathrooms because of the bathroom bills cropping up in state legislatures and public discourse. I started hormones that year, and while my bathroom anxieties should have decreased as my ability to pass as male increased, the fear that people were on the lookout for trans* people kept me on high alert. I had already begun avoiding going into women's bathrooms years before, except on the few occasions when I had no other option. In these situations, I would keep an eye out until no one was looking; if someone came in while I was in the stall, I'd stay there until they left, listening and waiting for a clearing when I could safely exit without encountering anyone. Of course, this is not only stressful, time-consuming, and inconvenient, but also not a fail-safe method: I had dozens of encounters with women whose reactions ranged from confusion to discomfort to outright hostility, though thankfully never resulted in violence. The impact of these transphobic policies was to engender fear in trans* people, an impact that harms regardless of who else might be in the bathroom. Though I pass consistently now, I am *still* never at ease using a public restroom unless it is gender neutral. Regardless of whether a bill becomes law, its introduction creates harmful environments for trans* people in spaces beyond its jurisdiction.

Harvey Molotch writes that "'public' and 'toilet' do not sit well together," because "the toilet involves doing the private in public and only loosely under the control of the actors directly involved."[21] As Isaac West notes, "to allay some of our anxieties, we invoke state-based protections to ensure that public bathrooms are places regulated by a variety of legal technologies," and these regulations have a particularly strong impact on transgender and gender nonconforming people.[22] The public bathroom is a site of gender norming and policing, Jack Halberstam writes, where "the sorting of bodies through seemingly irrefutable, inevitable, and obvious signage masked the intense enforcement of a set of gender norms through mechanisms that were cleverly folded into quotidian divisions of space."[23] In other words, a bathroom is not *just* a bathroom.

The current framing of transgender bathroom bills by conservatives borrows directly from past struggles for social equality. Arguments and ad campaigns in support of bathroom bills use fear tactics to scare up backers—most commonly harvesting terrors regarding male predators of female children.[24] The logic linking such fears to this legislation asserts that if transgender people are allowed to use any bathroom, then any man can say he is a woman to enter a women's bathroom. This dubious reasoning recalls anti-integration arguments from the 1950s and 1960s, which cited Black men as threats to white women's safety and sexual purity.[25] The reasoning also echoes 1970s arguments against the Equal Rights Amendment (ERA), which warned voters that if the ERA were passed, all institutions and facilities would become co-ed.[26] Sheila Cavanagh writes that public restrooms are designed to discipline gender: "Public restroom facilities stage gender so that non-experts—the general public—can decide if it is pure and intelligible or impure and indecipherable in heteronormative and cis-sexist landscapes."[27] The public bathroom is thus inherently a site of political struggle, and has been overtly the site of political organizing and protest.[28]

In the name of privacy and security: the cases of North Carolina and Texas

North Carolina's House Bill 2 (HB 2), titled the *Public Facilities Privacy & Security Act*, was adopted by the legislature and signed into law by Governor Pat McCrory in 2016. The law requires people to use the bathroom and changing room that align with their biological sex, defined as "the physical condition of being male or female, which is stated on a person's birth certificate."[29] The bill was passed in response to a Charlotte ordinance that protected transgender people from discrimination and allowed them to use the bathroom that aligns with their gender identity. After significant protest against HB 2, including costly corporate boycotts of the state by organizations ranging from the NBA to PayPal (the AP estimated the cost of the bill would be at least $3.7 billion over 12 years[30]) North Carolina repealed the bill but replaced it with a compromise that prevents municipalities from passing laws that protect LGBTQIA+ people until 2020. During his presidential campaign, Donald Trump made statements both in support of and against HB 2 in North Carolina, calling it a mistake because of the financial impact at one moment,[31] and a few months later, coming out in support of the law and Governor McCrory.[32] In the first few weeks of his presidency, Trump enacted a series of executive orders attacking transgender rights.[33] His actions have sent a strong message of support to his conservative base as his administration has sought to roll back recent legal gains for LGBTQIA+ people.

The bills proposed in Texas, unlike North Carolina, stipulated punishments for violations. HB 1747 would make it a crime of disorderly conduct for trans* people who have not changed the gender marker on their IDs to enter public restrooms aligning with their gender identities. HB 1748 went further, making it a Class A misdemeanor (punishable by up to a year in jail and a maximum fine of $4,000) for a person 13 years or older to use facilities that do not match the "gender established at the individual's birth or by the individual's chromosomes." Additionally, the law would make it a felony punishable by up to two years in state prison and a maximum $10,000 fine for "Texas building owners to allow any person seven years or older to use a restroom that does not fit his or her birth-assigned sex or chromosomes."[34] After the North Carolina bill was repealed in 2017, the Texas bill stalled in the legislature. In a debate leading up to the 2018 gubernatorial election, Governor Greg Abbott, who strongly supported the bill the year before, said it was no longer on his agenda.[35] His change in stance was also due, in no small part, to activist organizing and corporate pressure. The Texas Association of Business estimated that if it were to enact the proposed measures, the state could lose $5.6 billion through 2026.[36]

The question of cost is frequently used in different arguments regarding anti-trans* laws. When Trump initially argued against North Carolina's HB 2, he did so in the name of the significant financial losses it caused to the state. In the same interview, however, he also announced his opposition to requiring establishments to provide gender neutral restrooms because of the expense.[37] Later, he reversed course on the bill, and he has been undermining trans* rights since through a variety of measures. His administration opposed trans* inclusion in the military on the grounds that inclusion comes with "tremendous medical costs and disruption."[38] Yet a study commissioned by the Defense Department demonstrated that the military spends 10 times as much on erectile dysfunction drugs ($84 million) than the upper limit estimate for trans* health care ($8.4 million).[39] In April of 2019, the Health and Human Services Department (HHS) issued regulations that strengthen religious protections for health care workers, allowing them to deny treatment to LGBTQIA+ patients, or not provide services like abortions and sterilization. The administration justified the "Conscience Rule" that would legally allow denial of care by saying it would save $3.6 billion over five years by no longer requiring providers to post notices about discrimination or provide material in languages other than English, and by saving providers money, as they would no longer need to handle grievances.[40] The administration rationalized this as a cost-saving measure, yet also requested an additional $1 million for the newly created Conscience and Religious Freedom Division within the HHS Office for Civil Rights—a unit that supports religious protections for health care workers, including the right to not treat LGBTQIA+ patients. These are just two of many examples to show that financial cost is a rhetorical strategy rather than an economic one.

Economic pressure from the National Collegiate Athletic Association (NCAA) and National Basketball Association (NBA) against HB 2, while outwardly a measure of support for LGBTQIA+ people, should be considered with some skepticism, as the participation of trans athletes in professional sports is largely disallowed. Although Jason Collins came out as gay in 2014 as an NBA player, as of mid-2019, there are no out gay athletes currently playing in Major League Baseball, the National Football League, the National Basketball Association, or the National Hockey League.[41] Major League Soccer player Collin Martin is the only out gay male athlete in the United States who plays in a major professional league.[42] Halliburton and Exxon Mobil came out against the Texas bathroom bills, yet these companies have not been historically inclusive of LGBTQIA+ populations, and their transnational and human rights practices should compel us to ask, as Eli Massey and Yasmin Nair do, what are the costs of inclusion in atrocious institutions?[43] How do corporations' practices harm indigenous and trans* populations in the United States and globally, for example? What are the limits of inclusion justified as "good business policy"? As more companies and institutions have started offering health care benefits for married same-sex couples, they have stopped offering benefits for domestic partners, thus forcing people into the institution of marriage (often objectionable for a variety of reasons) to gain access to rights that should be guaranteed to all.[44] We should always be critical of inclusion alone within institutions and insist instead on accompanying structural changes to place those who are most affected and most vulnerable in positions of power and decision-making. And yet, under the Trump administration and in the state legislatures that are introducing anti-trans* laws, the fight is more fundamental: a fight for legal and discursive existence.

The rhetoric used by supporters of bathroom bills upholds a rigid and immutable gender binary and the common trope of trans* people as deceivers, rhetoric that enacts an erasure of trans* existence.[45] The popularly used Texas slogan for the bills was "no men in women's bathrooms." The 2015 Houston Equal Rights Ordinance (HERO), an antidiscrimination ordinance that would have banned discrimination in many categories, was defeated by its opponents' focus on the category of gender identity. In an anti-HERO advertisement, former Houston Astros baseball player Lance Berkman proclaims, "No men in women's bathrooms, no boys in girls' showers or locker rooms."[46] The strategies that defeated the equal rights protections of the HERO ordinance became a blueprint for how to go on the offensive by proposing anti-trans* bathroom bills.[47] Underscoring the binary logic of such campaigns with appeals to fear, Senator Ted Cruz argued, "men should not be going to the bathroom with little girls."[48] Trans* women do not exist within this framing, as they are reinvented as men "pretending" to be women. Speaking in April 2016 on *Meet the Press*, North Carolina Governor McCrory said "I don't think government should be telling the private sector what their restroom and shower law should be, to allow a man into a woman's restroom, or a shower facility at a YMCA, for example." Ratcheting up extreme appeals to fear, Lieutenant Governor Dan Forest claimed, "the loophole this ordinance created would have given pedophiles, sex offenders, and perverts free reign to watch women, boys, and girls undress and use the bathroom," thereby demonizing trans* people by willfully conflating us with sexual predators.[49]

In keeping with such rhetoric, far-right media outlet *Breitbart* published a story in February 2018 with the headline, "Walgreens Caves to Gay Agenda, Allows Men in Women's Restrooms." Perhaps in an attempt to create frightening images to accompany such headlines, conservative groups have actually encouraged men to enter women's bathrooms and locker rooms as a form of protest in support of bathroom bills.[50] This type of protest reaffirms supporters' claims that transgender women are "really" men, exploiting a well-worn transphobic trope. Mike Huckabee went so far as to use himself as an example for why bathroom bills are necessary:

> Now I wish that someone told me that when I was in high school that I could have felt like a woman when it came time to take showers in PE. I'm pretty sure that I would have found my feminine side and said, 'Coach, I think I'd rather shower with the girls today.'[51]

Reframing trans* bathroom access through his own heterosexist lens and fetishizing harassment is an abhorrent distortion of the issue. Inappropriate jokes like this ignore the real dangers and stigma that embodied femininity (both cis and trans*) faces in a culture of toxic masculinity. Huckabee's comments belie his understanding of transwomen as "really" heterosexual men who are posing as women to get closer to them. This strategy threatens uneducated audiences into fearing that failing to pass bathroom bills permits men to enter women's spaces, reasoning that any man can put on a wig and heels and call themselves trans*. Vilifying and negating trans* people is part of a broader conservative agenda that functions in conjunction with anti-immigrant policies to serve white nationalism.

Conservative attitudes and policies around gender, including rigid gender binaries, the policing of gender expression, and the notion that gender is immutable in the context of the law, are aligned with multiple modes of white settler colonialism and racist terrorism. As Scott Morgensen observes, "heteronormative national whiteness established its rule by sexualizing the racial, economic, and political subjugation of peoples of color in the name of protecting white settler society from queer endangerment by perversely racialized sexuality and gender."[52] It is no coincidence that the Trump administration's agenda is at once anti-immigrant and anti-transgender. Building a wall on the U.S.–Mexico border and preventing transgender people from serving in the military both fall under the nationalist and white supremacist mission to "Make America Great Again." They demonstrate actions by the state designed to whiten and straighten by means of exclusion, marginalization, imprisonment, and erasure. Shereen Yousuf and Bernadette Calafell note that "the striking overlaps with ICE separating families at the U.S.–Mexico border and placing them in detention facilities adhere to the same racialized logic that constructs Muslims as threats to the nation."[53] Though these discourses play out through different tropes, they argue, "the US is using the same policies and agencies to enact racial terror on both of these communities."[54] I am extending their argument to include a link between anti-trans*, anti-Muslim, and anti-immigrant rhetoric and policies.

In the history of race, immigration, and settler colonialism in the United States, gender and sexuality have been tools of both subjugation and exclusion. Trans* and gender nonconforming people who are impacted by anti-Muslim racism and anti-immigrant racism often take the full force of intersecting bigotry. MAGA white supremacy is heterosexist and gender-normative, invested in bolstering white male power and ostensibly protecting white femininity—a colonial act. Rod Ferguson explains it this way: "As a technology of race, U.S. citizenship has historically ascribed heteronormativity (universality) to certain subjects and nonheteronormativity (particularity) to others."[55] Margot Canaday, in her extensive history of U.S. federal regulation of homosexuality, details how immigrants arriving at Ellis Island and other checkpoints were screened for genital defects. If immigration inspectors found evidence of non-normative gender presentation or genitalia on a person's body, this typically led to their exclusion at points of entry: "Immigrants diagnosed with arrested sexual development or other related conditions were excluded under the public charge provision because immigration officials associated defective genitalia with perversion, and further viewed perversion as a likely cause of economic dependency."[56] Contemporary arguments against trans* inclusive policies share these two lines of argumentation: that to include trans* people (in the military, for example) would be a financial burden on the public, and that non-normatively gendered people are perverse and threatening and therefore should be excluded from public institutions.

Policing also links anti-Muslim, anti-immigrant, anti-Black, and anti-trans* policies and practices. A 2016 political cartoon by Steve Sack in the *Minneapolis Star Tribune* shows a uniformed gatekeeper outside the bathrooms, set up like a TSA agent at an airport ready to scan documents. A woman and a man stand in a hallway, holding hands a short distance from the women's bathroom. The woman says, "I wondered how North Carolina was going to enforce that new law," as the two look askance at the agent, who appears as a looming figure, a woman larger than the two of them combined. The sign on the front of her podium reads, "Halt! Hoo Hoo Inspector." Behind the inspector are signs that read "Birth Certificate Required" and "Present Photo I.D." She holds a magnifying glass and a flashlight, and leaning up against the podium is a mirror on wheels with a handle, presumably designed to look up potential patrons' skirts. This cartoon illustrates the impracticality of policing bathrooms more stringently than airplane passengers. It also uses normatively gendered white people in business attire to seed skepticism of the law's invasion of privacy and surveillance measures—it is not *only* trans* people who would be subjected to searches. This broadening of gender policing endangers *all* people who fall outside of societal gender norms. The cartoon points out the absurdity of enforcing such a law, but it also serves to connect two primary conservative issues: immigration and transgender rights. Some of the Trump administration's first actions were to target transgender people: to ban them from serving in the military, to rescind the Obama administration's Title IX guidance that transgender students should be able to use the bathroom matching their gender identity, and the October 2018 move to define gender as an immutable legal category—effectively negating the existence of transgender people in the eyes of the law. The political attacks on trans* people have been swift and far-reaching, as has the response from activists.

Social media activism and its limits

While resistance to bathroom bills took many forms, some trans* people's responses inadvertently reinscribed the rigid gender binary. The hashtag #wejustneedtopee emerged in 2016, and the social media posts that received the most attention and circulation showed normatively masculine transmen (with facial hair, cowboy hats, and large stature) and stereotypically feminine transwomen (wearing pencil skirts and lipstick) in the bathroom of their sex assigned at birth.[57] The subjects of these self-portraits are very clearly in the "wrong" bathroom. A caption from Michael Hughes, a bearded, tattooed man in a cowboy hat, reads

> Do I look like I belong in women's facilities? Republicans are trying to get legislation passed that would put me there, based on my gender at birth. Trans people aren't going into the bathroom to spy on you, or otherwise cause you harm.

James Sheffield took a selfie of his bearded, scowling face in a camo baseball cap and tweeted it at Pat McCrory with the caption, "It's now the law for me to share a restroom with your wife." Since these activists thoroughly embody masculine norms, no one would know they were trans* upon sight.

Such visual campaigns effectively demonstrate the absurdity of the bills by evidencing that many trans* people are not visibly trans; and in upholding the law, they would make people, both women and men, uncomfortable by appearing to be in the "wrong" bathroom. A stocky bearded cowboy clearly does not belong in the bathroom with women, these claims insist. Yet in proving the laws absurd and impractical, these kinds of messages simultaneously reaffirm gendered stereotypes of women as feminine and men as masculine. They leave no space for gender ambiguity or fluidity, for butch women who identify as women and would rather use the women's restroom, for example. Further, the hashtag "wejustneedtopee" sanitizes the debate, neglecting the other reasons trans* people might use the bathroom—to shit, cry, vomit, hook up, avoid the pain of dysphoria, or any number of other things. Besides washing one's hands, peeing is perhaps the most innocent of bathroom activities. By claiming it's "just" the need to pee, they argue that's all a trans* person would need the bathroom for rather than demanding full access, reducing a human rights issue to a mere practicality.

One trans woman, Ashley Smith, staged a social media protest using what Julia Serano has identified as one of the main archetypes for trans women in media representation: the deceiver. Deceivers have an ultrafeminine appearance, and "because deceivers successfully pass as women, they generally serve as unexpected plot twists, or play the role of sexual predators who fool innocent straight guys into falling for 'men.'"[58] Smith fits this model: she is young, ultrafeminine, and attractive. She posted a picture of herself with Texas Governor Greg Abbott, proponent of the state's bathroom bills with the caption, "How will the Potty Police know I'm transgender if the Governor doesn't?" She overlaid the photo with text, three lines reading #bathroombuddy at the top in red, white, and blue, and captions that read "Texas Governor/Greg Abbott and Ashley Smith/Trans-Woman." In this image, she appears to have indeed deceived him. In the photo, he is sitting and she is bending down, her hand on his back. Their faces are close, their bodies leaning toward one another. Abbott is grinning ear-to-ear, looking like he is thoroughly enjoying the moment.

While in media representations, the deceptive transwoman is typically played by a cismale actor, here, Smith is playing the role as a transwoman. This stereotype has a "recurring theme of deceptive trans women retaliating against men, often by seducing them," which, Serano writes, "seems to be an unconscious acknowledgement that both male and heterosexual privileges are threatened by transsexuals."[59] I cannot deny the momentary glee invited when seeing a politician who endorsed transphobic legislation being tricked by a transwoman in a public way. However, in deceiving Abbott and then outing herself, Smith's photo upholds the connection between transwomen and deception, which does not challenge the conservative argument that transwomen are not who they appear to be. Moreover, while this form of argumentation challenges the premise of bathroom bills, it does not address those who are most at risk in public bathrooms both before the introduction of these bills and afterwards—gender-nonconforming, androgynous, and nonbinary people, or those whose gender presentation marks them as trans*. Nonetheless, Smith's post went viral on Facebook and Instagram, and was picked up by Reuters and published by newspapers and media outlets around the globe. A likely reason that Smith's photo and the #wejustneedtopee images circulated so widely is the way they reinforce dominant transgender tropes in media representation even as they attempt to subvert them. They claim: We are just like you. We just have to pee. Although this approach does gain some political traction, it collapses trans* experience into its most normative form and does not address oppression at the structural level. Many trans* people do not pass—they do not slip into and out of public restrooms unnoticed. If the public faces of trans*-ness are normative and passing, this attention does nothing for those who fall somewhere in between, who are genderfluid, genderqueer, nonbinary, those who have socially but not medically transitioned (for a variety of reasons), and so on.

As access to medical transition increases, one danger we face is the limiting of who counts as trans* to those who have medically transitioned and/or present as gender normative and the concomitant further marginalization, or even invisibility, of those who cannot or do not transition or present in alignment these expectations. As Sara Ahmed warns, "To be included can thus be a way of sustaining and reproducing the politics of exclusion, where a life sentence for some is a death sentence for others."[60] When a stranger is brought into an institution, she argues, they become a subject, yet this subjecthood comes at the cost of being "willing to consent to the terms of inclusion."[61] For trans* people, this could codify who qualifies as trans* in very limiting, and white, middle- to upper-class, hetero- and gender-normative ways. We have seen this before. In Rod Ferguson's words, "As state and capital produced and utilized racial, gender, and erotic difference to exclude minoritized subjects from the rights and privileges of citizenship, the rhetoric of diversity concealed that use and production."[62]

With centrist liberal victories like gay marriage can come the illusion of inclusion and the increased invisibility of ongoing exclusion. Jack Halberstam warns,

> While we should undoubtedly celebrate recent attention to transgender rights, transgender visibility, and transgender recognition—not to mention trans* access to public facilities—we must also resist some of the streamlining effects of this recognition and ask what price will be paid and by whom for this new visibility.[63]

More radical forms of protest, following in a long history of radical queer activism (such as ACT-UP's die-ins and Ashes Action[64]) have included "shit ins" and a "Queer Dance Freakout" outside of the Texas governor's mansion in February 2017, modeled after a queer dance party outside of Vice President Mike Pence's D.C. home. While some musicians canceled performances in North Carolina to boycott the state, trans* punk singer Laura Jane Grace of the band Against Me! took a different approach. She refused to cancel her show, but she started it by burning her birth certificate on stage.[65] Actions like these both resist the dehumanization of the laws and the normalization of trans* experience. Her refusal to pass was a defiant act and an insistence on visibility in the face of erasure.

Conclusion

Bathroom bills and other transphobic legislation, while they have material consequences, also serve symbolic, performative functions, as part of enacting a conservative agenda. As numerous political cartoons and commentators point out, bathroom bills are wholly impractical, making trans* people out to be predators when we, in fact, are at a much higher risk of harm in public bathrooms than non-trans* people.[66] Dean Spade has argued extensively on the limits of legal solutions to oppression for trans* people, demonstrating that antidiscrimination and hate crime laws do not prevent anti-trans* violence (and can be undone), and that the law should not define who we are. He writes, "Trans people are told by legal systems, state agencies, employers, schools, and our families that we are impossible people who are not who we say we are, cannot exist, cannot be classified, and cannot fit anywhere."[67] This brings us to a central tension in the fight over bathroom access. When one loud strain of public discourse proclaims that trans* people are a burden, that we do not exist, that we are too expensive, too confusing, or too demanding, a clear logical response is to claim the opposite. We are not a burden! We do exist! We will not be erased! Such a position allows trans* people to claim space, name their existence, and commit to continued resistance. Yet at the same time, it is a reactionary position, responding to the terms set by a conservative agenda. What might we build if we refuse to accept these terms? What futures might we create if our existence and belonging were not the grounds of contention, but an absolute given? What if we refused to claim that we are just like anyone else? What if activism refused to limit itself to the claim that we just need to pee and instead built spaces of liberation and radical inclusion?

Notes

1 Jon Herskovitz, "Texas Governor Says 'Bathroom Bill' No Longer on His Agenda," *Reuters*, September 28, 2018, www.reuters.com/article/us-texas-lgbt/texas-governor-says-bathroom-bill-no-longer-on-his-agenda-idUSKCN1M901Y.
2 Joellen Kralik, "'Bathroom Bill' Legislative Tracking," *National Conference of State Legislatures*, July 28, 2017, www.ncsl.org/research/education/-bathroom-bill-legislative-tracking635951130.aspx.
3 For more on the legacy of *Lawrence v. Texas*, see Peter Odell Campbell, "The Procedural Queer: Substantive Due Process, Lawrence v. Texas, and Queer Rhetorical Futures." *Quarterly Journal of Speech* 98, no. 2 (May 2012): 203–29, https://doi.org/10.1080/00335630.2012.663923.
4 David A. Graham, "Red State, Blue City," *The Atlantic*, March 2017, www.theatlantic.com/magazine/archive/2017/03/red-state-blue-city/513857/.
5 Avery Tompkins, "Asterisk," *TSQ* 1, no. 1–2 (2014): 27, https://doi.org/10.1215/23289252-2399497.
6 Eva Haywood and Jami Weinstein, "Introduction: Tranimalities in the Age of Trans* Life," *TSQ: Transgender Studies Quarterly* 2, no. 2 (2015): 195–208, https://doi.org/10.1215/23289252-2867446.
7 Jack Halberstam, *Trans*: A Quick and Quirky Account of Gender Variability* (Oakland, CA: University of California Press, 2018), 52–53.

8 Kai M. Green, "Troubling the Waters: Mobilizing a Trans* Analytic," in *No Tea, No Shade: New Writings in Black Queer Studies,* ed. E. Patrick Johnson (Durham, NC: Duke University Press, 2016), 67.

9 Dean Spade, *Normal Life: Administrative Violence, Critical Trans Politics, and the Limits of Law* (Brooklyn: South End Press, 2011), 142.

10 Eithne Luibhéid and Lionel Cantú Jr., eds, *Queer Migrations: Sexuality, U.S. Citizenship, and Border Crossings* (Minneapolis, MN: University of Minnesota Press, 2005).

11 Dorothy Roberts, *Killing the Black Body: Race, Reproduction, and the Meaning of Liberty* (New York: Pantheon Books, 1997).

12 Marouf A. Hasian Jr., *Colonial Legacies in Postcolonial Contexts: A Critical Rhetorical Examination of Legal Histories* (New York: Peter Lang, 2002), 9.

13 See, for example, trans* work by queer scholars Karma Chávez, "Spatializing Gender Performativity: Ecstasy and Possibilities for Livable Life in the Tragic Case of Victoria Arellano," *Women's Studies in Communication* 33, no. 1 (May 2010): 1–15, https://doi.org/10.1080/07491401003669729; Roderick A. Ferguson, *Aberrations in Black: Toward a Queer of Color Critique* (Minneapolis, MN: University of Minnesota Press, 2004); and many others too numerous to name. In the field of communication, there are cisgender scholars who do transgender scholarship rooted in the work of trans* scholars, notably Isaac West, *Transforming Citizenships: Transgender Articulations of the Law* (New York: NYU Press, 2013); Leland Spencer, "Introduction to the Special Issue: Transcending the Acronym," *Women & Language* 41, no. 1 (2018): 7–15. There are also scholars like Finn Enke and Aaron Devor who started doing trans* scholarship before publicly identifying as trans*.

14 G. Patterson, "Entertaining a Healthy Cispicion of the Ally Industrial Complex in Transgender Studies," *Women & Language* 41, no. 1 (2018): 147.

15 Lisa Flores, "Between Abundance and Marginalization: The Imperative of Racial Rhetorical Criticism," *Review of Communication* 16, no. 1 (2016): 10, http://dx.doi.org/10.1080/15358593.2016.1183871.

16 Treva Ellison et al., "We Got Issues: Toward a Black Trans*/Studies," *Transgender Studies Quarterly* 4, no. 2 (May 2017): 162–169, https://doi.org/10.1215/23289252-3814949.

17 See C. Riley Snorton, *Black on Both Sides: A Racial History of Trans Identity* (Minneapolis, MN: University of Minnesota Press, 2017) and Tom Boellstorff et al., "Decolonizing Transgender: A Roundtable Discussion," *TSQ: Transgender Studies Quarterly* 1 (August 2014): 419–439, https://doi.org/10.1215/23289252-2685669.

18 Rob Kitchin and Robin Law, "The Socio-spatial Construction of (In)accessible Public Toilets," *Urban Studies* 38, no. 2 (February 2001): 287–298, https://doi.org/10.1080/00420980124395.

19 See West, *Transforming Citizenships.*

20 Rachel McKinnon, "Stereotype Threat and Attributional Ambiguity for Trans Women," *Hypatia* 29, no. 4 (Fall 2004): 866, https://doi.org/10.1111/hypa.12097.

21 Harvey Molotch, *Against Security: How We Go Wrong at Airports, Subways, and Other Sites of Ambiguous Danger* (Princeton, NJ: Princeton University Press, 2014), 25–26.

22 West, *Transforming Citizenships,* 73.

23 Halberstam, *Trans*,* 133.

24 Kristen Schilt and Laurel Westbrook, "Bathroom Battlegrounds and Penis Panics," *Contexts* 14, no. 3: (2015) 26–31, https://doi.org/10.1177/1536504215596943.

25 Maria L. LaGanga, "From Jim Crow to Transgender Ban: The Bathroom as Battleground for Civil Rights," *The Guardian,* March 30, 2016, www.theguardian.com/world/2016/mar/30/transgender-ban-bathrooms-north-carolina-civil-rights.

26 Sheila L. Cavanagh, *Queering Bathrooms: Gender, Sexuality, and the Hygienic Imagination* (Toronto: University of Toronto Press, 2010).

27 Cavanagh, *Queering Bathrooms,* 6.

28 West, *Transforming Citizens.*

29 Government of North Carolina, House of Representatives, *Public Facilities Privacy & Security Act of 2016,* § G.S. 115C-47. www.ncleg.net/Sessions/2015E2/Bills/House/PDF/H2v0.pdf.

30 Camila Domonoske, "AP Calculates North Carolina's 'Bathroom Bill' Will Cost More Than $3.7 Billion," *National Public Radio,* March 27, 2017, www.npr.org/sections/thetwo-way/2017/03/27/521676772/ap-calculates-north-carolinas-bathroom-bill-will-cost-more-than-3-7-billion.

31 *The Intelligencer,* "Donald Trump Comes Out Against North Carolina's 'Very Strong' Bathroom Bill, April 21, 2016.

32 HRC staff, "SHAME: Donald Trump Endorses Gov. McCrory's Vile HB2 Attaching LGBTQ People," *HRC,* July 6, 2016, www.hrc.org/press/shame-donald-trump-endorses-gov.-mccrorys-vile-hb2-law-attacking-lgbtq-peop.

33 On his inauguration day, mentions of LGBTQ people were removed from the White House, State Department, and Labor Department websites; on February 22, the Department of Justice the Department of Education withdrew the 2016 Title IX guidance protecting transgender students; on March 1, the Department of Justice declined to uphold the Affordable Care Act's protections for transgender people in the face of a nationwide preliminary court order stopping the law's enforcement of nondiscrimination protections. In the years since, there has been a systematic rollback of existing protections and codification of discrimination in the areas of housing, prisons, shelters, schools, the military, health care, employment, and asylum. The National Center for Transgender Equality maintains an updated list of the administration's actions (and inactions) that harm trans* people at https://transequality.org/the-discrimination-administration.

34 Demoya Gordon, "Anti-Transgender Bills Proposed in Texas," *Lambda Legal*, March 5, 2015, www.lambdalegal.org/blog/20150305_anti-trans-bathroom-bills-proposed-in-tx.

35 Herskovitz, "Texas Governor."

36 Herskovitz, "Texas Governor."

37 *The Intelligencer*, "Donald Trump."

38 *BBC*, "Trump Asks US Court for Review of Transgender Military Ban," November 23, 2018, www.bbc.com/news/world-us-canada-46321543.

39 Christopher Ingraham, "The Military Spends Five Times as Much on Viagra as it Would on Transgender Troops' Medical Care," *Washington Post*, July 26, 2017, www.washingtonpost.com/news/wonk/wp/2017/07/26/the-military-spends-five-times-as-much-on-viagra-as-it-would-on-transgender-troops-medical-care/?utm_term=.efb3cbfc6c35.

40 See Emmarie Huetteman, "Trump Administration Rule Would Undo Health Care Protections for LGBTQ Patients," *CT Mirror*, June 16, 2019, https://ctmirror.org/2019/06/16/trump-administration-rule-would-undo-health-care-protections-for-lgbtq-patients/; and United States, Center for Medicare and Medicaid Services, Office for Civil Rights, Office of the Secretary, Health and Human Services, Proposed Rule, "Nondiscrimination in Health and Health Education Programs or Activities," *Federal Register* 84 (June 14, 2019): 27846–27895, www.federalregister.gov/documents/2019/06/14/2019-11512/nondiscrimination-in-health-and-health-education-programs-or-activities.

41 Jim Buzinski, "Five Years After Jason Collins Came Out as Gay, Little Progress in Pro Team Sports," *Outsports*, April 30, 2018, www.outsports.com/2018/4/30/17299774/jason-collins-nba-gay-coming-out-anniversary.

42 Frank Pingue, "NBA: Barrier-breaking Collins Still Awaits Next Openly Gay Athlete," *Reuters*, February 28, 2019, https://ca.reuters.com/article/sportsNews/idCAKCN1QH21F-OCASP.

43 Eli Massey and Yasmin Nair, "Inclusion in the Atrocious," *Current Affairs*, March 22, 2018, www.currentaffairs.org/2018/03/inclusion-in-the-atrocious.

44 Massey and Nair, "Inclusion."

45 McKinnon, "Stereotype Threat."

46 Chuck Lindell, "Why Did the Bathroom Bill Become Such a Big Thing This Year?," *Statesman*, August 26, 2017, www.statesman.com/news/state–regional-govt–politics/why-did-the-bathroom-bill-become-such-big-thing-this-year/qBW03N4YMvZPyz9hMgsuyM/.

47 Lindell, "Why Did the Bathroom Bill."

48 *The Intelligencer*, "Donald Trump."

49 Zack Ford, "How North Carolina Became the Most Anti-LGBT State in Less Than a Day," *thinkprogress.com*, March 24, 2016, https://thinkprogress.org/how-north-carolina-became-the-most-anti-lgbt-state-in-less-than-a-day-d47a01e96120/.

50 Sydney Brownstone, "Conservative Trolls Have Been Suggesting Men Go into Women's Restrooms to Help Legislators Discriminate Against Trans People," *The Stranger*, February 17, 2016, www.thestranger.com/blogs/slog/2016/02/17/23584290/conservative-trolls-have-been-suggesting-men-go-into-womens-restrooms-to-help-legislators-discriminate-against-trans-people.

51 Mark Joseph Stern, "Mike Huckabee Says He Would Have Pretended to Be Trans to Shower with Girls in School," *Slate*, June 2, 2015, https://slate.com/human-interest/2015/06/mike-huckabee-says-he-would-have-pretended-to-be-trans-to-shower-with-girls-in-school-video.html.

52 Scott Lauria Morgensen, *Spaces Between Us: Queer Settler Colonialism and Indigenous Decolonization* (Minneapolis, MN: University of Minnesota Press, 2011), 43.

53 Shereen Yousuf and Bernadette Calafell, "The Imperative for Examining Anti-Muslim Racism in Rhetorical Studies," *Communication and Critical/Cultural Studies* 15, no. 4 (2018): 315, https://doi.org/10.1080/14791420.2018.1533641.

54 Yousuf and Calafell, "The Imperative," 315.

55 Ferguson, *Aberrations in Black*, 14.

56 Margot Canaday, *The Straight State: Sexuality and Citizenship in Twentieth-Century America* (Princeton, NJ: Princeton University Press, 2009), 34.

57 Dawn Ennis, "16 Images that Show #WeJustNeedToPee," *Advocate*, April 15, 2016, www.advocate.com/transgender/2016/4/15/19-images-show-wejustneedtopee.

58 Julia Serano, "Skirt Chasers: Why the Media Depicts the Trans Revolution in Lipstick & Heels," *Bitch Magazine* (Fall 2014): 41.

59 Serano, "Skirt Chasers," 42.

60 Sara Ahmed, *On Being Included: Racism and Diversity in Institutional Life* (Durham, NC: Duke University Press, 2012), 163.

61 Ahmed, *On Being Included*, 163.

62 Ferguson, *Aberrations in Black*, 71.

63 Halberstam, *Trans**, 46.

64 Phillip Picardi, "'George Bush, Serial Killer': ACT UP's Fight Against the President," *out.com*, December 1, 2018, www.out.com/out-exclusives/2018/12/01/george-bush-serial-killer-act-ups-fight-against-president.

65 Paul Blest, "Watch Laura Jane Grace of Against Me! Burn Her Birth Certificate on Stage in Durham," *Indy Week*, May 16, 2016, https://indyweek.com/music/archives/watch-laura-jane-grace-me-burn-birth-certificate-stage-durham.

66 Amira Hasenbush, Andrew R. Flores, and Jody L. Herman, "Gender Identity Nondiscrimination Laws in Public Accommodations: A Review of Evidence Regarding Safety and Privacy in Public Restrooms, Locker Rooms, and Changing Rooms," *Sexuality Research and Social Policy* 16, no. 1 (March 2019): 70–83, https://doi.org/10.1007/s13178-018-0335-z.

67 Spade, *Normal Life*, 209.

Bibliography

Ahmed, Sara. *On Being Included: Racism and Diversity in Institutional Life*. Durham, NC: Duke University Press, 2012.

BBC. "Trump Asks US Court for Review of Transgender Military Ban." November 23, 2018. www.bbc.com/news/world-us-canada-46321543.

Blest, Paul. "Watch Laura Jane Grace of Against Me! Burn Her Birth Certificate on Stage in Durham." *Indy Week*, May 16, 2016. https://indyweek.com/music/archives/watch-laura-jane-grace-me-burn-birth-certificate-stage-durham.

Boellstorff, Tom, Mauro Cabral, Micha Cárdenas, Trystan Cotten, Eric A. Stanley, Kalaniopua Young, Aren Z. Aizura. "Decolonizing Transgender: A Roundtable Discussion." *TSQ: Transgender Studies Quarterly* 1 (August 2014): 419–439. https://doi.org/10.1215/23289252-2685669.

Brownstone, Sydney. "Conservative Trolls Have Been Suggesting Men Go into Women's Restrooms to Help Legislators Discriminate Against Trans People." *The Stranger*, February 17, 2016. www.thestranger.com/blogs/slog/2016/02/17/23584290/conservative-trolls-have-been-suggesting-men-go-into-womens-restrooms-to-help-legislators-discriminate-against-trans-people.

Buzinski, Jim. "Five Years After Jason Collins Came Out as Gay, Little Progress in Pro Team Sports." *Outsports*, April 30, 2018. www.outsports.com/2018/4/30/17299774/jason-collins-nba-gay-coming-out-anniversary.

Campbell, Peter Odell. "The Procedural Queer: Substantive Due Process, Lawrence v. Texas, and Queer Rhetorical Futures." *Quarterly Journal of Speech* 98, no. 2 (May 2012): 203–229. https://doi.org/10.1080/00335630.2012.663923.

Canaday, Margot. *The Straight State: Sexuality and Citizenship in Twentieth-Century America*. Princeton, NJ: Princeton University Press, 2009.

Cavanagh, Sheila L. *Queering Bathrooms: Gender, Sexuality, and the Hygienic Imagination*. Toronto: University of Toronto Press, 2010.

Chávez, Karma. "Spatializing Gender Performativity: Ecstasy and Possibilities for Livable Life in the Tragic Case of Victoria Arellano." *Women's Studies in Communication* 33, no. 1 (May 2010): 1–15. https://doi.org/10.1080/07491401003669729.

Center for Medicare and Medicaid Services, Office for Civil Rights, Office of the Secretary, Health and Human Services. Proposed Rule. "Nondiscrimination in Health and Health Education

Programs or Activities." *Federal Register* 84 (June 14, 2019): 27846–95. www.federalregister.gov/documents/2019/06/14/2019-11512/nondiscrimination-in-health-and-health-education-programs-or-activities.

Domonoske, Camila. "AP Calculates North Carolina's 'Bathroom Bill' Will Cost More Than $3.7 Billion." *National Public Radio*, March 27, 2017. www.npr.org/sections/thetwo-way/2017/03/27/521676772/ap-calculates-north-carolinas-bathroom-bill-will-cost-more-than-3-7-billion.

Ellison, Treva, Kai M. Green, Matt Richardson, and C. Riley Snorton. "We Got Issues: Toward a Black Trans*/Studies." *Transgender Studies Quarterly* 4, no. 2 (May 2017): 162–169. https://doi.org/10.1215/23289252-3814949.

Ennis, Dawn. "16 Images that Show #WeJustNeedToPee." *Advocate*, April 15, 2016. www.advocate.com/transgender/2016/4/15/19-images-show-wejustneedtopee.

Ferguson, Roderick A. *Aberrations in Black: Toward a Queer of Color Critique*. Minneapolis, MN: University of Minnesota Press, 2004.

Flores, Lisa. "Between Abundance and Marginalization: The Imperative of Racial Rhetorical Criticism." *Review of Communication* 16, no. 1 (2016): 4–24. http://dx.doi.org/10.1080/15358593.2016.1183871.

Ford, Zack. "How North Carolina Became the Most Anti-LGBT State in Less Than a Day." *thinkprogress.com*, March 24, 2016. https://thinkprogress.org/how-north-carolina-became-the-most-anti-lgbt-state-in-less-than-a-day-d47a01e96120/.

Gordon, Demoya. "Anti-Transgender Bills Proposed in Texas." *Lambda Legal*, March 5, 2015). www.lambdalegal.org/blog/20150305_anti-trans-bathroom-bills-proposed-in-tx.

Graham, David A. "Red State, Blue City." *The Atlantic*, March 2017. www.theatlantic.com/magazine/archive/2017/03/red-state-blue-city/513857/.

Green, Kai M. "Troubling the Waters: Mobilizing a Trans* Analytic." In *No Tea, No Shade: New Writings in Black Queer Studies*, edited by E. Patrick Johnson, 65–82. Durham, NC: Duke University Press, 2016.

Halberstam, Jack. *Trans*: A Quick and Quirky Account of Gender Variability*. Oakland, CA: University of California Press, 2018.

Hasenbush, Amira, Andrew R. Flores, and Jody L. Herman. "Gender Identity Nondiscrimination Laws in Public Accommodations: A Review of Evidence Regarding Safety and Privacy in Public Restrooms, Locker Rooms, and Changing Rooms." *Sexuality Research and Social Policy* 16, no. 1 (March 2019): 70–83. https://doi.org/10.1007/s13178-018-0335-z.

Hasian, Marouf A. Jr. *Colonial Legacies in Postcolonial Contexts: A Critical Rhetorical Examination of Legal Histories*. New York: Peter Lang, 2002.

Haywood, Eva and Jami Weinstein. "Introduction: Tranimalities in the Age of Trans* Life." *TSQ: Transgender Studies Quarterly* 2, no. 2 (2015): 195–208. https://doi.org/10.1215/23289252-2867446.

Herskovitz, Jon. "Texas Governor Says 'Bathroom Bill' No Longer on His Agenda." *Reuters*, September 28, 2018. www.reuters.com/article/us-texas-lgbt/texas-governor-says-bathroom-bill-no-longer-on-his-agenda-idUSKCN1M901Y.

HRC staff. "SHAME: Donald Trump Endorses Gov. McCrory's Vile HB2 Attaching LGBTQ People." *HRC*, July 6, 2016. www.hrc.org/blog/shame-donald-trump-endorses-gov.-mccrorys-vile-hb2-attacking-lgbtq-peop.

Huetteman, Emmarie. "Trump Administration Rule Would Undo Health Care Protections for LGBTQ Patients." *CT Mirror*, June 16, 2019. https://ctmirror.org/2019/06/16/trump-administration-rule-would-undo-health-care-protections-for-lgbtq-patients/

Ingraham, Christopher. "The Military Spends Five Times as Much on Viagra as It Would on Transgender Troops' Medical Care." *Washington Post*, July 26, 2017. www.washingtonpost.com/news/wonk/wp/2017/07/26/the-military-spends-five-times-as-much-on-viagra-as-it-would-on-transgender-troops-medical-care/?utm_term=.efb3cbfc6c35.

The Intelligencer. "Donald Trump Comes Out Against North Carolina's 'Very Strong' Bathroom Bill." April 21, 2016.

Kitchin, Rob and Robin Law. "The Socio-spatial Construction of (In)accessible Public Toilets." *Urban Studies* 38, no. 2 (February 2001): 287–298. https://doi.org/10.1080/00420980124395.

Kralik, Joellen. "'Bathroom Bill' Legislative Tracking." *National Conference of State Legislatures*, July 28, 2017. www.ncsl.org/research/education/-bathroom-bill-legislative-tracking635951130.aspx.

LaGanga, Maria L. "From Jim Crow to Transgender Ban: The Bathroom as Battleground for Civil Rights." *The Guardian*, March 30, 2016. www.theguardian.com/world/2016/mar/30/transgender-ban-bathrooms-north-carolina-civil-rights.

Lindell, Chuck. "Why Did the Bathroom Bill Become Such a Big Thing This Year?" *Statesman*, August 26, 2017. www.statesman.com/news/state–regional–govt–politics/why-did-the-bathroom-bill-become-such-big-thing-this-year/qBW03N4YMvZPyz9hMgsuyM/.

Luibhéid, Eithne and Lionel Cantú Jr., eds. *Queer Migrations: Sexuality, U.S. Citizenship, and Border Crossings*. Minneapolis, MN: University of Minnesota Press, 2005.

Massey, Eli and Yasmin Nair. "Inclusion in the Atrocious." *Current Affairs*, March 22, 2018. www.currentaffairs.org/2018/03/inclusion-in-the-atrocious.

McKinnon, Rachel. "Stereotype Threat and Attributional Ambiguity for Trans Women." *Hypatia* 29, no. 4 (Fall 2004): 857–872. https://doi.org/10.1111/hypa.12097.

Molotch, Harvey. *Against Security: How We Go Wrong at Airports, Subways, and Other Sites of Ambiguous Danger*. Princeton, NJ: Princeton University Press, 2014.

Morgensen, Scott Lauria. *Spaces between Us: Queer Settler Colonialism and Indigenous Decolonization*. Minneapolis, MN: University of Minnesota Press, 2011.

Patterson, G. "Entertaining a Healthy Cispicion of the Ally Industrial Complex in Transgender Studies." *Women & Language* 41, no. 1 (2018): 146–151.

Picardi, Phillip. "'George Bush, Serial Killer': ACT UP's Fight Against the President." *Out.com*, December 1, 2018. www.out.com/out-exclusives/2018/12/01/george-bush-serial-killer-act-ups-fight-against-president.

Pingue, Frank. "NBA: Barrier-breaking Collins Still Awaits Next Openly Gay Athlete." *Reuters*, February 28, 2019. https://ca.reuters.com/article/sportsNews/idCAKCN1QH21F-OCASP.

Roberts, Dorothy. *Killing the Black Body: Race, Reproduction, and the Meaning of Liberty*. New York: Pantheon Books, 1997.

Schilt, Kristen and Laurel Westbrook. "Bathroom Battlegrounds and Penis Panics." *Contexts* 14, no. 3(2015): 26–31. https://doi.org/10.1177/1536504215596943.

Serano, Julia. "Skirt Chasers: Why the Media Depicts the Trans Revolution in Lipstick & Heels." *Bitch Magazine* (Fall 2014): 41–7.

Snorton, C. Riley. *Black on Both Sides: A Racial History of Trans Identity*. Minneapolis, MN: University of Minnesota Press, 2017.

Spade, Dean. *Normal Life: Administrative Violence, Critical Trans Politics, and the Limits of Law*. Brooklyn: South End Press, 2011.

Spencer, Leland. "Introduction to the Special Issue: Transcending the Acronym." *Women & Language* 41, no. 1 (2018): 7–15.

Stern, Mark Joseph. "Mike Huckabee Says He Would Have Pretended to Be Trans to Shower with Girls in School." *Slate*, June 2, 2015. https://slate.com/human-interest/2015/06/mike-huckabee-says-he-would-have-pretended-to-be-trans-to-shower-with-girls-in-school-video.html.

Tompkins, Avery. "Asterisk." *TSQ* 1, nos 1–2 (2014): 26–27. https://doi.org/10.1215/23289252-2399497.

West, Isaac. *Transforming Citizenships: Transgender Articulations of the Law*. New York: NYU Press, 2013.

Yousuf, Shereen and Bernadette Calafell. "The Imperative for Examining Anti-Muslim Racism in Rhetorical Studies." *Communication and Critical/Cultural Studies* 15, no. 4 (2018): 312–318. https://doi.org/10.1080/14791420.2018.1533641.

18

RESEARCH ON GENDER AND POLITICAL RHETORIC

Masculinity, ingenuity, and the double bind

Kristina Horn Sheeler, Serena Hawkins,
and Eline van den Bossche

Karine Jean-Peirre writes, "If 1992 was Year of the Woman, 2018 must be Year of Accountability."[1] She reminds readers of key events of 1992, including the successful campaign of Carol Moseley-Braun, the first Black woman elected to the Senate, who ran along with an unprecedented number of women the year following the Anita Hill hearings. Fast-forward 26 years and we have another Year of the Woman, fueled in part by reactions to Brett Kavanaugh's nomination to the Supreme Court, the #MeToo and #TimesUp movements, and Donald Trump's presidential election in 2016. While not a new phenomenon, gender and its relation to political rhetoric is a potent force for change in our current climate. However, we still have much work to do to achieve political parity if we are to answer Jean-Pierre's call to accountability.

This chapter assesses the state of the art in political rhetoric with a focus on gender and identifies three overlapping arguments: masculinity and bias, inventiveness of individual women, and confronting the double bind. We begin by reviewing seminal pieces as context for our essay as well as key frames through which scholars view women's political rhetoric. We continue by focusing on sites in which researchers assess gender and political rhetoric, including masculinity and the U.S. presidency. Gender as strategy is a rhetorical resource and constraint women candidates employ creatively, not only in U.S. presidential, congressional, and gubernatorial elections, but also in international political discourse. Finally, reactions to the 2016 U.S. presidential election provide rich scenarios through which to study the intersection of gender and political rhetoric, as evidenced by special journal issues focusing on events such as the Women's March.

We cast a wide net to identify the sources that populate this chapter. We started with databases such as Web of Knowledge, Communication and Mass Media Complete, and ProQuest, and followed the twitter feeds of @GenderWatch2018, @538politics, @GenderReport, @Women&PoliticsInstitute, and others. Starting broadly with search terms such as "gender and political rhetoric," we compiled a large number of academic articles, chapters, books, and blogs. From there, we filtered based on limiters that included election years, politicians, and terms such as "Women's March" and "#MeToo." Once we assembled a core group of relevant sources, we identified others based on those that cited our relevant sources and also mined bibliographies for other research. Using this process, we landed on the seminal pieces and concepts identified in this chapter, as well as an organizing structure that relies on chronology, case studies, and political offices in both U.S. and international contexts. While our timeframe features prominently the 2008 U.S. election cycle to the present, our chapter reaches further back and includes additional sources as important starting points for anyone who conducts research in the area of gender and political rhetoric from a rhetorical methodological perspective.

Feminist foundations

We must acknowledge research that informs our work and provides depth for any scholar hoping to add to the conversation surrounding gender and political rhetoric. Though not about political rhetoric specifically, previous research that assesses the growing and important body of feminist communication scholarship lays groundwork for this chapter. According to Susan Zaeske, research in the area of gender and political communication has its foundation in feminist rhetorical studies that sought to correct the historical record and recover women's voices and ingenuity.[2] For example, Bonnie J. Dow and Celeste M. Condit undertake a review of feminist communication scholarship in the years 1998–2003 that "is oriented toward the achievement of 'gender justice,' a goal that takes into account the ways that gender always already intersects with race, ethnicity, sexuality, and class."[3] They point out the theoretical rigor and sophistication of such scholarship, even concluding that the area "has been mainstreamed,"[4] while calling for more attention to "the unmarked race, class, sexuality, and geographical categories with which we work."[5] Their argument remains good advice for scholars today. Michaela D. E. Meyer, building on the work of Karen A. Foss[6] and Linda Aldoory and Elizabeth Toth,[7] articulates two contributions of feminist rhetoric, which she defines as "writing women in" to the canon and "challenging rhetorical standards."[8] Meyer concludes by offering advice for the future, advocating for an increased "sense of responsibility in constructing our own agency through theory and scholarly writing."[9] Focusing specifically on the power of oratory, Vanessa B. Beasley's historical assessment of political rhetoric is instructive even as it calls for more research "about the power of rhetoric across culture and nation" in response to the decidedly Americanist tradition of the current state of public address literature.[10] This chapter demonstrates Beasley's observation still applies seven years later and indicates where progress has been made: female political figures across the globe are speaking powerfully to call attention to sexist containment rhetoric that seeks to silence their efforts. Other sources, such as Susan A. Banducci's work on "Women as Political Communicators" provides still more breadth of research on gender and political communication by drawing on methodological traditions beyond rhetorical studies.[11]

Oxymoron, paradox, and the double bind emerge from the totality of scholarship on gender and political rhetoric as critical to understanding the arguments presented in this chapter. Indeed, scholarship on gender and political rhetoric relies on these terms,[12] noting the perceived incongruities that any woman must confront to advance public political change. Scholars often identify Karlyn Kohrs Campbell's "The Rhetoric of Women's Liberation: An Oxymoron"[13] as a starting point for understanding feminist public address. As Campbell notes, "the rhetoric of women's liberation appeals to *what are said to be* shared moral values, but forces recognition that those values are *not* shared, thereby creating the most intense of moral conflicts."[14] Similarly drawing on the notion of contradiction, Kathleen Hall Jamieson's "double binds"[15] are not new. Jamieson argues that "Western culture is riddled with evidence of traps for women that have forcefully curtailed their options,"[16] and identifies five double binds that plaque women in positions of leadership: womb/brain, silence/shame, sameness/difference, femininity/competence, and aging/invisibility. As Jamieson articulates throughout her work, and as many studies in this review demonstrate with increased clarity, while the binds are culturally constructed to silence women's contributions, they can be made to self-destruct as more and more women raise their voices and lay claim to public political power.[17] For example, Karrin Vasby Anderson and Kristina Horn Sheeler argue that U.S. first ladies and women governors in their study "managed to employ metaphor creatively, often capitalizing on its inventional potential to complicate the containment logic and confound the double bind that historically has challenged women's political agency."[18] The point of entry for scholars of gender and political rhetoric may indeed be a contradiction that demands further examination. One such contradiction is the dearth of female political figures who have inhabited, let alone sought, the highest office in U.S. political culture, the U.S. presidency.

Masculinity and the U.S. presidency

The normative way that the presidency has been fashioned discursively in U.S. political culture sets up obstacles for anyone perceived as lacking presidential masculinity. Jamieson's femininity/competence bind provides context. As she argues, women "who exercised their brains and brawn in public were thought to be tough, ... and masculine"; however, those women "deviated from the female norm of femininity while exceeding or falling short of the masculine norm of competence."[19] Caught in a political Catch-22, women who attempt to enter presidential politics as something other than voters or spouses are framed as "bitches" and unfeminine[20] on one hand and incompetent on the other.

The masculinity that functions as a primary barrier to entry for female political figures seeking executive level office is well documented and serves as a unifying theme in this essay.[21] As Suzanne Daughton writes, "First and foremost, the president is the national patriarch: the paradigmatic American man."[22] The rhetoric of fatherhood often features prominently in the image construction of presidential contenders, linking the performance of fatherhood to images of the president as the father of our country.[23] As Ann E. Burnette and Rebekah L. Fox argue, Mitt Romney, Rick Santorum, and Newt Gingrich in their 2012 presidential primary rhetoric "relied on enthymematic reasoning to conclude that, because a father is a strong protector and leader, these candidates' performances as fathers made them viable presidential candidates."[24] The authors go on to challenge rhetorical critics to question the invisibility of the norm of masculinity/fatherhood as presidential in order to "open the political process to a more diverse vision of leadership."[25]

Consistent with the logic of hegemonic masculinity[26] framing presidential frontrunners, less successful male candidates are framed in ways suggesting that anyone who might demonstrate characteristics deemed feminine runs afoul of the rigid gender norms of the presidency. For example, Anna Cornelia Fahey argues that John Kerry was feminized during the 2004 presidential election, which simultaneously constructed him as unfit for office.[27] Discussing the 2016 U.S. presidential election specifically, Jackson Katz argues that presidential elections are statements about manhood: "specifically, what kind of manhood is most exalted and should be in charge."[28] Katz goes on to describe manhood as embodying strength and stoicism that are performative. Thus, the president is a "pedagogue in chief. He literally teaches—by example—what one highly influential version of dominant masculinity looks like."[29] It is no wonder women (and even some men) have been disadvantaged by the normative nature of hegemonic masculinity that drives cultural assumptions about the presidency. The only examples we have witnessed to date are those casting a certain kind of man in the leading role.

Heightening the heteronormative masculine frame of the U.S. presidency, research on the rhetorical first lady demonstrates the extent to which tradition dictates she embody the ideal of womanhood of the culture at the time, serving as a feminine counter to the masculine president.[30] Some studies focus on traditional feminine framing of first ladies or their strategic use of gender in support of their issue positions.[31] As early as 1996, Karlyn Kohrs Campbell identified the office of U.S. president as "a two person career,"[32] placing the first lady firmly in the center of our understanding of what Trevor Parry-Giles and Shawn J. Parry-Giles more recently identify as "presidentiality."[33] Yet Campbell does not see the role of first lady as without challenge, noting in particular the difficulty Hillary Clinton faced as first lady because she did not conform to tradition.[34] Bucking custom, first ladies possess potential for agency and resistance. Shawn J. Parry-Giles and Diane M. Blair see the "rhetorical first lady" as functioning in both feminine and feminist ways[35] while Anderson argues that paradox best defines the role.[36] Further challenging the gendered assumptions of what it means to be a presidential candidate's spouse, Roseann M. Mandziuk sheds light on the contrast between what she argues is the silencing of female spouses in 2016 in contrast to the masculine ideology that defined the discourse of the potential first "first gentleman," Bill Clinton.[37]

Gender, 2008, and backlash

The 2008 election in particular demonstrated the extent to which hegemonic masculinity is a potent barrier for women in U.S. political culture. Sheeler and Anderson argue that the 2008 election reveals a cultural backlash against female presidentiality, fueled by a rhetoric of post-feminism.[38] Moreover, the election provided an important moment to assess such a backlash because Hillary Clinton and Sarah Palin each claimed a feminist identity, but performed that identity in very different ways. The backlash is evidenced in what Anderson terms pornification: sexist, misogynist, and violent rhetoric used to frame the candidates and undermine their credibility.[39] Lisa Glebatic Perks and Kevin A. Johnson also find evidence of aggressive sexualization of Sarah Palin in 2008, theorizing what they argue are burlesque binds of the MILF frame.[40] Not only was Palin herself hyper-sexualized, but her rhetoric served to heighten the gender norms of U.S. political culture, including her strategic deployment of particular metaphors in the service of frontier mythology[41] or hegemonic masculinity.[42] Not just media framing, but also the vice presidential candidate's rhetoric itself—her statements, performances, and campaign—played a role in reinforcing masculinity's lock on politics in 2008.

Assessment of 2008 political rhetoric demonstrated the various ways that Sarah Palin and Hillary Clinton confronted and were confronted by gender bias during their campaigns. For example, Erika Falk argues that the strength of the "playing the gender card" metaphor lies in its ability to serve as political shorthand for a host of unsupportable arguments, all of which discipline women like Hillary Clinton who dare to speak about bias and discrimination on the campaign trail.[43] Lindsey Meeks argues that *The New York Times* covered Hillary Clinton and Sarah Palin in stereotypical ways, using "novelty frames" to emphasize their candidacy as opposed to male candidates who received more substantive "masculinized" issue coverage.[44] Rebecca M. L. Curnalia and Dorian L. Mermer discuss Clinton's "emotional moment," arguing that it demonstrates the femininity/competence double bind faced by women candidates, while Ryan Shepherd finds that the incident became an important event that sparked conversations about gender bias in political campaigns.[45]

Hillary Clinton, authenticity, and paradox

Scholars have focused on Hillary Clinton in an ongoing area of research that draws attention to the political roles she has assumed and the challenges she has faced while first lady,[46] Senator, presidential candidate,[47] and nominee. Compelling studies uncovering sexist media framing and the cultural constraints of being a powerful political woman (as well as Hillary Clinton's resistance to those norms) have shared similar findings over and over.[48] Shawn J. Parry-Giles analyzes news coverage of Clinton from 1992 to 2008, pointing out the recurring narratives that frame Clinton as inauthentic.[49] Moreover, Parry-Giles documents the violent rhetoric with which

Clinton had to contend when she dared step outside traditional feminine gender norms to seek a Senate seat and later the Democratic nomination for president.

The double bind is often the theoretical frame through which Clinton's political performance is assessed.[50] In response to the 2016 election specifically, Anderson theorizes the "first-timer/frontrunner double bind" that plagued Clinton.[51] Unlike male candidates, female first-timers are positioned as symbolic and unproven rather than possessing credibility as political outsiders. Paradoxically, female candidates must garner significant political experience in order to be taken seriously, but as soon as they do so, political ambition is framed as an unnatural desire for power and outsider status can no longer be claimed. Anderson argues that those interested in changing these patterns should not focus on what women have to do to win the presidency, but on what we as voters have to do to change our cultural assumptions about women and leadership.

The research on masculinity and the U.S. presidency begins to articulate the three overlapping arguments we present in this essay. The normative masculinity of the U.S. presidency poses serious challenges to anyone who doesn't measure up to that unwritten standard. While women such as Hillary Clinton have tried to break the glass ceiling and demonstrated incredibly skillful rhetoric in the process, paradoxically they have not been able to achieve the ultimate goal. While Clinton as an area of study is rich and insightful, many other women who have aspired to and won positions of political leadership are the focus of research forming a solid body of work on the creativity of individual women in the face of cultural constraints. We will discuss these studies in the next section.

Beyond the U.S. presidency: case studies of resistance and change

A significant portion of the research on gender and political rhetoric takes the form of case studies of the discourse strategies of political women who have aspired toward offices including and in addition to the U.S. presidency. Some are positive in their assessment, focusing on the communication style of women such as Shirley Chisholm, Elizabeth Dole, or Hillary Clinton.[52] Others note the strategic rhetorical work behind any woman's attempt at public identity construction. Brenda DeVore Marshall and Molly A. Mayhead, in an edited collection, articulate the ways that women such as Barbara Jordan, Patricia Schroeder, Geraldine Ferraro, Wilma Mankiller, and Madeleine Albright use autobiography as a strategic political discourse to create "a personal and civic self situated within local, national, and/or international political communities."[53] Still others expose the resistance political women encounter and their work to change the face of leadership.

Research reveals the obstacles women face when enacting leadership at various levels (e.g., party politics, governorships, U.S. Congress, vice presidential and presidential roles), noting that the barriers and opportunities are different depending on the location and level of leadership.[54] Women in legislative bodies have more latitude than women who seek the presidency. Legislative women can subsume themselves to voters and work in collaboration with others in Congress, an expectation that is gendered feminine, whereas individuals seeking the presidency also seek executive level [masculine] authority.[55] Beyond gender and candidate discourse, other essays reveal challenges related to gender and voter issue expectations, illustrating the difficulties women face when national security or terrorism are high voter priorities.[56]

Strategies of resistance and bringing a broader perspective to what leadership looks and sounds like demonstrate the rhetorical inventiveness of female political contenders. For example, Marshall and Mayhead find that women governors "brought issues of caring, personal experience, empowerment, family, and inclusivity to the statehouse."[57] In another volume with chapters about women representatives, senators, and governors, Mayhead and Marshall conclude "that women have created and now operate in a rhetorical in-between space that values issues and perspectives traditionally seen as women's issues."[58] Moreover, they find evidence of "the dawning of a more collaborative approach to decision making" between women of all parties.[59] Similarly, in a comparison of male and female mayoral State of the City addresses, Mirya Holman argues that female mayors use more nurturing, inclusive frames in their speeches, even though male and female mayors discuss similar issues.[60] Kathryn Pearson and Logan Dancey make an analogous argument about women on the House floor, finding that they are more likely to discuss women during floor speeches and policy debates, regardless of political party identification, than are men.[61] Sheeler argues that Michigan Governor Jennifer Granholm confounded the beauty queen metaphor through which she was framed, using "the resources of the unruly woman within a context of gendered reporting" to capitalize on strategic visibility and ultimately win election.[62]

More recently, book-length studies consider the intersections of gender, political rhetoric, and public depictions, noting the increasing role that gendered popular culture plays in the ways that voters think about and participate in politics.[63] Perhaps one reason popular culture is becoming more influential in our understanding of politics is because we see a growing number of strong female characters on television and other venues. Anderson's edited collection considers programs such as *Madam Secretary*, *Scandal*, *Parks and Recreation*, and *VEEP*, examples of feminist comedians such as Amy Schumer and the women of *The Daily Show*, and other examples of feminist scholarship about popular culture.[64] While some of the programming illustrates the prevalence of stereotypical frames through which to view powerful women, Anderson argues, "feminist interventions in political pop culture are challenging these stereotypes, diversifying the frames through which we observe political women and moving the culture beyond a simplistic understanding of women's political identity."[65] According to Shawn J. Parry-Giles in her chapter on the "power paradox," the examples in Anderson's collection not only articulate "the continued anxieties that accompany women's empowerment" but also examine the various "ways in which women in popular culture speak back to power from inside and outside structures of power."[66] As Parry-Giles concludes, "Until we are able to reconcile the goal of empowerment with the attainment of power, women will continue to be restrained by this power paradox."[67] Importantly, viewers may envision a more inclusive politics as they grapple with the challenges confronted by these female characters.

Gender and international political rhetoric

Women around the world are constrained by the power paradox. Empirical research demonstrates that at the beginning of their careers, women are just as interested in political issues as are men and harbor similar ambitions as their male counterparts, but public perceptions, a legacy of discrimination, media bias, and family responsibilities curtail women's attempts at obtaining political power.[68] Karen Ross assesses media coverage of female political figures and argues that politics around the world is regarded as a masculine job, despite more women achieving the roles of president and prime minister in recent years. As her analysis demonstrates, novelty and inauthenticity dominate not only coverage of U.S. political women but female political candidates and leaders globally.

> What seems clear is that gender-differentiated coverage of politicians is a global phenomenon and a variety of factors are in play, including the circulation and routinization of gender-based stereotypes, the male-ordered nature of many newsroom environments, and the reliance on the 'usual suspects' as sources and subjects for news discourse.[69]

Numerous studies confirm Ross's conclusions. For example, research on Australia's first female Prime Minister Julia Gillard demonstrates gendered media framing[70] and negative public perceptions of an ambitious woman,[71] or what Jamieson would identify as the femininity/competence double bind.[72] Katrina Lee-Koo and Maria Maley identify a similarly gendered bind that confronted two Australian political women and that played out differently for each. They argue that the metaphor of the "Iron Butterfly" served to advance positive framings of Foreign Minister Julie Bishop's conservative femininity while Chief of Staff Peta Credlin's more aggressive warrior persona was at the root of negative public perceptions.[73] Battleground metaphors, which privilege masculinity and undermine a woman's competence, frame coverage of Canadian party leadership,[74] whereas frames through which female politicians are evaluated in the European press tend to focus on their roles as wives and mothers.[75] Australian Prime Minister Gillard went so far as to call out sexist and misogynist treatment in her widely publicized "Sexism and Misogyny" speech. Framing the speech as a strategic attack on men, an outpouring of emotions, and a hypocritical statement, media coverage served to undermine and minimize her charge,[76] vilify her sexually,[77] and facilitate the same gendered stereotypes that she brought to light.[78]

A few studies of international political women demonstrate their attempts to challenge stereotypes. For example, Sharon Halevi conducts a rhetorical analysis of the ways Tzipi Livni's "womanhood" was presented in the Israeli press in 2008–2009.[79] She finds that while resistance to strict gender ideologies was present, they did not alter masculine assumptions about leadership in Israel. Einat Lachover confirms this claim in an interpretive analysis of media frames, finding that national television coverage of women in local Israeli elections is becoming more complex: both conforming to patriarchal norms of news coverage and also giving voice to feminist concerns.[80] Content analysis demonstrates that even though Park Geun-hye, a female candidate in the 2007 Korean presidential primary, was presented to the public in stereotypical ways damaging her viability as a candidate,[81] she countered that coverage overtly on her candidate website.[82] More hopeful still, Middle Eastern and U.S. media visually framed Muslim women as active participants in political protest coverage of the Arab Spring.[83] Finally, research attempts to posit ways to close the gendered representation gap, such as more gender-friendly policies, childhood socialization, and changing traditional family norms,[84] but scholars argue that none of these changes will take hold until women are more prevalent in politics.[85]

Despite sexist coverage of women in international politics, young women around the world have long looked up to strong female leaders like Margaret Thatcher or Angela Merkel. Research on a new wave of women in right-wing parties, influenced by France's Marine LePen, considers gender and activism in populist movements, such as the Latvian National Front[86] and leaders such as Danish Pia Kjærsgaard.[87] Feminist activism in right-leaning parties has not always been common, as Sylvia Bashevkin argues in her empirical analysis of the feminist rhetoric of Canadian women candidates. She finds that centrist as well as opposition parties were more likely to voice feminist concerns than right-leaning or leading parties; although women in right leaning parties did not voice antifeminist positions.[88]

While challenging gendered perceptions may not be advantageous for some political women, aligning with stereotypes does not guarantee success. For example, research on campaign and leadership styles suggests that women are more inclusive in their campaign strategies;[89] however, this behavior may not be rewarded when it comes to public perceptions.[90] Yet, in combination with an ethnically diverse country, female leadership has been found to reap benefits.[91] Young-Im Lee conducts a compelling analysis of Park Geun-Hye's successful campaign for South Korean president in 2012. Lee argues that Geun-Hye relied on feminine stereotypes as well as political experience, campaigning as "the well-prepared female president."[92] Her case provides an interesting counter to Anderson's first-timer/frontrunner double bind.[93] Finally, a great deal of research exists on German Chancellor Angela Merkel's rhetorical style, assessing the extent to which gender features in her campaign and leadership discourse.[94]

These case studies demonstrate the challenges and originality political women around the world bring to their campaigns and leadership roles. Often written from the perspective of strategic choices by individual women running to counter culturally constructed stereotypes and biases, these cases are instructive for any woman aspiring to a position of political leadership. As the next section demonstrates, such strategies are necessary, for we are far from moving into an age in which stereotypes and bias are passé.

2016 and beyond: stereotypes and intersectional rhetorics

Stereotypes have long been a part of politics and the 2016 U.S. presidential election was no exception. The hypermasculinity associated with the presidency was on full display[95] along with attempts from female candidates to deploy these stereotypes to their advantage. Social media, a relatively new form of political communication, cemented its place in political discourse not only with regard to presidential contenders but gubernatorial candidates as well.[96]

Gendered assumptions remained the norm during the 2016 presidential race, as Kelly Dittmar reveals in her discussion of themes in the first year of the campaign.[97] Male candidates doubled down on masculinity, while female candidates did their best to combat negative stereotypes. Just as we have witnessed in previous election cycles, Republican candidates engaged in "locker room talk," questioning the masculinity of their fellow male candidates and reinforcing the supposition that the presidency is a heteronormative masculine role.[98] Sexist commentary is not party-bound. Bernie Sanders criticized Hillary Clinton's stance on gun control by saying "all the shouting in the world" would not end gun violence.[99] Republican Carly Fiorina was criticized not only for her speaking style but also her appearance. A journalist from *The Guardian*, Jeb Lund, described Fiorina's speaking style as peculiarly peevish, soft talking, and Grinchy.[100] Moreover, then candidate Donald Trump famously remarked: "Look at that face! … Would anyone vote for that?"[101]

Bernie Sanders' supporters, sometimes referred to as "Bernie Bros"—a derogatory nickname brought to popular usage by Robinson Meyer in *The Atlantic*[102]—made their own mark on stereotyping. More often, Democratic candidates align with feminist concerns, but the Sanders-Clinton democratic presidential primary race in 2016 complicated this assumption. According to Kelly Wilz, Sanders donned the mantle of street fighter against the wealthy upper class and corporations, channeling a vein of masculinity.[103] This street fighter mantra may have appealed to "a particular version of a political citizen whose investment in the campaign of Bernie Sanders straddled or crossed the line into a misogynist hatred toward Sanders' main political foe."[104] These "Bernie Bros," according to Michael Mario Albrecht, took to social media after Clinton won the nomination to spew gendered vitriol at delegates who voted for Clinton, many of whom were women.[105] The issue illustrates the significance that gender issues continue to play in our contemporary politics, an issue that scholars must understand in all its complexity if we are to move forward toward a more inclusive politics.

Intersectional rhetorics

In some ways, the 2016 election provided a watershed moment for research on the intersections of gender, race, sexuality, and politics. Dara Z. Strolovitch, Janelle Wong, and Andrew Procter consider the political significance of aligning with or resisting white heteropatriarchy[106] in 2016, while Donna Goldstein and Kira Hall see the election turning on a combination of "gendered and racialized nostalgia."[107] Paul Elliott Johnson focuses specifically on the demagogic strategies of Trump's campaign, finding that it relies on "victimized, White, toxic masculinity."[108]

When it comes to understanding electoral behavior, intersectionality is an increasingly important factor for scholars to consider. The Democratic Party in the U.S., with its long-term identification with southern whites and association with segregationist policies, saw change along race and gender lines in the mid-20th century. As early as the 1930s, white men moved toward the Republican Party and African American women aligned with the Democratic Party.[109] Gender is not the only determinant of voting behavior; race plays an important role as well. The 2016 election was a perfect example of this phenomenon. More than half of white female voters cast ballots for Trump while the majority of African American women and a plurality of African American men voted for Clinton.[110]

Along with a shifting gender gap, one study posits a change in assumptions about representation. Perhaps due to movement in the intersectional vote share, conservative candidates can be read as aligning with women's interests, as long as they include appeals and actions fitting categories of egalitarianism, inclusiveness, and responsiveness.[111] A further challenge to conservative female candidates in 2016 was the line between supporting and endorsing candidate Trump. Scott Smith argues that New Hampshire Senate candidate Kelly Ayotte navigated this challenge in her 2016 election rhetoric, although unsuccessfully. By examining Ayotte's response to the Access Hollywood tape scandal as well as her "discourse of renewal," Smith finds that she was able to delineate between supporting conservative values and condemning President Trump for his comments about women.[112]

One key reaction to the 2016 election and Trump's inauguration specifically was the first Women's March. Scholarship assessing the Women's March considered critically the agency of women who are not normally cast as political actors and responded to feminist calls for increased attention to the intersections of race, gender, class, religion, and other differences in advancing social change. Held one day after President Trump's inauguration, the first Women's March was a grassroots organization that exceeded expectations when nearly 600,000 demonstrators took to the National Mall in Washington, DC. Many thought that because the march was not focused on one issue and was largely organized via social media, attendance and news coverage would be below expectations. To the contrary, Kristine Nicolini and Sara Steffes Hansen review media coverage of the Women's March, finding that, for the most part, the event was conveyed to the public through four key frames in line with the organization itself: diversity, resistance, activation, and solidarity. Whereas previous protest coverage marginalized protestors, the Women's March coverage was different in that social media, website, and email communication was central to framing the supportive and intersectional coverage picked up by national media outlets, with the exception of FOX News.[113]

Not only was the Washington, DC march successful, but so were the marches in other cities.[114] However, some scholars criticized the event for emphasizing a narrow interpretation of femininity, arguing that the march marginalized women of color and transgender women.[115] Other scholars find both contradiction and resistance in the march as a form of protest. Anne Graefer, Allaina Kilby, and Inger-Lise Kalviknes Bore assess the "offensive humor" of the signs displayed at the marches, arguing "[p]rotesters and social media users attacked Trump's patriarchal and racist policies and practices through the use of gendered and raced insults that simultaneously reinforced established notions of ideal White masculinity."[116] Similarly, Banu Gökariksel and Sara Smith find in the protests an intersectional moment that considers difference in various forms, but that also may reinforce an "imperial feminism."[117] However, Rachel E. Presley and Alane L. Presswood argue that the march addressed the notion of "intersectional justice" and explore new forms of activism and resistance that emerged.[118]

Momentum continued with the 2018 primary and midterm elections. Female candidates not only performed as well as their male counterparts, but being a woman running for office was an advantage in many races.[119] Some of these advantages came from institutional factors. Women tend to lean further left than male candidates, so they have a built-in advantage in Democratic primaries that becomes a disadvantage in Republican primaries. There also tend to be more female candidates to pull from in Democratic races. Even though women are advantaged in Democratic races, to find success electorally, they often need more experience than the male candidates they run against.[120]

One strategy of the largely Democratic group of women new to Congress in 2018 is what Anderson calls "coalitional amplification.[121] Anderson described this as the process of "bolstering one another's messages … and forg[ing] an intersectional feminist practice that [brings] together feminists of different genders, sexualities, ethnicities, classes, and ages."[122] *The Washington Post* columnist Monica Hesse takes note of the strategy used by Alexandria Ocasio-Cortez and others "to lift one another up or highlight each other's accomplishments."[123] It remains to be seen whether the same sort of amplification will occur among the record number of women who declared their candidacies not only for the 2020 U.S. presidential election but for Congressional races as well. What is clear is that the media have not changed their playbook, providing women candidates not only less coverage in 2019 but less favorable coverage than the men running.[124]

Conclusion

Arguments in the field of political rhetoric with a focus on gender coalesce in three intersecting areas: barriers of masculinity and bias, inventive strategies of resistance, oxymoron, and double binds. Just like the feminist scholarship that informs this chapter, one strong argument thread here points to the persistence of masculine assumptions that inform our understanding of political office around the world. Normative masculinity functions as a barrier to entry for female candidates and for male candidates who may enact their candidacies in ways deemed insufficiently masculine. This obstacle can be witnessed not only in choice of candidate strategy but also in media frames and even in informing a backlash that disciplines strong political women. Of course, examples are plentiful of women candidates and leaders speaking in ways that not only make visible but also actively resist masculinity's stronghold on public perceptions of political office, our second intersecting theme. In particular, research on sexist and biased media coverage of women candidates and leaders remains an important thread that must continue not only to point out sexism but also to advocate for more gender-inclusive frames through which to present and evaluate political leadership. Confronting the double bind, our third intersecting thread for understanding research on gender and political communication, illustrates the difficulties of advancing a completely new understanding of political leadership. Like many of these studies demonstrate, stretching the boundaries in one direction may necessarily reinforce old assumptions in another.

Still, where do we go from here? The abundance of case studies on the rhetorical invention of individual women confronting stereotypes and deploying metaphors and binds in new ways suggests important work is being done to advance a more gender-inclusive understanding of leadership. That being said, what else are women to do? And perhaps that question is the problem. Individual women candidates and leaders are resisting and demanding they be taken seriously. We call for continued research that interrogates and seeks to dismantle the cultural constraints that make it difficult not only for women but people of color, and people of different sexualities, religions, classes, and ethnicities to achieve public positions of leadership. The advocacy and intersectional focus of research, especially since the 2016 U.S. presidential election, is heartening and much needed to advance this kind of cultural shift. Moreover, additional research on contexts beyond the presidency—and more specifically beyond the U.S. presidency—is much needed. Our focus on U.S. contexts may block us from considering other ways of thinking and doing leadership that could advance our own understanding.

Despite its limitations, the research on gender and political rhetoric is rich, compelling, sophisticated, and responsive to ways we may enact a more accountable political culture. As the 2018 U.S. midterm election demonstrated, the face of leadership is changing. And as some of the cases in this chapter reveal, these shifts are taking place across the globe. Still, we have much work to do to dismantle masculinity's hold on conventional understandings of leadership to bring forward a new chapter in the state of the art of research on gender and political rhetoric.

Notes

1 Karine Jean-Peirre, "It's not the 'Year of the Woman.' It's the 'Year of the Women,'" CNN, Opinion, November 4, 2018, www.cnn.com/2018/11/03/opinions/midterm-elections-year-of-woman-roundup/index.html.

2 Susan Zaeske, "History, Theory, and Method in Feminist Scholarship on Early American Discourse," *Rhetoric & Public Affairs* 6, no. 3 (2003): 567–579, www.jstor.org.uml.idm.oclc.org/stable/41939854.

3 Bonnie J. Dow and Celeste M. Condit, "The State of the Art in Feminist Scholarship in Communication," *Journal of Communication* 55, no. 3 (September 2005): 449, https://doi-org.uml.idm.oclc.org/10.1111/j.1460-2466.2005.tb02681.x.

4 Dow and Condit, "The State of the Art," 467.

5 Dow and Condit, "The State of the Art," 467.

6 Karen A. Foss, "Feminist Scholarship in Speech Communication: Contributions and Obstacles," *Women's Studies in Communication* 12, no. 1 (1989): 1–10, https://doi-org.uml.idm.oclc.org/10.1080/07491409.1989.11089729.

7 Linda Aldoory and Elizabeth L. Toth, "The Complexities of Feminism in Communication Scholarship Today," *Communication Yearbook* 24, no. 1 (2001): 345–361, https://doi-org.uml.idm.oclc.org/10.1080/23808985.2001.11678993.

8 Michaela D. E. Meyer, "Women Speak(ing): Forty Years of Feminist Contributions to Rhetoric and an Agenda for Feminist Rhetorical Studies," *Communication Quarterly* 55, no. 1 (2007): 2, https://doi-org.uml.idm.oclc.org/10.1080/01463370600998293.

9 Meyer, "Women Speak(ing)," 10.

10 Vanessa Beasley, "The Power of Rhetoric: Understanding Political Oratory," in *The Sage Handbook of Political Communication*, eds Holli A. Semetko and Margaret Scammell (Thousand Oaks, CA: Sage, 2012), 364.

11 Susan A. Banducci with Elisabeth Gidengil and Joanna Everitt, "Women as Political Communicators: Candidates and Campaigns," in *The Sage Handbook of Political Communication*, eds Holli A. Semetko and Margaret Scammell (Thousand Oaks, CA: Sage, 2012), 165–172.

12 See, for example, Karrin Vasby Anderson, "Every Woman Is the Wrong Woman: The Female Presidentiality Paradox," *Women's Studies in Communication* 40, no. 2 (2017): 132–135, https://doi-org.uml.idm.oclc.org/10.1080/07491409.2017.1302257; Karrin Vasby Anderson, "Hillary Rodham Clinton as 'Madonna': The Role of Metaphor and Oxymoron in Image Restoration," *Women's Studies in Communication* 25, no. 1 (2002): 1–24, https://doi-org.uml.idm.oclc.org/10.1080/07491409.2002.10162439; Karrin Vasby Anderson, "Presidential Pioneer or Campaign Queen? Hillary Clinton and the First-Timer/Frontrunner Double Bind," *Rhetoric & Public Affairs* 20, no. 3 (2017): 525–538, https://doi-org.uml.idm.oclc.org/10.1080/07491409.2002.10162439; Rebecca M. L. Curnalia and Dorian L. Mermer, "The 'Ice Queen' Melted and It Won Her the Primary: Evidence of Gender Stereotypes and the Double Bind in News Frames of Hillary Clinton's 'Emotional Moment'," *Qualitative Research Reports in Communication* 15, no. 1 (2014): 26–32, https://doi-org.uml.idm.oclc.org/10.1080/17459435.2014.955589; Dustin Harp, Jaime Loke, and Ingrid Bachmann, "Hillary Clinton's Benghazi Hearing Coverage: Political Competence, Authenticity, and the Persistence of the Double Bind," *Women's Studies in Communication* 39, no. 2 (2016): 193–210, https://doi-org.uml.idm.oclc.org/10.1080/07491409.2016.1171267; Lisa Glebatis Perks and Kevin A. Johnson, "Electile Dysfunction," *Feminist Media Studies* 14, no. 5 (2014): 775–790, https://doi-org.uml.idm.oclc.org/10.1080/14680777.2013.829860.

13 Karlyn Kohrs Campbell, "The Rhetoric of Women's Liberation: An Oxymoron," *Communication Studies* 50, no. 2 (1999): 125–137, https://doi-org.uml.idm.oclc.org/10.1080/10510979909388479.

14 Campbell, "The Rhetoric of Women's Liberation," 133.

15 Kathleen Hall Jamieson, *Beyond the Double Bind: Women and Leadership* (Oxford: Oxford University Press, 1995).

16 Jamieson, *Beyond the Double Bind*, 4.

17 Karrin Vasby Anderson and Kristina Horn Sheeler, *Governing Codes: Gender, Metaphor, and Political Identity* (Lanham, MD: Lexington Books, 2005); Karlyn Kohrs Campbell, "The Discursive Performance of Femininity: Hating Hillary," *Rhetoric & Public Affairs* 1, no. 1 (1998): 1–19, https://doi-org.uml.idm.oclc.org/10.1353/rap.2010.0172; Diana B. Carlin and Kelly L. Winfrey, "Have You Come a Long Way, Baby? Hillary Clinton, Sarah Palin, and Sexism in 2008 Campaign Coverage," *Communication Studies* 60, no. 4 (2009): 326–343, https://doi-org.uml.idm.oclc.org/10.1080/10510970903109904; Hinda Mandell, "Political Wives, Scandal, and the Double Bind: Press Construction of Silda Spitzer and Jenny Sanford Through a Gendered Lens," *Women's Studies in Communication* 38, no. 1 (2015): 57–77, https://doi-org.uml.idm.oclc.org/10.1080/07491409.2014.995327; Deborah Carol Robson, "Stereotypes and the Female Politician: A Case Study of Senator Barbara Mikulski," *Communication Quarterly* 48, no. 3 (2000): 205–222, https://doi-org.uml.idm.oclc.org/10.1080/01463370009385593; Kristina Horn Sheeler and Karrin Vasby Anderson, *Woman President: Confronting Postfeminist Political Culture* (College Station: Texas A&M University Press, 2013).

18 Anderson and Sheeler, *Governing Codes*.

19 Jamieson, *Beyond the Double Bind*, 120–121.

20 Karrin Vasby Anderson, "'Rhymes with Blunt': Pornification and US Political Culture," *Rhetoric & Public Affairs* 14, no. 2 (2011): 327–368, https://doi-org.uml.idm.oclc.org/10.1353/rap.2010.0228; Karrin Vasby Anderson, "'Rhymes with Rich': 'Bitch' as a Tool of Containment in Contemporary American Politics," *Rhetoric & Public Affairs* 2, no. 4 (1999): 599–623, https://doi-org.uml.idm.oclc.org/10.1353/rap.2010.0082; C. Robert Whaley and George Antonelli, "The Birds and the Beasts: Women as Animal," *Maledicta* 7 (1983) 219–29.

21 Kevin Coe, David Domke, Meredith M. Bagley, Sheryl Cunningham, and Nancy Van Leuven, "Masculinity as Political Strategy: George W. Bush, the 'War on Terrorism,' and an Echoing Press," *Journal of Women, Politics, and Policy* 29, no. 1 (2007): 31–55, https://doi-org.uml.idm.oclc.org/10.1300/J501v29n01_03; Suzanne Daughton, "Women's Issues, Women's Place," *Communication Quarterly* 42, no. 2 (1994): 106–119, https://doi-org.uml.idm.oclc.org/10.1080/01463379409369920; Greg Goodale, "The Presidential Sound: From Orotund to Instructional Speech, 1892–1912," *Quarterly Journal of Speech* 96, no. 2 (2010): 164–184, https://doi-org.uml.idm.oclc.org/10.1080/00335631003796651; Michael Kimmel, *Manhood in America: A Cultural History*, 3rd edn (New York: Oxford University Press, 2012); Clinton Rossiter, *The American Presidency* (Baltimore, MD: Johns Hopkins University Press, 1987); Sheeler and Anderson, *Woman President*; Mary Stuckey, "Rethinking the Rhetorical Presidency and Presidential Rhetoric," *Review of Communication* 10, no. 1 (2010): 38–52, https://doi-org.uml.idm.oclc.org/10.1080/15358590903248744.

22 Daughton, "Women's Issues, Women's Place," 114.

23 Ann E. Burnette and Rebekah L. Fox, "My Three Dads: The Rhetorical Construction of Fatherhood in the 2012 Republican Presidential Primary," *Journal of Contemporary Rhetoric* 2, nos 3/4 (2012): 80–91.

24 Burnette and Fox, "My Three Dads," 91.

25 Burnette and Fox, "My Three Dads," 91.

26 See R. W. Connell and James W. Messerschmidt, "Hegemonic Masculinity: Rethinking the Concept," *Gender & Society* 19, no. 6 (2005): 829–859, https://doi-org.uml.idm.oclc.org/10.1177/0891243205278639.

27 Anna Cornelia Fahey, "French and Feminine: Hegemonic Masculinity and the Emasculation of John Kerry in the 2004 Presidential Race," *Critical Studies in Media Communication* 24, no. 2 (2007): 132–150, https://doi-org.uml.idm.oclc.org/10.1080/07393180701262743.

28 Jackson Katz, *Man Enough? Donald Trump, Hillary Clinton and the Politics of Presidential Masculinity* (Northampton, MA: Interlink Pub Group, 2016).

29 Katz, *Man Enough?*

30 Karlyn Kohrs Campbell, "The Rhetorical Presidency: A Two-Person Career," in *Beyond the Rhetorical Presidency*, ed. Martin J. Medhurst (College Station, TX: Texas A&M University Press, 1996); Betty Boyd Caroli, *First Ladies: From Martha Washington to Michelle Obama* (New York: Oxford University Press, 2010); Molly Meijer Wertheimer, ed., *Inventing a Voice: The Rhetoric of American First Ladies of the Twentieth Century* (Lanham, MD: Rowman and Littlefield, 2004).

31 Tasha N. Dubriwny, "First Ladies and Feminism: Laura Bush as Advocate for Women's and Children's Rights," *Women's Studies in Communication* 28, no. 1 (2005): 84–114, https://doi-org.uml.idm.oclc.org/10.1080/07491409.2005.10162485; Keith V. Erickson and Stephanie Thomson, "First Lady International Diplomacy: Performing Gendered Roles on the World Stage," *Southern Communication Journal* 77, no. 3 (2012): 239–262, https://doi-org.uml.idm.oclc.org/10.1080/1041794X.2011.647502; Annette Madlock Gatison, "Michelle Obama and the Representation of Respectability," *Women & Language* 40, no. 1 (2017/2018): 101–110; Jenni M. Simon and Abby M. Brooks, "The 'First' Coverage of the First Lady: E-racing the Mom-in-Chief," *Women & Language* 40, no. 1 (2017/2018): 15–35.

32 Campbell, "The Rhetorical Presidency."

33 Trevor Parry-Giles and Shawn J. Parry-Giles, *The Prime-Time Presidency: "The West Wing" and US Nationalism* (Urbana, IL: University of Illinois Press, 2006), 210.

34 Campbell, "The Discursive Performance."

35 Shawn J. Parry-Giles and Diane M. Blair, "The Rise of the Rhetorical First Lady: Politics, Gender Ideology, and Women's Voice, 1789–2002," *Rhetoric and Public Affairs* 5, no. 4 (2002): 565–599, https://doi-org.uml.idm.oclc.org/10.1353/rap.2003.0011.

36 Karrin Vasby Anderson, "The First Lady: A Site of 'American Womanhood'," in *Inventing a Voice: The Rhetoric of American First Ladies of the Twentieth Century*, ed. Molly Meijer Wertheimer (Lanham, MD: Rowman and Littlefield, 2004), 17–30.

37 Roseann M. Mandziuk, "Whither the Good Wife? 2016 Presidential Candidate Spouses in the Gendered Spaces of Contemporary Politics," *Quarterly Journal of Speech* 103, nos 1/2 (2017): 136–159, https://doi-org.uml.idm.oclc.org/10.1080/00335630.2016.1233350.

38 Sheeler and Anderson, *Woman President*. For more on postfeminism, see Bonne J. Dow, *Prime-Time Feminism: Television, Media Culture, and the Women's Movement Since 1970* (Philadelphia, PA: University of Pennsylvania Press, 1996); Angela McRobbie, *The Aftermath of Feminism: Gender, Culture, and Social Change* (Thousand Oaks, CA: Sage, 2009); Mary Douglas Vavrus, *Postfeminist News: Political Women in Media Culture* (Albany, New York: State University of New York Press, 2002).

39 Anderson, "'Rhymes with Blunt'."

40 Perks and Johnson, "Electile Dysfunction."

41 Katie L. Gibson and Amy L. Heyse, "Depoliticizing Feminism: Frontier Mythology and Sarah Palin's 'The Rise of The Mama Grizzlies'," *Western Journal of Communication* 78, no. 1 (2014): 97–117, https://doi-org.uml.idm.oclc.org/10.1080/10570314.2013.812744.

42 Katie L. Gibson and Amy L. Heyse, "'The Difference between a Hockey Mom and a Pit Bull': Sarah Palin's Faux Maternal Persona and Performance of Hegemonic Masculinity at the 2008 Republican National Convention," *Communication Quarterly* 58, no. 3 (2010): 235–256, https://doi-org.uml.idm.oclc.org/10.1080/01463373.2010.503151.

43 Erika Falk, "Clinton and the Playing-the-Gender-Card Metaphor in Campaign News," *Feminist Media Studies* 13, no. 2 (2013): 192–207, https://doi-org.uml.idm.oclc.org/10.1080/14680777.2012.678074.

44 Lindsey Meeks, "All the Gender that's Fit to Print: How *The New York Times* Covered Hillary Clinton and Sarah Palin in 2008," *Journalism & Mass Communication Quarterly* 90, no. 3 (2013): 520–539, https://doi-org.uml.idm.oclc.org/10.1177/1077699013493791.

45 Curnalia and Mermer, "The 'Ice Queen' Melted," 26–32; Ryan Shepard, "Confronting Gender Bias, Finding a Voice: Hillary Clinton and the New Hampshire Crying Incident," *Argumentation & Advocacy* 46, no. 1 (2009): 64–77, https://doi-org.uml.idm.oclc.org/10.1080/00028533.2009.11821717.

46 Denise Oles-Acevedo, "Fixing the Hillary Factor: Examining the Trajectory of Hillary Clinton's Image Repair from Political Bumbler to Political Powerhouse," *American Communication Journal* 14, no. 1 (2012): 33–46; Shawn J. Parry-Giles, "Mediating Hillary Rodham Clinton: Television News Practices and Image-Making in the Postmodern Age," *Critical Studies in Media Communication* 17, no. 2 (2000): 205, https://doi-org.uml.idm.oclc.org/10.1080/15295030009388390.

47 David S. Kaufer and Shawn J. Parry-Giles, "Hillary Clinton's Presidential Campaign Memoirs: A Study in Contrasting Identities," *Quarterly Journal of Speech* 103, nos 1/2 (2017): 7–32, https://doi-org.uml.idm.oclc.org/10.1080/00335630.2016.1221529.

48 Ingrid Bachmann, Dustin Harp, and Jamie Loke, "Covering Clinton (2010–2015): Meaning-making Strategies in US Magazine Covers," *Feminist Media Studies* 18, no. 5 (2018): 793–809, https://doi-org.uml.idm.oclc.org/10.1080/14680777.2017.1358204; Regina G. Lawrence and Melody Rose, *Hillary Clinton's Race for the White House: Gender Politics and the Media on the Campaign Trail* (Boulder, CO: Lynne Rienner Publishers, 2010); Lindsey Meeks, "Is She 'Man Enough'? Women Candidates, Executive Political Offices, and News Coverage," *Journal of Communication* 62, no. 1 (2012): 175–93, https://doi-org.uml.

idm.oclc.org/10.1111/j.1460-2466.2011.01621.x; Shawn J. Parry-Giles, "Mediating Hillary Rodham Clinton," 205–226; Theodore F. Sheckels, ed., *Cracked but Not Shattered: Hillary Rodham Clinton's Unsuccessful Campaign for the Presidency* (Lanham, MD: Lexington Books, 2009); Kristina Horn Sheeler, "The Rhetoric of Impossible Expectations: Media Coverage of Hillary Clinton's 2016 General Election Campaign," in *An Unprecedented Election: Media, Communication, and the Electorate in the 2016 Election*, eds Benjamin R. Warner, Dianne G. Bystrom, Mitchell S. McKinney, and Mary C. Banwart (Santa Barbara, CA: ABC-Clio, 2018), 87–104; Diana Zulli, "The Changing Norms of Gendered News Coverage: Hillary Clinton in the *The New York Times*, 1969–2016," *Politics and Gender* 15, no. 3 (2019): 599–621, https://doi-org.uml.idm.oclc.org/10.1017/S1743923X18000466;

49 Shawn J. Parry-Giles, *Hillary Clinton in the News: Gender and Authenticity in American Politics* (Champaign-Urbana, IL: University of Illinois Press, 2014).

50 Anderson, "Every Woman Is the Wrong Woman," 132–132; Anderson, "Hillary Rodham Clinton as 'Madonna'," 1–24; Anderson, "Presidential Pioneer or Campaign Queen," 525–538; Harp, Loke, and Bachmann, "Hillary Clinton's Benghazi Hearing Coverage," 193–210.

51 Anderson, "Presidential Pioneer or Campaign Queen," 525–538.

52 Nichola Gutgold, *Almost Madam President: Why Hillary Clinton Won in 2008* (Lanham, MD: Lexington Books, 2009); Nichola Gutgold, *Paving the Way for Madam President* (Lanham, MD: Lexington Books, 2006).

53 Brenda Devore Marshall and Molly A. Mayhead, *Telling Political Lives: The Rhetorical Biographies of Women Leaders in the United States* (Lanham, MD: Lexington Books, 2008), 1.

54 Janis L. Edwards, ed., *Gender and Political Communication in America : Rhetoric, Representation, and Display* (Lanham, MD : Lexington Books, 2009); Michele Lockhart and Kathleen Mollick, eds, *Political Women: Language and Leadership* (Lanham, MD: Lexington Books, 2013); Sue Thomas and Clyde Wilcox, eds, *Women and Elective Office: Past, Present, and Future*, 3rd edn (Oxford: Oxford University Press, 2014).

55 See Anderson and Sheeler, *Governing Codes*.

56 Erika Falk and Kate Kenski, "Issue Saliency and Gender Stereotypes: Support for Women as Presidents in Times of War and Terrorism," *Social Science Quarterly* 87, no. 1 (2006): 1–18, www.jstor.org.uml.idm.oclc.org/stable/42956106; Ann Gordon and Jerry Miller, "Gender, Race, and the Oval Office," in *Anticipating Madam President*, eds Robert P. Watson and Ann Gordon (Boulder, CO: Lynne Rienner, 2003), 145–155; Jennifer Lawless, "Women, War, and Winning Elections: Gender Stereotyping in the Post-September 11th Era," *Political Research Quarterly* 57, no. 3 (2004): 479–490, https://doi-org.uml.idm.oclc.org/10.2307/3219857.

57 Brenda DeVore Marshall and Molly A. Mayhead, eds, *Navigating Boundaries: The Rhetoric of Women Governors* (Westport, CT: Praeger, 2000), 123–124.

58 Molly A. Mayhead and Brenda Devore Marshall, *Women's Political Discourse: A 21st Century Perspective* (Lanham, MD: Rowman and Littlefield, 2005), 209.

59 Mayhead and Marshall, *Women's Political Discourse*, 210.

60 Mirya R. Holman, "Gender, Political Rhetoric, and Moral Metaphors in State of the City Addresses," *Urban Affairs Review* 52, no. 4 (2016): 501–530, https://doi-org.uml.idm.oclc.org/10.1177/1078087415589191.

61 Kathryn Pearson and Logan Dancey, "Speaking for the Underrepresented in the House of Representatives: Voicing Women's Interests in a Partisan Era," *Politics & Gender* 7, no. 4 (2011): 493–519, https://doi-org.uml.idm.oclc.org/10.1017/S1743923X1100033X.

62 Kristina Horn Sheeler, "Beauty Queens and Unruly Women in the Year of the Woman Governor: Jennifer Granholm and the Visibility of Leadership," *Women's Studies in Communication* 33, no. 1 (2010): 34–53, https://doi-org.uml.idm.oclc.org/10.1080/07491401003669927.

63 Justin S. Vaughn and Lilly J. Goren, eds, *Women and the White House: Gender, Popular Culture, and Presidential Politics* (Lexington, KY: University of Kentucky Press, 2013).

64 Karrin Vasby Anderson, ed., *Women, Feminism, and Pop Politics: From "Bitch" to "Badass" and Beyond* (New York: Peter Lang, 2018).

65 Anderson, *Women, Feminism, and Pop Politics*, 23.

66 Shawn J. Parry-Giles, "Conclusion: Political Women and the Power Paradox: The Case of Hillary Clinton," in *Women, Feminism, and Pop Politics: From "Bitch" to "Badass" and Beyond*, ed. Karrin Vasby Anderson (New York: Peter Lang, 2018), 323.

67 Shawn J. Parry-Giles, "Conclusion: Political Women," 324.

68 Karina Kosiara-Pedersen, and Kasper M. Hansen, "Gender Differences in Assessments of Party Leaders," *Scandinavian Political Studies* 38, no. 1 (2015): 26–48, https://doi-org.uml.idm.oclc.org/10.1111/1467-9477.12033; Maria Helena Santos, Lígia Amâncio, and Hélder Alves, "Gender and Politics: The

Relevance of Gender on Judgements About the Merit of Candidates and the Fairness of Quotas," *Portuguese Journal of Social Science* 12, no. 2 (2013): 133–149, https://doi-org.uml.idm.oclc.org/10.1386/pjss.12.2.133_1; Conley Tyler, Melissa H., Emily Blizzard, and Bridget Crane, "Is International Affairs Too 'Hard' for Women? Explaining the Missing Women in Australia's International Affairs," *Australian Journal of International Affairs* 68, no. 2 (2014): 156–176, https://doi-org.uml.idm.oclc.org/10.1080/103 57718.2013.860582.

69 Karen Ross, *Gender, Politics, News: A Game of Three Sides* (Chichester: Wiley-Blackwell, 2017), 5.

70 Blair Williams, "A Gendered Media Analysis of the Prime Ministerial Ascension of Gillard and Turnbull: He's 'Taken Back the Reins' and She's 'A Backstabbing' Murderer," *Australian Journal of Political Science* 52, no. 4 (2017): 550–564, https://doi-org.uml.idm.oclc.org/10.1080/10361146.2017.1374347; Katharine A. M. Wright and Jack Holland, "Leadership and the Media: Gendered Framings of Julia Gillard's 'Sexism and Misogyny' Speech," *Australian Journal of Political Science* 49, no. 3 (2014): 455–468, https://doi-org.uml.idm.oclc.org/10.1080/10361146.2014.929089.

71 David Denemark, Ian Ward, and Clive Bean, "Gender and Leader Effects in the 2010 Australian Election," *Australian Journal of Political Science* 47, no. 4 (2012): 563–578, https://doi-org.uml.idm.oclc.org/10.1080/10361146.2012.731485.

72 Jamieson, *Beyond the Double Bind*.

73 Katrina Lee-Koo and Maria Maley, "The Iron Butterfly and the Political Warrior: Mobilizing Models of Femininity in the Australian Liberal Party, *Australian Journal of Political Science* 52, no. 3 (2017): 317–334, https://doi-org.uml.idm.oclc.org/10.1080/10361146.2017.1336202.

74 Bailey Gerrits, Linda Trimble, Angelia Wagner, Daisy Raphael, and Shannon Sampert, "Political Battlefield: Aggressive Metaphors, Gender, and Power in News Coverage of Canadian Party Leadership Contests," *Feminist Media Studies* 17, no. 6 (2017): 1088–1103, https://doi-org.uml.idm.oclc.org/10.1080/14680777.2017.1315734.

75 Iñaki Garcia-Blanco and Karin Wahl-Jorgensen, "The Discursive Construction of Women Politicians in the European Press," *Feminist Media Studies* 12, no. 3 (2012): 422–441, https://doi-org.uml.idm.oclc.org/10.1080/14680777.2011.615636.

76 Ngaire Donaghue, "Who Gets Played by 'The Gender Card'?" *Australian Feminist Studies* 30, no. 84 (2015): 161–178, https://doi-org.uml.idm.oclc.org/10.1080/08164649.2015.1038118; Anna Worth, Martha Augoustinos, and Brianne Hastie, "'Playing the Gender Card': Media Representations of Julia Gillard's Sexism and Misogyny Speech," *Feminism and Psychology* 26, no. 1 (2016): 52–72, https://doi-org.uml.idm.oclc.org/10.1177/0959353515605544.

77 Marian Sawer, "Misogyny and Misrepresentation," *Political Science* 65, no. 1 (2013): 105–117, https://doi-org.uml.idm.oclc.org/10.1177/0032318713488316.

78 Wright and Holland, "Leadership and the Media."

79 Sharon Halevi, "Damned If You Do, Damned If You Don't," *Feminist Media Studies* 12, no. 2 (2012): 195–213, https://doi-org.uml.idm.oclc.org/10.1080/14680777.2011.597100.

80 Einat Lachover, "Just Being a Woman Isn't Enough Any More," *Feminist Media Studies* 12, no. 3 (2012): 442–458, https://doi-org.uml.idm.oclc.org/10.1080/14680777.2011.615639.

81 Youngmin Yoon and Yoomin Lee, "A Cross-National Analysis of Election News Coverage of Female Candidates' Bids for Presidential Nomination in the United States and South Korea," *Asian Journal of Communication* 23, no. 4 (2013): 420–427, https://doi-org.uml.idm.oclc.org/10.1080/01292986.2012. 746995.

82 Yonghwan Kim, "Politics of Representation in the Digital Media Environment: Presentation of the Female Candidate between News Coverage and the Website in the 2007 Korean Presidential Primary," *Asian Journal of Communication* 22, no. 6 (2012): 601–620, https://doi-org.uml.idm.oclc.org/10.1080/01292986.2012.662513.

83 Shugofa Dastgeer and Peter J. Gade, "Visual Framing of Muslim Women in the Arab Spring: Prominent, Active, and Visible," *International Communication Gazette* 78, no. 5 (2016): 432–450, https://doi-org.uml.idm.oclc.org/10.1177/1748048516640204.

84 Marta Fraile and Raul Gomez, "Bridging the Enduring Gender Gap in Political Interest in Europe: The Relevance of Promoting Gender Equality," *European Journal of Political Research* 56, no. 3 (2017): 601–618, https://doi-org.uml.idm.oclc.org/10.1111/1475-6765.12200.

85 Yann P. Kerevel and Lonna Rae Atkeson, "Reducing Stereotypes of Female Political Leaders in Mexico," *Political Research Quarterly* 68, no. 4 (2015): 732–744, https://doi-org.uml.idm.oclc.org/10.1177/1065912915607637.

86   Anita Stasulane, "Female Leaders in a Radical Right Movement: The Latvian National Front," *Gender & Education* 29, no. 2 (2017): 182–198, https://doi-org.uml.idm.oclc.org/10.1080/09540253.2016.1274021.

87   Susi Meret, "Charismatic Female Leadership and Gender: Pia Kjærsgaard and the Danish People's Party," *Patterns of Prejudice* 49, nos 1/2 (2015): 81–102, https://doi-org.uml.idm.oclc.org/10.1080/0031322X.2015.1023657.

88   Sylvia Bashevkin, "Party Talk: Assessing the Feminist Rhetoric of Women Leadership Candidates in Canada," *Canadian Journal of Political Science/Revue Canadienne de Science Politique* 42, no. 2 (2009): 345–362, www.jstor.org.uml.idm.oclc.org/stable/27754471.

89   Kendall D. Funk, "Gendered Governing? Women's Leadership Styles and Participatory Institutions in Brazil," *Political Research Quarterly* 68, no. 3 (2015): 564–578, https://doi-org.uml.idm.oclc.org/10.1177/1065912915589130.

90   Renata Bongiorno, Paul G. Bain, and Barbara David, "If You're Going to Be a Leader, at Least Act Like It! Prejudice Towards Women Who Are Tentative in Leader Roles," *British Journal of Social Psychology* 53, no. 2 (2014): 217–234, https://doi-org.uml.idm.oclc.org/10.1111/bjso.12032.

91   Susan E. Perkins, Katherine W. Phillips, and Nicholas A. Pearce, "Ethnic Diversity, Gender, and National Leaders," *Journal of International Affairs* 67, no. 1 (2013): 85–104.

92   Young-Im Lee, "From First Daughter to First Lady to First Woman President: Park Geun-Hye's Path to the South Korean Presidency," *Feminist Media Studies* 17, no. 3 (June 2017): 377–391, https://doi-org.uml.idm.oclc.org/10.1080/14680777.2016.1213307.

93   Anderson, "Presidential Pioneer or Campaign Queen."

94   Joyce Marie Mushaben, "Kan-di(e)-dat? Unpacking Gender Images Across Angela Merkel's Four Campaigns for the Chancellorship, 2005-2017," *German Politics and Society* 36, no. 1 (2018): 31–51, https://doi-org.uml.idm.oclc.org/10.3167/gps.2018.360102; Kristina Horn Sheeler, and Karrin Vasby Anderson, "Gender, Rhetoric, and International Political Systems: Angela Merkel's Rhetorical Negotiation of Proportional Representation and Party Politics," *Communication Quarterly* 62, no. 4 (2014): 474–495, https://doi-org.uml.idm.oclc.org/10.1080/01463373.2014.922484.

95   Katz, *Man Enough?*

96   Shannon C. McGregor, Regina G. Lawrence, and Arielle Cardona, "Personalization, Gender, and Social Media: Gubernatorial Candidates' Social Media Strategies," *Information Communication & Society* 20, no. 2 (2017): 264–283, https://doi-org.uml.idm.oclc.org/10.1080/1369118X.2016.1167228.

97   Kelly Dittmar, "Watching Election 2016 with a Gender Lens," *PS-Political Science & Politics* 49, no. 4 (2016): 807–812, https://doi-org.uml.idm.oclc.org/10.1017/S1049096516001554.

98   Dittmar, "Watching Election 2016," 807–808.

99   Dittmar, "Watching Election 2016," 808.

100   Dittmar, "Watching Election 2016," 808–809.

101   Paul Solotaroff, "Trump Seriously: On the Trail with the GOP's Tough Guy," *Rolling Stone*, September 9, 2015, www.rollingstone.com/politics/politics-news/trump-seriously-on-the-trail-with-the-gops-tough-guy-41447/.

102   Robinson Meyer, "Here Comes the Berniebro," *The Atlantic*, October 17, 2015, www.theatlantic.com/politics/archive/2015/10/here-comes-the-berniebro-bernie-sanders/411070/.

103   Kelly Wilz, "Bernie Bros and Woman Cards: Rhetorics of Sexism, Misogyny, and Constructed Masculinity in the 2016 Election," *Women's Studies in Communication* 39, no. 4 (2016): 357–360, https://doi-org.uml.idm.oclc.org/10.1080/07491409.2016.1227178.

104   Michael Mario Albrecht, "Bernie Bros and the Gender Schism in the 2016 US Presidential Election," *Feminist Media Studies* 17, no. 3 (2017): 509–513, https://doi-org.uml.idm.oclc.org/10.1080/14680777.2017.1304715.

105   Albrecht, "Bernie Bros and the Gender Schism."

106   Dara Z. Strolovitch, Janelle S. Wong, and Andrew Proctor, "A Possessive Investment in White Heteropatriarchy? The 2016 Election and the Politics of Race, Gender, and Sexuality," *Politics, Groups, and Identities* 5, no. 2 (2017): 353–363, https://doi-org.uml.idm.oclc.org/10.1080/21565503.2017.1310659.

107   Donna M. Goldstein and Kira Hall, "Postelection Surrealism and Nostalgic Racism in the Hands of Donald Trump," *HAU: Journal of Ethnographic Theory* 7, no. 1 (2017): 397–406, https://doi-org.uml.idm.oclc.org/10.14318/hau7.1.026.

108   Paul Elliott Johnson, "The Art of Masculine Victimhood: Donald Trump's Demagoguery," *Women's Studies in Communication* 40, no. 3 (2017): 229–250, https://doi-org.uml.idm.oclc.org/10.1080/07491409.2017.1346533.

109 Liran Harsgor, "The Partisan Gender Gap in the United States a Generational Replacement?" *Public Opinion Quarterly* 82, no. 2 (2018): 231–251, https://doi-org.uml.idm.oclc.org/10.1093/poq/nfy013.

110 Jane Junn, "The Trump Majority: White Womanhood and the Making of Female Voters in the US," *Politics, Groups, and Identities* 5 (2017): 343–352, https://doi-org.uml.idm.oclc.org/10.1080/21565503. 2017.1304224.

111 Karen Celis and Sarah Childs, "Conservatism and Women's Political Representation," *Politics and Gender* 14, no. 1 (2018): 5–26, https://doi-org.uml.idm.oclc.org/10.1017/S1743923X17000575.

112 Scott J. Smith, "From the 'Ayotte Evasion' to Rejecting Trump: Senator Kelly Ayotte's Post-Crisis Discourse of Renewal," *Communication Quarterly* 66, no. 2 (2018): 117–137, https://doi-org.uml.idm. oclc.org/10.1080/01463373.2018.1438487.

113 Kristine M. Nicolini and Sara Steffes Hansen, "Framing the Women's March on Washington: Media Coverage and Organizational Messaging Alignment," *Public Relations Review* 44, no. 1 (2018): 1–10, https://doi-org.uml.idm.oclc.org/10.1016/j.pubrev.2017.12.005.

114 Carolyn Kitch, "'A Living Archive of Modern Protest': Memory-Making in the Women's March," *Political Communication* 16, no. 2 (2018): 119–127, https://doi-org.uml.idm.oclc.org/10.1080/1540570 2.2017.1388383.

115 Sydney Boothroyd, Rachelle Bowen, Alicia Cattermole, Kenda Chang-Swanson, Hanna Daltrop, Sasha Dwyer, Anna Gunn, Brydon Kramer, Delaney M. McCartan, Jasmine Nagra, Shereen Samimi, and Qwisun Yoon-Potkins, "(Re)producing Feminine Bodies: Emergent Spaces through Contestation in the Women's March on Washington," *Gender, Place & Culture* 24, no. 5 (2017): 711–721, https:// doi-org.uml.idm.oclc.org/10.1080/0966369X.2017.1339673.

116 Anne Graefer, Allaina Kilby, and Inger-Lise Kalviknes Bore, "Unruly Women and Carnivalesque Countercontrol: Offensive Humor in Mediated Social Protest," *Journal of Communication Inquiry* 43, no. 2 (2018): 171–193, https://doi-org.uml.idm.oclc.org/10.1177/0196859918800485.

117 Banu Gökarıksel and Sara Smith, "Intersectional Feminism Beyond U.S. Flag Hijab and Pussy Hats in Trump's America," *Gender, Place & Culture* 24, no. 5 (2017): 628–644, https://doi-org.uml.idm.oclc.org/ 10.1080/0966369X.2017.1343284.

118 Rachel E. Presley and Alane L. Presswood, "Pink, Brown, and Read All Over: Representation at the 2017 Women's March on Washington," *Cultural Studies ↔ Critical Methodologies* 18, no. 1 (2018): 61–71, https://doi-org.uml.idm.oclc.org/10.1177/1532708617735134.

119 Meredith Conroy, Mai Nguyen, and Nathaniel Rakich, "We Researched Hundreds of Races. Here's Who Democrats Are Nominating," *FiveThirtyEight*, August 10, 2018, https://fivethirtyeight.com/ features/democrats-primaries-candidates-demographics/.

120 Perry Bacon Jr., "Why the Republican Party Elects so Few Women," *FiveThirtyEight*, June 25, 2018, https://fivethirtyeight.com/features/why-the-republican-party-isnt-electing-more-women/.

121 Anderson, *Women, Feminism, and Pop Politics*, 3–4, 14.

122 Anderson, *Women, Feminism, and Pop Politics*, 14.

123 Monica Hesse, "The Revolutionary Strategy Hidden in Alexandria Ocasio-Cortez's Instagram Feed," *Washington Post*, November 16, 2018, www.washingtonpost.com/lifestyle/style/the-revolutionary-strategy-hidden-in-alexandria-ocasio-cortezs-instagram-feed/2018/11/15/d17ddd1c-e84d-11e8-a939-9469f1166f9d_story.html.

124 See, for example, Kara Alaimo, "What Woman Presidential Candidates Are Facing," CNN, Opinion, April 9, 2019, www.cnn.com/2019/04/09/opinions/women-candidates-voter-bias-alaimo/index. html; Kate Manne, "It's the Sexism, Stupid," *Politico*, April 11, 2019, www.politico.com/magazine/ story/2019/04/11/its-the-sexism-stupid-226620; Michelle Ruiz, "Women Candidates Have Fun Hobbies Too, They Just Don't Get to Talk About Them," *Vogue*, April 4, 2019, www.vogue.com/ article/2020-women-candidates-quirky-hobbies-intellect-too; Sabrina Siddiqui, "Why Women 2020 Candidates Face 'Likability' Question Even as They Make History," *The Guardian*, February 4, 2019, www.theguardian.com/us-news/2019/feb/04/why-the-likability-question-pursues-2020-female-candidates-even-as-they-make-history; Margaret Sullivan, "How Sexist Will the Media's Treatment of Female Candidates Be? Rule Out 'Not at All,'" *Washington Post*, February 17, 2019, www. washingtonpost.com/lifestyle/style/how-sexist-will-the-medias-treatment-of-female-candidates-be-rule-out-not-at-all/2019/02/15/117158e4-2fcb-11e9-8ad3-9a5b113ecd3c_story.html.

Bibliography

Alaimo, Kara. "What Woman Presidential Candidates are Facing." CNN [Opinion], April 9, 2019, www.cnn.com/2019/04/09/opinions/women-candidates-voter-bias-alaimo/index.html.

Albrecht, Michael Mario. "Bernie Bros and the Gender Schism in the 2016 US Presidential Election." *Feminist Media Studies* 17, no. 3 (2017): 509–513. https://doi-org.uml.idm.oclc.org/10.1080/14680777.2017.1304715.

Aldoory, Linda, and Elizabeth L. Toth. "The Complexities of Feminism in Communication Scholarship Today." *Communication Yearbook* 24, no. 1 (2001): 345–361. https://doi-org.uml.idm.oclc.org/10.1080/23808985.2001.11678993.

Anderson, Karrin Vasby. "Every Woman Is the Wrong Woman: The Female Presidentiality Paradox." *Women's Studies in Communication* 40, no. 2 (2017): 132–135. https://doi-org.uml.idm.oclc.org/10.1080/07491409.2017.1302257.

Anderson, Karrin Vasby. "The First Lady: A Site of 'American Womanhood'." In *Inventing a Voice: The Rhetoric of American First Ladies of the Twentieth Century*, edited by Molly Meijer Wertheimer, 17–30. Lanham, MD: Rowman and Littlefield, 2004.

Anderson, Karrin Vasby. "Hillary Rodham Clinton as 'Madonna': The Role of Metaphor and Oxymoron in Image Restoration." *Women's Studies in Communication* 25, no. 1 (2002): 1–24. https://doi-org.uml.idm.oclc.org/10.1080/07491409.2002.10162439.

Anderson, Karrin Vasby. "Presidential Pioneer or Campaign Queen? Hillary Clinton and the First-Timer/Frontrunner Double Bind." *Rhetoric & Public Affairs* 20, no. 3 (2017): 525–538. https://doi-org.uml.idm.oclc.org/10.14321/rhetpublaffa.20.3.0525.

Anderson, Karrin Vasby. "'Rhymes with Blunt': Pornification and US Political Culture." *Rhetoric & Public Affairs* 14, no. 2 (2011): 327–368. https://doi-org.uml.idm.oclc.org/10.1353/rap.2010.0228.

Anderson, Karrin Vasby. "'Rhymes with Rich': 'Bitch' as a Tool of Containment in Contemporary American Politics." *Rhetoric & Public Affairs* 2, no. 4 (1999): 599–623. https://doi-org.uml.idm.oclc.org/10.1353/rap.2010.0082.

Anderson, Karrin Vasby, ed. *Women, Feminism, and Pop Politics: From "Bitch" to "Badass" and Beyond*. New York: Peter Lang, 2018.

Anderson, Karrin Vasby, and Kristina Horn Sheeler. *Governing Codes: Gender, Metaphor, and Political Identity*. Lanham, MD: Lexington Books, 2005.

Bachmann, Ingrid, Dustin Harp, and Jamie Loke. "Covering Clinton (2010–2015): Meaning-making Strategies in US Magazine Covers." *Feminist Media Studies* 18, no. 5 (2018): 793–809. https://doi-org.uml.idm.oclc.org/10.1080/14680777.2017.1358204.

Bacon Jr., Perry. "Why the Republican Party Elects so Few Women." *FiveThirtyEight*, June 25, 2018, https://fivethirtyeight.com/features/why-the-republican-party-isnt-electing-more-women/.

Banducci, Susan A. with Elisabeth Gidengil, and Joanna Everitt. "Women as Political Communicators: Candidates and Campaigns." In *The Sage Handbook of Political Communication*, edited by Holli A. Semetko and Margaret Scammell, 165–172. Los Angeles, CA: Sage, 2012.

Bashevkin, Sylvia. "Party Talk: Assessing the Feminist Rhetoric of Women Leadership Candidates in Canada." *Canadian Journal of Political Science/Revue Canadienne de Science Politique*, 42, no. 2 (2009): 345–362. www.jstor.org.uml.idm.oclc.org/stable/27754471.

Beasley, Vanessa B. "The Power of Rhetoric: Understanding Political Oratory." In *The Sage Handbook of Political Communication*, edited by Holli A. Semetko and Margaret Scammell, 356–365. Los Angeles, CA: Sage, 2012.

Bongiorno, Renata, Paul G. Bain, and Barbara David. "If You're Going to Be a Leader, at Least Act Like It! Prejudice Towards Women Who Are Tentative in Leader Roles." *British Journal of Social Psychology* 53, no. 2 (2014): 217–234. https://doi-org.uml.idm.oclc.org/10.1111/bjso.12032.

Boothroyd, Sydney, Rachelle Bowen, Alicia Cattermole, Kenda Chang-Swanson, Hanna Daltrop, Sasha Dwyer, Anna Gunn, Brydon Kramer, Delaney M. McCartan, Jasmine Nagra, Shereen Samimi, and Qwisun Yoon-Potkins. "(Re)producing Feminine Bodies: Emergent Spaces Through Contestation in the Women's March on Washington." *Gender Place & Culture* 24, no. 5 (2017): 711–721. https://doi-org.uml.idm.oclc.org/10.1080/0966369X.2017.1339673.

Burnette, Ann E., and Rebekah L. Fox. "My Three Dads: The Rhetorical Construction of Fatherhood in the 2012 Republican Presidential Primary." *Journal of Contemporary Rhetoric* 2, nos 3/4 (2012): 80–91.

Campbell, Karlyn Kohrs. "The Discursive Performance of Femininity: Hating Hillary." *Rhetoric & Public Affairs* 1, no. 1 (1998): 1–19. https://doi-org.uml.idm.oclc.org/10.1353/rap.2010.0172.

Campbell, Karlyn Kohrs. "The Rhetoric of Women's Liberation: An Oxymoron." *Communication Studies* 50, no. 2 (1999): 125–137. https://doi-org.uml.idm.oclc.org/10.1080/10510979909388479.

Campbell, Karlyn Kohrs. "The Rhetorical Presidency: A Two-Person Career." In *Beyond the Rhetorical Presidency*, edited by Martin J. Medhurst, 179–195. College Station, TX: Texas A&M University Press, 1996.

Carlin, Diana B., and Kelly L. Winfrey. "Have You Come a Long Way, Baby? Hillary Clinton, Sarah Palin, and Sexism in 2008 Campaign Coverage." *Communication Studies* 60, no. 4 (2009): 326–343. https://doi-org.uml.idm.oclc.org/10.1080/10510970903109904.

Caroli, Betty Boyd. *First Ladies: From Martha Washington to Michelle Obama*. New York: Oxford University Press, 2010.

Celis, Karen, and Sarah Childs. "Conservatism and Women's Political Representation." *Politics and Gender* 14, no. 1 (2018): 5–26. https://doi-org.uml.idm.oclc.org/10.1017/S1743923X17000575.

Coe, Kevin, David Domke, Meredith M. Bagley, Sheryl Cunningham, and Nancy Van Leuven. "Masculinity as Political Strategy: George W. Bush, the 'War on Terrorism,' and an Echoing Press." *Journal of Women, Politics, and Policy* 29, no. 1 (2007): 31–55. https://doi-org.uml.idm.oclc.org/10.1300/J501v29n01_03.

Connell, R. W., and James W. Messerschmidt. "Hegemonic Masculinity: Rethinking the Concept." *Gender & Society* 19, no. 6 (2005): 829–859. https://doi-org.uml.idm.oclc.org/10.1177/0891243205278639.

Conroy, Meredith, Mai Nguyen, and Nathaniel Rakich. "We Researched Hundreds of Races. Here's Who Democrats Are Nominating." *FiveThirtyEight*, August 10, 2018, https://fivethirtyeight.com/features/democrats-primaries-candidates-demographics/.

Curnalia, Rebecca M. L., and Dorian L. Mermer. "The 'Ice Queen' Melted and It Won Her the Primary: Evidence of Gender Stereotypes and the Double Bind in News Frames of Hillary Clinton's 'Emotional Moment'." *Qualitative Research Reports in Communication* 15, no. 1 (2014): 26–32. https://doi-org.uml.idm.oclc.org/10.1080/17459435.2014.955589.

Dastgeer, Shugofa, and Peter J. Gade. "Visual Framing of Muslim Women in the Arab Spring: Prominent, Active, and Visible." *International Communication Gazette* 78, no. 5 (2016): 432–450. https://doi-org.uml.idm.oclc.org/10.1177/1748048516640204.

Daughton, Suzanne. "Women's Issues, Women's Place." *Communication Quarterly* 42, no. 2 (1994): 106–119. https://doi-org.uml.idm.oclc.org/10.1080/01463379409369920.

Denemark, David, Ian Ward, and Clive Bean. "Gender and Leader Effects in the 2010 Australian Election." *Australian Journal of Political Science* 47, no. 4 (2012): 563–578. https://doi-org.uml.idm.oclc.org/10.1080/10361146.2012.731485.

Dittmar, Kelly. "Watching Election 2016 with a Gender Lens." *PS-Political Science & Politics* 49 (2016): 807–812. https://doi-org.uml.idm.oclc.org/10.1017/S1049096516001554.

Donaghue, Ngaire. "Who Gets Played by 'The Gender Card'?" *Australian Feminist Studies* 30 (2015): 161–178. https://doi-org.uml.idm.oclc.org/10.1080/08164649.2015.1038118.

Dow, Bonne J. *Prime-Time Feminism: Television, Media Culture, and the Women's Movement Since 1970*. Philadelphia, PA: University of Pennsylvania Press, 1996.

Dow, Bonnie J., and Celeste M. Condit. "The State of the Art in Feminist Scholarship in Communication." *Journal of Communication* 55, no. 3 (2005): 448–478. https://doi-org.uml.idm.oclc.org/10.1111/j.1460-2466.2005.tb02681.x.

Dubriwny, Tasha N. "First Ladies and Feminism: Laura Bush as Advocate for Women's and Children's Rights." *Women's Studies in Communication* 28, no. 1 (2005): 84–114. https://doi-org.uml.idm.oclc.org/10.1080/07491409.2005.10162485.

Edwards, Janis L., ed. *Gender and Political Communication in America : Rhetoric, Representation, and Display*. Lanham, MD: Lexington Books, 2009.

Erickson, Keith V., and Stephanie Thomson. "First Lady International Diplomacy: Performing Gendered Roles on the World Stage." *Southern Communication Journal* 77, no. 3 (2012): 239–262. https://doi-org.uml.idm.oclc.org/10.1080/1041794X.2011.647502.

Fahey, Anna Cornelia. "French and Feminine: Hegemonic Masculinity and the Emasculation of John Kerry in the 2004 Presidential Race." *Critical Studies in Media Communication* 24, no. 2 (2007): 132–150. https://doi-org.uml.idm.oclc.org/10.1080/07393180701262743.

Falk, Erika. "Clinton and the Playing-the-Gender-Card Metaphor in Campaign News." *Feminist Media Studies* 13, no. 2 (2013): 192–207. https://doi-org.uml.idm.oclc.org/10.1080/14680777.2012.678074.

Falk, Erika, and Kate Kenski. "Issue Saliency and Gender Stereotypes: Support for Women as Presidents in Times of War and Terrorism." *Social Science Quarterly* 87, no. 1 (2006): 1–18. www.jstor.org.uml.idm.oclc.org/stable/42956106.

Foss, Karen A. "Feminist Scholarship in Speech Communication: Contributions and Obstacles." *Women's Studies in Communication* 12, no. 1 (1989): 1–10. https://doi-org.uml.idm.oclc.org/10.1080/074914 09.1989.11089729.

Fraile, Marta, and Raul Gomez. "Bridging the Enduring Gender Gap in Political Interest in Europe: The Relevance of Promoting Gender Equality." *European Journal of Political Research* 56, no. 3 (2017): 601–618. https://doi-org.uml.idm.oclc.org/10.1111/1475-6765.12200.

Funk, Kendall D. "Gendered Governing? Women's Leadership Styles and Participatory Institutions in Brazil." *Political Research Quarterly* 68, no. 3 (2015): 564–578. https://doi-org.uml.idm.oclc.org/10.1177/1065912915589130.

Garcia-Blanco, Iñaki, and Karin Wahl-Jorgensen. "The Discursive Construction of Women Politicians in the European Press." *Feminist Media Studies* 12, no. 3 (2012): 422–441. https://doi-org.uml.idm.oclc.org/10.1080/14680777.2011.615636.

Gatison, Annette Madlock. "Michelle Obama and the Representation of Respectability." *Women & Language* 40, no. 1 (2017/2018): 101–110.

Gerrits, Bailey, Linda Trimble, Angelia Wagner, Daisy Raphael, and Shannon Sampert. "Political Battlefield: Aggressive Metaphors, Gender, and Power in News Coverage of Canadian Party Leadership Contests." *Feminist Media Studies* 17, no. 6 (2017): 1088–1103. https://doi-org.uml.idm.oclc.org/10.1080/14680777.2017.1315734.

Gibson, Katie L., and Amy L. Heyse. "Depoliticizing Feminism: Frontier Mythology and Sarah Palin's 'The Rise of The Mama Grizzlies'." *Western Journal of Communication* 78, no. 1 (2014): 97–117. https://doi-org.uml.idm.oclc.org/10.1080/10570314.2013.812744.

Gibson, Katie L., and Amy L. Heyse. "'The Difference between a Hockey Mom and a Pit Bull': Sarah Palin's Faux Maternal Persona and Performance of Hegemonic Masculinity at the 2008 Republican National Convention." *Communication Quarterly* 58, no. 3 (2010): 235–256. https://doi-org.uml.idm.oclc.org/10.1080/01463373.2010.503151.

Gökarıksel, Banu, and Sara Smith. "Intersectional Feminism Beyond U.S. Flag Hijab and Pussy Hats in Trump's America." *Gender, Place & Culture* 24, no. 5 (2017): 628–644. https://doi-org.uml.idm.oclc.org/10.1080/0966369X.2017.1343284.

Goldstein, Donna M., and Kira Hall. "Postelection Surrealism and Nostalgic Racism in the Hands of Donald Trump." *HAU: Journal of Ethnographic Theory* 7, no. 1 (2017): 397–406. https://doi-org.uml.idm.oclc.org/10.14318/hau7.1.026.

Goodale, Greg. "The Presidential Sound: From Orotund to Instructional Speech, 1892–1912." *Quarterly Journal of Speech* 96, no. 2 (2010): 164–184. https://doi-org.uml.idm.oclc.org/10.1080/00335631003796651.

Gordon, Ann, and Jerry Miller. "Gender, Race, and the Oval Office." In *Anticipating Madam President*, edited by Robert P. Watson and Ann Gordon, 145–155. Boulder, CO: Lynne Rienner, 2003.

Graefer, Anne, Allaina Kilby, and Inger-Lise Kalviknes Bore. "Unruly Women and Carnivalesque Countercontrol: Offensive Humor in Mediated Social Protest." *Journal of Communication Inquiry* 43, no. 2 (2018): 171–193. https://doi-org.uml.idm.oclc.org/10.1177/0196859918800485.

Gutgold, Nichola. *Almost Madam President: Why Hillary Clinton Won in 2008*. Lanham, MD: Lexington Books, 2009.

Gutgold, Nichola. *Paving the Way for Madam President*. Lanham, MD: Lexington Books, 2006.

Halevi, Sharon. "Damned if You Do, Damned if You Don't: Tzipi Livni and the Debate on a 'Feminine' Leadership Style in the Israeli Press." *Feminist Media Studies* 12, no. 2 (2012): 195–213. https://doi-org.uml.idm.oclc.org/10.1080/14680777.2011.597100.

Harp, Dustin, Jaime Loke, and Ingrid Bachmann. "Hillary Clinton's Benghazi Hearing Coverage: Political Competence, Authenticity, and the Persistence of the Double Bind." *Women's Studies in Communication* 39, no. 2 (2016): 193–210. https://doi-org.uml.idm.oclc.org/10.1080/07491409.2016.1171267.

Harsgor, Liran. "The Partisan Gender Gap in the United States a Generational Replacement?" *Public Opinion Quarterly* 82, no. 2 (2018): 231–251. https://doi-org.uml.idm.oclc.org/10.1093/poq/nfy013.

Hesse, Monica. "The Revolutionary Strategy Hidden in Alexandria Ocasio-Cortez's Instagram Feed." *Washington Post*, November 16, 2018, www.washingtonpost.com/lifestyle/style/the-revolutionary-strategy-hidden-in-alexandria-ocasio-cortezs-instagram-feed/2018/11/15/d17ddd1c-e84d-11e8-a939-9469f1166f9d_story.html.

Holman, Mirya R. "Gender, Political Rhetoric, and Moral Metaphors in State of the City Addresses." *Urban Affairs Review* 52, no. 4 (2016): 501–530. https://doi-org.uml.idm.oclc.org/10.1177/1078087415589191.

Jamieson, Kathleen Hall. *Beyond the Double Bind: Women and Leadership*. Oxford: Oxford University Press, 1995.

Jean-Peirre, Karine. "It's Not the 'Year of the Woman.' It's the 'Year of the Women.'" CNN [Opinion], November 4, 2018, www.cnn.com/2018/11/03/opinions/midterm-elections-year-of-woman-roundup/index.html.

Johnson, Paul Elliott. "The Art of Masculine Victimhood: Donald Trump's Demagoguery." *Women's Studies in Communication* 40, no. 3 (2017): 229–250. https://doi-org.uml.idm.oclc.org/10.1080/07491409. 2017.1346533.

Junn, Jane. "The Trump Majority: White Womanhood and the Making of Female Voters in the U.S." *Politics, Groups, and Identities* 5 (2017): 343–352. https://doi-org.uml.idm.oclc.org/10.1080/21565503.2017 .1304224.

Katz, Jackson. *Man Enough? Donald Trump, Hillary Clinton and the Politics of Presidential Masculinity.* Northampton, MA: Interlink Pub Group, 2016.

Kaufer, David S., and Shawn J. Parry-Giles. "Hillary Clinton's Presidential Campaign Memoirs: A Study in Contrasting Identities." *Quarterly Journal of Speech* 103, nos 1/2 (2017): 7–32. https://doi-org.uml. idm.oclc.org/10.1080/00335630.2016.1221529.

Kerevel, Yann P., and Lonna Rae Atkeson. "Reducing Stereotypes of Female Political Leaders in Mexico." *Political Research Quarterly* 68, no. 4 (2015): 732–744. https://doi-org.uml.idm.oclc.org/ 10.1177/1065912915607637.

Kim, Yonghwan. "Politics of Representation in the Digital Media Environment: Presentation of the Female Candidate between News Coverage and the Website in the 2007 Korean Presidential Primary." *Asian Journal of Communication* 22, no. 6 (2012): 601–620. https://doi-org.uml.idm.oclc.org/10.1080/ 01292986.2012.662513.

Kimmel, Michael. *Manhood in America: A Cultural History*, 3rd edn. New York: Oxford University Press, 2012.

Kitch, Carolyn. "'A Living Archive of Modern Protest': Memory-making in the Women's March." *Popular Communication* 16, no. 2 (2018): 119–127. https://doi-org.uml.idm.oclc.org/10.1080/15405702. 2017.1388383.

Kosiara-Pedersen, Karina, and Kasper M. Hansen. "Gender Differences in Assessments of Party Leaders." *Scandinavian Political Studies* 38, no. 1 (2015): 26–48. https://doi-org.uml.idm.oclc.org/ 10.1111/1467-9477.12033.

Lachover, Einat. "Just Being a Woman Isn't Enough Any More: Israeli Television News of Women in Local Politics." *Feminist Media Studies* 12, no. 3 (2012): 442–458. https://doi-org.uml.idm.oclc.org/10. 1080/14680777.2011.615639.

Lawless, Jennifer. "Women, War, and Winning Elections: Gender Stereotyping in the Post-September 11th Era." *Political Research Quarterly* 57, no. 3 (2004): 479–90. https://doi-org.uml.idm.oclc. org/10.2307/3219857.

Lawrence, Regina G., and Melody Rose. *Hillary Clinton's Race for the White House: Gender Politics and the Media on the Campaign Trail.* Boulder, CO: Lynne Rienner Publishers, 2010.

Lee, Young-Im. "From First Daughter to First Lady to First Woman President: Park Geun-Hye's Path to the South Korean Presidency." *Feminist Media Studies* 17, no. 3 (2017): 377–391. https://doi-org.uml.idm. oclc.org/10.1080/14680777.2016.1213307.

Lee-Koo, Katrina, and Maria Maley. "The Iron Butterfly and the Political Warrior: Mobilising Models of Femininity in the Australian Liberal Party." *Australian Journal of Political Science* 52, no. 3 (2017): 317–334. https://doi-org.uml.idm.oclc.org/10.1080/10361146.2017.1336202.

Lockhart, Michele and Kathleen Mollick, eds. *Political Women: Language and Leadership.* Lanham, MD: Lexington Books, 2013.

Mandell, Hinda. "Political Wives, Scandal, and the Double Bind: Press Construction of Silda Spitzer and Jenny Sanford through a Gendered Lens." *Women's Studies in Communication* 38, no. 1 (2015): 57–77. https://doi-org.uml.idm.oclc.org/10.1080/07491409.2014.995327.

Mandziuk, Roseann M. "Whither the Good Wife? 2016 Presidential Candidate Spouses in the Gendered Spaces of Contemporary Politics." *Quarterly Journal of Speech* 103, nos 1/2 (2017): 136–159. https:// doi-org.uml.idm.oclc.org/10.1080/00335630.2016.1233350.

Manne, Kate. "It's the Sexism, Stupid." *Politico*, April 11, 2019, www.politico.com/magazine/story/ 2019/04/11/its-the-sexism-stupid-226620.

Marshall, Brenda DeVore and Molly A. Mayhead, eds. *Navigating Boundaries: The Rhetoric of Women Governors.* Westport, CT: Praeger, 2000.

Marshall, Brenda Devore, and Molly A. Mayhead. *Telling Political Lives: The Rhetorical Biographies of Women Leaders in the United States.* Lanham, MD: Lexington Books, 2008.

Mayhead, Molly, and Brenda Devore Marshall. *Women's Political Discourse: A 21st Century Perspective.* Lanham, MD: Rowman and Littlefield, 2005.

McGregor, Shannon C., Regina G. Lawrence, and Arielle Cardona. "Personalization, Gender, and Social Media: Gubernatorial Candidates' Social Media Strategies." *Information Communication & Society* 20, no. 2 (2017): 264–283. https://doi-org.uml.idm.oclc.org/10.1080/1369118X.2016.1167228.

McRobbie, Angela. *The Aftermath of Feminism: Gender, Culture, and Social Change.* Thousand Oaks, CA: Sage, 2009.

Meeks, Lindsey. "All the Gender That's Fit to Print: How the *New York Times* Covered Hillary Clinton and Sarah Palin in 2008." *Journalism & Mass Communication Quarterly* 90, no. 3 (2013): 520–539. https://doi-org.uml.idm.oclc.org/10.1177/1077699013493791.

Meeks, Lindsey. "Is She 'Man Enough'? Women Candidates, Executive Political Offices, and News Coverage." *Journal of Communication* 62, no. 1 (2012): 175–93. https://doi-org.uml.idm.oclc.org/10.1111/j.1460-2466.2011.01621.x.

Meret, Susi. "Charismatic Female Leadership and Gender: Pia Kjærsgaard and the Danish People's Party." *Patterns of Prejudice* 49, nos 1/2 (2015): 81–102. https://doi-org.uml.idm.oclc.org/10.1080/0031322X.2015.1023657.

Meyer, Michaela D. E. "Women Speak(ing): Forty Years of Feminist Contributions to Rhetoric and an Agenda for Feminist Rhetorical Studies." *Communication Quarterly* 55, no. 1 (2007): 1–17. https://doi-org.uml.idm.oclc.org/10.1080/01463370600998293.

Meyer, Robinson. "Here Comes the Berniebro." *The Atlantic*, October 17, 2015, www.theatlantic.com/politics/archive/2015/10/here-comes-the-berniebro-bernie-sanders/411070/.

Mushaben, Joyce Marie. "Kan-di(e)-dat? Unpacking Gender Images Across Angela Merkel's Four Campaigns for the Chancellorship, 2005-2017." *German Politics and Society* 36 no. 1 (2018): 31–51. https://doi-org.uml.idm.oclc.org/10.3167/gps.2018.360102.

Nicolini, Kristine M., and Sara Steffes Hansen. "Framing the Women's March on Washington: Media Coverage and Organizational Messaging Alignment." *Public Relations Review* 44, no. 1 (2018): 1–10. https://doi-org.uml.idm.oclc.org/10.1016/j.pubrev.2017.12.005.

Oles-Acevedo, Denise. "Fixing the Hillary Factor: Examining the Trajectory of Hillary Clinton's Image Repair from Political Bumbler to Political Powerhouse." *American Communication Journal* 14, no. 1 (2012): 33–46.

Parry-Giles, Shawn J. "Conclusion: Political Women and the Power Paradox: The Case of Hillary Clinton." In *Women, Feminism, and Pop Politics: From "Bitch" to "Badass" and Beyond*, edited by Karrin Vasby Anderson, 309–330. New York: Peter Lang, 2018.

Parry-Giles, Shawn J. *Hillary Clinton in the News: Gender and Authenticity in American Politics.* Champaign-Urbana, IL: University of Illinois Press, 2014.

Parry-Giles, Shawn J. "Mediating Hillary Rodham Clinton: Television News Practices and Image-Making in the Postmodern Age." *Critical Studies in Media Communication* 17, no. 2 (2000): 205–226. https://doi-org.uml.idm.oclc.org/10.1080/15295030009388390.

Parry-Giles, Shawn J., and Diane M. Blair. "The Rise of the Rhetorical First Lady: Politics, Gender Ideology, and Women's Voice, 1789-2002." *Rhetoric and Public Affairs* 5, no. 4 (2002): 565–599. https://doi-org.uml.idm.oclc.org/10.1353/rap.2003.0011.

Parry-Giles, Trevor, and Shawn J. Parry-Giles. *The Prime-Time Presidency: "The West Wing" and US Nationalism.* Urbana, IL: University of Illinois Press, 2006.

Pearson, Kathryn, and Logan Dancey. "Speaking for the Underrepresented in the House of Representatives: Voicing Women's Interests in a Partisan Era." *Politics & Gender* 7, no. 4 (2011): 493–519. https://doi-org.uml.idm.oclc.org/10.1017/S1743923X1100033X.

Perkins, Susan E., Katherine W. Phillips, and Nicholas A. Pearce. "Ethnic Diversity, Gender, and National Leaders." *Journal of International Affairs* 67, no. 1 (2013): 85–104.

Perks, Lisa Glebatis, and Kevin A. Johnson. "Electile Dysfunction: The Burlesque Binds of the Sarah Palin MILF Frame." *Feminist Media Studies* 14, no. 5 (2014): 775–790. https://doi-org.uml.idm.oclc.org/10.1080/14680777.2013.829860.

Presley, Rachel E., and Alane L. Presswood. "Pink, Brown, and Read All Over: Representation at the 2017 Women's March on Washington." *Cultural Studies ↔ Critical Methodologies* 18, no. 1 (2018): 61–71. https://doi-org.uml.idm.oclc.org/10.1177/1532708617735134.

Robson, Deborah Carol. "Stereotypes and the Female Politician: A Case Study of Senator Barbara Mikulski." *Communication Quarterly* 48, no. 3 (2000): 205–222. https://doi-org.uml.idm.oclc.org/10.1080/01463370009385593.

Ross, Karen. *Gender, Politics, News: A Game of Three Sides.* Chichester: Wiley-Blackwell, 2017.

Rossiter, Clinton. *The American Presidency.* Baltimore, MD: Johns Hopkins University Press, 1987.

Ruiz, Michelle. "Women Candidates Have Fun Hobbies Too, They Just Don't Get to Talk About Them." *Vogue,* April 4, 2019, www.vogue.com/article/2020-women-candidates-quirky-hobbies-intellect-too.

Santos, Maria Helena, Lígia Amâncio, and Hélder Alves. "Gender and Politics: The Relevance of Gender on Judgements about the Merit of Candidates and the Fairness of Quotas." *Portuguese Journal of Social Science* 12, no. 2 (2013): 133–149. https://doi-org.uml.idm.oclc.org/10.1386/pjss.12.2.133_1.

Sawer, Marian. "Misogyny and Misrepresentation: Women in Australian Parliaments." *Political Science* 65, no. 1 (2013): 105–117. https://doi-org.uml.idm.oclc.org/10.1177/0032318713488316.

Sheckels, Theodore F., ed. *Cracked but Not Shattered: Hillary Rodham Clinton's Unsuccessful Campaign for the Presidency.* Lanham, MD: Lexington Books, 2009.

Sheeler, Kristina Horn. "Beauty Queens and Unruly Women in the Year of the Woman Governor: Jennifer Granholm and the Visibility of Leadership." *Women's Studies in Communication* 33, no. 1 (2010): 34–53. https://doi-org.uml.idm.oclc.org/10.1080/07491401003669927.

Sheeler Kristina Horn. "The Rhetoric of Impossible Expectations: Media Coverage of Hillary Clinton's 2016 General Election Campaign." In *An Unprecedented Election: Media, Communication, and the Electorate in the 2016 Election,* edited by Benjamin R. Warner, Dianne G. Bystrom, Mitchell S. McKinney, and Mary C. Banwart, 87–104. Santa Barbara, CA: ABC-Clio, 2018.

Sheeler, Kristina Horn, and Karrin Vasby Anderson. "Gender, Rhetoric, and International Political Systems: Angela Merkel's Rhetorical Negotiation of Proportional Representation and Party Politics." *Communication Quarterly* 62, no. 4 (2014): 474–495. https://doi-org.uml.idm.oclc.org/10.1080/01463373.2014.922484.

Sheeler, Kristina Horn, and Karrin Vasby Anderson. *Woman President: Confronting Postfeminist Political Culture.* College Station: Texas A&M University Press, 2013.

Shepherd, Ryan. "Confronting Gender Bias, Finding a Voice: Hillary Clinton and the New Hampshire Crying Incident." *Argumentation & Advocacy* 46, no. 1 (2009): 64–77. https://doi-org.uml.idm.oclc.org/10.1080/00028533.2009.11821717.

Siddiqui, Sabrina. "Why Women 2020 Candidates Face 'Likability' Question Even as They Make History." *The Guardian,* February 4, 2019, www.theguardian.com/us-news/2019/feb/04/why-the-likability-question-pursues-2020-female-candidates-even-as-they-make-history.

Simon, Jenni M., and Abby M. Brooks. "The "First" Coverage of the First Lady: E-racing the Mom-in-Chief." *Women & Language* 40, no. 1 (2017/2018): 15–35.

Smith, Scott J. "From the 'Ayotte Evasion' to Rejecting Trump: Senator Kelly Ayotte's Post-Crisis Discourse of Renewal." *Communication Quarterly* 66, no. 2 (2018): 117–137. https://doi-org.uml.idm.oclc.org/10.1080/01463373.2018.1438487.

Solotaroff, Paul. "Trump Seriously: On the Trail with the GOP's Tough Guy." *Rolling Stone,* September 9, 2015, www.rollingstone.com/politics/politics-news/trump-seriously-on-the-trail-with-the-gops-tough-guy-41447/.

Stasulane, Anita. "Female Leaders in a Radical Right Movement: The Latvian National Front." *Gender & Education* 29, no. 2 (2017): 182–198. https://doi-org.uml.idm.oclc.org/10.1080/09540253.2016.1274021.

Strolovitch, Dara Z., Janelle S. Wong, and Andrew Proctor. "A Possessive Investment in White Heteropatriarchy? The 2016 Election and the Politics of Race, Gender, and Sexuality." *Politics, Groups, and Identities* 5, no. 2 (2017): 353–363. https://doi-org.uml.idm.oclc.org/10.1080/21565503.2017.1310659.

Stuckey, Mary. "Rethinking the Rhetorical Presidency and Presidential Rhetoric." *Review of Communication* 10, no. 1 (2010): 38–52. https://doi-org.uml.idm.oclc.org/10.1080/15358590903248744.

Sullivan, Margaret. "How Sexist Will the Media's Treatment of Female Candidates Be? Rule Out 'Not at All.'" *Washington Post,* February 17, 2019, www.washingtonpost.com/lifestyle/style/how-sexist-will-the-medias-treatment-of-female-candidates-be-rule-out-not-at-all/2019/02/15/117158e4-2fcb-11e9-8ad3-9a5b113ecd3c_story.html.

Thomas, Sue, and Clyde Wilcox, eds. *Women and Elective Office: Past, Present, and Future.* 3rd edn. Oxford: Oxford University Press, 2014.

Tyler, Conley, Melissa H., Emily Blizzard, and Bridget Crane. "Is International Affairs Too 'Hard' for Women? Explaining the Missing Women in Australia's International Affairs." *Australian Journal of International Affairs* 68, no. 2 (2014): 156–176. https://doi-org.uml.idm.oclc.org/10.1080/10357718.2013.860582.

Vaughn, Justin S. and Lilly J. Goren, eds. *Women and the White House: Gender, Popular Culture, and Presidential Politics.* Lexington, KY: University of Kentucky Press, 2013.

Vavrus, Mary Douglas. *Postfeminist News: Political Women in Media Culture.* Albany, New York: State University of New York Press, 2002.

Wertheimer, Molly Meijer, ed. *Inventing a Voice: The Rhetoric of American First Ladies of the Twentieth Century.* Lanham, MD: Rowman and Littlefield, 2004.

Whaley, C. Robert, and George Antonelli. "The Birds and the Beasts: Women as Animal." *Maledicta* 7 (1983): 219–229.

Williams, Blair. "A Gendered Media Analysis of the Prime Ministerial Ascension of Gillard and Turnbull: He's 'Taken Back the Reins' and She's 'A Backstabbing' Murderer.'" *Australian Journal of Political Science* 52, no. 4 (2017): 550–564. https://doi-org.uml.idm.oclc.org/10.1080/10361146.2017.1374347.

Wilz, Kelly. "Bernie Bros and Woman Cards: Rhetorics of Sexism, Misogyny, and Constructed Masculinity in the 2016 Election." *Women's Studies in Communication* 39, no. 4 (2016): 357–360. https://doi-org.uml.idm.oclc.org/10.1080/07491409.2016.1227178.

Worth, Anna, Martha Augoustinos, and Brianne Hastie. "'Playing the Gender Card': Media Representations of Julia Gillard's Sexism and Misogyny Speech." *Feminism and Psychology* 26, no. 1 (2016): 52–72. https://doi-org.uml.idm.oclc.org/10.1177/0959353515605544.

Wright, Katharine A.M., and Jack Holland. "Leadership and the Media: Gendered Framings of Julia Gillard's 'Sexism and Misogyny' Speech." *Australian Journal of Political Science* 49, no. 3 (2014): 455–468. https://doi-org.uml.idm.oclc.org/10.1080/10361146.2014.929089.

Yoon, Youngmin, and Yoomin Lee. "A Cross-National Analysis of Election News Coverage of Female Candidates' Bids for Presidential Nomination in the United States and South Korea." *Asian Journal of Communication* 23, no. 4 (2013): 420–427. https://doi-org.uml.idm.oclc.org/10.1080/01292986.2012.746995.

Zaeske, Susan. "History, Theory, and Method in Feminist Scholarship on Early American Discourse." *Rhetoric & Public Affairs* 6, no. 3 (2003): 567–579. https://doi-org.uml.idm.oclc.org/10.1353/rap.2003.0076.

Zulli, Diana. "The Changing Norms of Gendered News Coverage: Hillary Clinton in *The New York Times*, 1969–2016." *Politics and Gender* 15, no. 3 (2019): 599–621. https://doi-org.uml.idm.oclc.org/10.1017/S1743923X18000466.

19

RESISTING ORIENTALIST/ ISLAMOPHOBIC FEMINISMS

(Re)Framing the politics of difference

Fatima Zahrae Chrifi Alaoui and Shadee Abdi

Fatima Chrifi Alaoui dedicates this piece to her grandmother, Fatima El Alami (1930–2018), who taught me my first lessons in feminism and resilience, and continues to fill my soul with the strength to keep the fight for justice.

Their Muslimness is perceived as backward and oppressed, yet authentic and innate; their feminism is perceived as progressive and emancipated, yet corrupt and alien.[1]

The relationship between "Muslim," "Arab," and "Feminism" that has been deemed antonymic is currently a contested zone due to the resurgence of the popular Orientalist and Islamophobic stereotypical racial discourses on Arab and Muslim subjectivities. The liberal deployment of a rhetoric that conflates Arab and Muslim womanhood with victimization also positions a mainstream "American" feminist as the feminist subject par excellence. This intersection of Orientalism with Eurocentric imperial feminism traces the historical legacy of the Orientalist version of Arab and Muslim women and reifies certain events: in this case, September 11, 2001 as the central generator of "activism and visibility."

In this ontological pursuit, we choose to consider "9/11" as an event in the Deleuzian sense.[2,3] In *Terrorist Assemblages*, Puar argues that 9/11 reflects, "an assemblage of spatial and temporal intensities, coming together, dispersing, reconverging."[4] In other words, considering 9/11 as an event refuses a binary between past and present, between a "history-making moment" and a "history-vanishing moment," and between "politicality" versus "stability."[5] The eventness of 9/11 collapses past, present, and future and looks at 9/11 as a movement between the before and the after.[6] Situating 9/11 as an event frames Arab and Muslim feminism as an assemblage of theoretical interests—not as one specific stand, but rather as ideas that converge, diverge, and merge.[7] Such a treatment does not contend that a neonationalist feminism emerged because of 9/11 but apprehends it as a movement that is "always-becoming."[8] Foregrounding the political urgency of surveying the theoretical frameworks of Arab and Muslim scholars is imperative in creating new ways of understanding communication in postcolonial/neocolonial settings.[9]

In this essay, we first cover the literature on Orientalist/Islamophobic feminism that has "shaped" Arab and Muslim feminism as oxymoronic and nonexistent, emphasizing the events of 9/11. We also argue that Orientalist/Islamophobic feminism denies women's agencies and creates a framework of U.S. "gender exceptionalism." Second, we draw on theories of women of color to demonstrate how they forged an inclusive visibility for Arab and Muslim feminism. To elaborate on this argument, we use the anthologies of *This Bridge Called My Back*[10] to show their model of rethinking and rewriting the goals and strategies of feminist movements when new perspectives "emerge" and as historical events develop. Third, we theorize Muslim feminism as a movement that embarks to new spaces to understand the multiplicity and specificity of subjectivities within Muslim and Arab communities (*rethinking* feminism). We argue that Muslim feminism could be seen as an extension of theories of the flesh, where Muslim feminists theorize through their lived experiences and link the personal to the political. Through "fleshing," Muslim feminists reclaim their lived experiences and deconstruct the hegemonic universalist knowledge of feminism and struggle. They create a middle space "in between" the binaries to produce alternate discourses to challenge and disrupt the knowledge, the representation, and the discourse associated with them.

Defining Orientalist/Islamophobic feminism

Through Orientalist (European travelers to the Middle East) discourses, Arab and Muslim women have been constructed as oppressed, docile, silenced, and dominated by their counterparts: "Arab men." Ahmed, a Muslim feminist and a well-known researcher on Arab and Muslim women, has argued that "American women 'know' that Muslim women are overwhelmingly oppressed without being able to define the specific content of that oppression."[11] She adds that, "these are 'facts' manufactured in Western culture, by the same men who have littered the culture with 'facts' about Western women and how inferior and irrational they are."[12] Here, Ahmed is referring to the roots of the stereotype of Arab and Muslim women in the early 19th century. In this respect, Ahmed provides the example of Lord Cromer, the British Consul-General in early 20th-century Egypt.[13] He famously appropriated "feminist" arguments to supposedly save Egyptian women by unveiling them, while he himself opposed the suffragette movement and political enfranchisement of British women in his own home country.[14] In other words, for these European male travelers, the Orient and its women (metaphorically and physically) are the focal point of this male vision of the East.[15] However, those accepted narratives by mainstream feminists on Arab and Muslim women are fundamentally problematic as they support an epistemological and political domination of men's produced knowledge and history.[16] And they represent "a tokenistic apology that leaves uninterrogated a west/Islam binary."[17]

Scholars urge researchers to go beyond domestic women's issues and to invite transnational voices to participate in a conversation that they are seemingly already involved in. This is evidenced when Chowdhury writes,

> while there are important points of intersection between histories and struggles of US women of color and third world women, and therefore potential for powerful alliances, collapsing the two in to one category smudges over the necessity of analyses around nation as well as race.[18]

Moraga and Anzaldúa also highlight the noticeable lack of Arab and Muslim feminist voices within feminist discourse as problematic when (re)imagining a larger transnational feminist network:

> Thirty-five years ago, Egypt, Afghanistan, Nigeria seemed very far away. They are no longer so far… the prism of U.S. Third World Feminist consciousness has shifted as we turned our gaze *away* from a feminism prescribed by white women of privilege (even in opposition to them) and turned *toward* the process of discerning the multilayered and intersecting sites of identity and struggle—distinct and shared —among women of color across the globe.[19]

Instead, Arab and Muslim women continue to be presented as one singular, mono-lithic, undifferentiated, subordinate, and power-less group which primarily constitutes the contradictory "Eastern" pole of Western women. Western texts have commonly promoted fixed universal images of Muslim women and have presented them as poor, veiled, illiterate, victimized, sexually constrained, and docile housewives.[20]

(Sadiqi xvi-xvii)

In her analysis of the relationship between Western and Third World feminism, Nayaran critiques the misrepresentation of Third-World feminists as succumbing to Western ideology in order to seek gender equality:

Many Third-World feminists confront the attitude that our criticisms of our cultures are merely one more incarnation of a colonized consciousness, the views of "privileged native women in whiteface," seeking to attack their "non-Western culture" on the basis of "Western" values.[21]

Al-Ghabra furthers this notion, noting the ways in which colonialism continually impacts and implicates feminists in the Arab world.[22] Specifically, she writes that those who considered them-selves feminists had to adhere to the colonial narrative, or they risked being framed as anti-feminists. Thus, arguing that if Muslim women are only seen from the perspective of Western feminism, then their experiences will only be seen as relative to it and therefore always already starting from a position of inferiority. Such a limited approach justifies collapsing all other iden-tity markers into *muslimwoman,* a position doubly targeted by Islamophobia and misogyny.

Abu-Lughod has also written about the dangers of accepting the Islamophobic oppositional binary of Islam and the West as truth since,

those many people within Muslim countries who are trying to find alternatives to present injustices, those who might want to refuse the divide and take from different histories and cultures, who do not accept that being feminist means being Western, will be under pressure to choose, just as we are: Are you with us or against us?[23]

Accordingly, Arab and Muslim women continuously fight back against pervasive depictions of the region by (re)negotiating and reclaiming their subjective identities on their own terms. Simply put, to promote more inclusive feminist scholarship, Arab and Muslim women must be able to reclaim agency so that we might begin to understand the impact of patriarchy both locally and transnationally.

In the same vein, cultural scholars must not use the same yardstick to measure Arab and Muslim women that would be used to measure Western women, because, "Unlike western feminism, which has a history of being more individualistic, third-world feminism has a deep connection to family and nationalism."[24] The differentiation between Western and Muslim women is imperative to understanding complex and culturally nuanced women's lives, and to resisting "the assumption of a unity of women's interests on the basis of White experience."[25] To demonstrate, Sabbagh writes:

Western feminism, of course, is grounded in Western thought, ideology, and values. Arab women's struggle is equally grounded in the religious, cultural, and political norms of the Arab world. According to some Arab women, it is a difficult if not impossible task to write about Islamic feminism in a climate that assumes the universal supremacy of Western feminism. They believe that Western feminism is rejected by Muslim women because it calls for a form of cultural conversion at a time when the West is seen by them to be a dominating force.[26]

(Sabbagh xxiv)

Consequently, women of color urge that feminists "turn our attention from solely focusing on a particular 'culture' or 'tradition' to an examination of institutional, juridical and legislative practices of the state."[27] Simply put, feminist scholars are asked to look at the force of patriarchy and how it directly affects and impacts women of color globally.

Hence, Orientalist/Islamophobic feminism is a discursive strategy that appropriates feminist concepts for the purpose of domination, not liberation. The eventness of 9/11 has revitalized this form of feminism which deploys powerful Orientalist tropes to justify war and aggressive nationalism.[28] It has made Muslim and Arab women "visible" in mainstream discourses. However, it has done this through a dismissive and destructive visibility. During President George W. Bush's second term, the war on terror and the fight for women's rights became almost synonymous.[29] In her infamous radio address, Laura Bush conflates the two most explicitly, saying, "the fight against terrorism is also a fight for the rights and dignity of women."[30]

Protecting the rights of women became the most politically powerful rationale for invading Afghanistan. This resurgence of victimization discourses—to "save" Muslim women—as a justification for military intervention in Afghanistan employs the same logic and obscures endemic sexism within the United States.[31] The U.S. military intervention has little to do with actual concern for Muslim women's well-being; rather, Bush's neocolonial rhetoric about Islam's inferiority employs the century-old Orientalist trope that uses the status of women in Muslim societies as rationale for political domination and intervention in the Middle East.[32] The U.S. mainstream focus on Muslim women is usually limited to concerns and debates about burkas and veils (e.g., Afghan women) but overlooks the much more complex web of immediate and more urgent economic, political, and educational challenges Arab and Muslim women face daily. Further, this partial view obscures the role of U.S. foreign policy during the Cold War in co-creating misogynist regimes such as the Taliban, and the moral juncture of women's rights and imperialism divides the world into an easy grid of good versus evil.[33,34]

Building on these insights, McAlister argues that 9/11 and the succeeding moment of trauma in the U.S. enabled "a national amnesia" and a new narrative about the essentially "good" and benevolent nature of U.S. imperial power, manifested in the "popular" phrase, "Why do they hate us?," that effaces a long historic involvement of the United States in the Middle East.[35] With the changing power structures after decolonization and the current neoliberal trends in economic globalization, new classes of global citizens were created while degrading undesired humans to conditions of bare life.[36] Further, these neoliberal and neoimperial politics affect many people differently according to class and other "assets" that might provide them with what Ong has called "graduated sovereignty and flexible citizenship."[37] Mbembé lays out how the U.S. politics of a suspended state of emergency after 9/11 opened the gates for a wholesale perception of Muslim citizens as potential terrorists.[38] These perceptions "palimpsestically" register over Orientalist renderings of Muslim men as feminized, inferior, and queer, and yet simultaneously barbarically oppressing Muslim women. In our view, this serves as a screen for Western men to project and fantasize about "unrestrained," patriarchal "masculinity."[39] Interestingly, there was great feminist approval and appraisal of Bush's rhetoric and its neoliberal trends with regard to women's rights increased the misrepresentation and symbolic violence done to Muslim and Arab women.

Building on these points, we suggest that Orientalist feminism maintains two main ideologies that could be identified in what Puar calls "U.S. gender exceptionalism," as well as in the appropriation of execution of aggressive nationalist practices (under a feminist rhetoric). First, Puar defines gender exceptionalism as "a missionary discourse to rescue Muslim women from their oppressive males."[40] She goes on to say that this gender exceptionalism also proposes that, "in contrast to women in the United States, Muslim women are, at the end of the day, unsavable."[41] This rhetoric of gender exceptionalism locates the United States as the democratic empire and overlooks U.S.

abuse, violence, and policing of gender, racial, and sexual identities. It also denies Arab and Muslim women's agency and positions them as victims in need of rescue. In Puar's words, such discourses "posit America as the arbiter of appropriate ethics, human rights, and democratic behavior while exempting itself without hesitation from such universalizing mandates."[42]

The most recent/latest failure to interrogate the consolidation of mainstream (white) U.S. feminism with nationalistic and imperialistic agendas is problematic and dangerous. First, it denies women's agency and sees women only as passive victims in need of salvation, rather than as active political agents. For instance, in the case of Afghanistan in the aftermath of 9/11, media coverage focused on the role of Feminist Majority Foundation (a U.S.-based group) in "freeing Afghan women" and was praised for its efforts.[43] However, the media ignored the role of Afghan women acting on their own behalf, despite the fact that well-established women's groups like the Revolutionary Association of the Women of Afghanistan (RAWA) has been active in speaking out against oppressive laws in the country and against Western policies that damage women's livelihoods and communities.[44] Puar contends that the appropriation and erasure of the works of RAWA by the Feminist Majority Foundation was abusive. Furthermore, Puar describes the foundation as exemplifying a "hegemonic, U.S. centric, ego driven, corporate feminism."[45] Hence, the implication of this Orientalist/exceptional feminism is that it maintains that the solution to women's problems must come from the outside (mainly in the form of Western intervention).

Another vital implication of this Orientalist feminism is the misreading of women's concerns and demands. For instance, U.S. (white) feminist's preoccupation with the veiling and unveiling of women does not reflect what Muslim women consider to be the most pressing issues. As Sabbagh confirms of this discourse, "no Arab woman I know recognizes herself in it."[46] This misreading of Arab and Muslim women's rights advances the idea that inequality is merely cultural, ignoring key works on, and perspectives offered by, intersectional identity politics. The assumption that women's mistreatment is first and foremost grounded in an essentialized, monolithic Islamic or Arabic culture fails to recognize the plurality of gender regimes that coexist with Islamic and Arabic cultures, and the broader social and political factors shaping gender arrangements in Muslim and Arabic societies[47,48]

Furthermore, the "Muslim" woman as victim narrative is a clear example of the perceived opposition of feminism and multiculturalism that has marked dominant U.S. discourse about Muslim womanhood. As Volpp contends, "to posit feminism and multiculturalism as oppositional is to assume that minority women are victims of their cultures."[49] This argument opposes race to gender and provides a theoretical basis for Orientalist/Islamophobic feminism, because it renders certain cultures or religions as inherently violent against women...all while refusing to acknowledge Western culture's oppression of women.[50] Hence, it suggests that women will be better off without their respective cultures, which not only obscures the agency of women within patriarchal societies, but also condones and even encourages U.S. violent interventions to "save brown women" from brown men. Also, this view of women representing the fixed essence of women's culture keeps local feminist women trapped in a binary logic. Hence, we suggest that Orientalist/Islamophobic feminism misrepresents Arab and Muslim women and makes their visibility docile. In the substantive argument that follows, we advance theories of the flesh as models of rethinking and rewriting the goals and strategies of feminist movements seeking to become inclusive of different bodies. We also offer an analysis of the *Bridge* as an example of a theoretical framework that brings feminists together.

Theories of the flesh

Feminist history, especially in mainstream feminist discourse, presents patriarchy as a system of oppression on its own and feminist movements as part of an independent system of liberation and emancipation. As Mohanty argues, "to define feminism purely in gendered terms assumes that our consciousness of being 'women' has nothing to do with race, class, nation or sexuality, just with gender."[51] In that sense, critical feminism as a social theory and as a movement cannot disregard the interlocking points of the systems of oppression such as nationalism, racism, classism, and (hetero) sexism in order to question the relevance or usefulness of endeavors for liberation, emancipation, or critiques to those systems. For this reason, Chicana feminists have theorized a feminism that speaks to the needs of women of different backgrounds and struggles.

The construction of "woman of color" feminism responds to the exclusions of a white feminism that claims to represent all women. In Alarcon's essay, *The Theoretical Subject(s) of This Bridge Called My Back and Anglo- American Feminism,* she argues, "the fact that Anglo-feminism has appropriated the generic term [woman] for itself leaves many a woman in this country having to call herself otherwise, i.e., 'woman of color,' which is equally 'meaningless' without further specification."[52] Alarcon points to the way in which the term "woman" simultaneously fosters a problematic man/woman gender binary and subsumes racial difference.[53] The term "women of color" is a political strategy used by marginalized identities to problematize fixed definitions and challenge the binary that fails to acknowledge interracial and intercultural interactions.

In the collection of the *Bridge,* there is a demonstration of how bridging the diversity of genres is necessary to articulate the multiplicity of oppressions that women of color experience.[54] It also reveals the divergent concerns between groups, as well as the disparate issues of women within the same community. Such contradictions and conflicts point to the impossibility of formulating a fixed definition for "women of color" or "Third World feminism." In the introduction of the book, Moraga and Anzaldúa assert that *Bridge* is "a catalyst, not definitive statement of Third World feminism in the U.S."[55] They also make it clear that they do not intend to construct a rigid definition for "woman of color" feminism, nor do they expect their readership to take their essays as the final word on new and improved feminism. Their model is therefore exceptionally effective, as it promotes discussions rooted in tension and contradiction and, in turns, forces the readers and authors to confront and use conflict as a site for identity formation.

Moraga and Anzaldúa further ask people of color to examine the sources of knowledge and transform the process of theorizing. They challenge traditional interpretations of knowledge and encourage people of color to shift the research lens to one that recognizes their own experiences. They suggest that this negotiation occurs through and within the woman's body, which is the main site of conflict and oppression. They go on to outline a "theory of the flesh" in order to convey the significance of the individual female body confronting those challenges. Moraga and Anzaldúa state that "a theory of the flesh means one where physical realities of our lives—our skin color, the land or concrete we grew up on, our sexual longings—all fuse to create a politic born out of necessity."[56] Chicana feminists theorize a strategy of resistance that centers women of color's bodies because it simultaneously marks the site of women's victimization and the location where women must find their strength. Hence, theorizing through lived experience is a space for reclaiming agency and deconstructing oppression.[57]

Feminists of color encourage women of color to use their own voices to articulate their struggles; and in doing so, they work to legitimate women of color's voices and highlight the oppressions of racialized and gendered bodies.[58]

Feminists of color theoretical revisions: toward bridging

In the essays of *Bridge,* women of color situate their writings in opposition to white feminist theories and practices so that they may forge a feminism that characterizes challenges specific to women of color.[59] Anzaldúa articulates how this book collection fills gaps in white feminist discourses with multiple representations. She says, "I write to record what others erase when I speak, to rewrite the stories others have miswritten about me, about you…to show that I can and I will write."[60] She reveals that women of color refuse to be excluded from feminist discussions within academia. Moreover, she emphasizes the importance of writing the self into visibility.

Women's voices are significant, and should be heard; however, if those voices are not recorded, they can be erased. In the second edition of the *Bridge*, Arab and Arab American women were included as contributors to that theoretical framework. Moraga's foreword expands her feminist theoretical approach to formulate a more transnational *Bridge* that affirms connections between U.S. people of color and Third World feminism, as she claims: "thousands of unarmed people are slaughtered in the Middle East. The United States government daily drains us of nearly every political gain made by feminist, Third World, and anti-war work of the late '60s and early '70s."[61] Saldivar-Hull attests that Moraga expands a Chicana feminist agenda to express "solidarity with the Third World people struggling against the hegemony of the United States." and, in so doing, "Chicana feminist theories present material geopolitical issues and redirect feminist discourse, again pointing to a theory of feminism that addresses a multiplicity of experiences."[62] This reading of Moraga's theoretical revisions demonstrates an inclusive strategy to reach out to Muslim women by forging a site for visibility and inclusiveness. This revision also negates an Orientalist/Islamophobic rhetoric of salvation.

A revisionist model for inclusive visibility: the bridge

Groundbreaking efforts to widen and transform the various bridges linking different marginalized bodies have been sustained and developed in the more recent publication of *This Bridge We Call Home*.[63] This collection revisits and expands on *This Bridge Called My Back*, featuring previously excluded voices, and most particularly Arab and Arab American perspectives. In doing so, this edition complicates the understanding of identities and provides a revisionist model for undoing and redefining women of color's identities. Unlike the one-way approach created by colonial and Orientalist feminism, this book collection exemplifies the importance of rethinking and rewriting the goals and strategies of feminist movements when new perspectives are included, and as historical events alter the relationships between individuals and nations.

In the new collection, Anzaldúa and Keating add imagination and creativity as core factors within their revisionist standpoint. They say,

> imagination, a function of the soul, has the capacity to extend us beyond the confines of our skin, situation, and condition so we can choose our responses. It enables us to reimagine our lives, rewrite the self, and create guiding myths for our times.[64]

This revised standpoint positions women of color's struggles not only within the body but extends its perspective to mark the location of a woman's resistance in the mind and soul. It is important to note that Anzaldúa and Keating by no means erase the harsh reality that women's bodies are sites of oppression.[65] They do, however, emphasize the point that women need to contest violence through discourse. Another point of revision provided by the *Bridge* is the need to include other perspectives in order to develop and transform dialogue within feminists' discourses. Anzaldúa states, "diversity of perspectives expands and alters the dialogue, not in add-on fashion but through a multiplicity that's transformational…these inclusions challenge conventional identities and promote more expansive configurations of identity."[66]

It is these kinds of transformative feminism that reshape resistance in a move that imparts it with inclusive rather than exclusive characteristics. In this way, underrepresented groups such as Arab, Iranian, and South Asian voices are included. The essays by Evelyn Alsulatany, Nathalie

Handal, reem Abdelhadi, Rabab Abdulhadi, and Nadia Elia widen perspectives of intersectionality through the inclusion of Arab Americans' discussions of cultural multiplicity, exilic identities, and invisibility, carving out an inclusive platform opening dialogue between women of color.[67] In the *Bridge*, we see Arab and Muslim women speaking for themselves about their own struggles, unlike Orientalist feminists who assumed the role of speaking for them.

"9/11" as an event for converging and diverging

It is also intriguing that these Arab women's voices are included in *This Bridge We Call Home*, only after the events of 9/11 repositioned Arabs and Muslims at the forefront of both mainstream media and academia.[68] They were not included in anthologies created by other minorities even though they have shared similar experiences and have possessed rich contributions addressing struggles and resistance. It is unfortunate that it took a horrific tragedy on 9/11 to make Arab and Muslim women and their concerns visible; the eventness of 9/11 has also sparked urgency among Arab and Muslim women to make a space for them in those discourses. In addition, it has compelled other minorities in the United States to reach out to Arab and Muslim women and listen to their concerns. Furthermore, prior to 9/11, Arab and Muslim women's critical works were only published in volumes on Islam and the Middle East, but broader works on identity or intersectionality excluded their perspectives.[69] This is mainly due to the ambiguous location of Arab and Muslim identities within mainstream discourses as either "white" or perpetually "foreign." The relationship of "white" Arab and Muslim women to other women of color feminist groups had been tenuous before 9/11.[70] Neither of these types fall under the "typical" minority experience, which can explain, but not justify, their invisibility. While Arab and Muslim women have been a part of American society and discourses for many generations, and have grappled with issues of immigration, assimilation, and acculturation, they have remained unacknowledged as resistive voices.[71]

Despite the similarities between women of color feminism of the 1970s and 1980s and today's struggles of Arab and Muslim feminists, the post-9/11 decade is a different "post identity politics" context and the explicit juncture of war, neoimperial politics, and U.S. Orientalism poses specific challenges to Arab American feminists and their possible interventions.[72] Interestingly enough, the eventness of 9/11 had two intersecting movements. First, it brought Arab and Muslim women together with other women of color to theorize and advance their scholarship. Second, it established Islamic feminism as an oxymoron in mainstream discourses—a discourse that constituted a demand for Arab and Muslim feminists to legitimize their 'type' of feminism.

Muslim feminist thought

Within U.S. media depictions, Islam and feminism are two distinct ideologies that cannot coexist with each other since Islam is considered to be essentially misogynistic while to be feminist is to oppose misogyny. We take a stand for Muslim feminism by arguing that seeing Muslim feminism as an oxymoron can be considered an extension of seeing non-Western (white) movements with Western eyes. What is at stake in the discourse of critics of Muslim feminism is not Muslim women but women under Islam. That is, the discourse against Muslim feminism does not aim at a debate on feminism or women's movements, but allegedly proving that Islam is inherently misogynist.[73] In such critiques of Muslim feminism, Islam is configured as highly conflicting with women's liberation. This part of the debate implies Islam's character is despotic and barbaric (especially in terms of its views of women).

Thus, we argue that these ideas and assumptions should not be the starting point of a so-called feminist debate since the scope of those debates are not women or women's positionalities but an expression and representation of Orientalist and Islamophobic views of Arab and Muslim feminisms. As an alternative, Muslim feminism could be seen as an extension of theories of the flesh, where Muslim feminists theorize through their lived experiences and link the personal to the political. Muslim feminists reclaim their lives and deconstruct the hegemonic universalist knowledge of both Orientalism and patriarchy. Similar to Chicana feminist movements, Muslim feminism asks women to examine the sources of knowledge and transform the process of theorizing.

Theorizing Muslim feminism: between Orientalism and patriarchy

Muslim feminism can be described as a feminist movement that bases its methodology and epistemology on both post-colonial feminism and Islamic theology.[74] Najmabadi, one of the pioneering scholars in Muslim feminism, describes Muslim feminism as a "reform movement that opens up a dialogue between religious and secular feminists."[75] Scholars such as Najmabadi, Cooke, and Badran embrace Muslim feminism as an important and relevant movement to feminism. They argue that it critically approaches both Western feminist assumptions about Islam by challenging the male hegemonic domain of Islamic hermeneutics by presenting a middle ground between these two discourses.

Even if feminist or women's rights endeavors date back the early 20th century among Muslim women or in Muslim communities, the term Muslim feminism is relatively contemporary in both usage and circulation. The debate on Muslim feminism in academia started with Najmabadi's speech, delivered at the School of Oriental and African Studies at the University of London in 1994 when she described Muslim feminism "as a reform movement that opens up a dialogue between religious and secular feminists."[76] Najmabadi argued that Muslim feminism transcends the binary of the "secular" and "religious" through its critiques of unquestioned presuppositions of Western secular feminism regarding Muslim women. This important speech has come to mark and define how Muslim feminism is understood.

Fazaeli argues for the term Muslim feminism as a feminist response in Islam toward various social and political determinants.[77] Rhouni also embraces the movement or theorization of Muslim feminism, yet she problematizes the term "Islamic" or "Muslim" since, she argues, "it excludes both non-Muslims and secular scholars of Muslim background, who strive to contribute to the revitalization of Islamic thought through an approach that does not stigmatize Islam and recognizes its egalitarian scope."[78] In the case of Rhouni, then, Muslim feminism is a faith-oriented theory and movement.[79] This claim seems reasonable and it is one of the most common arguments among scholars who see Muslim feminism as an oxymoron.[80] For this statement, two points need clarification. First, Rhouni does not advocate for a specific alternative for "naming" feminists and/or scholars who interpret Islam through a more egalitarian lens, but she problematizes the adjective "Islamic/Muslim" just to show its dangers and traps.[81] Second, she does not offer possible adjectives for the particular kind of feminism or scholarship she endorses. We would argue that a Muslim feminism is a different articulation, which, we believe, serves only what Rhouni is being cautious about.[82] In other words, Muslim feminism only refers to self-described, pious practicing Muslims. That is why we position ourselves using the term "Muslim feminism" as we believe it is more inclusive.

However, Muslim feminism should not be confused with Islamist revivalist women who, as Mahmood demonstrates in *Politics of Piety*, also use their agency to actively be a part of and promote the Islamist revival of recent decades—but without challenging Islamic law and thought.[83] These very different perspectives on womanhood among various groups of practicing and secular

Muslim women are indicative of the current wider struggles over the meaning and practices of Islam within Muslim communities.

As for the methodological perspective of Muslim feminism, it derives its source of knowledge from both post-colonial feminist and classical Islamic epistemologies.[84] While Muslim feminism calls for gender equality in the social, political, and economic spheres, its methodology stems from reinterpretations or hermeneutics of the Quran and Islamic law. There are examples of feminist organizing that can be described as Muslim feminism in action, including the efforts of Iranian feminists for more gender-neutral laws, the demands of Egyptian feminists to participate in vocations which are currently not open to women such as the clergy, and the struggle of Turkish feminists to abolish the ban on veils in the public sector and on state premises.[85]

A feminist reinterpretation of Islam, as Badran argues, "renders compelling confirmation of gender equality in the Quran that was lost as male interpreters constructed a corpus of tafsir [explanation] promoting a doctrine of male superiority reflecting the mindset of the prevailing patriarchal cultures."[86] The aim of Muslim feminism as it utilizes reinterpretation as its methodological tool, therefore, is to interrupt and challenge patriarchal (and in some cases even misogynist) readings of sacred texts as well as the social formations constructed on those readings' approaches to the religion.

The deployment of women as reinterpreters, then, is a challenge to the orthodoxy of the religion for the sake of the equality of women. Second, Moghadam praises Muslim feminism since she sees the importance of this reformist movement as a common ground or a possible alliance with secular feminists in their efforts for gender equality.[87] In a similar vein, Badran argues that Muslim feminism is "Increasingly occupying a middle ground where the secular and religious meet or where the two collapse."[88] Therefore, Muslim feminism can be useful not only for building an alliance between the secular and the religious as two distinct ideologies but also for dismantling presumptions and assumptions of one regarding the other.

Besides the argument that Muslim feminism opens a space in between the dualism of the secular and the religious, there are two main arguments to support the relevance and usefulness of Muslim feminism in terms of theory and activism: one is, as Tohidi contends, that it's a step toward the secularization of state formations,[89] and the second is that it's a voice against the essentializing of the *muslimwoman* in that it requires reforming power relations both in and out of the communities from which Muslim feminism emerged, as Cooke and Mir-Hosseini claim.[90,91]

According to Badran, the relations between Muslim women in the Middle East and feminism emerged in the context of modernity and modernization in the late 19th century in response to nationalist, anti-colonialist, and/or Islamization discourses.[92] She points out that feminist movements have always been "discredited in the patriarchal mainstream as Western and a project of cultural colonialism and therefore were stigmatized as antithetical to Islam."[93] However, she asserts, the newly emerged movement of Muslim feminism offers a new path, a middle ground,[94] a "middle space of an independent site" between secular feminism and misogynist Islamism (or Islamic fundamentalism).[95] Furthermore, Moghadam argues that Muslim feminism is a part of the reform driven movement seen after the 1980s, which challenges patriarchal gender notions fueled by the Islamic fundamentalists.[96]

Writing on these developments, Mir-Hosseini says:

> By the late 1980s, there were clear signs of the emergence of a new consciousness, a new way of thinking, a gender discourse that was and is feminist in its aspiration and demands, yet Islamic in its language and sources of legitimacy. One version of this new discourse has come to be called Muslim feminism.[97]

In that sense, Muslim feminism's originality as a feminist movement and theory, according to Mir-Hosseini, stems from its double-agency as feminist and religious and from its task of bringing religion into the framework of feminism as well as making feminism legitimate within the religion:

> Muslim traditionalists and fundamentalists do this as a means of silencing other internal voices and abuse the authority of the text for authoritarian purposes. Secular fundamentalists follow the same pattern, but in the name of enlightenment, progress, and science—and as a means of showing the misogyny of Islam—while ignoring the contexts in which the texts were produced, as well as the existence of alternative texts. In doing so, they end up essentializing and perpetuating difference and reproducing a crude version of the orientalist narrative of Islam.[98]

Here, Mir-Hosseini raises the question of the "double exploitation" of feminist women in the Muslim world and Muslim communities. That is, she claims that women in Muslim communities, regardless of their feminist backgrounds from either Western or indigenous roots, have always been subjects of argument in terms of different parts of their identities. As Muslim, their identity is often questioned by secular fundamentalists and their feminism is viewed as suspicious by Muslim traditionalists and Islamic fundamentalists: "their Muslimness is perceived as backward and oppressed, yet authentic and innate; their feminism is perceived as progressive and emancipated, yet corrupt and alien."[99]

In making this argument, Mir-Hosseini is close to the position of Badran wherein Muslim feminism is viewed as a middle ground between secularist and non-secularist fundamentalism, and she adds:

> Though adhering to very different ideologies and scholarly traditions and following very different agendas, all these opponents of the feminist project in Islam share one thing—an essentialist and non-historical understanding of Islam and Islamic law. They fail to recognize that assumptions and laws about gender in Islam—as in any other religion—are socially constructed and thus historically changing and open to negotiation.[100]

What opponents of Muslim feminism miss is that religion (not limited to Islam) as a social phenomenon is not necessarily a series of dogmatic doctrines which are inevitably close to progress or change especially when it comes to the reforms in social orders including sex and gender orders. Instead, religion can be a dynamic way to cover what the contemporary requires with the help of constructive criticism.[101]

In fact, Muslim feminism, with its methodology of reinterpretation, is an example of the kind of constructive criticism that can push the traditional scholarship of Islam to meet the demands of Muslim women today. The secularist and Orientalist narrative of Islam is also discussed in the works of Cooke.[102] Even though it is not directly related to Muslim feminism, the term "muslimwoman" (coined by Cooke) is highly significant and reflects on the Orientalist point of view fueled after 9/11 to understand the opposition against Muslim feminism. The use of this term creates an image of a monolithic Muslim-woman or identity that assumes that being a Muslim woman is in essence something oppressive, and "muslimwoman," and all Muslim women, are victims of Islam's patriarchal essence and thus inevitably need liberation.[103]

Her understanding of Muslim feminism, then, is also related to her analysis of this image. That is to say, according to Cooke:

> Whenever Muslim women offer a critique of some aspect of Islamic history or hermeneutics, they do so with and/or on behalf of all Muslim women and their right to enjoy with men full participation in a just community, I call them Muslim feminists. This label is not rigid; rather it describes an attitude and intention to seek justice and citizenship for Muslim women.[104]

Therefore, for Cooke, *all* Muslim women would benefit from the critiques of (traditional) Islamic history and hermeneutics, as they provide positive changes in efforts to create a just community for Muslims.[105] At first glance, this argument may seem homogenizing. Yet Cooke asserts that multiple and different identities of Muslimhood (in terms of ethnicity, politics, and histories) can come together with Muslim feminism in order to claim "simultaneous and sometimes contradictory allegiances even as they resist globalization, local nationalisms, Islamization, and the pervasive patriarchal system."[106] Cooke's view appears to be in line with Mir-Hosseini and Badran's arguments on how Muslim feminism transcends the "limits" of both the inside and outside dimensions of a woman's movement, but by exceeding those limits, Cooke stands for "how a subalternized group can assume its essentialized representations and use them strategically against those who have ascribed them."[107]

In taking this position, Cooke follows critical cultural scholar Bhabha's re-reading of Orientalism in terms of power/knowledge relations and concludes that the subaltern takes up a middle space in between the binaries to produce alternate discourses to challenge and disturb the knowledge, representation, and discourse associated with and signifying the subaltern.[108] Muslim feminism is another example of this middle space where marginalized women reclaim a discursive space between the representations and assumptions about Muslim women.[109] According to Cooke, Muslim feminism, or Muslim women's critical attitude and intention, is in line with Bhabha's argument, as Muslim feminists, despite their diverse identities, produce a third space in between the binary of the secularist and the religious by disturbing the understanding of (Western) feminist ideals and pointing to the Orientalist values and images that render Islam misogynist.[110] Muslim feminists therefore challenge the traditional, orthodox readings of the religion of Islam for a more just sociopolitical order for women and men alike.[111]

Conclusion

Women of color writers and feminists have fought structurally similar fights and used the transformative power of writing and theorizing through their bodies and lived experiences against the pervasive racist and sexist hierarchies in hegemonic culture that leave imprints on women of color's selves. In this essay, we have extended this argument in the specific context of a Muslim feminism that theorizes outside an Orientalist/Islamophobic and a patriarchal frame of reference. We believe it is important to use revisionist models that rethink and rewrite the goals and strategies of feminist movements when new perspectives emerge. Moreover, we want to also acknowledge the important work that is being done in the field to transform how we come to understand Arab and Muslim feminism, particularly in relation to Muslim feminist activism and the use of social media for advocacy.[112,113,114] Finally, we argue that Muslim feminism could be seen as an extension of theories of the flesh, where Muslim feminists establish a middle space "in between" the binaries to produce alternate discourses to challenge and hegemonic structures and discourses. It is our sincere hope that this work can serve as one small step in that direction.

Notes

1 Ziba Mir-Hosseini, *Islam and Gender: The Religious Debate in Contemporary Iran* (Princeton, NJ: Princeton University Press, 1999), 9.
2 Jasbir K. Puar, *Terrorist Assemblages: Homonationalism in Queer Times* (Durham, NC: Duke University Press, 2007).
3 Fatima Zahrae Chrifi Alaoui, "Arabizing Vernacular Discourse: A Rhetorical Analysis of the Tunisian Revolutionary Graffiti," in *Rhetorics and Elsewhere and Otherwise: Contested Modernities, Decolonial Visions*, eds Damián Baca and Romeo Garcia (Urbana, IL: NCTE Press, 2019).
4 Puar, *Terrorist Assemblages*, xviii.
5 Puar, *Terrorist Assemblages*, xvii–xx.
6 Fatima Zahrae Chrifi Alaoui, "The Arab Spring between the Streets and the Tweets: Examining the Embodied (e)Resistance Through the Feminist Revolutionary Body," *Women of Color and Social Media Multitasking: Blogs, Timelines, Feeds, and Community*, eds Keisha Edwards Tassie and Sonja M. Brown Givens (Lanham, MD: Lexington Books, 2015), 35–68.
7 Chrifi Alaoui, "Arabizing Vernacular Discourse."
8 Puar, *Terrorist Assemblages*, xxiv.
9 Chrifi Alaoui, "Arabizing Vernacular Discourse."
10 Cherríe Moraga and Gloria Anzaldúa, *This Bridge Called My Back: Writings by Radical Women of Color* (Albany, NY: State University of New York Press, 2015).
11 Leila Ahmed, "Western Ethnocentrism and Perceptions of the Harem," *Feminist Studies* 8, no. 3 (1982): 522, doi: 10.2307/3177710.
12 Ahmed, "Western Ethnocentrism," 522.
13 Leila Ahmed, *A Quiet Revolution: The Veil's Resurgence, from the Middle East to America* (New Haven, CT: Yale University Press, 2012).
14 Ahmed, *A Quiet Revolution.*
15 Ahmed, *A Quiet Revolution.*
16 Ahmed, *A Quiet Revolution.*
17 Puar, *Terrorist Assemblages*, 7.
18 Elora Halim Chowdhury, "Locating Global Feminisms Elsewhere: Braiding US Women of Color and Transnational Feminisms," *Cultural Dynamics* 21, no. 1 (2009): 57, doi:10.1177/0921374008100407.
19 Moraga and Anzaldúa, *This Bridge Called My Back*, xvi.
20 Fatima Sadiqi, *Morocc.2+01an Feminist Discourses* (New York: Palgrave Macmillan, 2014), xvi–xvii.
21 Uma Narayan, *Dislocating Cultures: Identities, Traditions, and Third World Feminism* (New York: Routledge, 1997), 3.
22 Haneen Al-Ghabra, *Muslim Women and White Femininity: Reenactment and Resistance* (New York: Peter Lang, 2018).
23 Lila Abu-Lughod, "Do Muslim Women Really Need Saving? Anthropological Reflections on Cultural Relativism and Its Others," *American Anthropologist* 104, no. 3 (2002): 783–790.
24 Susan Muaddi Darraj, "Third World, Third Wave Feminism(s): The Evolution of Arab American Feminism." *Catching a Wave: Reclaiming Feminism for the 21st Century* (2003): 188–205.
25 Floya Anthias and Nira Yuval-Davis, *Racialized Boundaries: Race, Nation, Gender, Colour and Class and the Anti-Racist Struggle* (New York: Routledge, 2005), 71.
26 Suha Sabbagh, *Arab Women: Between Defiance and Restraint* (Northampton, MA: Olive Branch Press, 1996), xxiv–xxv.
27 Dicle Kogacioglu, "The Tradition Effect: Framing Honor Crimes in Turkey," *Differences: A Journal of Feminist Cultural Studies* 15, no. 2 (2004): 119.
28 Chrifi Alaoui, "Arabizing Vernacular Discourse."
29 Dana L. Cloud, "To Veil the Threat of Terror: Afghan Women and the Clash of Civilizations in the Imagery of the U.S. War on Terrorism," *Quarterly Journal of Speech* 90, no. 3 (2004), https://doi.org/10.1080/0033563042000270726.
30 Abu-Lughod "Do Muslim Women."
31 Amira Jarmakani, *Imagining Arab Womanhood: The Cultural Mythology of Veils, Harems, and Belly Dancers in the U.S.* (New York: Palgrave Macmillan, 2008), 159.
32 Abu-Lughod, "Do Muslim Women."
33 Puar, *Terrorist Assemblages*, 52.

34 Melani McAlister, *Epic Encounters: Culture, Media, and U.S. Interests in the Middle East Since 1945* (Berkeley: University of California Press, 2005), 282.

35 McAlister, *Epic Encounters*, 282.

36 J.-A Mbembé, and Libby Meintjes, "Necropolitics," *Public Culture* 15, no. 1 (2003): 12.

37 Aihwa Ong, "On the Edge of Empires: Flexible Citizenship Among Chinese in Diaspora," *Positions* 1, no. 3 (1993): 134–162.

38 Mbembé and Meintjes, Necropolitics.

39 Mbembé and Meintjes, Necropolitics 12.

40 Puar, *Terrorist Assemblages*, 5.

41 Puar, *Terrorist Assemblages*, 5.

42 Puar, *Terrorist Assemblages*, 8.

43 Puar, *Terrorist Assemblages*, 6.

44 Puar, *Terrorist Assemblages*, 6.

45 Puar, *Terrorist Assemblages*, 6.

46 Sabbagh, *Arab Women*, xi.

47 Maya Mikdashi, "How Not to Study Gender in the Middle East," *Jadliyya*, www.jadaliyya.com/pages/index/4775/how-not-to-study-gender-in-the-middle-east.

48 Chrifi Alaoui, "Arabizing Vernacular Discourse."

49 Leti Volpp, "Feminism Versus Multiculturalism," *Columbia Law Review* 101, no. 5 (2001): 1185.

50 Chrifi Alaoui, "Arabizing Vernacular Discourse."

51 Chandra Talpade Mohanty, *Feminism Without Borders: Decolonizing Theory, Practicing Solidarity* (Duke, NC: Duke University Press, 2003): 55.

52 Norma Alarcón, "The Theoretical Subject(s) of This Bridge Called My Back and Anglo-American Feminism," in *The Postmodern Turn*, ed. Steven Seidman (Cambridge: Cambridge University Press: 2010), 147.

53 Alarcón, "Theoretical Subject(s)".

54 Cherríe Moraga and Gloria Anzaldúa, *This Bridge Called My Back: Writings by Radical Women of Color* (New York: Kitchen Table/Women of Color Press: 1983).

55 Moraga and Anzaldúa, *This Bridge Called My Back*, xxvi.

56 Moraga and Anzaldúa, *This Bridge Called My Back*, 23.

57 Moraga and Anzaldúa, *This Bridge Called My Back*.

58 Moraga and Anzaldúa, *This Bridge Called My Back*.

59 Moraga and Anzaldúa, *This Bridge Called My Back*.

60 Moraga and Anzaldúa, *This Bridge Called My Back*, 169.

61 Moraga and Anzaldúa, *This Bridge Called My Back*, xix.

62 Sonia Saldívar-Hull. *Feminism on the Border: Chicana Gender Politics and Literature* (Berkeley, CA: University of California Press, 2000), 49.

63 Gloria Anzaldúa and Analouise Keating, *This Bridge We Call Home: Radical Visions for Transformation* (New York: Taylor and Francis, 2002).

64 Anzaldúa and Keating, *This Bridge We Call Home*, 5.

65 Anzaldúa and Keating, *This Bridge We Call Home*.

66 Anzaldúa and Keating, *This Bridge We Call Home*, 4.

67 Anzaldúa and Keating, *This Bridge We Call Home*.

68 Chrifi Alaoui, "The Arab Spring."

69 Miriam Cooke, "Deploying the Muslim," *Journal of Feminist Studies in Religion* 24, no. 1 (2008).

70 Anzaldúa and Keating, *This Bridge We Call Home*, 223.

71 Cooke, "Deploying the Muslim."

72 Cooke, "Deploying the Muslim."

73 Leila Ahmed, *Women and Gender in Islam: Historical Roots of a Modern Debate* (New Haven, CT: Yale University Press, 1993)

74 Chrifi Alaoui, "The Arab Spring."

75 Afsaneh Najmabadi, *Women with Mustaches and Men without Beards: Gender and Sexual Anxieties of Iranian Modernity* (Berkeley: University of California Press, 2002), 2.

76 Valentine M. Moghadam, "Islamic Feminism and Its Discontents: Toward a Resolution of the Debate," *Signs* 27, no. 4 (2002): 1143, doi: 10.1086/339639.

77 Roja Fazaeli, "Contemporary Iranian Feminism: Identity, Rights and Interpretations," *Muslim World Journal of Human Rights* 4, no. 1 (2007): 1–24.

78 Raja Rhouni, *Secular and Islamic Feminist Critiques in the Work of Fatima Mernissi* (Boston, MA: Brill, 2009), 33.
79 Rhouni, *Secular and Islamic Feminist.*
80 Rhouni, *Secular and Islamic Feminist.*
81 Rhouni, *Secular and Islamic Feminist.*
82 Rhouni, *Secular and Islamic Feminist.*
83 Saba Mahmood, *Politics of Piety: The Islamic Revival and the Feminist Subject* (Princeton, NJ: Princeton University Press, 2011).
84 Chrifi Alaoui, "Arabizing Vernacular Discourse."
85 Margot Badran, "Islamic Feminism: What's in a Name?," *Al-Ahram Weekly Online* (2002): 17–23, http://weekly.ahram.org.eg/2002/569/cu1.htm.
86 Badran, "Islamic Feminism."
87 Moghadam, "Islamic Feminism and Its Discontents."
88 Badran, "Islamic Feminism."
89 Nayereh Tohidi, "Islamic Feminism: Perils and Promises," *The Middle East Women's Studies Review* 16, nos 3–4 (2001).
90 Cooke, "Deploying the Muslim."
91 Ziba Mir-Hosseini, "The Quest for Gender Justice: Emerging Feminist Voices in Islam," *Islam* 21, no. 36 (2004): 1–5.
92 Badran, "Islamic Feminism."
93 Tohidi, "Perils and Promises."
94 Tohidi, "Perils and Promises."
95 Badran, "Islamic Feminism."
96 Moghadam, "Islamic Feminism and Its Discontents."
97 Mir-Hosseini, "The Quest for Gender Justice," 2–3.
98 Mir-Hosseini, "The Quest for Gender Justice," 3.
99 Tohidi, "Perils and Promises."
100 Mir-Hosseini, "The Quest for Gender Justice," 3.
101 Cooke, "Deploying the Muslim."
102 Cooke, "Deploying the Muslim."
103 Cooke, "Deploying the Muslim."
104 Miriam Cooke, "Multiple Critique: Islamic Feminist Rhetorical Strategies," *Nepantla: Views from South* 1, no. 1 (2000): 95.
105 Cooke, "Multiple Critique."
106 Cooke, "Multiple Critique," 95.
107 Cooke, "Multiple Critique," 100.
108 Homi K. Bhabha, *The Location of Culture* (New York: Routledge, 2004).
109 Chrifi Alaoui, "Arabizing Vernacular Discourse."
110 Cooke, "Multiple Critique."
111 Cooke, "Multiple Critique."
112 Al-Ghabra, Chrifi Alaoui, Abdi, and Calafell, *Voices in Middle Eastern and North African Communication and Cultural Studies: Thinking Transnationally* (Peter Lang Press: In Press).
113 Chrifi Alaoui, "The Arab Spring."
114 Chrifi Alaoui, "Arabizing Vernacular Discourse."

Bibliography

Abu-Lughod, Lila. "Do Muslim Women Really Need Saving? Anthropological Reflections on Cultural Relativism and Its Others." *American Anthropologist* 104, no. 3 (2002): 783–790.

Ahmed, Leila. *A Quiet Revolution: The Veil's Resurgence, from the Middle East to America.* New Haven, CT: Yale University Press, 2012.

Ahmed, Leila. "Western Ethnocentrism and Perceptions of the Harem." *Feminist Studies* 8, no. 3 (1982): 521–534. doi: 10.2307/3177710.

Ahmed, Leila. *Women and Gender in Islam: Historical Roots of a Modern Debate.* New Haven, CT: Yale University Press, 1993.

Alarcón, Norma. "The Theoretical Subject(s) of this Bridge Called My Back and Anglo-American Feminism." In *The Postmodern Turn*, edited by Steven Seidman, 140–152. Cambridge: Cambridge University Press, 2010.

Al-Ghabra, Haneen. *Muslim Women and White Femininity: Reenactment and Resistance*. New York: Peter Lang, 2018.

Al-Ghabra, Haneen, Fatima Zahrae Chrifi Alaoui, Shadee Abdi, and Bernadette Marie Calafell. *Voices in Middle Eastern and North African Communication and Cultural Studies: Thinking Transnationally*. New York: Peter Lang Press, In Press.

Anthias, Floya and Nira Yuval-Davis. *Racialized Boundaries: Race, Nation, Gender, Colour and Class and the Anti-Racist Struggle*. New York: Routledge, 2005.

Anzaldúa, Gloria and Analouise Keating. *This Bridge We Call Home: Radical Visions for Transformation*. New York: Taylor and Francis Group, 2002.

Badran, Margot. "Islamic Feminism: What's in a Name?" *Al-Ahram Weekly Online*, 2002, 17–23, http://weekly.ahram.org.eg/2002/569/cu1.htm.

Bhabha, Homi K. *The Location of Culture*. New York: Routledge, 2004.

Chowdhury, Elora Halim. "Locating Global Feminisms Elsewhere: Braiding US Women of Color and Transnational Feminisms," *Cultural Dynamics* 21, no. 1 (2009): 51–78. doi:10.1177/0921374008100407.

Chrifi Alaoui, Fatima Zahrae. "The Arab Spring between the Streets and the Tweets: Examining the Embodied (e)Resistance Through the Feminist Revolutionary Body." In *Women of Color and Social Media Multitasking: Blogs, Timelines, Feeds, and Community*, edited by Keisha Edwards Tassie and Sonja M. Brown Givens, 35–68. Lanham, MD: Lexington Books, 2015.

Chrifi Alaoui, Fatima Zahrae. "Arabizing Vernacular Discourse: A Rhetorical Analysis of the Tunisian Revolutionary Graffiti." In *Rhetorics and Elsewhere and Otherwise: Contested Modernities, Decolonial Visions*, edited by Damián Baca and Romeo Garcia, 112–140. Urbana, IL: NCTE Press, 2019.

Cooke, Miriam. "Multiple Critique: Islamic Feminist Rhetorical Strategies." *Nepantla: Views from South* 1, no. 1 (2000): 91–110.

Cooke, Miriam. "Deploying the Muslim." *Journal of Feminist Studies in Religion* 24, no. 1 (2008): 91–99.

Darraj, Susan Muaddi. "Third World, Third Wave Feminism(s): The Evolution of Arab American Feminism." In *Catching a Wave: Reclaiming Feminism for the 21st Century*, edited by Rory Dicker and Alison Piepmeier, 188–205. Lebanon, NH: University Press of New England, 2003.

Fazaeli, Roja. "Contemporary Iranian Feminism: Identity, Rights and Interpretations." *Muslim World Journal of Human Rights* 4, no. 1 (2007): 1–24. doi: 10.2202/1554-4419.1118.

Jarmakani, Amira. *Imagining Arab Womanhood: The Cultural Mythology of Veils, Harems, and Belly Dancers in the U.S.* New York: Palgrave Macmillan, 2008.

Kogacioglu, Dicle. "The Tradition Effect: Framing Honor Crimes in Turkey." *Differences: A Journal of Feminist Cultural Studies* 15, no. 2 (2004): 119–151. https://doi.org/10.1215/10407391-15-2-118.

Mahmood, Saba. *Politics of Piety: The Islamic Revival and the Feminist Subject*. Princeton, NJ: Princeton University Press, 2011.

Mbembé, J.-A., and Libby Meintjes. "Necropolitics." *Public Culture* 15, no. 1 (2003): 11–40. www.muse.jhu.edu/article/39984.

McAlister, Melani. *Epic Encounters: Culture, Media, and U.S. Interests in the Middle East Since 1945*. Berkeley, CA: University of California Press, 2005.

Mikdashi, Maya. "How Not to Study Gender in the Middle East." *Jadliyya*. www.jadaliyya.com/pages/index/4775/how-not-to-study-gender-in-the-middle-east

Mir-Hosseini, Ziba. *Islam and Gender: The Religious Debate in Contemporary Iran*. Princeton, NJ: Princeton University Press, 1999.

Mir-Hosseini, Ziba. "The Quest for Gender Justice: Emerging Feminist Voices in Islam." *Islam* 21, no. 36 (2004): 1–5.

Moghadam, Valentine M. "Islamic Feminism and Its Discontents: Toward a Resolution of the Debate." *Signs* 27, no. 4 (2002): 1135–1171. doi: 10.1086/339639.

Mohanty, Chandra Talpade. *Feminism without Borders: Decolonizing Theory, Practicing Solidarity*. Durham, NC: Duke University Press, 2003.

Moraga, Cherríe and Gloria Anzaldúa. *This Bridge Called My Back: Writings by Radical Women of Color*. New York: Kitchen Table/Women of Color Press, 1983.

Moraga, Cherríe and Gloria Anzaldúa. *This Bridge Called My Back: Writings by Radical Women of Color*. Albany, NY: State University of New York Press, 2015.

Narayan, Uma. *Dislocating Cultures: Identities, Traditions, and Third World Feminism*. New York: Routledge, 1997.

Ong, Aihwa. "Flexible Citizenship among Chinese Cosmopolitans." *Cultural Politics* 14 (1998): 134–162.

Puar, Jasbir K. *Terrorist Assemblages: Homonationalism in Queer Times*. Durham, NC: Duke University Press, 2007.

Rhouni, Raja. *Secular and Islamic Feminist Critiques in the Work of Fatima Mernissi*. Boston, MA: Brill, 2010.

Sabbagh, Suha. *Arab Women: Between Defiance and Restraint*. Northampton, MA: Olive Branch Press, 1996.

Sadiqi, Fatima. *Moroccan Feminist Discourses*. New York: Palgrave Macmillan, 2014.

Saldívar-Hull, Sonia. *Feminism on the Border: Chicana Gender Politics and Literature*. Berkeley, CA: University of California Press, 2000.

Tohidi, Nayereh. "Islamic Feminism: Perils and Promises." *The Middle East Women's Studies Review* 16, nos 3–4 (2001): 135–146

Volpp, Leti. "Feminism Versus Multiculturalism." *Columbia Law Review* 101, no. 5 (2001): 1181–1218. doi: 10.2307/1123774

20

NEGATIVE SPACES IN THE TRIANGLE OF GENDER, RELIGION, AND NEW MEDIA

A case study of the Ultra-Orthodox community in Israel

Rivka Neriya-Ben Shahar

Studying the Ultra-Orthodox community in Israel provides an opportunity to consider how gender, religion, and new media work together to form a complex system. This system can be seen as a triangle in which the sides represent gender, religion, and new media and the angles represent their interactions. This argument implies that a contemporary study of any one aspect should take the others and their interactions into account because all of these have multiple connections to and within the triangle. Alas Ruth offered a compelling rationale for the triangle model, arguing that it enables visibility of the connections between three critical ingredients to produce a systematic understanding of the interdependence of central elements representing ideas and processes.[1]

A triangular approach calls for working definitions of the key terms: gender, religion, and new media. For this study, I use the term gender to refer to a "social construct defining the attributes, behavior, and roles that are generally associated with men and women [...] constructed by the social and political implications."[2] My study makes use of a multifaceted account of religion as

(1) a system of symbols which acts to (2) establish powerful, pervasive, and long-lasting moods and motivations in men by (3) formulating conceptions of a general order of existence and (4) clothing these conceptions with such an aura of factuality that (5) the moods and motivations seem uniquely realistic.[3]

I understand new media to reference a network employing a variety of unique channels that "emerged on the contemporary landscape and offers new opportunities for social interaction, information sharing, and mediated communication."[4]

The definitions of these terms are already very complex. Nevertheless, I would like to add another level of complexity to my intersectional analysis. This new level is illustrated through the artistic concept of the "negative space" that can be located around or between subjects[5] or "the area surrounding a figure that makes the figure stand out"[6] as represented, for example, by Maurits Cornelis Escher's famous art. The crucial point this concept makes is that when the most visible element of the positive space occupied by the object itself becomes the focus, we neglect not only what is missing or absent or set apart but also the relationships *between* the spaces.

Supplementing triangulation with attention to negative spaces leads us to an approach that locates an intersection not only between the external forces but also between the internal relations. When conceptualized in this way, intersectionality[7] must understand gender by taking into account the relationships between social constructions for men and women. Similarly, careful attention to religion is not possible (even when including all of Clifford Geertz's ingredients—from symbols to motivations) without addressing secularization.[8] Likewise, for a deeper understanding of the contemporary opportunities offered by new media, we should carefully consider traditional media as well. Moreover, all of these aspects and oppositions create a full continuum that should also be part of the discussion. For example, between "woman" and "man" there are a variety of self and multiple identities that alter or augment these gender categories by complementing, contrasting, or even falling outside of them; people may define themselves as secular but also have some religious practices and beliefs, and new media users may employ the latest smartphone for some forms of communication and watch TV for others.

Therefore, I argue that (1) we cannot focus only on gender, religion, and new media as separated elements but must place them at an intersection. Moreover, (2) analyzing intersections of gender, religion, and new media requires attention to relations, continuums, and negative spaces that both create and are created by the pairing of the ostensibly oppositional categories of women/men, religion/secularization, and new/traditional media. From the starting point of these complicated relationships, this chapter asks: What can we learn about this system from these multiple points of view? How do gender issues affect new media and vice versa? How do religion and new media influence one another? What are the (power) relationships between gender issues and religious rules and values, and in what ways does this interplay intersect with new media? In the particular case study that follows, I explore complex relations between gender and new media among Ultra-Orthodox Israeli community members to answer these questions. Informed by the approach outlined above, I examine the sides and angles of this gender/religion/media triangle, taking into account both its complicated negative spaces and rich continuums. The next section will describe the case study chosen for these approaches.

The case study: the Ultra-Orthodox community in Israel

The Ultra-Orthodox Jews in Israel (hereafter: Ultra-Orthodox) are a minority group, constituting about 12% of Israel's population, or approximately 1 million people.[9] Their lives are bound by a commitment to the ongoing study of Torah and stringent interpretations of *Halakha,* religious Jewish law, through unquestioning faith, strict religious behaviors, and sincere obedience to the authority of its leaders.[10] An outsider might think of them as a singular "Ultra-Orthodox" group, but as in other communities, there are multiple differences, created by many groups and subgroups.[11]

Ultra-Orthodox community members are located mostly in urban eras with many living in Jerusalem, Bnei Brak (near Tel-Aviv), and Ashdod. They practice self-imposed isolation by living in separate neighborhoods, since they perceive their surrounding societies to be ideologically secular, Western, and modern enemies, embracing views and practices that run contrary to their religious values. This perception is promoted consistently by the rabbis, the leaders of the Ultra-Orthodox community. They use classes and harangues, including Bible verses and sages' articles to affirm the need for separation. The Ultra-Orthodox media, especially the newspapers that are controlled by the rabbis, echo these messages. Accordingly, Ultra-Orthodox communities have independent educational and legal systems and practice intra-community marriage. They also adopt a range of strategies involving dress (with stringent rules of modesty) and a purified discourse (without common Israeli slang and curses) to set themselves apart from mainstream Israeli culture.[12]

Nevertheless, recent studies point to community members' extensive personal, social, economic, and cultural ties with their broader contexts.[13] Many of these ties are created by women.[14] We can see that even among religious communities that try to separate themselves from the outside secular community, there are multiple connections made that are either necessary or by choice. These various connections form part of the evidence that this religion should not be studied as a unique phenomenon without taking into account the secular world around it.

The next part will illustrate the three sides and angels of the triangle: religion and gender, religion and media, and gender and media. The triangulation and the negative spaces will be demonstrated throughout the case study of the Israeli Ultra-Orthodox community. This discussion benefits from the perspective of the "third-person effect," which occurs when people perceive that media messages are significantly more influential on others than they are on themselves.[15]

Religion and gender among the Ultra-Orthodox community

I begin my triangular analysis in this case study by examining the side symbolizing the religion and gender connections. The Ultra-Orthodox community, characterized by strict boundaries between insiders and outsiders, enables another strong limitation: a rigorous separation between men and women and a focus on issues relating to modesty. This separation is implemented in all spheres and situations.[16] For example, the education system is segregated from the age of three years (some places begin segregation from the age of three months). Young boys and girls live in different rooms and generally are not allowed to play together. Special events (holidays, weddings, etc.) are celebrated separately with walls between gender groups. One of the most important places that excludes women is the space dedicated for Torah studies, where the men learn religious teachings and the rules for the entire community.

Paradoxically, Ultra-Orthodox women have surprising educational and economic power. Women tend to have around 14 years of schooling, but men stop their secular education at 7th grade. Moreover, sociopolitical circumstances have led to the unique model of a "society of scholars" wherein men predominantly spend their time pursuing religious studies[17] while women are the breadwinners: the primary earners at full-time jobs outside their homes and communities.[18] Ultra-Orthodox women might be considered superwomen, since they are responsible for their families of, on average, seven children, and also supply most of the social needs of their communities.

The negative space of this religion-gender triangle illustrates important complexities within these established roles and relations, enabling us to carefully consider its structures and stigmas. The requirement in Israel for citizens to serve in the army forces Ultra-Orthodox men into full-time Torah studies to be exempt from military service, thus pushing them to be much more religious then they might otherwise have been. While the common stigma regarding Ultra-Orthodox women is that they are obedient to the men, the reality is that their jobs provide higher salaries than their husbands' stipends from the Kollel (religious studies institutes for married men).

From a conventional perspective, we could look at the gender issues within this religious community by focusing on women's hardships imposed by tradition. But a complex triangular and negative space approach enables us to see that there are more actors in the game. Relationships between men and women within the Ultra-Orthodox community are not only a consequence of their tradition but also derived from political and social aspects of larger secular-religious relations in Israel.

My examination of gender and religion in this particular community requires a short personal note. Feminist research has demonstrated the value of self-reflexivity in research via incorporating reflections and addressing implications of the scholar's experiences and position.[19] Therefore, my own complex academic and cultural positions are relevant to this part of the study. I self-define as a radical feminist and commandment-observing (*shomeret mitzvot*) Jewish woman, using neither the term "religious" nor "orthodox." I have many Ultra-Orthodox family members and friends, and I was raised in a Nationalist Ultra-Orthodox family and community. This particular sect is similar to the larger Ultra-Orthodox community in its strict interpretations of religious law but different from other groups in its National-Zionist perspective and attitude. For example, the men study Torah for many years, similar to the Ultra-Orthodox community, but they also fulfill their National-Zionist commitments by serving in the Israeli army. During my academic studies, I became a feminist and joined "Shira Hadasha" [new song], which is a feminist Orthodox community that enables women to take part in the service by leading the prayers and reading the Torah. Women and men are equal participants in the committee leading Shira Hadasha.[20] The complexity of my own experiences and commitments echo throughout my studies of Ultra-Orthodox women. From a practical perspective, my relationships give me relatively easy access to this close community. My appearance is that of a religious woman, as I cover my hair and wear long skirts, so Ultra-Orthodox women allow me to come to their houses and open their hearts and minds to me during interviews and participant observations. However, my feminist academic training, identity, and views help me to pay critical attention to multiple gendered aspects of community practices and consider them through concepts drawn from feminist communication scholarship.

Religion and new media among the Ultra-Orthodox community

The side of the triangle model that represents relationships between religion and media shows the complicated interactions of these aspects in daily life. Religious communities' perceptions of media are comprised of a combination of acceptance, rejection, appropriation, and adaptation.[21] These complicated relationships derive from the fact that orthodox people negotiate daily between their religious expectations and the values of the larger secular society.[22] New media characterized by individualism, autonomy, personal empowerment, and networking pose higher challenges than traditional media to the core values of religious communities: traditionalism, cultural preservation, collective identity, hierarchy, patriarchy, authority, self-discipline, and censorship.[23] However, many religious communities use media and even new media technologies to spread their religious ideologies.[24]

The complexities and multiple communal ways of coping with communication technologies are represented by diverse perceptions of media among the variety of the Ultra-Orthodox subgroups.[25] Today, the Israeli Ultra-Orthodox media's map includes a market of four daily newspapers, many magazines, three radio (no television) channels, and many Internet websites. Ultra-Orthodox media use, as in any other society, is a reflection of users' socioreligious-demographic locations.[26]

Significant Ultra-Orthodox websites include: (1) religious websites that spread religious contents, studies, information, and activities, such as *Chabad.Org*; (2) business websites that target the Ultra-Orthodox community, such as *Bizzness* (a name combining English and Yiddish); (3) strictly censured news websites, such as *JDN*; (4) Ultra-Orthodox radio stations' websites; and (5) popular Ultra-Orthodox news websites like *Behadrei Haredim* and *Kikar Hashbat*, both of which have unique spaces for women.[27] The triangle theory enables us to see that the Ultra-Orthodox new media use secular but "kosher" mediums to transmit religious and secular content to men and women alike.

Ultra-Orthodox website managers, writers, and entrepreneurs face multiple challenges. Many of them are members of the Ultra-Orthodox community, while they are also heavy Internet users and content producers.[28] They locate their religious and professional identities in a liminal and sensitive space between the most modern secular platform and traditional religious culture.[29] The Ultra-Orthodox websites and their managers are busy focusing on the tensions between traditional versus new media as well as religious versus secular life. Chabad is a good example to represent this fraught tension, and Chabad websites' senders use smart strategies to cope with it.[30] Some try to sanctify the impure Internet through simultaneously bringing religious content to secular people and calling on secular people to become closer to the religion in practice.[31] Studies of Ultra-Orthodox websites have found that they cleverly combine the offline and online communities.[32] The Ultra-Orthodox social networks users[33] thus enable themselves to discuss various religious and secular subjects.[34]

Research shows that less than half of Ultra-Orthodox adults use the Internet. Only 43% of the Ultra-Orthodox (ages 20+) are online (47% of the women, compared to 39% of the men).[35] The relationships between the Ultra-Orthodox community and the Internet are complex. Their perceptions of it lie on a continuum, ranging from viewing it as a Satanic tool that will destroy community values and undermining rabbinic authority[36] to treating it as a tool necessary for work to valuing it as vital for achieving unlimited opportunities to reach out to people and give them access to necessary religious information. On one end we can find very strict individuals and groups that do not use the Internet at all, while the other shows us varied and frequent online media usage. However, along this continuum, there are many pragmatic practices and solutions, since the Internet is necessary for the employment of many people. Therefore, some subgroups enable minimal use (only for business needs), while others extend to other purposes for limited use.[37] It should be mentioned that the Ultra-Orthodox that live out of Israel have different attitudes toward the Internet. They are usually not part of the "society of scholars," and most of them have jobs.[38]

Even though the Ultra-Orthodox ideological voice is still firmly against the Internet, the daily reality, as always, is much more complex: not everyone takes up a set standpoint on the continuum according to the rules. For example, not all of the community members strictly separate their business needs and their private needs and pleasures; people who are supposed to use the Internet only away from home may have one important homework assignment, and from this point forward, the home Internet connection stays forever. Comparing the complex reality of online media usage to Ultra-Orthodox opposition to television is also helpful, as the Internet can be a work tool, but television is viewed as entertainment. This difference is part of the reason that rabbinic opposition to the Internet did not succeed entirely.[39] Moreover, the ability to control these mediums is different, because the Internet does not need an antenna outside the home, allowing people to hide their use. Another explanation for the differences is the technical solution offered by a "kosher Internet." Commercial companies joined with the rabbinic committee to create a strictly filtered version of online access. This creates some levels of blocking with different codes that enable different access for parents and children in order to meet various needs, such as business and connection with family abroad. Even though the filtered Internet has—like all computer software—some breaches in the fence, control and access is more complicated than the yes or no options as with installing a television at home.

Ultra-Orthodox community members try to keep the smartphone out of bounds due to the unique threat its unlimited options pose to religious restrictions. Since these devices place the Internet in one's pocket, they open boundaries to the world and reduce social control.[40] The smartphone established a new front in the war Ultra-Orthodox leaders waged against the cell phone. They were being threatened by its small size and availability, both of which reduce possibilities for policing its usage. The cell phone enables hidden connections between men and women, either within the community or with people outside it. The other central threats were inappropriate and immodest content, as well as wasting of precious time.[41] The small size of cell phones and smartphones make them easily escape detection,[42] creating a new phenomenon worthy of another study: Ultra-Orthodox people who own a visible kosher phone as well as a secret smartphone.[43]

In recent years a new device has appeared, a kosher smartphone, with kosher apps and a symbol indicating kosher supervision. This device allows not only calling and receiving calls but also limited access to GPS, diaries, email, and a filtered Internet that can be used for specific business needs. The procedure for acquiring a kosher smartphone includes coming before a rabbinic committee to prove that the smartphone is necessary for one's job and fully accepting the committee's jurisdiction and control over the filtered apps and Internet surfing.

Ultra-Orthodox women that responded to my study's questionnaires[44] perceive the smartphone as a threat to community values; to the question "Do you think that mobile Internet browsing is in keeping with Ultra-Orthodox values?" 98% (41) of the Ultra-Orthodox answered "no." They explained this position with statements such as "It is full of values and messages that contradict those which underlie the education of the Ultra-Orthodox community" and "It can destroy your spiritual world." Some women perceive the smartphone as a danger, saying, "The smartphone is the most dangerous device, it includes newspapers, radio, television, and mail. It has impure content, everything is there, obscene pictures, connections with strangers on Facebook." "Creating contacts is totally adultery," said another.

Triangulation and attention to negative space enable us to see, again, that issues surrounding new media are not simple, and relationships between a religious community and new media are not obvious. However, there are multiple complex junctions between the secular and religious influences that can be explored via a continuum of media and new media. We can conclude here that new media and religion, and the negative spaces created around traditional media and secularity are almost inseparable. The Ultra-Orthodox audience strongly reflects these tensions. The ideal of the "holy Jewish home" is key for understanding this complexity. This central value is based on the separation between the sanctified and religious home, compared to the profane and secular street. One of the common phrases I heard through my studies is "this media enters my home," compared to "this media can't enter my home." The media from the Ultra-Orthodox perspective is not judged according only to questions of new and modern technology, but from the perspective of the secular content that can penetrate the sanctified home from the street.

Gender and media among the Ultra-Orthodox community

Moving forward to the next side in the triangle enables us to focus on the connections between new media and gender. There is one comparative study about gender and media among the Ultra-Orthodox community in Israel that is important to consider for this analysis.[45] This part of my research compared approaches to reading Ultra-Orthodox newspapers and magazines and found that readers shared various strategies, from what Stuart Hall termed oppositional and critical interpretation to a dominant reading.[46] There are studies about Ultra-Orthodox men as the senders of new media[47] (as mentioned above), but we lack research examining male members of the community from an audience perspective for either traditional or new media.

Ultra-Orthodox traditional media excluded women as senders but included them as receivers. The main reason for this exclusion/inclusion is the strict dictates of modesty values. Women cannot talk on the community's radio because the men are not supposed to hear their voices.

Women can write in the Ultra-Orthodox press, but mostly do this under male pseudonyms. However, new media provides senders the chance to remain anonymous and hidden behind a keyboard, offering significant modes for change that enable Ultra-Orthodox women to voice their experiences and ideas.[48]

Content analysis of Ultra-Orthodox women's forums and blogs is an excellent way to see how they use new media as a means of empowerment. In these digital media, they are discussing issues that usually are taboo—such as the female body and purity rituals.[49] American Ultra-Orthodox women bloggers blur the limits between acceptable practices in public and private spaces, and their writing undermines traditional communal gender rules.[50]

The suggested triangulation and negative space shows that in contrast to the traditional expectation for men to control women's media consumption, the Ultra-Orthodox situation uniquely empowers these women. After a long period of exclusion from traditional media, women's out-of-home jobs enable them to use new media for various needs. Many of them not only get much higher salaries than their husbands, but also have higher technological skills that give them more power and control than in the past.

Gender, religion, and media among the Ultra-Orthodox community: triangulation and negative space

When looking at Ultra-Orthodox women as an audience exploring new media, it is important to attend to the ambivalence and multiplicity of their engagement with it. Five Haredi women who regularly use the Internet for work and leisure described their justification for the use of technology and the tension between their media usage and rabbinical views.[51] Members of closed online forums designed for Ultra-Orthodox women regard the Internet as a significant factor in their lives and a source of empowerment, but also note the danger it poses to their lifestyle traditions. The women use modern technology (whose legitimacy is questionable within their community) as a space in which they can communicate anonymously with similarly situated women about shared themes and concerns.[52] Their attitudes address the contrasts between Ultra-Orthodox ideology and daily practices, describe the opportunities and the dangers, and explore their means of coping with demanding jobs and vast families. Research participants responded in ways that illustrate key aspects of the relationships between this community and new media.

Ultra-Orthodox women employed in digital environments reported that on the ideological level, they are working to help support their families and keep their scholar–husbands within the boundaries of Ultra-Orthodox society. However, on the practical level, greater accessibility of these technologies (even only outside the home) widens the women's use patterns and enables them to see the technology's advantages.[53]

The Ultra-Orthodox women's attitudes and behaviors place them on a continuum between gatekeepers (trying hard to negotiate the positive and negative spaces and to glorify any non-use as a form of social and religious capital among the community) and change-agents (noting the practical advantages of the new media).

Living the ideology: the gatekeepers

Research indicates that most Ultra-Orthodox women perceive the Internet as a danger to their religiosity. Lev-On and Neriya-Ben Shahar showed that 92% (141) of the Ultra-Orthodox women they surveyed agreed with the statement: "I think that the Internet can weaken people religiously" (average agreement = 4.36, SD = 1.01).[54] Ninety percent (137) of the women agreed with the statement: "I think the Internet constitutes a danger to the Haredi lifestyle" (average agreement: 4.18, SD = 1.08). Seventy-five percent (116) agreed with the statement: "I think the Internet is as dangerous as the cell phone because it enables contacting other people" (average agreement = 3.74, SD = 1.41). Ninety-five percent (149) agreed with the statement: "I think the Internet is as dangerous as television because it enables people to view and hear prohibited content" (average agreement = 4.52, SD = 0.89).

This concern is also reflected in a related study by responses to the question "Do you think that browsing the Internet is in keeping with Ultra-Orthodox values?" 95% (40) of the Ultra-Orthodox women answered "No" to this query.[55] The women that responded to my questionnaires emphasized the role of social norms in their explanations with statements like "Internet use is not appropriate to ultra-Orthodox society" and "When the rabbis prohibit, we believe and do not ask questions."

Some examples from interviews with Ultra-Orthodox women clarify how access to online media is restricted. When a teacher and mother of 10 was asked this question, she stated,

> I heard about a woman that just got married, she had been accepted for a new job. She asked if they can block her internet connection, and the boss did not agree. She said that this is not her job and came back home.

Another teacher who has eight children reported,

> my husband took the new computer to a specialist because it came with an internet connection. I am not sure what the specialist did—but there is no option we can get access to the internet, without getting back to this specialist. He blocked it.

Three statements positioned on a continuum formulated from strict to permissive can summarize the Ultra-Orthodox women's perceptions of the rabbis' attitude toward the Internet.[56] Forty-eight percent (73) of respondents agreed with the moderate statement: "To the best of my knowledge, the rabbis in my social milieu permit the use of the Internet when needed" (average agreement = 2.61, SD = 1.28). Seventy-eight percent (120) of the respondents agreed with the intermediate statement (average agreement = 3.49, SD = 1.13), "From what I understand, the rabbis in my social milieu permit the use of the Internet for work purposes only." Thirty percent (46) of respondents agreed with the strictest statement: "To the best of my knowledge, the rabbis in my social milieu do not permit the use of the Internet for any purpose whatsoever" (average agreement = 2.37, SD = 1.27). These results represent a separated level of agreement for each Likert scale's statement. The overlapping of agreement among such divergent statements reflects the multiplicity and complexity of the women's perceptions.

The gatekeeping attitude also is illustrated in interviews with Ultra-Orthodox women who don't have Internet access. In an interview, a woman from the Gur Hassidic group who is a stay-at-home mother of five said:

> My husband and I had a real discussion and clarification for ourselves. This is not only interaction power […] We are Gur Hasidic, what does it mean? Why do we invest so much effort? […] we understood that the basis for life, for our way, is the obedience and discipline to the rabbis. […] before I got married, I had to think alone about all of these things. Now we are together, and it is clear for us that our basis for our home is obedience and submission to the rabbis […] If they said "no" for technology—it is no.

An Ultra-Orthodox woman interviewee who is a teacher and mother of five explained how the smartphone forces the gatekeepers to take a step farther.

> My husband said that the rabbi told that a person said to him that it is so easy to use the smartphone to get to the bank account, why should I spend half an hour to go to the bank? The rabbi asked how much time to get out of the phone after you entered it [the bank's application]. It is better to travel to the bank and back, and not to [go through] the phone, with everything [such as the potential for wasting time reading and replying to messages].

I asked my Ultra-Orthodox relative if she would like to see her parents, who live in Europe, through the Skype app in my smartphone. "They did not see your seven children for a long time," I said. "Here you have an opportunity to talk to them, and you can see them." Her eyes sparkled, then she said, "One moment, I have to ask my husband." She called him. "This is a Satan tool, and it is impure. We do not want to use it even to see the family," he said, so she gave up.

This event shows that while my relative is a gatekeeper who is invested in following the strictest rules of her community, she allowed herself to *consider* using the smartphone, even just for this one opportunity. The negative space of this forbidden yearning illustrates the borders these women must navigate, asking gently to move between tradition and new media via a gendered discussion between a husband and wife. The hard line they try to create between the secular "Satan tool" and the pure religious family won in this case.

A kindergarten teacher and mother of nine talked during an interview about her husband's strict gatekeeping rules and then about herself:

> My husband never looked on it [the smartphone]. Some week ago, came a plumber here to fix something. He wanted to show my husband a fire and firefighters by his device. My husband did not look. The plumbers said: Aren't you interested? My husband answered: It is very interesting, and I am curious, but I try not to look. The plumber said: you are fortunate! [...] It has also happened to me—I organized a summer camp, and somebody from the police came to see if I have all the documents. He asked me to sign on his smartphone; he wanted me to sign my name in this bomb! I said—I will not touch it! He said—I am a policeman, and you have to sign! I called the security guard of the neighborhood; I said to him: I gave you all the documents, I agree to sign, but on a paper. He said—the papers might be lost. Then he said that he would sign for me. The policemen signed for me and said: "you are fortunate and good for you"—and he runs away ....

This woman's stories illustrated the triangle interestingly. While the man and the women practiced gatekeeping by actively keeping to religious rules, a subtle feminist reading shows the woman as a hero. For her, the smartphone was a job requirement, and not just for her own curiosity. Moreover, there is a difference between negotiating the values with a policeman and with a plumber. The policeman represents the law and has the ability to inflict real punishment: he could take the license from the summer program if the leader refused to sign her name on his smartphone. The plumber is a simple worker that only wanted the men to see something interesting (but not important) on his phone.

Even though these women are gatekeepers, they have a rich discourse about new media. Mostly, they discuss issues and concerns with their husbands, within the context of religiosity and secularity, wherein the rabbi's voice echoes as a symbol of religious authority. Although they ultimately follow the rabbi's directions, the shared negotiation of new media and only eventual submission to leaders is something new that should perhaps be considered as a feminist development in this community.

Living the practice: ideology of non-use and practices of limited uses

Practically, most of the Ultra-Orthodox women use the Internet for their jobs and family needs. They use email, Skype, and websites for news, photos, shopping, searching, government information, recipes, paying bills, bus timetables, dictionaries, and so on.[57] Despite the conflict between these practices and Ultra-Orthodox views of new media, many of them do not define themselves as change agents but just as women trying to manage daily needs.

The next voices speak to how clear gender boundaries combine with new media needs and shared practices of life and work. A storekeeper and mother of 10 explained in an interview how she gets what she needs from the Internet, despite all of the limits: "My husband is a manager of a department, so he must use the Internet. He has the highest blocking. When I needed something, I went to his office, and he told me what to do." A teacher and mother of eight recounted that her job requires her to use the Internet for necessary documents but she doesn't want Internet access at home and her husband is a Torah scholar: "Near my husband's Yeshiva there is a computer for rent for 10 minutes [which is blocked—RNBS], he pays every 10 minutes. So, my husband runs my job's emails."

These two women described two of the three control mechanisms the Ultra-Orthodox community adhered to: (1) technical control afforded by content censorship and (2) out-of-the-home use—only in public places or the workplace. Other women, especially those who have Internet at home, address (3) self-control. Ultra-Orthodox community members use these tools to separate home from work needs, filtered from open access content, and work from personal usage. Their perceptions of the rabbis' directions enable them to navigate these distinctions without losing legitimation from the community.[58]

Paradoxically, looking closely at a community that tries very hard to build firm boundaries shows that there are no real boundaries in the contemporary world. There is no longer a real separation between home and work; the very strictly blocked Internet still enables people access to places they never could go previously, and traditional divisions between men and women have changed because of women's jobs out of the home and out of the community. Data from Lev-On and Neriya-Ben Shahar also show this complexity: Sixty-five percent (97) of the women agreed with the statement: "I think that I am capable of controlling the content to which I am exposed online" (average agreement = 3.12, SD = 1.31).[59] Still, 73% (108) agreed with the practical statement: "It happens that I am exposed to inappropriate content online (unintentionally)" (average agreement = 3.5, SD = 1.35).[60]

Additional responses from my research questionnaires described the inherent complexity of new media:

> The internet could be great and special, but it is too bad that they use this power to destroy the good and gentle in the human being. In blocked internet there is a control, but in practice we could see many that fall, and their head is not in the right place; It's such a useful tool, and simultaneously destroyer and ravager; It could be an efficient medium if it wouldn't be a garbage of filth and unlimited and non-filtered information.

These remarks show what Tova Hartman[61] called "strong multiplicity," defined as "insufficiently developed interpretive lexicon within postmodern narrative analysis [... for] finding legitimate expressions of identity among seemingly inconsistent self-representations." This multiplicity is one of the reasons for the existence of a "third-person effect" in Ultra-Orthodox women's attitudes toward perceived dangers originating from the Internet. As mentioned above, this effect occurs when people perceive that media messages are significantly more influential on others than they are on themselves.[62] In other words, these women perceive themselves as less vulnerable to the hazardous potential of the Internet than other people in their community.[63]

Taking triangulation and negative space into account, we can see that Ultra-Orthodox women fear the secular penetration of new media into their religious community. They perceive this new threat as higher than that of traditional media, and because of the third-person effect, they perceive themselves as strong enough to handle the danger. The role of gender on these perspectives is reflected when we asked about "other people" (either men or women). During the interviews I conducted with women, many of them mentioned that they want to defend the men, who they view as easily influenced by bad content. Therefore, this effect shows the women's perception that they control not only the traditional but also the new media, since they are stronger and perhaps even more religious than the men.

I should mention here that in contrast to these results, another study (currently in preparation) that I conducted regarding the third person effect among Old Order Amish and Ultra-Orthodox women found that they don't have the third person effect. There are two explanations for this contradiction. First, the published study[64] is based on data from 209 Ultra-Orthodox women, compared to the Amish/Ultra-Orthodox research that included only 42 Ultra-Orthodox women. The second (and maybe more important) explanation is that while the published study focused only on Ultra-Orthodox women that use computers and the Internet for their jobs and/or closed forums, the Amish/Ultra-Orthodox study included women from various occupations, many of whom work for the Ultra-Orthodox education system and don't use the Internet at all. Therefore, their non-use is important to note since they have no need to develop the third person effect.

Conclusion: gender-religion-new media ?

The Ultra-Orthodox community provides a case study that can contribute to our understanding of the triangular relationships between religion, gender, and media. By studying this community for more than 15 years, I found that the complexity of the suggested model allowed for an account of how their unique and paradoxical characteristics work together and in tension with one another.

In terms of religious practices, the Ultra-Orthodox are an ostensibly enclaved pious community that lives separately from modern secular culture. However, their multiple connections with their surroundings enable us to see that there is no real dichotomy between religiosity and secularity, even among members of the most traditional community.

The gender side of this community is also intriguingly complex, given the rare phenomenon (in the contemporary world) of a culture in which the women are the breadwinners and the men invest their time and power only in religious studies. The women tend to have higher levels of non-religious education than men: many of them have academic degrees, while the men usually finish their non-religious studies at about age 13. At the same time, this is a non-feminist community. The women claim that the men are the heads of the families while the women are only helping them to study Torah.[65]

Looking at the media angle, the strong religious community tries to control its members' uses of every sort of media but does not succeed with the most dangerous (from their perspective) technology: the Internet. Specific technological features, the ability to hide usage, and the necessities of employment in contemporary job markets can thus undermine key aspects of the entire

Ultra-Orthodox society. Men's efforts to limit women's media usage confronted the fact that many women use the Internet for their jobs, and due to this necessity, control becomes harder to maintain. These challenges as well as the creativity of the solutions found to meet changing needs make the influence of media in Ultra-Orthodox gender roles an interesting example that enables other scholars to draw new comparisons and distinctions.

The complicated relationships at the intersections between gender, religion, and new media necessarily entail changing boundaries between women and men, religion and secularization, traditional and new media. Triangulation of these factors demonstrates how the women take advantage of opportunities to use secular and religious media and new media, and they also make decisions about family consumption of a variety of new and traditional media.[66] A multilayered approach makes it clear that we cannot understand these women's lives without examining men's practices or religious isolation without tracking secular influences—and we must consider the role media plays in these relationships. Complex analysis also highlights the particular opportunities and forms of empowerment that Ultra-Orthodox women can create in this cultural and technological context.

Overall, my findings contribute to a broad field for emerging research. Even within the specific community studied, we can find numerous avenues for scholars to explore. For example, women's spaces in the popular Ultra-Orthodox websites, such as *Behadrei Haredim*, provide insight into busy women with high levels of education and various jobs that have diverse interests—from art to politics and from cooking to business. However, these spaces include particular commercial targeting of these women that has yet to attract much scholarly attention. Moreover, comparisons and contrasts between the Ultra-Orthodox community in Israel and those found in the United States and Europe (or other strict religious communities like the Amish as well as other Christian and Muslim groups around the world) are fertile fields for future studies.[67]

Shifting the spotlight from binary categories and moving forward toward multiple spheres and continuums may be difficult. Intersectionality in social science asks us to do something challenging. While we are accustomed to focusing on only one thing a time, training enables taking in multiple points of view simultaneously. One of the difficulties of complex analysis is the need to consider the full contexts for new and old phenomenon, such as social changes and power systems, but such understanding is both necessary and rewarding.

Taking such nuanced approaches, every new study that focuses on gender, religion, and new media will add to the sides and angles of this triangle, filling in its negative spaces and elaborating its continuums. On a personal note, I have found that my unique position as a strictly religious and radical feminist who studies media enables me to live within this challenging multiplicity. This chapter is an invitation for every scholar first to see the surprising complexities (whether they be represented as triangles, squares, octagons—and the negative spaces surrounding them) and then to implement challenging tensions into their studies, moving research forward with unexpected insights. I hope that one day the complex relations and tensions of gender-religion-new-media (or: *gendereligionewmedia*) will be indivisible and inseparable just like scholarship linking previously distinct inquiries into geopolitics and socioeconomic studies and scholarship.

Notes

1 Ruth Alas, "The Triangle Model for Dealing with Organization Change," *Journal of Change Management* 7, nos 3–4 (2007): 255–71; Simon Bernd and Bert Klandermans, "Politicized Collective Identity: A Social Psychological Analysis," *American Psychologist* 56, no. 4 (2001): 319–31.

2 Mia Lovheim, "Introduction: Gender – a Blind Spot in Media, Religion, and Culture?" in *Media, Religion and Gender: Key Issues and New Challenges*, ed. Mia Lovheim (London: Routledge, 2013), 16.

3 Clifford Geertz, *The Interpretation of Cultures* (New York: Basic Books, 1973), 90.

4 Heidi Campbell, *When Religion Meets New Media* (London: Routledge, 2010), 9.

5 Stephen A. Buetow, "Something in Nothing: Negative Space in the Clinician-Patient Relationship," *Annals of Family Medicine* 7, no. 10 (2009): 80–83.

6 Elizabeth Rosenblatt, "Intellectual Property's Negative Space: Beyond the Utilitarian," *Florida State University Law Review* 40, no. 3 (2013): 447.

7 Orit Avishai, Afshan Jafar, and Rachel Rinaldo, "A Gender Lens on Religion", *Gender and Society* 29, no. 1 (2015): 5–25.

8 José Casanova, "Rethinking Secularization: A Global Comparative Perspective after Secularization," *Hedgehog Review: Critical Reflections on Contemporary Culture* 8, nos 1–2 (2006): 7–22; José Casanova, "Religion, Secularizations and Modernities," *European Journal of Sociology* 52, no. 3 (2011): 425–45.

9 Lee Cahaner, Gilad Malach, and Maya Choshen, *The Yearbook of Ultra-Orthodox society in Israel* (Jerusalem: The Israel Democracy Institute, 2017).

10 Kimmy Caplan, *The Internal Popular Discourse in Israeli Haredi Society* (Jerusalem: Zalman Shazar, 2007) [Hebrew]; Michael K. Silber, "The Emergence of Ultra-Orthodoxy: The Invention of a Tradition," in *The Uses of Tradition: Jewish Continuity in the Modern Era*, ed. Jack Wertheimer (New York-Jerusalem: JTS, 1992), 23–84.

11 Benjamin Brown, *The Haredim* (Tel Aviv: Am Oved, 2017) [Hebrew].

12 Simeon D. Baumel, *Sacred Speakers: Language and Culture Among the Haredim in Israel* (New York: Berghahn Books, 2006); Samuel C. Heilman and Menachem Friedman, "Religious Fundamentalism and Religious Jews: The Case of the Haredim," in *Fundamentalisms Observed*, eds Martin E. Marty and Scott Appleby (Chicago: The University of Chicago Press, 1991), 197–264; Menachem Friedman, *Haredi Society* (Jerusalem: Jerusalem Institute for Israel Studies, 1991); Emanuel Sivan, "Enclaved Culture," *Alpayim Magazine* 4 (1991): 45–99 [Hebrew]; Orit Yafeh, "The Time in the Body: Cultural Construction of Femininity in Ultra-Orthodox Kindergartens for Girls," *Ethos* 35, no. 4 (2007): 516–33; Sima Zalcberg, "Grace is Deceitful, and Beauty is Vain: How Hasidic Women Cope with the Requirement of Shaving One's Head and Wearing a Black Kerchief," *Gender Issues* 24, no. 3 (2007): 13–34.

13 Lee Cahaner and Haim Zicherman, *Modern Ultra-Orthodoxy: The Emergence of the Haredi Middle Class in Israel* (Jerusalem: The Israel Democracy Institute, 2012); Kimmy Caplan and Nurit Stadler, "Introduction," in *From Survival to Consolidation: Changes in Israeli Haredi Society and its Scholarly Study*, eds Kimmy Caplan and Nurit Stadler (Jerusalem: Van Leer Institute, 2012) [Hebrew]; Emanuel Sivan and Kimmy Caplan, "Introduction," in *Israeli Haredim: Integration Without Assimilation?*, eds Emanuel Sivan and Kimmy Caplan (Jerusalem: Jerusalem Institute for Israel Studies, 2003) [Hebrew].

14 Rivka Neriya-Ben Shahar, "*Haredi (Ultra-Orthodox) Women and Mass Media in Israel – Exposure Patterns & Reading Strategies*," (PhD diss., The Hebrew University of Jerusalem, 2008); Rivka Neriya-Ben Shahar, "The Learners' Society: Continuity and Change in Characteristics of Education and Employment among Ultra-Orthodox (Haredi) Women," *Sociological Papers* 14 (2009): 1–15.

15 Phillips W. Davison, "The Third-Person Effect in Communication," *Public Opinion Quarterly* 47, no. 1 (1983): 1–15.

16 Lea Taragin-Zeller, "Between Modesty and Beauty: Reinterpreting Female Piety in the Israeli Haredi Community," in ed. Fishman Barack, *Sylvia, Love, Marriage, and Jewish Families Today: Paradoxes of the Gender Revolution* (Waltham, MA: Brandeis University Press, 2015), 308–326; Lea Taragin-Zeller, "Modesty for Heaven's Sake: Authority and Creativity among Female Ultra-Orthodox Teenagers in Israel, Nashim," *A Journal of Jewish Women's Studies & Gender Issues* 26, Spring 5774 (2014):75–96.

17 Friedman, *Haredi Society*; Nurit Stadler, *Yeshiva Fundamentalism* (New York: New York University Press, 2009); Nurit Stadler, *A Well-Worn Tallis for a New Ceremony: Trends in Israeli Haredi Culture* (Brighton, MA: Academic Studies Press, 2012).

18 Tamar El-Or, *Educated and Ignorant: Ultra-orthodox Jewish Women and Their World* (Boulder, CO: Lynne Rienner Publishers, 1994); Tamar El-Or, "Ultra-Orthodox Jewish Women," in *Israeli Judaism: The Sociology of Religion in Israel*, ed. Shlomo Deshen (New Brunswick: Transaction, 1995), 149–69; Tamar El-Or, "Visibility and Possibilities: Ultra-Orthodox Jewish Women between the Domestic and Public Spheres," *Women's Studies International Forum* 20, no. 5 (1997): 665–73; Neriya-Ben Shahar, "*Haredi Women*".

19 Anna Piela, "Claiming Religious Authority: Muslim Women and New Media," in *Media, Religion and Gender: Key Issues and New Challenges*, ed. Mia Lovheim (London: Routledge, 2013), 252–82; Mary E. Hess, "Digital Storytelling: Empowering Feminist and Womanist Faith Formation with Young Women," in *Media, Religion and Gender: Key Issues and New Challenges*, ed. Mia Lovheim (London: Routledge, 2013), 336–62.

20 "Shira Hadasha".Accessed January 20, 2020, https://shirahadasha.org;Tova Hartman, *Feminism Encounters Traditional Judaism: Resistance and Accommodation* (Waltham, MA: Brandeis University Press, 2007).

21 Campbell, *When Religion*; Stuart M. Hoover and Knut Lundby, "Introduction," in *Rethinking Media, Religion and Culture*, eds Stuart M. Hoover and Knut Lundby (Thousand Oaks, CA: Sage, 1997), 3–14; Donald B., Kraybill, Karen M. Johnson-Weiner, and Steve M. Nolt, *The Amish* (Baltimore, MD: Johns Hopkins University Press, 2013); Daniel A. Stout, *Media and Religion: Foundation of an Emerging Field* (New York: Routledge, 2012); Daniel A. Stout, "Secularization and the Religious Audience: A study of Mormons and Las Vegas Media," *Mass Communication & Society* 7, no. 1 (2004): 61–75; Daniel A. Stout, *Media and Religion* (New York: Routledge, 2012).

22 Peter Berger, *The Sacred Canopy: Elements of a Sociological Theory of Religion* (Garden City, NJ: Anchor Books, 1967).

23 Heidi Campbell, "Challenges Created by Online Religious Networks," *Journal of Media and Religion*, 3, no. 2 (2004): 81–99; Heidi Campbell, "'What Hath God Wrought': Considering How Religious Communities' Culture (or Kosher) the Cell Phone," *Continuum: Journal of Media and Cultural Studies* 2, no. 2 (2007): 191–203; Heidi Campbell, "Who's Got the Power? The Question of Religious Authority and the Internet," *Journal of Computer-Mediated Communication* 12, no. 3 (2007): 1043–62; Heidi Campbell, "Religion and the Internet in the Israeli Orthodox Context," *Israel Affairs* 17, no. 3 (2011): 364–83; Mia Lovheim and Heidi Campbell, "Considering Critical Methods and Theoretical Lenses in Digital Religion Studies," *New Media and Society* 19, no. 1 (2017): 5–14.

24 Heidi Campbell, *Digital Religion: Understanding Religious Practice in New Media Worlds* (London: Routledge, 2013); Heidi Campbell, "Studying Jewish Engagement with Digital Media and Culture," in *Digital Judaism: Jewish Negotiations with Digital Media and Technology*, ed. Heidi Campbell (New York: Routledge, 2015), 1–15; Orly Tzorfati and Dotan Bleis, "Between 'Enclave Culture' and 'Virtual Enclave': The Ultra-Orthodox Community and the Digital Media," *Kesher* 32 (2002): 47–56 [Hebrew].

25 Menachem Blondheim, "The Jewish Communication Tradition and its Encounter with New Media," in *Digital Judaism: Jewish Negotiations with Digital Media and Technology*, ed. Heidi Campbell (New York: Routledge, 2015), 16–39; Caplan, *The Internal*.

26 Neriya-Ben Shahar, "*Haredi Women*".

27 Rephael Mann, "The Ultra-Orthodox Media 2016: Between Conservative and Modernization," in *the Annual Report: The Mass Media in Israel*, ed. Rephael Mann and Azi Lev-On (Ariel: Ariel University Press, 2016), 91–105.

28 Nakhi Mishol-Shauli, *Fundamentalist Knowledge Brokers: New Media Journalists as Agents of Informal Education Among the Ultra-Orthodox Community in Israel* (Master's diss., Haifa University, 2016) [Hebrew].

29 Campbell, "Religion and the Internet"; Oren Golan, "Re-Constructing Religious Boundaries Online: The Emergence of the Denominational Jewish Internet," *La Temps Des Medias* no. 17 (2011): 118–19.

30 Oren Golan, "Charting Frontiers of Online Religious Communities: The Case of Chabad Jews," in *Digital Religion: Understanding Religious Practice in New Media Worlds*, ed. Heidi Campbell (London: Routledge, 2013), 155–63; Oren Golan, "Legitimation of New Media and Community Building Amongst Jewish Denominations in the US," in *Digital Judaism: Jewish Negotiations with Digital Media and Technology*, ed. Heidi Campbell (New York: Routledge, 2015), 125–44; Oren Golan and Nurit Stadler, "Building the Sacred Community Online: The Dual Use of the Internet by Chabad," *Media, Culture and Society* 38, no. 1 (2016): 71–88; Oren Golan and Heidi Campbell, "Strategic Management of Religious Websites: The Case of Israel's Orthodox Communities," *Journal of Computer-mediated Communication* 20, no. 4 (2015): 467–86; Tsuriel Rashi and Maxwell McCombs, "Agenda Setting, Religion, and New Media: The Chabad Case Study," *Journal of Religion, Media and Digital Culture* 4, no. 1 (2015): 128–53.

31 Heidi Campbell and Wendi Bellar, "Sanctifying the Internet: Aish's Use of the Internet for Digital Outreach," in *Digital Judaism: Jewish Negotiations with Digital Media and Technology*, ed. Heidi Campbell (New York: Routledge, 2015), 74–90; Sharrona Pearl, "Exceptions to the Rule: Chabad-Lubavitch and the Digital Sphere," *Journal of Media and Religion* 13, no. 3 (2014): 123–37.

32 Yoel Cohen, "Media Events, Jewish Religious Holidays, and the Israeli Press," in *Global Perspectives on Media Events in Contemporary Society*, ed. Andrew Fox (Hershey, PA: IGI Global, 2016), 122–31; Yoel Cohen, "Holy Days, News Media, & Religious Identity: A Case Study in Jewish Holy Days and the Israeli Press," in *Spiritual News: Reporting Religion Around the World*, ed. Yoel Cohen (New York: Peter Lang, 2017), 303–22.

33 Karine Barzilai-Nahon and Gad Barzilai, "Cultured Technology: The Internet and Religious Fundamentalism," *The Information Society* 21, no. 1 (2005): 25–40.

34 Sarit Okun, "Exploring the Characteristics of an Online Ultra-Orthodox Jewish Community," (Master's thesis), Ben-Gurion University of the Negev, 2015; Sarit Okun, "Identity Games in the Online Ultra-Orthodox Religious Community," *Heker Hahevra Ha'Haredit. [The Journal for the Haredi Society Research]*, 3 (2016): 56–77; Sarit Okun and Galit Nimrod, "Online Ultra-Orthodox Religious Communities as a Third Space: A Netnographic Study," *International Journal of Communication* 11 (2017): 2825–41.

35 Cahaner, Malach, and Choshen, *Yearbook*.

36 Blondheim, "Jewish Communication"; Campbell, *When Religion*; Campbell, "Religious Engagement"; Campbell, "Religion and the Internet"; Campbell, "Studying Jewish"; Heidi Campbell and Oren Golan, "Creating Digital Enclaves: Negotiation of the Internet Amongst Bounded Religious Communities," *Media, Culture & Society* 33, no. 5: 709–24.

37 Campbell, "Religion and the Internet"; Campbell, *When Religion*.

38 Ayala Fader, "Nonliberal Jewish Women's Audiocassette Lectures in Brooklyn: A Crisis of Faith and the Morality of Media," *American Anthropologist* 115, no. 1 (2013): 72–84; Nethanel Deutsch, "The Forbidden Fork, the Cell Phone Holocaust, and Other Haredi Encounters with Technology," *Contemporary Jewry* 29, no. 1 (2009): 3–19.

39 Yoel Cohen, "Israeli Rabbis and the Internet," in *Digital Judaism: Jewish Negotiations with Digital Media and Technology*, ed. Heidi Campbell (New York and London: Routledge, 2015), 183–204; Yoel Cohen, "Judaism in the Computer-Mediated Era," in *Changing Cultures and Religious Practices in Asia,* ed. B. Agrawal (Manila: University of Santo Tomas, 2015), 131–40.

40 Deutsch, "The Forbidden Fork".

41 Campbell, "What Hath"; Campbell, "Who's Got"; Campbell, *When Religion*; Akiba A. Cohen, Dafna Lemish, and Amit Schejter, *The Wonder Phone in the Land of Miracles* (New York, NY: Hampton Press, 2008); Deutsch, "The Forbidden Fork"; Tsuriel Rashi, "The Kosher Cell Phone in Ultra-Orthodox Society: A Technological Ghetto Within the Global Village?" in *Digital Religion: Understanding Religious Practice in New Media Worlds,* ed. Heidi Campbell (Philadelphia: Routledge, 2013), 173–81; Hananel Rosenberg, Menachem Blondheim, and Elihu Katz, "The Wall-Keepers: Social Enclaves, Social Control and the Kosher Cellphone Campaign in Jewish Ultra-Orthodox Society," *Israeli Sociology* 17, no. 2 (2016): 116–37 [Hebrew]; Hananel Rosenberg and Tsuriel Rashi, "Pashkevillim in Campaigns Against New Media: What Can Pashkevillim Accomplish that Newspapers Cannot?" in *Digital Judaism: Jewish Negotiations with Digital Media and Technology*, ed. Heidi Campbell (London: Routledge, 2015), 161–82.

42 Campbell, *When Religion*.

43 Campbell, *When Religion*; Deutsch, "The Forbidden Fork".

44 Rivka Neriya-Ben Shahar, "'Mobile Internet is Worse than the Internet; it Can Destroy Our Community': Old Order Amish and Ultra-Orthodox Women Responses to Cellphone and Smartphone Use," *The Information Society* 36, no. 1 (2020): 1–18.

45 Neriya-Ben Shahar, "Haredi (Ultra-Orthodox) Women".

46 Stuart Hall, "Encoding/Decoding," in *Culture, Media, Language: Working Papers in the Cultural Studies 1972-79*, eds S. Hall, D. Hobson, A. Lowe, and P. Willis (London: Hutchinson & Co, 1981), 128–38.

47 Golan, "Charting Frontiers"; Golan, "Legitimation of New Media".

48 Hess, "Digital Storytelling".

49 Judy Tydor Baumel-Schwartz, "Frum Surfing: Orthodox Jewish Women's Internet Forums as a Historical and Cultural Phenomenon," *Journal of Jewish Identities* 2, no. 1 (2009): 1–30.

50 Andrea Lieber, "A Virtual Veibershul: Blogging and the Blurring of Public and Private Among Orthodox Jewish Women," *College English* 72, no. 6 (2010): 621–37.

51 Oren Livio and Karen Tenenboim-Weinblatt, "Discursive Legitimation of a Controversial Technology: Ultra-Orthodox Jewish Women in Israel and the Internet," *The Communication Review* 10, no. 1 (2007): 29–56.

52 Azi Lev-On and Rivka Neriya-Ben Shahar, "A Forum of Their Own: Views About the Internet Among Ultra-Orthodox Jewish Women Who Browse Designated Closed Forums," *First Monday* 16, no. 4 (2011).

53 Lev-On and Neriya Ben-Shahar, "A Forum of Their Own".

54 Lev-On and Neriya Ben-Shahar, "A Forum of Their Own".

55 Rivka Neriya-Ben Shahar, "Negotiating Agency: Amish and Ultra-Orthodox Women's Responses to New Media," *New Media and Society* 19, no. 1 (2017): 81–95.

56 Lev-On and Neriya-Ben Shahar, "A Forum of Their Own".

57 Neriya-Ben Shahar, "Negotiating Agency".

58 Neriya-Ben Shahar, "Negotiating Agency".

59 Lev-On and Neriya Ben-Shahar, "A Forum of Their Own".
60 Lev-On and Neriya Ben-Shahar, "A Forum of Their Own".
61 Tova Hartman, "Strong Multiplicity: An Interpretive Lens in the Analysis of Qualitative Interview Narratives," *Qualitative Research* 15, no. 1 (2015), 22.
62 Davison, "The Third-Person Effect".
63 Azi Lev-On and Rivka Neriya-Ben Shahar, "To Browse, or Not to Browse? Perceptions of the Danger of the Internet by Ultra-Orthodox Jewish Women," in *New Media and Intercultural Communication*, eds Pauline Hope-Cheong, Judith N. Martin, and Lea P. Macfadyen (New York: Peter Lang, 2012), 223–36.
64 Lev-On and Neriya-Ben Shahar, "To Browse".
65 Neriya-Ben Shahar, "*Haredi Women*".
66 Neriya-Ben Shahar, "*Haredi Women*"; Neriya-Ben Shahar, "Negotiating Agency"; Rivka Neriya-Ben Shahar, "The Medium Is the Danger: Discourse About Television Among Amish and Ultra-Orthodox (Haredi) Women," *Journal of Media and Religion* 16, no. 1 (2017): 27–38.
67 Neriya-Ben Shahar, "Negotiating Agency"; Neriya-Ben Shahar, "The Medium".

Bibliography

Alas, Ruth. "The Triangle Model for Dealing with Organization Change." *Journal of Change Management* 7, nos 3–4 (2007): 255–71.

Avishai, Orit, Jafar, Afshan, and Rinaldo, Rachel. "A Gender Lens on Religion." *Gender and Society* 29, no. 1 (2015): 5–25.

Barzilai-Nahon, Karine, and Barzilai, Gad. "Cultured Technology: The Internet and Religious Fundamentalism." *The Information Society* 21, no. 1 (2005): 25–40.

Baumel, Simeon D. *Sacred Speakers: Language and Culture among the Haredim in Israel.* New York: Berghahn Books, 2006.

Baumel-Schwartz, Judy T. "Frum Surfing: Orthodox Jewish Women's Internet Forums as a Historical and Cultural Phenomenon." *Journal of Jewish Identities* 2, no. 1 (2009): 1–30.

Berger, Peter. *The Sacred Canopy: Elements of a Sociological Theory of Religion.* Garden City: Doubleday, 1967.

Blondheim, Menachem. "The Jewish Communication Tradition and Its Encounter with New Media." In *Digital Judaism: Jewish Negotiations with Digital Media and Technology*, edited by Heidi Campbell, 16–39. New York: Routledge, 2015.

Brown, Benjamin. *The Haredim: A Guide to Their Beliefs and Sectors.* Tel Aviv: Am Oved, 2017. [Hebrew].

Buetow, Stephen A. "Something in Nothing: Negative Space in the Clinician-Patient Relationship." *Annals of Family Medicine* 7, no. 10 (2009): 80–83.

Cahaner, Lee, Malach, Gilad, and Choshen, Maya. *The Yearbook of Ultra-Orthodox Society in Israel.* Jerusalem: The Israel Democracy Institute, 2017. [Hebrew].

Cahaner, Lee, and Zicherman, Haim. *Modern Ultra-Orthodoxy: The Emergence of the Haredi Middle Class in Israel.* Jerusalem: The Israel Democracy Institute, 2012 [Hebrew].

Campbell, Heidi. "Challenges Created by Online Religious Networks." *Journal of Media and Religion* 3, no. 2 (2004): 81–99.

Campbell, Heidi. "Religion and the Internet." *Communication Research Trends* 25, nos 1–3 (2006): 1–24.

Campbell, Heidi. "What Hath God Wrought': Considering How Religious Communities Culture (or Kosher) the Cell Phone." *Continuum: Journal of Media and Cultural Studies* 2, no. 2 (2007): 191–203.

Campbell, Heidi. "Who's Got the Power? The Question of Religious Authority and the Internet." *Journal of Computer-Mediated Communication* 12, no. 3 (2007): 1043–62.

Campbell, Heidi. *When Religion Meets New Media.* London: Routledge, 2010.

Campbell, Heidi. "Religion and the Internet in the Israeli Orthodox Context." *Israel Affairs* 17, no. 3 (2011): 364–83.

Campbell, Heidi (ed.). *Digital Religion: Understanding Religious Practice in New Media Worlds.* London: Routledge, 2013.

Campbell, Heidi. "Studying Jewish Engagement with Digital Media and Culture." In *Digital Judaism: Jewish Negotiations with Digital Media and Technology*, edited by Heidi Campbell, 1–15. New York: Routledge, 2015.

Campbell, Heidi, and Wendi, Bellar. "Sanctifying the Internet: Aish's Use of the Internet for Digital Outreach." In *Digital Judaism: Jewish Negotiations with Digital Media and Technology*, edited by Heidi Campbell, 74–90. New York: Routledge, 2015.

Campbell, Heidi, and Oren, Golan. "Creating Digital Enclaves: Negotiation of the Internet Amongst Bounded Religious Communities." *Media, Culture & Society* 33, no. 5 (2011): 709–24.

Caplan, Kimmy. *The Internal Popular Discourse in Israeli Haredi Society.* Jerusalem: Zalman Shazar, 2007 [Hebrew].

Caplan, Kimmy, and Stadler Nurit, eds. *From Survival to Consolidation: Changes in Israeli Haredi Society and its Scholarly Study.* Jerusalem: Van Leer Institute, 2012. [Hebrew].

Casanova, José. "Rethinking Secularization: A Global Comparative Perspective After Secularization." *Hedgehog Review: Critical Reflections on Contemporary Culture* 8, nos 1–2 (2006): 7–22.

Casanova, José. "Religion, Secularizations and Modernities." *European Journal of Sociology* 52, no. 3 (2011): 425–45.

Cohen, Yoel. "Israeli Rabbis and the Internet." In *Digital Judaism: Jewish Negotiations with Digital Media and Technology,* edited by Heidi Campbell, 183–204. New York and London: Routledge, 2015.

Cohen, Yoel. "Judaism in the Computer-Mediated Era." In *Changing Cultures and Religious Practices in Asia,* edited by B. Agrawal, 131–40. Manila: University of Santo Tomas, 2015.

Cohen, Yoel. "Media Events, Jewish Religious Holidays, and the Israeli Press." In *Global Perspectives on Media Events in Contemporary Society,* edited by A. Fox, 122–31. Hershey, PA: IGI Global, 2016.

Cohen, Yoel. "Holy Days, News Media, & Religious Identity: A Case Study in Jewish Holy Days and the Israeli Press." In *Spiritual News: Reporting Religion Around the World,* edited by Cohen Yoel, 303–22. New York: Peter Lang, 2017.

Cohen, Yoel, ed. *Spiritual News: Reporting Religion Around the World.* New York: Peter Lang, 2017.

Cohen, Akiba A., Lemish, Dafna, and Schejter, Amit. *The Wonder Phone in the Land of Miracles.* New York: Hampton Press, 2008.

Davison, W. Phillips. "The Third-Person Effect in Communication." *Public Opinion Quarterly* 47, no. 1 (1983): 1–15.

Deutsch, Nethanel. "The Forbidden Fork, the Cell Phone Holocaust, and Other Haredi Encounters with Technology." *Contemporary Jewry* 29, no. 1 (2009): 3–19.

El-Or, Tamar. *Educated and Ignorant: Ultra-Orthodox Jewish Women and Their World.* Boulder, CO: Lynne Rienner Publishers, 1994.

El-Or, Tamar. "Ultra-Orthodox Jewish Women." In *Israeli Judaism: The Sociology of Religion in Israel,* edited by Shlomo Deshen, 149–69. New Brunswick: Transaction, 1995.

El-Or, Tamar. "Visibility and Possibilities: Ultra-Orthodox Jewish Women Between the Domestic and Public Spheres." *Women's Studies International Forum* 20, no. 5 (1997): 665–73.

Fader, Ayala. "Nonliberal Jewish Women's Audiocassette Lectures in Brooklyn: A Crisis of Faith and the Morality of Media." *American Anthropologist* 115, no. 1 (2013): 72–84.

Friedman, Menachem. *Haredi Society: Sources, Trends and Processes.* Jerusalem: Jerusalem Institute for Israel Studies, 1991 [Hebrew].

Geertz, Clifford. *The Interpretation of Cultures.* New York: Basic Books, 1973.

Golan Oren. "Re-Constructing Religious Boundaries Online: The Emergence of the Denominational Jewish Internet." *La Temps Des Medias* 17, no. 3 (2011): 118–19.

Golan, Oren. "Charting Frontiers of Online Religious Communities: The Case of Chabad Jews." In *Digital Religion,* edited by Heidi Campbell, 155–63. London: Routledge, 2013.

Golan, Oren. "Legitimation of New Media and Community Building Amongst Jewish Denominations in the US." In *Digital Judaism: Jewish Negotiations with Digital Media and Technology,* edited by Heidi Campbell, 125–44. New York: Routledge, 2015.

Golan, Oren, and Campbell, Heidi. "Strategic Management of Religious Websites: The Case of Israel's Orthodox Communities." *Journal of Computer-Mediated Communication* 20, no. 4 (2015): 467–86.

Golan, Oren, and Stadler, Nurit. "Building the Sacred Community Online: The Dual Use of the Internet by Chabad." *Media, Culture and Society* 38, no. 1 (2016): 71–88.

Hall, Stuart. "Encoding/Decoding." In *Culture, Media, Language: Working Papers in the Cultural Studies 1972–79,* edited by S. Hall, D. Hobson, A. Lowe, and P. Willis, 128–38. London: Hutchinson & Co, 1981.

Hartman, Tova. *Feminism Encounters Traditional Judaism: Resistance and Accommodation.* Waltham: Brandeis University Press, 2007.

Hartman, Tova. "Strong Multiplicity: An Interpretive Lens in the Analysis of Qualitative Interview Narratives." *Qualitative Research* 15, no. 1 (2015): 22–38.

Heilman, Samuel C., and Friedman, Menachem. "Religious Fundamentalism and Religious Jews: The Case of the Haredim." In *Fundamentalisms Observed,* edited by Martin E. Marty and Scott Appleby, 197–264. Chicago: The University of Chicago Press, 1991.

Hess, Mary E. "Digital Storytelling: Empowering Feminist and Womanist Faith Formation with Young Women." In *Media, Religion and Gender: Key Issues and New Challenges*, edited by Mia Lovheim, 336–62. London: Routledge, 2013.

Hoover, Stuart M., and Lundby, Knut. "Introduction." In *Rethinking Media, Religion and Culture*, edited by Stuart M. Hoover and Knut Lundby, 3–14. Thousand Oaks, CA: Sage, 1997.

Kraybill, Donald B., Johnson-Weiner, Karen M., and Nolt, Steve M. *The Amish*. Baltimore, MD: Johns Hopkins University Press, 2013.

Lev-On, Azi, and Neriya-Ben Shahar, Rivka. "A Forum of Their Own: Views About the Internet Among Ultra-Orthodox Jewish Women Who Browse Designated Closed Forums." *First Monday* 16, no. 4 (2011).

Lev-On, Azi, and Neriya-Ben Shahar, Rivka. "To Browse, or Not to Browse? Perceptions of the Danger of the Internet by Ultra-Orthodox Jewish Women." In *New Media and Intercultural Communication*, edited by Pauline Hope-Cheong, Judith N. Martin, and Lea P. Macfadyen, 223–36. New York: Peter Lang, 2012.

Lieber, Andrea. "A Virtual Veibershul: Blogging and the Blurring of Public and Private Among Orthodox Jewish Women." *College English* 72, no. 6 (2010): 621–37.

Livio, Oren, and Tenenboim-Weinblatt, Karen. "Discursive Legitimation of a Controversial Technology: Ultra-Orthodox Jewish Women in Israel and the Internet." *The Communication Review* 10, no. 1 (2007): 29–56.

Lovheim, Mia. "Introduction: Gender – a Blind Spot in Media, Religion, and Culture?," In *Media, Religion and Gender: Key Issues and New Challenges*, edited by Mia Lovheim, 15–42. London: Routledge, 2013.

Lovheim, Mia, and Campbell, Heidi. "Considering Critical Methods and Theoretical Lenses in Digital Religion Studies." *New Media and Society* 19, no. 1 (2017): 5–14.

Mann, Rephael. "The Ultra-Orthodox Media 2016: Between Conservative and Modernization." In *The Annual Report: The Mass Media in Israel*, edited by Rephael Mann and Azi Lev-On, 91–105. Ariel: Ariel University Press, 2016. [Hebrew].

Mishol-Shauli, Nakhi. "*Fundamentalist Knowledge Brokers: New Media Journalists as Agents of Informal Education Among the Ultra-Orthodox Community in Israel*." Master's diss., Haifa University, 2016. [Hebrew].

Neriya-Ben Shahar, Rivka. "*Haredi (Ultra-Orthodox) Women and Mass Media in Israel - Exposure Patterns & Reading Strategies*." PhD diss., The Hebrew University of Jerusalem, 2008 [Hebrew].

Neriya-Ben Shahar, Rivka. "The Learners' Society: Continuity and Change in Characteristics of Education and Employment Among Ultra-Orthodox (Haredi) Women." *Sociological Papers* 14 (2009): 1–15.

Neriya-Ben Shahar, Rivka. "Negotiating Agency: Amish and Ultra-Orthodox Women's Responses to New Media." *New Media and Society* 19, no. 1 (2017): 81–95.

Neriya-Ben Shahar, Rivka. "The Medium Is the Danger: Discourse About Television Among Amish and Ultra-Orthodox (Haredi) Women." *Journal of Media and Religion* 16, no. 1 (2017): 27–38.

Neriya-Ben Shahar, Rivka. "Mobile Internet Is Worse Than the Internet; It Can Destroy Our Community': Old Order Amish and Ultra-Orthodox Women Responses to Cellphone and Smartphone Use." *The Information Society* 36, no. 1 (2020): 1–18.

Neriya-Ben Shahar, Rivka, and Lev-On Azi. "Gender, Religion and New Media: Attitudes and Behaviours Related to the Internet Among Ultra-Orthodox Women Employed in Computerized Environments." *International Journal of Communication* 5 (2011): 875–95.

Okun, Sarit. "*Exploring the Characteristics of an Online Ultra-Orthodox Jewish Community*." Master's thesis., Ben-Gurion University of the Negev, 2015.

Okun, Sarit. "Identity Games in the Online Ultra-Orthodox Religious Community." *Heker Hahevra Ha'Haredit [The Journal for the Haredi Society Research]* (2016): 56–77.

Okun, Sarit and Nimrod, Galit. "Online Ultra-Orthodox Religious Communities as a Third Space: A Netnographic Study." *International Journal of Communication* 11 (2017): 2825–41.

Pearl, Sharrona. "Exceptions to the Rule: Chabad-Lubavitch and the Digital Sphere." *Journal of Media and Religion* 13, no. 3 (2014): 123–37.

Piela, Anna. "Claiming Religious Authority: Muslim Women and New Media." In *Media, Religion and Gender: Key Issues and New Challenges*, edited by Mia Lovheim, 252–82. London: Routledge, 2013.

Rashi, Tsuriel. "The Kosher Cell Phone in Ultra-Orthodox Society: A Technological Ghetto Within the Global Village?" In *Digital Religion: Understanding Religious Practice in New Media Worlds*, edited by Heidi Campbell, 173–81. Philadelphia, PA: Routledge, 2013.

Rashi, Tsuriel, and McCombs, Maxwell. "Agenda Setting, Religion, and New Media: The Chabad Case Study." *Journal of Religion, Media and Digital Culture* 4, no. 1 (2015): 128–53.

Rosenberg, Hananel and Rashi, Tsuriel. "Pashkevillim in Campaigns Against New Media: What Can Pashkevillim Accomplish That Newspapers Cannot?" In *Digital Judaism: Jewish Negotiations with Digital Media and Technology*, edited by Heidi A. Campbell, 161–82. London: Routledge, 2015.

Rosenberg, Hananel, Blondheim, Menachem, and Katz, Elihu. "The Wall-Keepers: Social Enclaves, Social Control and the Kosher Cellphone Campaign in Jewish Ultra-Orthodox Society." *Israeli Sociology* 17, no. 2 (2016): 116–37 [Hebrew].

Rosenblatt, Elizabeth. "Intellectual Property's Negative Space: Beyond the Utilitarian." *Florida State University Law Review* 40, no. 3 (2013): 441–86.

Silber, Michael K. "The Emergence of Ultra-Orthodoxy: The Invention of a Tradition." In *The Uses of Tradition: Jewish Continuity in the Modern Era*, edited by Jack Wertheimer, 23–84. New York-Jerusalem: JTS, 1992.

Simon, Bernd, and Klandermans, Bert. "Politicized Collective Identity: A Social Psychological Analysis." *American Psychologist* 56, no. 4 (2001): 319–31.

Sivan, Emanuel. "Enclaved Culture." *Alpayim Magazine* 4 (1991): 45–99 [Hebrew].

Sivan, Emanuel, and Kimmy Caplan, eds. *Israeli Haredim: Integration Without Assimilation?* Jerusalem: Jerusalem Institute for Israel Studies, 2003 [Hebrew].

Stadler, Nurit. *Yeshiva Fundamentalism: Piety, Gender, and Resistance in the Ultra-Orthodox World.* New York: New York University Press, 2009.

Stadler, Nurit. *A Well-Worn Tallis for a New Ceremony: Trends in Israeli Haredi Culture.* Brighton, MA: Academic Studies Press, 2012.

Stout, Daniel A. "Secularization and the Religious Audience: A Study of Mormons and Las Vegas Media." *Mass Communication & Society* 7, no. 1 (2004): 61–75.

Stout, Daniel A. *Media and Religion: Foundation of an Emerging Field.* New York: Routledge, 2012.

Tzorfati, Orly, and Bleis, Dotan. "Between 'Enclave Culture' and 'Virtual Enclave': The Ultra-Orthodox Community and the Digital Media." *Kesher* 32 (2002): 47–56 [Hebrew].

Yafeh, Orit. "The Time in the Body: Cultural Construction of Femininity in Ultra-Orthodox Kindergartens for Girls." *Ethos* 35, no. 4 (2007): 516–33.

Zalcberg, Sima. "Grace is Deceitful, and Beauty is Vain: How Hasidic Women Cope with the Requirement of Shaving One's Head and Wearing a Black Kerchief." *Gender Issues* 24, no. 3 (2007): 13–34.

21

INVISIBLE IN/HUMANITY

Feminist epistemic ethics and rhetorical studies[1]

Kundai Chirindo

Three days before *Black Panther's* wide release, in February 2018, Ann Hornaday, *The Washington Post's* lead movie critic, wrote a review of the Marvel blockbuster. Rather than comment only on *Black Panther's* internal dynamics and other cinematic esoterica, Hornaday's review of *Black Panther* focused on the movie's broader cultural significance. Specifically, Hornaday explained how *Black Panther* epitomized a gaping hole in 21st-century cultural, racial, and epistemic politics; it spoke to a deficiency endemic to American and global public psyches. Here is part of what she wrote:

> When "Black Panther" arrives in theaters this weekend, one thing is sure: There will be tears ... Some tears will be of pride, at the sight of a movie dominated by strong, smart, funny, beautiful characters and actors of African descent. Others will be a spontaneous emotional response to "Black Panther's" story and subtext, which include moments of personal betrayal and loss, as well as reflections on a painful legacy of colonialism and dispossession. But some tears might also be of grief. Because just as palpable as the celebratory joy and sheer artistry that define "Black Panther," there lies an unspoken absence, the costs of which are no less real for being diffuse and virtually unfathomable.[2]

For Hornaday, as productive as *Black Panther* was (and *it was* productive) the film had additional cultural significance because of the lacuna it addressed. To Hornaday, *Black Panther* highlighted a hole in the range of cultural ideas to which moviegoers around the world have been exposed. For her, as much as *Black Panther's* release was rightly understood by many as a moment of celebration and exhilaration at the widespread dissemination and consumption of symbols and practices putatively drawn from the African world, it was also a moment of deep regret at the recognition of just how narrowly our collective cultural palates have been restricted. At the time of its release, the film's central affirmation was "that the world [including that of cinematography] is plural"[3] as Achille Mbembé puts it. By basing the film's symbolic, cultural, and cinematic grammar in a decidedly non-white world, director Ryan Coogler explicitly demonstrated the commonsense point that the Euro-American world is not the only cultural palate on which auteurs can draw to great success. The regret Hornaday's review noted was because *Black Panther's* release was the first time in a long time that the filmgoing audience, at least in the United States, had experienced this with a major film.

This chapter is not about *Black Panther*. It is about the epistemic and hermeneutic ecologies of communication studies broadly, and of rhetorical studies specifically. There is a parallel between the crux of Hornaday's critique and the general point I want to make here. Just as Hornaday's response to *Black Panther* draws attention to parochialism in film, I want to make explicit the *epistemic injustice* manifest in the narrow range of cognitive and hermeneutic resources reflected and drawn upon in a vast majority of rhetorical scholarship. My argument relies on what some have labeled the "deliberative turn" in epistemology. Over the last decade or so, epistemologists have returned to deliberation's constitutive role in the production of knowledge. Specifically, I take feminist philosopher Miranda Fricker and her 2007 book, *Epistemic Injustice* (as well the work it has inspired), as my guide.

Rather than presume a normative standard as the baseline for her analysis, Fricker begins by identifying specific practices that result in epistemic injustice. By so doing, she renders epistemic practice morally ambivalent, capable of both good and evil. We can thus appreciate that for epistemic practices to be virtuous or just, they should be free of unjust processes and outcomes. There are two distinct types of wrongs that constitute epistemic injustice according to Fricker: (i) *testimonial injustice* wherein a listener ascribes undue credibility excess or deficit to a giver of knowledge[4] or (ii) *hermeneutical injustice* wherein "some significant area of one's social experience [is] obscured from collective understanding owing to persistent wide and wide-ranging hermeneutical marginalization."[5] Borrowing these two concepts from the feminist critical lexicon, this chapter argues that masculinist, hetero- and cis-, and Euro- and U.S. American-centered heuristic schemas dominate rhetorical studies. As one among many Obama scholars in rhetorical studies, I'm going to use our discipline's encounter with the Obama phenomenon as a representative case.

The emergence and presidency of Obama inspired numerous dissertations, classes, articles, books, conferences, and more in rhetorical studies. Given Obama's cosmopolitan biography, one would expect his popularity as a subject of study to have brought to the fore a similarly worldly (i.e., beyond traditionally androcentric, Euro-American, hetero- and cis- normative) palate of rhetorical concepts and concerns. Yet, rhetorical studies of Obama have thus far amounted to little more than, "attempts to explain…Obama's politics as either an instantiation of conventional wisdom or as an apparition that retrieves some long-lost and undervalued wisdom [one, we might note, that is andro-, hetero-, cis-, and Euro-American-centric] presenting it as (re) new(ed)."[6] Given this, I attempt to answer the following questions in the remaining pages: (i) how can a feminist concern with epistemic injustice help us understand this puzzle?; (ii) what are the processes and outcomes that effect epistemic injustice in rhetorical studies?; and (iii) what can we do about it? To explore these questions, I begin by defining epistemic injustice with particular attention to the two types I just mentioned—testimonial injustice and hermeneutical injustice. After that, I briefly survey some of the rhetorical scholarship on Obama in light of what we will then understand as the call to epistemic justice. I conclude with a few thoughts about why rhetoricians would do well to care about combatting epistemic injustice and how we might facilitate and promote epistemic justice in the field. But first, what is epistemic injustice?

Epistemic injustice

Put simply, the concern for epistemic injustice is a concern with "a wrong done to someone specifically in their capacity as a knower."[7] Epistemic injustice assumes that *every individual is a knower*; one who is capable of attesting to what s/he knows, and one whose lived experiences condition the knowledge the individual possesses. Notice, though, that Fricker's definitional phrasing here is at odds with the object of inquiry in epistemology. Epistemology is not typically concerned with the knower *per se*, but with knowledge as such.[8] The case is different with epistemic injustice, which takes as its point of departure the ontological fact that the subject in epistemology is *always* an ethical subject because they are *always* human beings. This means that epistemic injustice is concerned with those practices that have the peculiar quality of being "at once epistemically bad, and ethically bad."[9] The question, then, is by what means can knowledge, it's production and pursuit, simultaneously wrong the individual *qua* human being, and wrong the individual *qua* knower? If, that is to say, the human is a knower, how might knowledge production violate the individual's humanness (violating the code of ethics) and status as a knower (sullying the pursuit of knowledge)?

Fricker posits two primary conduits of epistemic injustice[10]: *testimonial injustice* and *hermeneutical injustice*. Before I go any further, let me observe that those two categories signal relevance to rhetorical scholars on the basis of their names alone—they both suggest that epistemic events are *always* discursive or communicative events where meaning is made/exchanged, or where salience is cultivated. Here is Fricker commenting on testimonial injustice, for instance: "few would deny that an enormous amount of what we know is, at root, testimonially acquired."[11] *Pace* Aristotle's suggestion that testimonial evidence is an "inartistic proof" for the art of rhetoric, I think precisely the opposite: because of the ubiquity of testimonial practice, I agree with Brad Vivian that testimonial sites and testimonial practices belong squarely within the ambit of rhetorical theory.[12] This is what the deliberative turn in epistemology means.

Testimonial injustice results from the unwarranted excess or deficit attribution of credibility to an individual making knowledge claims.[13] It occurs when "a speaker receives a deflated [or inflated] degree of credibility owing to prejudice on the hearer's part."[14] Testimonial injustice occurs when "the speaker," argues Fricker, "is misjudged and perceived as epistemically lesser [or more]."[15]

Fricker cites two fictional examples—drawn from novels—which both are widely acclaimed for their commentary on nonfictional gender and racial politics to illustrate testimonial injustice. The first is from a scene in Patricia Highsmith's 1955 novel, *The Talented Mr. Ripley*, in which Herbert Greenleaf dismisses Marge Sherwood's suspicions about who murdered her fiancé (Herbert's son), saying, "there is female intuition, and then there are facts."[16] Greenleaf's derision is premised on his judgment of Marge's gendered identity, *and* at the gendered knowledge practices—"female intuition"—upon which he presumes her conjecture is based. Fricker's second example of testimonial injustice is drawn from the trial of Tom Robinson in Harper Lee's *To Kill a Mockingbird*. Here, she keys in on the scene when the jury is unable to overcome their racial animus toward Robinson in their adjudication of his innocence despite compelling evidence. With this second fictional case, Fricker demonstrates the important point that testimonial injustice can be socially and systemically embedded. These two examples make clear that "when someone suffers testimonial injustice, they are degraded *qua* knower, and they are symbolically degraded *qua* human."[17] Their claims to knowledge, or the testaments to what they know are dismissed *prima facie* because of who they are. The point of the dismissals is a commentary on their *humanity*: in the dismissal, they are taken not to be a *knowledge-capable* human. The denial of the other's knowledge-ability constitutes a degradation of their humanity; this is how epistemic injustice amounts to inhumanity.

Testimonial injustice can also occur absent intentional malice on the part of the individual actors that enact it. It is not simply about people acting badly; the seductions of testimonial injustice can be baked into our everyday processes and social practices. Here's why: we routinely employ various schemas as shortcuts as we go about the business of processing information. For instance, as hearers of information we "often make use of stereotypes as heuristics to facilitate [t]his judgment of a speaker's credibility."[18] This is something that the likes of Claude M. Steele[19] in social psychology, Glenn C. Loury[20] in economics, and others have demonstrated in their work on implicit bias and on stereotype threat. At stake in these snap judgments is nothing short of the evaluation and analysis of one's "competence and sincerity."[21] These, of course, are two crucial dimensions of what rhetorical scholars recognize as *ethos*. The stakes, where testimonial injustice is concerned, could not be higher:

> ...for prejudice presents an obstacle to truth, either directly by causing the hearer to miss out on a particular truth, or indirectly by creating blockages in the circulation of critical ideas[...] the fact that prejudice can prevent speakers from successfully putting knowledge into the public domain reveals testimonial injustice as a serious form of *unfreedom in our collective speech situation.*[22]

Fricker goes further still; she argues that since prejudice can preclude individuals from "trustful conversation"[23] which, as we know from George Herbert Mead, is fundamental to the formation of "mind, self, and society,"[24] then "persistent testimonial injustice can indeed inhibit the very formation of the self [and following Mead, I would add, mind and society]."[25] To that extent, this form of injustice is an *invisible form of inhumanity*. Failing to acknowledge the individual as a knower is failing to acknowledge (in full) the humanity of the other. That is what is at stake with testimonial injustice.

As for hermeneutical injustice, it happens

> when someone is trying to make sense of a social experience but is handicapped in this by a certain sort of gap in collective understanding—a hermeneutical lacuna whose existence is owing to the relative powerlessness of a social group to which the subject belongs.[26]

Before we had names for things like sexual harassment, racism, anti-gay discrimination, or transphobia, people victimized by these practices literally could not speak about their experiences because they did not have a vocabulary specific to the wrong-doings they endured. The insidiousness of hermeneutic marginalization extends to what we now consider mundane knowledge: like postpartum depression, microaggressions, stereotype threat, or even imposter syndrome. Before we developed trenchant labels that made each of these nameable, the experiences, though familiar to many, remained unrecognized by our shared pool of knowledge. Those who go through these ineffable experiences suffer injury twice over: first, they are victimized by the specific breach of their general well-being (we can blame persons or nonhuman actors for the infringement). They suffer, in the second position, the injustice of an imposed aphasia because the cultural lexicon to which they have access renders their lived experience moot and indescribable. That is hermeneutical injustice. Hermeneutical injustice, Fricker insists, is experienced as an effect of prior hermeneutic marginalization. The aggrieved individual is "at an unfair disadvantage when it comes to making sense of their social experiences"[27] because the dominant "grid[s] of intelligibility"[28] are incongruent with and detached from their lived experiences. Hermeneutical injustice, like testimonial injustice, works to dehumanize people by enforcing a collective silence(ing) around the human experiences of those who are marginalized, discounting the veracity of their claims, and foreclosing the inclusion of their experiences in what is considered as part of human experience.

As post-/decolonial, and Foucauldian analysis have demonstrated, the imbrications of power-knowledge,

> skew shared hermeneutical resources so that the powerful tend to have appropriate understandings of their experiences ready to draw on as they make sense of their social experiences, whereas the powerless are more likely to find themselves having some social experiences through a glass darkly, with at best ill-fitting meanings to draw on in the effort to render them intelligible.[29]

Whereas testimonial injustice can be perpetrated by individual actors (though not necessarily owing from malicious intent), hermeneutical injustice is wholly embedded in social structures, practices, and institutions. Hermeneutical injustice is a barrier that limits what is deemed worth knowing and what is deemed unworthy of contributing to a culture's knowledge base.

Epistemic justice

Having outlined the nature of epistemic injustice, we are now in a position to inquire after testimonial and hermeneutic justice. First, testimonial justice: since it derives from the fact that "the hearer perceives the speaker in an epistemically loaded way,"[30] and since we learn a great deal of epistemic evaluation via social training, one way to guard against misperceptions of testimonial quality is by recovering the contingency and partiality of our schemas. Hearers of testimonial claims, "must be alert to the impact not only of the speaker's social identity but also the impact of their *own* social identity on their credibility judgment."[31] In other words, the road to testimonial justice is paved through a recognition of the contingency of our final vocabularies, to borrow an expression from Richard Rorty.[32] Such a recognition is only possible if and when we maintain reflexivity in our epistemic practices because "testimonial responsibility requires a distinctly *reflexive* critical social awareness."[33]

As a reminder, hermeneutical injustice occurs when the lived experiences of some individuals cannot be understood according to the cultural vocabularies that dominate discourse in the society to which they belong. There are two possible remedies Fricker directs our attention to for this problem. The first one is going to be a hard one for us as writing teachers, journal editors, and manuscript reviewers to swallow: it is reminding oneself that the other's inarticulacy or incomprehensibility can reflect something other than banality or incompetence on their part. Instead, we should be open to the possibility that it is *our heuristics, our sense making apparatus* that occasion the aphasia. This means that we need to constantly (re)evaluate our heuristics and remain open to alternative epistemes, to different ways of understanding what counts as rhetoric/communication. At the very least, those who enjoy influential positions like undergraduate and graduate mentors, journal editors, and reviewers, as well as others with relative security should refrain from overzealously policing disciplinary boundaries by asking, for example, "where is the rhetoric in that?" or "what has that got to do with rhetoric?" Rather than reacting negatively to novel and unfamiliar research topics or objects of study (by enterprising graduate students, for example, or in the publication process), our shared obligation to hermeneutic justice requires that we remain receptive to rhetoric's and communication's multiple ontologies, *always*. Pursuing hermeneutic justice implies that the ideal posture for those already accepted as "part of the field" is one of curiosity, one of recursively (re)learning rhetorical (and communicative) polythetics[34] from what we don't (yet) understand as part of our scholarly purview. If we aren't already, we should be uncomfortable (i) that a broad swath of human communicative situations get little or no attention in discipline-sanctioned conversation spaces—conferences and journals, for instance—and (ii) to participate in any epistemic enterprise (especially one putatively concerned with "human communication") that is so parochially circumscribed that it mundanely ignores the very real practices of our fellow humans and our non-human cohabitants on this planet.

Fricker's second antidote for hermeneutical injustice follows from the first. She urges the adoption of what she calls "epistemic affirmative action," explained as when "any member of a stereotyped group says something anomalous, [we] should assume that it is *us* who don't understand."[35] The virtuous interlocutor enacts hermeneutical virtue by forestalling "*the impact of structural identity prejudice on one's credibility judgment*."[36] Critics might argue that the cost of this hermeneutical virtue is too high, as it severely damages epistemic standards by which what is worth knowing is determined while overemphasizing the role that identity plays in knowledge production. I, for one, reject the false choice implied in a critique contrasting the pursuit of "pure knowledge" with attention to processes of identification. As Stuart Hall observed, "cultural identities matter not because they fix us into place politically but because *they are what is at stake*—what is won or lost—in cultural politics."[37] Likewise, identity is precisely what is at stake in epistemic activity.

Feminist epistemic ethics, Obama, and rhetorical studies

It is now time to consider how these ideas about epistemic injustice relate to rhetorical studies. How can we extrapolate these insights to *our* disciplinary practice? I'm going to focus my response to these questions on the pursuit of hermeneutic justice because I am primarily concerned with the narrowness of our disciplinary lexicon, a narrowness which results in far too many human experiences being left out of the annals of rhetorical studies as has been recently documented.[38] As I explained above, the concern for hermeneutic justice compels us to ask: can our scholarship account for an even broader sampling of human experiences than we currently do? What social experiences, what epistemic perspectives are yet to manifest in our scholarly conversations?

As one of many people who have written about him over the last decade or so, my thinking on this question brought me to the 44th president of the United States. Here, I want to recall Hornaday's review of *Black Panther*. When I think about Hornaday's gloss on the *Black Panther* moment and how it might relate to our discipline, our field's encounter with the "Obama phenomenon" comes to mind. I think that the "Obama phenomenon" was to rhetorical studies what *Black Panther* was to fans of the Marvel universe and movies more generally. After all, Obama's emergence was marked by virtuoso rhetorical performances both in print and speech. We could even say he prefigured the Afrofuturism and Afro-optimism to which the leading figures in Wakanda would later attest. And, as with Prince T'Challa (the movie's protagonist), Obama's odyssey was marked, too, by its own "moments of personal betrayal and loss, as well as reflections on a painful legacy of colonialism and dispossession."[39] As Hornaday realizes about film, I think rhetorical studies has our own "unspoken absence, the costs of which are no less real for being diffuse and virtually unfathomable"?[40] How do we turn our epistemic lacunae into heuristic opportunities? And, importantly, how do we counter, pause, and reverse the parochialism entrenched in our disciplinary practices?

There are two relevant observations we can make about Obama's ascent to the presidency. First, the rise of Obama was a boon to rhetorical scholarship, particularly for those studying the rhetorical presidency and presidential rhetoric. Second, Obama's rise refigured or recast the connections between American politics and the world's communities. Since its founding, the United States has stood apart—as the new world, as the city on a hill, the world's last best hope, and by other familiar metaphors of U.S.-American exceptionalism. Yet in Obama's personal story, those relationships of exteriority to the world became relationships of interiority. There is consensus between Obama's first memoir and his biographers that his early stint in Indonesia was crucial to the formation of his awareness that his racial identity had vast implications. His quest for a sense of racial belonging advanced during and after his first trip to Kenya. The roots of his political involvement were in the South African divestment campaign he joined when he was student at Occidental College. And, as I've argued elsewhere, in Selma, Obama's "references to Kenya and Africa established African life as parallel and connected, not opposite and detached from the American saga."[41] Obama's were cosmopolitan influences.

Likewise, Obama's social formation was undoubtedly influenced by his mother, grand-mother, sister Auma, and countless others. His was hardly a stereotypically masculine childhood. Yet when we turn to our scholarship—and I want to be clear that I include the work I've done on Obama here—rhetoricians have treated Obama largely in terms drawn from a familiar canon with relatively few exceptions—by people like Molefe Asante, Mark and Roger McPhail, Darrel Wanzer-Serrano, Kristen Hoerl, Dorthy Pennington, David Cisneros, and Vince Pham—into ideas drawn from beyond the andro- and Euro American-centric orthodoxy in rhetorical theory.[42] I cannot enumerate every single rhetorical study of Obama. But I think you will allow the generalized observation that Obama has been explained in terms of the hallmark *topoi* of American public address—American Exceptionalism,[43] the American Dream,[44] the American prophetic tradition,[45] American pragmatism,[46] and civil rights.[47] While some have interrogated Obama's rhetoric in the context of global discourses of post-/neo-colonialism, immigration, and white supremacy,[48] inquiries into the interactions between Obama's rhetorical practices and the social practices of the Indonesian and Kenyan communities *he credits as major influences early in his life*, are conspicuous by their absences in our discipline's official record. This collection of approaches to Obama in rhetorical studies is remarkable, too, for only recognizing Obama's masculine influences. Reading this scholarship we are led to believe that even though the social ontology of Obama was incredibly diverse, when it came to the "important" work of politics, only his U.S. American male influences are worth considering. Fricker's feminist epistemology makes plain that these absences constitute a hermeneutical injustice in rhetorical scholarship. The opportunity waits for us to turn these sociocultural perspectives into heuristics that explore "the imaginative possibilities that [Obama's] campaign and administration offer to us a way of examining our assumptions about the rhetorical presidency and about presidential rhetoric"[49] among other things. I think what has happened in this corner of our discipline might be a microcosm of the vast heuristic opportunities that lay in wait for those who will write more human experiences—including but not limited to sexual violence, transphobia, and workplace and marketplace gender discrimination—into our disciplinary conversations.

Conclusion

To conclude, let me turn now to why feminist rhetoricians committed to critiquing injustices based on gender, sexuality, race, class, and national origin should care about epistemic justice. You might be wondering: Why should we care about expanding the range of theories we have avail-able in our toolkit?" Specifically, why concern ourselves with diversifying the geocultural origins of the theories we bring to bear on the study of communication and rhetorical discourses that constitute identity? The first thing we might note about this line of questioning is the underlying and unspoken assumption that authorizes the angst in the first place. There is, in asking either why ought we care about theories and practices beyond what is already known and familiar, an assumption, that pure knowledge, pure rhetoric, and reason are disembodied and dislocated, points which feminist scholars such as Elizabeth Grosz,[50] Kyla Schuller,[51] and Doreen Massey[52] have thoroughly critiqued. This assumption holds true for the even more powerful means of both silencing and shutting down knowledge claims from the margins, which goes like this: "why bother if what is claimed is ultimately similar to what is already known?" The underlying idea is that the realm of knowledge production and knowledge itself is free from influences of the geographical, historical, and cultural contingencies that attend to its emergence. Linda Alcoff has labeled this desire for decontextualized knowledge a "transcendentalist delusion."[53] The decontextualization and despatialization aspired to in the transcendentalist delusion pose serious problems for us rhetoricians. As we know well, from the origins of the systemic study of rhetorical theory in Athens to the present, rhetorical theory is thoroughly sensitive to spatiality and emplacement.[54] Rhetoric is often *topo*-centric. Our obligation to this tradition of caring about rhetoric's topographies is one reason why rhetorical scholars should care about epistemic justice. The incitement to epistemic justice of the hermeneutical kind might stem, for us rhetoricians, from the desire to recover in full the places or *topoi* of rhetoric, including bodies in their diverse materialities.

The second and final reason why we should care is implicated in the recent materialist turn in rhetorical theory. Diane Davis has argued that responsivity—the corporeal condition of being exposed to external stimuli, including to rhetoric itself—is the *sine qua non* of rhetoric.[55] In thus defining rhetoric, Davis enjoins rhetoricians to probe corporeality as rhetoric's precondition.[56] What I would like to note about Davis' injunction here is her shift in focus from words to bodies, from speech and speaking to responses, from speaker and audiences as rhetorical agents to the human species and human experiences. This shift in emphasis highlights what is at stake in the critique of rhetorical scholarship from the standpoint of epistemic injustice. Writes Davis, "Rhetoric is not first of an all an essence or property "*in* the speaker" (a natural function of biology) but an underivable obligation to respond that issues from an irreducible relationality."[57] To practice rhetoric is take up the mantle of relationality that *always* entails the relatability of parties that are rhetorically involved. It is therefore incumbent on scholars of rhetoric to attend to the relational matrix as fully as possible. Focusing on just one actor or on one side of the relational ledger might produce analyses that are informative and insightful, but it might also make us culpable of epistemic injustice. True, polling data and other contextual factors that we routinely take into account usefully texture our analyses, but stopping there (as we sometimes do) leaves too much of the story untold, only recognizing rhetoric as addressed without probing the *addressivity* that licenses rhetorical address to begin with. For Fricker, such an account enacts epistemic injustice—it violates the rhetorical other's (i.e., the public's) status as knowers. As gender activists have long known, recognizing and drawing attention to injustice is a crucial step toward repairing our broken humanity. Scholars of gender, race, and sexuality in rhetorical and communication studies have an important role to play in enabling communication about more human experiences than the field presently accounts for.

Notes

1 This chapter is drawn from a paper first presented at the 2018 Public Address Conference. I am indebted to Lisa Flores, Phaedra Pezzullo and the College of Media, Communication and Information at UC Boulder for the invitation. Thanks to Joan Faber McAlister for her longstanding generosity and for the opportunity to contribute to this volume. I am grateful, too, to Lindsay Harroff, Vince Pham, G. Mitchell Reyes, David Schulz, and Mary Stuckey for reading and responding to previous drafts.
2 Ann Hornaday, "'Black Panther' Is a Revelation but Also a Reminder of What We've Been Missing," *The Washington Post*, February 15, 2018.
3 Joseph-Achille Mbembe and Laurent Dubois, *Critique of Black Reason* (Durham, NC: Duke University Press, 2017), 157.
4 Miranda Fricker, *Epistemic Injustice: Power and the Ethics of Knowing* (Oxford, UK: Oxford University Press, 2007/2010), 17.

5 Fricker, *Epistemic Injustice*, 154.

6 Kundai Chirindo, "Paradigmatic and Syntagmatic Approaches to the Obama Presidency," *Rhetoric & Public Affairs* 19, no. 3 (2016): 492.

7 Fricker, *Epistemic Injustice*, 1.

8 As far back as Plato's separation of knowledge from the knower, most accounts of epistemology are only tangentially concerned with human subjects. Some, like Descartes, go so far as to make the veracity of human subjectivity contingent upon the capacity to know.

9 Miranda Fricker, "Replies to Critics," *Theoria: an International Journal for Theory, History an Foundations of Science* 23, no. 1 (2008): 84.

10 David Coady subsequently reclassified epistemic injustice into two types: unjust ignorance or error and unjust credibility or intelligibility deficits. He placed Fricker's work under the latter category. See Coady "Two Concepts of Epistemic Injustice," *Episteme* 7, no. 2 (2010): 101–13.

11 Miranda Fricker, "Précis," *Theoria: An International Journal for Theory, History and Foundations of Science* 23, no. 1 (2008): 70.

12 Bradford Vivian, *Commonplace Witnessing: Rhetorical Invention, Historical Remembrance, and Public Culture* (New York: Oxford University Press, 2017).

13 Fricker, *Epistemic Injustice*, 17.

14 Fricker, "Précis," 70.

15 Ian James Kidd, José Medina, and Gaile Pohlhaus Jr, *The Routledge Handbook of Epistemic Injustice* (New York: Routledge, 2017), 94.

16 Quoted in Fricker, *Epistemic Injustice*, 9.

17 Fricker, *Epistemic Injustice*, 44.

18 Fricker, *Epistemic Injustice*, 36.

19 Claude Steele, *Whistling Vivaldi: And Other Clues to How Stereotypes Affect Us* (New York: W.W. Norton & Company, 2011).

20 C. Loury Glenn, *The Anatomy of Racial Inequality* (Cambridge, MA: Harvard University Press, 2003).

21 Fricker, *Epistemic Injustice*, 32.

22 Fricker, *Epistemic Injustice*, 43. Italics added.

23 Fricker, *Epistemic Injustice*, 52.

24 George Herbert Mead and Charles W. Morris, *Mind, Self & Society from the Standpoint of a Social Behaviorist* (Chicago: The University of Chicago Press, 1934).

25 Fricker, *Epistemic Injustice*, 55.

26 Fricker, "Précis," 70.

27 Fricker, *Epistemic Injustice*, 1.

28 Michel Foucault and Robert Hurley, *The History of Sexuality. Volume I* (New York: Random House, 1978), 93.

29 Fricker, *Epistemic Injustice*, 148.

30 Fricker, *Epistemic Injustice*, 86.

31 Fricker, *Epistemic Injustice*, 90.

32 Richard Rorty, *Contingency, Irony, and Solidarity* (Cambridge; New York: Cambridge University Press, 1989).

33 Fricker, *Epistemic Injustice*, 91. Italics in original.

34 See Nathan Stormer, "Rhetoric's Diverse Materiality: Polythetic Ontology and Genealogy," *Review of Communication* 16 (2016); "Recursivity: A Working Paper on Rhetoric and Mnesis," *Quarterly Journal of Speech* 99, no. 1 (2012).

35 Fricker, *Epistemic Injustice*, 170–71.

36 Fricker, *Epistemic Injustice*, 173. Italics in original.

37 Stuart Hall, Kobena Mercer, and Henry Louis Gates, "The Fateful Triangle: Race, Ethnicity, Nation," (Cambridge, MA: Harvard University Press, 2017): 130. Italics in original.

38 See, for example, Paula Chakravartty et al., "#Communicationsowhite," *Journal of Communication* 68, no. 2 (2018); Darrel Wanzer-Serrano, "Rhetoric's Rac(E/Ist) Problems," *Quarterly Journal of Speech* 105, no. 4 (2019).

39 Hornaday.

40 Hornaday.

41 Kundai Chirindo, "Barack Obama, Tropology, and Ideas of Africa," in *Rhetorics Change/Rhetoric's Change*, eds Jenny Rice, Chelsea Graham, and Eric Detweiler (Anderson, SC: Parlor Press, 2018), 351.

42 Molefi Kete Asante, "Barack Obama and the Dilemma of Power," *Journal of Black Studies* 38, no. 1 (2007); J. David Cisneros, "A Nation of Immigrants and a Nation of Laws: Race, Multiculturalism, and

Neoliberal Exception in Barack Obama's Immigration Discourse," *Communication, Culture & Critique* 8, no. 3 (2015); Darrel Enck-Wanzer, "Barack Obama, the Tea Party, and the Threat of Race: On Racial Neoliberalism and Born Again Racism," *Communication, Culture & Critique* 4, no. 1 (2011): 23; Kristen Hoerl, "Selective Amnesia and Racial Transcendence in News Coverage of President Obama's Inauguration," *Quarterly Journal of Speech* 98, no. 2 (2012): 178; Mark Lawrence McPhail and Roger McPhail, "(E)Raced Men: Complicity and Responsibility in the Rhetorics of Barack Obama," *Rhetoric & Public Affairs* 14 (2011): 673; Vincent N. Pham, "Our Foreign President Barack Obama: The Racial Logics of Birther Discourses," *Journal of International & Intercultural Communication* 8, no. 2 (2015): 86.

43 Kundai Chirindo and Ryan Neville-Shepard, "Obama's 'New Beginning': U.S. Foreign Policy and Comic Exceptionalism," *Argumentation & Advocacy* 51, no. 4 (2015): 215; Robert L. Ivie and Oscar Giner, "American Exceptionalism in a Democratic Idiom: Transacting the Mythos of Change in the 2008 Presidential Campaign," *Communication Studies* 60, no. 4 (2009): 359.

44 Robert C. Rowland, "The Fierce Urgency of Now: Barack Obama and the 2008 Presidential Election," (2010): 203; Robert C. Rowland and John M. Jones, "Recasting the American Dream and American Politics: Barack Obama's Keynote Address to the 2004 Democratic National Convention," *Quarterly Journal of Speech* 93, no. 4 (2007): 425; Rowland and Jones, "One Dream: Barack Obama, Race and the American Dream," *Rhetoric & Public Affairs* 14 (2011): 125.

45 David A. Frank, "The Prophetic Voice and the Face of the Other in Barack Obama's 'A More Perfect Union' Address, March 18, 2008," *Rhetoric & Public Affairs* 12, no. 2 (2009): 167; John M. Murphy, "Barack Obama, the Exodus Tradition, and the Joshua Generation," *Quarterly Journal of Speech* 97, no. 4 (2011): 387; Dorthy Pennington, "Barack Obama's "Authentic Self" as a Spiritual Warrior: Discipleship to the Apostolic," *Howard Journal of Communications* 26, no. 1 (2015): 74.

46 Robert Danisch, "The Roots and Form of Obama's Rhetorical Pragmatism," *Rhetoric Review* 31, no. 2 (2012): 148; James T. Kloppenberg, *Reading Obama : Dreams, Hope, and the American Political Tradition* (Princeton, NJ: Princeton University Press, 2011).

47 David A. Frank and Mark Lawrence McPhail, "Barack Obama's Address to the 2004 Democratic National Convention: Trauma, Compromise, Consilience, and the (Im)Possibility of Racial Reconciliation," *Rhetoric & Public Affairs* 8, no. 4 (2005): 571; McPhail and McPhail; David Remnick, *The Bridge : The Life and Rise of Barack Obama* (New York: Vintage Books, 2011); Robert E. Terrill, "Unity and Duality in Barack Obama's "a More Perfect Union," *Quarterly Journal of Speech* 95, no. 4 (2009): 363; Ron Walters, "Barack Obama and the Politics of Blackness," *Journal of Black Studies* 38, no. 1 (2007): 7.

48 Pham; Cisneros; Enck-Wanzer.

49 Mary E. Stuckey, "Rethinking the Rhetorical Presidency and Presidential Rhetoric," *Review of Communication* 10, no. 1 (2010): 39.

50 See, for instance, Elizabeth Grosz, *Volatile Bodies Toward a Corporeal Feminism* (Bloomington, IN: Indiana University Press, 2011).

51 Kyla Schuller, *The Biopolitics of Feeling: Race, Sex, and Science in the Nineteenth Century* (Durham, NC: Duke University Press, 2018).

52 Doreen Massey, *Space, Place and Gender* (Cambridge: Polity Press, 2007).

53 Linda Martín Alcoff, "Philosophy and Philosophical Practice: Eurocentrism as an Epistemology of Ignorance," in *The Routledge Handbook of Epistemic Injustice*, eds Ian James Kidd, José Medina, and Gaile Pohlhaus Jr (New York: Routledge, 2017).

54 Kundai Chirindo, "A (Hetero)Topology of Rhetoric and Obama's African Dreams," *Advances in the History of Rhetoric* 19, no. 1 (2016): 50; Kundai Chirindo, "Rhetorical Places: From Classical Topologies to Prospects for Post-Westphalian Spatialities," *Women's Studies in Communication* 39, no. 2 (2016): 127.

55 Diane Davis, "Creaturely Rhetorics," *Philosophy & Rhetoric* 44, no. 1 (2011): 88.

56 Others have called for an even wider aperture, one that includes all "spacetimemattering," as Karen Barad puts it, in laying out the conditions of possibility that authorize rhetorics. See Karen Michelle Barad, *Meeting the Universe Halfway: Quantum Physics and the Entanglement of Matter and Meaning* (Durham, NC: Duke University Press, 2007) especiallly Chapters 4 and after; Stephen H. Browne, "Rhetorical Criticism and the Challenges of Bilateral Argument," *Philosophy & Rhetoric* 40, no. 1 (2007): 108; Nathan Stormer, "Articulation: A Working Paper on Rhetoric and Taxis," *Quarterly Journal of Speech*, 90, no. 3 (2004): 257; Nathan Stormer, "Rhetoric's Diverse Materiality: Polythetic Ontology and Genealogy," *Review of Communication* 16, no. 4 (2016): 299.

57 Davis, *Philosophy & Rhetoric*, 84.

Bibliography

Alcoff, Linda Martín. "Philosophy and Philosophical Practice: Eurocentrism as an Epistemology of Ignorance," Chap. 37 In *The Routledge Handbook of Epistemic Injustice*, edited by Ian James Kidd, José Medina, and Gaile Pohlhaus Jr, 397–408. New York: Routledge, 2017.

Asante, Molefi Kete. "Barack Obama and the Dilemma of Power." *Journal of Black Studies* 38, no. 1 (2007): 105–15.

Barad, Karen Michelle. *Meeting the Universe Halfway: Quantum Physics and the Entanglement of Matter and Meaning*. Durham, NC: Duke University Press, 2007.

Browne, Stephen H. "Rhetorical Criticism and the Challenges of Bilateral Argument." *Philosophy & Rhetoric* 40, no. 1 (2007): 108–18.

Chakravartty, Paula, Rachel Kuo, Victoria Grubbs, and Charlton McIlwain. "#Communicationsowhite." *Journal of Communication* 68, no. 2 (2018): 254–66.

Chirindo, Kundai. "Barack Obama, Tropology, and Ideas of Africa." In *Rhetorics Change/Rhetoric's Change*, edited by Jenny Rice, Chelsea Graham, and Eric Detweiler, 214–23. Anderson, SC: Parlor Press, 2018.

———. "A (Hetero)Topology of Rhetoric and Obama's African Dreams." *Advances in the History of Rhetoric* 19, no. 1 (2016): 50–70.

———. "Paradigmatic and Syntagmatic Approaches to the Obama Presidency." *Rhetoric & Public Affairs* 19, no. 3 (2016): 491–504.

———. "Rhetorical Places: From Classical Topologies to Prospects for Post-Westphalian Spatialities." *Women's Studies in Communication* 39, no. 2 (2016): 127–31.

Chirindo, Kundai, and Ryan Neville-Shepard. "Obama's 'New Beginning': U.S. Foreign Policy and Comic Exceptionalism." *Argumentation & Advocacy* 51, no. 4 (2015): 215–30.

Cisneros, J. David. "A Nation of Immigrants and a Nation of Laws: Race, Multiculturalism, and Neoliberal Exception in Barack Obama's Immigration Discourse." *Communication, Culture & Critique* 8, no. 3 (2015): 356–75.

Coady, D. "Two Concepts of Epistemic Injustice." *Episteme* 7, no. 2 (2010): 101–13.

Danisch, Robert. "The Roots and Form of Obama's Rhetorical Pragmatism." *Rhetoric Review* 31, no. 2 (2012): 148–68.

Davis, Diane. "Creaturely Rhetorics." *Philosophy & Rhetoric* 44, no. 1 (2011): 88–94.

Enck-Wanzer, Darrel. "Barack Obama, the Tea Party, and the Threat of Race: On Racial Neoliberalism and Born Again Racism." *Communication, Culture & Critique* 4, no. 1 (2011): 23–30.

Foucault, Michel, and Robert Hurley. *The History of Sexuality.* Volume I. New York: Random House, 1978.

Frank, David A. "The Prophetic Voice and the Face of the Other in Barack Obama's "A More Perfect Union" Address, March 18, 2008." *Rhetoric & Public Affairs* 12, no. 2 (2009): 167–94.

Frank, David A., and Mark Lawrence McPhail. "Barack Obama's Address to the 2004 Democratic National Convention: Trauma, Compromise, Consilience, and the (Im)Possibility of Racial Reconciliation." *Rhetoric & Public Affairs* 8, no. 4 (2005): 571–93.

Fricker, Miranda. *Epistemic Injustice: Power and the Ethics of Knowing*. Oxford: Oxford University Press, 2007/2010.

———. "Précis." *Theoria: An International Journal for Theory, History and Foundations of Science* 23, no. 1 (2008): 69–71.

———. "Replies to Critics." *Theoria: An International Journal for Theory, History and Foundations of Science* 23, no. 1 (2008): 81–86.

Glenn, C. Loury. *The Anatomy of Racial Inequality* [in English]. Cambridge, MA: Harvard University Press, 2003.

Grosz, Elizabeth. *Volatile Bodies toward a Corporeal Feminism*. Bloomington, IN: Indiana University Press, 2011.

Hall, Stuart, Mercer, Kobena and Gates, Henry Louis. *The Fateful Triangle: Race, Ethnicity, Nation*. Cambridge, MA: Harvard University Press, 2017.

Hoerl, Kristen. "Selective Amnesia and Racial Transcendence in News Coverage of President Obama's Inauguration." *Quarterly Journal of Speech* 98, no. 2 (2012): 178–202.

Hornaday, Ann. "'Black Panther' Is a Revelation but Also a Reminder of What We've Been Missing." *The Washington Post*, February 15, 2018, http://link.galegroup.com.library.lcproxy.org/apps/doc/A527687143/AONE?u=lcc&sid=AONE&xid=6f8cfdc9 [Accessed 6/8/2018].

Ivie, Robert L., and Giner, Oscar. "American Exceptionalism in a Democratic Idiom: Transacting the Mythos of Change in the 2008 Presidential Campaign." *Communication Studies* 60, no. 4 (2009): 359–75.

Kidd, Ian James, Medina, José, and Pohlhaus Gaile, Jr. *The Routledge Handbook of Epistemic Injustice*. New York: Routledge, 2017.

Kloppenberg, James T. *Reading Obama: Dreams, Hope, and the American Political Tradition* [in English]. Princeton, NJ: Princeton University Press, 2011.

Massey, Doreen. *Space, Place and Gender*. Cambridge, UK: Polity Press, 2007.

Mbembe, Joseph-Achille, and Dubois, Laurent. *Critique of Black Reason*. Durham, NC: Duke University Press, 2017.

McPhail, Mark Lawrence, and McPhail, Roger. "(E)Raced Men: Complicity and Responsibility in the Rhetorics of Barack Obama." *Rhetoric & Public Affairs* 14 (2011): 673–91.

Mead, George Herbert, and Morris, Charles W. *Mind, Self & Society from the Standpoint of a Social Behaviorist*. Chicago, IL: The University of Chicago Press, 1934.

Murphy, John M. "Barack Obama, the Exodus Tradition, and the Joshua Generation." *Quarterly Journal of Speech* 97, no. 4 (2011): 387–410.

Pennington, Dorthy. "Barack Obama's "Authentic Self" as a Spiritual Warrior: Discipleship to the Apostolic." *Howard Journal of Communications* 26, no. 1 (2015): 74–94.

Pham, Vincent N. "Our Foreign President Barack Obama: The Racial Logics of Birther Discourses." *Journal of International & Intercultural Communication* 8, no. 2 (2015): 86–107.

Remnick, David. *The Bridge: The Life and Rise of Barack Obama*. New York: Vintage Books, 2011.

Rorty, Richard. *Contingency, Irony, and Solidarity*. Cambridge: Cambridge University Press, 1989.

Rowland, Robert C. "The Fierce Urgency of Now: Barack Obama and the 2008 Presidential Election." *American Behavioral Scientist* 54, no. 3 (2010): 203–21.

Rowland, Robert C., and Jones, John M. "One Dream: Barack Obama, Race and the American Dream." *Rhetoric & Public Affairs* 14 (2011): 125–54.

———. "Recasting the American Dream and American Politics: Barack Obama's Keynote Address to the 2004 Democratic National Convention." *Quarterly Journal of Speech* 93, no. 4 (2007): 425–48.

Schuller, Kyla. *The Biopolitics of Feeling: Race, Sex, and Science in the Nineteenth Century*. Durham, NC: Duke University Press, 2018.

Steele, Claude. *Whistling Vivaldi: And Other Clues to How Stereotypes Affect Us*. New York: W.W. Norton & Company, 2011.

Stormer, Nathan. "Articulation: A Working Paper on Rhetoric and Taxis." *Quarterly Journal of Speech* 90, no. 3 (2004): 257–84.

———. "Recursivity: A Working Paper on Rhetoric and Mnesis." *Quarterly Journal of Speech* 99, no. 1 (2012): 27–50.

———. "Rhetoric's Diverse Materiality: Polythetic Ontology and Genealogy." *Review of Communication* 16, no. 4 (2016): 299–316.

Stuckey, Mary E. "Rethinking the Rhetorical Presidency and Presidential Rhetoric." *Review of Communication* 10, no. 1 (2010): 38–52.

Terrill, Robert E. "Unity and Duality in Barack Obama's 'a More Perfect Union'." *Quarterly Journal of Speech* 95, no. 4 (2009): 363–86.

Vivian, Bradford. *Commonplace Witnessing: Rhetorical Invention, Historical Remembrance, and Public Culture*. New York: Oxford University Press, 2017.

Walters, Ron. "Barack Obama and the Politics of Blackness." *Journal of Black Studies* 38, no. 1 (2007): 7–29.

Wanzer-Serrano, Darrel. "Rhetoric's Rac(E/Ist) Problems." *Quarterly Journal of Speech* 105, no. 4 (2019): 465–76.

PART IV

Gendered contexts and strategies

Introduction to Part IV

In this section, contributors examine how gender and sexuality affect and are affected by their changing contexts. These contexts are extensive, varied, and represent male-dominated organizations, sport, settings where health decisions are made such as hospitals and hospice-care facilities, dating applications, and fandom. Using a variety of theories and frameworks, the authors discuss communication strategies used to manage gender imbalance in these contexts and provide suggestions for coping with, and often resisting, dominant power structures.

Consistent in these contexts is that homophobia, sexism, and racism represent prevalent and powerful values. Further, narratives of what represents the "ideal" (worker, player, partner, patient, character, etc.) are framed around maleness, heterosexuality, and whiteness. As a result, those who are "not ideal" suffer in the performances of their identities. For example, Kallia O. Wright and Kesha Morant Williams discuss Black women's physical pain and how its severity is frequently not believed by health-care workers. As a result, Black women have higher rates of death, particularly during and after childbirth, in the United States. In sport settings, overt acts of masculinity are privileged, with those who behave in feminine ways excluded using verbal and nonverbal interactions. The ideal worker, player, partner, patient, and character are frequently the ones leading the charge toward the exclusion of other groups, yet the authors in the section find that traditionally marginalized groups may also engage in similar oppressive behaviors (Mel Stanfill, for example, finds that straight white women frequently feed into gendered stereotypes in slash fan fiction).

However, strategies for resistance against oppressive and gendered narratives are evident in the chapters in this section, occurring at both micro and macro levels and in overt and covert ways, though it is notable that some of the strategies are more effective than others and can potentially uphold gendered power structures. Resistance strategies offered by the authors include flaunting one's femininity and being one's own advocate. The authors also provide strategies for those who represent the "ideal" to use to implode oppressive structures, such as training and listening to the narratives of those who are historically marginalized. Additional recommendations are suggested for both those being oppressed and those in power, including speaking verbally against acts of sexism, racism, and homophobia.

The first chapter in this section, "Organizational discourse and sexuality in male-dominated organizational settings," is written by Clifton Scott, Aly Stetyick, and Jaime Bochantin and points out the dearth of research on sexuality in male-dominated organizations. The authors thoughtfully review Karen Ashcraft's frames of studying gendered organizing: sexuality-based difference at work and discourse; hegemonic sexuality and organizational performance; organizations and occupations as sexualized institutional forms; and sexuality discourse as societal narrative. Using Ashcraft's frames, Scott, Stetyick, and Bochantin provide a thoughtful literature review of sexuality research and find that the ideal worker is typically a white, straight male; those who are outside of this norm do not obtain leadership positions at the same rate and are often pressured to conform to the norms of the ideal worker. The authors argue that the body itself is an organizational performance and the ideal professional body has the same physical characteristics of the ideal worker: white, straight, young, and male. Those who are excluded include women, sexual minorities, and aging workers. However, similar to other chapters in this section and book, the authors offer strategies to resist power structures, including subtle acts, like hidden transcripts, and more overt responses, such as flaunting one's femaleness.

The realm of sport is another gendered context discussed in this section. In their chapter, "Shifting sands and moving goalposts: communicating gender in sport," Kitrina Douglas and David Carless argue that narrative theory plays a pivotal role in our understanding of sport in that stories allow us to analyze situations, understand behavior, and remove artificial boundaries. Stories can also be used as a pedagogical tool to help repair the damage of abuse. The chapter highlights gender inequality in sport, specifically noting the "boys will be boys" narrative that frequently serves as an excuse for sexist behavior and attitudes. Honestly sharing their own stories and discussing firsthand stories from others, Douglas and Carless disclose to their readers the misogyny and homophobia that flourishes within the culture of sport.

Homophobia and misogyny extend beyond the context of sport into other industries, including the health industry. Kallia O. Wright and Kesha Morant Williams detail the impact of gender and sexuality in health-care settings, paying special attention to the illness-related experiences of patients. Their chapter, "Gender, sexuality, and health communication during the illness experience," argues that women's health needs are framed around sexual and reproductive functions and that doctors minimize and resist the concerns of patients, particularly women of color. Men, on the other hand, often avoid health-care settings, viewing sickness as weakness. This avoidance extends to men's mental health around the world, where men frequently fear shame if a mental diagnosis is health or shared. Wright and Morant Williams also find that while a number of physicians state that they are comfortable treating LGBTQ patients, they are not educated on how best to communicate with the patients. The chapter concludes with clear strategies for patients (including knowledge of one's own health and self-advocacy) and practitioners (training and asking questions of patients) that can reduce the impact of gender stereotypes in health-care settings.

Online contexts represent another area where gendered imbalance occurs. In their chapter, "Women first: Bumble™ as a model for managing online gendered conflict," Sean Eddington and Patrice M. Buzzanell identify mobile dating applications as a technoculture that amplifies gendered conflict and harassment. Using gender structure theory, they argue that organizations are far from gender-neutral and normalize the privileging of the masculine over the feminine. The authors use the online dating platform Bumble as a case study of macro- and micro-level conflict management strategies that subvert traditional power dynamics. Bumble espouses feminist values and, as examples, encourages its women users to make the first move, has rules regarding the types of photos that can be shown, and offers a zero-tolerance policy for harassment and disrespect. Such strategies for managing organizational conflict represent a model to be used in other organizational settings.

Mel Stanfill's chapter, "Straight (white) women writing about men bonking? Complicating our understanding of gender and sexuality in fandom," discusses fan studies and slash fan fiction in the conclusion to this section. The author defines slash fan fiction as stories about men in relationships with other men, and femslash as inclusive of slash between men, but with writers representing their own identities. Stanfill thoughtfully argues that fandoms are predominantly white and that men and women fans feed into gendered stereotypes. That is, slash between men is often homophobic, with women's roles being to advance the relationship between men. In addition, marriage and family are valued as goals, furthering heteronormative values. Ultimately, while fandom breaks down norms, it also upholds numerous power structures.

While the contexts discussed in this section are varied, it is evident that historically marginalized communities continue to suffer because of imbalanced and gendered power structures that highlight an ideal worker who represents straight white maleness. The suffering can be physical, such as in health-care settings, but it can also be more covert, such as in online contexts. Even contexts that claim to effectively reduce gender stereotypes in numerous ways, like fandom, have the ability to reproduce dominant structures that privilege maleness, whiteness, and heterosexuality.

Strategies of resistance, such as sharing narratives, modeling organizations that espouse feminist values, and training those in positions of power are useful in centralizing those that are at the margins of these contexts. Some strategies of resistance occur at micro levels, allowing one individual or small group to survive varied contexts of discomfort and violence. Yet other strategies can occur at macro levels, with the ability to create safe spaces on a larger scale. The chapters in this section do not privilege one strategy over others, and it is worth noting that survival and change are the most significant and meaningful outcomes.

22
ORGANIZATIONAL DISCOURSE AND SEXUALITY IN MALE-DOMINATED ORGANIZATIONAL SETTINGS

Clifton Scott, Aly Stetyick, and Jaime Bochantin

The idea that most organizational communication processes reflect and sustain hegemonic masculinity has been well demonstrated in scholarship for decades, but the more specific role of sexuality in organizational life has received far less attention.[1] In a sense, the emerging "#MeToo" social movement has resulted in hegemonic masculinity being taken more seriously in a manner that begins to conform to what critical and postmodern gender theorists have been saying about organizational life for decades. But in another sense, those of us who study gender can point to a relatively small number of studies that examine hegemonic masculinity in the very setting such movements might suggest examination is most badly needed: male-dominated organizations and occupations.

For example, during the time we were writing this chapter, approximately 20,000 employees in one male-dominated domain, high-technology organizations, staged a global walkout. They walked away from their work for the day to demonstrate against cultures (organizational, occupational, popular) that not only fail to sanction abuse but in some ways actively encourage it through attraction, selection, and attrition processes that sustain employment of these bad actors.[2] These and other public protests, along with many of the high-profile corporate terminations of the accused, take an openly critical view of hegemonic sexuality in the workplace, one that has ironically become more mainstream among practitioners than the scholars who study them. This is particularly true of sexuality in male-dominated organizations and occupations where most research remains normative, ahistorical, and preoccupied with individual differences (e.g., gender-based differences in attitude or behavior assumed to be static, natural, normal, and absolute) rather than the ongoing discursive processes that constitute them. This oversight is particularly extreme when it comes to sexuality, an apparent element of diversity that organizational research tends to ignore, perhaps under the assumption that sexuality can or should be neutralized in the workplace. In other words, although broad movements like "#MeToo" are still in their early stages, a relatively critical and progressive stance on hegemonic masculinity has become more mainstream in the world of practice than in scholarship. Practitioners increasingly recognize what most scholarship does not: that organizations and occupations are not neutral when it comes to gender *and* sexuality and historically have reified normative practices and policies that privilege already dominant groups (cisgender men, heterosexuals, management, etc.), a diversity issue that not only damages careers and generates lawsuits but also inhibits organizational effectiveness.

In this chapter we provide an updated review of the literature on sexuality in male- dominated organizations and attempt to show how the study of gender and sexuality in the workplace in *general* may benefit from an expanded understanding of sexuality in the *particular* context of male-dominated organizations and occupations, one that views this sexuality as both process and product of organizational and institutional discourse, the situated language use in and between organizations that makes communication possible.[3] We begin by defining key terms and describing the Ashcraft framework around which we organize this review.[4] The remainder of the chapter discusses the topic in terms of sex-based differences, workplace sexuality as performance, and strategies of resistance.

Key terms and assumptions

To begin, let us briefly describe our working assumptions about key terms. Following the lead of most gender and sexuality research, we conceive of sex and gender as separate but highly related phenomena. Sex type is at least initially related to anatomical differences apparent at birth, and gender is behavioral and socially constructed and reified, though still highly predicted by sex type.[5] Sexuality, as a term, is generally defined in one or both of two ways. In one sense, sexuality is about type of sexual orientation, whether sexual orientation is understood in binary terms (either heterosexual or homosexual) or in more fluid terms (e.g., pansexuality). In another sense, sexuality may be understood more in terms of activity, mutual or individual behavior, that is related to perceived sex differences and/or sexual acts. This second definition may include a wide range of sexual behavior that can be thought of as having "message value," (i.e., behavior that potentially communicates something—positive, negative, or otherwise—to an audience intentionally or unintentionally), whether it be how closely one worker stands next to another during a job task, what interviewers assume about the sexuality of job candidates they interview, or overt harassment that would legally constitute what is understood at least in the United States as quid pro quo or hostile work environment harassment.[6]

Although we have pointed to some conceptual differences between gender and sexuality, it is important to note that the two constructs are often intentionally or unintentionally conflated in the organization studies literature, including organizational communication studies. Indeed, if we assume, as the vast majority of communication scholarship does, that gender norms are not fixed but rather socialized, constituted, and reconstituted over time in formal and informal interaction, then the enactment of sexuality in the workplace has significant implications for gender norms and vice versa.[7] Gender and sexuality play out in organizational and occupational environments that are always and already gendered, so enactments of gendered sexuality by individuals are always process and product of prototypically masculine organizational environments that enable and constrain (bureaucratically and symbolically) how sexuality is understood and carried out.[8] Although we do not believe that sexuality and gender can be neatly disentangled in the course of a handbook chapter, we attempt wherever possible to highlight sexuality as figure in contrast to a gendered organizational ground.

We generally view sexuality in work life as constituted over time through discourse, everyday talk situated in a given cultural setting that makes communication possible.[9] In other words, the normative state of sexuality in a given organization or occupation is a direct consequence of how we talk about sex and gender. As such, sexuality is normative but ambiguous, meaning that although sexuality in a given organization may reflect and reinforce particular themes, the content of communication about sexuality is "neither arbitrary nor absolute" and open to inference and interpretation.[10] For example, as Clair demonstrates, stories of sexual harassment and the experiences that inspire them may be characterized discursively in ways that enable or constrain hegemonic interpretations (e.g., "boys will be boys").[11]

A final note about our guiding assumptions regards the impacts of hegemonic male sexuality in organizational life. Although discursive and behavioral norms around sexuality in male-dominated organizations are certainly a product of dominant, privileged social actors, we assume that they also impact members of dominant groups, in addition to the marginalized. In other words, sexuality in male-dominated organizations is not just about masculinity or impacts of gendered performances of sexuality on men. It is about all of us. Thus, since gender and sexuality are highly relational phenomena, the topic of sexuality in male-dominated organizations includes not only discourse of and about the sexuality of white men but also the relevant discourse of a broad range of identity groups whose communication also contributes to local, often contested, meanings for various sexual subject positions related not only to sexuality but other categories of social identity (e.g., working-class lesbian sexuality, transsexuality, pansexuality, manly sexuality, masculine sexuality, gay male sexuality, straight female sexuality, and cisgender sexuality).

Ashcraft's four frames and male-dominated organizations

We organize this review mainly around three of four categories described in Karen Ashcraft's "four frames" approach to the study of gendered organizing.[12] The purpose of these frames has been to categorize the ways in which gender can be assumed to relate to organizational discourse and organizing.[13] The Ashcraft framework has brought some order to a generally fragmented and disordered gender and organization literature that included multiple streams of research conceptualizing gender differently depending on the assumptions and goals of the research.

Although Ashcraft developed this framework for the study of gender, we believe it offers just as much promise in its capacity to serve as a useful organizing scheme in the study of sexuality in male-dominated organizational contexts for at least three reasons. First, as scholars of identity continue to call for intersectional approaches, using a framework developed with gender in mind to make sense of findings about sexuality enables scholars to assess the literature with a value for understanding conceptual variations the two literatures have in common. Second, although we note a growing consensus that gender and sexuality are two independent phenomena (i.e., one's gender identity does not automatically predict their sexuality), the two are shaped by very similar discursive forces and ideological positions. Finally, although gender is increasingly understood in less categorical, more fluid terms, the distinction between the two remains unclear in many areas of scholarship.[14] As readers will see from our synthesis, the distinctions that comprise Ashcraft's frames highlight competing approaches to the study of gender that are as useful to the study of sexuality with some minor conceptual adaptation.

Frame one: sexuality-based difference at work → discourse

One of the most popular and palatable ways of thinking about the relationship between sexuality and discourse is to begin with the assumption that there are stable, predictable differences in the discursive habits of individuals (i.e., individual differences) that function systematically as effects of relatively fixed categories of sexuality. From this perspective, straight men, for example, tend to have a deep-seated personal communication style that corresponds to their sexual orientation in a way that they typically believe is qualitatively different from that of gay men. This frame generally relies on a representational view of language, in that it assumes discourse is not generative but rather a reflection of "baked in" differences (genetic and sociological) in sex and gender identities.[15] In this view, sexuality is understood as more behavioral than discursive, where discourse becomes a secondary, relatively unimportant reflection of sex-based differences in behavior. This "difference" view characterizes gendered and sexualized preferences and practices as relatively stable and predictable. For example, the leadership practices and linguistic choices of executives are understood as "individual differences" that are merely reflected in discourse, not always and already constituted by discourse. From this point of view, gender and sexuality are enacted "in organizations" and do not fundamentally characterize the organization or its culture but rather its individual inhabitants and their "engrained personal communication style."[16] This is by far the most mainstream way of framing gender and sexuality, popularized in numerous airport bookstores and departments of psychology.

Frame two: hegemonic sexuality → organizational performance

A second frame considers sexuality as identity work that is performed more or less credibly as impression management or cultural performance in mundane interaction for local organizational or occupational audiences.[17] Frame two focuses on the way that sex and gender identities are embodied and performed, constituting multiple, competing subjectivities or discourses. From this perspective, gendered enactments of sexuality occur not because of fixed, individual-level differences but rather because of how sexuality and gendered roles are normatively performed in a given organizational culture. The meaning of these performances is co-constructed via interaction with various audiences in the course of impression management that is more or less effective in accomplishing its goals with various outcomes, intentional and unintentional. Research on the "doing" or performance of gendered sexuality exemplifies this work.

Frame three: organizations and occupations as sexualized institutional forms

Moving up to an institutional or occupational level of analysis, this frame conceives of sexuality as a discursive "narrative of gender and power relations that gels into organizational design and thereby fosters gendered interaction among members."[18] From this perspective, discourses of sexuality are the products of an ongoing dialectic between abstract, latent texts (e.g., the idea that allegations of harassment are personal rather than organizational problems) and concrete, conversational enactments of those texts (e.g., "he said, she said").[19] Although this frame is partially based on performance-oriented elements of the previous frame, it goes a step further by including not just the figure of performed identities of individual employees but also the discursive ground of the organization and/or occupation as text.[20] Organizational discourse constitutes not just the identities of members but also the organization itself, enabling the emergence, sedimentation, and occasional transformation of a textual repertoire shared by members. Organizational communication from this perspective is an ongoing interplay between local conversation and the prevailing texts discourse typically draws upon and reproduces. As Ashcraft summarizes, "Organization produces gendered discourse, even as it is also a product of such discourse."[21] Over time, a shared narrative about gender develops and shapes the organization by influencing organizational structure and design (e.g., military policy on sexuality).[22] Given space limitations and the limited research on the institutionalization of sexuality, especially at the occupational level of analysis, this chapter does not focus on research reflecting this frame.

Frame four: sexuality discourse as societal narrative

Finally, the fourth frame assumes that macro culture, more than meso-level institutional activity, shapes the narratives that comprise organizational and occupational discourses. Here, culture is seen as transcending organizations and occupations such that the locus of control for sexuality is macro cultural discourses, not organizational or institutional (e.g., occupational) forces. From this perspective, for example, military policies about sexual harassment or transgendered personnel are shaped primarily by the broader culture more than the discursive structures (texts) and (conversational) practices that comprise the military as institution or a particular armed force as an organization. Organizational discourse is significant only to the extent that it does the work of drawing upon and sustaining cultural discourses.

The discourse of sexuality-based difference at work

Although we attempt to limit the current review to differences related to sexuality rather than gender and sex, it will be difficult to disentangle related and systematic oppressions based on sexuality and those based on gender and sex, as they are "inherently linked."[23] People view gender, sex, and sexuality as highly salient in most contexts, and the workplace is no exception. Social, cognitive, and evolutionary psychologists have found that the human brain is remarkably adept (and eager) to identify the sex of human stimuli.[24] Others have argued that socialization patterns lead us to disproportionately weight sex-related information against other available signals or cues.

Although there has been almost no empirical evidence found to support differences in job *performance* between white, heterosexual males and outgroups in male-dominated organizations, "common (non)sense" beliefs about such comparisons have been particularly difficult to extinguish. The "ideal worker" norm favored throughout culture and media is typically white, straight, and male, which leads to an implicit othering and devaluing of those who do not perfectly fit the bill.[25] Women's bodies are seen as too unruly and sexual to comply with supposedly universal standards of job performance, leading to discrimination and harassment in hiring, promotion, and other areas of life at work. Meanwhile, concerns about reactions from coworkers lead many homosexual workers to cover their status by "passing," often to the detriment of their mental and physical health.[26] Academia has largely shied away from producing research that can (a) dispute these claims and (b) be translated and distributed for public consumption, likely due in part to the politicization and privatization of scholarship.

Though this literature is lacking, there is a considerable body of research on more distal predictors of job performance, including individuals' internal experiences in and around work, job attitudes, and both systematic and isolated (in)actions that disproportionately affect minority groups, including discrimination and harassment. From here, we will review the "not-so-objective standards" applied differentially across groups in the workplace, the effects of harassment, discrimination, and related events, and the resultant pressure to conform to available norms of the ideal worker.

(Not so) objective standards and policies

Those who fail to acknowledge the differences in treatment between white, heterosexual, and masculine employees and traditionally "othered" groups in the workplace typically cite extant policies as catchall defenses against systematic disadvantages; however, this proposition has received little empirical support.[27] Moreover, such policies are often difficult to enforce (e.g., assembling enough evidence to substantiate "antidiscrimination" claims) or applied differentially across individuals (e.g., mid-level supervisors using their discretion to judge the legitimacy of requests for time off). As such, policies alone are rarely able to "level the playing field."

The discursive reach of the "ideal worker" extends even into public policy around employee benefits. The United States' somewhat outdated healthcare system, which links access to medical services with employment status, lends employers a disproportionate amount of control over employees' sexual autonomy. In a landmark case, Hobby Lobby, a closely held Christian craft store, won the right to deny its female employees birth control and abortion coverage by citing company leadership's objections to such policies, which were based in religion.[28] Interestingly, there were no similar comments or complaints regarding erectile dysfunction drugs. Similarly, medical care for those transitioning between genders was recently sensationalized by the national news media as an undue strain on the armed forces' (notoriously expansive) budget following an attempt from the President to block their service altogether. Insiders and medical professionals both insisted that the care of such individuals was *not* an undue cost. An amendment to cut funding for transitions was rejected, as was the attempted ban on transgender enrollment in the armed forces; however, reports of delayed processing of applications—an informal ban—have surfaced.[29]

Almost all women experience the consequences of inaction, biased attributions, and outright discrimination, regardless of their status outside of the employment environment. Given the dominant discourse of women as "ideal mothers," and Western society's pseudo-reverence for motherhood, one might reasonably expect that mothers, at least, would be well protected by their employers. Instead, mothers have been repeatedly found to be targets of discriminatory organizational practices. Hebl et al. found that women who appeared pregnant were subject to increased hostility during job interviews, especially in male-dominated organizations; however, they also found that those same women received increased positive attention when posing as customers rather than job applicants.[30] Similarly, Correll, Benard, and Paik found that women whose resumes contained signals of motherhood received callbacks half as often as women whose résumés did not.[31] Pregnancy is also often overlooked as a legitimate medical condition by employers looking to maximize employee value and minimize cost. For example, the United Parcel Service (UPS) has refused disability payments to pregnant women and instead forced them to keep working or go without pay well into their pregnancies. The company argued that its disability policies were not envisioned as solutions to "the elective problem of pregnancy."[32] Similarly, XPO, a leading shipping and fulfillment company, refused to honor doctors' notes for pregnant warehouse workers, resulting in a wave of miscarriages in a Tennessee warehouse.[33]

Should a working woman manage to have children, the accommodations necessary to raise them adequately are still difficult to find or to use in many organizations. In the United States, paid maternity leave remains extremely controversial, and even female managers have noted that they are uncertain about whether paid maternity leave is necessary and justifiable.[34] Such resistance is unfortunate for both individuals and their employers.

Some large companies seem to have embraced work/life balance accommodating policies more readily than their smaller counterparts, which may face significant financial constraints. Companies facing tight talent markets are also apt to enact more policies aimed at "work/life balance" to attract applicants. Recent years have seen an increase in the number of companies offering paid paternal leave, paid leave for couples who have recently adopted children, and "flex-time."[35] Despite these advances, employees who negotiate work-from-home arrangements to accommodate childcare still face significant stigma.[36] These accommodations are even more difficult to obtain for those who arguably need them most: shift workers whose low wages preclude them from affording childcare.

People that fall outside the narrow ideal worker norm are systematically blocked from leadership positions. Scholars commonly cite a "pipeline problem" that limits the number of women entering a given field—especially those typically associated with masculinity (e.g., science, technology, engineering, and mathematics [STEM])—and thus the pool of individuals available for promotion. However, the pipeline problem does not adequately explain the lack of promotion for women already inside such organizations. Glass ceiling effects still stall advancement of women in many fields. Women accounted for just 22% of C-suite executives across the 329 companies that participated in McKinsey & Company's 2019 Women in the Workplace survey.[37] Traits that are perceived as feminine also systematically depress performance raters' perceptions of person-job fit in masculine organizations, leading to fewer promotions for women.[38] Interestingly, this effect holds constant even when women receive better performance ratings than men.[39] Over time, as women are systematically passed over for promotions, their satisfaction with promotion and trust in their organizations erodes.[40]

Even at the top, expectations of "nontraditional" leaders are still the result of stereotype bias. Despite an increased interest in feminine leadership styles, tropes associated with the fetishization of feminine leadership have received little empirical support. Emergent research on the "glass cliff" shows that women and other minorities are often promoted to executive positions in times of organizational crisis. Boards expect nontraditional leaders to "shake things up" and let them go with little thought when such expectations are not translated into reality.[41] Furthermore, Diefendorff et al. found that despite the dominant narrative of the "altruistic woman," organizational citizenship behaviors were distributed evenly across male and female employees.[42]

As the above cases show, policies often fall short of preventing discrimination in practice; in fact, they are sometimes intentionally written in vague or abstract terms that preserve actors' ability to *police gender* at their discretion. As such, organizations discursively and practically privilege masculinity, subtly reinforce heterosexual gender roles, and systematically disadvantage those deemed "less than ideal."

In many cases, the hegemony associated with policing gender can escalate to physical and/or sexual harassment in the workplace, or even battery and assault. Physical harassment encompasses all unwanted, threatening, or belittling behaviors aimed at an individual or group. Sexual harassment is "unwanted sex-related behavior at work that is appraised by the recipient as offensive, exceeding [one's] resources, or threatening [to one's] wellbeing."[43] Available data suggests that in the United States, one in two women will be harassed at least once during their time in academia and at work and that such events can have significant long-term effects on wellbeing at work.[44] Waldo, Berdahl, and Fitzgerald suggest that men also experience some of the same categories of sexual harassment as women (unwanted sexual attention, sexual coercion) in addition to "lewd comments," "negative remarks about men," and "enforcement of the heterosexual male gender role."[45] Sexual minorities are one of the most targeted groups with regard to harassment, assault, and discrimination. In fact, 25–66% of sexual minorities have been harassed at work or experienced discrimination based on their sexual identities.[46] We are unaware of studies that track the frequency of such problems in male-dominated organizations.

In addition to issues faced by their heterosexual counterparts, LGBTQ individuals also commonly face heterosexism at work. Heterosexism is classified as a minority stress and is defined as inappropriate sexuality-based behaviors at work used to reinforce traditional gender structures and their related hierarchy.[47] Such behavior is a less extreme, yet related, method of policing both sexuality and gender. The negative effects of heterosexism include distress, health problems, decreased job satisfaction, higher turnover intentions, lower health satisfaction, increased withdrawal behaviors, and decreased perceptions of person-organization fit.[48] These are all especially likely to affect employees in male-dominated organizations, which are linked to increased incidence of direct heterosexism.[49]

While prevalent, some heterosexist behaviors can be difficult for management to detect and prevent without an initial incident of victimization and subsequent appeal for intervention by the affected individual. This is especially true of implicit and/or indirect heterosexist behaviors (e.g., failure to promote LGBT employees, conversational assumptions that LGBT employees are in heterosexual relationships).[50] Unfortunately, the events needed to trigger protective policies eventually take a toll on many LGBT employees. The mere presence of antidiscrimination policy does little, if anything, to prevent heterosexism, encourage disclosure behaviors, improve affective commitment and job satisfaction, or reduce conflict between work and home.[51] Moreover, those who believe their companies are tolerant of heterosexism are more likely to experience it.[52]

Pressure to conform

Although many organizations offer at least minimal protection from harassment at work, there are notable exceptions that may push employees to attempt to cover, rather than protect, their identities. For example, the United States has yet to add LGBT individuals to the list of protected classes that can legally take action against organizations for harassment and discrimination. Furthermore, attempts to do so at a local level have been challenged and struck down by higher courts (e.g., North Carolina's now [in] famous House Bill 2, which sought to remove city-wide protections for members of the city of Charlotte's local LGBT community).[53] In order to avoid harassment and discrimination, many resort to attempts to emulate normative heterosexuality and masculinity by altering their appearances, patterns of emotional display, and social and task-based behaviors at work. In instances in which this is not explicitly endorsed by organizational leadership, policies like the United States' Armed Forces now-defunct "Don't Ask, Don't Tell" push employees to internalize and manage their deviations from the norm by refusing to acknowledge such deviations exist.

The pressure to conform takes its toll on seemingly cisgender[54] employees who choose to remain closeted at work. Being open about one's sexuality (i.e., disclosure behavior) is positively related to affective commitment, harmony between home life and work life, lower role ambiguity, continuance commitment (i.e., commitment out of economic necessity), and lower role conflict.[55] Individuals are more likely to engage in disclosure behaviors if they believe their company is supportive of LGBT individuals, and if top management supports antidiscrimination policies.[56] Coworker reactions to these behaviors mediate the relationship between being "out" and reduced job anxiety as well, evidencing a need for holistic support and a feeling of person-organization fit, rather than mere policy protection.[57] While job stress remains the same for individuals both "in" and "out," the other benefits of open and comfortable work environments are less clear.[58]

As we will cover in the next section, organizations' efforts to cover femininity and other deviations from the ideal worker norm can significantly alter employees' interpretations of their duties, roles, relationships, and cultural or organizational norms and expectations.

Sexuality and discourses of organizational performance

Gender and sexuality have become central features of organizational culture through the continual performance and institutionalization of key behaviors and rituals.[59] However, research tends to focus on gender to the exclusion of concerns for sexuality. But organizations are wrought with issues related to both, as gender and sexuality inform one another. Brenda Allen has said that when organizational members do not directly refer to sexuality at work, sexuality is always an "absent presence"—thus the focus on performance.[60] As a result of experiencing gender and sexuality at work and participating in organizational discourse, shared understandings, beliefs, behaviors, and symbols develop through interactions among organizational actors. Scholars find that workers in a wide range of occupations and organizations "do gender" in particular ways, based on assumptions about what customers, peers, and supervisors like, motivations, and "normal" interactive behaviors.[61] It is easy to see how issues related to sexuality also inform our occupations. Particularly in service-oriented occupations that privilege femininity, women work as "traditional women" and are instructed (more or less directly) not to deviate from the script. Thus, femininity is constructed and reified in ways that reinforce heterosexuality and male dominance and "naturalize" stereotypical images of women.[62] In this section, we review how gender and sexuality performances are ongoing aspects of organizational life, how the "professional" body gets performed at work, and the implications of emotional labor.

Gender and sexuality as ongoing organizational performances

We know that organizations are anything but gender neutral.[63] That said, gender and sexuality performances are ongoing aspects of organizational life revealed through rites, rituals, practices, policies, and other aspects of culture. Allen describes these as performances that embody issues of power, control, resistance, and change.[64] It also suggests certain value systems related to "performing" one's sexuality (or sometimes, hypersexuality). One of those value systems is hegemonic masculinity, which often includes men's sexual domination of women in order to ensure continuation of the gender hierarchy.[65] This can be seen in highly masculine, male-only environments such as fraternities. Fraternities employ sexual symbolism in the socialization of newcomers, asserting heterosexuality with and for other men.[66] While men perform hetero-masculinity through "sex talk" and other heterosexual posturing, the fraternity men also expect each other to pursue women at house parties, bars and clubs, and many other social gatherings.[67] In fact, Sweeney found that men who "perform" heterosexual success are rewarded with peer attention and status improvements.[68]

Just as in fraternities and other male-dominated, highly masculine environments, it is through everyday organizational processes and practices that social discourses of power are appropriated, reproduced, and/or transformed in ways that enable and constrain how members enact and embody professional identities.[69] Buzzanell states: "Gender organizes every aspect of our social and work lives including how we formally and informally communicate in organizational settings."[70] Thus, our language and discourse help to reproduce stereotypical masculine and feminine identities and ideologies in association with sexuality. This seeps into various work contexts. For example, pink-, blue-, and white-collar job designations help to preserve asymmetrical power relations and reproduce hegemonic discourses of sexuality and culturally "appropriate" roles both men and women should play. Blue-collar workers tend to use more explicit sexual language and are generally permitted to be more vulgar. In fact, some would say it is an expectation.[71] Sexualizing the workplace may even instantiate a form of resistance against the rigors and repetition of their work.[72] Unfortunately, this only furthers women's subordination and men's oversexualization of masculinity. Much of the blue-collar workforce is male-dominated and tends to espouse masculine norms. In this environment, women often have to choose between the "defeminization" of their role (an overemphasis of the job expectations and gender norms) or they must "deprofessionalize" (an obligation of meeting sex role norms while on the job) for the sake of the workplace.[73] As a result, they are often trapped within the duality of unfeminine woman versus incompetent worker. Thus, many women end up exiting blue-collar jobs as a result of an overly sexualized environment.

"Doing gender": performing the "professional" body

While gender and sexuality get enacted via organizational rites, rituals, vocabularies, and practices, the body itself (both male and female) also get implicated in the organizational performance.[74] Furthermore, the body is a "text" that others read and interpret in line with divergent expectations for men and women when it comes to identity and the body.[75]

First, there exists the professional, heterosexual male body. This represents the status quo to which all other bodies are measured. Images of men's bodies and masculinity pervade organizational processes, marginalizing alternate perspectives. In contrast, the professional, heterosexual, female body is marginalized against the male standard. Women are told to "control" and discipline their apparently unruly bodies.[76] To get ahead in organizations, they need to consciously tone down the ascribed sexuality of their bodies, lest they risk disrupting the (masculine) organizational environment.[77] Women are constantly trapped in the double bind of not dressing too feminine while at the same time not dressing too masculine, as that would be a deviation from their gender stereotype.

Additionally, an unfit body equals lack of endurance, while a fit body communicates to others that one has the ability and endurance to complete tasks.[78] Thus, the goal for women at work is to mask any "unfit" tendencies, any "leakages" due to pregnancy, menstruation, lactation, or menopause, and to control excessive bulging that the physical body may possess and communicate back to colleagues a lack of competency or stamina for work. Dellinger and Williams' study of the use of makeup at work found that some heterosexual women wear makeup as a performative indicator of their heterosexuality to avoid being labeled as a lesbian. In addition, some gay women were noted to wear makeup in order to "pass" as heterosexual by conforming to the institutionalized organizational norms of what femininity should look like.[79]

This notion of "passing" has been discussed in past research and is something that many members of the gay community participate in actively in order to conceal their sexuality. They participate in passing strategies/techniques to avoid performing anything aside from heteronormativity. In fact, Spradlin describes six communication strategies that lesbians and gay men use at work: distancing from colleagues, dissociating from other known gay employees, dodging certain topics, distracting, denial, and deceiving or delivering dishonest messages of heterosexuality as opposed to homosexuality.[80] More recent work continues to document the mostly negative consequences associated with gay and lesbian employees (e.g., stigma, discrimination), especially lesbians in male-dominated organizations, working "out" in the workplace, and not attempting to pass.[81] Unfortunately, many women help reinforce heteronormativity by adhering to heterosexual norms. This is due to the latent sexual systems that are modeled on the male-female relationship.

Sexuality discourse as cultural narrative

As a social institution, the modern workplace appears designed to erase sexuality from the organizational context.[82] This is done to varying degrees of success, as the complete erasure of sexuality from the workplace is functionally impossible. In contrast to these attempts, the media paints a different portrait of the organization, one dripping with gender tropes and sexualized relationships. In some cases, these accounts seem to be fairly representative; in others, they are exaggerated to the point of caricature. In this section we will review sexualized work-based tropes in media accounts of organizational life. Here we will condense these tropes, patterns, and characters into three main categories to uncover deeper meanings: reproductions of heterosexual relationships, exclusion from sex as a proxy for powerlessness, and the "othering" or erasure of certain individuals and groups.

Reproduction of heterosexual relationships

Many of the most iconic professions and occupations portrayed in the media are built around sexualized male-female pairings that mimic heteronormative and traditional relationships, with masculine "principal" workers and feminized assistants. Examples include executives and secretaries, doctors and nurses, and principals and teachers. As Ashcraft et al. proposed, this is not accidental—masculine traits associated with high status jobs in media likely entered into public discourse as a result of image-building campaigns, which simultaneously mirror and are mirrored by culture.[83] Even in the absence of such formalized systems, the "work spouse" trope, for example, encourages the formation of exclusive opposite-gender bonds that quite literally reflect existing sexual relationships.

The above rests on an implicit understanding of men as career-centered and successful and women as subservient and domestic. Men are typically portrayed as the image of corporate success: forceful, aggressive, overpowering, technically skilled, and largely emotionless (save for "righteous" anger).[84] Successful women in the workplace are portrayed as defiant and intimidating.[85] Caught in a double bind, they must at once be extremely feminine, young, and conventionally attractive in appearance, yet distinctly anti-feminine in attitude.[86] Women's control over their lives at work also seems to be under a timer—while young women may be able to excel in the corporate world, women who are too young are ignored, while women who are too old are viewed as having failed to establish families or even lusting after much younger men. No such age restrictions exist for men. Similarly, women are expected to talk exclusively about men, compete over men, and manipulate men with their wardrobes. Most of women's character development in television and cinema revolves around their male counterparts. Men, meanwhile, are given complex and varied storylines with neatly compartmentalized romantic elements.

Even constructions of fictional characters' work reflect latent cultural stereotype bias. Much contemporary writing and film reflect the gender segregation of work, which have been described elsewhere extensively.[87] As this segregation is maintained in popular discourse, it reifies itself at the occupational, organizational, and individual level.[88] More technical and "worthy" creative work is typically associated with men, while more supportive, domestic, and "unworthy" artistry work is reserved for female characters. These characters also notably perform the "heavy lifting" of emotional labor for the men around them, including customers, male friends, and romantic partners.

This division can even be observed in the ways that workspaces are portrayed. Most "technical" spaces appear quite masculine, with sharp edges, cool colors, and metallic surfaces; "creative" and "comforting" spaces, meanwhile, appear quite feminine with their rounded edges and warm colors. This is not only propagated by media but reified by interior designers themselves. Interesting alternative understandings are possible—for example, laboratories "giving birth" to inventions, microorganisms, and medicines could be portrayed as feminine.

Sex(lessness) as power(lessness)

The problem of the masculine "ideal worker" extends into the intersections of gender and age. While sexuality seems to have permeated every aspect of organizational life, the intersections of gender and age present (possible) boundaries to its scope and span. Trethewey's interviews with aging women at work found that the sexlessness associated with aging women could be interpreted as either erasure or empowerment.[89] Unfortunately, the former is much preferred by current media narratives. Older women who escaped the double bind by relinquishing powerful work identities are presented as grandmotherly and nonthreatening, while women who retained their power at work by rejecting family obligations are portrayed as lonely spinsters, full of regret, and lacking meaningful relationships. Menopause is seen as a relinquishing of the power, attention, and influence associated with the sexuality of younger people by the media. Indeed, it is rare to find someone who "has it all" on television.

Sexlessness and the exclusion from power also extend to men and women who do not adhere to traditional notions of femininity and masculinity. Fat or ugly women are portrayed as ineffective, careless, desperate, or, at best, *funny*—a trend certainly intensified by niche comedians who have taken ownership of the space, playing a seemingly uniform character across various media. Men who deviate from heteronormative masculinity by breaking emotional display rules are also portrayed as pandering, ineffectual, effeminate, and/or overly sensitive. Though the physical standards men are held to are somewhat less strict than those for women, those whose bodies are deemed too large, too small, or too old are similarly plagued. As such, they are denied access to the hegemonic masculinity of their more traditional counterparts and used as comic relief. Even the days of the "fat cat" boss are over, as he is replaced by more coolly intimidating and conventionally attractive figures.

Normalizing and othering sexuality

Despite progress made toward representation of various genders, sexualities, races, and ethnic groups in recent years, widespread tokenism persists in films, television, and advertising. Rather than normalizing the sexuality of typically "othered" groups, these representations provide nominal representation without allowing for the depth and variability of experience typically enjoyed by majority groups (namely, white heterosexual men and women). As such, there is little room in much mainstream media for sexuality involving racial minorities, and that which does appear is often fetishized, caricatured, and co-opted by members of majority groups. The aging are similarly ignored, with an exception made for older men who have married much younger women. There appears to be even less room for sexual minorities. Asexuality, lesbianism, pansexuality, bisexuality, and polyamory are still mostly missing from media accounts of organizational life. The exception here seems to be gay men, who, nonetheless, are reduced to the "sassy gay friend" trope and again stripped of the substance of their sexuality.

Taking into account Ashcraft's theory of the present other, heteronormativity and its elevated status are also held up implicitly by practices that quietly exclude nontraditional couples and families.[90] These include organizational events that assume male-female pairings, like company dinners and "bring your child to work" days which favor traditional family units and employees of an ever-less-specific age range.

Strategies of resistance and sexuality discourse

Resisting the status quo can be difficult. How then, do the "powerless" carve out discursive spaces for influence, control, and resistance? In this section, we discuss different approaches to resistance including both overt and covert resistance tactics that get used in the context of sexuality at work.

Power and resistance

Contemporary approaches to power embrace a relationally negotiated meaning where power is not an attribute of an individual but is a relationship negotiated between individuals. Power can also be understood as domination—as such, it is far from a neutral feature of contemporary society. Constructions of "power-as-domination" both privilege and marginalize individuals or entire groups, which necessarily create social inequalities and serve organizational interests.[91] Power-as-domination is also a primarily masculine construction of power.[92] According to research, women conceptualize power as a relationally negotiated experience (i.e., power with) as opposed to the hierarchy and domination described by their male counterparts.[93] Marshall notes that men tend to value both boundaries and hierarchy while women tend to value networking and more personalized, flowing communities.[94] Further, men tend to view power productively as hierarchical authority.[95]

There are many approaches to the concept of power. Like power-as-domination, the critical-interpretive approach sheds light on how the less powerful may seek to retake control. The basic tenet of this approach is that reality is socially constructed and constituted (and reconstituted) through the language we use and the relationships we cultivate. Thus, the world we are born into exists in tension against the world we create.

Communication is central to understanding how organizational realities are created, reproduced, and altered.[96] This is potentially enabling and constraining.[97] Power relations are reproduced through discursive formations.[98] Dominant organizational meanings permeate our organizations, but where there is domination, there is resistance. This can enable even the most oppressed individual to experience fleeting moments of power and dominance.[99] Deetz has found that humans are bound to resist (both effectively and ineffectively).[100] Moreover, the literature shows that most people resist covertly through subtle action (which will be described shortly) although overt resistance is also possible.[101]

Many employees have historically engaged in a form of covert resistance through subtle and tacit acts that serve to empower them at the individual level. These types of resistance strategies are not usually done to create any sort of large-scale change at an organization, but rather they are done so employees can make it through their day feeling vindicated, even if only in a small way.[102] Many feminists believe women and other marginalized groups can resist specific instances of patriarchal power or engage in small-scale resistance, either individually or with coworkers. Examples include participating in rituals or practices beyond the direct observation of oppressive power holders that reject organizational rules such as stealing office supplies or breaking dress code requirements.[103]

Scott argues that all subordinate groups employ strategies of resistance that go unnoticed by superordinate groups. These are encapsulated in his concept of "hidden transcripts"—interactions, stories, myths, and rituals that employees participate in offstage, or beyond the direct surveillance of power holders and the public.[104] In his analysis, the open, public interactions between dominators and the oppressed are contrasted as "public transcripts." Thus, marginalized or dominated groups cannot be understood by their public actions alone. While the oppressed may appear to accept their domination in public, they may also question that domination offstage. Lexa Murphy provides a relevant organizational example of this duality in her study of flight attendants, who use hidden transcripts to carve out spaces of resistance in an otherwise rigid, bureaucratic environment.[105] Other relevant organizational studies include Bell and Forbes' examination of covert resistance through office folklore. They drew particular attention to cartoons and pictures posted in workspaces by employees that reveal individual-level differences and celebrate employees' diversity.[106] Bochantin and Cowan, meanwhile, describe ways in which female police officers resist the masculine, male-dominated nature of policing through "flaunting their femaleness."[107] Examples of resistance included taking advantage of male supervisors' squeamishness by using sick days for menstrual cramps and commissioning custom maternity uniforms to avoid light duty, since the rule-of-thumb was to move pregnant officers to light duty once they could no longer fit in their uniforms.

Conclusion

Our review of the literature on sexuality in male-dominated organizations suggests that, for better or worse, contemporary scholarly understandings of organizational discourse remain grounded in the ontological assumptions of gender theory. Although the organization of this literature review around Ashcraft's gender framework certainly facilitates such a conclusion, we were surprised at the ease with which research on the discourse of sexuality in male-dominated work settings could be sorted into these same categories.[108] After all, gender scholarship, especially in the realm of feminist organization studies, has been particularly careful in recent decades not to conflate gender and sexuality. Nevertheless, Muehlenhard and Peterson note that although there is some consensus that gender is more cultural and sex more biological, the two terms are still often used interchangeably in scholarship.[109] Furthermore, their review describes considerable variability in conceptual definitions, even among those scholars who do attempt to distinguish sex and gender.

If the object of gender discourse is so cultural and the object of sexuality discourse so biological, how can our research make better theoretical use of these conceptual differences while at the same time capturing important similarities? Certainly, more research is needed, especially in the domain of sexuality, that addresses the ontological uniqueness of discursive practices around human sexuality in the workplace. Based on the work of Michele Foucault and other foundational social theorists, organizational discourse studies have, for decades, considered the body a topic of fundamental significance for the enactment of both gender and sexuality.[110] But if there is consensus that sexuality is more prone to the influences of biology and gender more a product of culture and communication, then the relationships between sexuality and discourse should, to at least some extent, diverge even while at the same time acknowledging that elements of sexuality are also socially constructed. In this way, feminist and postmodern scholarship on gender and sexuality runs the risk of reproducing a problematic binary distinction between phenomena that have biological versus social bases. Future research should do more to consider the mutual influence of both, as it is arguably true that neither gender nor sexuality is fully cultural or fully biological.

The development and implementation of organizational policies related to sexuality via organizational discourse presents us with a significant opportunity to identify elements of sexuality discourse that are distinct from that of gender. Just as studies of gender policy enactment have provided unique insights about organizational discourse, research on the discursive practices organizational actors develop and refine in the course of developing and implementing additional policies related to sexuality may also provide distinct contributions to our understanding of sexuality discourse.[111] Occupational and organizational contexts that have typically been dominated by heterosexual men but now appear to struggle with accommodating a range of sexualities (e.g., military and paramilitary organizations) may offer especially potent research opportunities in this regard.

The social and linguistic dynamics of gender discourse have captured a great deal of attention from organizational communication scholars over the brief history of this subfield. Sexuality in male-dominated organizations has much to teach us about the prevailing discourses from which a range of workplace identities are produced, reproduced, and occasionally transformed. Even though plenty of work on gender itself remains to be done, there are also considerable opportunities on the horizon to examine other organizational discourses that are simultaneously distinct from and overlapping with gender.

Notes

1 Angela Trethewey, Clifton Scott, and Marianne LeGreco. "Constructing Embodied Organizational Identities: Commodifying, Securing, and Servicing Professional Bodies," in *Handbook of Gender and Communication*, eds Bonnie Dow and Julia Wood (Thousand Oaks, CA: Sage, 2006), 123–141.

2 James, G., "Why the Google Walkout Terrifies the Tech Moguls," *Inc*, November 08, 2018, www.inc. com/geoffrey-james/why-google-walkout-terrifies-tech-moguls.html.

3 Gail Fairhurst and Linda Putnam, "Organizations as Discursive Constructions," *Communication Theory* 14, no. 1 (2004): 5–26, https://doi.org/10.1111/j.1468-2885.2004.tb00301.x.

4 Karen Ashcraft, "Gender, Discourse, and Organization: Framing a Shifting Relationship," in *The Sage Handbook of Organizational Discourse*, eds David Grant, Cynthia Hardy, Clifford Oswick, and Linda Putnam (Thousand Oaks, CA: Sage, 2004), 275–291.

5 Michael Monsour, "Communication and Gender among Adult Friends," in *The Sage Handbook of Gender and Communication*, eds Bonnie Dow and Julia Wood (Thousand Oaks, CA: Sage, 2006), 57–69.

6 Shereen Bingham, "Communication Strategies for Managing Sexual Harassment in Organizations: Understanding Message Options and Their Effects," *Journal of Applied Communication Research* 19 (1991): 88–115, https://doi.org/10.1080/00909889109365294.

7 Dennis Mumby, "Organizing Men: Power, Discourse, and the Social Construction of Masculinity(s) in the Workplace," *Communication Theory* 8, no. 2 (1998): 164–183, https://doi.org/10.1111/j.1468-2885.1998. tb00216.x; Carol Spitzack, "The Production of Masculinity in Interpersonal Communication," *Communication Theory* 8, no. 2 (1998): 143–164, https://doi.org/10.1111/j.1468-2885.1998.tb00215.x.

8 Joan Acker, "Hierarchies, Jobs, Bodies: A Theory of Gendered Organizations," *Gender and Society* 4, no. 2 (1990): 139–158, https://doi.org/10.1177/089124390004002002; Jaime Bochantin and Renee Cowan, "On Being 'One of the Guys': How Female Police Officers Manage Tensions and Contradictions in Their Work and Their Lives," *The Ohio Communication Journal* 46 (2008): 145–170; Putnam, L. L. and J. E. Bochantin, "Gendered Bodies: Negotiating Normalcy and Support," *Negotiation and Conflict Management Research* 2, no. 1 (2009): 57–73; Trethewey et. al., "Embodied Organizational Identities," 123–141.

9 Ashcraft, "Gender, Discourse, and Organization," 275–291; Fairhurst and Putnam, "Organizations," 5–26.

10 Ashcraft, "Gender, Discourse, and Organization," 275–291.

11 Robin Clair, "The Use of Framing Devices to Sequester Organizational Narratives: Hegemony and Harassment," *Communications Monographs* 60, no. 2 (1993): 113–136, https://doi. org/10.1080/03637759309376304.

12 Ashcraft, "Gender, Discourse, and Organization," 275–291.

13 Karen Ashcraft and Dennis Mumby, "Organizing a Critical Communicology of Gender and Work," *International Journal of the Sociology of Language* 166 (2004): 19–44, https://doi.org/10.1515/ijsl.2004.012.

14 Laurel Westbrook, L. and Karen Schilt. "Doing Gender, Determining Gender: Transgender People, Gender Panics, and the Maintenance of the Sex/Gender/Sexuality System." *Gender & Society* 28, no. 1 (2014): 32–57, https://doi.org/10.1177/0891243213503203.

15 James Barker and George Cheney, "The Concept and Practices of Discipline in Contemporary Organizations," *Communications Monographs* 61, no. 1 (1994): 19–43, https://doi. org/10.1080/03637759409376321.

16 Ashcraft, "Gender, Discourse, and Organization," 276.

17 Michael Pacanowsky and Nick O'Donnell-Trujillo, "Organizational Communication as Cultural Performance," *Communications Monographs* 50, no. 2 (1983): 126–147, https://doi. org/10.1080/03637758309390158.

18 Ashcraft, "Gender, Discourse, and Organization," 276.

19 Robin Clair, "Hegemony and Harassment: A Discursive Practice," in *Conceptualizing Sexual Harassment as Discursive Process*, ed. Shereen Bingham (Westport, CT: Praeger, 1994), 59–70.

20 James Taylor and Elizabeth Van Every, *The Emergent Organization: Communication as Its Site and Surface* (Mahwah, NJ: Lawrence Erlbaum, 2000).

21 Ashcraft, "Gender, Discourse, and Organization," 282.

22 Dan Brouwer, "Corps/Corpse: The U. S. Military and Homosexuality," *Western Journal of Communication* 68 (2009): 411–430, https://doi.org/10.1080/10570310409374811.

23 Julie Konik and Lilia Cortina, "Policing Gender at Work: Intersections of Harassment Based on Sex and Sexuality," *Social Justice Research* 21, no. 3 (2008): 313–337, https://doi.org/10.1007/s11211-008-0074-z.

24 Juan Contreras, Mahzarin Banaji, and Jason Mitchell, "Multivoxel Patterns in Fusiform Face Area Differentiate Faces by Sex and Race," *PloS ONE* 8, no. 7 (2013), https://doi.org/10.1371/journal. pone.0069684.

25 Joan Williams, *Unbending Gender* (New York: Oxford University Press, 2000).

26 Waldo, C. R., "Working in a Majority Context: A Structural Model of Heterosexism as Minority Stress," *Journal of Counseling Psychology* 46, no. 2 (1999): 218–232, https://doi.org/10.1037/0022-0167.46.2.218.

27 Nancy Day and Patricia Schoenrade, "The Relationship among Reported Disclosure of Sexual Orientation, Anti-Discrimination Policies, Top Management Support, and Work Attitudes of Gay and Lesbian Employees," *Personnel Review* 29, no. 3 (2000): 346–363, https://doi.org/10.1108/00483480010324706; Michele Hebl, Eden King, Peter Glick, Sarah Singletary, and Stephanie Kazama, "Hostile and Benevolent Reactions Toward Pregnant Women: Complementary Interpersonal Punishments and Rewards That Maintain Traditional Roles," *Journal of Applied Psychology* 92, no. 6 (2007): 1499–1511, https://doi.org/10.1037/0021-9010.92.6.1499; Waldo, "Heterosexism," 218–232.

28 Burwell v. Hobby Lobby Stores, Inc., 573 U.S. ___ (2014)

29 Stark, L., "Hartzler: Transgender Service Members 'Costly' to Military," *CNN*, www.cnn.com/2017/07/26/politics/transgender-military-ban-vicky-hartzler-cnntv/index.html; Philipps, D., "Ban was Lifted, but Transgender Recruits Still Can't Join Up," *The New York Times*, www.nytimes.com/2018/07/05/us/military-transgender-recruits.html.

30 Hebl et al., "Pregnant Women," 1499–1511.

31 Shelley Correll, Stephen Benard, and In Paik, "Getting a Job: Is There a Motherhood Penalty?," *American Journal of Sociology* 112, no. 5 (2007): 1297–1338, https://doi.org/10.1086/511799.

32 N. Kitroeff and J. Silver-Greenberg, "Pregnancy Discrimination is Rampant inside America's Biggest Companies," *The New York Times*, www.nytimes.com/interactive/2018/06/15/business/pregnancy-discrimination.html.

33 N. Kitroeff and J. Silver-Greenberg, "Miscarrying at Work: The Physical Toll of Pregnancy Discrimination," *The New York Times*, www.nytimes.com/interactive/2018/10/21/business/pregnancy-discrimination-miscarriages.html.

34 Patrice Buzzanell and Meina Liu, "Struggling with Maternity Leave Policies and Practices: A Poststructuralist Feminist Analysis of Gendered Organizing," *Journal of Applied Communication Research* 33, no. 1 (2005): 1–25, https://doi.org/10.1086/511799.

35 Society for Human Resource Management (SHRM), "2018 Employee Benefits: The Evolution of Benefits," *SHRM*, www.shrm.org/hr-today/trends-and-forecasting/research-and-surveys/Documents/2018%20Employee%20Benefits%20Report.pdf.

36 Erin Cech and Mary Blair-Loy, "Consequences of Flexibility Stigma among Academic Scientists and Engineers," *Work and Occupations* 41, no. 1 (2014): 86–110, https://doi.org/10.1177/0730888413515497.

37 McKinsey & Company, "Women in the Workplace – 2019," (2019), https://wiw-report.s3.amazonaws.com/Women_in_the_Workplace_2019.pdf.

38 M. J. Budig and P. England, "The Wage Penalty for Motherhood," *American Sociological Review* 66, no. 2 (2001): 204–225, https://doi.org/10.2307/2657415.

39 Philip Roth, Kristen Purvis, and Philip Bobko, "A Meta-Analysis of Gender Group Differences for Measures of Job Performance," *Journal of Management* 38, no. 2 (2012): 719–739, https://doi.org/10.1177/0149206310374774.

40 Thomas Ng and Daniel Feldman, "The Relationships of Age with Job Attitudes: A Meta-Analysis," *Personnel Psychology* 63, no. 3 (2010): 677–718, https://doi.org/10.1111/j.1744-6570.2010.01184.x.

41 Alexander Haslam and Michele Ryan, "The Road to the Glass Cliff: Differences in the Perceived Suitability of Men and Women in Succeeding and Failing Organizations," *The Leadership Quarterly* 19, no. 5 (2008): 530–546, https://doi.org/10.1016/j.leaqua.2008.07.011; Michele Ryan and Alexander Haslam, "The Glass Cliff: Evidence that Women are Over-Represented in Precarious Leadership Positions," *British Journal of Management* 16, no. 2 (2005): 81–90, https://doi.org/10.1111/j.1467-8551.2005.00433.x; Michele Ryan and Alexander Haslam, "The Glass Cliff: Exploring the Dynamics Surrounding the Appointment of Women to Precarious Leadership Positions," *Academy of Management Review* 32, no. 2 (2007): 549–572, https://doi.org/10.5465/amr.2007.24351856; Michele Ryan and Alexander Haslam, Metta Hersby, and Renata Bongiorno, "Think Crisis-Think Female: The Glass Cliff and Contextual Variation in the Think Manager- Think Male Stereotype," *Journal of Applied Psychology* 96, no. 3 (2011): 470, https://doi.org/10.1037/a0022133.

42 James Diefendorff, Douglas Brown, Allen Kamin, and Robert Lord, "Examining the Roles of Job Involvement and Work Centrality in Predicting Organizational Citizenship Behaviors and Job Performance," *Journal of Organizational Behavior* 23, no. 1 (2002): 93–108, https://doi.org/10.1002/job.123.

43 Louise Fitzgerald, Suzanne Swan, and Vicki J. Magley, "But Was It Really Sexual Harassment?: Legal, Behavioral, and Psychological Definitions of the Workplace Victimization of Women," in *Sexual Harassment: Theory, Research, and Treatment,* ed. W. O'Donohue (Needham Heights, MA: Allyn & Bacon, 1997): 5–28.

44 Fitzgerald, L. F., "Sexual Harassment: Violence against Women in the Workplace," *American Psychologist* 48, no. 10 (1993): 1070–1076, https://doi.org/10.1037/0003-066X.48.10.1070.

45 Craig Waldo, Jennifer Berdahl, and Louise Fitzgerald, "Are Men Sexually Harassed? If so, by Whom?" *Law and Human Behavior* 22, no. 1 (1998): 60, https://doi.org/10.1023/A:1025776705629.

46 Konik and Cortina, "Policing Gender," 313–337.

47 Nathan Smith and Kathleen Ingram, "Workplace Heterosexism and Adjustment Among Lesbian, Gay, and Bisexual Individuals: The Role of Unsupportive Social Interactions," *Journal of Counseling Psychology* 54, no. 1 (2004): 57, https://doi.org/10.1037/0022-0167.51.1.57; Jennifer Berdahl, Vicki Magley, and Craig Waldo, "The Sexual Harassment of Men? Exploring the Concept with Theory and Data," *Psychology of Women Quarterly* 20, no. 4 (1996): 527–547, https://doi.org/10.1111/j.1471-6402.1996.tb00320.x; Katherine Franke, "What's Wrong with Sexual Harassment?" *Stanford Law Review* 49 (1997): 691–772, https://doi.org/10.2307/1229336; Margaret Stockdale, Michelle Visio, and Leena Batra, "The Sexual Harassment of Men: Evidence for a Broader Theory of Sexual Harassment and Sex Discrimination," *Psychology, Public Policy, & Law* 5 (1999): 630–664, https://doi.org/10.1037/1076-8971.5.3.630.

48 Waldo, "Heterosexism," 218–232.

49 Heather Lyons, Bradley Brenner, and Ruth Fassinger, "A Multicultural Tests of the Theory of Work Adjustment: Investigating the Role of Heterosexism and Its Perceptions in the Job Satisfaction of Lesbian, Gay, and Bisexual Employees," *Journal of Counseling Psychology* 52, no. 4 (2005): 537–548, https://doi.org/10.1037/0022-0167.52.4.537.

50 Waldo, "Heterosexism," 218–232; James Croteau, "Research on the Work Experiences of Lesbian, Gay, and Bisexual People: An Integrative Review of Methodology and Findings," *Journal of Vocational Behavior* 48 (1996): 195–209, https://doi.org/10.1006/jvbe.1996.0018; Martin Levine and Robin Leonard, "Discrimination against Lesbians in the Work Force," *Journal of Women in Culture and Society* 9 (1984): 700–710, https://doi.org/10.1086/494094; Bella Ragins and John Cornwell, "Pink Triangles: Antecedents and Consequences of Perceived Workplace Discrimination against Gay and Lesbian Employees," *Journal of Applied Psychology* 86 (2001): 1244–1261, https://doi.org/10.1037/0021-9010.86.6.1244; Bella Ragins, John Cornwell, and Janice Miller, "Heterosexism in the Workplace: Do Race and Gender Matter?" *Group & Organization Management* 28 (2003): 45–74, https://doi.org/10.1177/1059601102250018.

51 Waldo, "Heterosexism," 218–232; Kristen Griffith and Michelle Hebl, "The Disclosure Dilemma for Gay Men and Lesbians: 'Coming Out' at Work," *Journal of Applied Psychology* 87, no. 6 (2002): 1191–1199, https://doi.org/10.1037/0021-9010.87.6.1191; Day and Schoenrade, "Disclosure," 346–363.

52 Waldo, "Heterosexism," 218–232.

53 Gordon, M., M. S. Price, and K. Peralta, "Understanding HB2: North Carolina's Newest Law Solidifies State's Role in Defining Discrimination," *The Charlotte Observer,* www.charlotteobserver.com/news/politics-government/article68401147.html.

54 Anna Spradlin, "The price of 'passing': A lesbian perspective on authenticity in organizations," *Management Communication Quarterly* 11, no. 4 (1998): 598–605.

55 Nancy Day and Patricia Schoenrade, "Staying in the Closet versus Coming Out: Relationships Between Communication about Sexual Orientation and Work Attitudes," *Personnel Psychology* 50, no. 1 (1997): 147–163, https://doi.org/10.1111/j.1744-6570.1997.tb00904.x; Day and Schoenrade, "Disclosure," 346–363.

56 Griffith and Hebl, "Disclosure Dilemma," 1191–1199; Day and Schoenrade, "Disclosure," 346–363.

57 Griffith and Hebl, "Disclosure Dilemma," 1191–1199; Lyons et al., "Work Adjustment," 537–548; Day and Schoenrade, "Disclosure," 346–363.

58 Day and Schoenrade, "Staying in the Closet," 147–163; Day and Schoenrade, "Disclosure," 346–363.

59 Nick Trujillo, "Organizational Communication as Cultural Performance: Some Managerial Considerations," *Southern Journal of Communication* 50, no. 3 (1985): 201–224, https://doi.org/10.1080/10417948509372632; Mary Nell Trautner, "Doing Gender, Doing Class: The Performance of Sexuality in Exotic Dance Clubs," *Gender & Society* 19, no. 6 (2005): 771–788, https://doi.org/10.1177/0891243205277253.

60 Brenda Allen, *Difference Matters: Communicating Social Identity* (Long Grove, IL: Waveland Press, 2010), 131.

61 Kirsten Dellinger and Christine Williams, "The Locker Room and the Dorm Room: Workplace Norms of Sexual Harassment in Magazine Editing," *Social Problems* 42, no. 2 (2002): 242–257, https://doi.org/10.1525/sp.2002.49.2.242; Sylvia Gherardi, *Gender, Symbolism, and Organizational Cultures* (Thousand Oaks, CA: Sage, 1995); Harrison Miller Trice, *Occupational Subcultures in the Workplace* (Ithaca, NY: Cornell University Press, 1993).

62 Dellinger and Williams, "The Locker Room and the Dorm Room," 242–257; Robin Leidner, "Serving Hamburgers and Selling Insurance: Gender, Work, and Identity in Interactive Service Jobs," *Gender & Society* 5, no. 2 (1991): 154–177, https://doi.org/10.1177/089124391005002002; Christine Williams, Patti Giuffre, and Kirsten Dellinger, "Sexuality in the Workplace: Organizational Control, Sexual Harassment, and the Pursuit of Pleasure," *Annual Review of Sociology* 25, no. 1 (1999): 73–93, https://doi.org/10.1146/annurev.soc.25.1.73.

63 Acker, "Gendered Organizations," 139–158; Putnam and Bochantin, "Gendered Bodies," 57–73; Trethewey et al., "Embodied Organizational Identities," 123–141; Allen, *Difference Matters*.

64 Raewyn Connell, "Masculinities and Globalization," *Men and Masculinities* 1, no. 1 (1998): 3–23, https://doi.org/10.1177/1097184X98001001001.

65 Tristan Bridges and Cheri Pascoe, "Hybrid Masculinities: New Directions in the Sociology of Men and Masculinities," *Sociology Compass* 8, no. 3 (2014): 246–258, https://doi.org/10.1111/soc4.12134.

66 Donna Eder, *School Talk: Gender and Adolescent Culture* (New Brunswick, NJ: Rutgers University Press, 1995).

67 Mary Jane Kehily, "Sexing the Subject: Teachers, Pedagogies, and Sex Education," *Sex Education: Sexuality, Society, and Learning* 2, no. 3 (2002): 215–231, https://doi.org/10.1080/1468181022000025785; Brian Sweeney, "Performance Anxieties: Undoing Sexist Masculinities Among College Men." *Culture, Society & Masculinities* 5, no. 2 (2013): 208–218, https://doi.org/10.3149/CSM.0502.208.

68 Trethewey et al., "Embodied Organizational Identities," 123–141.

69 Patrice Buzzanell, "Employment Interviewing Research: Ways We Can Study Underrepresented Group Members' Experiences as Applicants," *The Journal of Business Communication* 39, no. 2 (2002): 257–275, https://doi.org/10.1177/002194360203900206.

70 Jaime Bochantin, "'Ambulance Thieves, Clowns, and Naked Grandfathers': How PSEs and Their Families Use Humorous Communication as a Sensemaking Device," *Management Communication Quarterly* 31, no. 2 (2017): 278–296. https://doi.org/10.1177/0893318916687650.

71 Amy Denissen, "The Right Tools for the Job: Constructing Gender Meanings and Identities in the Male-Dominated Building Trades," *Human Relations* 63, no. 7 (2010): 1051–1069, https://doi.org/10.1177/0018726709349922; Allen, *Difference Matters*.

72 Denissen, "The Right Tools," 1051–1069.

73 Majia Nadesan and Angela Trethewey, "Performing the Enterprising Subject: Gendered Strategies for Success," *Text and Performance Quarterly* 20, no. 3 (2000): 223–250, https://doi.org/10.1080/10462930009366299.

74 Bochantin, J. E., "'Long Live the Mensi-Mob': Communicating Support Online with Regards to Experiencing Menopause in the Workplace," *Communication Studies* 65, no. 3 (2014): 260–280, https://doi.org/10.1080/10510974.2013.811433.

75 Nadesan and Trethewey, "Gendered Strategies for Success," 223–250; Trethewey et al., "Embodied Organizational Identities," 123–141; Bochantin, "Menopause," 260–280.

76 Kirsten Dellinger and Christine Williams, "Makeup at Work: Negotiating Appearance Rules in the Workplace," *Gender & Society* 11, no. 2 (1997): 151–177, https://doi.org/10.1177/089124397011002002; Nadesan and Trethewey, "Gendered Strategies for Success," 223–250; Trethewey, A., "Disciplined Bodies: Women's Embodied Identities at Work," *Organization Studies* 20, no. 3 (1999): 423–450, https://doi.org/10.1177/0170840699203003.

77 Trethewey, "Disciplined Bodies," 423–450.

78 Trethewey, "Disciplined Bodies," 423–450.

79 Dellinger and Williams, "The Locker Room and the Dorm Room," 242–257.

80 Spradlin, A.L., "The Price of 'Passing': A Lesbian Perspective on Authenticity in Organizations," *Management Communication Quarterly* 11, no. 4 (1998): 598–605, https://doi.org/10.1177/0893318998114006.

81 Andrea Lewis, "Destructive Organizational Communication and LGBT Workers' Experiences," in *Destructive Organizational Communication: Processes, Consequences, and Constructive Ways of Organizing*, eds Pamela Lutgen-Sandvik and Brenda Sypher (New York: Routledge/Taylor & Francis Group, 2009), 184–202.

82 Giuffre, P. and C. Caviness, "Desexualization," in *The Wiley-Blackwell Encyclopedia of Gender and Sexuality Studies* (2016): 1–3.

83 Karen Ashcraft, Sara Muhr, Jens Rennstam, and Katie Sullivan. "Professionalization as a Branding Activity: Occupational Identity and the Dialectic of Inclusivity-Exclusivity." Gender, *Work & Organization* 19, no. 5 (2012): 467–488, https://doi.org/10.1111/j.1468-0432.2012.00600.x.

84 Jean Lipman-Blumen, "Female Leadership in Formal Organizations: Must the Female Leader Go Formal?" *Readings in Managerial Psychology* 54, no. 21 (1980): 341–362.; Moss Kanter, R., "Women and the Structure of Organizations: Explorations in Theory and Behavior," *Sociological Inquiry* 45 (1975): 34–74, https://doi.org/10.1111/j.1475-682X.1975.tb00331.x.

85 Trethewey, A., "Reproducing and Resisting the Master Narrative of Decline: Midlife Professional Women's Experiences of Aging," *Management Communication Quarterly* 15, no. 2 (2001): 183–226, https://doi.org/10.1177/0893318901152002.

86 Acker, "Gendered Organizations," 139–158.

87 Richard Anker, *Gender and Jobs: Sex Segregation of Occupations in the World* (Geneva: International Labour Organization, 1998); Cha, Youngjoo. "Overwork and the Persistence of Gender Segregation in Occupations." *Gender & Society* 27, no. 2 (2013): 158–184, https://doi.org/10.1177/0891243212470510; Maria Charles and David Grusky, *Occupational Ghettos: The Worldwide Segregation of Women and Men* (Stanford, CA: Stanford University Press, 2004).

88 Karen Ashcraft, "Appreciating the 'Work' of Discourse: Occupational Identity and Difference as Organizing Mechanisms in the Case of Commercial Airline Pilots," *Discourse & Communication* 1, no. 1 (2007): 9–36, https://doi.org/10.1177/1750481307071982.

89 Trethewey, "Master Narrative of Decline," 183–226.

90 Ashcraft et al., "Professionalization," 467–488.

91 Tamyra Pierce and Debbie Dougherty, "The Construction, Enactment, and Maintenance of Power-as-Domination Through an Acquisition: The Case of TWA and Ozark Airlines," *Management Communication Quarterly* 16, no. 2 (2002): 129–164, https://doi.org/10.1177/089331802237232.

92 Dougherty, D. S. "Toward a Theoretical Understanding of Feminist Standpoint Process in Organizations: The Case of Sexual Harassment." Paper presented at The National Communication Association Convention, Atlanta, GA, November, 2001.

93 Dougherty, "Feminist Standpoint Process."; Marlene Fine, "New Voices in Organizational Communication: A Feminist Commentary and Critique," in *Transforming Visions: Feminist Critiques in Communication Studies*, eds S. Perlmutter Bowen and N. Wyatt (Cresskill, NJ: Hampton Press, 1993), 125–166; Judi Marshall, "Viewing Organizational Communication from a Feminist Perspective," in *Communication Yearbook,* ed. S. Deetz (Newbury Park, CA: Sage, 1993), 122–143; Parker, P. S., "African American Women Executives' Leadership Communication within Dominant Culture Organizations: (Re)Conceptualizing Notions of Collaboration and Instrumentality," *Management Communication Quarterly* 15 (2001): 42–48, https://doi.org/10.1177/0893318901151002.

94 Marshall, "Viewing Organizational Communication," 122–143.

95 Dougherty, "Feminist Standpoint Process."

96 Deetz, S., *Democracy in an Age of Corporate Colonization* (Albany, NY: SUNY Press, 1992); Murphy, A. G., "Hidden Transcripts of Flight Attendant Resistance," *Management Communication Quarterly* 11, no. 4 (1998): 499–535, https://doi.org/10.1177/0893318998114001.

97 Anthony Giddens, *Central Problems in Social Theory* (London: Macmillan, 1979); Anthony Giddens, *The Constitution of Society* (Cambridge: Polity, 1984); Anthony Giddens, *New Rules of Sociological Method*, 2nd edn (Stanford, CA: Stanford University Press, 1993).

98 Michel Foucault, *Power/Knowledge: Selected Interviews and Other Writings, 1972–1977* (New York: Pantheon, 1980).

99 Murphy, "Hidden Transcripts," 499–535.

100 Deetz, *Democracy.*

101 Kathy Ferguson, *The Feminist Case Against Bureaucracy* (Philadelphia, PA: Temple University Press, 1984).

102 Pfafman, T. M. and J. E. Bochantin, "Negotiating Power Paradoxes: Contradictions in Women's Constructions of Organizational Power," *Communication Studies* 63, no. 5 (2012): 574–592.

103 Ferguson, *The Feminist Case against Bureaucracy*; Murphy, "Hidden Transcripts," 499–535.

104 James Scott, *Domination and the Arts of Resistance: Hidden Transcripts* (New Haven, CT: Yale University Press, 1990).

105 Murphy, "Hidden Transcripts," 499–535.

106 Elizabeth Bell and Linda Forbes, "Office Folklore in the Academic Paperwork Empire: The Interstitial Space of Gendered (Con)Texts," *Text and Performance Quarterly* 14, no. 3 (1994): 181–196, https://doi.org/10.1080/10462939409366082.

107 Bochantin and Cowan, "One of the Guys," 162.

108 Ashcraft, "Gender, Discourse, and Organization," 275–291.

109 Charlene Muehlenhard and Zoe Peterson, "Distinguishing Between Sex and Gender: History, Current Conceptualizations, and Implications," *Sex Roles* 64 nos 11–12 (2011): 791–803, https://doi.org/10.1007/s11199-011-9932-5.
110 Trethewey et al., "Embodied Organizational Identities," 123–141.
111 Brouwer, "Corps/Corpse," 411–430; Erika Kirby and Kathleen Krone, "'The Policy Exists but You Can't Really Use It': Communication and the Structuration of Work-Family Policies," *Journal of Applied Communication Research* 30, no. 1 (2002): 50–77, https://doi.org/10.1080/00909880216577.

Bibliography

Acker, Joan. "Hierarchies, Jobs, Bodies: A Theory of Gendered Organizations." *Gender & Society* 4, no. 2 (1990): 139–58. doi:10.1177/089124390004002002.

Allen, Brenda J. *Difference Matters: Communicating Social Identity*. Long Grove, IL: Waveland Press, 2011.

Anker, Richard. *Gender and Jobs: Sex Segregation of Occupations in the World*. Geneva: International Labor Organization, 2001.

Ashcraft, Karen Lee. "Appreciating the 'Work' of Discourse: Occupational Identity and Difference as Organizing Mechanisms in the Case of Commercial Airline Pilots." *Discourse & Communication* 1, no. 1 (2007): 9–36. doi:10.1177/1750481307071982.

Ashcraft, Karen Lee. "Gender, Discourse, and Organization: Framing a Shifting Relationship." *The SAGE Handbook of Organizational Discourse*. Thousand Oaks, CA: Sage, 275–98. doi:10.4135/9781848608122.n13.

Ashcraft, Karen Lee, and Dennis K. Mumby. "Organizing a Critical Communicology of Gender and Work." *International Journal of the Sociology of Language* 166 (2004). doi:10.1515/ijsl.2004.012.

Ashcraft, Karen Lee, Sara Louise Muhr, Jens Rennstam, and Katie Sullivan. "Professionalization as a Branding Activity: Occupational Identity and the Dialectic of Inclusivity-Exclusivity." *Gender, Work & Organization* 19, no. 5 (2012): 467–88. doi:10.1111/j.1468-0432.2012.00600.x.

Barker, James R., and George Cheney. "The Concept and the Practices of Discipline in Contemporary Organizational Life." *Communication Monographs* 61, no. 1 (1994): 19–43. doi:10.1080/03637759409376321.

Bell, Elizabeth, and Linda C. Forbes. "Office Folklore in the Academic Paperwork Empire: The Interstitial Space of Gendered (con)texts." *Text and Performance Quarterly* 14, no. 3 (1994): 181–96. doi:10.1080/10462939409366082.

Berdahl, Jennifer L., Vicki J. Magley, and Craig R. Waldo. "The Sexual Harassment of Men?: Exploring the Concept with Theory and Data." *Psychology of Women Quarterly* 20, no. 4 (1996): 527–47. doi:10.1111/j.1471-6402.1996.tb00320.x.

Bingham, Shereen G. "Communication Strategies for Managing Sexual Harassment in Organizations: Understanding Message Options and Their Effects." *Journal of Applied Communication Research* 19, nos 1–2 (1991): 88–115. doi:10.1080/00909889109365294.

Bingham, Shereen G. *Conceptualizing Sexual Harassment as Discursive Practice*. Westport, CT: Praeger, 1994.

Bochantin, Jaime E. "Ambulance Thieves, Clowns, and Naked Grandfathers." *Management Communication Quarterly* 31, no. 2 (2017): 278–96. doi:10.1177/0893318916687650.

Bochantin, Jaime. "'Long Live the Mensi-Mob': Communicating Support Online with Regards to Experiencing Menopause in the Workplace." *Communication Studies* 65, no. 3 (2014): 260–80. doi:10.1080/10510974.2013.811433.

Bridges, Tristan, and C. J. Pascoe. "Hybrid Masculinities: New Directions in the Sociology of Men and Masculinities." *Sociology Compass* 8, no. 3 (2014): 246–58. doi:10.1111/soc4.12134.

Brouwer, Daniel C. "Corps/corpse: The U.S. Military and Homosexuality." *Western Journal of Communication* 68, no. 4 (2004): 411–30. doi:10.1080/10570310409374811.

Budig, Michelle J., and Paula England. "The Wage Penalty for Motherhood." *American Sociological Review* 66, no. 2 (2001): 204. doi:10.2307/2657415.

Buzzanell, P. M. "Employment Interviewing Research: Ways We can Study Underrepresented Group Members' Experiences as Applicants." *Journal of Business Communication* 39, no. 2 (2002): 257–75. doi:10.1177/002194360203900206.

Buzzanell, Patrice M., and Meina Liu. "Struggling with Maternity Leave Policies and Practices: A Poststructuralist Feminist Analysis of Gendered Organizing." *Journal of Applied Communication Research* 33, no. 1 (2005): 1–25. doi:10.1080/0090988042000318495.

Cech, Erin A., and Mary Blair-Loy. "Consequences of Flexibility Stigma among Academic Scientists and Engineers." *Work and Occupations* 41, no. 1 (2014): 86–110. doi:10.1177/0730888413515497.

Cha, Youngjoo. "Overwork and the Persistence of Gender Segregation in Occupations." *Gender & Society* 27, no. 2 (2013): 158–84. doi:10.1177/0891243212470510.

Clair, Robin Patric. "The Use of Framing Devices to Sequester Organizational Narratives: Hegemony and Harassment." *Communication Monographs* 60, no. 2 (1993): 113–36. doi:10.1080/03637759309376304.

Connell, R. W. "Globalization, Imperialism, and Masculinities." *Handbook of Studies on Men & Masculinities.* Thousand Oaks, CA: Sage, 71–89. doi:10.4135/9781452233833.n5.

Contreras, Juan Manuel, Mahzarin R. Banaji, and Jason P. Mitchell. "Multivoxel Patterns in Fusiform Face Area Differentiate Faces by Sex and Race." *PLoS One* 8, no. 7 (2013). doi:10.1371/journal.pone.0069684.

Correll, Shelley J., Stephen Benard, and In Paik. "Getting a Job: Is There a Motherhood Penalty?" *American Journal of Sociology* 112, no. 5 (2007): 1297–339. doi:10.1086/511799.

Croteau, James M. "Research on the Work Experiences of Lesbian, Gay, and Bisexual People: An Integrative Review of Methodology and Findings." *Journal of Vocational Behavior* 48, no. 2 (1996): 195–209. doi:10.1006/jvbe.1996.0018.

Day, Nancy E., and Patricia Schoenrade. "Staying in the Closet Versus Coming Out: Relationships between Communication about Sexual Orientation and Work Attitudes." *Personnel Psychology* 50, no. 1 (1997): 147–63. doi:10.1111/j.1744-6570.1997.tb00904.x.

Day, Nancy E., and Patricia Schoenrade. "The Relationship among Reported Disclosure of Sexual Orientation, Anti-discrimination Policies, Top Management Support and Work Attitudes of Gay and Lesbian Employees." *Personnel Review* 29, no. 3 (2000): 346–63. doi:10.1108/00483480010324706.

Deetz, Stanley A. *Democracy in an Age of Corporate Colonization: Developments in Communication and the Politics of Everyday Life.* Albany, NY: State University of New York, 1992.

Dellinger, Kirsten, and Christine L. Williams. "The Locker Room and the Dorm Room: Workplace Norms and the Boundaries of Sexual Harassment in Magazine Editing." *Social Problems* 49, no. 2 (2002): 242–57. doi:10.1525/sp.2002.49.2.242.

Dellinger, Kirsten, and Christine L. Williams. "Makeup at Work." *Gender & Society* 11, no. 2 (1997): 151–77. doi:10.1177/089124397011002002.

Denissen, Amy M. "The Right Tools for the Job: Constructing Gender Meanings and Identities in the Male-dominated Building Trades." *Human Relations* 63, no. 7 (2010): 1051–069. doi:10.1177/0018726709349922.

Diefendorff, James M., Douglas J. Brown, Allen M. Kamin, and Robert G. Lord. "Examining the Roles of Job Involvement and Work Centrality in Predicting Organizational Citizenship Behaviors and Job Performance." *Journal of Organizational Behavior* 23, no. 1 (2002): 93–108. doi:10.1002/job.123.

Fairhurst, Gail T., and Linda Putnam. "Organizations as Discursive Constructions." *Communication Theory* 14, no. 1 (2004): 5–26. doi:10.1111/j.1468-2885.2004.tb00301.x.

Farkas, George. *Occupational Ghettos: The Worldwide Segregation of Women and Men. By Maria Charles and David B. Grusky.* Stanford, CA: Stanford University Press, 2004, Xvii 381. *American Journal of Sociology* 111, no. 2 (2005): 621–23. doi:10.1086/499003.

Ferguson, Ann Arnett, Donna Eder, Catherine Colleen Evans, and Stephen Parker. "School Talk: Gender and Adolescent Culture." *Contemporary Sociology* 25, no. 5 (1996): 614. doi:10.2307/2077545.

Ferguson, Kathy E. *The Feminist Case against Bureaucracy.* Philadelphia, PA: Temple University Press, 1984.

Fitzgerald, Louise F. "Sexual Harassment: Violence against Women in the Workplace." *American Psychologist* 48, no. 10 (1993): 1070–076. doi:10.1037/0003-066x.48.10.1070.

Foucault, Michel, and Colin Gordon. *Power/knowledge: Selected Interviews and Other Writings 1972–1977.* Sussex, UK: Harvester Press, 1980.

Franke, Katherine M. "What's Wrong with Sexual Harassment." *Directions in Sexual Harassment Law*, 2003, 169–79. doi:10.12987/yale/9780300098006.003.0013.

Gherardi, Silvia. *Gender, Symbolism and Organizational Cultures.* Thousand Oaks, CA: Sage, 2000.

Giddens, Anthony. *Central Problems in Social Theory.* Berkeley, CA: University of California Press, 1979.

Giddens, Anthony. *New Rules of Sociological Method: A Positive Critique of Interpretive Sociologies.* Cambridge: Polity Press, 1993.

Giddens, Anthony. *The Constitution of Society.* Berkeley, CA: University of California Press, 1984.

Gordon, Michael, Mark S. Price, and Katie Peralta. "Understanding HB2: North Carolina's Newest Law Solidifies State's Role in Defining Discrimination." *Charlotte Observer.* March 26, 2016. www.charlotteobserver.com/news/politics-government/article68401147.html.

Griffith, Kristin H., and Michelle R. Hebl. "The Disclosure Dilemma for Gay Men and Lesbians: 'Coming Out' at Work." *Journal of Applied Psychology* 87, no. 6 (2002): 1191–199. doi:10.1037/0021-9010.87.6.1191.

Haslam, S. Alexander, and Michelle K. Ryan. "The Road to the Glass Cliff: Differences in the Perceived Suitability of Men and Women for Leadership Positions in Succeeding and failing Organizations." *The Leadership Quarterly* 19, no. 5 (2008): 530–46. doi:10.1016/j.leaqua.2008.07.011.

Hebl, Michelle R., Eden B. King, Peter Glick, Sarah L. Singletary, and Stephanie Kazama. "Hostile and Benevolent Reactions toward Pregnant Women: Complementary Interpersonal Punishments and Rewards That Maintain Traditional Roles." *Journal of Applied Psychology* 92, no. 6 (2007): 1499–511. doi:10.1037/0021-9010.92.6.1499.

James, Geoffrey. "Why the Google Walkout Terrifies the Tech Moguls." *Inc.com.* November 08, 2018. www.inc.com/geoffrey-james/why-google-walkout-terrifies-tech-moguls.html.

Kanter, Rosabeth Moss. "Women and the Structure of Organizations: Explorations in Theory and Behavior." *Sociological Inquiry* 45, no. 2–3 (1975): 34–74. doi:10.1111/j.1475-682x.1975.tb00331.x.

Kehily, Mary Jane. "Sexing the Subject: Teachers, Pedagogies and Sex Education." *Sex Education* 2, no. 3 (2002): 215–31. doi:10.1080/1468181022000025785.

Kirby, Erika, and Kathleen Krone. "'The Policy Exists but You can't Really Use It': Communication and the Structuration of Work-family Policies." *Journal of Applied Communication Research* 30, no. 1 (2002): 50–77. doi:10.1080/00909880216577.

Kitroeff, Natalie, and Jessica Silver-Greenberg. "Pregnancy Discrimination Is Rampant Inside America's Biggest Companies." *The New York Times.* June 15, 2018. www.nytimes.com/interactive/2018/06/15/business/pregnancy-discrimination.html.

Konik, Julie, and Lilia M. Cortina. "Policing Gender at Work: Intersections of Harassment Based on Sex and Sexuality." *Social Justice Research* 21, no. 3 (2008): 313–37. doi:10.1007/s11211-008-0074-z.

Leidner, Robin. "Serving Hamburgers and Selling Insurance: Gender, Work, and Identity in Interactive Service Jobs." *Gender & Society* 5, no. 2 (1991): 154–77. doi:10.1177/089124391005002002.

Levine, Martin P., and Robin Leonard. "Discrimination against Lesbians in the Work Force." *Signs: Journal of Women in Culture and Society* 9, no. 4 (1984): 700–10. doi:10.1086/494094.

Lutgen-Sandvik, Pamela, and Beverly Davenport Sypher. *Destructive Organizational Communication: Processes, Consequences, and Constructive Ways of Organizing.* New York: Routledge, 2009.

Lyons, Heather Z., Bradley R. Brenner, and Ruth E. Fassinger. "A Multicultural Test of the Theory of Work Adjustment: Investigating the Role of Heterosexism and Fit Perceptions in the Job Satisfaction of Lesbian, Gay, and Bisexual Employees." *Journal of Counseling Psychology* 52, no. 4 (2005): 537–48. doi:10.1037/0022-0167.52.4.537.

Monsour, Michael. "Communication and Gender among Adult Friends." In *The SAGE Handbook of Gender and Communication.* Thousand Oaks, CA: Sage, 57–70. doi:10.4135/9781412976053.n4.

Muehlenhard, Charlene L., and Zoe D. Peterson. "Distinguishing between Sex and Gender: History, Current Conceptualizations, and Implications." *Sex Roles* 64, no. 11–12 (2011): 791–803. doi:10.1007/s11199-011-9932-5.

Mumby, Dennis K. "Organizing Men: Power, Discourse, and the Social Construction of Masculinity(s) in the Workplace." *Communication Theory* 8, no. 2 (1998): 164–83. doi:10.1111/j.1468-2885.1998.tb00216.x.

Murphy, Alexandra G. "Hidden Transcripts of Flight Attendant Resistance." *Management Communication Quarterly* 11, no. 4 (1998): 499–535. doi:10.1177/0893318998114001.

Nadesan, Majia Holmer, and Angela Trethewey. "Performing the Enterprising Subject: Gendered Strategies for Success (?)." *Text and Performance Quarterly* 20, no. 3 (2000): 223–50. doi:10.1080/10462930009366299.

Naples, Nancy A., Renée Carine Hoogland, Maithree Wickramasinghe, and Wai Ching Angela Wong. *The Wiley Blackwell Encyclopedia of Gender and Sexuality Studies.* Hoboken, NJ: Wiley Blackwell, 2016.

Ng, Thomas W. H., and Daniel C. Feldman. "The Relationships of Age with Job Attitudes: A Meta-Analysis." *Personnel Psychology* 63, no. 3 (2010): 677–718. doi:10.1111/j.1744-6570.2010.01184.x.

O'Donohue, William T. *Sexual Harassment: Theory, Research, and Treatment.* Boston, MA: Allyn and Bacon, 1997.

Pacanowsky, Michael E., and Nick O'Donnell-Trujillo. "Organizational Communication as Cultural Performance." *Communication Monographs* 50, no. 2 (1983): 126–47. doi:10.1080/03637758309390158.

Parker, Patricia S. "African American Women Executives' Leadership Communication Within Dominant-Culture Organizations." *Management Communication Quarterly* 15, no. 1 (2001): 42–82. doi:10.1177/0893318901151002.

Pfafman, Tessa M., and Jaime E. Bochantin. "Negotiating Power Paradoxes: Contradictions in Women's Constructions of Organizational Power." *Communication Studies* 63, no. 5 (2012): 574–92. doi:10.108 0/10510974.2012.681100.

Philipps, Dave. "Ban Was Lifted, but Transgender Recruits Still can't Join Up." *The New York Times.* July 05, 2018. www.nytimes.com/2018/07/05/us/military-transgender-recruits.html.

Pierce, Tamyra, and Debbie S. Dougherty. "The Construction, Enactment, and Maintenance of Power-As-Domination through an Acquisition." *Management Communication Quarterly* 16, no. 2 (2002): 129–64. doi:10.1177/089331802237232.

Putnam, Linda L., and Jaime Bochantin. "Gendered Bodies: Negotiating Normalcy and Support." *Negotiation and Conflict Management Research* 2, no. 1 (2009): 57–73. doi:10.1111/j.1750-4716.2008.00028.x.

Ragins, Belle Rose, John M. Cornwell, and Janice S. Miller. "Heterosexism in the Workplace." *Group & Organization Management* 28, no. 1 (2003): 45–74. doi:10.1177/1059601102250018.

Ragins, Belle Rose, and John M. Cornwell. "Pink Triangles: Antecedents and Consequences of Perceived Workplace Discrimination Against Gay and Lesbian Employees." *Journal of Applied Psychology* 86, no. 6 (2001): 1244–261. doi:10.1037/0021-9010.86.6.1244.

Roth, Philip L., Kristen L. Purvis, and Philip Bobko. "A Meta-Analysis of Gender Group Differences for Measures of Job Performance in Field Studies." *Journal of Management* 38, no. 2 (2010): 719–39. doi:10.1177/0149206310374774.

Ryan, Michelle K., and S. Alexander Haslam. "The Glass Cliff: Evidence That Women Are Over-Represented in Precarious Leadership Positions." *British Journal of Management* 16, no. 2 (2005): 81–90. doi:10.1111/j.1467-8551.2005.00433.x.

Ryan, Michelle K., and S. Alexander Haslam. "The Glass Cliff: Exploring the Dynamics Surrounding the Appointment of Women to Precarious Leadership Positions." *Academy of Management Review* 32, no. 2 (2007): 549–72. doi:10.5465/amr.2007.24351856.

Ryan, Michelle K., S. Alexander Haslam, Mette D. Hersby, and Renata Bongiorno. "Think Crisis–Think Female: The Glass Cliff and Contextual Variation in the Think Manager–Think Male Stereotype." *Journal of Applied Psychology* 96, no. 3 (2011): 470–84. doi:10.1037/a0022133.

Scott, James C. *Arts of Resistance: Hidden Transcript of Subordinate Groups.* New Haven, CT: Yale University Press, 1990.

Silver-Greenberg, Jessica, and Natalie Kitroeff. "Miscarrying at Work: The Physical Toll of Pregnancy Discrimination." *The New York Times.* October 21, 2018. www.nytimes.com/interactive/2018/10/21/business/pregnancy-discrimination-miscarriages.html.

Smith, Nathan Grant, and Kathleen M. Ingram. "Workplace Heterosexism and Adjustment Among Lesbian, Gay, and Bisexual Individuals: The Role of Unsupportive Social Interactions." *Journal of Counseling Psychology* 51, no. 1 (2004): 57–67. doi:10.1037/0022-0167.51.1.57.

Spitzack, Carole. "The Production of Masculinity in Interpersonal Communication." *Communication Theory* 8, no. 2 (1998): 143–64. doi:10.1111/j.1468-2885.1998.tb00215.x.

Spradlin, Anna L. "The Price of 'Passing'." *Management Communication Quarterly* 11, no. 4 (1998): 598–605. doi:10.1177/0893318998114006.

Stark, Liz. "Hartzler: Transgender Service Members 'Costly' to Military." *CNN.* July 26, 2017. www.cnn.com/2017/07/26/politics/transgender-military-ban-vicky-hartzler-cnntv/index.html.

Stockdale, Margaret S., Michelle Visio, and Leena Batra. "The Sexual Harassment of Men: Evidence for a Broader Theory of Sexual Harassment and Sex Discrimination." *Psychology, Public Policy, and Law* 5, no. 3 (1999): 630–64. doi:10.1037/1076-8971.5.3.630.

Sweeney, Brian. "Performance Anxieties: Undoing Sexist Masculinities Among College Men." *Culture, Society and Masculinities* 5, no. 2 (2013): 208–18. doi:10.3149/csm.0502.208.

Taylor, James R., and Van Every Elizabeth J. *The Emergent Organization: Communication as Its Site and Surface.* New York; Hove, UK: Psychology Press, 2008.

Trautner, Mary Nell. "Doing Gender, Doing Class." *Gender & Society* 19, no. 6 (2005): 771–88. doi:10.1177/0891243205277253.

Trethewey, Angela, Cliff Scott, and Marianne Legreco. "Constructing Embodied Organizational Identities: Commodifying, Securing, and Servicing Professional Bodies." *The SAGE Handbook of Gender and Communication*: 123–42. doi:10.4135/9781412976053.n7.

Trethewey, Angela. "Disciplined Bodies: Women's Embodied Identities at Work." *Organization Studies* 20, no. 3 (1999): 423–50. doi:10.1177/0170840699203003.

Trethewey, Angela. "Reproducing and Resisting the Master Narrative of Decline." *Management Communication Quarterly* 15, no. 2 (2001): 183–226. doi:10.1177/0893318901152002.

Trice, Harrison Miller. *Occupational Subcultures in the Workplace*. Ithaca, NY: ILR Press, 1993.

Trujillo, Nick. "Organizational Communication as Cultural Performance: Some Managerial Considerations." *Southern Speech Communication Journal* 50, no. 3 (1985): 201–24. doi:10.1080/10417948509372632.

Waldo, Craig R., Jennifer L. Berdahl, and Louise F. Fitzgerald. "Are Men Sexually Harassed? If So, by Whom?" *Law and Human Behavior* 22, no. 1 (1998): 59–79. doi:10.1023/a:1025776705629.

Waldo, Craig R. "Working in a Majority Context: A Structural Model of Heterosexism as Minority Stress in the Workplace." *Journal of Counseling Psychology* 46, no. 2 (1999): 218–32. doi:10.1037/0022-0167.46.2.218.

Westbrook, Laurel, and Kristen Schilt. "Doing Gender, Determining Gender." *Gender & Society* 28, no. 1 (2013): 32–57. doi:10.1177/0891243213503203.

Williams, Christine L., Patti A. Giuffre, and Kirsten Dellinger. "Sexuality in the Workplace: Organizational Control, Sexual Harassment, and the Pursuit of Pleasure." *Annual Review of Sociology* 25, no. 1 (1999): 73–93. doi:10.1146/annurev.soc.25.1.73.

Williams, Joan. *Unbending Gender: Market Work and Family Work in the 21st Century*. Oxford: Oxford University Press, 1999.

23

SHIFTING SANDS AND MOVING GOALPOSTS

Communicating gender in sport

Kitrina Douglas and David Carless

It was a beautiful day and the walk to the park took barely five minutes. But well before we had reached the park gate, the sunshine had already begun to warm our bodies and conversation. And it was usually this way. We'd step outside for a walk or a surf, and in doing so, the "what we were going to write about" would begin to take shape.

As we came to the park gates, so intense and engaged were we in our conversation, neither of us really gave much attention to a group of footballers training. In fact, we might have almost failed to register the group as they ran past, but for a comment made by one of the men with COACH written in capital letters on the back of his tracksuit, who shouted, "C'mon girls, put your backs into it!"

And then they were gone, as quickly as they'd appeared. The dregs of the sexist message, in contrast, hung in the air. I looked at David, he looked back, each wondering what we could have said that would make a difference.

Take the most prestigious tournament

to a golf course with no women members

Put flowers in the urinals

A screen down the middle of the locker room

Men on the left-Ladies to the right

And finally, hammer a "Championship" marker on the men's tees

And make out everything is OK

We let them play!

"Jeremy, we can't play today," said a young male returning to the open trunk of a car in the car park, "there's some *l-a-d-i-e-s* event on."

"Bloody women!" his mate replied, slamming the door.[1]

Adapted from Douglas and Carless, 2008

389

A call to stories

Like many qualitative researchers, we have often found traditional research methods are not well suited to exploring and/or representing emotional connection, empathic understanding, or embodied knowledge.[2,3,4,5] Yet, sport is renowned for displays of emotion and passion; its axis is physical, sensual, and fleshy. We have thus been challenged to find ways to preserve the corporeality of sport without diminishing, devaluing, or diluting its complexity during the research process. Arts-based methodologies, and particularly storytelling, has provided us one route forward. It is not our intention to describe here all the benefits storytelling *per se*, but given we began this chapter with two short stories and will later use dialogical and storied representations, we feel some background to narrative, stories, and storytelling would be useful.

Narrative theory and stories

Underscoring our use of storytelling is narrative theory, of which the central tenet is that communication through stories, is a basic human activity.[6,7,8] In order to make sense, bring meaning and create order, people create and share stories. Through creating and sharing stories, an individual is able to create an identity, a sense of place, belonging, and community. Narrative theory provides an understanding about the "work" that these stories "do," making clear how stories are implicated in their influence on human behavior, action, and relationships. An important concept underpinning narrative theory is that stories can only be created through the narrative resources that are available to an individual. This being the case, we learn a great deal about a culture or "people" by exploring the stories that are told and/or silenced within that culture.

Important to our narrative research in sport has been recognizing the way stories cluster together and give rise to recognizable plots. *Dominant narratives* are clusters of stories that have moral force so that some actions and behaviors are justified over others. The dominant narrative in sport has been called the *performance narrative*,[9] a hegemonically masculine type of story,[10] where winning is the only accepted outcome and being tough, strong, and powerful are the only routes to success. If stories about emotional engagement, loss, weakness, shame, empathic awareness, or sensitivity do surface, they are often subsequently silenced, hidden, devalued, and/or rendered taboo. One consequence is that there is a dearth of stories to validate alternative ways of living and being in sport, yet these alternatives are needed to help people negotiate their lives in sport, for example, when the performance narrative fails to fit or align with individual experience. At these times, *counter narratives* are needed to oppose, resist, or undermine the authority of a dominant narrative, bringing to light morally relevant details that have been missed or misrepresented by the dominant narrative.[11,12]

Under these circumstances, storytelling methodologies do work that moves beyond analyzing stories. A story can be created by a researcher in order interrogate a scene, to better understand behavior, and/or to communicate research findings. A well-told story, for example, includes physicality and emotions. In recent years, we have found reading or performing stories to students evokes a very different response compared with traditional types of [re]presentation.[13]

From an ethical and moral standpoint, stories also have the potential to remove artificial boundaries while preserving and revealing the intimacy we develop with a participant that is usually obscured in research findings. Stories also allow us to stand with participants on the page and in the unfolding action, thus showing the philosophy underpinning our research. Before providing some examples, we first revisit some of the issues that are relevant to understanding how hegemonic masculinity has been constructed and maintained in sport.

Constructing hegemonic masculinity through sport

In 1989, when Australian feminist, activist, and scholar Lois Bryson wrote that sport is a powerful institution through which male hegemony is constructed and maintained,[14] she may possibly have hoped that by 2020 things might have changed. So too might have Bennett, Whitaker, Wooley Smith, and Sablove,[15] who said of sport:

> Sport perpetuates domination and submission. Sport is built on a capitalistic model of competition and survival of the fittest. Sport uses people and discards them. The very language of sport is the language of assault and dehumanization. One team penetrates the other's defences; seeks another's weakness; wipes out the opponent. Men who err are sissies (women). Men who perform well are studs. The recruitment pool is a meat market; injured players are put out to pasture.

In the intervening years, the number of women and girls participating in sport and the types of activities now accessible to women have increased.[16] In numerous professional sports women can earn sizeable incomes through prize money, salaries, and endorsements.[17] There are now more women coaching sport, more women taking up leadership roles in sport, and more opportunities for women to work in sport media and television. These types of outcomes often lead people to assume that there is equity and equality in sport.

Against this, it is important to note, women in sport are still being subjected to "gender verification" testing, which is invasive and infringes on their human rights, dignity, and privacy. There remains a perception that women's sport requires rule and regulation changes to make up for a lack of strength, stamina, and power or to reduce risk of injury.[18] In professional sports, women play for lower prize money compared to their male counterparts, are given less media and TV coverage, and have fewer endorsements.[19] Added to this, for women who move into coaching and leadership roles, their progression is blocked by a ceiling made of concrete. [20] [21]

While there have been changes in sport, the culture that underpins sport and the ideology that sustains it has been—and can still be—a bastion of sexism, heterosexism, heteronormativity, and homophobia.[22] [23] [24] [25] [26] [27] There are, of course, many reasons for this. Anderson describes sport culture as a "closed-loop system"[28]—that is, athletes, coaches, and officials are likely to have grown up within a sporting culture. As such, sensitivities around gender issues along with an exclusive focus on winning means it becomes difficult to recognize, let alone challenge, oppression, homophobia, and sexism. These processes can particularly affect diversity and inclusion around gender and gender identities, as well as sexuality and sexual identities. Thus, sport continues to resemble the world of sport that Bennett et al. described.

In 1978, the United Nations Educational, Scientific and Cultural Organization published the "International Charter of Physical Education, Physical Activity and Sport," the first rights-based document establishing physical education and sport as a fundamental right for all. The document was widely criticized for failing to represent women and girls. In response, "The Berlin Declaration" made amendments in an attempt to include women and girls' physical activity, as did a revised version of the International Charter. Following these revisions, the Women's Sport Foundation noted:

> Unfortunately, despite these well-written and comprehensive documents, it is plain to see that inequalities in sport persist unaddressed even though nearly all the countries of the world have promised otherwise. Just as we see with Title IX in the United States, without vigorous monitoring and accountability, pretty words on paper do not produce changes—people do.
>
> (Oglesby, 2017)

In saying "pretty words on paper" do not bring about change, whereas "people do," the Women's Sports Foundation brings into sharp focus the necessity for individual accountability. That is, we all share a responsibility to understand what constitutes sexist, racist, abusive, and homophobic behaviors—it cannot be left to a change of policy or to well-intentioned charters alone. In order to bring about change within society, it is important that each of us begin to contest oppressive forces within abusive power systems.[29] But how do we recognize those forces that oppress us if, as Anderson suggests, the closed-loop systems in sport means we may be blind to those forces and if they are the sea within which we swim?[30] And, if we do recognize oppression and subjugation, by what means might it be challenged so it results in social change?

Moral accountability

A useful starting place is with an example of sexism and abuse that took place in 2016 and involved Donald Trump in the run-up to the U.S. presidential election. After making abusive and predatory comments about women, he stated that his comments were "locker room talk."[31] It was the use of this particular context, the male changing room, that brought it into the horizon of us in sport. However, what made this incident particularly interesting was the ideological challenge to Trump's suggestion that such talk is typical and not unusual in locker rooms. The following, published in the Telegraph online,[32] provides some of the responses:

"As an athlete, I've been in locker rooms my entire adult life and uh, that's not locker room talk." Baseball All Star pitcher Sean Doolittle of the Oakland Athletics on Twitter, October 10, 2016

"Don't throw us in there—we have nothing to do with it … that ain't our locker room talk. I don't know what locker room he's been in." Basketball player Udonis Haslem of the Miami Heat

"Claiming Trump's comments are 'locker room banter'…is to suggest they are somehow acceptable. They aren't." Basketball player Dahntay Jones of the Cleveland Cavaliers, Twitter, October 9, 2016

"I'm offended as an athlete that @realDonaldTrump keeps using this 'locker room talk' as an excuse." Soccer player Robbie Rogers of LA Galaxy, the first openly gay athlete to play in a top North American sport league, Twitter, October 10, 2016

There are a number of processes at play in the above that seem important to consider more closely. From a narrative theoretical position, dominant narratives can damage an individual's moral agency through hiding, naturalizing, and normalizing processes. They *naturalize* (the oppression and abuse of women by men) through suggestions that these actions are part of male biology or in other words, a masculine trait where "boys will be boys." By identifying himself within such a narrative script, Trump uses the narrative to deflect attention away from his personal accountability toward a male biology, hormones, and DNA, and sport culture allows or even encourages it. Secondly, dominant narratives *normalize* sexist behavior, presenting the story as if all male athletes talk like this in a locker room. This also deflects attention away from the behavior so that "nothing morally objectionable appears to be going on."[33] In doing so, sexist and homophobic comments and behavior hide the way such talk subjugates people: in this case, men (because all men are implicated if it is natural and normal) as well as women (because it condones such behavior and talk towards women) are subjugated. These types of actions and behavior also keep alive myths about how men in sport behave, and what is acceptable in a sport context or between men. It is no wonder the general public often misunderstands athletes.

A problem for male athletes, therefore, are those clusters of stories that fan the "boys will be boys" flame also recognize narratives like, "there is no smoke without fire." In other words, although perhaps few people believed that all men are as abusive and predatory as Trump, these types of stories "are notoriously evidence-resistant"[34] and leave a question mark in some people's minds. Our identities are created around the aspects of our lives and relationships that matter to us, yet they are also constituted by the stories other people create about us.[35] Thus, the "boys will be boys" type of narrative omit problematic information or evidence (as illustrated previously) that undermines the truth of their claims and, at the same time, "undermines the cognitive authority of people who are in a position to point out those inconvenient facts."[36]

Hilde Lindemann Nelson provides some theoretical background insights that shed light on this issue. She suggests that if an identity has been morally damaged, one route towards repair is through acts of purposeful self-definition, such as developing or creating a counterstory.

> Counterstories, which root out the master narratives in the tissue of stories that constitute an oppressive identity and replace them with stories that depict the person as morally worthy, supply the necessary means of resistance. Here, resistance amounts to *repair*: the damaged identity is made whole. Through their function of narrative repair, counterstories thus open up the possibility that the person could attain, regain, or extend her freedom of moral agency.[37]

By publicly distancing themselves through comments like: "That's not locker room talk," "Don't throw us in there," and "Bragging about getting away with sexually assaulting women because you're famous is not locker room talk!" these athletes help create a reservoir of counterstories that set out to cause a shift[38] in cultural understanding. Their actions make a clear and public statement about individual moral identities, about accountability towards women and sport, and about what is acceptable. While these acts help reinstate the moral agency of these athletes, they also show solidarity and advocacy for women, which are equally important in terms of challenging sexist behavior and attitudes.

While this is a very public example, sexist and homophobic behavior also exists under the radar, in more covert messages, and through signs and signals that are more difficult to decode and identify as sexism or homophobia. For example, the coach who said "C'mon girls" to a group of males might believe the comment is just a bit of ribbing and not recognize that the roots of the comment (and arguably why it motivates some men) are embedded deeply within a much larger narrative. Thus, many people—who do not see themselves as sexist and or homophobic—may be unaware of their sexist and homophobic comments and behavior.

As researchers, we have a responsibility not just to point out or document what is happening. While this task has arguably been necessary, perhaps the priority now is to make more explicit and understandable how and in what ways people need to become more aware, and more sensitive and behave differently. Along with this, we also need to imagine and envisage better alternatives.

Using stories in sport

In what follows, we draw on dialogue and stories to explore how stories might challenge hegemonic masculinity, sexism, heterosexism, and homophobia in sport. Our dialogical approach takes the form of a conversation around particular stories we have written from researching three different sport contexts. These are same sex attraction in school sport; high performance sport; and sport and adventurous training for wounded/injured soldiers.

Example one: researching same sex attraction in school sport

KITRINA: David, the stories you often write seem to go beyond talking about or showing narrative theory. Why is this?

DAVID: Denzin makes the point that the "performance tale is a utopian tale of self and social redemption, a tale that brings a moral compass back into the reader's (and the writer's) life."[39] Performance texts *act on the world in order to change it,* so stories, vignettes, and other arts-based methods provide a way for people to think and understand the world differently. Along with that, there is so much of the human experience that lies at the borders of what can be said;[40,41,42,43,44] we can't just leave it out because it's difficult to communicate. Creating a story (or at times a song, poem, or ethnodrama) helps us to communicate emotionally significant information and some of the things that go unsaid, or can't be put into words. It also makes it possible to take the reader into unfamiliar scenes, or make the familiar seem strange, and that helps make more obvious how people are hurt by what, on the surface, might seem innocuous comments. The following extract is from a story I wrote based on school experiences of same sex attraction.

Adapted from "A story in ten fragments"[45]

"Ahhhh! You two are always in here!" A loud, abrasive voice calls out from the corner of the tiled shower area. It's Marcus, a team-mate from the under-15 rugby team, and he's leaning round the entrance to the showers, in his army uniform grinning and staring at us. Simultaneously, Robert and I turn away from Marcus to face the wall. "*Oh brilliant!*" I whisper to Rob sarcastically, jolted from my relaxed state. Since giving up rugby, Robert has mostly been spared Marcus's presence. I have not.

"So, let me have a look now … which one of you is the giver and which is the taker?" Marcus shouts out, chuckling to himself. "C'mon, turn around girls, don't be shy!" Both Robert and I stand where we are, facing the wall. Neither Rob nor I look at each other and neither of us speaks. "Oh yeah! You are definitely the taker!" Marcus laughs, pointing his finger. "You've definitely got the arse for it."

KITRINA: I've never been into a male changing room, nor been in the shower at a boys school, or travelled on a coach to a rugby match, but your stories allowed me to get a feel of what it may have been like to be a young man attracted to another young man in a culture that is hegemonically masculine and homophobic. It helped me to understand things in an emotional and connected way, and also see how gender is policed through verbal abuse. The story made it clear how hegemonic masculinity silences gay and bisexual men, while and at the same time, it degrades and devalues women.

DAVID: Storytelling makes it possible for the reader to feel and experience with the characters in the story what is emotionally significant for them and why. If it's done well, the writer gains new insights, as well as the reader, so there is potential to develop a much more sophisticated level of ethical and moral reflection.[46]

Example two: researching high performance sport

KITRINA: I have only learned through the process of writing stories that it is possible, as a writer, to learn, be surprised, or even shocked, when interrogating a scene and characters/dialogue as I write. At these moments it becomes possible to find connections between a throw-away comment like "turn around girls, don't be shy!" and an ideology that damages women and men. In the following story, where I was attempting to take the reader into my world of professional golf, I drew on my experiences and observations, as well as research interviews with female tour players. The story is a fiction based on fact; there are no people called Bernard, Ken, or Monica, but these types of behaviors unfold each week. The story is based on the contractual event that occurs before each tournament, the pro-am, in which women professional golfers have to play. The main character is Bernard, a type of powerful and successful businessman you are likely to find at women's golf tournaments. The following extract picks up the story after Monica has hit her tee shot.

"The Pro-am"[47]

"On the tee Bernard Brasco!" The starter announced, but Bernard wasn't ready. He was still looking down the fairway to where Monica's ball lay…then it dawned on him, as he looked again towards her ball in the middle of the fairway, a long, long way away, that he must hit it further than this girl.

A bead of sweat formed on his brow, it gained weight and size before dripping onto his shaky grip. He stood in the hitting position, muscles tense, stance too wide, and going for the big shot, he swung too quickly—it was all over in a flash. A huge divot flew up in the air and the ball squirted off into bushes 50 yards away. Bernard didn't look up.

"Nice one Mary!" came a voice from the crowd along with muffled chuckles, "Does your husband play? I think you forgot your makeup and yellow handbag!"

Paul, the third member of the group recognized the voice and nodded toward Julian, a slightly inebriated client of theirs. Bernard piped up.

"It's this Italian, Julian—she's got me in a fluster!"

"Oooh! Now I see why you hit in the bushes! Maybe she'll help you look for it!" Julian and Bernard both laughed loudly.

Monica smiled. The other professionals rolled their eyes; they'd heard it all before.

DAVID: Even from the short extract above, I get a sense of the ways women can be infantilized by using the term *girl*, and the way some men feel shame by not being able to hit the ball further than a woman. Then they can be mocked by others, as Julian does when he calls Bernard "Mary," and through other abusive comments that question the hegemonically masculine gender order. Both men then attempt to restore a hegemonic, heterosexist gender order by making predatory sexual innuendos—that Bernard hit into the bushes on purpose in order to lure Monica to look for his lost ball, and then, he would have sex with her which would provide evidence of his manhood and allow him to save face. When women are thus disempowered publicly, perhaps the only dignified thing to do is stay silent. The Pro-am is a light-hearted story, as opposed to "That Night," another story from your research which I always find difficult to listen to.

Example three: researching taboo issues in high performance sport

KITRINA: I hate the "That Night" story, but I hate even more that people get raped, and how often I hear some men say, "She was asking for it," without any idea about the person, the circumstances or how being raped can bring a level of trauma that can change a person's life forever.

DAVID: Why did you choose to write about rape through a story about rape?

KITRINA: The most honest answer is a participant I'd been interviewing for six years as part of my PhD contacted me for help. She'd recently retired from professional sport and the process had provoked a lot of emotional distress and triggers to trauma. She'd never told her Mum she had been sexually assaulted when she was a child, nor had she felt able to recount her experience of rape. She lived with these in silence, feeling shame. Well, in fact she was only able to tell me a few facts. I spent the whole day just witnessing her distress and handing her tissues. I wasn't researching rape; I didn't invite this story and I felt inadequate. I didn't know what to do, but I also felt a responsibility to do something. When I decided to try and write the story, there was a lot I didn't know, but through drawing on my imagination and experiences on tour as a container for her story, I was able to use the facts that I had and set them into motion through a story plot that I was familiar with, thinking *this is how it might go down*. The story begins with a group of pro golfers and caddies watching football on TV one evening in one of the player's hotel rooms; nothing unusual about that. The game ends and people disperse, two are left. This happens all the time, again, nothing unusual. One of them, Graham, is a horny male and wants sex, the participant didn't tell me who this person was. She would say, he was just one of the "boys." The character in the story, Val, represents the participant. While confident and resourceful on a golf course, she was extremely vulnerable off the golf course due to her childhood experiences. She wants love, affection, a relationship, and acceptance (don't we all?), but she was struggling with her sexuality. Because the culture is heterosexist and she wants to fit in, she doesn't want to be labeled a lesbian. But nor does she want her experience of being sexually assaulted as a child to keep ruining and impinging on her life. Slowly, through emotional and psychological coercion, Graham has sex with her.

DAVID: I don't think it is possible to communicate the horrors of rape, or the emotional, physical, spiritual and psychological harm that is done, or how this is such a taboo subject in sport, as powerfully, by just reporting the occurrence, or presenting "data" and stats. We need to create, from our research, accounts that are visceral, evocative, compelling, and unflinching. Stories can do that, and we also need to share stories that resist cleaning up, sterilizing, or diminishing the brutal realities of discrimination, misogyny, abuse and trauma. I remember film director Steve McQueen said something similar on his Oscar winning film *12 Years A Slave*. He said, "I didn't want to censor myself on anything. I decided I'm going to show everything … How can I make a movie about slavery and not show some aspects of it? I cannot. For my sisters, and for other people, it would be a travesty … It was a very costly, emotionally charged story to tell on everybody's part."[48] Likewise with your story, I find the final scene especially damning and depressing. In the following extract, we take up the story as Graham leaves Val's room.

"That night"[49]

Val wanted to curl up into a ball and die, but she needed to vomit and so she staggered awkwardly to the bathroom where she threw up immediately, grabbing the toilet bowl to steady herself as she wretched out her guts, her pain, her stupidity, her worthlessness, her hopelessness, and then, she collapsed on the floor and sobbed and sobbed and sobbed.

Meanwhile, Graham ran into Dino and Phil in the lobby.

"Ah, Graham," Phil said putting his arm around his friend's shoulder as much to steady himself as it was to get close to his mate. "What you been up to? Coming for a game of pool?"

Graham's hand, in an obvious fashion, went directly to his crotch where he massaged his genitals. Simultaneously, his facial expression produced a huge smile and raised eyebrows.

"You bastard, you haven't?" asked Dino.

"Nahhh, course not," he said, laughing and raising both hands in mock surrender.

"What? With Val?" Phil laughed, and joined the others in their walk to the snooker room, "You fucker!"

"Yeah, I know," Graham chuckled, "and you owe me ten euros, Phil."

"What, is she?" Phil asked setting the balls up on the table.

"Yeaaah," Graham replied laughing while sorting a snooker cue from the rack and quickly chalking its tip, "Definitely a member of the finger in the wall club." He bent over the table, steadied the cue in his left hand, pulled it back, and hammered its tip into the white ball. As balls exploded in all directions, he turned to his mates.

"She's a dyke, gentlemen, a lesbo, and not a very good shag."

KITRINA: So much for stories helping us to imagine a better world eh?! But, I imagine for every person who has been raped, male or female, and for those who have been sexually abused, it *is* bleak. And sadly, I have heard some men talk like Graham, in a predatory and abusive ways.

DAVID: But, in terms of this type of research it is needed. When you performed the story in lectures and conferences, often to sport science students, and then invited group discussions and written feedback, it's made it very clear how important these types of stories are.[46] Class discussions provoked students to think about their ethical behavior, for example, when the male students publicly sided with Val. On the feedback sheets from female students, many noted how they found the responses of their male colleagues to be empowering and generated feelings of solidarity. But secondly, in group conversations after you read the story, the male students distanced themselves from Graham. By making their remarks (that Graham was wrong, and that he had no right), "in public" in front of both their male and female peers, the male students' declarations become a moral act. This is much like the athletes who publicly challenged Trump's "locker room" remarks, sending a clear message that this type of talk/behavior is wrong, and not how they behave.

KITRINA: And as opposed to rape and sexual abuses being "taboo" and a woman being silenced, the story and discussion brought these issues and the behavior of men into the open in a way that allowed sport students (many of whom will go into high level sport teams) to claim a moral identity and understanding in a deeper way about how some types of "talk" or "banter" are part of a dominant and damaging narrative. Several students also wrote in their feedback that before the lecture, if they had read in the news media that a star athlete had been accused of rape, they would have thought it was the woman's fault, but following the story and the discussion, they had changed their minds.

Example four: researching a sport intervention for soldiers returning from war

DAVID: Researching injured soldiers returning from wars in Afghanistan and Iraq during a sport and outdoor program has been a completely different environment. For some, "sport" is playing an important role in their recovery. The following story, based on ethnographic field research, gives a glimpse of what this sport recovery means to the soldiers who participate in wheelchair basketball.

Wheelchair basketball[50]

Have you ever seen a man in a wheelchair perform a pirouette? Spin on the spot, 360-degrees, at speed and then stop, as they say, on a dime. Then, roll backward, tip the front up, and balance on two wheels, laugh, and spin again. Now, give him a ball, and put him on a basketball court.

Imagine that same man—dirt in his mouth, his eyes, his teeth and hands bloody, legs missing, genitals exposed through torn fatigues, broken skin, lying motionless on an Arab desert floor, thousands of miles from his home, unable to move—when the thought flashes across his mind: "I CAN'T FUCKING MOVE!" Fast forward. Through the months of rehab, pain, lies, humiliation, lost hope, fear. Sat at home, alone, drugged up, angry, lost. And no one comes to visit. A month passes, 2 months, 3 months, 4, a year ... apathy ... another year.

The clash of iron, the noise, the exhilaration, the possibility, fun, laughter, blokes ribbing each other, speed down the court. Lungs bursting, perspiration flooding, adrenaline pumping, a body that recognizes, remembers, the feelings, they were good, weren't they? They were lost, weren't they? And here they are—again. Dare a man ... believe again?

KITRINA: We both found that individually soldiers were really respectful and valued an opportunity to talk, both about their experiences of war and life changing injuries, but also how this particular sport project was helping them. That said, we both found military culture sexist and homophobic. The following story, "How do you take it?" is one example, and one way we have attempted to challenge sexism.

How do you take it? (Based on field notes, May 2015)

Needing to backup my interviews I bypassed the refreshments trolley, sat down and opened my laptop. Alan and Norman, two of the civilian coaches came in, sat nearby and passed the time joking and exchanging stories while watching YouTube clips on Norm's phone. Slowly, one-by-one, the soldiers trickled in; most went straight for the coffee and cake.

"Have you been to Northern Spain then?" Alan asked, picking up on a conversation we'd been having earlier.

"Yes, a few times," I said looking up as he stood and walked towards the coffee.

"Where was that?" he continued.

"Oh, Santander, Somo, Bilbao." I watched him pour his coffee.

"Do you want a coffee, Norm?" He asked.

"Oh – yes, thanks," Norm replied, not looking up from his phone.

"Kitrina?" he asked, smiling and holding up another cup.

"Yes, thanks." I shut my laptop and walked over, glad to be asked, glad to be included. I was dying for a coffee.

"How do you take it?" he asked.

I thought Alan's question was aimed at me, but before I could answer, Norm spoke up: "I like mine 18 inches," he said laughing and then, "Are we talking about the same thing?"

I wished I hadn't heard, and didn't laugh. I felt everyone's gaze. Whatever I had been going to say to Alan about Spain was now lost to me, not knowing what to say to Norm, I just answered Alan.

"I don't need milk or sugar thanks."

Alan passed me the coffee. He wasn't smiling now but carried on pouring coffees, unwrapping the cake, and offering pieces round. Everyone in the room seemed to be suddenly quieter than before the comment. Perhaps they were waiting to see what happened next.

A moment later, not looking up from his phone, Norm carried on.

"We have this joke don't we, Alan?" there was a pause while his fingers danced over the phone again, then, "Sorry Kitrina, I shouldn't say that," he laughed.

"No!" I said, in an offhand way, trying not to make too much of it. "You shouldn't have."

"But it's the army in me coming out, I've only been out a couple of years."

"It doesn't matter, you shouldn't say that." I repeated, putting my coffee by the chair, and getting on with my work. Alan passed Norman a coffee.

"Thanks, mate," he said, without taking his eyes off of his phone, "That's great."

In a whirl of high energy and enthusiasm the head coach burst into the room. "Right everybody," he announced in a loud voice, rubbing his hands and smiling. Any remaining tension was lost. I was lost in thought, remembering the stories some of the female soldiers had told me that week. Stories of sexual abuse and rape, and I wondered if Norm or the other coaches had any idea what it was like being a female in this type of culture, one where, to get on, a woman feels she has to become 'one of the boys.' I hated that I was the one who noticed, hated it had to be me who spoke out and that, alongside all the very excellent coaching, there were moments like *this*.

DAVID: The previous story illustrates how difficult, or impossible, it can be, to "call out" sexist, homophobic behavior/talk when no one else in the group will stand with you. Norm knows his comment, which refers to the size of his penis, is wrong, and offers a (lame) apology. But, in the same way that Trump did, he attempts to deflect attention away from personal account-ability by including another male coach, "We have this joke don't we, Alan?" and then, to garner further support he suggests this is normal behavior in the army. Unlike the athletes, none of the other soldiers or Alan choose to distance themselves from the comment. Your strategy, saying, "No, you shouldn't," marks his behavior as wrong, but that is the end of it. There is no counterstory, no acts or behaviors that reinstate the moral agency of Alan, or other soldiers— thus they are all tarnished with the same brush. There is no show of solidarity and advocacy for you and/or other women.

KITRINA: Ultimately, there is little change from this approach. In our final story however, you explore a different strategy. This story is also based on ethnographic field research during the sport/adventure course for soldiers.[51] Here you (David) narrate the story, while Billy (pseudonym) is the civilian coach/course leader. In contrast to the above story, "Three Seconds Flat" shows how storytelling can provoke moral understanding along with behavior change.

Dining room, day 3 (adapted from "Three seconds flat")

It's a demanding job, being course leader, and one that I can see Billy takes seriously. I imagine he would rather be spending the evening preparing for tomorrow's sessions, so I appreciate him making time for me to interview him.

I've brought along drafts of two stories I've written which foreground moments of sexism and homophobia this week. I want to talk with him about these, but don't feel comfortable just bringing them up in general conversation. I suppose I want him to have a chance to feel what I've felt being here. So I begin our interview by asking if he would read the stories as a way to kick-start our dialogue. He agrees and begins to read while I switch on my recorder, and wait.

Billy finished reading, looks up and sits back in his chair. "Good piece of writing that," he says as we both laugh, a little nervously. He rubs his hands together, pausing again.

I'm finding the silence uncomfortable, so begin to speak, "Well, it's trying to get …" then force myself to stop, and remind myself, *leave space for him*. "No, I should let you speak." The seconds tick over one by one on the recorder's digital display. Fifteen seconds pass. Billy leans forward, places his elbows on the table. Twenty seconds. Billy shifts back in his chair. Thirty seconds. I'm drawn to fill the silence, but hold off.

Eventually, Billy continues. "That," he pauses again, "is a good piece of writing, a fly on the wall." He halts. "Fly on the wall is the device used by" I feel that he is filling time again, searching for words, "… And it probably takes a person like you who is outside of the main." He hesitates again, leaning forward once more, elbows on knees.

Where is he going with this? Is he working towards positioning me as an outsider, just the researcher?

He begins to speak again: "Um, errrr, I …" then halts, stumbles. "It does make me think, actually, because I still use inappropriate things sometimes. But I'm aware of … I'm hoping to get a laugh. But I'm not including …." He stops again, sits back in the chair, rubs the back of his head.

"Women in the Services are an interesting thing …"

Now where is he going with THIS?

He stops. Thinks some more. "They can either sacrifice all of their femininity to fit in, or … But it's not just women—could be all sorts of things."

I sense Billy filling time again, trying to decide what is okay to say.

"It does make me think we need to be more … One of us might be cracking a joke about picking up soap in the shower or whatever, which is referring to homosexuality, and of course we may have a … We're not going to get feedback on that are we?"

"Well," I begin, "my partner is male. I think that's one of the reasons I've often felt like an outsider, particularly in sport." Billy is listening—intently. "I'm sensitive to exclusion because of sexuality. But I don't want to just state in a report or article that the culture of the course is homophobic or sexist. The reason I am recounting these events as stories is to engage with the complexities behind that, explore how it unfolds, how it came to be that way, ask questions, invite dialogue about how it could be different."

Billy nods. "I think that's really valuable. I would never want to be in a situation where I was hurting or excluding anybody. Perhaps I should have said, 'Are you a gay guy?' I almost don't even know how to approach that. I know quite a few people who are gay but I'm still …" He pauses again, rubbing the back of his neck, again, before continuing: "I'm as nervous about sexuality. I need to be careful because if what we do offends people then the course is not doing what it's supposed to do. It's as applicable if the guy is disabled, gay, black, you know, we may be giving him a hard time without even knowing it. We need to be more careful."

Sports hall, the next afternoon

Billy, Paul, and I are laying out the mats for a seated volleyball session that will begin shortly. We haven't got long to cover the floor with the simple geometric arrangement of mats that will provide the court for the two teams. The phrase "military precision" might be a cliché, but right now it's appropriate: three men swiftly manoeuvring 40-odd gym mats from the storage room into a tight arrangement on the sports hall floor. We're chatting, joking, and calling to each other as we work.

"I got this one, Billy," Paul shouts.

"Speed m-a-c-h-i-n-e!" Billy echoes.

"This one's stuck, gimmie me a hand, Paul?" I ask.

"Sure thing," Paul replies, giving the other end of the mat a kick.

"I've got the left side covered," Billy says to us both.

"OK, I'm coming in behind you, Billy," I reply.

"Uh oh! Better watch out, Billy!" Paul shouts, laughing.

Did he just say that? Was that an anal sex "joke"? My heart jumps, I feel an impulse to freeze, but keep on working. Maybe I'm being hypersensitive?

Before there's time to decide, Billy drops his mat, turns to Paul and says: "Hey, mate. Come on. We don't need any of that kind of chat around here."

Then I know right away. Billy is on it. He gets it. Advocacy. Solidarity.

Concluding thoughts: shifting sands and moving goalposts

Sport remains a bastion of hegemonic masculinity, homophobia, and misogyny. That is to say that sport remains an arena that reproduces a desire for the toughest form of masculinity, an attitude in which "men are men"; an arena in which homosexuality, femininity, and other assumed "weaknesses" are not perceived as being conducive to the ultimate quest for victory.[52]

The shifting sands which support gender discussion and awareness are changing, and that's a good thing. But there is a great need to reflect on these changes, both how we understand them, as well how we document and communicate them.

Historically, sport culture has made it easy for misogyny, homophobia, and abuse not only to take place, but often, to flourish. At the same time, the consequences were often disregarded or hidden. Those who experienced abuse, along with those who perhaps wanted to advocate for social justice, had no vocabulary or narrative maps to challenge the dominant narrative or chart a better an alternative course. However, the frame through which we now view such behavior and talk has changed.

While it is still difficult to weed out and link a comment or action that is presented as "just a bit of fun" or "just my nature" with the underpinning ideology of hegemonic masculinity, we now have more accessible strategies and devices to make sexism and homophobia visible, one of these being through using stories. Stories help people find their identities within a storyline with which they are familiar. Such stories make clear the "roots" of hegemonic masculinity that are still hidden in the *tissues of stories*,[53] at a micro level. As a pedagogical tool, as well as a strategy for transformative learning, stories therefore help reorient our moral compass while also having the potential to narratively repair damage. To this end we hope the examples used here provide some insights into how such devices might help move the goal posts when communicating and interrogating issues around gender in the world of sport.

Notes

1 Kitrina Douglas and David Carless, "The team are off: Getting inside women's experiences in professional sport," *Aethlon: The Journal of Sport Literature* XXV, I (2008): 241–251.

2 David Carless, "Negotiating sexuality and masculinity in school sport: An autoethnography," *Sport, Education and Society* 17, 5 (2012): 607–625.

3 Kitrina Douglas, "Signals and Signs," *Qualitative Inquiry* 18, 6 (2012): 525–532.

4 Kitrina Douglas, "Challenging interpretive privilege in elite and professional sport: One [athlete's] story, revised, reshaped, reclaimed," *Qualitative Research in Sport, Exercise and Health* 6, 2 (2014): 220–243.

5 Kerry McGannon, "Am 'I' a work of art(?): Understanding exercise and the self through critical self-awareness and aesthetic self-stylization," *Athletic Insight* 4 (2012): 79–95.

6 Dan P. McAdams, *The Stories We Live By* (New York: Guildford Press, 1993).

7 Arthur W. Frank, *Letting Stories Breathe: A Socio-Narratology* (Chicago: The University of Chicago Press, 2010).

8 John McLeod, *Narrative and Psychotherapy* (London: Sage, 1997).

9 David Carless and Kitrina Douglas, "Living, resisting, and playing the part of athlete: Narrative tensions in elite sport," *Psychology of Sport and Exercise* 14, 5 (2013): 701–708.

10 Eric Anderson, *In the Game* (Albany, NY: State University of New York Press, 2005).

11 Kitrina Douglas, "Storying my self: Negotiating a relational identity in professional sport," *Qualitative Research in Sport, Exercise & Health* 1, 2 (2009): 176–190.

12 Hilde Lindemann Nelson, *Damaged Identities, Narrative Repair* (Ithaca, NY: Cornell University Press, 2001).

13 David Carless and Kitrina Douglas, "Performance ethnography as an approach to health-related education," *Educational Action Research* 18, 3 (2010): 373–388.

14 Lois Bryson, "Sport and the maintenance of masculine hegemony," *Women's Studies International Forum* 10, 4 (1987): 349–360.

15 Roberta S. Bennett, K. Gail Whitaker, Nina Jo Woolley Smith, and Anne, Sablove, "Changing the rules of the game: Reflections toward a feminist analysis of sport," *Women's Studies International Forum* 10, 4 (1987): 369–379.

16 For example, since 2000, women can compete in weightlifting, modern pentathlon, taekwondo, triathlon, bobsleigh, wrestling, golf, rugby, and boxing at Olympic levels.

17 For example, basketball, NASCAR, football, cricket, golf, tennis, badminton, and surfing.

18 Examples include Olympic-level female boxers who compete over two-minute rounds, as opposed to three-minute rounds for male competitors. In swimming, there is no women's 1500 m freestyle. In gymnastics there are six disciplines for male competitors and four for women—who do not perform on the pommel horse or hoops and neither do male gymnasts compete to music on the floor exercise.

19 Janet Fink, "Female athletes, women's sport, and the sport media commercial complex: Have we really 'come a long way, baby?'" *Sport Management Review* 18 (2014): 331–342.

20 Leanne Norman, Alexandra Rankin-Wright, and Wayne Allison, "'It's a concrete ceiling; it's not even glass': Understanding tenets of organizational culture that supports the progression of women as coaches and coach developers," *Journal of Sport and Social Issues* 42, 5 (2018): 393–414.

21 Gregory Cranmer, Maria Brann and Nicholas Bowman, "Male athletes, female aesthetics: The continued ambivalence toward female athletes in ESPN's The Body issue," *International Journal of Sport Communication* 7 (2014): 145–165.

22 David Carless, "Negotiating sexuality and masculinity in school sport: An autoethnography," *Sport, Education and Society* 17, 5 (2012): 607–625.

23 Beth Fielding-Lloyd and Lindsey Mean, "'I don't think I can catch it': Women, confidence and responsibility in football coach education," *Soccer & Society* 12 (2018): 345–364.

24 Vicky Krane, "We can be athletic and feminine, but do we want to? Challenging hegemonic femininity in women's sport," *Quest* 53 (2001): 115–133.

25 Shawn Ladda, "Where Are the Female Coaches?," *The Journal of Physical Education, Recreation & Dance* 86, 4 (2015): 3–4.

26 Kelly Poniatowski, "You're not allowed body checking in women's hockey: Preserving gendered and nationalistic hegemonies in the 2006 Olympic ice hockey tournament," *Women in Sport & Physical Activity Journal* 20 (2011): 39–52.

27 Sally Shaw and Larena Hoeber, "'A strong man is direct and a direct woman is a bitch': Gendered discourses and their influence on employment roles in sports organizations." *Journal of Sport Management* 10 (2003): 347–375.

28 Eric Anderson, *In the Game.*

29 Hilde Lindemann Nelson, *Damaged Identities.*

30 Arthur W. Frank, *Letting Stories Breathe.*

31 Chira Pilazzo, "Professional athletes to Donald Trump: Sexual assault is not 'locker room talk,'" *Telegraph Online*, accessed June 1, 2018, www.telegraph.co.uk/news/2016/10/11/professional-athletes-to-donald-trump-sexual-assault-is-not-lock/.

32 Clara Eroukhmanoff, "A feminist reading of foreign policy under Trump: Mother of all bombs, wall and the "locker room banter," *Critical Studies on Security* 5, 2 (2017): 177–181.

33 Lindemann Nelson, *Damaged Identities*, 162.

34 Lindemann Nelson, 167.

35 Lindemann Nelson.

36 Lindemann Nelson, 167.

37 Lindemann Nelson, 150.

38 Lindemann Nelson, 156.

39 Norman Denzin, *Performance Ethnography* (Thousand Oaks, CA: Sage, 2003).

40 Judith Butler, *Excitable Speech: A Politics of the Performative* (New York: Routledge, 1997).

41 Jerome Bruner, *Acts of Meaning* (Cambridge, MA: Harvard University Press, 1990).

42 John Dewey, "The Quest for Certainty", in *John Dewey: The Latter Works 1925–1953* (Vol. 10, pp. 1–444), ed. J. Boydstone (Carbondale, IL: Southern Illinois University Press, 1989[1934]).

43 Eugene Gendlin, "The Wider Role of Bodily Sense in Thought and Language," in *Giving the Body Its Due*, ed. M. Sheets-Johnstone (Albany, NY: State University of New York Press, 1992), 192–207.

44 Mark Freeman, *Hindsight: The Promise and Peril of Looking Backward* (New York: Oxford University Press, 2010).

45 David Carless, "Negotiating sexuality and masculinity in school sport."

46 Frank, *Letting Stories Breathe.*

47 Kitrina Douglas and David Carless, *The Team Are Off.*

48 Steve McQueen, "And the academy award goes to … 12 Years a Slave," *BBC Radio Four*, accessed January 16, 2019, www.bbc.co.uk/programmes/b08g2mgc.

49 Kitrina Douglas and David Carless, "Exploring taboo issues in high performance sport through a fictional approach," *Reflective Practice* 10, 3 (2009): 311–323.

50 David Carless and Kitrina Douglas, "When two worlds collide: A story about collaboration, witnessing and life story research with soldiers returning from war," *Qualitative Inquiry* 23, 5 (2017): 375–383.

51 David Carless, "Three Seconds Flat: Autoethnography Within Commissioned Research and Evaluation Projects," in *International Perspectives on Autoethnographic Research and Practice*, ed. Lydia Turner, Nigel P. Short, Alec Grant and Tony E. Adams (London: Routledge, 2018), 123–132.

52 Anderson, *In the Game*, 7.

53 Lindemann Nelson.

Bibliography

Anderson, Eric. *In the Game*. Albany, NY: State University of New York Press, 2005.

Bennett, Roberta S., K. Gail Whitaker, Nina Jo Woolley Smith and Anne Sablove. "Changing the rules of the game: Reflections toward a feminist analysis of sport," *Women's Studies International Forum* 10, no. 4 (1987): 369–79.

Bruner, Jerome. *Acts of Meaning*. Cambridge, MA: Harvard University Press, 1990.

Bryson, Lois. "Sport and the maintenance of masculine hegemony," *Women's Studies International Forum* 10, no. 4 (1987): 349–60.

Butler, Judith. *Excitable Speech: A Politics of the Performative*. New York: Routledge, 1997.

Carless, David. "Negotiating sexuality and masculinity in school sport: An autoethnography," *Sport, Education and Society* 17, no. 5 (2012): 607–25.

———. "Three Seconds Flat: Autoethnography within Commissioned Research and Evaluation Projects," in *International Perspectives on Autoethnographic Research and Practice* ed. Lydia Turner, Nigel P. Short, Alec Grant and Tony E. Adams (London: Routledge, 2018), 123–32.

Carless, David and Kitrina Douglas. "Performance ethnography as an approach to health-related education," *Educational Action Research* 18, no. 3 (2010): 373–88.

———. "Living, resisting, and playing the part of athlete: Narrative tensions in elite sport," *Psychology of Sport and Exercise* 14, no. 5 (2013): 701–8.

———. "When two worlds collide: A story about collaboration, witnessing and life story research with soldiers returning from war," *Qualitative Inquiry* 23, no. 5 (2017): 375–83.

Cranmer, Gregory, Maria Brann, and Nicholas Bowman. "Male athletes, female aesthetics: The continued ambivalence toward female athletes in ESPN's The Body issue," *International Journal of Sport Communication* 7 (2014): 145–65.

Dewey, John. "The Quest for Certainty", in *John Dewey: The Latter Works 1925-1953* (Vol. 10), ed. J. Boydstone (Carbondale, IL: Southern Illinois University Press, 1989[1934]), 1–444.

Douglas, Kitrina. "Storying my self: Negotiating a relational identity in professional sport," *Qualitative Research in Sport, Exercise & Health* 1, 2 (2009): 176–90.

———. "Signals and Signs," *Qualitative Inquiry* 18, no. 6 (2012): 525–32.

———. "Challenging interpretive privilege in elite and professional sport: One [athlete's] story, revised, reshaped, reclaimed," *Qualitative Research in Sport, Exercise and Health* 6, no. 2 (2014): 220–43.

Douglas, Kitrina and David Carless. "The team are off: Getting inside women's experiences in professional sport," *Aethlon: The Journal of Sport Literature* XXV, I (2008): 241–51.

———. "Exploring taboo issues in high performance sport through a fictional approach," *Reflective Practice* 10, no. 3 (2009): 311–23.

Denzin, Norman. *Performance Ethnography*. Thousand Oaks, CA: Sage, 2003.

Eroukhmanoff, Clara. "A feminist reading of foreign policy under Trump: Mother of all bombs, wall and the "locker room banter," *Critical Studies on Security* 5, no. 2 (2017): 177–81.

Fielding-Lloyd, Beth and Lindsey Mean. "'I don't think I can catch it': Women, confidence and responsibility in football coach education," *Soccer & Society* 12 (2018): 345–64.

Fink, Janet. "Female athletes, women's sport, and the sport media commercial complex: Have we really 'come a long way, baby?'" *Sport Management Review* 18 (2014): 331–42.

Frank, Arthur W. *Letting Stories Breathe: A Socio-narratology*. Chicago, IL: The University of Chicago Press, 2010.

Freeman, Mark. *Hindsight: The Promise and Peril of Looking Backward*. New York: Oxford University Press, 2010.

Gendlin, Eugene. "The Wider Role of Bodily Sense in Thought and Language", in *Giving the Body its Due*, ed. M. Sheets-Johnstone (Albany, NY: State University of New York Press, 1992), 192–207.

Krane, Vicky. "We can be athletic and feminine, but do we want to? Challenging hegemonic femininity in women's sport." *Quest* 53 (2001): 115–33.

Nelson, Hilde Lindemann. *Damaged Identities, Narrative Repair*. New York: Cornell University Press, 2001.

Ladda, Shawn. "Where Are the Female Coaches?" *The Journal of Physical Education, Recreation & Dance* 86, no. 4 (2015): 3–4.

McAdams, Dan P. *The Stories We Live By*. New York: Guildford Press, 1993.

McGannon, Kerry. "Am 'I' a work of art(?): Understanding exercise and the self through critical self-awareness and aesthetic self-stylization," *Athletic Insight* 4 (2012): 79–95.

McLeod, John. *Narrative and Psychotherapy*. London: Sage, 1997.

McQueen, Steve. "And the academy award goes to … 12 Years a Slave," BBC Radio Four, accessed January 16, 2019, www.bbc.co.uk/programmes/b08g2mgc

Norman, Leanne, Alexandra Rankin-Wright and Wayne Allison. "'It's a Concrete Ceiling; It's Not Even Glass': Understanding Tenets of Organizational Culture That Supports the Progression of Women as Coaches and Coach Developers," *Journal of Sport and Social Issues* 42, no. 5 (2018): 393–414.

Pilazzo, Chira. "Professional athletes to Donald Trump: Sexual assault is not 'locker room talk,'" Telegraph Online, accessed June 1, 2018. www.telegraph.co.uk/news/2016/10/11/professional-athletes-to-donald-trump-sexual-assault-is-not-lock/

Poniatowski, Kelly. "'You're not allowed body checking in women's hockey': Preserving gendered and nationalistic hegemonies in the 2006 Olympic ice hockey tournament," *Women in Sport & Physical Activity Journal* 20 (2011): 39–52.

Shaw, Sally and Larena Hoeber. "'A Strong Man Is Direct and a Direct Woman Is a Bitch': Gendered Discourses and Their Influence on Employment Roles in Sports Organizations." *Journal of Sport Management* 10 (2003): 347–75.

24

GENDER, SEXUALITY, AND HEALTH COMMUNICATION DURING THE ILLNESS EXPERIENCE

Kallia O. Wright and Kesha Morant Williams

Introduction

The health-care context demands a greater consideration of the impact of gendered communication. This is especially important because interactions between the patient and physician can affect a patient's health and healing. Gendered distinctions are evident in perceptions of health and illness, with gender a verified social determinant of health outcomes.[1] Gendered attributes in the health-care setting need to be identified, monitored, and continuously assessed[2] so that the well-being of patients is not negatively affected and that gender-related stereotypes or expectations in the health-care context are eliminated.

Additionally, while the terms are often interchanged, there is a clear distinction between the disease and illness experience. The disease experience is objective relating to the pathology—that is, the facts, figures, categories, and other areas doctors are trained to identify and manage. The disease experience does not consider the broader impact on a patient's ability to exist. This is problematic because the patient typically focuses on the subjective illness experience. The illness experience is contextual and shaped by personal circumstances connected to a person's ability to function, interact, and maintain relationships.[3] If physicians and patients are privileging different components of the health experience, their interaction begins at a disadvantage because it is built on different assumptions. These conflicting vantage points likely add to patients' fear of being stigmatized or rejected, making patients reluctant to share their health experiences.[4]

This chapter focuses on the impact of gender and sexuality during the illness experience. Organized according to the experiences of heterosexual female-identifying patients, heterosexual cisgender male patients, and then, members of the Lesbian, Gay, Bisexual, Transgender, and Queer/Questioning (LGBTQ) community, each section provides an overview of current issues that emerge as a result of communication influenced by the patient's gender and sexuality. The prominent causes of prejudicial communication and the effects of that form of communication on the health of the patient are highlighted. Specifically, this chapter underscores the continued disregard for women's pain, the silence concerning the mental health challenges men face, and the ignorance on the part of practitioners in the treatment of LGBTQ patients. This review also notes that the impact of gender and sexuality cannot be considered in isolation, but in tandem with race, ethnicity, and socioeconomic factors. Additionally, the chapter presents strategies that patients and practitioners can adopt to cope with, manage, or resist dominant structures that sustain harmful gendered communication. The chapter concludes with suggestions for areas of future research in this field.

Gender and sexuality-related constraints and effects in health care

There are several factors related to gender and sexuality that constrain the experiences of patients. For many, the illness experience is fraught with complexities linked to the diagnosis experience, stigma, interpersonal interactions, and gendered assumptions or expectations. Some prominent challenges include prejudiced expectations regarding the health of patients, the delegitimization of the illness experience, socioeconomic forces, and fear of disclosure.

Constraints in women's health

The illness experience of cisgender female-identifying patients is varied and intertwined with several factors. Across the globe, merely being born a woman can be a negative implication for health. Among the constraints experienced are an overemphasis on the reproductive health of women and a struggle to gain credibility as a patient. Global approaches to women's health have been influenced by their perceived social role as child bearers; that is, health concerns are often treated as secondary to preserve the women's ability to procreate.[5] The health needs of women are typically categorized as those related to potentially harming sexual and reproductive function, general diseases that impact women, and finally, negative social and cultural experiences, such as female genital mutilation, sexual abuse, and domestic violence. In referring to reproductive health, the United Nations' Division for the Advancement of Women asserted that:

> …the reproductive system, in function, dysfunction, and disease, plays a central role in women's health…A major burden of the disease in females is related to their reproductive function and reproductive system, and the way society treats or mistreats them because of their gender. While more men die because of what one may call their "vices," women often suffer because of their nature-assigned physiological duty for the survival of the species, and the tasks related to it.[6]

In some instances, women are considered a "means in the process of reproduction…"[7] Therefore, the health needs of women tend to come second to their roles or potential roles as incubators.

Another challenge facing women's health is the delegitimization of their illnesses.[8] This cynicism or disbelief in *their* account of *their* experiences often leads to feelings of frustration, isolation, and self-doubt as patients begin to wonder if the illness is, in fact, a figment of their imaginations.[9] Tennis star Serena Williams brought national attention to issues related to women's health in a 2018 *Vogue* magazine cover story.[10] In it, she reveals that shortly after her emergency c-section she experienced a pulmonary embolism. Williams, who has a history of blood clots, was not taking her prescribed medication because of the emergency procedure. Shortly after delivering her daughter, she experienced shortness of breath. She notified medical personnel and requested a CT scan and a blood thinner, having experienced similar symptoms in the past. According to the article, Williams' concerns were dismissed as confusion which was perhaps a reaction to pain medication. Instead of the scan she received an ultrasound on her leg that revealed nothing. Several small clots had settled in her lungs when she finally received the requested CT scan and blood thinner.[11] Williams' experience is directly in line with delegitimization research which highlights instances where women are met with resistance when self-advocating during illness experiences. Coupled with race/ethnicity, Williams and other Black women face maternal health disparities equal to those found in developing countries.[12]

Similarly, Shalon Irving, a Black Lieutenant Commander in the U.S. Public Health Service Commissioned Corps and an epidemiologist at the Centers for Disease Control (CDC) who spent her career working to improve the health of marginalized groups, died suddenly because of complications just three weeks after giving birth. Irving, who, following the delivery of her daughter made repeated visits to her physician, joins the disproportionate number of Black women who die from pregnancy-related complications.[13] According to the CDC, Black mothers in the United States die at three to four times the rate of white mothers, and while a medical explanation for this enormous disparity is unclear, many would assert that unconscious bias adds to this troubling statistic.[14]

Further exacerbating the disparity is the variable of income. Low-income people are more likely to experience barriers to health care and have poorer health outcomes than their affluent counterparts. Women tend to have fewer financial resources[15] and less access to health resources, such as information or care for treatment interventions.[16] In particular, indigenous women[17] and undocumented female immigrants are less likely to have access to quality health resources because of poverty.[18] Furthermore, the health of women in rural areas globally are impacted by a range of factors including food insecurity, less access to medical practitioners, a reduction in insurance funding, and stringent medical policy and laws, such as for maternity or family leave.[19]

More specifically, low-income Black people experience greater risk factors that lead to poorer health than those who are affluent. Equally troubling, while social and economic advantages are typically considered protective factors, affluent Black people still experience health inequalities outpacing European Americans with similar socioeconomic backgrounds.[20] Therefore, while Williams and Irving had social and economic advantages, they are a reflection of the enormous disparity linked to Black women and maternal health. Raegan McDonald-Mosley, the chief medical director for Planned Parenthood Federation of America, asserts,

> It tells you that you can't educate your way out of this problem. You can't health care-access your way out of this problem. There's something inherently wrong with the system that's not valuing the lives of Black women equally to white women.[21]

An additional example of delegitimizing women's illness narratives is the enduring and differing expectations regarding experiences with pain. Thyberg explored how it is socially acceptable for women to verbally express pain and distress during health incidents, however, men are conditioned to display strength and tolerance.[22] This perspective assumes that women are less tolerant of pain and therefore exaggerate the severity of their illness experience.[23] Conversely, other research suggests that women are treated for pain less aggressively than men because some medical professionals assume a woman's capacity to bear children makes her more equipped to tolerate pain.[24] This leads some physicians to believe women's pain is caused by emotional rather than physical distress.[25] Black women are doubly impacted, as research on racial bias and health-care support that the pain of Black women is often underestimated and undertreated by medical professionals.[26]

Additionally, cultural beliefs can act as barriers to health care. Stigmatized illnesses can place both men and women on the margins of families and society, and even the health-care con-text.[27] For example, cultural beliefs, such as religious fatalism, have caused Latinas to delay cancer screenings.[28] Stringent religious beliefs about interactions between opposite sexes[29] have influenced some immigrant and Muslim women to request treatment only from female physicians.[30] Also, women begin to enter into medical encounters with reduced trust due to years of poor treatment caused by racial discrimination.[31] In summary, the treatment of women's health is influenced by social assumptions and gender norms that lead to irresponsible practices which place women at greater risk for health inequities.

Constraints in men's health

Similar assumptions about social and health expectations of gender impact men's health. Unfortunately, many of these expectations privilege stereotypical depictions of masculine identity and therefore aid in poor health decision making. Men are half as likely to seek medical attention as women[32] and are considerably more likely to delay help-seeking for concerns related to reproductive health.[33] This framework is particularly challenging when coupled with factors that impact the outcomes of illness experiences such as race, ethnicity, socioeconomic class,[34] and the performance of masculinity.[35]

Though men define health differently, there are significant commonalities in terms of what constitutes a "healthy man." For instance, Black men may define health and masculinity in terms of social roles.[36] Hooker et al. found that Black men, considered the following attributes as "manly": leading a family/household, being a provider, having a strong work ethic, and having a masculine physique. The men also identified a "typical man" as being responsible, principled, and of good character.[37] Therefore, an unhealthy or an unwell man is perceived as one who cannot perform these roles. As a result, Black men are reluctant to admit that they are unwell or seek medical assistance because this would indicate that they have failed at fulfilling the role of the "manly man."

Relatedly, while the experiences of male-identifying patients are varied, a cross-cultural analysis of experiences reveals several common themes, such as a delay in seeking medical help, fear of the loss of the masculine identity, and loss of control. In Australia, patients revealed that their idealized constructions of masculinity were challenged as they moved through the diagnosis and treatment stages for prostate cancer.[38] In other studies, male patients stated that they were often reluctant to disclose illnesses for fear of being viewed differently by their peers[39] and when they had to, used humor to dispel tension and elicit sympathy.[40] For research participants in the United Kingdom, cancer also caused an uncomfortable sense of vulnerability[41] and embarrassment that threatened the cultural characteristic of strength associated with men.[42] The threat of being perceived as weak, inadequate, out of control, incapable of providing for one's family, and lacking in resilience or self-reliance are prominent causes for the delay in seeking and accessing health care. As was observed in Malawi,[43] South Africa,[44] and Zimbabwe,[45] the delay in seeking help is compounded by a fear of being associated with stigmatized illnesses, such as HIV/AIDS and tuberculosis.[46]

Men and mental health

Mental health is another gendered health experience that profoundly impacts personal identities, relationships, and performance in society. The social implications are acute, and in some instances fatal, when stigmatized mental illness is revealed. While heterosexual cisgender women and members of the LGBTQ community also encounter challenges with mental health, heterosexual cisgender men in particular face social obstacles in identifying, acknowledging, and treating mental illness. More significant for men is the overwhelming silence and shame that surround men's struggles with mental health. In India, men with schizophrenia endure ridicule and shame and live in fear of losing their job if their diagnosis is revealed. These men engage in self-silencing because once the diagnosis is public, they also risk losing both their immediate and extended family.[47] Black men in the United States with depression exemplify the compounding relationship that race, gender, and sociocultural background has with health. Daily, insidious, personal and institutional racism, and prejudice are cultural experiences in the United States that heavily affect the social performance of Black men,[48] the roles they play in the family as mentioned earlier,[49] and ultimately, their health.

Masculine expressions of underlying emotional distress and mental illness come in various forms and include "acting out" behaviors[50] such as, substance abuse, gambling, physical violence, sexual risk, anger, emotional stoicism, workaholism, and social isolation.[51] Many men mask their psychological distress and go under the radar especially if they work in a highly masculine environment, as a study on farmers in Australia revealed.[52] In fact, for Muslim American men, cultural distrust, cultural beliefs about mental health, the threat of imposed isolation from their cultural and religious community, and a distrust of the medical system prevent them from reaching out for medical help.[53]

Emotional containment and reticence are common coping choices among men struggling with mental illness. High-profile news stories, such as the suicides of actor Robin Williams,[54] television host Chef Anthony Bourdain,[55] and grunge musician Chris Cornell,[56] have highlighted that men struggle silently and unsuccessfully with mental illness. In writing about his anxiety and panic attacks, Kevin Love, a Cleveland Cavaliers player in the National Basketball Association League in the United States, stated that when he was younger, he learned the appropriate behavior norms for a boy: "You learn what it takes to 'be a man.' It's like a playbook: *Be strong. Don't talk about your feelings. Get through it on your own*."[57] His remarks echo research that has found that men are not used to talking about personal or family health on a day-to-day basis.[58] Nevertheless, Love's story demonstrates that there is a slow, but increasingly open, conversation about mental illness and men. For example, Olympic swimmer Michael Phelps has publicly disclosed his battle with depression and suicide at a mental health conference.[59] Also, in the United Kingdom, Prince William and Prince Harry have thrown attention on mental health challenges faced by men by discussing their own personal struggles with depression, especially after the tragic death of their mother.[60] As these stories about men's mental illness experiences indicate, men's health is complex and multilayered.

However, as the previous news stories and research also reveal, men can add more positive health behaviors to their perception of manhood by accepting more inclusive definitions of masculinity and being more health conscious.[61] In fact, there are instances when male-identifying patients have resisted traditional gendered constructs. Rather than exhibit emotional or verbal reticence or distance,[62] some seek emotional support from and connection with others experiencing a similar illness.[63] Additionally, as more men become involved in single parenting, same-sex relationships, and cohabiting stepfamilies, they are adopting traditional health responsibilities usually ascribed to women. Men are taking on more health monitoring, providing more social support, and are becoming primary caregivers.[64] In essence, the sociocultural environment in which men exist cannot be ignored because it impacts the complete lived illness experience. Discourses of masculinity do not work in isolation. They often work with other notions of personal identity connected to culture, race, ethnicity, sexual identity, sexual orientation, disability status, and geography.[65]

Constraints in the health of the LGBTQ community

Research on the health-care experiences of members of the LGBTQ community has revealed themes of fear of disclosure, discrimination, and poor practitioner training on how to communicate with a patient about sexuality. One of the main challenges within a patient-practitioner interaction is the disclosure of one's gender and sexuality. For members of the LGBTQ community, a concern is that their sexuality may not always be obvious to the practitioner.[66] Fear of judgment is the main reason LGBTQ patients avoid disclosing their health status[67] or even seek health care for family members.[68] Barriers to health care for lesbian patients include witnessing poor treatment of the partner and not finding a gay-friendly practitioner.[69] As research on the health experiences of lesbian women in Norway demonstrated, many refrained from divulging their sexuality to avoid risking the poor handling of a partner or encountering a hostile practitioner who may refuse them or provide treatment.[70] Due to fear of homophobia, lesbian women have also avoided or underutilized health-care resources. These patients' reticence continues even while receiving care,[71] because practitioners do not ask them about their sexual orientation and have assumed that they were heterosexual.[72]

The same fear of judgment and of poor medical treatment is present in the research on gay men and their experiences with health care. Men have reported that their interactions have been impacted by a historical distrust of the medical system (early studies regarded homosexuality as a pathology to be cured), sexual orientation stigma, medical miscommunication, and for men who are part of racial minorities, racial stigma.[73] These internal and external challenges span cultures. For instance, in the United States, Black men who have sex with men (MSM) are the least likely to reveal their sexual status and this decision is exacerbated by various factors including the lack of LGBTQ-friendly signs in a doctor's office.[74] A study on HIV-infected Ghanaian MSM revealed that these men will often not receive care, or they will delay care due to illness stigma, such as a fear of being seen at a medical facility, financial costs, or inadequacies in the medical facility itself.[75] A mistrust of the system is also understandable when practitioners have expressed prejudice against homosexuality. For instance, medical students in Hong Kong stated that after graduation they would refuse to touch homosexual patients for fear of contracting an illness.[76] In a more recent survey in India, 28% of medical students reported that they believed that homosexuals were promiscuous.[77]

A majority of physicians express comfort in treating LGBTQ patients, but many admit that they are not well educated about the communicative needs of this specific population. Some practitioners are uncomfortable bringing up the topic, forget to address it, or do not think it is important or relevant to the issue at hand,[78] and some do not think they have the skills to talk about sexual orientation.[79] Additionally, some physicians question the necessity of highlighting the differences between LGBTQ patients and heterosexual patients and prefer a broader approach.[80] For example, a gastrointestinal oncology provider stated, "I see a lot of in my practice, but I don't know that knowing their sexual behaviors would change the way I would treat their cancer."[81] More recently, a survey of oncologists revealed that knowledge, attitudinal, and institutional practices toward members of the LGBTQ community are still lacking. While about 95% of oncologists felt comfortable in treating LGBTQ patients, the percentage fell to 53% feeling confident that they knew of their patients' unique health needs. The percentages were lower for the treatment of transgender patients, with about 82% of the oncologists feeling comfortable treating transgender patients and only about 37% feeling confident that they knew their specific health needs.[82]

The transgender community struggles to receive nondiscriminatory health care. Research has found that practitioners are resistant to treat transgender patients,[83] refuse to treat them,[84] or are prejudiced when they administer treatment.[85] Transgender patients have reported that they have had to educate medical practitioners about transgenderism and how to care for these patients, even as they have suffered abuse in the health-care setting.[86] As a result, transgender individuals also avoid disclosing for as long as they can, like other LGBTQ members. Unfortunately, the lack of proper health-care treatment for this group leads to higher rates of suicide, depression, and drug abuse.[87] For instance, in Australia, long waiting lists and legal obstacles have created an increased risk of attempted suicide particularly among transgender youth.[88] Advocate and prominent member of the transgender community, U.S. American actress, Laverne Cox, adds that the deadnaming of or misgendering of transgender people (which means refusing to use a person's new name, preferred pronoun, and embraced gender) can be fatal because these actions create psychological and emotional injury that could lead to suicide.[89]

While attempts are being made to improve the experiences of transgender people within health care, challenges still exist. Transgender patients present unique medical requirements, such as hormone therapy due to sex-change surgery and mental health challenges exacerbated by social rejection. Historically, treatments for transgender individuals, such as gender reassignment surgery, were not covered by insurance. Organizations such as the World Professional Organization of Transgender Health and the Mount Sinai Center for Transgender Medicine and Surgery offer alternatives to mainstream medical organizations by focusing on trans-specific medical care.[90] In the United States of America, for instance, the passing of the Affordable Care Act in 2010 helped to facilitate the availability of more funds and protection against discrimination.[91] In light of this, the U.S. Centers for Medicare and Medicaid provided terminology and guidelines for transgender people when they applied for health care and reviewed insurance options, as some insurance companies cover trans-specific health-care issues and others do not.[92] While these were positive steps to provide health care for this community, opposing viewpoints and policies persist. For example, in 2017, U.S. President Donald Trump banned most transgender people with a history of or a diagnosis of gender dysmorphia from joining the military.[93] The justification was that health-care costs for this group were too much for the military to bear. Although the policy was subsequently challenged, the U.S. Supreme Court upheld the ban in January 2019, effectively prohibiting not only transgender people's access to health care but also access to a particular occupation.[94]

Strategies to manage dominant structures and behaviors in health care

The illness experience calls for a number of interactions. In these health-care spaces, it becomes necessary for patients to determine how to communicate effectively with medical practitioners and family members. In the same manner, it is also important for practitioners to engage effectively and ethically with patients. In the following section, strategies for harnessing adverse behaviors and assumptions in the health-care context are suggested.

Strategies for patients

While challenging, there are several strategies patients can enact to resist or confront unconscious bias and intentional prejudice within health care. Several studies have emphasized and recommended strategies to empower patients as they communicate in contexts that are characterized by gender and sexuality. Among the recommendations are that patients must ensure that they are prepared, advocate for themselves, and receive appropriate social support.

For any health appointment, patients need to be prepared.[95] Patients must know the purpose of the meeting and they should also be aware of the underlying social and cultural factors that could influence the interaction. Preparation includes knowing one's health history and having a list of questions ready for the practitioner. For members of the LGBTQ community, knowing the practitioner's personal views regarding gender and sexuality is highly recommended to prevent any negative experiences from occurring.[96]

Patients also need to advocate for themselves in any health context. They can do so by asking clarifying questions, paraphrasing, clearly describing symptoms and feelings, and insisting on clearer communication from the practitioner, such as written instructions at the end of appointments.[97] Additionally, patients are encouraged to disclose their sexuality early in the visit.[98] Disclosure becomes even more important when the concern is linked to genital or reproductive health. Disclosure to general practitioners is equally important, as it provides a more wholistic view of the patient and intimate partnerships that could impact the patient's health. Furthermore, due to a history of discrimination, transgender patients need to not only disclose their medical history but also educate themselves about their rights, such as the right to receive medical treatment, and advocate for those rights if they are being violated.[99]

It is also important that patients have an advocate with them during health appointments. That person can act as an extra listener, ask preplanned questions, legitimize symptoms, demand clarification, or even call out gender-related stereotypical behavior during the visit.[100] An advocate is especially valuable for women, who often must work harder to achieve the sick role, particularly when that illness is an invisible one. In this manner, the patient attempts to restructure the balance of power by becoming an active participant in the experience.

Strong social support is key to effectively managing any health experience. Research shows that families provide instrumental support such as driving to appointments, providing family members with a place to stay, and taking care of children and ill family members.[101] However, emotional support was less likely.[102] Patients may need to elicit instrumental and emotional support from family and friends when resisting and coping with dominant or unconscious structures of prejudice. Support can also come in the form of groups that provide a protected space for patients to release emotions, celebrate milestones, redefine gendered expectations, and ask for help.[103] This is significant for men who encounter and struggle with dominant gendered notions of health and who embody negative perceptions of masculinity that delay the healing or management of illness.

Patients can teach loved ones how to help them during an illness. Since men tend to be less silenced than women during the health experience, female-identifying patients should articulate the specific types of instrumental and emotional support they desire. Women who often privilege family over their health needs should renegotiate the assumed responsibilities of their roles. Articulation of women's desire for support can lessen the ambiguity of the shifting role while allowing for a clearer understanding of their needs. Social support for women is essential, but it must be balanced. As a study on Black women with breast cancer revealed, seemingly harmless and complimentary statements such as "You are a strong Black woman" can be immensely injurious to a woman's health. While loved ones think they are being supportive with the descriptor, the label instead heaps guilt on the struggling patient and results in even fewer offers of tangible support from loved ones.[104] Therefore, patients also need to beware of the impact of sociocultural perceptions and beliefs on their lives. The more educated a patient and family members are about an illness, the more they can help mitigate against the impact of culture on health outcomes.

Strategies for practitioners

Overall, practitioners need to acknowledge their own gendered communication and gendered beliefs about patients when they encounter each other. Clinicians must realize that gender impacts how a patient perceives an illness reality, as well as the decisions and the communication that occur during this experience. Additionally, practitioners must acknowledge that masculine-feminine behaviors do not always align with society's assigned expectations for biological sex.

Therefore, practitioners must check or be mindful of their expectations before and during communication with patients. Several suggestions are provided for ethically and effectively communicating with patients, including becoming more educated about gender and sexuality, building trust, being more culturally sensitive, and creating a more inclusive environment.

Practitioners should educate themselves about the relationship between gender and communication in the health-care context. Specifically, practitioners should learn about the experiences and needs of intersex or non-gender normative patients. A major concern of patients is that too often it is the patient who is tasked with the responsibility of educating the practitioner.[105] Practitioner training has beneficial results. Helitzer et al. found that when practitioners undergo brief training, their communication competence increases and even persists after two years.[106]

Practitioners' knowledge of gendered behaviors will impact how they share information with patients, how they build trust, and how they collaborate with patients in the treatment process.[107] Gendered communicative styles may impact a patient's attitude, approach to treatment,[108] and ultimately, the patient-practitioner relationship.[109] Once a practitioner assesses the patient's communicative behaviors (e.g., a desire for communication that emphasizes directness, independence, control, and status, or communication that is nurturing and builds relationships), it will help the practitioner determine the best approach in the interaction.[110] This purposeful consideration is essential because of the intimate medical information that needs to be shared during these medical conversations.[111] A more engaging patient-centered communication[112] may ultimately lead to better health outcomes.[113]

Beyond increasing one's knowledge, practitioners are also encouraged to sharpen their communication skills, particularly with individuals who are part of historically underrepresented communities. At some point, practitioners need to acknowledge that they may have implicit biases about gender and sexuality and that these may have unconsciously developed during professional training or during their personal socialization process in the larger society. The more comfortable a practitioner is talking about gender and sexuality, the greater the likelihood of a favorable interaction with a patient.[114] In fact, when provided with a model of how to initiate conversation about sexuality and sexual health, practitioners stated that they felt not only more comfortable but also more competent when communicating with patients about sexuality.[115]

While expanding the conversation with patients is a good step, the practitioner also needs to acknowledge larger and more conspicuous structures that impact relationships with patients. For instance, clinicians must know that various cultures have gendered roles and responsibilities that are different from mainstream culture, and these can impact perceptions of illness and health.[116] It is important that practitioners determine these culturally specific expectations and use that information to assist the patient.[117] Another act of cultural inclusivity involves changing the overall heteronormative and gendered environment that patients encounter, such as the medical documents patients must complete.[118] The recommendations include removing a request to check boxes for a specific gender identity and replacing that with blank spaces that allow patients to self-identify. Additionally, forms need to briefly explain why practitioners are requesting information about one's sexual history. Forms should also ask for preferred pronouns and remind patients that their privacy is protected and that they can discuss their privacy concerns with practitioners. Finally, forms need to remind patients that they have a right to refuse to answer questions.[119]

Practitioners are encouraged to ask about the partner's sexual orientation and to use gender-neutral language to encourage disclosure from the patient.[120] It is important for practitioners to recognize that the patient is taking a risk when disclosing sexual orientation or a non-normalized gendered preference.[121] They risk being only regarded as that orientation (e.g., gay man), and they risk being mistreated by the system. Therefore, it is important that practitioners help create an open and welcoming, nonjudgmental, and safe atmosphere.[122] Practitioners are encouraged to take advantage of the early stages of the visit (when small talk and medical history taking are being conducted), because these are the key moments for persons to reveal and discuss their sexuality.[123] In an effort to affirm patients, practitioners can even inquire if the patient is in the early stages of the "coming out" process. Knowing this will further help the practitioner engage in more empathetic communication with the patient.

Beyond interpersonal strategies, health practitioners need to be mindful of the gendered norms in the larger society and in health care. For instance, in trying to encourage early health-seeking behaviors among men, practitioners should note that strategies that overtly threaten identities of masculinity (such as emphasizing losing control, inability to provide or protect one's family or weakness) are not effective. Instead, strategies that reinforce more positive forms of masculinity or provide alternative forms of masculinity (such as showing one's emotions equals strength) are preferred.[124] Additionally, practitioners can appoint cultural practices rather than avoid them. For instance, rather than balk at social perceptions of women as the main caregiver, it may be possible to work within the norm so that it becomes an asset for women and their families. All cultural practices do not have to be hindrances; health practitioners can seek to determine how to use them as advantages and as tools for personal change.

Recommendations for future research and practices

Based on the preceding research, this section presents recommendations for future research and practices concerning the role of gender and sexuality during the illness experience. Female identifying patients have a myriad of experiences that are compounded by sociocultural expectations grounded in patriarchy, economic disenfranchisement, and lack of access to information and education about illnesses and health outcomes. Researchers need to purposefully incorporate concepts of sex and gender into their work, so that the lived experiences of women are captured accurately. Specifically, more effort should be made to unearth those accounts of when women's health challenges were placed secondary to reproductive health.[125] Additionally, research can highlight those positive stories where women felt that they were treated as more than their ability to bear children. Similarly, practitioners must engage in holistic care and make every effort to treat the specific concerns of the female patient rather than focus solely on reproductive health. Connected to this approach, research also needs to highlight the effects of the delegitimization of the illness experiences of women.[126] More importantly, practitioners must be mindful of their own implicit biases and tendencies to dismiss the complaints of these patients, particularly women of color. As was illustrated through the Williams'[127] and Irving's[128] stories, not listening to women can lead to fatal consequences.[129] Finally, researchers and practitioners should note the impact of the socioeconomic culture on women's attempts to maintain their health. Economic factors influence one's health, and cultural roles and expectations also have a significant impact.

Performances of masculinity and its impact on men's health is a necessary area of future research. For instance, in the United States the racial, economic, social, and cultural disparities among groups of men[130] are so great that their health experiences will differ significantly.[131] This is true for men of color, and even more so for men of color who belong to sexual minority groups.[132] Future research should also evaluate how the performance of masculinity differs according to health outcomes and communication regarding one's health. Additionally, more research is needed on which aspect of masculinity is dominant or most likely associated with a particular illness, health behavior, or outcome.[133] Importantly, while the conversations about mental health are increasing, more research is needed into men's various experiences with mental illness and stigma. While mentioned peripherally in this chapter, there is value in researching men's reproductive health experience, with questions being asked about stigma, patient credibility, and impact on identity to determine best and negative practices undertaken by practitioners. Finally, there is a call for men's health research to take on a more intersectional lens. Research needs to explore how men's health is affected simultaneously by multiple intertwined factors such as race, age, ethnicity, sexual orientation, and institutional and social practices; practitioners must also be mindful of the impact of these factors when interacting with the patient.[134]

While much research has covered the gendered prejudice and discrimination that LGBTQ people encounter in health care, more attention ought to be given to communicative interactions that are effective, ethical, and sensitive. These best practices may act as models for future medical practitioners and may operate as teaching tools for patients to advocate for themselves more strongly. Attention needs to be given to practitioners and institutions that demonstrate that they are LGBTQ friendly and that they are competent communicators who provide a safe zone for this group of patients. This is especially needed for the youth who are vulnerable to suicide tendencies and depression. Additional research is still needed on the interactions of the transgender community and people who exist on the margins of the LGBTQ community, such as those who identify as bisexual and intersex (individuals born with unique reproduction or sexual genitalia that cannot be classified as either male or female).[135]

Conclusion

In summary, the social constructs of gender and sexuality become even more amplified within the context of health and illness. The examples articulated in this chapter support the notion that illness experiences are compounded by sociocultural expectations. Though implicit, the societal narrative connected to women's health suggests that nothing, including her own life, is more important than a woman's ability to bear children. This crippling narrative creates an environment where reproductive care is privileged more than other health concerns that might more directly impact the woman's health. Equally troubling are irresponsible decisions surrounding the assessment and management of pain in women. Assumptions surrounding a woman's natural ability to manage pain because of childbirth, as well as those suggesting that women are emotional and therefore exaggerate pain, negatively impact the care women receive.

Gendered assumptions also impact men's health by focusing on sociocultural expectations of masculinity rather than health needs. Expressions of vulnerability are treated as weaknesses that prevent many men from seeking the medical attention they need. This is especially true when exploring mental health concerns. Men, who are expected to display resilience and self-reliance, are often concerned with the stigma that may arise from seeking help. Furthermore, members of the LGBTQ community are met with significant levels of ignorance, as well as discrimination and prejudice. While some physicians grapple with the utility of working to service this population differently, others admit that while open to addressing the needs of the LGBTQ community, they are unaware of what the needs are and how to do so.

Researchers and practitioners are encouraged to adopt an intersectional lens when examining the illness experiences of individuals. Society's consistent privileging of the disease experience has created an environment where patients struggle to give voice to their experiences. Patients do not live within a vacuum, but within powerful institutional, cultural, and social structures that impact health outcomes. The gendered interactions of women, men, and members of the LGBTQ community highlight the necessity for health-related self-advocacy.

Notes

1 Renata Schiavo, *Health Communication: From Theory to Practice* (San Francisco, CA: Jossey-Bass, 2014), 87.
2 Schiavo, *Health Communication: From Theory to Practice*, 94.
3 Antonio Ventriglio, Julio Torales and Dinesh Bhugra, "Disease versus Illness: What Do Clinicians Need to Know?" *International Journal of Social Psychiatry* 63, no. 1 (February 2017): 3.
4 Katy Gallop et al., "Development of a Conceptual Model of Health-related Quality of Life for Systemic Lupus Erythematosus from the Patient's Perspective," *Lupus* 21, no. 9 (2012): 942.
5 Ruth Macklin, "Global Inequalities in Women's Health: Who Is Responsible for Doing What?," *Philosophical Topics* 37, no. 2 (2009): 97.
6 Mahmoud Fathalla, "Health and Being a Woman." Lectures, Speeches and Statements, UN Commission on the Status of Women United Nations, New York, March 3, 1999, para 2.
7 Fathalla, "Health and Being a Woman," para 4.
8 Benjamin J. Newton et al., "A Narrative Review of the Impact of Disbelief in Chronic Pain," *Pain Management Nursing* 14, no. 3 (2013): 162.
9 Newton et al., "A Narrative Review" 166.
10 Rob Haskell, "Serena Williams on Motherhood, Marriage, and Making Her Comeback," *Vogue*, January 10, 2018, accessed January 14, 2019, www.vogue.com/article/serena-williams-vogue-cover-interview-february-2018 para. 1
11 Haskell, "Serena Williams," para. 11.
12 Centers for Disease Control and Prevention, "Pregnancy Mortality Surveillance System," August 7, 2018, accessed January 14, 2019, www.cdc.gov/reproductivehealth/maternalinfanthealth/pregnancy-mortality-surveillance-system.htm, para. 11.
13 Nina Martin and Renee Montagne, "Black Mothers Keep Dying After Giving Birth. Shalon Irving's Story Explains Why," *NPR*, December 7, 2017, accessed January 19, 2019, www.npr.org/2017/12/07/568948782/black-mothers-keep-dying-after-giving-birth-shalon-irvings-story-explains-why, para. 6.
14 Centers for Disease Control and Prevention, para. 12.
15 Abiodun Obayelu and I. Ogunlade, "Analysis of the Uses of Information Communication Technology (ICT) for Gender Empowerment and Sustainable Poverty Alleviation in Nigeria," *International Journal of Education and Development using Information and Communication Technology* 2, no. 3 (2006): 53.
16 Schiavo, *Health Communication: From Theory to Practice*, 93.
17 Jeffrey E. Holm et al., "Assessing Health Status, Behavioral Risks, and Health Disparities in American Indians Living on the Northern Plains of the US," *Public Health Reports* 125, no. 1 (2010): 76.
18 Kinsey Hasstedt et al., "Immigrant Women's Access to Sexual and Reproductive Health Coverage and Care in the United States," *Commonwealth Fund*, November 2018, accessed July 1, 2019, https://doi.org/10.26099/2dcc-mh04.
19 Kate Sinclair et al., "Rural Women: The Most Vulnerable to Food Insecurity and Poor Health in the Global South," *The FASEB Journal* 31, no. 1_supplement (2017): 791.
20 Dhruv Khullar and Dave A. Chokshi, "Health, Income, and Poverty: Where We Are and What Could Help," *Health Affairs*, October 4, 2018, accessed May 9, 2019, www.healthaffairs.org/do/10.1377/hpb20180817.901935/full/.
21 Martin and Montagne, para. 16.
22 Hannah Thyberg, "A Semblance of Normalcy: Social Isolation and the Burden of Looking Well," in *Reifying Women's Experiences with Invisible Illness*, ed. Kesha Morant Williams and Frances Selena Morant (Lanham, MD: Lexington, 2018), 33.
23 Kallia Wright, ""You Have Endometriosis": Making Menstruation-Related Pain Legitimate in a Biomedical World," *Health Communication* (2018): 2.
24 Diane E. Hoffmann and Anita J. Tarzian, "The Girl Who Cried Pain: A Bias Against Women in the Treatment of Pain," *Journal of Law, Medicine, & Ethics* 29 (2001): 16.
25 Hoffmann and Tarzian, "The Girl Who Cried Pain," 19.
26 Kelly N. Hoffman et al., "Racial Bias in Pain Assessment and Treatment Recommendations, and False Beliefs About Biological Differences Between Blacks and Whites," *Proceedings of the National Academy of Sciences of the United States of America* 113, no. 16 (2016): 4296–4301.
27 Schiavo, *Health Communication: From Theory to Practice*, 93.
28 Silvia Tejeda et al., "Breast Cancer Delay in Latinas: The Role of Cultural Beliefs and Acculturation," *Journal of Behavioral Medicine* 40, no. 2 (2017): 347.

29 Mahzabin Ferdous et al., "Barriers to Cervical Cancer Screening Faced by Immigrant Women in Canada: A Systematic Scoping Review," *BMC Women's Health* 18, no. 1 (2018): 170.

30 Milkie Vu et al., "Predictors of Delayed Healthcare Seeking Among American Muslim Women," *Journal of Women's Health* 25, no. 6 (2016): 591.

31 Adolfo Cuevas et al., "African American Experiences in Healthcare: 'I Always Feel like I'm Getting Skipped Over,'" *Health Psychology* 35, no. 9 (2016): 991.

32 U.S. Department of Health and Human Services, Centers for Disease Control and Prevention, & National Center for Health Statistics, "Summary Health Statistics: National Health Interview Survey," accessed June 9, 2019, https://ftp.cdc.gov/pub/Health_Statistics/NCHS/NHIS/SHS/2014_SHS_Table_A-18.pdf.

33 Alan Dolan et al., "'It's like taking a bit of Masculinity away from you': Towards a Theoretical Understanding of Men's Experiences of Infertility," *Sociology of Health & Illness* 39, no. 6 (2017): 882.

34 Will Courtenay, "A Global Perspective on the Field of Men's Health: An Editorial," *International Journal of Men's Health* 1, no. 1 (2002): 6.

35 Clare Moynihan, "Theories in Health Care and Research: Theories of Masculinity," *BMJ* 317, no. 7165 (1998): 1072.

36 Joseph Ravenell et al., "African American Men's Perceptions of Health: A Focus Group Study," *Journal of the National Medical Association* 98 (2006): 546.

37 Steven P. Hooker et al., "The Potential Influence of Masculine Identity on Health-Improving Behavior in Midlife and Older African American Men," *Journal of Men's Health* 9, no. 2 (2012): 84.

38 Alex Broom, "Prostate Cancer and Masculinity in Australian Society: A Case of Stolen Identity?" *International Journal of Men's Health* 3, no. 2 (2004): 87.

39 Mark Jeffries and Sarah Grogan, "'Oh, I'm Just, You Know, a Little Bit Weak Because I'm Going to the Doctor's': Young Men's Talk of Self-Referral to Primary Healthcare Services" *Psychology & Health* 27, no. 8 (2012): 903.

40 Shona Hilton et al., "Disclosing a Cancer Diagnosis to Friends and Family: A Gendered Analysis of Young Men's and Women's Experiences." *Qualitative Health Research* 19, no. 6 (2009): 754.

41 Shona Hilton et al., "Have Men Been Overlooked? A Comparison of Young Men and Women's Experiences of Chemotherapy-Induced Alopecia." *Psycho-Oncology* 17, no. 6 (2008): 579.

42 Jeffries and Grogan, "Young Men's Talk of Self-Referral," 907.

43 Jeremiah Chikovore et al., "Control, Struggle, and Emergent Masculinities: A Qualitative Study of Men's Care-Seeking Determinants for Chronic Cough and Tuberculosis Symptoms in Blantyre, Malawi," *BMC Public Health* 14, no. 1 (2014): 1053.

44 Ingrid Lynch, Pierre W. Brouard and Maretha J. Visser, "Constructions of Masculinity Among a Group of South African Men Living with HIV/AIDS: Reflections on Resistance and Change," *Culture, Health & Sexuality* 12, no. 1 (2010): 15.

45 Stephen Pearson and Makadzange Panganai, "Help-Seeking Behaviour for Sexual-Health Concerns: A Qualitative Study of Men in Zimbabwe," *Culture, Health & Sexuality* 10, no. 4 (2008): 361.

46 Nicoli Nattrass, "Gender and Access to Antiretroviral Treatment in South Africa," *Feminist Economics* 14, no. 4 (2008): 19.

47 Santosh Loganathan and R. Srinivasa Murthy, "Living with Schizophrenia in India: Gender Perspectives," *Transcultural Psychiatry* 48, no. 5 (2011): 569.

48 Daphne C. Watkins et al., "'Their Depression is Something Different … It Would Have to Be': Findings from a Qualitative Study of Black Women's Perceptions of Depression in Black Men," *American Journal of Men's Health* 7, no. 4 (2013): 50S.

49 Daphne C. Watkins et al., "A Meta-study of Black Male Mental Health and Well-being," *Journal of Black Psychology* 36, no. 3 (2010): 321.

50 Christopher Kilmartin, "Depression in Men: Communication, Diagnosis and Therapy," *Journal of Men's Health and Gender* 2, no. 1 (2005): 96.

51 Phillipe Roy et al., "Male Farmers with Mental Health Disorders: A Scoping Review," *Australian Journal of Rural Health* 21 (2013): 5.

52 Roy et al., "Male Farmers with Mental Health Disorders," 5.

53 Saara Amri and Fred Bemak, "Mental Health Help-seeking Behaviors of Muslim Immigrants in the United States: Overcoming Social Stigma and Cultural Mistrust," *Journal of Muslim Mental Health* 7, no. 1 (2013): 59.

54 Susan Schneider Williams, "The Terrorist Inside My Husband's Brain," *Neurology* 87, no. 13 (2016): 1310.

55 Peter Sblendorio, "Anthony Bourdain Discussed His Mental Health Struggles during Therapy Session on 'Parts Unknown,'" *New York Daily News*, June 8, 2018, accessed January 19, 2019, www.nydailynews. com/entertainment/tv/ny-ent-anthony-bourdain-parts-unknown-therapy-20180608-story.html, para. 5.

56 Mike Zimmerman, "What Chris Cornell Told Me About His Depression Years Before His Suicide," *Men's Health*, May 19, 2017, accessed January 19, 2019, www.menshealth.com/trending-news/a19516660/ chris-cornell-death-depression-suicide-interview/, para. 1.

57 Kevin Love, "Everyone is Going Through Something," *The Players' Tribune*, March 6, 2018, accessed January 19, 2019, www.theplayerstribune.com/en-us/articles/kevin-love-everyone-is-going-through-something, para. 4.

58 Sally Brown, "What Makes Men Talk About Health?" *Journal of Gender Studies* 10, no. 2 (2001): 192.

59 Melissa Matthews, "Michael Phelps Opens Up About His Struggles with Depression and Thoughts of Suicide," *Men's Health*, October 26, 2018, accessed January 19, 2019, www.menshealth.com/health/ a24268441/michael-phelps-depression/, para. 1.

60 Caroline Davies, "Prince William: Suicide Callout Shed Light on Men's Mental Health," *The Guardian*, April 17, 2017, accessed May 15, 2019, www.theguardian.com/society/2017/apr/18/ prince-william-duke-cambridge-call-out-suicide-men-mental-health.

61 Hooker, "The Potential Influence of Masculine Identity," 84–85.

62 Will Courtenay, "Constructions of Masculinity and Their Influence on Men's Well-Being: A Theory of Gender and Health," *Social Science & Medicine* 50 (2000): 1389.

63 Stephen Trapp et al., "Male Coping Processes as Demonstrated in the Context of a Cancer-Related Social Support Group," *Supportive Care in Cancer* 21, no. 2 (2013): 625.

64 Kathleen M. Galvin, "Gender and Family Interaction," in *The Sage Handbook of Gender and Communication*, eds Bonnie J. Dow and Julia T. Wood, 41–55 (Thousand Oaks, CA: Sage, 2006), 50.

65 Derek Griffith, "An Intersectional Approach to Men's Health," *Journal of Men's Health* 9, no. 2 (2012): 108.

66 Mari Bjorkman and Kirsti Malterud, "Being Lesbian–Does the Doctor Need to Know? A Qualitative Study About the Significance of Disclosure in General Practice," *Scandinavian Journal of Primary Health Care* 25, no. 1 (2007): 59.

67 Nicole C. Hudak and Benjamin R. Bates, "In Pursuit of 'Queer-friendly' Healthcare: An Interview Study of How Queer Individuals Select Care Providers," *Health Communication* (2018): 3.

68 Mary E. Clark et al., "The GLBT Health Access Project: A State-funded Effort to Improve Access to Care," *American Journal of Public Health* 91, no. 6 (2001): 896.

69 Angela Barbara, Sara A. Quandt, and Roger T. Anderson, "Experiences of Lesbians in the Health Care Environment," *Women & Health* 34, no. 1 (2001): 51.

70 Bjorkman and Malterud, "Being Lesbian–Does the Doctor Need to Know?", 59.

71 Katharine A. O'Hanlan et al., "Advocacy for Women's Health Should Include Lesbian Health," *Women's Health* 13, no. 2 (2004): 229.

72 Mary Ann A. van Dam, Audrey S. Koh and Suzanne L. Dibble, "Lesbian Disclosure to Health Care Providers and Delay of Care," *Journal of the Gay and Lesbian Medical Association* 5, no. 1 (2001): 17.

73 David J. Malebranche et al., "Race and Sexual Identity: Perceptions about Medical Culture and Healthcare Among Black Men Who Have Sex with Men," *Journal of the National Medical Association* 96, no. 1 (2004): 104.

74 Shan Qiao et al., "Disclosure of Same-Sex Behaviors to Health-Care Providers and Uptake of HIV Testing for Men Who Have Sex with Men: A Systematic Review," *American Journal of Men's Health* 12, no. 5 (2018): 1197. https://doi.org/10.1177/1557988318784149.

75 Adedotun Ogunbajo et al., "Barriers, Motivators, and Facilitators to Engagement in HIV Care Among HIV-Infected Ghanaian Men Who Have Sex with Men (MSM)," *AIDS & Behavior* 22, no. 3 (2018): 832.

76 R.W.M. Kan et al., "Homophobia in Medical Students of the University of Hong Kong," *Sex Education* 9, no. 1 (2009): 72.

77 A. Kar et al., "Attitude of Indian Medical Students Towards Homosexuality," *East Asian Archives of Psychiatry* 28, no. 2 (2018): 59.

78 Robert L. Kitts, "Barriers to Optimal Care between Physicians and Lesbian, Gay, Bisexual, Transgender, and Questioning Adolescent Patients," *Journal of Homosexuality* 57 (2010): 737.

79 Chalmer E. Labig and Tim O. Peterson, "Sexual Minorities and Selection of a Primary Care Physician in a Midwestern U.S. City," *Journal of Homosexuality* 51 (2006): 4.

80 Christina L. Tamargo et al., "Cancer and the LGBTQ Population: Quantitative and Qualitative Results from an Oncology Providers' Survey on Knowledge, Attitudes, and Practice Behaviors," *Journal of Clinical Medicine* 6, no. 93 (2017): 8.

81 Tamargo et al., "Cancer and the LGBTQ Population," 8.

82 Matthew B. Schabath et al., "National Survey of Oncologists at National Cancer Institute – Designated Comprehensive Cancer Centers: Attitudes, Knowledge, and Practice Behaviors About LGBTQ Patients with Cancer," *Journal of Clinical Oncology* (2019): 4.

83 Heather L. Corliss et al., "An Evaluation of Service Utilization Among Male to Female Transgender Youth: Qualitative Study of a Clinic Based Sample," *Journal of LGBT Health Research* 3 (2007): 55.

84 Amaya Perez-Brumer et al., "'We Don't Treat Your Kind': Assessing HIV Health Needs Holistically Among Transgender People in Jackson, Mississippi," *PLoS ONE* 11 (2018): 5.

85 Jodi Sperber, Stewart Landers, and Susan Lawrence, "Access to Health Care for Transgendered Persons: Results of a Needs Assessment in Boston," *International Journal of Transgenderism* 8 (2005): 85.

86 Jaime Grant et al., *Executive Summary. Injustice at Every Turn*, 2011, 5.

87 Redfern and Sinclair, "Improving Health Care Encounters," 30.

88 Michelle Telfer et al., "Transformation of Health-Care and Legal Systems for the Transgender Population: The Need for Change in Australia," *Journal of Paediatrics and Child Health* 51, no. 11 (2015): 2.

89 Anna Papachristos, "Laverne Cox Shares Struggle with Mental Health to Promote Respect for Trans Identities," *A Plus*, August 14, 2018, accessed January 14, 2019, https://aplus.com/a/laverne-cox-on-deadnaming-and-mental-health-of-trans-community?, para. 5.

90 "Mount Sinai Center for Center for Transgender Medicine and Surgery," accessed January 19, 2019, www.mountsinai.org/locations/center-transgender-medicine-surgery.

91 Jan S. Redfern and Bill Sinclair, "Improving Health Care Encounters and Communication with Transgender Patients," *Journal of Communication in Healthcare* 7, no. 1 (2014): 35.

92 U.S. Centers for Medicare and Medicaid Services, "Transgender Health Care," accessed January 21, 2019, www.healthcare.gov/transgender-health-care/.

93 Sophie Tatum, "White House announces Policy to Ban Most Transgender People from Serving in the Military," *CNN*, March 24, 2018, accessed January 19, 2019, www.cnn.com/2018/03/23/politics/transgender-white-house/index.html, para. 1.

94 Ariane de Vogue, "Supreme Court Allows Transgender Military Ban to Go into Effect," *CNN*, January 22, 2019, accessed January 22, 2019, www.cnn.com/2019/01/22/politics/scotus-transgender-ban/index.html, para. 1.

95 Linda C. Lederman et al., *Health Communication in Everyday Life* (Dubuque, IA: Kendall Hunt, 2017), 66.

96 Bjorkman and Malterud, "Being Lesbian," 59.

97 Lederman et al., *Health Communication in Everyday Life*, 66.

98 Bjorkman and Malterud, "Being Lesbian," 59.

99 National Center for Transgender Equality, "Know Your Rights: Healthcare," para 7.

100 Lederman et al., *Health Communication in Everyday Life*, 66.

101 Christina J. Cross et al., "Instrumental Social Support Exchanges in African American Extended Families," *Journal of Family Issues* 39, no. 13 (2018): 3535.

102 Rukmalie Jayakody et al., "Family Support to Single and Married African American Mothers: The Provision of Financial, Emotional, and Child Care Assistance," *Journal of Marriage and Family* 55, no. 2 (1993): 261.

103 John L. Oliffe et al., "Connecting Humor, Health, and Masculinities at Prostate Cancer Support Groups," *Psycho-Oncology* 18, no. 9 (2009): 920.

104 Annette Madlock Gatison, *Health Communication and Breast Cancer Among Black Women: Culture, Identity, Spirituality, and Strength* (Lanham, MD: Lexington, 2016), 11.

105 Damien W. Riggs and Clemence Due, "Mapping the Health Experiences of Australians Who Were Female Assigned at Birth but Who Now Identify with a Different Gender Identity," *Lambda Nordica* no. 3–4 (2013): 69.

106 Deborah L. Helitzer et al., "A Randomized Controlled Trial of Communication Training with Primary Care Providers to Improve Patient Centeredness and Health Risk Communication," *Patient Education and Counseling* 82 (2011): 28.

107 Michael P. Pagano, *Health Communication for Health Care Professionals: An Applied Approach* (New York, NY: Springer Publishing, 2017), 53.

108 Donald J. Cegala et al., "The Impact of Patient Participation on Physicians' Information Provision During a Primary Care Medical Interview," *Health Communication* 21, no. 2 (2007): 181.

109 Schiavo, 91.
110 Donald J. Cegala, "The Impact of Patients' Communication Style on Physicians' Discourse: Implications for Better Health Outcomes," in *Applied Interpersonal Communication Matters: Family, Health, and Community Relations*, ed. Rene M. Daily and Beth A. Le Poire, 201–217 (New York: Peter Lang, 2006), 213.
111 Pagano, *Health Communication for Health Care Professionals: An Applied Approach*.
112 Lederman et al., *Health Communication in Everyday Life*, 63–64.
113 Richard L. Street et al., "How Does Communication Heal? Pathways Linking Clinician-Patient Communication to Health Outcomes," *Patient Education Counseling* 74 (2009): 297–299.
114 Ryan Gamlin, "Sexuality: A Challenge for Nursing Practice," *Nursing Times* 95, no. 7 (1999): 50.
115 Chris Quinn and Brenda Happell, "Talking About Sexuality with Consumers of Mental Health Services," *Perspectives in Psychiatric Care* 49, no. 1 (2013): 17.
116 Ruthbeth Finerman, "The Burden of Responsibility: Duty, Depression, and Nervios in Andean Ecuador," *Health Care for Women International* 10, nos 2–3 (1989): 154.
117 Vlassoff and Manderson, "Anthropology of Infectious Diseases," 1018.
118 Elizabeth S. Goins and Danee Pye, "Check the Box that Best Describes You: Reflexively Managing Theory and Praxis in LGBTQ Health Communication Research," *Health Communication* 28 (2013): 404.
119 Goins and Pye, "Check the Box That Best Describes You," 406.
120 Bjorkman and Malterud, "Being Lesbian," 60.
121 Bjorkman and Malterud, "Being Lesbian," 59.
122 Hudak and Bates, "In Pursuit of 'Queer-friendly' Healthcare," 4.
123 Maria K. Venetis et al., "Characterizing Sexual Orientation Disclosure to Health Care Providers: Lesbian, Gay, and Bisexual Perspectives," *Health Communication* 32, no. 5 (2017): 581.
124 Chikovore et al., "Control, Struggle, and Emergent Masculinities," 1060.
125 Macklin, "Global Inequalities in Women's Health," 97.
126 Newton et al., "A Narrative Review," 162.
127 Haskell, "Serena Williams," para. 11.
128 Martin and Montagne, "Black Mothers Keep Dying," para. 6.
129 Centers for Disease Control and Prevention, para. 12.
130 Joan Evans et al., "Health, Illness, Men and Masculinities (H1MM): A Theoretical Framework for Understanding Men and Their Health," *Journal of Men's Health* 8, no. 1 (2011): 9.
131 Derek Griffith et al., "Measuring Masculinity in Research on Men of Color: Findings and Future Directions," *American Journal of Public Health* 102, no. S2 (2012): S187.
132 Joan Evans et al., "Health, Illness, Men and Masculinities (H1MM)," 9.
133 Griffith, "An Intersectional Approach to Men's Health," 108.
134 Griffith, "An Intersectional Approach to Men's Health," 110.
135 Alison B. Alpert et al., "What Lesbian, Gay, Bisexual, Transgender, Queer, and Intersex Patients Say Doctors Should Know and Do: A Qualitative Study," *Journal of Homosexuality* 64, no. 10 (2017): 1377.

Bibliography

Alpert, Alison B., Cichoski Kelly, Eileen M., and Fox, Aaron D. "What Lesbian, Gay, Bisexual, Transgender, Queer, and Intersex Patients Say Doctors Should Know and Do: A Qualitative Study." *Journal of Homosexuality* 64, no. 10 (2017): 1368–89. doi:10.1080/00918369.2017.1321376.
Amri, Saara, and Bemak, Fred. "Mental Health Help-seeking Behaviors of Muslim Immigrants in the United States: Overcoming Social Stigma and Cultural Mistrust." *Journal of Muslim Mental Health* 7, no. 1 (2013): 43–63. http://dx.doi.org/10.3998/jmmh.10381607.0007.104.
Barbara, Angela M., Quandt, Sara A., and Anderson, Roger T. "Experiences of Lesbians in the Health Care Environment." *Women & Health* 34, no. 1 (2001): 45–62.
Bjorkman, Mari, and Malterud, Kirsti. "Being Lesbian–Does the Doctor Need to Know? A Qualitative Study about the Significance of Disclosure in General Practice." *Scandinavian Journal of Primary Health Care* 25, no. 1 (2007): 58–62. https://doi.org/10.1080/02813430601086178.
Broom, Alex. "Prostate Cancer and Masculinity in Australian Society: A Case of Stolen Identity?" *International Journal of Men's Health* 3, no. 2 (2004): 73–91. doi:10.3149/jmh.0302.73.
Brown, Sally. "What Makes Men Talk About Health?" *Journal of Gender Studies* 10, no. 2 (2001): 187–95. doi:10.1080/09589230120053300.

Cegala, Donald J. "The Impact of Patients' Communication Style on Physicians' Discourse: Implications for Better Health Outcomes." In *Applied Interpersonal Communication Matters: Family, Health, and Community Relations*, edited by Rene M. Daily and Beth A. Le Poire, 201–17, New York: Peter Lang, 2006.

Cegala, Donald J., Street Jr., Richard L. and Clinch, C. Randall. "The Impact of Patient Participation on Physicians' Information Provision During a Primary Care Medical Interview." *Health Communication* 21, no. 2 (2007): 177–85.

Centers for Disease Control and Prevention. "Pregnancy Mortality Surveillance System." August 7, 2018. Accessed January 14, 2019. www.cdc.gov/reproductivehealth/maternalinfanthealth/pregnancy-mortality-surveillance-system.htm.

Chikovore, Jeremiah, Hart, Graham, Kumwenda, Moses, Chipungu, Geoffrey A., Desmond, Nicola, and Corbett, Liz. "Control, Struggle, and Emergent Masculinities: A Qualitative Study of Men's Care-Seeking Determinants for Chronic Cough and Tuberculosis Symptoms in Blantyre, Malawi." *BMC Public Health* 14, no. 1 (2014): 1053–65. doi:10.1186/1471-2458-14-1053.

Clark, Mary E., Landers, Stewart, Linde, Rhonda, and Sperber, Jodi. "The GLBT Health Access Project: A State-funded Effort to Improve Access to Care." *American Journal of Public Health* 91, no. 6 (2001): 895–96.

Corliss, Heather L., Belzer, Marvin, Forbes, Catherine, and Wilson, Erin C. "An Evaluation of Service Utilization among Male to Female Transgender Youth: Qualitative Study of a Clinic Based Sample." *Journal of LGBT Health Research* 3 (2007): 49–61. https://doi.org/10.1300/J463v03n02_06.

Courtenay, Will. "A Global Perspective on the Field of Men's Health: An Editorial." *International Journal of Men's Health* 1, no. 1 (2002): 1–13.

Courtenay, Will. "Constructions of Masculinity and Their Influence on Men's Well-Being: A Theory of Gender and Health." *Social Science & Medicine* 50 (2000): 1385–401.

Cross, Christina J., Nguyen, Ann W., Chatters, Linda M., and Taylor, Robert Joseph. "Instrumental Social Support Exchanges in African American Extended Families." *Journal of Family Issues* 39, no. 13 (2018): 3535–63.

Cuevas, Adolfo G., O'Brien, Kerth, and Saha, Somnath. "African American Experiences in Healthcare: 'I Always Feel like I'm Getting Skipped Over'." *Health Psychology* 35, no. 9 (2016): 987–95. doi:10.1037/hea0000368.

Davies, Caroline. "Prince William: Suicide Callout Shed Light on Men's Mental Health." *The Guardian*, April 17, 2017. Accessed May 15, 2019, www.theguardian.com/society/2017/apr/18/prince-william-duke-cambridge-call-out-suicide-men-mental-health.

de Vogue, Ariane. "Supreme Court Allows Transgender Military Ban to Go into Effect." *CNN*. January 22, 2019. Accessed January 22, 2019, www.cnn.com/2019/01/22/politics/scotus-transgender-ban/index.html, para. 1.

Dolan, Alan, Lomas, Tim, Ghobara, Tarek, and Hartshorne, Geraldine. "'It's like taking a bit of masculinity away from you': Towards a Theoretical Understanding of Men's Experiences of Infertility." *Sociology of Health & Illness* 39, no. 6 (2017): 878–92.

Evans, Joan, Frank, Blye, Oliffe, John L., and Gregory, David. "Health, Illness, Men and Masculinities (H1MM): A Theoretical Framework for Understanding Men and their Health." *Journal of Men's Health* 8, no. 1 (2011): 7–15. https://doi.org/10.1016/j.jomh.2010.09.227.

Fathalla, Mahmoud. "Health and Being a Woman." Lectures, Speeches and Statements, UN Commission on the Status of Women United Nations, New York, March 3, 1999.

Ferdous, Mahzabin, Lee, Sonya, Goopy, Suzanne, Yang, Huiming, Rumana, Nahid, Abedin, Tasnima, and Turin, Tanvir C. "Barriers to Cervical Cancer Screening Faced by Immigrant Women in Canada: A Systematic Scoping Review." *BMC Women's Health* 18, no. 1 (2018): 165–178. doi:10.1186/s12905-018-0654-5.

Finerman, Ruthbeth. "The Burden of Responsibility: Duty, Depression, and Nervios in Andean Ecuador." *Health Care for Women International* 10, no. 2–3 (1989): 141–57. https://doi.org/10.1080/07399338909515846.

Gallop, K., Nixon, A., Swinburn, P., Sterling, K. L., Naegeli, A. N., and Silk, M. E. T. "Development of a Conceptual Model of Health-related Quality of Life for Systemic Lupus Erythematosus from the Patient's Perspective." *Lupus* 21, no. 9 (2012): 934–43. doi:10.1177/0961203312441980.

Galvin, Kathleen M. "Gender and Family Interaction." In *The Sage Handbook of Gender and Communication*, edited by Bonnie J. Dow and Julia T. Wood, 41–55, Thousand Oaks, CA: Sage, 2006.

Gamlin, Ryan. "Sexuality: A Challenge for Nursing Practice." *Nursing Times* 95, no. 7 (1999): 48–50.

Gatison, Annette Madlock. *Health Communication and Breast Cancer Among Black Women: Culture, Identity, Spirituality, and Strength*. Lanham, MD: Lexington, 2016.

Goins, Elizabeth S., and Pye, Danee. "Check the Box that Best Describes You: Reflexively Managing Theory and Praxis in LGBTQ Health Communication Research." *Health Communication* 28 (2013): 397–407. doi:10.1080/10410236.2012.690505.

Grant, Jaime M., Mottet, Lisa A., Tanis, Justin, Harrison, Jack, Herman, Jody L., and Keisling, Mara. *Executive Summary. Injustice at Every Turn*, 2011. https://transequality.org/sites/default/files/docs/resources/NTDS_Exec_Summary.pdf.

Griffith, Derek. "An Intersectional Approach to Men's Health." *Journal of Men's Health* 9, no. 2 (2012): 106–12. http://health-equity.lib.umd.edu/4064/1/Intersectional_Approach_to_Men's_Health.pdf.

Griffith, Derek M., Gunter, Katie, and Watkins, Daphne C. "Measuring Masculinity in Research on Men of Color: Findings and Future Directions." *American Journal of Public Health* 102, no. S2 (2012): S187–94. doi:10.2105/AJPH.2012.300715.

Haskell, Rob. "Serena Williams on Motherhood, Marriage, and Making Her Comeback." *Vogue.* January 10, 2018. Accessed January 14, 2019. www.vogue.com/article/serena-williams-vogue-cover-interview-february-2018.

Hasstedt, Kinsey, Desai, Sheila, and Ansari-Thomas, Zohra. "Immigrant Women's Access to Sexual and Reproductive Health Coverage and Care in the United States." *Commonwealth Fund* (November 2018). https://doi.org/10.26099/2dcc-mh04.

Helitzer, Deborah L., LaNoue, Marianna, Wilson, Bronwyn, de Hernandez, Brisa Urquieta, Warner, Teddy, and Roter, Debra. "A Randomized Controlled Trial of Communication Training with Primary Care Providers to Improve Patient Centeredness and Health Risk Communication." *Patient Education and Counseling* 82 (2011): 21–9.

Hilton, Shona, Emslie, Carol, Hunt, Kate, Chapple, Alison, and Ziebland, Sue. "Disclosing a Cancer Diagnosis to Friends and Family: A Gendered Analysis of Young Men's and Women's Experiences." *Qualitative Health Research* 19, no. 6 (2009): 744–754. https://doi.org/10.1177/1049732309334737.

Hilton, Shona, Hunt, Kate, Emslie, Carol, Salinas, Maria, and Ziebland, Sue. "Have Men been Overlooked? A Comparison of Young Men and Women's Experiences of Chemotherapy-induced Alopecia." *Psycho-Oncology: Journal of the Psychological, Social and Behavioral Dimensions of Cancer* 17, no. 6 (2008): 577–583. https://doi.org/10.1002/pon.1272.

Hoffmann, Diane E., and Tarzian, Anita J. "The Girl Who Cried Pain: A Bias Against Women in the Treatment of Pain." *Journal of Law, Medicine, & Ethics*, 29 (2001): 13–27. doi:10.1111/j.1748-720X.2001.tb00037.x.

Hoffman, Kelly N., Trawalter, Sophie, Axt, Jordon R., and Oliver, M. Norman. "Racial Bias in Pain Assessment and Treatment Recommendations, and False Beliefs about Biological Differences between Blacks and Whites." *Proceedings of the National Academy of Sciences of the United States of America*, 113, no. 16 (2016): 4296–301. doi:10.1073/pnas.1516047113.

Holm, Jeffrey E., Vogeltanz-Holm, Nancy, Poltavski, Dmitri, and McDonald, Leander. "Assessing Health Status, Behavioral Risks, and Health Disparities in American Indians Living on the Northern Plains of the US." *Public Health Reports* 125, no. 1 (2010): 68–78. https://doi.org/10.1177/003335491012500110.

Hooker, Steven P., Wilcox, Sara, Burroughs, Ericka, Rheaume, Carol E., and Courtenay, Will. "The Potential Influence of Masculine Identity on Health-Improving Behavior in Midlife and Older African American Men." *Journal of Men's Health* 9, no. 2 (2012): 79–88. doi:10.1016/j.jomh.2012.02.001.

Hudak, Nicole C., and Bates, Benjamin R. "In Pursuit of 'Queer-friendly' Healthcare: An Interview Study of how Queer Individuals Select Care Providers." *Health Communication* (2018): 1–7. doi:10.1080/10410236.2018.1437525

Jayakody, Rukmalie, Chatters, Linda M., and Taylor, Robert Joseph. "Family Support to Single and Married African American Mothers: The Provision of Financial, Emotional, and Child Care Assistance." *Journal of Marriage and Family* 55, no. 2 (1993): 261–76. doi:10.2307/352800.

Jeffries, Mark and Grogan, Sarah. "'Oh, I'm Just, You Know, a Little Bit Weak Because I'm Going to the Doctor's': Young Men's Talk of Self-Referral to Primary Health Care Services." *Psychology & Health* 27, no. 8 (2012): 898–915. doi:10.1080/08870446.2011.631542.

Kan, R.W.M., Au, K. P., Chan, W. K., Cheung, L. W. M., Lam, C.Y.Y., Liu, H. H.W., Ng, L.Y., Wong, M.Y., and Wong, W. C. "Homophobia in Medical Students of the University of Hong Kong." *Sex Education* 9, no. 1 (2009): 65–80. doi:10.1080/14681810802639848.

Kar, A., Mukherjee, S., Ventriglio, A., and Bhugra, D. "Attitude of Indian Medical Students Towards Homosexuality." *East Asian Archives of Psychiatry* 28, no. 2 (2018): 59–63. doi:10.12809/eaap181728.

Khullar, Dhruv and Chokshi, Dave A. "Health, Income, and Poverty: Where We Are and What Could Help." *Health Affairs*. October 4, 2018. Accessed May 9, 2019. www.healthaffairs.org/do/10.1377/hpb20180817.901935/full/.

Kilmartin, Christopher. "Depression in Men: Communication, Diagnosis and Therapy." *Journal of Men's Health and Gender* 2, no. 1 (2005): 95–99.

Kitts, Robert L. "Barriers to Optimal Care Between Physicians and Lesbian, Gay, Bisexual, Transgender, and Questioning Adolescent Patients." *Journal of Homosexuality* 57 (2010): 730–47. https://doi.org/10.1080/00918369.2010.485872.

Labig, Chalmer E., and Peterson, Tim O. "Sexual Minorities and Selection of a Primary Care Physician in a Midwestern U.S. City." *Journal of Homosexuality* 51 (2006): 1–5. https://doi.org/10.1300/J082v51n03_01.

Lederman, Linda C., Kreps, Gary L., and Roberto, Anthony J. *Health Communication in Everyday Life.* Dubuque, IA: Kendall Hunt, 2017.

Loganathan, Santosh and Murthy, R. Srinivasa. "Living with Schizophrenia in India: Gender Perspectives." *Transcultural Psychiatry* 48, no. 5 (2011): 569–84. doi:10.1177/1363461511418872.

Love, Kevin. "Everyone Is Going Through Something." *The Players' Tribune.* March 6, 2018. Accessed January 19, 2019. www.theplayerstribune.com/en-us/articles/kevin-love-everyone-is-going-through-something (2018, March 6).

Lynch, Ingrid, Brouard, Pierre W., and Visser, Maretha J. "Constructions of Masculinity Among a Group of South African Men Living with HIV/AIDS: Reflections on Resistance and Change." *Culture, Health & Sexuality* 12, no. 1 (2010): 15–27. https://doi-org.illinoiscollege.idm.oclc.org/10.1080/13691050903082461.

Macklin, Ruth. "Global Inequalities in Women's Health: Who Is Responsible for Doing What?" *Philosophical Topics* 37, no. 2 (2009): 93–108. www.jstor.org/stable/43154558.

Malebranche, David J., Peterson, John L., Fullilove, Robert E., and Stackhouse, Richard W. "Race and Sexual Identity: Perceptions About Medical Culture and Healthcare Among Black Men Who Have Sex with Men." *Journal of the National Medical Association* 96, no. 1 (2004): 97–107.

Martin, Nina and Montagne, Renee. "Black Mothers Keep Dying After Giving Birth. Shalon Irving's Story Explains Why." *NPR.* December 7, 2017. Accessed January 19, 2019, www.npr.org/2017/12/07/568948782/black-mothers-keep-dying-after-giving-birth-shalon-irvings-story-explains-why.

Matthews, Melissa. "Michael Phelps Opens Up About His Struggles with Depression and Thoughts of Suicide." *Men's Health.* October 26, 2018. Accessed January 19, 2019. www.menshealth.com/health/a24268441/michael-phelps-depression/.

Mount Sinai Center for Center for Transgender Medicine and Surgery. Accessed January 19, 2019, www.mountsinai.org/locations/center-transgender-medicine-surgery.

Moynihan, Clare. "Theories in Health Care and Research: Theories of Masculinity." *BMJ* 317, no. 7165 (1998): 1072–75. Accessed December 22, 2018. www.ncbi.nlm.nih.gov/pmc/articles/PMC1114070/.

National Center for Transgender Equality. "Know Your Rights: Healthcare." Accessed January 8, 2019. https://transequality.org/know-your-rights/healthcare.

Nattrass, Nicoli. "Gender and Access to Antiretroviral Treatment in South Africa." *Feminist Economics* 14, no. 4 (2008): 19–36. doi:10.1080/13545700802266452.

Newton, Benjamin J., Southall, Jane L., Raphael, Jon H., Ashford, Robert L., and LeMarchand, Karen. "A Narrative Review of the Impact of Disbelief in Chronic Pain." *Pain Management Nursing* 14, no. 3 (2013): 161–71. doi:10.1016/j.pmn.2010.09.001.

Obayelu, Abiodun and Ogunlade, I. "Analysis of the Uses of Information Communication Technology (ICT) for Gender Empowerment and Sustainable Poverty Alleviation in Nigeria." *International Journal of Education and Development using Information and Communication Technology* 2, no. 3 (2006): 45–69. Accessed December 22, 2018. http://ijedict.dec.uwi.edu/viewarticle.php?id=172&layout=html.

Ogunbajo, Adedotun, Kershaw, Trace, Kushwaha, Sameer, Boakye, Francis, Wallace-Atiapah, Nii-Dromo, and Nelson, Laron E. "Barriers, Motivators, and Facilitators to Engagement in HIV Care Among HIV-Infected Ghanaian Men Who Have Sex with Men (MSM)." *AIDS & Behavior* 22, no. 3 (2018): 829–39. doi:10.1007/s10461-017-1806-6.

O'Hanlan, Katharine A., Dibble, Suzanne L., Hagan, H. Jennifer, and Davids, Rachel. "Advocacy for Women's Health Should Include Lesbian Health." *Women's Health* 13, no. 2 (2004): 227–34. https://doi.org/10.1089/15409990432966218.

Oliffe John L., Ogrodniczuk, John, Bottorff, Joan L., Hislop, T. Gregory, and Halpin, Michael. "Connecting Humor, Health, and Masculinities at Prostate Cancer Support Groups." *Psycho-Oncology* 18, no. 9 (2009): 916–26. doi:10.1002/pon.1415.

Pagano, Michael P. *Health Communication for Health Care Professionals: An Applied Approach.* New York, NY: Springer Publishing, 2017.

Papachristos, Anna. "Laverne Cox Shares Struggle with Mental Health to Promote Respect for Trans Identities." *A Plus.* August 14, 2018. Accessed January 14, 2019. https://aplus.com/a/laverne-cox-on-deadnaming-and-mental-health-of-trans-community?

Pearson, Stephen and Makadzange, Panganai. "Help-Seeking Behaviour for Sexual-Health Concerns: A Qualitative Study of Men in Zimbabwe." *Culture, Health & Sexuality* 10, no. 4 (2008): 361–76. doi:10.1080/13691050801894819.

Perez-Brumer, Amaya, Nunn, Amy, Hsiang, Elaine, Oldenburg, Catherine, Bnder, Melverta, Beauchamps, Laura, Mena, Leandro, and MacCarthy, Sarah. "'We Don't Treat Your Kind': Assessing HIV Health Needs Holistically among Transgender People in Jackson, Mississippi." *PLoS One* 13, no. 11 (2018): e0202389. https://doi.org/10.1371/journal.pone.0202389.

Qiao, Shan, Zhou, Guangyu, and Li, Xiaoming. "Disclosure of Same-Sex Behaviors to Health-Care Providers and Uptake of HIV Testing for Men Who Have Sex with Men: A Systematic Review." *American Journal of Men's Health* 12, no. 5 (2018): 1197–1214. https://doi.org/10.1177/1557988318784149.

Quinn, Chris, and Happell, Brenda. "Talking About Sexuality with Consumers of Mental Health Services." *Perspectives in Psychiatric Care* 49, no.1 (2013): 13–20. https://doi.org/10.1111/j.1744-6163.2012.00334.x.

Ravenell, Joseph E., Johnson, Jr. Waldo E., and Whitaker, Eric E. "African American Men's Perceptions of Health: A Focus Group Study." *Journal of the National Medical Association* 98 (2006): 544–50.

Redfern, Jan S. and Sinclair, Bill. "Improving Health Care Encounters and Communication with Transgender Patients." *Journal of Communication in Healthcare* 7, no. 1 (2014): 25–40. doi:10.1179/1753807614Y.0000000045.

Riggs, Damien W., and Due, Clemence. "Mapping the Health Experiences of Australians Who Were Female Assigned at Birth but Who Now Identify with a Different Gender Identity." *Lambda Nordica* nos 3–4 (2013): 54–76. Accessed December 18, 2018. https://digital.library.adelaide.edu.au/dspace/bitstream/2440/89383/2/hdl_89383.pdf.

Roy, Phillipe, Tremblay, Giles, Oliffe, John L., Jbliou, Jalila, and Robertson, Steve. "Male Farmers with Mental Health Disorders: A Scoping Review." *Australian Journal of Rural Health* 21 (2013): 3–7.

Sblendorio, Peter. "Anthony Bourdain Discussed his Mental Health Struggles During Therapy Session on 'Parts Unknown.'" *New York Daily News.* June 8, 2018. Accessed January 19, 2019. www.nydailynews.com/entertainment/tv/ny-ent-anthony-bourdain-parts-unknown-therapy-20180608-story.html.

Schabath, Matthew B., Blackburn, Catherine A., Sutter, Megan E., Kanetsky, Peter A., Vadaparampil, Susan T., Simmons, Vani N., Sanchez, Julian A., Sutton, Steven K., and Quinn, Gwendolyn P. "National Survey of Oncologists at National Cancer Institute—Designated Comprehensive Cancer Centers: Attitudes, Knowledge, and Practice Behaviors About LGBTQ Patients with Cancer." *Journal of Clinical Oncology* (2019): 1–12. doi:10.1200/JCO.18.00551.

Schiavo, Renata. *Health Communication: From Theory to Practice.* San Francisco, CA: Jossey-Bass, 2014.

Sinclair, Kate, Ahmadigheidari, Davod, Dallmann, Diana, and Melgar-Quiñonez, Hugo. "Rural Women: The Most Vulnerable to Food Insecurity and Poor Health in the Global South." *The FASEB Journal* 31, no. 1_supplement (2017): 791–28.

Sperber, Jodi, Landers, Stewart, and Lawrence, Susan. "Access to Health Care for Transgendered Persons: Results of a Needs Assessment in Boston." *International Journal of Transgenderism* 8 (2005): 75–91. https://doi.org/10.1300/J485v08n02_08.

Street, Richard L., Makoul, Gregory, Arora, Neeraj K., and Epstein, Ronald M. "How Does Communication Heal? Pathways Linking Clinician-Patient Communication to Health Outcomes." *Patient Education Counseling* 74 (2009): 295–301. doi: 10.1016/j.pec.2008.11.015.

Tamargo, Christina L., Quinn, Gwendolyn P., Sanchez, Julian A., and Schabath, Matthew B. "Cancer and the LGBTQ Population: Quantitative and Qualitative Results from an Oncology Providers' Survey on Knowledge, Attitudes, and Practice Behaviors." *Journal of Clinical Medicine* 6, no. 93 (2017): 1–13. doi:10.3390/jcm6100093.

Tatum, Sophie. "White House announces Policy to Ban Most Transgender People from Serving in the Military." *CNN.* March 24, 2018. Accessed January 19, 2019, www.cnn.com/2018/03/23/politics/transgender-white-house/index.html, para. 1.

Tejeda, Silvia, Gallardo, Rani, Ferrans, Carol, and Rauscher, Garth. "Breast Cancer Delay in Latinas: The Role of Cultural Beliefs and Acculturation." *Journal of Behavioral Medicine* 40, no. 2 (2017): 343–51. doi:10.1007/s10865-016-9789-8.

Telfer, Michelle, Tollit, Michelle, and Feldman, Debi. "Transformation of Health-Care and Legal Systems for the Transgender Population: The Need for Change in Australia." *Journal of Paediatrics and Child Health* 51, no. 11 (2015): 1051–53. doi:10.1111/jpc.12994.

Thyberg, Hannah. "A Semblance of Normalcy: Social Isolation and the Burden of Looking Well." In *Reifying Women's Experiences with Invisible Illness*, edited by Kesha Morant Williams and Frances Selena Morant, 29–40, Lanham, MD: Lexington Books, 2018.

Trapp, Stephen K, Woods, Jacqueline D., Grove, Alicia, and Stern, Marilyn. "Male Coping Processes as Demonstrated in the Context of a Cancer-Related Social Support Group." *Supportive Care in Cancer* 21, no. 2 (2013): 619–627. doi:10.1007/s00520-012-1565-x.

U.S. Centers for Medicare and Medicaid Services. "Transgender Health Care." Accessed January 21, 2019, www.healthcare.gov/transgender-health-care/.

U.S. Department of Health and Human Services, Centers for Disease Control and Prevention, & National Center for Health Statistics. "Summary Health Statistics: National Health Interview Survey." Accessed June 9, 2019, https://ftp.cdc.gov/pub/Health_Statistics/NCHS/NHIS/SHS/2014_SHS_Table_A-18.pdf.

Utamsingh, Pooja D., Richman, Laura S., Martin, Julie L., Lattanner, Micah R., and Chaikind, Jeremy R. "Heteronormativity and Practitioner–Patient Interaction." *Health Communication* 31, no. 5 (2016): 566–74. https://doi.org/10.1080/10410236.2014.979975.

van Dam, Mary Ann A., Koh, Audrey S., and Dibble, Suzanne L. "Lesbian Disclosure to Health Care Providers and Delay of Care." *Journal of the Gay and Lesbian Medical Association* 5, no. 1 (2001): 11–19. https://doi.org/10.1023/A:1009534015823.

Venetis, Maria K., Meyerson, Beth E., Friley, L. Brooke, Gillespie, Anthony, Ohmit, Anita, and Shields, Cleveland G. "Characterizing Sexual Orientation Disclosure to Health Care Providers: Lesbian, Gay, and Bisexual Perspectives." *Health Communication* 32, no. 5 (2017): 578–586. https://doi.org/10.1080/10410236.2016.1144147.

Ventriglio, Antonio, Torales, Julio, Bhugra, Dinesh. "Disease versus Illness: What Do Clinicians Need to Know?" *International Journal of Social Psychiatry* 63, no. 1 (February 2017): 3–4. doi:10.1177/0020764016658677.

Vlassoff, Carol, and Manderson, Lenore. "Incorporating Gender in the Anthropology of Infectious Diseases." *Tropical Medicine and International Health* 3, no. 12 (1998):1011–1019.

Vu, Milkie, Azmat, Alia, Radejko, Tala, and Padela, Aasim I. "Predictors of Delayed Healthcare Seeking among American Muslim Women." *Journal of Women's Health* 25, no. 6 (2016): 586–93. doi:10.1089/jwh.2015.5517.

Watkins, Daphne C., Abelson, Jamie M., and Jefferson, S. Olivia. "'Their Depression Is Something Different… It Would Have to Be': Findings from a Qualitative Study of Black Women's Perceptions of Depression in Black Men." *American Journal of Men's Health* 7, no. 4 (2013): 45S–57S. doi:10.1177/1557988313493697.

Watkins, Daphne C., Walker, Rheeda L., and Griffith, Derek M. "A Meta-study of Black Male Mental Health and Well-being." *Journal of Black Psychology* 36, no. 3 (2010): 303–30. doi:10.1177/0095798409353756.

Williams, Kesha. "Introduction." In *Reifying Women's Experiences with Invisible Illness*, edited by Kesha Morant Williams and Frances Selena Morant, xiii–xviii, Lanham, MD: Lexington, 2018.

Williams, Susan Schneider. "The Terrorist Inside My Husband's Brain." *Neurology* 87, no. 13 (2016): 1308–11. https://doi.org/10.1212/WNL.0000000000003162.

Wright, Kallia. "'You Have Endometriosis': Making Menstruation-Related Pain Legitimate in a Biomedical World." *Health Communication* (2018): 1–4. doi:10.1080/10410236.2018.1440504.

Zimmerman, Mike. "What Chris Cornell Told Me about His Depression Years before His Suicide." *Men's Health*. May 19, 2017. Accessed January 19, 2019. www.menshealth.com/trending-news/a19516660/chris-cornell-death-depression-suicide-interview/.

25

WOMEN FIRST

Bumble™ as a model for managing online gendered conflict

Sean Eddington and Patrice M. Buzzanell

Scholarship on conflict in organizational contexts has examined the ways in which people organize their relationships and resources based on gendered power relations and structures. In general, this scholarship attempts to sort through contexts in which women and men interact to reproduce and disrupt expectations and experiences that systematically disadvantage one group, typically women, over the more dominant group, men. From experiments to ethnographic studies, researchers have demonstrated that there are sex and gender differences (women, men; feminine, masculine) in gendered conflict style preferences, negotiation patterns, framings of experiences, and other aspects of workplace through organizing.[1] However, such research also indicates points of contention, contradiction, and possibility for equitable gender processes and practices in offline and online contexts.

In this chapter, we present an example of how the internet dating platform Bumble™ has managed online gendered conflict. Bumble addresses gendered conflict by promoting relational and structural strategies online. That is, Bumble subverts technocultural norms by advancing policies that deter and practices that minimize conflict and harassment on both micro (e.g., individual) and macro (e.g., institutional and platform) levels. We note that offline and online contexts share many similarities as triggers for and responses to conflict, yet we focus on online contexts, as online organizing is enabled by technical affordances. After this introduction, we discuss prominent gender, conflict, and communication theories (giving specific attention to online contexts), then return to Bumble to develop the case in depth. Using Bumble, we note conflict management practices that other organizations can adapt to their own contexts. We conclude our chapter by presenting strategies that move beyond "fix the woman" or "fix the problem" to engage in broader gendered organizing relationships and structures.[2]

To begin, consider the following fictionalized scenario: Jan is a single, 20-something engineer and is interested in dating. Jan's friends have told her that she ought to try one of the popular mobile dating applications, as they have had success matching with potential partners. Jan downloads the application Tinder™ and creates her profile, complete with details that she believes will present her authentically but also in an interesting way for potential partners (e.g., she loves most major sports, reading spy novels, is involved in her local sketch comedy troupe, and has a bucket list for visiting local breweries in all 50 states). Her profile also includes several photos from her life that shows her doing various activities (camping, traveling, and walking her two-year old Golden Retriever, Frank). Jan believes that her profile is dating-ready, and she

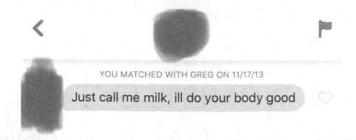

YOU MATCHED WITH GREG ON 11/17/13

Just call me milk, ill do your body good

Figure 25.1   Actual message sent to a user on Tinder (TinderNightmares, 2018)[3]

begins swiping left and right looking for love. One Friday night, Jan matches with Greg, who responds quickly with the message (see Figure 25.1):

Jan's quickly saves the screenshot, unmatches with Greg, and sends it to her girlfriends with the message, "What kind of app did you tell me download?!" Her friends respond, "Oh no! We forgot to tell you that there are those types of guys out there—you just learn to ignore them!"

Although Jan's message was "tame" compared to what others have received, women often receive unwanted explicit photos and harassing sexist messages on mobile dating applications and in an online world that continues to objectify and harm women.[4] Mobile dating applications like Tinder are dominated by men; they outnumber women two to one.[5] Moreover, these applications often are perpetuated through toxic technocultures enabled through various communicative affordances that support and normalize egregious forms of sexist communication— what Jane dubbed as "e-bile."[6]

Online technocultures are an emerging area of inquiry considering conflict management. Technocultures are the customs, values, traditions, and symbols that exist within an online setting.[7] Like offline contexts, technocultures promote gender neutrality while simultaneously ignoring the masculine default that structure and organize many online spaces.[8] Within technocultures, gendered conflict can be exaggerated, amplified, and even more pervasive than such behaviors in offline settings, as evidenced by an increase in behaviors such as trolling, gender-based harassment, and rape threats.[9] For individuals who are targets of this behavior, there are way to mitigate these conflicts: speak out, fight back, or exit platforms altogether.[10] (Some of these ways for addressing conflict online through subversion of and challenges toward toxic, masculine technocultures are addressed at the end of this chapter.[11])

Mobile dating applications like Bumble have tried to limit the amount of toxic communication that occur in online spaces. We view Bumble as an example of an online, alternative organization that attempts to cultivate a unique, inclusive technoculture. To demonstrate these possibilities, we present a case study of the online, feminist dating application Bumble, which was created and launched by founder and CEO Whitney Wolfe Herd in December 2014. Bumble, as Bennett describes,

> aims to be a little less agonizing for women. It features photo verification that assuages users' fears that they might be getting catfished (lured into an online relationship with a false identity) and security that makes it easy to report harassment. The company says its abuse report rate is among the lowest of its competitors, at 0.005 percent.[12]

Moreover, Wolf Herd's Bumble is a unique case in that it subverts traditional male-dominated dating sites by placing power and agency to control communication at the fingers of women by creating a contra-technocultural ecosystem.[13] Additionally, Wolfe Herd's own professional experiences and challenges working in the male-dominated technology industry were reported through her high-profile gender discrimination lawsuit against her former employer and rival dating application Tinder.[14]

Gender in organizational contexts

In what follows, we discuss gender structure theory as it impacts and organizes its impact on social life. From there, we extend these theories, concepts, and arguments into online technocultures. We conclude by providing a brief overview of conflict management theorizing pertaining to our case study of Bumble.

Gender structure theory

Although organizational scholars have described and theorized the central role that gender plays in organizing, organizational communication scholarship has yet to richly explore and embrace gender as a form of organizing that spans both micro (e.g., individuals) and macro levels (e.g., organizations).[15] As such, we draw from the sociological theory of gender structure that describes gender organizing as both a social structure and institution, as it is "embedded not only in individuals but throughout social life."[16] Risman's gender structure theory contends that gender has been institutionalized throughout time through micro, meso, and macro levels that continually enact difference through the structure's privileging the masculine over feminine norms, values, and identities in society. That is, gender is both relationally enacted through individual interaction and normalized and legitimized through institutionalization. For example, when people talk and interact, they socially construct what is appropriate (and not) through gendered, as well as other, lenses of difference.[17] Arguing that "social structures not only act on people, people act on social structures," Risman articulated the importance of individuals' interactions and accountability in constituting gender and gendered behaviors/practices.[18] Over time, as individuals perform gendered identities, their own focus on policing and enacting norms or idealized identities (i.e., hegemonic masculinity and femininity) institutionalize what and how identities emerge as privileged and normed.

Gendered social structures emerge in workplaces and organizations, as they *do* gender in a variety of ways.[19] Although organizations consider themselves gender-neutral, they are far from it, as relational aspects of gender enact heteronormative, masculine organizational norms within organizations.[20] These norms are evident in labor markets, leadership (i.e., who makes up senior leadership), ideas on work ethic and identities (i.e., The Ideal Worker Norm), and organizational hierarchies (i.e., tiered hierarchies versus flat organizations). These norms are both a product of individuals' constant enactment of gender but also institutionalize and reify masculine norms of competition, individualization, power, and control. Seeing these masculine norms as prevalent throughout the social world, gender, as both d/Discourse, creates and enacts difference through its privileging the masculine over feminine.[21] Masculine norms become an invisible default in many contexts, meaning they are normalized unless challenged and contested. Masculine norms pervade social structures such as family, communities, organizations, and institutions and influence what scholars study and do not study.[22]

Building upon structurational approaches to gender, gender Discourses also organize, constitute, and reinforce much within organizational life from organizational culture, to industry norms, and physical spaces. Organizational cultures are gendered insomuch as they often are values-laden and promote masculine ideals.[23] Organizational rituals and symbols often undergird organizational life and performances of gender and are made visible in workplace interactions.[24]

Interactions from women's conversational style, their deference to compliments from male bosses or peers, and even the use of subordinated styles of communication are symbolic negotiations that women often undertake as they engage in organizational life.[25]

Gender Discourses also organize pathways for both career and career choice. Feminist organizational communication scholars have argued that careers are gendered as masculine with career patterns and pathways often described in linear, sequential ways.[26] Others have argued that career choices are inherently gendered. For example, Ashcraft's "glass slipper" metaphor refers to how careers are configured through idealized discursive assumptions about who and which social identities (e.g., gender, race, class) "fit" for certain careers.[27] Building upon this metaphor, Trotter's ethnographic study of two cohorts of nursing professionals found that the nursing field reproduced gender roles in two ways.[28] First, the nursing field's flexible work arrangements allowed women to balance their attention to both work and family life (e.g., parental responsibilities). The promise of a flexible career to balance a home and work life often provided an attractive career for some women; however, flexible work arrangements also continued to promote gender stereotypes. The promise of flexible work arrangements in nursing further endorsed "mommy tracks" in careers as the women in Trotter's study often were the only ones to utilize these organizational policies. Over time, the use of these policies disadvantage women's economic potential and long-term career goals, which further serve to enact gender inequality in organizations.

Other studies have examined notions of space as gendered. For example, Craig and Liberti's study of a women-only gym showcased organizational processes that promoted a safe space for women to work out, lose weight, and accept one another; however, the authors argued that "the women appear to be doing just what girls do—waiting for a turn, supporting each other's weight loss, hating exercise and the complications of machines, and measuring their fitness progress through shopping expeditions."[29] These instances and examples demonstrate several ways that the gym encouraged traditional gender roles and continued to support gendered hierarchies as the women continued to promote stereotypical and traditional roles for women in society. As Craig and Liberti argued, despite the possibilities for transgression from gender roles, these types of spaces reinforced normative ideas and performances of womanhood—that of women concerned with body image, lamenting the struggle of exercise, and discussing shopping successes with one another.

Workplaces also highlight how gendered identity negotiations take place. In male-dominated industries like technology or engineering, male identities are often legitimized and privileged over other identities.[30] For example, in an intersectional analysis of 18 women in the technology industry, Alfrey and Twine uncovered various experiences of participants as they negotiated and gained insider status in their all-male workplaces.[31] Tensions emerged regarding feminine and masculine gender performances and, oftentimes, the participants noted that feminine performances created liabilities and further exacerbated outsider status. For several of the LGBTQ, white, and Asian workers, gender fluidity became an asset in cultivating belonging with the male-dominated workplaces. More specifically, for the white and Asian workers who identified as gender fluid, they were able to assimilate into the workplaces as they performed more masculine behaviors. As such, these gender performances aided in avoiding microaggressions that impacted other women of color *and* women who adopted more feminine gender performances.

To this point, the prior overview of literature has showcased a few examples of how gender structure theory occurs in offline settings through the interrelated concepts of space, discourse, and organizational culture. The interconnections between space, discourse, and culture are a central part of online technocultures. In the next section, we build upon Risman's gender structure theory by arguing that online spaces, too, institutionalize gender from micro- to macro-level communicative interactions.

Gendered online technocultures

Online spaces enact organizational cultures over time. Whereas in offline spaces, cultures are constituted through verbal and nonverbal forms of communication, in online spaces, language plays an integral role in organizing technocultures.[32] Technocultures are constituted through the "relationship between technology and culture and the expression of that relationship in patterns of social life, economic structures, politics, art, literature and popular culture."[33] In online spaces, technocultures are enacted through continual user engagement—that is, through their text-based interactions with one another. Language shared by users in online spaces and communities shape vernaculars, slang terms, topics, values, traditions, and symbols within spaces.[34]

As scholars have contended, there are often intersections between gender, race, and online technologies as "the design and deployment of technology can, knowingly or unknowingly, perpetrate sexist or exclusionary gender politics."[35] Throughout the history and evolution of the internet, white men have dominated technocultures—both as creators and users.[36] Additionally, like offline contexts, technocultures, too, promote gender and racial neutrality while simultaneously ignoring the white masculine default that structure and organize many online spaces.[37] Online spaces promote specific ideals about who should be—often perpetuating digital forms of white masculinity. For instance, Eddington drew upon the online comment sections of the men's rights organization The Red Pill to examine the discursive and gendered construction of organizational identity.[38] The organizational identity of The Red Pill was continually constituted by and through contradictions pertaining to hegemonic and subordinated ideals of geek masculinities.

Other scholars have called for investigations that shed light on how "claims to online space, made through the affordances of digital infrastructures are gendered, material, and embodied."[39] In their study of Tinder and Bumble's affordances, MacLeod and McArthur identified specific constraints with the applications that reinforce limited forms of gender expression and promote gender binaries.[40] While individuals have some freedom in creating and performing identities in their profiles, users are limited to "male" and "female" categories of whom they are shown in the apps. Although the role of gender categories may be a necessary component of Tinder and Bumble's sorting algorithm, they reinforce the limitations of an individual's nuanced and unique expressions of their gender identities. In doing so, specific and traditional forms of gender identity become standardized and normalized in online spaces. Thus, these dating applications can reinforce traditional performances and norms surrounding gender identity.

In order to subvert online norms, users, collectivities, and social movements often employ alternative strategies (e.g., naming conventions) to exist and convene online.[41] For instance, the Cybergrrl women's movement employed the use of the word "grrl" instead of "girl," which was useful for two reasons. First, according to creator of the movement, Aliza Sherman, it was both a sign of strength and attitude. Sherman noted that she was inspired by the Riot Grrrl movement of the 1990s, which was a "young, bristling feminist subset of the alternative punk rock music scene."[42] Second, the use of "grrls" helped to limit the visibility of Cybergrrls, as "searching on the net for 'girls' mainly produces sex sites and very little relevant material for women."[43] Although online spaces are often characterized as utopian and egalitarian marketplaces of ideas and identities, the promise of equality is often denied as individuals perform identities online that challenge the existing white, male, heterosexual norms of cyberspace.[44]

When there are challenges to technocultural norms, the resulting backlash is often vitriolic and increasingly toxic. Ging described the backlash as a form of digital hegemonic masculinity that seeks to reestablish dominance of masculine ideologies that pervade online spaces.[45] Scholars have detailed the manifestations of the backlashes through events like Gamergate and the Fappening, and through the proliferation of online networks like extremist groups and the Manosphere.[46] Additionally, drawing on the organizing nature of language online, scholars like Jane have studied and developed language generators that predict how misogyny online will be communicated.[47] Jane's *Random Rape Threat Generator* evolved from observing linguistic patterns (what she dubbed, "Rapeglish") that emerged from several of the events, groups, and spaces online. Rapeglish and other types of communication behaviors are found, encouraged, and normalized on some social network sites. For example, the social news site Reddit has been examined as a hotbed of and nexus for many of these issues.[48] Bolstered by the technological affordances of Reddit (e.g., the ability to create pseudonymous identities), users often are racialized as white and adopt masculine gender performances.[49] In spaces like Reddit, women and other underrepresented minorities who perform identities that run counter to platform norms and even speak out against social issues online are cast as faceless "Others"—often being characterized as villains, fanatics, or social justice warriors.[50] Given these trends in online spaces, the question, then, is how do individuals begin to navigate and negotiate gendered conflict in and throughout technocultures?

Conflict management strategies

Conflict is "the interaction of interdependent people who perceive incompatible goals and interference from one another in achieving those goals," and its presence in organizational life often is experienced through emotional responses to incompatibility.[51] Gender organizes how conflict is managed in a variety of ways.[52] Masculine forms of conflict often adopt and privilege the role of exchange throughout the process. In exchange models, individuals adopt *dominating* approaches that are "characterized by the use of forceful tactics such as threats and put-downs, an unwillingness to move from one's initial position, and a focus on defeating the opponent."[53] Exchange models engender these types of responses through the emphasis on and role of competition and strategy. Given the masculine values that undergird organizations and organizational activities, conflict is framed in terms of competition for scarce resources.[54] Competition engenders win-lose dynamics, linear thinking, and rationality as essential parts of the both masculine organizations and thus as traditional forms of conflict management strategies.[55] Although there has been scholarly focus on conflict types *or* its management in interpersonal and organizational relationships, these strands rarely investigate the intersections between type and management or include various understandings of conflict.[56]

Considering the latter, understanding of conflict from various perspectives can highlight power imbalances that exist in and during conflict processes. For example, gender and other forms of difference (e.g., cultural, ethnic, and religious) shape and are shaped by traditional forms of conflict management strategies through their focus on "neutral" (or masculine) hierarchical resolutions and distributive bargaining.[57] With competitive individualism, gendered patterns are often replicated, even exacerbated, during short-term conflict episodes and longer-term negotiations where there are mixed motives and cultural role differences.[58] Moreover, traditional forms of conflict management often ignore complexities of gender, culture, and power relations that exist in organizations.[59]

Traditional models of conflict management are often limited in addressing gendered conflict online. Given these realities, how do researchers engage with the complexities of conflict management in online spaces wherein these complexities are often made more visible, widespread, and amplified through explicit and problematic content shared online that target women and underrepresented minorities? Additionally, how might conflict management strategies create opportunities for change at micro and macro levels? That is, how might conflict management strategies enact change at both individual and organizational levels in technocultures? In what follows we begin to answer these questions referring back to, and expanding, our case study of \Bumble.

Challenging technocultural norms: Whitney Wolfe Herd and Bumble

Bumble started in late 2014. CEO and founder Whitney Wolfe Herd left rival dating application Tinder in 2014 and filed a sexual harassment lawsuit against the company, "alleging that her ex-boss and ex-boyfriend Justin Mateen called her a 'whore' and 'gold digger' and bombarded her with threatening and derogatory text messages…[and] that Tinder…had wrongly stripped her of a cofounder title."[60] The suit was settled for an estimated one million dollars; however, Tinder did not express wrongdoing, although Justin Mateen resigned from Tinder.[61] As the news of the harassment suit spread throughout tech and business circles, online commenters and journalists speculated and characterized Wolfe Herd in a variety of negative and unfavorable ways. In her own words, Wolfe Herd described the experience in the following way:

> Online strangers, journalists and trolls threatened me in article comment sections and called me hateful names on Twitter. Gossip pages published story after story of speculation meant to discredit me. People who'd never met me took cruel pleasure in dissecting and distorting my private life, reducing me to a caricature I did not know or recognize. Still, if you had Googled me back in 2014, that was the only Whitney you'd see.

> For months, it was hard for me not to feel like all that ugliness was stamped across my forehead. I sank into a deep depression. I became an insomniac and drank too much in weak attempt to numb the pain and fear I was experiencing. At my lowest point, I wanted to die. I was only 24, and already I felt like I was finished. That's the poisonous power of online harassment and abuse—especially when it lands on your phone every morning and follows you everywhere you go.[62]

Wolfe Herd's description further sheds light on vitriolic and problematic communication patterns and behaviors that abound in online spaces. Driven by a desire to challenge these online, technocultural norms, Wolfe Herd partnered with Andrey Andreev, the CEO of the European dating application Badoo to launch Bumble in December 2014.[63] In describing the advent of Bumble as a response to the online harassment and misogyny that is often perpetuated online, Wolfe Herd noted, "You have to start a business to solve something that's a personal pain point. That's where the best businesses come from."[64] In four years of operation, Bumble is one of the fastest growing mobile dating applications and currently boasts a user network of 44.5 million; however, Wolfe Herd's story and Bumble's growth and evolution provide pathways for how gendered organizational conflict can be mitigated online.[65] In what follows, we discuss a variety of artifacts, policies, and media reports that showcase both Wolfe Herd and Bumble as exemplars in confronting, challenging, and mitigating gendered conflict online.

Conflict management strategies: organizational structure

Conflict is ever-present online; however, Wolfe Herd and Bumble borrowed and extended Robbins' caution that the proper orientation toward conflict is in managing rather than resolving conflict.[66] Given our focus on managing gendered conflict, Bumble's macro-level organizational strategies are present in a variety of ways that foster forms of both social control and change.[67] Given the context of Bumble, the organizational structure engenders inclusion and feminist values through a myriad of ways that reverse traditional gender roles and norms. In considering conflict management and change processes, Risman argued that creating gender equality must recursively be enacted from macro to micro levels and occurs through the "undoing" of gender.[68] If gender can be undone, then conflict, as a gendered act, can also adopt strategies wherein undoing gender occurs through a variety of organizational and individual communicative acts.

Bumble's subversion of traditional power dynamics within both romantic relationships and technocultural communications are bolstered by its organizing and espousal of feminist values. Bumble enacts and performs its organizational mission and vision in a variety of ways that challenge traditional gender norms. Additionally, Bumble adopts an explicit feminist sensibility about its organizing, and enacts feminist values (e.g., kindness, respect, equality) in policy creation and enforcement, reinforcement of its values through its social media platforms, engagement in larger social issues, and organizational expansion into other social networking markets—namely, business (BumbleBizz) and friendship (BumbleBFF) social networking platforms. These examples provide insights into Bumble's desire to manage gendered conflict that exists online.

At its inception, Bumble's organizational mission that was adopted and enacted provided a framework for managing and countering gendered conflict. Specifically, its mission states, "Bumble is a platform and community that creates empowering connections in love, life, and work. We promote accountability, equality, and kindness to end misogyny and re-write archaic gender roles. On Bumble, women always make the first move."[69] By adopting and creating an online technoculture where women hold the power, Bumble subverts traditional ideas of masculine leadership in romantic relationships. This transgressive act is echoed by Bumble's vision, which is driven by a desire to challenge online misogyny and promote an online technoculture "where all relationships are equal."[70] Here, Bumble makes explicit its organizational desire to (re)negotiate both technocultural norms (e.g., online misogyny) through a promotion of inclusionary relationships. Bumble's vision is continually reinforced throughout the Bumble application and its web presence. For example, on its webpage, "About Bumble," it states, "We prioritize kindness and respect, providing a safe online community for users to build new relationships."[71]

Conflict management strategies: individual actions

On the micro, interactional level, Bumble encourages and promotes three specific conflict management strategies for its users. Bumble's technocultural environment – from its application ecosystem to how users design their profile and communicate with one another – was built around concerns for user safety. First, Bumble's application disrupts the traditional online dating spaces by giving the power to women to initiate communication. That is, once a woman and man mutually match with one another, no communication between the two occurs until the woman makes the first communicative move. This strategy and policy are part of Bumble's "Women First" initiative. Bumble representative and tennis champion Serena Williams described this affordance and mission in the following way: "Society has taught us as women to kind of sit back and not necessarily be the first one to speak up. We want to take that and flip the story."[72]

Second, Bumble users have specific requirements for how they create profiles. Like rival Tinder, profiles are linked to Facebook accounts; however, Bumble adds additional measures to protect its users. For instance, users have an option to opt in for a verified photo, which can help identify internet trolls and other fake accounts.[73] Additionally, there are community limitations on what photos can be part of users' profiles. For example, Bumble banned firearms from appearing in profile photos.[74] These guidelines help contribute to and cultivate a safe and inclusive community.

Third, Bumble's terms of service offer a zero-tolerance policy for harassment and disrespect. Specifically, the terms of service explicitly state that speech containing "language or imagery which could be deemed offensive or is likely to harass, upset, embarrass, alarm or annoy any other person" is prohibited as is language that is "abusive, insulting or threatening, discriminatory or which promotes or encourages racism, sexism, hatred or bigotry."[75] All users (not just men) who violate Bumble's terms of service are sanctioned or banned from the application, as individuals are encouraged to report harassment through the application when it occurs. Alexandra Williamson el-Effendi, Bumble's head of brand, described,

> We rely on our users. We have a block and report feature in the app where [users] can report [other users] on the app, or we have a lot of people who reach out [to us] via social media, too.[76]

Furthermore, el-Effendi noted that Bumble seriously considers all harassment reports and is unafraid to ban users as it helps sustain a positive community and maintain high user expectations.

From time to time, Bumble takes active steps to highlight its actions through social media and its organizational blog.[77] For example, on June 3, 2016, Bumble Headquarters posted, "An Open Letter to Connor," in response to a user ("Connor") who berated a user ("Ashley") that asked about his career stating, "I don't have time for entitled, gold-digging whores…I don't…prescribe to this neo-liberal, Beyoncé, feminist cancer which plagues society and says a guy can't as so much as give constructive criticism to and call a girl out on her bullshit."[78] Connor's message was shared in screenshots to Bumble administrators (called the "Bumble Hive"), which prompted the following response from the Bumble Hive:

> While you [Connor] may view this as a "neo-liberal, Beyoncé, feminist-cancer" and rant about the personal wounds you've sustained from "entitled gold digging whores," we're going to keep working. We're going to expand our reach and make sure women everywhere receive the message that they're just as empowered in their personal lives as they are in the workplace. We're going to continue to build a world that makes small-minded, misogynist boys like you outdated.
>
> We're hoping one day you come around. We hope the hate and resentment welling up inside of you will subside and you'll be able to engage in everyday conversations with women without being afraid of their power to the extent you feel the need to lash out at them. But until that day comes, Connor, consider yourself blocked from Bumble. #LaterConner

The Bumble Hive's response and banning of Connor offer two insights. First, their blog post demonstrated their commitment to feminist organizing and establishment of a community wherein all individuals communicate respectfully with one another. Second, their blog post (and subsequent posts) end with #LaterConnor. The use of this hashtag encourages conversation and attention on other social networking sites like Twitter, Facebook, or Instagram. These forms of boundary spanning across social media again serve as an extension and reinforcement of Bumble's organizational mission.

Bumble as a model for managing online gendered conflict

Using Bumble as an exemplar of how gendered conflict is managed in a counter-technocultural online space, we have demonstrated how gendered conflict is managed in online spaces. The connection between the individual members of the Bumble community and the organization of Bumble help to create a safer, online technoculture that challenges online misogyny. Because of Bumble's efforts, the organization boasts a 0.008 percent abuse report rate.[79] Although the threat of conflict in online spaces is ever-present, organizations like Bumble have adopted and enacted both organizational commitments and individual-level policies and practices that both challenge traditional forms of gendered organizing and promote gender equity and healthy relationships. Throughout this chapter, we have offered a number of conflict management strategies and use this section to pull them together with commentary.

We begin with individual-oriented conflict management strategies. Opening with the fictional but all-too-common case of Jan and Greg, one bit of advice is to ignore those people who initiate and perpetuate problematic interactions online, as Jan's friends advised ("Oh no! We forgot to tell you that there are those types of guys out there—you just learn to ignore them!"). Another strategy (albeit riskier) is counter-disciplining by "calling out" or "naming and shaming" individuals who send lewd and sexist content online.[80] Counter-disciplining also occurs through publicizing or sharing screenshots of the awkward, lewd, and perverted interactions (as evidenced in Figure 25.2).[81] Screenshots like the one in Figure 25.2 are shared through various social media platforms and, at times, can go viral through sites like *TinderNightmares*. An integral part of viral photos on sites like *TinderNightmares* is the use of humor. Humor can serve as an additional facet of counter-disciplining for women as it "provides women with a form of discursive agency through the showcasing of witty replies."[82]

Counter-disciplining also has the potential to challenge men's hypermasculine and sexist behaviors with the potential for illuminating and identifying men's misogynistic content online. The specter of awkward moments shared in online spaces does serve as a reminder that one's presentation and performance online can always be seen, made visible, and shared throughout networks.[83] That is, there are consequences and social ramifications for behaving like a jerk through viral shaming. Dating applications like Tinder often see these forms of "digilantism" because the technocultural values and norms promote "hook-up culture" as a key part of why users are on the application.[84] Whereas in feminist technocultures like Bumble, the promotion and enactment of values like respect and kindness create expectations that communicate how users should interact with one another.

Counter-disciplining practices can also be extended to users' profiles. Although scholarship has explored how individuals construct, perform, and communicate their identities online, it is possible that conflict can be proactively addressed by communicating expectations on individual's

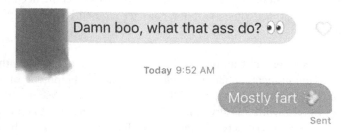

Figure 25.2 A screenshot of one woman's humorous response on Tinder that garnered 48,559 likes on Instagram

profiles.[85] For instance, a user could enact guidelines on their profile for (1) why one should initiate communication (e.g., "swipe right if you're into feminism"); (2) how to communicate (e.g., ground rules for messaging); and (3) what can be communicated (e.g., what topics are and are not off-limits; expectations for continuing conversation on other social media platforms or offline, etc.). Although such responses might be more easily accomplished on some internet sites, the individual response does not engage the gendered cultural context that promotes such offensive displays. While these strategies might feel immediately satisfying, conflict research indicates that such strategies are not effective in the long run and might escalate to stalking and other forms of harassment.

Less individual-oriented strategies involve engaging the larger technocultural spaces in which the users engage. In the instance of Bumble, these relational and structural tactics can include confrontation of problematic relationships and resource allocations through use of and change in organizational policies, standard operating procedures, reward structures, and other means that align with our expanded version of the Bumble Hive's reactive and proactive responses to gendered conflict, as noted earlier. At the heart of these strategies lies the recognition that gendered conflict is complex, complicated, and sometimes normative in situations saturated with masculine values and behaviors. Simple remedies would be inadequate. Instead, multifaceted actions designed to address the varied and deep-seated bases of human action can attend to the tensions and contradictions that individual through organizational change would require.[86]

Other recommendations for engaging in organizational change adopt an action orientation.[87] Action orientations can encourage organizations to balance proactivity and reactivity regarding issues of harassment and gender-based conflict on online platforms. That is, organizations' leaders and leadership teams (i.e., Wolfe Herd, Williamson el-Effendi, and members of the Bumble Hive) can brainstorm potential responses to gender-based conflict and harassment if there are no local or federal policies or guidelines for online harassment (often, there are not), and "look for productive alignment among ideas and regulations, and craft policy and protocol from there."[88] Adopting an action-oriented approach situates the organization as a key partner in mitigating conflict, thus potentially materializing the organization (and its actors) as an entity involved in the conflict management process.[89] Furthermore, this conceptualization of the material organization challenges traditional organizational structures that are structured and institutionalized as masculine. As Harris argued, traditional communicative approaches to conflict tend to ignore the complicity of organizations in promotion and maintenance of conflict; however, when viewed from a gendered, action-orientation perspective, the organization becomes a central actor and part of the broader communicative conflict process.[90] In this sense, it is imperative that organizations take an active part in managing and solving conflict when it occurs.

Concluding thoughts

In this chapter, we extend gender structure theorizing into online gendered technocultures. We highlight scholarship that challenges assumptions of gender neutrality by arguing that online, like offline, spaces are gendered through the promotion of masculine values and norms. Scholarship on conflict in organizational contexts adopts a similar gender-neutral positioning; however, conflict, too, is a gendered process. Given our focus on online technocultures, conflict is an emerging area of inquiry in online spaces. Considering the rise of gender-based threats and violence online that are cultivated vis-à-vis toxic technocultures, we use Bumble as a case of both a technoculture and organization that is actively engaged in addressing gendered conflict online through the promotion and cultivation of gender equitable strategies and processes. Through organizing and creating an alternative technoculture that reverses gender roles and norms for communication,

Bumble's promotion of "Women First" serves as a crucial action-orientation that positioned the organization as a key stakeholder in addressing and reducing online gendered conflict. This positioning aids in promoting gender equity at interpersonal, organizational, and industrywide levels.

Notes

1 John G. Oetzel and Stella Ting-Toomey, *The SAGE Handbook of Conflict Communication: Integrating Theory, Research, and Practice* (Thousand Oaks, CA: Sage, 2013).

2 Patrice M. Buzzanell, "Gender and Feminist Theory" in *Origins and Traditions of Organizational Communication: A Comprehensive Introduction to the Field*, ed. Anne Nicotera (New York: Routledge 2020), 250–269.

3 Tindernightmares, "Unspirational on Instagram: 'VOM'," *Instagram*, accessed May 08, 2019, www.instagram.com/p/BqvApTmHE2U/.

4 Aaron Hess and Carlos Flores, "Simply More than Swiping Left: A Critical Analysis of Toxic Masculine Performances on Tinder Nightmares," *New Media & Society* 20, no. 3 (2016): 1085–1102, https://doi.org/10.1177/1461444816681540; Emma A. Jane, "'Your a Ugly, Whorish, Slut,'" *Feminist Media Studies* 14, no. 4 (2014): 531–46, https://doi.org/10.1080/14680777.2012.741073.

5 Kate Hakala, "This Is Why Men Outnumber Women Two-to-One on Tinder," Mic, May 07, 2019, accessed May 08, 2019, www.mic.com/articles/110774/two-thirds-of-tinder-users-are-men-here-s-why.

6 Emma Alice Jane, "'Back to the Kitchen, Cunt': Speaking the Unspeakable about Online Misogyny," *Continuum* 28, no. 4 (2014): 558–570, https://doi.org/10.1080/10304312.2014.924479.

7 Andrea Braithwaite, "It's about Ethics in Games Journalism? Gamergaters and Geek Masculinity," *Social Media + Society* 2, no. 4 (2016): 1–10, https://doi.org/10.1177%2F2056305116672484; Adrienne Massanari, "#Gamergate and the Fappening: How Reddit's Algorithm, Governance, and Culture Support Toxic Technocultures," *New Media & Society* 19, no. 3 (2015): 329–346, https://doi.org/10.1177%2F1461444815608807.

8 Michael Salter, "Privates in the Online Public: Sex(Ting) and Reputation on Social Media," *New Media & Society* 18, no. 11 (2016): 117, https://doi.org/10.1177%2F1461444815604133.

9 Whitney Phillips, *This Is Why We Can't Have Nice Things: Mapping the Relationship Between Online Trolling and Mainstream Culture* (Cambridge, MA: MIT Press, 2015); Karla Mantilla, *Gendertrolling: How Misogyny Went Viral* (Santa Barbara, CA: Praeger, 2015).

10 Bridget Blodgett and Anastasia Salter, "*Ghostbusters* Is for Boys: Understanding Geek Masculinity's Role in the Alt-Right," *Communication Culture & Critique* 11, no. 1 (2018): 133–146, https://doi.org/10.1093/ccc/tcx003; Caitlin Lawson, "Platform Vulnerabilities: Harassment and Misogynoir in the Digital Attack on Leslie Jones," *Information, Communication & Society* 21, no. 6 (2018): 818–833, https://doi.org/10.1080/1369118X.2018.1437203.

11 Aristea Fotopoulou, "Digital and Networked by Default? Women's Organisations and the Social Imaginary of Networked Feminism," *New Media & Society* 18, no. 6 (2016): 1–17, https://doi.org/10.1177%2F1461444814552264.

12 Jessica Bennett, "With Her Dating App, Women Are in Control," *The New York Times*, March 18, 2017, accessed May 08, 2019, www.nytimes.com/2017/03/18/fashion/bumble-feminist-dating-app-whitney-wolfe.html.

13 Bennett, "With Her Dating App," para 4.

14 Bennett, "With Her Dating App," para 5; Leora Yashari, "Meet the Tinder Co-Founder Trying to Change Online Dating Forever," *Vanity Fair*, August 07, 2015, accessed May 08, 2019, www.vanityfair.com/culture/2015/08/bumble-app-whitney-wolfe.

15 Suzy D'Enbeau. "Gender and Organizing" in *The International Encyclopedia of Organizational Communication*, eds Craig R. Scott and Laurie K. Lewis (Malden, MA: Wiley-Blackwell, 2017a), 1–13, https://doi.org/10.1002/9781118955567.wbieoc085; Jessica A. Pauly and Patrice M. Buzzanell, "Gender and sexuality," in *Movements in Organizational Communication Research: Current Issues and Future Directions*, eds Jamie McDonald and Rahul Mitra (New York: Routledge, 2019), 116–134.

16 Barbara J. Risman, "Gender as a Social Structure," *Gender & Society* 18, no. 4 (2004): 431, www.jstor.org/stable/4149444.

17 Candace West and Don H. Zimmerman, "Doing Gender," *Gender & Society* 1, no. 2 (1987): 125–151, www.jstor.org/stable/189945.

18 Risman, "Gender as a Social Structure," 432.

19 Patrice M. Buzzanell, "Gaining a Voice: Feminist Organizational Communication," *Management Communication Quarterly* 7, no. 4 (1994): 339–383, https://doi.org/10.1177%2F0893318994007004001; West and Zimmerman, "Doing Gender," 128.

20 Joan Acker, "Hierarchies, Jobs, Bodies: A Theory of Gendered Organizations," *Gender & Society* 4, no. 2 (1990): 139–58, https://doi.org/10.1177%2F089124390004002002.

21 Risman, "Gender as a Social Structure," 436; Barbara J. Risman, "From Doing to Undoing: Gender as We Know It," *Gender & Society* 23, no. 1 (2009): 81–84, www.jstor.org/stable/20676752.

22 Silvia Gherardi, "The Gender We Think, the Gender We Do in Our Everyday Organizational Lives," *Human Relations* 47, no. 6 (1994): 591–610, https://doi.org/10.1177%2F001872679404700602; Buzzanell, "Gaining a Voice," 342.

23 Acker, "Hierarchies, Jobs, Bodies," 149–150; Gherardi, "The Gender We Think," 599.

24 Buzzanell, "Gaining a Voice," 340; Gherardi, "The Gender We Think," 599.

25 Gherardi, "The Gender We Think," 605.

26 Patrice. M. Buzzanell, and Kristen Lucas, "Gendered Stories of Career: Unfolding Discourses of Time, Space, and Identity," in *SAGE Handbook of Gender and Communication*, eds Bonnie J. Dow, and Julia T. Wood (Thousand Oaks, CA: Sage, 2006), 161–178.

27 Karen Lee Ashcraft, "The Glass Slipper: 'Incorporating' Occupational Identity in Management Studies," *The Academy of Management Review* 38, no. 1 (2013): 6–31, www.jstor.org/stable/23416300.

28 Latonya J. Trotter, "Making A Career: Reproducing Gender within a Predominately Female Profession," *Gender & Society* 31, no. 4 (2017): 503–525, https://doi.org/10.1177%2F0891243217716115.

29 Maxine Leeds Craig and Rita Liberti, "'Cause that's What Girls Do,'" *Gender & Society* 21, no. 5 (2007): 697, www.jstor.org/stable/27641005.

30 Lauren Alfrey and France Winddance Twine, "Gender-Fluid Geek Girls," *Gender & Society* 31, no. 1 (2016): 28–50, https://doi.org/10.1177%2F0891243216680590; Wendy Faulkner, "Doing Gender in Engineering Workplace Cultures. I. Observations from the Field," *Engineering Studies* 1, no. 1 (2009): 3–18, https://doi.org/10.1080/19378620902721322; Wendy Faulkner, "Doing Gender in Engineering Workplace Cultures. II. Gender In/authenticity and the In/visibility Paradox," *Engineering Studies* 1, no. 3 (2009): 169–189, https://doi.org/10.1080/19378620903225059.

31 Alfrey and Twine, "Gender-Fluid Geek Girls," 42–43.

32 Andrew Pilny and Jeffrey Proulx, "The Influence of Prototypical Communication in Dark Online Organizations: How to Speak Like a Monger," *Journal of Language and Social Psychology* 37, no. 2 (2017): 249–261, https://doi.org/10.1177%2F0261927X17722581.

33 Debra Benita Shaw, *Technoculture: The key concepts* (New York: Berg, 2008), 4.

34 Bryan Pfaffenberger, "'If I Want It, It's OK': Usenet and the (Outer) Limits of Free Speech," *The Information Society* 12, no. 4 (1996): 365–386, https://doi.org/10.1080/019722496129350; Massanari, "#Gamergate and the Fappening," 8.

35 Alice Marwick, "Gender, Sexuality, and Social Media," in *The Social Media Handbook*, eds Jeremy Hunsinger, and Theresa M. Senft (New York: Routledge, 2013), 66; Marwick, Alice E. and Robyn Caplan, "Drinking Male Tears: Language, the Manosphere, and Networked Harassment," *Feminist Media Studies* 18, no. 4, 2018: 543–559. https://doi.org/10.1080/14680777.2018.1450568.

36 Liesbet van Zoonen, "Feminist Internet Studies," *Feminist Media Studies* 1, no. 1 (2001): 67–72, https://doi.org/10.1080/14680770120042864.

37 Salter, "Ghostbusters Is for Boys" 13.

38 Sean M. Eddington, "Organizing and Identification within r/TheRedPill: The Communicative Constitution of Organizational Identity Online." (Unpublished PhD diss., Purdue University, 2019).

39 Brandee Easter, "'Feminist_Brevity_in_Light_of_Masculine_Long-Windedness:' Code, Space, and Online Misogyny," *Feminist Media Studies* 18, no. 4 (2018): 677, https://doi.org/10.1080/14680777.2018.1447335.

40 Caitlin Macleod and Victoria Mcarthur, "The Construction of Gender in Dating Apps: An Interface Analysis of Tinder and Bumble," *Feminist Media Studies* 2018, 1–19, https://doi.org/10.1080/14680777. 2018.1494618.

41 van Zoonen, "Feminist Internet Studies," 68.

42 Laura Lambert, *The Internet: A Historical Encyclopedia* (Santa Barbara, CA: ABC-CLIO, 2005), 210.

43 van Zoonen, "Feminist Internet Studies," 68.

44 Mantilla, *Gendertrolling*, 170; Phillips, *This Is Why*, 158.

45 Debbie Ging, "Alphas, Betas, and Incels: Theorizing the Masculinities of the Manosphere," *Men and Masculinities* 2017, https://doi.org/10.1177%2F1097184X17706401.

46 Mantilla describes Gamergate in the following way: "a campaign of harassment against women in the video game industry that received widespread media attention in August 2014" (46). Similarly, Massanari noted that "The Fappening" was an online event wherein a large cache of private photographs from female celebrities were posted online on sites like 4chan and Reddit (1–2). The Manosphere is a loose network of blogs, forums, and online communities wherein men's rights activists and right-wing extremists converge online (Ging 2017); Sean M. Eddington, "The Communicative Constitution of Hate Organizations Online: A Semantic Network Analysis of 'Make America Great Again,'" *Social Media Society* 4, no. 3 (2018): 1–12, https://doi.org/10.1177%2F2056305118790763; Sean M. Eddington, "Alt-resilience: A Semantic Network Analysis of Identity (Re)construction in an Online Men's Rights Community," *Journal of Applied Communication Research* 48, no. 1 (2020): 114–135, https://doi.org/10.1080/00909882.2019.1706099.

47 Emma A. Jane, "Systemic Misogyny Exposed: Translating Rapeglish from the Manosphere with a Random Rape Threat Generator," *International Journal of Cultural Studies* 21, no. 6 (2018): 661–681, https://doi.org/10.1177%2F1367877917734042.

48 Christine Lagorio-Chafkin, *We Are the Nerds: The Birth and Tumultuous Life of Reddit, the Internet's Culture Laboratory* (New York: Hachette Books, 2018).

49 Phillips, *This Is Why*, 76.

50 Eddington, "Organizing and Identity," 120; Massanari, "#Gamergate and the Fappening," 7.

51 Linda L. Putnam and Marshall Scott Poole, "Conflict and Negotiation" in *Handbook of Organizational Communication: An Interdisciplinary Perspective*, eds Fredric M. Jablin, Linda L. Putnam, Karlene Roberts, and Lyman Porter (Thousand Oaks, CA: Sage, 1987), 552.

52 Linda L. Putnam and Deborah Kolb, "Rethinking Negotiation: Feminist Views of Communication and Exchange," in *Rethinking Organizational and Managerial Conflict from Feminist Perspectives*, ed. Patrice M. Buzzanell (Thousand Oaks, CA: Sage, 2000), 76–104.

53 Deborah Cai and Edward Fink, "Conflict Style Differences between Individualists and Collectivists," *Communication Monographs* 69, no. 1 (2002): 69, https://doi.org/10.1080/03637750216536.

54 Patrice M. Buzzanell and Meina Liu, "It's a 'Give and Take': Maternity Leave as a Conflict Management Process," *Human Relations* 60, no. 3 (2007): 563–595, https://doi.org/10.1177%2F0018726707076688; Min Jiang and Patrice M. Buzzanell, "Qualitative Approaches to Conflict" in *The SAGE Handbook of Qualitative Communication*, eds John G. Oetzel, and Stella Ting-Toomey (Thousand Oaks, CA: Sage, 2013), 67–98.

55 Acker, "Hierarchies, Jobs, Bodies," 147; Buzzanell, "Gaining a Voice," 344.

56 Cai and Fink, "Conflict Style Differences," 82–83; Elisabeth Naima Mikkelsen and Stewart Clegg, "Unpacking the Meaning of Conflict in Organizational Conflict Research," *Negotiation and Conflict Management Research* 11, no. 3 (2018): 185–203, https://doi.org/10.1111/ncmr.12127; James A. Wall Jr. and Ronda Roberts Callister, "Conflict and Its Management: Discover It!," *Journal of Management* 21, no. 3 (1995): 515–558, https://psycnet.apa.org/doi/10.1177/014920639502100306.

57 Cai and Fink, "Conflict Style Differences," 70–71; Marjaana Gunkel, Christopher Schlaegel, and Vas Taras, "Cultural Values, Emotional Intelligence, and Conflict Handling Styles: A Global Study," *Journal of World Business* 51, no. 4 (2016): 568–585, https://doi.org/10.1016/j.jwb.2016.02.001.

58 Meina Liu, "Cultural Differences in Goal-directed Interaction Patterns in Negotiation," *Negotiation and Conflict Management Research* 4, no. 3 (2011): 178–199, https://doi.org/10.1111/j.1750-4716.2011.00079.x; Meina Liu and Steven R. Wilson, "The Effects of Interaction Goals on Negotiation Tactics and Outcomes: A Dyad-Level Analysis Across Two Cultures," *Communication Research* 38, no. 2 (2010): 248–277, https://doi.org/10.1177%2F0093650210362680.

59 Buzzanell and Liu, "It's a 'Give and Take,'" 467.

60 Clare O'Connor, "Billion-Dollar Bumble: How Whitney Wolfe Herd Built America's Fastest-Growing Dating App," *Forbes*, November 17, 2017, para 7, accessed May 08, 2019, www.forbes.com/sites/clareoconnor/2017/11/14/billion-dollar-bumble-how-whitney-wolfe-herd-built-americas-fastest-growing-dating-app/.

61 O'Connor, para. "Billion-Dollar Bumble," 8.

62 Whitney Wolfe Herd, "You Reported Sexual Harassment, Now What? Bumble's Whitney Wolfe Herd Offers Advice," Medium, September 26, 2018, para 2–3, accessed May 08, 2019, https://medium.com/harpers-bazaar/you-reported-sexual-harassment-now-what-bumbles-whitney-wolfe-herd-offers-advice-653036a400a8.

63 Ahiza Garcia, "Bumble Founder Created the App after Experiencing Online Harassment," *CNNMoney*, accessed May 08, 2019, https://money.cnn.com/2017/09/14/technology/business/bumble-whitney-wolfe-fresh-money/index.html.

64 Garcia, para "Bumble Founder," 6.

65 Stephen P. Robbins, "'Conflict Management' and 'Conflict Resolution' Are Not Synonymous Terms." *California Management Review* 21, no. 2 (1978): 67–75, https://doi.org/10.2307%2F41164809.

66 Robbins, "'Conflict Management,'" 69.

67 Putnam and Kolb, "Rethinking Negotiation," 84.

68 Barbara J. Risman, *Where the Millennials Will Take Us* (New York: Oxford University Press, 2018).

69 "Bumble Mission, Vision and Values," *Comparably*, accessed May 08, 2019, www.comparably.com/companies/bumble/mission.

70 "Bumble Mission, Vision, & Values," para 2.

71 "About," Bumble, accessed May 08, 2019, https://bumble.com/en-us/about.

72 Lisa Richwine, "Serena Williams to Take Bumble's Woman-first Message to Super Bowl," Reuters, January 03, 2019, accessed May 08, 2019, www.reuters.com/article/us-people-serenawilliams/serena-williams-to-take-bumbles-woman-first-message-to-super-bowl-idUSKCN1OX0V4.

73 "Privacy," Bumble, accessed May 08, 2019, https://bumble.com/privacy/.

74 Shauna Dillavou, "Women-run Bumble to Silicon Valley: User Safety Is the New Cool," Women's Media Center, accessed May 08, 2019, www.womensmediacenter.com/news-features/women-run-bumble-to-silicon-valley-user-safety-is-the-new-cool.

75 "Terms." Bumble. Accessed January 28, 2019. https://bumble.com/en-us/terms.

76 Candice Jalili, "Bumble Bans Awful Fat-Shaming Male User from the App, So Now We Can Celebrate," Elite Daily, May 08, 2019, para 10, accessed May 08, 2019, www.elitedaily.com/dating/fat-shaming-bumble/1755334.

77 Sara Ashley O'Brien, "Dating App Bumble Slams—and Bans—Misogynist User," *CNNMoney*, accessed May 08, 2019, https://money.cnn.com/2016/06/10/technology/bumble-app-imwithashley/index.html.

78 "An Open Letter to Connor," Bumble, accessed May 08, 2019, http://thebeehive.bumble.com/bumbleblog/an-open-letter-to-connor.

79 Dillavou, "Women-run Bumble," para 7.

80 Emma A. Jane, "Online Misogyny and Feminist Digilantism," *Continuum* 30, no. 3 (2016): 284–297, https://doi.org/10.1080/10304312.2016.1166560.

81 Tindernightmares, "Unspirational on Instagram: 'Makes Sense,'" *Instagram*, accessed May 08, 2019, www.instagram.com/p/Bm8_rZ0BxR8/.

82 Hess and Flores, "Simply More than Swiping Left," 1099.

83 Taina Bucher, "Want to Be on the Top? Algorithmic Power and the Threat of Invisibility on Facebook," *New Media & Society* 14, no. 7 (2012): 1164–1180, https://doi.org/10.1177%2F1461444812440159.

84 Hess and Flores, "Simply More than Swiping Left," 1099; Jane, "Online Misogyny," 287–288.

85 Nicole Ellison, Rebecca Heino, and Jennifer Gibbs, "Managing Impressions Online: Self-Presentation Processes in the Online Dating Environment," *Journal of Computer-Mediated Communication* 11, no. 2 (2006): 415–441, https://doi.org/10.1111/j.1083-6101.2006.00020.x; Jennifer L. Gibbs, Nicole B. Ellison, and Chih-Hui Lai, "First Comes Love, Then Comes Google: An Investigation of Uncertainty Reduction Strategies and Self-Disclosure in Online Dating," *Communication Research* 38, no. 1 (2010): 70–100, https://doi.org/10.1177%2F0093650210377091.

86 Buzzanell, "Gaining a Voice," 367.

87 Suzy D'Enbeau, "Unpacking the Dimensions of Organizational Tension: The Case of Sexual Violence Response and Prevention Among College Students," *Journal of Applied Communication Research* 45, no. 3 (2017): 237–255, https://doi.org/10.1080/00909882.2017.1320568.

88 D'Enbeau, "Unpacking the Dimensions," 251.

89 Karen Lee Ashcraft, Timothy R. Kuhn, and François Cooren, "1 Constitutional Amendments: "Materializing" Organizational Communication," *The Academy of Management Annals* 3, no. 1 (2009): 1–64, www.researchgate.net/publication/247527086_1_Constitutional_Amendments_Materializing_Organizational_Communication.

90 Kate Lockwood Harris, "Re-situating Organizational Knowledge: Violence, Intersectionality and the Privilege of Partial Perspective," *Human Relations* 70, no. 3 (2017): 263–85, https://doi.org/10.1177%2F0018726716654745.

Bibliography

"About," Bumble, accessed May 08, 2019, https://bumble.com/en-us/about.

Acker, Joan. "Hierarchies, Jobs, Bodies: A Theory of Gendered Organizations." *Gender & Society* 4, no. 2 (1990): 139–158, https://doi.org/10.1177%2F089124390004002002.

Alfrey, Lauren, and France Winddance Twine. "Gender-Fluid Geek Girls: Negotiating Inequality Regimes in the Tech Industry." *Gender & Society* 31, no. 1 (2017): 28–50, https://doi.org/10.1177%2F0891243216680590.

"An Open Letter to Connor." *Bumble*. Accessed March 11, 2020. https://bumble.com/the-buzz/an-open-letter-to-connor.

Ashcraft, Karen Lee. "The Glass Slipper: 'Incorporating' Occupational Identity in Management Studies." *Academy of Management Review* 38, no. 1 (2013): 6–31, www.jstor.org/stable/23416300.

Ashcraft, Karen Lee, Timothy R. Kuhn, and François Cooren. "1 Constitutional Amendments: 'Materializing' Organizational Communication." *Academy of Management Annals* 3, no. 1 (2009): 1–64, www.researchgate.net/publication/247527086_1_Constitutional_Amendments_Materializing_Organizational_Communication.

Bennett, Jessica. "With Her Dating App, Women Are in Control." *The New York Times*, March 18, 2017. www.nytimes.com/2017/03/18/fashion/bumble-feminist-dating-app-whitney-wolfe.html.

Blodgett, Bridget, and Anastasia Salter. "Ghostbusters Is for Boys: Understanding Geek Masculinity's Role in the Alt-Right." *Communication Culture & Critique* 11, no. 1 (2018): 133–146, https://doi.org/10.1093/ccc/tcx003.

Braithwaite, Andrea. "It's About Ethics in Games Journalism? Gamergaters and Geek Masculinity." *Social Media+ Society* 2, no. 4 (2016): 1–10. https://doi.org/10.1177%2F2056305116672484.

Bucher, Taina. "Want to Be on the Top? Algorithmic Power and the Threat of Invisibility on Facebook." *New Media & Society* 14, no. 7 (2012): 1164–1180, https://doi.org/10.1177%2F1461444812440159.

"Bumble Mission, Vision & Values." *Comparably*. Accessed March 11, 2020. www.comparably.com/companies/bumble/mission.

Buzzanell, Patrice M. "Gaining a Voice: Feminist Organizational Communication Theorizing." *Management Communication Quarterly* 7, no. 4 (1994): 339–383, https://doi.org/10.1177%2F0893318994007004001.

Buzzanell, Patrice M. "Gender and Feminist Theory." In *Origins and Traditions of Organizational Communication: A Comprehensive Introduction to the Field*, edited by Anne Nicotera, 250–269, New York: Routledge, 2020.

Buzzanell, Patrice, and Meina Liu. "It's a 'Give and Take': Maternity Leave as a Conflict Management Process." *Human Relations* 60, no. 3 (2007): 463–495, https://doi.org/10.1177%2F0018726707076688.

Buzzanell, Patrice M., and Kristen Lucas, "Gendered Stories of Career: Unfolding Discourses of Time, Space, and Identity," in *The SAGE Handbook of Gender and Communication*, edited by Bonnie J. Dow and Julia T. Wood, 161–178, Thousand Oaks, CA: Sage, 2006.

Cai, Deborah, and Edward Fink. "Conflict Style Differences between Individualists and Collectivists." *Communication Monographs* 69, no. 1 (2002): 67–87, https://doi.org/10.1080/03637750216536.

Craig, Maxine Leeds, and Rita Liberti. "'Cause that's What Girls Do'; The Making of a Feminized Gym." *Gender & Society* 21, no. 5 (2007): 676–699, www.jstor.org/stable/27641005.

D'Enbeau, Suzy. "Gender and Organizing." In *The International Encyclopedia of Organizational Communication*, edited by Craig R. Scott and Laurie K. Lewis, 1–13. Malden, MA: Wiley-Blackwell, 2017.

D'Enbeau, Suzy. "Unpacking the Dimensions of Organizational Tension: The Case of Sexual Violence Response and Prevention among College Students." *Journal of Applied Communication Research* 45, no. 3 (2017): 237–255, https://doi.org/10.1080/00909882.2017.1320568.

Dillavou, Shauna, "Women-Run Bumble to Silicon Valley: User Safety Is the New Cool," Women's Media Center, accessed May 08, 2019, www.womensmediacenter.com/news-features/women-run-bumble-to-silicon-valley-user-safety-is-the-new-cool.

Easter, Brandee. "'Feminist_Brevity_in_Light_of_Masculine_Long-Windedness': Code, Space, and Online Misogyny." *Feminist Media Studies* 18, no. 4 (2018): 675–685, https://doi.org/10.1080/14680777.2018.1447335.

Eddington, Sean M. "The Communicative Constitution of Hate Organizations Online: A Semantic Network Analysis of 'Make America Great Again.'" *Social Media+ Society* 4, no. 3 (2018): 1–12, https://doi.org/10.1177%2F2056305118790763.

Eddington, Sean M. "Organizing and Identification within /r/TheRedPill: The Communicative Constitution of Organizational Identity Online." PhD diss., Purdue University Graduate School, 2019.

Eddington, Sean M. "Alt-Resilience: A Semantic Network Analysis of Identity (Re)Construction in an Online Men's Rights Community." *Journal of Applied Communication Research* 48, no. 1 (2020): 114–135, https://doi.org/10.1080/00909882.2019.1706099.

Ellison, Nicole, Rebecca Heino, and Jennifer Gibbs. "Managing Impressions Online: Self-Presentation Processes in the Online Dating Environment." *Journal of Computer-Mediated Communication* 11, no. 2 (2006): 415–441, https://doi.org/10.1111/j.1083-6101.2006.00020.x.

Faulkner, Wendy. "Doing Gender in Engineering Workplace Cultures. I. Observations from the Field." *Engineering Studies* 1, no. 1 (2009): 3–18, https://doi.org/10.1080/19378620902721322.

Faulkner, Wendy. "Doing Gender in Engineering Workplace Cultures. II. Gender In/Authenticity and the In/Visibility Paradox." *Engineering Studies* 1, no. 3 (2009): 169–189, https://doi.org/10.1080/19378620903225059.

Fotopoulou, Aristea. "Digital and Networked by Default? Women's Organisations and the Social Imaginary of Networked Feminism." *New Media & Society* 18, no. 6 (2016): 989–1005, https://doi.org/10.1177%2F1461444814552264.

Garcia, Ahiza. "Bumble Founder Created the App after Experiencing Online Harassment." CNNMoney. Cable News Network. Accessed March 11, 2020. https://money.cnn.com/2017/09/14/technology/business/bumble-whitney-wolfe-fresh-money/index.html.

Gherardi, Silvia. "The Gender We Think, the Gender We Do in Our Everyday Organizational Lives." *Human Relations* 47, no. 6 (1994): 591–610, https://doi.org/10.1177%2F001872679404700602.

Gibbs, Jennifer L., Nicole B. Ellison, and Chih-Hui Lai. "First Comes Love, then Comes Google: An Investigation of Uncertainty Reduction Strategies and Self-Disclosure in Online Dating." *Communication Research* 38, no. 1 (2011): 70–100, https://doi.org/10.1177%2F0093650210377091.

Ging, Debbie. "Alphas, Betas, and Incels: Theorizing the Masculinities of the Manosphere." *Men and Masculinities* 22, no. 4 (2019): 638–657, https://doi.org/10.1177%2F1097184X17706401.

Gunkel, Marjaana, Christopher Schlaegel, and Vas Taras. "Cultural Values, Emotional Intelligence, and Conflict Handling Styles: A Global Study." *Journal of World Business* 51, no. 4 (2016): 568–585, https://doi.org/10.1016/j.jwb.2016.02.001.

Hakala, Kate. "This Is Why Men Outnumber Women Two-to-One on Tinder." Mic. Mic, February 18, 2015. www.mic.com/articles/110774/two-thirds-of-tinder-users-are-men-here-s-why.

Harris, Kate Lockwood. "Re-situating Organizational Knowledge: Violence, Intersectionality and the Privilege of Partial Perspective." *Human Relations* 70, no. 3 (2017): 263–285, https://doi.org/10.1177%2F0018726716654745.

Hess, Aaron, and Carlos Flores. "Simply More than Swiping Left: A Critical Analysis of Toxic Masculine Performances on Tinder Nightmares." *New Media and Society* 20, no. 3 (2018): 1085–1102. https://doi.org/10.1177/1461444816681540.

Jalili, Candice. "Bumble Bans Awful Fat-Shaming Male User from the App, So Now We can Celebrate." Elite Daily. Elite Daily, January 17, 2017. www.elitedaily.com/dating/fat-shaming-bumble/1755334.

Jane, Emma A. "'Back to the Kitchen, Cunt': Speaking the Unspeakable about Online Misogyny," *Continuum* 28, no. 4 (2014): 558–570, doi:10.1080/10304312.2014.924479.

Jane, Emma A. "'Your a Ugly, Whorish, Slut': Understanding E-Bile." *Feminist Media Studies* 14, no. 4 (2014): 531–546. https://doi.org/10.1080/14680777.2012.741073.

Jane, Emma A. "Online Misogyny and Feminist Digilantism." *Continuum* 30, no. 3 (2016): 284–297. https://doi.org/10.1080/10304312.2016.1166560.

Jane, Emma A. "Systemic Misogyny Exposed: Translating Rapeglish from the Manosphere with a Random Rape Threat Generator." *International Journal of Cultural Studies* 21, no. 6 (2018): 661–680. https://doi.org/10.1177%2F1367877917734042.

Jiang, Min, and Patrice M. Buzzanell, "Qualitative Approaches to Conflict," in *The SAGE Handbook of Qualitative Communication*, edited by John G. Oetzel and Stella Ting-Toomey, 67–98, Thousand Oaks, CA: Sage, 2013.

Lagorio-Chafkin, Christine. *We Are the Nerds: The Birth and Tumultuous Life of Reddit, the Internet's Culture Laboratory*. Hachette, UK: Hachette Books, 2018.

Lambert, Laura. *The Internet: A Historical Encyclopedia*. Santa Barbara, CA: ABC-CLIO, 2005.

Lawson, Caitlin E. "Platform Vulnerabilities: Harassment and Misogynoir in the Digital Attack on Leslie Jones." *Information, Communication & Society* 21, no. 6 (2018): 818–833, https://doi.org/10.1080/1369118X.2018.1437203.

Liu, Meina. "Cultural Differences in Goal-Directed Interaction Patterns in Negotiation." *Negotiation and Conflict Management Research* 4, no. 3 (2011): 178–199, https://doi.org/10.1111/j.1750-4716.2011.00079.x.

Liu, Meina, and Steven R. Wilson. "The Effects of Interaction Goals on Negotiation Tactics and Outcomes: A Dyad-Level Analysis Across Two Cultures." *Communication Research* 38, no. 2 (2011): 248–277, https://doi.org/10.1177%2F0093650210362680.

MacLeod, Caitlin, and Victoria McArthur. "The Construction of Gender in Dating Apps: An Interface Analysis of Tinder and Bumble." *Feminist Media Studies* 19, no. 6 (2019): 822–840, https://doi.org/10.1080/14680777.2018.1494618.

Mantilla, Karla. *Gendertrolling: How Misogyny Went Viral: How Misogyny Went Viral.* Santa Barbara, CA: Praeger. 2015.

Marwick, Alice. "Gender, Sexuality, and Social Media." In *The Social Media Handbook*, edited by Jeremy Hunsinger and Theresa M. Senft, 67–83. New York: Routledge, 2013.

Marwick, Alice E., and Robyn Caplan. "Drinking Male Tears: Language, the Manosphere, and Networked Harassment." *Feminist Media Studies* 18, no. 4 (2018): 543–559, https://doi.org/10.1080/14680777.2018.1450568.

Massanari, Adrienne. "#Gamergate and the Fappening: How Reddit's Algorithm, Governance, and Culture Support Toxic Technocultures." *New Media & Society* 19, no. 3 (2017): 329–346, https://doi.org/10.1177%2F1461444815608807.

Mikkelsen, Elisabeth Naima, and Stewart Clegg. "Unpacking the Meaning of Conflict in Organizational Conflict Research." *Negotiation and Conflict Management Research* 11, no. 3 (2018): 185–203, https://doi.org/10.1111/ncmr.12127.

O'Connor, Clare. "Billion-Dollar Bumble: How Whitney Wolfe Herd Built America's Fastest-Growing Dating App." *Forbes Magazine*, November 17, 2017. www.forbes.com/sites/clareoconnor/2017/11/14/billion-dollar-bumble-how-whitney-wolfe-herd-built-americas-fastest-growing-dating-app/#22f6410248b3.

O'Brien, Sara Ashley. "Dating App Bumble Slams – and Bans – Misogynist User." CNNMoney. Cable News Network. Accessed March 11, 2020. https://money.cnn.com/2016/06/10/technology/bumble-app-imwithashley/index.html.

Oetzel, John G. and Ting-Toomey, Stella, eds. *The SAGE Handbook of Conflict Communication: Integrating Theory, Research, and Practice.* Thousand Oaks CA: Sage, 2013.

Pauly, Jessica, A., and Patrice M. Buzzanell. "Gender and Sexuality," in *Movements in Organizational Communication Research: Current Issues and Future Directions*, edited by Jamie McDonald and Rahul Mitra, 116–134, New York: Routledge, 2019.

Phillips, Whitney. *This Is Why We can't Have Nice Things: Mapping the Relationship between Online Trolling and Mainstream Culture.* Cambridge, MA: MIT Press, 2015.

Pfaffenberger, Bryan. "'If I Want It, It's OK': Usenet and the (Outer) Limits of Free Speech." *The Information Society* 12, no. 4 (1996): 365–386, https://doi.org/10.1080/019722496129350.

Pilny, Andrew, and Jeffrey Proulx. "The Influence of Prototypical Communication in Dark Online Organizations: How to Speak Like a Monger." *Journal of Language and Social Psychology* 37, no. 2 (2018): 249–261, https://doi.org/10.1177%2F0261927X17722581.

"Privacy," *Bumble*, accessed May 08, 2019, https://bumble.com/privacy/.

Putnam, Linda, L., and Deborah Kolb, "Rethinking Negotiation: Feminist Views of Communication and Exchange," in *Rethinking Organizational and Managerial Conflict from Feminist Perspectives*, edited by Patrice M. Buzzanell, 76–104, Thousand Oaks, CA: Sage, 2000.

Putnam, Linda, L., and Marshall Scott Poole, "Conflict and Negotiation" in *Handbook of Organizational Communication: An Interdisciplinary Perspective*, edited by Frederic M. Jablin, Linda L. Putnam, Karlene Roberts, and Lyman Porter, 549–599, Thousand Oaks, CA: Sage, 1987.

Richwine, Lisa, "Serena Williams to Take Bumble's Woman-first Message to Super Bowl," Reuters, January 03, 2019, accessed May 08, 2019, www.reuters.com/article/us-people-serenawilliams/serena-williams-to-take-bumbles-woman-first-message-to-super-bowl-idUSKCN1OX0V4.

Risman, Barbara J. "Gender as a Social Structure: Theory Wrestling with Activism." *Gender & Society* 18, no. 4 (2004): 429–450, www.jstor.org/stable/4149444.

Risman, Barbara J. "From Doing to Undoing: Gender as We Know It." *Gender & Society* 23, no. 1 (2009): 81–84, www.jstor.org/stable/20676752.

Risman, Barbara J. *Where the Millennials Will Take Us: A New Generation Wrestles with the Gender Structure.* Oxford: Oxford University Press, 2018.

Robbins, Stephen P. "'Conflict Management' and 'Conflict Resolution' Are Not Synonymous Terms." *California Management Review* 21, no. 2 (1978): 67–75, https://doi.org/10.2307%2F41164809.

Salter, Michael. "Privates in the Online Public: Sex (Ting) and Reputation on Social Media." *New Media & Society* 18, no. 11 (2016): 2723–2739, https://doi.org/10.1177%2F1461444815604133.

Shaw, Debra Benita. *Technoculture: The Key Concepts*. New York: Berg, 2008.

"Terms." *Bumble*. Accessed January 28, 2019. https://bumble.com/en-us/terms.

Tindernightmares. "Unspirational on Instagram: 'Makes Sense.'" *Instagram*. Accessed March 11, 2020. www.instagram.com/p/Bm8_rZ0BxR8/.

Tindernightmares. "Unspirational on Instagram: 'VOM.'" *Instagram*. Accessed March 11, 2020. www.instagram.com/p/BqvApTmHE2U/.

Trotter, Latonya J. "Making a Career: Reproducing Gender within a Predominately Female Profession." *Gender & Society* 31, no. 4 (2017): 503–525, https://doi.org/10.1177%2F0891243217716115.

Wall Jr, James A., and Ronda Roberts Callister. "Conflict and Its Management." *Journal of Management* 21, no. 3 (1995): 515–558, https://psycnet.apa.org/doi/10.1177/014920639502100306.

West, Candace, and Don H. Zimmerman. "Doing Gender." *Gender & Society* 1, no. 2 (1987): 125–151, www.jstor.org/stable/189945.

Wolfe Herd, Whitney. "You Reported Sexual Harassment, Now What? Bumble's Whitney Wolfe Herd Offers Advice." Medium. Harper's Bazaar, September 26, 2018. https://medium.com/harpers-bazaar/you-reported-sexual-harassment-now-what-bumbles-whitney-wolfe-herd-offers-advice-653036a400a8.

Yashari, Leora. "Meet the Tinder Co-Founder Trying to Change Online Dating Forever." Vanity Fair. Accessed August 7, 2015. www.vanityfair.com/culture/2015/08/bumble-app-whitney-wolfe.

Zoonen, Liesbet van. "Feminist Internet Studies." *Feminist Media Studies* 1, no. 1 (2001): 67–72, https://doi.org/10.1080/14680770120042864.

26

STRAIGHT (WHITE) WOMEN WRITING ABOUT MEN BONKING?

Complicating our understanding of gender and sexuality in fandom

Mel Stanfill

The field of fan studies is now more than 30 years old. Over its history, it has amassed a body of knowledge with particularly rich analysis of gender and sexuality. One of the key findings at the beginning of fan studies was both the existence of slash fan fiction that romantically pairs men who are not gay in the source texts, named for the practice of separating character names with the slash punctuation mark when labeling stories, and the fact that these stories were written by heterosexual women. This was seen as a way both for people ill-served by media to rework media texts to meet their needs and for those discouraged from sexual agency to claim it and explore desire, and therefore interpreted as progressive. As this chapter will show, each component of this narrative has been complicated by subsequent research. Fans are not all women, which has been explored by studies of men and masculinity in fandom—a term referring either to the collective body of fans or to the condition of being a fan. Fans are not necessarily heterosexual, either because they hold an identity as lesbian, bisexual, or queer or because writing fan fiction is a practice of eroticism between women. Fan fiction is also not exclusively written about pairs of men, as research into femslash, or stories featuring same-sex relationships between women, has shown. Further, recent research has complicated the ways that the formulation "straight women" in early fan studies had an implicit modifier of "white," with scholars both interrogating unexamined whiteness in fandom and pushing to explore the experiences of fans of color. Finally, research is beginning to unravel how fandom is not necessarily progressive, including how slash fiction and its writers can be misogynistic and homophobic; the role of homonormativity in producing desires for domesticity, childbearing, and property ownership as the goal of fan fiction and media representation alike; and a trend toward seeing industry approval as the purpose of fandom.

Early understandings of gender and sexuality in fandom

When fans and fandom first became subject to academic inquiry, the framework for understanding them was what Jonathan Gray, Cornel Sandvoss, and C. Lee Harrington call "Fandom is Beautiful."[1] In this research, focused primarily on U.S.-based communities of fans of television shows like *Star Trek*, fan practices were seen as "the guerilla-style tactics of those with lesser resources" in the struggle over mass media meaning, "a communal effort to form

446

interpretive communities that in their subcultural cohesion evaded the preferred and intended meanings."[2] Specifically, fandom was understood as resistant to power structures. Often, foundational research spoke of fandom as "progressive,"[3] "liberal" or "left,"[4] or even "utopian."[5] Of course, this drive to establish fandom as progressive and, particularly, resistant is understandable, rooted in these scholars' own resistance to pathologizing discourses of fans as passive, uncritical consumers mesmerized by media that were in wide circulation by the early 1990s.[6] Given pathologized activities that seemed to pay too much attention to media, early research on fans sought to "redeem them as creative, thoughtful, and productive."[7]

What the field lauded was specifically a gendered kind of progressiveness and transgression. Henry Jenkins contended that while men were certainly fans, "male media fans […] have learned to play according to the interpretive conventions" of a community comprised primarily of women.[8] While there is no definitive explanation, the ways television has historically been seen as feminized may account for why media fandom has historically been primarily women, as opposed to science fiction book fandom being dominated by men. Fan fiction—writing new stories based on media texts—was a distinctive practice of media fandom seeking to "reclaim female experiences from the margins of male-centered texts, offering readers the kinds of heroic women still rarely available elsewhere in popular culture; their stories address feminist concerns about female autonomy, authority, and ambition."[9] Camille Bacon-Smith and Constance Penley similarly highlighted that fandom was women's space to tell their own stories: the former was excited to find in fandom "a conceptual space where women can come together and create—to investigate new forms for their art and for their living outside the restrictive boundaries men have placed on women's public behavior!"[10] while the latter described fans as "fiercely proud of having created a comfortable yet stimulating social space in which women can manipulate the products of mass-produced culture to stage a popular debate around issues of technology fantasy, and everyday life."[11] Bacon-Smith emphasized media fandom's "freedom of expression," while Jenkins called it specifically "a progressive reformulation" of cultural tropes.[12] Penley and Bacon-Smith both emphasized the subversiveness of fan writing;[13] Bacon-Smith also went further to identify fans as "rebels in the cause of a women's art/communication system."[14]

The practice of women taking storytelling into their own hands, reworking mass media in fan fiction, challenged gender norms. Moreover, the content of these stories itself challenged gender norms. Jenkins contended that fans worked "toward a gradual redefinition of existing gender roles."[15] This included not only women having more options than previously but also what Penley calls "retooling masculinity itself" and Jenkins more strongly characterizes as "an explicit critique of traditional masculinity."[16] One important reworking of masculinity in fan fiction was that it placed "emotional responsibility on men for sustaining relationships while men in reality frequently dodge such responsibility."[17] In this way, it is clear why later authors characterized this as writing "(straight) men the way (straight) women want them or want them to behave."[18]

In particular, these were often stories about men in romantic and sexual relationships with other men, known as slash fiction. Jenkins noted that these same-sex relationships were "egalitarian," while Penley noted their "democratic equality."[19] This then served as "an alternative to repressive and hierarchical male sexuality."[20] Slash stories tended to center on men who in source texts had relationships only with women; Jenkins specifically identified this "transcendence of rigidly defined categories of gender and sexual identity" as one of "many progressive elements" of slash.[21] Another key aspect of fandom and sexuality in this early work is identifying fan fiction as a space of sexual desire and pleasure. Penley noted that fans took "pride in having created a unique, hybridized genre that ingeniously blends romance, pornography, and utopian science fiction," calling it a "guerrilla erotics."[22] Much of this early work is focused on anglophone fandom, but Sharon Kinsella similarly identified amateur manga (Japanese graphic novel) writers as young women drawing homoerotic stories about men.[23] Importantly, this was specifically women's desire for and pleasure in men—in the language of the title of the 1998 academic paper that my own title here riffs on, "normal female interest in men bonking."[24] While such women may of course have various sexual identities, typically the interpretation has been (rightly or wrongly) that they are heterosexual,[25] as in Penley's report of the quip "slash fandom—as close as two straight women can get!"[26] In all of these ways, then, the baseline understanding of fandom generated by the first wave of research is that it is progressive, specifically because women transgressed and reworked gender and sexuality norms.

Men and masculinity in fandom

However, as tends to happen as fields develop, subsequent fan studies work has nuanced these early understandings. Each component of the above characterization of fandom was complicated by later research. First, there has been research on men and masculinity in fandom. Men tend to engage in different practices than women, notably fan filmmaking[27] and game modding, in which nonprofessionals add to or modify video games,[28] as opposed to the fan fiction discussed previously, and as Julie Levin Russo notes, the kind of work "disproportionately produced by men" falls into "the 'original' genres that enjoy legal and corporate sanction."[29] That is, men's fan work is more often approved by the same media industries that are wary of fan fiction and especially slash. One important factor is that men's fan works tend to be affirmational rather than transformational[30]—that is, to stick closely to the text rather than change it substantially.[31] Men are also, not coincidentally, the fans who tend to be able to commercialize their works and leverage fandom into a career.[32]

There has also been research exploring fandoms where, by contrast to women's usual numerical dominance, men are most prevalent. Anne Gilbert's work on Bronies, men who are fans of television show *My Little Pony: Friendship Is Magic*, has shown that they both "profess to deliberately rewrite codes of masculine behavior that valorize competition and aggression" and justify their affection for "*My Little Pony* by likening it to artifacts from traditionally masculine geek culture."[33] This upholding of hegemonic masculinity has also been dramatically displayed in fan backlash movements protesting the inclusion of white women, people of color, and LGBTQ+ people in fannish media, such as Gamergate, the Sad and Rabid Puppies campaigns, and the 2016 harassment of Black woman actor Leslie Jones.[34] Bethan Jones argues that while there might be a tendency to separate the two, "the fact that Gamergaters seem to be protesting progressive moves toward inclusivity does not make them any less part of a fan community, or their behavior any less antifandom against progressiveness."[35] Thus, this toxic geek masculinity is within the purview of fan studies.[36] These campaigns are related to Katie Wilson's argument that "in 2014 and 2015, fan communities both online and offline were infiltrated by the growing grassroots men's rights and neomasculine activist groups"; this infiltration was successful because the men's rights movement "operates on the premise that men have be[en] oppressed by the feminist movement," thereby providing feminism as a scapegoat for "the historical bullying of male fans, both in real-life situations and in media representations."[37] In these ways, we have gained greater understanding of gender in fandom through attention to men and masculinity.

Sexual diversity in fandom

Similarly, further research has complicated the assumption that fans are heterosexual, whether because they identify as lesbian, bisexual, or queer or because writing fan fiction is an erotic practice between women. First, many fans hold identities other than heterosexual. In Katherine

E. Morrissey's 2008 survey, 23% of her respondents identified themselves as bisexual, 4% as homosexual, 3% as asexual, and 2% as none of the above.[38] Non-heterosexual people in fandom have likely both numerically increased over time and become increasingly visible over time as social norms change and technology enables marginalized sexualities to build community across distance. Moreover, as Alexis Lothian, Kristina Busse, and Robin Anne Reid argue, "participation in electronic social networks can induct us into new and unusual narratives of identity and sexuality, calling into question familiar identifications and assumptions."[39] Sexuality, that is, is more complex in online fandom than it might be offline, and perhaps more complex than even in fans' own offline lives. Eden Lackner, Barbara Lynn Lucas, and Robin Anne Reid put a finer point on it, contending that

> even the "straight" women are doing something that can arguably be seen as pretty "queer" (producing writing designed to give sexual pleasure to other women, whether the texts are called erotica, pornography, slash, or smut, whether the texts are defined as het or queer or bi).[40]

This compounds the non-normativity of women desiring as they give pleasure to one another. Through this line of research, we now understand sexuality in fandom more expansively.

Not just men bonking: understanding femslash

Related to this, though early work focused on slash between men, subsequent research expanded to include femslash. Femslash is still not well-studied—though, as these stories represent only approximately 4% of fiction, this phenomenon may not be *under*studied.[41] Though it may seem the same as slash between men, femslash is fundamentally different because of the usually close correspondence of writer and/or audience identity with the relationship explored. Ling Yang and Hongwei Bao describe what is called *Girls' Love* fan fiction in China as enabling various forms of intimacy and nonnormative sexuality between women.[42] Russo argues that

> it is the fusion that femslash presumes between fans and characters in terms of sexual and gender identities—its primary difference from the way male slash has been defined—that affords it this powerful platform for literal campaigns of resistance to heteronormative structures.[43]

That is, because femslash writers tend to represent their own identities, they are more often overtly political. Indeed, Russo is deeply critical of any argument that "positions not necessarily queer-identified women writing about not necessarily queer-identified men as *more* queer than people who are queer."[44] However, such resistance is deeply contextual. It is, for example, more possible in Anglo-American contexts than the Hong Kong context Cheuk Yin Li describes, which is freighted with "residual Chinese ethics, which value social harmony and family unity; the British colonial legacy; and the growing influence of rightist Christian influences and non-governmental organizations."[45]

The "fusion" between fiction and identity also makes femslash more personally powerful. Rosalind Hanmer describes how, with historical fantasy television show *Xena: Warrior Princess* (1995-2001), fans' coming out narratives reflected the show's role in not only their "lesbian subjectivity" but even their recognition of their own same-sex desire.[46] *Xena* was the first major femslash fandom, and that distinctiveness invited investigation.[47] Sara Gwenllian Jones noted that "lesbian and bisexual women dominate *Xena: Warrior Princess* slash, but have much less presence in most other slash fandoms."[48] Russo attributed the rise of femslash in the mid-1990s in part to more shows with "two or more strong and complex central female characters";[49] elsewhere I have added that media that tend to be femslash-heavy are those where women's relationships to each other are central, in line with Rich's theorization of the lesbian continuum.[50] However, Russo also contends that technological change has influenced the growth of femslash; she highlights both "the internet's role in expanding the accessibility and diversity of fan fiction" and "how femslash fanworks are one instantiation of an emerging technological configuration that makes it increasingly difficult to contain audience desire and use within economically and normatively dominant bounds."[51] In fact, Russo identifies the post-*Xena* period as "laying the groundwork for femslash to become a unifying investment that transcends any particular show or couple," as has subsequently become the case.[52] In these ways, our understanding of fandom has been enriched through studying femslash.

Decentering whiteness and making space for fans of color

Some of the most important recent research has emphasized how the formulation of "straight women" in early fan studies had an implicit modifier of "white." Contemporary scholarship both interrogates unexamined whiteness in fandom and explores the experiences of fans of color. A 2009 roundtable session gathered fans of color together to discuss their experiences, noting that "debates about racism and other forms of global structural inequality […] have increasingly shaped the public online landscapes of some sectors of science fiction and media fandom in recent years."[53] One particular flashpoint became known as RaceFail '09, which

> refers to a series of blog posts written by SF/F [science fiction/fantasy] fans in response to SF/F author Elizabeth Bear's (2008) advice about "writing the other" in fiction. These posts pointed out both Bear's apparent hypocrisy, critiquing her record of portraying people of color, and encompassed the failings of the SF/F genre as a whole on the issue of race.[54]

Such debates help illuminate how, as Rukmini Pande points out, "media fandom spaces, theorized as inclusive and liberating, are not immune to hierarchies structured by privilege accruing to income, class, racial, ethnic, and cultural identity," which has recently become better understood.[55]

This body of scholarship calls attention to Predominantly White Fandoms[56]—on the model of "Predominantly White Institutions" in higher education, which does not allow such colleges to go unmarked. Thinking in terms of PWFs disrupts the neutrality or naturalness of the whiteness of such fandoms—fan identity is often assumed to take precedence, effectively asking fans of color to be fans first and people of color second.[57] It is not neutral that discussing gender and/ or sexuality in fan spaces is seen as vital work, but discussing race is "drama" or inappropriate for the space.[58] Similarly, as one fan in the 2009 discussion noted,

> fandom, as encompassing women, immediately understands the power of the phrase *character rape*, while it does not so swiftly regard the words *colonial* or *Oriental* with equal visceral horror, even though the effects of rape and colonialism have been equally scarring on women of color.[59]

Fans of color also experience tone policing for discussing racism.[60] This is a tendency to center white feelings—feelings about being called to account for their racism or the perceived injustice of being falsely accused of racism—and not grappling with the harms of racism itself.[61] Hurt white feelings are frequently held up as why white fans refuse to write about characters of color.[62]

An important recent body of research investigates how fans of color claim their right to exist in fan spaces. Jessica Seymour notes that "fans of media that fail to reflect their world beyond the harmful stereotypes traditionally perpetuated in mainstream media have begun combatting this lack of diversity by producing their own works" that include more people of color.[63] Pande argues that contemporary platforms like Tumblr and Twitter "offer greater visibility, both in terms of a willingness of individual fans to 'claim' a non-white identity within a fannish space and, in doing so, find others who share or understand their experiences."[64] A growing body of literature centers the experiences of fans of color. Rebecca Wanzo argues that some of the foundational assumptions of fan studies, like that fans are outsiders, are particular to white fandom, whereas "African American fans make hypervisible the ways in which fandom is expected or demanded of some socially disadvantaged groups as a show of economic force and ideological combat."[65] Kristen J. Warner describes real person shipping[66]—desiring a romance between the lead actors in *Scandal* and not just their characters—as resulting from Black women's desire for the possibility, in the face of persistent devaluation, that women like them are desirable to powerful men in real life as they are in the show.[67] In all of these ways, we see that understanding gender and sexuality in fandom is incomplete without attention to race.

On *not* transgressing and reworking norms

Finally, research is beginning to unravel the ways fandom is not necessarily progressive, including (in addition to the racism discussed previously) how slash fiction and its writers can be misogynistic and homophobic; the role of homonormativity in producing fan desires for domesticity, childbearing, and property ownership as the goal of fan fiction; and notions of industry approval as the purpose of fandom. First, despite the contention that women in fandom transgress gender norms or roles, slash fiction often upholds gender hierarchies through being implicitly or explicitly misogynistic. Christine Scodari calls attention to slash's "misogynistic nuances and roots in heterosexual desire."[68] Women in such stories often exist purely to further the men's relationship or cement them as bisexually available for women's desire despite their relationship with each other.[69] As Scodari notes, "if canonical homosociality between men is a catalyst for traditional slash, it is not threatened merely by male/female romance but by a female character's centrality in the narrative."[70] Thus, frequently women in slash have to be removed as obstacles. While this sometimes-violent disappearing of women is often a continuation of a misogynistic source text,[71] it nevertheless upholds norms.

Additionally, slash between men is surprisingly often vaguely homophobic. The idea that men in slash aren't gay,[72] on one hand, much like including women in the story to demonstrate the character's attraction to women, insists that they are still available to the slasher's gaze as an object. On the other hand, it is shot through with, and inextricable from, the broader homophobic insistence that it is bad to be gay. However, just making the characters gay men does not necessarily resolve all concerns. Such stories could be described as

> having a motivation comparable to that associated with male-targeted pornography featuring lesbian encounters—namely, removal of the competition and the desire to frame both attractive characters of the opposite sex as performing for and serving only the individual indulging in the fantasy.[73]

451

It also raises the question of what it might mean, ethically, for heterosexual women to use gay men as playthings for their own sexual satisfaction. In these ways, slash fiction often reproduces heterosexist devaluation of LGBTQ+ lives.

Third, at the level of content, fan fiction often upholds the normative valuation of marriage, family, and property. This is what Lisa Duggan calls homonormativity, "a politics that does not contest dominant heteronormative assumptions and institutions, but upholds and sustains them, while promising the possibility of a demobilized gay constituency and a privatized, depoliticized gay culture anchored in domesticity and consumption."[74] For example Catherine Tosenberger found characters in one genre of fan fiction to be "portrayed as firmly middle class or wealthy and holding middle class values."[75] Berit Åström similarly noted that male pregnancy (mpreg) fiction—a seemingly dramatic transgression of norms in which a cisgender man becomes pregnant—tends to extensively use normative tropes "like buying their first house, deciding which room should be the nursery, buying baby clothes, deciding on names and so forth," and may not be so transgressive after all.[76] Ultimately, Tosenberger identifies these tendencies as "containment within Western romantic tropes that are relentlessly heterosexist"[77]—though I'd call it homonormative. These are "conventional themes of love, trust, and homemaking"[78] in slash stories that often "rewrite romantic comedies, Harlequin romances, and Disney movies."[79] While fiction writers and readers are of course free to enjoy any trope, the prevalence of these particular tropes shows the limits of norm transgression in slash.

Finally, though fan fiction is often seen as contesting media texts, in recent years fans increasingly value media industry approval and official, or "canon," representation of their preferred slash relationships. Eve Ng offers an analysis of "queerbaiting," typically defined as

> situations where those officially associated with a media text court viewers interested in LGBT narratives—or become aware of such viewers—and encourage their interest in the media text without the text ever definitively confirming the nonheterosexuality of the relevant characters.[80]

Ng argues that "the crucial element is not a lack of canonicity, but how satisfactorily queerness plays out in the canonical text relative to viewer expectations."[81] That is, people feel queerbaited when they have higher expectations for canon representation than materialize; as Ng's comparison of *Xena* to contemporary TV demonstrates, expectations for representation have increased over time. Seeking this sort of validation means desiring "a grounding for slash consumption practices in the objective attributes of what is consumed, and, by the same token, a legitimisation of those practices."[82]

Demands for same-sex romance to be unambiguously present in media are understandable—it does important work to be visible—but it comes at a cost. Victoria Gonzalez describes the ways one fandom's centering of canonicity "insinuates that abiding by canon is the primary way in which normal or sane shippers function within fandom."[83] Similarly, Neta Yodovich finds that Israeli women fans repeat much the same stigmatizing stereotypes used against them, directed toward other fans.[84] The further implication is that "canon and TPTB [The Powers That Be; media production personnel] are infallible, inflexible, and not something that fans control."[85] Despite resistance from scholars to the idea that texts are straight by default,[86] tendencies to chase canon demonstrate beliefs that texts *are* straight unless there are overt LGBTQ+ characters. As Ng points out, "the fact that in such critiques a text is deemed queer enough only when two characters are in a canonically romantic relationship tends to obscure other nonnormative modes of being and relating."[87] Certainly, such practices put the existence of queerness in the hands of

TV makers, giving away power fans formerly exercised for decades to find it where they wanted. Similarly, Li finds that close ties to industry specifically inhibit activism in the Hong Kong-based fandom she analyzes.[88]

I argue that these ongoing and new attachments to normativity in slash fandom reinforce, rather than disrupt, norms and their attendant power structures. Drawing on the distinction queer and disability theorist Robert McRuer—borrowing from and extending Michael Warner and Judith Butler—makes between "virtually" crip or queer and "critically" crip or queer,[89] which I reframe as a distinction between "nonnormative" and "queer," I argue that the presence or absence of "queer" in the sense of norm breaking and transgression in fandom needs fuller attention. Fandom surely breaks norms, but we must ask: Which norms are disrupted and which upheld?

Conclusion

Over the course of more than 30 years of studying fans and fandom, research has discovered much about gender and sexuality identities and practices in this community. The baseline understanding of fandom generated by the first wave of research was that it is progressive, specifically because women transgressed and reworked norms around gender and sexuality. Subsequently, we gained greater understanding of gender in fandom through attention to men and masculinity. Second, further research shows that fans are not necessarily heterosexual, either identifying as lesbian, bisexual, or queer or because they give erotic pleasure to other women. Third, though early work focused on slash between men, subsequent analysis increased our knowledge of femslash. Fourth, it is clear that understanding of gender and sexuality in fandom is incomplete without careful attention to race. Finally, I've called attention here to the role of homonormativity in undermining progressive and transgressive traditions in fandom. Future research in the field can continue to help us deepen our understandings of how popular culture is a site of conflict and exclusion as much as of pleasure, desire, and identity.

Notes

1 Jonathan Gray, Cornel Sandvoss, and C. Lee Harrington "Introduction: Why Study Fans?," in *Fandom: Identities and Communities in a Mediated World*, ed. Jonathan Gray, Cornel Sandvoss, and C. Lee Harrington (New York, NY: New York University Press, 2007), 1–16.
2 Gray, Sandvoss, and Harrington, *Fandom*, 2.
3 Henry Jenkins, "Star Trek Rerun, Reread, Rewritten: Fan Writing as Textual Poaching," *Critical Studies in Mass Communication* 5, no. 2 (1988): 99, https://doi.org/10.1080/15295038809366691.
4 Constance Penley, *NASA/Trek: Popular Science and Sex in America* (New York, NY: Verso, 1997), 99.
5 Henry Jenkins, *Textual Poachers: Television Fans and Participatory Culture* (New York, NY: Routledge, 1992), 189; Penley, *NASA/Trek*, 2.
6 For analysis of these stereotypes, see: Jenkins, *Textual Poachers*; Joli Jensen, "Fandom as Pathology: The Consequences of Characterization," in *The Adoring Audience: Fan Culture and Popular Media*, ed. Lisa A. Lewis (London, UK: Routledge, 1992), 9–29; Lisa A. Lewis, "Introduction," in *The Adoring Audience: Fan Culture and Popular Media*, ed. Lisa A. Lewis (London, UK: Routledge, 1992), 1–6; Lisa A. Lewis, "'Something More Than Love': Fan Stories on Film," in *The Adoring Audience: Fan Culture and Popular Media*, ed. Lisa A. Lewis (London: Routledge, 1992), 135–59.
7 Gray, Sandvoss, and Harrington, "Introduction," 3.
8 Jenkins, *Textual Poachers*, 6.
9 Jenkins, *Textual Poachers*, 167.
10 Camille Bacon-Smith, *Enterprising Women: Television Fandom and the Creation of Popular Myth* (Philadelphia, PA: University of Pennsylvania Press, 1991), 3.
11 Penley, *NASA/Trek*, 101.
12 Bacon-Smith, *Enterprising Women*, 3; Jenkins, "Star Trek Rerun, Reread, Rewritten," 99.

13 Penley, *NASA/Trek*; Bacon-Smith, *Enterprising Women*.

14 Bacon-Smith, *Enterprising Women*, 3.

15 Jenkins, "Star Trek Rerun, Reread, Rewritten," 99.

16 Penley, *NASA/Trek*, 127; Jenkins, *Textual Poachers*, 219.

17 Shoshanna Green, Cynthia Jenkins, and Henry Jenkins, "Normal Female Interest in Men Bonking: Selections from the Terra Nostra Underground and Strange Bedfellows," in *Theorizing Fandom: Fans, Subculture and Identity*, ed. Cheryl Harris and Alison Alexander (Creskill, NJ: Hampton Press, 1998), 19.

18 Eden Lackner, Barbara Lynn Lucas, and Robin Anne Reid, "Cunning Linguists: The Bisexual Erotics of Words/Silence/Flesh," in *Fan Fiction and Fan Communities in the Age of the Internet: New Essays*, eds Karen Hellekson and Kristina Busse (Jefferson, NC: McFarland, 2006), 194.

19 Jenkins, *Textual Poachers*, 219; Penley, *NASA/Trek*, 145.

20 Jenkins, *Textual Poachers*, 219.

21 Jenkins, *Textual Poachers*, 219.

22 Penley, *NASA/Trek*, 101.

23 Sharon Kinsella, "Japanese Subculture in the 1990s: Otaku and the Amateur Manga Movement," *Journal of Japanese Studies* 24, no. 2 (1998): 289–316.

24 Green, Jenkins, and Jenkins, "Normal Female Interest."

25 Lackner, Lucas, and Reid, "Cunning Linguists"; Christine Scodari, "'Nyota Uhura Is Not a White Girl': Gender, Intersectionality, and Star Trek 2009's Alternate Romantic Universes," *Feminist Media Studies* 12, no. 3 (2012): 335–351, https://doi.org/10.1080/14680777.2011.615605.

26 Penley, *NASA/Trek*, 132.

27 John Walliss, "Fan Filmmaking and Copyright in a Global World: 'Warhammer 40,000' Fan Films and the Case of 'Damnatus,'" *Transformative Works and Cultures*, 5 (2010): n.p., https://doi.org/10.3983/twc.v5i0.178; Abigail De Kosnik, "Should Fan Fiction Be Free?" *Cinema Journal* 48, no. 4 (2009): 118–24, https://doi.org/10.1353/cj.0.0144.

28 De Kosnik, "Should Fan Fiction Be Free?"

29 Julie Levin Russo, "User-Penetrated Content: Fan Video in the Age of Convergence," *Cinema Journal* 48, no. 4 (2009): 128, https://doi.org/10.1353/cj.0.0147.

30 obsession_inc, "Affirmational Fandom vs. Transformational Fandom," June 1, 2009, https://obsession-inc.dreamwidth.org/82589.html.

31 Russo, "User-Penetrated Content"; Walliss, "Fan Filmmaking and Copyright in a Global World."

32 Kristina Busse, "Introduction," *Cinema Journal* 48, no. 4 (2009): 104–7, https://doi.org/10.1353/cj.0.0131; De Kosnik, "Should Fan Fiction Be Free?"

33 Anne Gilbert, "What We Talk about When We Talk about Bronies," *Transformative Works and Cultures* 20 (2015): 4.3, https://doi.org/0.3983/twc.2015.0666.

34 Gamergate, which took place in 2014, was a fan backlash movement that especially contested women in video games. The Sad and Rabid Puppies campaigns resisted the alleged exclusion of heterosexual white men from science fiction awards. Leslie Jones was harassed as the focal point for racist and misogynist backlash against the all-women reboot of Ghostbusters.

35 Bethan Jones, "#AskELJames, Ghostbusters, and #Gamergate: Digital Dislike and Damage Control," in *A Companion to Fandom and Fan Studies*, ed. Paul Booth (Oxford, UK: Wiley-Blackwell, 2018), 424.

36 Anastasia Salter and Bridget Blodgett, *Toxic Geek Masculinity in Media: Sexism, Trolling, and Identity Policing* (New York, NY: Palgrave Macmillan, 2017).

37 Katie Wilson, "Red Pillers, Sad Puppies, and Gamergaters: The State of Male Privilege in Internet Fan Communities," in *A Companion to Fandom and Fan Studies*, ed. Paul Booth (Oxford, UK: Wiley-Blackwell, 2018), 431, 443.

38 Katherine E. Morrissey, "Fandom Then/Now: A Participatory Project on Fan Fiction," 2014, http://katiedidnt.net/fandomthennow/pages/surveydemographics.html.

39 Alexis Lothian, Kristina Busse, and Robin Anne Reid, "'Yearning Void and Infinite Potential': Online Slash Fandom as Queer Female Space," *English Language Notes* 45, no. 2 (2007): 103.

40 Lackner, Lucas, and Reid, "Cunning Linguists," 201.

41 centrumlumina, "AO3 Ship Stats 2017," The Slow Dance of the Infinite Stars, August 3, 2017, http://centrumlumina.tumblr.com/post/163750676579/now-presenting-the-fifth-annual-ao3-ship-stats-top.

42 Ling Yang and Hongwei Bao, "Queerly Intimate: Friends, Fans and Affective Communication in a Super Girl Fan Fiction Community," *Cultural Studies* 26, no. 6 (2012): 842–71, https://doi.org/10.1080/0950 2386.2012.679286.

43 Julie Levin Russo, "The Queer Politics of Femslash," in *The Routledge Companion to Media Fandom*, ed. Melissa A. Click and Suzanne Scott (New York: Routledge, 2017), 161.

44 Russo, "The Queer Politics of Femslash," 160.

45 Cheuk Yin Li, "The Absence of Fan Activism in the Queer Fandom of Ho Denise Wan See (HOCC) in Hong Kong," *Transformative Works and Cultures* 10 (2012): para. 7.2, https://doi.org/10.3983/twc.2012.0325.

46 Rosalind Hanmer, "Lesbian Subtext Talk: Experiences of the Internet Chat," *International Journal of Sociology and Social Policy* 23, nos 1–2 (2003): 80–106, https://doi.org/10.1108/01443330310790453.

47 Julie Levin Russo, "Textual Orientation: Queer Female Fandom Online," in *The Routledge Companion to Media and Gender*, ed. Cynthia Carter, Linda Steiner, and Lisa McLaughlin (New York: Routledge, 2013), 450–60.

48 Sara Gwenllian Jones, "The Sex Lives of Cult Television Characters," *Screen* 43, no. 1 (2002): 80, https://doi.org/10.1093/screen/43.1.79.

49 Julie Levin Russo, "Indiscrete Media: Television/Digital Convergence and Economies of Online Lesbian Fan Communities" (Dissertation, Providence, RI, Brown University, 2010), 2, http://j-l-r.org/diss.

50 Adrienne Rich, "Compulsory Heterosexuality and Lesbian Existence," *Signs* 5, no. 4 (1980): 631–60; Mel Stanfill, "Where the Femslashers Are: Media on the Lesbian Continuum," *Transformative Works and Cultures* 24 (2017), http://journal.transformativeworks.org/index.php/twc/article/view/959.

51 Russo, "Indiscrete Media: Television/Digital Convergence and Economies of Online Lesbian Fan Communities," 2, 267.

52 Russo, "The Queer Politics of Femslash," 156.

53 TWC Editor, "Pattern Recognition: A Dialogue on Racism in Fan Communities," *Transformative Works and Cultures* 3 (2009): 1.2, https://doi.org/10.3983/twc.2009.0172.

54 Rukmini Pande, "Squee from the Margins: Racial/Cultural/Ethnic Identity in Global Media Fandom," in *Seeing Fans: Representations of Fandom in Media and Popular Culture*, eds Lucy Bennett and Paul Booth (New York: Bloomsbury Academic, 2016), 214. For an example of Bear's posts, see Elizabeth Bear, "Whatever You're Doing, You're Probably Wrong," *LiveJournal*, January 12, 2009, https://matociquala.livejournal.com/1544111.html.

55 Pande, "Squee from the Margins," 210.

56 Mel Stanfill, "The Unbearable Whiteness of Fandom and Fan Studies," in *A Companion to Fandom and Fan Studies*, ed. Paul Booth (Oxford, UK: Wiley-Blackwell, 2018), 305–17.

57 Sarah N. Gatson and Robin Anne Reid, "Editorial: Race and Ethnicity in Fandom," *Transformative Works and Cultures* 8 (2011): n.p.; Stanfill, "The Unbearable Whiteness of Fandom and Fan Studies."

58 TWC Editor, "Pattern Recognition: A Dialogue on Racism in Fan Communities"; Pande, "Squee from the Margins."

59 TWC Editor, "Pattern Recognition: A Dialogue on Racism in Fan Communities," 5.1, original emphasis.

60 Pande, "Squee from the Margins"; Stanfill, "The Unbearable Whiteness."

61 TWC Editor, "Pattern Recognition"; Stanfill, "The Unbearable Whiteness"; Mel Stanfill, "Fans of Color in Femslash," *Transformative Works and Cultures* 29 (2019), https://doi.org/10.3983/twc.2019.1528.

62 TWC Editor, "Pattern Recognition"; Stanfill, "The Unbearable Whiteness."

63 Jessica Seymour, "Racebending and Prosumer Fanart Practices in Harry Potter Fandom," in *A Companion to Fandom and Fan Studies*, ed. Paul Booth (Oxford, UK: Wiley-Blackwell, 2018), 334.

64 Pande, "Squee from the Margins," 214.

65 Rebecca Wanzo, "African American Acafandom and Other Strangers: New Genealogies of Fan Studies," *Transformative Works and Cultures* 20 (2015): sec. 2.1.

66 Shipping, from "relationship," refers to advocating for a romance between two or more characters.

67 Kristen J. Warner, "If Loving Olitz Is Wrong, I Don't Wanna Be Right," *The Black Scholar* 45, no. 1 (2015): 16–20, https://doi.org/10.1080/00064246.2014.997599.

68 Scodari, "'Nyota Uhura Is Not a White Girl'," 3.

69 Berit Åström, "'Let's Get Those Winchesters Pregnant': Male Pregnancy in 'Supernatural' Fan Fiction," *Transformative Works and Cultures,* 4 (2010): n.p., https://doi.org/10.3983/twc.v4i0.135.

70 Scodari, "'Nyota Uhura Is Not a White Girl'," 9.

71 Åström, "Let's Get Those Winchesters Pregnant."

72 Jenkins, *Textual Poachers*; Christine Scodari, "Resistance Re-Examined: Gender, Fan Practices, and Science Fiction Television," *Popular Communication* 1, no. 2 (2003): 111–30, https://doi.org/10.1207/S15405710PC0102_3.

73 Scodari, "Resistance Re-Examined," 114.

74 Lisa Duggan, *The Twilight of Equality? Neoliberalism, Cultural Politics, and the Attack on Democracy* (Boston, MA: Beacon Press, 2004), 50.

75 Catherine Tosenberger, "'The Epic Love Story of Sam and Dean': 'Supernatural,' Queer Readings, and the Romance of Incestuous Fan Fiction," *Transformative Works and Cultures* 1 (2008): 4.2, https://doi.org/10.3983/twc.v1i0.30.

76 Åström, "Let's Get Those Winchesters Pregnant," 3.4.

77 Tosenberger, "The Epic Love Story of Sam and Dean," 4.2.

78 Åström, "Let's Get Those Winchesters Pregnant," 1.1.

79 Tosenberger, "The Epic Love Story of Sam and Dean," 4.4.

80 Eve Ng, "Between Text, Paratext, and Context: Queerbaiting and the Contemporary Media Landscape," *Transformative Works and Cultures* 24 (2017): para. 1.2, http://journal.transformativeworks.org/index.php/twc/article/view/917.

81 Ng, "Between Text, Paratext, and Context," para. 2.8.

82 Daniel Allington, "'How Come Most People Don't See It?': Slashing the Lord of the Rings," *Social Semiotics* 17, no. 1 (2007): 49, https://doi.org/10.1080/10350330601124650.

83 Victoria M. Gonzalez, "Swan Queen, Shipping, and Boundary Regulation in Fandom," *Transformative Works and Cultures* 22 (2016): 4.2, http://journal.transformativeworks.org/index.php/twc/article/view/669.

84 Neta Yodovich, "'A Little Costumed Girl at a Sci-Fi Convention': Boundary Work as a Main Destigmatization Strategy Among Women Fans," *Women's Studies in Communication* 39, no. 3 (2016): 289–307, https://doi.org/10.1080/07491409.2016.1193781.

85 Gonzalez, "Swan Queen, Shipping, and Boundary Regulation in Fandom," 4.9.

86 Alexander Doty, *Making Things Perfectly Queer: Interpreting Mass Culture* (Minneapolis: University of Minnesota Press, 1993); Tosenberger, "The Epic Love Story of Sam and Dean"; Ng, "Between Text, Paratext, and Context."

87 Ng, "Between Text, Paratext, and Context," 9.8.

88 Li, "The Absence of Fan Activism."

89 Robert McRuer, *Crip Theory: Cultural Signs of Queerness and Disability* (New York, NY: New York University Press, 2006); Michael Warner, "Normal and Normaller: Beyond Gay Marriage," *GLQ: A Journal of Lesbian & Gay Studies* 5, no. 2 (1999): 119–71; Judith Butler, *Bodies That Matter: On the Discursive Limits of "Sex"* (New York, NY: Routledge, 1993).

Bibliography

Allington, Daniel. "'How Come Most People Don't See It?': Slashing the Lord of the Rings." *Social Semiotics* 17, no. 1 (2007): 43–62. https://doi.org/10.1080/10350330601124650.

Åström, Berit. "'Let's Get Those Winchesters Pregnant': Male Pregnancy in 'Supernatural' Fan Fiction." *Transformative Works and Cultures* 4 (2010): n.p. https://doi.org/10.3983/twc.v4i0.135.

Bacon-Smith, Camille. *Enterprising Women: Television Fandom and the Creation of Popular Myth.* Philadelphia, PA: University of Pennsylvania Press, 1991.

Bear, Elizabeth. "Whatever You're Doing, You're Probably Wrong." *LiveJournal*, January 12, 2009. https://matociquala.livejournal.com/1544111.html.

Busse, Kristina. "Introduction." *Cinema Journal* 48, no. 4 (2009): 104–7. https://doi.org/10.1353/cj.0.0131.

Butler, Judith. *Bodies That Matter: On the Discursive Limits of "Sex."* New York: Routledge, 1993.

centrumlumina. "AO3 Ship Stats 2017." The Slow Dance of the Infinite Stars, August 3, 2017. http://centrumlumina.tumblr.com/post/163750676579/now-presenting-the-fifth-annual-ao3-ship-stats-top.

De Kosnik, Abigail. "Should Fan Fiction Be Free?" *Cinema Journal* 48, no. 4 (2009): 118–24. https://doi.org/10.1353/cj.0.0144.

Doty, Alexander. *Making Things Perfectly Queer: Interpreting Mass Culture.* Minneapolis, MN: University of Minnesota Press, 1993.

Duggan, Lisa. *The Twilight of Equality? Neoliberalism, Cultural Politics, and the Attack on Democracy.* Boston, MA: Beacon Press, 2004.

Gatson, Sarah N., and Reid, Robin Anne. "Editorial: Race and Ethnicity in Fandom." *Transformative Works and Cultures* 8 (2011): n.p.

Gilbert, Anne. "What We Talk About When We Talk About Bronies." *Transformative Works and Cultures* 20 (2015): 15. https://doi.org/0.3983/twc.2015.0666.

Gonzalez, Victoria M. "Swan Queen, Shipping, and Boundary Regulation in Fandom." *Transformative Works and Cultures* 22 (2016). http://journal.transformativeworks.org/index.php/twc/article/view/669.

Gray, Jonathan, Sandvoss, Cornel, and Harrington, C. Lee. "Introduction: Why Study Fans?" In *Fandom: Identities and Communities in a Mediated World*, edited by Jonathan Gray, Cornel Sandvoss, and C. Lee Harrington, 1–16, New York: New York University Press, 2007.

Green, Shoshanna, Jenkins, Cynthia, and Jenkins Henry. "Normal Female Interest in Men Bonking: Selections from the Terra Nostra Underground and Strange Bedfellows." In *Theorizing Fandom: Fans, Subculture and Identity*, edited by Cheryl Harris and Alison Alexander, 9–38, Creskill, NJ: Hampton Press, 1998.

Hanmer, Rosalind. "Lesbian Subtext Talk: Experiences of the Internet Chat." *International Journal of Sociology and Social Policy* 23, no. 1–2 (2003): 80–106. https://doi.org/10.1108/01443330310790453.

Jenkins, Henry. "Star Trek Rerun, Reread, Rewritten: Fan Writing as Textual Poaching." *Critical Studies in Mass Communication* 5, no. 2 (1988): 85–107. https://doi.org/10.1080/15295038809366691.

———. *Textual Poachers: Television Fans and Participatory Culture*. New York: Routledge, 1992.

Jensen, Joli. "Fandom as Pathology: The Consequences of Characterization." In *The Adoring Audience: Fan Culture and Popular Media*, edited by Lisa A. Lewis, 9–29, London: Routledge, 1992.

Jones, Bethan. "#AskELJames, Ghostbusters, and #Gamergate: Digital Dislike and Damage Control." In *A Companion to Fandom and Fan Studies*, edited by Paul Booth, 415–30, Oxford, UK: Wiley-Blackwell, 2018.

Jones, Sara Gwenllian. "The Sex Lives of Cult Television Characters." *Screen* 43, no. 1 (2002): 79–90. https://doi.org/10.1093/screen/43.1.79.

Kinsella, Sharon. "Japanese Subculture in the 1990s: Otaku and the Amateur Manga Movement." *Journal of Japanese Studies* 24, no. 2 (1998): 289–316.

Lackner, Eden, Lucas, Barbara Lynn, and Reid, Robin Anne. "Cunning Linguists: The Bisexual Erotics of Words/Silence/Flesh." In *Fan Fiction and Fan Communities in the Age of the Internet: New Essays*, edited by Karen Hellekson and Kristina Busse, 189–206, Jefferson, NC: McFarland, 2006.

Lewis, Lisa A. "Introduction." In *The Adoring Audience: Fan Culture and Popular Media*, edited by Lisa A. Lewis, 1–6, London: Routledge, 1992.

———. "'Something More than Love': Fan Stories on Film." In *The Adoring Audience: Fan Culture and Popular Media*, edited by Lisa A. Lewis, 135–59, London: Routledge, 1992.

Li, Cheuk Yin. "The Absence of Fan Activism in the Queer Fandom of Ho Denise Wan See (HOCC) in Hong Kong." *Transformative Works and Cultures* 10 (2012): n.p. https://doi.org/10.3983/twc.2012.0325.

Lothian, Alexis, Busse, Kristina, and Reid, Robin Anne. "'Yearning Void and Infinite Potential': Online Slash Fandom as Queer Female Space." *English Language Notes* 45, no. 2 (2007): 103–11.

McRuer, Robert. *Crip Theory: Cultural Signs of Queerness and Disability*. New York: New York University Press, 2006.

Morrissey, Katherine E. "Fandom Then/Now: A Participatory Project on Fan Fiction," 2014. http://katiedidnt.net/fandomthennow/pages/surveydemographics.html.

Ng, Eve. "Between Text, Paratext, and Context: Queerbaiting and the Contemporary Media Landscape." *Transformative Works and Cultures* 24 (2017). http://journal.transformativeworks.org/index.php/twc/article/view/917.

obsession_inc. "Affirmational Fandom vs. Transformational Fandom," June 1, 2009. https://obsession-inc.dreamwidth.org/82589.html.

Pande, Rukmini. "Squee from the Margins: Racial/Cultural/Ethnic Identity in Global Media Fandom." In *Seeing Fans: Representations of Fandom in Media and Popular Culture*, edited by Lucy Bennett and Paul Booth, 209–20, New York: Bloomsbury Academic, 2016.

Penley, Constance. *NASA/Trek: Popular Science and Sex in America*. New York: Verso, 1997.

Rich, Adrienne. "Compulsory Heterosexuality and Lesbian Existence." *Signs* 5, no. 4 (1980): 631–60.

Russo, Julie Levin. "Indiscrete Media: Television/Digital Convergence and Economies of Online Lesbian Fan Communities." Dissertation, Brown University, 2010. http://j-l-r.org/diss.

———. "Textual Orientation: Queer Female Fandom Online." In *The Routledge Companion to Media & Gender*, edited by Cynthia Carter, Linda Steiner, and Lisa McLaughlin, 450–60, New York: Routledge, 2013.

———. "The Queer Politics of Femslash." In *The Routledge Companion to Media Fandom*, edited by Melissa A. Click and Suzanne Scott, 164–55, New York: Routledge, 2017.

———. "User-Penetrated Content: Fan Video in the Age of Convergence." *Cinema Journal* 48, no. 4 (2009): 125–30. https://doi.org/10.1353/cj.0.0147.

Salter, Anastasia, and Blodgett, Bridget. *Toxic Geek Masculinity in Media: Sexism, Trolling, and Identity Policing.* New York: Palgrave Macmillan, 2017.

Scodari, Christine. "'Nyota Uhura Is Not a White Girl': Gender, Intersectionality, and Star Trek 2009's Alternate Romantic Universes." *Feminist Media Studies* 12, no. 3 (2012): 335–51. https://doi.org/10.1080/14680777.2011.615605.

———. "Resistance Re-Examined: Gender, Fan Practices, and Science Fiction Television." *Popular Communication* 1, no. 2 (2003): 111–30. https://doi.org/10.1207/S15405710PC0102_3.

Seymour, Jessica. "Racebending and Prosumer Fanart Practices in Harry Potter Fandom." In *A Companion to Fandom and Fan Studies*, edited by Paul Booth, 333–48, Oxford, UK: Wiley-Blackwell, 2018.

Stanfill, Mel. "Fans of Color in Femslash." *Transformative Works and Cultures* 29 (2019). https://doi.org/10.3983/twc.2019.1528.

———. "The Unbearable Whiteness of Fandom and Fan Studies." In *A Companion to Fandom and Fan Studies*, edited by Paul Booth, 305–17, Oxford, UK: Wiley-Blackwell, 2018.

———. "Where the Femslashers Are: Media on the Lesbian Continuum." *Transformative Works and Cultures* 24 (2017). http://journal.transformativeworks.org/index.php/twc/article/view/959.

Tosenberger, Catherine. "'The Epic Love Story of Sam and Dean': 'Supernatural,' Queer Readings, and the Romance of Incestuous Fan Fiction." *Transformative Works and Cultures* 1 (2008): n.p. https://doi.org/10.3983/twc.v1i0.30.

TWC Editor. "Pattern Recognition: A Dialogue on Racism in Fan Communities." *Transformative Works and Cultures* 3 (2009): n.p. https://doi.org/10.3983/twc.2009.0172.

Walliss, John. "Fan Filmmaking and Copyright in a Global World: 'Warhammer 40,000' Fan Films and the Case of 'Damnatus.'" *Transformative Works and Cultures* 5 (2010): n.p. https://doi.org/10.3983/twc.v5i0.178.

Wanzo, Rebecca. "African American Acafandom and Other Strangers: New Genealogies of Fan Studies." *Transformative Works and Cultures* 20 (2015): n.p.

Warner, Kristen J. "If Loving Olitz Is Wrong, I Don't Wanna Be Right." *The Black Scholar* 45, no. 1 (2015): 16–20. https://doi.org/10.1080/00064246.2014.997599.

Warner, Michael. "Normal and Normaller: Beyond Gay Marriage." *GLQ: A Journal of Lesbian & Gay Studies* 5, no. 2 (1999): 119–71.

Wilson, Katie. "Red Pillers, Sad Puppies, and Gamergaters: The State of Male Privilege in Internet Fan Communities." In *A Companion to Fandom and Fan Studies*, edited by Paul Booth, 431–46, Oxford, UK: Wiley-Blackwell, 2018.

Yang, Ling, and Bao, Hongwei. "Queerly Intimate: Friends, Fans and Affective Communication in a Super Girl Fan Fiction Community." *Cultural Studies* 26, no. 6 (2012): 842–71. https://doi.org/10.1080/09502386.2012.679286.

Yodovich, Neta. "'A Little Costumed Girl at a Sci-Fi Convention': Boundary Work as a Main Destigmatization Strategy among Women Fans." *Women's Studies in Communication* 39, no. 3 (2016): 289–307. https://doi.org/10.1080/07491409.2016.1193781.

Gendered violence and communication

Introduction to Part V

The pervasive problem of gendered violence is an urgent subject to take up in this volume, as it requires careful attention to sociocultural forces that contribute to—or even constitute—matters of life and death. The research in this section examines the role of communication in fostering, identifying, or resisting many types of violent attacks animated by bias and bigotry forming around gender and sexuality. As the included chapters illustrate, these often operate in consort with racial, ethnic, and economic discrimination and dehumanization. However, our featured authors also locate resources that offer some hope for countering the damage and destruction wrought by gendered logics and social hierarchies.

The central themes woven through this section of writings on gendered violence and communication include the function of media in social change movements, the material force of symbol systems, and the crucial connections between how bodies are made objects or agents via discourses of difference. From meticulous case studies to provocative concepts, the following chapters demand disciplinary change in how the communicative dimensions of violence and the violent dimensions of communication are researched and theorized. We want to preface this section by highlighting how these studies foreground neglected populations and problems while offering new frameworks and approaches for grappling with the complex connections between gender, communication, and violence. The careful attention to race, class, sexuality, age, femininity, and masculinity in the studies previewed below enables their authors to challenge communication scholarship in ways that should not be missed.

The first chapter is Sunera Thobani's "Imaging rape, imagining woman in popular Indian cinema: victim, vigilante, or Goddess?" In a carefully contextualized yet broadly conceived profile of troubling themes in Indian film, Thobani investigates the danger and promise of recent cinematic sexual violence narratives. The analysis takes in eroticized and empowering depictions, exploring how numerous portrayals reinforce or upset conventional tropes and hierarchies. The author approaches controversial films and difficult topics such as suicide, victim blaming, and vengeance with care and nuance, helping readers to hone their attention to specific tools of the medium. Moreover, Thobani links critical cultural insights to urgent problems stemming from neonationalist movements and persistent socioeconomic inequities.

The next chapter, Tia C. M. Tyree's "Speak Up, Sis: black women, race, and news coverage of the Me Too movement," expands on the theme of timely social movement research by shifting the focus to online journalism. In a study that untangles key strands in the thick web of U.S. media representations surrounding racialized sexual violence, Tyree profiles the role of Black media sources in centering Black women's voices and experiences. Tyree's discourse analysis examines framing devices evident in close to 50 stories that appeared early in the Me Too movement's emergence in mass media. By directing scholarly attention to understudied outlets, Tyree shows that news media can inadvertently maintain some facets of traumatizing events and stigmatizing systems, even as they create rare opportunities to address complexities of both.

A related but distinct study of media and the Me Too movement follows in "Digital *testimonios* and witnessing of Salma Hayek and America Ferrera's disclosures of sexual harassment and assault" by Raisa F. Alvarado and Michelle A. Holling. This chapter confronts the tendency for Me Too coverage to fixate on famous women in Hollywood by focusing on two Latina actresses who used online media to represent their own experiences with sexual assault. With an approach that details the possibilities and limits of testifying and witnessing via online platforms, Alvarado and Holling bring much-needed attention to understudied figures and their publics. The authors sift meticulously through anonymous online comments in response to Salma Hayek's story of Harvey Weinstein's predatory assaults as well as Instagram posts reacting to America Ferrera's account of childhood sexual grooming and abuse. In doing so, Alvarado and Holling explicate complex forms of agency and advocacy that survivors and supporters may employ in and around digital disclosures regarding firsthand encounters with sexual violence.

Taking the theme of agency in the Me Too movement in a radical new direction, Elena Elías Krell's "From innocents to experts: queer and trans of color interventions into #MeToo" offers a new frame for attending to the marginalized survivors of sexual violence so often silenced by racist, ageist, and cis-centric heterosexism. Krell probes the frightening connections between vulnerability and violence and the relationships between theory and politics that facilitate hidden abuse and exploitation in and through social institutions such as the family. This urgent and unique intervention into anti-abuse advocacy shows readers how activism that fails to pay attention to the wisdom and power of queer of color survivors of childhood sexual abuse will reinforce the ideals of innocence and victimhood that erase the experiences and perspectives of queer and trans of color children and youth. Moreover, Krell models artistic and autoethnographic approaches to advocacy and representation that respect the expertise of survivors who are quite capable of speaking for themselves and have much to teach those who would be allies.

The connections between communication and practices in enabling and excusing violence touched on in earlier chapters get thorough attention in "Symbolic erasure as gendered violence: the link between verbal and physical harm," by Kate Lockwood Harris. Through a careful overview of interdisciplinary scholarship positing relations between representations and behaviors, Harris shows how and why communication scholars are uniquely positioned to provide a more complex understanding of these dynamics in ways consistent with both contemporary theory and empirical research. Harris examines the threads and tensions between symbolic and material dimensions of violence along multiple vectors, finding promise in pathways opened up by scholarship pursuing administrative violence, critical race theory, epistemic violence, internalized oppression, organizational violence, symbolic violence, and trauma studies.

The final chapter in this section on gendered violence is Ashley Morgan's "Sherlock Holmes and the case for toxic masculinity." Departing from the focus in the previous studies on representations of violence and its victims, Morgan traces this urgent social problem back to a seemingly harmless and wildly influential mythic male idol: the iconic detective whose intellectual and heroic qualities have won him countless fans through literature, film, television, and many other media. Morgan takes a close look at recent television portrayals of Sir Arthur Conan Doyle's famous character to demonstrate how subtle messages of masculine agency permeate popular cultural discourses. The resulting discoveries concern performances of verbal denigration and interpersonal dominance, and combine to give readers a detailed exposition of understated communicative modes of manhood embodied by beloved figures. In addition to challenging contemporary celebrations of Holmes as a potentially feminist- and queer-friendly figure, Morgan expands the time frame and sources conventionally considered when charting the origins of dangerous norms of hegemonic hetero-masculinity.

Taken together, the studies of communication and violence collected in this section help to direct scholars into new territory by drawing their attention to understudied topics, shared assumptions, and subtle influences. Each chapter presents bold arguments backed by detailed analyses that contribute new texts to examine, new perspectives to take, and new concepts to theorize the role of communication in violence and violence in communication. Grounded in extensive conversations both within and beyond the field, these studies make lasting contributions to communication scholarship. They will be valuable sources of critique and inspiration for readers interested in better understanding urgent problems relating to communication and gendered violence.

27

IMAGING RAPE, IMAGINING WOMAN IN POPULAR INDIAN CINEMA

Victim, vigilante, or Goddess?

Sunera Thobani

A sisterhood of women—who have either been raped themselves or have relatives who have been raped—make a plan to entice, kidnap, and castrate their rapists in *Zakhmi Aurat* (Wounded Woman, Bhogal 1988). Set in Bombay, the film's heroine is a police officer who is gang-raped and then retraumatized as the rapists are all acquitted by the courts. Enraged by this doubled injustice, she initiates the sisterhood and works with the other women to track down one rapist after another. Not surprising, the film has been read as "one of the most daring films from Hindi cinema."[1]

Since *Zakhmi Aurat* opens with a montage of newspaper and magazine clippings about rape, the film is self-consciously inserted into then-unfolding public debates about sexual and gender violence in India.[2] By casting the heroine as a police officer, the film is also interjected into international feminist debates of the time: the film's heroine Kiran Dutt was a namesake of India's first woman police officer, Kiran Bedi, who joined the police force in 1972 and quickly garnered an international reputation for her work on violent crimes against women. Featuring a police officer as rape victim also recalls the 1985 opening of the first woman-only police station to deal with crimes against women in Brazil following the feminist organizing to that end.[3] Breaking with earlier cinematic depictions of the raped woman as helpless victim (*Aap Aye Bahar Aayee*, Kumar 1971) and of the rape victim-turned-vigilante as "lawless" (*Insaf ka Tarazu* 1980), *Zakhmi Aurat* was also notable for its depiction of a woman police officer breaking the law explicitly on behalf of women to end sexual violence in their lives. Because the state's law fails to protect women, the heroine argued, the women were compelled to call on a greater "law"—that of Justice—as they planned and carried out their "un-manning" of the rapists.

Yet by depicting the heroine as an upper caste and upper middle class educated woman, the film reproduced older gendered tropes of "woman as nation," and more particularly, of the Hindu woman as heroic "daughter of the nation." Moreover, the film's (admittedly few) titillating scenes and song sequences objectified the female body to ensure its box office attraction. Nevertheless, its explicit condemnation of the law, and of the corrupt lawyers and politicians who stand for the state, packed a powerful punch. The film's end does not render a "legal" verdict on the violence committed by the women, nor does it redeem the justice system. Instead, the film's rendering of its "moral" verdict redeems the heroine as she is embraced by her prospective mother-in-law. The other members of her sisterhood, however, find no such social validation. The film's drawing of an equivalence between female rape and male castration in a libidinal economy of exchange went beyond the pale for contemporaneous mainstream cinema; the death of the rapists through such violence produced the trope of the "angry young woman" after the Emergency (imposed by Indira Gandhi in the 1970s[4]), a time when the iconic image of the "angry young man" was reshaping popular understandings of the post-independence state (ineffectual), and politicians, policemen, and lawyers (corrupt). Yet few among the rape-revenge films that followed *Zakhmi Aurat*, including *Dushman* (1998), *Daman* (2001), and *Mom* (2017), have picked up the earlier film's message of women's collective resistance, or of the possibility of women's cross-class, cross-caste, and cross-religion solidarity. Limited as the message of feminist solidarity proved to be in *Zakhmi Aurat*, its filmic themes would thereafter appear only sporadically, and even then in a more truncated manner, in popular films (*Gulaab Gang* 2014).

The question of "woman" has been mediated in no small way through the prism of rape in the popular cinema of post-independence India. The spectacular convergence of gender, nation, and rape—as allegory, metaphor, or empirical fact—in filmic representation shapes public understandings of 'woman' (chaste, violate-able, degraded) sexual violence (the conditions of its possibility, extent, and consequences) as well as of ideas about the nation (insiders, outsiders, threats). Whether in the foundational text of nationhood, *Mother India* (Khan 1957), the "social issue" films of the 1970s (*Aap Aaye Bahar Ayee* 1971; *Roti Kapda Aur Makaan* 1974; *Prem Rog* 1982), the rape revenge films of the 1980s (*Insaf Ka Tarazu* 1980; *Zakhmi Aurat* 1988), or the proto-feminist films of the 1990s and 2000s (*Damini* 1993; *Dushman* 1998; *Daman* 2001), rape remains an omnipresent threat to the female characters. Moreover, rape and sexual assault also feature centrally in the seminal works of "new wave" or "art" cinema (*Arth* 1982; *Bhumika* 1977; *Salaam Bombay!* 1988; *Bandit Queen* 1994), in South Asian diasporic films (*Heaven on Earth* 2008; *Monsoon Wedding* 2001; *Water* 2005); in the recent crop of post-Gujarat (2002) "communalism" films (*Firaaq* 2008; *Parzania* 2005); and in the films sparked by the Delhi gang rape case of 2012 (*Mom* 2017 and *Pink* 2016, among others). If Projansky notes in her study of sexual violence in film and television that rape is "one of contemporary U.S. popular culture's compulsory citations," the most cursory of readings demonstrates this is no less the case in popular Indian film culture.[5] In the latter case, however, the conditions of post-coloniality are crucial to decoding the gender-sexual politics of rape.

Fashioned in an economy of exchange with related forms of gender exploitation and oppression in a caste-ridden, communal, and capitalist-modernizing society, and addressing the "counter-hegemonic" politics of feminist, Dalit, queer, minority rights, and other social justice movements in India, the films studied here have consistently engaged audiences through the rubric of sexual violence in public as well as private spaces. These films have pointed to —sometimes explicitly, sometimes implicitly—the intrinsic instability of the private/public divide. If rape functions within the family to organize male power, control, and access to women's sexuality (*Prem Rog*; *Damini, Daman*), this violence intersects with its public spectacularization to organize caste, class, and religious domination within the larger social order (*Bandit Queen, Parzania, Dil Se*). Further, the filmic treatment of sexual violence also speaks to the pressures of the globalization of the mediascape, particularly the entry of powerful cultural competitors into the terrain over which Bollywood once held uncontested sway (*Slumdog Millionaire*, for example). Yet despite the powerful hold of sexual imagery over the popular cultural consciousness, and despite the performative and prescriptive aspects of cinematic depiction of sexual violence, rape is arguably amongst the most under-researched topics in South Asian film and media studies.

The following sections introduce the reader to a general overview of some of the key representational strategies and practices at work in films such as those mentioned above. My analysis of the major themes in their depictions of rape situates them in the political and cultural shifts that mark the development of the Indian nation-state, beginning with the early post-independence period, the economic crisis of the late 1970s, the advent of neoliberalism in the 1990s, and the consolidation of Hindu extremism in national politics during the 2000s. Using specific examples from representative films, I illustrate what they reveal about the role of sexual violence in the construction of femininities and masculinities in the changing discourses of nation, identity, and belonging. I take a historicizing approach in my study, not because I mean to suggest that an inevitable linearity bending toward the arc of gender justice is to be found in these depictions, but because these representations of rape shift quite dramatically in a symbiotic relation with social and political development in South Asia, and in other parts of the world. Put differently, this chapter shows how media depictions of sexual violence and their deployment of particular representational strategies cannot be taken as outside the relations of power they tap into and, in turn, affectively reshape.

Theorizing rape, reading representation

In her study of sexual violence, Sunder Rajan critiques narratives that posit rape as the core truth of patriarchy; her point is that this leads to the essentialization of feminist politics. Such an approach reduces sexual and gender politics to the male/female binary, and in the process, obfuscates the complex power relations and hierarchies at work among and between differently positioned women. Within film studies in general, as well as in South Asian film studies, however, the critical perspective advanced by Sunder Rajan remains marginal.

The work of early Western feminist scholars on sexual violence shaped what are even now the dominant approaches in feminist and other critical traditions.[6] The view that sexual violence is central to the construction of patriarchy and the reproduction of women's oppression has become a truism in most critical studies, such that the relation of rape to caste, race, and class relations, and to the overall forms of post/colonial subjugation is generally overlooked. Correspondingly, the early tendency in feminist film and media studies to focus on how film texts as well as the cinematic apparatus foster male power and privilege homogenized the category "woman." So, for example, feminists addressed the gendered structure of cinema, arguing this produces and organizes the privileged stance of the 'male gaze' to generate masculine pleasure by objectifying and sexualizing the figure of the woman.[7] Tied to the politics of feminist movements, these feminist scholars challenged the idea that sexual violence could be explained away as the personal misfortune of individual women or caused by the personal pathologies of individual men. Instead, they deconstructed the sociocultural practices and legal-political norms that constructed rape as a purely interpersonal matter and unconnected to the political order.[8] This relation of the personal to the social was captured in the phenomenally successful second-wave feminist slogan "the personal is political."

Such feminist scholarship was soon enough critiqued for its neglect of the mutually constitutive relation of gender to race, class, ethnicity, and nation, among other relations of power.[9] Consequently, more recent Western feminist scholarship has integrated intersectional approaches to some extent, advancing the earlier understanding of rape as "the originary moment of the social contract" to account for other social axis of power.[10] Representation of sexual violence, for example, is often redefined as fueling "cultural fantasies of power and domination, gender and sexuality, and class and ethnicity."[11] Naming the representation of such violence "public rape," Horeck has argued that representations of rape enable the making of social bonds and the mapping out of public spaces to inform collective ideas about sexual difference. Yet theoretical and methodological approaches grounded in critical race, cross-cultural and queer of color scholarly traditions remain outside the 'core' of feminist film and media studies.[12]

South Asian film and media studies draw heavily from Western feminist traditions but the former's development has also been strongly informed by left and post-colonial feminist traditions. Although the early feminist scholarship adopted homogenizing approaches to the category of woman, as pointed out by Sunder Rajan, more recent work is informed by intersectionality,[13] and by post-structural as well as post-colonial approaches, including Subaltern Studies (Ansari, 2008; Kabir, 2003). Yet there is a dearth of studies focusing on how sexual violence is related to the condition of post-coloniality, to the construction of the nation-state and its religio-cultural, class- and caste-based formations of gender and sexuality. While the disproportionate effects of sexual violence in the lives of different communities of women are garnering more attention, much work remains to be done in this regard. As Virdi has pointed out, feminists in South Asia remain divided over how to approach representations of rape: they are caught in debates over whether its depiction serves the male gaze, sensationalizes and eroticizes the violence, or simply titillates the viewer. But, as Virdi also argues, not attending to the depiction of sexual violence cannot be an option for feminists, for this furthers the erasure of rape within public debate and discussion, and has a silencing effect on its profound consequences in women's lives.

South Asian feminists who study how sex and violence come together in the assertion of gendered, caste, class, and religious domination also note that rape organizes women's access to public space, shapes their relations to community, and extends over these the political power of the state.[14] In this context, "The taking of embattled positions around a raped woman's cause often marks an identity crisis for a group…," Sunder Rajan argues, for the concerns of women who are raped disappear when rape functions as "an emotional war-cry."[15] In contrast, a woman who survives rape becomes a subject through this experience, as well as her responses in its aftermath, within critical feminist depictions of sexual violence, Sunder Rajan demonstrates. In such feminist interventions, the woman's post-rape strategies of survival—not the rape itself—shape the narrative structure and flow. These narratives highlight the operations of systems of oppression rather than the "romantic" heterosexual relationships that go awry following the rape. Moreover, there is a literal depiction of rape and *not* its mystification in such narratives, and the actual costs of the violence to the woman are centered. Importantly, these narratives do not end in the woman's death.

If rape is an "expression of male power," it is also sanctioned by the larger structures of power, explains Sunder Rajan, as she points out that most reported cases of rape in the country fall under the category of "institutional" rape, in that they are perpetrated by state forces, landlords, and so on; women of the "oppressed classes" are the majority of those violated and brutalized.[16] Let me now turn to a discussion of how this complex array of feminist approaches help us identify and locate key themes in the treatment of rape in popular Indian films.

Contexts and constructs

Popular Indian cinema is a dynamic and complex, if highly productive, field to study. This film industry is the world's largest; its films have been hugely popular not only in South Asia but across Asia, Africa, Latin America, and among their diasporas in the west. The Bombay film industry (now also known as Bollywood) is dominant within this cinema; its Hindi/Urdu films are an amalgam of various genres (drama, romance, action, comedy, musical, etc.) stitched together by overgenerous doses of melodrama. Popular films were generally treated by audiences, scholars, as well as critics until fairly recently as unrelenting celebrations of traditional and cultural "Indian" values, with certain exceptions, of course. However, as I show in this section, these films produced for mass consumption are neither simplistic reproductions of dominant ideologies nor are the "traditions" and "values" they depict static. Indeed, even as these films engage mainstream ideologies and hegemonic norms and customs, these are interrogated, contested, and reworked, as are their underlying structures of power. This is the case even when the conflicts depicted are resolved in ways that contribute to upholding the prevailing status quo, for the ruptures and fissures that need to be sutured over to get to happy endings reveal problems, conflicts, and tensions that are deeply embedded within the social and political fabric.

Film and film culture (both representations and practices) have undergone striking changes in India. If the classic nation-building texts of the 1950s (*Shree 420* 1955; *Mother India* 1957; *Navrang* 1959, etc.) are renowned for their portrayals of strong female leads and concern about socio-economic inequalities within the newly independent society, the films of the 1970s reflected the political pessimism of the time. The trope of the "angry young man" that dominates the landmark films of this period provide important glimpses into the widespread public disillusionment with the corruptions of the elites as well as the state (*Deewaar* 1975; *Trishul,* 1978); these films were also marked by a shift in depictions of female characters that relegated them to a secondary role, while the seething and righteous hero dominated the text. By the 1990s, a neoliberal ethos took center-stage in the celebration of the middle-class lifestyle in family-oriented films as globalization, with its consumerist cultures, took hold in the country (*Dilwale Dulhaniya Le Jayenge* 1995; *Hum Aapke Hain Koun* 1994; *Pardes* 1997). These films were soon followed by the "terrorist" genre in the late 1990s and 2000s, which reflected the post-Babri Masjid/Ayodhya decades of resurgent Hindu nationalism, communal violence, and Islamophobic discourses (*Bombay* 1995; *Mission Kashmir* 2000; *Fanaa* 2006, etc.). The point here is that popular films have not just been reflections, but integral components, of the social and political shifts underway within the nation-state, as they were within regional and global politics. These thematic developments in narrative plots as well aesthetic styles mirrored, negotiated, reworked, and sometimes contested the changing ideas about nationhood and its dilemmas as encountered by the films' communities of address.

Gender, sexuality, violence, exploitation—these are the key themes that shape the core conflicts, albeit in different ways, in these popular films. The characterization of women to signify "good" or "bad," "traditional" or "fallen," individuals, families, and communities set them up as either rape-able or not. Many of these characterizations were formulaic, but once the researcher moves beyond caricatured images of virtuous and virile heroes (slim, trim, and amazingly dexterous), hapless tradition-bound and/or modern female leads (even slimmer and trimmer, impossibly "fair and lovely"), ruthless villains (handlebar mustaches, diabolical laughter, maniacal grins) that even now shape the depiction of sexual violence, she can begin to unpack how their spectacularization of anti-woman violence interacts with the web of fantasies, desires, and power hierarchies that constitute the body politic itself. Tracking which "woman" is construed as violate-able, as victim, and which "woman" is depicted as respectable or empowered is key to reading the representational strategy and the sexual-national politics at work in these texts, in their "encoding" of particular ideologies. Attending to the caste, class, religion, and ethnicity of the "woman" as reading practice reveals how women are marked as conforming to the "national," "cultural," and "communal" norm, and hence how gender functions to mark and police these borders. The question of how sexual violence functions to transgress, secure and negotiate the limits of the social can thus be opened up for study.

A number of scholars of this cinema and its treatment of gender relations have argued that "… popular Indian films play a role in sexual violence by teaching men to equate sex with force as part of a normal script for expressing love and developing relationships with women."[17] Critical feminist scholars, however, have adopted more nuanced reading practices to note that these films are also important sites that open up transgressive possibilities through their deployment of women's agency in resistance to such violence.[18]

A study of the popular films of the 1980s and early 1990s finds that 56% of the sample showed women being sexually assaulted and harassed, and 40% of the sexual scenes in the films of the late 1990s included some form of violence.[19] Moreover, in their study of film culture during 2000–2012, Manohar and Kline argue that this socializes young people "… into gender roles and scripts that produces misperceptions about violence against women…."[20] Identifying a "sexual assault script" at work in the films, these researchers describe their depictions as typically featuring single males as perpetrators and "traditional" women as victims. The women are blamed for the attacks they experience and the films end with either the death of the perpetrator at the hands of the women or/and their families, or with his getting away with the violence. Unfortunately, the analytic frame used in this study relies on neo-Orientalist constructs of "ancient" Hindu civilization and modern progressivism (so, for example, the authors identify "existing cultural beliefs" as informing the representation of sexual violence but they do not interrogate how these "cultural beliefs" are actually constructed. It is therefore significant that the study nevertheless finds women's active resistance to assault and violence to be a "cultural expectation," which suggests that these films demonstrate the potential for the emergence of "new beliefs" about women and violence. It is also significant that this study identifies the perpetrators as usually single men of the upper castes. In the following section, I show how many of the popular films explicitly centered on the theme of rape complicate dominant ideas about "tradition" and 'culture', they also raise important questions about specific forms of women's agency and empowerment.

Victims, killers, or goddesses?

The most significant films of the "rape" genre, all box office hits, offer stinging critiques of the failure of the family, community, state, and law to protect women from sexual violence. In most of these texts, the raped woman either stands in for the nation or, depending on her class, caste, and religion, is rescued by a subject who does. *Insaaf ka Tarazu* (The Scales of Justice, Chopra 1980), *Prem Rog* (Love Sickness, Kapoor 1982), and *Damini* (Lightening, Satoshi 1993) are all melodramas with plots centered on rape. As these films are made by some of the industry's most notable filmmakers and their female characters are essayed by the top actresses of their day, they provide good examples for studying how the trope of rape functions in the nation's politics of belonging.

Insaaf ka Tarazu, predating yet in many ways anticipating *Zakhmi Aurat*, critiques the evidence-based criminal justice system, that is, its insistence on eye witnesses to sexual violence in order to convict the perpetrators. The film also challenges the assumption that the "degraded" character of a woman (the heroine works as a model) makes her rape an impossibility. Confronting this stereotype, the heroine's claim to justice is based on the truth of her experience, on the violation of her bodily and sexual integrity. Her insistence on this truth opens up her understanding of the power relations and structures of oppression at work in her life, and in the lives of other women, that make them vulnerable to the men who prey on them. The woman lawyer who defends the heroine (named Bharti, a synonym for the nation) warns her that the rape trial will be yet another form of "rape," as she bases her legal strategy on the idea of consent. The question of consent would reframe the feminist anti-violence strategy during the 1990s, popularized in the slogan, "No Means No." The film itself defines the law as an impossible site for the emergence of woman as a violate-able subject, and thus sets the stage for *Zakhmi Aurat*, in which women take the law into their own hands. Paradoxically however, both films' presentation of women's agency conceives of this in terms of vigilantism, a "feminist" parallel to the violent "angry young man" who would dominate the cinema of the period.

Prem Rog and *Damini*, for their part, critique the family as the site of sexual violence, and the community for enabling—if not outright sanctioning—this. The elite and middle-class families featured here are unable to protect women from being raped by their men, whether these men are situated higher (*Prem Rog*) or lower (*Damini*) within its family hierarchy. Fathers and mothers are unable to protect daughters; male and female employers are unable to protect women domestic workers. Whereas *Damini* critiques the middle-class upper-caste family for protecting its honor by silencing women who are raped in the home, *Prem Rog* exposes the multi-level networks within the upper-caste Thakurs that facilitate the sexual violence in their community. The film identifies widows as particularly vulnerable and condemns their treatment within the family and larger society. Both films also expose the hypocrisy of the other castes/classes, who uphold the social ostracism of widows and the stigmatized status of raped women due to their own aspiration for upward social mobility.

While *Insaaf Ka Tarazu* presents the rape of women as an attack on the nation itself, *Zakhmi Aurat* shows how powerful men manipulate law and the criminal justice system. *Damini* implicates the medical establishment for pathologizing unruly women, locking them up in mental asylums when they resist family pressure, and *Prem Rog* exposes how the concept of "honor" functions to make women sexually available to male kin. Certainly these films reproduce problematic constructs of woman-as-nation, they redeem "modern" husbands and male lovers who stand with the heroines, but they also imagine cross-class, cross-caste, cross-religious alliances among women that emphasize female solidarity in their acts of resistance. Such resistance counters, albeit in ways that are contained at the films' end, the negation of women's agency and their presentation as unthinking and willing to sacrifice their integrity—in sexual terms—for either family or nation. Indeed, these heroines' condemnation of family relationships, law, the juridical system, the state, when read in tandem with their call on a higher justice carries considerable subversive potential. I do not mean to suggest these films should be read as "feminist" texts. What I am arguing here is that they engage with feminist critiques in ways that demand attention, so that despite their too hasty closing down of their own transgressive possibilities, they do play a significant role in disseminating feminist ideas and ideologies.

The "modern" man represented in these films usually has access to masculine, caste, and class privilege, but his depiction complicates this masculinity in very specific ways. He either upholds the family or social power structure and is confronted by the women for doing so (*Damini*; *Prem Rog*), plays second-fiddle to the resisting heroine (*Insaf ka Tarazu*; *Zakhmi Aurat*), or is himself in a position of relative disadvantage within the male caste/class/patriarchal hierarchy (*Prem Rog*). Indeed, in *Zakhmi Aurat*, the sympathetic judge resigns his position at the end of the film, signaling the inability of the masculine power structure and patriarchal state to deliver justice to women. My point here is that to read this genre of films as unrealistic rape-revenge fantasies is to miss the often trenchant critiques they offer of masculinity and the existing social order, notwithstanding their numerous limitations.

The more recent crop of sexual violence films (*Pink*, *Mom*) rework in interesting ways the critiques of the state, law, and male power proffered by the rape films of the 1980s and 1990s. In *Pink*, for example, the older male lawyer who represents the three young women who are sexually assaulted by the sons of rich and powerful politicians becomes a father figure to them, symbolizing the wise patriarch who stands in for the state, as the narrative unfolds. In this character, male power and privilege are reinscribed; the lawyer is the heroic savior of the women's honor. The hero is thus a vindication of the law and its male-dominated structure and institutions. Likewise, *Mom* empties out the subversive potential of the woman who avenges her daughter's rape. The titular mother kills the young men who have gang-raped her daughter, but she does so under the watchful eye of a sympathetic male police officer, even using his gun to kill the last of the rapist gang. Moreover, the mother's main motivation throughout is to win her stepdaughter's affection, for the young woman refuses to accept her as stepmother. It is only after her rape is fully avenged that the daughter accepts her stepmother. There is no notion of feminist solidarity here, as the mother's sole objective is to end the alienation that pains her husband and herself, and in the process, keep her upper-middle-class family intact. Equally problematic is the film's appropriation of, and reworking, of class politics and struggle. The filmic rape is based on the actual rape of a young student/intern, daughter of a village family who sold their land to pay for her education. The gang rape led to the death of the young woman and provoked outrage across India and around the world. The young woman's pseudonym was "Nirbhaya" (Fearless) and the incident became known as the "Delhi gang rape" (2012). The fictional rape in the film *Mom* reproduces pertinent details of the actual gang rape, which took place on a moving bus. However, the film inserts the upper-class/-caste daughter as the victim, displacing "Nirbhaya," who was from a marginalized community of modest background. The film's rape victim is an elite and privileged young woman and the film thus rewrites the well-known case, erasing the compounded threat to single women from a lower economic stratum to replace her with a more "worthy" victim. Where the real-life family publicly struggled for justice for their daughter, the fictional stepmother's actions have the depoliticized and sentimentalized objective of winning the daughter's love.

Mom's exploitation of the Delhi rape case to uphold the sanctity of the elite, professional family differs dramatically from *Prem Rog*, a film that addressed women's complicity in reproducing class and caste prejudices and violence. Although the young Manorama loves Devdhar, her caste/class investments preclude her from recognizing her own desire. It is only after her upper-caste/-class brother-in-law rapes her that Devdhar, from a lower caste/class, stands a chance at winning her love. The upper-class/-caste women in the family, Manorama's mother and sister-in-law, try to protect her from the brother-in-law, but their own class/caste loyalties prohibit this (the sister-in-law covers up the rape committed by her own husband; the mother counsels her daughter to keep silent in order to preserve her own family). Ultimately, the mother reveals the rape to the family patriarch…but only when the rapist brother-in-law attempts to return Manorama to the fold of his family, where he can secure sexual access to her. That the mother and the sister-in-law are unable to protect the heroine because their own lives are subject to caste/class constraints is evident in their struggle to reconcile the rape with maintaining their own privileged status within the family. The film ends "happily" with Manorama's union with the lower-class self-made 'modern man, Devdhar, who "saves" this upper-caste/-class rape survivor from a lifetime of sexual exploitation. His democratic and secular ideals, however, remain in uneasy tension with his willing subservience to the "good" patriarch of Manorama's Thakur family.

Some of the rape films with the most affective power present their victims as Devi, or Goddess. They draw on popular Hindu religious iconography to present the nation as Hindu and the rape victim as its daughter. These films typically construct the assaulted woman—or her bene-factress—as personally chaste and pure of spirit. She is thereby rendered worthy of symbolizing the Goddess, of being possessed by the spiritual strength and supernatural power, the Shakti, of the Devi. Drawing on the Hindu Epics, the Mother Goddess is manifested here in her incarnation as either Durga or Kali. The violated woman, whether "traditional" in her orientation (*Daman*) or aspiring of modernity (*Damini*), becomes the embodiment of the Devi as she metes out her vengeance, transforming the rapist into a sacrificial demonic object whose bloody killing placates the awakened Goddess. Symbolizing the powerful female energy of the Goddess, this rape survivor's violence restabilizes a cosmic order based on the unending struggle between good and evil. In those films where the woman regains her lover's/husband's acceptance after the rape, she usually undergoes an agni pariksha (trial by fire) to get to her "happy ending." These films thus also draw on the figure of Sita, the ideal of wifely chastity in the Ramayana, who was compelled by husband, family, and community to prove her purity in a trial by fire. Once the woman's virtuous nature is publicly confirmed, she can be folded back into the family to resume her position as worthy companion of the Ram-like husband figure (*Prem Rog, Prem Granth*). The narrative structure and characters in popular films make ample "use of traditional elements from the epics," Bloom argues, this explains their "ongoing popularity" (1995: 172). This is certainly the case with many of the rape films referenced in this chapter. For example, Devdhar, the hero of *Prem Rog*, is depicted as a Krishna figure (playing the flute) and he recounts the Radha/Krishna narrative to challenge the villagers who accuse him and Manorama of immorality. This hero's casting as a secular, modern man does not contradict his representation as a Krishna figure, whose playfulness is seen in the hero's attempt to bring some comfort into Manorama's austere life as a widow. Indeed, the figure of this Krishna becomes coupled with that of Ram as he redeems 'his' own Sita.

The male heroes of the rape revenge films *Zakhmi Aurat, Insaf ka Tarazu, Damini*, and *Prem Rog* all stand by the women they love, often against the wishes of their family and the castigation of their community. They defy social norms by standing up to the public insults that attempt to humiliate them. Gupta-Casalle reads the male figure in such films (in particular, *Zakhmi Aurat* and *Damini*) as witnesses; she defines them as the "new" Indian man, modern and liberal, who uses his authority as male to sanction the woman's resistance since he too recognizes the state to be ineffectual and society to disavow women's legal and social agency (2000). My reading of *Zakhmi Aurat* and *Insaf ka Tarazu*, however, suggests these men are themselves quite ineffectual, pale shadows of the heroic women whose courage and strength shines through in the most banal of film genres. Indeed, the men hardly feature as central characters within the narrative, they become less and less relevant to the women's post-rape lives in which they move beyond their original desire for legal and social acceptance. These heroines reach a stage where they attach little or no significance to romantic love, or to the possibility of marriage. In films where the woman protagonist is an independent and confident decision maker in her own right before she is raped, her character becomes stronger and more determined, less likely to compromise in unfolding the post-rape events. The offer of marriage at these films' end seems superfluous, if not utterly trite. It is only in films such as *Prem Rog* or *Mom*, where the woman is strongly family-bound in her sense of self, that the character's achievement of her desire does not depend on her own independent action. Here, a male lover/husband/confidante becomes central to the action, it is the men who facilitate her advancement and bring her actions to fruition. This is not the case in *Insaf ka Tarazu* or *Zakhmi Aurat*; the distinction is no small matter in thinking through depictions of women's agency in popular culture.

To sum up this section, by the time the film *Mom* is made, the rape revenge film is no longer an unambiguous call for justice by the women who are raped. Instead, the prime motivation for this Mom-turned-killer is to save her family from disintegration when her stepdaughter blames her for the gang rape—the mother, a teacher, disciplines one of her students who then rapes the stepdaughter. There is no substantive critique of either the law or the family here; instead, the stepmother's desire for the estranged daughter's acceptance is what leads her to exact revenge. Thus, restoring the upper-middle-class nuclear family to "normalcy" drives the post-rape events in *Mom*. While the Delhi rape case, on which the film's narrative hinges, galvanized national and international feminist mobilization against rape and sexual violence, the gang-rape in *Mom* remains a purely private affair. It is resolved quietly. Moreover, the mother is successful in her revenge only as a result of the intervention of not one, but two men, a police officer and a private detective.

Directions for future research

The treatment of rape in popular films opens a window into understanding the organization of social relations, power hierarchies, and the formation of the self and community at particular historical junctures. Just as critical race and post-colonial feminists have debunked the idea of a universal "woman," there can be no unidimensional theory of rape, no single all-encompassing 'explanation' for the motive, meaning, and consequence of rape, or of the conditions of its perpetration. Critical feminist scholars and activists have long argued that rape and sexual violence are neither individual aberrations nor private encounters, but widespread cultural-political practices that help organize the family, community, and larger social order within which these institutions function. Sexual violence is thus always contextual and best approached contextually in order to unpack the specificity of the power relations at work.

Cinematic depictions of raped women as chaste but helpless victims, exotic and sexually available, or enraged (proto)feminists who kill in the name of justice are to be found across various genres, from melodrama to action and art films. These representations inform ideas about "woman" and femininity, as well as "man" and masculinity. Taking a broad historical approach to the treatment of rape across these genres allows the researcher to situate sexual violence in the particularities of the sociocultural and political moment of its perpetration. Using an intersectional approach that attends to how various axis of power crosscut to shape the lives of "women" demonstrates that the constitution of gender, sex, and sexuality (like caste, class, religion, and nation) is never static. The challenge to film and media scholars is how to conceptualize and study these linkages without losing sight of their nuances and complexities. Moving beyond essentialist approaches to gender is a necessary requirement for such careful readings.

Sexual violence remains vastly underreported and under-prosecuted, with its consequences under-theorized in India as elsewhere. The representation of rape in one of the most powerful cultural industries, film, is thus a crucial area for study. How is rape treated in popular culture? What does the depiction of rape accomplish? What do particular representational strategies of rape reveal about dominant conceptions of masculinity and femininity, about the efficacy of legal responses, about the family and the larger community? Can rape revenge films be considered "feminist" responses to the sexual violation of women? What do the rape revenge films reveal about the present high level of support among young women in India for capital punishment for sexual offenders, about women's public celebration of the notorious "encounters" in which police officers kill alleged rapists, as was recently the case? What methodologies are useful for comparative research in cross-cultural contexts? In what ways does the representation of sexual violence against gender/sexual minorities differ from that against middle class women? Moreover, feminists have been particularly incensed by the song and dance numbers in popular Indian films for their objectification of women, particularly the "item songs" that are now ubiquitous. This development requires close attention, particularly with regard to what these performances reveal about the psychic structure of sexual desire. What representational strategies disrupt the pleasures generated by the seeming glorification of rape in many of these particular song sequences? What are the social consequences of celebrating vigilante acts of violence by women characters as revenge for sexual violence? What are the ranges of options available to women as they attempt to end sexual violence? What do these options reveal about popular conceptions of female sexual agency? Clearly, much work remains to be done on these, and many other, questions in the study of popular Indian cinema.

Notes

1 Sing, *Bobby Talks Cinema.*
2 Bedi, *Dare to Do!*

3 Hautzinger, *Criminalising Male Violence.*
4 Rudisill, *Encyclopedia of Social Movement Media.*
5 Projansky, *Watching Rape*, 2.
6 Brownmiller, *Against Our Will*; Daly, *Gyn/Ecology*; Dworkin, *Intercourse*; MacKinnon, *Towards a Feminist Theory*; Whisnant, *Feminist Perspectives on Rape.*
7 Mulvey, *Visual Pleasure*; Oliver, *The Male Gaze is More Relevant.*
8 Brownmiller, *Against Our Will*; MacKinnon, *Towards a Feminist Theory*; Horeck, *Public Rape.*
9 Davis, *Women, Race and Class*; Mohanty, *Third World Women*; Crenshaw, 1991; Hill Collins, *Black Feminist Thought*; Spivak, 1988.
10 Horeck, *Public Rape*, 10.
11 Horeck, *Public Rape*, 3.
12 Shohat and Stam, *Unthinking Eurocentrism*; Mercer, 1991; Fung, 1991; hooks, 1992; Rony, 1996; Arya, 2012.
13 Loomba and Lukoose, *South Asian Feminisms.*
14 Basu, *Sexual Property.*
15 Sunder Rajan, *Life After Rape*, 72.
16 Sunder Rajan, *Life After Rape*, 78.
17 Manohar and Kline, *Sexual Assault Portrayals*, 235.
18 Virdi, *Reverence, Rape – and then Revenge.*
19 Manohar and Kline, *Sexual Assault Portrayals.*
20 Manohar and Kline, *Sexual Assault Portrayals*, 233.

References

Aap Aye Bahar Aayee (As Soon as You Arrived, Spring Arrived). Directed by Mohan Kumar, 1971 (India).

Ansari, Usamah. "There are Thousands Drunk by the Passion of These Eyes," *South Asia: Journal of South Asian Studies*, XXXI (2): 290–316, 2008.

Arth (Meaning). Directed by Mahesh Bhatt, 1982 (India).

Bandit Queen. Directed by Shekhar Kapoor, 1994 (India).

Basu, Srimati. "Sexual property: staging rape and marriage in Indian law and feminist theory," *Feminist Studies*, 37(1): 185–211, Spring, 2011.

Bhumika (Role). Directed by Shyam Benegal, 1977 (India).

Bedi, Kiran. *Dare to Do! For the New Generation*. Hay House, 2012.

Bombay. Directed by Mani Ratnam, 1995 (India).

Brownmiller, Susan. *Against Our Will: Women and Rape*. New York: Simon & Schuster, 1975.

Daly, Mary. *Gyn/Ecology: The Metaethics of Radical Feminism*. Boston, MA: Beacon Books, 1978.

Daman: A Victim of Marital Violence. Directed by Kalpana Lajmi, 2001 (India).

Damini (Lightening). Directed by Rajkumar Santoshi, 1993 (India).

Davis, Angela. *Women, Race and Class*. New York: Random House, 1981.

Deewaar (The Wall). Directed by Yash Chopra, 1975 (India).

Dil Se (From the Heart). Directed by Mani Ratnam, 1998 (India).

Dilwale Dulhaniya Le Jayenge. Directed by Aditya Chopra, 1995 (India).

Dushman (Enemy). Directed by Tanuja Chandra, 1998 (India).

Dworkin, Andrea. *Intercourse*. New York: Basic Books, 1987.

Fanaa. Directed by Kunal Kohli, 2006 (India).

Firaaq. Directed by Nandita Das, 2008 (India).

Gulaab Gang (Rose Gang). Directed by Soumik Sen, 2014 (India).

Hautzinger, Sarah. "Criminalizing male violence in Brazil's women's stations: from flawed essentialism to imagined communities," *Journal of Gender Studies*, 11(3): 243–251, Aug, 2010.

Heaven on Earth aka *Videsh*, 2008 (Canada and India).

Hill Collins, Patricia. *Black Feminist Thought: Knowledge, Consciousness and the Politics of Empowerment*. New York: Routledge, 2009.

hooks, bell. *Black Looks: Race and Representation*. New York: Routledge, 1992.

Horeck, Tanya. *Public Rape: Representing Violation in Fiction and Film*. New York: Routledge, 2004.

Hum Aapke Hain Koun (Who Am I to You?) Directed by Sooraj Barjatya, 1994 (India).

Insaf ka Tarazu (The Scales of Justice). Directed by Baldev Raj Chopra, 1980 (India).

Kabir, Jahanara Ananya. "Allegories of Alienation and Politics of Bargaining," *South Asian Popular Culture*, 1(2): 141–159, 2003.

Loomba, Ania and Lukose Ritty A. (eds). *South Asian Feminisms*. Durham, NC: Duke University Press, 2012.

MacKinnon, Katherine. *Toward a Feminist Theory of the State*. Cambridge, MA: Harvard University Press, 1991.

Manohar, Uttara and Kline Susan L. "Sexual assault portrayals in Hindi cinema," *Sex Roles*, 71:233–245, 2014.

Mission Kashmir. Directed by Vidhu Vinod Chopra, 2000 (India).

Mohanty, Chandra Talpade. "Under western eyes: feminist scholarship and colonial discourses," *Boundary 2*, 12: 3, 333–358, Spring-Autumn, 1984.

Mom. 2017. Directed by Ravi Udyawar, 2017 (India).

Monsoon Wedding. Directed by Meera Nair, 2001 (India/US).

Mother India. Directed by Mehboob Khan, 1957 (India).

Mulvey, Laura. "Visual pleasure and narrative cinema," *Screen*, 16:6–19, 1975.

Navrang. Directed by V. Shantaram, 1959 (India).

Oliver, Kelly. "The male gaze is more relevant, and more dangerous, than ever," *New Review of Film and Television Studies*, 15(4): 451–455, 2017.

Pardes. Directed by Subhash Ghai, 1997 (India).

Parzania (Heaven and Hell on Earth). Directed by Rahul Dholakia, 2005 (US/India).

Pink. Directed by Aniruddha Roy Chowdhury, 2016 (India).

Prem Rog (Love Sickness). Directed by Raj Kapoor, 1982 (India).

Projansky, Sarah. *Watching Rape: Film and Television in Postfeminist Culture*. New York: New York University Press, 2001.

Rudisill, Kristen. Social Movement Media in the Emergency (India), in Downing, John D. H. (ed.) *Encyclopedia of Social Movement Media*. Sage Publications, Inc., 2020, 499–500.

Roti, Kapda Aur Makaan (Food, Clothing and Shelter). Directed by Manoj Kapoor, 1974 (India).

Salaam Bombay! Directed by Mira Nair, 1988 (India/US).

Shree 420. Directed by Raj Kapoor, 1955 (India).

Shohat, Ella and Stam Robert. *Unthinking Eurocentrism: Multiculturalism and the Media*. New York: Routledge, 1994.

Sing, Bobby. "Zakhmi Aurat," *Bobby Talks Cinema*. 2012, September 9. www.bobbytalkscinema.com/recentpost/Zakhmi-Aurat-1988-One-of-t-963

Sunder Rajan, Rajeshwari. "Life After Rape: Narrative, Rape and Feminism," *Real and Imagined Women: Gender, Culture and Postcolonialism*. New York: Routledge, 1993, 64–82.

Trishul. Directed by Yash Chopra, 1978 (India).

Virdi, Jyotika. "Reverence, Rape – and then Revenge: Popular Hindi Cinema's "Women's Films," in Burfoot, Annette (ed.) *Killing Women*. Waterloo, Ontario: Wilfred Laurier University Press, 2006, 251–272.

Water. Directed by Deepa Mehta, 2005 (Canada, India, US).

Whisnant, Rebecca. "Feminist Perspectives on Rape," *Stanford Encyclopedia of Philosophy*. June 21, 2017. https://plato.stanford.edu/entries/feminism-rape/

Zakhmi Aurat (Wounded Woman). Directed by Avtar Bhogal, 1988 (India).

28

SPEAK UP, SIS

Black women, race, and news coverage of the Me Too movement

Tia C. M. Tyree

Introduction

January 2019 marked an important moment for sexual abuse allegations in the Black community. On that date, the *Lifetime* television network began airing a three-night documentary series entitled "Surviving R. Kelly." Drawing in about two million viewers each night and garnering an incredible amount of "chatter" across social media, it explored the abuse allegations against R&B singer Robert (R. Kelly) Kelly.[1] According to the network's official website, women were "emerging from the shadows and uniting their voices to share their stories" and "shedding light on the R&B star whose history of alleged abuse of underage African American girls has, until recently, been largely ignored by mainstream media."[2] While rumors and news reports about R. Kelly were in the public eye for decades, the climate long tolerant of these types of situations had changed, and it all started with one single tweet.

In October 2017, Alyssa Milano, a white actress, sent a tweet encouraging women to speak out about sexual harassment and assault. She wrote, "If you've been sexually harassed or assaulted write 'Me Too' as a reply to this tweet." The tweet referenced a suggestion "from a friend" and noted, "if all women who have been sexually harassed or assaulted wrote 'Me Too.' as a status, we might give people a sense of the magnitude of the problem."[3] A firestorm of responses eventually sparked the Me Too movement known today. In fact, in about one day, there were over 12 million #MeToo posts on social media, and despite the tweet coming late in the year, *Time* magazine honored "The Silence Breakers" as their Person of the Year 2017, signaling a celebration of those who had now "had it" with their mistreatment and started a revolution "simmering for years, decades, centuries."[4]

The Me Too movement is now a true social movement that is said to be a "narrative movement by people, primarily women, telling their stories of sexual harassment or assault."[5] Yet, almost immediately after its development, activists had to advocate for acknowledgment of the Black woman who deserved credit for originating the Me Too movement: Tarana Burke. Burke launched the campaign 10 years before Milano's 2017 tweet. It was a part of a grassroots organization created to address the lack of services for Black women and girls in underprivileged communities who experienced abuse, especially sexual assault.[6] Unlike the widespread usage of "Me Too" on social media after 2017, the original phrase was previously only privately shared among advocates and survivors of sexual assault.[7] According to Burke,

It wasn't built to be a viral campaign or a hashtag that is here today and forgotten tomorrow. It was a catchphrase to be used from survivor to survivor to let folks know that they were not alone and that a movement for radical healing was happening and possible.[8]

According to Burke's website, her movement is about "empowerment through empathy and community based action," and since its start, the guiding vision was "to address both the dearth in resources for survivors of sexual violence and to build a community of advocates, driven by survivors, who will be at the forefront of creating solutions to interrupt sexual violence in their communities." Further, Burke's website acknowledged how the viral #MeToo hashtag "thrust" the conversation of sexual violence not just into the national dialogue, but also into conversations around the world. The grassroots campaign's message was now reaching survivors across the globe from varying communities and aiding in the destigmatization of surviving sexual violence by bringing attention to its scope and impact. What, too, was now acknowledged on Burke's website was the goal to "reframe and expand the global conversation" to address "young people, queer, trans, and disabled folks, Black women and girls, and all communities of color."[9]

Quickly, or "six months" later as the website notes, the movement had shifted Me Too efforts *away* from centering on Black women and girls. This is not unique, as Burke herself noted. She said, "many times when White women want our support, they use an umbrella of 'women supporting women' and forget that they didn't lend the same kind of support."[10] However, Burke added, "I think that women of color use social media to make our voices heard with or without the amplification of White women."[11] This begs the answers to several important questions: why and how do Black women and girls get ignored and left behind in larger conversations and issues in U.S. society, and how do individuals like R. Kelly with long lists of accusers remain unpunished for their behavior?

This idea grounds the chapter's inquiry, as there is an undeniable need to investigate how Black women and sexual violence and assaults against them are framed in news media, including those focused on providing news about and to the Black community. Taking into consideration the news-making process and interconnectedness of race, class, and gender, this much-needed analysis fills a gap in mass media studies by investigating Black women—an often-overlooked group within scholarly literature—as well as Black news outlets. Following the groundbreaking work of Marian Meyers in her study of news coverage of Black women, a discourse analysis was utilized to examine how sexual violence, assaults, and perpetrators of violence against Black women were framed in Me Too movement news coverage. The goal was to assess whether longstanding practices of stereotyping, exclusion and the dismissal of Black women's voices and experiences occurred during the early framing of the Me Too movement in U.S. online Black news sources.

While scholars like Michelle Rodino-Colocino argue the #MeToo movement works to challenge systems of power that establish the foundations for harassment, discrimination, and assault by promoting empathy from the ground up, systems in the United States have not treated Black people, including Black women, fairly.[12] Race and gender-based discrimination against Black women is well documented in every significant social system, including mass media.[13] While people of all ages and races experience sexual violence, the problem is racialized at a social scale in the United States, with privilege being afforded to whites. Historically, sexual violence on U.S. soil has been a mechanism to control African Americans, women, poor people, and others, and with respect to Black women, it has carried no public name, attracted no notable large-scale public disapproval, and has been compartmentalized as a "gender issue," diverting attention away from Black political efforts to fight racism.[14] By virtue of their treatment at the hands of white slave owners, Black women were sexual violence victims "deemed inherently rapable" by both white male slave masters and other Black men who were often forced to rape them.[15] Further, according to Gerda Lerner, it was Black women's femaleness that made them sexually vulnerable to racist domination, but their Blackness excluded them from protections afforded to white women.[16] This colonial ideology has not disappeared and is still seen in popular culture and in the rates of violence against Black women in society.[17]

White privilege is real in the United States. In fact, the logical rationalizing of white domination is so embedded in the collective consciousness of U.S. society that is difficult to be freed of it.[18] Cheryl Harris asserts it is embedded in the benefits, assumptions, and privileges that pair with the status of being white, and whiteness has become an asset white people have worked to protect.[19] Further, whites expect and rely on those benefits, which are supported, legitimized, and protected by the law. Because of the domination of whites over Blacks, white identity became the basis of racialized privilege, which was confirmed and legitimized through laws, such as formalized segregation. Further, since Black women emerged from slavery as collective rape victims, they have historically been encouraged to keep quiet in ways that promote the longstanding belief they somehow encouraged or wanted such sexual assaults.[20] For centuries, Black women have endured systematic mistreatment, both sexually and socially, in the United States, and the Me Too movement created a place where these issues collided in public spaces.

The beginning of the Me Too movement was fueled by reactions and responses to sexual harassment incidents involving high-profile men, whether confirmed or alleged, against women celebrities or women in other industries and contexts.[21] It is not uncommon for those in more privileged gendered classes or with a higher status, high visibility, and high-profile occupations to receive different treatment, and studies of women entering male-dominated employment areas have long reported more harassment.[22] Yet, many of the abusers and accusers stories garnering media attention as a part of the Me Too movement were white celebrities and public figures, and as asserted by Leigh Gilmore, the "he said/she said" must also be disrupted for women of color. This is especially the case because those environments where abuse of power by men thrives must be recognized as spaces in which similar forms of violent abuse and discrimination, such as racism and homophobia, also exist.

U.S. news, news production, and framing

U.S. media narratives often perpetuate the historical, political, and cultural marginalization of Black women's experiences and Black women's pain. The problematic portrayals of Black women in mass media are multifaceted and longstanding, so all cannot be addressed in this one chapter (and other authors in this book investigate several related aspects of these larger issues). Thus, this study will focus on uncovering how Black women were framed specifically in Me Too movement stories appearing in Black online news sources as well as the roles women played in creating those news narratives.

First, understanding the structure of media and news is critical. By definition, news is "new information about a subject of some public interest that is shared with some portion of the public."[23] News is a tying element in societies of diverse people, as it works to establish a shared sense of current information by which the community operates, and those who write for mass media have important jobs with tremendous impact on the direction and shape of their communities.[24]

Despite newspapers' dwindling readerships and the varied options for news gathering in social media, newspapers are still major sources for understanding how issues and events are presented in our society. In fact, a Pew Research Center survey in December 2018 showed news websites and print newspapers were among the top five ways U.S. citizens obtained news; others were television, radio, and social media.[25] Media are powerfully influential in shaping what people think about or how they think about specific issues or topics based on their ability to select the stories that gain media attention, also known as "agenda setting."[26]

Newspapers must serve their readers to continue to publish. While the shared aim of providing news to inform society is central to a press outlet's existence, how it is done, the tone it uses, and the sources it taps to provide voices and expertise in stories considerably differ. These are tailored to ensure profitability for each newspaper publisher. In fact, newspapers have latitude in what they cover, and several factors influence their reporting, including multiple biases that can lead to the omission of certain stories, limits of coverage based on economic necessity, and the activities of their competitors.[27] In many ways, this is still reflective of Johan Galtung and Mari Holmboe Ruge's research from more than 50 years ago outlining key factors that influenced news value, which were prominence, impact, proximity, frequency, currency, continuity, uniqueness, simplicity, personality, predictability, exclusivity, and negativity.[28] Several of these are relevant to the Me Too movement, including the impact of the story, as the movement connected to literally millions of people, and the prominence of the actors (i.e., people involved), since they were initially elites who were powerful in the United States.

News is not created in a perfectly sterile and unbiased environment. Media owners with power can work to manipulate information for their own goals.[29] This can happen in several different ways, but most relevant to this analysis is the manner in which reporters are allowed to have a specific news slant (i.e., bias). In 1940, Samuel Hayakawa created the term "slanting," defined as "the process of selecting details that are favorable or unfavorable to the subject being described." News slant has been well documented in communication research, as it relates to news creation (e.g., Groseclose and Milyo).[30] Yet, researchers have also seen how consumers may prefer slanted coverage to support their cognitive biases, and they, too, may trust sources more that report information conforming to their personal beliefs.[31]

In terms of what consumers like, Western audiences are intrigued by crime and justice, and mass media play an important role in establishing the public's knowledge base through the construction of criminals, victims, law enforcement officials, and deviants.[32] Crime stories often provide very intimate details of the private lives of individuals, and when violence and sex are included in stories, they provoke "an instinctual need to remain alert to the whereabouts of potential threats and potential mates."[33] Further, by reporting on the punishment connected to criminal activities, the public becomes aware of the association of crime with punishment, helping to clarify and reinforce what is acceptable and unacceptable behavior in society.[34] Media arguably also play a role in the development of the public's opposition to or support of social movements, as people gain information from the media to make their decisions.[35] The Me Too movement is unique in that it is one of the few recent social movements, similar to the Black Lives Matter movement, bringing these together via social media activism. Yet, the critical focus of this chapter is on understanding how media portray Black women when these issues collide.

How a story is framed in the media can determine how viewers or readers think and feel as well as create blame or establish responsibility for an issue, and frame manipulations can make certain facts about an issue seem more important.[36] The frame is the group of facts packaged to create a story, and an issue frame involves the theme, storyline, or label used to suggest a favorable interpretation of a policy question.[37] Elements of a story frame include language, presentation of the story's beginning and overall focus, as well as sources and opinions.[38] Ultimately, as noted by Hans Magnus Enzenberger, no writing, broadcasting, or filming can be considered without attention to manipulation, and the larger question should not be *whether* media products are manipulated, but *who* is doing the manipulation.[39] One answer to this question can be reporters. This is largely based on the ability of reporters to display several representations through the selection of which part of an issue to "include and emphasize" or to "ignore or downplay" in a story.[40]

Negative frames of Blacks in the media stem from the racial and structural makeup of those who both own media and work as journalists. U.S. society is majority white, and "white ethnocentrism prevails, attributing a positive image to the whites and a negative one to other racial groups."[41] The media are a predominate way in which this occurs. As Hall states, "The media serve, in societies like ours, ceaselessly to perform the critical ideological work or 'classifying out the world' within the discourses of the dominant ideology."[42]

With respect to feminist thinking, language reflects experiences of men and its categories do not align with women's lives.[43] Voices and words matters in the media. Who is quoted and seen as an authority makes a difference to those who are reading news. Journalists in mainstream media tend to interview "knowns" as sources, which is a problematic as it shifts power, authority, and knowledge to a small minority and can lead to the exclusion of others both outside of and even within a minority group.[44] The dependence on what is often a small network of sources or "insiders" can contribute to an unquestionable consensus viewpoint held by journalists who hold power over audience members.[45] Yet, multiple studies have drawn attention to the scarcity of women as sources in news, and it has been noted that women reporters have a greater tendency to cite women as sources.[46] Some scholars assert this is a conscious effort and may even be linked to them feeling more comfortable interviewing other women.[47] This possibility can be very important to consider when investigating the coverage of sexual harassment by women in the media, as women reporters may be more inclined to include women as sources in reporting.

Sources are also important in relation to the Black community and reporting by and for Black women about sexual harassment. As noted by Teresa Heinz,

> ...the politics of representation underlie the performance of community, identity, and Blackness. In articulating the authenticity of experience within a particular community, the quoted representatives are always speaking from a certain subject position and viewpoint that perhaps silences other voices. Within this process, both authority and power are also performed through constructed categories, such as Blackness, which imply separateness from other communities.[48]

For Black women in the United States, race and gender are intimately connected; race develops the way they experience gender, while gender constructs the way they experience race.[49] Within the dominant cultural ideology of U.S. society, Black women's experiences can be conceptualized as intersectional. Being both Black and women, they are often treated in ways that do not take into consideration their unique positioning in the United States, and thus their experiences can either be seen as the same as other women and Black people's experiences and "absorbed" into their collective experiences or ignored because they are too different.[50] Further, their particularly unique location within America's established social relations makes them the subject of "double discrimination—the combined effects of practices which discriminate on the basis of race, and on the basis of sex."[51]

Past news research about sexual violence tended to blame women in some way for their own abuse.[52] In fact, in a news study about violence against Black women, Marian Meyers concluded that media often took the blame away from men and, instead, made women responsible for their own victimization, which worked to establish and reinforce the guidelines for what is appropriate behavior for women in public. She also determined that Black men were more often criminalized for their damage to property than their abuse of Black women, a signaling of their value—or lack thereof—in society.[53]

Blacks and Black female representations in media

A pressing concern is that little news research focuses specifically on Black women, and when research has focused on Black people over the last 25 years, they were rarely used as news sources and primarily seen as problems, threats, or deviants who strain the social fabric that connects the United States.[54] Herman Gray's critique of media illustrates how deep structural problems within mass media create the negative racial depictions of Black people and stand as a form of ideology designed to keep whiteness connected to dominance and power and Black and brown minorities related to inferior social roles.[55] In fact, Gray notes that narratives in media

> ...presume and then fix in representation the purported natural affinity between Black criminality and threats to the nation. By fixing the blame, legitimating the propriety of related moral panics, these representations (and the assumptions on which they are based) help form the discursive logic through which policy proscriptions for restoring order...are fashioned.[56]

Robert Entman and Andrew Rojecki concluded the news media overrepresented Blacks as criminals and underrepresented them as victims.[57] A study of sex crimes concluded racism and class impacted coverage, with crimes against white victims more likely to be covered than crimes against Blacks, and rapes involving Black perpetrators and white victims were covered with "exaggerated frequency, class prejudice and racist stereotypes."[58] It is not just Black men being seen as violent either. Black women are also portrayed as aggressive or dangerous.[59] Historically, reporting of gang- and drug-related crime was often done as a "matter of public service," and by doing so, news accounts helped Black people and their neighborhoods become places framed as crime and violence ridden and thus naturalized their status as contrary to the public's best interests.[60] Further, stereotypes, such as the Black thug and crazy Black woman, can be deadly, when they are coupled with police powers within a society that is structured by intersectional forms of racial and gender dominance.[61] The falsehoods associated with the stereotypical images of Black people can cause problematic or violent responses toward them by law enforcement and citizens, often including unwarranted and even deadly force.

The differences between Black and white women are consistently positioned in a "good woman/bad woman" frame that has negative consequences in U.S. society.[62] This is frequently done by framing Black women as "Jezebels" and "naïfs," with the Jezebel's bad behavior and lapses in moral judgement related to being Black and poor.[63] Historically, sex is often the main theme connecting poverty and blackness, putting the Jezebel into sharper perspective, as her behavior links her to cultural assumptions about who she is and what she will do as a Black underclass woman.[64] The good woman is often portrayed through the Black lady stereotype. Patricia Hill Collins argues that the Black lady sharply contrasts images of poor and working class Blacks, and in being favored as respectable, the Black lady avoids the negative trappings associated with behaviors of the working class by "negotiating the complicated politics that accompany this triad of bitchiness, promiscuity, and fertility."[65] The adoption of the Black lady is directly related to the desire to avoid "deviance" and to claim the femininity white patriarchy refuses to bestow on Black women.[66]

In addition to the hypersexual Jezebel and the Black lady, multiple other stereotypes of Black women exist in news coverage, including the emasculating sapphire, Black bitch, and the powerful Black bitch.[67] Issues with reporting on Blacks and positioning their existence through stereotypes date very far back, and Black feminists argue male supremacist and white supremacist ideologies are reflected in existing stereotypes of Black women.[68] Patricia Hill Collins argues that sexism, racism, and poverty are naturalized and normalized by defining Black women as "stereotypical mammies, matriarchs, welfare recipients and hot mammas."[69] Black women have long been seen as sexually promiscuous, mammies, and welfare cheaters, and in U.S. iconography, Black women are "whores" and white women are placed on pedestals as Madonnas.[70]

Hope Landrine found Black women were more likely to be stereotyped as "dirty, hostile, and superstitious," while white women were more likely to be stereotyped as "competent, dependent, emotional, intelligent, passive...and warm."[71] In addition, as summarized by Pauline K. Brennan and Abby L. Vandenberg, four media studies (i.e., Bond-Maupin; Farr, "Aggravating and Differentiating Factors," "Defeminizing and Dehumanizing Female Murderers"; and Huckerby) focusing on comparative portrayals shared a similar main conclusion: "White women are more likely than minority women to have their behavior excused in some way."[72] Scholars also conducted research regarding portrayals of white female offenders versus Black female offenders in the media, and they found minority women were depicted less favorably than white women in front-page news stories from the *Los Angeles Times* and *New York Times*.[73]

Media coverage of violence against women rarely contains stories about violence against Black women, unless there is an unusual or sensational angle.[74] Yet, the issue of ignoring violence against Black women is not just reflected in newspapers; it is deeply embedded in Black communities and culture. Discussing violence against Black women in the Black community is controversial for several reasons. First, Anima Mama argues that since racists presume all Black people are violent, Blacks would prefer not to address Black violence.[75] If they do, it can perpetuate damaging racist representations of Black people and cause some who are aware of the dominant culture's portrayal of them to take a defensive posture. Second, when Black men are violent against Black women, Black ideologies point to racism as the cause, instead of Black gender relations. Mama exposes a misconstrued and highly problematic rationalization of violence in the Black community, suggesting white men beat Black men who then beat Black women who then beat their children. This all creates a warped myth based on "racial fiction" and not "social fact."[76]

There are a few highly publicized and relatively recent sexual harassment cases involving Black women and Black men that garnered tremendous U.S. media attention. These include coverage of U.S. Supreme Court Justice nominee Clarence Thomas, who Anita Hill accused of sexual harassment, and championship boxer Mike Tyson, who was eventually convicted of raping Desiree Washington (after many questioned Tyson's guilt based on Washington's voluntary visit to his hotel room).[77] Media stereotypes and relationships between Black community members caused many questions about Hill and Washington. The image of the hypersexualized Black woman was "successfully used" to discredit Hill, and she, too, was said to be harmed by the "myth" of the strong Black woman stereotype.[78] This occurred because many Black men and women struggled to understand why she did not immediately report him or "just lash back," which is tied to "That myth, based on the reality of generations of Black women straining their resources to make better lives for their families, leads some women today to drastically overestimate their individual strength relative to men."[79] In short, both cases point to the "continuing power" of the misrecognition of Black women as "promiscuous and sexually immoral," and showcase how even those in the Black community accept gendered and racist stereotypes created and perpetuated by "White social, political and economic institutions" that allow problematic depictions of Black women to infiltrate internal politics within Black communities.[80]

Ultimately, through media narratives, crime signifies Blackness, and Blackness signifies crime.[81] Thus, it can be problematic to frame Black women as victims of crimes in the news, when historically the media have largely ignored Black female issues, negatively portrayed Black women, and often framed violence against women, including Black women, as deserved or their own fault. This analysis investigated exactly how Black women and their allegations, including those involving sexual violence, sexual harassment, and sexual assault, were framed in the highly visible beginning of the Me Too movement.

Methodology

To examine representations of Black women in Me Too movement news articles, this analysis borrowed heavily from the methodologies of Pauline Brennan and Abby Vandenberg's study investigating the way race and ethnicity influenced how women were portrayed in the media and Marian Meyers's study analyzing representations of violence against Black women in news coverage.[82] Using discourse analysis, this study explored how Black women and violence were portrayed in news published in the six months following Milano's tweet. The time frame is significant, because it is a milestone marked on Burke's website as the turning point carrying her work to an audience beyond Black women and girls.

There are existing patterns of Black discourse, and Black discursive practices have an influence on the ways Black people read and respond to the social world.[83] Discourse is closely tied to ideology and the reproduction of social hierarchies, and analyzing it provides a mechanism to examine ideologies present within visual, spoken, and written texts.[84] Texts are important elements of social events, and they can create changes in social relations, the material world, and people's knowledge, beliefs, attitudes, and values.[85] Analyzing discourse is important, because it allows researchers to dissect power in media texts. As proposed by John Fiske, discourse analysis provides a framework to examine how language and representation create meaning and to understand relationships between power, meaning, and representation as well as investigate the construction of identities and subjectivities.[86] Further, discourse analysis, according to Teun Adrianus van Dijk, provides the ability to analyze headlines and story leads; schematic structures that express the meaning of the text through conventional categories, such as layout and sentence structure; local meanings, which contextualize stories based on a community's knowledge and beliefs about society; and semantic strategies, which are the goal-directed properties of discourse.[87] Discourse analysis was coupled with an intertexual analysis in this study. Combining these helped to make connections between the dependence of texts on history and society as well as the relationships between the discursive practices themselves.[88]

Three popular and longstanding Black-owned online news sites were selected for this study: *TheGrio.com*, *NewsOne.com*, and the *Afro.com*, which is the official website for the national Afro-American newspaper. The foundation of the Black press is rooted in goals to uplift, educate, and encourage readers, develop readers' sense of cultural identity, and help them fight for social justice.[89] According to Clint C. Wilson II, Armistead Pride noted four characteristics of the Black press: (1) African-American ownership, (2) Black journalists producing the editorial content, (3) media working primarily as an advocate for its audience, and (4) outlets acting as opinion forums concerning issues relevant to individuals who share an African diasporic heritage.[90] It is important to note there are many news sites with Black-centered coverage, like *The Root* and *Huffington Post*'s *Black Voices*. Yet, by this definition, they would not be considered part of the Black press. As noted, ownership matters in news media production. The assumption is that the three under investigation would have stories independent of white ownership structures that have historically influenced news coverage of Black women.

In total, 47 news stories were obtained from the three online news sources. Several key words, which were respective of past research involving the subject matter, were used to assist with the collection of the stories: "Black," "Black women," "Black woman," "Black men," "Black man," "Black community," "race," "sexual violence," "sexual assault," and "sexual harassment." Only stories mentioning the Me Too movement; sexual violence, assault, or allegations that mentioned a Black accuser or accused; quoted an identifiable Black woman as a source related to sexual violence, assault, or allegations; or centered on Black women or the Black community as a collective group as it related to sexual violence, assault, or allegations were included in the study. Stories were analyzed for the presence of a racial frame (e.g., race was specifically mentioned or used as a qualifier); existence of Black women stereotypes and past dominant themes in news coverage about Black women, which Meyers noted were victimization, poverty, crime, dysfunction and violence; and an overall favorable or unfavorable tone.[91] With respect to the existence of a particular tone toward Black women, the characteristics associated with the women offenders in the Brennan and Vandenberg study were reversed, as this study focuses on Black women as victims. To assess an unfavorable tone toward Black women, Sykes and Matza's "techniques of neutralizers" were used, which are denial of responsibility, denial of injury, denial of victims, condemnation of the condemners, and appeal to higher loyalties.[92] To assess a favorable tone toward Black women regarding violence against them, "exacerbation" characteristics were examined, which were guilt attributed, real injury noted, and real victim assumed. Further, the presences of van Dijk's semantic structures, which are blaming the victim, admission, comparisons, as well as contrast and division, were analyzed.

News portrayals of Black women and Me Too movement

While Black women's voices are historically largely muted in traditional news outlets, the Black press privileged their voices by quoting them at a rate of about 7 to 1 compared to white men, white women, and others as well as at a rate of 3.5 to 1 compared to Black men. Stories related to the Me Too movement and sexual allegations covered a wide range of abuses and types of violence (see Box 28.1) and provided a space for Black women to offer information, opinions and personal stories. Black women, too, were often the only ones quoted in stories, which positioned them as legitimate sources and showcased their ability to be reputable resources for reporters to create narratives for readers. In terms of Black press byline credits, most articles reflected title lines of wire services or multiple staff contributions, such as "TheGrio," "NewsOne Staff," and "Special to the Afro," which made identifying reporters impossible. When present, women did slightly outnumber men. This supports Craft, Wanta, and Lee's claim that women tend to interview other women more often.

Both Black women and Black reporters discourse spoke of the "collective," which is language closely aligned with Black Feminist Thought. In their discussions of the U.S. sexual climate and ways to dismantle the structures causing the mistreatment of Black women, they discussed their "collective power," "collective action," "collective apprehension," "collective quiet," and "collective wisdom." All worked to provide readers with a sense of community and understanding of the work Black women can do together to change the current sexual politics, violence, and abuse in the United States.

Box 28.1 Allegations named in Me Too movement coverage in Black press

Rape	Verbal abuse	Molestation
Groping	Abuse	Sexual advances
Fondling	Physical abuse	Sexual misconduct
Hitting	Sexual harassment	Assault
Sexual abuse	Sexual violence	Sexual assault

Stories also provided a chance to uplift Black women not only in the current "fight" against sexual misconduct, but in relation to others who worked hard in the past to bring to light civil rights issues or other problems facing Black women, such as Recy Taylor, a Black woman who fought for justice after being abducted and raped by six white men.[93] Others mentioned included Barbara Jordan, Ida B. Wells, Rosa Parks, and Fannie Lou Hammer. These stories support the longstanding practice in the Black community of valuing storytelling or the telling of our stories. By evoking previous leaders in the context of the modern-day struggles of Black women, the discourse surrounding Black women's fights for equality were recollected, reimagined, and rejuvenated.

Hill was the most highlighted in the Black press regarding her past efforts to share her experience of sexual harassment. Unlike her coverage in past mainstream media, Black press framed her in largely celebratory and supportive ways, applauding her speaking up as well as her efforts for justice. Yet, there was no one more heralded than Burke, a sexual violence survivor herself. She was described as a "recognizable face all over the country when it comes to women's rights." From stories mentioning her attendance at the Golden Globes, suggesting she could "get Trump out of the White House" in 2020, and being selected to push the button for the ceremonial ball drop on New Year's Eve in New York City, she was the subject of multiple stories and discourse in the Black press. However, most stories worked to credit her for starting the Me Too movement, condemn those who did not initially recognize her work, and position her as a credible and valuable resource.

Burke also became a spokesperson advocating for getting credit for her work and ensuring others knew she did not agree to its usage. She was quoted as saying, "Initially, I panicked. I felt a sense of dread because something that was a part of my life's work was going to be coopted and taken from me and used for a purpose that I hadn't originally intended." She was also quoted declaring, "The point…over the last decade with the 'me too movement' is to let women, particularly young women of color know that they are not alone—it's a movement." She also worked to maintain and sustain the discourse needed to ensure Americans understood that Black girls' and women's experiences with sexual politics were different from those of other women. For example, she was quoted as stating, "We're socialized to not believe Black women. We're socialized to believe that we are fast and sexually promiscuous and things of that nature."

The Me Too movement itself was described in the Black press in a multitude of ways (see Box 28.2). However, it was largely seen as a movement *not* for Black women. The news narrative surrounding the absence of Black women in the movement provided several reasons for this, including the lack of Black women willing to tell their stories, the general disbelief of those Black women who do tell their stories, and the historic and systematic structures and people within the United States—especially white women—who hindered Black women's recognition as victims of sexual violence and assault. For example, one Black woman was quoted, saying, "White women in media are being ushered to safety and communities of empathy and care, Black women and girls have a history of sexual assault in this country, and simultaneously, no one caring." One reporter even called out white women's behavior directly: "If White women were to ever speak loudly for all women as they do for themselves—perhaps there would be a change in the way abusers of black girls and women are held accountable."

Perhaps the two women most noted in the Black press during the period of analysis for not being embraced or believed after sharing their stories were Lupita Nyong'o, who alleged a sexual assault by Harvey Weinstein after which he "quickly denied doing anything inappropriate with Nyong'o, after days of silence following similar accusations by famous White accusers" and Aurora Perrineau, who accused "Girls" writer Murray Miller of raping her when she was 17.[94] He denied

---

Box 28.2 Select Me Too movement descriptions present in Black press

TheGrio

"[The Me Too] movement is meant to spotlight how pervasive the problem is, and also to give women a voice for the problems often swept under the rug."

"Women speaking out about sexual assault."

"The #MeToo movement inspired millions of women to come forward and tell their stories as well, detailing the everyday struggle faced by so many women in a culture steeped in normalizing sexual harassment."

"women began sharing their own harassment and assault experiences using the hashtag #MeToo."

NewsOne

"Prompted in large part by the #MeToo and #TimesUp movements, fans are less forgiving of displays of toxic masculinity."

"speaks against systematic sexism"

"…the #MeToo hashtag flourished on social media as women recounted their personal experiences with sexual assault to show their solidarity with the victims who came forward."

"…Tarana Burke, a Black activist who created the #MeToo movement on Twitter in 2006 to raise awareness around sexual violence"

"#MeToo movement associated with bringing attention to and ending assault, harassment and overall misconduct."

"#MeToo movement, initial begun by Tarana Burke, who is Black, has crossed cultural divides."

"MeTOO founder and leader created her movement fighting against sexual harassment and abuse for women more than a decade ago."

Afro-American

"The success of the #MeToo campaign proved a watershed in opening dialogue, including issues surrounding intersectionality, the space where issues facing women are framed by White women, whose concerns and objectives may be decidedly distinct from those of women of color."

"…Where millions have shared their stories about being sexually harassed and assaulted."

"…#MeToo movement that has engulfed Hollywood and spread into the culture at large with astonishing speed."

"Metoo does not validate our experiences."

"#MeToo, the social media movement denouncing sexual assault."

---

the accusations, and so, too, did Lena Dunham, creator, writer, and star of the same show. Dunham later apologized for not believing Perrineau and called her own actions "inexcusable."[95]

While race may have played a role in how Black women were connected to the coverage, it did not seem to matter with the men accused of wrongdoings. Men named in Black press stories were of all races, ages, and industries, including rappers, singers, former athletes, news anchors, politicians, media moguls, and actors (see Box 28.3). Considering the time frame under

---

Box 28.3 Men named in Me Too movement news stories

Curt Anderson	Charlie Rose
Bill Cosby	Warren Sapp
Heath Evans	Russell Simmons
Marshall Faulk	Tavis Smiley
Mark Halperin	Kevin Spacey
Robert Kelly (R. Kelly)	Louis Székely (Louis C.K.)
Matt Lauer	Ike Taylor
Floyd Mayweather	James Toback
Murray Miller	Donald Trump
Cameron Mitchell	Terdema Ussery
Tremaine Neverson (Trey Songz)	Harvey Weinstein
Brett Ratner	Jeffrey Williams (Young Thug)

---

investigation, several men dominated coverage. Weinstein, Kelly, Bill Cosby, Russell Simmons, and Donald Trump received the most news mentions.

Past news representations and frames of Black women

Although the stories under investigation came from the Black press, not all had racial frames. It is important to note that all three online sources often relied on other news entities for stories, including *CNN*, *Associated Press*, and *New York Times*, with the Afro-American and TheGrio relying on them more than *NewsOne*. Thirty-four percent of the stories framed race as a central part of the narrative, and nearly all of the stories came from *NewsOne*, which was the only online news site with an identifiable, dedicated Black female reporter, Clarissa Hamilton, covering issues related to the Me Too Movement and Black women's issues regarding race, class, and sexual politics in the United States.[96] She, along with a few other named writers, created stories with headlines, such as "WTF? Tavis Smiley, Accused of Sexual Misconduct, Holds Town Hall on Harassment," "Russell Simmons, The Cost of Brothers' Keepers and Black Women Reporting Our Own," and "Recent Sex Abuse Scandals Highlight Problem of Ignoring and Blaming Black Women" as well as "Here's Why Only A Few Women of Color Report Sexual Assault." These headlines, and the stories that accompanied them, showed the focus was on keeping Black women and their stories at the forefront of coverage.

Yet, in the coverage overall, a contrasting and ironic representation of Black women was presented in the Black press. Black women were framed as both the stereotypical "Superwoman" and a "vulnerable victim." Michelle Wallace notes the superwoman is an intelligent, independent, articulate, strong, professional, assertive, and extremely talented Black women who can handle large amounts of distasteful work and, unlike other women, she does not have the same insecurities, weaknesses, and fears.[97] As the superwoman, the Black woman was strong, despite being admittedly harmed, and ready to work for the cause, regardless of what it took. These traits were said to be manifested in women like Former Secretary of State Condoleezza Rice and media mogul Oprah Winfrey, both of whom were present in news articles under investigation.[98]

In the rare instances when women did chronicle details of sexual misconduct, two women were portrayed stronger or better than others: Daisy Morgan and Cynthia Marshall. The articles featuring them were the only two in which Black women were in the headlines and also remained the subjects of the story. Daisy Morgan, a Black woman who served in the U.S. Navy during the Vietnam War, noted she endured "verbal, emotional and physical abuse and I experienced some inappropriate touching" by the sailors. However, she was then quoted stating, "None of this abuse set me back, but it made me stronger and more determined to persevere." The super-woman narrative was also strongly seen in the coverage of the National Basketball Association's Dallas Mavericks' hiring Cynthia Marshall after the former team president, Terdema Ussery, was associated with a "pattern of misogyny and predatory sexual misconduct." This article stated: "Marshall's hiring proves that when the ish hits the fan, it's a good look to call in a Black woman to save the day. Yes, she will put on her cape!"[99]

When being framed as the "vulnerable victim," Black women were real victims with real injuries resulting from the accused men's behaviors in a country that had longed ignored them. One reporter noted there was a strategy deployed "by abusers time and time again: attack and blame the victim." He noted this strategy was "particularly successful when that person is a Black woman, as she is typically unlikely to receive much empathy and or support in the first place."[100] The longstanding news themes of victimization and dysfunction were present, which came from stories chronicling Black women being associated with numerous stories of sexual violence, abuse, and other misconduct as well as the dysfunction of living in the United States while within a community that often did not support their accusations. One reporter wrote,

> Characteristically, it has not been until rich Anglo (and importantly) American women have come forward to share their stories, that people have taken notice. But Black women have long known the lewd behaviors of powerful men (and women) who could and did use fear and force to push sexual advances and postures of power.

Still another *NewsOne* reporter wrote, "A demonizing pattern of social conditioning, based on racism, has tainted the images of women of color, especially Black females."

The accused-accused narrative was present in Black press coverage. In fact, 12 of the 47 (25 percent) articles directly dealt with accusations. However, of those articles, only three focused solely on the woman's story, of which one was primarily centered on a white model accusing Simmons of rape. Considering the entire sample, only 6 percent of stories quoted women providing personal accounts of their accusations. Therefore, reporters were left to create narratives largely absent of Black women's own words. Using specific language, Black women were framed as victims by first actually describing them as victims, without any rebuttal from accusers or others. Second, accused men were specifically framed in negative ways, such as being described as "shunned," "disgraced," "slammed on social media," or even having "lured her." This discourse provided readers with cues as to how to understand the Black woman's position and victimization. Further, accused white men were overwhelmingly portrayed as "powerful," and Black men were consistently portrayed as sexually aggressive and dangerous. Both characterizations of men are common and longstanding media representations. Framing men in these ways further positioned Black women as vulnerable within the news coverage.

Reporters also described accused men's past histories of allegations or abuse, which helped to portray women accusers as victims. For example, men were described as having a "past of violently abusing Black women," "a man with a long and well-documented history of disrespecting women," and a "history of previous incidents involving law enforcement." In fact, there was only one instance in the coverage of Sykes' and Matza's denial of injury and denial of the victim. In most cases, reports noted the accused claimed the interactions were consensual or denied the incident occurred. While this may have neutralized the accusation within the article by providing a response contrary to the accuser, it did not overwhelmingly change the favorable tone in which the Black press portrayed Black women accusers.

The narrative construction of a Black women's sexual victimization conundrum and cycle of silence

Black women are locked in a cycle of silence deeply rooted in problematic systematic structures that exist both within the United States and the Black community. The semantic structures within Black press coverage uncovered why Black women were in such a conundrum. Black women's existence in the United States was portrayed as being rife with divisions and differences that both legitimized and undermined their abilities to name their abusers or detail their experiences. Articles depicted Black women as largely silenced by living within the United States and the Black community, both of which were not open to accepting them as victims. Further, Black women were constantly compared to white women, and not in a positive way. Black women were largely portrayed as not having support and not being believed, while white women had both support and believability.

By far the most compelling sematic structure present was the use of contrasts and divisions. News narratives contrasted Black women and Black men, Black women and white women, white men and Black men, U.S. culture and the Black community, U.S. elite and "everyday" people, U.S. rich and U.S. poor as well as modern times and historical time frames with questionable U.S. laws and social practices. The divisive environment described within the Black press included reasons why Black women often remained silent in their victimization. The Black women's sexual violence cycle of silence included the factors present in the United States and the Black community existing within it (see Figure 28.1).

As noted, there was clearly an overall lopsided representation within the Black press of stories about men being accused of sexual misconduct. On the surface, it may seem victims were not worthy of news coverage since stories about those accused were quadruple the number of those primarily about victims, but news narratives worked to explain this issue. Statistics were used in news coverage to substantiate the Black woman's position, including "nearly one in five Black women are the victims of sexual assault in the United States," "for every Black woman who reports her rape, at least 15 Black women maintain silence," and "approximately 40 percent of Black women report coercive sexual contact by age 18." These and other facts and statistics

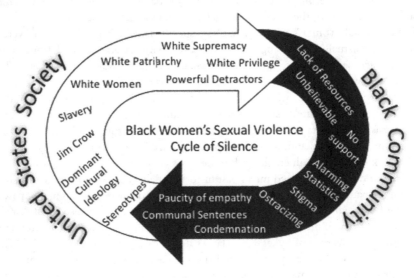

Figure 28.1 Factors outlined in the Black press contributing to Black women remaining silent

Source: author

validated the existence of sexual misconduct and violence against Black women and offered evidence as to why their stories are often not told.

Reporters not only used Black women as credible sources to speak for other Black women in society, but they, too, did the same. Overwhelmingly, both blamed and problematized the Black community for not supporting Black women. In one article, Burke was quoted regarding the Me Too movement and the Black community. She said,

> But I think a deeper issue that we haven't scratched the surface on is within the Black community—and it's so nuanced. While we are constantly living with white supremacy and oppression and how it plays out in every part of our lives, we still have to be accountable for who we are to each other. One of the primary reasons Black men haven't been swept up (yet) in this moment is because Black women aren't coming forward. Black women aren't coming forward because we have a harder time being believed both outside and INSIDE our own community.

In another article, a criminal justice activist said, "the lens through which Black women are viewed has to shift to encourage Black girls and women to speak up and the Black community to support them when they do." Yet, another article quoted Deborah Gray White, author of *Ar'n't I a Woman?: Female Slaves in the Plantation South*, stating, "Black women's bodies, from Day One, have been available to all men." This was immediately followed by the reporter's narrative that "In this harmful pattern, doubt has been cast on Black women who dare to say they have been victims of sexual exploitation. As to avoid condemnation from detractors, a silence has grown among females." Still another reporter wrote, "Each time we laugh at a (Jerrika) Karlae or ignore a Nyong'o, we perpetuate the same White supremacist patriarchy that we, ourselves, are victimized by our entire lives."

The shared narrative in the Black press is the "stakes is high" for Black women who experience sexual assault or violence to "publicly say #metoo." Thus, they often remain silent. The question becomes why? Answers were provided. Overall, there was a problematic and dangerous proposition that "Black women have to consider entire communities, even when victimized." More specifically, it was noted that "abusers not only remain in (the) community, but often 'to elevated statuses." Plus, in the past, "many" Black women were said to be "ostracized into keeping quiet for fear of 'ruining a good brother's life' or being blamed for the abuse because of her clothing or deportment." Finally, there was the "communal sentencing" associated with being vocal. This was said to be "anything from psycho-terrorism, a violent cussing out, threats, physical harm, isolation, accusations around betraying the race, demanding to know why they didn't tell, loss of wages, fear, shaming." The issue of communal sentencing is that even when Black women do not go to law enforcement for assistance with their abusers, there still could be very dangerous, disturbing, and distressing consequences they or others might endure within the Black community. Thus, these communal sentences are likely adding to the pressures of Black women to remain silent.

Conclusion

This analysis investigated how Black women and their allegations, including those involving sexual violence, sexual harassment, and sexual assault, were framed in the Black press during the highly visible beginning of the Me Too movement. Coverage in the Black press showed the three online news sites under investigation stayed true to the original goals of advocating for their audience (in this case Black women), and working as an opinion forum concerning an important issue in the Black community. Unlike the mainstream press, the Black press privileged Black women, allowing them to speak for and about themselves in ways usually unseen in traditional media. Staying true to its mission, Black press news articles provided readers with real actions to use to create change. Readers were offered solutions to stop the problem, and they were even pointed to those people and structures at the root of the problem.

The Me Too movement coverage of Black women allowed them a space to discuss what was occurring in various industries in the United States, but it was largely framed as a movement not for and about Black women. Yet, what was clear in the coverage within the Black press was how the privilege afforded to white women and accused white and Black men often left Black women questioning their ability to be safeguarded and believed. This showed that in today's U.S. society whiteness is still privileged and gender inequality remains problematic, and the Black press continued to create a space for this to be highlighted.

Yet, what was clear in the coverage was the existence of a troubling pattern of behavior by Black women of not speaking up or out publicly about their sexual allegations. Reporters, utilizing other Black women and their own biases, worked to contextualize Black women and their bodies for readers. Black women were seen as either the Superwoman able to handle their sexual abuse or vulnerable victims unable to speak up and get support and help. Their victimization was closely tied to the dysfunction surrounding the Black woman's everyday lived experience in the United States, which moved from her historic past of being a slave to being muted in today's society. Black women's sexual violence cycle of silence was framed as devastating to them, and it was blamed not only on white supremacy and patriarchy, but also the Black community of which they were a part. The stories underscored systemic and problematic forces that can envelope a Black woman's thoughts and experiences and push her into a cycle of silence, instead of being liberated into speaking her truth.

The Black press, its reporters, and its sources advocated for Black women and framed the issues of sexual violence and sexual assault as problematic. Discourse within stories showed the layers of stigma and trauma associated with Black women being unbelievable, unrespectable, disrespected, and discredited, which pushed many into the proverbial shadows along with their allegations. As noted, despite not having the supportive umbrella structures and resources within the Me Too movement or U.S. society, Black women deserve and have the right to speak up and name their abusers and violators. Absent those external support structures, the Black community must become the support structure Black women need to deal with any ramifications that might occur, when they decide to name their abusers.

#TimesUp

Notes

1 Tracy Swartz, "Millions Watched New TV Docuseries about Abuse allegations Against R. Kelly," *Chicago Tribune*, January 9, 2019, www.chicagotribune.com/entertainment/tv/ct-ent-r-kelly-docuseries-20190108-story.html.

2 "Surviving R. Kelly," My Lifetime, accessed May 16, 2019, www.mylifetime.com/shows/surviving-r-kelly/about.

3 Allyssa Milano (@Alyssa_Milano), "If You've Been Sexually Harassed or Assaulted Write 'Me Too' as a Reply to This Tweet," Twitter, October 15, 2017, 1:21 p.m., https://twitter.com/Alyssa_Milano/status/919659438700670976/photo/1.

4 "More than 12M 'Me Too' Facebook Posts, Comments, Reactions in 24 Hours," *CBS*, October 17, 2017, www.cbsnews.com/news/metoo-more-than-12-million-facebook-posts-comments-reactions-24-hours/; Stephanie Zacharek, Eliana Dockterman, and Sweetland Edwards Haley, "Person of the Year 2017," *Time*, accessed May 16, 2019, http://time.com/time-person-of-the-year-2017-silence-breakers.

5 Margaret E. Johnson, "Feminist Judgments & #MeToo," *Notre Dame Law Review Online* 94 (2018): 51.

6 Jeff Hearn, "Author's Copy: This Is a So-Called Personal Version (Author's Manuscript as Accepted for Publishing after the Review Process but Prior to Final Layout and Copyediting) of the Article," *European Journal of Women's Studies* 25, no. 2 (2018): 228–35.

7 Leigh Gilmore, "He Said/She Said: Truth-Telling and# MeToo," *University of Edinburgh Postgraduate Journal of Culture & the Arts* 25 (2017), www.forumjournal.org/article/view/2559.

8 Zahara Hill, "A Black Woman Created the 'Me Too' Campaign Against Sexual Assault 10 Years Ago," *Ebony*, October 18, 2017, www.ebony.com/news-views/black-woman-me-too-movement-tarana-burke-alyssa-milano 11/30/2018.

9 Me Too Movement, "History & Vision," accessed May 16, 2019, https://metoomvmt.org/about/.

10 Hill, "A Black Woman."

11 Ibid.

12 Michelle Rodino-Colocino, "Me Too, #MeToo: Countering Cruelty with Empathy," *Communication and Critical/Cultural Studies* 15, no. 1 (2018): 96.

13 Frances Beale, "Double Jeopardy: To Be Black and Female," in *The Black Woman: An Anthology*, ed. Toni Cade (New York: New American Library, 1979), 90–100; Deborah K. King, "Multiple Jeopardy, Multiple Consciousness: The Context of a Black Feminist Ideology," in *Race, Gender and Class*, ed. Bart Landry (New York: Routledge, 2016), 36–57.

14 Patricia Hill Collins, *Black Sexual Politics: African Americans, Gender, and the New Racism* (New York: Routledge, 2004).

15 Andrea Smith, *Conquest: Sexual Violence and American Indian Genocide* (Durham, NC: Duke University Press, 2015).

16 Gerda Lerner, ed., *Black Women in White America: A Documentary History* (New York: Vintage, 1992).

17 Traci C. West, *Wounds of the Spirit: Black Women, Violence, and Resistance Ethics* (New York: NYU Press, 1999).

18 Ibid.

19 Cheryl I. Harris, "Whiteness as Property," *Harvard Law Review* 106 (1992): 1707.

20 Collins, *Black Sexual Politics*.

21 Hearn.

22 Barbara A. Gutek and Bruce Morasch, "Sex-Ratios, Sex-Role Spillover, and Sexual Harassment of Women at Work," *Journal of Social Issues* 38, no. 4 (1982): 55–74.

23 Mitchell Stephens, *A History of News*, 3rd edn (New York: Oxford University Press, 2007), 4.

24 James Glen Stovall, *Writing for the Mass Media* (Upper Saddle River, NJ: Prentice-Hall, 1985).

25 Elisa Shearer, "Social Media Outpaces Print Newspapers in the U.S. as a News Source," *Pew Research Center*, December 10, 2018, www.pewresearch.org/fact-tank/2018/12/10/social-media-outpaces-print-newspapers-in-the-u-s-as-a-news-source/.

26 Maxwell E. McCombs and Donald L. Shaw, "The Agenda-Setting Function of Mass Media," *Public Opinion Quarterly* 36, no. 2 (1972): 176–87.

27 Pamela Ban, Alexander Fouirnaies, Andrew B. Hall, and James M. Snyder, "How Newspapers Reveal Political Power," *Political Science Research and Methods* (2018): 1–18.

28 Johan Galtung and Mari Holmboe Ruge, "The Structure of Foreign News: The Presentation of the Congo, Cuba and Cyprus Crises in Four Norwegian Newspapers," *Journal of Peace Research* 2, no. 1 (1965): 64–90.

29 Andrea Prat, "Media Power," *Journal of Political Economy* 126, no. 4 (2018): 1747–83.

30 Tim Groseclose and Jeffrey Milyo, "A Measure of Media Bias," *Quarterly Journal of Economics* 120, no. 4 (2005): 1191–237.

31 Sendhil Mullainathan and Andrei Shleifer, "The Market for News," *American Economic Review* 95, no. 4 (2005): 1031–53; Matthew Gentzkow and Jesse M. Shapiro, "Media Bias and Reputation," *Journal of Political Economy* 114, no. 2 (2006): 280–316.

32 Kenneth Dowler, "Media Consumption and Public Attitudes Toward Crime and Justice: The Relationship between Fear of Crime, Punitive Attitudes, and Perceived Police Effectiveness," *Journal of Criminal Justice and Popular Culture* 10, no. 2 (2003): 109–26.

33 Stephens, *A History of News*, 101.

34 Kai Ericson, "Notes on the Sociology of Deviance," *The Other Side: Perspectives on Deviance* (1964): 9–21.

35 Kaitlynn Mendes, "Framing Feminism: News Coverage of the Women's Movement in British and American Newspapers, 1968–1982," *Social Movement Studies* 10, no. 1 (2011): 81–98.

36 Shanto Iyengar, *Is Anyone Responsible?: How Television Frames Political Issues* (Chicago: The University of Chicago Press, 1994); Thomas E. Nelson and Zoe M. Oxley, "Issue Framing Effects on Belief Importance

and Opinion," *Journal of Politics* 61 (1999):1040–67; Thomas E. Nelson, Zoe M. Oxley, and Rosalee A. Clawson, "Toward a Psychology of Framing Effects," *Political Behavior* 19 (1997): 221–46.

37 Lawrence Wallack, Lori Dorfman, David Jernigan, and Makani Themba-Nixon, *Media Advocacy and Public Health: Power for Prevention* (Thousand Oaks, CA: Sage, 1993); John D. Richardson and Karen M. Lancendorfer, "Framing Affirmative Action: The Influence of Race on Newspaper Editorial Responses to the University of Michigan cases," *Harvard International Journal of Press/Politics* 9, no. 4 (2004): 74–94.

38 Catherine A. Taylor and Susan B. Sorenson, "The Nature of Newspaper Coverage of Homicide," *Injury Prevention* 8, no. 2 (2002): 121–27.

39 Hans Magnus Enzenberger, "Constituents of a Theory of the Media," in *The New Media Reader*, eds N. Wardrip-Fruin and N. Montfort (Cambridge, MA: MIT Press, 2003), 261–75.

40 Richardson and Lancendorfer, 75–76.

41 Minako Kurokawa, "Mutual Perceptions of Racial Images: White, Black, and Japanese Americans 1," *Journal of Social Issues* 27, no. 4 (1971): 214.

42 Stuart Hall, "The Work of Representation," in *Representation: Cultural Representations and Signifying Practices*, ed. Stuart Hall, vol. 2 (Sage, 1997), 346.

43 Marjorie L. DeVault, "Talking and Listening from Women's Standpoint: Feminist Strategies for Interviewing and Analysis," *Social Problems* 37, no. 1 (1990): 96–116.

44 Teresa L. Heinz, "From Civil Rights to Environmental Rights: Constructions of Race, Community, and Identity in Three African American Newspapers' Coverage of the Environmental Justice Movement," *Journal of Communication Inquiry* 29, no. 1 (2005): 59; Gina Dent, "Black Pleasure, Black Joy," in *Black Popular Culture*, ed. Gina Dent (Seattle, WA: Bay, 1995), 8.

45 Stephen D. Reese, August Grant, and Lucig H. Danielian, "The Structure of News Sources on Television: A Network Analysis of 'CBS News,' 'Nightline,' 'MacNeil/Lehrer,' and 'This Week with David Brinkley,'" *Journal of Communication* 44, no. 2 (1994): 84–107.

46 Eric Freedman and Frederick Fico, "Male and Female Sources in Newspaper Coverage of Male and Female Candidates in Open Races for Governor in 2002," *Mass Communication & Society* 8, no. 3 (2005): 257–72; David D. Kurpius, "Sources and Civic Journalism: Changing Patterns of Reporting?," *Journalism & Mass Communication Quarterly* 79, no. 4 (2002): 853–66.

47 Stephanie Craft, Wayne Wanta, and Cheolan Lee, "A Comparative Analysis of Source and Reporter Gender in Newsrooms Managed by Men and Women," in *National Conference of the Association for Education in Journalism and Mass Communication, Kansas City, Missouri* (2003).

48 Heinz, "From Civil Rights," 60.

49 Jane Mansbridge and Katherine Tate, "Race Trumps Gender: The Thomas Nomination in the Black Community," *PS: Political Science & Politics* 25, no. 3 (1992): 488–92.

50 Kimberle Crenshaw, "Demarginalizing the Intersection of Race and Sex: A Black Feminist Critique of Antidiscrimination Doctrine, Feminist Theory, and Antiracist Politics [1989]," in *Feminist Legal Theory*, eds Katherine Bartlett and Rosanne Kennedy (New York: Routledge, 2018), 150.

51 Ibid., 149.

52 Helen Benedict, *Virgin or Vamp: How the Press Covers Sex Crimes* (Oxford: Oxford University Press on Demand, 1993); Marian Meyers, *News Coverage of Violence Against Women: Engendering Blame* (Thousand Oaks, CA: Sage Publications, 1997), "African American Women and Violence: Gender, Race, and Class in the News," *Critical Studies in Media Communication* 21, no. 2 (2004): 95–118.

53 Marian Meyers, "African American Women and Violence Gender, Race, and Class in the News," *Critical Studies in Media Communication* 21, no. 2 (2004): 95–118.

54 Marian Meyers, *African American Women in the News: Gender, Race, and Class in Journalism* (New York: Routledge, 2013).

55 Herman Gray, "Black Masculinity and Visual Culture," *Callaloo* 18, no. 2 (1995): 401–05.

56 Herman Gray, *Cultural Moves: African Americans and the Politics of Representation* (Berkeley, CA: University of California Press, 2005), 24–25.

57 Robert M. Entman and Andrew Rojecki, *The Black Image in the White Mind: Media and Race in America* (Chicago: The University of Chicago Press, 2001).

58 Benedict, *Virgin or Vamp*, 251.

59 Pauline Katherine Brennan, *Women Sentenced to Jail in New York City* (El Paso: LFB Scholarly Pub. LLC, 2002); "Sentencing Female Misdemeanants: An Examination of the Direct and Indirect Effects of Race/Ethnicity," *Justice Quarterly* 23, no. 1 (2006): 60–95; Kathryn Ann Farr, "Aggravating and Differentiating Factors in the Cases of White and Minority Women on Death Row," *Crime & Delinquency* 43, no. 3 (1997): 260–78.

60 Ruth Rosen, "Who Gets Polluted? The Movement for Environmental Justice," *Dissent* 41 (1994): 223–30.

61 Patricia Hill Collins and Sirma Bilge, *Intersectionality* (Malden, MA: Polity Press, 2016).

62 Carole J. Sheffield, *Sexual Terrorism: The Social Control of Women* (Thousand Oaks, CA: Sage Publications, Inc, 1987), 173.

63 Meyers, "African American Women and Violence."

64 Nell Irvin Painter, "Hill, Thomas, and the Use of Racial Stereotype," in *Race-ing Justice, En-gendering Power: Essays on Anita Hill, Clarence Thomas, and the Construction of Social Reality*, ed. Toni Morrison (Pantheon, 1992), 200–14; Patricia Hill Collins, *Black Feminist Thought* (New York: Routledge, 1991).

65 Collins, *Black Sexual Politics*, 139.

66 Verna L. Williams, "The First (Black) Lady," *Denver University Law Review* 86 (2008): 841.

67 Meyers, *African American Women in the News.*

68 Meyers, "African American Women and Violence."

69 Collins, *Black Feminist Thought*, 67.

70 Painter, "Hill, Thomas," 207.

71 Hope Landrine, "Race X Class Stereotypes of Women," *Sex Roles* 13, nos 1/2 (1985): 71–72.

72 Pauline K. Brennan and Abby L. Vandenberg, "Depictions of Female Offenders in Front-Page Newspaper Stories: The Importance of Race/Ethnicity," *International Journal of Social Inquiry* 2, no. 2 (2009): 148; See also, Lisa Bond-Maupin, "'That Wasn't Even Me They Showed': Women as Criminals on America's Most Wanted," *Violence Against Women* 4, no. 1 (1998): 30–44; Farr, "Aggravating and Differentiating Factors," "Defeminizing and Dehumanizing Female Murderers: Depictions of Lesbians on Death Row," *Women & Criminal Justice* 11, no. 1 (2000): 49–66; Jayne Huckerby, "Women Who Kill Their Children: Case Study and Conclusions Concerning the Differences in the Fall from Maternal Grace by Khoua Her and Andrea Yates," *Duke Journal of Gender Law & Policy* 10 (2003): 149.

73 Brennan and Vandenberg, "Depictions of Female Offenders".

74 Meyers, *News Coverage of Violence against Women*; Benedict.

75 Anima Mama, "Violence against Black Women in the Home," in *Home Truths About Domestic Violence: Feminist Influences on Policy and Practices: A Reader*, eds Jalner Hanmer, Catharine Itzin, Sheila Quaid, and Debra Wigglesworth (New York: Routledge, 2000), 44–56.

76 Ibid., 44.

77 Melissa V. Harris-Perry, *Sister Citizen: Shame, Stereotypes, and Black Women in America* (New Haven, CT: Yale University Press, 2011).

78 John Fiske, *Media Matters: Everyday Culture and Political Change* (Minneapolis: University of Minnesota Press, 1994); Mansbridge and Tate, "Race Trumps Gender," 490.

79 Mansbridge and Tate, "Race Trumps Gender," 490.

80 Harris-Perry, *Sister Citizen*, 54–55.

81 David J. Leonard, "The Real Color of Money: Controlling Black Bodies in the NBA," *Journal of Sport and Social Issues* 30, no. 2 (2006): 158–79.

82 Pauline K. Brennan and Abby L. Vandenberg, "Depictions of Female Offenders in Front-Page Newspaper Stories: The Importance of Race/Ethnicity," *International Journal of Social Inquiry* 2, no. 2 (2009): 141–75; Meyers, "African American Women and Violence."

83 Elaine Richardson, "'She Was Workin Like Foreal': Critical Literacy and Discourse Practices of African American Females in the Age of Hip Hop," *Discourse & Society* 18, no. 6 (2007): 789–809.

84 Meyers, "African American Women and Violence."

85 Norman Fairclough, *Analysing Discourse: Textual Analysis for Social Research* (New York: Routledge, 2003).

86 Hall, "The Work of Representation."

87 Teun Adrianus van Dijk, *Racism and the Press* (New York: Routledge, 1991).

88 Norman Fairclough, "Discourse and Text: Linguistic and Intertextual Analysis within Discourse Analysis," *Discourse & Society* 3, no. 2 (1992): 193–217, "Critical Discourse Analysis and the Marketization of Public Discourse: The Universities," *Discourse & Society* 4, no. 2 (1993): 133–68.

89 Clint C. Wilson, II, *Whither the Black Press?: Glorious Past, Uncertain Future* (Bloomington, IN: Xlibris Corporation, 2014).

90 Armistead Scott Pride and Clint C. Wilson, *A History of the Black Press* (Washington, DC: Howard University Press, 1997).

91 Meyers, *African American Women in the News.*

92 Gresham M. Sykes and David Matza, "Theory of Neutralization: A Theory of Delinquency," *American Sociological Review* 22, no. 6 (1957): 664–70.

93   Sewell Chan, "Recy Taylor, Who Fought for Justice after a 1944 Rape, Dies at 97," *New York Times*, December 29, 2017, www.nytimes.com/2017/12/29/obituaries/recy-taylor-alabama-rape-victim-dead.html.

94   "Why Few Women of Color in Wave of Accusers? 'Stakes Higher,'" November 18, 2017, https://afro.com/women-color-wave-accusers-stakes-higher/.

95   Leah Dunham, "Lena Dunham: My Apology to Aurora," *Hollywood Reporter*, December 5, 2018, www.hollywoodreporter.com/news/lena-dunham-my-apology-aurora-perrineau-1165614.

96   Clarissa Hamilton, "Mavericks Hire Black Woman CEO Amid Ex-Team President's Sexual Harassment Scandal: Cynthia Marshall Is the Latest Black Woman to the Rescue," NewsOne, February 27, 2018, https://newsone.com/3777445/dallas-mavericks-news-ceo-cynthia-marshall-scandal-sexual-misconduct-controversy/.

97   Michele Wallace, *Black Macho and the Myth of the Superwoman* (New York: Verso, 1999).

98   Ibid.

99   Hamilton, "Mavericks Hire Black Woman CEO."

100 Austin Williams, "Recent Sex Abuse Scandals Highlight Problem of Ignoring and Blaming Black Women: Harvey Weinstein and R. Kelly's Scandals Remind Us that Black Women and Girls just Aren't Valued Enough," *NewsOne*, November 2, 2017, https://newsone.com/3756837/harvey-weinstein-update-Black-women-lupita-nyongo-sexual-assault-abuse-controversy/.

29

DIGITAL *TESTIMONIOS* AND WITNESSING OF SALMA HAYEK AND AMERICA FERRERA'S DISCLOSURES OF SEXUAL HARASSMENT AND ASSAULT

Raisa F. Alvarado and Michelle A. Holling

In 2017, the #MeToo hashtag trended globally, bringing heightened awareness to the prominence of sexual assault by Hollywood elites. Survivors came forward in droves to condemn the actions of powerful figures such as Harvey Weinstein to Kevin Spacey with vast records of abuse. Yet, far from solely operating in elitist circles, the #MeToo hashtag circulated in diverse networks, bringing invigorated attention to the prevalence of sexual assault. Popularized and made viral by actress Alyssa Milano, the origins of the hashtag were far removed from the Hollywood platform that would eventually bolster it. Rather, Tarana Burke, a youth-camp director, conceived of the phrase two decades earlier as a unifying call for people who had experienced sexual assault. Burke hoped the "me too" catchphrase and hashtag "could release survivors from the shame they felt and … empower them—especially in minority communities."[1] Initially issuing an intersectional call for unity and solidarity, Burke would eventually launch the "Me Too" campaign, an initiative that engages the varied obstacles survivors of color negotiate as they decide whether to disclose.

The 2017 #MeToo viral hashtag, however, was distanced from Burke's early iteration as Hollywood disclosures "whitewashed" the phrase[2] without consideration of how identity standpoints complicate a person's ability and willingness to disclose.[3] "Woman," as a homogenous category, became the war-cry and attention was overwhelmingly given to white survivors and the "individual bogeymen" who commit sexual assault.[4] The structural conditions that place women of color at higher risk of being assaulted were largely ignored. Furthermore, positionality as a factor that shapes an individual's willingness to disclose was not part of mainstream conversations.[5] In the spirit of Burke's original intersectional hailing of the catchphrase and the Hollywood platform that would later popularize its use, we center the #MeToo disclosures of two prominent Latina thespians, Salma Hayek and America Ferrera, as well as subsequent electronic responses to their disclosures. These responses employed reframing, deflecting, and counter-disclosing in important ways worth analyzing.

We focus on the disclosures of Hayek and Ferrera for two reasons. First, Hayek and Ferrera are prominent Latina actors and activists with far-reaching audiences. The distinct platforms in which they chose to disclose their experiences and the considerable responses they received offers a significant sample of our socio-technological context relative to understanding sexual harassment and assault disclosures by two high-profile women generally, and two women of color specifically.[6] Hayek's descriptive opinion piece, entitled "Harvey Weinstein Is My Monster Too," in *The New York Times* enabled those who wanted to engage her admission to maintain their anonymity with a site-specific handle in the "comment" section, disconnected from their social media profiles.[7] Conversely, Ferrera's brief disclosure via Instagram reached a younger audience who would (potentially) contend with their personal accounts being linked to their engagement.[8] Due to these differences, responding or "liking" the statements came with different stakes, respectively. Second, Hayek and Ferrera's disclosures represent distinct levels of raced and gendered violence that highlight dynamics of professional and interfamilial contexts. Hayek's disclosure is firmly situated within a Hollywood landscape replete with power dynamics. While Hayek calls out her abuser, Harvey Weinstein, and details the abuse inflicted upon her, her disclosure also explicitly addresses unique struggles surrounding Mexican American artistry. Through an account of how she struggled to produce, fund, and act in the 2002 film *Frida*, Hayek accentuates the white gatekeeping that forced her to prioritize her project over disclosing sexual harassment. Notably, at the height of the #MeToo movement, Hayek and Lupita Nyong'o, a prominent Kenyan Mexican actor, were the singular actors of color who Weinstein publicly discredited. In an interview, Hayek responded to Weinstein singling her and Nyong'o out: "We [Hayek and Nyong'o] are the easiest to get discredited … it is a well-known fact. So he went back, attacking the two women of color, in hopes that he could discredit us."[9] Hayek's recognition that women of color are more likely to be disparaged when they disclose abuse accentuates the necessity of centering identity standpoints. In contrast, Ferrera's disclosure focused on the abuse she endured at nine years of age from a person closely connected to her family. Through references to the grooming that often takes place in assaults by friends and family members,[10] Ferrera's disclosure represents proximal abuse from people to whom one is closely connected. As survivors of sexual harassment and assault, united under the #MeToo hashtag, each disclosure represents how women can resist being reduced to a homogenized category.

While Ferrera's disclosure, via Instagram, does not make explicit reference to her Honduran American ethnicity, we argue that her exhaustive Latin@[11] oriented activism, presence in Latin@ inspired programing,[12] and viral opinion piece via the Huffington Post, "Thank you, Donald Trump"—wherein she argued that Trump's xenophobic rhetoric would motivate "Latinos" to go to the voting polls—situate her as a prominent Latina figurehead.[13] We read her sexual assault disclosure through her Latina identity and note the manner in which she encouraged other women of color to mark identity standpoints relative to sexual assault disclosures. Although Hayek and Ferrera represent a niche Hollywood enclave, we argue that their disclosures function as digital *testimonios*, a unique rhetorical structure that broadens *testimonio* and the act of witnessing. Moreover, given the unique digital platforms used and the responses accrued, we look toward the way digital *testimonios* provide new options for centering one's positionality. We additionally consider the nuanced forms of witnessing technology provides. Witnessing as a feature of *testimonio* can refer to an individual recounting their story and thereby bearing witness[14] to their experiences as well as to an audience who is the recipient of the *testimonio*. For the purposes of this chapter, we examine the digital audience of Hayek and Ferrera's disclosures and identify three thematic forms of witnessing that take place: reframing, deflecting, and counter-disclosing.

Digital *testimonios* and digital disclosures

Testimonios, as rhetorical structures, are narratives of urgency that work toward generating wisdom, consciousness, and action against structural and institutional oppression.[15] Through the use of individual stories of discrimination, violence, and subjugation that also represent collective experience and struggle, these accounts promote societal transformation. A signature tool of

Latin@/x-oriented epistemologies, a *testimonio* as rhetorical structure gives urgency to "social injustices"[16] for the purposes of inspiring collaborative engagement and action. Amplifying the act of recounting one's truth, a *testimonio* relies heavily on the practices of bearing witness and, in particular, transforming "listeners-to-witnesses."[17] In Michelle Holling's analysis of feminicidio *testimonios*, she argues that the rhetorical crafting of positionality, prior to, throughout, and following delivery of a *testimonio* has the capacity to transform listeners-to-witnesses through a recognition that speakers and witnesses are "gathered in the flesh."[18] More specifically, in the context of feminicidio *testimonio*, scenes of address that prime audiences combine with the act of sharing horrifying and detailed accounts of violence and sexual assault, augmented by the display of deceased victims' portraits, interpolated audiences into witnesses with the capacity to act. Expanding upon the notion of listener-to-witness theorized by Holling, this chapter looks at how a technological landscape alters or affects the act of expressing positionality and witnessing. We explore this dynamic twofold as we consider how Hayek and Ferrera craft their Latina positionalities in relation to their sexual harassment and assault disclosures and examine the responses each *testimonios* garnered. Given the added dimensions of optional anonymity and the popular culture prominence of both figures, we also consider how technology can expand upon a *testimonio*.

Digital *testimonios* are largely understudied with respect to how their rhetorical structure changes, if at all, and how a digital landscape alters audiences' witnessing capabilities. Among the small number of studies analyzing digital *testimonios*, Rina Benmayor's[19] work provides particular insight into how digital landscapes expand productions and theorizations of *testimonios*. Given the overwhelming narratives available online, Benmayor clarifies that all digital *testimonios* are digital stories yet the opposite is not the case. More specifically, a digital *testimonio* focuses on social injustice for the purpose of resisting oppression and transforming the world. A rhetorical structure that "moves beyond narrative, biography or oral history," a digital *testimonio* remains grounded in a transformative political agenda.[20] Benmayor further notes the "broader, more democratic authorship, dissemination, and reception" that technology enables provides indefinite and more accessible ways (e.g., blogs, social media, chat rooms) to testimoniar.[21] Unfettered by constraints of space, time, or a material platform, digital *testimonios* are able to reach audiences through nuanced and creative modes of expression, as authors can include, for example, images, videos, and hashtags for the purposes of promoting identification and action. We argue that digital *testimonios* further transcend the confines of academic inquiry, as authors can write and disseminate information without a researcher acting as translator.[22] In the case of Hayek and Ferrera, each digital *testimonio* was authored by each actress and shared on unique media platforms that impacted their reception and the manner audiences (when inclined) witnessed them.

We read Hayek and Ferrera's sexual harassment and assault disclosures as digital *testimonios* as each account was shared digitally, used personal experience to note structural oppression, and presented readers with technologically informed ways of witnessing. Both admissions recalled personal experiences with sexual harassment and assault for the sake of bringing urgency to the pervasiveness of sexual assault and to further the #MeToo social movement. Yet, while similar in both disclosing abuse, each actress' disclosure represented a distinct dynamic of sexual harassment and assault. Hayek brought attention to the prevalence of sexual harassment and sexual coercion in an overwhelmingly white Hollywood landscape, and Ferrera spoke of familial dynamics that impede one's willingness and ability to disclose. Further, while Hayek and Ferrera centered their individual experiences, both emphasized the social significance behind their disclosures, proclaiming a desire to bring awareness and mobilize other survivors. Hayek ended her digital *testimonio* by referencing the institutional implications of her individuated account, saying, "I hope that adding my voice to the chorus of those who are finally speaking out will shed light on why it is so difficult, and why so many of us have waited too long."[23] Similarly, Ferrera's digital *testimonio* ended with the following call to arms: "Ladies, let's break the silence so the next generation of girls won't have to live with this bullshit."[24] In what follows, we ana-lyze each digital *testimonio* twofold: first we examine how Hayek and Ferrera crafted their Latina positionalities within a digital landscape and then we investigate the types of witnessing each *testimonio* accrued. We provide three classifications of witnessing that Hayek and Ferrera over-whelmingly received (reframing, deflecting, and counter-disclosures) and conclude by noting how counter-disclosures fit and expand upon Moors and Webber's[25] identified types of digital sexual assault disclosures.

Digital *testimonios*: crafting Latina positionality

Benmayor's work acknowledges how technology provides those who are testimoniando diverse tools to craft their "social identities, positionalities, and inequalities."[26] With its wide array of visual and auditory capabilities, technology can accentuate aspects of the *testimonio* that may not be evident in other forms of discourse. We first focus on the technological crafting of each *testimonio*, as strategic choices were made to distinctly and similarly emphasize Hayek and Ferrera's Latina positionalities.

Hayek's first-person digital *testimonio* was published in *The New York Times* website under the opinion column section, a portion of the editorial that visually changes based on the published piece. In the case of Hayek's opinion piece, the stylized elements visually framed the digital *testimonio* in a manner that accentuated her Latina positionality. First, upon accessing the art-icle, audiences are confronted with an image of Hayek staring intently. The picture is consid-erable in size and takes up the entire screen, so readers must scroll past her image to read her disclosure. This is not a heavily doctored image of Hayek, as gray strands frame her face, she wears subtle makeup, and emotes a concerned expression. In sum, the image visually beckons readers to view her as in individual who exists outside of her manufactured Hollywood image. Here, too, audiences are confronted with her Latina physical features: her olive complexion, her dark hair and eyebrows. At the side of her image, descriptive quotes from the article alterna-tively flash, such as "I don't think he hated anything more than the word no."[27] The visual focus on Hayek's portrait in conjunction with flashing text produced a layout encouraging varied degrees of meaning and interpretation. Based on the amount of responses that alluded to Hayek's "Mexican American" identity, "strength," "beauty," and "resilience," one such interpretation is that the tone was candid and vulnerable. A digital landscape allowed Hayek to communicate beyond the wording of the *testimonio* itself. In this case, readers are confronted with Hayek's Latina image and the harm inflicted upon her prior to reading her account. Having engaged her *testimonio*, user HS "wonder[s] how much more abuse she [Hayek] experienced as a result of being Mexican," demonstrating recognition that factors beyond gender contributed to Hayek's experience with sexual abuse. The importance of centering complex positionality relative to sexual assault becomes further evident in the discursive portion of Hayek's digital *testimonio*, as she continuously foregrounds her Mexican identity in relation to her sexual harassment and unwillingness to disclose.

At the start of the digital *testimonio*, Hayek provides an account of the contextual environment in which she operated. Particularly, Hayek notes the considerable obstacles that impeded her career as a Latina actress. During "the 14 years that [she] stumbled from schoolgirl to Mexican soap star to an extra in a few American films to catching a couple of lucky breaks," Hayek writes that "Harvey Weinstein had become the wizard of a new wave of cinema that took original content into the mainstream."[28] Hayek is careful to frame her Mexican identity in relation to her "lucky breaks,"[29] as her positionality was a significant factor that shaped her interactions with Weinstein as well as his subsequent abuse of her. Weinstein was an especially powerful representative of white, Hollywood gatekeeping with considerable influence over media representations and artistic projects. To this point, Hayek opines that at the time she worked under Weinstein to produce and star in *Frida*, "it was unimaginable for a Mexican actress to aspire to a place in Hollywood."[30] Moreover, drawing on her experiences as a Latina performer struggling to cross over successfully to mainstream American cinema, Hayek references systemic practices (e.g., producing, funding, and casting choices) that informed her decision not to disclose initially. Hayek specifically notes that "between 2007 and 2016, only 4 percent of directors were female and 80 percent of those [women] got the chance to make only one film," an observation recognizing how her gender and race worked doubly to exclude her from Hollywood.[31]

Deploying a key feature of *testimonio*, Hayek then connected her individual experience with larger systems of gendered and raced oppression. Continuing to draw attention to the stark absence of nonstereotypical roles for Latinas in Hollywood, Hayek shared, "[i]t became my mission to portray the life of this extraordinary artist and to show my native Mexico in a way that combated stereotypes."[32] There are numerous reasons why survivors of sexual assault do not disclose, ranging from the belief that no one will believe them[33] to fear of retaliation.[34] While those reasons may have also impacted Hayek, the added commitment to represent Latinas in dynamic and different roles was at the forefront of her silence. Her commitment to Latina artistry in a white-dominated landscape shaped her experiences with sexual harassment, silence, and eventual digital disclosure.

Next, we turn to discerning Ferrera's positionality via her disclosure on her verified Instagram account.[35] The choice to disclose on Instagram offered a platform-specific means of connoting her Latina identity. First, Ferrera's digital disclosure[36] (see Figure 29.1) is brief and is visually

Figure 29.1 Image of Ferrera's sexual assault disclosure via Instagram

499

represented through an image of text as opposed to a photograph, the photo being typical of an Instagram post. While one can speculate on Ferrera's choice to disclose via Instagram, we note the manner of her usage of the site promotes her Latina artistry and activism. This is evident in the posts immediately prior to and following her October 16 disclosure. On October 15, Ferrera shared a picture of herself with other prominent Latina actresses (e.g., Eva Longoria, Gina Rodriguez, Rosario Dawson, Isabella Gomez, Andrea Navedo, Justina Machado, Yara Martinez, Gloria Kellett, Jaina Ortiz, Stephanie Beatriz, and Eiza Gonzalez) with the caption "Latinas Who Lunch. Giving me life. #FiercelyLatina."[37] Similarly, on the day following her digital *testimonio*, Ferrera posted[38] an image of her first film, *Real Women Have Curves*, accompanied by a caption expressing her gratitude for staring in the role: "once in a lifetime chance to portray Ana … a Latina woman who still hungers to see more authentic and complex portrayals of women from all walks of life."[39] The posts that surround Ferrera's disclosure are significant, as Instagram operates within a matrix of chronological continuity, each post existing in relation to the ones around it. Instagram allowed Ferrera to indirectly center her Latina positionality outside of the actual statement disclosing sexual abuse.

Discursively, Ferrera's digital *testimonio*, in spite of its brevity, represents a dynamic of sexual violence wherein the assaulter is someone the survivor knows. Ferrera starts by recounting her first experience of sexual assault at "9 years old" and the "shame and guilt" she felt at her perceived responsibility for her own suffering.[40] As a survivor who "had to see this man [her abuser] on a daily basis," Ferrera notes the psychological and emotional trauma that leads many survivors not to disclose: "my guts carrying the burden of what only he & I knew—that he expected me to shut my mouth and smile."[41] While the disclosure does not explicitly reference familial factors, the proximity of Ferrera's attacker represents a form of sexual violence known all too well: assaults perpetrated by someone one knows. Through reference to the expectation that she should "shut" her mouth, Ferrera further implicates gendered and age-related power dynamics that dissuade childhood survivors from coming forward. Significantly (and unlike Hayek), Ferrera does not name her assailant, a decision that became a point of contention discussed further in relation to witnessing below. Yet, as rhetorically constructed, Ferrera's decision not to expose her assailant also represents familial and societal dynamics that frame women, and particularly women of color, as unreliable sources. For reasons unknown, Ferrera chose to maintain the anonymity of her attacker and instead briefly recounted her sexual assault for the purpose of promoting and forwarding the #MeToo social movement.

Hayek and Ferrera chose distinct platforms that worked in site-specific ways to frame their positionality in relation to their digital *testimonio*, and while the differences in rhetorical structure are considerable, also important to note are the commonalities between each disclosure. First, each actress placed regulated Latina artistry at the forefront of their politics. Ferrera's account may have represented proximal sexual assault, but the images and captions that surround and work in collaboration with her disclosure confront media as an industry that reduces Latina identity to stereotypical representations. Hayek and Ferrera both use their Latina positionalities to bring heightened awareness to the need for more diverse representations of Latina identity. Additionally, both Hayek and Ferrera use personal experience to promote action and resistance beyond Hollywood. As Latinas who operate in privileged positions with respect to socioeconomic status and influence, they are careful to frame their disclosures within an ethic of inclusion and solidarity. When Hayek recognizes that she adds her "voice to the chorus of those who are finally speaking out"[42] and Ferrera encourages activism ("Ladies, Let's break the silence"[43]), each actress moves beyond their personal experience to promote a collective response.

Witnessing in digital disclosures

Digital *testimonios* expand upon the concept of witnessing, a central attribute of *testimonio*. First, we note that witnessing occurs twofold within *testimonio*, as the author acts as witness to their own experiences and the audience witnesses the "experiences and struggles"[44] of the individual testimoniando. In this section, we focus on the audience of a digital *testimonio* and the manner audiences are moved to act through collaborative engagement.

Technology and social media provide audiences with an array of tools to engage digital *testimonios*. In the case of Hayek, *The New York Times* offers readers the option of anonymously creating an account or linking their Facebook accounts to comment. The site also gives witnesses the option of commenting or replying to other posters on the site. In contrast, Ferrera's digital *testimonio* via Instagram provides audiences the option of "liking" or commenting through their linked accounts. This is a noteworthy difference, as Instagram accounts are typically users' personal profiles. While temporary accounts can be created to engage a digital *testimonio*, in the case of Ferrera's disclosure, users chose overwhelmingly to respond with an account that is not anonymous. Given the distinct demographics of each digital platform, the types of responses garnered varied considerably. All of the responses to Hayek's digital *testimonio* were discursive, formal, and (for the most part) adhered to proper grammar. Conversely, the responses to Ferrera were more informal and a significant amount of comments were expressing support via emojis or the #MeToo hashtag. For the scope of this project, we focused on written responses to each digital *testimonio* (hence excluded emojis). In total, the digital *testimonios* received considerable response (Hayek's generated 994 comments and Ferrera's accrued 54,098 likes and 1,171 comments). After reading through responses and coding them for thematic similarities, three types of witnessing emerged across the two digital *testimonios*.

First, we noticed a considerable amount of reframing as witnesses would focus on singular aspects of the disclosures without an interrogation of the wider reaching implications of sexual assault referenced in the original statements. Second, we noted deflecting, a rhetorical shifting wherein witnesses would turn the focus away from the sexual harassment and assault. Finally, the most common form of witnessing we identified was counter-disclosure of a sexual assault, as survivors sought to express solidarity and promote the movement through accounts of their own abuse. In order to classify and better understand the types of counter-disclosures Hayek and Ferrera received, we turn to the digital sexual assault classifications identified by Rosetta Moors and Ruth Webber.[45]

Reframing can be understood as a strategic pivoting wherein witnesses focused on individuated aspects of Hayek and Ferrera's digital *testimonios* as opposed to the structural implications of sexual harassment relative to widespread patterns of it and further redirected attention from discriminatory practices in Hollywood. Specifically, although Hayek draws attention to the "chorus" of women she is joining in revealing her me too moment, witnesses overwhelmingly fixated on her[46] physical appearance and filmography. For example, references to physical appearance are evident in comments by user CFM ("Hayek first came to my attention as a startlingly beautiful screen presence in small roles") and user sandhillgarden ("Salma, You are one of the most beautiful women in the world, intelligent, and talented"), only two of the many responses that exemplified this pattern. Similarly, one Instagram user[47] responded to Ferrera by posting, "Sick what happened to you. U have been BEAUTIFUL both inside & out." Here, the larger implications of sexual harassment and assault for women of color in a regulated Hollywood landscape are ignored in favor of individualized aspects of the disclosers that maintain the hyper-fetishization of Latinas. One comment standing apart from this pattern appears on Hayek's digital *testimonio* from user JKR, who remarked, "Can we start by not emphasizing 'beauty' when we talk about women and girls?" This singular user made note of the overwhelming focus on Hayek's (and similarly on Ferrera's) physical appearance, but the uniqueness of the observation combined with a failure to engage it suggests that fixation on women's appearance hampers individuals' grasp of sexual harassment or assault as enmeshed in issues of power and control.

Likewise, comments focused on each actress's filmography rather than the social problems they were addressing. A user identified as Nicole on Hayek's digital *testimonio* remarked, "I deeply love Frida's work as an artist and your portrayal of her in the film" and a user on Ferrera's disclosure commented, "You are a beautiful brave woman who I've adored watching on TV." Such reactions exemplify the inclination to highlight each actress's work instead of structural oppression as in the lack of representation of Latinas in Hollywood as well as the inattention to Latina survivors of sexual assault who are not famous actresses. The inclination to reframe the cultural and societal implications of each disclosure accentuates markers of witnessing relative to digital *testimonio*. We note that these comments, although reductive and problematic, may have functioned as acts of recognition and support. Particularly, in lieu of the nonverbal support present when one provides a *testimonio* in person, witnesses might have felt inclined to validate Hayek and Ferrera through what they know best about them: their appearances and filmographies.

Where the structural and/or systemic elements of sexual harassment and assault remain present albeit overshadowed by a focus on beauty in reframing, the next manifestation of witnessing—deflection—commits harm by implicitly negating Hayek's and Ferrera's disclosures. Deflection surfaced when witnesses changed the themes (e.g., sexual harassment-assault and the underwhelming representation of Latinas in Hollywood) in an effort to draw attention away from inequitable gendered practices that make men more likely to sexually assault. In this case, witnesses focused on men who don't sexually assault, such as user Ms: "As a woman, I also want to thank the men who supported Ms. Hayek and the other women on her team" or user Jim O'Brien: "Yours is a very important voice, not only for all women, but also for the millions of decent men who recoil from the rash of stories about the vulgar, ugly and nauseating treatment of women." Here, witnesses deflect from Hayek's sexual harassment and, on a larger scale, the prominence and prevalence of sexual assault by positioning men as saviors and protectors. While these responses drew attention away from the raced, power dynamics that further complicate assault(s) on women of color, other users employed explicitly racist pivots. Deflecting attention from Ferrera's disclosure digitally, some users made Islamophobic comments, such as one who wrote:

> Imagine how all the little girls in UK, Netherlands, Sweden, France, Germany etc etc etc feel about being raped by Muslim refugees and perm residents. I'm sure they blame themselves too. Unfortunately they have no voice. And Hollywood and MSM is silent about them too.

As exemplified in the comment, some respondents chose to express xenophobic sentiments that further promote a white supremacist narrative, white victimhood, and fear of the Other.

An alternate way that deflecting is present appears when users minimize the act of speaking out, drawing attention away from sexual assault as a systemic and hence cultural problem. In Hayek's digital *testimonio* and users' (i.e., Olivia; Kcgrantz; cheerful dramatist; Erica; Connie Moffit; Gina Albert) disclosures, sexual assault in other industries such as technology, theater, and film is acknowledged. In response to Hayek's digital *testimonio*, some users derisively mock her outspokenness as with user James Wilton, who remarks, "… this essay is much easier to write when married to a French billionaire" or user david henry, who opined, "It takes no courage to jump on the bandwagon after so many came before." Both users, identified by male names, negate the significance of the appalling acts of sexual assault recounted by Hayek and instead deflect attention by emphasizing the protection of gendered, socioeconomic disparity. Similarly, and in response to another user, user Michael Vincent questions, "Strength? Sounds like she's pretty weak to me. No strength is demonstrated in being the 91st person to speak out against Weinstein." The most prominent deflecting that took place was on Hayek's disclosure, a type of witnessing that we speculate is encouraged by the anonymity *The New York Times* comment section allowed. As users were able to create pseudonyms untethered to their identities and relationships, they likely felt a degree of safety (in terms of protection from retaliation) in vocalizing their opinions. While Ferrera did receive some deflecting, the majority of users were posting from personal accounts, a factor that likely impacted the types of comments made.

The final and most prominent theme identified from responses to Hayek and Ferrera's digital *testimonios* were counter-disclosures of sexual assault. Technology has presented survivors of sexual assault with safe and wide-reaching platforms to discuss the prevalence of assault and potentially to disclose their own experiences. Given the accessibility of online forums in the 21st century and the option for anonymity, survivors are turning to social media in droves.[48] Yet, the types of digital disclosures taking place are largely understudied. For this reason, we turn to the work of Moors and Webber as their qualitative study on digital sexual assault disclosures via Yahoo! Answers provides a working framework to classify the types of counter-disclosures Hayek and Ferrera received. More specifically, Moors and Weber identified four types of disclosures: "'naming' the incident as sexual assault, 'unburdening' or recounting ones' story, expressing emotions ('emoting'), and 'help-seeking.'"[49] Each type of disclosure appears in response to Hayek and Ferrera's digital *testimonios*, but most evident were witnesses who chose to unburden and emote or even express a desire for vengeance, an important theme that was not identified in Moors and Webber's work.

With regard to unburdening, Moors and Webber note that a significant amount of users treat online platform(s) as a confessional wherein they may release "themselves through writing."[50] Contrary to a question and answer dialogic, those digitally unburdening adhered to comments of a "circuitous nature" where emphasis is placed on the age and act of sexual assault for the sake of disclosing rather than seeking support.[51] We observed a considerable amount of unburdening in response to Ferrera's digital *testimonio*. These include brief posts, such as from one user stating, "Me too, ages 4, 16, 17, 24 …. Me too" to more detailed responses, such as,

> Thank you America! I was 4 … I didn't tell my parent until I was 14 as I had blocked it from my mind but was fondled by a stranger on a class trip to the museum and it brought it all back.

Through such remarks, users were able to simultaneously unburden and express support. Extending solidarity across gender-sex lines were counter-disclosures by males, which were posted in response to Ferrera. One user recounts his experience while in a "group home" while another laments "#metoo I didn't think it could happen to boys, but it did. Thanks for giving me the courage." Although male users' disclosures are few, they nonetheless open a space for others to support them and, importantly, to help debunk common associations of sexual assault, molestation, or other sexual violence with female survivors only. Finally, the counter-disclosures often crossed linguistic lines as several commenters unburdened in Spanish, as evidenced in one such disclosure: "yo sufri un abuso a 5 anitos no pude hablar hasta los 40 años."[52] Perhaps the safety of anonymity[53] in online counter-disclosures enables survivors, domestic or international, to voice experiences, thus extending unburdening to exceed implicit solicitations for support. Ferrera's digital *testimonio* presented witnesses with an opportunity to share their own sexual assaults, in acts that cumulatively emphasized the prevalence of sexual violence.

Moors and Webber identify emoting as the expression of "a complex web of emotions experienced by [a] survivor."[54] While survivors often emote in various ways, Moors and Webber note that "internalized emotions such as shame, guilt, and fear [are] more frequently [expressed] than externalized emotions such as anger."[55] Within the counter-disclosures analyzed, emoting was a prominent theme amongst witnesses who expressed feelings for Hayek and Ferrera but also noted their own emotions regarding experiences with harassment and assault. One such example of emoting is evident in a counter-disclosure to Hayek in which user Olivia writes, "As someone who had her share of harassment in and out of Hollywood, I felt exactly the same way. Disheartened, disregarded, humiliated, threatened, abused, demeaned, embarrassed, and disembodied." Notably, the counter-disclosure references harassment, not sexual assault, yet Hayek's *testimonio* offered this user an opportunity to emote about similar experiences in Hollywood. We recognize the manner in which the user, Olivia, was able to articulate clearly feelings about discriminatory practices and how this clarity was not the case for all those emoting within counter-disclosures. Instead, emotive responses to Ferrera's post would often code their emotions through descriptive characterizations. For example, one user wrote that seeing her uncle at family events made her want to "vomit" and another user described her heart as "racing" due to intimately knowing the feeling expressed by Ferrera. Though explicit terms such as disgust or anger were not used to describe emotions relative to sexual assault, diverse descriptive words functioned in a similar capacity as users were able to negotiate the visceral and consuming feelings that follow a sexual assault.

The final theme we found in responses, a desire for vengeance, was not identified by Moors and Webber. However, this category composed a considerable number of the comments accrued. This desire for vengeance was distinct for each disclosure, as Hayek revealed her assailant (Weinstein) and Ferrera did not. As a result, there were vastly different approaches to vengeance seeking. At the time of Hayek's opinion piece, other prominent actresses had already come forward with allegations against Weinstein.[56] This being the case, Hayek's sexual harassment disclosure operated within a cultural context that explicitly marked Weinstein as a perpetrator. Thus, most responses to Hayek made clear that Weinstein needed to be punished for his "crimes and abjection" as expressed by user SQ. In contrast to these focused responses, counter-disclosures to Ferrera expressed a desire for vengeance on their own perpetrators but also urged Ferrera as well as other users to expose her/their assailants. The first theme is evident when a user recalled that her father "raped" her and her sister. She ends her counter-disclosure by saying, "I wish him a slow painful death." The second theme is variously expressed by urging survivors to "name him," to "#namenames," or more emphatically, "Name the son of a bitch!!! Name him! If you really wanna help these girls, name the son of a bitch!!!! Don't be just another story! Name him!!!!" Outwardly, demanding names of perpetrators does not convey vengeance, but revealing names will likely bring forth accountability such as "let[ting] him bare some shame!!!" Finally, users' calls for exposing assailants are often linked to saving the next victim, as seen in the following examples: "Name Him. He's not done. Save the next victim"; "Call them out, expose them prevent someone else being abused!"; and "If you are really coming forward then you need to name them. Doubtless you are not the only one they abused and they will go on abusing unless you out them." In the absence of names of abusers, some reactions chide those who counter-disclose and/or Ferrera as failing to break through silence. One user comments to Ferrera: "What silence did you break? Tell everybody his name and report him to the authorities." The conflicting and layered revenge-seeking counter-disclosures speak to the prominence of anger for survivors of sexual assault and the varied styles of coping survivors employ. Overall, the responses represented a vast spectrum wherein some witnesses disclosed vaguely, some survivors used euphemisms, and others demanded transparency via naming the assailant. While distinct and diverse, we draw attention to how each digital disclosure presented witnesses with an opportunity to speak back to the systemic prevalence of sexual assault for the purpose of promoting change.

Conclusion

Analysis of Salma Hayek's and America Ferrera's disclosures of sexual harassment and sexual assault, respectively, provided an opportunity to examine what transpires when survivors of gendered violence disclose on online platforms. Their digital *testimonios* evince a rhetorical structure of first-person online disclosures that are augmented by visuals highlighting racial/ethnic and gendered positionalities. Important to recall is that "the first person 'I' of their *testimonios*" stands for shared experience.[57] Inasmuch as Hayek and Ferrera each come forth and relay their individual experiences that then provoked a myriad of responses, there is nonetheless a communal "we" that is crafted by and through the counter-disclosures. This community is accentuated not only in the collaborative hailing of sexual assault survivors by Hayek and Ferrera but by the multitude of responders who acknowledged the larger implications of their individuated disclosures via references to the #MeToo social movement.

Unlike other types of *testimonios* studied[58] with a reach or impact that is difficult to assess, the rhetorical structure of digital *testimonios* produces discourse that shows how they facilitate heightened participation by audiences, subsequently expanding a circle of resistance. Audiences of both Hayek's and Ferrera's digital *testimonios* engaged with extended responses that subsequently received "recommendations," "likes," and hashtags. Moreover, digital *testimonios* expand the terrain for witnessing, as comments are not bound to a spatial and temporal moment but digitally persist, promoting repeated engagement. Expanding upon the digital *testimonio* work of Benmayor, technology widens the scope and types of witnessing that take place. Particularly, a digital landscape presents those that are testimoniando diverse tools by which an audience can "bear witness to experiences and struggles ... beyond [their] own realities."[59]

A possible limitation of witnessing via technology is the potentially condescending and pejorative responses one opens themselves up to via public disclosure, particularly for women of color, who are already subjected to racism and sexism. Not all who engage are doing so for the transformative intent of digital *testimonios*—quite the opposite. In sharing an experience of oppression or subjugation, one becomes vulnerable to an audience who may comment with the intent of maintaining inequitable power relations (in this case, concerning both gender and race). Yet, in spite of this potential, we affirm (as evidenced through the counter-disclosures prompted by statements by Hayek and Ferrera) the substantive, transformative potential of recounting one's *testimonio* digitally. In each case, digital *testimonio* added breadth to the #MeToo movement, providing space for people to commiserate and in some cases reveal their own experiences with sexual assault.

Notes

1 Elizabeth Adetiba and Tarana Burke, "Tarana Burke Says #MeToo Should Center Marginalized Communities," in *Where Freedom Starts: Sex Power Violence #MeToo* (N.p.: Verso, 2017), 15.
2 During a question and answer session in a public talk, Burke expressed her disagreement with individuals who accuse white women of appropriating the catchphrase and hashtag, noting they "didn't take

anything." Instead, Burke identified mass media as culprits who seek to "pit" women against each other and who fail to recognize that there are "vulnerable white women." Tarana Burke, "Founder of the 'Me Too' Movement and Social Justice Activist." February 5, 2019. Arts and Lecture Series. San Marcos, CA: California State University San Marcos.

3 Burke, "Founder of the."

4 Adetiba and Burke, "Tarana Burke Says," 16.

5 Melissa Villarreal, "Latinas' Experience of Sexual Assault Disclosure," *Psychology* 5, no. 10 (2014): 1285–1300.

6 Studies of sexual harassment as experienced by women of color is sparse within communication; rectifying that from an intersectional approach are Brian K. Richardson and Juandalynn Taylor, "Sexual Harassment at the Intersection of Race and Gender: A Theoretical Model of the Sexual Harassment Experiences of Women of Color," *Western Journal of Communication* 73, no. 3 (2009): 248–272.

7 Salma Hayek, "Harvey Weinstein Is My Monster Too," *New York Times*, December 12, 2017, www. nytimes.com/interactive/2017/12/13/opinion/contributors/salma-hayek-harvey-weinstein.html.

8 americaferrera, "metoo" Instagram, 16 Oct 2017, www.instagram.com/p/BaVNRL5DtQm/.

9 Stuart Oldham, "Salma Hayek Says Harvey Weinstein Only Responded to Her and Lupita Nyong'o's Harassment Claims because Women of Color Are Easier to Discredit," *Variety*, May 13, 2018, https:// variety.com/2018/film/news/salma-hayek-says-harvey-weinstein-only-responded-to-her-and-lupita-nyongos-harassment-claims-because-women-of-color-are-easier-to-discredit-1202808828/.

10 Ann Wolbert Burgess and Carol R Hartman. "On the Origin of Grooming," *Journal of Interpersonal Violence* 33, no. 1 (2018): 17–23; Anne-Marie Mcalinden, "'Setting 'Em Up': Personal, Familial and Institutional Grooming in the Sexual Abuse of Children." *Social & Legal Studies* 15, no. 3 (September 2006): 339–362; Nicole Van Zyl, "Sexual Grooming of Young Girls: The Promise and Limits of Law," *Agenda* 31, no. 2 (2017): 1–10.

11 We adopt "Latin@" as gender inclusive, expressive of Latina and Latino subjects, and symbolic of alliances (consult Calafell and Holling 2011, xvi).

12 Examples include the television series *Ugly Betty* (consult Sowards and Pineda 2011) or the acclaimed film *Real Women Have Curves* (2002).

13 America Ferrera, "Thank You, Donald Trump!," *Huffington Post,* July 2, 2016, www.huffingtonpost.com/ america-ferrera/thank-you-donald-trump_b_7709126.html.

14 John Beverley, *Testimonio: On the Politics of Truth* (Minneapolis: University of Minnesota Press, 2004).

15 Roberto Avant-Mier and Marouf Hasian Jr., "Communicating 'Truth': Testimonio, Vernacular Voices, and the Rigoberto Menchú Controversy," *Communication Review* 11 (2008): 323–345; Claudia G. Cervantes-Soon, "Testimonios of Life and Learning the Borderlands: Subaltern Juárez Girls Speak," *Equity & Excellence in Education* 45, no. 3 (2012): 373–391; Fernando Delgado, "Rigoberto Menchú and Testimonial Discourse: Collectivist Rhetoric and Rhetorical Criticism," *World Communication* 28 (1999): 17–29; Michelle A. Holling, "So My Name is Alma. I Am the Sister of…": A Feminicidio Testimonio of Violence and Violent Identifications," *Women's Studies in Communication* 37, no. 3 (2014): 313–338. doi: 10.1080/07491409.2014.944733; Lindsay Pérez Huber, "Disrupting Apartheid of Knowledge: Testimonio as Methodology in Latina/o Critical Race Research in Education," *International Journal of Qualitative Studies in Education* 22, no. 6 (2009): 639–654.

16 Rina Benmayor, "Digital Testimonio as a Signature Pedagogy for Latin@ Studies," *Equity & Excellence in Education* 45, no. 3 (2012): 507–524.

17 Holling, "So My Name."

18 Holling, "So My Name," 331.

19 Benmayor, "Digital Testimonio."

20 Kalina Brabeck, "Testimonio: A Strategy for Collective Resistance, Cultural Survival and Building Solidarity," *Feminism & Psychology* 13, no. 2 (2003): 253.

21 Benmayor, "Digital Testimonio," 508.

22 In this regard, we have in mind subalterns' production of *testimonios* that relied upon an interlocutor, usually a researcher, to record, transcribe, and disseminate a *testimonio* that scholars later study and analyze (see, for example, work completed by Scholz).

23 Hayek, "Harvey Weinstein Is My."

24 americaferrera, "metoo."

25 Rosetta Moors and Ruth Webber, "The Dance of Disclosure: Online Self-Disclosure of Sexual Assault," *Qualitative Social Work* 12, no. 6 (2013): 799–815.
26 Benmayor, "Digital Testimonio," 508.
27 Hayek, "Harvey Weinstein is My."
28 Hayek, "Harvey Weinstein is My."
29 Hayek, "Harvey Weinstein is My."
30 Hayek, "Harvey Weinstein is My."
31 Hayek, "Harvey Weinstein is My."
32 Hayek, "Harvey Weinstein is My."
33 Bonnie S. Fisher, Leah E. Daigle, Francis T. Cullen, and Michael G. Turner, "Reporting Sexual Victimization to the Police and Others: Results from a National-Level Study of College Women," *Criminal Justice and Behavior* 30, no. 1 (2003): 6–38.
34 Marjorie R. Sable, Fran Danis, Denise L. Mauzy, and Sarah K. Gallagher, "Barriers to Reporting Sexual Assault for Women and Men: Perspectives of College Students," *Journal of American College Health* 55, no. 3 (2006): 157–162.
35 Verified within social media refers to a blue checkmark that appears next to a celebrity or public figure to mark their "authentic presence," in this case, her verified account substantiates her identity (@ americaferrera). While one has to create an Instagram account to like or comment, Ferrera's account is "public," meaning anyone can access her images and subsequent captions or comments.
36 americaferrera, "metoo."
37 americaferrera, "Latinas who lunch. Giving me life" Instagram, 15 Oct 2017, www.instagram.com/p/ BaSUTNQDK9e/.
38 americaferrera, "15 years ago, I began my career with this beautiful film, REAL WOMEN HAVE CURVES." Instagram, 17 Oct 2017, www.instagram.com/p/BaWuZZWDPyu/.
39 americaferrera, "15 years ago."
40 americaferrera, "metoo."
41 americaferrera, "metoo."
42 Hayek, "Harvey Weinstein Is My."
43 americaferrera, "metoo."
44 Pérez Huber, "Disrupting Apartheid of," 649.
45 Moors and Webber, "The Dance of Disclosure."
46 Witnesses' attention to physical beauty and filmography was also apparent in responses to Ferrera's disclosure.
47 We omit Instagram usernames to preserve anonymity as most accounts are linked to witnesses' identities.
48 Kaitlynn Mendes, Jessica Ringrose, and Jessalynn Keller, "#MeToo and the Promise and Pitfalls of Challenging Rape Culture through Digital Feminist Activism," *European Journal of Women's Studies* 25, no. 2 (2018): 236–246: 10.1177/1350506818765318; Richardson and Taylor, "Sexual Harassment at the." Hayek, "Harvey Weinstein is My."
49 Moors and Webber, "The Dance of Disclosure," 812.
50 Moors and Webber, "The Dance of Disclosure," 810.
51 Moors and Webber, "The Dance of Disclosure," 810.
52 English translation: I suffered an abuse at 5 years old, I couldn't speak until I was 40 years old.
53 Anonymity not only in the sense of the user's identity but also the anonymity of being able to be one of thousands posting to Ferrera and the strong unlikeliness that one's identity would be known.
54 Moors and Webber, "The Dance of Disclosure," 810.
55 Moors and Webber, "The Dance of Disclosure," 810.
56 Actresses such as Ashley Judd, Gwyneth Paltrow, and Rose McGowan.
57 Benmayor, "Digital Testimonio," 509.
58 Holling, "So My Name"; T. M. Linda Scholz, "The Rhetorical Power of Testimonio and Ocupación: Creating a Conceptual Framework for Analyzing Subaltern Rhetorical Agency," PhD thesis, University of Colorado Boulder, 2007; T. M. Linda Scholz, "Hablando Por (Nos)Otros, Speaking for Ourselves: Exploring the Possibilities of 'Speaking Por' Family and Pueblo in the Bolivian Testimonio '*Si Me Permiten Hablar*'," in *Latina/o Discourse in Vernacular Spaces: Somos de Una Voz?*, ed. Michelle A. Holling and Bernadette Marie Calafell (Lanham, MD: Lexington Press, 2011), 203–222.
59 Pérez Huber, "Disrupting Apartheid of," 649.

Bibliography

Adetiba, Elizabeth and Tarana Burke. "Tarana Burke Says #MeToo Should Center Marginalized Communities." In *Where Freedom Starts: Sex Power Violence #MeToo*, 15–18, n.p.: Verso, 2017.

americaferrera. "Latinas who lunch. Giving me life." *Instagram*, October 15, 2017. www.instagram.com/p/BaSUTNQDK9e/.

americaferrera. "15 years ago, I began my career with this beautiful film, REAL WOMEN HAVE CURVES." *Instagram*, October 17, 2017. www.instagram.com/p/BaWuZZWDPyu/.

Avant-Mier, Roberto and Marouf Hasian Jr. "Communicating 'Truth': Testimonio, Vernacular Voices, and the Rigoberto Menchú Controversy." *Communication Review* 11 (2008): 323–45.

Benmayor, Rina. "Digital Testimonio as a Signature Pedagogy for Latin@ Studies." *Equity & Excellence in Education* 45, no. 3 (2012): 507–24. DOI: 10.1080/10665684.2012.698180

Beverley, John. *Testimonio: On the Politics of Truth*. Minneapolis: University of Minnesota Press, 2004.

Brabeck, Kalina. "Testimonio: A Strategy for Collective Resistance, Cultural Survival and Building Solidarity." *Feminism & Psychology* 13, no. 2 (2003); 252–8.

Burgess, Ann Wolbert, and Carol R. Hartman. "On the Origin of Grooming." *Journal of Interpersonal Violence* 33, no. 1 (2018): 17–23.

Burke, Tarana. "Founder of the 'Me Too' Movement and Social Justice Activist." *Arts and Lecture Series*, San Marcos, CA: California State University San Marcos. February 5, 2019.

Calafell, Bernadette M. and Michelle A. Holling "Introduction." In *Latina/o Discourse in Vernacular Spaces: Somos de Una Voz?*, edited by Michelle A. Holling and Bernadette Marie Calafell, xv–xxv. Lanham, MD: Lexington Press, 2011.

Cervantes-Soon, Claudia G. "Testimonios of Life and Learning the Borderlands: Subaltern Juárez Girls Speak." *Equity & Excellence in Education* 45, no. 3 (2012): 373–91. DOI: 10.1080/10665684.2012698182.

Delgado, Fernando. "Rigoberto Menchú and Testimonial Discourse: Collectivist Rhetoric and Rhetorical Criticism." *World Communication* 28 (1999): 17–29.

Ferrera, America. "Thank You, Donald Trump!" *Huffington Post*, July 2, 2016. www.huffingtonpost.com/america-ferrera/thank-you-donald-trump_b_7709126.html.

Fisher, Bonnie S., Leah E. Daigle, Francis T. Cullen, and Michael G. Turner. "Reporting Sexual Victimization to the Police and Others: Results from a National-Level Study of College Women." *Criminal Justice and Behavior* 30, no. 1 (2003): 6–38.

Hayek, Salma. "Harvey Weinstein Is My Monster Too." *New York Times*, December 12, 2017. www.nytimes.com/interactive/2017/12/13/opinion/contributors/salma-hayek-harvey-weinstein.html.

Holling, Michelle A. "'So My Name Is Alma. I Am the Sister of …': A Feminicidio Testimonio of Violence and Violent Identifications." *Women's Studies in Communication* 37, no. 3 (2014): 313–38. DOI: 10.1080/07491409.2014.944733.

Lokot, Tetyana. "#IamNotAfraidToSayIt: Stories of Sexual Violence as Everyday Political Speech on Facebook." *Information, Communication & Society* 21, no. 6 (2018): 802–17.

Mcalinden, Anne-Marie. "'Setting 'Em Up': Personal, Familial and Institutional Grooming in the Sexual Abuse of Children." *Social & Legal Studies* 15, no. 3 (September 2006): 339–62.

Moors, Rosetta, and Ruth Webber. "The Dance of Disclosure: Online Self-disclosure of Sexual Assault." *Qualitative Social Work* 12, no. 6 (2012): 799–815. DOI: 10.1177/1473325012464383

Oldham, Stuart. "Salma Hayek Says Harvey Weinstein Only Responded to Her and Lupita Nyong'o's Harassment Claims because Women of Color Are Easier to Discredit." *Variety*, May 13, 2018. https://variety.com/2018/film/news/salma-hayek-says-harvey-weinstein-only-responded-to-her-and-lupita-nyongos-harassment-claims-because-women-of-color-are-easier-to-discredit-1202808828/.

Pérez Huber, Lindsay. "Disrupting Apartheid of Knowledge: Testimonio as Methodology in Latina/o Critical Race Research in Education." *International Journal of Qualitative Studies in Education* 22, no. 6 (2009): 639–54.

Reyes, Kathryn Blackmer, and Julia E. Curry Rodríguez. "Testimonio: Origins, Terms, and Resources." *Equity & Excellence in Education* 45, no. 3 (2012): 525–38. DOI: 10.1080/10665684.2012.698571

Richardson, Brian K., and Juandalynn Taylor. "Sexual Harassment at the Intersection of Race and Gender: A Theoretical Model of the Sexual Harassment Experiences of Women of Color." *Western Journal of Communication* 73, no. 3 (2009): 248–72.

Sable, Marjorie R., Fran Danis, Denise L. Mauzy, and Sarah K. Gallagher. "Barriers to Reporting Sexual Assault for Women and Men: Perspectives of College Students." *Journal of American College Health* 55, no. 3 (2006): 157–62.

Scholz, T. M. Linda. "The Rhetorical Power of Testimonio and Ocupación: Creating a Conceptual Framework for Analyzing Subaltern Rhetorical Agency." PhD thesis, University of Colorado Boulder, 2007.

Scholz, T. M. Linda "Hablando Por (Nos)Otros, Speaking for Ourselves: Exploring the Possibilities of 'Speaking Por' Family and Pueblo in the Bolivian Testimonio '*Si Me Permiten Hablar*'." In *Latina/o Discourse in Vernacular Spaces: Somos de Una Voz?*, edited by Michelle A. Holling and Bernadette Marie Calafell, 203–22. Lanham, MD: Lexington Press, 2011.

Sowards, Stacey, and Richard D. Pineda. "Latinidad in Ugly Betty: Authenticity and the Paradox of Representation." In *Latina/o Discourse in Vernacular Spaces: Somos de Una Voz?*, edited by Michelle A. Holling and Bernadette Marie Calafell, 123–43. Lanham, MD: Lexington Press, 2011.

Van Zyl, Nicole. "Sexual Grooming of Young Girls: The Promise and Limits of Law." *Agenda* 31, no. 2 (2017): 1–10.

Villarreal, Melissa. "Latinas' Experience of Sexual Assault Disclosure." *Psychology* 5, no. 10 (2014): 1285–300.

30

FROM INNOCENTS TO EXPERTS

Queer and trans of color interventions into #MeToo

Elena Elías Krell

This essay investigates the theoretical–political links between sexual abuse and structural oppression, including the quotidian labor of "difference" that queer people of color (QPOC) perform in everyday life. I argue that anti-sexual assault campaigns on college campuses and elsewhere, as well as efforts to end childhood sexual abuse, would benefit from centering QPOC analytics. Doing so reveals the racialization of "innocence" and its limited availability to certain subject positions, as well as its dubious usefulness even for those to whom it does accrue. Instead, we could rebrand survivors from innocent victims to *experts* in CSA prevention. Drawing from Ignacio Rivera, Janet Mock, and Ann Cvetkovich, I demonstrate that QPOC activists theorize abuse-prevention extensively and, crucially, lay bare the historical links between intimate violence and systemic modes of oppression. The essay offers a case study of Ignacio Rivera's "The Heal Project," a multimodal undertaking that uses an intersectional framework and aims to end childhood sexual abuse through education, discussion, and performance. I discuss my own work with this organization, offering an autoethnographic view into a QTPOC-centric approach to the sexual abuse epidemic. Centering queer and trans of color voices offers a survivor/expert-centered approach to CSA, and helps move us closer to a world without systemic sexual abuse. The essay ultimately suggests that incest and sexual abuse must be thought alongside state-sanctioned vulnerability.

> Accountability is love and love is accountability; you can't have one without the other.
>
> Najma Johnson

In her groundbreaking transfeminist memoir, Janet Mock frames research on queer youth of color and childhood sexual abuse (CSA) through the lens of the systemic quotidian productions of difference. Specifically, she cites that queer and trans people of color (QTPOC) disproportionately experience isolation from friends and family, and that isolation can lead to heightened vulnerability:

> Being or feeling different, child sexual abuse research states, can result in social isolation and exclusion, which in turn leads to a child being more vulnerable to the instigation and continuation of abuse. Abusers often take advantage of a child's uncertainties and insecurities about their identity and body.[1]

Theorizing difference, Mock moves beyond statistics that demonstrate heightened QTPOC vulnerability to CSA and shows how everyday life and its cisheteronormative structure produces this disproportionate vulnerability. In other words, youth can experience isolation due to their queerness, in the original meaning of being odd or different, from a community that otherwise would provide a level of protection against would-be abusers, and thus everyday life of queer youth is linked to with systemic (sexual) violence.

I take this opening Mock provides for thinking CSA structurally and, in this essay, contribute to the growing body of literature addressing the lack of intersectionality in conversations about #MeToo and CSA. First, I consider the relationship between queerness and sexual abuse in cis and trans feminist scholarship on sexual trauma. I then move to an ethnographic case study of Ignacio Rivera's "Heal Project." Third, I suggest rebranding survivors from innocents to theorists of power and difference. Centering queer and trans theorizers of color, I reveal the structural scaffolding buttressing childhood and adult sexual violence.[2] Rather than focusing on individual perpetrators and victims, a QTPOC approach to ending CSA and sexual assault frames sexual abuse as systemic violence that is both separate from and also imbricated with other state-sanctioned forms of oppression, such as racism, sexism, and homo/transphobia.

Centering QPOC theorizers of CSA not only helps us understand sexual abuse as systemic but also provides us with actual strategies and tools to end sexual violence. After sharing a case study of Ignacio Rivera, an activist who is at the center of the QTPOC movement to end childhood sexual abuse, I suggest that the discourse of "survivorship" needs to be reconfigured. Even though we have moved from the language of "victim" to "survivor," the dominant affect sympathetic non-survivors show toward survivors is pity. I suggest we start thinking of survivors as experts with unique vantage points into our society, that have much to contribute to the movement to end CSA and other forms of sexual violence.

I want to add one, for which I am grateful to a close friend, who, upon hearing my thesis, nervously asked, "If survivors are experts, is sexual abuse their training?" I return to this excellent question at the end of the essay and, throughout, listen for the resonances between #MeToo, childhood sexual abuse, and queer and trans of color critique. I aim to amplify the muted connections between sexual abuse and the "difference" that QTPOC are perceived to be performing in their everyday lives.

Cis/trans feminism and the queerness of sexual trauma

In *An Archive of Feelings*, especially her third chapter "Sexual Trauma/Queer Memory," Ann Cvetkovich courageously warns that queer and feminist resistance to any relationship between queerness and childhood sexual abuse (i.e., "you're gay because you were molested") has led feminists to place a chasm between conversations on queerness and childhood sexual abuse. The (white) women who wrote the leading books on CSA never mention their queerness, even though the majority of them are openly queer in other (con)texts.

She even goes so far as to say that there might be *generative* ways queerness and sexual abuse might be linked[3]:

> ... why can't saying that "sexual abuse causes homosexuality" just as easily be based on the assumption that there's something right, rather than something wrong, with being lesbian or gay? As someone who would go so far as to claim lesbianism as one of the *welcome* effects of sexual abuse, I am happy to contemplate the therapeutic process by which sexual abuse turns girls queer ....The rejection of causality ... need not preclude the value of exploring the productive and dense relations among these terms.[4]

In her chapter, Cvetkovich disavows any neat causality between queerness and CSA and resists the idea that a change in object choice (i.e., from male to female) can resolve trauma in any simplistic way. However, she productively insists that queer studies engage with CSA. We miss out on a lot, she observes, if we allow the homophobic/antifeminist idea that abuse causes queerness to frame the conversation.

Rather than framing them as innocent victims in the way most self-help literature does, Cvetkovich shows that queer women have queer modes of survivorship through performance, writing, and other means of artistic expression. The performances of Tribe 8 that Cvetkovich analyzes show how survivors castrate themselves by chopping off dildos they are wearing in performance and sing about sexual abuse, survivorship from a survivor-centered, empowered perspective. By cutting off their own phalluses, they are not simply saying "F★ the patriarchy!" but are involving themselves as part of the matter to be lobotomized, compromised, and questioned relating to power and consent.

Returning to the quotation that opened my discussion, Mock provides a framework for reading Ann Cvetkovich's theorization of the fraught relationship between queerness and sexual trauma. Cvetkovich spends the majority of her chapter showing how (white and cis) queer people create particular strategies of survivorship after abuse, while Janet Mock thinks closely about the factors that make QTPOC disproportionately vulnerable to CSA in the first place.

Cvetkovich's text opens theoretical avenues of connection, and I find Mock useful for grounding those theories in everyday realities. Mock's insistence that queerness-as-difference makes children more vulnerable shows how queerness *can* be causally related to sexual abuse—but in the reverse order that the antifeminists suppose. Mock and Cvetkovich together help me think about CSA in its broader sociocultural, political, and economic context because we are not equally made vulnerable to sexual violence.

Undergirding my understanding of sexual abuse as systemic is women of color feminist texts that provide a critical context for the current #MeToo conversation. In the United States, the use of rape as a tool of genocide in the formation of the nation state, the rape of Black women's bodies as a mode of production under chattel slavery, as Angela Davis, Hortense Spillers, and others have written, and sexual violence women of color experience due to imperialism and war around the world.[5] Andrea Smith's *Conquest* catalogues the diaries, juridical texts, and print media that targeted sexual and gender deviants as a central tactic of colonialism. Smith writes, "sexual violence [is] a primary tool of genocide … by which certain peoples become marked as inherently 'rapable.'"[6]

Conversations around #MeToo have a great deal to learn from theories of decolonization and Black and indigenous feminisms, such as Saidiya Hartman's theorization of the ways in which repulsion and desire, fear and joy are paradoxically both involved in the construction of race under white supremacy.[7] She shows that pleasure, as much as pain, were constitutive elements of cultivating the hegemony of anti-blackness under slavery, as today, and the institutionally sanctioned rape of Black women was one of its most significant manifestations. In the United States, the sexual vulnerability of female populations has been characterized by the subjugation of specific ethnic and racial categories. As Smith suggests, "the demonization of Native women can be seen as a strategy of white men to maintain control over white women" (21). She explains, the construction of women of color as "inherently rapable" is a disciplinary force that co-constitutes the taboo on violating white women (who nevertheless remain sexually vulnerable to white men), demonizes men of color, and positions white men as "saviors" of both white women and women of color, a position whose violence becomes evident when centering women of color perspectives.[8]

Rather than advocating for replacing white women with women of color at the center of #MeToo, I suggest we think #MeToo through the prism of the mutual imbrication of white women, women of color, *and* men of color as vulnerable to sexual violence in the afterlife of colonialism and slavery. While no one is arguing white women are not targets of violence, these works make it clear that centering women of color's voices uncovers a history of sex as a form of racialized violence, including setting up the innocent victim as white woman. If we choose not to grapple with the histories of racism that undergird sexual assault today, we run the very likely risk of reproducing the systems of oppression that make all women and some men targets, while pretending to protect (some in) those groups.

Isolation is caused by difference because "difference" is vilified and underappreciated in our society. As Audre Lorde famously theorized, the problem is not difference itself, but that we have such uncreative ways of navigating it.[9] It may seem obvious but I think is worth stating that difference transmits a lower status to the perceived carrier of that difference in our country. Lorde implores us to move beyond the white liberal dictum that "we are all the same" and learn how to see difference as a good thing. Further, her intervention forces liberal people to recognize the underbelly of the superficially palliative idea that we are more similar than different: why is similarity a necessary precursor for respect?

The answer, of course, is that this violent paradigm was necessary for the justification of the violent means by which the modern post/colonial world was invented, and this is why women of color feminism is so useful and necessary for #MeToo and CSA movements today.

Our inability to respect and celebrate difference is not just a theoretical problem: it creates what Lorde and Mock together help us see as the violence that "different" children disproportionately experience.

The Combahee River Collective offered a groundbreaking politico-theoretical tactic of centering the most vulnerable in social justice efforts.[10] They insisted that doing so was necessary in order to effectively and comprehensively account for all the many arms of power (my paraphrase). Centering queer and trans of color lives in #MeToo discourse and on conversations on sexual abuse on college campuses is one implementation of this politico-theoretical tactic. We can and must apply this to #MeToo and CSA discourse or miss the big picture of how and why sexual assault is a global epidemic.

As b. binaohan has shown, *all* people of color have been targets of cissexism and homophobia because gender and sexuality were primary means by which POC have been historically marked as aberrant.[11] The future of #MeToo has a radical potential of tracking within and alongside movements for racial justice because "… sexual violence is not simply a tool of patriarchy but also a tool of colonialism and racism … [since] entire communities of color are the victims of sexual violence."[12]

Centering QTPOC in the conversation on sexual abuse is useful for another practical reason: QTPOC have theories and strategies for resisting and preventing CSA! I now turn to the case of Ignacio Rivera, an organizer who is centering racism, ableism, xenophobia, U.S. exceptionalism, and other manifestations of whitecisheteropatriarchy in/as CSA, in The HEAL Project.

Gimme that "structural healing" (with apologies to Marvin Gaye)

When National Public Radio and other major media stations broke news of Harvey Weinstein's legacy of abuse, Ignacio Rivera had been working to end childhood sexual abuse through education and outreach for over 10 years. The "Hidden Encounters Altered Lives" (HEAL) Project is Rivera's most recent and most comprehensive campaign to end childhood sexual abuse.[13] Supported by a two-year fellowship from the Just Beginnings Collaborative,[14] initially, and currently sponsored by the Effing Foundation for Sex-Positivity,[15] the HEAL Project is characterized by four prongs of action to end CSA: "building community," "critical analysis," "social media campaigns," and "mobilization and education."

As part of HEAL, Rivera gives workshops and lectures around the country, maintains an active blog called "Heal2end" and a vlog entitled "Pure Love," performs theatrical acts on the topic, physically tours, and launches media campaigns regularly.[16] Rivera insists that the problem of and solution to CSA are both interpersonal and structural. Therefore, Rivera teaches people how to talk about sex and sexuality with their children, modeling such conversations with their daughter on the Pure Love vlog.

Rivera's *Guatu* tour took HEAL to 14 cities in 10 weeks, where they workshopped with organizers, students, parents, and concerned citizens in order to share their work on CSA prevention. Rivera has produced creative and scholarly work—including a parent/guardian toolkit, poetry, and performances—that explores the resonances between kink, polyamory, consent, love, CSA, substance abuse, sobriety, structural racism, and LGBT communities.

Over the course of a year from 2016 to 2017, I had the privilege of being on the advisory board of the HEAL Project along with a rotating group of queer of color activists and organizers. We were a sounding board for Rivera as they provided updates, shared challenges and successes, and dreamed up new manifestations in the project. It became clear on day one that Rivera centers the ways sexual abuse intersects with other forms of systemic oppression, including racism, homo/transphobia, poverty, and ableism.

Under the "Building Community" section of the website, Rivera writes that it is a goal of the project to "Utilize longstanding and new connections within the LGBT, Anti-Violence, Sex positive/Liberation, Women's Rights, Sex worker movement, Anti-rape, economic justice movements …"[17] In one of our first conversations, Rivera and I spoke about their work and they said "my activities as an organizer, performer, poet, and educator are all connected. Everything is connected. Structural oppression, racism, transphobia, CSA, that's why an intersectional framework is at the core of everything I do."[18]

I invited Rivera to perform in a series I curated at Vassar College in 2014–2015. For two hours, one evening in October 2015, they performed poetry to a full house in the little black box theater on campus. Their performance included themes of internalized trauma around raising a daughter as a survivor of incest and CSA, of finding power in their Taino/Puerto Rican ancestry and two spirit identities, exploring kink and BDSM (bondage, discipline, submission, and dominance) as forms of reclaiming and healing (which recalls Cvetcovich's work), and being a survivor of CSA. The HEAL Project is one manifestation of Rivera's commitment to holistic organizing that is diverse methodologically: a multipronged matrix of liberation, or to adapt Patricia Hill Collins' term, "matrix of domination."[19]

When they formed the advisory board of HEAL, Rivera invited only other QTPOC people, going out of their way to make sure that trans women were present in order to counteract the dominance of trans masculine people. This is how you center the most marginalized.

When they invited me to join the advisory board of HEAL, I offered a hesitant "yes." I admired their integrative approach to activism, performance, and writing, as well as its deep intersectionality, and I had become disillusioned by the whiteness of the conversations on sexual assault on college campuses.[20] By taking into account structural inequality, HEAL shows us that intersectionality is a practice that #MeToo ignores at its own peril.

On the campus where I teach, conversations about racial justice center around men of color. Conversations on sexual assault, on the other hand, are dominated by white staff, faculty, and survivors. Women of color students do not trust the support systems available to them around sexual assault enough to even report crimes against them. Buffeting the cases involving cisgender white female students is the history I briefly outlined in the previous section, a history in which sexual innocence is tied to white femininity. Recently, some journalists have started to address the whiteness of the conversations around #MeToo and sexual assault on and off college campuses.[21] Overall, I am stunned by the lack of awareness of how white-centric the conversations on sexual assault are on my campus. It is almost as if women of color are assumed not to experience sexual abuse, when we know scientifically that they almost certainly are experiencing it more.

In pushing against what Rivera calls the "culture of silence" around sexual abuse in childhood in particular, The HEAL Project explicitly names the violence that silence engenders. Rivera speaks to the ways that the price of silence, the price of non-survivors' comfortability in not having to talk or think about abuse, is the altered lives of people who do not have the privilege of not talking or thinking about it. The HEAL Project is survivor-centric in its end goal of *ending* childhood sexual abuse. While an individualist model might focus on putting particular perpetrators in jail, Rivera's holistic intersectional one understands sexual violence as systemic and as part of the settler colonial project foundational to the modern nation state. With this holistic and complex view of the problem, solutions must likewise be structural in approach.

The labor QTPOC perform to navigate their everyday lives is not unlike that involved in reclaiming one's body after a history of interpersonal abuse, since much of QTPOC existence entails self-determining one's identity and body in the face of hostility and societal abuse. Indeed, the two are not completely separable, as Mock has theorized. In their essay, "Accountability to Ourselves and Our Children," Ignacio Rivera writes:

> The silencing of sexual, physical, psychological, and economic violence within … family structures is anchored to our experiences of oppression …. Not ruffling any feathers while trying to survive a law enforcement system that is homophobic, transphobic, racist, and sexist comes at a cost. This is where the cultural, historical, and community-driven measures of addressing CSA become a necessity.[22]

In this passage, Rivera explicitly links structural oppression to sexual violence and shows how the two must be thought alongside one another.

I suggest that survivors of CSA ought to "rebrand" ourselves as experts of energetic and physical boundary making, of healing, and of imagining other ways of parenting, of interacting with others. This is what my epigraph "Love is accountability—and accountability is love" speaks to. Rivera theorizes that accountability should not only be about retribution or justice after harm has been done. Since the vast majority of sexual abuse cases involve people who are friends and family, "we know the harm doers … if we acknowledge that reality, we have work to do to interact more intentionally with those around us; if we do not acknowledge it, we are left to rely only on reactive measures."[23] It is the proactive work that HEAL offers to the world. The information Rivera shares in their workshops, lectures, vlog, and writing goes beyond good/bad touch and the sex talk:

> The HEAL Project picks up where CSA prevention has left off. It pushes parents to engage with their children around sex and sexuality and helps create well-informed young people. The HEAL Project aids children in finding their voice and agency … it eliminates shame and uncovers secrecy—the very places where abuse breeds.[24]

This is just one example of the kinds of expertise that survivors of sexual abuse offer. Survivors have experience with the specific form of violence we wish to end, and have often theorized it more than anyone else. This was the case with Rivera, who, upon having their own child experienced intense paranoia around abusing their own child, investigated their own fears until they healed themselves. They now share their expertise on accountable love with others.

From innocents to experts

It seems sad to me now that I hesitated before joining Rivera's board and this lacuna in my self-awareness that I had a great deal to offer such a project was part of the impetus for this essay. If we reorient survivorship around expertise rather than innocence, survivors may feel more empowered to speak from their experience of survivorship as a form of expertise. Once people experience sexual abuse, the effects can and often do last a lifetime. At the same time, viewing victims as only that is to miss a tremendous opportunity to engage the people most directly affected by it in the effort to rid the world of sex as a tool of violence. The upside of the reality of the profound effect that CSA has on most people is that survivors develop effective strategies for navigating the world—the same world that allowed the abuse to happen—and thus have unique insights into what allows abuse to occur and what would have made it less possible. We are, for example, many of us, experts at healing. As Brené Brown discusses in her 2012 TED talk, sharing and speaking on trauma has been demonstrated to be an antidote to the shame that drives survivors of sexual abuse to not come forward to accuse their abusers.[25]

My sense of inadequacy as part of Rivera's board is glaring not only because my of status as someone with a doctorate in studying power and gender and an Assistant Professor of Women Studies, but also because I so closely associated being a survivor with being a victim (what is a survivor if not someone marked by trauma?) that I couldn't imagine occupying a place of power or expertise. It had never occurred to me that I had things to say about CSA from a QTPOC perspective.[26] I attribute this in large part to the culture of shame around survivorship and the fact that no one expects (or wants) us to talk about it—and also that my "innocence" and its preservation have never been things I could take for granted as valued commodities in any of the places I have lived (Greece, England, Mexico, Germany, Honduras, United States, and Canada).

Our current conceptions of childhood stem from Victorian-era discourses of innocence that were part of a series of campaigns to institute child labor laws, as Michaela di Leonardo observes.[27] This narrative emerged from a particular political economic landscape where Victorian British society was trying to differentiate itself from its colonial Others across the globe, and where child labor laws were just beginning to take form. A primary way that child labor advocates won rights for young people, who at that point were working longer workdays than adults in contemporary Western society, was to argue that they were innocent and in need of care-taking.[28]

Understanding ourselves as innocents/victims/survivors is *especially* unhelpful for women of color, and queer women of color in particular, I suggest. Innocence is coded as white, as José E. Muñoz (2006) and others have argued, and thus women of color do not have the same structural access to being perceived as innocent in the first place. As children, we are taught that we are who and how people treat us. Women of color who experience CSA are thus taught that their bodies exist for other people to do with them what they wish.[29]

As Sara Ahmed (2004) writes, the un/conscious feelings we have toward a phenomenon play a critical role in the ways we think and act toward that phenomenon.[30] Affect is not simply feeling, but a structure that constitutes the space between ourselves and other people and things. If women of color, especially queer feminine youth, do not have the protection of the affect of innocence, what other affects and effects take its place? The structure of innocence occludes other more empowering structures of feeling that women of color need in order to have agency over their own bodies.

Further, the victim-as-innocent-in-need-of-saving distracts us from the critical conversation about what anti-CSA and #MeToo movements have to gain from an intersectional structural lens. It centers non-survivors and puts their view at the center because it is their view of us as innocent or in need of protecting, rather than centering the often much more complex reality of survivorship. Understanding victims as innocents needing to be saved withholds possibilities for agency and action, thus primarily benefiting those who (want to) imagine themselves as saviors. Ending sexual violence necessitates centering survivors in our conversations about survivorship.

A glaring example of how we center non-survivors in dominant media is that there exist so few representations of abusers. Rather than being vilified, the trope of the creepy uncle or neighbor is all but lauded in popular culture.[31] The term "daddy issues" protects men who use women as objects upon which to project their subjectivity and desire by erasing their responsibility altogether. We have very little language for the "dad" who causes these issues, ironically, though he is named. These examples demonstrate the double-edged sword aspect of the trope of innocence: it can be taken away just as easily as it is given.

Lastly, "survivorship," albeit an improvement from "victim," creates a behemoth boundary between survivors and those who have not experienced sexual abuse—a boundary that reality does not support. Many of us have experienced uncomfortable crossings of boundaries that are laced or wrought through with sexual energy and power dynamics, and there is no strict boundary between abuse of various kinds. QTPOC survivor-experts are uniquely positioned to see CSA as an inevitable extension of a white cis heteropatriarchy.

Conclusion

In this essay, I have tried to show how the trope of innocent victim disallows the connections between CSA and racism, the latter defined by Ruth Wilson Gilmore as the production of heightened vulnerability to premature death for specific populations.[32] We should certainly rally around survivors and give them time to heal, but healing also means being able to have the choice and chance to respond to trauma. Any project investing in ending sexual assault must grant survivors the agency and collective capability to change the conditions under which they were abused.

Centering women of color makes apparent the need for us to address the cognitive absence in what survivors think they have to offer these conversations besides the risky confession.[33] For this to change, a structural elision in the affective paradigm of survivorship must shift. For Raymond Williams affect is a "structure of feeling" that permeates an entire historical era.[34] Ironically, "innocence" attached to survivorship actually contributes to the heightened vulnerability of women who do and do not approximate our cultural picture of innocence. The second definition of violence in the Merriam Webster Dictionary is as a descriptor of something "powerful, intense, and potent." A queer reading of "violence" recasts survivors as powerful mobilizers around ending CSA.[35]

Rebuilding one's sense of self from the ground up after sexual abuse in youth is a project that requires expertise of various kinds. As mentioned, in an early sharing of this essay, a friend queried, "If survivors are experts, is sexual abuse their training?" I am grateful for her positing this concern because it helps me clarify that when I say that survivors are experts, I am not suggesting that we condone the violence itself. Survivorship constitutes everything that happens after a trauma, not only the trauma itself. We have extensive experience in surviving and healing, and daily practice in how to navigate the same world that created the conditions of our abuse. To affirm this is not to laud abuse in any way. To say that people of color have developed ingenious ways of dealing with racism is not to say that racism should continue to exist. I mean only that

CSA survivors have skills and knowledge that comprise a vast resource for our conversations on sexual violence. Someone like Ignacio Rivera, a QTPOC expert, is a daily theorizer of the links between structural power, CSA, and communication, and should be at the center of the #MeToo movement and movements against CSA.

The #MeToo movement has brought critical attention to the issue of sexual harassment. Centering queer and trans people of color in this discussion helps different valences of violence to emerge. This approach explores how we are not just vulnerable in places of work but also in our families, neighborhoods, and very homes—all places that can be sites of violence, as *The Revolution Starts at Home* attests.[36] Thus, we learn different things when we center queer and trans youth of color in discussions of sexual abuse.

Rivera's HEAL Project exemplifies Kimberlé Crenshaw's intersectionality as a praxis, without which we actively exclude some of the most marginalized students.[37] Relatedly, centering women of color in conversations around sexual abuse also allows us to learn from the strategies of survival that WOC and/or queer students mobilize, which can only serve to enrich conversations around #MeToo.

This discussion has profiled Ignacio Rivera's formidably intersectional HEAL project to see how survivors can operate as experts. Rivera's work and Mock's writing underscore not just the possibility but also the urgency of linking queerness and sexual abuse. Rethinking the relationships between queerness and sexual abuse, not only through causality but also through power and difference, allows for QTPOC lives to be centered in #MeToo. My training in performance studies helps me think about performance as a praxis and way of doing theory. It is time that the experts be celebrated for their expertise as living lives before, during, and after CSA and #MeToo moments. We should be at the center of the #MeToo and CSA conversation.

Notes

1 Janet Mock, *Redefining Realness: My Path to Womanhood, Identity, Love & So Much More* (New York: Atria Books, 2014), 47.
2 I am indebted to R. Gilmore for her formulation of racism as the disproportionate production of heightened vulnerabilities for certain populations. "Seeing: The Problem or the Infrastructure of Feeling." Public Lecture, Dartmouth College, Hanover, NH, May 8, 2018.
3 Ann Cvetkovich, *An Archive of Feelings: Trauma, Sexuality, and Lesbian Public Cultures* (Durham: Duke University Press, 2003), 88–90.
4 Cvetkovich, *An Archive of Feelings*, 90, emphasis in original.
5 Angela Davis, *Women, Race, and Class* (Vintage, 2011); Hortense Spillers, "Mama's Baby, Papa's Maybe: An American Grammar Book," *Diacritics* 17, no. 2 (Summer, 1987): 64–81.
6 Andrea Smith, *Conquest: Sexual Violence and American Indian Genocide* (Cambridge: South End Press, 2005), 3.
7 Saidiya Hartman, *Scenes of Subjection: Terror, Slavery, and Self-Making in Nineteenth-Century America* (New York: Oxford University Press, 1997), 28–29, 44.
8 My thanks to Joan McAlister for the suggestion to include and framing of this point.
9 Audre Lorde, "Age, Race, Class, and Sex: Women Redefining Difference," in *Sister Outsider: Essays and Speeches* (Trumansburg, NY: Crossing Press, 1984), 111–113.
10 Combahee River Collective, *The Combahee River Collective Statement: Black Feminist Organizing in the Seventies and Eighties*, 1986.
11 B. Binaohan, *Decolonizing Transgender 101* (Biyuti Press), 2014.
12 Smith, *Conquest,* 8.
13 www.heal2end.com
14 Justbeginnings.org
15 www.effingfoundation.org
16 www.igrivera.com/watch-episodes.html

17 www.igrivera.com/the-heal-project.html
18 Rivera Ignacio, *Personal Skype Conversation* (Vassar College, Poughkeepsie, NY, May 9, 2014).
19 Patricia Hill Collins' term "matrix of domination" is theorized in her book *Black Feminist Thought: Knowledge, Consciousness, and the Politics of Empowerment* (pp. 225–227) (1991).
20 By intersectional, I mean a commitment to engaging the importance of the imbrication of vectors of power such as race, gender, sexuality, class, and ability. Cf. Kimberlé Crenshaw, "Mapping the Margins: Intersectionality, Identity Politics, and Violence against Women of Color," *Stanford Law Review* 43.6 (1991): 1241–1299.
21 See www.latimes.com/politics/la-na-pol-black-women-sexual-assault-20170828-story.html and www.insidehighered.com/advice/2017/09/01/sexual-violence-prevention-requires-focusing-how-multiple-forms-oppression
22 Ignacio Rivera. "Accountability to Ourselves and Our Children," in *Love with Accountability: Digging Up the Roots of Childhood Sexual Abuse*, ed Aishah Shahidah Simmon (Chico, CA: AK Press, 2019), 279.
23 Rivera, "Accountability," 281.
24 Rivera, "Accountability," 283.
25 www.ted.com/talks/brene_brown_listening_to_shame?language=en
26 I was also concerned that I might take up too much space and/or say the wrong thing (I tend to open up quickly and offer a lot). White guilt and internalized racism work in tandem, as my former colleague Treva Ellison theorizes through a phenomenon called *vanta* power (personal conversation May 2018, Dartmouth College, Hanover, NH).
27 di Leonardo, In-class lecture. "Gender and Race." Anthropology course, Northwestern University, 2011.
28 See Michaela di Leonardo, *Exotics at Home: Anthropologies, Others, American Modernity* (Chicago, IL: The University of Chicago Press, 1998).
29 José Estéban Muñoz, "Feeling Brown, Feeling Down: Latina Affect, the Performativity of Race, and the Depressive Position," *Signs* 31, no. 3 (2006): 675–88.
30 Ahmed writes: "Emotions are not 'in' the individual or the social, but produce the very surfaces and boundaries that allow the individual and the social to be delineated as if they are objects." Sara Ahmed, *The Cultural Politics of Emotion* (New York: Routledge), 2004, 10.
31 We can think of Kevin Spacey's character in *American Beauty* and "Creed" in the American version of *The Office*.
32 Gilmore, *Golden Gulag: Prisons, Surplus, Crisis, and Opposition in Globalizing California* (Berkeley, CA: University of California Press, 2007).
33 Cvetkovich discusses the fraught nature of confession and the ways it is dependent on the public who receives it (citation).
34 Raymond Williams, *Marxism and Literature*. Marxist Introductions (Oxford: Oxford UP, 1977).
35 This framing, while on the surface seeming to be about the victim, centers those who are not victims, and even perpetrators. The discomfort non-survivors feel around discussing the issue refracts the stigma and shame of abuse on to the survivor.
36 Edited by Ching-In Chen, Jai Dulani, Leah Lakshmi Piepzna-Samarasinha; preface by Andrea Smith, *The Revolution Starts at Home: Confronting Intimate Violence within Activist Communities* (Brooklyn, NY: South End Press, 2011).
37 Crenshaw, "Mapping the Margins."

Bibliography

Ahmed, Sara. *The Cultural Politics of Emotion*. New York: Routledge, 2004.
binaohan, b. *decolonizing transgender 101*. biyuti press, 2014.
Brown, Brené. "Listening to Shame." TED. Accessed January 12, 2020. www.ted.com/talks/brene_brown_listening_to_shame?language=en.
Chen, Ching-In, Dulani, Jai, and Piepzna-Samarasinha, Leah Lakshmi, eds. *The Revolution Starts at Home: Confronting Intimate Violence within Activist Communities*. Brooklyn, NY: South End Press, 2011.
Collins, Patricia Hill. *Black Sexual Politics: African Americans, Gender, and the New Racism*. New York: Routledge, 2006.
Crenshaw, Kimberlé. "Mapping the Margins: Intersectionality, Identity Politics, and Violence against Women of Color." *Stanford Law Review* 43, no. 6 (1991): 1241. https://doi.org/10.2307/1229039.

Cvetkovich, Ann. *An Archive of Feelings: Trauma, Sexuality, and Lesbian Public Cultures*. Durham, NC: Duke University Press, 2013.

Davis, Angela. *Women, Race, and Class*. New York: Vintage Press, 2011.

di Leonardo, Michaela. *Exotics at Home: Anthropologies, Others, American Modernity*. Chicago, IL: The University of Chicago Press, 1998.

Gilmore, Ruth Wilson. "Seeing: The Problem or the Infrastructure of Feeling." *Public Lecture at Dartmouth College*. Hanover, NH, May 8, 2018.

—. *Golden Gulag: Prisons, Surplus, Crisis, and Opposition in Globalizing California*. Berkeley, CA: University of California Press, 2007.

Hartman, Saidiya V. *Scenes of Subjection: Terror, Slavery, and Self-Making in Nineteenth-Century America*. New York: Oxford University Press, 2010.

"HEAL2END." *IGNACIO G RIVERA*. Accessed January 12, 2020. www.igrivera.com/heal2end.

"Inside Higher Ed." *Sexual Violence Prevention Requires Focusing on How Multiple Forms of Oppression Intersect with Sexism (essay)*. Accessed January 12, 2020. www.insidehighered.com/advice/2017/09/01/sexual-violence-prevention-requires-focusing-how-multiple-forms-oppression.

Just Beginnings Collaborative. Accessed January 12, 2020. http://justbeginnings.org/.

Lorde, Audre. "Age, Race, Class, and Sex: Women Redefining Difference." In *Sister Outsider: Essays and Speeches*. Trumansburg, NY: Crossing Press, 1984.

Mock, Janet. *Redefining Realness: My Path to Womanhood, Identity, Love & So Much More*. New York: Atria Books, 2014.

Muñoz, José Esteban. "Feeling Brown, Feeling Down: Latina Affect, the Performativity of Race, and the Depressive Position." *Signs: Journal of Women in Culture and Society* 31, no. 3 (2006): 675–88. https://doi.org/10.1086/499080.

"Q&A: Why It's Harder for African American Women to Report Campus Sexual Assaults, Even at Mostly Black Schools." *Los Angeles Times*, August 28, 2017. www.latimes.com/politics/la-na-pol-black-women-sexual-assault-20170828-story.html.

Rivera, Ignacio. "Accountability to Ourselves and Our Children," in *Love with Accountability*. Ed. Aishah Shahidah Simmons. Chico, CA: AK Press, 2019.

Smith, Andrea. *Conquest: Sexual Violence and American Indian Genocide*. Cambridge: South End Press, 2005.

Snorton, C. Riley. *Nobody Is Supposed to Know: Black Sexuality on the Down Low*. Minneapolis, MN: University of Minnesota Press, 2014.

Spillers, Hortense J. "Mama's Baby, Papa's Maybe: An American Grammar Book." *Diacritics* 17, no. 2 (1987): 64. https://doi.org/10.2307/464747.

The Combahee River Collective Statement: Black Feminist Organizing in the Seventies and Eighties. Albany, NY: Kitchen Table, 1986.

"The Effing Foundation for Sex-Positivity." The Effing Foundation for Sex-Positivity. Accessed January 12, 2020. www.effingfoundation.org/.

Williams Raymond. *Marxism and Literature*. Marxist Introductions Series. Oxford: Oxford University Press, 1977.

31

SYMBOLIC ERASURE AS GENDERED VIOLENCE

The link between verbal and physical harm

Kate Lockwood Harris

In popular discourse, two contradictory ideas circulate regarding the relationship between symbolic and material violence. First, some people say that words do no harm and have no real impact on the world. From this perspective, people who are "offended" by others' statements should toughen up and be less sensitive. Furthermore, instances of violence bear no connection to how people speak or the contents of their remarks. Violence, in this account, is purely physical and has no symbolic components. Words are inert and merely describe or explain the world. By contrast, other people say that words and thoughts can directly transform the world, as in the bestselling book *The Secret*. This approach encourages people to imagine words in their minds, positing that this act alone will change the conditions around them. Words become so powerful that any distinction between symbols and violence collapses: A simple phrase can be shattering, and words have magical powers. Whereas in the first perspective, communication and violence have no relationship, in the second, they are synonymous. This chapter shows that neither of these ideas adequately explains how physical and symbolic factors intertwine in gendered violence. In place of simplistic but popular assumptions, this chapter provides background on several intellectual traditions that offer more nuanced accounts of the relationship between verbal and physical harm.

The importance of a verbal-physical link in research on gendered violence

For scholars who research gendered violence, the first perspective (i.e., words are inert) is inadequate because it misses key aspects of violence. Gendered violence almost always encompasses more than physical brutality. For example, although intimate partner violence (IPV) often involves one person bruising, breaking, bloodying, or otherwise injuring another individual's body, it is not limited to these behaviors. IPV may also involve financial control, social isolation, and verbal abuse, none of which are immediately observable on a person's body. Given its varied components, IPV is neither simply physical assault nor simply words. Instead, relational abuse involves connections between the physical and symbolic aspects of violence. As Kate Lockwood Harris and Jenna Hanchey argued, sexual and gendered violence is "a constellation of action, meaning, institutions, and relations" such that "rape, other sexual assault, and intimate partner violence" as well as "the structures and discourses that enable those acts" are included in definitions of these phenomena.[1] To assume a total separation between verbal and physical harm is to deny how systems of meaning support interpersonal violence.

By rejecting the perspective in which words are inert, scholars also show that violence is more than a singular, physical event between two people. They establish parallels between relational abuse (such as that which occurs in IPV) and abusive social structures and systems. Andrea Smith, for example, asserted that sexual violence is "a tool by which certain peoples become marked as inherently 'rapable.' These people then are violated, not only through direct or sexual assault, but through a wide variety of state policies, ranging from environmental racism to sterilization abuse."[2] For Smith, violence occurs not only during acute, physical trauma. Instead, slower processes—such as those whereby a group of people are exposed to toxic chemicals over years and, as a result, have higher cancer rates and shorter lifespans—are also included. This environmental racism operates through complex decision-making, policies, struggles over property rights, and the circulation of social and financial capital. Smith incorporates these contributing factors into a definition of sexual violence to underscore the mechanisms and processes that denigrate marginalized groups and thereby make it possible to violate people. Scholars can consider these enabling patterns of symbol use to be an aspect of sexual and gendered violence only if physical and verbal harms are related. Without some tie between the two, nothing but torn and damaged flesh would be noticeable, and gradual, collective violence would not be intelligible as such.

Furthermore, by assuming that physical and verbal processes are linked, scholars can theorize the ties among multiple levels of violence. Similar to Smith, the collective called Incite! asserted that to address gendered violence, people must "make connections between interpersonal violence, the violence inflicted by domestic state institutions (such as prisons, detention centers, mental hospitals, and child protective services), and international violence (such as war, military base prostitution, and nuclear testing)."[3] Incite! foregrounded the associations between harmful acts in intimate relationships and broader social and global political patterns. Neither Smith nor Incite! define gendered or sexual violence as one-time events that have a clear beginning and end. Instead, gendered and sexual violence are implicated in the historical and contemporary distribution of resources such as security, health, and access to basic rights. When considered within this framework, a nuanced account of the relationships between gender, violence, and communication begins to emerge: Words and language never exist in a vacuum, and human actions are not independent of words and language. Symbols are imbricated in the material and social processes that maintain ongoing gendered violence. Although a word alone cannot draw blood, words are connected to the myriad systems that enact and perpetuate violence. Similarly, violence entails meaning and symbolism.

In sum, scholars have several reasons for rejecting the idea that words are inert: It conflicts with empirical studies of relational abuse, it narrows definitions of violence to include only immediate and interpersonal harms, and it ignores myriad symbolic processes without which violence could not occur. Explaining and intervening in gendered violence requires assuming *some* relationship between physical and verbal harms, but scholars have conceptualized that relationship in many different ways. The remainder of this chapter details several of these conceptualizations. Because an inert view of words aligns with popular understandings of communication,[4] any departure from that assumption can be misunderstood as the magical perspective outlined in the introduction. Importantly, none of the perspectives that follow *eradicate* a distinction between symbols and violence. Instead, each exists on a continuum somewhere in between inert and magical assumptions about what words can do.

Words as weapons: symbols can maintain historical and cultural domination

At first glance, scholars who theorize hate speech appear to argue that words can cause direct injury, much like the magical stance I outlined in the introduction. For example, in the edited volume *Words that Wound,* many authors borrow terms used to describe physical injury and suggest that speech can be assaultive.[5] Although the categories *word* and *weapon* collapse in a cursory interpretation of their work, these scholars do not actually assert that symbolic and physical violence are indistinguishable. Instead, they demonstrate that symbols become violent only in specific contexts that encompass more than a single utterance. One word *can* have immediate, bodily effects. When a person uses hate speech against another—that is, employs words as weapons—the target may experience physiological responses: "rapid breathing, headaches, raised blood pressure, dizziness, [and] rapid pulse rate."[6] Despite these short-term effects, the critical race and gender theorists I discuss in this section are not principally concerned with any particular collection of syllables. Instead, they show that over time, and with repeated exposure to messages that devalue groups to which an individual belongs, these instantaneous impacts can become lasting and lead to depression, other life-threatening health conditions, and suicide. By itself, no word can commit violence; however, a word along with its history and the cumulative impact of its deployment can.

Words enact violence when they elicit and perpetuate historical and cultural systems of domination. Explicating this line of thinking, Charles Lawrence argued,

> There is a great difference between the offensiveness of words that you would rather not hear because they are labeled dirty, impolite, or personally demeaning and the *injury* inflicted by words that remind the world that you are fair game for physical attack, that evoke in you all of the millions of cultural lessons regarding your inferiority that you have so painstakingly repressed.[7]

In distinguishing offense from injury, Lawrence makes a specific link between verbal and physical harm. Rather than eradicate a split between symbols and violence, he suggests that only some utterances lead to imprisonment, rape, other sexual assaults, murder, and additional unjust deaths. Additionally, Richard Delgado and Jean Stefancic argued that people who use words as weapons leave those they target less able to excel in their workplaces, thrive in school, and be involved in democratic processes.[8] Said differently, people in dominant groups use particular words to exclude and systematically shut others out of full participation in society.

Words that sustain domination can circulate independent of people who articulate them. Describing "the injurious action of names," Judith Butler said that words invoking systemic racism or homophobia (or, I will add, transphobia) can "arrive without a speaker—on bureaucratic forms, the census, adoption papers, employment applications…. [The] individuals who write and distribute the forms … are not the originators of the discourse they convey."[9] In this claim, Butler is not focused on a single human who repeats a phrase or a term. Instead, she argues that the harm and violence of a name emerge from its circulation across many different events and locations. Butler thus disassociates the violence of words from any individual who wishes to cause harm. Verbal injuries can be inflicted without intent because the force of words depends upon discourse, related statements that cohere across time and place to form a system of meaning.

Though scholars who study hate speech have broad consensus that weaponized words have physiological, psychological, and sociopolitical impacts, they do not agree about what should be done in response. Critical race theorists such as Delgado and Stefancic advocate for legal restrictions on racist hate speech. They aim to interrupt the white supremacy of the United States' juridical system from within. To support this position, scholars point out that U.S. law already limits freedom of speech: One cannot defame another, invade another's privacy, commit fraud, or use fighting words. By contrast, one can use words to invoke and bolster racism, a system that operates through financial control, barriers to property ownership, educational segregation, resource and health disparities, and other processes. Critical race theorists conclude that the current pattern of allowed and regulated speech upholds white supremacy.[10] Because of this, they argue that the United States should apply existing restrictions on speech to racist hate speech and suggest that this change would not rework current law. Instead, it would apply existing laws in ways that uphold rather than undermine the principles of the country. In seeming opposition to this position, Judith Butler argued against increasing the state's ability to arbitrate injuries associated with hate speech, noting that the state has often inflicted such injuries.[11] Through analysis of several cases, Butler contended that the collapse of speech into conduct—which is related but not identical to the magical collapse of communication into violence I detailed in the introduction—supports state intervention, regardless of political ideology. Butler asserted that arguments whereby communication *is* violence (i.e., words inflict direct harm) have reinforced a puritanical and violently heteronormative society that represses representations of nonnormative sex and sexuality. Charles Lawrence offered a rejoinder to the type of argument Butler advanced:

> Whenever we decide that racist hate speech must be tolerated because of the import-ance of tolerating unpopular speech, we ask Blacks and other subordinated groups to bear a burden for the good of society—to pay the price for the societal benefit of cre-ating more room for speech. And we assign this burden to them without seeking their advice or consent. This amounts to white domination, pure and simple.[12]

Though they differ in their arguments about what should be done to address the problem, these authors each offer theoretical approaches that explain how words participate in, contribute to, and enact violence via reference to and recreation of historical and cultural domination.

Each of the scholars mentioned in this section suggests that words build meanings that have material impacts. Because of the lessons they teach about who matters, words support systems whereby groups accrue inequitable benefits and access to resources. Even so, none of these scholars argue simply that a single word should be banned in all situations. They do not believe in incantations or spells. Indeed, it would be hard to find a communication scholar who identifies a "correct" thing to say in any given scenario, or who asserts that a person can avoid doing violence by simply choosing one word over another. The arguments in this section are far more complex than that, and they offer theoretically informed assertions that meanings form over generations and embolden contemporary action.

Trauma studies: erasure is evidence of violence

Whereas the scholarship overviewed in the last section explained how words can enact violence, scholars who study trauma explain how violence can erase words. Per this set of ideas, generally called *trauma studies*, an absence of particular symbols, language, and stories is evidence that vio-lence has already occurred. Traumatic events can make it difficult if not impossible for individuals and groups to provide descriptions of violent events. Because it can be so challenging for people to talk about violence, and because some forms of violence actively undermine representations of that violence, trauma studies suggests that violence is present in what appear to be absences in language, symbols, and stories. Both the inert and magical stances outlined in the introduction address whether words can be physically violent, but this intellectual tradition focuses on how physical violence delimits what words can be said or written.

Trauma studies proceeds from two approaches. The psychological approach demonstrates that experiences of trauma reconfigure people's brain chemistry such that they may be unable to recall events in chronological order. After acute trauma ends, some stimuli, including certain symbols, can return a person's brain to the state it was in during the original trauma. Because of this process, the person may re-experience the traumatic event, even though the external threat is no longer immediately present. If they are not aware of how violence can impact a person's ability to recall and speak about trauma, members of the public may dismiss the testimonies of people who have experienced horrors. In particular, people who have been raped often cannot recall certain details or provide to friends and investigators a "coherent" account of what happened. A second approach to trauma studies emerges from literary and cultural studies and is influenced by psychodynamic theories. This scholarship suggests that individuals and societies repress knowledge about violence and wrongdoing because they are unable to cope with those events. This difficulty with processing terror could lead people who have experienced violence to avoid reminders of those experiences. It could also lead to more pernicious outcomes, such as societies suggesting that specific instances of mass violence, even genocide, never happened. Trauma studies thereby illuminates how individuals respond to the aftermath of violence. It also provides one explanation for why societies routinely institutionalize amnesia, fail to acknowledge egregious harms, and defer accountability.

These ideas from trauma studies have been influential in communication studies. For example, echoing the notion that some symbols can return bodies to the immediate moment of trauma, Sarah Projansky argued that depictions or descriptions of rape, even when designed with anti-rape aims, actually "expand, heighten, and perpetuate the experience of rape."[13] Given the availability of few, if any, representational strategies that disrupt rape logics, Projansky decided not to detail any scenes of assault in her own writing, acknowledging that the choice to omit such accounts might, paradoxically, contribute to the seeming absence of sexual violence from texts. Relying on neuroscience and literary studies, Lydia Huerta Moreno showed that when readers encounter narratives about *Los Feminicidios,* a spate of murdered women in Ciudad Juárez, the stories stimulate neurons that would be active if the reader was directly experiencing assault.[14] This claim is consistent with the trauma studies principle that talking or writing about violence can produce bodily responses similar to those that occur during physical violence. Kate Lockwood Harris and James Fortney surfaced a similar theme in their critique of traditional approaches to reflexivity, arguing that if one considers both trauma studies and disability studies together, asking scholars simply to reflect on their encounters with violence or trauma can reinforce the original violence.[15] They described academics who urge a war veteran to consider how her combat experiences influence her research design and analysis, but doing so triggers her flashbacks and exacerbates her post-traumatic stress disorder. Instead of asking people to describe their experiences of trauma and disability, Harris and Fortney suggested that paying attention to the absences in language and communication may be more appropriate. Joëlle Cruz also adopted ideas from trauma studies when she showed how Liberian women's informal credit organizations became more closed off after their members survived a war.[16] On the whole, such communication studies scholarship on trauma challenges the notion that people can describe or speak easily about violence without also experiencing or participating in violent dynamics. Verbal accounts, whether written or spoken, can replicate humans' neurological and chemical experiences of violence.

Trauma studies also demonstrates that trauma crosses multiple levels of social life. For example, English literature scholar Ann Cvetkovich maintained that trauma is the "hinge between systemic structures of exploitation and oppression and the felt experience of them."[17] In this argument, any individual experience of trauma is connected to society's exploitation, abuse, and marginalization of particular groups of people. Stephanie Houston Grey, in an essay tracing the uptake of trauma studies in communication scholarship, stressed that trauma is "an individual *and* social concern," a "private *and* public event."[18] Grey emphasized that trauma, and the violence that often produces it, operates across various levels and forms of communication: It "reverberates throughout individual lives, local peer networks, institutions, and mass mediated narratives."[19] She suggested that violence and trauma "produce ripples"[20] across relationships, time, and place as they "circulate throughout a culture for decades via art, film, literature, music, and politics."[21] Trauma studies, then, resists reducing violence to local episodes that are bound to particular bodies or eras. It also challenges the idea that people can simply and explicitly speak about violent events. Instead, it emphasizes that violence changes communication and that, through absences in symbolic systems, trauma travels across generations and throughout cultures.

Epistemic violence: erasure enables harm

Whereas the previous section suggested that people's experiences of trauma can leave gaps in their own language, this section focuses on how dominant groups and abusive individuals can actively create gaps in communication in order to perpetrate violence. These gaps are often produced when powerful people deny that particular violent and abusive events have occurred. Over time, these denials and refusals sometimes alter reality by erasing evidence of egregious wrongdoing or by convincing people who have suffered violence that they were mistaken about their experiences. According to Gayatri Chakravorty Spivak, this process can occur when "*one* explanation and narrative of reality" is "established as the normative one."[22] The result is called *epistemic violence*. Kristie Dotson elaborated that epistemic violence is "a refusal, intentional or unintentional, of an audience to communicatively reciprocate a linguistic exchange owing to pernicious ignorance … that, in a given context, harms another person (or set of persons)."[23] This refusal can occur when dominant groups discount the knowledge of minoritized groups.[24] To illustrate this idea, several examples from the academy follow. Epistemic violence is rife in this institution. It shows up in how schools respond to racial and gendered violence, in the systems and processes used to produce and evaluate scholarship, and in lived experiences of people who build careers in academia. Similar to trauma studies, the scholarship discussed in this section focuses on the violent erasure of certain words and stories. It adds analytic tools that highlight how perpetrators of violence actively make and use these erasures to continue a cycle of violence.

Epistemic violence includes discrediting and ignoring evidence of harassment and other gendered violence on campuses. This process can occur even when universities attempt to eliminate sexual assault: The actions they take often reinforce existing power structures that contribute to violence.[25] For example, Kate Lockwood Harris showed that, in U.S. universities, the policies designed to respond to racial and gender discrimination (which includes sexual assault) enact epistemic violence.[26] When members of marginalized groups in those schools provide testimony about their experiences of violence, administrators often do not take meaningful action in response, citing a lack of evidence to support these stories. The universities are unable to find evidence, as Harris showed, because people with the most bureaucratic power are able to be ignorant of that evidence. Given the so-called absence of evidence, the universities take no action to revise their practices and therefore do not respond to existing violence that continues to harm marginalized and minoritized groups. As Joshua Daniel Phillips and Rachel Alicia Griffin argued, these institutional processes can silence people who have experienced violence.[27] Using Black feminist theory, they analyzed the media coverage surrounding Crystal Mangum, a Black sex worker who reported that white male lacrosse players at Duke University assaulted her. Phillips and Griffin argued that the media repressed Mangum's voice, erased her subjectivity, and cast her as a "denigrated object."[28] Tracing the ways in which privileged groups dismiss the accounts of the people they marginalize, the authors said, "Popular perceptions of guilt and innocence in sexual violence cases continue to favor the narratives of upper-class White men and remain suspicious of, and even hostile toward, the narratives of Black women."[29] In this case, the prioritized reality was that of wealthy white men, and the public refused to find Mangum's story credible. In both of these examples, people with considerable power and influence dismissed ample evidence of harm. They therefore saw no need for responsiveness and allowed violence to continue.

Epistemic violence also operates in the academy when scholarship is produced, circulated, and evaluated. In an article demonstrating these processes, Paula Chakravartty et al. showed that white communication scholars systematically over-cite the publications of other white people while overlooking the contributions of scholars of color, especially women of color.[30] Other scholarly practices are similarly infused with whiteness. For example, Alexandra Moffett-Bateau argued that disciplinary boundaries and borders erase Black feminist methodologies, "to the detriment of social sciences."[31] These practices, which include the over-citation of white scholars, maintenance of homogenous networks, and defense of disciplinary borders, uphold systemic racism in the academy. As a consequence of these practices and many others, white scholars continue to move up in rank, earn raises, and become tenured based on artificially inflated citation counts and related metrics. Meanwhile, scholars of color continue to be underrepresented at every level of higher education, and especially so at the highest ranks. These raced and gendered inequities are one result of epistemic violence.

Unsurprisingly, epistemic violence also shapes the experiences of scholars who work in academia. Joëlle Cruz identified processes related to race, nation, and immigration status that marginalize particular bodies:

- Have you ever wondered
- How discursive closure
- Feels on tender immigrant bodies?
- Feels like a high speed train
- Hurling itself across a spine
- Metal sizzling against flesh
- …
- Erasure
- They look through you
- You are perfectly transparent
- When you leave, nobody says goodbye[32]

With these passages, Cruz evokes the lived, felt dynamics of being an immigrant in the U.S. academy, ones that involve erasure. White citizens in the U.S. academy who speak with standard American accents fail to be responsive to these experiences and reflective about their own actions, and, in so doing, maintain control by shutting down challenges to their privileges. Aileen Moreton-Robinson, an aboriginal Australian feminist scholar, illustrated a related form of epistemic violence that a prominent white male academic enacted. He engaged with Aboriginal scholars' publications only to critique them and to reinforce the seeming universality and neutrality of his own perspective. Moreton-Robinson showed that this move reinforced white patriarchy.[33] These forms of epistemic violence pattern both everyday encounters and scholarly production processes in higher education.

In each of these examples from the academy, erasure contributes to violence. Under the dominant regime, evidence of violence becomes unintelligible, not because it does not exist, but because systems of power conspire not to notice it. Whiteness, too, operates in this way when individuals and systems are rewarded for and buffered from challenges to their incompetence. When reality gets codified and solidified as one thing—especially when that one thing includes repeated denials of oppression—it operates in the service of existing power structures that are able to continue enacting violence while people in power pretend not to notice. Indeed, this willful disavowal of particular experiences and realities is a common technique used in abusive relationships. In that context, this form of epistemic violence is often called *gaslighting*, and it is discussed more in the next section.

Internalized oppression and symbolic violence: verbal harms occupy psyches

The last section detailed how epistemic violence—a refusal to acknowledge another person's reality—can make certain people and events less noticeable, or erase them altogether. In the context of interpersonal relationships, an abuser will often accomplish this by gaslighting, a process of invalidating someone else's feelings and observations about the world.[34] Over time, gaslighting can lead the people experiencing abuse or assault to doubt their own perceptions. The abuser's goal is to produce this doubt because it allows him to substitute his own reality for that of the person he abuses. As a consequence, the abused person must rely on the abuser to know the truth. In this way, violent systems and structures in the external world can begin to shape people's inner landscapes. Scholars have theorized the processes through which this internalization occurs, and their work is the focus of this section.

Many traditions problematize an easy division between violence directed outward toward other humans and violence directed internally toward oneself. They do so by linking physical–symbolic systems to self-perception, mental health, and embodied habits. In one of the earliest theorizations, Frantz Fanon suggested that people who are colonized may internalize the colonizer's reality and sometimes mimic the colonizer.[35] This idea echoes through generations of scholarship, and it has been particularly important in theories about racism and gendered violence in the United States. When describing his experiences in early 20th-century America, W.E.B. DuBois prefaced such an argument, writing, "It is a peculiar sensation, this double-consciousness, this sense of always looking at one's self through the eyes of others, of measuring one's soul by the tape of a world that looks on in amused contempt and pity."[36] Decades later, in a volume first published in the 1980s, Audre Lorde wrote:

> America's measurement of me has lain like a barrier across the realization of my own powers. It was a barrier which I had to examine and dismantle, piece by painful piece, in order to use my energies fully and creatively. It is easier to deal with the external manifestations of racism and sexism than it is to deal with the results of those distortions internalized within our consciousness of ourselves and one another.[37]

As is evident in the accounts above, DuBois and Lorde asserted that raced (and gendered) domination and violence operate both internally and externally. Repeated messages from society can create interior pain and obstacles in a person's psyche. Building upon Fanon's discussion of "psychic alienation," Sandra Bartky described the dehumanizing processes associated with femininity.[38] This scholar showed that repeated sexual objectification is a form of psychological oppression, one linked to gendered violence. Other scholars have demonstrated that heteronormativity, homophobia, and transphobia, when internalized, can negatively impact LGBTQ people's mental health, promote relational violence, and lead to discriminatory behavior.[39] Through these varied processes of internalization, an oppressor's views and prejudices can threaten or even erase a marginalized group's positive self-regard.

These ideas have been taken up in communication research. For example, Raka Shome described growing up in postcolonial India, where the educational system relied on Western authors, textbooks, and ideologies.[40] She mentioned that if she spoke Bengali rather than English in school, her teachers would penalize her. She recounted how teachers and others in her social network "all caught up in a neocolonial desire to 'Westernize' and modernize their sons and daughters and to bring about a generation of 'progress,' enacted an ideological violence whose structures they had inherited from history."[41] Like others who take up Fanon's work, Shome linked internalized oppressions to geopolitics, histories, and economies. Each of these dynamics involve both verbal and symbolic components.

In communication scholarship, researchers rely less often on Fanon and more often on Pierre Bourdieu. Bourdieu lived in France, the country that colonized Martinique, where Fanon was born. That Bourdieu is cited more often illustrates Shome's claim that the "very machinery of knowledge production [is] caught up in the intricacies and complexities of ... neocolonial global capitalism,"[42] and it implicates communication scholarship, including this chapter, in those dynamics. Bourdieu coined the term *symbolic violence*, which refers to "violence which is exercised upon a social agent with his or her complicity."[43] This idea has generated useful critiques of gender and class formations (though these critiques often omit analyses of racism and colonialism). So, for example, Angela McRobbie drew upon Bourdieu's theory of embodiment to analyze popular television makeover shows.[44] McRobbie detailed a process whereby the show derides and shames the (mostly) female subjects of the makeovers to conscript them into the dominant social order. The makeover subjects participate in this process by emulating the clothing, bodily habits, and mannerisms of the dominant group.

When encountering these ideas, some readers suggest that the oppressed could end their domination by simply deciding to reject the beliefs of the colonizer or oppressor. This "solution" stems from the assumption that individuals have total control over their own psyches. It also makes the oppressed responsible for their own oppression, thereby obscuring the actions of dominant groups. This misreading overlooks a core idea: Internalized oppression, symbolic violence, and related phenomena are not simply self-inflicted harms, nor are they matters of individual psychology. Instead, internalized oppression emerges from cultural processes that secure social stratification and inequity, processes external to any one mind. By problematizing an internal/external opposition, this tradition challenges easy distinctions among force, coercion, and consent. Consequently, people's actions and worldviews are never entirely independent from social systems, and the concept of *choice* has limited utility. Despite making force and choice more complicated, this perspective does not eradicate the agency of marginalized groups—as so-called power feminists have argued. Instead, scholars who draw upon some aspect of this tradition contend that power emerges from relational engagement and coalition building, not through isolated decision-making.[45]

In sum, work on internalized oppression and symbolic violence theorizes how dominant social orders, which are predicated on misogynist, colonialist, and racist processes, operate both outside and inside of people's bodies. Such scholarship problematizes a conceptual separation between self and society, and it highlights how systems of domination feed on the participation of those who are dominated. It undermines and also nuances the "free will" about which white boys philosophize. Most importantly, for the purposes of this chapter, it calls attention to a specific link between verbal and physical harm. A person cannot literally internalize the punches another human throws, but the symbolic systems that make the punch possible can inhabit that person and occupy their mind. Those who benefit from oppression depend on this peculiar power of words.

Organizational and administrative violence: symbols institutionalize harm

Research on organizations and institutions challenges individualized understandings of violence, as does the scholarship on internalized oppression discussed in the previous section. Building on the notion that violence is dispersed throughout systems, scholarship on organizational violence further complicates the connections between physical and verbal harm. Google, Netflix, Enron, Harvard, and the Catholic Church have no genitalia, no fingers, and no biceps (other than in a metaphorical sense). So if violence occurs only when human bodies either directly cause physical harm to someone (e.g., a fist punches a person's face) or employ material tools to cause injury (e.g., a hand grips a gun that fires a shot at a person's chest), then organizations cannot enact violence. Yet organizations have been involved in atrocities that devastate human bodies, and these facts challenge the notion that organizations can do no violence. In 1984, a Union Carbide plant in Bhopal, India, leaked a toxic gas that killed thousands of people and caused more than half a million injuries. Earlier, in the 1940s and 1950s, the United States government conducted tests of atomic bombs on the Marshall Islands, and the impact on people's bodies continues today. If violence is located purely in the physical realm, and if it also requires a physical body to enact it (as is often presumed in research on sexual assault), then it would be difficult if not impossible to connect organizations to these kinds of violence.

Scholars who study organizations render the relationship between physical and verbal harms in ways that make an analysis of organizations' role in enacting violence possible. Dean Spade discussed *administrative violence*, which highlights how norms—specifically gender binaries encoded in law—"create structured insecurity and (mal)distribute life chances across populations" leaving trans people particularly subject to assault, incarceration, poverty, death, and murder.[46] For Spade, words do not simply and straightforwardly spur people to violence, nor do words independently do violence (as in the magical stance outlined in the introduction). Instead, words influence the U.S. justice system in ways that prompt complex decisions, processes, and institutional practices whereby people who fit cisnormative gender binaries are better able to thrive. Similarly, Jeff Hearn and Wendy Parkin identified *organizational violations* as those gendered organizational practices that cause harm through "structural presence, operation, and social enactment."[47] None of these elements involves purely physical harm. They elaborated,

> once violence has been done, including being threatened … mere reference to that violence (verbally, by a look, or a slight movement or some other cue or clue) may be enough to invoke and connote violence, and thus the modification of material behavior.[48]

For Hearn and Parkin, symbolic suggestions of violence shift physical organizational practices. In this way, violence becomes ingrained in an organization. For example, in 2015 at the University of Missouri, someone made an anonymous social media post-threatening to shoot students of color. To preserve their safety, many students of color chose not to attend classes that day. Professors who did not take the threat or students' responses seriously penalized individuals who were absent, exacerbating structures that produce uneven access to education. In each of these examples, verbal harms shape organizational policies and decisions that in turn solidify both slow and acute violence.

Many scholars who identify the organizational components of gendered violence rely on intersectionality, a theoretical framework that Kimberlé Crenshaw named and developed.[49] Though intersectionality is now often applied to individual humans' identities, the original publications emphasized raced and gendered organizational processes, not identity categories alone. Specifically, Crenshaw noted that legal, political, and social service institutions presume an artificial separation between race and gender that allows them to ignore women of color's sexually violent experiences. In a comprehensive assessment of intersectional scholarship, Sumi Cho, Kimberlé Crenshaw, and Leslie McCall reiterated this point, stating, "We emphasize an understanding of intersectionality that is not exclusively or even primarily preoccupied with categories, identities, and subjectivities. Rather, the intersectional analysis foregrounded here emphasizes political and structural inequalities."[50] Underscoring the importance of this perspective, Patricia Hill Collins observed, "the U.S. is awash with individuals who perpetrate violent acts that, in the absence of intersectional analyses, can seem random, individualized and senseless."[51] In other words, intersectional analyses make the organizational and administrative aspects of gender, race, and violence apparent. They do so by providing tools that link micro, meso, and macro levels of practice and by illustrating the harmful, physical consequences of institutionalized verbal systems that separate interconnected facets of human social categories.

Recently, many scholars interested in organization and organizing have taken up new materialist theory. This emerging tradition asserts that physical objects and things (not only human subjects) influence the world. New materialism undermines the idea that intentional change is exclusive to human beings. By tracing how discursive and material systems are intertwined, this perspective can show how gendered and raced violence becomes sedimented and organized. For example, in scholarship on #SayHerName and #BlackLivesMatter, Armond Towns asserted that links between symbolic and physical forms of communication are patterned by gender and race.[52] Outlining Black feminist new materialism, he critiqued scholars interested in prevailing approaches to the power of objects: "How is thing-power not Black Power, if Black bodies are historically thought of as things?"[53] In a resonant argument, Nina Maria Lozano suggested that new materialist theories are inadequate for understanding and stopping *feminicidio* at and near the U.S. border because they do not grapple with "political power structures and larger cultural hegemonic discourses."[54] Similarly, in an argument about rape at U.S. universities, Kate Lockwood Harris cautioned that new materialism glosses histories whereby certain people have been cast as inactive, nonhuman objects, a move that enables sexual violence.[55] These scholars who study gendered violence assert that physical and verbal harms are not simply linked, but that specific renderings of those linkages are themselves part of violent systems. Despite these risks, some iterations of new materialist thinking may usefully dismantle the organization of violence if they are attentive to power structures already sedimented along lines of difference.

Conclusion

In the perspectives discussed in this chapter—from critical race theory, postcolonial theory, trauma theory, and organizational theory—words are not simply things that describe violence, nor are they always violent. Instead, verbal actions have complex relationships to physical violence, and those relationships emerge from intricate social, historical, and political processes that shape who accrues the most power and influence. Though each intellectual tradition has a distinct lineage, purpose, and set of guiding assumptions, these ways of thinking share the following overlapping ideas regarding the relationship between physical and verbal harm:

a. Violence can occur within a person, in dyadic encounters, in institutions, and across societies. It is not limited merely to relationships, and it need not be enacted by a single human. Communication connects each of these sites of violence, and it provides a mechanism whereby violence can occur independent of a single, embodied perpetrator.
b. Societal patterns and individual psychology are connected. Because of this relationship, when analyses focus merely on choice and individual responsibility, they miss or erase structural and systemic aspects of violence.
c. Violence often erases aspects of language and memory, thereby altering communication. Moreover, when individuals and groups create those erasures intentionally, or when they strategically forget certain stories and ways of knowing, they use communication to accomplish violence.

d. Communication passes meaning across generations and thereby links historical actions with contemporary ones. It is a means by which past violence continues in the present, but these linkages are often difficult to notice.

e. Violence is organized through repetition. Patterns of statements and symbols sediment violence in processes, practices, policies, and procedures. Once institutionalized, violence coordinates the actions of large groups of people.

Together, these perspectives discard both notions introduced at the outset of the chapter. Violence cannot be considered to be only physical, disconnected from symbols. Additionally, no single utterance, stripped of context and culture, can enact violence on its own. Communication is one consequential phenomenon in a complex set of relations that can secure, reinforce, or undermine brutality and inequity. Researchers who utilize and extend these theories and analytic tools draw attention to those symbols and lives that have been erased. They also continue to develop means for resisting and dismantling ongoing systems of violence.

Notes

1 Kate Lockwood Harris and Jenna N. Hanchey, "(De)stabilizing Sexual Violence Discourse: Masculinization of Victimhood, Organizational Blame, and Labile Imperialism," *Communication and Critical/Cultural Studies* 11, no. 4 (2014): 323, https://doi-org.uml.idm.oclc.org/10.1080/14791420. 2014.972421.

2 Andrea Smith, *Conquest: Sexual Violence and American Indian Genocide* (Cambridge, MA: South End Press, 2005), 3.

3 INCITE! Women of Color Against Violence, eds, *Color of Violence: The INCITE! Anthology* (Durham, NC: Duke University Press, 2016), 110.

4 For an analysis of how communication scholarship resonates with and complicates commonplace understandings of communication, see Robert T. Craig, "Communication Theory as a Field," *Communication Theory* 9, no. 2 (1999): 119–161, https://doi-org.uml.idm.oclc.org/10.1111/j.1468-2885.1999. tb00355.x.

5 Mari J. Matsuda et al., *Words that Wound: Critical Race Theory, Assaultive Speech, and the First Amendment* (Boulder, CO: Westview Press, 2013).

6 Richard Delgado and Jean Stefancic, *Understanding Words that Wound* (Boulder, CO: Westview Press, 2004), 13.

7 Matsuda et al., *Words that Wound,* 74.

8 Delgado and Stefancic, *Understanding Words that Wound.*

9 Judith Butler, *Excitable Speech: A Politics of the Performative* (New York: Routledge, 1997), 32.

10 For examples of this argument, see Cheryl I. Harris, "Whiteness as Property," *Harvard Law Review* 106, no. 8 (1993): 1707–1791, https://doi-org.uml.idm.oclc.org/10.2307/1341787; Matsuda et al., *Words that Wound*; and Delgado and Stefancic, *Understanding Words that Wound.*

11 Butler, *Excitable Speech.*

12 Matsuda et al., *Words that Wound,* 80.

13 Sarah Projansky, *Watching Rape: Film and Television in Postfeminist Culture* (New York: New York University Press, 2001), 218.

14 Lydia Cristina Huerta Moreno, *Affecting Violence: Narratives of Los Feminicidios and Their Ethical and Political Reception* (PhD diss., University of Texas at Austin, 2012).

15 Kate Lockwood Harris and James Fortney, "Reflexive Caring: Rethinking Reflexivity through Trauma and Disability," *Text and Performance Quarterly* 37, no. 1 (2017): 20–34, https://doi-org.uml.idm.oclc.org/ 10.1080/10462937.2016.1273543.

16 Joëlle Cruz, "Memories of Trauma and Organizing: Market Women's Susu Groups in Postconflict Liberia," *Organization* 21, no. 4 (2014): 447–462, https://doi-org.uml.idm.oclc.org/10.1177/1350508814527254.

17 Ann Cvetkovich, *An Archive of Feelings: Trauma, Sexuality, and Lesbian Public Cultures* (Durham, NC: Duke University Press, 2003), 12.

18 Stephanie Houston Grey, "Wounds Not Easily Healed: Exploring Trauma in Communication Studies," *Annals of the International Communication Association* 31, no. 1 (2007): 177 [emphasis added], https://doi-org.uml.idm.oclc.org/10.1080/23808985.2007.11679067.

19 Grey, "Wounds Not Easily Healed," 175.

20 Grey, "Wounds Not Easily Healed," 179.

21 Grey, "Wounds Not Easily Healed," 211.

22 Gayatri Chakravorty Spivak, *Critique of Postcolonial Reason: Toward a History of the Vanishing Present* (Cambridge, MA: Harvard University Press, 1999), 267.

23 Kristie Dotson, "Tracking Epistemic Violence, Tracking Practices of Silencing," *Hypatia* 26, no. 2 (2011): 238, https://doi-org.uml.idm.oclc.org/10.1111/j.1527-2001.2011.01177.x.

24 See Delgado Bernal, "Critical Race Theory, Latino Critical Theory, and Critical Race-Gendered Epistemologies: Recognizing Students of Color as Holders and Creators of Knowledge," *Qualitative Inquiry* 8, no. 1 (2002): 105–126, https://doi-org.uml.idm.oclc.org/10.1177/107780040200800107.

25 For an extended argument about how this process works, see Sara Ahmed, *On Being Included: Racism and Diversity in Institutional Life* (Durham, NC: Duke University Press, 2012).

26 Kate Lockwood Harris, "Re-situating Organizational Knowledge: Violence, Intersectionality, and the Privilege of Partial Perspective," *Human Relations* 70, no. 3 (2017): 263–285, https://doi-org.uml.idm.oclc.org/10.1177/0018726716654745; Kate Lockwood Harris, *Beyond the Rapist: Title IX and Sexual Violence on US Campuses* (New York: Oxford University Press, 2019).

27 Joshua Daniel Phillips and Rachel Alicia Griffin, "Crystal Mangum as Hypervisible Object and Invisible Subject: Black Feminist Thought, Sexual Violence, and the Pedagogical Repercussions of the Duke Lacrosse Rape Case," *Women's Studies in Communication* 38, no. 1 (2015): 36–56, https://doi-org.uml.idm.oclc.org/10.1080/07491409.2014.964896.

28 Phillips and Griffin, "Crystal Mangum," 37.

29 Phillips and Griffin, "Crystal Mangum," 50.

30 Paula Chakravartty et al., "#CommunicationSoWhite," *Journal of Communication* 68, no. 2 (2018): 254–266, https://doi-org.uml.idm.oclc.org/10.1093/joc/jqy003.

31 Alexandra Moffett-Bateau, "Feminist Erasures: The Development of a Black Feminist Methodological Theory," in *Feminist Erasures: Challenging Backlash Culture*, ed. Kumi Silva and Kaitlynn Mendes (New York: Palgrave Macmillan, 2015), 69.

32 Joëlle Cruz, "A Brown Body of Knowledge: A Tale of Erasure," *Cultural Studies ⇔ Critical Methodologies* 18, no. 5 (2018): 365, https://doi-org.uml.idm.oclc.org/10.1177/1532708617735131.

33 Aileen Moreton-Robinson, "The White Man's Burden: Patriarchal White Epistemic Violence and Aboriginal Women's Knowledges within the Academy," *Australian Feminist Studies* 26, no. 70 (2011): 413–431, https://doi-org.uml.idm.oclc.org/10.1080/08164649.2011.621175.

34 According to the *Oxford English Dictionary,* the term *gaslighting* emerged from a play in the 1930s, *Gaslight,* which was later adapted for a movie. The protagonist changes the environment in the home he and his wife share, for example, by moving furniture. When his wife notices, he denies anything is different. At one point he makes the lights flicker, but convinces her that she is imagining the change. Eventually, he successfully manipulates her into believing that she is crazy.

35 Frantz Fanon, *Black Skin, White Masks* (New York: Grove Press, 1967).

36 W.E.B. DuBois, *The Souls of Black Folk*, ed. Brent Hayes Edwards (New York: Oxford University Press, 2007), 8.

37 Audre Lorde, *Sister Outsider: Essays and Speeches* (Berkeley, CA: Crossing Press, 2007), 147.

38 Sandra Bartky, *Femininity and Domination: Studies in the Phenomenology of Oppression* (New York: Routledge, 1990).

39 For an example of such work, see Nabina Liebow, "Internalized Oppression and Its Varied Moral Harms: Self-Perceptions of Reduced Agency and Criminality," *Hypatia* 31, no. 4 (2016): 713–729, https://doi-org.uml.idm.oclc.org/10.1111/hypa.12265.

40 Raka Shome, "Whiteness and the Politics of Location: Postcolonial Reflections," in *Whiteness: The Communication of Social Identity*, eds Thomas K. Nakayama and Judith N. Martin (Thousand Oaks, CA: Sage, 1999).

41 Shome, "Whiteness," 116.

42 Shome, "Whiteness," 114.

43 Pierre Bourdieu and Loïc J. D. Wacquant, *An Invitation to Reflexive Sociology* (Cambridge, UK: Polity, 2002), 167.

44 Angela McRobbie, "Notes on 'What not to Wear' and Post-Feminist Symbolic Violence," *The Sociological Review* 52, no. 2_suppl (2004): 99–109, https://doi-org.uml.idm.oclc.org/10.1111/j.1467-954X.2005.00526.x.

45 For examples, see Robin Bauer, *Queer BDSM Intimacies: Critical Consent and Pushing Boundaries* (New York: Palgrave Macmillan, 2014); and Sheena Malhotra and Aimee Carillo Rowe, eds, *Silence, Feminism, Power: Reflections at the Edges of Sound* (New York: Palgrave Macmillan, 2013).

46 Dean Spade, *Normal Life: Administrative Violence, Critical Trans Politics, and the Limits of the Law* (Brooklyn, NY: South End Press, 2011), 29.

47 Jeff Hearn and Wendy Parkin, *Gender, Sexuality and Violence in Organizations: The Unspoken Forces of Organization Violations* (Thousand Oaks, CA: Sage, 2001), 73.

48 Hearn and Parkin, *Violence in Organizations*, xii.

49 Kimberlé Crenshaw, "Demarginalizing the Intersection of Race and Sex: A Black Feminist Critique of Antidiscrimination Doctrine, Feminist Theory and Antiracist Politics," *The University of Chicago Legal Forum* (1989), 139–167; Kimberlé Crenshaw, "Mapping the Margins: Intersectionality, Identity Politics, and Violence against Women of Color," *Stanford Law Review* 43, no. 6 (1991): 12241–1299, https://doi-org.uml.idm.oclc.org/10.2307/1229039.

50 Sumi Cho, Kimberlé Crenshaw, and Leslie McCall, "Toward a Field of Intersectionality Studies: Theory, Applications, and Praxis," *Signs* 38, no. 4 (2013): 797.

51 Patricia Hill Collins, "On Violence, Intersectionality and Transversal Politics," *Ethnic and Racial Studies* 40, no. 9 (2017): 1463.

52 Armond R. Towns, "Geographies of Pain: #SayHerName and the Fear of Black Women's Mobility," *Women's Studies in Communication* 39, no. 2 (2016): 122–126, https://doi-org.uml.idm.oclc.org/10.1080/07491409.2016.1176807; Armond R. Towns, "Black 'Matter' Lives," *Women's Studies in Communication* 41, no. 4 (2018): 349–358, https://doi-org.uml.idm.oclc.org/10.1080/07491409.2018.1551985.

53 Towns, "Black 'Matter' Lives," 350.

54 Nina Maria Lozano, *Not One More! Feminicidio on the Border* (Columbus, OH: The Ohio State University Press, 2019), 69.

55 Kate Lockwood Harris, *Beyond the Rapist: Title IX and Sexual Violence on US Campuses* (New York: Oxford University Press, 2019).

Bibliography

Ahmed, Sara. *On Being Included: Racism and Diversity in Institutional Life*. Durham, NC: Duke University Press, 2012.

Bartky, Sandra. *Femininity and Domination: Studies in the Phenomenology of Oppression*. New York: Routledge, 1990.

Bauer, Robin. *Queer BDSM Intimacies: Critical Consent and Pushing Boundaries*. New York: Palgrave Macmillan, 2014.

Berman, Elise. "Force Signs: Ideologies of Corporal Discipline in Academia and the Marshall Islands." *Linguistic Anthropology* 28, no. 1 (2018): 22–42. https://doi-org.uml.idm.oclc.org/10.1111/jola.12175.

Bourdieu, Pierre, and Loïc J. D. Wacquant. *An Invitation to Reflexive Sociology*. Cambridge: Polity, 2002.

Butler, Judith. *Excitable Speech: A Politics of the Performative*. New York: Routledge, 1997.

Campbell, Rebecca. *The Neurobiology of Sexual Assault*. Washington, DC: National Institute of Justice, 2012. https://nij.gov/multimedia/presenter/presenter-campbell/pages/presenter-campbell-transcript.aspx.

Campbell, Rebecca, Emily Dworkin, and Giannina Cabral. "An Ecological Model of the Impact of Sexual Assault on Women's Mental Health." *Trauma, Violence, & Abuse* 10, no. 3 (2009): 225–246. https://doi-org.uml.idm.oclc.org/10.1177/1524838009334456.

Caruth, Cathy. *Unclaimed Experience: Trauma, Narrative, and History*. Baltimore, MD: Johns Hopkins University Press, 1996.

Chakravartty, Paula, Rachel Kuo, Victoria Grubbs, and Charlton McIlwain. "#CommunicationSoWhite." *Journal of Communication* 68, no. 2 (2018): 254–266. https://doi-org.uml.idm.oclc.org/10.1093/joc/jqy003.

Cho, Sumi, Kimberlé Williams Crenshaw, and Leslie McCall. "Toward a Field of Intersectionality Studies: Theory, Applications, and Praxis." *Signs* 38, no. 4 (2013): 785–810.

Collins, Patricia Hill. "On Violence, Intersectionality and Transversal Politics." *Ethnic and Racial Studies* 40, no. 9 (2017): 1460–1473.

Craig, Robert T. "Communication Theory as a Field." *Communication Theory* 9, no. 2 (1999): 119–161. https://doi-org.uml.idm.oclc.org/10.1111/j.1468-2885.1999.tb00355.x.

Crenshaw, Kimberlé. "Demarginalizing the Intersection of Race and Sex: A Black Feminist Critique of Antidiscrimination Doctrine, Feminist Theory and Antiracist Politics." *The University of Chicago Legal Forum* (1989): 139–167.

Crenshaw, Kimberlé. "Mapping the Margins: Intersectionality, Identity Politics, and Violence Against Women of Color." *Stanford Law Review* 43, no. 6 (1991): 1241–1299. https://doi-org.uml.idm.oclc.org/10.2307/1229039.

Cruz, Joëlle M. "A Brown Body of Knowledge: A Tale of Erasure." *Cultural Studies ⇔ Critical Methodologies* 18, no. 5 (2018): 363–365. https://doi-org.uml.idm.oclc.org/10.1177/1532708617735131.

Cruz, Joëlle M. "Memories of Trauma and Organizing: Market Women's Susu Groups in Postconflict Liberia." *Organization* 21, no. 4 (2014): 447–462. https://doi-org.uml.idm.oclc.org/10.1177/1350508414527254.

Cvetkovich, Ann. *An Archive of Feelings: Trauma, Sexuality, and Lesbian Public Cultures.* Durham, NC: Duke University Press, 2003.

DasGupta, Debanuj. "Rescripting Trauma: Trans/Gender Detention Politics and Desire in the United States." *Women's Studies in Communication* 41, no. 4 (2018): 324–328. https://doi-org.uml.idm.oclc.org/10.1080/07491409.2018.1544011.

Delgado, Richard, and Jean Stefancic. *Understanding Words that Wound.* Boulder, CO: Westview Press, 2004.

Delgado Bernal, Dolores. "Critical Race Theory, Latino Critical Theory, and Critical Race-Gendered Epistemologies: Recognizing Students of Color as Holders and Creators of Knowledge." *Qualitative Inquiry* 8, no. 1 (2002): 105–126. https://doi-org.uml.idm.oclc.org/10.1177/107780040200800107.

Dotson, Kristie. "Tracking Epistemic Violence, Tracking Practices of Silencing." *Hypatia* 26, no. 2 (2011): 236–257. https://doi-org.uml.idm.oclc.org/10.1111/j.1527-2001.2011.01177.x.

DuBois, W. E. B. *The Souls of Black Folk,* edited by Brent Hayes Edwards. New York: Oxford University Press, 2007.

Fanon, Frantz. *Black Skin, White Masks.* New York: Grove Press, 1967.

Grey, Stephanie Houston. "Wounds Not Easily Healed: Exploring Trauma in Communication Studies." *Annals of the International Communication Association* 31, no. 1 (2007): 174–222. https://doi-org.uml.idm.oclc.org/10.1080/23808985.2007.11679067.

Harris, Cheryl I. "Whiteness as Property." *Harvard Law Review* 106, no. 8 (1993): 1707–1791. https://doi-org.uml.idm.oclc.org/10.2307/1341787.

Harris, Kate Lockwood. *Beyond the Rapist: Title IX and Sexual Violence on US Campuses.* New York: Oxford University Press, 2019.

Harris, Kate Lockwood. "Re-situating Organizational Knowledge: Violence, Intersectionality, and the Privilege of Partial Perspective." *Human Relations* 70, no. 3 (2017): 263–285. https://doi-org.uml.idm.oclc.org/10.1177/0018726716654745.

Harris, Kate Lockwood, and James Michael Fortney. "Reflexive Caring: Rethinking Reflexivity through Trauma and Disability." *Text and Performance Quarterly* 37, no. 1 (2017): 20–34. https://doi-org.uml.idm.oclc.org/10.1080/10462937.2016.1273543.

Harris, Kate Lockwood, and Jenna N. Hanchey. "(De)stabilizing Sexual Violence Discourse: Masculinization of Victimhood, Organizational Blame, and Labile Imperialism." *Communication and Critical/Cultural Studies* 11, no. 4 (2014): 322–341. https://doi-org.uml.idm.oclc.org/10.1080/14791420.2014.972421.

Harris, Kate Lockwood, Megan McFarlane, and Valerie Wieskamp. "The Promise and Peril of Agency as Motion: A Feminist New Materialist Approach to Sexual Violence and Sexual Harassment." *Organization.* Advance online publication (2019). https://doi-org.uml.idm.oclc.org/10.1177/1350508419838697.

Hearn, Jeff, and Wendy Parkin. *Gender, Sexuality and Violence in Organizations: The Unspoken Forces of Organization Violations.* Thousand Oaks, CA: Sage, 2001.

Herman, Judith. *Trauma and Recovery: The Aftermath of Violence, from Domestic Abuse to Political Terror.* New York: Basic Books, 1997.

Huerta Moreno, Lydia Cristina. "Affecting Violence: Narratives of *Los Feminicidios* and Their Ethical and Political Reception." PhD diss., University of Texas at Austin, 2012.

INCITE! Women of Color against Violence, eds. *Color of Violence: The INCITE! Anthology.* Durham, NC: Duke University Press, 2016.

Liebow, Nabina. "Internalized Oppression and Its Varied Moral Harms: Self-Perceptions of Reduced Agency and Criminality." *Hypatia* 31, no. 4 (2016): 713–729. https://doi-org.uml.idm.oclc.org/10.1111/hypa.12265.

Lorde, Audre. *Sister Outsider: Essays and Speeches.* Berkeley, CA: Crossing Press, 2007.

Lozano, Nina Maria. *Not One More! Feminicidio on the Border.* Columbus, OH: The Ohio State University Press, 2019.

Malhotra, Sheena, and Aimee Carrillo Rowe, eds. *Silence, Feminism, Power: Reflections at the Edges of Sound.* New York: Palgrave Macmillan, 2013.

Matsuda, Mari J., Charles R. Lawrence III, Richard Delgado, and Kimberlé Williams Crenshaw. *Words that Wound: Critical Race Theory, Assaultive Speech, and the First Amendment.* Boulder, CO: Westview Press, 2013.

McRobbie, Angela. "Notes on 'What not to Wear' and Post-Feminist Symbolic Violence." *The Sociological Review* 52, no. 2_suppl (2004): 99–109. https://doi-org.uml.idm.oclc.org/10.1111/j.1467-954X.2005.00526.x.

Moffett-Bateau, Alexandra. "Feminist Erasures: The Development of a Black Feminist Methodological Theory." In *Feminist Erasures: Challenging Backlash Culture,* edited by Kumi Silva and Kaitlynn Mendes, 54–71. New York: Palgrave Macmillan, 2015.

Moreton-Robinson, Aileen. "The White Man's Burden: Patriarchal White Epistemic Violence and Aboriginal Women's Knowledges within the Academy." *Australian Feminist Studies* 26, no. 70 (2011): 413–431. https://doi-org.uml.idm.oclc.org/10.1080/08164649.2011.621175.

Phillips, Joshua Daniel, and Rachel Alicia Griffin. "Crystal Mangum as Hypervisible Object and Invisible Subject: Black Feminist Thought, Sexual Violence, and the Pedagogical Repercussions of the Duke Lacrosse Rape Case." *Women's Studies in Communication* 38, no. 1 (2015): 36–56. https://doi-org.uml.idm.oclc.org/10.1080/07491409.2014.964896.

Projansky, Sarah. *Watching Rape: Film and Television in Postfeminist Culture.* New York: New York University Press, 2001.

Shome, Raka. "Whiteness and the Politics of Location: Postcolonial Reflections." In *Whiteness: The Communication of Social Identity,* edited by Thomas K. Nakayama and Judith N. Martin, 107–128. Thousand Oaks, CA: Sage, 1999.

Smith, Andrea. *Conquest: Sexual Violence and American Indian Genocide.* Cambridge, MA: South End Press, 2005.

Spade, Dean. *Normal Life: Administrative Violence, Critical Trans Politics, and the Limits of Law.* Brooklyn, NY: South End Press, 2011.

Spivak, Gayatri Chakravorty. *A Critique of Postcolonial Reason: Toward a History of the Vanishing Present.* Cambridge, MA: Harvard University Press, 1999.

Towns, Armond R. "Black 'Matter' Lives." *Women's Studies in Communication* 41, no. 4 (2018): 349–358. https://doi-org.uml.idm.oclc.org/10.1080/07491409.2018.1551985.

Towns, Armond R. "Geographies of Pain: #SayHerName and the Fear of Black Women's Mobility." *Women's Studies in Communication* 39, no. 2 (2016): 122–126. https://doi-org.uml.idm.oclc.org/10.1080/07491409.2016.1176807.

Walters, Karina L., Selina A. Mohammed, Teresa Evans-Campbell, Ramona E. Beltrán, David H. Chae, and Bonnie Duran. "Bodies Don't Just Tell Stories, They Tell Histories: Embodiment of Historical Trauma Among American Indians and Alaska Natives." *DuBois Review: Social Science Research on Race* 8, no. 1 (2011): 179–189. https://doi-org.uml.idm.oclc.org/10.1017/S1742058X1100018X.

WHO Department of Reproductive Health. "Understanding and Addressing Violence Against Women: Intimate Partner Violence." Geneva: World Health Organization, 2012. https://apps.who.int/iris/bitstream/handle/10665/77432/WHO_RHR_12.36_eng.pdf;jsessionid=82947E963EF98DB0EFE1EAE08C892155?sequence=1.

32

SHERLOCK HOLMES AND THE CASE FOR TOXIC MASCULINITY

Ashley Morgan

Introduction

Post-millennial depictions of Sherlock Holmes have generated a considerable amount of new fandom and academic scrutiny.[1] Much of this attention is concerned with the way in which the character's identity has been presented in a contemporary format, ostensibly portraying a different kind of masculinity than in previous iterations. Yet these celebrations of a "new" Sherlock lack attention to the character's potential to perform and reiterate toxic masculinity rather than to challenge it. This chapter seeks to return to the first representation of the character in the original stories by Arthur Conan Doyle as a lens by which to analyze the concept of toxic masculinity in the BBC's *Sherlock* (2010–2017)[2] and CBS's *Elementary* (2012–2019).[3] These media texts have been chosen because of their phenomenal global popularity, and to address the gap in academic analysis of toxic masculinity in these texts.[4] Despite the apparent differences in these shows, which will be addressed further in the chapter, I argue that the way in which Sherlock Holmes employs language serves to denigrate others and elevate his own position as a hegemonic man. In contemporary popular discourse, there is much analysis and critique of male behavior, both in media representations and in everyday life, and this chapter will draw upon these critiques, drawing similarities of toxicity to the much-loved character of Sherlock Holmes. Moreover, I examine the ways in which behaviors considered suitable for enacting masculinity have been updated and refigured for contemporary audiences.

Arthur Conan Doyle created the character of Sherlock Holmes in 1887, and such was the subsequent popularity and continuing longevity of his stories that I contend that Sherlock might be considered one of the first figures of popular culture. Hence, he is laden with mythology surrounding his brilliance. Yet, rather than focus on his intellectualism, this chapter concentrates on his use of language and masculinity. In one of the early stories, *A Scandal in Bohemia*, Doyle describes Holmes's perspective of women: "He never spoke of the softer passions, save with a gibe and a sneer."[5] Despite having been written in 1891, this sentence has much in common with contemporary toxic discursive practices of men toward both women and other men.[6] This chapter opens with a working definition of toxic masculinity, focusing on the way in which it appears discursively, and is perpetuated through popular culture texts, and demonstrates that toxic masculinity has been present and emerges from the concept of hyper masculinity as normative male behavior. Arguably, toxic masculinity was present in the original texts, and now appears as a form of censorship in language and affect, through the updated character of Sherlock Holmes in Sherlock and Elementary.

Why "toxic" masculinity?

Toxic masculinity is the kind of masculinity that is predicated on aggressive tendencies (either verbal or physical), emotional coldness, and competitive behavior that leads to the suppression and domination of women and other men.[7] This conception of masculinity is a common norm in real life and is often portrayed in media representations and is maintained through social media. Overt examples of random male aggression such as road rage and mass shootings appear to be modern phenomenon, and arguably may be a backlash against feminism and shifts in social expectation of men's roles.[8]

Raewyn Connell was one of the first people to use the word toxic to designate a performance contingent on power gained through violence, part of the "toxic practices including physical violence, with a strong relationship to hegemonic masculinity" that defined the behaviors of many men, especially those in power.[9] While the concept of hegemonic masculinity itself is loose, each decade appears to bring with it the apex of masculinity to which all masculinity should aspire. In the 1980s and 1990s Connell referred to "transnational business masculinity," apparent in countries such as Australia, the United Kingdom, and the United States, as being the hegemonic apex, with a focus on masculinity embodied through the trappings of aggressive financial acumen and business expertise.[10] I have previously argued that since the economic crisis of 2008, a shift has occurred in popular culture representations of masculine hegemonic identity, with the transnational business masculine identity (embodied in be suited Wall Street characters) falling out of favor in media representations. The new ideal focused on masculine hegemony gained through specific knowledge, and a greater emphasis on more academically inclined, scholastic, and scientific expertise.[11]

Arguably, male power has been enacted through domination,[12] and Bliss describes violent domination as "poison" that contributes to the insidious nature of certain types of dangerous masculinity.[13] Moreover, the portrait of Sherlock Holmes appears to correspond with Kupers' 20th-century description of toxic masculinity present in the prison system, and can be applied to behaviors in wider society: "Toxic masculinity is the constellation of socially regressive male traits that serve to foster domination, the devaluation of women, homophobia, and wanton violence."[14] The concepts of domination, especially through discursive means, through verbal censorship will be addressed in the representations of Sherlock Holmes in this chapter. On a micro and everyday level, common displays of toxic masculinity include deriding others in order to suppress them, while putting oneself forward as an overbearing and sole source of information, often coupled with social ineptitude and nonchalance. Arguably, the character of Sherlock Holmes in the TV texts under scrutiny here is written in a manner that deploys the original character's self-aggrandizement and verbal veracity (often to the detriment of the feelings of other characters) and updates it to more commonly used language.

Prior to joining post-millennial common parlance, men's behaviors that now are considered "toxic" were framed using the term "hyper" (as in "hyper masculinity"), a popular term in the 1980s and 1990s. Mosher and Sirkin have suggested that hyper masculinity was a combination of the following qualities: "callous sexual attitudes towards women, belief that violence is manly and experience of danger as exciting."[15] Such elements framed hyper masculinity as part of a "macho personality constellation."[16] The belief that violence is manly (and even admirable) and somehow naturally produced appears as a strong theme here. This may in part explain the ways in which commonly violent behaviors enacted by some men are condoned and championed as being necessary in order for men to compete with each other at the top—whether this is in the military, sports, or in work organizations, against other men or women.

Arguably, the notion of hyper masculinity appears to have more positive associations when compared with the concept of toxic masculinity. Media research of the 1990s focused on the provenance of hyper masculinity and found that violence was usually represented contextually and therefore appeared as an appropriate response to certain situations. For example, Horrocks refers to some genres of film, such as the Western, in which violence was desirable in order to delineate between good and bad elements and acted as a vehicle through which good could trounce bad.[17] Katz's research on advertising and white masculinity found that violent masculinity was condoned and reproduced through popular culture tropes, which were then emulated.[18] Katz and Earp examined the concepts of "tough" and "macho" in media representations, but none of these words could be considered as problematic as the word "toxic."[19] Moreover, hooks has argued that in the United States, toxic masculinity appears specifically as part of racialized male behavior, and that "violent behavior is standard" among men.[20] Black men are stereotypically portrayed as hyper masculine, and are even expected to enact violence, like bombs waiting to go off at the slightest provocation. Such racial profiling has dire consequences for many American Black men, not least in the form of increased discrimination from police.[21]

Physical violence therefore seems to be the main driver for considering hyper masculinity and appears to be a normative and affirmative value of masculinity, suggesting a level of social acclimatization to physical violence amongst men. Yet there is increasing media attention being paid to discourses of toxicity which are being enacted by men in prominent mediated positions, such as Piers Morgan in the United Kingdom and Ben Shapiro and Donald Trump in the United States.[22] The kind of masculinity perpetuated by these prominent figures resonates with that described earlier: of a repudiation of any kind of masculinity that is different to their own, which is based on disparaging others, and putting themselves forward as the sole source of information. Rather than the "positivity" of physical hyper masculinity, verbal toxic masculinity is always negative, as it based on an obvious desire to censor others.[23] While public critiques of such masculinity are available on social and news media, equally, there is support for such behaviors.[24] This suggests that censorship by some men of others is considered an eminently suitable facet of masculinity, especially, but not limited to, ideals of hegemonic masculinity.

Representations of toxic masculinity in the media

Recent media research focuses specifically on representations of male violence against women. For example, scholars have explored violence against women in video games as well as a perceived lack of empathy for women amongst men who play such games in everyday life.[25] Studies also examine the relationship between violence in video games and rape.[26] Yet media research has been slow to acquire the term "toxic masculinity" to describe such behaviors and has focused more firmly on hyper masculinity. For example, Vokey et al. analyze representations of hyper masculinity in adverts and their subsequent impact on certain social class masculinities.[27] Salter and Blodgett spend some time attempting to investigate toxic masculinity in geek identities in popular culture films, television, video games, and fandom, but the analysis stops short of adequately defining what is meant by toxic masculinity in the first instance, and instead relies heavily on the outmoded concept of hyper masculinity—which the authors relate to heroism.[28] Moreover, as mentioned above, the term "hyper masculinity" has an aura of positivity about it due to contextualized portrayals, and therefore makes it seem more tolerable. In terms of violence and toxic masculinity on TV, Park has highlighted the popularity of Korean television dramas regularly imported into other countries in the east (such as Cambodia) and the ways they portray physical, sexual, and emotional violence by men against women with regularity in order to shock and thrill audiences.[29]

In wider popular literature, Bealer considers masculinity in the Twilight books by Stephanie Meyer and suggests that vampires have been considered "hyper masculine" because of the ways in which they penetrate the bodies of their victims. Moreover, the character of Edward, particularly, embodies "phallic masculinity" as the physical and emotional desire to dominate—a power with which he wields ambivalently against main love interest, Bella Swan.[30]

In order to discourage the apparently inexorable growth of toxic masculinity, which some have argued begins in school,[31] both Biddulph and Winton call for a change in parenting boy children. They suggest that parenting should move away from naturalizing aggressive tendencies in boys, away from the "boys don't cry" trope and toward a gentler and more nurturing approach to raising them.[32] Moreover, this idea has emerged in concert with Bridges and Pascoe's research on hybrid masculinity, finding that some men are choosing to engage with traditionally feminized qualities, once rejected due to marginalization, in order to distance themselves from the damaging connotations of masculine hegemony and to create a more hybridized masculinity.[33]

Victorian toxicity

There are considerable commonalties between the contemporary concept of toxic masculinity and Victorian hegemonic masculinity as the apex of manliness.[34] Moreover, examples of toxic masculinity are also reflected in Victorian literature, as found in Kipling and Charles Kingsley, for example.[35] In Victorian times, there was a strong belief that physical prowess would lead to mental strength and that "manliness" should be demonstrated.[36] This, coupled with a strong belief in Christian religion, spawned the Victorian version of masculine ideals encompassing hegemonic values: "muscular Christianity."[37] This masculinity is evident in the desire for physical strength, engendered through outdoor exertions in order to develop camaraderie and teamwork (such as in the scouting movement as advocated by Baden Powell), along with domination of others through social class and gender discourse. Doyle's famous character emerged from this same historical and cultural context.

The stories of Sherlock Holmes are popular from a global perspective, with arguably every decade having their own iteration of the famous detective. Portrayals frequently refer to Victorian times, usually through iconography incorporating a pipe, cloak, and a deerstalker hat, or foggy London (which has become synonymous with the character), while keeping some basic principles of the distinctive character as described originally by Doyle. Sherlock Homes is a mythopoetic character, in that new versions of him evidence "re-mythologizing, not merely repeating the old stories,"[38] and iterations of this character have lasting global appeal, as deducing and detection appear to be appealingly western narratives.

Unpacking the toxicity of Sherlock Holmes

It is to a scrutiny of Sherlock Holmes as a recent TV series that this chapter now turns. In order to demonstrate the ways in which Doyle's' description of this popular character has been mythopoeticized and reimagined through the contemporary lens of toxic masculinity,[39] this section will examine discursive elements in two recent iterations of Sherlock Holmes: the BBC's *Sherlock* and CBS's *Elementary*. The popularity of both shows is significant. In 2014, *Sherlock* was reported to be the biggest selling television show to emerge from Britain and was "licenced in 224 territories across the globe."[40] Moreover, because of the show's reach, sales of the Sherlock Holmes books increased by 53%. While there is considerably less fanfare regarding *Elementary*, it lasted from 2012 until August 2019 and remains a much-loved police procedural.[41]

On first examination, all episodes of *Sherlock* and *Elementary* were to be analyzed, but due to the sheer number of episodes in *Elementary*,[42] it was not possible to conduct a more comprehensive content analysis. Therefore, a mixed-method approach to analysis was developed through employing popular discourse in addition to content analysis. Each episode in each first season of both shows was viewed and notes were made.[43] Quotations were gathered from episodes when Sherlock's identity is either being described by other characters or by himself. This was in order to grasp the way in which Sherlock's masculinity may be defined differently to other men. For example, Sherlock variously describes himself as "I'm a high-functioning sociopath" (from *Sherlock*) and "I find sex repellent" (from *Elementary*). In order to supplement the viewings, main quotes from the series were scanned on the Internet using quotation websites,[44] several of which provide key dialogue from both series. Moreover, they demonstrate the phrases from *Sherlock* that have become popular with fans through Internet memes and repeated through popular GIFs, and T-shirts such as "Brainy Is the New Sexy," "I Am Sher-Locked," and "The Game Is On."

Therefore, following Rose, a method of discourse analysis emergent from Foucault was employed. In this approach, discourse "refers to groups of statements which structure the way a thing is thought, and the way we act on the basis of that thinking."[45] Here, the discursive power of toxic masculinity and the consequence of the discourse will be examined, making what Sherlock says, and to whom, of prime importance.

The first seasons of both shows were useful to watch in entirety. From this initial viewing, three key themes were identified in concert with the original description of Holmes' perspective of women, outlined at the beginning of this chapter. These were sentences or phrases that were delivered with a "gibe and sneer," "coldness" which is played out in indifference towards the feelings or even the presence of other people, and the inability to engage socially. Additionally, characters were assessed for their violent tendencies, which Kupers describes as being part of a toxic identity. Notes were made while watching the shows whenever the character of Sherlock offers either a gibe (derisory and personal comments, especially against women) or a sneer (censoring others, belittling or disparaging personal remarks at either gender) or calls the intelligence of others into question.

In terms of length and type of series, *Sherlock* and *Elementary* are extremely different, raising concern that they may not comparable. For example, between 2010 and 2017, four seasons of *Sherlock* were produced, with an average of four 90-minute long episodes per series for a total of 15 episodes. Conversely, from 2012 until the series finale in August 2019, there are seven seasons of *Elementary* with an average of 23 40-minute episodes, producing 174 episodes. While *Sherlock* was a rare, annual series, with a Christmas or New Year Special, *Elementary* appeared as a weekly police procedural. Both iterations successfully appeared within two years of each other, which suggests a public appetite for the character, and they share an English actor playing the part of Sherlock Holmes, albeit in very different locations; moreover both iterations fundamentally refer to key elements of the original Sherlock Holmes canonical tales. For example, Moriarty is the villain, Watson is still a medical doctor and Sherlock's sidekick, and Sherlock himself is resolutely both a socially awkward and verbally toxic Englishman.

BBC's Sherlock

Sherlock stars the formally moderately successful film and television actor Benedict Cumberbatch in the main role and Martin Freeman, who played Tim Canterbury in the original British version of *The Office* (2001–2003), as Watson. These actors play youthful versions of the popular characters and display elements of sex appeal, which are much removed from the famous illustrations of Sherlock Holmes by Sydney Padget in the 19th century, which portray him as a gaunt, middle-aged, imperious man with a receding hairline, a pipe, and a deerstalker hat.

A great deal has been written about the BBC's version of *Sherlock* —considerably more than Arthur Conan Doyle ever wrote about the character himself. Much of the academic writing on this show is organized around themes of queer and gender issues apparent in its representations;[46] the relationship to this iteration and fans;[47] the portrayal of London;[48] and Sherlock as a modern adaptation.[49] Alternatively, I argue that Holmes eschews sex in favor of deduction, and rather than offering a queer interpretation, I suggest that Sherlock is sexually ascetic.[50] Mark Gatiss, one of the main writers, plays Mycroft Holmes with similar austerity. However, it is the physically handsome nature of Cumberbatch, with his "Byronic" hair[51] as he swaggers around in a long coat to denote gravitas[52] as well as the palpable chemistry between Cumberbatch as Holmes and Freeman as Watson in which fans seem to delight. Moreover, the series is set in contemporary London and pays homage to the city with its many obvious external locations.

A case of inflated self-importance

Part of the trope of toxic masculinity is that those who perpetuate it need to keep reinforcing their position by undermining others. In each iteration of Sherlock Holmes, he is frequently portrayed as subverting the knowledge and expertise of men who are at the pinnacle of the hierarchical system of the police force, such as Detective Inspector Lestrade (Rupert Graves) in *Sherlock*. Men who have titles such as these are usually considered to be authoritative figures, as they have earned a place at the top through their performance. Sherlock, however, is prone to ignore social hierarchies and frequently suppresses the expertise of others in favor of inflating his own sense of importance. In the following quotation, when Holmes and Watson are getting to know each other in *Sherlock*, Holmes is at pains to reinforce his own importance by undermining others and referring to his uniqueness.

HOLMES: "I'm a Consulting Detective. Only one in the world. I invented the job".
WATSON: "What does that mean?"
HOLMES: "It means that whenever the police are out of their depth, which is always, they consult me".
WATSON: "The police don't consult amateurs"[53]

Here, Sherlock ably demonstrates one of the key principles of toxic masculinity: that of putting himself forward as the sole source of information. In this quotation, it is suggested that Sherlock has created a role that the police had no idea that they needed, that only he could see, and which role only he could fill. In this example, the sneering and self-inflation does not go unnoticed by other characters, as Watson is skeptical of Sherlock's derisory pronouncements that the police are clueless and have come to rely on him.

While Watson is correct in his assertions that "the police don't consult amateurs", the word amateur is misleading here, as it suggests someone who lacks professional qualifications and expertise. Yet, Doyle sets Holmes up as a learned maverick who has circumvented the traditional routes of professionals and experts. The dialogue above suggests that it is Holmes's innate diligence, his identity as a genius, and his nonconformity to rules that conveys expertise. Nonconformity often appears as a masculine trope of genius, exemplified in the figures of innovator Steve Jobs and businessman Richard Branson, who both undermined the more traditional aspects of education and business to reach the top.

Moreover, I have argued elsewhere that what Holmes' so-called "amateur status" means is that he is free from the more pedestrian issues that often arise from professional roles. These include the paperwork, management, and accountability that customarily focus on solving the crime rather than taking any glory.[54]

Another common way that Holmes exerts his toxic masculinity is to censor those around him, either through discursive means, bodily gestures, or actions. He will often hold up his hand to prevent people from speaking and shuts doors in people's faces. He verbally censors speaking, especially potential clients, by shouting "bored!" and waving his hands about. In the briefest and most visceral of manners, he tells people around him to "shut up" to allow him to concentrate, or simply because he finds them annoying. This includes DI Lestrade and Forensics Officer Anderson (Jonathan Aris). Moreover, it is clear that Sherlock believes that Anderson is not particularly intelligent and takes every opportunity to deride him. Feeling superior to others is both a classic narcissistic and toxically masculine trait. An exemplar Sherlock quote that illustrates this attitude is, "Dear God what is it like in your tiny little brains? It must be *so* boring!"[55]

Moreover, toxic masculinity is also displayed by Sherlock's brother, Mycroft, when he is being chastised for allowing Sherlock to engage in risky behaviors by Sherlock's landlady, Mrs. Hudson (Una Stubbs).

MRS. HUDSON: "It's a disgrace, sending your little brother into danger like that. Family is all we have in the end, Mycroft Holmes"
MYCROFT HOLMES: "Oh, shut up, Mrs. Hudson".
SHERLOCK HOLMES: "Mycroft!"
MYCROFT HOLMES: "Apologies".
MRS. HUDSON: "Thank you"
SHERLOCK HOLMES: "Though do, in fact, shut up".[56]

Arguably the term "shut up" is the ultimate censorial phrase, used by people from all walks of life. Most notably, it has become the go-to phrase of men in positions of high privilege and power, such as U.S. President Donald Trump and English TV Broadcaster Piers Morgan, who commonly use this phrase against everyone who appears to disagree with them, from journalists to Gillette adverts.[57]

In one of the first scenes of *Sherlock*, Molly the medical examiner (Louise Brearley) seems smitten with Sherlock. While he appears to be unaware of personal interactions, he uses her to conduct experiments on dead bodies. Yet because of his inability or unwillingness to read the social signs of potential affection (as a consequence of his toxicity) and employ normal social discourse, his response to her question of whether he would like to meet for a coffee (a common courtship overture) is: "Yes, black, no milk two sugars." Despite her professional position as a doctor, he treats her as the woman who brings drinks, undermining her authority and asserting his own. She is an easy target for his toxicity, and as a man who prides himself on the art of deduction from the tiniest of clues, he must be aware of her longing, but he goes out of his way to deride and insult Molly. Arguably, his poor social skills are part of his toxicity and desire to ensure that his knowledge trounces everyone else's. While some of the characters such as Anderson and Donovan (Vinette Robinson) suggest he is a "psychopath," apparently due to his delight at the macabre elements of the work, his lack of emotion and empathy, as well as his inability to work with others, these behaviors are excused by Sherlock himself as characteristic of a "high functioning sociopath," suggesting that he is aware of his behaviors but that they are necessary in order for him to excel at deduction.

While it might be argued that his poor social skills are played for laughs in the show, in one rather poignant scene with Mycroft, Sherlock wonders whether it is right to be so apparently emotionless:

SHERLOCK HOLMES: "Look at them, they all care so much. Do you ever wonder if there's something wrong with us?"
MYCROFT HOLMES: "All lives end. All hearts are broken. Caring is not an advantage, Sherlock."[58]

Acts of caring, apart from self-care, do not compute in toxic masculinity, given that one of its central tenets is a deficiency of care for others, in lieu of looking after and promoting themselves. There are elements of toxicity present in the character of Watson; he misses the thrill of violence in warfare, yet he is a considerably more fragile yet more sociable and a greater man "of the world" than Sherlock.[59] This makes him a suitable companion to the often-unpredictable Sherlock. When Sherlock asks Watson to join him in a dangerous escapade, Watson's relief is palpable. Sherlock: "Want to see some more [trouble]?" Watson: "Oh God, yes!"[60] While Watson misses the war, his toxicity is much less present than Sherlock's, and he often acts as the foil to make Sherlock's behavior seem even more acute by being shocked by or even by rebuking him.

The BBC's *Sherlock* has become a much-loved television series, experiencing global success. While the characters appear to have been "updated" for the 21st century, the central character of Sherlock Holmes arguably reflects elements of Victorian toxic masculinity, present in the original text, especially in terms of verbal discourse. The following section examines the verbal discourse of Sherlock Holmes in *Elementary,* and I argue that while the series is distinct from the original texts and *Sherlock* in terms of its contemporary American feel, there are distinct commonalities in terms of toxic verbal discourse and a sense of punctilious correctness.

Toxicity in CBS's Elementary

In *Elementary*, we see a very different Holmes to the mannered one that Cumberbatch creates. Despite Cumberbatch and Johnny Lee Miller being contemporaries, Miller plays a humbler, less physically present character. This is in the sense that his appearance is unexceptional (no long coat or Victorian iconography for him), and he looks like everyone else in contemporary New York, with his buzz-cut hair and clean-shaven face. Yet his speech is quaint, and Anglicized, and he delights in verbal interplay as a quintessentially Englishman in stark contrast with New York and its many American accents. Curiously, very little has been written about *Elementary* from an academic perspective.[61] This may be because it is a police procedural, and therefore quotidian, in this iteration, Sherlock is likely to be seen eating and having sex. Moreover, there appear to be commonalities between TV series such as *Lie to Me*[62] and *The Mentalist*.[63] Such shows feature characters who work with the police and engage in detection as consultants with special skills, rather than being employed as detectives. *Elementary*'s Sherlock uses discursive means to deploy toxic masculinity and there is considerably more talking and less action in this iteration than in *Sherlock,* and they are clearly different shows. One of the first things to note is Sherlock's verbal dominance: he is both loquacious and verbose, especially in contrast to the considerably more laconic Watson.

WATSON: "Any luck?"
HOLMES: "Luck is an offensive, abhorrent concept. The idea that there is a force in the universe tilting events in your favor or against it is ridiculous. Idiots rely on luck".
WATSON: "So that'd be a no".[64]

Here, Sherlock painstakingly defines the correct use of the work "luck" and disparages its common misuse while demonstrating his superiority over Watson. Arguably, in this iteration, some of Sherlock's toxicity might be attributable to his Englishness and therefore "foreign" demeanor amongst die hard New Yorkers.

At the beginning of the show, Sherlock treats Watson in very much the same way as he does other women, as an object of derision.

HOLMES : "Tell me, how do clients typically introduce you?"
WATSON : "What do you mean?"
HOLMES : "I mean, I find it hard to believe they'd actually tell someone they've been assigned a glorified helper monkey".

While this changes considerably through the seasons, Sherlock's approach to women other than Watson appears to be based on the original text's "gibes, and sneers." He uses words to describe women that most American people would never have heard before: "strumpet, dolly mops, joie de fille," and reduces women to things on which to have sex or conduct experiments. Using words such as these increases his relationship to the character in the original texts, as these words are not commonly used in everyday contemporary language. This may be appealing to American audiences, perhaps especially to those who have engaged with traditional British humor (such as *Monty Python*, for example), and those who might find Sherlock's "Englishman out of England" speech exotically amusing rather than toxic.

In the first episode, Sherlock tells Watson that the woman she passes on the steps to his house is a prostitute who fulfils his "needs," saying, "I actually find sex repellent. All those fluids and odd sounds. My brain and my body require it to function at optimum levels, so I feed them as needed. Moreover, he frequently uses language that describes women a sexual object, emphasizing the way he might use them. They are: "soft and pillowy – a means to an end". In such statements, the cold aloofness of the original character seeps through, although arguably *Elementary* is the only televisual iteration of Sherlock Holmes that shows him doing regular social and domestic things (such as eating and having sex).

HOLMES: "I'm not sleeping. I'm just reviewing the details of the case in my mind".
WATSON: "I'm sorry. Were you talking to me? Because I thought I was just a cavernous expanse between two ears".
HOLMES: "You mustn't be so sensitive, Watson. The service you're providing is quite valuable. For a brief stretch in London I talked only to a phrenology bust I kept in my study. I named him Angus. It wasn't the same. I realized that when it came to listeners I preferred animates to inanimate. Quite a breakthrough, really".
WATSON: "Angus. I'm glad I made it to the "animate" category".

One of the differences between this media portrayal of Sherlock Holmes and the original books is how toxic Sherlock's performance of his manhood is, as he is at constant pains to elevate himself over others. While in *Elementary*, there is a considerable amount of humility from Sherlock, the net effect of derision is the same, as evident in the following dialogue:

HOLMES: "I am smarter than everyone I meet, Watson. I know it's bad form to say it, but in my case, it's a fact".[65]

Sherlock demonstrates at least a scintilla of wider knowledge of social etiquette. While it might be "bad form to say it," it is part of his habitual toxically discursive practices that he speaks his mind fully, and his masculine identity creates expectations that his assertions are correct, in ways that feed his confident authority. Yet unlike Cumberbatch's Sherlock, Miller plays him in a flawed manner, and learns that his actions affect others, so he changes. The grandiloquence becomes dialed down, and the toxic masculinity moves beyond Sherlock's gibes and sneering to actual violence, gang warfare, and skirmishes with the mafia. The language that Sherlock uses mimics that of a solider engaged in warfare, such as in his frequent use of the term "Lieutenants" to describe people acting out the wishes of the "enemy" (especially Moriarty). Such scenes imply that Sherlock is fighting against forces of evil in contemporary New York. This shift of toxic masculinity from the domestic, cozy sphere of Baker Street in *Sherlock* to the rough streets of New York makes *Elementary* have more in common with a usual police procedural where there are elements of hyper masculinity, employing nefarious means in order for good to trounce evil, for example, in scenes of violence between the police and the public.

Why is toxic masculinity still prevalent in media representations?

Recent popular television shows featuring Sherlock Holmes represent power enacted through toxic masculinity in very similar ways, such as deploying discourse rather than overt violence. Both depictions of the character reflect the masculinity that is present in the very first description of him as being cold, lacking in emotion, and disparaging toward others.[66] This is consistent with the antisocial account of toxic masculinity offered by Kupers.[67] The successes of both *Sherlock* and *Elementary* suggest that there is still an audience appetite for representations of toxic masculinity and underscores the enduring popularity of the character of Sherlock Holmes himself. Moreover, the more mundane, but no less domineering, discursive elements of toxic masculinity (such as derision) are much more prevalent in these television shows than in other examples of post-millennial Sherlock,[68] such as in the films *Sherlock Holmes*[69] and *Sherlock Holmes Game of Shadows*[70] by Guy Ritchie, or in the recent *Mr. Holmes* starring Ian Mckellen.[71] Both *Sherlock* and *Elementary* are very positive (albeit very different) versions of a mythopoetic character. Both have been remythologized, in terms of their impact on popular culture, although Miller's portrayal much less so than Cumberbatch's.[72]

Yet, the tide appears to be turning against instances of toxic masculinity. A phenomenon growing in concert with increased interest in toxic masculinity is that of women and men speaking up and fighting back against male toxicity en masse.[73] The past few years has seen a rejection of the normalization of men's commonplace toxic behaviors that are perceived as being detrimental to women—from catcalling in the street to "manspreading" to inappropriate touching at music concerts. Activists are now attempting to address such behavior and reject it. For example, prominent alleged harassment cases against purportedly toxic men, such as film producer Harvey Weinstein, actor Kevin Spacey, and businessman-turned-president, Donald

Trump, has spawned #TimesUp, a charitable organization set up to support women who have faced harassment by men in the workplace, and supported by male and female celebrities.[74] The #MeToo movement is also a prominent, celebrity-endorsed agitation group calling out toxic male behaviors. Therefore, it is not just simply entrenched and random violence that activists are campaigning against; it is also the everyday sexism and the kinds of behaviors and use of language that would have in the past been excused as "banter"—especially in parts of the United Kingdom. This chapter has examined the ways in which portrayals of the character of Sherlock Holmes in popular television shows positions him as means by which toxic masculinity might be articulated through discursive means, in the way that he uses language to both censor (*Sherlock*) and mock (*Elementary*) other characters. This type of censorship has the net effect of the character positioning himself as the sole source of information. Despite the shift in masculinity toward greater hybridity away from hegemonic values of dominance and oppression, the displays of toxic masculinity analyzed in this chapter have much in concert with prevailing contemporary political, economic, and social masculine ideologies, especially the opportunities to enact power through the suppression of others. Such actions are now being held up as toxic and therefore unacceptable.[75]

Notes

1 Matt Hills, "Sherlock 'Content' Onscreen: Digital Holmes and the Fannish Imagination," *Journal of Popular Film and Television* 45, no. 2 (2017): 69. https://doi.org/10.1080/01956051.2017.1319200; in *Sherlock and Transmedia Fandom*, eds Louise Ellen Stein and Kristina Busse (Jefferson, NC: McFarland and Company, 2012).

2 *Sherlock*, seasons 1–4, episodes 1–15, directed by Paul Mcguigan et al. (first aired July 25, 2010, BBC, UK).

3 *Elementary*, seasons 1–7, directed by Guy Ferland et al. (first aired October 20, 2012, CBS, USA).

4 Zack Handelin, "It's Elementary, Sherlock: How the CBS Procedural Surpassed the BBC Drama," *tv. avclub.com*, January 20, 2014. https://tv.avclub.com/it-s-elementary-sherlock-how-the-cbs-procedural-surpa-1798265525; Troy Nankervis, "Sherlock's Benedict Cumberbatch 'Honoured' to Be Named Most Popular TV Character Across the Globe," *Metro*, February 15, 2017. https://metro.co.uk/2017/02/15/sherlocks-benedict-cumberbatch-honoured-to-be-named-most-popular-tv-character-across-the-globe-6449335/?ito=cbshare.

5 Arthur Conan Doyle, *Sherlock Holmes: The Complete Stories* (London: Wordsworth Editions, 1996).

6 Sarah Young, "Piers Morgan Calls Daniel Craig an 'Emasculated Bond' for Using a Baby Carrier," *The Independent*, October 16, 2018. www.independent.co.uk/life-style/piers-morgan-daniel-craig-baby-carrier-james-bond-emasculated-papoose-twitter-reaction-a8585836.html.

7 Raewyn Connell and James Messerschmidt, "Hegemonic Masculinity: Rethinking the Concept," *Gender and Society* 19 (2005): 840. doi:10.1177/0891243205278639.

8 Syed Haider, "The Shooting in Orlando Terrorism, or Toxic Masculinity (or Both)? *Men and Masculinities* 19, no. 5 (2016): 557. doi:abs/10.1177/1097184X16664952.

9 Raewyn Connell, "Masculinities and Globalization," *Men and Masculinities* 1, no. 1 (1998): 5. doi: 10.1177/1097184X98001001001.

10 Connell, "Masculinities and Globalization," 6.

11 Ashley Morgan, "The Rise of the Geek: Exploring Masculine Identity in the Big Bang Theory," *Masculinities: A Journal of Identity and Culture* 2 (2014): 35.

12 Demetrakis A. Demetriou. "Connell's Concept of Hegemonic Masculinity: A Critique," *Theory and Society* 30, no. 3 (2001): 378. doi: 10.1023/A: 1017596718715.

13 Shepherd Bliss, "Mythopoetic Men's Movements," in *The Politics of Manhood: Profeminist Men Respond to the Mythopoetic Men's Movement (and the Mythopoetic Leaders Answer)*, ed. Michael S. Kimmel (Philadelphia, PA: Temple University Press, 2009): 32.

14 Terry A. Kupers, "Toxic Masculinity as a Barrier to Mental Health Treatment in Prison," *Journal of Clinical Psychology* 61, no. 6 (2005): 714. doi:abs/10.1002/jclp.20105.

15 Donald Mosher and Mark Sirkin, "Measuring a Macho Personality Constellation," *Journal of Research in Personality* 18, no. 2 (1984): 151. https://doi.org/10.1016/0092-6566(84)90026-6.

16 Mosher and Sirkin, "Macho Personality," 150.

17 Roger Horrocks, *Male Myths and Icons* (Basingstoke: Palgrave Macmillan, 1995): 15–16.

18 Jackson Katz, "Advertising and the Construction of Violent White Masculinity," in *Gender, Race and Class in Media*, eds Gail Dines and Jean Humez (Thousand Oaks, CA: Sage, 2011): 261.

19 Jackson Katz and Jeremy Earp, *Tough Guise* (Northampton, MA: Media Education, 1999).

20 bell hooks, *We Real Cool* (London/New York: Routledge 2004): 16.

21 Joscha Legewie, "Racial Profiling and Use of Force in Police Stops: How Local Events Trigger Periods of Increased Discrimination," *American Journal of Sociology* 22, no. 2 (2016): 379. doi: 10.1086/687518.

22 Jared Yates Sexton, "Donald Trump's Toxic Masculinity," *The New York Times*, October 13, 2016. www.nytimes.com/2016/10/13/opinion/donald-trumps-toxic-masculinity.html.

23 Ashley Morgan, "The Real Problem with Toxic Masculinity Is That Is Assumes There Is Only One Way of Being a Man," *The Conversation*, February 7, 2019. http://theconversation.com/the-real-problem-with-toxic-masculinity-is-that-it-assumes-there-is-only-one-way-of-being-a-man-110305.

24 BBC News, "Geoffrey Boycott Doesn't Give a 'Toss' About Knighthood Criticism," *BBC News*, September 10, 2019. www.bbc.co.uk/news/uk-49639222.

25 Alessandro Gabbiadini, Brad J. Bushman, Paolo Riva, Luca Andrighetto, and Chiara Volpato, "Acting Like a Tough Guy: Violent-Sexist Video Games, Identification with Game Characters, Masculine Beliefs, and Empathy for Female Violence Victims," *Plos One*, April 13 (2013): 2. doi: 10.1371/journal.Pone.015212.

26 Victoria Simpson Beck, Stephanie Boys, Christopher Rose, and Eric Beck, "Violence Against Women in Video Games: A Prequel or Sequel to Rape Myth Acceptance," *Journal of Interpersonal Violence* 27, no. 5 (2012): 3016–3031. doi: 10.1177/0886260512441078.

27 Megan Vokey, Bruce Tefft, and Chris Tysiaczny, "Analysis of Hyper-Masculinity in Magazine Adverts," *Sex Roles: A Journal of Research* 68, nos 9–10 (2013): 562–576. doi:10.1007/s11199-013-0268-1.

28 Anastasia Salter and Bridget Blodgett, *Toxic Geek Masculinity in Media: Sexism, Trolling, and Identity Policing* (London: Palgrave Macmillan, 2017).

29 Chami Park, "Violence against Women in Korean Television Dramas," *The Asia Foundation* (2016): 1–2. https://asiafoundation.org/publication/violence-women-korean-television-dramas/.

30 Tracey L. Bealer, "Of Monsters and Men: Toxic Masculinity and the 21st Century Vampire in the Twilight Saga," in *Bringing Light to the Twilight*, ed. G.L. Anatol (New York: Palgrave Macmillan, 2011): 140.

31 Kathleen Elliott, "Challenging Toxic Masculinity in Schools and Society," *On the Horizon* 26, no. 1 (2018): 17. https:doi/10.1108/OTH-11-2017-0088.

32 Steve Biddulph, *Raising Boys* (London: Thorsons Classics, 2015); Tim Winton, "About the Boys: Tim Winton on How Toxic Masculinity Is Shackling Men to Misogyny," *The Guardian*, April 6, 2018. www.theguardian.com/books/2018/apr/09/about-the-boys-tim-winton-on-how-toxic-masculinity-is-shackling-men-to-misogyny.

33 Tristan Bridges and C.J. Pascoe, "Hybrid Masculinities: New Directions in the Sociology of Men and Masculinities," *Sociology Compass* March 18, 8, no. 3 (2014): 250. https://doi.org/10.1111/soc4.12134.

34 Biddulph, *Raising Boys*, 151.

35 Josephine Jobbins, "Man-Up: The Victorian Origins of Toxic Masculinity," *The Historian* 2016. https://projects.history.qmul.ac.uk/thehistorian/2017/05/12/man-up-the-victorian-origins-of-toxic-masculinity/.

36 Jobbins, "Man-Up," 2016.

37 Donald E. Hall, *Muscular Christianity* (Cambridge: Cambridge University Press, 1994).

38 Bliss, "Mythopoetic Men," 300.

39 Bliss, "Mythopoetic Men."

40 Paul Jones, "Benedict Cumberbatch's Sherlock Boosts Book Conan Doyle Book Sales," *Radio Times*, January 1, 2012. www.radiotimes.com/news/2012-01-17/benedict-cumberbatchs-sherlock-boosts-conan-doyle-book-sales/.

41 Stuart Heritage, "Sick of Sherlock? Elementary Has All the Holmes Comfort You Need," *The Guardian*, January 14, 2017. www.theguardian.com/tv-and-radio/2017/jan/14/elementary-jonny-lee-miller.

42 There are 154 episodes of *Elementary* across seven seasons: Imdb.com.

43 There are 24 episodes in *Elementary* season 1, and three episodes in *Sherlock* season 1.

44 Wikiquotes. https://en.wikiquote.org/wiki/Sherlock_%28TV_series%29 and imdb.com.

45 Gillian Rose, *Visual Methodologies* (London: Sage, 2001): 136.

46 Francesca Coppa, "Sherlock as Cyborg: Bridging Mind and Body," in *Sherlock and Transmedia Fandom*, eds Louise Ellen Stein and Kristina Busse (Jefferson, NC: McFarland and Company, 2012); Stephen Greer, "Queer (Mis)recognition in the BBC's Sherlock," *Adaptation* 8, no. 1 (2015): 50–67. https://doi.org/10.1093/adaptation/apu039; Judith Fathallah, "Moriarty's Ghost: Or the Queer Disruption of the BBC's Sherlock," *Television and New Media* 15, no. 5 (2014): 490–500. doi/10.1177/1527476414543528.

47 Hills, "Sherlock 'Content' Onscreen," 2017.

48 Benjamin Lee, "When I Say Sherlock, You Think Cumberbatch," *The Cumberbatch Watch*, *cumberbatchwatch.com*, 2014. https://sites.uci.edu/cumberbatchwatch/2014/05/23/when-i-say-sherlock-you-think-cumberbatch/.

49 Ashley Polasek, "Winning the Grand Game: Sherlock and the Fragmentation of Fan Discourse," in *Transmedia Fandom*, eds Stein and Busse (2012): 41–55.

50 Ashley Morgan, "'Sex Doesn't Alarm Me': Exploring Heterosexual Male Identity in BBC's Sherlock," *Journal of Popular Television* 7, no. 3 (2019): 337–335. doi: 10.1386/jptv_00004_1.

51 Coppa, "Sherlock as Cyborg," 2012: 212.

52 Morgan, "Sex Doesn't Alarm Me," 2019.

53 *Sherlock*, season 1, episode 1, "A Study in Pink," directed by Paul Mcguigan, aired July 25, 2010, BBC, UK.

54 Morgan, "Sex Doesn't Alarm Me," 2019.

55 *Sherlock*, "A Study in Pink," 2010.

56 *Sherlock*, season 2, episode 2, "A Scandal in Belgravia," directed by Paul Mcguigan, aired January 1, 2012, BBC, UK.

57 Ashley Morgan, "The Real Problem with Toxic Masculinity Is That It Assumes There Is Only One Way of Being a Man," *The Conversation*, February 7, 2019. https://theconversation.com/the-real-problem-with-toxic-masculinity-is-that-it-assumes-there-is-only-one-way-of-being-a-man-110305.

58 *Sherlock*, "A Scandal in Belgravia," 2012.

59 Morgan, "Sex Doesn't Alarm Me," 2019: 14.

60 *Sherlock*, "A Study in Pink," 2012.

61 Roberta Pearson, "A Case of Identity, Sherlock, Elementary and Their National Broadcasting Systems," in *Storytelling in the Media Convergence Age*, eds Roberta Pearson and Anthony N. Smith (London: Palgrave Macmillan, 2015): 122–148.

62 *Lie to Me*, directed by Daniel Sackheim (2009–2011) Fox, USA.

63 *The Mentalist*, directed by Chris Long (2009–2015) Warner Bros TV, USA.

64 *Elementary*, season 1, episode 5, "Lesser Evils," directed by Colin Bucksey, aired November 1, 2012, CBS, USA.

65 *Elementary*, season 1, episode 14, "The Deductionist," directed by John Polson, aired February 3, 2013, CBS, USA.

66 Doyle, "Sherlock Holmes," 1996.

67 Kupers, "Toxic Masculinity," 2015.

68 Polasek, "Winning the Grand Game," 2012.

69 *Sherlock Holmes*, directed by Guy Ritchie (2009) Warner Bros, USA.

70 *Sherlock Holmes, A Game of Shadows*, directed by Guy Ritchie (2011) Warner Bros, USA.

71 *Mr. Holmes*, directed by Bill Condon (2015) Miramax, UK/USA.

72 Bliss, "Mythopoetic Men."

73 #TheWomensMarch. https://womensmarch.com/2019-march/.

74 Time's Up. www.timesupnow.com/.

75 The Everyday Sexism Project. https://everydaysexism.com/.

Bibliography

Anderson, Eric. n.d. "Toxic Masculinity, the Problem with the Phrase." *The Book of Man*. Accessed March 13, 2019. https://thebookofman.com/mind/masculinity/what-is-toxic-masculinity/.

Bates, Laura. n.d. The Everyday Sexism Project. https://everydaysexism.com.

BBC News, "Australian Senator Leyonhjelm Criticised for 'Sexist Slurs'." *BBC News*, July 22, 2018. Accessed September 28, 2018. www.bbc.co.uk/news/world-australia-44678148.

BBC News, "Geoffrey Boycott Doesn't Give a 'Toss' about Knighthood Criticism." *BBC News*, September 10, 2019. Accessed September 10, 2019. www.bbc.co.uk/news/uk-49639222.

Bealer, Tracey L. "Of Monsters and Men: Toxic Masculinity and the 21st Century Vampire in the Twilight Saga." In *Bringing Light to the Twilight*, edited by G. L. Anatol, 139–152, New York: Palgrave Macmillan, 2011.

Beck, Victoria Simpson, Boys, Stephanie, Beck, Christopher, and Rose, Eric. "Violence against Women in Video Games: A Prequel or Sequel to Rape Myth Acceptance." *Journal of Interpersonal Violence* 27, no. 5 (2012): 3016–3031. doi:10.1177/0886260512441078.

Biddulph, Steve. *Raising Boys*. London: Thorsons Classics, 2015.

Bliss, Shepherd. "Mythopoetic Men's Movements." In *The Politics of Manhood: Profeminist Men Respond to the Mythopoetic Men's Movement (and the Mythopoetic Leaders Answer)*, edited by Michael. S. Kimmel, 292–309, Philadelphia, PA: Temple University Press, 2009.

Bridges, Tristan, and Pascoe, C.J. "Hybrid Masculinities: New Directions in the Sociology of Men and Masculinities." *Sociological Compass* 8, no. 3 (2014): 246–258.

Busari, Stephanie. "UNICEF: Boko Haram Has Kidnapped More than 1000 Children in Nigeria," *CNN*, April 3, 2018. Accessed December 11, 2018. https://edition.cnn.com/2018/04/13/africa/boko-haram-children-abduction-intl/index.html.

Condon, Bill, dir. *Mr. Holmes*. United States: Miramax, 2015.

Connell, Raewyn, and Messerschmidt, James. "Hegemonic Masculinity: Rethinking the Concept." *Gender and Society* 19 (2005): 829–859. https://doi.org/10.1177/0891243205278639.

Connell, Raewyn. "Masculinities and Globalization." *Men and Masculinities* 1, no. 1 (1998): 3–23. doi: https://doi.org/10.1177/1097184X98001001001.

Coppa, Francesca. "Sherlock as Cyborg: Bridging Mind and Body," In *Sherlock and Transmedia Fandom*, edited by Louisa Ellen Stein and Kristina Busse, 210–223, Jefferson, NC: McFarland and Co, 2012.

Demetriou, A. Demetrakis. "Connell's Concept of Hegemonic Masculinity: A Critique." *Theory and Society* 30, no. 3 (2001): 337–361. doi: 10.1023/A: 1017596718715.

Doyle, Arthur Conan. *Sherlock Holmes: The Complete Stories*. London: Wordsworth Editions, 1996.

Dreardon, Lizzie. "Knife Crime Hits Highest Levels Since Records Began," *The Independent*, April 25, 2019. Accessed June 28, 2019. www.independent.co.uk/news/uk/crime/knife-crime-stabbings-london-gang-police-england-wales-uk-a8885486.html.

Elliot, Kathleen. "Challenging Toxic Masculinity in Schools and Society." *On the Horizon* 26, no. 1 (2018): 17–22. https:doi/10.1108/OTH-11-2017-0088.

Fathallah, Judith. "Moriarty's Ghost: Or the Queer Disruption of the BBC's Sherlock." *Television and New Media* 15, no. 5 (2015): 490–500. doi: 10.1177/1527476414543528.

Ferland, Guy et al. dirs. *Elementary*, seasons 1–7. Aired October 20, 2012 on CBS.

Gabbiadini, Allessandro, Brad J. Bushman, Paola Riva, Luca Andrighetto, and Chiara Volpato. "Acting Like a Tough Guy: Violent-Sexist Video Games, Identification with Game Characters, Masculine Beliefs, and Empathy for Female Violence Victims." *PLoS One* 11, no. 4 (2013): 1–14. doi: 10.1371/journal.pone.015212.

Greer, Stephen. "Queer (Mis)recognition in the BBC's Sherlock." *Adaptation* 8, no. 1 (2015): 50–67. https://doi.org/10.1093/adaptation/apu039.

Haider, Syed. "The Shooting in Orlando, Terrorism, Toxic Masculinity (or Both)?" *Men and Masculinities* 19, no. 5 (2016): 555–565. doi:10.1177/1097184X16664952.

Hall, Donald E. *Muscular Christianity*. Cambridge: Cambridge University Press, 1994.

Handelin, Zack. "It's Elementary, Sherlock: How the CBS Procedural Surpassed the BBC Drama," *TV—AV Club*, January 20, 2014. Accessed June 14, 2016. https://tv.avclub.com?it-s-elementary-sherlock-how-the-cbs-procedural-surpa-1798265525.

Heritage, Stuart. "Sick of Sherlock? Elementary Has All the Holmes Comfort You Need," *The Guardian*, January 14, 2017. Accessed March 22, 2017. www.theguardian.com/tv-and-radio/2017/jan/14/elementary-jonny-lee-miller.

Hills, Matt. "Sherlock 'Content' Onscreen: Digital Holmes and the Fannish Imagination." *Journal of Popular Film and Television* 45, no. 2 (2017): 68–78. https://doi.org/10.1080/01956051.2017.1319200.

Hills, Matt. "Sherlock's Epistemological Economy and the Value of 'Fan' Knowledge." In *Sherlock and Transmedia Fandom*, edited by Louise Ellen Stein and Kristina Busse, 27–40, London: I.B. Taurus, 2012.

hooks, bell. *We Real Cool*. London/New York: Routledge, 2004.

Horrocks, Roger. *Male Myths and Icons*. Basingstoke: Palgrave Macmillan, 1995.

Jobbins, Josephine. "Man-Up: The Victorian Origins of Toxic Masculinity," *The Historian* (2016). Accessed November 11, 2018. https://projects.history.qmul.ac.uk/thehistorian/2017/05/12/man-up-the-victorian-origins-of-toxic-masculinity/.

Jones, Paul. "Benedict Cumberbatch's Sherlock Boosts Conan Doyle Book Sales," *Radio Times*, January 17, 2012. Accessed February 7, 2016. www.radiotimes.com.news/2012-01-17/benedictcumberbatchs-sherlock-boots-conan-doyle-book-sales/.

Kabeer, Nalia. "Grief and Rage in India: Making Violence against Women History," *Open Democracy* 5 March 2015. Accessed August 19, 2019. www.opendemocracy.net/en/5050/grief-and-rage-in-india-making-violence-against-women-history/.

Katz, Jackson. "Advertising and the Construction of Violent White Masculinity." In *Gender, Race and Class in Media*, edited by Gail Dines and Jean Humez, 260–268, Thousand Oaks CA: Sage, 2011.

Katz, Jackson and Jeremy Earp. *Tough Guise*. Northampton, MA: Media Education, 1999.

Kupers, Terry A. "Toxic Masculinity as a Barrier to Mental Health Treatment in Prison." *Journal of Clinical Psychology* 61, no. 6 (2005): 713–724. doi:10.1002/jclp.20105.

Lee, Benjamin. "When I Say Sherlock, You Think Cumberbatch," *The Cumberbatch Watch*, 2014. Accessed March 21, 2015. https://sites.uci.edu/cumberbatchwatch/2014/05/23/when-i-say-sherlock-you-think-cumberbatch/.

Legewie, Joscha. "Racial Profiling and Use of Force in Police Stops: How Local Events Trigger Periods of Increased Discrimination." *American Journal of Sociology* 22, no. 2 (2016): 379–424. doi:10.1086/687518.

Mcguigan, Paul, et al., dir. *Sherlock*, seasons 1–4, episodes 1–15. Aired July 25, 2010 on BBC, UK.

Morgan, Ashley. "'Sex Doesn't Alarm Me': Exploring Heterosexual Male Identity in BBC's Sherlock." *Journal of Popular Television* 7, no. 3 (2019): 337–335. doi: 10.1386/jptv_00004_1.

———. "The Real Problem with Toxic Masculinity is that it Assumes there is Only One Way of Being a Man," *The Conversation*, February 7, 2019. Accessed May 21, 2019. https://theconversation.com/the-real-problem-with-toxic-masculinity-is-that-it-assumes-there-is-only-one-way-of-being-a-man110305.

———. "The Rise of the Geek: Exploring Masculine Identity in the Big Bang Theory (2007)." *Masculinities: A Journal of Identity and Culture* 2 (2014): 31–35.

Mosher, Donald, and Mark Sikin. "Measuring a Macho Personality Constellation." *Journal of Research in Personality* 18, no. 2 (1984): 150–163. doi:10.1016/0092-6566(84)90026-6.

Nankervis, Troy. "Sherlock's Benedict Cumberbatch 'Honoured' to Be Named Most Popular TV Character Across the Globe," *Metro*, February 15, 2017. Accessed June 13, 2016. https://metro.co.uk/2017/02/15/sherlocks-benedict-cumberbatch-honoured-to-be-named-most-popular-tv-character-across-the-globe-6449335/?ito=cbshare.

Park, Chami. "Violence Against Women in Korean Television Dramas," *The Asia Foundation*, 2016. Accessed November 12, 2018. https://asiafoundation.org/publication/violence-women-korean-television-dramas/.

Pearson, Roberta. "A Case of Identity, Sherlock, Elementary and Their National Broadcasting Systems." In *Storytelling in the Media Convergence Age*, edited by Roberta Pearson and Anthony N. Smith, 122–148, London: Palgrave Macmillan, 2015.

Polasek, Ashley. "Winning the Grand Game: Sherlock and the Fragmentation of Fan Discourse," In *Sherlock and Transmedia Fandom*, edited by Louise Ellen Stein and Kristina Busse, 41–55, Jefferson, NC: McFarland and Company, 2012.

Ritchie, Guy, dir. *Sherlock Holmes: A Game of Shadows*. Warner Bros, USA, 2009.

Rose, Gillian. *Visual Methodologies*. London: Sage, 2001.

Salter, Anastasia, and Blodgett Bridget. *Toxic Geek Masculinity in Media*. London: Palgrave Macmillan, 2017.

Sexton, Jared Yates. "Donald Trump's Toxic Masculinity," *The New York Times*, October 13, 2016. Accessed August 5, 2019. www.nytimes.com/2016/10/13/opinion/donald-trumps-toxic-masculinity.html.

Time's Up. www.timesupnow.com/.

Vokey, Megan, Bruce Tefft, and Chris Tysiaczny. "Analysis of Hyper-Masculinity in Magazine Adverts." *Sex Roles: A Journal of Research* 68, nos 9–10 (2013): 562–576. doi:10.1007/s11199-013-0268-1.

Winton, Tim. "About the Boys: Tim Winton on How Toxic Masculinity Is Shackling Men to Mysogyny," *The Guardian*, April 9, 2018. Accessed August 6, 2019. www.theguardian.com/books/2018/apr/09/about-the-boys-tim-winton-on-how-toxic-masculinity-is-shackling-men-to-misogyny.

PART VI

Gender advocacy in action

Introduction to Part VI

We have reached a critical moment in history: a time in which issues of gender, gender determination, gender identification, and gender reassignment have been uncloistered and are no longer private but are now public. And because the private and personal have become necessarily public, this period in time has become more political as a liberatory project—not merely political as in governmental interest in public affairs, though that is a continued concern, or political as in related to the strategies of one's political party or group, though issues of gender identity and rights are clearly grounded in certain political camps linked to the propriety of bodies, self-determination, military service, as well as rights and privileges linked with particularity and desire. Instead, we refer to political in the everyday sense of partisan politics that guide decision-making and the social construction of acceptability in the world—and the need for progressive positive activism—and political as an unflinching defense of the very ground we stand on, the queer bodies and sensibilities we live in, to claim our right to be fully alive and fully present in the world.

The notion of "gender advocacy in action" that frames this section is about varying methods, modes, and modalities of overt partisan politics that promote the free expression of gendered identities toward social justice. Thus, the chapters featured in this segment address the possibilities for empowering the innate dignity of all human beings in and through the performance of their complex gendered identities (both singular and plural, both normative and the presumptions of its opposite). The final section of this volume links academic research on communication and gender to activism and advocacy beyond the academy. Entries explore practical pathways for scholars' critical insights to shape community outreach, coalitional movements, and pragmatic institutional intervention. Moreover, contributors in the closing segment of the volume challenge current readers and future researchers to reflect on the gendered dynamics and communicative ethics of their own practices in pedagogical, interpersonal, and communal encounters to reflect backwards on those inhibiting factors on gendered beings and to project forward the potentialities of becoming a just society.

In the first chapter, "Queer praxis: the daily labors of love and agitation," Dustin Bradley Goltz and Jason Zingsheim (scholars, gendered subjects, and politically public intimate partners) interrogate the personal and relational work of queering in daily life as labor, loving, agitation, and as praxis. They engage this work with a passion that speaks to a queer praxis that is not just that place where theory and practice meet but a centralized location of lived experience, a livable location of queer relationality. They approach queerness as an action, an ethic, and a relational project owned by no one and differently accessible to everyone. In the chapter, they first challenge the tendency for queerness to linger in the theoretical by positioning queerness in the material, lived, embodied, and relational—thus, the performative—dimensions of queer lives and practices. Then, the chapter interrogates patterns that reduce the work of queer worldmaking to either gender queer or categorical queer postures, re-centering white, cis, masculine frameworks, while calling for constant reflection, decentering, and accountability for who and how queerness is framed, what stories are centralized, and to whose benefit or exclusion. Finally, the chapter looks to queer praxis as a daily and moment-by-moment resistance, as well as an opportunity and radically contextualized action, rather than an identity or fixed position.

In the second chapter, the notion of a *fixed liquidity* comes to mind in the work of Sarah Jane Blithe and Mackenna Neal as they reapproach the sometimes fixed, and yet always fluid, construction of feminism as a social movement that has oscillated between periods of high activity and lulls in activist work, depicted through the familiar wave metaphor in their chapter, "Communicating gender advocacy: riding the fourth wave of feminism." The notion of *fixed liquidity* in this chapter is not about a stationary and anchored logic of feminism. The authors allude to a deep political commitment that has staked territory—even as it pivots, shifts, flows, ripples, and expands to become more mindfully inclusive and historically relevant. Based on a recent surge in feminist advocacy, they argue that gender advocates have created a fourth wave of feminism, which presupposes intersectionality but also calls for justice for "women" broadly defined, prioritizes social justice, relies on social media, and has significantly influenced the political scene in the United States. Grown through the ashes of postfeminism and from the seeds of third-wave feminists, fourth-wave feminism has taken a particular turn to focus on reducing sexual harassment and gendered violence in multiple organizational spaces. In this chapter, they review relevant advocacy in the first three waves of feminism and post-feminism, arguing that each wave has taken arguments and efforts from earlier advocacy. Therefore, fourth-wave feminism is deeply connected to other periods of advocacy and crests from the strength of previous feminist work.

In the next chapter in this section, "The oppositional gaze as spectacle: feminist visual protest movements in China," Nickesia S. Gordon and Yuhan Huang explore the power of the gaze and the spectacle that it generates as resistive forces in certain visual protest movements in China. The oppositional gaze serves as a type of looking relation that involves a political rebellion of not just seeing, or an intense fixed focus but an intentional act of being seen looking back—that breaks, breaches, and inevitably bridges the social expectations and emergent resistance of raced and gendered subjects. Such particular acts of protest, inspired by similar actions in the United States, bring attention to and challenge normative ideals about female identity, power, and oppression that occur across the globe. These visual protest movements also rely heavily on social media for organizing and galvanizing efforts. In their analysis, the authors use an extended version of bell hooks' concept of the oppositional gaze to make the case for the rebellious nature of the visual practices that these movements internalize and invite. Using China as a case study, they conduct a discursive analysis of how the oppositional gaze unfolds when actors respond to misogyny and patriarchy by using visual or nonverbal forms of protest. Gordon and Huang also consider how movements in the United States (such as #BlackLivesMatter, #MeToo, Occupy, #WhyIStayed, #SayHerName, and #PussyHat, among many more) help frame sister movements in China.

Taking the notion of oppositional resistance in a different direction but still firmly grounded in activism, Benny LeMaster and Deanna L. Fassett, in their chapter "Refusing mastery, mastering refusal: critical communication pedagogy and gender," articulate the state of the art of critical communication pedagogy (CCP) as it does and does not speak to gender. In turn, they advance two pillars that serve as pedagogical anchors for theorizing, embodying, and enacting gender futurity grounded in criticality. The pillars include the dialectic coupling of "refusing mastery" and "mastering refusal." They argue that CCP and gender teach us that to *refuse mastery* is to *embrace failure*; that is, refusing mastery means to engage and critique the structural means by which one is rendered a "failure" as a result of their embodied difference. To embrace failure is to reject hegemonic expectancies and to affirm self and/as other in terms that refuse cisheterosexism, white supremacy, ableism, and classism, for instance. CCP in/of gender additionally teaches us that to *master refusal* is to *embrace uncertainty*. While the former commitment challenges epistemological certainty through embracing failure, the latter commitment centers the affective registers that seek to shift the status quo through a willful refusal to find calm in the midst of uncertainty. To master refusal is to bask in the improvisational and playful space of unknowing gender so as to better engage gender on its emergent, creative, constrained, and relational intersectional terms.

In their chapter, "Gender futurity at the intersection of Black Lives Matter and Afrofuturism," Amber Johnson explores the intersections of race and gender in social movement contexts and offers alternative structures for creating a gender-free future via intersectional social movements, gender futurity, and transfuturism. The chapter begins with a discussion of intersectionality and Black Lives Matter and Afrofuturism, two social movements that center the lived experiences of the Black diaspora in an attempt to create systemic change, while also falling short of producing gender freedom and inclusion. Then, using Black Lives Matter and Afrofuturism as departure points, the chapter argues for gender futurity using a case study in Transfuturism to unpack how the promoted praxis can push the boundaries of gender embodiment and freedom. The chapter offers a kind of performative utopianism with strategies of engagement that make possible a future unshackled in the binary, making way for a fluid future of possibilities as an activist sensibility.

Capitalizing on the impulse of a resistant particularity, and within a clear historizing of still current challenges, in the essay "Lantinx feminist activism for the safety of women journalists," Aimée Vega Montiel references how gender-based violence against women journalists has increased dramatically, both offline and online. These professionals face more barriers to performing their roles in safe conditions. Slow progress has been made by governments and news media industries to protect women journalists. Most of the progressive actions to address this widespread problem have been achieved by journalists, nongovernmental organizations (NGOs), and scholars. Thanks to the work done by these actors, violence against women journalists has been unveiled and caught the attention of human rights courts. Grounded in both communication studies and feminist theory, this chapter analyzes the structural conditions enabling gender-based violence against women journalists in Latin America, explores how it affects the collective right to freedom of expression, and considers responses to this crisis from activist movements. The chapter does some heavy lifting in documenting these challenges but provides an opening for new possibility and serves as a call to attend to how the vulnerability of socially constructed gender expectations has resistive potential that provides perspective.

The final chapter in this section is the perfect bookend, as it focuses on "Pushing boundaries: toward the development of a model for transing communication in (inter)cultural contexts." Gust A. Yep, Sage E. Russo, and Jace Allen ask the question: How do individuals and groups create and maintain identities, enhance and sustain cultural intelligibility, cultivate and increase a sense of belonging, and negotiate and contest social meanings—indeed communicate—within and across these complex boundaries? To begin to answer this question and to tap into the potential of the negotiation, crossing, and (re)definition of such boundaries, their chapter provides sketches of a model for *transing communication* in (inter)cultural contexts. To do so, the chapter first discusses the process of transing communication, outlining four domains of a model. Using the domains, the authors identify and review past research on transgender communication. They conclude the chapter by providing an exploration of potential implications and new directions for transing communication research in (inter)cultural contexts.

As a whole, this section on "Gender advocacy in action" takes the reader through queer praxis as daily resistance in queer worldmaking to transing communication as calls for a transrelationality of visibility and communication across trans people, thus making possible a transfuturity. The section moves from a fluidity of impact between waves of feminism to a gender futurity as a promoted praxis toward freedom, and from operationalizing the oppositional gaze as a form of resistance to an overt refusing of mastery. At the end of the volume, these chapters offer empowering notions of *mastering refusal* and *refusing masters,* as well as developing new models for *transing communication.* The latter two chapters speak powerfully to the intention of the whole volume: establishing and encouraging new templates of sociality, new models of resistance, and expansive frameworks of viewing oneself in relation to society. Ultimately, this final section offers strategies for changing society: debunking the notion of a monolithic culture and reconstructing it as a constellation of social orientations and cultural practices that can be remade anew, with broader potentials of liberation for expansively defined gendered subjects.

33

QUEER PRAXIS

The daily labors of love and agitation

Dustin Bradley Goltz and Jason Zingsheim

Queerness, while taking some root in the abstract idea, is given shape, necessity, urgency, utility and contestation in the lived experience.

Queer Praxis authors[1]

Introduction: queer interconnections

Queer praxis is the working, doing, living, and everyday processes of queering. It is the work of hope, anticipation, resistance, and imagining outside the violences of social categorizations—of longing sideways[2] and gesturing other-wards and otherwise. In short, it is the work of worldmaking. Thus, it is a creative, collaborative, and ever-shifting project for a multiracial, multigendered, multi-everything social imaginary. It is about interrogating legacies of privilege around our relations, our bodies, and our communities in order to disrupt and challenge the exclusions, limitations, and violences of our patterned ways of being, knowing, and doing. Queer praxis is about a constant and arduous process to be better in relation and connection to a range of differences too often devalued. It is about interrogating the places we each may have had voice, position, comfort, and access. It is about undoing, redoing, and continually learning to make room, surrender space, redirect our gaze, and halt our defensive dances. Queer praxis hopes and labors for more humane, more durable ways to be and exist in our world across difference. It is work. It is process. And, it is ever-fallible.

This chapter interrogates the personal and relational work of queering in daily life—as labor, loving, agitation, and praxis. Thus, we frame queerness as an action, an ethic, and a relational project owned by no one and differently accessible to everyone. Specifically, we examine faulty tensions in this work—competing dimensions of queer work that discursively create divisions and oppositions that paradoxically mobilize *and* impede—both somewhat literally and meta-phorically akin to the Cartesian mind/body split, where a wholly enmeshed and inter-reliant system is severed into a false binary. Queer praxis sits at the intersections and reintegrations of the lived and the theoretical, the individual and the relational, and the global and immediately situated. This chapter is a meditation upon these interconnections and false binaries that shape, foster, and impede queer praxis. First, this chapter challenges the tendency for queerness to linger in the abstract and theoretical. This, in turn, demands the necessity to position queerness in the material, lived, embodied, and relational—thus the performative—dimensions of queer lives and practices. Second, the chapter explores resistance to minoritizing[3] patterns that reduce the work of queer worldmaking to either gender queer or categorical queer[4] postures. Resisting normalizing tendencies to cast queerness in terms that privilege white, cis, masculine frameworks, this resistance calls for constant reflection, decentering, and accountability for who and how queerness is being framed, what stories are being centralized, and to whose benefit and exclusion. Third, the chapter looks to queer praxis as a daily and moment-by-moment resistance, an opportunity and radically contextualized action, rather than an identity or fixed position. These resistances are elaborated through a series of mundane questions, statements, and encounters, exploring both the strategies and risks of queering (even as these shift and change).

Tension #1: the theoretical/personal split

My mind and body agree on one thing only: I have nothing to lose. What-the-fuck: caution to the wind. Become confessional. Get queer.

Frederick Corey[5]

Our personal narratives rooted in the experience of an embodied, complex repertoire offer a queer potentiality for silenced and/or not yet imagined queer of color worlds.

Robert Gutierrez-Perez and Luis Andrade[6]

The developing anti-assimilationist/anti-identity project of queer theory, at least by name, emerged in the academy in the early 1990s as activists in the LGBTQ community were literally fighting for their lives. As a broad project of inquiry and activism, queer theory was a project of "remapping the terrain of gender, identity, and cultural studies."[7] Yet, while a contemporaneous exertion and extension of AIDS activism, activist groups like Queer Nation, and art movements such as Queercore and New Queer Cinema, queer theoretical work in the academy was often criticized for being abstract, elitist, needlessly convoluted, and of questionable practical use to the vulnerable queer lives it sought to examine and support. It claimed radicalism, and yet often appeared inaccessible outside the walls of academia. What's more, its (intended) playful subversion—is it a noun, a verb, or an adjective (the answers being a sly "yes")—carried an arrogance. The term "queer," and the definition and scope of queer theory—at the outset—was famously slippery. To say queerness defied, resisted, rejected, and/or evaded definition became, itself, an exhausting cliché. However, over the next few decades, the project of queer theory and queer worldmaking found footings for its self-proclaimed abstractions (and, perhaps, tempering some of its scope and fire). In turn, these growing pains of queer theory (where, it seemed, someone was always declaring its demise) pushed queer theoretical projects that sought out modes, approaches, and grounded strategies to *do* queer work.

Within the field of communication and performance studies, building and extending traditions from third-wave feminism's "theories of the flesh,"[8] queer theorists looked to the personal story and embodied experience as a space to examine the doings of queer work.[9] This marked one place of concerted attention and commitment to executing queer *praxis*, the embodiment and enactment of queer theory into action, into realization, in the lived and practiced. Performative writing, personal narrative, and autoethnography took hold as practices to resist hegemonic modes of knowledge production[10] and disrupt patterns of LGBTQ erasure and silencing.[11] After all, what could testify in the brashest, loudest, and most unapologetic voice than the contested assertion of the queer "I"—demanding visibility for one's body, intelligibility of one's life, and the politics of one's experience? At its most vital, its most uproarious, the queer "I" worked to defix and dismantle master narratives[12] of queer punishment, queer shame, queer isolation, the violences of heteronormativity,[13] and the always intersecting systems of oppression that shape the lives of queer bodies. "The abstractions of theory work to make sense of the situated, contextual and embodied experiences of the personal. The personal, in turn, dialogues with the theory. They frustrate and challenge each other with questions."[14] In tension and dialogue with the theoretical—which is inherently abstract—the personal voice, the performing body, and one's individual experience mark a space of theory building[15] that can speak, howl, and whisper back to queer theories with urgency, messiness, love, and complication.

A performative framework to interrogate, theorize, uncover, and anticipate queerness locates the queer body, the workings of queerness in the body, in the lived, and traces it through the moment-by-moment negotiations of power, history, patterns of behavior, and cultural discourse. Queerness is framed as located and situated action—a space between the immediate past and the not- yet-solidified future where resistance, disruption, and rearticulation linger in potentiality. We are driven by norms and patterns, and yet, with each passing moment, a space is opened up where the possibility of otherwise is present, is lingering, is just outside the margins of normativity and intelligibility. This is where performance and queerness rub against one another as conceptual partners, as kin, as relation. Performative theories' attention to the liminal, the second-by-second negotiation of preceding norms with an eye on breaking the sedimented routine, or challenging the hegemony, or considering the otherwise is conceptually tied to queerness. Queerness is about disrupting the moment-by-moment systems, patterns, and discourses that foreclose lives, futures, identities, and relations beyond the white, cis, neoliberal heteronormative mandate. Queerness and performance linger on the act of doing—rather than on an ontology of being—not fixed in space and time but ever-open to disruption and rearticulation. Joined as interrogations and disruptors of patterns, of historical inertia, queerness finds a footing, a praxis in the performative.

The personal and the performative also provide an alternative entrée point for students to engage with queer theories. Let us begin with the notion that lives are queer, bodies are queer, desires and relations are queer. Theoretical work, driven by abstracted language of categories and systems, often works to hold and discipline. And yet our lives, bodies, and desires remain stubbornly undisciplined. The personal presses back against the abstractions of the norms of academic writing. The *Queer Praxis* collection asks, "what happens if we hold up the abstractions and interventions of queer theory to the scrutiny and experience of our daily lives,"[16] embracing the messy and unruly dimensions of the personal while maintaining suspicions. As Martinez explains, the queer "I" is fraught, contested, requires critical frustration, as the personal always runs the risk of being wrong, being narcissistic.[17]

And yet it is to the personal we must turn, as the very (theoretical) language used to make this argument is mired in abstractions, losing heat and affect and grounding in order to generalize its claim. It loses the rage and play of the personal. It misses the queered limits of decorum, as Pérez and Brouwer examine their experiences of sitting in straight wedding spaces, erased, bodies still and disciplined, irate, immobile, alongside their love and intimacy in sharing these experiences. "I stayed for the dinner reception, pushing the food into my mouth while bleeding, trying to heal."[18] It neglects the anger, frustration, and isolating burst that Fox recounts, trying to get his colleagues to see how, every day and always, their pedagogy is about sexuality (it's just their vanilla-ass heterosexuality).[19] It loses both the deep love, disappointment, and struggle of BROWN DYKE GIRL and WHITE JEW GAY GUY attempting to nurture lines of connection, while honoring all the messy and devastating ways they cannot ever fully know one another, who they are and are not to one another, and how each of them are never but one thing—always shifting in relation, shifting in context, severed and sliced across multiple lines of difference.[20] It omits the demand, celebration, and burn that Gutierrez-Perez and Andrade poetically craft in demanding their queer love, their queer brown marriages, be etched into the archive that seeks to straighten and whitewash their story.[21] These stories enflesh, push, tear, and stretch the limiting boundaries of our theoretical abstractions that function to cloak or diminish the very queer seams of our experiences.

Thus, queer praxis insists upon a resistance to the abstract, for in generalization, there is a violence, a normalizing that should never be left unchecked—never be left to roam freely. With a continual lean into the material, the lived, the embodied, and the relational, the queer "I" carries power to disrupt, intervene, and dislodge. However, the queer "I" is also fraught, fallible, and limited in its illusions and blindness. The queer "I" requires frustration, demanding it be placed into dialogue, in relation—never final or without contest.[22] Queerness, that undefined vista or horizon of theory[23] maintains its openness, its eye toward the elsewhere. Yet it is grounded in the relational process, the work of reflexively storying and re-storying and de-storying as a queer effort, as queer action. This is the queer praxis through which we engage in queer worldmaking.

Tension #2: normative pollution/relational accountability

Blessed are the beloved who I didn't describe, I couldn't describe, will learn to describe and respect and love.

Mark Aguhar[24]

Prompted by one another, we continually deconstruct, but not in any purely intellectual sense. We throw our shoulders into the labor of our own and each other's becoming. Growing sideways, the hinges of our planters cannot contain us.

Aimee Carrillo-Rowe[25]

As the project of queerness extended beyond statements of abstraction, several (perhaps predictable) assimilationist trends worked to file down, lay claim to, and reduce the radical scope of queer work. Queerness is always and forever in tension with multiple corrosive normativities—working against an ever-present force, an always already polluting presence in any articulation. Queerness exists in the resistance to domestication, to use Madison's phrase.[26]

> It means that she will use the resources, skills, and privileges available to her to make accessible—to penetrate the borders and break through the confines in defense of—the voices and experiences of subjects whose stories are otherwise restrained or out of reach.[27]

Just as Madison calls for the critical ethnographer to contribute to "emancipatory knowledge and discourses of social justice," these aims run through the queer project.[28]

It hardly took long for the anti-identity reappropriation of the term "queer" to emerge as an option for clicking one's identity on gay.com (in the early days of the Internet, long before the days of social media). From there, in the 1990s we moved to a generic and mainstreamed usage, as demonstrated in *Queer Eye for the Straight Guy* and *Queer as Folk*.[29] Queer, in all its anti-normative rustlings, started to become co-opted, mainstream, and marketable (which also meant cis, white, middle-classed, able-bodied, Western, and male-centric). Susan Stryker details one of these primary reductions through a discussion of *homonormativity* and the different roots of this conceptual term.[30] As a concept, homonormativity is often traced through the sexualized dimensions of neoliberalism based on the critical work of Lisa Duggan—who challenges the emergence of mainstreamed, moderate, privatized, and "sensible" gays at "the reasonable center" of political life.[31] Stryker traces a different usage of the term predating Duggan's work in the early

2000s. Dating back to the early 1990s, homonormativity was enlisted to speak to the "double sense of marginalization and displacement experienced within transgender political and cultural activism."[32] Specifically, she refers to the cis-centric and gender-normative commitments and leanings that, even within more gay and lesbian spaces, worked to isolate and disparage trans and gender queer persons. This homonormativity, as a term, worked to name a tension in how "homosexuality, as a sexual orientation category based on constructions of gender it shared with dominant culture, sometimes had more in common with the straight world than it did with us."[33] Extending the need to place this tension into language, to name this rift and experience, Stryker asserts a distinction in the conception of categorical queerness or *categorical queers* from gender queerness or *gender queers*.

Much queer theory focused its energies on one facet of heteronormativity, interrogating the rigid categorization and solidification of sexual identity (gay/straight/bisexual), working to dismantle these boxes. Queerness was the work of making strange the categorization of sexuality through object choice, and highlighting the negative impact of this categorization to all sexual subjects in a universalizing (rather than minoritizing) project.[34] The work demanded we call into question established norms, values, and their effects. It invited us to flip over the tables of our commitments, viewing such disruption as a radical act of inquiry. Yet, too often, normative gender configurations remain intact and unchallenged in these interrogations into categorical queerness. The categorical queers, in Stryker's articulation, failed to tackle or meaningfully engage how systems of gender were central to this project, and how the radical queer potentials of transgenderism and gender queerness are often circumvented or skirted in much "queer" work. Categorical queerness, in this frame, claims a queer politic and project that refuses to investigate its many normative leanings and foundations, often leaving cis-normativity, patriarchy, and an underlying stink of whiteness untroubled in its efforts.

The second tension examined in this chapter is the struggle to hold ourselves and each other accountable within the constant, ongoing pull toward normative longings and commitments, such as—but not limited to—those Stryker identifies. This is a constant struggle and tension in queer praxis. The earliest conceptions that sought to place queerness at the margins of discourse,[35] as an ideality not yet here,[36] worked to capture the forward-driving hope that queerness might carry. Yet they also understood the absolute dangers of positioning queerness in the here and now. The now is stifling and bankrupt, as Muñoz reminds us; it is weighed down in the gravitational pull and the immobilizing web of normativities. Stryker's articulation of the limitations of categorical queerness mark one such space where normativity burrows and drives an articulation of queerness. Yet there are many, operating across a broad range of intersectionalities. Commitments to whiteness and civility mark additional modes of normative infestation and hidden allegiances, always enmeshed with classism, ableism, and systems of nationality and belonging. Ageism, and the centralization of youth and youthism mark different modes of normativity that take root and infest queer work. What could it look like to imagine alternative modes of future and hope that are not confined to the logics of Edelman's reproductive futurism?[37] Where the future is not only about discourses of youth (be they heteronormative or homonormative) but open up vistas for imagining and celebrating the playfulness and expansiveness of a broad range of expressions, identities, and ages? Is queerness aged? Additionally, queer theory still needs to contend more directly with the privileges and queer exclusions tied to academic spaces, academic performativity, and the very theoretical tools enlisted to drive the project, this chapter, our own capacity to name our fallibilities and limitations—for it all stinks of complex systems of power and discipline.[38] Sure enough, we have theories to help us see the co-presence of agency within these structures. Yet queer praxis demands an orientation to critical reflexivity that directs an eye inward in the spirit of the collective.

Queer praxis calls for love and agitation, as well as reflexivity and suspicions. We ought to be suspicious when one declares their queerness above others. We ought to be suspicious when one demands the reflexivity of others, claiming to have done their work, and builds their careers and identity and theory through heroic stories of disciplining others. In trust and accountability to each other, there is, there must be, accountability for ourselves. This resistance to defining what queer is, who is queer—or determining who is "more" queer and the "queerest of the queer"[39]— seems laughable from a distance. Yet the disciplinary postures of calling out others repeated in nonprofit organizations, protests, classrooms, and academic conferences are fraught and stinking with power and hypocrisies. Like Sedgwick's calling out of the paranoid read, the critical orientation toward negative effects, such a posture will always find the thing it seeks—failure. But what is gained in our quest to claim and name the failures of others without also implicating our always implicated selves? When we call out and slap down, bolstering our own superiority (and rhetorically constructing our own innocence), what could be more normative? More rooted in legacies of violence and oppression? What story of self is being asserted in those moments of disciplining? I the hero? I the victim? I the innocent? I the infallible? The queer "I" is fallible, just as the queer "we" is fraught and contested. Failure, especially in the now, is a given. But, for the process of queer worldmaking, the work of the "we" must still remain as a structure, a mechanism, a mode, and a hope.

Situating the relational dimensions of queerness is at the fore of queer praxis, examining what queerness does and how queerness is done in our lives, in community, and in relation. This is the daily work of holding ourselves accountable to one another, that messy and fraught work that resists the solo voice, the singular I, the uncontested account, the heroic and tragic tale, and the illusion of innocence. In dialogue, in the ongoing ethical work of love/agitation,[40] queer praxis is about the hope and disruption, the elsewhere and potentiality of queered modes of intimacy. Tied to the notion of relationality, and grounded in the work of scholars such as Carrillo-Rowe, Ann Russo's 2019 text[41] articulates queer worldmaking through a project of deep and processual accountability. In each moment, in each interaction, queer praxis is the work of understanding the complex ways that we all, each and every one of us, is complicit in a range of systems that seek to benefit us at the expense of others. It is challenging work that mandates an ethic of fallibility and forgiveness. It adamantly rejects tragic models of thinking and call-outs that appoint uncomplicated heroes and victims, while condemning simplified villains.

The work of this collective accountability of coalition does not require friendship or liking or similarity to one another. Rather, according to Corey and Nakayama, it requires anger; it requires people being really pissed.[42]

> The [AIDS activist] coalitions were neither permanent nor lasting—as nothing ever is. We came together over certain injustices that arose, like whack-a-mole… Perhaps we can borrow the term "crowd-sourcing" to build alliances that are focused on defeating a specific bill, or pushing for new laws, or so on.[43]

Herein, once again, the shift is from a communal identity of who "we" are and what "we" do, and the "we" is wholly determined in the moment by the people working together—if only for a moment. Coalition is, at its root, an agreement to stand together in times of consequence and resistance. It means when those who seek to demean or annihilate us show themselves, they face our collective, united, and pissed resistance in the moment, the room, the streets, the web. In those spaces, a "we" emerges based on what Cohen calls "the queer politics of positionality" to work together.[44] Queer praxis, for us, is not about cuddles and adulation. It's not about liking and it's not even always about listening. As cultural theorists have routinely argued, the balance of holding each other accountable and/while allowing space for growth is delicate, situated, relational work without a linear map. The real honest self-reflexive work of unpacking our shit, takes love and trust and time and mutual respect. In the face of a clear and present danger, when we are bodies standing next to other bodies, because rights, and safety, and humanity, and dignity are on the line, it is also about considering the time and function of your critique. Sometimes it is just about who shows up, helps make a sign, can offer a hand, a space, a donation, or recognition.

It is commonly said you cannot change other people but you can change yourself—shift your own behavior, do better. Starting with a little humility and reflexivity that we all have a complex and toxic load of normativity streaming through our veins and that we, each of us, are always inside the Matrix means we are always fallible, polluted, and corruptible, but we are also always connected. We share this culture of toxicity. We are all sick. We will all fail, in differing degrees and manners. Queer praxis, the ongoing work of putting this theory into practice within the context of relation through our daily performances means wrestling with this unpleasant fact as often and frequently as we allow ourselves. It is resisting the idea of queer as a noun, as something fixed and stabilized, where one becomes certified as being "queer." Instead, queer praxis calls us to continually seek out what it is to queer, to do queer work, and to activate the workings of queerness in our worlds.

Tension #3: queer worldmaking: in time and place

To me, queer relations of belonging have to do with choices. And choices have to be made within the paradoxical ambiguities of our lives. Sometimes, the tyranny of consistency becomes the problem. We are all human beings in struggle. We all fuck up… How about allowing not only for our differences/individualities, but also for our inconsistencies?

Sheena Malhotra[45]

Go then, there are other worlds than these.

Stephen King[46]

In the *Queer Praxis* volume, we brought together 29 scholars from across the field of queer communication to consider what it means to approach queer worldmaking as a relational project. "Worldmaking is not … a definitive alternative, or a substitute …. Worldmaking seeks an elsewhere, a disruption, and a rejection of the legitimized and routinized conventions of normativity."[47] Consequently, we also frame queerness as liminal and performative. It "takes root in the body, is negotiated within space, in time, through relations, and is practiced/resisted/negotiated moment-by-moment."[48] The political operations of sexual and gendered discourses are ever-present in our social worlds, and thus the opportunity for modes of queer resistance is always and everywhere. Yet, the ability and mobility to enact these differing modes of resistance will be tempered and contingent upon a broad range of intersectional and contextual factors. In some spaces for some bodies queer work is tolerated, considered and/or welcomed, while in other spaces, for other bodies, it can court anger, violence, and even death. Yet large and small, loud and soft, varying modes, volumes, and voices of queer resistance are the seeds, the soil, the impossible hope, and the daily practice of queer worldmaking. This section of the chapter briefly surveys a range of concrete worldmaking strategies, which present modes and performances of differing queerness that inform and gesture to the heterogeneous work of queer praxis.

In classrooms, we sometimes have students engage a very mundane practice of queering. When faced with a series of commonplace, yet heavily loaded questions, we talk about what it means to be hailed and interpellated into heteronormative logics. "When are you getting married?" "When are you going to have kids?" "When are you going to give up the 'single thing?'" "Are you gay or straight?" "A boy or a girl?" The pedestrian, and often well-intentioned (or at least thought to be neutral or acceptable) hailings work to trap and discipline queer (and other) persons into a hegemonic *Cosmo* quiz, with only two or three imaginable answers. Thus, we liberate the question as training for facing the forces outside (and sadly too often inside) the classroom walls. How to queer the question is inspired by the work of Kate Bornstein and marks a form of queer worldmaking—a project that Fox suggests is "the presentation of alternate worldviews that run alongside, rather than replace, master narratives."[49]

"Are you going to get married?"
"Oh sure, as often as I can."
"When are you going to have kids?"
"Never, unless you die and decide to throw me into a *Beaches* scenario with your offspring."
"When are you going to give up the 'single thing?'"
"For about two lovely hours tonight, assuming all three of them make it."
"Are you gay or straight?"
"Hmmm. Depends. What day is today?"
"Are you a boy or a girl?"
"It all depends on the lighting."

The daily encounters with the hailing of heterosexuality, for some, however, cannot risk such overt resistance. Systems, pressures, and violences force queers to adopt a broad range of strategies in the face of heteronormative pulls. For some, bold and playful resistance is a strategy. For others, when survival is on the line, resistances takes different forms, approaches, and volumes. When positioned within systems of power and control "the glance is sometimes all we got."[50] Barrett and Killen theorize the worldmaking potential of a glance, a side eye, as captured by the photograph of Doug Jones' out gay son, as his father is sworn into the Senate by the homophobic Ken doll VP Mike Pence. That glance, captured in the photo, upset a system and norm, offering a defiant opening to another voice, community and world.

Beyond the daily questions and utterances we encounter on streets, in classrooms, and at dinner tables, the hailings of heteronormative discourse emerge in the ads we are bombarded with, the pop culture we consume and artists we allow into our ear buds. Thus, our very reading practices mark an important space of queer resistances, as queer spectatorship and queer engagements with mainstream—and assumed (by many) to be heterosexual—texts mark a space of contest, invention, connection, and creation. From Mary Rogers calling for us to see how Barbie is a drag queen, to Doty's read of the *Wizard of Oz* as lesbian fantasia, the queer read of popular culture is not only a playful call to see the world through queer eyes (though it is surely that). It is also the work of survival, of insistence that we are here, and we see one another, speak to and with one another, offering hope and support and community—even in the most isolating spaces.

Yet the work of queering is not only about deliberate and conscious choices of resistance. Some bodies, in some spaces, always already carry a threat to normative systems by queering existing structures. The body can queer. Dress, makeup, and gender play can queer. Bodily gesture, movement, attitude, and affect can queer, thus challenging systems and resulting in threat and/or violence. Muñoz's highly influential theorization of disidentifications traces such negotiated strategies for queers of color who, in the face of normative hailing and command, locate spaces, modes, and performances of resistance. Extending Muñoz's project, and Chávez's call to continue queering the work of intercultural communication,[51] Eguchi and Asante examine how disidentification strategies are enlisted in liminal spaces through the embodied performances of queer of color transnational immigrants, presenting "complex, contested, contradictory, and dynamic intersections and interplays between assimilation and resistance."[52]

How to queer, how queerness is performed and enacted, and what the project of queer worldmaking is and can be shifts and moves across different cultural spaces. It is important to practice cultural humility when considering queer practices across global contexts,[53] in order to actively resist and disrupt "epistemic narcissism in which global queering is all about or only happening in relation to white men or the Western gay male gaze."[54] Rather, Lim calls for a *glocal* approach that regularly attends to the animated interplay between global and local histories, logics, practices, and tensions.

In our research in Nairobi, the term queer, itself, as well as the acronym SOGIEM (Sexual Orientation, Gender Identity and Expression Minorities) marked an important term of inclusivity for many women in Kenya who expressed strong alienation and invisibility in the LGBT community, as well as a resistance to LGBTQ as a Western label.[55] Much of the queer work participants discussed revolved around challenging and upsetting sedimented mythologies about homosexuality (itself being thought of as a choice, a Western form of indoctrination or recruitment). For the lesbian-identified women we worked with, many found the problematic and oppressive discourses surrounding feminine (homo)sexuality to provide spaces of negotiated queer resistance. For example, we spoke with queer women who used personal storytelling to construct themselves as agents, sources of queer power, and cultural tricksters. Some women talked about how pervasive norms of lesbian invisibility and erasure actually allowed and enabled the existence of same-sex relationships without raising family suspicions. For women in Kenya, where same-sex behaviors are commonly demonized and identity reduced to choice, the very statement of speaking desire, speaking one's worth, and declaring one's value marks a radical act of queer resistance.[56]

Other glocal approaches are described in Huang and Brouwer's interview study of queer Chinese interviewees, where they found a cultural tension between coming out and coming home. Wherein coming out marks the simultaneous embrace of one's sexuality, this act and gesture culturally works in immediate tension with one's family and the notion of home. Thus, these two ideas sit in binary tension. As a queer negotiation in this space, the concept of *coming with* "combines the preservation of space for one's queer sexuality with tactics that stay with the family either by cultivating parental harmony or actively interrogating heteronormative family structures."[57] It is a mode of disidentification, contextualized negotiation, and a queering of existing cultural and discursive systems in order to craft liveable lives.[58] Facing similar but distinct familial pressures in Vietnam, gay men have developed distinctive linguistic features that serve to create community, allow them to speak freely when among non-LGBTQ folks, and foster playfully queer resignification.[59]

Conclusion: always a question

> This project begins and ends with questions, working to trouble foundations rather than necessarily solidify new ones.
>
> Queer Praxis authors[60]

Queer praxis means different things, to different persons, at different moments in time. The *Queer Praxis* book project began prior to the election of Barack Obama as the 44th president of the United States, before the passing of Prop 8 in California. Our initial concerns were tied to the dangers of marriage equality being an all-encompassing focus for LGBT rights movements, and yet these concerns were tempered and troubled through differing voices, positions, experiences, and histories. While writing this chapter in 2019, with Trump as president of the United States, the threats of the active practices of government-sanctioned trans erasure, and refugee abuse at the southern border, we witnessed different contexts for LGBTQ worldmaking emerge across space and time. Still, the tensions of anger/decorum remain. Still, the struggles of coalition are real, and the power of normativity to shame and discipline queer lives is ever-present. Many of the questions we started with remain—questions about building community, building coalition, holding a community into accountability, questioning how we hear, honoring the present with our pasts, demanding we interrogate our longings of future, our meanings of age, and how we negotiate the many institutions, roles, relations, and systems of decorum that rub up against our queer angers, queer rage, and queer desires. Statements of coherence remain fraught and suspicious, except for the need to stay open, look outward, and resist the lure to stop resisting.

In the end, this chapter argues that any attempts to explain what queer is, or what queer does, are better served by offering radically contextualized questions, frustrations, and concerns. What is queer praxis? What do we need it to be in this moment, in this place? What can it be? How might being queer be "both an identity and a position relative to oppressive state power?"[61] A body in motion? A moment? A relation? A glance? An opportunity seized or lost? A risk? A temporal orientation? A question? Queerness moves and shifts and functions in different spaces, through different bodies, in different relations in ways that, from its conception, cannot be tailor-made for me, or an "I," or even an assumed "Us." The "I" is fallible, and those who work to draw the lines of "Us" run into equally fallible and exclusionary trappings of limited vision and scope. Queer praxis, like Muñoz's conception of queer futurity, requires space to move, shift, play, and refuse fixity. Paradoxically abstracted in theory and grounded in specifics, queer praxis must also continually shift and expand to account for a precious, precarious, and ever-present frustration, a contested and fleeting potentiality, an imaginary—a "we" and a then and a there longed for, gestured to, but never—absolutely never—possessed.

Notes

1 Dustin Goltz, et al., *Queer Praxis: Questions for LGBTQ Worldmaking* (New York: Peter Lang, 2015), 12.
2 Aimee Carrillo-Rowe in Goltz et al., *Queer Praxis*, 4.
3 Eve Kosofsky Sedgwick, *Epistemology of the Closet* (Berkeley, CA: University of California Press, 1990).
4 Susan Stryker, "Transgender History, Homonormativity, and Disciplinarity," *Radical History Review* 100 (2018): 145–157.
5 Frederick C. Corey, "Gay Life/Queer Art," *Canadian Journal of Political and Social Theory* 16, nos 1–3 (1993): 80.
6 Robert Gutierrez-Perez and Luis Andrade, "Queer of Color Worldmaking: <Marriage> in the Archive and the Embodied Repertoire," *Text & Performance Quarterly* 38, no. 1–2 (2018): 1–18.
7 Bryant Keith Alexander, "Queer(y)ing the Postcolonial Through the West(ern)," in *Handbook of Critical and Indigenous Methodologies*, eds Norman K. Denzin, Yvonna S. Lincoln, and Linda T. Smith (Thousand Oaks, CA: Sage, 2008), 101–133, 108.
8 Cherie Moraga and Gloria Anzaldúa, *This Bridge Called My Back: Writings by Radical Women of Color* (2nd edn) (New York: Kitchen Table/Women of Color Press, 1983).
9 Ragan Fox, "'Homo'-Work: Queering Academic Communication and Communicating Queer in Academia," *Text & Performance Quarterly* 33, no. 1 (2013): 64.
10 Dwight Conquergood, "Performance Studies: Interventions and Radical Research," *The Drama Review* 46, no. 2 (2002): 145–56.
11 Fox, "'Homo'-Work," 68.
12 Frederick C. Corey, "The Personal: Against the Master Narrative," In *The Future of Performance Studies: Visions and Revisions,* ed. Sheron J. Dailey (Annandale, VA: National Communication Association, 1998), 249–253.

13 Gust Yep, "The Violence of Heteronormativity in Communication Studies: Notes on Injury, Healing and Queer World-Making," in *Queer Theory and Communication: From Disciplining Queers to Queering the Discipline(s)*, eds Gust A. Yep, Karen E. Lovaas and John P. Elia (New York: Harrington, 2003), 11–60.

14 Goltz et al., *Queer Praxis*, 13.

15 Moraga and Anzaldúa, *This Bridge*.

16 Goltz et al., *Queer Praxis*, 1.

17 Goltz et al., *Queer Praxis*, xi.

18 Kimberlee Pérez and Daniel C. Brouwer, "Potentialities and Ambivalences in the Performance of Queer Decorum," *Text & Performance Quarterly* 30, no. 3 (2010): 319.

19 Fox, "'Homo'-Work."

20 Kimberlee Pérez and Dustin B. Goltz, "Treading Across Our 'Lines in the Sand': Performing Bodies in Coalitional Subjectivity," *Text & Performance Quarterly* 30, no. 3 (2010): 247–268.

21 Robert Gutierrez-Perez and Luis Andrade, "Queer of Color Worldmaking: <Marriage> in the Archive and the Embodied Repertoire," *Text & Performance Quarterly* 38, no. 1–2 (2018): 1–18.

22 Dustin B. Goltz, "Frustrating the 'I': Critical Dialogic Reflexivity with Personal Voice," *Text & Performance Quarterly* 31 (2011): 386–405.

23 Judith Butler, *Bodies That Matter: On the Discursive Limits of "Sex"* (New York: Routledge, 1993).

24 Mark Aguhar, "Litanies to My Heavenly Brown Body", March 13, 2012, http://culturaldisruptions. blogspot.com/2012/03/litanies-to-my-heavenly-brown-body.html.

25 Aimee Carrillo-Rowe, *Queer Praxis*, 5.

26 Soyini D. Madison, *Critical Ethnography: Methods, Ethics, and Performance* (2nd edn). (Thousand Oaks, CA: Sage, 2012), 6.

27 Madison, *Critical Ethnography*, 6.

28 Madison, *Critical Ethnography*, 6

29 *Queer Eye. Queer as Folk*. See also Jay Clarkson, "Contesting Masculinity's Makeover: Queer Eye, Consumer Masculinity, and 'Straight-Acting' Gays," *Journal of Communication Inquiry* 29, no. 3 (2005): 235–255.; Dustin B. Goltz, *Queer Temporalities in Gay Male Representation: Tragedy, Normativity, and Futurity* (New York: Routledge, 2010); Wendy Peters, "Pink Dollars, White Collars: *Queer as Folk*, Valuable Viewers, and the Price of Gay TV," *Critical Studies in Media Communication* 28, no. 3 (2011): 193–212.; Katherine Sender, "Queens for a Day: *Queer Eye for the Straight Guy* and the Neoliberal Project," *Critical Studies in Media Communication* 23, no. 2 (2006): 131–151.; Robert Westerfelhaus and Celeste Lacroix "Seeing 'Straight' Through *Queer Eye*: Exposing the Strategic Rhetoric of Heteronormativity in a Mediated Ritual of Gay Rebellion."

30 Stryker, "Transgender."

31 Lisa Duggan, "The New Homonormativity: The Sexual Politics of Neoliberalism," in *Materializing Democracy: Towards a Revitalized Cultural Politics*, eds Russ Castronovo and Dana D. Nelson (Durham, NC: Duke University Press, 2002), 174–194.

32 Stryker, "Transgender," 145.

33 Stryker, "Transgender," 146.

34 Sedgwick, *Epistemology*.

35 Butler, *Bodies*.

36 Jose Esteban Muñoz, *Cruising Utopia: The then and there of Queer Futurity* (New York: New York University Press, 2009).

37 Lee Edelman, *No Future: Queer Theory and the Death Drive* (Durham, NC: Duke University Press, 2004).

38 Alexander in Goltz et al., *Queer Praxis*, 208.

39 Gingrich-Philbrook and Gray in Goltz et al., *Queer Praxis*, 241–245.

40 Goltz et al., *Queer Praxis*, 13.

41 Ann Russo, *Feminist Accountability: Disrupting Violence and Transforming Power* (New York: New York University Press, 2018).

42 Goltz et al., *Queer Praxis*, 154.

43 Goltz et al., *Queer Praxis*, 164.

44 Cathy Cohen, "The Radical Potential of Queer? Twenty Years Later," *GLQ: A Journal of Lesbian and Gay Studies* 24, no. 1 (2019): 143.

45 Malhotra in Goltz et al., *Queer Praxis*, 138–139.

46 Stephen King, *The Dark Tower I: The Gunslinger* (New York: Signet, 1982), 269.

47 Goltz et al., *Queer Praxis*, 12.

48 Goltz et al., *Queer Praxis*, 12.

49 Fox, "'Homo'-Work," 62.

50 Joshua Trey Barrett and Brandon S. Killen, "Catching Site: Queer Worldmaking in a Glance," *QED: A Journal in GLBTQ Worldmaking* 5, no. 2 (2018): 30.

51 Karma Chávez, "Pushing Boundaries: Queer Intercultural Communication," *Journal of International and Intercultural Communication* 6, no. 2 (2013): 84.

52 Shinsuke Eguchi and Godfried Asante, "Disidentifications Revisited: Queer(y)ing Intercultural Communication," *Communication Theory* 26 (2016): 183.

53 Goltz et al., "Discursive Negotiations."

54 Lim, "Glocalqueering," 386.

55 Dustin Goltz et al., "Discursive Negotiations of Kenyan LGBTI Identities: Cautions in Cultural Humility," *Journal of International and Intercultural Communication* 9, no. 2 (2016): 104–121; Jason Zingsheim et al., "Narrating Sexual Identities in Kenya: 'Choice,' Value, and Visibility," *Journal of Lesbian Studies* 21, no. 2 (2017): 151–168.

56 Zingsheim et al., "Narrating Sexual Identities."

57 Shuzan Huang and Daniel C. Brouwer, "Coming Out, Coming Home, Coming with: Models of Queers Sexuality in Contemporary China," *Journal of International and Intercultural Communication* 11, no. 2 (2018): 107.

58 Huang and Brouwer, "Coming Out," 110.

59 Tri Hoang Dang, "A Preliminary Study of Gay Spoken Language in Ho Chi Minh City," *Language in India* 13, no. 9 (2013): 464.

60 Goltz et al., *Queer Praxis*, 12.

61 Cohen, "Radical Potential," 143.

Bibliography

Alexander, Bryant K. "Queer(y)ing the Postcolonial through the West(ern)." In *Handbook of Critical and Indigenous Methodologies*, edited by Norman K. Denzin, Yvonna S. Lincoln, and Linda T. Smith, 101–133. Thousand Oaks, CA: Sage, 2008.

Aguhar, Mark. "Litanies to My Heavenly Brown Body." http://culturaldisruptions.blogspot.com/2012/03/litanies-to-my-heavenly-brown-body.html March 13, 2012.

Barrett, Joshua Trey, and Brandon S. Killen, Brandon S. "Catching Site: Queer Worldmaking in a Glance." *QED: A Journal in GLBTQ Worldmaking* 5, no. 2 (2018): 23–31.

Bornstein, Kate. *My Gender Workbook*. New York: Routledge, 1998.

Butler, Judith. *Bodies that Matter: On the Discursive Limits of "Sex"*. New York: Routledge, 1993.

Chávez, Karma. "Pushing Boundaries: Queer Intercultural Communication." *Journal of International and Intercultural Communication* 6, no. 2 (2013): 83–95.

Clarkson, Jay. "Contesting Masculinity's Makeover: Queer Eye, Consumer Masculinity, and 'Straight-Acting' Gays." *Journal of Communication Inquiry* 29, no. 3 (2005): 235–255.

Cohen, Cathy. "The Radical Potential of Queer? Twenty Years Later." *GLQ: A Journal of Lesbian and Gay Studies* 24, no. 1 (2019): 140–144.

Corey, Frederick C. "The Personal: Against the Master Narrative." In *The Future of Performance Studies: Visions and Revisions*, edited by Sheron J. Dailey. Annandale, VA: National Communication Association, 1998, 249–253.

Corey, Frederick C. "Gay Life/Queer Art." *Canadian Journal of Political and Social Theory* 16, no. 1–3 (1993): 67–81.

Conquergood, Dwight. "Performance Studies: Interventions and Radical Research." *The Drama Review* 46, no. 2 (2002): 145–156.

Dang, Tri Hoang. "A Preliminary Study of Gay Spoken Language in Ho Chi Minh City." *Language in India* 13, no. 9 (2013): 448–467.

Doty, Alexander. *Making Things Perfectly Queer*. Minneapolis, MN: University of Minnesota Press. 1993.

Duggan, Lisa. "The New Homonormativity: The Sexual Politics of Neoliberalism." In *Materializing Democracy: Towards a Revitalized Cultural Politics*, edited by Russ Castronovo and Dana D. Nelson, 175–194, Durham, NC: Duke University Press, 2002.

Edelman, Lee. *No Future: Queer Theory and the Death Drive*. Durham, NC: Duke University Press. 2004.

Eguchi, Shinsuke, and Godfried Asante. "Disidentifications Revisited: Queer(y)ing Intercultural Communication." *Communication Theory* 26 (2016): 171–189.

Fox, Ragan. "'Homo'-Work: Queering Academic Communication and Communicating Queer in Academia." *Text & Performance Quarterly* 33, no. 1 (2013): 58–76.

Goltz, Dustin B. "Frustrating the 'I': Critical Dialogic Reflexivity with Personal Voice." *Text & Performance Quarterly* 31 (2011): 386–405.

Goltz, Dustin B. *Queer Temporalities in Gay Male Representation: Tragedy, Normativity, and Futurity*. New York: Routledge, 2010.

Goltz, Dustin B., Jason Zingsheim, Teresa Mastin, and Alexandra G. Murphy. "Discursive Negotiations of Kenyan LGBTI Identities: Cautions in Cultural Humility." *Journal of International and Intercultural Communication* 9, no. 2 (2016):104–121.

Goltz, Dustin. B., Jason Zingsheim, Kimberlee Pérez, Aimee Carrillo-Rowe, Raechel Tiffe, and Meredith Bagley. *Queer Praxis: Questions of LGBTQ Worldmaking*, edited by Dustin B. Goltz and Jason Zingsheim, New York: Peter Lang, 2015.

Gutierrez-Perez, Robert, and Luis Andrade. "Queer of Color Worldmaking: <Marriage> in the Archive and the Embodied Repertoire." *Text & Performance Quarterly* 38, no. 1-2 (2018): 1–18.

Huang, Shuzan, and Daniel C. Brouwer. "Coming Out, Coming Home, Coming with: Models of Queers Sexuality in Contemporary China." *Journal of International and Intercultural Communication* 11, no 2 (2018): 97–116.

King, Stephen. *The Dark Tower I: The Gunslinger*. New York: Signet, 1982.

Lim, Eng-Beng. "Glocalqueering in New Asia: The Politics and Performing Gay in Singapore." *Theatre Journal* 57, no 3 (2005): 383–405.

Madison, Soyini D. *Critical Ethnography: Methods, Ethics, and Performance* (2nd edn). Thousand Oaks, CA: Sage, 2012.

Moraga, Cherie, and Gloria Anzaldúa. *This Bridge Called My Back: Writings by Radical Women of Color* (2nd edn). New York: Kitchen Table: Women of Color Press, 1983.

Muñoz, Jose. E. *Cruising Utopia: The then and there of Queer Futurity*. New York: NY University Press, 2009.

Pérez, Kimberlee, and Daniel C. Brouwer. "Potentialities and Ambivalences in the Performance of Queer Decorum." *Text & Performance Quarterly* 30, no. 3 (2010): 317–323.

Pérez, Kimberlee, and Dustin B. Goltz. "Treading Across Our 'Lines in the Sand': Performing Bodies in Coalitional Subjectivity." *Text & Performance Quarterly* 30, no. 3 (2010): 247–268.

Peters, Wendy. "Pink Dollars, White Collars: *Queer as Folk*, Valuable Viewers, and the Price of Gay TV." *Critical Studies in Media Communication* 28, no. 3 (2011): 193–212.

Queer as Folk. Created by Ron Cowen and Daniel Lipman. 2000. Showtime.

Queer Eye for the Straight Guy. Created by David Collins. 2003. Bravo.

Rogers, Mary F. *Barbie Culture*. London: Sage, 1999.

Russo, Ann. *Feminist Accountability: Disrupting Violence and Transforming Power*. New York: New York University Press, 2018.

Sedgwick, Eve Kosofsky. *Epistemology of the Closet*. Berkeley, CA: University of California Press, 1990.

Sedgwick, Eve Kosofsky. "Paranoid Reading and Reparative Reading: Or, You're So Paranoid, You Probably Think this Introduction Is About You." In *Novel Gazing: Queer Readings in Fiction*, 1–37, Durham: Duke University Press, 1997.

Sender, Katherine. "Queens for a Day: Queer Eye for the Straight Guy and the Neoliberal Project." *Critical Studies in Media Communication* 23, no. 2 (2006): 131–151.

Stryker, Susan. "Transgender History, Homonormativity, and Disciplinarity." *Radical History Review* 100 (2018): 145–57.

Westerfelhaus, Robert, and Celeste Lacroix. "Seeing 'Straight' through Queer Eye: Exposing the Strategic Rhetoric of Heteronormativity in a Mediated Ritual of Gay Rebellion." *Critical Studies in Media Communication* 23, no. 5 (2006): 426–444.

Yep, Gust A. "The Violence of Heteronormativity in Communication Studies: Notes on Injury, Healing and Queer World-Making." In *Queer Theory and Communication: From Disciplining Queers to Queering the Discipline(s)*, edited by Gust A. Yep, Karen E. Lovaas and John P. Elia, 11–60, New York: Harrington, 2003.

Zingsheim, Jason, Dustin. B. Goltz, Alexandra G. Murphy, and Teresa Mastin. "Narrating Sexual Identities in Kenya: 'Choice,' Value, and Visibility." *Journal of Lesbian Studies* 21, no. 2 (2017): 151–168.

34

COMMUNICATING GENDER ADVOCACY

Riding the fourth wave of feminism

Sarah Jane Blithe and Mackenna Neal

Feminism is for everybody.

bell hooks[1]

Feminism is having a new moment. Merriam-Webster declared feminism as the word of the year in 2017,[2] and across political affiliations, high-profile women have screamed their identity as feminists. This moment has grown through the ashes of postfeminism and from the seeds of third-wave feminists, who advocated for intersectional feminisms and multiple versions of woman/personhood. In this chapter, we discuss the recent upsurge in gender justice projects and argue that these action-oriented movements are indicative of a different kind of feminism; a fourth wave of feminist advocacy that presupposes intersectionality, prioritizes both targeted and comprehensive social justice agendas, *and yet* also calls for progress for "women" broadly defined. We argue that gender advocates in the fourth wave have collaborative agendas around sexual harassment and gendered violence, and a renewed momentum toward social justice.

Feminist waves

The term "feminism" dates back to 1895, but gender advocacy has existed for centuries.[3] Feminist advocacy has morphed and changed through its history, organizing around particular issues and moments in time, dependent on other societal and cultural ideologies and events, and influenced by technology. Continuous throughout is the concept of feminism in waves. The metaphor has been critiqued as matrophobic and colorblind, representing mostly white feminism.[4] As we will argue, postfeminism and third-wave feminism both grappled extensively with broadening the scope of concerns with which feminist advocates must grapple. While the postfeminists certainly tried to distance themselves from second wave, primarily white feminism, we argue that fourth-wave feminists largely assume that differences among humans exist, are important, and should be represented in the scope of feminism writ large. At the same time, fourth-wave feminist activism is still concerned with equality and representation.

When a wave crashes, it does not dissipate into thin air; rather, it recedes and becomes part of the next wave. It is in this way that postfeminist, postcolonial, and third-wave critiques of second-wave feminism are still relevant for fourth-wave advocates. Joined with sophisticated critiques of the latter movements as well, fourth-wave activism seems to embrace inclusivity and social justice while still grappling with exclusion and privilege. This tension is highlighted in the 2019 Women's March, which was canceled in some cities because the participants were too white[5] or anti-sematic.[6] Much work remains in decolonizing white feminism, and most fourth-wave advocates have retained the importance of this lineage in their work.

First wave

Although the first wave of feminism is often characterized as the movement to get voting rights for women, advocates in the first wave also worked tirelessly to get women recognized as humans worthy of consideration beyond their roles as daughters and wives. In 1792, Mary Wollstonecraft wrote *Vindication of the Rights of Woman*, which advocated for a woman's right to education.[7] She appealed to the patriarchy by saying, "If woman isn't fitted by education to become man's companion, she will stop the progress of knowledge."[8] Wollstonecraft marketed educating women as a benefit to men. By using this communicative strategy, she lessened the threat to men.[9] Although Wollstonecraft's reputation was diminished through her advocacy, her legacy and feminist beliefs lived on through the work of many upcoming scholars.[10]

Feminism in the United States is typically marked as beginning at the Seneca Falls Convention in 1848.[11] Activists Elizabeth Cady Stanton, Fredrick Douglass, Lucretia and James Mott, and Martha Wright, among others, came together with 300 people to discuss "social, civil, and religious conditions and rights of woman" and the right to vote.[12] At the convention, the Declaration of Sentiments put into writing that men and women are created equal and are entitled to the same basic rights, serving as a catalyst for the first wave of feminism in the United States.[13]

Influential women during this first wave advocated to change laws about divorce, rights for married women to own property, and, of course, the right to vote.[14] The battle of suffrage began in the 1830s and continued for almost 100 years.[15] It was not an easy battle: women were divided over joining the fight for gender equality versus preserving traditional gender role differences—a division that has stayed constant through all waves of feminism and has prolonged the end to gender inequality.

The first wave of feminism came from primarily privileged white women. It focused heavily on education for women so that they could help contribute to society. Women were eventually granted the right to vote, but this did not mean that every woman was able to vote or wanted to. Black women during the turn of the early 20th century were enslaved and were considered second-class citizens. Black feminists like Sojourner Truth, Fredrick Douglass, Maria Stewart, Frances E. W. Harper, Anna Julia Cooper, Ida B. Wells, and Mary Church Terrell advocated for women of color during the 19th century, but most history books do not recount their advocacy into the history of first-wave feminism.[16] It is important to say the names of these feminists and recognize their contributions to feminism as the fourth wave crests.

Also excluded from the narrative of first-wave feminism were Indigenous women, despite some strong feminist contributions. For example, the Navajo Origin story *The Changing Woman*[17] describes the celebration of the eternal and ever-changing strength and beauty of women. The story remains important for young girls as they become women and is itself activist writing. Reverend Anna Shaw's 1891 paper "Woman Versus the Indian" argued that the rights of women and people of color would come to fruition together,[18] and Waheenee's 1921 "An Indian Girl's Story Told by Herself to Gilbert L. Wilson" accounts for a way of life, gone forever at the expense of white settlers.[19] White suffragists advocating for the rights of Indigenous women have also been overlooked in mainstream reporting of feminist works at this time. For example, Matilda Joslyn Gage protested the forced citizenship and subsequent sale of tribal lands in 1878,[20] yet these efforts are often overlooked in the discourse of suffrage. Rather than advocating for the right to vote per se, Indigenous advocates during this time focused on human rights to life and land.

The first wave of feminism faded after women were legally allowed to vote, and advocacy efforts remained relatively quiet for a few decades. World War II and the economic status of the United States are generally credited with suppressing feminist efforts during this time. Nine years after women gained the right to vote, the United States faced "conditions that were among the most difficult and chaotic in its history," the Great Depression, which lasted from 1929 to the start of World War II in 1939.[21] World War II gradually pulled the United States out of its economic downturn, but with the war came an "all-hands-on-deck" approach. With war looming over the world, issues of gender inequality were put on the back burner.

The popularity of Rosie the Riveter during World War II is sometimes credited as a brief uptick in feminist activity during the war years. In actuality, the notion of Rosie the Riveter, the strong, iconic woman worker, is a "modern day myth" for feminism.[22] Rosie the Riveter was simply a recruitment strategy to help get women into the workforce while men were away at war. Although this "historical accident" did empower women to enter the workforce, the second wave of feminism did not arrive until the 1960s.[23]

Second wave

In 1963, a new book swept through the United States, reigniting feminism. Betty Friedman's (1963) *Feminine Mystique* captivated upper-middle-class white women. In the 15 years after World War II, many privileged women took on the role of housewives: cooking, cleaning, raising children, and supporting their husbands' careers. As Friedman described, women seemed to lose their identities and purpose during this postwar "bliss." The *Feminine Mystique* described this general malaise as the *problem with no name*[24]— women's experiences in their roles as supporters—and inspired women to find identities beyond their roles as wives and mothers. Like the first wave, the second wave emerged out of the need to leave the domestic lifestyle.

Although advocacy in the second wave of feminism is colloquially known for focusing on the right to work, feminists were remarkably successful in a number of gender justice areas. In particular, feminist advocacy in the second wave focused on disrupting the patriarchy by improving political representation, defying gender roles, increasing reproductive rights, and resisting the subordination, subservience, and exploitation of women.[25] However, advocacy for workplace rights was particularly successful.

The first half of the second wave of feminism was successful in promoting equality for white women in the workplace, but was focused almost exclusively on the issues of white, middle-class women. However, at the turn of the decade, women of color were not only participating in the mainstream feminist organizations like the National Organization of Women (NOW) but actively created their own organizations that took on the intersection of race and feminism.[26] Although the term *intersectionality* was not yet introduced, identifying the advocacy of second-wave feminists beyond the white, heterosexual, middle-class feminists, is, of course, imperative to understanding how third-wave feminism developed. Hijas de Cuauhtemoc, The National Black Feminist Organization, Third World Women's Alliance, National Alliance of Black Feminists, Black Women Organized for Action, and many more women of color oriented organizations started during this second wave.[27]

The Third World Women's Alliance (TWWA) was founded in 1968 with the intentions of advocating for a feminism that supported women of color.[28] They incorporated social justice issues into their mission and educated others on the second layer of issues women of color face.

The Third World Women's Alliance "brought the struggles, condition, and statues of women in Latin American, Asia, Africa, and the Middle East to the forefront."[29] This organization fought to deconstruct feminism and sought to meet the needs of women of color. It is organizations like TWWA that worked to include all voices during the second wave of feminism.

Advocacy in the second wave focused around reproductive rights, equal access to work, and equal opportunities in educational spaces. The longest and hardest fight during the second wave of feminism was the Equal Rights Amendment. This amendment to the U.S. Constitution would have ensured legal gender equality.[30] In 1972, Congress passed the Equal Rights Amendment, leaving ratification to the individual states.[31] After 10 years of trying to get at least 75% of the 50 states to ratify this amendment, second-wave feminists came up three states short. It was, perhaps, the heartbreak of not seeing this amendment come to fruition, combined with serious critiques of second-wave feminism ideology, that ultimately ended the second wave of feminism.

Postfeminism and the third wave

After the success of the second wave of feminism, gender equality advocates again experienced a dramatic lull in activism from the late 1970s until the early 1990s. Eventually, two related camps of feminist thought emerged: postfeminism and third-wave feminism.[32] Around 1990, postcolonial feminists such as Gayatri Spivak, T. Minha Trinh, and Chandra T. Mohanty, and other feminist theorists such as Judith Butler and Donna Haraway fully interrogated the essentialist nature of the second wave and moved feminist thought toward a more inclusive representation of women of different races, sexualities, and bodies.[33] Theorists Julia Kristeva, Hélène Cixous, Catherina Clément, and Laura Mulvey are often labeled as postfeminists as well, in part for their conceptual work with gender deconstruction and identity. Naomi Wolf is perhaps the most identifiable face of postfeminism. Her understanding of a postfeminist position sees failure in earlier feminist movements to capitalize on these gains, and in her later work, she called for women to seize their own power.[34]

British scholars Imelda Whelehan, Angela McRobbie, Ros Gill, and U.S. journalist Susan Faludi described postfeminism as backlash against the achievements of the first and second waves of feminism, and saw postfeminist agendas as undermining gender equity progress and hypersexualizing the images of thin women.[35] Ringrose defined postfeminism as "part backlash, part cultural diffusion, part repressed anxiety over shifting gender orders."[36] Uneasy about the direction of the feminist movement, many activists found themselves in an uncomfortable limbo. Had they achieved all they wanted in the pursuit of gender justice? As Germaine Greer put it in the introduction of her book, "The future is female, we are told. Feminism has served its purpose and should now eff off. Feminism was long hair, dungarees and dangling earrings; postfeminism was business suits, big hair and lipstick;...ostentatious sluttishness and disorderly behavior."[37] Generally speaking, the postfeminist debate centers around images of victimization, autonomy, and responsibility, assuming that women, newly empowered through second-wave feminism, act as independent agents.

Increasingly, young women started to distance themselves from the politics of postfeminism and instead embraced third-wave feminism. Third-wave feminism is distinctly different from the essentialism of the second wave, and uniquely individual. Third-wave feminism is more optimistic than postfeminism and characterized by high beauty standards, consumer culture, and consumption, and the assumption that women are individuals, in charge of their own destiny. Postfeminism is markedly different, critical of the progress third-wave feminists too easily claimed.[38] Thus, while postfeminists worried that flippant commercials would undo the hard work of the first and second waves, third-wave feminists often embraced (and found power in) hyper-feminine identities.

Third-wave feminism accepted pluralism as a given. As Sarah Gamble described,

> We know that what oppresses me may not oppress you, that what oppresses you may be something I participate in, and that what oppresses me may be something you partici-
> pate in. Even as different strands of feminism and activism sometimes directly contradict
> each other, they are all part of our third wave lives, our thinking, and our praxes.[39]

Both the postfeminist and the third-wave feminist movement are rife with contradictions. They are at once concerned with continued violence against women yet resistant to cast women as "victims" of assault and harassment; grappling with inequities in the paid and unpaid labor forces yet insistent that women *choose* their work-life destiny; against objectification of women yet reveling in visible body positivity and hyper-sexualized confidence; and heteronormative but sexually freer than previous generations. Perhaps the most enduring contradiction to emerge from these social movements is the acute focus on individual problems and issues, which do not require collective action.[40] As Julia Schuster argued,

> Everyday feminism and related branches of third wave feminism, such as DIY-feminism,
> lipstick-feminism, feminist online activism and power-feminism all focus on the pri-
> vate rather than the public sphere of women's lives and therefore are less suited to the
> broader mobilization of a women's movement.[41]

It is here that differences between the activist of the third wave and the fourth wave of feminism are most obvious.

The fourth wave of feminism

In recent years, there has been a noticeable uptick in collective feminist movements.[42] Unlike the third-wave feminists, who built political alliances primarily for issue-specific or individual concerns rather than long-lasting coalitions, fourth-wave feminism is much more collaborative. Gender activism is different now than in previous waves in three primary ways: (1) the mode of organizing and message sharing through social media has drastically changed the face of feminist activism; (2) fourth-wave feminists are particularly concerned with reducing gender-based violence and sexual harassment and assault; and (3) fourth-wave feminists presuppose intersectionality and advocate for different groups of individuals. Flowing through social media, and with powerful supporters in high-profile positions, gender activism in the last few years is loud and collective, and demands accountability for gender-based violence.

Because the third wave focused on diverse experiences and was informed by Black feminism, postcolonial feminism, queer and intersectionality theories, international and transnational fem-inism, and anti-essentialist understandings of "women," fourth-wave feminists begin with the understanding that gender identities are fluid, multiple, intersectional, and situated in systems of power.[43] Fourth-wave feminist advocacy presupposes intersections and embraces differences among women.[44] They have demonstrated a renewed momentum toward social justice for the everyday lives of individual women. At the same time, these feminists simultaneously collaborate in order to make progress for "women" broadly defined.

Previous waves of feminism laid much legal groundwork for gender equality, yet discrimin-ation and oppression continue. Most women are still paid less than men, mothers face continued discrimination in the workplace, and sexual assaults on college campuses and at work remain rampant. Women of color, LGBTQ+ individuals, women with disabilities, and low-income white women experience discrimination at incredibly high rates. Thus, while previous waves have made some progress toward greater equality, renewed passion to drive forward gender advocacy is welcome. The tides of the fourth wave are increasingly visible around the world, making both tidal and incremental changes in the name of social justice. Increasingly, this advocacy often occurs through social media.

Fourth-wave feminism and social media

Advocacy groups increasingly use social media sites to improve their outreach and messaging efforts.[45] Social media platforms, especially Facebook and Twitter, have facilitated a number of international social movements in recent years, changing the means through which advocacy groups function. For example, antigovernment protests in Brazil, Indonesia, Bulgaria, and Turkey; revolutionary protests during the Arab Spring; the It Gets Better Campaign for LGBTQ+ youth empowerment; and Occupy Wall Street and #MeToo in the United States all used social media to mobilize participants, frame news and events, and incite people to join the movements.[46]

However, how social media is used for advocacy is still in the early stages of research, and remains quite contested.[47] Some scholars see social media as an effective tool for grassroots advocacy, with the ability to connect and empower people, foster faster communication, and reach dispersed populations. At the same time, others suggest that social media campaigns rarely generate action, and worse, foster false sense of participation, or *clicktivism*.[48]

In their study of how advocacy groups use social media, Jonathan Obar, Paul Zube, and Clifford Lampe found that most organizations place great importance on social media platforms and use either volunteers or paid employees to manage mobilization, message framing, and daily communication.[49] The participants in their study found that Facebook and Twitter were the most effective for civic engagement and collective action. Chao Guo and Gregory Saxton also found Facebook and Twitter were the most used social media sites by advocacy organizations.[50] They found that advocacy groups used these platforms to communicate informational messages, community messages, action messages, support messages, and strategic messages—highlighting the many ways and reasons advocacy groups can use social media.

Fourth-wave feminist activism has evolved to deeply embrace social media.[51] Like other advocacy groups, feminist activists use social media sites to organize, discuss, share information, protest, coordinate, and network.[52] Social media has enabled a major increase in feminism, breathing new life into how women share their stories, join together, and shine light on important feminist issues. Fourth wavers are incredibly savvy in their use of social media, creating campaigns to solve both micro and macro gender-based problems.[53]

Social media has enabled fourth-wave feminism to connect across national borders, and thus audiences for feminist issues are global.[54] When organizing happens at an international scale, and in view of large publics, action and problem solving can happen faster.[55] Put simply, social media has sped up the process of feminist activism, broadened the audience for feminist messages, connected multiple groups of women, and yet also made feminism accessible for individual everyday women.

While most scholarship and popular press coverage positions social media as a benefit to advocacy, feminist activism on social media is particularly vulnerable to harassment, flaming, trolling, and e-bile.[56] These can include threats of violence, rape, and accusations of ugliness or a lack of intelligence, all designed to silence feminist activism through social media. Larisa Mann cautions that the publicity, visibility, and connectedness promised from social media may not always be beneficial and can make it difficult to manage images and messages. She argues, "Traditional mass media channels (and academia) have historically excluded, silenced, or heavily mediated/edited the words of Black women and others on the wrong side of hegemonic power."[57] Thus, while social media does open space for wider audiences and provides opportunities for women of color to bring their voices into important conversations with mainstream feminists, it is important to be mindful of the ways in which power and representation operate through the transparency and openness of social media.

Sexual harassment and assault advocacy

Recent feminist activism has sprung up around a variety of issues, but as the fourth wave takes shape, activism to prevent and reduce gendered violence, sexual harassment, and sexual assault is coalescing into primary focus. This focus is evident through the global movement #MeToo, and repercussions of the movement are particularly obvious in the political scene of the United States. As these cases demonstrate, activism to reduce sexual assault and harassment moved through social media, reaching wide audiences and resulting in social justice action. Activism against sexual harassment and assault has empowered the pursuit of gender equality and continuously fueled the fourth wave of feminism.

Intersectional and decolonial activism

Having learned from their postfeminist and third-wave predecessors, fourth-wave feminists presuppose the importance of intersectionality and are moving toward decolonizing feminism. A number of campaigns emerged to support women from marginalized groups around the world, and advocacy for women previously left out of mainstream feminism is a priority in the fourth wave. As mentioned earlier in this chapter, decolonizing white feminism is far from complete, and women of color, women with disabilities, and LBTQ+ individuals are continually marginalized, oppressed, and left out of many advocacy campaigns, organizations, and the media. However, advocacy for inclusion and equality is heightened, and, as evidenced in the recent 2016 U.S. presidential election, finally gaining traction. A few high-profile movements highlight this aspect of fourth-wave feminism.

Some recent intersectional movements specifically focus on race (discussed here as it is constructed in the United States). For example, Black Lives Matter began in response to the acquittal of George Zimmerman, who murdered Trayvon Martin in 2012, during a time of multiple shootings of young Black men in the United States. The movement is feminist and intentionally intersectional, insisting on inclusivity for Black women, Black LGBTQ+ individuals, Black people with disabilities, or undocumented status.[58]

In another example, the Lakota People's Law Project has a number of advocacy campaigns. First, Empowering Native Voters is geared toward increasing voter rights for Indigenous people. Second, Reuniting Children & Families protects families from state separation and the removal of children from tribal lands. Finally, Defending Dissent protects native rights.[59]

In addition to advocacy for particular race/ethnic groups around the world, the fourth wave has also seen a significant increase in advocacy for LGBTQ+ people. Although the Human Rights Campaign (HRC) originated in 1980, it has expanded its advocacy to support a wide variety of issues in recent years, and is the largest advocacy group for LGBTQ+ rights.[60] The organization is responsible for iconic campaigns #LoveConquersHate (supporting LGBTQ Russian Olympians during the 2014 Games) and #AsktheGays (to identify political agendas, in response to President Trump's challenge to "ask the gays" about where he stood on LGBTQ equality). The HRC was also instrumental in the fight for marriage equality, and now continues to work to reduce discrimination for LGBTQ+ people in workplaces, family law, and politics. Another campaign, the It Gets Better Project, seeks to empower and connect LGBTQ+ youth through uplifting stories of acceptance and welcome.[61] In another example, the National Center for Transgender Equality advocates for transgender and nonbinary people across a multitude of issues.[62]

Other fourth-wave feminist movements specifically advocate around particular issues that affect women from marginalized groups. For example, Daughters of Eve creates awareness and provides support for women at risk of female genital mutilation.[63] Free a Girl advocates to free minor girls from prostitution, particularly in Bangladesh, Brazil, India, Iraq, Laos, the Netherlands, Nepal, and Thailand.[64] At the same time, the Red Umbrella Project advocates for sex workers rights.[65] Girl Up focuses on a multitude of issues and is region specific so that it can best aid local women around the world. It aims to create equality for girls in education, health, and economic viability. While it began as a movement for girls in the United States, it is now borderless and has specific initiatives to support girls in Liberia, India, Guatemala, Ethiopia, Uganda, and Malawi. It supports refugee programs, leadership education, violence prevention, ending child marriage and pregnancy, improved nutrition, sustainable development, and documentation.[66] Ni Una Mas is a movement to prevent femicides in Mexico,[67] and United We Dream provides support and advocacy for undocumented youth in the United States.[68]

Significant momentum has occurred in the effort to decolonize feminism and in decentering white feminism. Fourth-wave feminists take an intersectional lens and apply it often to unjust conditions in society. While goals of intersectional feminism are not yet met, fourth-wave activists continue to make women of color, women with disabilities, and LBTQ+ individuals part of feminist campaigns.

Fourth-wave backlash

As is true for all social justice movements, fourth-wave feminism has experienced some backlash. Some people have critiqued the continued use of the wave metaphor. Others have accused fourth-wave feminism as slacktivism and for continuing to exclude groups of people who should be included. Conservative backlash, concerns about call-out culture, and a belief that society no longer needs feminism are also critiques leveled at fourth-wave feminism.

As discussed earlier, the wave metaphor is highly critiqued as matraphobic,[69] unclear,[70] and reductive.[71] Often, waves are considered to be monolithic, discrete periods, in which newer wave feminists critique previous waves and fight for united causes. This conceptualization of feminism and its movements is inaccurate, and overlooks how waves continue and blend into each other. It also overlooks the multitude of feminisms and projects occurring within each wave. Feminist arguments exist simultaneously, yet disagreements about if there is a fourth wave (or a third wave, for that matter) and convoluted dates and definitions complicate understanding feminist activism as waves.[72] There is much debate among Anglo-Saxon academics and activists regarding the true definition of the third wave of feminism. In addition, it can be difficult to apply the wave metaphor on a global scale because waves did not occur at the same time in different regions, nor did they focus on the same project.[73]

As mentioned, fourth-wave feminism has been ridiculed for *clicktivism*[74] or *slacktivism*.[75] Because it is defined in large part by social media, concerns about whether or not change is actually accomplished is often part of the discourse against fourth-wave feminism. It may seem like less of a commitment to action that people are willing to show support to a cause by signing a petition, sharing a post, or wearing a bracelet. There is little cost or sacrifice in this sort of activism, and critics worry that it reduces the occurrence of more engaged activist agendas.

Although inclusivity and intersectionality are cornerstones of fourth-wave feminism, the movement has been critiqued for persistent exclusion. Some critiques have pointed out that fourth-wave feminism is only inclusive in Western feminist movements and that it still focuses primarily in the global north, ignoring other regions.[76] Ragna Rök Jóns argued that disproportionate access to social media devices creates an inherently classist, ableist, and privileged space.[77] Some politicians, public figures, and feminists have expressed particularly transphobic rhetoric, claiming that feminism is only for "real" women. This faction pushes against the inclusivity of the fourth wave and seeks to refocus efforts on cis, heterosexual, white women.[78]

Conservative backlash to fourth-wave feminism often focuses on the inclusion of LGBTQ+ individuals. In some geographic regions, trans-exclusionary activists reject queer feminism. Others reject gender studies altogether. From this conservative backlash is a strong contingent of men's rights—a carryover from previous feminist movements. Fourth-wave men's rights activists claim that men are forgotten when there is a focus on women's advancement.[79] Others have pointed out that men are suffering from false sexual assault accusations and from a lack of opportunity when women take their jobs.[80]

Another issue with fourth-wave feminism lies in *call-out culture*. In part based on social media, the tendency to aggressively call people out on missteps is part of the fabric of fourth- wave feminism.[81] The emergence of *privilege checking*, open letters, high-profile media, and other forms of conscious raising has been critiqued as too aggressive to be effective. These critiques are often in defense of white feminism, patriarchy, men, and people engaging in so-called innocuous, everyday sexism. Other critics claim that while white feminism and patriarchy should be critiqued, strong allies are lost when they fear public shaming for accidental missteps.[82]

Finally, fourth-wave feminism has experienced some backlash because people believe women have achieved equality. Those drawing on this critique claim that girls outperform boys in school at all levels and that women hold many positions of power in all sectors.[83] This argument has plagued many feminist movements, in multiple iterations, but cries that feminism is over have only intensified in recent years.

Conclusion

To conclude, there are clear demarcations in recent feminist advocacy that are different from previous waves. First, the means through which people advocate are strikingly different. Social media has completely changed how people organize, frame messages, and share information. Second, the new wave of feminism is particularly concerned with reducing gender-based violence and sexual harassment and assault, as evident in the #MeToo movement. Finally, because fourth-wave feminists already assume intersectionality, advocacy for myriad groups of women has emerged to significantly broaden the scope and mission of feminist advocacy. The fervor with which fourth wavers advocate is inspiring; yet through their passion and reach, it is clear there remains much work to do in the quest for gender equality.

Looking forward, we expect to see continued activism against sexual harassment and assault, and in political representation. We imagine that feminist activists will pursue justice for women who have not been sufficiently represented by earlier waves of feminism, including women of color, LGBTQ+ individuals, sex workers, and immigrant women. We also envision a global impact of feminist activism, as women rally around issues in local contexts that both differ and connect with simultaneous feminist movements. It is likely that greater attention to toxic masculinity and how men are implicated in patriarchy and gender norms will help include men in feminist movements, and how environmental impacts affect people in a variety of ways. This list is at once global and local, expansive yet promising, as the seeds of these movements have been planted and are beginning to grow. While we remain optimistic that fourth-wave feminism can overcome some of the challenges of earlier waves, addressing multiple audiences and issues, feminist activists must continue to be vigilant in efforts of inclusivity and engagement.

Notes

1 bell hooks, *Feminism Is for Everybody: Passionate Politics* (Cambridge: South End Press, 2000), 1.
2 "Merriam-Webster's 2017 Words of the Year." Merriam-Webster, n.d., accessed March 10, 2019, www.merriam-webster.com/words-at-play/word-of-the-year-2017-feminism.
3 Valerie Sanders, "First Wave Feminism," in *The Routledge Companion to Feminism and Post Feminism*, ed. Sarah Gamble (New York: Routledge), 2006, 15–24; Ian Buchanan, *A Dictionary of Critical Theory* (Oxford: Oxford University Press, 2010).
4 Lynn O'Brien Hallstein, *White Feminists and Contemporary Maternity: Purging Matrophobia* (Springer, 2010).
5 Michael Brice-Saddler, "California Women's March Rally Canceled Over Concerns that It Would Be 'Overwhelmingly White,'" *Washington Post*, January 5, 2019, accessed January 6, 2019, www.denverpost.com/2019/01/05/california-womens-march-rally-canceled-overwhelmingly-white/.
6 Valerie Richardson, "Organizers Cancel Women's March in New Orleans as Support Drops 'Drastically,'" *The Washington Times*, December 31, 2018, accessed January 6, 2019, www.washingtontimes.com/news/2018/dec/31/organizers-cancel-womens-march-new-orleans/.
7 Valarie Sanders, *First Wave Feminism* (New York: Routledge, 2006), 15.
8 Ibid., 16.
9 Sanders, "First Wave Feminism," 2006.
10 Ibid., 17.
11 Ibid., 17.
12 Judith Wellman, "The Seneca Falls Women's Rights Convention: A Study of Social Networks," *Journal of Women's History* 3 (1991): 9, accessed December 2019.
13 Elizabeth Stanton, "The Declaration of Sentiments," *National Park Service*, February 2015, 2, accessed April 4, 2019.
14 Ibid., 20.
15 Sanders, "First Wave Feminism," 2006; Paula Baker, "The Domestication of Politics: Women and American Political Society, 1780-1920," *The American Historical Review* 89, no. 3 (1984): 620–647.
16 Elizabeth Alexander, "'We Must Be About Out Father's Business': Anna Julia Cooper and the Incorporation of the Nineteenth-Century African-American Woman Intellectual," *Signs: Journal of Women in Culture and Society* 20, no. 2(1995): 336–365; Wilma Peebles-Wilkins and E. Aracelis Francis, "Two Outstanding Black Women in Social Welfare History: Mary Church Terrell and Ida B. Wells," *Women and Social Work* 5, no. 4(1990): 87–100; Hazel V. Carby, "'On the Threshold of Woman's Era': Lynching, Empire, and Sexuality in Black Feminist Theory," *Critical Inquiry* 12, no. 1 (1985): 262–277.
17 Anonymous, "The Changing Woman" in Kolmar, Wendy K., and Frances Bartkowski. "Feminist Theory: A Reader." (1999).
18 Mann, Susan Archer, and Ashly Suzanne Patterson, *Reading Feminist Theory: From Modernity to Postmodernity* (New York: Oxford University Press, 2016).
19 Ibid., 422.
20 Ibid., 409.
21 Ben S. Bernanke, "Non-Monetary Effects of the Financial Crisis in the Propagation of the Great Depression," *The American Economic Review*, 73, no. 3 (1983): 257–276.
22 James Kimble and Lester Olson, "Visual Rhetoric Representing Rosie the Riveter: Myth and misconception," in J. Howard Miller's "We Can Do It!" poster," *Rhetoric and Public Affairs* 9 (2006): 533–569.
23 Sanders, "First Wave Feminism," 2006; Urszula M. Prunchniewska, "'A Crash Course in Herstory": Remembering the Women's Movement in MAKERS: Women Who Make America," *Southern Communication Journal* 82, no. 4 (2017): 228–238.
24 Betty Friedman, *The Feminine Mystique* (New York: Norton, 1963).
25 Mel Gray and Jennifer Boddy, "Making Sense of the Waves: Wipeout or Still Riding High?," *Affilia* (2010); Julia Schuster. "Why the Personal Remained Political: Comparing Second and Third Wave Perspectives on Everyday Feminism." *Social Movement Studies* 16 (2017): 647–659.
26 Becky Thompson, "Multiracial Feminism: Recasting the Chronology of Second Wave Feminism," *Feminist Studies* 28 (2002): 336–360.
27 Kimberley Springer, "The Interstitial Politics of Black Feminist Organizations," *Meridians* 1, no. 2 (2001): 155–191.

28 Smith College Archive, Third World Women's Alliance records, accessed April 7, 2019, https://asteria. fivecolleges.edu/findaids/sophiasmith/mnsss527.html.

29 Ibid., para 2.

30 Barbara Brown, Thomas Emerson, Gail Falk, and Ann Freedman, "The Equal Right Amendment: A Constitutional Basis for Equal Right for Women," *The Yale Law Journal* 80 (1971): 871–981.

31 Sarah Soule and Susan Olzak, "When Do Movements Matter? The Politics of Contingency and the Equal Rights Amendment," *American Sociological Review* 69 (2004): 473–497.

32 Sarah Gamble, *The Routledge Companion to Feminism and Postfeminism* (New York: Routledge, 2004).

33 Gayatri Spivak, "'Can the Subaltern Speak?" in *Marxism and the Interpretation of Culture*, eds C. Nelson L. Grossberg (Chicago, IL: University of Illinois Press, 1988), 271–317; Minha T. Trinh, *Woman Native Other* (Bloomington, IN: Indiana University Press, 1989); Chandra T. Mohanty, "Under Western Eyes," in *The Postcolonial Studies Reader*, eds B. Ashcroft, G. Griffiths, and H. Tiffin (London: Routledge, 1995); Judith Butler, *Gender Trouble* (New York: Routledge, 1990); Donna Haraway, *Simians Cyborgs and Women* (London: Free Association Books, 1991); Angela McRobbie "Post-feminism and Popular Culture," *Feminist Media Studies* 4, no. 3 (2004): 255–264.

34 Gamble, *The Routledge Companion to Feminism and Postfeminism*.

35 Imelda Whelehan, *Modern Feminist Thought: From the Second Wave to Post-Feminism* (New York: New York University Press, 1995; Angela McRobbie, *The Aftermath of Feminism: Gender, Culture and Social Change* (Thousand Oaks, CA: Sage, 2009); Ros Gill, "From Sexual Objectification to Sexual Subjectification: The Resexualization of Women's Bodies in the Media," *Feminist Media Studies* 3, no. 1(2007): 100–105; Ros Gill, "Post-Feminist Media Culture: Elements of a Sensibility," *European Journal of Cultural Studies* 10, no. 2 (2003): 147–166; Andrea L. Press, "Feminism and Media in the Post-Feminist Era: What to Make of the 'Feminist,'" *Feminist Media Studies* 11 (2011): 107–113.

36 Jessica Ringrose, "A New Universal Mean Girl: Examining the Discursive Construction and Social Regulation of a New Feminine Pathology," *Feminism & Psychology* 16, no. 4 (2006): 2.

37 Germaine Greer, *The Whole Woman* (London: Doubleday, 1999).

38 Press, Feminism and Media in the Post-Feminist Era.

39 Sarah Gamble, ed. *The Routledge Companion to Feminism and Postfeminism* (London/New York: Routledge, 2004), 43.

40 Natalie Fixmer and Julia T. Wood, "The Personal Is Still Political: Embodied Politics in Third Wave Feminism," *Women's Studies in Communication* 28, no. 2 (2005): 235–257; Maura Kelly, "Feminist Identity, Collective Action, and Individual Resistance Among Contemporary U.S. Feminists," in *Women's Studies International Forum*, vol. 48 (Oxford: Pergamon, 2015), 81–92.

41 Julia Schuster, "Why the Personal Remained Political: Comparing Second and Third Waves Perspectives on Everyday Feminism", *Social Movement Studies* 16 (2017): 650.

42 Kira Cochrane, *All the Rebel Women: The Rise of the Fourth Wave of Feminism*, vol. 8 (London: Guardian Books, 2013); Ruxandra Looft. "#Girlgaze: Photography, Fourth Wave Feminism, and Social Media Advocacy," *Continuum* 31, no. 6 (2017): 892–902.

43 Ibid.

44 Cochrane, All the Rebel Women, 2013.

45 Adam Chalmers and Paul Shotton, "Changing the Face of Advocacy? Explaining Interest Organizations' Use of Social Media Strategies," *Political Communication* 33, no. 3 (2016): 374–391; Erica L. Ciszek, "Advocacy Communication and Social Identity: An Exploration of Social Media Outreach," *Journal of Homosexuality* 64, no. 14 (2017): 1993–2010; Chao Guo and Gregory Saxton, "Tweeting Social Change: How Social Media are Changing Nonprofit Advocacy," *Nonprofit and Voluntary Sector Quarterly* 43, no. 1 (2014): 57–79.

46 Chalmers and Shotton, "Changing the Face," 2016; Ciszek, "Advocacy Communication," 2017; Jonathan Obar, Paul Zúbe, and Clifford Lampe, "Advocacy 2.0: An Analysis of How Advocacy Groups in the United States' Perceive and Use Social Media as Tools for Facilitating Civic Engagement and Collective Action," *Journal of Information Policy* 2 (2012): 1–25.

47 Chalmers and Shotton, 2016; Ella McPhearson, "Advocacy Organizations' Evaluation of Social Media Information for NGO Journalism: The Evidence and Engagement Models," *American Behavioral Scientist* 59, no. 1 (2015).

48 Chalmers and Shotton, "Changing the Face," 2016; Ciszek, "Advocacy Communication," 2017; Obar, Zube, and Lampe, "Advocacy 2.0," 2012; Jeffrey W. Treem, and Paul M. Leonardi, "Social Media Use in Organizations: Exploring the Affordances of Visibility, Editability, Persistence, and Association," *Annals of the International Communication Association* 36, no. 1 (2013): 143–189.

49 Obar, Zube, and Lampe, "Advocacy 2.0," 2012.

50 Guo and Saxton, "Tweeting Social Change," 2014.

51 Cochrane, *All the Rebel Women*, 2013; Looft, 2017.

52 Jessamy Gleeson, "'(Not) working 9-5': The Consequences of Contemporary Australian-Based Online Feminist Campaigns as Digital Labour," *Media International, Australia* 121, no. 1 (2016); Jessica Megarry, "Under the Watchful Eyes of Men: Theorizing the Implications of Male Surveillance Practices for Feminist Activism on Social Media," *Feminist Media Studies* 18, 6 (2018): 1070–1085.

53 Cochrane, 2013; Looft, "#Girlgaze," 2017.

54 Jennifer Baumgardner, "Is There a Fourth Wave? Does it Matter?" *Feminist.com*, 2011, accessed January 2, 2019, www.feminist.com/resources/artspeech/genwom/baumgardner2011.html.

55 Megarry, "Under the Watchful Eyes of Men," 2018.

56 Gleeson, "'(Not) working 9-5,'" 2016.

57 Larissa K. Mann, "What Can Feminism Learn from New Media?," *Communication and Critical/Cultural Studies* 11 (2014), 124.

58 Alicia Garza, "A Herstory of the# BlackLivesMatter Movement," *Are All the Women Still White* (2014): 23–28; Sarah J. Jackson. "(Re) Imagining Intersectional Democracy from Black Feminism to Hashtag Activism," *Women's Studies in Communication* 39, no. 4 (2016): 375–379.

59 Lakota's People Law Project, accessed December 20, 2018, www.lakotalaw.org/our-campaigns.

60 Human Rights Campaign, accessed December 20, 2018, www.hrc.org/hrc-story/about-us.

61 It Gets Better Project, accessed December 20, 2018, https://itgetsbetter.org/.

62 National Center for Transgender Equality, accessed December 20, 2018, https://transequality.org/.

63 Daughters of Eve, accessed December 20, 2018, www.dofeve.org/.

64 Free a Girl, accessed December 20, 2018, www.freeagirl.nl/en/.

65 NSWP Global Network for Sex Work Projects, accessed December 20, 2018, www.nswp.org/members/red-umbrella-project.

66 Girl Up, accessed December 20, 2018, www.girlup.org/#sthash.WTKc0Zxq.dpbs.

67 Justice in Mexico, accessed December 20, 2018, https://justiceinmexico.org/femicidesinmexico/.

68 United We Dream, accessed December 20, 2018, https://unitedwedream.org/about/.

69 Hallstein, *White Feminists and Contemporary Maternity*, 2010.

70 Ragna Rök Jóns, "Is the '4th Wave' of Feminism Digital?," *Bluestockings Magazine*, August 19, 2013, accessed December 9, 2019, http://bluestockingsmag.com/2013/08/19/is-the-4th-wave-of-feminism-digital/.

71 Constance Grady, "The Waves of Feminism, and Why People Keep Fighting Over Them, Explained," *Vox,* July 20, 2018, accessed December 9, 2019, www.vox.com/2018/3/20/16955588/feminism-waves-explained-first-second-third-fourth.

72 Elizabeth Evans and Prudence Chamberlain, "Critical Waves: Exploring Feminist Identity, Discourse and Praxis in Western Feminism," *Social Movement Studies* 14, no. 4 (2015): 396–409; Stacy Gillis and Rebecca Munford, "Genealogies and Generations: The Politics and Praxis of Third Wave Feminism," *Women's History Review* 13, no. 2 (2004): 165–182.; Grady, 2018.

73 Jóns, "Is the '4th Wave'," 2019.

74 Jeffrey Treem and Paul M. Leonardi, "Social Media Use in Organizations," 2013.

75 Kristofferson, Kirk, Katherine White, and John Peloza, "The Nature of Slacktivism: How the Social Observability of an Initial Act of Token Support Affects Subsequent Prosocial Action," *Journal of Consumer Research* 40, no. 6 (2013): 1149–1166.

76 Ealasaid Munro, "Feminism: A Fourth Wave?" *Political Studies Association*, September 5, 2013, accessed December 9, 2019, www.psa.ac.uk/psa/news/feminism-fourth-wave; Evans and Chamberlain, 2015.

77 Jóns, "Is the '4th Wave'," 2019.

78 Jessica Abrahams, "Everything You Wanted to Know About Fourth Wave Feminism—But Were Afraid to Ask," *Prospect Magazine*, August 14, 2017, accessed December 9, 2019, /www.prospectmagazine.co.uk/magazine/everything-wanted-know-fourth-wave-feminism.

79 Ibid.

80 Joanna Williams, "Fourth Wave Feminism: Why No One Escapes," *The American Conservative*, September 4, 2018, accessed December 9, 2019, www.theamericanconservative.com/articles/fourth-wave-feminismwhy-no-one-escapes/.

81 Munro, "Feminism," 2013.

82 Ibid.

83 Williams, "Fourth Wave Feminism," 2018; Jóns, "Is the '4th Wave'," 2019.

Bibliography

Abrahams, Jessica. "Everything You Wanted to Know About Fourth Wave Feminism—But Were Afraid to Ask," *Prospect Magazine*, August 14, 2017. Accessed on December 9, 2019. www.prospectmagazine. co.uk/magazine/everything-wanted-know-fourth-wave-feminism.

Alexander, Elizabeth. "'We Must Be about Our Father's Business': Anna Julia Cooper and the Incorporation of the Nineteenth-Century African-American Woman Intellectual." *Signs: Journal of Women in Culture and Society* 20, no. 2 (1995): 336–365.

Baker, Paula. "The Domestication of Politics: Women and American Political Society, 1780- 1920." *The American Historical Review* 89, no. 3 (1984): 620–647.

Baumgardner, Jennifer. "Is There a Fourth Wave? Does It Matter?" *Feminist.com* (2011). Accessed on January 2, 2019. www.feminist.com/resources/artspeech/genwom/baumgardner2011.html.

Brice-Saddler, Michael. "California Women's March Rally Canceled Over Concerns That It Would Be 'Overwhelmingly White.'" *Washington Post*, January 5, 2019. Accessed on January 6, 2019. www. denverpost.com/2019/01/05/california-womens-march-rally-canceled-overwhelmingly-white/.

Brown, Barbara, Emerson, Thomas, Falk, Gail, and Freedman, Ann. "The Equal Right Amendment: A Constitutional Basis for Equal Rights for Women." *The Yale Law Journal* 80, no. 5 (1971): 871–981.

Buchanan, Ian. *A Dictionary of Critical Theory*. Oxford: Oxford University Press, 2010.

Butler, Judith. *Gender Trouble*. New York: Routledge, 1990.

Carby, Hazel V. "'On the Threshold of Woman's Era': Lynching, Empire, and Sexuality in Black Feminist Theory." *Critical Inquiry* 12, no. 1 (1985): 262–277.

Carillo-Rowe, Amy. *Powerlines: On the Subject of Feminist Alliances*. Durham, NC/London: Duke University Press, 2008.

Chalmers, Adam, and Shotton, Paul. "Changing the Face of Advocacy? Explaining Interest Organizations' Use of Social Media Strategies." *Political Communication* 33, no. 3 (2016): 374–391.

Cixous, Hélène, and Clément, Catherine. *The Newly Born Woman*, trans. Betsy Wing. Manchester: Manchester University Press, 1986.

Cizek, Erica L. "Advocacy Communication and Social Identity: An Exploration of Social Media Outreach." *Journal of Homosexuality* 64, no. 14 (2017): 1993–2010.

Cobb, Shelly, and Horeck, Tanya. "Post-Weinstein: Gendered Power and Harassment in the Media Industries." *Feminist Media Studies 18*, no. 3 (2018): 489–508.

Cochrane, Kira. *All the Rebel Women: The Rise of the Fourth Wave of Feminism*. Vol. 8. London: Guardian Books, 2013.

Cooney, Samantha. "Here Are All the Public Figures Who've Been Accused of Sexual Misconduct After Harvey Weinstein." *Time Magazine* (2018). Accessed on December 9, 2019. http://time. com/5015204/harvey-weinstein-scandal/.

Daughters of Eve (2018). Accessed on December 20, 2018. www.dofeve.org/.

Deslippe, Dennis. "Organized Labor, National Politics, and Second-Wave Feminism in the United State, 1965–1975." *International Labor and Working-Class History*, no. 49 (1996): 143–165. www.jstor.org/stable/27672282.

Evans, Elizabeth, and Prudence Chamberlain. "Critical Waves: Exploring Feminist Identity, Discourse and Praxis in Western Feminism." *Social Movement Studies* 14, no. 4 (2015): 396–409.

Faludi, Susan. *Backlash: The Undeclared War Against Women*. London: Vintage, 1992.

Fixmer, Natalie, and Julia T. Wood. "The Personal Is Still Political: Embodied Politics in Third Wave Feminism." *Women's Studies in Communication* 28, no. 2 (2005): 235–257.

Friedan, Betty. *The Feminine Mystique*. New York: Norton, 1963.

Gamble, Sarah, ed. *The Routledge Companion to Feminism and Postfeminism*. London/New York: Routledge, 2004.

Garza, Alicia. "A Herstory of the# BlackLivesMatter Movement." *Are All the Women Still White* (2014): 23–28.

Gelb, Joyce. *Feminism and Politics: A Comparative Perspective*. Berkeley, CA: University of California Press, 1989.

Gibson, Katie. "In Defense of Women's Rights: A Rhetorical Analysis of Judicial Dissent." *Women's Studies in Communication* (2012): 123–137.

Gill, Ros. "From Sexual Objectification to Sexual Subjectification: The Resexualization of Women's Bodies in the Media." *Feminist Media Studies* 3, no. 1 (2003): 100–105.

Gill, Ros. "Post-Feminist Media Culture: Elements of a Sensibility." *European Journal of Cultural Studies* 10, no. 2 (2003): 147–166.

Gillis, Stacy, and Rebecca Munford. "Genealogies and Generations: The Politics and Praxis of Third Wave Feminism." *Women's History Review* 13, no. 2 (2004): 165–182.

Gleeson, Jessamy. "'(Not) Working 9–5': The Consequences of Contemporary Australian-Based Online Feminist Campaigns as Digital Labour." *Media International, Australia* 121, no. 1 (2016): 77–85.

Grady, Constance. "The Waves of Feminism, and Why People Keep Fighting Over Them, Explained." *Vox*, July 20, 2018. Accessed on December 9, 2019. www.vox.com/2018/3/20/16955588/feminism-waves-explained-first-second-third-fourth.

Gray, Mel, and Boddy, Jennifer. "Making Sense of the Waves: Wipeout or Still Riding High?" *Affilia* (2010).

Greensburg, K. "Still Hidden in the Closet: Trans Women and Domestic Violence." *Berkeley Journal of Gender, Law, and Justice* 27, no. 2 (2012): 198–251.

Greer, Germaine. *The Whole Woman.* London: Doubleday, 1999.

Guo, Chao, and Saxton, Gregory. "Tweeting Social Change: How Social Media Are Changing Nonprofit Advocacy." *Nonprofit and Voluntary Sector Quarterly* 43, no. 1 (2014): 57–79,

Hallstein, Lynn O'Brien. *White Feminists and Contemporary Maternity: Purging Matrophobia.* Springer, 2010.

Haraway, Donna. *Simians Cyborgs and Women.* London: Free Association Books, 1991.

Harris, Aisha. "She Founded Me Too. Now She Wants to Move Past the Trauma." *New York Times.* Accessed on April 6, 2019. www.nytimes.com/2018/10/15/arts/tarana-burke-metoo-anniversary.html.

hooks, bell. *Feminism Is for Everybody: Passionate Politics.* Cambridge, MA: South End Press, 2000.

Jackson, Sarah J. "(Re) Imagining Intersectional Democracy from Black Feminism to Hashtag Activism." *Women's Studies in Communication* 39, no. 4 (2016): 375–379.

Jóns, Ragna Rök. "Is the '4th Wave' of Feminism Digital?" *Bluestockings Magazine*, August 19, 2013. Accessed on December 9, 2019. http://bluestockingsmag.com/2013/08/19/is-the-4th-wave-of-feminism-digital/.

Jodi, Kantor, and Twohey, Megan. "Harvey Weinstein Paid Off Sexual Harassment Accusers for Decades." *New York Times.* Accessed on December 19, 2018. www.nytimes.com/2017/10/05/us/harvey-weinstein-harassment-allegations.html

Kelly, Maura. "Feminist Identity, Collective Action, and Individual Resistance Among Contemporary U.S. Feminists." In *Women's Studies International Forum*, vol. 48, Oxford: Pergamon, 2015, 81–92,

Kimble, James, and Olson, Lester. "Visual Rhetoric Representing Rosie the Riveter: Myth and Misconception in J. Howard Miller's 'We Can Do It!' Poster." *Rhetoric and Public Affairs* 9 (2006): 533–569.

Kristofferson, Kirk, Katherine White, and John Peloza. "The Nature of Slacktivism: How the Social Observability of an Initial Act of Token Support Affects Subsequent Prosocial Action." *Journal of Consumer Research* 40, no. 6 (2013): 1149–1166.

Lakota People's Law Project (2018). Accessed on December 20, 2018. www.lakotalaw.org/our-campaigns.

Looft, Ruxandra. "#Girlgaze: Photography, Fourth Wave Feminism, and Social Media Advocacy." *Continuum* 31, no. 6 (2017): 892–902.

Mack, Ashely Noel, and McCann, Bryan, J. "Critiquing State and Gendered Violence in the Age of #Metoo." *Quarterly Journal of Speech* 194, no. 3 (2018): 329–344.

Mann, Larisa K. "What Can Feminism Learn from New Media?" *Communication and Critical/Cultural Studies* 11, no. 3 (2014): 293–297.

McPherson, Ella. "Advocacy Organizations' Evaluation of Social Media Information for NGO Journalism: The Evidence and Engagement Models." *American Behavioral Scientist* 59, no. 1 (2015): 124–148.

McRobbie, Angela. "Post-Feminism and Popular Culture." *Feminist Media Studies* 4, no. 3 (2004): 255–264.

McRobbie, Angela. *The Aftermath of Feminism: Gender, Culture and Social Change.* Los Angeles: Sage, 2009.

Media Diversified (2018). Accessed on December 9, 2019. https://mediadiversified.org/.

Megarry, Jessica. "Under the Watchful Eyes of Men: Theorizing the Implications of Male Surveillance Practices for Feminist Activism on Social Media." *Feminist Media Studies* 18, no. 6 (2018): 1070–1085, doi: 10.1080/14680777.2017.1387584.

Me Too Movement (2018). "History & Vision." Accessed on May 16, 2019, https://metoomvmt.org/about/. https://metoomvmt.org/about/.

Mohanty, Chandra T. "Under Western Eyes." In *The Postcolonial Studies Reader*, edited by Bill Ashcroft, Gareth Griffiths, and Helen Tiffin, 259–265, London: Routledge, 1995.

Munro, Ealasaid. "Feminism: A Fourth Wave?" Political Studies Association, September 5, 2013. Accessed on December 9, 2019. www.psa.ac.uk/psa/news/feminism-fourth-wave.

Obar, Jonathan, Zube, Paul, and Lampe, Clifford. "Advocacy 2.0: An Analysis of How Advocacy Groups in the United States Perceive and Use Social Media as Tools for Facilitating Civic Engagement and Collective Action." *Journal of Information Policy* 2 (2012): 1–25.

Peebles-Wilkins, Wilma, and Aracelis Francis, E. "Two Outstanding Black Women in Social Welfare History: Mary Church Terrell and Ida B. Wells." *Women and Social Work* 5, no. 4 (1990): 87–100.

Press, Andrea L. "Feminism and Media in the Post-Feminist Era: What to Make of the 'Feminist' in Feminist Media Studies." *Feminist Media Studies* 11, no. 1 (2011): 107–113.

Prunchniewska, Urszula M. "'A Crash Course in Herstory': Remembering the Women's Movement in MAKERS: Women Who Make America." *Southern Communication Journal* 82, no. 4 (2017): 228–238.

Richardson, Valerie. "Organizers Cancel Women's March in New Orleans as Support Drops 'Drastically.'" *The Washington Times*, December 31, 2018. Accessed on January 6, 2019. www.washingtontimes.com/news/2018/dec/31/organizers-cancel-womens-march-new-orleans/.

Ringrose, Jessica. "A New Universal Mean Girl: Examining the Discursive Construction and Social Regulation of a New Feminine Pathology." *Feminism & Psychology* 16, no. 4 (2006): 405–424.

Rodino-Colocino, Michelle. "Man Up, Woman Down: Mama Grizzlies and Anti-Feminist Feminism During the Year of the (Conservative) Woman and Beyond." *Women & Language* 35, no. 1 (2012): 79–95.

Sanders, Valerie. "First Wave Feminism." In *The Routledge Companion to Feminism and Post Feminism*, edited by Sarah Gamble. New York: Routledge, 2006.

Schuster, Julia. "Why the Personal Remained Political: Comparing Second and Third Waves Perspectives on Everyday Feminism." *Social Movement Studies* 16 (2017): 647–659.

Soule, Sarah A., and Olzak, Susan. "When Do Movements Matter? The Politics of Contingency and the Equal Rights Amendment." *American Sociological Review* 69 (2004): 473–497.

Spivak, Gayatri. "Can the Subaltern Speak?" In *Marxism and the Interpretation of Culture*, edited by Cary Nelson and Lawrence Grossberg, 271–317, Chicago, IL: University of Illinois Press, 1998.

Springer, Kimberley. "The Interstitial Politics of Black Feminist Organizations." *Meridians* 1, no. 2 (2001): 155–191.

Stanton, Elizabeth. "Declaration of Sentiments – Seneca Falls." Accessed on December 19, 2018. https://liberalarts.utexas.edu/coretexts/_files/resources/texts/1848DeclarationofSentiments.pdf.

Thompson, Becky. "Multiracial Feminism: Recasting the Chronology of Second Wave Feminism." *Feminist Studies* 28 (2002): 336–360.

Treem, Jeffrey W., and Paul M. Leonardi. "Social Media Use in Organizations: Exploring the Affordances of Visibility, Editability, Persistence, and Association." *Annals of the International Communication Association* 36, no. 1 (2013): 143–189.

Trinh, T. Minha. *Woman Native Other*. Bloomington, IN: Indiana University Press, 1989.

U.S. Equal Employment Opportunity Commission (b). "Title VII of the Civil Rights Act of 1964." Accessed on December 19, 2018. EEOC.gov. www.eeoc.gov/laws/statutes/titlevii.cfm.

U.S. Equal Employment Opportunity Commission (a). "The Equal Pay Act of 1963." Accessed on December 19, 2018. EEOC.gov. www.eeoc.gov/laws/statutes/epa.cfm.

Whelehan, Imelda. *Modern Feminist Thought: From the Second Wave to 'Post-Feminism'*. New York: New York University Press, 1995.

Weaver, Hilary. N. "The Colonial Context of Violence: Reflections on Violence in the Lives of Native American Women." *Journal of Interpersonal Violence* 24, no. 9 (2009): 1552–1563.

Wellman, Judith. "The Seneca Falls Women's Rights Convention: A Study of Social Networks." *Journal of Women's History* 3 (1991): 9–37.

Williams, Joanna. "Fourth Wave Feminism: Why No One Escapes," *The American Conservative*, September 4, 2018. Accessed December 9, 2019. www.theamericanconservative.com/articles/fourth-wave-feminismwhy-no-one-escapes.

Wolf, Naomi. *The Beauty Myth*. London: Vintage, 1990. *Fire with Fire: The New Female Power and How It Will Change the 21st Century* (London: Chatto & Windus, 1993).

Zernike, Kate, and Lu, Denise. "A Surge of Women Candidates, but Crowded Primaries and Tough Races Await. *New York Times* (2018). Accessed on April 6, 2019. www.nytimes.com/interactive/2018/05/12/us/politics/women-midterm-elections.html.

35

THE OPPOSITIONAL GAZE AS SPECTACLE

Feminist visual protest movements in China

Nickesia S. Gordon and Yuhan Huang

[W]oman always has a problematic relation to the visible, to form, to structures of seeing…

Mary Ann Doane[1]

Introduction

Doane's above observation of women's relationship to looking illuminates the patriarchal practice of associating womanhood with in/visibility. It also highlights the historical construction of female identity as being somehow, naturally, dissociated from gazing. Through this disconnection, a woman becomes positioned as the recipient of the gaze, or, as John Berger (1972) observed, "an object of vision: a sight."[2] Men, on the other hand, according to this logic, are subjects and therefore instigators of the gaze. They "act" while women "appear,"[3] and subsequently, it is men who look at women while women "watch themselves being looked at."[4] Women are thus construed as passive objects of the male gaze and, unsurprisingly, as non-agentic when it comes to visuality. Recalling Doane's above reflection, it would appear that women have traditionally been "constructed differently in relation to processes of looking."[5] This gendered construction of relations to looking (hooks 1996; Mulvey 2001; Russo 1986; Shugart and Waggoner 2005) speaks to a tradition of trying to control female subjectivity, corporeally as well as intellectually. But what is it about a look, particularly from a woman, that requires such surveillance? What makes our gazes so dangerous? bell hooks points us to a simple but compelling answer: "[t]here is power in looking."[6] Looking implies knowledge, awareness, agency, activity rather than passivity, and the ability to be creative. A look can arrest, transfix, or transform. A look can be the site of a thousand rebellions. It is not surprising, therefore, that within a system of patriarchy, a woman's ability to look would be constrained. In such a system, she is mainly permitted only (if at all) to engage the gaze in certain ways—not directly, not into, not at, not back—since penetrating looks are dangerous and have historically been punished, demonized, and feared. Indeed, the influence of Sigmund Freud (1941) is never far from these gendered relations of power as it pertains to looking. A woman's gaze, according to this sensibility, is death (to a man) since it encapsulates the quintessential male fear of castration. Accordingly, a woman's ability to look can be Medusa-like, a "symbol of horror" and danger to male subjects.[7]

The gendered politics of looking becomes heightened by the scene it creates when one dares to do it. Looking signals, as Mary Russo (1986) surmises, a breech that induces the spectacular:

> For a woman, making a spectacle out of herself had more to do with a kind of inadvertency and loss of boundaries: the possessors of large, aging, and dimpled thighs displayed at the public beach, of overly rouged cheeks, of a voice shrill in laughter, or of a sliding bra strap…were at once caught out by fate and blameworthy. It was my impression that these women had done something wrong, had stepped, as it were, into the limelight out of turn—too young or too old, too early or too late—and yet anyone, any woman, could make a spectacle out of herself if she was not careful.[8]

What Russo describes above is the perceptual side of the act of looking (Zsolt Gyenge 2016), a self-reflexive response that at once recognizes the power of spectacle and the constitutive nature of an oppositional gaze. Instead of constituting the public display of "aging" and "dimpled thighs" as "wrong," spectacle arouses a new vision of woman in relation to herself. That is to say, woman is the surveyor of herself, not man. She thus turns herself into subject through the vision of her unruly, spectacular self as object of the gaze. She sees that she transgresses boundaries and makes a spectacle of herself. A woman who looks or who assumes the gaze is therefore a disorderly but visible woman. Just as there is power in looking, there is also power in the spectacle created for the gaze. Although looking is agentic, creating a spectacle may be deliberate and serve specific aims. Therein lies the critical potential of the gaze, of a woman's look, namely, "in the blatant positioning of oneself as spectacle."[9] In this regard, the gaze invokes oppositionality through attendant spectacle, a powerful site of resistance.

In this chapter, we explore how the power of the gaze and the spectacle that it generates operate as resistive forces in certain visual protest movements in China that were inspired by similar actions in the United States. For the purposes of this chapter, we interpret the gaze as mirrorlike, an inward look that interpolates patriarchal renditions of female subjectivity—and then spits them out as a reinvented, restored vision of the female self. In this sense not only does the gaze create the spectacular through the "disorder" it creates, the spectacle created in turn captures the gaze, transfixing the onlooker through this interplay. This idea is not unlike the gaze discussed in Foucault's (2002) "Las Meninas," wherein the binary between the spectator's gaze and the gaze of the art collapses into a ceaseless co-creation of intersubjectivity. This conceptualization spectacularly challenges traditional notions of female identity, particularly in the way that it draws attention to itself, or in the way that it makes the invisible visible. This idea is applied to an examination of popular protest movements and public embodiments of resistance performed by women in China. These acts of protest bring attention to and challenge normative ideals about female identity, power, and oppression that occur across the globe. They also rely heavily on social media for organizing and galvanizing efforts. But beyond the mechanics of the logistics with which social media helps, the latter also has the added effect of heightening the spectacular nature of their protests. Social media also assists in creating participatory sites of rebellion that offer alternative "mediatized" accounts of gender norms regarding both looking and making a spectacle of oneself.

In our analysis, we use an extended version of bell hooks' concept of the oppositional gaze to make the case for the rebellious nature of the visual practices that these movements internalize and also invite. Using China as a case study, we conduct a discursive analysis of how the oppositional gaze unfolds when actors respond to misogyny and patriarchy (often perpetuated through mass media) by using visual or nonverbal forms of protest. We make reference to social protests and performances such as "the wounded brides" and "occupying men's toilet" events in

2012, and other actions against gender bias and sexual harassment that consequently led to the detainment of Chinese activists later known as the "Feminist Five" in 2015. We also consider how movements in the United States such as the #MeToo movement help frame sister movements in China. Binding these different protest movements together across geographic divides is social media, which has gained a reputation in recent times for playing a key role in popular collective political actions. When investigating the unique nature of visual demonstrations in China inspired by U.S. activism, we pay attention to how the politics of visual resistance is shaped by the particularities and politics of each location, especially as it pertains to social media surveillance.

Oppositional gaze and resistance

In his seminal work, *Ways of Seeing*, John Berger (1972) describes the act of gazing as a particularly male phenomenon. According to Berger, "[m]en look at women" and "[w]omen watch themselves being looked at."[10] Through this process, women become objectified and essentially stripped of any agency, making them passive observers of their own dehumanization. Berger is rightly drawing attention to the codes embedded in visual culture that invite particular readings of women's bodies and by extension, female subjectivity, which position women as passive, sexualized, and mute. Laura Mulvey (2001) makes a similar point about the gaze in her seminal article from 1973, "Visual Pleasure and Narrative Cinema," where she argues that the "scopophilic" nature of the cinematic lens renders "[t]he image of woman as (passive) raw material for the (active) gaze of man."[11] Both Berger's and Mulvey's discussions of the gaze offer important understandings of how gender, or more precisely patriarchy, configures relations of looking. They lay bare the voyeurism, fetishism, and exploitation that characterize the male gaze. Notwithstanding these concerns, other accounts of the gaze such as that offered by bell hooks (1996) provide an alternative perspective that posit looking as an act of rebellion from the position of the dominated other. Herein lies the idea of the oppositional gaze, a look that is full of ferment and that aims to "change reality."[12] This concept not only takes up the potential of "looking back" as an act of defiance, but also critically assesses relations of power and agency entailed in this resistant practice.":

> Subordinates in relations of power learn experientially that there is a critical gaze, one that "looks" to document, one that is oppositional. In resistance struggle, the power of the dominated to assert agency by claiming and cultivating "awareness" politicizes "looking" relations—one learns to look a certain way in order to resist.[13]

In her essay "The Oppositional Gaze," from which the above excerpt derives, hooks was specifically writing about what she calls "the oppositional black gaze."[14] The latter term describes Black women's relationship to cinematic spectatorship, which hooks presents as potentially critical of and resistive to the predominant "whiteness" of Hollywood films. hooks argues that "[a]s critical spectators, Black women participate in a broad range of looking relations that contest, resist, revise, interrogate, and invent on multiple levels."[15] While acknowledging this very essential focus of hooks' work, the writers of this study also see an opportunity to extend this idea of the oppositional gaze to an examination of how visual protest movements that are female driven operate as sites of resistance in other places and other times. The gaze, as appropriated by these actors, seeks to push back against forms of patriarchal domination that invite sexual, psychological, physical, and other forms of gender-based violence. The "pleasure of interrogation"[16] that the oppositional gaze facilitates, is, in these instances, being leveled at certain state and cultural institutions that perpetuate misogyny.

Going back to hooks' statement that there is power in the act of looking, we would like to add that there is even more power in "looking *back*." That power is heightened by the fact that for women, looking back creates a particular danger. This is because "looking back" always holds the potential for punishment, via the surveillance it implies, perhaps because of the spectacle that it creates. Since looking back draws attention to oneself (thereby making the invisible visible), it produces a spotlight that the onlooker has little power to evade. It forces one to look at and to look into the once negated presence of the oppressed or invisible, whose specular presence now demands attention and also exposes them to the danger that Russo (1986) describes in her essay. The particular visual protest movements cited in this essay point to the perils women face when they "make a spectacle of themselves" in the manner described above—dangers that may include being sexually harassed on and offline or even being jailed as a result of their visual protests.

Despite the dangerous responses they may generate, these acts represent women's resistance to how they are traditionally viewed or "looked at." This is the nexus between the oppositional gaze and feminist political activism from the perspective of this essay: women daring to look at themselves in a way that challenges traditional constructions of their gender identities. The implied reflexivity of internalizing the gaze as a gateway to self-knowledge is echoed by Jean-Paul Sartre (1956) when he argued that the gaze is the frontline for the self "to define and redefine itself" since "we become aware of our self as subject only when confronted with the gaze of the Other and become aware of our self as object."[17] Just as the Black women hooks writes about grappled with and ultimately rejected identification with cinematic representations of themselves, so too do these visual protests critically interrogate patriarchal ideas about gender and invent new ways to communicate these lived experiences. This contestation plays itself out via the spectacle of the protests, which range from scenes of bloody wedding dresses to shaved heads, in the streets and social media spaces of China and the United States. In this sense, spectacle emerges as a form of political resistance that disrupts normative ideas about female identity via its denaturalization of gender (Butler 1990) and its redefinition of who can be a subject with the gaze.

Spectacle and resistance

Spectacular performances in popular culture and beyond have received their fair share of derision from scholars over time. Notable among those criticisms is Guy Debord's (2001) scathing evaluation of the role of spectacle in modern consumer culture. Not only does he view spectacular performances as ideological tools used by the capitalist elites to maintain the status quo, he also sees them as deceptive tactics used to sow disunity among ordinary citizens:

> The spectacle itself presents itself simultaneously as all of society, as part of society, and as *instrument of unification*. As a part of society it is specifically the sector which concentrates all gazing and all consciousness. Due to the very fact that this sector is *separate*, it is the common ground of the deceived gaze and of false consciousness, and the unification it achieves is nothing but an official language of generalized separation.[18]

Debord's caution has particular implications for visual culture, a domain of popular culture long associated with spectacle and mediated representations. Feminists are especially wary (and rightly so) of the hegemonic propensities of spectacle as it pertains to discourses of gender (Shugart and Waggoner 2005). This concern is not unlike Peggy Phelan's (2003) critique of contemporary culture's reliance on image or representation to make visible the "hitherto unseen."[19] This is because, as she argues, representation "fails to reproduce the real exactly,"[20] thereby creating gaps and potentially reifying image. However, as other scholars have noted, there have always been disruptions created by spectacular performances of marginalized others that display the transgressive possibilities of spectacle (Doane 1982; Russo 1986; Shugart and Waggoner 2005). It is this rendering of spectacle that the writers draw upon to illustrate the resistive power of the visual protests being examined for this essay. As Debord correctly points out, the spectacle concentrates all gazing and in doing so, we argue, manifests itself as a resistive force given women's traditional relations to looking. There is evident intent involved in the spectacle created though organized protests designed to grab media attention and capture public viewing. By deliberately making themselves the concentration of all gazing and all consciousness, women display their agency and confront patriarchal powers that assert that they should not look nor become visible by making spectacles of themselves. The spectacular performances rendered through the visual protest movements under examination thus shine a spotlight on misogyny while simultaneously resisting the "imposition of dominant ways of knowing and looking."[21]

One of the ways in which the oppositional gaze operates in these visual protest movements is through bodily resistance. That is to say, the body becomes the primary vehicle for the specular in certain events. This corporeal resistance is in line with hooks' and Foucault's observation that for oppressed groups, the body is a primary site where agency may be located. Our essay is specifically interested in the agency enacted and displayed by the unruly body. When "set loose in the public sphere,"[22] the bodies of marginalized groups such as women become unruly and disorderly. They disrupt and threaten the status quo through their spectacular presence.

The second way in which the oppositional gaze functions in these movements is through social media amplification of the specular. Most, if not all, of the visual protest movements being studied utilize social media as part of their arsenal of resistance. In doing so, visual protests manage to create what Hsiao and Yang (2018) describe as an augmented reality "where the power of digitality meets the politics of physicality."[23] Not only do social media transform the communication of these movements from a network perspective (Poell and van Dijck 2018), they also amplify the spectacle of the unruly body and intensify the oppositional gaze. We now turn to an analysis of how these phenomena unfold in the very different context of China as compared to the United States.

Digital feminist activism: China versus the United States

Information communication technologies, including social media and other digital spaces, have given rise to what scholars such as Julianne Guillard (2016) see as a fourth wave of feminist activities and practices. Human rights movements in general have been growing correlative to the rapid global expansion of the Internet. Digital technologies promise to make such movements more powerful and formidable not only through facilitating protest and political agitation among citizens but also through the rapid exportation of ideas that activate resistance to oppression among people in different parts of the world. Digital platforms such as social media sites also make solidarity for human rights issues accessible and palpable because of their strong networking capabilities, while also giving visibility to traditionally disenfranchised groups. For example, in China, as Yalan Huang (2016) writes, "[w]ith the fact that feminists are often excluded or demonized by mainstream mass media, the Internet, especially social media, is expected to be a useful medium for feminists to disseminate information, organize activities, and interact with the public."[24] As a result, feminist activists in China (as elsewhere) have taken their political agitation to the "highways" of the Internet, spaces that allow them to "deploy the precarious female body to make visible contradictions of contemporary social reality."[25] In other words, digital platforms such as social media sites facilitate the spectacle of looking back. This is a particularly dangerous enterprise for feminists in China, where digital spaces are subject to state-sanctioned surveillance and violations can lead to imprisonment…or worse. Under such circumstances, the oppositionality of the gaze is heightened given the perils associated with this digital and less-than-subtle medium for resistance. In the United States, while activists may not face jail time, they are subject to misogynist threats of violence and cyberbullying. The message behind such threats is that women, whether they are in China or the United States, do not belong in the public sphere. Women look back nonetheless.

As Turley and Fisher (2018) observe, there has been an explosion of feminists using social media and other digital spaces (such as blogs) to raise awareness and galvanize support for pushback against misogyny, gender-based violence, sexism, and a range of other issues related to patriarchy. In the United States, Twitter has been a central campaign tool for many digital feminist movements. Harnessing the power of the hashtag, many campaigns have exposed mis-ogyny and paved the way for global engagement with such issues. Examples of these include such movements as #BlackLivesMatter, #MeToo (perhaps one of the most well-known hashtags), Occupy, #WhyIStayed, #SayHerName, and #PussyHat, among many more. Other important digital feminist activism predates the rise of Twitter. For example, the annual Gladys Ricart & Victims of Domestic Violence Memorial Walk, aka the Brides' March, relied on Facebook in its early days for much of its digital activism, and more recently has used Instagram. Other movements founded by women, while not being distinctly feminist in nature, have also inspired activism on a global scale. The most prominent of these is the #Occupy movement. For the purposes of this chapter, we focus on the inspiration of the #MeToo, Brides' March, and #Occupy movements as major catalysts that fed important social movements in China. They seem to have had an especially strong impact in sparking namesake protests in China, such as the "Wounded Brides," "Occupy Men's Toilets," and the "Feminist Five." These events also starkly illuminate the power of the oppositional gaze in the spectacle they generate both on and offline, through the deploy-ment of disorderly bodies as public sites of resistance.

The oppositional gaze and unruly bodies: Chinese feminists in action

Feminism in China is found to be perceived as morally deviant, foreign-rooted, and intertwined with issues like nationalism, social polarization, and a modernization-tradition contradiction.[26]

In her book *Betraying Big Brother: The Feminist Awakening in China* (2018), advocate and scholar Leta Hong Fincher recorded a moment of resistance and self-empowerment from an interview with a detained feminist activist who looked back at her male interrogators. Li Tingting (also known as Li Maizi) was arrested by the police in 2015 at the age of 25 and taken away from her home, along with her girlfriend Teresa. At the station, one of the first things the police did was deprive them of any belongings they had, including the glasses of near-sighted Teresa. Then Li was put in a separate room and two men questioned her about the motives and funding for her activism without identifying themselves nor explaining the charges against her. In the process of this interrogation, Li, although in an extremely disadvantageous position without any kind of support, derived a sense of power and resistance by using her good eyesight to stare at the male questioners, saying, "The good thing about not being blind [being near-sighted and deprived of glasses] is that you can see the interrogators' faces [...] That's kind of a threat, because they're afraid you'll remember them."[27] Li's realization echoes what bell hooks points out about the power in looking.[28] Li Tingting's case epitomizes the struggles and strategies of Chinese feminists in the face of government censorship and control. When street protests and mass movements are not possible, the act of looking back—be it literal or metaphorical—becomes the site of resistance and self-empowerment.

Li Tingting belongs to a group of young feminists who created the most recent wave of feminist movements in China since 2012. The group of young women organized a series of regional protests including the "Wounded Bride," "Occupying Men's Toilet," and "Bald Sisters" events, focusing on social issues related to women's welfare in different aspects of life such as domestic violence, the availability of public bathrooms for women, and gender discrimination in university admissions. The activists demonstrated creative spontaneity by engaging with current events and generating striking images for newsworthy stories. At the center of these movements was a strategy that combines the use of online and offline resources, embodied in the creation and dissemination of a set of visually powerful images of women-in-action. The images circulating from the "Wounded Bride" and the "Bald Sisters" protests are not only sensational but also created a spectacle that disrupted normative gender assumptions.

Visual images play an important role in extending the influence of these Chinese feminist movements and in engaging the general public. Compared with protests in the West, most offline feminist demonstrations in China were very limited in their scale, consisting of only a small group of participants due to heavy political restraints on mass public gatherings. However, images of the feminist protests generated instant interest, receiving coverage by both the traditional news outlets and independent outlets in social media. While public protests of social problems in China are highly sensitive and heavily censored, some mainstream and official news agencies reported the visual protests staged by feminist activists in the early stages of the movements. For example, both *China Daily* and *People.com* reported on the "Wounded Bride" protests that took place at various points in 2012 with about 10 or fewer participants at each of the street gatherings, depoliticizing them as a "street performance art against domestic violence." The fact that these small-scale street protests were being reported by major official mouthpieces allowed feminist activism to be seen and taken into consideration by decision makers—and this was particularly important since the first "Wounded Bride" protest was launched to support a legal case involving domestic violence.

Inspired by the Brides' March, a few Chinese activists employed the striking visual combination of wedding and wounds of abuse to raise public awareness of domestic violence, expressing their support for Kim Lee. Kim was a victim of domestic violence who filed a legal case against her husband Li Yang, a billionaire and the founder of a highly successful company for English education. The case caused a sensation in Chinese society where the issue of domestic abuse had been often neglected. The activists wore white bridal gowns tainted with blood outside the courthouse in Chaoyang district, where Kim Lee's case was being deliberated. The group also organized further "Wounded Bride" campaigns for Valentine's Day and during the United Nations' End Violence Against Women campaign in 2012, and even started a public petition, collecting more than 12,000 signatures for introducing an anti-domestic violence law in China, the largest women's rights action to date.[29] In 2013, the final jurisdiction for the case against Li Yang ruled in favor of the plaintiff Kim Lee, granting her a divorce on the grounds of domestic violence, full custody of her child, and a sum of $1.9 million in child support. The case not only established a landmark precedent in China but also contributed to a longtime and concerted effort that led to the approval of China's first domestic violence law in 2015.

Along with persistent lobbying and campaigning by established women's rights groups, a new generation of feminists started to play an important role in Chinese feminist movements, mobilizing social media and online resources to support offline efforts. The photo of the "Wounded Bride" protest in Beijing, featuring women activists Li Maizi, Wei Tingting, and Xiao Meili in white wedding gowns tainted with blood, became an iconic image for feminist movements around this time, creating a vision of feminist women in action. In the photograph, the three activists stand in the middle of the street in Beijing, their faces showing bruises and bloodstains, holding up posters that reads, "Love is not an excuse for violence." To some audiences, these Chinese feminist actions may appear much less radical than their Western counterparts in terms of the scale and the extent of issues engaged. However, when we situate them in a society of heightened political control and censorship, it becomes clear that what these women achieved entailed immense risk. Unlike in the West, it is impossible for Chinese feminists to organize a large-scale parade. The only form of mobilization available is through a staged photo shooting with a group of limited and trusted participants, and even such cases of small-scale gatherings could be disbanded by the police or security forces. The very action of standing in the street of a country where public protests are not allowed and advocating for feminism in itself embodied the spirit of "looking back"—resisting both the patriarchal culture that condones domestic violence and the political control of anyone who dares to challenge the existing social and cultural structure.

This is why offline activism had very limited space for impact, and Chinese feminists used it mainly as a means to "make the news" (Li and Li 2017[30]; Lü 2018d). In reviewing Chinese feminist movements from 2012 to 2015, Lü Pin (a key member who worked mostly behind the scenes of many feminist initiatives) pointed out that one of their strategies was to increase visibility and reportage of women's issues in mass media: "the aim of every action was to make the news, following the logic of mass media and occupying its content."[31] A visually striking image served as the best medium to attract attention, provoke debate, and take an oppositional stance without triggering the mechanisms of censorship. By resorting to the power of images and the influence of both social and mass media, Chinese feminist movements entered a new stage of "media activism" (Wang 2018[32]; Han 2018[33]; Li and Li 2017).

Using image and social media as the backbone of these movements, new Chinese feminists invariably participated in the creation of visual spectacles, but of a different and subversive type that challenged dominant ways of representing and looking at female bodies. The images they produced in movements between the years of 2012 and 2015 such as the "Bald Sisters," "nude selfies," and "armpit hair photo contest" not only resisted normative and photogenic portrayals of female bodies, but also brought attention to important and practical issues relevant to women's welfare. An illustrative example is the "Bald Sisters" campaign, centered on the problem of gender discrimination in Chinese university admissions requiring female applicants to achieve higher scores to qualify for top universities. To protest this phenomenon and to make it more visible in the public sphere, a group of four feminists started a campaign by shaving their hair at the foot of the Confucianist Temple in Guangzhou. This was a potent protest, because hair has rich symbolic meanings in Chinese culture. In traditional China, Confucianism sees hair as a part of the physical body received from parents, so tradition dictates that it should be cared for carefully. Given this association, shaving one's hair off constitutes a gesture of total disobedience. In addition, the Chinese idiom "long-haired women have little insights" betrays blatant sexism. Aware of the cultural implications of hair in Chinese culture, the protestors chose to become bald, performing "a complete break with the traditional social female image of a woman."[34] The Guangzhou campaign was echoed by feminists in Beijing and also by anonymous online supporters who shaved their own hair and shared their "selfies" in support. According to Xiao Meili, one of the first who shaved her hair, the campaign not only exceeded expectations by receiving the support of so many strangers but also attracted attention far beyond initial imagination.[35] The women with bald heads created a powerful image of resistance, returning the male gaze that often casts and shapes what a female should look like. Similarly, the nude selfies campaign subverted the erotic image of female nudity and used the body as a site for protesting domestic violence and the "Armpit Hair Photo contest" challenged the social norm of shaving and created a set of provocative images of unruly female bodies. By voluntarily rendering themselves as spectacles and engaging in transgressive acts that sought to redefine standards for evaluating the female body, Chinese feminists brought problematic gender norms to the forefront of social attention and discussion.

This series of feminist efforts culminated in a nationwide campaign against sexual harassment on public transportation, which was planned as an event celebrating International Women's Day in 2015. However, before the campaign could take place, many feminist activists were arrested on March 6, 2015 for engaging in "subversive activities." Among them were Wei Tingting, Zheng Churan, Li Tingting, Wu Rongrong, and Wang Man (who later became known as the "Feminist Five"), detained by the Chinese government for 37 days.[36] They were accused of taking part in movements and having connections to foreign powers subversive to the country. This major crackdown led to a diminishing space for organized visual protests and online feminist movements. Nonetheless, these movements have left behind a legacy and created significant impacts vital to the development of more spontaneous feminist movements such as #MeToo in China in the years that followed.

The rise of a new generation of feminists and their bold practices engaging visual images and social media needs to be put into perspective within both historical and transnational contexts. The development of feminism in China, marked by progress and setbacks in a time of political caprice, provides a telling case of the global circulation of ideas and images related to women's activism. Representations of women and women's bodies have been central to the emergence of female subjectivities and feminist movements ever since the late 19th century when women's magazines played an essential role in fostering the image of modern women in China. In the communist era, the image of proletarian women was brought to the forefront and promoted in revolutionary propaganda posters and films while liberal feminist movements were replaced by top-down state feminism woven into the socialist nationalist discourse. The Reform era witnessed a gradual process of reintroducing femininity in an age of fast-paced globalization and commercialization. The current wave of feminist movements was thus part of a long social process, which women responded to by working to denaturalize gender images and to reverse the power relations between the gaze and the spectator.

Social media: sites of spectacular performances

Social media can be considered as a form of public space that facilitates social, political, and cultural encounters. Thus conceived, social media spaces lend themselves to "public occupation" by resistance movements. As Hsiao and Yang (2018) put it, forms of digital media opening social media platforms have "brought forth unprecedented opportunities for citizens to autonomously and collectively participate in democracy"[37] in an augmented reality wherein "the power of digitality meets the politics of physicality." Not surprisingly, many resistance movements have utilized social media to mobilize their activities, both on and offline. This has certainly been the case with the protest movements discussed in this essay, in which tweeting and posting connected seamlessly with activities on the ground, creating a "non-linear relationship between social media activity and offline participation."[38] Digitally enabled protests are dynamic in nature since they utilize virtual spaces that "constantly grow, interconnect, disjoin, or break down [and], which may or may not be connected with physical locations."[39] This dynamism creates a nonlinear relationship between on and offline participation, as there are no clear demarcations among or between how online and offline activities influence each other. Coordination of actions is decentralized and loose at best.

For feminist activists in China, as in the United States, social media public spaces and their non-linear connections to non-digital spaces created opportunities to resist invisibility. These activists literally and digitally inserted themselves into the public sphere through their photographs, tweets, hashtags, blogs, and other types of communication, making a spectacle of themselves through exposure and in defiance of the cultural politics that define who has rights to certain kinds of visibility in public spaces. Perhaps because of their mashable digital infrastructure, social media intrinsically lends itself to spectacle and also creates new possibilities for "looking back." These platforms allowed activists to circulate, in spectacular scale, anything and everything, "from protest hashtags, second-hand rumors and photoshopped images, to first-hand eyewitness reports and video evidence,"[40] all while creating networks of outrage (Castells 2012). An example of this can be seen in how the #Occupy and #MeToo movements spread from the United States to China.

The #MeToo movement in China that started in early 2018 was not only a feminist battle but also a fight against media censorship. An evident and critical difference in the role of media in Western and Chinese social movements lies in degrees of government interference and censorship. Without freedom of the press and freedom of speech, Chinese feminists had to continually fight against political erasure (Lü 2018b; Han 2018[41]). As the feminist activist Lü Pin puts it, "it's not that #MeToo finally came to China but rather that [...] #MeToo has appeared and re-appeared in China."[42] The appearance, disappearance, and reappearance of feminism in social media records this intermittent, but persistent, history of Chinese women's movements.

The year 2015 was a turning point for Chinese feminist activism, as it began a period when the government further confined the already limited public space for civil and social movements. With key activists detained, mass media withdrew from reporting feminist issues due to censorship. The formerly gray zone that allowed feminist "performance art" had now become forbidden. While the detainment of the Chinese Feminist Five received increasing international attention and support from global feminist non-government organizations (NGOs),[43] relevant information was blocked in Chinese media. From then on, offline action became even harder to organize, and online content was heavily censored. The year 2015 therefore marked an end to an era of active and frequent feminist actions.

Nonetheless, the sudden disruption of collective efforts did not stop feminist activism in China. Many continued their feminist initiatives on their own. When denied the space for a public advertisement against sexual harassment, Zhang Leilei, a feminist based in the southern city of Guangzhou, continued her campaign by wearing a billboard in May of 2017. Walking around in public transportation with the billboard, Zhang made herself a moving site of feminist advocacy. Her single-handed effort moved many in the country to echo her, wearing anti-harassment signs in public spaces.[44] The spontaneous response and support Zhang received shows that a broader but loosely connected feminist community has been fostered in China, which forms the foundation for the Chinese #MeToo movement to take place.

Inspired by the #MeToo movement in the United States, Luo Xixi, a Beijing University of Aeronautics and Astronautics alumna, broke her silence by writing an open letter in January of 2018 that detailed the sexual harassment of her former adviser, Chen Xiaowu, The letter soon spread on the Chinese microblogging site Weibo and the social communication platform Wechat. Following her case, students across the country exposed similar incidents of harassment they had suffered. Many signed petitions calling for universities to take measures to prevent such abuse. Meanwhile, the movement expanded beyond higher education with celebrities and leading figures of various industries getting exposed publicly. Among the many cases, two caused the most stir on social media. Zhu Jun, the host of China Central Television's (CCTV) Spring Gala and one of the best-known personalities in the country, as well as Lei Chuang, founder of a prominent non-government charity who worked on anti-harassment campaigns, were both accused of sexual harassment. While posts about Zhu Jun were quickly erased, Lei Chuang promptly responded to the allegation.

The #MeToo Movement in China won an initial victory when Beijing University of Aeronautics and Astronautics fired Luo Xixi's adviser, Chen Xiaowu, and the Chinese Ministry of Education reversed Chen's award as a Yangtze River Scholar, a prestigious academic title. However, as the movement increased in scope, it was met with relentless censorship. The case that received both wide attention and also extensive censorship was the case involving Zhu Jun, the famous TV personality who has also served multiple times on national committees affiliated with the government, in accusations made by a 25-year-old screenwriter and an intern at the state-run Chinese Central Television network, Zhou Xiaoxuan. Not long after getting exposure, social media posts about the accusation were promptly deleted, and the latest mentions of Zhu's name were censored (Hernández). It was under such circumstances that the #MeToo Movement in China became a relay race that shared and replicated posts before they disappeared in social media—when one post was deleted, more emerged in the form of reposts; when a certain word was censored, a substitute was used; when text posts would not work, images that contained the censored texts were created and circulated. At the center of this movement was the phrase #MeToo, which ultimately became a politically sensitive term and was censored. Standing in place of #MeToo was #MiTu (米兔), which is the homophone of "Me Too" in Mandarin and literally means "rice rabbit." Images of a bowl of rice and the head of a rabbit also became a popular hashtag in resistance to censorship of #MeToo in China.

Why did the Chinese government censor #MeToo? Some may find it surprising that feminism is politically sensitive in China, where Mao Zedong had once famously said, "Women hold up half the sky." Additionally, the state openly supported gender equality. The crackdown on feminist activism partly results from shrinking public spaces in China after the ascendance of the current president, Xi Jinping. Additionally, recent Chinese feminist movements targeted unequal power structures in Chinese society and voiced criticism against existing political and legal systems that fail to address issues such as domestic violence and sexual harassment.

While social media serves a vital role for Chinese feminist action and the Chinese #MeToo movement exists largely online, the fact that information can be erased quickly and online groups can be easily disbanded also means that it takes a lot more to keep up the momentum. Firstly, breaking out of censorship requires activists to keep the public interested, always producing newer and more shocking content; secondly, it also takes more time and energy for a feminist cohort to reestablish the online community over and over again after the initial accounts that produce content are banned. In addition, the lack of an effective support system in real life for those who speak out has led to a serious backlash. For example, TV celebrity Zhu Jun later sued Zhou Xiaoxuan for defamation. With very limited space for organizational efforts and resources to help and support the people who spoke out, the #MeToo movement in China is slow-moving and gradually loses its visibility, heightening a power imbalance that prevents more people from coming forward.

The #MeToo movement in China demonstrates both the potential and the limitation of using social media as the core of a social movement. Social media can attract popular support, short-term enthusiasm, and wide mobilization. It can circumvent and overcome censorship, but at the same time, it has limited long-term appeal and cannot always support an agenda needed for lasting political and social change in real life.

Conclusion

So that the spectacle may wound and move the audience member, Teatro Valdoca has at its disposal two powerful vehicles: the body and the voice…with faces and bodies garishly covered in make-up, the actor manifests, above all, a knowledge of the body that makes [them] capable of being at one with [their] own emotions.[45]

The above quote, from an analysis of Teatro Valdoca's spectacular performances, sums up the connection between spectacle and the oppositional gaze in feminist resistance movements. Not only does it allude to the power of the gaze to arrest one's attention, but it also speaks to the way in which this attention is captivated. It is through pure spectacle, resisting domination by bodily insistence, that critiquing and returning a look will change reality or "wound and move" the audience. Such action also affirms hooks' (1996) assertion that it is on and through the body that agency resides. Through unruly bodies occupying public spaces both digitally and in visually spectacular ways, feminist activists in China, as in the United States, push back against misogyny. In doing so, they have created new visual narratives that may allow women's voices to be heard and women to be seen (and to see) differently. There are of course inherent dangers that accompany such brazenness, especially given the increased surveillance to which public spaces are now subject and the harsh challenges to enacting certain kinds of political spectacle in public places.[46] However, this is what makes the resistance even more spectacular and transformative, especially in Chinese contexts. The use of social media has been at the core of contemporary Chinese feminist movements since 2012, a conscious move by the activists who engaged the general public in their actions. Visually striking images, deriving from feminist "performance art," served as an effective way for them to attract attention, provoke debate, and take an oppositional stand without triggering censorship. Photos of bruised, hurt, and unusual female bodies not only brought attention to social issues like domestic violence but also denaturalized the idealized image of the female body. In staging these photos, activists exercise subjective control over their own bodies, thereby reversing the conventionally gendered power of looking.

Notes

1 Mary Ann Doane, "Film and the Masquerade: Theorizing the Female Spectator," *Screen* 23, nos 3–4 (1982): 80, https://doi.org/10.1093/screen/23.3-4.74.
2 John Berger, *Ways of Seeing* (London: Penguin, 1972), 47.
3 Berger, *Ways of Seeing*, 47.
4 Berger, *Ways of Seeing*, 47.
5 Mary Ann Doane, "Film and the Masquerade," 80.
6 bell hooks, "The Oppositional Gaze: Black Female Spectators," in *Movies and Mass Culture*, ed. John Belton (New Brunswick, NJ: Rutgers University Press, 1996), 247.
7 Sigmund Freud, "Medusa's Head," *International Journal of Psychoanalysis* 22 (1941): 273, www.freud2lacan. com/docs/Medusa%27s_Head.pdf.
8 Mary Russo, "Female Grotesques: Carnival and Theory," in *Feminist Studies/Critical Studies (Language, Discourse, Society)*, ed. Teresa de Lauretis (London: Palgrave Macmillan, 1996), 213.
9 Helene A. Shugart and Catherine Egley Waggoner, "A Bit Much: Spectacle as Discursive Resistance," *Feminist Media Studies* 5, no. 1 (2005): 67, https://doi.org/10.1080/14680770500058215.
10 Berger, *Ways of Seeing*, 47.
11 Laura Mulvey, "Visual Pleasure and Narrative Cinema," in *Media and Cultural Studies: Keyworks*, eds Meenakshi Gigi Durham and Douglas M. Kellner (Malden, MA: Blackwell Publishers, 2001), 402.
12 hooks, "Oppositional Gaze," 248.
13 hooks, "Oppositional Gaze," 248.
14 hooks, "Oppositional Gaze," 249.

15 hooks, "Oppositional Gaze," 261.

16 hooks, "Oppositional Gaze," 258.

17 Jean-Paul Sartre, "The Look," in *Being and Nothingness* (New York: Philosophical Library), 110.

18 Guy Debord, "The Commodity as Spectacle," in *Media and Cultural Studies: Keyworks*, eds Meenakshi Gigi Durham and Douglas M. Kellner, Malden (Malden, MA: Blackwell Publishers, 2001), 139.

19 Peggy Phelan, *Unmarked: The Politics of Performance* (London: Routledge, 2003), 1.

20 Phelan, *Unmarked*, 2.

21 hooks, "The Oppositional Gaze," 261.

22 Russo, "Female Grotesques," 214.

23 Yuan Hsiao and Yunkang Yang, "Commitment to the Cloud? Social Media Participation in the Sunflower Movement," *Information, Communication & Society* 21, no.7 (2018): 996, https://doi.org/10.1080/13691 18X.2018.1450434.

24 Yalan Huang, "War on Women: Interlocking Conflicts Within The Vagina Monologues in China," *Asian Journal of Communication* 26, no. 5 (2016): 466, https://doi.org/10.1080/01292986.2016.1202988.

25 Hester Baer, "Redoing Feminism: Digital Activism, Body Politics, and Neoliberalism," *Feminist Media Studies* 16, no. 1 (2016): 30, http://dx.doi.org/10.1080/14680777.2015.1093070.

26 Huang, "War on Women," 466.

27 Leta Hong-Fincher, *Betraying Big Brother: The Feminist Awakening in China* (New York: Verso, 2018), 28.

28 hooks, "Oppositional Gaze," 247.

29 Lü Pin, "Two Years On: Is China's Domestic Violence Law Working?," *Amnesty International*, published March 7, 2018, accessed December 5, 2018, www.amnesty.org/en/latest/campaigns/2018/03/is\ chinadomestic-violence-lawworking/.

30 Jun Li and Xiaoquin Li, "Media as a Core Political Resource: The Young Feminist Movements in China," *Chinese Journal of Communication* 10, no. 1 (2017): 61, https://doi.org/10.1080/17544750.2016.1274265.

31 Lü Pin, "How Could Change Happen: Strategy and Transformation of the Contemporary Feminist Movement in China," *Online Talk* presented at the Chinese Feminist Collective, December 15, 2018.

32 Qi Wang, "Young Feminist Activists in Present-Day China: A New Feminist Generation?" *China Perspectives* no. 3 (2018): 59.

33 Xiao Han, "Searching for an Online Space for Feminism? The Chinese Feminist Group Gender Watch Women's Voice and It's Changing Approaches to Online Misogyny," *Feminist Media Studies* 18, no. 4 (2018): 734–749, https://doi.org/10.1080/14680777.2018.1447430.

34 Vera Penêda, "Taking It to the Street," *Feminism in China*, accessed December 5, 2018, www. feminisminchina.com/.

35 Penêda, "Taking It to the Street."

36 Hong-Fincher, *Betraying Big Brother*, 15.

37 Yuan Hsiao and Yunkang Yang, "Commitment to the Cloud," 996.

38 Yuan Hsiao and Yunkang Yang, "Commitment to the Cloud," 997.

39 Yuan Hsiao and Yunkang Yang, "Commitment to the Cloud," 997.

40 Thomas Poell and Jose van Dijck, "Social Media and New Protest Movements," in *The Sage Handbook of Social Media*, eds Jean Burgess, Alice Marwick, and Thomas Poell (London: Sage, 2018), 1.

41 Han, "Searching for an Online Space," 743.

42 Lü Pin, "'MeToo': From Butterflies to Hurricanes," trans. Emile Dirks, *China Digital Times (CDT)*, September 2018. https://chinadigitaltimes.net/2018/09/lu-pin-metoo-from-butterflies-to-hurricanes/.

43 Zheng Wang, "Detention of the Feminist Five in China," *Feminist Studies* 41, no. 2 (2015): 481.

44 Lü Pin, "'#MeToo': From Butterflies."

45 Valentina Valentini, "A Spectacle of Resistance and Exhortation," *PAJ: A Journal of Performance and Art* 32, no. 1 (2010): 86, www.jstor.org/stable/20627961.

46 Jorge La Barre and Blagovesta Momchedjikova, "Introduction: Between Spectacle and Resistance: Some Thoughts on Public Space Today," *Streetnotes* 25, no. 0 (2016.0): 1–12, https://escholarship.org/uc/item/1zg1280p.

Bibliography

Baer, Hester. 2016. "Redoing Feminism: Digital Activism, Body Politics, and Neoliberalism." *Feminist Media Studies* 16(1): 17–34. http://dx.doi.org/10.1080/14680777.2015.1093070.

Berger, John. 1972. *Ways of Seeing*. London: Penguin.

Butler, Judith. 1990. *Gender Trouble: Feminism and the Subversion of Identity*. New York: Routledge.

Castells, Manuel. 2012. *Networks of Outrage and Hope: Social Movements in the Internet Age*, 1st Edition. Massachusetts: Polity Press.

China Daily. 2012. "Beijing nü daxuesheng qingrenjie faqi xingwei yishu fandui banlü baoli [Beijing Female Students Stage Performance Art to Oppose Domestic Violence]." *China Daily*. February 14, 2012. www.chinadaily.com.cn/hqzx/2012-02/14/content_14607836.htm.

Debord, Guy. 2001. "The Commodity as Spectacle." In *Media and Cultural Studies: Keyworks*, edited by Meenakshi Gigi Durham and Douglas M. Kellner, 139–143. Malden, MA: Blackwell Publishers.

Doane, Mary Ann. 1982. "Film and the Masquerade: Theorizing the Female Spectator." *Screen* 23(3-4): 74–88. https://doi.org/10.1093/screen/23.3-4.74.

Foucault, Michel. 2002. "Las Meninas." In *The Order of Things*. London: Routledge Classics.

Freud, Sigmund. 1941. "Medusa's Head." *International Journal of Psychoanalysis* 22: 273–274. www.freud2lacan. com/docs/Medusa%27s_Head.pdf

Guillard, Julianne. 2016. "Is Feminism Trending? Pedagogical Approaches to Countering (Sl)Activism." *Gender and Education* 28(5): 609–626. https://doi.org/10.1080/09540253.2015.1123227.

Gyenge, Zsolt. 2016. "Subjects and Objects of the Embodied Gaze: Abbas Kiarostami and the Real of the Individual Perspective." *Acta Univ. Sapientiae, Film and Media Studies* 13: 127–141. DOI: 10.1515/ausfm-2016-0018.

Han, Xiao. 2018. "Searching for an Online Space for Feminism? The Chinese Feminist Group Gender Watch Women's Voice and It's Changing Approaches to Online Misogyny." *Feminist Media Studies* 18(4): 734–749. https://doi.org/10.1080/14680777.2018.1447430.

Hernández, Javier C. 2019. "She's on a #MeToo Mission in China, Battling Censors and Lawsuits." *The New York Times*, January 7, 2019, sec. World. www.nytimes.com/2019/01/04/world/asia/china-zhou-xiaoxuan-metoo.html.

Hong-Fincher, Leta. 2016. "China's Feminist Five." *Dissent Magazine*, Fall 2016. www.dissentmagazine.org/article/china-feminist-five.

———. 2018. *Betraying Big Brother: The Feminist Awakening in China*. New York: Verso.

hooks, bell. 1996. "The Oppositional Gaze: Black Female Spectators." In *Movies and Mass Culture*, edited by John Belton, 247–264. New Brunswick, NJ: Rutgers University Press.

Hsiao, Yuan and Yunkang Yang. 2018. "Commitment to the Cloud? Social Media Participation in the Sunflower Movement." *Information, Communication & Society* 21(7): 996–1013. https://doi.org/10.1080/1369118X.2018.1450434.

Hu, Alice. 2016. "Half the Sky, but Not Yet Equal: China's Feminist Movement." *Harvard International Review* 37(3): 15–18.

Huang, Yalan. 2016. "War on Women: Interlocking Conflicts Within *The Vagina Monologues* in China." *Asian Journal of Communication* 26(5): 466–484. https://doi.org/10.1080/01292986.2016.1202988.

Huang, Yujie. 2012. "Shoushang xinniang jietou shangyan xingwei yishu xuanchuan fanjiabao [Wounded Bride Stage Performance Art on the Street, Opposing Domestic Violence]." People.com. December 3, 2012. http://legal.people.com.cn/n/2012/1203/c42510-19775286.html.

Jacobs, Katrien. 2016. "Disorderly Conduct: Feminist Nudity in Chinese Protest Movements." *Sexualities* 19(7): 819–35. https://doi.org/10.1177/1363460715624456.

Kaun, Anne. 2017. "'Our Time to Act Has Come': Desynchronization, Social Media Time and Protest Movements." *Media Culture and Society* 39(4): 469–486. https://doi.org/10.1177/0163443716646178.

La Barre, Jorge and Blagovesta Momchedjikova. 2016. "Introduction: Between Spectacle and Resistance: Some Thoughts on Public Space Today." *Streetnotes* 25(0): 1–12. https://escholarship.org/uc/item/1zg1280p.

Li, Jun and Xiaoquin Li. 2017. "Media as a Core Political Resource: The Young Feminist Movements in China." *Chinese Journal of Communication* 10(1): 54–71. https://doi.org/10.1080/17544750.2016.1274265.

Lü, Pin. 2018a. "Two Years On: Is China's Domestic Violence Law Working?" Amnesty International. Published March 7, 2018. Accessed December 5, 2018. www.amnesty.org/en/latest/campaigns/2018/03/is-china-domestic-violence-law-working/.

———. 2018b. "Gone but Not Forgotten." *China Channel* (blog). Accessed December 5, 2018. https://chinachannel.org/2018/04/13/gone-but-not-forgotten/.

———. 2018c. "'MeToo': From Butterflies to Hurricanes." Translated by Emile Dirks. *China Digital Times (CDT)*, September 2018. https://chinadigitaltimes.net/2018/09/lu-pin-metoo-from-butterflies-to-hurricanes/.

————. 2018d. "How Could Change Happen: Strategy and Transformation of the Contemporary Feminist Movement in China." Online Talk presented at the Chinese Feminist Collective, December 15.

Mulvey, Laura. 2001. "Visual Pleasure and Narrative Cinema." In *Media and Cultural Studies: Keyworks*, edited by Meenakshi Gigi Durham and Douglas M. Kellner, 393–404. Malden, Malden, MA: Blackwell Publishers.

Penêda, Vera. "Taking It to the Street." Feminism in China. Accessed December 5, 2018. www. feminisminchina.com/.

Phelan, Peggy. 2003. *Unmarked: The Politics of Performance*. London: Routledge.

Poell, Thomas and Jose van Dijck. 2018. "Social media and New Protest Movements." In *The Sage Handbook of Social Media* edited by Jean Burgess, Alice Marwick, and Thomas Poell, 546–561. London: Sage.

Russo, Mary. 1986. "Female Grotesques: Carnival and Theory." In *Feminist Studies/Critical Studies (Language, Discourse, Society)*, edited by Teresa de Lauretis, 213–229. London: Palgrave Macmillan.

Sartre, Jean-Paul. 1956. "The Look." In *Being and Nothingness*. New York: Philosophical Library.

Shugart, Helene A. and Catherine Egley Waggoner. 2005. "A Bit Much: Spectacle as Discursive Resistance." *Feminist Media Studies* 5(1): 65–81. https://doi.org/10.1080/14680770500058215.

Turley, Emma and Jenny Fisher. 2018. "Tweeting Back While Shouting Back: Social Media and Feminist Activism." *Feminism & Psychology* 20(1): 128–132. https://doi.org/10.1177/0959353517715875.

Valentini, Valentina. 2010. "A Spectacle of Resistance and Exhortation." *PAJ: A Journal of Performance and Art* 32(1): 83–89. www.jstor.org/stable/20627961.

Wang, Qi. 2018. "Young Feminist Activists in Present-Day China: A New Feminist Generation?" *China Perspectives* (3): 59–68.

Wang, Zheng. 2015. "Detention of the Feminist Five in China." *Feminist Studies* 41(2): 476–485.

Wang, Zheng and Ying Zhang. 2010. "Global Concepts, Local Practices: Chinese Feminism since the Fourth UN Conference on Women." *Feminist Studies* 36(1): 233–234.

36

REFUSING MASTERY, MASTERING REFUSAL

Critical communication pedagogy and gender

Benny LeMaster and Deanna L. Fassett

Gender is an unstable signifier shifting across time and through space. In their bid for a trans-affirming, intersectional approach to critical communication pedagogy, Benny LeMaster and Amber Johnson characterize gender as *ineffable*, *elusive*, and *particular* rendering claims of epistemological certainty with regards to gender questionable at best.[1] Moreover, the popular, the institutional, and the individual perform respective—although systemically derived—enactments of gender that are often in contradiction and rarely defined through consensus. Said differently, popular representations of gender chafe against individual means of living/becoming gender(ed) across intersections of identity as institutions define and constrain gender potentiality through administrative means.[2] Lisa Flores unpacks gender through a critical intercultural communication framework suggesting gender is more the result of intersectional "conflicting and complementary narratives" than it is particularized groups.[3] Thus to learn gender, one must *unlearn* gender—at least a typified framework that presumes mastery or categorical certainty in favor of an emerging relational cultural phenomenon. Gender is, in its broadest and most materially profound sense, a colonial refrain: a racialized relic grounded in taxonomic flights of fancy and fiction made manifest through historically contextualized material consequence that presumes an allegedly self-evident binary structure.[4] In short, teaching and (un)learning gender is complicated. Critical communication pedagogy (CCP) provides a framework through which to "unlearn" gender in its perpetual—and productive—instability as it shifts through time and across space even as it is anchored in historically significant ways.[5]

CCP as a disciplinary framework opens a context and offers a vocabulary for scholars to explore communication as it is constitutive of learning in settings ranging from the formal classroom to the mundane and interpersonal or familial. Deanna Fassett and Kyle Rudick clarify CCP marks a paradigmatic approach to the study of communication and instruction that centers "analysis of culture and power in the service of social justice."[6] In their original articulation of CCP, Deanna Fassett and John Warren explore how researchers may intertwine a Freirean critical pedagogy and its emphasis on reflexivity and praxis with a fine-grained analysis of how everyday communication practices constitute and may, therefore, alter oppressive social forces.[7] Paulo Freire notably critiqued what he termed the "banking" approach to education in which the teacher-student relationship is animated through an oppressive vertical relational dynamic. He uses the metaphor of banking to illustrate approaches to education that cast knowledge as a "gift bestowed by those who consider themselves knowledgeable upon those whom they consider to know nothing."[8] This approach hinders the development of a critical consciousness—what Freire termed *conscientização*—in which students and teachers come to understand themselves as historically, materially, and relationally situated in, through, and across culture(s). In contrast, Freire proposes "problem-posing" education[9] that, in David Kahl's words, seeks to "heighten students' awareness of hegemony and to help them learn to respond to its presence, both in the classroom and in society."[10] The path to consciousness is through dialogue, which Freire asserts is not a technique but rather an "epistemological relationship" that "presents itself as an indispensable component of the process of both learning and knowing."[11]

In effect, dialogue reconfigures the dichotomous vertical teacher/student relationship in the banking model to a horizontal orientation in which teacher/student is reconfigured as "teacher-student" and "students-teacher."[12] This repositioning of student-and/as-teacher marks what Dwight Conquergood terms a "dialogical stance" in which the entities are brought "together even while it holds them apart. It is more like a hyphen than a period."[13] Returning to the question of gender, a banking approach to the communicative study of gender may include imparting knowledge *about* gender derived of empirical studies that seek to understand gender through binaristic and mimetic terms (e.g., "male" *or* "female") and/or biological imperatives (e.g., gendered communicative expectancies based on sex assigned at birth). This model equally risks universalizing gender experience such that what is "learned" (e.g., gendered communication) is presumed to be generalizable across gender experience just as it asserts a particularized way to *be* gender(ed); in this regard, gender is not that which one embodies but rather that to which one acquiesces. Conversely, a problem-posing approach to the study of gender explores gender as performative and embodied and is, as a result, inevitably at odds with claims to gender mastery or categorical certainty. Indeed, a problem-posing approach can begin with the assumption that gender is an *imposed* colonial refrain that contains and thus delimits gender potentiality.[14] The material means by which individuals and collectives navigate neocolonial structural impositions, including gender, render valuable insight about the performative constitution and maintenance of culture across a variety of intersecting identities in a given context.

This chapter articulates the state of the art of critical communication pedagogy as it does and does not speak to gender, providing readers with a process that supports and identifies possible directions for continued reflexivity and praxis in and beyond the classroom. To begin, we introduce the constitutive elements that comprise CCP paying particular attention to the ways in which the CCP framework has opened spaces for novel insights about communication and gender in instructional contexts, institutional intervention, and public pedagogy (including coalitional movements).[15] Thereafter, we nuance the relationship between CCP and gender and offer the reader two pillars set in dialectic tension that may guide critical work at the intersections of CCP and gender. These pillars serve as pedagogical anchors for theorizing, embodying, and enacting gender futurity grounded in criticality. They include the dialectic coupling of "refusing mastery" and "mastering refusal."

We argue CCP and gender teach us that to *refuse mastery* is to *embrace failure*; that is, to engage and critique the structural means by which one is rendered a "failure" as a result of their embodied difference across intersections of identity.[16] To embrace failure is to reject hegemonic expectancies and to affirm self and/as other in terms that refuse to conform to hegemonic orders including cisheterosexism, white supremacy, ableism, and classism, for instance. CCP in/of gender additionally teaches us that to *master refusal* is to *embrace uncertainty*. While the former commitment challenges epistemological certainty through embracing failure, the latter commitment centers the affective registers that seek to shift the status quo through a willful refusal to find calm in the midst of uncertainty.[17] Rather, to master refusal is to bask in the improvisational and playful space of unknowing—to unknow gender so as to better engage gender on its emergent, creative, constrained, and relational intersectional terms. With that, we turn to CCP's constitutive elements.

Gender in and against critical communication pedagogy

Critical communication pedagogy, as paradigm and practice, evolves from and makes possible extensions of communication scholarship in and of gender, identity, teaching, and learning. Fassett and Warren articulate CCP via a series of ten commitments.[18] In this initial iteration of CCP, the commitments serve to draw contemporary communication theory (including work in critical theory and cultural studies) into instructional communication scholarship—as such, the commitments are didactic. More recent articulations of CCP, including Fassett and Nainby[19] and Fassett and Rudick,[20] are less insular or intra-disciplinary and more elastic, allowing for more scholars and practitioners in and beyond communication studies to engage in and expand this social justice agenda. These works explore, expand, and contextualize the commitments of CCP into meaningful trajectories of investigation, such as discerning, destabilizing, and resisting neoliberalism, toxic masculinity, white nationalism, and other forces at play in dehumanizing educators and learners. Like Fassett and Rudick, here we find it helpful to group CCP's original 10 commitments into three broad themes or overarching areas of understanding and action: (1) communication is constitutive, (2) social justice as process, and (3) teaching and learning environments are meaningful sites of activism and interpersonal justice.[21]

Communication is constitutive of what matters. CCP is grounded in and functions to nurture and sustain understandings of human identity as fluid, emergent, and evolving. In this definition, it is communication that gives rise to identities and relationships of all sorts, including organizations and cultures. The first of Fassett and Warren's commitments—"identity is constituted in communication"[22]—asserts that identities are not inherent but rather emergent *through* communication, in ways that are always already shaped by relationships of power and privilege. To illustrate this, we have found it helpful to discuss with our students the elaborate and often unreflective practices associated with gendering children, ranging from decisions about baby names and the colors to paint nurseries to the perniciousness of gender reveal parties. That familial assumptions about a child's gender precede the arrival of that very particular child helps to illustrate not only the power of communication to delimit and complicate a child's possibilities but also the power of our families, our teachers, our doctors, and other significant others to (often unreflectively, in the name of tradition) wield that power to shape our lives.

A CCP approach to power departs from traditional instructional communication research (derived from social scientific, post-positivist approaches) that frames power as a "tool" or a "skill set."[23] Drawing from Jo Sprague's critical articulation of power in education,[24] Fassett and Warren observe that when power is framed as such, student and/or teacher resistance is (mis)interpreted as "deviance, as decontextualized (mis)behaviors that, at first blush, may seem arbitrary or irrational."[25] This leads them to explore how CCP is premised on power as both fluid and material—their second commitment.[26] In this regard, power is dialectically animated.[27] When scholars and educators understand power as dialectical, as fluid or moving in and through us in ways we often do not realize, we may (re)interpret resistance as a valid response to contextual material conditions. Drawing on the works of Antonio Gramsci and Raymond Williams, Susana Martínez Guillem articulates power in dialectical terms that link everyday micro cultural practices to macro cultural flows of power. Guillem argues a dialectical approach highlights the material conditions that enliven power relations while offering an "explanatory dimension" that "distinguish[es] between acceptable and unacceptable forms" of power, which in turn maps means for "concrete political intervention."[28] As such, one possible research trajectory may be exploring with students their experiences of resisting gendering, as well as inviting their own discomfort with such resistance to illuminate how power (as individuals, institutions, and ideologies) moves in and through them.

Notably, CCP also explores all aspects of identity and culture as created in and through communication, opening space for intersectional analyses of identity and power that illuminate conflict, contradiction, and paradox. Fassett and Warren challenge CCP scholars to recognize culture as a meaningful and foundational component of any exploration of teaching and learning contexts. While this is a stark departure from mainstream instructional communication research, this commitment does not adequately capture the possibilities of intersectional analysis. For instance, Patricia Hill Collins and Sirma Bilge note that power relations are intersectionally animated across structural, disciplinary, cultural, and interpersonal domains.[29] First, intersectionality names the "cultural synergy" that is created as a result of our various identities interacting through time and across space.[30] Intersectionality illuminates this synergy of how one is granted and denied material access across a variety of domains. Second, the domains that Collins and Bilge name (structural, disciplinary, cultural, and interpersonal) may function as a heuristic device to examine power relations including the means by which to distinguish between oppressive and empowering enactments of power.[31] They clarify, "[l]ooking at how power works in *each* domain can shed light on the dynamics of a larger social phenomenon."[32] Critical communication educators may work with their students toward critical attention to power relations "*via their intersections*" and "*across domains of power*" to describe and navigate power as a fluid force with very real material effects.[33]

As always, our present sociocultural moment affords educators ample opportunity for such investigation. See, for example, actor Viola Davis's speech at the 2018 Women's March in Los Angeles where she challenges listeners to not only call out and call to account toxic masculinity but also the ways in which the initial rootedness of the #MeToo movement in the experiences of Black women became more diffuse via its amplification through social media platforms. As bell hooks importantly highlights, feminism requires political commitment; unless one actively works to unlearn *and* combat sexism and racism, for instance, one risks reifying them in subtle unconscious ways.[34] In this regard, women can perform and perpetuate the sexist oppression of women just as racialized and/or racially ambiguous folks can perform and perpetuate whiteness and in turn white supremacy. Indebted to these critical insights, we propose that CCP "unhinge" cissexism from non-trans, cisgender, and cissexual subjects so as to critically engage the multifarious ways in which bodies of all genders can perform and perpetuate cissexism as well as sexism across intersecting lines of flight in a variety of domains. Assuming a dialectic framing of power implores scholars at work in the intersections of teaching, learning, and identity to name the material conditions that enable and restrict access to relationally derived power across intersecting lines of flight in structural, disciplinary, cultural, and interpersonal domains. CCP work must continue to explore and respond to the ways sexism, cissexism, and cisheterosexism— as intersectionally constituted structures—materially organize the lived experiences of different bodies, identities, and subjectivities across varying domains of power.

Social justice as inherently communicative. Fassett and Warren's next three commitments define social justice as explicitly communicative, as a process that requires intentional reflection and practice. These three commitments explore how CCP focuses on "mundane communication practices as constitutive of large social structural systems,"[35] "social structural critique as it places mundane communication practices in a meaningful context,"[36] and the centrality of "language and analysis of language as constitutive of social phenomena."[37] These commitments lean on and open space to productively explore Freire's assertion that "changing language is part of the process of changing the world."[38] To contextualize this in CCP, Fassett and Rudick add, "communication, and the production of meaning, is never simply a self-contained, isolable, and dyadic event, but rather draws upon and (re)produces the norms, rules, and identities that make meaning possible."[39] Gender pronouns offer a timely and insightful example.

In a Western context, gender pronouns provide a means to grammatically categorize cultural phenomena and to subjectively identify a sense of self as belonging to culturally specific gender categories. Transnational students may lead with language(s) that do not draw on distinctive gender pronouns nor that distinguish an individual sense of self rendering the demand for a pronoun a manifestation of English hegemony.[40] Typically framed in binary terms (e.g., masculine [he/him/his] *or* feminine [she/her/hers]), gender pronouns performatively constitute and in turn reveal a localized and gendered understanding of culture. In articulating a nonbinary pedagogy, Finn Enke frames binary gender pronouns as a "technology" that at once reflects binary gender as it "create[s], teach[es], and enforce[s] binary gender."[41] For instance, students who identify as nonbinary reveal the limits of a binary framing when implored to identify their pronouns in public space including a classroom. Forced to "choose one," the nonbinary student may set a gendered course of interactions with peers and faculty that does not align with their nonbinary sense of self. Moreover, some students are uncertain of their gender identity in a particular moment or would rather evade being marked as "the" trans person in a given space. This realization is exacerbated by the very real understanding that gender-variant students of color are already marked as "deviant" in a white supremacist culture, suggesting that pronoun disclosure may aggravate an already stressful reality for racialized students specifically.

Conversely, many nonbinary students have turned to gender-neutral pronouns as alternatives (e.g., xe, zie, hir). The "singular they" pronoun has experienced a recent increase in circulation as an alternative to binary gender pronouns, though its use dates back to at least the 16th century when Shakespeare's Antipholus of Syracuse opened Act 4, Scene 3 of *The Comedy of Errors* thusly: "There's not a man I meet but doth salute me//As if I were *their* well-acquainted friend."[42] Still, critics persist, suggesting the singular "they" pronoun is grammatically improper; these critics tend to identify themselves in binary terms, effectively asserting their own gendered legitimacy at the dismissal of variation.[43] Fassett and Warren remind us "[I]t is in our communication practices that we produce knowledge, define how identities are negotiated and maintained, and imply that power is something only the powerful possess."[44] In this regard, the way in which instructors approach gender pronouns in the classroom is always and already power laden, a discussion linked to broader cultural processes that directly implicate the performative constitution of binary gender.

Teaching and learning environments are meaningful sites of activism and interpersonal justice. Fassett and Warren's final commitments set an agenda for CCP scholars and practitioners to advance: The classroom as a meaningful site of interpersonal engagement, social change, and personal justice. It is important to note that they do not minimize other possible sites of activism and justice even as they work to reclaim classrooms and other teaching and learning settings as more than preparation for "the real world." The final four commitments include: "reflexivity [as] an essential condition for critical communication pedagogy,"[45] "pedagogy and research [are] praxis,"[46] "critical communication educators embrace…a nuanced understanding of human subjectivity and agency,"[47] and "dialogue [is] both metaphor and method for our relationships."[48]

Freire described praxis as "reflection and action upon the world in order to transform it."[49] In his view, oppression is a "domesticating" force and praxis marks the "critical intervention" that enables transformation.[50] Praxis is often marked as the coupling of theory with action such that theory guides action. For Freire, critical pedagogy facilitates reflection upon the ways power orders everyday lived experience so that one is equipped to name and resist the ways power delimits subjective potentiality in fluid and material ways. Within CCP, praxis marks collaborative efforts to "locate and name, reflexively, the taken-for-granted" in both pedagogical and research contexts so as to "decenter [(our own)] normative readings" of cultural phenomena.[51] In so doing, students and researchers engage the ways "we build our identities in contest and collusion with one another," which Fassett and Warren argue holds the potential to "interrupt those processes and shape them anew in meaningful ways."[52] Rudick, Golsan, and Cheesewright describe praxis as the dialogic means by which instructors and students as well as researchers and research participants collaboratively set guidelines that are open to change in community with the processual goal of "improving their [/our] craft."[53] Let's consider a contemporary example of praxis in the context of research to help illustrate contemporary directions.

Satoshi Toyosaki develops Calvin Schrag's philosophy of subjectivity as a communicative phenomenon so as to advocate a "praxis-oriented" approach to the study of whiteness. For Schrag, the "praxis-oriented self" is "defined by its communicative practices, oriented toward an understanding of itself" in its discourse, action, and relational ties."[54] This reflexive approach implicates intrapersonal dimensions and implores researchers to be cognizant of the paradoxical and fluid ways whiteness persists even in the critique of whiteness. Guillem responds to Toyosaki, arguing praxis must attend to power as a dialectic force that is both fluid and material. As such, Guillem envisions praxis as "the product of the relationship between human will and different constraining and enabling structures."[55] For Guillem praxis can only be a powerful force when it places in tension "objective social conditions" with "subjective communicative interventions" because:

> As much as it hurts, we need to accept that willingness, self-awareness, and acknowledgement of "others" do not really alter the fact that some people systematically enjoy more access to resources than others, that some people are terrorized by others, that some do not enjoy the emancipating possibilities of human communication because they are not even considered human.[56]

Taken together, Toyosaki and Guillem illustrate praxis as a force that desires cultural transformation. Reflexivity and implication guide transformation for Toyosaki while materiality guides transformation for Guillem. In the context of gender and communication, praxis implores critical pedagogues to work with students in community to envision the transformation and destruction of systems including sexism, cissexism, and cisheterosexism at the intersections of difference while centering the lived experiences of those most materially disadvantaged (e.g., trans women of color and Black and Indigenous trans women in particular, for instance). Failing to center the lived experiences of those most materially disadvantaged risks reifying the very structures we purport to challenge.

Emerging commitments: refusing mastery, mastering refusal

Thus far, we have illustrated CCP as a disciplinary framework that enables scholars and educators to explore the communicative constitution of cultural phenomena including gender. Social justice as an axiological commitment guides CCP praxis. As a result, (un)learning gender through CCP entails a commitment to socially just gender being and becoming at the intersections of difference. For us, this commitment requires centering pedagogical inquiry and praxis around those most materially disadvantaged by what María Lugones terms the *coloniality of gender*.[57] In a 21st-century U.S. context—in which the terrorism of slavery and genocide persists through the carceral state[58]—this includes trans women, trans feminine, and nonbinary folks of color. Black and Indigenous trans women are disproportionately affected due to intersectional stigmatization, dehumanization, and criminalization of blackness, indigeneity, transness, and womanness.[59]

In their 2015 report "Violence Against LGBTI Persons," the Inter-American Commission on Human Rights names a number of factors exacerbating racialized cisheterosexist violence across the Americas including "exclusion, discrimination and violence within the family, regarding education, and in society at large; lack of recognition of their gender identity; involvement in [criminalized] occupations that put them at a higher risk for violence; and high rates of criminalization."[60] In turn, we pause to remember and name the known[61] trans women and trans feminine subjects who were brutally murdered—at the time of our writing—in the United States in 2018: Christa Leigh Steele-Knudslien (42), Viccky Gutierrez (33), Celine Walker (36), Tonya Harvey (35), Zakaria Fry (28), Phylicia Mitchell (45), Amia Tyrae Berryman (28), Sasha Wall (29), Karla Patricia Flores-Pavón (26), Gigi Pierce (28), Antash'a Devine Sherrington English (38), Diamond Stephens (39), Cathalina Christina James (24), Keisha Wells (54), Sasha Garden (27), Vontashia Bell (18), Dejanay Stanton (24), Shantee Tucker (30), Londonn Moore (20), Nikki Enriquez (28), Ciara Minaj Carter Frazier (31), and Tydie (unknown age). At the same time, we affirm the multiple mundane ways in which transgender and gender non-conforming (TGNC) folks "survive, thrive, and fight multiple systems of oppression every day"[62] effectively shifting the "symbolic burden"[63] placed on trans women of color as exclusively consistent victims. To affirm such requires an epistemological shift in the ways we approach the question of gender in communication classrooms and beyond such that trans women of color are understood as more than statistics—as humans living livable lives.[64]

In his bid for Blacktransfeminist Thought (BTFT), Marquis Bey notes Black and/or trans bodies disrupt normativity as they are framed as "nonlives—nonbeings whose deaths are not and cannot be mournable since their claims to humanity have been revoked" as a result of existing outside of intersecting matrices of racialized, gendered, and classed intelligibility.[65] Said differently, cisheteronormativity constitutes Black trans women as *non-human* and *fugitive* under racial capitalism.[66] Informed by Bey's BTFT, a CCP social justice approach to gender radically validates Black trans women in their "nonnormativity, the result of which will yield the disruption of the gender binary structuring the terms and conditions of 'life.'"[67] To accomplish this, we (un)learn gender through the dialectic reversal refusing mastery and mastering refusal to which we turn in this section, which maps a CCP approach to the study of gender.

Refusing mastery marks a decolonial labor that intervenes in the performative sedimentation of modern cultural formations like gender. In turn, decolonization refuses the hegemony of gender as a universal condition. It accomplishes this by "colliding with the limits of knowledge structures ... derived from, and enabled by, various imperial and national modernities."[68] Evident here are constraints (always and already) animating what is commonly termed "gender" in a Western educational context, which Julia Wood and Natalie Fixmer-Oraiz attribute to the "social meaning assigned to sex" that is "not innate."[69] In turn, sex is—in their terms—a "biological classification based on external genitalia (penis and testes in males [sic], clitoris and vagina in females [sic])."[70] To put it frankly, these criteria (for both gender *and* sex) concurrently erase TGNC subjectivity and embodiment as it normalizes non-trans subjectivity and embodiment just as it centers whiteness through colonial taxonomy.

The categorical criteria for genitalia, and thus sex, are left unquestioned, presuming neutrality in the structure itself. That is to say, these criteria communicatively *sex* genitals, which in turn normalizes the sort of body that is expected to have particular genitalia. More than that, the presumption that gender is not innate counters the administrative and narrative constitution of, specifically, white transsexual subjectivity in a Western medicalized context.[71] For a non-trans person, gender may not feel innate, while for some transsexual folks that internal essence is what they are implored to routinely "prove" when *bargaining* for care each time they visit a health-care provider (given they have access to health care).[72] More than that, that it is not innate positions gender as an external phenomenon distinct from the body; a point the authors punctuate when they privilege scientific "evidence" as grounds to affirm "'true' sex"—criteria that is not applied to non-trans subjects in the same text.[73] This internal/external dualism emerged out of Western modernity.

Transpinay activist b. binaohan centers Indigenous ontology when she reminds us: "we cannot talk about an 'inner' conception of gender vs. an 'outer' physical body. it is incoherent, since there is no 'inner' and no 'outer,' simply one complete, unified mind/body."[74] To insist otherwise is to assert a hegemonic framing of sex and gender that is incommensurate with epistemologies and ontologies many of our students embody as they enter formalized learning environments (and increasingly teachers and scholars as *we* advance through academic ranks). To press the point further, Wood and Fixmer-Oraiz go on to use intersex embodiment as a "rhetorical device that points to the crisis of category and the category of crisis"—an exploitative and dehumanizing gesture found in postmodern thought experiments that reduce sexual and gender non-normativity to figures used to ironically increase understanding about normative embodiment and subjectivity.[75] Wood and Fixmer-Oraiz's oft-assigned text serves as a synecdoche for a broad array of gender texts published in the discipline that center Western ideological formations.[76] Our observations highlight the effects of what Chakravartty, Kuo, Grubbs, and McIlwan term "citational segregation" or the tendency to cite authors with whom we share a group affiliation.[77] Non-Western-centric scholarship that decenters whiteness is underrepresented throughout the communication discipline, leading to reductive and insular discussions regarding gender and communication. Indeed, that upwards of 80% of academic research is published in English effectively turns gender understanding in favor of Western ideological formations, specifically emerging out of a U.S. context.[78] As a result, global gender knowledge privileges U.S. formations that draw on colonial binary structures and violent histories that are always and already racialized.[79] CCP's critical posture implores us to diversify our citational commitments and to affirm a multiplicity of genders. To accomplish this, we begin by theorizing through queer failure.

We theorize queer failure as an embodied and resistant posture that refuses to acquiesce to normative formations—even if performance(s) of self appear to concede[80]—positioning the individual as "continually resist[ing]" the performative (re-)constitution of administrative and institutional structures.[81] In this regard—and drawing on Fassett and Warren's original articulation of CCP—embracing failure facilitates learning environments in which students are "unwilling and unable to reproduce a given ideology."[82] José Esteban Muñoz theorizes queer failure as a mundane "mode of escape" for minoritarian Subjects that affirms the embodied means by which they/we refuse a "dominant order and its systemic violence."[83] In Muñoz's framing, failure is constitutive of minoritarian subjectivity—queerness at the intersections of race and class in particular—as a result of being defined in relation to normative criteria. LeMaster clarifies, the normative constructs non-normativity as "failing to meet (its own) dominant cultural values."[84] However, this is *not* failure as we are theorizing it here, as *queer failure* is understood from the vantage of the non-normative. From this positionality, the Subject has not failed normative criteria but rather "normative culture has failed to make room for and to affirm non-normative embodiment, identification, and subjectivity."[85] Thus, to embrace failure in this regard is to resist mastery or to engage and critique the structural means by which one is rendered a "failure" as a result of their embodied difference across intersections of identity. It is to conceptualize of modern constructs like gender as a colonial technology used to categorize bodies, which in turn shapes how we interpret bodies at all. Queer failure thus implores educators to, drawing on Gust Yep's de-subjugation of racialized knowledges, *provincialize* normative cultural machinations (e.g., whiteness, cisheteronormativity) and to, in turn, theorize mundane and embodied means of navigating and surviving the same.[86] This labor implicates "the systemic means by which normative formations fail to affirm non-normative cultural difference."[87] Our task, then, includes naming the normative means by which gender becomes intelligible and, in turn, theorizing embodied means by which those who fail to meet said criteria survive and thrive nonetheless. For this task, we turn to Maria Lugones, who helps us to refuse mastering colonial categories including gender.

Lugones theorizes gender as a "colonial imposition" that sought to shift not the identity but the very nature of those who were colonized.[88] Similarly, binaohan argues what we know as the gender binary coupled with transmisogyny in concert with white supremacy emerged as tools of colonialism used to justify the destruction of gender and sexual formations that were unintelligible to the colonizer's gaze.[89] Colonization thus references the coercive and compulsory categorization of people that was justified by the "process of active reduction of people."[90] That is to say—prior to "gender"—Indigenous difference was distinguished from a presumed Western core understood as the marker of civilization and thus humanity. To put it bluntly: Indigeneity was/is framed as non-human—in a category entirely distinct from that of the human Western subject. Decolonization, at the core then, is a praxiological project that reformulates the "genre of human" constituted over 600 years of colonization.[91] In the colonial framework, gender categories distinguished human "men" from their sub-human form "woman" whose sole task it was to reproduce the human-man form. In turn, sex categories were used to determine the labor potential of non-human colonized subjects. Lugones writes, "As primitive, wild, not quite human, the colonized were [...] understood sexually as males and females, the female the inferior, inverted male."[92] Distinct from "females" were what we might term gender and sexual non-normative formations, which were categorized as "deviations from [non-human] male perfection."[93] Both "females" and non-normative sexual and gender formations were distinguished from "males"—one an inversion the other a deviation—while *all* were rendered non-human under white supremacy. The coloniality of gender reveals that racist ideology always and already underpins sex and gender intelligibility. As a result, critical communication educators and theorists cannot affirm Black and Indigenous trans women in discussions of gender without engaging the ways in which white supremacy has historically undergirded sex/gender distinctions.

The coloniality of gender is not a relic of the past but the foundation for contemporary articulations of sex and gender intelligibility evidenced in Bey's BTFT in which our goal is to radically validate Black and Indigenous trans subjectivity, which is a project humanizing those who have been sociopolitically, historically, and materially dehumanized. To refuse mastery and to embrace failure is to perpetually resist the coloniality of gender as a hegemonic structure organizing life at macro, meso, and micro levels. Refusing mastery can be conceptualized in bell hooks' words as a "practice of freedom" that marks the limits of knowledge so as to re-envision it differently and in tension with colonialization's persistence; doing otherwise "merely strives to reinforce domination."[94] Lugones writes, "The long process of subjectification of the colonized toward adoption/internalization of the men/women dichotomy as a normative construction of the social, a mark of civilization, citizenship, membership in civil society was and is constantly renewed."[95] This reframing of performativity in terms that prioritize race through the project of humanization highlights the need to reflexively grapple with the coloniality of gender as the ground on which gender and sex can ever be known in any meaningful way. Indeed, it is precisely because trans women of color are dehumanized that they are disregarded as inconsequential to the maintenance of white manhood and white womanhood centered in much gender and communication thought and discourse. Naming this in blunt terms is implicative and thus requires reflexivity, which points us to our dialectical reversal: Mastering refusal. Whereas refusing mastery resists the colonizer's gaze through engaging and reimagining the coloniality of gender, mastering refusal engages and preempts the affective registers that seek to refuse transformation. This is to say, mastering refusal is about mastering or embracing the potential in categorical uncertainty—of our own and others. Whereas refusing mastery marks an epistemological commitment, mastering refusal marks an ontological shift.

For instance, reading gender on its own terms, through the "cosmologies that inform it," without translating said vernacular gender through a colonial binary structure begins the work necessary to decolonize gender.[96] Doing so, however, is uneasy for those who refuse to decenter the (colonial) self. It requires self-implication. Richard Jones theorizes "intersectional reflexivity" as an embodied means of holding ourselves accountable to our concurrent privileges and disadvantages as they emerge and shift across communicative contexts "even as we work to subvert them."[97] He writes elsewhere, this work "cuts to the bone. It implicates you. Reflexivity is uncomfortable because it forces you to acknowledge that you are complicit in the perpetuation of oppression."[98] Thus, to master refusal is to engage the improvisational space of unknowing[99] so as to affirm gender at the intersections of race on its emergent, creative, constrained, and relational intersectional terms. In this regard, white nonbinary students cannot universally claim that binary people of color oppress them. For instance, I (Benny) am a nonbinary trans person who is also mixed-race Asian/white, though I pass as white. As a result, it would be self-serving and racist to presume a binary trans woman of color has more power than me simply because I am nonbinary; I have access to white privilege *even as* a nonbinary and mixed-race person. In this way, white TGNC folks always and already have access to white privilege and thus always and already embody genders that are more intelligible to white supremacy relative to TGNC folks of color, whose genders are initially filtered through racist tropes. To affirm trans women of color, and Black and Indigenous trans women in particular, is to theorize gender in terms that affirm the materiality of race, racism, and coloniality.

A pedagogical means of accomplishing this can include extending critiques on the coloniality of gender through distinguishing between gender *attribution* (the gendered meaning we attribute to bodies based on our presumptions about bodies including ideological scripts that animate bodily attributions like white supremacy), gender *identity* (a gender sense of self that may not acquiesce to Western discourses of intelligibility), *assigned* sex/gender (administrative and performative means by which our bodies are *sexed* and in turn *gendered* based on white and non-trans criteria), and gender *expression* (the means by which we express our gender(s); emergent on terms that may not acquiesce to Western discourses of intelligibility). Gender scholars have long shown us that there is far more intragroup variability than there is intergroup variability among gender categories. That is, there is more variance among "men" and among "women" than between "men" and "women"—this is made especially evident when we center the racialized terms set under colonization that always and already distinguish types of men and types of women in relation to normative notions of white manhood and white womanhood. This revelation is not new and instead marks the core of women of color and transnational feminisms that refute white and thus universal framings of "woman."

The history of colonization cannot be unwritten. We cannot return to a precolonial moment. As a result, decolonization engages the hybrid forms that emerge out of colonization including the creative, mundane, and embodied ways by which minoritarian subjects navigate the coloniality of gender.[100] In her bid to decolonize the communication classroom in a time of rising white nationalism and xenophobia, Devika Chawla implores educators to explore identity as "racially, culturally, and ethnically" hybrid and thus intersectional while questioning hegemonic cultural narratives and grappling with agency and resistance in nuanced and embodied ways through storytelling.[101] Incorporating stories by TGNC folks of color—Indigenous and Black trans women in particular—living livable lives begins the important work of rehumanizing difference under the coloniality of gender. Storying variation from normativity humanizes difference as it highlights our relative and intersubjective navigation of the coloniality of gender. This point of connection is useful as it implores students and pedagogues to engage the embodied means by which we are rendered non-human as a result of our embodied differences with concerted focus granted to the ways race undergirds these intersectional and monstrous renderings. This can include subtleties (e.g., disciplining emotional men or sexual women as informed through racist scripts) just as it can include relative intensities (e.g., racialized sexual and physical violence against TGNC folks of color). The goal is to decenter the presumption that decolonization only benefits a few people and to instead engage the multifarious impact the coloniality of gender plays in all of our gender lives as subjects navigating colonization's wake with particular focus granted to those who remain most materially impacted by the coloniality of gender. bell hooks reminds us pedagogy can be liberatory only when "everyone claims knowledge as a field in which we all labor."[102]

Contrary to antifeminist beliefs, gender is a cultural phenomenon with a rich history impacting more than women and trans folks. Gender is, as we have exhibited here, communicatively constituted through colonial logics that binaristically dehumanize blackness and indigeneity as it exemplifies and humanizes whiteness. To refuse mastery is to resist the colonial logics perpetually pitting gender against race, which inevitably ends with the erasure of those most materially impacted by the coloniality of gender: TGNC folks of color and Black and Indigenous trans women in particular. This requires an epistemological shift if we are to affirm trans women of color as living livable lives under the imposition of coloniality. Conversely, mastering refusal marks an ontological shift in which the individual is implicated in their perpetuation of the coloniality of gender in unconscious ways. For Freire, *conscientização* marks our critical pedagogical goal. That goal requires that we explore gender, in this case, as a historically, materially, and relationally situated cultural phenomenon. More than that, affect anchors *conscientização* through the willful embracing of uncertainty in the face of unlearning what we presume to have known all along. This work is necessarily uncomfortable. And it should be.

Notes

1 Benny LeMaster and Amber L. Johnson, "Unlearning Gender," *Communication Teacher*, 33, no. 4 (2018): 189, https://doi.org/10.1080/17404622.2018.1467566.
2 Dean Spade, *Normal Life* (Durham, NC: Duke University Press, 2015).
3 Lisa Flores, "Gender with/out Borders," in *The SAGE Handbook of Gender and Communication*, eds Bonnie J. Dow and Julia T. Wood (Thousand Oaks, CA: Sage, 2006), 386.
4 María Lugones, "Toward a Decolonial Feminism," *Hypatia* 25, no. 4 (2010): 742–759, www.jstor.org/stable/40928654.
5 LeMaster and Johnson, "Unlearning."
6 Deanna L. Fassett and C. Kyle Rudick, "Critical Communication Pedagogy," in *Communication and Learning*, ed. Paul Witt (Berlin: De Gruyter Mouton, 2016), 579.
7 Deanna L. Fassett and John T. Warren, *Critical Communication Pedagogy* (Thousand Oaks, CA: Sage, 2007).
8 Paolo Freire, *Pedagogy of the Oppressed*, trans. Myra Bergman Ramos (New York: Continuum, 2005), 72.
9 Freire, *Oppressed*, 79.
10 David H. Kahl Jr., "Addressing the Challenges of Critical Communication Pedagogy Scholarship," *Journal of Applied Communication Research* 45, no. 1 (2017): 116–120, https://doi.org/10.1080/00909882.2016.1248468.
11 Paolo Freire and Donaldo P. Macedo, "A Dialogue," *Harvard Educational Review* 65, no. 3 (1995): 379, https://doi.org/10.17763/haer.65.3.12g1923330p1xhj8.
12 Freire, *Oppressed*, 80.

13 Dwight Conquergood, "Performing as a Moral Act," *Literature in Performance* 85, no. 5 (1985): 9, https://doi.org/10.1080/10462938509391578.

14 María Lugones, "Methodological Notes toward a Decolonial Feminism," in *Decolonizing Epistemologies*, eds Ada Mara Isasi-Daz and Eduardo Mendieta (New York: Fordham University Press, 2011), 68–86.

15 Fassett and Warren, *Critical*.

16 J. Halberstam, *The Queer Art of Failure* (Durham, NC: Duke University Press, 2011); Benny LeMaster, "Pedagogies of Failure," in *Critical Intercultural Communication Pedagogy*, eds Ahmet Atay and Satoshi Toyosaki (Lanham, MD: Lexington, 2018), 81–96.

17 Benny LeMaster, "Embracing Failure: Improvisational Performance as Critical Intercultural Praxis," *Liminalities* 14, no. 4 (2018), http://liminalities.net/14-4/embracing.pdf.

18 Fassett and Warren, *Critical*.

19 Deanna L. Fassett and Keith Nainby, "Critical Communication Pedagogy," in *Handbook of Instructional Communication*, eds Marian Houser and Angela Hosek (New York: Routledge, 2018), 248–259.

20 Deanna L. Fassett and C. Kyle Rudick, "Critical Communication Pedagogy: Toward 'Hope in Action,'" in *Oxford Research Encyclopedia of Communication* (New York: Oxford University Press, 2018), doi 10.1093/acrefore/9780190228613.013.628.

21 Fassett and Rudick, "Hope."

22 Fassett and Warren, *Critical*, 39.

23 Fassett and Rudick, *Critical*, 576.

24 Jo Sprague, "Ontology, Politics, and Instructional Communication Research," *Communication Education* 43, no. 4 (1994): 273–290, https://doi.org/10.1080/03634529409378986.

25 Fassett and Warren, *Critical*, 42.

26 In their original articulation, Fassett and Warren describe power as "fluid and complex" (xxv). While power is indeed complex, a material shift encourages practitioners to ground their CCP praxis in the materiality of lived experience as intersectional subjects navigating both oppressive and destructive as well as empowering and productive manifestations of power. That is, name the complexity that animates power as a cultural force.

27 Susana Martínez Guillem, "Rethinking Power Relations in Critical/Cultural Studies," *The Review of Communication* 13, no. 3 (2013): 184–204, http://dx.doi.org/10.1080/15358593.2013.843716.

28 Guillem, "Rethinking," 192, 190, 192. We are equally indebted to Martin and Nakayama's groundbreaking work in dialectic thought and praxis. Judith N. Martin and Thomas K. Nakayama, "Intercultural Communication and Dialectics Revisited," in *The Handbook of Critical Intercultural Communication*, eds Thomas K. Nakayama and Rona Tamiko Halualani (Malden, MA: Blackwell Publishing, 2010), 59–83; Judith N. Martin and Thomas K. Nakayama, "Thinking Dialectically about Culture and Communication," *Communication Theory* 9 (1999): 1–25, https://doi.org/10.1111/j.1468-2885.1999.tb00160.x.

29 Patricia Hill Collins and Sirma Bilge, *Intersectionality* (Cambridge: Polity, 2016), 15–18.

30 Robin M. Boylorn and Mark P. Orbe, "Critical Autoethnography as Method of Choice," in *Critical Autoethnography*, eds Robin M. Boylorn and Mark P. Orbe (New York: Routledge, 2013), 16.

31 For a feminist discussion on power see Karma R. Chávez and Cindy L. Griffin, "Power, Feminisms, and Coalitional Agency," *Women's Studies in Communication* 32, no. 1 (2009): 1–11, https://doi.org/10.1080/07491409.2009.10162378; Aimee Carrillo Rowe, "Subject to Power—Feminism without Victims," *Women's Studies in Communication* 32, no. 1 (2009): 12–35, https://doi.org/10.1080/07491409.2009.10162379.

32 Collins and Bilge, *Intersectionality*, 25.

33 Collins and Bilge, *Intersectionality*, 25, emphasis original.

34 bell hooks, *Feminism is for Everybody* (New York: Routledge, 2015). See also Gloria Anzaldúa, "(Un)natural Bridges, (Un)safe Spaces," in *This Bridge We Call Home*, eds Gloria E. Anzaldúa and AnaLouise Keating (New York: Routledge, 2002), 1–5; AnaLouise Keating, *Transformation Now!* (Urbana: University of Illinois Press, 2013).

35 Fassett and Warren, *Critical*, 43.

36 Fassett and Warren, *Critical*, 45.

37 Fassett and Warren, *Critical*, 48.

38 Paolo Freire, *Pedagogy of Hope*, trans. Robert R. Barr (New York: Continuum, 1994), 56.

39 Fassett and Rudick, "Critical," 578.

40 Yukio Tsuda, "Speaking against the Hegemony of English," in *The Handbook of Critical Intercultural Communication*, eds Thomas K. Nakayama and Rona Tamiko Halualani (Malden, MA: Blackwell Publishing, 2010), 248–269.

41 Finn Enke, "Stick Figures and Little Bits," in *Trans Studies: The Challenge to Hetero/Homo Normativities*, eds Yolanda Martínez and Sarah Tobias (New Brunswick, NJ: Rutgers University Press, 2016), 218.

42 William Shakespeare, *The Plays of Shakespeare: The Comedy of Errors* (London: William Heinemann, 1904), 45, emphasis added.

43 Canadian professor Jordan Peterson has infamously refused to use gender-neutral pronouns for students on the grounds of "academic freedom." See Jessica Murphy, "Toronto Professor Jordan Peterson Takes on Gender-Neutral Pronouns," *BBC News* (Toronto), November 4, 2016, www.bbc.com/news/world-us-canada-37875695.

44 Fassett and Warren, *Critical*, 45.

45 Fassett and Warren, *Critical*, 50.

46 Fassett and Warren, *Critical*, 50.

47 Fassett and Warren, *Critical*, 52.

48 Fassett and Warren, *Critical*, 54.

49 Freire, *Oppressed*, 51.

50 Freire, *Oppressed*, 52, 51.

51 Fassett and Warren, *Critical*, 51.

52 Fassett and Warren, *Critical*, 51.

53 C. Kyle Rudick, Kathryn B. Golsan, and Kyle Cheesewright, *Teaching from the Heart* (San Diego: Cognella, 2017), 25.

54 Calvin O. Schrag, *The Self after Postmodernity* (New Haven, CT: Yale University Press, 1997), 9.

55 Guillem, "Rethinking," 266.

56 Guillem, "Rethinking," 266.

57 Lugones, "Methodological."

58 Michelle Alexander, *The New Jim Crow* (New York: The New Press, 2012).

59 Inter-American Commission on Human Rights, *Violence against LGBTI Persons* (Washington, DC: Organization of American States, 2015), www.oas.org/en/iachr/reports/pdfs/violencelgbtipersons.pdf; See also Elías Cosenze Krell, "Is Transmisogyny Killing Trans Women of Color?" *Transgender Studies Quarterly* 4, no. 2 (2017): 226–242, https://doi.org/10.1215/23289252-3815033.

60 Inter-American Commission on Human Rights, *Violence*, 157–158.

61 These data reflect persons known to be or to identify as transgender and/or gender nonconforming and whose deaths are known to have been motivated by racialized cisheterosexism.

62 Aren Z. Aizura, Trystan Cotton, Carsten Balzer/Carla LaGata, Marcia Ochoa, and Salvador Vidal-Ortiz, "Introduction," *Transgender Studies Quarterly* 1, no. 3 (2014): 312, https://doi.org/10.1215/23289252-2685606.

63 Aren Z. Aizura, "Affective Vulnerability and Transgender Exceptionalism," in *Trans Studies*, eds Yolanda Martínez-San Miguel and Sarah Tobias (New Brunswick, NJ: Rutgers University Press, 2016), 133.

64 Vivane K. Namaste, *Invisible Lives* (Chicago, IL: The University of Chicago Press, 2000).

65 Marquis Bey, "The Shape of Angels' Teeth," *Departures in Critical Qualitative Research* 5, no. 3 (2016): 43, https://doi.org/10.1525/dcqr.2016.5.3.33.

66 Bey, "The Shape," 44. See also C. Riley Snorton, *Black on Both Sides* (Minneapolis, MN: University of Minnesota Press, 2017).

67 Bey, "The Shape," 45.

68 Raka Shome and Radha S. Hegde, "Postcolonial Approaches to Communication," *Communication Theory* 12, no. 3 (2002): 250, https://doi.org/10.1111/j.1468-2885.2002.tb00269.x.

69 Julia T. Wood and Natalie Fixmer-Oraiz, *Gendered Lives* (Boston, MA: Cengage, 2017), 20.

70 Wood and Fixmer-Oraiz, *Gendered*, 19.

71 Julia Serano, *Excluded* (Berkeley, CA: Seal Press, 2013); Jay Prosser, *Second Skins* (Chichester: Columbia University Press, 1998).

72 b. binaohan, *decolonizing trans/gender 101* (Toronto: Biyuti publishing, 2014): 68; see also Spade, *Normal*.

73 Wood and Fixmer-Oraiz, *Gendered*, 26.

74 binaohan, *decolonizing*, 35. Quotations by this author appear as published—free of edits and in keeping with binaohan's wishes.

75 Namaste, *Invisible*, 15.

76 For a text that shifts the traditional articulation of gender and communication, see Catherine Helen Palczewski, Victoria Pruin DeFrancisco, and Danielle Dick McGeough, *Gender in Communication*, 3rd edn (Boston: Cengage, 2017).

77 Paula Chakravartty, Rachel Kuo, Victoria Grubbs, and Charlton McIlwan, "#CommunicationSoWhite," *Journal of Communication* 68, no. 2 (2018): 254–266, https://doi.org/10.1093/joc/jqy003.

78 Yukio Tsuda, "The Hegemony of English and Strategies for Linguistic Pluralism," in *The Global Intercultural Communication Reader*, eds Moledi Kete Asante, Yoshitaka Miike, and Jing Yin (New York: Routledge, 2013), 445–456.

79 Aizura, Cotton, Balzer/LaGata, Ochoa, and Vidal-Ortiz, "Introduction."

80 Devika Chawla and Ahmet Atay, "Introduction," *Cultural Studies ⇔ Critical Methodologies* 18, no. 1 (2018): 3–8, https://doi.org/10.1177/1532708617728955.

81 Lugones, "Methodological," 78.

82 Fassett and Warren, 25. See also Benny LeMaster, "Embracing."

83 José Esteban Muñoz, *Cruising Utopia* (New York: New York University Press, 2009), 172.

84 Benny LeMaster, "Pedagogies of Failure," in *Critical Intercultural Communication Pedagogy*, eds Ahmet Atay and Satoshi Toyosaki (Lanham, MD: Lexington Press, 2018), 85.

85 LeMaster, "Pedagogies," 86.

86 Gust Yep, "Toward the De-subjugation of Racially Marked Knowledges in Communication," *Southern Communication Journal* 75, no. 2 (2010): 171–175, https://doi.org/10.1080/10417941003613263.

87 LeMaster, "Pedagogies," 94.

88 Lugones, "Toward," 748.

89 binaohan, *decolonizing*, 122.

90 Lugones, "Methodological," 75.

91 Aman Sium, Chandni Desai, and Eric Ritskes, "Towards the 'Tangible Unknown,'" *Decolonization* 1, no. 1 (2012): xi, https://jps.library.utoronto.ca/index.php/des/article/view/18638/15564.

92 Lugones, "Methodological," 73.

93 Lugones, "Methodological," 73.

94 bell hooks, *Teaching to Transgress* (New York: Routledge, 1994), 4.

95 Lugones, "Methodology," 79.

96 Lugones, "Methodology," 79; see also Vic Muñoz, "Gender Sovereignty," in *Transfeminist Perspectives*, ed. Finn Enke (Philadelphia, PA: Temple University Press, 2012), 23–33.

97 Richard G. Jones Jr., "Queering the Body Politic," *Qualitative Inquiry* 21, no. 9 (2015): 767, https://doi.org/10.1177/1077800415569782.

98 Richard G. Jones Jr., "Putting Privilege into Practice Through 'Intersectional Reflexivity,'" *Faculty Research and Creative Activity* 3 (2010): 124, https://thekeep.eiu.edu/commstudies_fac/3; see also Dawn Marie D. McIntosh and Kathryn Hobson, "Reflexive Engagement," *Liminalities* 9, no. 4 (2013), http://liminalities.net/9-4/reflexive.pdf; Haneen Ghabra and Bernadette Marie Calafell, "From Failure to Allyship to Feminist Solidarities," *Text and Performance Quarterly* 38, nos 1–2 (2018): 38–54, https://doi.org/10.1080/10462937.2018.1457173.

99 LeMaster, "Embracing."

100 Chawla and Atay, "Introduction."

101 Devika Chawla, "Contours of a Storied Decolonial Pedagogy," *Communication Education* 67, no. 1 (2018): 116, https://doi.org/10.1080/03634523.2017.1388528.

102 hooks, *Teaching*, 14.

Bibliography

Aizura, Aren Z. "Affective Vulnerability and Transgender Exceptionalism." In *Trans Studies*, edited by Yolanda Martínez-San Miguel and Sarah Tobias, 122–140, New Brunswick, NJ: Rutgers University Press, 2016.

Aizura, Aren Z., Trystan Cotton, Carsten Balzer/Carla LaGata, Marcia Ochoa, and Salvador Vidal-Ortiz. "Introduction." *Transgender Studies Quarterly* 1, no. 3 (2014): 308–19. https://doi.org/10.1215/23289252-2685606.

Alexander, Michelle. *The New Jim Crow*. New York: The New Press, 2012.

Anzaldúa, Gloria. "(Un)natural Bridges, (Un)safe Spaces." In *This Bridge We Call Home*, edited by Gloria E. Anzaldúa and AnaLouise Keating, 1–5, New York: Routledge, 2002.

Bey, Marquis. "The Shape of Angels' Teeth." *Departures in Critical Qualitative Research* 5, no. 3 (2016): 33–54. https://doi.org/10.1525/dcqr.2016.5.3.33.

binaohan, b. *Decolonizing Trans/Gender 101*. Toronto: Biyuti publishing, 2014.

Boylorn, Robin M. and Mark P. Orbe. "Critical Autoethnography as Method of Choice." In *Critical Autoethnography*, edited by Robin M. Boylorn and Mark P. Orbe, 13–26. New York: Routledge, 2013.

Chakravartty, Paula, Rachel Kuo, Victoria Grubbs, and Charlton McIlwan, "#CommunicationSoWhite." *Journal of Communication* 68, no. 2 (2018): 254–66. https://doi.org/10.1093/joc/jqy003.

Chávez, Karma R. and Cindy L. Griffin. "Power, Feminisms, and Coalitional Agency." *Women's Studies in Communication* 32, no. 1 (2009): 1–11. https://doi.org/10.1080/07491409.2009.10162378.

Chawla, Devika. "Contours of a Storied Decolonial Pedagogy." *Communication Education* 67, no. 1 (2018): 115–20. https://doi.org/10.1080/03634523.2017.1388528.

Chawla, Devika and Ahmet Atay. "Introduction." *Cultural Studies ⇔ Critical Methodologies* 18, no. 1 (2018): 3–8. https://doi.org/10.1177/1532708617728955.

Collins, Patricia Hill and Sirma Bilge. *Intersectionality*. Cambridge: Polity, 2016.

Conquergood, Dwight. "Performing as a Moral Act." *Literature in Performance* 85, no. 5 (1985): 1–13. https://doi.org/10.1080/10462938509391578.

Enke, A. Finn. "Stick Figures and Little Bits," In *Trans Studies*, edited by Yolanda Martínez and Sarah Tobias, 215–29, New Brunswick, NJ: Rutgers University Press, 2016.

Fassett, Deanna L. and C. Kyle Rudick. "Critical Communication Pedagogy." In *Oxford Research Encyclopedia of Communication*, New York: Oxford University Press, 2018. doi 10.1093/acrefore/9780190228613.013.628.

Fassett, Deanna L. and C. Kyle Rudick. "Critical Communication Pedagogy." In *Communication and Learning*, edited by Paul Witt, 573–98, Berlin: De Gruyter Mouton, 2016.

Fassett, Deanna L. and John T. Warren. *Critical Communication Pedagogy*. Thousand Oaks, CA: Sage, 2006.

Fassett, Deanna L. and Keith Nainby. "Critical Communication Pedagogy." In *Handbook of Instructional Communication*, edited by Marian Houser and Angela Hosek, 248–59, New York: Routledge, 2018.

Freire, Paolo. *Pedagogy of Hope*, translated by Robert R. Barr. New York: Continuum, 1994.

Freire, Paolo. *Pedagogy of the Oppressed*, translated by Myra Bergman Ramos. New York: Continuum, 2005.

Freire, Paolo and Donaldo P. Macedo. "A Dialogue." *Harvard Educational Review* 65, no. 3 (1995): 377–403. https://doi.org/10.17763/haer.65.3.12g1923330p1xhj8.

Flores, Lisa. "Gender with/out Borders." In *The SAGE Handbook of Gender and Communication*, edited by Bonnie J. Dow and Julia T. Wood, Thousand Oaks, CA: Sage, 2006.

Ghabra, Haneen and Bernadette Marie Calafell. "From Failure to Allyship to Feminist Solidarities." *Text and Performance Quarterly* 38, no. 1–2 (2018): 38–54. https://doi.org/10.1080/10462937.2018.1457173.

Guillem, Susana Martínez. "Rethinking Power Relations in Critical/Cultural Studies." *The Review of Communication* 13, no. 3 (2013): 184–204. http://dx.doi.org/10.1080/15358593.2013.843716.

Halberstam, J. *The Queer Art of Failure*. Durham, NC: Duke University Press, 2011.

hooks, bell. *Feminism Is for Everybody*. New York: Routledge, 2015.

hooks, bell. *Teaching to Transgress*. New York: Routledge, 1994.

Inter-American Commission on Human Rights. Violence against LGBTI Persons. Washington, DC: Organization of American States, 2015. www.oas.org/en/iachr/reports/pdfs/violencelgbtipersons.pdf.

Jones, Richard G., Jr., "Putting Privilege into Practice Through 'Intersectional Reflexivity.'" *Faculty Research and Creative Activity* 3 (2010): 122–25. https://thekeep.eiu.edu/commstudies_fac/3.

Jones, Richard G., Jr., "Queering the Body Politic." *Qualitative Inquiry* 21, no. 9 (2015): 766–75. https://doi.org/10.1177/1077800415569782.

Kahl, David H., Jr., "Addressing the Challenges of Critical Communication Pedagogy Scholarship." *Journal of Applied Communication Research* 45, no. 1 (2017): 116–20. https://doi.org/10.1080/00909882.2016.1248468.

Keating, AnaLouise. *Transformation Now!* Urbana, IL: University of Illinois Press, 2013.

Krell, Elías Cosenze. "Is Transmisogyny Killing Trans Women of Color?" *Transgender Studies Quarterly* 4, no. 2 (2017): 226–42. https://doi.org/10.1215/23289252-3815033.

LeMaster, Benny. "Embracing Failure: Improvisational Performance as Critical Intercultural Praxis." *Liminalities* 14, no. 4 (2018): http://liminalities.net/14-4/embracing.pdf.

LeMaster, Benny. "Pedagogies of Failure," In *Critical Intercultural Communication Pedagogy*, edited by Ahmet Atay and Satoshi Toyosaki, 81–96, Lanham: Lexington, 2018.

LeMaster, Benny and Amber L. Johnson. "Unlearning Gender." *Communication Teacher* 33, no. 4 (2018): 189–198. https://doi.org/10.1080/17404622.2018.1467566.

Lugones, María. "Methodological Notes Toward a Decolonial Feminism." In *Decolonizing Epistemologies*, edited by Ada Mara Isasi-Daz and Eduardo Mendieta, 68–86, New York: Fordham University Press, 2011.

Lugones, María. "Toward a Decolonial Feminism." *Hypatia* 25, no. 4 (2010): 742–59. www.jstor.org/stable/40928654.

Martin, Judith N. and Thomas K. Nakayama. "Intercultural Communication and Dialectics Revisited." In *The Handbook of Critical Intercultural Communication*, edited by Thomas K. Nakayama and Rona Tamiko Halualani, 59–83, Malden, MA: Blackwell Publishing, 2010.

Martin, Judith N. and Thomas K. Nakayama. "Thinking Dialectically about Culture and Communication." *Communication Theory* 9 (1999): 1–25. https://doi.org/10.1111/j.1468-2885.1999.tb00160.x.

McIntosh, Dawn Marie D. and Kathryn Hobson. "Reflexive Engagement." *Liminalities* 9, no. 4 (2013): http://liminalities.net/9-4/reflexive.pdf.

Muñoz, José Esteban. *Cruising Utopia*. New York: New York University Press, 2009.

Muñoz, Vic. "Gender Sovereignty." In *Transfeminist Perspectives*, edited by Finn Enke, 23–33. Philadelphia, PA: Temple University Press, 2012.

Murphy, Jessica. "Toronto Professor Jordan Peterson Takes on Gender-Neutral Pronouns." *BBC News* (Toronto). November 4, 2016. www.bbc.com/news/world-us-canada-37875695.

Namaste, Vivane K. *Invisible Lives*. Chicago, IL: The University of Chicago Press, 2000.

Palczewski, Catherine Helen, Victoria Pruin DeFrancisco, and Danielle Dick McGeough. *Gender in Communication*, 3rd edition, Boston, MA: Cengage, 2017.

Prosser, Jay. *Second Skins*. Chichester, UK: Columbia University Press, 1998.

Rowe, Aimee Carrillo. "Subject to Power—Feminism without Victims." *Women's Studies in Communication* 32, no. 1 (2009): 12–35. https://doi.org/10.1080/07491409.2009.10162379.

Rudick, C. Kyle, Kathryn B. Golsan, and Kyle Cheesewright. *Teaching from the Heart*. San Diego, CA: Cognella, 2017.

Schrag, Calvin O. *The Self After Postmodernity*. New Haven, CT: Yale University Press, 1997.

Serano, Julia. *Excluded*. Berkeley, CA: Seal Press, 2013.

Shakespeare, William. *The Plays of Shakespeare: The Comedy of Errors*. London: William Heinemann, 1904.

Shome, Raka and Radha S. Hegde. "Postcolonial Approaches to Communication." *Communication Theory* 12, no. 3 (2002): 249–70. https://doi.org/10.1111/j.1468-2885.2002.tb00269.x.

Sium, Aman, Chandni Desai, and Eric Ritskes. "Towards the 'Tangible Unknown.'" *Decolonization* 1, no. 1 (2012): i–xiii. https://jps.library.utoronto.ca/index.php/des/article/view/18638/15564.

Snorton, C. Riley. *Black on Both Sides*. Minneapolis, MN: University of Minnesota Press, 2017.

Spade, Dean. *Normal Life*. Durham, NC: Duke University Press, 2015.

Sprague, Jo. "Ontology, Politics, and Instructional Communication Research." *Communication Education* 43, no. 4 (1994): 273–90. https://doi.org/10.1080/03634529409378986.

Tsuda, Yukio. "Speaking Against the Hegemony of English," In *The Handbook of Critical Intercultural Communication*, edited by Thomas K. Nakayama and Rona Tamiko Halualani, 248–69, Malden, CA: Blackwell Publishing, 2010.

Tsuda, Yukio. "The Hegemony of English and Strategies for Linguistic Pluralism," In *The Global Intercultural Communication Reader*, edited by Moledi Kete Asante, Yoshitaka Miike, and Jing Yin, 445–56, New York: Routledge, 2013.

Wood, Julia T. and Natalie Fixmer-Oraiz. *Gendered Lives*, 12th edition, Boston, MA: Cengage, 2017.

Yep, Gust. "Toward the De-Subjugation of Racially Marked Knowledges in Communication." *Southern Communication Journal* 75, no. 2 (2010): 171–175. https://doi.org/10.1080/10417941003613263.

37

GENDER FUTURITY AT THE INTERSECTION OF BLACK LIVES MATTER AND AFROFUTURISM

Amber Johnson

In this chapter, I explore the intersections of race and gender in social movement contexts and offer alternative structures for creating a gender-free future via intersectional social movements, gender futurity, and Transfuturism. This chapter begins with a discussion of intersectionality and Black Lives Matter and Afrofuturism, two social movements that center the lived experiences of the Black diaspora in an attempt to create systemic change, while also falling short of producing gender freedom and inclusion. Then, using Black Lives Matter and Afrofuturism as departure points, this chapter argues for gender futurity using a case study in Transfuturism to unpack how our praxis can push the boundaries of gender embodiment and freedom.

Intersectionality

Intersectionality is both a normative theoretical argument and an approach to conducting empirical research ... that considers the interaction of race, gender, class, and other organizing structures of society a key component influencing political access, equality and the potential for any form of justice.[1]

Theorists who use intersectionality as a theoretical framework are interested in unpacking the cumulative impact of social systems on humans versus separating systems as if they are distinct and easily separated. For instance, in her groundbreaking article, Kimberlé Crenshaw delineated the qualitative differences between cisgender women's experiences with sexual assault and domestic abuse across intersections of race and class. She argues that women experience assault and, hopefully, eventual liberation differently depending on their race, class, and documentation status. Undocumented women do not have access to the same resources as documented women who are English speaking or employed women who have access to emergency funds. She puts forth three categories to map out the multiple dimensions of women's experiences regarding race, class, gender, and a host of other social identity categories.[2]

Structural intersectionality refers to location and the ways in which women of color experience differences due to situational circumstances like unemployment, immigrant status, lack of job skills, race and gender oppression, or lack of resources and information. Approaching structural intersectionality from a nonbinary point, we can attend to the ways in which Black trans women experience structural hardships at an even greater degree due to a lack of cisgender identity, often experiencing physical and discursive violence from both cismen and ciswomen. *Political intersectionality* refers to the ways in which women of color are often silenced and marginalized by contradicting political agendas like antiracist versus antisexist groups. As groups attempt to fuse intersectional difference across race, class, and gender, those moves are still largely targeting cisgender experiences. For instance, the Women's March attempted to be inclusive, but the pink pussy hats identified womanhood based on genitalia and thus excluded a lot of trans women and nonbinary femmes.[3] *Representational intersectionality* refers to the cultural and social construction of women of color within popular cultural contexts, which often paint Black ciswomen stereotypically, while trans and nonbinary people are largely absent from mediated representation.

What Crenshaw's work contributes to communication scholarship is a delineation of the multiple and overlapping ways in which bodies not only perform as political entities through discourse but endure the systematic attempts to maintain power through representation, politics, and institutions.[4] The labels attached to our bodies mimic more than arbitrary words designed to highlight difference; they also have the ability to (re)inforce and challenge power structures as divisive systems of oppression. As Judith Lorber contends,

> the margin and the center, the insider and the outsider, the conformist and the deviant are two sides of the same concept. Introducing even one more term, such as bisexuality, forces a rethinking of the oppositeness of heterosexuality and homosexuality.[5]

While Crenshaw's work falls shorts because of her centering of cisgender women, structural, representational, and political intersectionality are important entry points to this conversation, but gender variation must also be included as an intersection for femme identified people within the three areas. Unpacking the discourse and turning the reflexive onto our own praxis is a beginning step toward de-privileging cisgender identity at the intersections and creating a more inclusive social movement aimed at liberation for all. As Cromwell reminds us, "disruption occurs because an individual is capable of articulating an identity founded upon both/and as well as neither/or and either/or."[6] Our gendered bodies are complex and require nuance. When we unpack the nuance of gender justice in the movement for Black Lives and Afrofuturism, it becomes very apparent that our movements, while advancing racial equity, often fail to uplift those in the margins of the margins due to non-normative gender, ability, and a host of other social categories. The next sections interrogate Black Lives Matters and Afrofuturism for the gender oppression despite radical racial potential.

The movement for Black Lives

Black Lives Matter espouses to be an intersectional movement because it attends to the needs of Black women and was led by three Black queer ciswomen; however, its shortcomings in terms of valuing the work and lives of trans and nonbinary folks are multifold. Black Lives Matter, as a movement, can be traced to mediated responses to Trayvon Martin's death. On February 26, 2012, George Zimmerman murdered Trayvon Martin, a 17-year-old African American, in Sanford, Florida. Massive protests and marches followed, prompting Black and progressive America to try and answer the question:

> What makes it possible for a 17-year-old boy, a teen, someone's son, someone's friend, someone's student to be shot dead by a vigilante, after fighting for his life, and no one truly *hears* the fear, the desire to survive, the anguish and angst, encapsulated in his scream?[7]

Don't our lives matter at all? Alicia Garza, Opal Tometi, and Patrisse Cullors answered that question with the hashtag #BlackLivesMatter. While the hashtag was created by the three women, the social movement surrounding the hashtag is an international phenomenon seeded in several tragic incidents inspiring momentum.

On August 9, 2014, Darren Wilson murdered Michael Brown, another unarmed, African American teenager in Ferguson, Missouri. A preemptive answer to Garza, Tometi, and Cullors' question, "Don't our lives matter at all?" resounded in a loud and angry, "NO!" Black life does not matter. Black people can be killed in the streets while the world continues to live. Shortly after Michael Brown's death, Eric Garner, VonDerrit Myers, John Crawford, and Tamir Rice's murders gained traction in social media. These five men and children represent only 0.4% of the staggering 1,112 victims murdered by police officers in 2014 alone.[8] With these new events emerging and begging for the attention of national, international, and social media, tweets turned into protests, a social media hashtag turned into social momentum, and a movement was birthed.

The Black Lives Matter hashtag and subsequent social movement mark a resurgence in social action. Bailey and Leonard acknowledge the twofolded nature of the movement: it not only challenges white supremacy, systemic racial oppression, and state-sanctioned violence against Black bodies, the hashtag also articulates Black love, community, and possibility.[9] With every mention of the hashtag, mouths form the words that proclaim Black people do, in fact, matter. "Black Lives Matter is an act of collective imagination [that] envisions … [the] conditions that guarantee [our] voices, humanity, and lives … are protected, valued, and embraced."[10] The hashtag represents more than a "successful discursive tool"; it is a powerful rhetorical tool that calls for justice, the end of systemic disregard for our bodies, and the exigency for community movement.[11]

Black Lives Matter strives to be intersectional and intentional with naming, especially as its platform is published. One such passage reads,

> We believe in elevating the experiences and leadership of the most marginalized Black people, including but not limited to those who are women, queer, trans, femmes, gender nonconforming, Muslim, formerly and currently incarcerated, cash poor and working class, disabled, undocumented, and immigrant. We are intentional about amplifying the particular experience of state and gendered violence that Black queer, trans, gender nonconforming, women, and intersex people face. There can be no liberation for all Black people if we do not center and fight for those who have been marginalized. It is our hope that by working together to create and amplify a shared agenda, we can continue to move towards a world in which the full humanity and dignity of all people is recognized.[12]

This passage explicates an intersectional understanding of how the marginalized move within the margins of the margins and why those voices and bodies cannot afford to be silenced or erased. Albeit poverty, disability, non-normative gender identity, or queer sexuality, it is imperative that we keep in mind those most vulnerable to white supremacy, capitalism, and imperialism. However, despite this rhetorical move to be intersectional by including different types of Black people due to religion, ability, sexuality, and gender identity, the ways in which activists embody Black Lives Matter has been far from inclusive. While Black Lives Matter garnered the support of millions nationally and internationally, there are Black identities missing from the structural, political, and representational conversations covering social change. Two important moments illustrate whose bodies are left out of massive communication efforts.

First, if we reference the aforementioned names of people who pushed the hashtag to become an international social movement, we would be remiss to not notice all of the names belong to cisgender Black men and boys. The majority of #BlackLivesMatter organizing occurs for cisgender men and boys who fall victim to state-sanctioned violence. One glaring example of omitting women occurred after Sandra Bland, an African American cisgender female, died and her name failed to go viral until #sayhername was created. When Sandra Bland was found dead hanging in her jail cell, her story and name did not immediately go viral. Eventually, #sayhername was created to question why her case had not gone viral and establish more support for ciswomen victims. #sayhername was both a rhetorical plea and a critique. Activists wanted her name to go viral, but also wanted to question why it hadn't. The hashtag points to an overwhelming issue, which is a lack of naming when it comes to ciswomen. The hashtag #sayhername would go on to cover all ciswomen victims of state-sanctioned violence, with only two names going viral on their own after the hashtag was created, Korryn Gaines and Sandra Bland.

Another thing to note of the victims whose names went viral is the physical support. From national marches and protests to camping outside of the police station where Sandra Bland died, people showed up en masse to protest against police brutality and state-sanctioned violence against Black cismen and ciswomen. However, 23 nonbinary and trans Black people were killed in 2016, and 29 trans and nonbinary Black people were killed in 2017.[13] Their names did not go viral, there was very little physical support outside of small private crowdsourcing, and vigils were more popular than protests for these victims, who were all Black, but not cisgender.

From a structural standpoint, Black Lives Matter has attended to the lives of trans and nonbinary people through their platform, describing their organization as one that is "guided by the fact that all Black lives matter, regardless of actual or perceived sexual identity, gender identity, [or] gender expression," and dedicated to making space for transgender brothers and sisters to participate and lead."[14] Additionally, the organization states they are "self-reflexive and do the work required to dismantle cisgender privilege and uplift Black trans folk, especially Black trans women who continue to be disproportionately impacted by trans-antagonistic violence."[15] However, the movement has fallen short in its representational and political support of trans people, whose stories largely go unnoticed in media.

In order for current social movements to be intersectional, movement practitioners must attend to the healing and imagination required to move, think, and do beyond our current patri-archal constraints. If people who believe #blacklivesmatter have not attended to the ways social injustices show up in intra-racial spaces, then we will continue to traumatize queer and ciswomen, trans and non-binary people, people currently incarcerated, people who are un-housed, and people with disabilities. How do we promote healing from trauma while also fighting against systems of oppression? One way to combat erasure in reactive justice movements based on those murdered by state-sanctioned violence is to radically reimagine how we can move through the world, which is the work of Afrofuturism.

Afrofuturism

Birthed from a nexus of social movement, technology, transnational capital, and artistic expression, Afrofuturism is an aesthetic manifestation of storytelling critically aware of possibility.[16] Designed to project the mind and body into a future free from colonialism, Afrofuturistic artists, activists, and scholars look toward the critical embodiment of Afrocentric imagination in art forms such as film, music, visual art, fashion, and literature as a means of replacing presumed whiteness as authority. Afrocentric in practice, Afrofuturism "emerged as a means to understand the trans-formation of African peoples as they dealt with the oppressive forces of discrimination, and the complexities of modern urban life and postmodernity."[17] "Astroblackness is an Afrofuturistic concept in which a person's Black state of consciousness, released from the confining and crippling slave or colonial mentality, becomes aware of the multitude and varied possibilities and probabilities within the universe."[18] The precise moment of imagined possibilities wherein the Black body seeks a future that centers its Blackness instead of whiteness is Afrofuturism. In short, Afrofuturism is the umbrella term for both an ideology and body of work that speculates what a future free from colonialism can look like. It is a future that relies on technological advancement, critical imagination, and the centering of Blackness above any other racial category. Famous Afrofuturistic artists include jazz musician Sun Ra's, novelists Octavia Butler and Samuel Delaney, and painter Jean-Michel Basquiat.

While the language of Afrofuturism creates a level of unification that renders the critical imagination of Blackness visible, the ways in which the intersections of gender and race have been highlighted leave much to be desired. Feminine identified characters fall on a binary between strong and powerful or androgynous. In superhero narratives, for example, Black and femme identified characters are often powerful and strong. While this may look like progress for combatting sexism from a white perspective because brute strength and power contradict dominant narratives historically attached to white ciswomen's bodies, Black ciswomen's bodies have long been viewed as anti-feminine due to stereotypes associated with being too loud, too independent, too strong, too surly, too masculine, and too emasculating.[19] The strong Black woman stereotype has plagued Black ciswomen's existence for decades and impacts social desirability, health outcomes, and familial structures.[20] Black ciswomen are often deemed unworthy of partnership,[21] hypertolerant to pain via pain perception and discriminated against in the medical profession,[22] and often assumed to be matriarchs vis-à-vis emasculation. In recent big data studies, Black ciswomen were not deemed human or identified as masculine over 90% of the time by racist algorithms.[23]

Shifting away from the stereotypically strong Black women, some Afrofuturists employ androgyny as an answer, rendering bodies genderless. For examples, look to the work of Octavia Butler's *Xenogenesis* or Janelle Monae's *Dirty Computer.*[24] Both texts eliminate the human form altogether for their nonbinary characters, thereby eliminating the gender binary. *Xenogenesis* turns to aliens that can procreate and merge with any human gender, while *Dirty Computer* turns to computers and robots to challenge the gender binary. While androgynous android narratives create a particular kind of gender freedom, or freedom from gender in futuristic imaginings, not all bodies want to be genderless, and not all bodies can transgress into genderless embodiment due to various aesthetic, genetic, capitalistic, and/or cultural reasons. Instead of employing gender stereotypes or removing gender altogether, artists, scholars, and activists must radically reimagine how we understand and employ gender at the intersection of race. I attempt to do that with a photography, oral history, and art-activism project called Transfuturism.

Transfuturism

The principal argument of Transfuturism is that trans and nonbinary people embody what Afrofuturism only speculates in terms of gender freedom. Transbodies have embraced the critical embodiment of imagination as push-back, refusing to accept cisgender performances as normal, natural, and preferred—not just in art, but in the flesh. As such, Transfuturism renders the lives of Black trans folk complex and visible; it creates discursive and physical spaces to live and breathe in the critical body, or the body whose mere presence acts as a form of resistance in a culture predicated on assimilation to and simulation of gender standards. The following three sections—The Transfuturism Project, Centering Trans Embodiment in the Afrofuture, and Transfutures beyond Technology and in the Flesh—define the Transfuturism Project and offer two areas to explore Transfuturism as an ideological manifestation of gender futurity at the intersection of Black racial identity.

The Transfuturism Project

The Transfuturism Project exemplifies transfutures through photography, oral history storytelling, and Afrofuturistic art. I photographed and spoke with trans and gender nonconforming (GNC) people and Wriply Bennet, a Transfuturism artist in residence and a Black trans woman, then used the stories and photographs to create a superhero sketch that honored their gender embodiment. After this, I painted each superhero character on a large canvas and created an environment that built from their interview. The artistic renderings of each informant pushed the boundaries of what we know and where we are going, as it relates to gender at the intersection of race (to see the photographs, interviews, sketches, and paintings, visit www.thejsuticefleet.com or www.transfuturism.com).

The conversations and comments from those with whom we spoke were insightful. Kei Williams, one of the nonbinary Transfuturism participants, stated,

> When it comes to trans and GNC folks, most folks are looking for you to pass in some type of way. And if you don't pass, you are a considered a threat. Being able to stand in truth in one's self is bravery.[25]

Essentially, as a Black body, we are already at once predisposed to a culture of simulation: "The 'culture of simulation' is no different from 'the culture' for people of color in this country, who have been 'inventing' themselves, their multiple selves, as they go along, and 'constructing the world, too.'"[26] This is testament to our survival as a mode of invention. Raquel Willis, a Black transwoman and Transfuturism Project participant, echoed the idea of invention when she said:

> My liberation stems from the journey of actually claiming my identity on my own terms. Trans folks and GNC folks, we are architects of this world that we are envisioning where people aren't held back based on what they were told they were supposed to be at birth or some point in their lives.[27]

Eb Brown, a nonbinary participant in this project, expanded on Raquel's sentiment stating,

> The thing that makes me feel liberated is the fact that I can actually morph my body into different shapes depending what I am feeling for the day. It makes me feel like I am deconstructing and reconstructing gender on a daily basis. I think it challenges my idea of liberation because in that deconstruction and reconstruction, it makes me think about myself in relationship to the world and others every day.[28]

Eb continues with thoughts on what liberation can ultimately be:

> A liberated future looks like one where people can stand fully in themselves in every moment of the day. The interstellar future is the interlocking of my orbit and your orbit. So all that you are inhabiting and all that I am inhabiting circulate around each other all the time and vibing with each other and learning from each other. But there is no competition because you are fully you and I am fully me. I don't need to be you and you don't need to be me. And that is what liberation truly is.[29]

What these responses all have in common is the necessity to claim our own identities and reinvent our gendered selves while also being cautious of the insidious ways the gender binary can reemerge, even in non-normative spaces. The next section exemplifies how the gender binary is erased via nonconformity while also reveling in the possibility of trans embodiment in Afrofuturistic renderings.

Centering trans embodiment in the Afrofuture

Unfortunately, transgender embodiment has not been centered enough in the work of Afrofuturist scholars. However, Afrofuturistic artforms have taken up nonconformity beautifully. For examples, we can visit the work of Octavia Butler, Jimi Hendrix, Prince, and other artists who cross gender norms, in their own work and/or personal lives. As Reynaldo Anderson reminds us, "The Black queer futuristic performance of Sylvester James not only demonstrated the ability of the artist to reimagine and influence popular culture and the political sphere, impacting internal and external communities, but simultaneously created new discursive spaces."[30] Ayai Nikos, a transwoman participant in the Transfuturism Project speaks volumes when she says, "My pro-Black, femme body is a testament to Octavia Butler's theorizing about a new sun when she writes, 'There is nothing new under the sun, but there are new suns.'"[31]

We can look to the trans body as a "new sun" or site of new discursive and physical space. In their recent article, Karen Jaime interrogates "Chasing Rainbows" by Black Cracker to locate the convergence of transgender identity and Afrofuturism.[32] They argue for Trans Afrofuturity as a site of lyrical, visual, and sonic possibility that pushes for inclusivity with the use of the word *we* in a repetitious loop that signals a lack of arrival. They write, "Ultimately, Glenn as Black Cracker presents transness as unsettling to configurations of Afrofuturism as a politics of arrival: The future is not about arrival. The future operates as an outside, an elsewhere ..."[33] Evolve, another nonbinary participant in the Transfuturism Project, speaks to the notion of constant evolution versus a perceived destination when they say,

> The liberated future looks like a place where we can change our minds, our gender, our names, or how we describe ourselves and not be ostracized or pushed out of different communities. It is a place where we're valued for evolving and not punished. That would be a beautiful future.[34]

It is this new discursive and liminal space that Transfuturism explores, specifically as it relates to gender liberation and gender performance in Black communities. Black trans bodies signify a critical resistance toward colonialism and gender-confining rhetoric. Jae Shephard, a nonbinary participant, speaks of authenticity and the ruptures of social constructs when they said, "Being my authentic self is liberating because I am able to exist beyond the confines of socially constructed gender binaries."[35] Black trans bodies use skin as a semi-structured, blank canvas, embarking on a journey of re-scripting[36] the body as an act of critical imagination. Black trans bodies use voice, hair, fashion, walk, sway, swag, and presence to resurrect possibility. The Black trans body imagines a future self with the aid of external and internal resources without a preexisting model for existence. Every body that transitions becomes their own entity, not a duplication of a trans body that has already transitioned or is transitioning. Therefore, with every iteration, the trans body becomes a new body, or a new sun, thus relying on critical imagination to begin the process, in whatever iteration that may take. The new body, then, becomes a performative signifier of manifestation.

Kai Green and Treva Ellison speak to this notion of manifestation. They call the Black trans body a tranifesting body.[37] "Tranifesting (transformative manifesting) calls attention to the epistemologies, sites of struggle, rituals, and modes of consciousness, representation, and embodiment that summons into being flexible collectivities … capable of operating across normativizing and volatile configurations of race, gender, class, sex, and sexuality."[38] "Tranifesting is a form of radical political and intellectual production that takes place at the crossroads of trauma, injury, and the potential for material transformation and healing."[39] Tranifesting is a futuristic and critical performance of embodied social identity that creates strategic space for transformation. Trans bodies are like transformers for social justice. The very act of disrupting the static gender continuum is an act of critical transformation that creates space for disrupting various normativities pertaining to a multitude of social identities, including being Black, queer, and/or gender nonconforming.

Shifting from *all* disciplinarian regimes requires a summoning of manifested power and imagination. The Black transbody is more than a Black body; it is more than a gendered body; it is more than a trans body. It is a body that meets at specific, fluid, always changing intersections. It is a body that has to constantly navigate those intersections with precise care and critical attention because it is a body that is not always welcome, appreciated, or loved. It is a body that has to reinvent itself over and over again at the sites of critical embodiment. And it is in the flesh of critical embodiment that trans bodies extend beyond the current scope of Afrofuturism. If Afrofuturism is about the speculation of future and possibility in cultural artistic production, then Transfuturism is the critical, lived embodiment of that production. The Black trans body becomes a physical and live disruptor and builder of social change with regard to intersectional oppression. Black trans bodies live physically in the world artistically crafted by Afrofuturism. As Kali Tal poignantly posits, "We need, as a culture, to pay attention to the theory and literature of those among us who have long been wrestling with multiplicity."[40]

Transfutures beyond technology and in the flesh

Several scholars have wrestled with technology, gender, and multiplicity as a form of theorizing. Ras Mashramani discusses the power of the Internet to create safe spaces and live beyond the watchful eyes of those who control, violate, and bruise our bodies.[41] In these safe spaces, we are able to experiment with our identities and grow beyond our binaries in a hypermodal setting where we are constantly folding over and into ourselves and others.

> We are living in a science-fiction reality, and if science fiction has taught as anything, it's that a mastery of technology is integral to survival in a plugged-in world … our identities hinge on our ability to create and manipulate data in the cybersphere to affect change in real life.[42]

The Internet literally becomes a space "for us freaks and outcasts, whose existences are politicized by overpowering mainstream media that tries its best to distract the masses."[43]

Haraway offers a different account that utilizes the cyborg as a genderless analogy.[44] "Haraway embraces technology as a way of moving away from humanism built on normative binaries and dualisms in order to create a regenerated world without gender."[45] The cyborg uses technology to negate binary systems much like trans bodies use technology to alter the physical characteristics of the body, resulting in a new identity that disrupts binaries. While moving along the continuum might reinforce the binary, the mere ability to move along the continuum also serves as a disruptor of the binary. If our hope is for a monstrous world without gender, then trans bodies are the growing seedlings of that vision. And while Haraway's cyborg vision has been met with warranted critique regarding the "reinscriptions, *not* regenerations, of the same old meanings of race and gender in cyborg imaginary,"[46] trans identities have been met with criticism regarding the perpetuation of the gender binary. However, we must remember that "We remake and even exceed language, but we do not escape it."[47] That is, language is finite and limits our ability to grow beyond its boundaries. Therefore, we create new language, but must always be attentive to the language that already exists, especially in the realm of technological advancement.

Science fiction, which relies on technology to explore scientific futurity, offers an important opportunity to wrestle with the contradictions surfacing between technologies, scientific thought, and lived experience.[48] When referencing the cyborg, Maggie Eighteen reminds us, science fiction is not a simple process of joining science with fiction or humanity and technology, it is extremely fluid, complex, and often contradictory.[49] Nestled within the contradictions, however limiting or creative, is the fact that to attempt to shift across rigid binaries, however reifying, still points to the possibility for disruption. "Gender becomes legible through acts of translation that betray disciplinary success and failure simultaneously."[50] One of those failures lies in the inability to escape the language that binds our bodies to binary performances. "Transgender highlights the labors of translation, inhering an implied 'before' and 'from which.' The present moment does not tell the story, only that there is one worth telling."[51] And there is a story worth telling here that is replete with future possibility.

Gender futurity

As evidenced in this handbook, gender is complicated and eludes definition.[52] Gender grows and becomes something new in everyday expressions of identity, proving a certain materiality and particularity, yet how it grows, consumes, overlaps, and comes back again signals an ineffable uncertainty. How do we harvest the capacity to understand gender, while also freeing ourselves from the messy constraints if we cannot wrap our minds and bodies around the many complex and interconnected links? Binary gender assumptions posit that there are only two genders, masculine and feminine, and that those gender identities are directly attached to genitalia and genetics. People who are cisgender identify and embody the gender they were assigned at birth based on genitalia. Nonbinary, trans, and intersex people do not fit the binary and are subject to discursive violence (e.g., name calling, dead naming, and micro-aggressions), physical violence (e.g., assault and murder), and legislative violence (e.g., discriminatory laws and policies that justify the maltreatment of trans, nonbinary, and intersex people).[53] While physical, legislative, and discursive violence should be enough to warrant the overhaul of gender systems that cause harm, there are not enough measures to interpolate cisgender people into the fight against the gender binary.

Gender futurity is an intervention that asks cisgender, transgender, and nonbinary gender people to use their imagination to begin thinking about alternative ways to embody gender that doesn't control, oppress, or limit the functions of expression. The term *futurity* signals a future environment free from oppression, trauma, violence, and discrimination. It is a fictional future that assumes humans have done the work to rid their communities of violence, albeit physical, discursive, or legislative. It is the imaginative work required to create the space to even wonder if a liberated future is possible. It is the moment that exists beyond the horizons of our current constraints. Imagination and futurity have a complex and critical relationship to social justice. They are tools designed to help people stuck in systems of oppression build the futures they want to live in as an act of activism. It is not enough to deconstruct systems of power; we have to build the world we want in the wake of deconstruction, and the world must be critical, imaginative, and rooted in a futurity not predicated on current systems of injustice.

In his groundbreaking text, *Freedom Dreams: The Black Radical Imagination,* Robin D. G. Kelley offers a historical outline of Black radical imagination in the United States that follows how Black intellectuals and freedom fighters envisioned freedom. He wanted to understand how radical thinkers imagined their life after revolution and where their ideas came from. He posed the question, "What kind of world [do] we want to struggle for?"[54] To tack gender to the understanding of radical imagination and futurity is to think through what it means for bodies to be free from the constraints of binary gender. It is to ask the question, "What kind of gender freedom do we want to struggle for?" Transfuturism pushes us to wrestle with other people's struggles with gender while also recognizing the beauty in gender freedom and expression and pushing us to embark on beautiful struggles of our own for the sake of liberation.

Conclusion

When thinking about the future of the communication discipline, our world, and the need for reimagining gender beyond current constraints, I think of the importance of researching gender at the intersection of race, the different interventions attempting to alleviate the stress of systemic oppression, and ways to evoke the critical and radical imagination in our everyday lives and research. Gender and race are inextricably linked, as this essay posits. Afrofuturism pushes us to intervene into racialized marginalization by actively thinking beyond the constraints of racism and colonialism. Transfuturism pushes our discipline and our bodies to create space for gender equity and generative co-creation.

I offer the term Transfuturism as a theoretical disruption of critical gender binaries within Black communities, while simultaneously disrupting post-colonial and neoliberal thought predicated on identity binaries, demonization, and dehumanization. The term generates visibility for Black transbodies that separates their unique struggles from other bodies that do not carry the weight of these specific intersections, while also highlighting the possibility in critical imagination. The term also renders visible the multiple ways in which transbodies have begun laying the groundwork for enfleshed possibilities, or the possibility of performing futurism in the now body, free from speculative fictitious accounts and into lived reality. In thinking about Transfuturism, transbodies have the ability to transcend gender binaries in multiple ways that inform freedom from gender in the body, or what is labeled gender futurity. Gender futurity, as an intervention, asks all of us to do this imaginative work to create a space of liberation and freedom for all.

The Transfuturism Project celebrates gender and Blackness in all of their fullness and nuance. The project asks witnesses to think about their own gender embodiment and ways they can use the project as a springboard toward freedom. As participants are able to see themselves in their own power and generate language to render visible their experiences, Transfuturism opens space for witnesses to move beyond critical deconstruction and into reconstruction, or the reimagining of structures to be more liberatory and just. Participants are able to simultaneously offer critique and analyses of current social issues related to gender and race while also privileging their own voices in the reimagination of those spaces and modes of being. This empowers participants and witnesses and asks researchers and scholars to keep pushing the boundaries of our discipline to be more inclusive and equitable, but also more imaginative and generative, specifically in intersectional ways that liberate our endlessly complex and growing social identities.

Notes

1 Ange-Marie Hancock, "When Multiplication Doesn't Equal Quick Addition: Examining Intersectionality as a Research Paradigm," *Perspectives on Politics* 5, no. 1 (2007): 75.

2 Kimberly Crenshaw, "Mapping the Margins: Intersectionality, Identity Politics, and Violence Against Women of Color," *Stanford Law Review* 43, no. 1 (1993): 1241–99.

3 CindyAnn Rose-Redwood and Reuben Rose-Redwood, "'It definitely felt very white': Race, Gender, and the Performative Politics of Assembly at the Women's March in Victoria, British Columbia," *Gender, Place & Culture* 24, no. 5 (2017): 645–54.

4 Crenshaw, "Mapping the Margins," 1256–57.

5 Judith Lorber, "Beyond the Binaries: Depolarizing the Categories of Sex, Sexuality, and Gender," *Sociological Inquiry* 66, no. 2 (1996): 155.

6 Jason Cromwell, "Queering the Binaries: Transsituated Identities, Bodies and Sexualities," Chap. 35 in *Transgender Studies Reader*, eds Susane Stryker and Stephen Whittle (New York: Routledge, 2006).

7 Nyle Fort and Darnell L. Moore, "Last words: A Black Theological Response to Ferguson and Anti-Blackness," *QED: A Journal in GLBTQ Worldmaking* 2, no. 2 (2015): 213.

8 KBP. "Killed by Police." *Killed by Police*, last modified 2019, www.killedbypolice.net/kbp2014.html.

9 Julius Bailey and David J. Leonard, "#BlackLivesMatter: Post-Nihilistic Freedom Dreams," *Journal of Contemporary Rhetoric* 5, nos 3–4 (2015): 67–77.

10 Bailey and Leonard, "Post-Nihilistic Freedom Dreams," 68.

11 Bailey and Leonard, "Post-Nihilistic Freedom Dreams," 76.

12 "Platform," *M4BL*, last modified December 20, 2018, https://policy.m4bl.org/platform/.

13 "Violence Against the Transgender Community in 2017," *HRC*, last modified December 20, 2018, www.hrc.org/resources/violence-against-the-transgender-community-in-2017.

14 "Platform," *M4BL*.

15 "What We Believe," *BLM*, last modified 2019, https://blacklivesmatter.com/about/what-we-believe/.

16 Reynaldo Anderson and John Jennings, "Afrofuturism: The Digital Turn and the Visual Art of Kanye West," Chap. 3 in *The Cultural Impact of Kanye West*, ed. Julius Bailey (New York: Palgrave Macmillan, 2014).

17 Anderson and Jennings, "Afrofuturism," 35.

18 Andrew Rollins, "Afrofuturism and Our Old Ship of Zion: The Black Church in Post Modernity," Chap. 7 in *Afrofuturism 2.0: The Rise of Astro-Blackness*, eds Reynaldo Anderson and Charles Jones (Lanham, MD: Lexington Books, 2015).

19 P. H. Collins, *Black Feminist Thought: Knowledge, Consciousness, and the Politics of Empowerment* (Boston, MA: Unwin Hyman, 1990).

20 Robin Boylorn, "A Story and a Stereotype," Chap. 7 in *Critical Autoethnography: Intersecting Cultural Identities in Everyday Life*, eds Robin Boylorn and Mark Orbe (Walnut Creek, CA: Left Coast Press, 2013).

21 Erica Chito Childs, "Looking Behind the Stereotypes of the 'Angry Black Woman': An Exploration of Black Women's Responses to Interracial Relationships," *Gender and Society* 19, no. 4 (2005): 544–61, www.jstor.org/stable/30044616.

22 Kelly Hoffman et al. "Racial Bias in Pain Assessment and Treatment Recommendations, and False Beliefs About Biological Differences Between Blacks and Whites," *Proceedings of the National Academy of Sciences of the United States of America* 113, no. 16 (2016): 4296–301, doi:10.1073/pnas.1516047113.

23 Joy Adowaa Buolamwini, "Gender Shades: Intersectional Phenotypic and Demographic Evaluation of Face Datasets and Gender Classifiers," Master's thesis, *MIT*, 2017, http://hdl.handle.net/1721.1/114068.

24 For examples, see Octavia Butler, *Xenogensis* (New York: Grand Central Publishing, 1989), or Janelle Monae, "Dirty Computer," *Wondaland Records*, 2018.

25 Kei Williams (Activist) in discussion with the author, June 2016.

26 Kali Tal, "The Unbearable Whiteness of Being: African American Critical Theory and Cyberculture," *Wired Magazine*, October 1996.

27 Raquel Willis (activist) in discussion with the author, July 2018.

28 Eb Brown (activist and community healer) in discussion with the author, June 2016.

29 Brown, June 2016.

30 Reynaldo Anderson, "Fabulous: Sylvester James, Black Queer Afrofuturism and the Blackfantastic," *Dancecult: Journal of Electronic Dance Music Culture* 5, no. 2 (2013): n.p.

31 Ayai Nikos (teacher) in discussion with author, April 2016.

32 Karen Jaime, "Chasing Rainbows: Black Cracker and Queer, Trans Afrofuturity," *Transgender Studies Quarterly* 4, no. 2 (2017): 208–18.

33 Jaime, "Chasing Rainbows," 216.

34 Evolve (activist and poet) in discussion with the author, July 2018.

35 Jae Shepherd (activist) in discussion with the author, April 2016.
36 Ronald L. Jackson, *Scripting the Black Masculine Body: Identity, Discourse, and Racial Popular Media* (Albany, NY: State University of New York Press, 2006).
37 Kai M. Green and Treva Ellison, "Tranifest," *Transgender Studies Quarterly* 1, nos 1–2 (2014): 222.
38 Green and Ellison, "Tranifest," 222.
39 Green and Ellison, "Tranifest," 223.
40 Kali Tal, "The Unbearable Whiteness of Being: African American Critical Theory and Cyberculture," *Wired Magazine*, October 1996.
41 Ras Mashramani, "Science Fiction, the Political, Metropolarity Crew," *Journal of Speculative Vision & Critical Liberation Technologies: Future Now Edition* 1, no. 1 (2013): 6.
42 Mashramani, "Science Fiction," 6.
43 Mashramani, "Science Fiction," 6.
44 Donna J. Haraway, "A Cyborg Manifesto: Science, Technology, and Socialist-Feminism in the Late Twentieth Century," Chap. 8 in *Simians, Cyborgs and Women: The Reinvention of Nature*, ed. Donna Haraway (Abingdon, England: Routledge, 1991).
45 Tiffany Barber, "Cyborg Grammar? Reading Wangechi Mutu's *Non je ne regrette rien* through *Kindred*," Chap. 1 in *Afrofuturism 2.0: The Rise of Astro-Blackness*, eds Reynaldo Anderson and Charles Jones (Lanham, MD: Lexington Books, 2015).
46 Janell Hobson, *Body as Evidence: Mediating Race, Globalizing Gender* (New York: SUNY Press, 2012).
47 A. F. Enke, "Translation," *Transgender Studies Quarterly* 1, nos 1–2 (2014): 242.
48 Riley C. Snorton, "An Ambiguous Heterotopia: On the Past of Black Studies' Future," *The Black Scholar* 44, no. 2 (2014): 29–36.
49 Maggie Eighteen, "Science Fiction, the Political, Metropolarity Crew," *Journal of Speculative Vision & Critical Liberation Technologies: Future Now Edition* 1, no. 1 (2013): 5.
50 Enke, "Translation," 242.
51 Enke, "Translation," 243.
52 Benny LeMaster and Amber L. Johnson, "Unlearning Gender—Toward a Critical Communication Trans Pedagogy," *Communication Teacher* (2018).
53 Serena Nanda, "The Hijras of India: Cultural and Individual Dimensions of an Institutionalized Gender Role," *Journal of Homosexuality* 11, nos 3–4 (1986): 35–54.
54 Robin D. G. Kelley, *Freedom Dreams: The Black Radical Imagination* (Boston, MA: Beacon Press, 2002): 7.

Bibliography

Anderson, Reynaldo. "Fabulous: Sylvester James, Black Queer Afrofuturism and the Blackfantastic." *Dancecult: Journal of Electronic Dance Music Culture* 5, no. 2 (2013): n.p.

Anderson, Reynaldo, and John Jennings. "Afrofuturism: The Digital Turn and the Visual Art of Kanye West," Chap. 3 in *The Cultural Impact of Kanye West*, ed. Julius Bailey (New York: Palgrave Macmillan, 2014).

Bailey, Julius, and David J. Leonard. "#BlackLivesMatter: Post-Nihilistic Freedom Dreams." *Journal of Contemporary Rhetoric* 5, no. 3–4 (2015): 67–77.

Barber, Tiffany. "Cyborg Grammar? Reading Wangechi Mutu's *Non je ne regrette rien* through *Kindred*," Chap. 1 in *Afrofuturism 2.0: The Rise of Astro-Blackness*, eds Reynaldo Anderson and Charles Jones (Lanham, MD: Lexington Books, 2015).

Black Lives Matter. "What We Believe," *BLM*, last modified 2019, https://blacklivesmatter.com/about/what-we-believe/.

Boylorn, Robin. "A Story and a Stereotype," Chap. 7 in *Critical Autoethnography: Intersecting Cultural Identities in Everyday Life*, eds Robin Boylorn and Mark Orbe (Walnut Creek, CA: Left Coast Press, 2013).

Brown, Eb (activist and community healer) in discussion with the author, June 2016.

Buolamwini, Joy Adowaa. "Gender Shades: Intersectional Phenotypic and Demographic Evaluation of Face Datasets and Gender Classifiers," Master's thesis, *MIT*, 2017. http://hdl.handle.net/1721.1/114068.

Childs, Erica Chito. "Looking Behind the Stereotypes of the 'Angry Black Woman': An Exploration of Black Women's Responses to Interracial Relationships." *Gender and Society* 19, no. 4 (2005): 544–61.

Collins, Patricia Hill. *Black Feminist Thought: Knowledge, Consciousness, and the Politics of Empowerment* (Boston, MA: Unwin Hyman, 1990).

Crenshaw, Kimberly. "Mapping the Margins: Intersectionality, Identity Politics, and Violence against Women of Color." *Stanford Law Review* 43, no. 1 (1993): 1241–99.

Cromwell, Jason. "Queering the Binaries: Transsituated Identities, Bodies and Sexualities," Chap. 35 in *Transgender Studies Reader*, eds Susane Stryker and Stephen Whittle (New York: Routledge, 2006).

Evolve (activist and poet) in discussion with the author, July 2018.

Enke, A. F. "Translation." *Transgender Studies Quarterly* 1, no. 1–2 (2014): 242.

Fort, Nyle, and Darnell L. Moore. "Last Words: A Black Theological Response to Ferguson and Anti-Blackness." *QED: A Journal in GLBTQ Worldmaking* 2, no. 2 (2015): 213.

Green, Kai M., and Treva Ellison, "Tranifest." *Transgender Studies Quarterly* 1, no. 1–2 (2014): 222.

Hancock, Ange-Marie. "When Multiplication Doesn't Equal Quick Addition: Examining Intersectionality as a Research Paradigm." *Perspectives on Politics* 5, no. 1 (2007): 75.

Haraway, Donna, J. "A Cyborg Manifesto: Science, Technology, and Socialist-Feminism in the Late Twentieth Century," Chap. 8 in *Simians, Cyborgs and Women: The Reinvention of Nature*, ed. Donna Haraway (Abingdon, UK: Routledge, 1991).

Hobson, Janell. *Body as Evidence: Mediating Race, Globalizing Gender* (New York: State University at New York Press, 2012).

Hoffman, Kelly. "Racial Bias in Pain Assessment and Treatment Recommendations, and False Beliefs about Biological Differences between Blacks and Whites." *Proceedings of the National Academy of Sciences of the United States of America* 113, no. 16 (2016): 4296–301. doi:10.1073/pnas.1516047113.

Human Rights Campaign. "Violence Against the Transgender Community in 2017," *HRC*, last modified December 20, 2018. www.hrc.org/resources/violence-against-the-transgender-community-in-2017.

Jackson, Ronald L. Scripting the Black Masculine Body: Identity, *Discourse, and Racial Popular Media* (Albany, NY: State University of New York Press, 2006).

Jaime, Karen. "Chasing Rainbows: Black Cracker and Queer, Trans Afrofuturity." *Transgender Studies Quarterly* 4, no. 2 (2017): 208–18.

Kelley, Robin D.G. *Freedom Dreams: The Black Radical Imagination* (Boston, MA: Beacon Press, 2002).

KBP. "Killed by Police." *Killed by Police.* Last modified 2019. www.killedbypolice.net/kbp2014.html.

LeMaster, Benny, and Amber L. Johnson, "Unlearning Gender—Toward a Critical Communication Trans Pedagogy," *Communication Teacher* (2018).

Lorber, Judith. "Beyond the Binaries: Depolarizing the Categories of Sex, Sexuality, and Gender." *Sociological Inquiry* 66, no. 2 (1996): 155.

Maggie Eighteen. "Science Fiction, the Political, Metropolarity Crew." *Journal of Speculative Vision & Critical Liberation Technologies: Future Now Edition* 1, no. 1 (2013): 5.

Mashramani, Ras. "Science Fiction, the Political, Metropolarity Crew." *Journal of Speculative Vision & Critical Liberation Technologies: Future Now Edition* 1, no. 1 (2013): 6.

Movement for Black Lives. "Platform," M4BL, last modified December 20, 2018. https://policy.m4bl.org/platform/.

Nanda, Serena. "The Hijras of India: Cultural and Individual Dimensions of an Institutionalized Gender Role." *Journal of Homosexuality* 11, no. 3-4 (1986): 35–54.

Nikos, Ayai (teacher) in discussion with author, April 2016.

Rollins, Andrew. "Afrofuturism and Our Old Ship of Zion: The Black Church in Post Modernity," Chap. 7 in *Afrofuturism 2.0: The Rise of Astro-Blackness*, eds Reynaldo Anderson and Charles Jones (Lanham, MD: Lexington Books, 2015).

Rose-Redwood, CindyAnn, and Reuben Rose-Redwood. "'It definitely felt very white': Race, Gender, and the Performative Politics of Assembly at the Women's March in Victoria, British Columbia." *Gender, Place & Culture* 24, no 5 (2017): 645–654.

Shepherd, Jae (activist) in discussion with the author, April 2016.

Tal, Kali. "The Unbearable Whiteness of Being: African American Critical Theory and Cyberculture," *Wired Magazine*, October 1996.

Snorton, Riley, C. "An Ambiguous Heterotopia: On the Past of Black Studies' Future." *The Black Scholar* 44, no. 2 (2014): 29–36.

Williams, Kei (activist) in discussion with the author, June 2016.

Willis, Raquel (activist) in discussion with the author, July 2018.

38

LATINX FEMINIST ACTIVISM FOR THE SAFETY OF WOMEN JOURNALISTS

Aimée Vega Montiel

There has been an increase in violence, threats and harassment against women journalists. Women journalists are subjected to the same wide range of human rights violations as are directed against men journalists [...] However, they also experience workplace and employment related discrimination and gender-based violence, including threats of violence, abuse and harassment. Both are symptomatic of the gender-based inequality, discrimination and violence experienced by women globally across many aspects of their lives.

UN General Secretary, 2017:3.

Introduction

There has been a severe increase of violence against women journalists, both offline and online. According to the 2018 UNESCO's Report on the Safety of Journalists and the Danger of Impunity, 2017 was the year with the highest number of murdered women journalists: 11 victims. This is happening in a number of nations, but mostly in conflict and post-conflict countries where the human rights of women journalists have become more precarious. Three regions recorded the highest rates: the Arab States, Asia and the Pacific, and Latin America and the Caribbean. The third of these has become paradigmatic as one of the most dangerous regions in the world for women to practice journalism.

Violence against women journalists is a form of censorship that threatens the vital role of a free press: "not only it is a grave violation of human rights, it also represents a broader attack on the collective right to freedom of expression and access to information."[1] That this often happens with the consent of governments results in a cycle of impunity in contexts where news media companies are not ensuring safe working conditions for these professionals.

The advocacy work of women journalists, like the efforts of activists and academics, has become a key strategy to make visible the structural nature and systemic dimensions of aggressions against women with the aim of putting this issue on national agendas.

Grounded in both communication studies and feminist theory, the purpose of this chapter is to analyze the structural conditions enabling gender-based violence against women journalists in Latin America, to explore how it affects the collective right to freedom of expression, and to consider responses to this crisis from activist movements. The final goal is to contribute to a constructive debate regarding the specific dangers and larger problems that can be addressed by research on communication and gendered attacks on journalists.

Gender-based violence

Gender-based violence against women is defined as "any act…that results in physical, sexual or psychological harm or suffering to women, including threats of such acts, coercion or arbitrary deprivation of liberty, whether occurring in public or private life" (UN, 1994). Violence against women (VAW) is a manifestation of unequal gender relations that have resulted in domination of and discrimination against women.

International and regional organizations have taken significant steps to call on Member States[2] to be accountable for and to eradicate VAW. The Convention on the Elimination of All Forms of Discrimination Against Women (CEDAW) was the first international instrument that defined VAW and set an agenda for national action to end such discrimination. The Inter-American Convention to Prevent, Punish and Eradicate Violence against Women, Belem do Parà, was launched in 1994 by the Organization of American States (OAS). It recognizes all kinds of violence against women as a violation of women's human rights. These institutions and events have influenced the creation of legal frameworks at the national level.

The definitions of gender-based violence included in such international advocacy are grounded in feminist theory, which initially focused on sexual violence to call attention to gendered discrimination targeting women. In the 1980s, feminist scholars moved to the concept of domestic violence to unveil dangerously unequal relations between men and women in the private sphere. Thanks to those initial definitions, contemporary activists and scholars can take a more holistic approach to VAW that identifies its various types and modalities. In this context, typologies identify the forms that VAW takes (physical, psychological, sexual, economic, and femicide/murder), while analysis of modalities examines the social spheres where gender-based violence is executed (institutions, schools, workplaces, communities, and homes). Viewed from this broader perspective, the victimized of a particular woman is considered in relation to different forms and locations of violence enacted against her. For example, femicide (the murder of a woman) is viewed as only the tip of the iceberg, as it is often the final outcome of physical, psychological, and often sexual violence too.[3]

As gender-based violence is linked to widespread gendered social conditions for women, it is experienced across specific contexts and cultures. VAW will increase and will be more tolerated in those societies where women's disadvantaged position is exacerbated by other forms of discrimination. This is often the case for indigenous women (who may be ethnically marginalized and less privileged than white women) as well as for poor and younger women (who can be more vulnerable and less protected than middle-class and adult women).[4]

Violence against women journalists

In 2017, in honor of International Women's Day, the United Nations Special Rapporteur on the Promotion and Protection of the Right of Opinion and Expression warned about the increase in violence against women journalists.[5]

Women journalists are vulnerable to attacks, not only by those actors who attempt to silence their work, but also from their colleagues and bosses. In 2014, INSI and IWMF, with the support of UNESCO, jointly published the results of the first international survey on violence against women journalists. Based in the responses of 1,000 professionals, results showed that two-thirds of women journalists had been victims of sexual harassment in the workplace, mainly perpetrated by their coworkers and supervisors. More than 25% were victims of online violence. Most women journalists do not report having been victims of sexual violence for fear of discrimination and of retaliation from their bosses. The effects of VAW targeting journalists include fear, self-censorship, and silence.[6]

Contemporary VAW in journalism includes online violence such as cyberbullying, trolling, doxxing, hate speech, public shaming, intimidation, threats, and cyber-stalking.[7] This type of violence "ranges from pernicious, gendered online harassment to overt, targeted attacks that frequently involve threats of sexual violence."[8] Although the types of such violence vary, research shows more attacks that "include digital security breaches from exposure of identifying information (exacerbating the offline risks) to malicious misrepresentation using Artificial Intelligence technologies."[9]

The UN General Secretary Report (2017) identifies gender inequality as a root cause of discrimination against women in the professional journalistic arena, since "the freedom of movement of women journalists can be restricted based on discriminatory laws, overt threats against their safety and cultural norms and stereotypes regarding women's conduct."[10] These conditions link to structural barriers: women are often paid less than men, fewer women hold decision-making positions in the news media industry, and women's work is less visible than men's.[11]

In most of the countries, the national legal framework adopts international standards for fighting gender-based violence. However, there is an inadequate implementation of measures to tackle violence against women journalists by Member States. This problem is replicated in online environments. Although most countries formally recognize that online harassment has become a problem, governments are doing little to combat it, a failure that combines with the inaction of private companies to worsen the problem.[12]

In 2012, the UN adopted a Plan of Action on the Safety of Journalists and the Issue of Impunity, which calls attention to the need to make visible the gendered dimensions of violence against journalists. In 2017, the UN General Assembly adopted a resolution to condemn attacks on women journalists, including intimidation and harassment taking place online as well as offline. In summary, recommendations to Member States included:

- Ensure safety conditions for women journalists to perform their work.
- Monitor and report attacks against women journalists.
- Build data on VAW journalists.
- Publicly condemn attacks.
- Guarantee women journalists access to justice and combat impunity.
- Put in place safe gender-sensitive investigative measures and provide protection and support to them.

The Inter-American Commission on Human Rights Special Rappporteur on Freedom of Expression specifically denounced the increasing violence against women journalists in Latin America and the Caribbean.[13] UNESCO decried the killings of 38 women journalists worldwide from 2012 to 2016, representing 7% of all the journalists murdered, both men and women. Alarmingly, this percentage more than doubled in Latin America and the Caribbean during the following year (2016–2017), where the average increased to 16% in the period.[14] A key finding of the report is that most of the women journalists murdered did not work in conflict zones.

In the Latin American region, violence against women journalists is combined with corruption and impunity.[15] According to the Committee to Protect Journalists (CPJ), Mexico, Colombia, and Brazil are the Latin American countries with the highest rates of impunity in cases involving killed journalists.

Acts of violence against journalists constitute the most serious form of attack against freedom of expression at the individual and social level, because these attacks violate the access a society has to information. Crimes against journalists cannot be treated casually or dismissed as spontaneous or personal. These must be considered crimes that function to silence voices and prevent investigations that can reveal corruption.[16]

One of the critical barriers to understanding violence against women journalists in Latin America is the lack of existing data available. Violence against women journalists is not widely visible within the region because it is displaced by the increasing rates of murders and violence against male journalists. Both governments and NGOs fail to adopt a gender-sensitive approach when analyzing violence against these professionals. The few studies that can be found are prepared by journalistic organizations, as with the CIMAC's report, or by scholars such as Julie Posetti and Hannah Storm.[17]

According to the UNESCO Chair of Communication of the University of Malaga, Argentina, Colombia and Mexico are the countries with the highest rates of femicides of women journalists. In Colombia, the Foundation for the Freedom of Press (FLIP) reported aggressions against 33 women journalists in 2015, 14 more cases than in 2014. Eight femicides happened between 1987 and 2015.[18] In Mexico, the NGO CIMAC published the first report on violence against women journalists in 2012. According to this organization, from 2002 to 2013 there were 184 women journalist victims of gender-based violence. Between 2014 and 2015 alone, there were 147—marking a dramatic increase. More than 15 women journalists have been victims of femicide in the country. Smaller but still significant numbers are reported from Brazil, where 86% of women journalists have been victims of gender-based violence.[19]

However, countries in the Central American region are also among the most dangerous.[20] The Civitas Center reports that in Guatemala alone in 2015 there were 24 documented incidents of violence against women journalists—all of them linked to their profession, and most of a sexual nature committed by their peers and their information sources. Indeed, the perpetrators include a presidential candidate and a deputy in the Leader Party.[21] In Honduras, the case of journalist Lourdes Ramírez gained public attention, as she was threatened in 2005 for denouncing violations to the human rights of women workers and for unveiling the corruption of authorities in her country.[22]

These existing reports and cases show that women journalists face challenges in their profession linked to gender discrimination that male journalists do not have to face. In summary, according to those reports, main types of gender-based violence against women journalists in the region are psychological, physical, and sexual and include harassment and rape, economic precarity in working conditions, patrimonial theft of work materials and of personal objects, forced entry into their homes, and institutional omissions. These problems are heightened by the lack of the rule of law, by corruption, and by impunity. Most of the women journalists attacked covered corruption and unveiled links between governments and criminal groups. According to CIMAC, in Mexico the main perpetrators of violence against women journalists are government officials (comprising 60% of aggressors). Gender-based violence is evidently a way to stop female journalists from reporting on corruption.

Common forms of violence used by governments against women journalists are re-victimization, which may include accusing them of having links to criminals attributing attacks to personal conflicts or denying that the violence is related to their professional work. Another form of governmental violence is the use of espionage instruments against women journalists.

A critical form of institutional violence is impunity. Impunity happens because of a lack of action in following investigations, weak judicial systems, corruption, and negligence. According to UNESCO, only 15% of cases have been classified as solved (31 out of 204 cases) in Latin America and the Caribbean.[23] However, in Mexico the average reaches only 1%.[24] Linked to impunity, access to justice is a major challenge for these professionals. The lack of gender experts in the judicial system in the region is one reason why sexist stereotypes are reproduced in judicial processes.

Criminal organizations are the second main aggressors attacking women journalists. In Mexico, most of the local news media are constrained by criminal groups that force journalists to remove unfavorable coverage. Those journalists who insist on unveiling links between governments and organized crime often become victims of attacks. This was the case of Mexican journalist Miroslava Breach, who was murdered in 2016.[25]

Although such attacks against women journalists are linked to their work, news media industries do not take responsibility for ensuring safe working conditions for them.[26]

Legal conditions

Women journalists have built their own protective mechanisms by creating professional connections through organizations such as the "Women Journalists Network with a Gender Perspective," which has representation in Mexico and Colombia. Through this and other groups, they have compelled governments to adopt measures to prevent attacks. Nevertheless, they still face obstacles.

The legal framework for the protection of journalists in Latin America is particularly weak. Mexico, Colombia, and Honduras rely on national mechanisms for the protection of human rights defenders and journalists, yet they have failed in their duty to ensure safe conditions for these professionals (as evidenced by the continuing murder of journalists in the region).

According to the report from the organization WOLA PBI, main problems with current mechanisms to combat the problem are an absence of funds to protect journalists who are in risky situations, limited and unqualified staff, a lack of gender-focused protocols, delays in risk analysis to determine whether a journalist needs protecting and the kind of measures that need to be taken, and slow responses to requests for protection. Since 2013, Freedom House (a U.S. NGO) has provided technical assistance to some national efforts to prevent violence targeting journalists in Latin America; however, its quantitative methodology does not include a broader contextual analysis and this impedes fully evaluating risk situations.[27]

In summary, a variety of legal and social conditions in Latin American shows that the protection of women journalists is not a priority for governments in the region.

The effects of violence against women journalists on freedom of expression

Violence against women journalists threatens freedom of expression. Freedom of Expression is stated in Article 19 of the Universal Declaration of Human Rights. It declares that *everyone has the right to freedom of opinion and expression; this right includes freedom to hold opinions without interference and to seek, receive and impart information and ideas through any media regardless of frontiers.* The promotion of freedom of expression starts with governmental commitments to the rule of law, followed by the protection of both regional and international human rights courts.

According to UNESCO, a free press and free access to information are corollaries of the human right to freedom of expression. "Press freedom covers the freedom of all individuals or institutions to use media platforms in order that their expression may reach the public."[28]

For this UN agency, gender equality is required throughout all dimensions of press freedom:

> Women have the right to be equally involved in all dimensions of press freedom. Press freedom in its holistic and gender-sensitive conceptualization is particularly relevant to the production of journalism, which is a public exercise of freedom of expression [...] The freedom to participate in media, the rights of expression and access to and production of media content are all issues that can be fully understood only by considering their gender equality dimensions.[29]

One of the key components of press freedom is the safety surrounding public expression. "Safety issues point specially to the responsibility of the State in protecting media freedom and ensuring that there is not impunity for crimes against the people who do journalism."[30] When a woman journalist is victimized, this right is violated.

Violence against journalists is a form of censorship, as it violates not only the human rights listed above but also attacks individual and collective rights to communicate: it "censors the voices of individuals, works to intimidate others, and encourages that use of self-censorship with a 'chilling-effect' on freedom of expression."[31].

Activism for the safety of women journalists

As shown above, women journalists, NGOs, academics, and intergovernmental organizations in Latin America have taken active roles in combating violence against women journalists. It is fair to say that these actors are accomplishing most of the tasks that the governments are not undertaking to guarantee safe conditions for these professionals. The forms of intervention are diverse and linked to varied areas of expertise. This does not mean the problem is solved, but these individuals and groups have succeeded in drawing public attention to this issue.

Reports on types and modalities of violence against women journalists are usually produced by women journalists and scholars. That is the case with the organization CIMAC (Comunicación e Información de la Mujer) in Mexico, which in 2012 completed the first national report on violence against women journalists. This document unveiled the individual experiences of the victims as well as their links to structural barriers. Such reports serve to inform policies like the adoption of a gender-sensitive approach in national mechanisms offered for the protection of journalists, which constitutes another type of activism.

In tandem with governmental efforts, journalistic organizations have established strategies to minimize the risk of women journalists, online and offline. Both Rapporteurs Sans Frontiers (RSF) and the Committee for the Protection of Journalists (CPJ) have published guides to increase the safety of journalists in the region.

Financial and physical protection is another form of activism in Latin America. Organizations such as the RSF provide financial support to displaced journalists and their families.

More recently, the NGO Article 19 has started focusing its analysis and activities on the gendered dimension of violence. The organization provides security training to women journalists in high-risk countries such as Mexico, Brazil, and Honduras. In 2015, Article 19 designated a thematic lead on gender issues in order to create a cross-cutting approach to gender, sexuality, and its intersection with the right to freedom of expression. The increasing focus of Article 19 entailed specific changes within the organization. For example, the research methodology and the model for responding to individual cases were reviewed by a group of experts on women's rights that made concrete recommendations for planning and operational changes. This included training on gender sensitive techniques for monitoring and documenting, as well as on how to address cases related to sexual violence.[32]

This organization has played a key role in countries such as Mexico and Colombia concerning legal protections for women journalists, which include reporting to human rights courts regarding violations of the rights of these professionals. On July 31, 2018, the United Nations Human Rights Committee made history with a resolution against the Mexican government stating its responsibility for violations of the human rights of journalist Lydia Cacho (after her arbitrary detention in 2005). Legally represented by the organization Article 19, this journalist reported attacks to her human rights before the Committee in 2014, where she argued violations to her rights to gender equality, freedom of expression, security, and the prohibition to get tortured.

The Global Alliance on Media and Gender (GAMAG), an initiative launched in 2013 by UNESCO and more than 500 organizations, presented a Shadow Report on Violence Against Women in Media and ICT before the CEDAW Committee on the occasion of the 9th Periodic Report of the Government of Mexico in July 2018. The report prepared by the Alliance called attention to the increasing violence against women journalists in the country and outlined the danger of impunity. Thanks to this effort, the recommendations to that government published by the Committee included a section on the protection of these professionals.

Other groups, such as Women Journalists for the Life and Freedom of Expression in Honduras, have taken legal defense into their own hands, calling for access to justice.

Finally, campaigns to raise public awareness are another strategy to combat the problem. In 2018, a national campaign started in Brazil to report on violence against women journalists in sports. Under the hashtag #DeixaElaTrabalhar (#LetHerWork), female sports journalists highlighted the gender-based violence with which they are often victimized.[33]

Conclusion

These are critical times for women journalists in Latin America and the Caribbean. As gender inequality deepens rather than lessens, they face more barriers to performing their professional roles in safe conditions.

Gender-based violence against women journalists encompasses a continuum of different types of attacks, both offline and online. The forms are diverse (kidnapping, sexual assault, sexual harassment, threats, surveillance, and sexist stereotyping), but all of them are linked to gendered social conditions. The effects of violence against women journalists are both individual and collective, which places freedom of expression at risk.

Although some progress has been made at the legal level, these laws have not been effective and the evidence supports an increase in murders and other forms of attacks, as is shown at the beginning of this chapter.[34] The UN has called on both governments and media institutions to improve mechanisms for the protection of women journalists, but they have not done enough to prevent violence. On the contrary, some Latin American governments are part of the problem, as their officials are the main aggressors and most of the crimes go unpunished.

Many of the progressive actions to address this widespread problem have been achieved by women journalists, NGOs, and scholars. Thanks to the work done by these actors, violence against women journalists has been unveiled and caught the attention of human rights courts. The work done by these advocates includes building data; providing legal assistance as well as physical and financial support; promoting policies, guidelines, and training; launching public education campaigns; and improving protocols to combat gender-based violence online and offline.

Activism has become a way to promote and protect the human rights of women journalists. However, activists place themselves at risk by doing this work. The effectiveness of their activities will depend mostly on the accountability of both governments and news media industries. These institutional components are part of the problem, but they can become part of the solution: as allies in the fight to protect women journalists in their vital role to create a free press and freedom of speech for all.

Notes

1 UNESCO, *UNESCO Director-General's Report on the Safety of Journalists and the Danger of Impunity* (Paris: UNESCO, 2018a), 134.
2 A Member State is a state that is a member of an international organization, such as the UN.
3 Aimée Vega Montiel, "Violence Against Women in Media and Digital Contents," in *Setting the Gender for Communication Policy: New Proposals from the Global Alliance on Media and Gender*, eds Aimée Vega Montiel and Sarah Macharia (Paris: UNESCO, 2019), 67–74.
4 Vega Montiel, "Violence Against Women," 67–74.
5 UNESCO, Montiel, "Violence Against Women," 2018a: 63.
6 Alana Barton and Hannah Storm, *Violence and Harassment Against Women in the News Media: A Global Picture* (London: IWMF/INSI, 2014).
7 UNESCO, *UNESCO Director-General's Report*, 2018a.
8 Julie Posetti and Hannah Storm, "Violence Against Women Journalists—Online and Offline," in *Setting the Gender Agenda for Communication Policy. New Proposals from the Global Alliance on Media and Gender*, eds Aimée Vega Montiel and Sarah Macharia (Paris: UNESCO, 2019), 75.
9 Posetti and Storm "Violence Against Women Journalists," 75.
10 UN General Secretary, *The Safety of Journalists and the Issue of Impunity Report*, http://undocs.org/A/72/290, 2017, 3.
11 Aimée Vega Montiel, "Mujeres en las industrias de comunicación en México," in *Revista Ciencia y Desarrollo* (Mexico: Consejo Nacional de Ciencia y Tecnología, 2014), 56–61.
12 Anita Gurumurthy, "A Feminist Perspective on Gender, Media and Communication Rights in Digital Times," *Setting the Gender Agenda for Communication Policy: New Proposals from the Global Alliance on Media and Gender*, eds Aimée Vega Montiel and Sarah Macharia (Paris: UNESCO, 2019), 103–118.
13 UNESCO, *UNESCO Director-General's Report*, 2018a.
14 According to the 2018 UNESCO Report on the Safety of Journalists and the Danger of Impunity (UNESCO, 2018b), Latin America is one of the regions where more murders of journalists happened between 2016 and 2017. The two countries with the highest numbers were Mexico (13) and Afghanistan (11). Six more Latin American countries are on the list of the 24 most dangerous countries for journalists: Honduras, Dominican Republic, Brazil, Colombia, Guatemala, and Peru.
15 Impunity refers to the exemption from punishment and the failure to bringing victims the right to justice and redress.
16 Edilson Lanza, "La seguridad de periodistas y la lucha contra la impunidad," Paper presented at the UNESCO Forum on Safety of Journalists and the Fight Against Impunity, San Salvador, August 31, 2017.
17 Julie Posetti and Hannah Storm, "Violence Against Women Journalists," 2019.
18 Foundation for the Freedom of the Press. 2016. "Así es la violencia contra las mujeres periodistas," http://pacifista.co/asi-es-la-violencia-contra-las-mujeres-periodistas/.
19 Associaçao Brasileira de Jornalismo Investigativo and Genero e Número, 2017, *Mulheres No Jornalismo Brasileiro*. http://gnjb-spa.surge.sh.
20 Ruth De Frutos, "Mujeres periodistas: violencia aumentada," *Infoamérica Revista Iberoamericana de Comunicación* 10 (2016): 69–84.
21 Ligia Flores, "El riesgo de ser periodista en Guatemala," *Sala de Redacción*, November 4, 2015.
22 International Consortium for Investigative Journalists, "Lourdes Ramírez," July 2018, www.icij.org/journalists/lourdes-ramirez/.
23 UNESCO, 2018b.

24 Article 19. "Article 19 Submission to OHCHR for the Report on Safety of Journalists," February 2016, www.ohchr.org/Documents/Issues/Journalists/GA72/Article19.docx.
25 Further information is available at this link: www.bbc.com/news/world-latin-america-39376061.
26 Vega Montiel, Aimée, "Violencia contra mujeres periodistas," *Revista Interdisciplina* 7, no. 17 (Enero-Abril 2019): 57–68.
27 WOLA PBI. *Mexico's Mechanism to Protect Human Rights Defenders and Journalists: Progress and Continued Challenges* (London: WOLA PBI, 2016).
28 UNESCO, *UNESCO Director-General's Report*, 2018a: 20–21.
29 UNESCO, *UNESCO Director-General's Report*, 2018a: 20–21, 62.
30 UNESCO, *UNESCO Director-General's Report*, 2018a: 23.
31 UNESCO, *UNESCO Director-General's Report*, 2018a: 134.
32 Article 19, "Article 19 Submission," 2016.
33 This initiative is available at its Twitter account: https://twitter.com/deixaelatrab?s=17.
34 UNESCO, *UNESCO Director-General's Report*, 2018a.

References

Article 19. "Article 19 submission to OHCHR for the report on Safety of Journalists. February 2016." Accessed March 12, 2017: www.ohchr.org/Documents/Issues/Journalists/GA72/Article19.docx.

Associaçao Brasileira de Jornalismo Investigativo and Genero e Número. "Mulheres No Jornalismo Brasileiro." Accessed December 26, 2018: http://gnjb-spa.surge.sh.

Barton, Alana and Hannah Storm. *Violence and Harassment Against Women in the News Media: A Global Picture.* Report. London: IWMF/INSI, 2014.

CIMAC. *Impunity and Violence Against Female Journalists in Mexico.* México: CIMAC, 2012. Accessed July 17, 2017: www.cimacnoticias.com.mx/sites/default/files/Impunity%20and%20Violence%20Against%20Female%20Journalists%20in%20Mex.%20Inglés.pdf.

De Frutos, Ruth, "Mujeres periodistas: violencia aumentada." *Infoamérica Revista Iberoamericana de Comunicación* 10 (2016): 69–84.

Flores, Ligia, "El riesgo de ser periodista en Guatemala." *Sala de Redacción.* November 4, 2015.

Foundation for the Freedom of the Press, "Así es la violencia contra las mujeres periodistas." Accessed December 25, 2018: http://pacifista.co/asi-es-la-violencia-contra-las-mujeres-periodistas/.

Gurumurthy, Anita. "A Feminist Perspective on Gender, Media and Communication Rights in Digital Times" In *Setting the Gender Agenda for Communication Policy: New Proposals from the Global Alliance on Media and Gender*, edited by Aimée Vega Montiel and Sarah Macharia, 103–118, Paris: UNESCO, 2019.

International Consortium for Investigative Journalists. "Lourdes Ramírez." Accessed July 10, 2018: www.icij.org/journalists/lourdes-ramirez/.

Lanza, Edilson. 2017. "La seguridad de periodistas y la lucha contra la impunidad." Paper presented at the UNESCO Forum on Safety of Journalists and the Fight Against Impunity, San Salvador, August 31, 2017.

Posetti, Julie and Hanna Storm, "Violence Against Women Journalists—Online and Offline." In *Setting the Gender Agenda for Communication Policy: New Proposals from the Global Alliance on Media and Gender*, edited by Aimée Vega Montiel and Sarah Macharia, 75–86, Paris: UNESCO, 2019.

UN. *Declaration on the Elimination of Violence Against Women.* December 20, 1993. Accessed October 2, 2005: www.ohchr.org/en/professionalinterest/pages/violenceagainstwomen.aspx.

UN General Secretary. *The Safety of Journalists and the Issue of Impunity Report.* Accessed December 10, 2018: http://undocs.org/A/72/290.

UNESCO. *World Trends in Freedom of Expression and Media Development. Global Report 2017–2018.* Paris: UNESCO, 2018.

———. *2018 UNESCO Director-General's Report on the Safety of Journalists and the Danger of Impunity.* Paris: UNESCO, 2018.

Vega Montiel, Aimée. "Violencia contra mujeres periodistas." *Revista Interdisciplina* 7, no. 17 (Enero-Abril 2019): 57–68.

————. "Violence Against Women in Media and Digital Contents." In *Setting the Gender Agenda for Communication Policy: New Proposals from the Global Alliance on Media and Gender,* edited by Aimée Vega Montiel & Sarah Macharia, 67–74, Paris: UNESCO, 2019.

————. "Mujeres en las industrias de comunicación en México." *Revista Ciencia y Desarrollo,* Guadalajara, Mexico: Consejo Nacional de Ciencia y Tecnología, 2014: 56–61.

Vega Montiel, Aimée and Sarah Macharia (Eds). *Setting the Gender for Communication Policy: New Proposals from the Global Alliance on Media and Gender,* Paris: UNESCO, 2019.

WOLA PBI. *Mexico's Mechanism to Protect Human Rights Defenders and Journalists: Progress and Continued Challenges.* London: WOLA PBI, 2016.

39

PUSHING BOUNDARIES

Toward the development of a model for transing communication in (inter)cultural contexts

Gust A. Yep, Sage E. Russo, and Jace Allen

Transgender has, in the twenty-first century, become a global assemblage.[1, 2] Reminding us of the complex, evolving, productive, and contested nature of such assemblage, Stryker writes,

> "Transgender" is, without a doubt, a category of First World origin that is currently being exported for Third World consumption. Recently, however, engagements between ... "transgender" ... that [circulate] globally with Eurocentric privilege, and various non-European, colonized, and diasporic communities whose members configure gender in ways that are marginalized within Eurocentric contexts, have begun to produce entirely new genres of analysis. Such encounters mark the geo-spatial, discursive, and cultural boundaries of transgender ... but also point toward [its] untapped potential.[3]

As transgender embodiments, practices, subjectivities, communities, and politics traverse a wide range of geographical spaces and cultural systems, such boundaries are maintained and challenged, drawn and redrawn, perpetuated and transformed. Indeed, social meanings of transgender have evolved within and between cultural spaces and geopolitical systems over time.[4] How do individuals and groups create and maintain identities, enhance and sustain cultural intelligibility, cultivate and increase a sense of belonging, and negotiate and contest social meanings—indeed communicate—within and across these complex boundaries? To begin to answer this question and to tap into the potential of the negotiation, crossing, and (re)definition of such boundaries, our chapter provides sketches of a model for transing communication in (inter)cultural contexts. To do so, our chapter consists of three sections. First, we discuss the process of transing communication and provide a brief description of our model. Next, using the four domains outlined in our model, we identify and review past research on transgender communication. We conclude by exploring the potential implications and new directions for transing communication research in (inter)cultural contexts.

Transing communication

Transing, introduced by Stryker, Currah, and Moore, is a deconstructive tool that can be used within, across, and between gendered spaces and configurations.[5, 6] More specifically, it is a practice that examines how gender is contingently assembled and reassembled with other structures

639

and attributes—for example, race, nation—of bodily being. Similar to processes of queering advanced by contemporary queer studies, transing is a critical practice that unpacks underlying relations of power within specific cultural, geopolitical, and historical contexts from a universalizing perspective which maintains that gender is a critical concern for *all* individuals inhabiting various positions in a gender system (e.g., gender normative and gender non-normative people in a given culture). In this sense, transing focuses on all gender embodiments and subjectivities across a broad cultural spectrum.

Transing communication is based on four fundamental premises. First, gender is understood and rendered meaningful in relation to other vectors of social and bodily difference (e.g., race, sexuality) within a specific cultural system. For example, one cannot fully and accurately understand the category of "woman" without examining how gender intersects with race, class, sexuality, ability, nation, and culture, among other social categories.[7] Second, gender is simultaneously a performative iteration and an administrative structure. In other words, it is both a set of repetitive acts that gives gender the illusion of substance[8] and a form of governance for individual and collective action that becomes institutionalized as "natural" (e.g., gender classification systems) in a social structure.[9] Third, gender is characterized by multiplicity rather than duality (i.e., we prefer to think in terms of a gender galaxy rather than a gender binary). There are multiple genders that transcend the simplistic cultural classification of "woman" and "man" in various societies and historical periods.[10] Finally, transing communication highlights the centrality of the subjectivity of individuals inhabiting different genders rather than the social and cultural imposition of gender meanings and categories on such individuals. To put it differently, it prioritizes the experiences of people in their own gendered bodies as they engage with the social world—for example, how a trans person navigates a gender-oppressive cultural system.[11] Taken together, transing communication provides a powerful tool for the examination of the relationship between gender and power, both microscopically (e.g., how gender influences identity and interpersonal relationships) and macroscopically (e.g., how gender is administered in social institutions such as media, education, and law, among many others).

Adhering to the above premises, we sketch a model for transing communication (see Figure 39.1) that consists of two interdependent and orthogonal axes—degree of difference and degree of mediatedness. Degree of difference focuses on how cultural, social, and geopolitical systems inform and constitute conceptions of gender and sexuality between individuals and groups. As such, it outlines a continuum based on cultural distinctiveness ranging from low (e.g., individuals from the same cultural system interacting with each other) to high (e.g., individuals from very different cultural systems interacting with one another).[12] Degree of mediatedness focuses on the extent through which various technologies mediate communication. This axis outlines another continuum based on the qualities of communication influenced by technology ranging from low (e.g., two individuals engaged in face-to-face interaction without technological devices) to high (e.g., an individual or group communicating exclusively through multiple communication technologies with a potentially large audience). Together, the two axes intersect to form four domains of communication (see Figure 39.1): (1) low degree of difference and low degree of mediatedness (e.g., face-to-face interactions between gender normative and non-normative people from similar cultural systems); (2) low degree of difference and high degree of mediatedness (e.g., mediated representations of gender non-normative people within a cultural system); (3) high degree of difference and low degree of mediatedness (e.g., face-to-face interactions between gender normative and non-normative people from different cultural systems); and (4) high degree of difference and high degree of mediatedness (e.g., popular discourses of gender non-normativity from two distinctive cultural systems). Characterized by broken boundary lines and potentially separate and sometimes overlapping areas (see Figure 39.1),

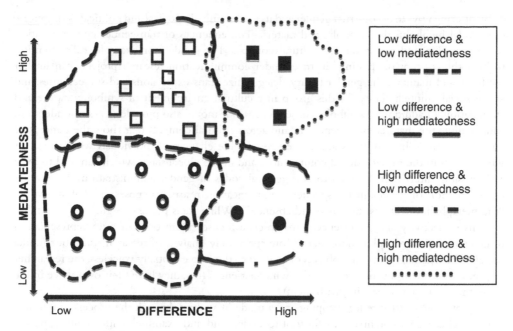

Figure 39.1   A model for transing communication in (inter)cultural contexts

Source: author.

these domains are fluid and dynamic as communication shifts and changes along the continua of difference and mediatedness.

Our model provides both a structure for mapping research on transgender communication and a mode of analysis of such research. Given the recent surge of research on transgender communication and its increasing theoretical and methodological diversity, our model offers a structure and process for understanding these complexities. More specifically, by identifying past research using our four research domains, we can more closely examine the construction of difference and cultural viewpoint in the study (degree of difference) as well as the nature and texture of the communication under investigation (degree of mediatedness). Such examination can provide a clearer perspective of cultural communication across the intracultural-intercultural continuum, as defined by degrees of difference, in a variety of settings featuring various degrees of mediatedness. In addition, our model provides a mode of analysis of the conceptualization and construction of gender in transgender communication research. Indeed, transing highlights the process of gender construction itself in research—for example, a study may conceptualize gender unidimensionally (i.e., gender by itself) or intersectionally (i.e., gender in relation to other vectors of difference), microscopically (i.e., gender as a set of individual traits or performances) and/or structurally (i.e., gender as an administrative structure and a form of collective govern-ance), dualistically (i.e., gender as a binary) or multiplicatively (i.e., gender as a galaxy). Through the process of highlighting the construction of gender in research, our model provides ways to deconstruct gender and its underlying assumptions. For example, if a study constructs gender as an inherent, stable, and universal set of characteristics (i.e., gender as essential traits), the model, using the process of transing, calls attention to the ideologies and politics of the project. As such, it deconstructs the intricate relationship between gender and power within specific geopolitical and historical contexts (e.g., an essential and universal gender binary that transcends culture and history erases trans existence and subjectivity).

By offering ways to deconstruct gender and its underlying ideological and political assumptions, our model highlights the symbolic and material consequences of transgender communication research within and across various cultural contexts. The model, for example, calls attention to how the concept of "passing" in transgender communication research projects can further reinforce and maintain the gender binary, delegitimize trans expressions and subjectivities, and create social unintelligibility for this group in a cultural environment of symbolic erasure and annihilation. In addition to symbolic consequences, our model also provides ways to understand and examine the materiality of gender in and across cultural contexts (e.g., how the concept of "passing" harms—indeed limits and decreases—the life chances of individuals who cannot or will not pass in their own cultural communities and in the global stage, which can lead to local and, possibly, international efforts to change legal codes of gender administration). Finally, the deconstruction of gender in transgender communication research exposes, in Rubin's words, "the metaphysics of (trans) presence and absence."[13] While trans presence signals the centrality of trans subjectivity and experience, trans absence focuses more on trans as a representation of gender enactment and achievement. More specifically, analysis of research using our model reveals whether trans people are objects of study (i.e., they are essentially voiceless, the researcher is speaking for them) or subjects of their own experiences (i.e., they have their own voice in the study, the researcher is speaking with them).

By providing a structure for mapping and a mode of analysis through the process of transing, our model can be a potentially useful tool to understand and examine gender, culture, power, and communication in transgender communication research. To illustrate, we use it to review and analyze past research before we explore new directions for transing communication in (inter) cultural contexts.

Reviewing research on transgender communication

Transgender communication research is becoming increasingly visible. After providing a more detailed conceptualization of our model, we provide in this section a review of representative research associated with each domain.[14]

The first domain: low difference, low mediatedness

This domain explores interpersonal interactions within similar cultural systems. Talk show interviews (while highly mediated to a larger audience) are interpersonal interactions with varying degrees of low cultural difference. For example, an interview with Laverne Cox (star of the popular Netflix series *Orange Is the New Black*) and Carmen Carrera (TV personality and model) by Katie Couric has perhaps a slightly lower degree of cultural difference (similar avowed gender identity) than an interview between Janet Mock (author and former *People* magazine editor) and Piers Morgan (different avowed gender identity). To understand these interactions, we highlight three themes within the research in this area: (1) *issues surrounding language,*[15] (2) *construction of identity,*[16] and (3) *exploration of relationships.*[17]

Language

In this domain, language refers to the creation, construction, and production of meanings in and through various linguistic systems within similar cultural contexts. Language constructs, affirms, and invalidates identities. When a language, such as English (both U.S. American and British), only has two socially accepted pronouns (i.e., him/he, her/she) to describe two legible genders (i.e., man, woman), those who do not fit the dichotomy become unintelligible. Unintelligibility produces invisible, invalid, and impossible identities and lives.[18] For trans individuals, the issues surrounding language are threefold: the conflation of the terms *sex* and *gender*, the lack of a consistent definition of trans, and language as a source of regulation of trans identities.

Of the numerous issues surrounding language and the trans community, the conflation of the terms *sex* and *gender* is vital to examine. Much of the research in this domain addresses the fact that in Western cultures, colloquially using *sex* and *gender* as interchangeable terms invalidates trans identities by the inherent naturalization that occurs from the notion that biological sex (i.e., a specific phenotypic presentation of genitalia) should and must be concurrent with gender expression and identity.[19] Acknowledging gender as separate from biological features would create a space where trans identities could be as valid as genders that are already biologically celebrated by the hegemonic binary. Even with the acknowledgement that sex and gender should be separate terms, the consequences of their conflation remain. Most of the research conducted in Western cultures for this domain continues to address issues among "transmen" and "transwomen" or describe multiple trans identities as "transmasculine" and "transfeminine." Due to the lack of readily available, socially digestible language that acknowledges gender as a galaxy, research ultimately ends up perpetuating and reinforcing the dominant social systems.

As with sex and gender, there is some disagreement among researchers as to the definition of *transgender*, which is culturally specific. Some research does not acknowledge any differentiation of the multiple identities within the transgender "umbrella."[20] Others specifically conflate *transgender* and *transsexual* to produce their own terms (e.g., "transpeople"[21] or "gender variance" and "gender nonconformity").[22] Still others simply condemn the use of *transgender* as an oversimplified umbrella term and call for a linguistic modification to the concept altogether.[23] This lack of consensus and understanding are preserving the confusing climate engulfing a perpetually muted and grossly misunderstood community.

The language we utilize in our everyday lives functions to validate and exclude certain people and identities in a culture. As such, it affects everyone (usually beginning at a young age), including those who are consciously working to deconstruct existing gender systems. Social scripts do this work without our conscious knowledge. Typical examples of Western social scripts include, "boy or girl?" as well as the prevalent focus on transitioning and genitalia in conversations with trans people.[24] The reinforcement of the Western gender binary system, particularly through the English language, serves to maintain the gender hierarchy, which labels trans identities (or anything other than boy/man or girl/woman) as "unnatural."[25] The notion of unnaturalness easily leads to the pathologization of trans bodies, namely that trans people in U.S. American and English cultures (among others) must be diagnosed with gender identity disorder (GID) to be able to receive hormone therapy, surgery, or other forms of medical intervention.[26]

Identity

In this domain, identity refers to the creation, construction, and production of self and others in and through interactions within cultural contexts. Based on the research in this domain, identity is examined in terms of the construction and maintenance of self through language, discourse, and narrative; interaction with power; and interaction with borders, specifically the gender binary.

Language, discourse, and narratives of trans voices that circulate in a culture influence how trans people construct their own identities. Because language and discourse in U.S. culture, for example, (re)create the gender binary, trans people are forced to conceive of their identities in terms of these two gender boxes. Much of the literature pointedly describes the ideas that trans people want or feel pressured to "pass" as either a stereotypically presenting man or woman. The "success" and "authenticity" of a trans person can be measured by the ability to pass as a particular gender for social legibility, perceived productivity, or safety.[27]

Constructions of self are intricately connected to power, including those of labeling and discourses of cultural intelligibility—that is, to be named, recognized, seen, and registered in the cultural imaginary. Although passing can be favorable in terms of access to privileges afforded to normative cisgender identities in a culture,[28] pressure to pass pertains to power relations. This pressure is inherently problematic because it reinforces the oppressive gender binary, invalidates trans identities (particularly for those who cannot or will not pass), and serves to protect the hegemonic cultural hierarchy of sex and gender through which heterosexuality is founded. In other words, the Western hegemonic sexual system based on gender object choice that produces the homosexual-heterosexual binary depends on the rigid separation of genders into "proper" (i.e., socially intelligible) women and men.[29]

Trans identities work with and against the gender borders of a culture (i.e., gender binary). This is manifested in a number of ways, including the health-care establishment and sexual minority communities, among others. For a trans person to receive medical intervention in the United States or United Kingdom, for example, they are forced to gain social, cultural, and medical approval from their provider, giving the doctor the power to control and define the trans person's identity.[30] If a trans person does receive medical intervention (i.e., sex reassignment surgery or hormonal therapy), their shift in identity can be staggering.

Shifts in identity are inevitably coupled with shifts in power dynamics in intracultural contexts. Circulating within a cultural system, some trans people report gaining access to privileges that were not afforded to their previous gender (generally those who have specifically transitioned to "passing" masculine identities), while others were surprised by the apparent loss of privilege that was coupled with their transitions (generally those who transitioned to feminine identities or are unable/unwilling to pass as cisgender).[31]

Relationships

In this domain, relationships refer to the creation, maintenance, and negotiation of relational meanings in and through interpersonal interactions. Relationships with trans identities, based on published research, are explored through two lenses—interpersonal relationships and relationships to and within communities.

Interpersonal relationships with trans people can be complicated simply because of the gender systems trans people are forced to occupy. As previously stated, sexuality in most Western cultures is based on a strict gender structure that forces participants in a relationship to identify themselves and the relationship in particular ways.[32] For instance, if a person identifies as a man and is in a heterosexual relationship with a woman, but decides to transition to become a woman, do they become a lesbian couple? Their individual and collective sexual identities have inevitably shifted, regardless of the couple's willingness or desire to reconfirm the sexual and gender binary of the culture.[33] Due to the constraints and pressures of the current sex and gender systems, the couple will likely shift their sexual identities to be congruent with that of their gender identities.

The communal ties of trans people can be unfortunately complicated for a multitude of reasons. Experiences within these relationships can range from a sense of belonging and understanding to discomfort and violence.[34] Transphobia and genderism from the outside and within the LGBTQ[35] community are unfortunate realities for trans individuals.[36] Internalized transphobia and genderism are likely the result of the Western social ideologies that are cultivated from (before) birth.[37] Take, for example, a selfidentified lesbian couple, in a Western culture, with a female-to-male (FTM) transitioning partner who has been shunned from lesbian communities for no longer fitting communal identity expectations of lesbianism. Trans narratives deem these instances of discrimination and violence as the most hurtful due to the fact that in these cases "the oppressed become oppressors."[38] In another trusted and seemingly safe relationship, parents grappling with their child's "gender nonconformity," unwilling to acknowledge their children as agents of their own identities, can end up harming the child's self-image by steering them toward cisgender identities that may not be the healthiest option for the child.[39]

The second domain: low difference, high mediatedness

This domain explores communication based on cultural homogeneity (i.e., low difference) in the context of mass communication technologies (i.e., high mediatedness). This often comes in the form of media portrayals of transgender people within the context of their own culture. *Self Evident Truths*, a U.S.-based photojournalism project depicting headshots of people who identify as anything outside of the culture's gender and sexual binaries, is an example of highly mediated communication with a relatively low degree of difference because of the U.S. participants and audience. Grounded in the research within this area, we highlight three emerging themes: (1) *construction of identity*,[40] (2) *sexual rights*,[41, 42] and (3) *sexual activism*.[43, 44]

Identity

In this domain, identity refers to the construction of self and others in and through public discourses within cultural contexts. Subthemes that arise surrounding identity include the conflation of the terms *gender* and *sexuality*, othering of trans identities, and visibility of trans identities.

While we have acknowledged that the conflation of *sex* and *gender* is inherently problematic and harmful, it is important to note that the conflation of *gender* and *sexuality* is also dangerous, but regrettably common. For example, U.S. and Thai media representations of trans people tend to conflate non-normative gender expression with an assumed homosexual identity. This assumption not only perpetuates the gender binary by presuming that all trans people will fall into conventional gender expressions that reify heteronormative understandings of sexuality but also reduces the transgender person's gender and sexual identity to that of their gender expression.[45] Though Brandon Teena identifies as a man in the 1999 film *Boys Don't Cry*, he is declared a lesbian by the intolerant townspeople, forcing him into inaccurate gender and sexual identities based solely on his genitalia.[46] Similarly, media representations of trans people generally depict trans identities as a threat and betrayal to the cisgender and heterosexual populations. For example, representations of the *tom-dee* population (women who do not adhere to the socially acceptable feminine dress and performance codes) in Thai media have implied that not only is gender expression indicative of sexual identity, but that diverse representations of gender (and therefore sexuality) are a threat to Thai culture as a whole.[47] In the U.S. reality television show *Dancing with the Stars*, Chaz Bono's gender identity is so confusing for the mainstream public that his sexuality is stripped from him altogether. While the spouses and partners of his fellow competitors are featured during the show, Bono's fiancé is rarely mentioned and never in a sexual context.[48]

In order to preserve the prevalent heteronormative structures through media representation, both Thai and U.S. media consciously portray trans identities as inherently separate from the cultural mainstream—that is, normative cisgender. The creation of the "other" creates a space of uncertainty and anxiety about trans identities that further reinforces and solidifies the cultural gender hierarchy. For example, Thai media also portray the *tom-dees* as "brash" and "consumer-oriented," insinuating that masculine gender expression in women is intrinsically tied to unrelated personality characteristics that are viewed as divergent from ideal Thai culture.[49] Within these media contexts, trans identities are marked in obvious and debilitating ways. As a self-identified trans man in the beginning of the show, Bono, in *Dancing with the Stars,* is constantly separated from the other contestants. Though Bono attempts to perform heteronormative masculinity through song choice, the show undercuts his attempts with choreography that is generally aligned with the women on the show. While most dances on the show tend to be romantic and/or sexual, culminating in a close embrace, in six of eight dances that Bono performed, he and his partner were separated in the final pose.[50]

In recent years, there has been some trans visibility in U.S. media. As people cultivate understandings of self through social comparison and self-reflection, it is vital that they see depictions of their identities within media contexts in order to validate their existence, legitimacy, and societal value. Keeping certain identities and their voices out of mainstream media insinuates that they are wrong, invalid, and inferior in the culture. This can be detrimental to the development of a healthy self-image for trans individuals and only serves to maintain the hierarchy of sex and gender systems that work to oppress non-normative identities. With such limited media representations of trans identities, it is crucial that trans individuals are presented consciously as multidimensional and humanistic. Unfortunately, U.S. popular media continue to present trans identities with simultaneously positive and negative valences. A promotional poster for *TransAmerica,* a 2005 film starring Felicity Huffman, depicts a trans individual's struggle with everyday activities, such as bathroom use, providing visibility for trans identities and shedding light on the cisgender privilege afforded in those seemingly mundane acts. However, by depicting the main character, Bree, wearing all blue standing in front of an entirely pink background, the poster works to reify the established gender binary and highlight Bree's gender non-normativity in her culture.[51]

Sexual rights

In this domain, sexual rights refer to the construction, regulation, and contestation of various sex and gender systems in and through public discourses within a singular cultural context. Perhaps not surprisingly, the research mostly focuses on laws.

Laws not only govern the ways we interact with the social world, but they also serve as a primary mechanism of the construction of cultural ideologies, including gender.[52] According to Vitulli,[53] sexual laws in the United States serve to govern people's bodies, partially through the validation of gender and sexual identities. If people's genders are not aligned with their sex (and therefore do not adhere to the established Western cultural gender binary), they can be deemed inhuman and unworthy of the rights that are afforded to cisgender individuals.[54] Within Western constructions of normative bodies, cisgender people are implicitly constructed as human and worthy of protection by law and further validated by social law (i.e., interpersonal and self-regulation of societal norms). On the other hand, non-normative bodies are pathologized as unhealthy and unstable, which might serve as justification for regulation and discrimination by both law and social law.[55]

Sexual activism

In this domain, sexual activism refers to the contestation, deconstruction, and reconstruction of various sex and gender systems in and through discourses within cultural contexts. The identified research on sexual activism focuses mostly on advocacy for trans communities and increasing the visibility of trans identities in mediated contexts.

As sexual activism becomes a more pressing issue throughout the world—including, for example, the "decline to state" option on birth certificates in Germany and the legal recognition of *hijra* identity in India—advocacy for sexual rights attempts to increase awareness to produce social change. South African trans activists, for example, have created a sundry approach to advocacy by encouraging solidarity among trans communities, addressing social law and the way that discrimination functions in everyday life, and deconstructing public discourses to break down the hegemonic forces plaguing sexual rights.[56] Politics and legislation are a pressing concern for the trans community, as trans people are routinely denied access to basic human rights and subjected to discrimination within public services including medical care, among many others. However, trans people simply speaking publicly about their lived experiences and being represented through various media sources is a form of advocacy in a world that refuses to recognize and validate their existence.[57]

The third domain: high difference, low mediatedness

This domain is conceptualized through interpersonal interactions where there is a high degree of cultural difference. Dana International, an Israeli male-to-female (MTF) transsexual, and Guildo Horn, a German cisgender male, competed in the highly popular Eurovision Song Contest in 1998. Interpersonal interactions between Dana and Guildo during the event would exemplify this domain—face-to-face communication between two different individuals attempting to understand and relate to each other through their own linguistic and cultural filters. Research in this area generally takes the form of ethnographic accounts of cultural contexts with which the researcher does not identify. Three themes emerged from these cross-cultural interactions: (1) *use of language,* [58](2) *construction of identity,*[59] and (3) *construction and maintenance of relationships.*[60]

Language

In this domain, language refers to the creation, construction, and production of meanings in and through various linguistic systems across cultural contexts. When different languages are involved, such meanings become much more complex. As we have established, language holds an immense power in the construction and regulation of identity and relationships. When people attempt to understand the same concept across cultural contexts (e.g., gender in, say, Andean, Indonesian, Norwegian, South African, and U.S. American), meanings can be confounding, especially across diverse geopolitical regions and historical periods.

Identity

In this domain, identity is referred to as the creation, construction, and production of self and others in and through interactions across cultural contexts. It would be highly ethnocentric and inaccurate to assume that non-Western cultures would (and could) readily adapt Western versions of sex and gender systems. However, these portrayals are available to anyone with access to the Internet and have a potential influence on other cultures throughout the world. Samoan culture, for example, has been negatively affected by Western trans ideologies. As reported by Roen, a Samoan *fa'afafine* (i.e., a highly respected male who takes on feminine presentations of dress and familial duties) who visited the United States was referred to as a "cock in a frock,"[61] demonstrating an ethnocentric projection of U.S. culture (i.e., that "cross dressing" is a joke or source of entertainment as opposed to a respected gender identity). On the other hand, Blackwood noted that some Indonesian identities derive their meanings from Western understandings of gender and sexuality, which gives them recognition and validation on a global scale.[62]

Relationships

In this domain, we refer to relationships as the creation, maintenance, and negotiation of relational meanings in and through interactions across cultural contexts. Blackwood's experience with the *lesbis* and *tombois* of Indonesia serves as an example of Western influence on another culture's language systems that works to define and construct relationships through a Western lens.[63] More specifically, Blackwood proposed that the word *lesbi* (used colloquially and in Indonesian media) is derived from the term *lesbian* and describes two females (note the reference to biology) in a relationship with each other.[64] A *lesbi* relationship is comprised of a *tomboi* and a *girlfriend*, that is, a masculine presenting female and a feminine presenting female. *Tombois* do not consider themselves women, but rather feel that their performance of everyday life (e.g., sleeping wherever they choose, engaging in physical labor for work) falls more in line with Indonesian men, which is preferable considering the privileges afforded to men in Indonesian society. *Girlfriends* are feminine women. While the description of *tomboi* refers to gender expression and identity, in this case, sexuality is also implied. Due to the fact that *tombois* consider themselves to be closer to men than to women, they refer to themselves as heterosexual. Similarly, *girlfriends* do not consider themselves homosexual, though it is acknowledged that a *lesbi* relationship is a relationship between two females. Similarly, in the case of the Samoan *fa'afafine*, Roen observes that gender generally implies a homosexual identity and the terms gender and sexuality are practically interchangeable.[65]

According to Blackwood, the *lesbi* community will only acknowledge relationships where there is one masculine identified partner and one feminine identified partner, actively upholding a similar structure to the Western idea of heteronormativity.[66] Adhering to the gender binary and acknowledging that there is one partner on either side is a crucial aspect of belonging to the *lesbi* community. However, these very particular understandings of gender and sexuality create a community where *lesbis* and *tombois* can feel safe, free, and proud to express themselves and their love for one another.

The fourth domain: high difference, high mediatedness

This domain explores communication based on cultural heterogeneity (high difference) in mass mediated contexts (high mediatedness). To put it differently, it focuses on media representations of trans bodies from different cultural systems. An example of a highly mediated artifact that has spread across the globe is Thai pop singer Vid Hiper R Siam's music video "Ladyboy Never Cheats" (title loosely translated). Although problematic in various ways, the video has, as of 2014, reached well over 13 million viewers across the globe, including non-Thai speaking audiences, cisgender audiences, and audiences that know very little about Thai culture or transgender subjectivities (exemplifying this domain) as well as Thai audiences (exemplifying the second domain). Research in this domain highlights three central themes: (1) *identity*,[67] (2) *sexual rights*,[68, 69] and (3) *sexual activism*.[70, 71]

Identity

In this domain, identity refers to the construction of self and others in and through public discourses, such as media representations, across cultural contexts. Research in this domain focuses on various conceptions of identity and conflation between gay and trans identities leading to a range of intelligibility of transgenderism.

Trans identity has been conceptualized in different ways ranging from stable to fluid. Patel describes *hijras*, individuals in Indian culture who are neither male nor female, as people who were born male but identify as female by performing femininity through bodily deportment, appearance, and dress.[72] Often referred to as "third gender," which Patel translates into the Western conception of "transgender," the identity of a *hijra* is fairly fixed and stable.[73] On the other hand, identity has been characterized as the interplay between stability and closure as well as fluidity and ambiguity. Focusing on Dana International, the first Israeli MTF singer to represent her country and to win the Eurovision Song Contest, Maurey examines Dana's performance, choice of songs, and lyrics.[74] The analysis suggests that Dana engages in a constant play with her identity ranging from reification of stable womanhood as passive and consumer-oriented to evolving performances of female agency as transgressive and highly sexual.

As stated previously, the literature we examined suggests that there is a pervasive conflation between *gender* and *sexuality*. In discourses across different contexts, we continue to see trans identity associated with homosexuality. In Japan, for example, McLelland notes that transgenderism, in the form of cross-dressing, is often conflated with homosexual desire and attraction.[75] This often results in elevating homosexual identity, with its accompanying markers of cross-dressing, entertainment, sex work, and effeminacy, at the expense of the intelligibility of trans subjectivity and identity. On the other hand, *hijras* have a trans identity that is distinctively different from homosexuality. Patel notes that there are two types of men who have sex with men—what the Western sexual system would call "homosexual"—in Indian culture: *panthis* (those who are masculine in gender performance and penetrators in sexual practice) and *kothis* (those who are feminine in gender performance and are sexually passive in practice).[76] Although both *kothis* and *hijras* perform traditional femininity in their culture, *kothis* refers to a sexual identity while *hijras* adopt a culturally and religiously structured feminine way of living, which Patel equates to a transgender identity. In this cultural context, *hijra* identity is highly intelligible.[77]

Sexual rights

In this domain, sexual rights refer to the construction, regulation, and recognition of various sex and gender systems in and through public discourses across cultural contexts. Previous research continues to observe the conflation of sex and gender and different conceptualizations of trans identity, as reported above, and focuses on transgender rights as human rights. In India, for example, Patel notes that in spite of social acceptance, *hijras* struggle with their basic human rights, such as the lack of basic access to health care, sexual health information, and general political rights; employment and police discrimination; and ongoing gender-based violence and police brutality, among others.[78] Such struggles have been placed in an international context and covered by various cultural media outlets as a battle for transgender rights.

Sexual activism

In this domain, sexual activism refers to the contestation, deconstruction, and reconstruction of various sex and gender systems across cultural contexts. As such, it is both a symbolic and material extension of what researchers call "sexual rights." Not surprisingly, work in this area focuses on political and legislative change. Returning to the case of the *hijras* in India, a lot of focus was placed on Indian Penal Code (IPC) Section 377—imposed in 1860 by the United Kingdom, India's former colonizers, to criminalize sodomy. IPC 377, in many ways, serves as a

classic example of how distinct cultural systems of sex, gender, and sexuality come into conflict through the process of colonialism and demonstrate how the colonizers—in this case, the
British—used their power to impose, formalize, and enforce their cultural values on the native
community. Legal discrimination against homosexuality and *hijras* in India is one of the results.
Given the rampant employment discrimination and other forms of social regulation surrounding
the workplace, most *hijras* turn to sex work to survive, and IPC 377 is used to intimidate, harass,
target, and punish them.[79] Using the language of transgender rights to fight legal discrimination
against *hijras*, a legislative victory was achieved in April 2014, giving Indian *hijras* the right to be
recognized as a "third gender."

Transing communication in (inter)cultural contexts: exploring new directions for research

Without a doubt, our research indicates that there is a need to acknowledge the vast invisibility
and invalidation of trans lives across cultural contexts. Utilizing the deconstructive method of
transing to recognize and redefine the dominant sex, gender, and sexual systems within and
between cultural spaces is the first step to dismantling the pervasive oppression of trans identities. By prioritizing the experiences of trans individuals, we can begin to construct communication that serves to challenge current social and ideological systems, validate diverse identities
regardless of normativity, and produce research that reflects the rich complexities of trans lives
within and across cultural systems. Recognizing that communication is always already cultural, our model offers a way to think about cultures in terms of difference, such as meanings
and frameworks of intelligibility, views of self and others, embodiments and practices, as well
as structures and social institutions, among others. Further, acknowledging the unprecedented
changes and uses of technology in the last few decades, our model provides a way to consider its
impact on communication and discourse, such as individual and collective identity, representation within and between cultures, construction of rights, and collective mobilization for social
change, among others. Taken together, the model provides ways to map and examine current
research and suggests new directions for transgender communication scholarship within and
across cultures.

Using the model we developed and focusing on the studies we identified for each domain,
we now discuss some implications for transing communication in (inter)cultural contexts and
suggest new directions for research in these arenas. Our discussion focuses on two broad areas:
(1) issues that emerged from the research *within* the four domains, which might be viewed as
a "deep look from within," and (2) patterns that emerged from the research *between* the four
domains, which might be characterized as a "panoramic view from above."

Transing communication, as stated earlier, demands that we examine how gender is assembled
and reassembled with other qualities (e.g., race, sexuality) and structures (e.g., culture, systems
of representation) of bodily being within geopolitical and historical contexts. A "deep look
from within" the four domains—individually and collectively—suggests a number of potential
implications and new directions for research.

First, researchers must be mindful of the ways they construct gender in their projects. More
specifically, a number of the studies we examined, perhaps inadvertently, reified the gender binary
by forcing trans people to conform to a system of gender duality rather than affirming multiple,
even infinite, gender expressions, embodiments, configurations, and practices. In the process, trans
people are othered and, in doing so, they will continue to be, as Suess, Espineira, and Walters
forcefully remind us, "defined as pathological."[80] In addition to being symbolically and materially violent to trans people, such pathologization serves to reify gender norms (manifestations

of societal gender standards), normativity (the culturally accepted belief that gender norms are "natural"), and ideals of normality (the creation of "natural" as distinct and separate from "unnatural" gendered bodies) in a culture. When researchers conceptualize normality as a dynamic relation rather than a static set of qualities (i.e., gender normality, such as the binary, is defined in relation to gender abnormalities, such as those who do not fit the binary, rather than defining gender normality as a "natural" or "given" group of traits), they expose the constructed, unstable, contingent, and improvisational nature of gender.[81] This is a critical move in future transgender communication research. Another critical move in this research is to examine gender inclusively through the trope of gender galaxy by having individuals and groups self-define—in other words, gender self-determination[82]—rather than forcing them into existing categories.

Second, researchers must be attentive to how sex, gender, and sexuality systems are configured within a cultural context. Several studies point to the conflation of sex and gender; gender and sexuality; and sex, gender, and sexuality. In many Western cultures, for example, sex is always already gendered and normative gendered performances are intricately connected to the institutions of heterosexuality and patriarchy.[83] In such a system, what is the potential of a trans body to reify and/or disrupt heteropatriarchy? Attention to this question as well as the relationship between microscopic interactions—for example, everyday conversations—and macroscopic structures and forces—for example, cultural ideology of cisgenderism that denigrates trans identities, bodies, genders, and sexualities[84]—are important in future work in transgender communication.

Third, researchers must examine how gender is defined and constituted through other salient cultural categories, such as race, class, sexuality, body, among others.[85] In other words, gender must be examined intersectionally, which to us, means attending to both marked (i.e., non-normative identities within a culture such as non-white, non-middle class, non-heterosexual, etc.) and unmarked (i.e., normative identities within a culture such as white, middle-class, heterosexual, etc.) intersections.[86] Unfortunately, much of the research we identified focused mostly on gender and sexuality with little attention to race, class, or the body.

Fourth, studies that examined trans bodies and identities in the United States did not generally focus on culture while research on bodies and identities outside the United States—particularly non-Western societies—had a much greater emphasis on culture. As a result, Western cultures become an invisible and normative center through which "other" cultures are described, measured, and declared to deviate. A potentially productive direction in future transgender communication research is to examine how culture—visibly or invisibly—constructs gender, enforces gender normativity, and produces trans bodies and identities at particular moments in time and in various geopolitical contexts.[87]

Moving away from a focus on research *within* the domains and turning to the examination of research *between* the four domains of our model, a "panoramic view from above" emerges. It is clear that transgender communication research of various degrees of mediatedness in intracultural settings are much more common than intercultural ones. Given the additional complexities and demands of intercultural research, this is, of course, hardly surprising. Intercultural communication research requires exploration into different ways of seeing, knowing, and living; constructing, reconstructing, and narrating bodies and subjectivities; and their subsequent translation into other systems of meaning, legibility, and cultural representation.[88] This process of knowledge production exists in a complex field of power relations as Enke reminds us, "gay, queer, and transgender all comingle with imperialist institutions;" in this sense, intercultural researchers, who act like translators, carry "the burden of destruction and creation."[89] With sensitivity, care, and hope, the burden of creating new knowledge systems can be met in future transgender (inter)cultural communication research partly through "intersectional reflexivity,"[90] which refers to the researchers' acknowledgment of and reflection on their own intersectional identity and their own self-implication in systems of privilege and marginalization in relation to their own culture as well as the culture of the group they are studying.

One obvious direction for future transgender communication research is to examine the production of gender—and its accompanying intersections—in cultural contexts beyond the United States and the global West. A perhaps less obvious future research direction is the examination of gendering processes in the West that focuses on whiteness, body normativity, and cisgender privileges.[91] Such work is, in many ways, about unpacking invisible and unmarked systems of power and privilege, an important area of future transgender communication research. A final implication for researchers is to be mindful that in intercultural settings, identity, relationships, gender/sexual rights, and gender/sexual activism may have very different meanings for the individuals and cultural communities involved. Gender/sexual rights, for example, may look different from one cultural context to another depending on geopolitics, history, and the interplay between the local and the global.[92]

"Transgender" is itself an imperfect translation across cultural contexts, one that, as Enke points out, "carries institutional and imperial discipline: to be named and to name oneself transgender is to enter into disciplinary regimes that distribute recognition and resources according to imperial logics," that is, systems of cultural intelligibility, value, and protection.[93] As an imperfect translation, Enke further notes, "transgender demands above all the need for more context, more story, and thus the translation into transgender never arrives and rests."[94] Transing communication in (inter)cultural contexts is, in many ways, an invitation to engage with the boundaries and untapped potential of transgender and to participate in the necessary, risky, and hopeful act of unending, creative, and imperfect translation. We hope our model serves as a tool and a beginning of a journey to start thinking about communication and gender, within and across cultures, more critically, politically, expansively, and inclusively.

Postscript 2019

Since the original publication of this article in 2015, transgender, as a global assemblage, has continued to grow and expand. In the academy, Trans Studies has become more widely recognized, as demonstrated by increasing number of conferences, publications, and university courses focusing on the topic. Although the process has not been necessarily seamless and without conflict, Trans Studies has transformed and enriched work in Feminist and Queer Studies through an increased attention to questions of embodiment and cultural normativity.[95] In the cultural domain, transgender, as a word, has shifted and changed (e.g., transgender as gender variance and gender nonconformity; transgender as an umbrella term for various modes of gender transgression). More recently, Halberstam calls for the use of "trans*" to "open the term up to unfolding categories of being organized around but not confined to forms of gender variance."[96] In addition, transgender, in terms of representation, has seemingly exploded in popular culture. Although such representations have increased the visibility of trans experiences and lives in mainstream culture, they are not necessarily unproblematic.[97] For example, Yep, Russo, Allen, and Chivers[98] note that seemingly transgressive representations, such as the character of Unique—a fat African American gender non-normative person—in the popular U.S. television series *Glee*, ultimately end up reinforcing and perpetuating heteronormative ideologies and structures. Further, Yep, Chrifi Alaoui, and Lescure[99] point out that mainstream U.S. trans representations tend to erase the subjectivities and experiences of trans women of color, particularly if they are poor. However, there are cultural spaces where trans and gender nonconforming people can claim and validate their own subjectivities and experiences.[100] Taken together, the study of gender through the framework of transing has become an increasingly useful and important critical theoretical tool.

Current research on transgender communication

To provide an update of the research published since our original article, we conducted a literature search of scholarly, peer-reviewed works from the beginning of 2014 to the beginning of 2019. Using the same parameters and search terms from our initial project, we located research and collectively discussed and placed them in the four quadrants of our model. Similar to what we did in the original project, we highlight, in this section, some of the most current findings.

Generally, we found three trends in current research on transgender communication. First, studies are examining current struggles that trans people are facing at this historical moment, including trans rights,[101] immigration and asylum,[102] trans mobilizing,[103] gender and racial oppression,[104] and trans activism,[105] among others. Second, there is the emergence of a "new transnormativity," which, according to Yep, Russo, and Gigi, refers to a "new set of standards through which trans identities, experiences, expressions, and aesthetics—in short, trans lives—are to be measured and judged in the cultural domain."[106] Such standards include fixing the trans body through consumption of products (e.g., cosmetics for trans women) and services (e.g., gender reassignment surgery, voice therapy, facial feminization surgery for trans women, hormone therapy); concern about physical appearance that strictly adhere to the gender binary (e.g., focus on "passing" as masculine men for trans men); and centering the individual at the expense of community and politics (e.g., trans identity as a personal choice rather than a cultural and political category to produce social change). Finally, as in our original project, more intracultural research on various degrees of mediatedness (first and second domains of our model) than intercultural projects (third and fourth domains) continue to be published, arguably for the same reasons we discussed earlier. We now turn to highlight some of the current research in each one of the four domains.

The first domain: low difference, low mediatedness

Current research in this domain has maintained a focus on the resounding themes of *language*, *identity*, and *relationships* identified in the first iteration. However, the research has grown significantly more complex in the ways scholars are analyzing gender in relation to intersecting oppressions (e.g., culture, race, class, gender, immigration status). Further, these intersections mirror the current struggles facing trans people much more than our last search. Take, for instance, Cheney et al.'s work confronting the issues that transgender Mexican asylum seekers face on both sides of the border as they attempt to immigrate to the United States, focusing on the interlocking ways in which already marginalized people are forced to navigate multiple cultural contexts at once (e.g., Mexico, United States).[107] The research presents self-disclosed experiences of gender discrimination and violence through the contexts of parents, community, school, police, and the workplace. Participants in the study reported an overwhelming sense of fear, hopelessness, and anxiety about gender, immigration, and the loss of family and community.

In another example of research that could be viewed as a direct response to the current cultural climate is Edelman and Zimman's analysis of how trans men circumvent hegemonic embodiment by utilizing recuperative language surrounding their genitals.[108] As the mainstream conversation about "the trans experience" has settled into an overwhelming trope about being "born in the wrong body," the authors explore a subversive narrative that refuses to consider trans genitals as solely a site of conflict. By invoking the concept of "social flesh," the authors are able to illuminate the nuances of personal gender identities and ideologies are co-constructed and maintained as an ongoing conversation between the person and the social fabric they are being (forcibly) draped in.

The second domain: low difference, high mediatedness

Our current review suggests that the second domain is the most populous of the four, providing numerous studies that fall into the quadrant. The research in this domain is perhaps the most easily accessible considering that researchers are analyzing cultural artifacts within cultural systems they are familiar with, focusing on the ways that *identity* is constructed in relation to *sexual rights* and *sexual activism*. For example, Loh's work brings us back to the *hijra* communities of India, discussing the legalization of gender minorities and the ways in which non-Western cultures are influenced by the Western assumption that gender and sexuality are universal experiences.[109] Loh provides a critique of the rhetoric used within mainstream U.S. LGBTQ movements, noting that the universalization of LGBTQ experiences serves to satiate the general populace and yet still remain superior to those most intersectionally marginalized (e.g., rural, non-U.S. cultural members). More specifically, Loh examines the ways in which the legalization of the *hijra* population has affected related arenas such as a national definition of gender, access to health care, and privacy rights. This legalization has, in some cases, contributed to a limiting scope of transgender needs and desires.

In contrast, Jackson and associates offer a thorough analysis of #GirlsLikeUs—a movement by trans women, for trans women—and the potentiality of trans people getting to lead their own movements.[110] When trans communities are allowed the space to define and advocate for themselves, it becomes clear that they are more careful about making space for trans identities beyond their own to participate and feel validated, rather than rely on essentialist narratives to only allow the palatable identities, such as trans women who pass in terms of cultural standards of femininity and beauty, be recognized. Most importantly, this work primarily builds community from within while centering trans women's experiences and issues, and thus is able to trickle out into the general population providing self-constructed visibility. The authors expand the idea of trans worldmaking within a reality that often considers the trans critique a "disruption" to the larger LGBTQ narrative.[111]

The third domain: high difference, low mediatedness

The third domain remains the least populated quadrant, as we were only able to locate one article that came out since the time of our last publication. Kerry examines the discourse and rhetoric across research and conferences in order to understand the issues facing indigenous and non-indigenous transgender Australians.[112] Indigenous and non-indigenous communities report discrimination in multiple settings and experience economic instability, social exclusion, limited access to comprehensive medical care, abuse, and lack of community engagement. Though the overarching issues that face trans Australians are shared by both communities, the research points to the ways that racism intersects with trans identities, creating unique experiences for indigenous and non-indigenous trans Australians. Specifically, there is much less research that directly examines the unique issues faced by indigenous trans Australians, which contributes to the sweeping erasure of the particularities of trans identities outside of whiteness. Further, there is even less research about the nuances of identity construction of nonconforming genders within indigenous communities done by people who are able to truly understand those communities' desires and struggles.

The fourth domain: high difference, high mediatedness

The fourth domain continues to have a very limited amount of research. However, the two recent studies are relatively similar in their framework and scope as the research in our original iteration. More specifically, they revolve around the same themes of *sexual rights, sexual activism*, and *identity*. Émon and Garlough examine case studies of South Asian women in Madison, Wisconsin, attempting to weave LGBTQ identities and activism into their communities through a queer, feminist live performance event called "Yoni Ki Baat."[113] The example of using live performance to explore the intersection of South Asian gender and sexual identity in the middle of the Midwest provides a unique experience for both the audience and the performer. Hackl, Becker, and Todd's quantitative analysis of the management of Chelsea Manning's gender and pronoun change across international media offers a less subjective take on identity construction.[114] Unlike the performers of Yoni Ki Baat, Manning is unable to control the way in which her identity is being shaped to a world audience. It is worth noting that media outlets outside of the United States used Manning's correct name and pronouns faster and more often that domestic U.S. outlets. These findings are in line with our original research, which shows that cultures outside of the United States have more holistically embraced gender and sexuality diversity.

Further transing communication: additional directions for future research

Although the suggestions for future research in the original article continue, in our view, to be relevant in transgender communication projects, we add three potential new directions—*transsubjectivity, transrelationality*, and *transfuturity*—that highlight and focus on the experiences and lives of individuals inhabiting their gender in a cultural system, the fourth and last fundamental premise of transing communication described at the beginning of this chapter. Since the experiences of gender normative individuals (e.g., cis women, cis men) are widely represented in research and in popular culture, we make an urgent call to center the experiences—indeed the lives—of people who are considered gender non-normative in their cultural system. To put it differently, there is a critical need to create spaces for gender non-normative individuals to speak for themselves. Three theoretical concepts might be useful for this endeavor.

Transsubjectivity, according to Yep et al., refers to "modalities of feeling and sense-making that (re)center the lived experience and perspectives of trans individuals in a cultural context of liminality and ongoing tension between cisnormativity and trans performativity."[115] To put it differently, transsubjectivity involves making the feelings and perspectives of trans people—surviving, coping, living, and thriving in an unfriendly and hostile cisnormative world—legible, vital, and invaluable. For example, Yep and associates, analyzing the internationally acclaimed film *Ma Vie en Rose*, report on how Ludovic, the seven-year-old protagonist, manages to communicate a unique, clear, and deliberate sense of identity that transgresses normative ideals of boyhood masculinity in Ludovic's culture.[116] Transsubjectivity highlights these communicative processes of identity construction from the viewpoint of the gender non-normative subject.

Transrelationality refers to "modes of affinity, connection, and affective expression that enable and materialize transsubjectivities governed by symbolic and material cultural conditions to imagine and actualize networks of possibilities for alliance and potential transformation."[117] In other words, transrelationality makes legible relationships and possibilities that are outside of culturally recognized and sanctioned relations and networks—for example, creating new conceptualizations of family that challenge traditional legal definitions of the institution. In their analysis of the highly acclaimed independent film *Tangerine*, Yep and colleagues document how the two central characters—Sin-dee and Alexandra, two poor trans sex workers of color—forge a friendship and create a familial system of love, solidarity, and support in the hostile, and sometimes violent, physical and social environment for sex workers in Los Angeles.[118] Transrelationality makes such modes of affinity and connection visible and recognizable.

Through transsubjectivity and transrelationality, transfuturity could emerge. By hearing the voices and perspectives of trans people and by recognizing their affinities, affects, and connections, new possibilities, visions, relations, communities, and worlds can be imagined and created without the trappings of cisnormativity and constraints of gender binarism. We call this novel and emerging horizon transfuturity—an open, fluid, indeterminate, and inclusive process of worldmaking guided by the gender galaxy.

Notes

1 Gust thanks Wenshu Lee and Philip Wander, John Elia and Gina Bloom, and Amy Taira, my "soul friends," for their ongoing inspiration and support; and Yogi Enzo and Pierre Lucas, my affectionate Pomeranian companions, for their sweet presence and unconditional love. Finally, Gust acknowledges his co-authors, Sage and Jace, for a collaborative project full of positive energy and mutual support; we make a great team.

2 For purposes of clarity and continuity, we use "trans people" and "trans identities" to describe the galaxy of gender expressions and identities that challenges the gender binary system (i.e., woman/man). See Stryker and Currah's "Introduction" and Williams' "Transgender" for a brief historical overview of the evolving meanings of the term.

3 Susan Stryker, "(De)Subjugated Knowledges: An Introduction to Transgender Studies," in *The Transgender Studies Reader*, eds Stephen Whittle and Susan Stryker (New York: Routledge, 2006), 14.

4 For example: Giorgia Aiello et al., "Here, and Not Yet Here: A Dialogue at the Intersection of Queer, Trans, and Culture," *Journal of International and Intercultural Communication* 6, no. 2 (2013): 96–117, https://doi.org/10.1080/17513057.2013.778155; Nael Bhanji, "Trans/scriptions: Homing Desires (Trans) Sexual Citizenship, and Racialized Bodies," in *The Transgender Studies Reader 2*, eds Susan Stryker and Aren Z. Aizura (New York: Routledge, 2013), 512–526; Michael J. Horswell, *Decolonizing the Sodomite: Queer Tropes of Sexuality in Colonial Andean Culture* (Austin, TX: University of Texas Press, 2005); Susan Stryker and Paisley Currah, "Introduction," *TSQ: Transgender Studies Quarterly* 1, no. 1–2 (2014): 1–18, https://doi.org/10.1215/23289252-2398540; David Valentine, *Imagining Transgender: An Ethnography of a Category* (Durham, NC: Duke University Press, 2007); Cristan Williams, "Transgender," *TSQ: Transgender Studies Quarterly* 1, nos 1–2 (January 2014): 232–234, https://doi.org/10.1215/23289252-2400136.

5 Susan Stryker, Paisley Currah, and Lisa Jean Moore, "Introduction: Trans-, Trans, or Transgender?," *WSQ: Womens Studies Quarterly* 36, nos 3–4 (2008): 11–22, https://doi.org/10.1353/wsq.0.0112.

6 We are using the gerund in our headings to highlight that these processes—transing, reviewing the literature, and exploring implications and future directions for research—are active, ongoing, and evolving. For example, our model is in a process of evolution and change as new modes of communication emerge and transform and (inter)cultural encounters increase and magnify in our current era of neoliberal globalization.

7 Chandra Talpade Mohanty, *Feminism Without Borders: Decolonizing Theory, Practicing Solidarity* (Durham, NC: Duke University Press, 2003); Gust A. Yep, "Queering/Quaring/Kauering/Crippin'/Transing 'Other Bodies' in Intercultural Communication," *Journal of International and Intercultural Communication* 6, no. 2 (2013): 118–126, https://doi.org/10.1080/17513057.2013.777087; Gust A. Yep, "Toward Thick(er) Intersectionalities: Theorizing, Researching, and Activating the Complexities of Communication and Identities," in *Globalizing Intercultural Communication: A Reader*, eds Kathryn Sorrells and Sachi Sekimoto (Thousand Oaks, CA: Sage, 2016), 86–94.

8 Judith Butler, *Gender Trouble: Feminism and the Subversion of Identity* (New York: Routledge, 1990).

9 Dean Spade, *Normal Life: Administrative Violence, Critical Trans Politics, and the Limits of Law* (Durham, NC: Duke University Press, 2015).

10 For example: Kate Bornstein, *My New Gender Workbook* (London: Routledge, 2013); Horswell, *Decolonizing the Sodomite*.

11 Spade, *Normal Life*.

12 Consistent with a critical approach to intercultural communication, our definition of culture is not restricted to the nation-state. As Yep stated elsewhere, "culture is a contested conceptual, discursive, and material terrain of meanings, practices, and human activities within a particular social, political, and historical context,"[119] which recognizes power in the construction and maintenance of differences. For further discussion of the definitions of culture in intercultural communication, see Moon's work.[120]

13 Henry S. Rubin, "Trans Studies: Between a Metaphysics of Presence and Absence," in *Reclaiming Genders: Transsexual Grammars at the Fin De Siècle*, eds Kate More and Stephen Whittle (London: Cassell, 1999), 173–192, 178.

14 To illustrate the potential contours and content of each domain, we review the literature and provide examples of interdisciplinary research. Such review focuses more on qualities and patterns of representative research and should not be read as a comprehensive assessment of the literature.

15 For example: Mollie V. Blackburn, "Agency in Borderland Discourses: Examining Language Use in a Community Center with Black Queer Youth," *Teachers College Record* 107, no. 1 (2005): 89–113, https://doi.org/10.1111/j.1467-9620.2005.00458.x; Sally Hines, "Queerly Situated? Exploring Negotiations of Trans Queer Subjectivities at Work and within Community Spaces in the UK," *Gender, Place & Culture* 17, no. 5 (2010): 597–613, https://doi.org/10.1080/0966369x.2010.503116; Alex Iantaffi and Walter O. Bockting, "Views from Both Sides of the Bridge? Gender, Sexual Legitimacy and Transgender Peoples Experiences of Relationships," *Culture, Health & Sexuality* 13, no. 3 (2011): 355–370, https://doi.org/10.1080/13691058.2010.537770; Dan Irving, "Normalized Transgressions: Legitimizing the Transsexual Body as Productive," *Radical History Review* 2008, no. 100 (2008): 38–59, https://doi.org/10.1215/01636545-2007-021; Heidi M. Levitt and Maria R. Ippolito, "Being Transgender: Navigating Minority Stressors and Developing Authentic Self-Presentation," *Psychology of Women Quarterly* 38, no. 1 (2013): 46–64, https://doi.org/10.1177/0361684313501644; Elizabeth Anne Riley et al., "The Needs of Gender-Variant Children and Their Parents: A Parent Survey," *International Journal of Sexual Health* 23, no. 3 (2011): 181–195, https://doi.org/10.1080/19317611.2011.593932; Susan Saltzburg and Tamara S. Davis, "Co-Authoring Gender-Queer Youth Identities: Discursive Tellings and Retellings," *Journal of Ethnic and Cultural Diversity in Social Work* 19, no. 2 (2010): 87–108, https://doi.org/10.1080/15313200903124028; Tam Sanger, "Trans Governmentality: the Production and Regulation of Gendered Subjectivities," *Journal of Gender Studies* 17, no. 1 (2008): 41–53, https://doi.org/10.1080/09589230701838396; Spade, *Normal Life*; Mel Wiseman and Sarah Davidson, "Problems with Binary Gender Discourse: Using Context to Promote Flexibility and Connection in Gender Identity," *Clinical Child Psychology and Psychiatry* 17, no. 4 (2011): 528–537, https://doi.org/10.1177/1359104511424991.

16 For example: Butler, *Gender Trouble*; Adrienne Hancock and Lauren Helenius, "Adolescent Male-to-Female Transgender Voice and Communication Therapy," *Journal of Communication Disorders* 45, no. 5 (2012): 313–324, https://doi.org/10.1016/j.jcomdis.2012.06.008; Griffin Hansbury, "King Kong & Goldilocks: Imagining Transmasculinities Through the Trans–Trans Dyad," *Psychoanalytic Dialogues* 21, no. 2 (2011): 210–220, https://doi.org/10.1080/10481885.2011.562846; Hines, "Queerly Situated?"; Iantaffi and Bockting, "Views from Both Sides"; Laura E. Kuper, Laurel Wright, and Brian Mustanski, "Stud Identity Among Female-Born Youth of Color: Joint Conceptualizations of Gender Variance and Same-Sex Sexuality," *Journal of Homosexuality* 61, no. 5 (2014): 714–731, https://doi.org/10.1080/00918369.2014.870443; Levitt and Ippolito, "Being Transgender"; Eric Darnell Pritchard, "'This Is Not an Empty-Headed Man in a Dress': Literacy Misused, Reread and Rewritten in Soulopoliz," *Southern Communication Journal* 74, no. 3 (2009): 278–299, https://doi.org/10.1080/10417940903061094; Riley, et al., "Needs of Gender-Variant Children"; Saltzburg and Davis, "Co-Authoring Gender-Queer Youth Identities"; Sanger, "Trans Governmentality"; Wiseman and Davidson, "Problems with Binary Gender Discourse."

17 For example: C. Aramburu Alegría, "Relationship Challenges and Relationship Maintenance Activities Following Disclosure of Transsexualism," *Journal of Psychiatric and Mental Health Nursing* 17, no. 10 (2010): 909–916, https://doi.org/10.1111/j.1365-2850.2010.01624.x; Talia Mae Bettcher, "When Selves Have Sex: What the Phenomenology of Trans Sexuality Can Teach Us About Sexual Orientation," *Journal of Homosexuality* 61, no. 5 (2014): 605–620, https://doi.org/10.1080/00918396.2014.865472; Hansbury, "King Kong & Goldilocks"; Hines, "Queerly Situated?"; Iantaffi and Bockting, "Views from Both

Sides"; Levitt and Ippolito, "Being Transgender"; Riley et al., "Needs of Gender-Variant Children"; Avery Brooks Tompkins, "'There's No Chasing Involved': Cis/Trans Relationships, 'Tranny Chasers,' and the Future of a Sex-Positive Trans Politics," *Journal of Homosexuality* 61, no. 5 (2014): 766–780, https://doi.org/10.1080/00918369.2014.870448; Yuenmei Wong, "Islam, Sexuality, and the Marginal Positioning of *Pengkids* and Their Girlfriends in Malaysia," *Journal of Lesbian Studies* 16, no. 4 (2012): 435–448, https://doi.org/10.1080/10894160.2012.681267.

18 Spade, *Normal Life.*

19 Hines, "Queerly Situated?"; Iantaffi and Bockting, "Views from Both Sides"; Levitt and Ippolito, "Being Transgender"; Sanger, "Trans Governmentality"; Wiseman and Davidson, "Problems with Binary Gender Discourse".

20 Levitt and Ippolito, "Being Transgender".

21 Sanger, "Trans Governmentality".

22 Riley et al., "Needs of Gender-Variant Children."

23 Hines, "Queerly Situated?"; Iantaffi and Bockting, "Views from Both Sides."

24 Iantaffi and Bockting, "Views from Both Sides."

25 Wiseman and Davidson, "Problems with Binary Gender Discourse."

26 Hines, "Queerly Situated?"; Iantaffi and Bockting, "Views from Both Sides"; Levitt and Ippolito, "Being Transgender"; Riley et al., "Needs of Gender-Variant Children"; Wiseman and Davidson, "Problems with Binary Gender Discourse."

27 Hines, "Queerly Situated?"; Iantaffi and Bockting, "Views from Both Sides"; Levitt and Ippolito, "Being Transgender"; Riley et al., "Needs of Gender-Variant Children"; Sanger, "Trans Governmentality"; Wiseman and Davidson, "Problems with Binary Gender Discourse."

28 Levitt and Ippolito, "Being Transgender."

29 Butler, *Gender Trouble.*

30 Levitt and Ippolito, "Being Transgender"; Riley et al., "Needs of Gender-Variant Children."

31 Levitt and Ippolito, "Being Transgender"; Kristen Schilt, "Just One of the Guys?," *Gender & Society* 20, no. 4 (2006): 465–490, https://doi.org/10.1177/0891243206288077.

32 Levitt and Ippolito, "Being Transgender."

33 Iantaffi and Bockting, "Views from Both Sides"; Levitt and Ippolito, "Being Transgender."

34 Iantaffi and Bockting, "Views from Both Sides"; Levitt and Ippolito, "Being Transgender"; Riley et al., "Needs of Gender-Variant Children."

35 LGBQT (lesbian, gay, bisexual, queer, transgender) is an umbrella term used to describe non-heteronormative sexual and gender identities. LGBQT has become the trademark of a community that mostly serves lesbian and gay identities.

36 Talia Mae Bettcher, "Transphobia," *TSQ: Transgender Studies Quarterly* 1, nos 1–2 (2014): 249–251, https://doi.org/10.1215/23289252-2400181; Hines, "Queerly Situated?"; Iantaffi and Bockting, "Views from Both Sides"; Levitt and Ippolito, "Being Transgender"; Riley et al., "Needs of Gender-Variant Children."

37 Iantaffi and Bockting, "Views from Both Sides."

38 Levitt and Ippolito, "Being Transgender," 57.

39 Riley et al., "Needs of Gender-Variant Children."

40 For example: Andre Cavalcante, "Centering Transgender Identity via the Textual Periphery: *TransAmerica* and the 'Double Work' of Paratexts," *Critical Studies in Media Communication* 30, no. 2 (2013): 85–101, https://doi.org/10.1080/15295036.2012.694077; Richard Mocarski et al., "'A Different Kind of Man': Mediated Transgendered Subjectivity, Chaz Bono on *Dancing with the Stars*," *Journal of Communication Inquiry* 37, no. 3 (2013): 249–264, https://doi.org/10.1177/0196859913489572; Megan Sinnott, "The Semiotics of Transgendered Sexual Identity in the Thai Print Media: Imagery and Discourse of the Sexual Other," *Culture, Health & Sexuality* 2, no. 4 (2000): 425–440, https://doi.org/10.1080/13691050050174431; Eliza Steinbock, "On the Affective Force of 'Nasty Love,'" *Journal of Homosexuality* 61, no. 5 (2014): 749–765, https://doi.org/10.1080/00918369.2014.870446; Tompkins, "There's no Chasing Involved"; Jan Wickman, "Masculinity and Female Bodies," *NORA—Nordic Journal of Feminist and Gender Research* 11, no. 1 (2003): 40–54, https://doi.org/10.1080/08038740307272.

41 Although it might be more accurately labeled as "gender/sexual rights," we are keeping the terms used by the researchers in this area.

42 For example: Leigh Goodmark, "Transgender People, Intimate Partner Abuse, and the Legal System," *Harvard Civil Rights-Civil Liberties Law Review* 48, no. 1 (2013): 51–104; Randi Gressgård, "When Trans Translates into Tolerance—Or Was It Monstrous? Transsexual and Transgender Identity in Liberal

Humanist Discourse," *Sexualities* 13, no. 5 (2010): 539–561, https://doi.org/10.1177/1363460710375569; Spade, *Normal Life*; Elias Vitulli, "Racialized Criminality and the Imprisoned Trans Body: Adjudicating Access to Gender-Related Medical Treatment in Prisons," *Social Justice* 37, no. 1 (2010): 53–68.

43 Although it might be more accurate to label this theme as "gender/sexual activism," we are, once again, remaining close to the original terms used in the research we identified.

44 For example: Darryl B. Hill, "Coming to Terms: Using Technology to Know Identity," *Sexuality and Culture* 9, no. 3 (2005): 24–52, https://doi.org/10.1007/s12119-005-1013-x; Ryan Thoreson, "Beyond Equality: The Post-Apartheid Counternarrative of Trans and Intersex Movements in South Africa," *African Affairs* 112, no. 449 (July 2013): 646–665, https://doi.org/10.1093/afraf/adt043.

45 Sinnott, "Semiotics of Transgendered Sexual Identity."

46 Cavalcante, "Centering Transgender Identity via the Textual Periphery."

47 Sinnott, "Semiotics of Transgendered Sexual Identity."

48 Mocarski et al., "A Different Kind of Man."

49 Sinnott, "Semiotics of Transgendered Sexual Identity," 43.

50 Mocarski et al., "A Different Kind of Man."

51 Cavalcante, "Centering Transgender Identity via the Textual Periphery."

52 Spade, *Normal Life*.

53 Vitulli, "Racialized Criminality."

54 Gressgård, "When Trans Translates into Tolerance."

55 Gressgård, "When Trans Translates into Tolerance"; Vitulli, "Racialized Criminality."

56 Thoreson, "Beyond Equality."

57 Hill, "Coming to Terms."

58 For example: Evelyn Blackwood, "Transnational Discourses and Circuits of Queer Knowledge in Indonesia," *GLQ: A Journal of Lesbian and Gay Studies* 14, no. 4 (January 2008): 481–507, https://doi.org/10.1215/10642684-2008-002; Katrina Roen, "Transgender Theory and Embodiment: The Risk of Racial Marginalisation," *Journal of Gender Studies* 10, no. 3 (2001): 253–263, https://doi.org/10.1080/09589230120086467.

59 For example: Blackwood, "Transnational Discourses in Indonesia"; Sara Davidmann, "Imag(in)ing Trans Partnerships: Collaborative Photography and Intimacy," *Journal of Homosexuality* 61, no. 5 (2014): 636–653, https://doi.org/10.1080/00918369.2014.865481; Roen, "Transgender Theory and Embodiment."

60 For example: Blackwood, "Transnational Discourses in Indonesia"; Davidmann, "Imag(in)ing Trans Partnerships"; Roen, "Transgender Theory and Embodiment."

61 Roen, "Transgender Theory and Embodiment," 258.

62 Blackwood, "Transnational Discourses in Indonesia."

63 Blackwood, "Transnational Discourses in Indonesia."

64 Blackwood, "Transnational Discourses in Indonesia."

65 Roen, "Transgender Theory and Embodiment."

66 Blackwood, "Transnational Discourses in Indonesia."

67 For example: Yossi Maurey, "Dana International and the Politics of Nostalgia," *Popular Music* 28, no. 1 (2009): 85–103, https://doi.org/10.1017/s0261143008001608; Mark McLelland, "Is There a Japanese Gay Identity?" *Culture, Health & Sexuality* 2, no. 4 (2000): 59–472, https://doi.org/10.1080/13691050050174459; Amisha R. Patel, "India's Hijras: The Case for Transgender Rights," *George Washington International Law Review* 42, no. 4 (2010): 835–863; Eliza Steinbock, "On the Affective Force of 'Nasty Love,'" *Journal of Homosexuality* 61, no. 5 (September 2014): pp. 749–765, https://doi.org/10.1080/00918369.2014.870446; Tompkins, "There's No Chasing Love."

68 Although it might be more accurately labeled as "gender/sexual rights," we are keeping the terms used by the researchers in this area.

69 For example: Mark McLelland and Katsuhiko Suganuma, "Sexual Minorities and Human Rights in Japan: An Historical Perspective," *The International Journal of Human Rights* 13, no. 2-3 (2009): 329–343, https://doi.org/10.1080/13642980902758176; Patel, "India's Hijras."

70 Although it might be more accurate to label this theme as "gender/sexual activism," we are, once again, remaining close to the original terms used in the research we identified.

71 For example: McLelland and Suganuma, "Sexual Minorities and Human Rights in Japan"; Patel, "India's Hijras."

72 Patel, "India's Hijras."

73 Patel, "India's Hijras," 836.

74 Maurey, "Dana International."

75 McLelland, "Japanese Gay Identity?"
76 Patel, "India's Hijras."
77 Patel, "India's Hijras."
78 Patel, "India's Hijras."
79 Patel, "India's Hijras."
80 Amets Suess, Karine Espineira, and Pau Crego Walters, "Depathologization," *TSQ: Transgender Studies Quarterly* 1, nos 1–2 (2014): 73–77, https://doi.org/10.1215/23289252-2399650, 73.
81 Elizabeth Stephens, "Normal," *TSQ: Transgender Studies Quarterly* 1, nos 1–2 (2014): 141–145, https://doi.org/10.1215/23289252-2399848.
82 Eric A. Stanley, "Gender Self-Determination," *TSQ: Transgender Studies Quarterly* 1, nos 1–2 (2014): 89–91, https://doi.org/10.1215/23289252-2399695.
83 Butler, *Gender Trouble.*
84 Erica Lennon and Brian J. Mistler, "Cisgenderism," *TSQ: Transgender Studies Quarterly* 1, nos 1–2 (2014): 63–64, https://doi.org/10.1215/23289252-2399623.
85 Julia R. Johnson, "Cisgender Privilege, Intersectionality, and the Criminalization of CeCe McDonald: Why Intercultural Communication Needs Transgender Studies," *Journal of International and Intercultural Communication* 6, no. 2 (2013): 135–144, https://doi.org/10.1080/17513057.2013.776094; Yep, "Queering/Quaring/Kauering/Crippin'/Transing"; Yep, "Toward Thick(er) Intersectionalities."
86 Yep, "Toward Thick(er) Intersectionalities."
87 Valentine, *Imagining Transgender;* Yep, "Queering/Quaring/Kauering/Crippin/Transing."
88 Yep, "Queering/Quaring/Kauering/Crippin'/Transing."
89 A. Finn Enke, "Translation," *TSQ: Transgender Studies Quarterly* 1, nos 1–2 (2014): 241–244, https://doi.org/10.1215/23289252-2400163, 242.
90 Richard G. Jones and Bernadette Marie Calafell, "Contesting Neoliberalism Through Critical Pedagogy, Intersectional Reflexivity, and Personal Narrative: Queer Tales of Academia," *Journal of Homosexuality* 59, no. 7 (2012): 957–981, https://doi.org/10.1080/00918369.2012.699835, 963.
91 Johnson, "Criminalization of CeCe McDonald"; Salvador Vidal-Ortiz, "Whiteness," *TSQ: Transgender Studies Quarterly* 1, nos 1–2 (2014): 264–266, https://doi.org/10.1215/23289252-2400217.
92 For example: Carsten Balzer and Carla Lagata, "Human Rights," *TSQ: Transgender Studies Quarterly* 1, nos 1–2 (2014): 99–103, https://doi.org/10.1215/23289252-2399731; McLelland and Suganuma, "Sexual Minorities and Human Rights in Japan"; Spade, *Normal Life.*
93 Enke, "Translation," 243.
94 Enke, "Translation," 243.
95 Yolanda Martínez-San Miguel and Sarah Tobias, "Introduction: Thinking Beyond Hetero/Homo Normativities," in *Trans Studies: The Challenge to Hetero/Homo Normativities,* ed. Yolanda Martínez-San Miguel and Sarah Tobias (New Brunswick, NJ: Rutgers University Press, 2016), 1–17.
96 Jack Halberstam, *Trans*: A Quick and Quirky Account of Gender Variability* (Oakland, CA: University of California Press, 2018), 4.
97 Halberstam, *Trans*;* Gust A. Yep, Sage E. Russo, and Rebecca N. Gigi, "The New Transnormativity? Reading Mainstream Representations of Caitlyn Jenner in the University Classroom," in *Leadership, Equity, and Social Justice in American Higher Education: A Reader,* ed. Charles P. Gause (New York: Peter Lang, 2017), 149–159.
98 Gust A. Yep et al., "Uniquely *Glee*: Transing Racialized Gender," in *Race and Gender in Electronic Media: Content, Context, Culture,* ed. Rebecca Ann Lind (New York: Routledge, 2017), 55–71.
99 Gust A. Yep, Fatima Zahrae Chrifi Alaoui, and Ryan M. Lescure, "Representations of Trans Women of Color in Sean Baker's *Tangerine,*" in *Race/Gender/Class/Media 4.0: Considering Diversity across Content, Audiences and Production,* ed. Rebecca Ann Lind (New York: Routledge, 2019), 154–158.
100 For example: Joshua Trey Barnett, "Fleshy Metamorphosis: Temporal Pedagogies of Transsexual Counterpublics," in *Transgender Communication Studies: Histories, Trends, and Trajectories,* eds Leland G. Spencer and Jamie C. Capuzza (Lanham, MD: Lexington Books, 2015), 155–169; Gust A. Yep, Sage E. Russo, and Ryan M. Lescure, "Transing Normative Boyhood Masculinity in Alain Berliner's Ma Vie En Rose," *Boyhood Studies* 8, no. 2 (2015): 43–60, https://doi.org/10.3167/bhs.2015.080204; Yep, Chrifi Alaoui, and Lescure, "Representations of Trans Women in *Tangerine.*"
101 Jennifer Ung Loh, "Transgender Identity, Sexual versus Gender 'Rights' and the Tools of the Indian State," *Feminist Review* 119, no. 1 (2018): 39–55, https://doi.org/10.1057/s41305-018-0124-9.

102 Marshall K. Cheney et al., "Living Outside the Gender Box in Mexico: Testimony of Transgender Mexican Asylum Seekers," *American Journal of Public Health* 107, no. 10 (2017): 1646–1652, https://doi.org/10.2105/ajph.2017.303961.

103 Sarah J. Jackson, Moya Foucault Bailey, and Brooke Welles, "#GirlsLikeUs: Trans Advocacy and Community Building Online," *New Media & Society* 20, no. 5 (2017): 1868–1888, https://doi.org/10.1177/1461444817709276.

104 Stephen Craig Kerry, "Sistergirls/Brotherboys: The Status of Indigenous Transgender Australians," *International Journal of Transgenderism* 15, nos 3–4 (2014): 173–186, https://doi.org/10.1080/15532739.2014.995262.

105 Elijah Adiv Edelman and Lal Zimman, "Boycunts and Bonus Holes: Trans Men's Bodies, Neoliberalism, and the Sexual Productivity of Genitals," *Journal of Homosexuality* 61, no. 5 (2014): 673–690, https://doi.org/10.1080/00918369.2014.870438; Ayeshah Émon and Christine Garlough, "Refiguring the South Asian American Tradition Bearer: Performing the 'Third Gender' in *Yoni Ki Baat*," *The Journal of American Folklore* 128, no. 510 (2015): 412–437, https://doi.org/10.5406/jamerfolk.128.510.0412.

106 Yep, Russo, and Gigi, "The New Transnormativity?" 156.

107 Cheney et al., "Living Outside the Gender Box."

108 Edelman and Zimman, "Boycunts and Bonus Holes."

109 Loh, "Transgender Identity and the Tools of the Indian State."

110 Jackson et al., "#GirlsLikeUs."

111 Jackson et al., "#GirlsLikeUs," 1870.

112 Kerry, "Sistergirls/Brotherboys."

113 Émon and Garlough, "Performing the 'Third Gender.'"

114 Andrea M. Hackl, Amy B. Becker, and Maureen E. Todd, "'I Am Chelsea Manning': Comparison of Gendered Representation of Private Manning in U.S. and International News Media," *Journal of Homosexuality* 63, no. 4 (2015): 467–486, https://doi.org/10.1080/00918369.2015.1088316.

115 Yep, Chrifi Alaoui, and Lescure, "Representations of Trans Women in *Tangerine*," 155–156.

116 Yep, Russo, and Lescure, "Transing Normative Boyhood Masculinity."

117 Yep, Chrifi Alaoui, and Lescure, "Representations of Trans Women in *Tangerine*," 156.

118 Yep, Chrifi Alaoui, and Lescure, "Representations of Trans Women in *Tangerine*."

Bibliography

Aiello, Giorgia, Sandeep Bakshi, Sirma Bilge, Lisa Kahaleole Hall, Lynda Johnston, Kimberlee Pérez, and Karma Chávez. "Here, and Not Yet Here: A Dialogue at the Intersection of Queer, Trans, and Culture." *Journal of International and Intercultural Communication* 6, no. 2 (2013): 96–117. https://doi.org/10.1080/17513057.2013.778155.

Alegría, C. Aramburu. "Relationship Challenges and Relationship Maintenance Activities Following Disclosure of Transsexualism." *Journal of Psychiatric and Mental Health Nursing* 17, no. 10 (2010): 909–16. https://doi.org/10.1111/j.1365-2850.2010.01624.x.

Balzer, Carsten, and Carla Lagata. "Human Rights." *TSQ: Transgender Studies Quarterly* 1, no. 1–2 (2014): 99–103. https://doi.org/10.1215/23289252-2399731.

Barnett, Joshua Trey. "Fleshy Metamorphosis: Temporal Pedagogies of Transsexual Counterpublics." In *Transgender Communication Studies: Histories, Trends, and Trajectories*, edited by Leland G. Spencer and Jamie C. Capuzza, 155–69. Lanham, MD: Lexington Books, 2015.

Bettcher, Talia Mae. "Transphobia." *TSQ: Transgender Studies Quarterly* 1, no. 1–2 (2014): 249–51. https://doi.org/10.1215/23289252-2400181.

Bettcher, Talia Mae. "When Selves Have Sex: What the Phenomenology of Trans Sexuality Can Teach Us About Sexual Orientation." *Journal of Homosexuality* 61, no. 5 (2014): 605–20. https://doi.org/10.1080/00918396.2014.865472.

Bhanji, Nael. "Trans/scriptions: Homing Desires (Trans)Sexual Citizenship, and Racialized Bodies." In *The Transgender Studies Reader 2*, edited by Susan Stryker and Aren Z. Aizura, 512–26. New York: Routledge, 2013.

Blackburn, Mollie V. "Agency in Borderland Discourses: Examining Language Use in a Community Center with Black Queer Youth." *Teachers College Record* 107, no. 1 (2005): 89–113. https://doi.org/10.1111/j.1467-9620.2005.00458.x.

Blackwood, Evelyn. "Transnational Discourses and Circuits of Queer Knowledge in Indonesia." *GLQ: A Journal of Lesbian and Gay Studies* 14, no. 4 (2008): 481–507. https://doi.org/10.1215/10642684-2008-002.

Bornstein, Kate. *My New Gender Workbook*. London: Routledge, 2013.

Butler, Judith. *Gender Trouble: Feminism and the Subversion of Identity*. New York: Routledge, 1990.

Cavalcante, Andre. "Centering Transgender Identity via the Textual Periphery: *TransAmerica* and the 'Double Work' of Paratexts." *Critical Studies in Media Communication* 30, no. 2 (2013): 85–101. https://doi.org/10.1080/15295036.2012.694077.

Cheney, Marshall K., Mary J. Gowin, E. Laurette Taylor, Melissa Frey, Jamie Dunnington, Ghadah Alshuwaiyer, J. Kathleen Huber, Mary Camero Garcia, and Grady C. Wray. "Living Outside the Gender Box in Mexico: Testimony of Transgender Mexican Asylum Seekers." *American Journal of Public Health* 107, no. 10 (2017): 1646–52. https://doi.org/10.2105/ajph.2017.303961.

Davidmann, Sara. "Imag(in)ing Trans Partnerships: Collaborative Photography and Intimacy." *Journal of Homosexuality* 61, no. 5 (2014): 636–53. https://doi.org/10.1080/00918369.2014.865481.

Edelman, Elijah Adiv, and Lal Zimman. "Boycunts and Bonus Holes: Trans Men's Bodies, Neoliberalism, and the Sexual Productivity of Genitals." *Journal of Homosexuality* 61, no. 5 (2014): 673–90. https://doi.org/10.1080/00918369.2014.870438.

Émon, Ayeshah, and Christine Garlough. "Refiguring the South Asian American Tradition Bearer: Performing the 'Third Gender' in *Yoni Ki Baat*." *The Journal of American Folklore* 128, no. 510 (2015): 412–37. https://doi.org/10.5406/jamerfolk.128.510.0412.

Enke, A. Finn. "Translation." *TSQ: Transgender Studies Quarterly* 1, no. 1–2 (2014): 241–44. https://doi.org/10.1215/23289252-2400163.

Goodmark, Leigh. "Transgender People, Intimate Partner Abuse, and the Legal System." *Harvard Civil Rights-Civil Liberties Law Review* 48, no. 1 (2013): 51–104.

Gressgård, Randi. "When Trans Translates into Tolerance—Or Was It Monstrous? Transsexual and Transgender Identity in Liberal Humanist Discourse." *Sexualities* 13, no. 5 (2010): 539–61. https://doi.org/10.1177/1363460710375569.

Hackl, Andrea M., Amy B. Becker, and Maureen E. Todd. "'I Am Chelsea Manning': Comparison of Gendered Representation of Private Manning in U.S. and International News Media." *Journal of Homosexuality* 63, no. 4 (2015): 467–86. https://doi.org/10.1080/00918369.2015.1088316.

Halberstam, Jack. *Trans*: A Quick and Quirky Account of Gender Variability*. Berkeley, CA: University of California Press, 2018.

Hancock, Adrienne, and Lauren Helenius. "Adolescent Male-to-Female Transgender Voice and Communication Therapy." *Journal of Communication Disorders* 45, no. 5 (2012): 313–24. https://doi.org/10.1016/j.jcomdis.2012.06.008.

Hansbury, Griffin. "King Kong & Goldilocks: Imagining Transmasculinities Through the Trans–Trans Dyad." *Psychoanalytic Dialogues* 21, no. 2 (2011): 210–20. https://doi.org/10.1080/10481885.2011.562846.

Hill, Darryl B. "Coming to Terms: Using Technology to Know Identity." *Sexuality and Culture* 9, no. 3 (2005): 24–52. https://doi.org/10.1007/s12119-005-1013-x.

Hines, Sally. "Queerly Situated? Exploring Negotiations of Trans Queer Subjectivities at Work and within Community Spaces in the UK." *Gender, Place & Culture* 17, no. 5 (2010): 597–613. https://doi.org/10.1080/0966369x.2010.503116.

Horswell, Michael J. *Decolonizing the Sodomite: Queer Tropes of Sexuality in Colonial Andean Culture*. Austin: University of Texas Press, 2005.

Iantaffi, Alex, and Walter O. Bockting. "Views from Both Sides of the Bridge? Gender, Sexual Legitimacy and Transgender Peoples Experiences of Relationships." *Culture, Health & Sexuality* 13, no. 3 (2011): 355–70. https://doi.org/10.1080/13691058.2010.537770.

Irving, Dan. "Normalized Transgressions: Legitimizing the Transsexual Body as Productive." *Radical History Review* 2008, no. 100 (2008): 38–59. https://doi.org/10.1215/01636545-2007-021.

Jackson, Sarah J., Moya Foucault Bailey, and Brooke Welles. "#GirlsLikeUs: Trans Advocacy and Community Building Online." *New Media & Society* 20, no. 5 (2017): 1868–88. https://doi.org/10.1177/1461444817709276.

Johnson, Julia R. "Cisgender Privilege, Intersectionality, and the Criminalization of CeCe McDonald: Why Intercultural Communication Needs Transgender Studies." *Journal of International and Intercultural Communication* 6, no. 2 (2013): 135–44. https://doi.org/10.1080/17513057.2013.776094.

Jones, Richard G., and Bernadette Marie Calafell. "Contesting Neoliberalism Through Critical Pedagogy, Intersectional Reflexivity, and Personal Narrative: Queer Tales of Academia." *Journal of Homosexuality* 59, no. 7 (2012): 957–81. https://doi.org/10.1080/00918369.2012.699835.

Kerry, Stephen Craig. "Sistergirls/Brotherboys: The Status of Indigenous Transgender Australians." *International Journal of Transgenderism* 15, no. 3-4 (2014): 173–86. https://doi.org/10.1080/15532739.2014.995262.

Kuper, Laura E., Laurel Wright, and Brian Mustanski. "Stud Identity Among Female-Born Youth of Color: Joint Conceptualizations of Gender Variance and Same-Sex Sexuality." *Journal of Homosexuality* 61, no. 5 (2014): 714–31. https://doi.org/10.1080/00918369.2014.870443.

Lennon, Erica, and Brian J. Mistler. "Cisgenderism." *TSQ: Transgender Studies Quarterly* 1, no. 1–2 (2014): 63–64. https://doi.org/10.1215/23289252-2399623.

Levitt, Heidi M., and Maria R. Ippolito. "Being Transgender: Navigating Minority Stressors and Developing Authentic Self-Presentation." *Psychology of Women Quarterly* 38, no. 1 (2013): 46–64. https://doi.org/10.1177/0361684313501644.

Loh, Jennifer Ung. "Transgender Identity, Sexual versus Gender 'Rights' and the Tools of the Indian State." *Feminist Review* 119, no. 1 (2018): 39–55. https://doi.org/10.1057/s41305-018-0124-9.

Martínez-San Miguel, Yolanda, and Sarah Tobias. "Introduction: Thinking Beyond Hetero/Homo Normativities." In *Trans Studies: The Challenge to Hetero/Homo Normativities*, edited by Yolanda Martínez-San Miguel and Sarah Tobias, 1–17. New Brunswick, NJ: Rutgers University Press, 2016.

Maurey, Yossi. "Dana International and the Politics of Nostalgia." *Popular Music* 28, no. 1 (2009): 85–103. https://doi.org/10.1017/s0261143008001608.

McLelland, Mark. "Is There a Japanese Gay Identity?" *Culture, Health & Sexuality* 2, no. 4 (2000): 459–72. https://doi.org/10.1080/13691050050174459.

McLelland, Mark, and Katsuhiko Suganuma. "Sexual Minorities and Human Rights in Japan: An Historical Perspective." *The International Journal of Human Rights* 13, no. 2–3 (2009): 329–43. https://doi.org/10.1080/13642980902758176.

Mocarski, Richard, Sim Butler, Betsy Emmons, and Rachael Smallwood. "'A Different Kind of Man': Mediated Transgendered Subjectivity, Chaz Bono on *Dancing with the Stars*." *Journal of Communication Inquiry* 37, no. 3 (2013): 249–64. https://doi.org/10.1177/0196859913489572.

Mohanty, Chandra Talpade. *Feminism without Borders: Decolonizing Theory, Practicing Solidarity*. Durham, NC: Duke University Press, 2003.

Patel, Amisha R. "India's *Hijras*: The Case for Transgender Rights." *George Washington International Law Review* 42, no. 4 (2010): 835–63.

Pritchard, Eric Darnell. "'This Is Not an Empty-Headed Man in a Dress': Literacy Misused, Reread and Rewritten in Soulopoliz." *Southern Communication Journal* 74, no. 3 (2009): 278–99. https://doi.org/10.1080/10417940903061094.

Riley, Elizabeth Anne, Gomathi Sitharthan, Lindy Clemson, and Milton Diamond. "The Needs of Gender-Variant Children and Their Parents: A Parent Survey." *International Journal of Sexual Health* 23, no. 3 (2011): 181–95. https://doi.org/10.1080/19317611.2011.593932.

Roen, Katrina. "Transgender Theory and Embodiment: The Risk of Racial Marginalisation." *Journal of Gender Studies* 10, no. 3 (2001): 253–63. https://doi.org/10.1080/09589230120086467.

Rubin, Henry S. "Trans Studies: Between a Metaphysics of Presence and Absence." In *Reclaiming Genders: Transsexual Grammars at the Fin De Siècle*, edited by Kate More and Stephen Whittle, 173–92. London: Cassell, 1999.

Saltzburg, Susan, and Tamara S. Davis. "Co-Authoring Gender-Queer Youth Identities: Discursive Tellings and Retellings." *Journal of Ethnic and Cultural Diversity in Social Work* 19, no. 2 (2010): 87–108. https://doi.org/10.1080/15313200903124028.

Sanger, Tam. "Trans Governmentality: The Production and Regulation of Gendered Subjectivities." *Journal of Gender Studies* 17, no. 1 (2008): 41–53. https://doi.org/10.1080/09589230701838396.

Schilt, Kristen. "Just One of the Guys?" *Gender & Society* 20, no. 4 (2006): 465–90. https://doi.org/10.1177/0891243206288077.

Sinnott, Megan. "The Semiotics of Transgendered Sexual Identity in the Thai Print Media: Imagery and Discourse of the Sexual Other." *Culture, Health & Sexuality* 2, no. 4 (2000): 425–40. https://doi.org/10.1080/13691050050174431.

Spade, Dean. *Normal Life: Administrative Violence, Critical Trans Politics, and the Limits of Law*. Durham, NC: Duke University Press, 2015.

Stanley, E. A. "Gender Self-Determination." *TSQ: Transgender Studies Quarterly* 1, no. 1–2 (2014): 89–91. https://doi.org/10.1215/23289252-2399695.

Steinbock, Eliza. "On the Affective Force of 'Nasty Love.'" *Journal of Homosexuality* 61, no. 5 (2014): 749–65. https://doi.org/10.1080/00918369.2014.870446.

Stephens, E. "Normal." *TSQ: Transgender Studies Quarterly* 1, no. 1–2 (2014): 141–45. https://doi.org/10.1215/23289252-2399848.

Stryker, Susan. "(De)Subjugated Knowledges: An Introduction to Transgender Studies." In *The Transgender Studies Reader*, edited by Stephen Whittle and Susan Stryker, 1–17. New York: Routledge, 2006.

Stryker, Susan, Paisley Currah, and Lisa Jean Moore. "Introduction: Trans-, Trans, or Transgender?" *WSQ: Womens Studies Quarterly* 36, no. 3–4 (2008): 11–22. https://doi.org/10.1353/wsq.0.0112.

Stryker, Susan, and Paisley Currah. "Introduction." *TSQ: Transgender Studies Quarterly* 1, no. 1–2 (2014): 1–18. https://doi.org/10.1215/23289252-2398540.

Suess, Amets, Karine Espineira, and Pau Crego Walters. "Depathologization." *TSQ: Transgender Studies Quarterly* 1, no. 1–2 (2014): 73–77. https://doi.org/10.1215/23289252-2399650.

Thoreson, Ryan. "Beyond Equality: The Post-Apartheid Counternarrative of Trans and Intersex Movements in South Africa." *African Affairs* 112, no. 449 (2013): 646–65. https://doi.org/10.1093/afraf/adt043.

Tompkins, Avery Brooks. "'There's No Chasing Involved': Cis/Trans Relationships, 'Tranny Chasers,' and the Future of a Sex-Positive Trans Politics." *Journal of Homosexuality* 61, no. 5 (2014): 766–80. https://doi.org/10.1080/00918369.2014.870448.

Valentine, David. *Imagining Transgender: An Ethnography of a Category.* Durham, NC: Duke University Press, 2007.

Vidal-Ortiz, Salvador. "Whiteness." *TSQ: Transgender Studies Quarterly* 1, no. 1–2 (2014): 264–66. https://doi.org/10.1215/23289252-2400217.

Vitulli, Elias. "Racialized Criminality and the Imprisoned Tran Body: Adjudicating Access to Gender-Related Medical Treatment in Prisons." *Social Justice* 37, no. 1 (2010): 53–68.

Wickman, Jan. "Masculinity and Female Bodies." *NORA—Nordic Journal of Feminist and Gender Research* 11, no. 1 (2003): 40–54. https://doi.org/10.1080/08038740307272.

Williams, Cristan. "Transgender." *TSQ: Transgender Studies Quarterly* 1, no. 1–2 (2014): 232–34. https://doi.org/10.1215/23289252-2400136.

Wiseman, Mel, and Sarah Davidson. "Problems with Binary Gender Discourse: Using Context to Promote Flexibility and Connection in Gender Identity." *Clinical Child Psychology and Psychiatry* 17, no. 4 (2011): 528–37. https://doi.org/10.1177/1359104511424991.

Wong, Yuenmei. "Islam, Sexuality, and the Marginal Positioning of *Pengkids* and Their Girlfriends in Malaysia." *Journal of Lesbian Studies* 16, no. 4 (2012): 435–48. https://doi.org/10.1080/10894160.2012.681267.

Yep, Gust A. "Queering/Quaring/Kauering/Crippin'/Transing 'Other Bodies' in Intercultural Communication." *Journal of International and Intercultural Communication* 6, no. 2 (2013): 118–26. https://doi.org/10.1080/17513057.2013.777087.

Yep, Gust A., Sage E. Russo, and Ryan M. Lescure. "Transing Normative Boyhood Masculinity in Alain Berliner's *Ma Vie En Rose.*" *Boyhood Studies* 8, no. 2 (2015): 43–60. https://doi.org/10.3167/bhs.2015.080204.

Yep, Gust A. "Toward Thick(er) Intersectionalities: Theorizing, Researching, and Activating the Complexities of Communication and Identities." In *Globalizing Intercultural Communication: A Reader*, edited by Kathryn Sorrells and Sachi Sekimoto, 86–94. Thousand Oaks, CA: Sage, 2016.

Yep, Gust A., Sage E. Russo, Jace Allen, and Nicholas T. Chivers. "Uniquely *Glee*: Transing Racialized Gender." In *Race and Gender in Electronic Media: Content, Context, Culture*, edited by Rebecca Ann Lind, 55–71. New York: Routledge, 2017.

Yep, Gust A., Sage E. Russo, and Rebecca N. Gigi. "The New Transnormativity? Reading Mainstream Representations of Caitlyn Jenner in the University Classroom." In *Leadership, Equity, and Social Justice in American Higher Education: A Reader*, edited by Charles P. Gause, 149–59. New York: Peter Lang, 2017.

Yep, Gust A., Fatima Zahrae Chrifi Alaoui, and Ryan M. Lescure. "Representations of Trans Women of Color in Sean Baker's *Tangerine.*" In *Race/Gender/Class/Media 4.0: Considering Diversity across Content, Audiences and Production*, edited by Rebecca Ann Lind, 154–58. New York: Routledge, 2019.

INDEX